COMMUNICATION SYSTEM ENGINEERING HANDBOOK

OTHER McGRAW-HILL HANDBOOKS OF INTEREST

COMMUNICATION SYSTEM ENGINEERING HANDBOOK

DONALD H. HAMSHER, *Editor-in-Chief*

U. S. Army Electronics Command

McGRAW-HILL BOOK COMPANY

New York   St. Louis   San Francisco   Düsseldorf

Johannesburg   Kuala Lumpur   London   Mexico

Montreal   New Delhi   Panama   Rio de Janeiro

Singapore   Sydney   Toronto

COMMUNICATION SYSTEM ENGINEERING HANDBOOK

07-025960-7

4 5 6 7 8 9 0 K P K P 7 9 8 7 6 5 4 3

CONTRIBUTORS

ROY K. ANDRES, *Manager of Advanced Planning, RCA Communications, Inc.*

E. D. BECKEN, *Vice President and Chief Engineer, RCA Communications, Inc.*

EDWARD W. BROWN, JR., *Raytheon Company; now with Page Communications Engineers*

HENRY F. BURKHARD, *Supervisory Electronic Engineer, U. S. Army Electronics Command*

JOSEPH E. CLARE, *Engineer, New Jersey Bell Telephone Company*

MAURICE E. COOKSON, *Lenkurt Electric Company, Inc.; now with Electro-Mechanical Research, Inc.*

H. I. DOBSON, *Staff Engineer, Tennessee Valley Authority*

JAMES M. DUGUID, *Consulting Engineer; retired from Bell Telephone Laboratories*

C. H. ELDER, *Vice President and Chief Engineer, Reliable Electric Company*

THEODOR FRANKEL, *Engineering Scientist, Radio Corporation of America*

ENRIQUE FURTH, *Radio Corporation of America; now Chief Scientist of Dunlap and Associates, Inc.*

L. F. GOELLER, Jr., *Member of the Technical Staff, Bell Telephone Laboratories*

R. GUENTHER, *Manager, Advanced Communications Technology, Radio Corporation of America*

C. JULIAN GUSTAFSON, *Consultant; retired from Western Electric Company, Inc.*

NORMAN HOVAGIMYAN, *Senior Project Member, Technical Staff, Radio Corporation of America*

KENNETH D. HOWATT, *Assistant Vice President, American Telephone and Telegraph Company*

ARDEN C. JOHNSON, *American Telephone and Telegraph Company*

WALTER LEVY, *Radio Corporation of America; now with Pennsylvania Research Associates, Inc.*

M. A. LISH, *Engineering Director—Cost Analyses, American Telephone and Telegraph Company*

HENRY MAGNUSKI, *Associate Director of Research, Motorola, Inc.*

WILLARD F. MEEKER, *Engineer, Radio Corporation of America*

S. J. MEHLMAN, *Manager, Advanced Transmission Techniques, Radio Corporation of America*

PIERRE MERTZ, *Consultant; retired from Bell Telephone Laboratories*

J. Z. MILLAR, *Assistant Vice President, Western Union Telegraph Company*

LUIS F. NANNI, *Professor of Industrial Engineering, Rutgers, the State University*

DANIEL E. NOBLE, *Vice Chairman of the Board and Chief Technical Officer in charge of four technical divisions, Motorola, Inc.*

PHILIP SCHNEIDER, *Manager, Digital System Engineering, Radio Corporation of America*

B. SHEFFIELD, *Senior Project Member, Technical Staff, Radio Corporation of America*

NEAL H. SHEPHERD, *Communication Products Department, General Electric Company*

PHILIP F. SILING, *Consultant; retired from Director, RCA Frequency Bureau*

LEON STAMBLER, *Leader, Advanced Switching Techniques, Radio Corporation of America*

T. M. SWINGLE, *Chief, Communication Engineering and Design Branch, Tennessee Valley Authority*

EWELL M. THOMPSON, *Technical Promotion Manager, Lenkurt Electric Company, Inc.*

JACK H. WOLFF, *Engineering Leader, Communications Systems Division, Radio Corporation of America*

FOREWORD

In recent years, definitions covering the engineering and science functions have become confused, and it is not uncommon to discover public references to engineers as scientists, and to scientists as engineers. A generic definition for engineering could be "the application of natural law for the service of mankind." In other words, any processing of either materials or energy which is carried out to produce a result useful to man, is engineering. If engineering is concerned with the problem of making something which is useful to man, then systems engineering is the particular professional subdivision concerned with finding the approach to an optimum systems pattern which will produce the maximum practical usefulness for each application, within the limits of the overall controlling parameter, cost.

Shakespeare was one of the first to define systems engineering procedures with the following explicit statement:[1]

> "When we mean to build
> We first survey the plot, then draw the model;
> And when we see the figure of the house,
> Then must we rate the cost of the erection."

The systems engineering function has always been with us, but many years ago when the problems of electrical communication related to the transmission, over a few miles of wire, of either telegraph digital signals or electrical voice analog signals, the systems-engineering problems were readily incorporated in the total engineering job of installing the required communications means between two designated points. Today, an input of high-speed digital and video signals may be added to electrical analog signals for transmission through wires, along coaxial cables, between microwave stations, and even halfway across the earth by satellite relay, before they are picked up again and delivered through more wires and cables to the ultimate user. For this elaborate processing pattern, the problems of achieving compatibility at the interfaces of each of the subsystems and the problems of minimizing the signal deterioration produced by accumulating noise and distortion, become so complex that there is

[1] *King Henry IV*, Part 2, Act 1, Scene 3, Line 4.

justification for the extraction of systems engineering as a separate and very necessary professional function dedicated to the analysis and solution of total systems-design problems. The identification of systems engineering as a special branch of the engineering profession is, therefore, a product of the rising complexity of systems. It is no longer an acceptable procedure to assign complex-systems design to the equipment-design engineer as a secondary responsibility. It is understood, of course, that all engineering is concerned with "trade-offs," but systems engineering emphasizes the search for optimum trade-offs to achieve the best balance among all of the factors relating to costs, technical performance, compatibility with other systems, and the anticipation of future extensions and improvements of the systems.

The rising complexity of systems problems has forced upon us the necessity for developing formalized systems-engineering techniques for analysis and modeling. The use of formal procedures tends to prevent the possibility of overlooking important elements of design which must include customer requirements considered in relation to feasibility, costs, and equipment design, in addition to the practical equipment and systems specifications, which will integrate the final systems design.

Systems design and equipment design must not be treated as isolated problems. Not only does the systems engineer concern himself with the problems of compatibility among the elements of the system; he must also stimulate the use of both direct and feedback paths of communications among the several disciplines involved in the solution of any complex application problem. The need for the feedback requirement is obvious, but as an example, consider the fact that the introduction of integrated circuits to equipment design gives new emphasis to the need for effective communications, even among the engineers working on equipment or subsystems design.

All engineering branches must deal with systems-engineering problems, but communications engineering, involving, as it does, analog signals, digital signals, spectrum utilization, bandwidth, noise, and distortion, must deal with systems-engineering problems of a very high level of complexity. This book on communications systems engineering develops the nature of technical interfaces between the elements of the communication machine, including machine-to-machine communications, as well as the man-machine relationships, and therefore it should prove to be of substantial value to communications systems designers and equipment engineers.

Dr. Daniel E. Noble
Vice Chairman of the Board
and Chief Technical Officer
in charge of four technical
divisions, Motorola, Inc.

PREFACE

This handbook has been prepared to satisfy a need for information on all aspects of communication system engineering. As is true of many other large systems, it is characteristic of a communication system that each part of it has considerable influence on the other parts. The specialist working on a problem therefore must have reference material at hand that enables him to relate the effect of his solution on other equipment in the system and on the performance of the system as a whole.

The book is intended for engineers working in communication systems or on equipment that utilizes communication systems. It will also be useful to those users of communication systems who need to express their requirements for service in terms that are understandable to the communication engineer. Finally, the engineering graduate student may find it useful to broaden his understanding of applied electronic equipment design.

Communication system engineering requires criteria for such activities as the selection of techniques or the siting of facilities from among alternative choices, as well as data on specific characteristics. This handbook provides this information in a digest form based upon experience and current practices. In any particular chapter, a topic is treated from a point of view and in a depth that depend upon its relation to:

The information needed prior to system design
The development of the system plan
The variety of transmission and switching components available
The essentials of power, facility layout, operation, and costing

System standards, practices, and principles have been included, and attention has been given to important foreign divergence from United States practice. Carefully selected references to supplement the text are provided and keyed to the appropriate passages.

Since no comparable book on this subject has been published there were no guidelines to follow. The experience of the contributors who are all active in practical communication system engineering has strongly influenced the selection of the material presented. As editor I am indebted to them for the talent and interest they have brought to bear in producing this handbook.

Donald H. Hamsher

ix

CONTENTS

COMMUNICATION SYSTEM ENGINEERING HANDBOOK

Chapter 1

SYSTEM DESIGN REQUIREMENTS

ROY K. ANDRES, *RCA Communications, Inc.*

Historically, the communications field is rich in imagination, starting with one of the first optical telegraph systems proposed in 1684 by Robert Hooke, an English physicist, to the Royal Society in London. More than a hundred years elapsed before Claude Chappe, a French engineer, constructed over 500 semaphore stations in France which became the first organized communications network.

The next 60 years were studded with experimentation with electric and electro-static telegraphs by many illustrious names culminating with Samuel F. B. Morse's first operating telegraph line in 1845 between Washington and Baltimore. The next century brought many profound changes in the communications art commencing with the transatlantic telegraph cables, followed by the telephone, wireless telegraphy, radio broadcasting, radar, television, and the first voice transmission over a synchronous satellite in 1963. In 1965, the first commercial satellite, "Early-bird," was placed in synchronous orbit, positioned midway between Europe and North America providing over 4 Mc of bandwidth for voice, video, and all forms of record communications.

A complete history of communications may be found in *From Semaphore To Satellite.*[1]

PRINCIPLES

1. The Problem of System Design

Communication-system design is perhaps more of an art than any other branch of engineering when we consider the entire range of disciplines involved in designing and engineering an effective communication system.

If it were possible to write a general equation as a starting point for all communication-system designs, it would most likely contain an unlimited number of terms with many of the terms having a range of apparently satisfactory values rather than a precise single value. We can attempt to reduce this hypothetical equation to one containing the most important elements of the problem and then, by borrowing the flow-charting techniques of computer programming, attempt to reduce each factor to a finite number of possible paths or branch decisions so that we can reach a logical conclusion. While this is the usual intuitive process when evaluating the possible solutions to any problem, the increasing complexity and sophistication of modern communication systems require that an orderly process be set down during the problem-definition stage.

After the communications requirements have been defined and the possible solutions outlined, the system engineer is then faced with the task of selecting one of the various alternatives in each segment of the system. He must evaluate the expected performance through each portion of each communication link, adding the deleterious effects of noise, distortion, and diminishing signal-to-noise ratios while at the same time considering the probabilities of equipment and path failure and the alternate routing available in a complex network.   This process of trading off technical advantages and disadvantages vs. cost may require compromising some of the original requirements in whole or in part, which may or may not be acceptable to the user.

If, for example, the service demands high reliability in a long-haul system with several tandem links, the degrading factors introduced by the equipment and transmission medium of each link should be assumed to add fortuitously in an algebraic fashion to arrive at a proper calculation of the maximum total end-to-end distortion. This method, carried out rigorously, will result in a system design capable of successfully offsetting the worst case condition occurring simultaneously in each communication link.   The process of comparison and compromise in trading one factor for another while weighing all the possibilities affecting total performance is the "art" in communication-system engineering.

2. Characteristics of Communication Systems

The most general definition states that a communication system is one containing an information source, an information sink or user, and a communication link for moving the information or intelligence between the source and the sink.

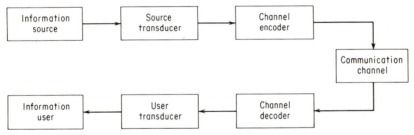

FIG. 1. A basic communication system.

Information or intelligence originates from two basic sources and a quasi third source:

1. Random or formulated thoughts in the brain of man (example—spontaneous speech)

2. Random or predictable changes in the physical world (examples—temperature, pressure, velocity, flow, volume)

3. Information generated as a result of man's stimulus to a physical system (example—unique information generated by a computer or information-processing device as a result of combining known data and programming from man with random data from the physical world)

The basic problem of transmitting intelligence and successfully recovering it from an environment disturbed by noise has been expressed in terms of mathematical models by the work of Claude Shannon in two classical papers, "A Mathematical Theory of Communications" and "Communications in the Presence of Noise."[2,3] Modern communication theory is based very heavily on these mathematical models and, in particular, on the postulate that intelligence can be processed to advantage both before and after transmission through a channel corrupted by noise.   This fundamental precept of manipulating intelligence in terms of the probability of achieving a successful recovery is the foundation of modern "information theory."

Figure 1 shows a block diagram of a basic communication system.   The blocks may be defined in terms of the functions to be performed and the equipment or devices used to carry out these functional requirements.

The "information source" represents the person or device generating the original information to be transmitted by the communication system to the user. It may be the human voice, the scene in a television studio, a message from a teleprinter keyboard, the sound of a violin, information stored in a computer's high-speed memory, or a photograph from a TIROS satellite revealing the earth's cloud cover for a weather map. The "source transducer" is any device capable of being actuated by waves or a source of energy and of supplying waves or energy related to the input. The waves or energy entering the transducer may produce an output which is of the same or different types (e.g., electric, acoustic or mechanical, etc.). Some examples of source transducers are the microphone in a telephone handset, a video camera, a perforated-tape reader, a magnetic-tape read head, and the photoelectric cell scanning a photograph.

The "channel encoder" may be defined as a device which can accept the electrical signals from the source transducer and transform or condition these signals into a form suitable for transmission through the selected communication channel. The channel encoder may be a simple relay repeater, an amplitude or frequency-shift modulator, a voice multiplex and microwave modulator, or a synchronous telegraph time-division multiplex complete with code translator and error-detection-and-correction logic.

Several important differences can be recognized at this point between the transducers and the encoder-decoder blocks in Fig. 1. The transducers are subject to and dependent upon the characteristics of the "information" at the input and output of the system, while the encoder-decoder pair must meet the demands and characteristics of the "communication channel." A second difference, not apparent from Fig. 1, is the frequent requirement for the communication channel to carry several signals simultaneously, each generated by a separate information source and directed to separate "information users." In this case, the channel encoder must include the facilities and equipment for accepting a number of similar or dissimilar signals from a number of source transducers while the channel decoder recovers the signals and delivers them to the proper "user transducer." The channel encoder must modify the signals from each source so that there is a distinctive and detectable difference among the several signals at the receiving end of the communication channel. This coding process, which identifies subchannels within the whole communication channel, must not in any way alter or destroy the sense of the original information from each source. Similarly, the channel decoder must be designed to include the additional function of selectively filtering out each signal intended for a discrete information user as well as transforming each signal back into a form suitable to the particular "user transducer." This process of "modulation" and "demodulation" is covered in subsequent chapters.

If the sending and receiving points (information source and user) are reasonably close (the distance depending upon the information rate), the transducer-encoder as well as the transducer-decoder functions can be performed by a single device between the source and the channel and between the channel and the user.

For example, a telephone or telegraph circuit transmitting over metallic wires can operate without a channel encoder-decoder if the line resistance and source potential will produce a signal within the lower limits of sensitivity of the user transducer. In this case, the demands of the source and user can be satisfied without elaborate signal conditioning if the degradation of the communication channel will not render the signal unusable.

The channel decoder must be able to accept the signal delivered by the communication channel and transform or condition this signal into a form suitable to the user transducer. The channel decoder may be a simple repeating device such as a biased polar relay; a voice channel separation network; an amplitude, phase, or frequency detector; or a receiving multiplex terminal complete with code translator, error-detection-and-correction logic, and synchronous regeneration. The user transducer must transform the signal delivered by the channel decoder and be able to deliver electrical, optical, acoustical, mechanical, or any other form of energy acceptable to and called for by the information user. The user transducer might be a loudspeaker, teleprinter television receiver, magnetic-tape recorder, an analogue signal for input to a computer, or a solenoid for operating a control valve in a pipeline.

While the user transducer usually supplies a physical replica of the original information, the final intelligence delivered to the user may be intentionally converted to a different form by the user transducer. For example, coded information read from punched cards at the source may produce magnetic tape or perforated tape or may pass as electrical signals directly into a computer memory by utilizing a different user transducer and/or a different code converter in the channel decoder. A more complex example made possible by the full exploitation of information theory and the integration of communications and stored-program computers is a communication system having the human voice as the information source, a voice encoder which digitizes the voice so that the computer can recognize the words and translate them into a different language. The translated words are again digitized for transmission through the communication channel and are processed by similar language-recognition equipment at the receiving end. In the decoding process, the words may be translated again to

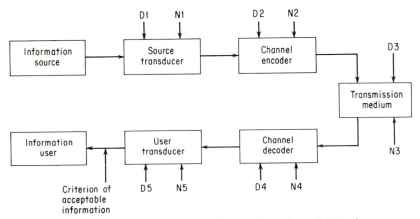

Fig. 2. A communication system disturbed by noise and distortion.

some other digital code and cause a typesetting machine to prepare a press to print words in one language which originally entered the communication system as spoken words of another language.

3. Criterion of Acceptable Information

Now that the basic communication system has been defined and examined, some consideration should be given to the principal factors which can disturb, distort, or otherwise corrupt a typical communication system. Figure 2 shows a communication system disturbed by noise and distortion from all possible sources. Before examining the various types and sources of noise and distortion, we should consider the criterion of performance expected from a communication system. The "criterion of acceptable information" is the minimum amount of information that will satisfy the user or conversely the maximum amount of degradation which the original information can be permitted to suffer and still satisfy the user. The establishment of this criterion depends largely on the type and purpose of the communication service being rendered. The absolute values of acceptability will vary widely from one communication requirement to another, which, in turn, will profoundly influence the demands placed on the terminal equipment and the performance of the communication channel. In the case of speech communication, the system acceptability may be judged differently in accordance with the user's purpose, as shown in Table 1. The transmission characteristics required to satisfy the criterion of acceptability for the various voice services shown will vary widely with respect to the standards of amplitude and frequency distortion, acceptable volume stability, background and impulse noise, the need for echo suppression and frequency discrimination. If we considered the range and discriminatory capabilities of human hearing as the standard for acceptable voice transmission, modern technology would be hard pressed to furnish this fidelity at double the present service cost. Table 2 indicates some of the characteristics of human hearing which would be technically difficult or economically impractical to provide in a worldwide

Table 1. Range of Acceptability in Speech Communication

Voice requirement	*Criterion of acceptability*
Public and private telephone	User expects "discrete word and sentence intelligibility," including preservation of sufficient tonal quality to permit recognition of the particular voice and the transient emotions of the speaker
Public utility and industrial voice control dispatch, police, fire, mobile, fleet control, talker channels for communicatains maintenance	"Discrete word intelligibility" is desirable but "discrete sentence intelligibility" may be acceptable. Recognition of a particular voice or tonal quality not essential
Radio broadcast (program transmission)	The voice channel must pass the full range of articulation of the human voice as well as the frequencies required to reproduce music. This requires from two to three times the bandwith necessary for discrete word intelligibility

Table 2. Some Characteristics of Human Hearing

Characteristic	*Range or discriminatory power*
Volume	0 to 120 db, a range of 10^{12} in sound power.
	0 db = a mosquito buzzing at 6 ft
	90 db = the noisiest spot at Niagara Falls
	115 db = hammering on steel plate at 2 ft
	120 db = threshold of pain
Detectable volume variation	At 1,000 cycles and 5 db above threshold, the least detectable change is 3 db. At 1,000 cycles and 100 db above threshold, the least detectable change is 0.25 db
Frequency discrimination (pure tones)	At 20 to 30 db above threshold and frequencies below 1,000 cycles, a change of 3 cps is detectable
Number of discrete detectable frequencies	1,400 discriminably different pitches of sound

Distortion perception	Transmission bandwidth	% sine-wave distortion	
		Perceptible	Objectionable
	15,000	0.7	2.6
	10,000	1.0	4.0
	5,000	1.2	8.0
	3,000	1.4	18–20

telephone network. It is fortunate for the communication engineer that the percentage of distortion which is objectionable to the average listener increases to 20 per cent when the bandwidth of the speech channel is reduced to 3,000 cycles. This anomaly in human hearing along with many others uncovered by psychoacoustic research has led to the conclusion that "we hear what we expect to hear." A complete treatise on human hearing will be found in Ref. 4. Voice-transmission systems and standards will be found in later chapters as well as Ref. 5. The principal acoustic effects relating to voice transmission are described in Chap. 3.

Similar ranges of the criterion of acceptable information can be established for each basic type of communication. In the transmission of record information by telegraphic techniques, the criterion of acceptable error rates depends on the type of traffic, the end use of the message, and the degree of redundancy in the text. A social message conveying anniversary greetings in plain language contains sufficient redundancy within the words and sentences to permit an error rate of 1 character in every 6 without losing the sense of the message. On the other hand, a business message directing the sale of securities or the transfer of funds requires an error rate of 3 characters in 100,000 and a collation by repetition of the key alphanumeric characters to ensure the accuracy of the resulting transaction. The acceptability of a half-tone photograph transmitted by a communication system involves subjective judgment, which may vary from one individual to another. The acceptability of a black-and-white weather map, however, can be measured precisely in terms of the legibility of the lines and the readability of the alphanumeric and special weather symbols.

4. Distortion and Noise

Referring again to Fig. 2, we can examine in some detail the nature of the various types and sources of distortion and noise and their probable effect on the overall communication-system performance.

The word "noise" in the field of communication has such a broad base and variety of manifestations that it is difficult to affix a generalized definition. One of the many standard definitions[6] states that "noise is any unwanted disturbance within a useful frequency band, such as undesired electric waves in any transmission channel or device." Distortion may be broadly defined as "any undesired change in wave form." Figure 2 indicates a noise factor entering each portion of the system labeled N_1 through N_5 and an associated distortion factor labeled D_1 through D_5. If we consider all the noise and distortion factors within each portion of the system, we can proceed from block to block through the system and determine the probability of meeting the criterion of acceptable information established by the user.

If the system is without any form of selective amplification, regeneration, or signal-restoration devices, the distortion factors can be added algebraically and the noise factors combined in accordance with their fortuitous or periodic occurrence. This summation will determine the probability of recovering acceptable information when discriminating between the signal and noise power at the receiving end of the system. Since selective amplification and regeneration techniques are freely employed in all but the most elementary systems, we shall consider a system which includes successive amplification, wave shaping, and regeneration techniques.

The information source of Fig. 2 is assumed to inject distortionless and noise-free intelligence into the source transducer, although, in some cases, an information source device may deliver distorted information via a noisy input circuit. Since the information source is usually completely under the control of the system designer, every attempt should be made to deliver a perfect signal to the input transducer, as subsequent compensation and correction are more difficult and costly.

The first determination of noise and distortion is usually an assessment of the quality of the transmission medium and the probable distortion and noise which will be contributed by the selected mode of transmission. The difference between the distortion and noise contributed by the transmission channel and the maximum acceptable levels of noise and distortion at the input to the channel decoder is equal to the maximum values of distortion and noise which can leave the channel encoder and enter the transmission medium. Since the source transducer factors D_1 and N_1 are usually dictated by the requirements of the information source and the characteristics of given devices at the input, the signal quality entering the channel encoder is substantially known. If the channel encoder equipment cannot accept the output signal of the source transducer, then amplification, wave shaping, or regeneration must be employed at the input to the channel encoder. If the channel encoder cannot be made to deliver the required signal fidelity upon entering the transmission medium, then an alternate transmission path or the provision of an intermediate relay point must be considered. Similarly, if the channel decoder can successfully recover the intelligence but not deliver an acceptable signal to the user transducer, amplification, wave shaping, or regeneration must be added to the output of the channel decoder.

When considering the location of any signal-restoration device such as amplifiers, wave shapers, or regenerative repeating equipment, it is a cardinal rule to insert such devices as close as possible to the source of distortion or noise. For example, if the signal entering the transmission medium is substantially degraded, it is far better to locate the signal-restoration equipment at the entrance to the transmission channel rather than require such devices to correct the cumulative distortion and noise appearing at the output of the transmission channel.

In many cases, the proper location of amplifiers and signal-restoration equipment can mean the difference between a marginal communication system and one which is

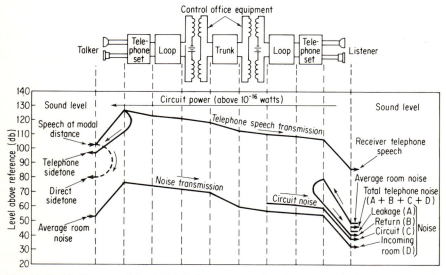

FIG. 3. Level diagram of typical telephone connection (302-type telephone set.) (*Bell Telephone Laboratories, Transmission Systems for Communications*, p. 57, *Bell Telephone Laboratories, Inc.*, 1964. *By permission.*)

considered highly reliable. This is particularly true in binary transmission systems using pulse sampling techniques to test the signal for one of two possible conditions. If the margin at the sampling point is very narrow owing to uncontrollable factors such as propagation multipath, the addition of 2 or 3 per cent distortion because of an improperly located signal-restoration device can increase the bit error rate by several orders of magnitude, rendering the system unusable.

In voice transmission, the elimination of noise is becoming steadily more difficult because of the increasing level of background room noise at both ends of the circuit, the steady improvement in transmission quality, and the overall reduction in circuit losses as newer equipment is added to the network.

The total "message circuit noise" which reaches the listener's ear as either an annoyance factor or reduced intelligibility can be controlled by the introduction of repeaters or amplifying devices. If they are introduced at the proper points in the system, the signals will be of sufficient amplitude to override the noise at the points where noise is introduced. Similarly, power-supply noise and battery noise can be more effectively controlled by filtering at the source rather than attempting to eliminate such noise after it has entered the voice channel and mixed with the intelligence. Whenever noise can be traced to the point of entry in any system, it is far less costly and technically more feasible to eliminate it at the source or point of entry rather than permit the noise to be amplified and possibly propagated over multiple paths in the system.

Figure 3 shows a speech and noise level diagram of a typical telephone connection. The reduction of noise by repeaters which discriminate against it and the natural filtering of high-frequency noise by the frequency cutoff characteristics of the system are evident from the level diagram. Construction of level diagrams which include the worst path through a switched network permits a final assessment of the overall signal-to-noise ratio expected at the listener's location. A more detailed description of the various forms of noise and distortion, including their definitions and methods of correction, will be found in the section Noise and Distortion Types.

DEFINITION OF THE COMMUNICATION REQUIREMENT

5. A Flow Chart for Communication-system Design

The "definition of the communication requirement" is the fundamental first step in system design and, in many cases, the most difficult task. Since most communication-system designs involve a great many factors, involved technical evaluations, and time-consuming cost comparisons, the importance of starting with a precise definition of the user's requirements and communication objectives cannot be overemphasized.

Earlier in this chapter, we considered that it would not be practical to write a general equation as a starting point for all communication-system designs; however, we can establish a procedure which will cover the essential steps in an orderly fashion. Figure 4a and b shows a "flow chart for communication-system design" which establishes a procedure applicable to most problems.

Starting at the top of Fig. 4a, the fundamental question is posed "What is desired?" This is the problem-definition stage, requiring a clear understanding of the form of "intelligence to be moved" (i.e., voice, telegraph, facsimile, video, etc.) and a statement of the overall "communication objectives" in terms of the characteristics of the "source" of this intelligence and what use the "sink," or end user, intends to make of the intelligence. If the communication requirement is part of a corporate or government plan, the policies of the organization and the short- and long-range plans for future communications become a vital part of the problem-definition stage, since they affect the final system requirements for flexibility, growth, and new services. When the requirements and objectives have not been clearly defined, the chart shows a return to management or the user for clarification until the system designer is satisfied that the basic criteria for the design are sound and defined as far as possible. If certain factors cannot be adequately defined, the known requirements can be carried through the evaluation process while allowing a range of values to be applied to the unknown factors. This will complicate the evaluations and may, in some cases, make them inconclusive; however, a range of solutions or a qualified solution may be useful to the user by assisting him in defining the final communication requirement.

The next question to be asked is "how much" communication is required. This step requires the accumulation of traffic statistics and load distributions over the applicable time period and predictions of future requirements by extrapolating past growth or other influential factors. In a complex communications network involving large quantities of users, lines, switching points, and interconnecting trunks, the traffic analysis is often so time consuming that it proceeds continuously long after the system design is finished and is thus setting the parameters for the next stage or subsequent expansion of the overall network. In this event, the present system design or expansion of an existing system must start with the known present traffic volume and correlate the expected volume growth during the desired period.

At this point, the "objectives" have been defined and the "quantity" of communications determined, which permits an examination of the more detailed question of

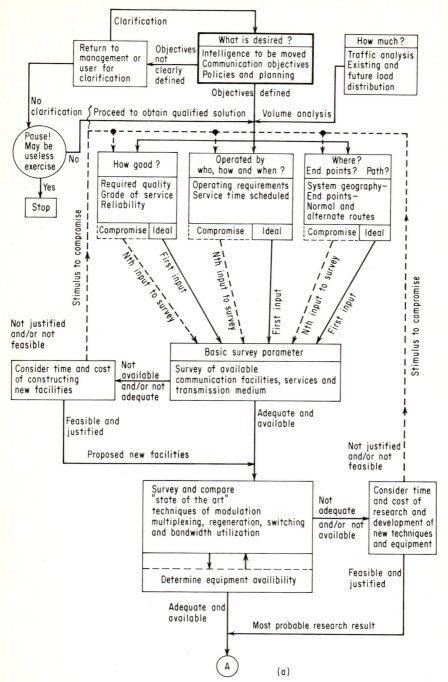

Fig. 4. Flow chart for communication-system design.

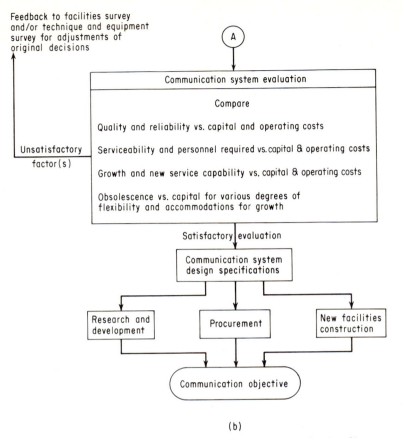

(b)

Fig. 4. Flow chart for communication-system design. (*Continued.*)

"how good" the service must be in a qualitative sense. This stage of the process involves setting down the original objectives with the known volume requirements in terms of the expected or desirable technical performance. It is also possible at this stage to establish the operating requirements for the system and to establish the desired service schedule as well as the system geography. Consideration of each of these parameters will give rise to ideal values or most desirable objectives which are carried forth into the first survey of available facilities, services, and transmission mediums. Figure 4a indicates that, if this first survey shows that existing facilities are not adequate or not available, the system designer must consider the time and cost of constructing new facilities. If this is not justified or feasible, then the original ideal parameters must be compromised in a fashion which will match the available facilities without sacrificing the original objectives. This process may involve several compromises and deviations from the original objectives but all justified by avoiding the time and cost of constructing new facilities.

When the question of facilities has been decided, the original parameters or compromised parameters, as the case may be, are carried into the next stage, that of surveying the "state of the art" techniques of transmission, reception, and signal processing to determine if existing techniques are adequate. If the techniques are not adequate or the equipment is not available, consideration must be given to the time and cost of research and development of new techniques and equipment.   Again, if this is not justified, the basic parameters may have to be compromised still further in order to remain within the time allotted to provide the service or the budget allocated to the project.   Finally, when the techniques have been decided and the equipment has been selected, all factors are examined in a final "communication-system evaluation."   In this final evaluation, all factors are examined again by comparing the quantitative and qualitative aspects to the initial capital cost requirements and the costs of maintaining satisfactory service.   The factor of obsolescence is examined to determine in the last analysis what additional capital, time, and manpower can be justified to add a given degree of flexibility and accommodations for growth and new service requirements.   With these factors satisfactorily evaluated, the "communication-system design specification" can be completed and the "communication objective" reached.

6. The Parameters of System Design

The flow chart of Fig. 4a and b requires certain fundamental parameters to be established to permit the evaluation and determination of facilities and equipment.   These parameters, like the terms of any equation, must be evaluated if the process is to lead to a determinate result.

The following fundamental parameters of system design are listed in the usual order of consideration:

1. Intelligence to be moved, communication objectives
2. Traffic analysis—volume and loading requirements
3. Criterion of acceptability—quality and grade of service
4. Operating requirements—service schedule
5. Initial survey—geography of the system
6. Facilities survey—available services, facilities, and transmission mediums
7. Survey of communication techniques and equipment
8. Service quality and reliability vs. cost
9. Serviceability and personnel requirements vs. cost
10. Volume growth, new services, and future prospects vs. cost
11. Obsolescence

7. Intelligence to Be Moved, Communication Objectives

To establish the type and variety of "intelligence to be moved" and the "communication objectives," the system engineer must establish the following facts:

1. The reason for the communication in terms of what useful purpose will be made of the information at the receiving end.

2. Will the successful transmission and reception of information A result in a second transmission of information B in the reverse direction, either manually or automatically?   If not, a simplex or one-direction communication system can be considered unless the desired accuracy and reliability dictate the use of error-detection and -correction techniques.   If the error-detection and -correction technique utilizes repetitive transmission by request from the receiving end, a reverse communication path or duplex facility must be provided.

3. If voice service is not required, what minimum amount of intelligence, using coding techniques, will satisfy the purpose of the user without jeopardizing the correct interpretation of the intelligence at the receiving terminal?

4. What is the maximum amount of time which can be tolerated to transmit and receive successfully the given amount of intelligence in 3 above?   Defining the minimum useful intelligence in the maximum tolerable time will, in general, determine the bandwidth of the required system.

5. What time of the day must the intelligence be moved, and can the intelligence be stored for a period prior to transmission, or must it be moved as fast as it is generated on a continuous basis?   These factors will determine if the system must operate on a "real-time" basis or if a "store and forward" system will meet the objectives.

6. If the intelligence must be moved in real time or quasi real time, can the periodic delays encountered in the public switched networks be tolerated? If not, leased dedicated channels in the public network or private communication facilities must be considered.

7. Can the new communication requirement being considered be combined with existing communications to effect a combined service at lower cost and/or an improvement in the technical and operating performance of both services? If a combination of several communication services such as voice and data, telegraph and data, etc., is technically and operationally feasible, do the tariff provisions of the public domestic services and/or the international record common carriers permit such combinations?

8. If the communication service extends to overseas points, will the available facilities and networks of the overseas PTT (Post Telephone & Telegraph) or private operating agencies abroad match the facilities of the United States international common carrier? This determination is best carried out by the international common carrier who is in a position to negotiate and secure the necessary facilities and interconnections with other governments and private operating agencies.

After considering these basic factors and the capabilities of the available public and private facilities, the user may wish to consider altering the communication requirement to match the available facilities, services, and transmission mediums.

8. Traffic Analysis

In order to determine "how much" intelligence the system must handle, it is essential to make a thorough analysis of the traffic volume and patterns of distribution. After the communication objectives have been clearly defined, the immediate and future traffic-handling requirements can be determined from the following traffic analysis factors and procedures:

1. Establish a suitable "communication unit" (usually termed a "message") which can be used throughout the traffic analysis. If several types of communication are required, it may be necessary to establish separate communication units for each distinct type of service.
2. Define the message as completely as possible by considering the structure of the communication event in terms of:
   a. The total time required to move a message from the source to the sink or user. If the public network is involved, the time to dial and establish the connection, the time to carry out acknowledgement procedures, and the disconnect time should be included.
   b. If portions of the message are fixed by traffic procedures or established format conventions define the message in terms of the fixed and the variable areas of intelligence. A typical message in telegraph communication might be defined as follows:

Fixed Areas

Header   Start-of-message indicator, channel sequence number, origin reference number, destination code, priority symbol, chargeable class-of-service symbol, origin code, customer accounting code plus standard spacing and paper positioning functions = 52 character intervals.

Closure   End-of-text symbol, fixed length collation or parity code, message separation functions, end-of-message symbol = 26 character intervals.

Semifixed Areas

Address   A clear language address not to exceed three full lines, multiple-address codes not to exceed 24 letter codes, alternate routing codes, and instructions not to exceed 60 character intervals.

Variable Areas

Text   The text may be completely variable with no restrictions on length or may require separation into pages of n words, blocks of

m characters, or transmission units of t sec. When text separation is required, there are usually separation codes between sections and sometimes a repetition of certain key information from the original header between each group of x sections.

3. Determine the distribution of message volume over various periods of time, and plot a complete profile of the traffic load by sampling several average days, peak days, and below-normal days. If the traffic patterns show distinct peak periods which appear repetitive or cyclic in occurrence, examine a number of such peaks to establish the start-of-peak and end-of-peak time, the maximum demand for service during the peak, and the probability of occurrence. Traffic load profiles in the international public networks clearly show the effects of overlapping and nonoverlapping office hours when the communication service spans many world time zones.

4. If priority classifications are used, establish the desired speed of service and maximum permissible delay for each priority. The time factors associated with various message priorities are particularly important in the design of large multiple-trunk, central-office switching systems. If a superhigh priority exists, determine if it is to be handled on a priority interrupt basis (seizing the line and interrupting the message in progress) or by jumping the message over all others in the queue and transmitting it immediately following the message in progress.

5. Determine the growth pattern from previous traffic statistics, and extrapolate forward in time, adding or subtracting from the normal growth pattern according to special circumstances or changes in the state of the art which are expected in the years ahead.

6. Consider the changes in traffic patterns which may result from adding or subtracting relay points or transit switching centers in the overall network. Similarly, consider the effects of new transmission facilities such as satellite networks which may bypass existing cable routes, permitting a redistribution of traffic loads within the networks.

7. Examine the possibility that the standard communication unit, or message, may evolve into a substantially different size or structure and thus affect the philosophy of the overall system design.

9. Criterion of Acceptability—Quality and Grade of Service

In order to determine "how good" the communication service must be, the minimum acceptable level of intelligence and grade of service must be established. Although some of the factors which determine the criterion of acceptability, as well as several examples, appear elsewhere in this chapter, there are additional factors to be considered when defining the required quality and grade of service.

Each basic type of communication service has a set of established quality factors and standards of measurement which should be examined to determine the degree of applicability to the system under consideration. A complete treatise on this subject is beyond the scope of this chapter; however, as a general guide, the following list of prime factors will lend direction to the task of establishing the required service quality assisted by the references, standards, and bibliography at the end of the chapter.

1. With the communication objectives firmly fixed, the required criterion of acceptability can be specified by referring to the existing standards of performance (Table 3) which pertain to the particular class of communication under consideration and modifying these to suit the given objectives. In each of the classifications given, the prime factors are listed with their CCITT recommendations and ASA definition numbers. These references will lead, by cross-referencing, to subclassifications and secondary factors within the definitions of terms and standards of measurement found in Refs. 5 and 6 and elsewhere in this Handbook.

2. The term "grade of service" has several connotations in communication systems, varying in definition from networks using line switching to those using store-and-forward or message-switching techniques. The grade of service in telephone systems usually refers to the percentage of completed calls which will fall within a certain

Table 3. Appropriate Existing Technical Standards

	CCITT recommendation	ASA definition
Voice Communication		
Relative interfering effect	Vols. III, V	65.08.189
Noise	Vols. III, V	65.08.192
Disturbance	Vols. III, V, VII	65.08.224
Cross modulation	Vols. III, V	65.08.235
Crosstalk	Vols. V, VII	65.08.252
Fidelity	Vol. V	65.08.318
Distortion	Vols. III, V, VII	65.08.321
Master reference system for telephone transmission	Vols. III, V	65.08.500
Articulation (intelligibility)	Vol. V	65.08.533
Record Communication—Telegraph and Data Transmission		
Gross information content	Vol. VII	65.02.240
Redundancy	Vol. VII	65.02.246
Resolution	Vol. VII	65.02.348
Telegraph signal distortion	Vol. VII, Part I	65.02.348
Range	Vol. VII, Part I	65.26.140
Margin	Vol. VII, Part I	65.26.141
Error rate	Vol. VII, p. 45, pp. 173–179, pp. 466–488	
Reliability of the domestic and international public networks	Vol. VII, Part I	
Record Communication—Facsimile and Phototelegraph		
Speed equalization	Vol. VII, T.1	
Index of cooperation	Vol. VII, T.1	
Drum factor	Vol. VII, T.1	
Drum and picture standards	Vol. VII, T.1	
Drum rotation speed	Vol. VII, T.1	
Equalization of speeds	Vol. VII, T.1	
Phasing	Vol. VII, T.1 and 11	
Resolution, fidelity	Vol. VII, Recommendation, T.20, Test Chart, p. 112	
Television		
Geometric distortion	Vols. III, IV	65.32.150
Keystone distortion	Vols. III, IV	65.32.153
Contrast	Vols. III, IV	65.32.162
Gray scale	Vols. III, IV	65.32.164
Resolution wedge	Vols. III, IV	65.32.170
Signal-to-noise ratio	Vols. III, IV	65.32.190

quality specification with respect to received volume, distortion, noise, and fidelity. The term is also commonly used in association with line-switching systems to define the probability factor that, at any given time t, point A will be able to call point B and obtain a satisfactory connection. Thus, in the telephone network, the grade of service refers to the quality of an established connection as well as the probability of making the connection in a given switched network.

In record communication systems, the term grade of service in store-and-forward or message-switching systems usually denotes the average time to forward a message through the system, measured from the time the end of the message enters the system at the sending point until the message is completely received by the addressee. It may also be expressed as a probability of achieving delivery in less than time t, where t equals the average time for a message to move from the sender through the network to the addressee.

When the term is used in association with line-switching systems carrying record communications such as the international telex network (telegraphic conversation by direct interconnection of teleprinters), it usually refers to the probability of establishing a successful connection through the network at a given time.

The grade of service or probability of lost calls in a line-switching system can be predicted by utilizing the classical formulas of Erlang, which permit the calculation of loss probability when the traffic to be carried (in erlangs) and the number of available circuits is known. See also Chap. 7. Table 4, which includes Erlang's formula, and Figs. 5 and 6 illustrate the method of calculating the number of circuits required

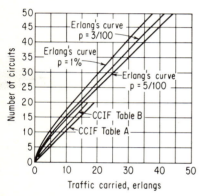

FIG. 5. Relation between the number of circuits and the traffic in erlangs which can be carried by these circuits, in the case of the Erlang formula ($p = 1$, 3, and 5 per cent), the CCIF, Tables A and B. (*CCITT Red Book, Vol. II bis, Geneva, 1961.*)

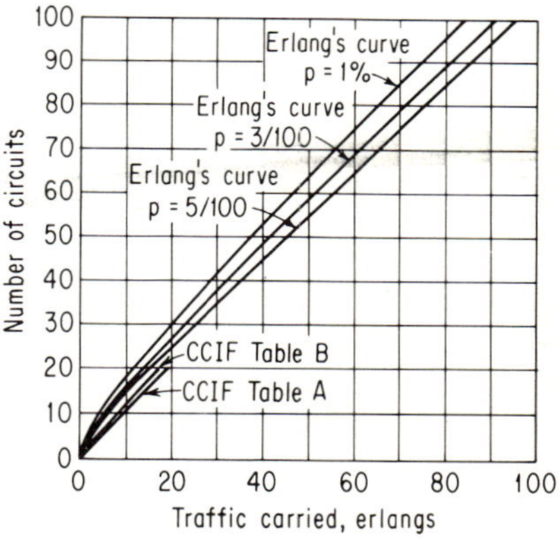

FIG. 6. Relation between the number of circuits and the traffic in erlangs which can be carried by these circuits, in the case of the Erlang formula ($p = 1$, 3, and 5 per cent), the CCIF, Tables A and B. (*CCITT Red Book, Vol. II bis, Geneva, 1961.*)

in a typical telephone-line switching network to render successful connections in 99, 97, and 95 per cent of the first attempts or lost-call probabilities of not achieving a connection of 1, 3, and 5 per cent. The method can be used for any type of communication system having a large number of subscribers sharing a smaller number of circuits, providing the characteristics of the average communication unit, or message, is known. See Ref. 5, Vol. II bis, Sec. 5, pp. 133–156 for additional information on the use of the Erlang formula and a second method, termed the "Swedish method," which includes:

Determination of the number of direct circuits for the most economical arrangement
Calculation of the number of circuits in the overflow group
Calculation of the number of circuits required of the alternate route(s) when the primary route is unable to accept additional calls

10. Operating Requirements—Service Schedule

Following the determination of communication objectives, traffic loading, and required grade of service, the operating requirements and service schedules can be

Table 4. Erlang No. 1 Formula for Loss Probabilities of 1, 3, and 5 Per Cent*

Formula:

Let p = loss probability
   y = the traffic to be carried (in erlangs)
   n = the number of circuits

$$E_{1,n}(y) = p = \frac{y^n/n!}{1 + y/1 + y^2/2! + \cdots + n^n/n!}$$

n	$p = 1\%$	$p = 3\%$	$p = 5\%$	n	$p = 1\%$	$p = 3\%$	$p = 5\%$
1	0.01	0.03	0.05	51	38.80	42.89	45.52
2	0.15	0.28	0.38	52	39.70	43.84	46.52
3	0.46	0.715	0.90	53	40.60	44.80	47.53
4	0.87	1.26	1.52	54	41.50	45.77	48.53
5	1.36	1.875	2.22	55	42.41	46.73	49.53
6	1.91	2.54	2.96	56	43.31	47.69	50.52
7	2.50	3.25	3.74	57	44.22	48.66	51.52
8	3.13	3.99	4.54	58	45.13	49.62	52.50
9	3.78	4.75	5.37	59	46.04	50.6	53.5
10	4.46	5.53	6.22	60	46.95	51.5	54.5
11	5.16	6.33	7.08	61	47.86	52.5	55.5
12	5.88	7.14	7.95	62	48.77	53.4	56.5
13	6.61	7.97	8.83	63	49.69	54.4	57.5
14	7.35	8.80	9.73	64	50.60	55.4	58.5
15	8.11	9.65	10.63	65	51.52	56.3	59.5
16	8.87	10.505	11.54	66	52.44	57.3	60.5
17	9.65	11.37	12.46	67	53.35	58.3	61.5
18	10.44	12.24	13.38	68	54.27	59.2	62.5
19	11.23	13.115	14.31	69	55.19	60.2	63.6
20	12.03	14.00	15.25	70	56.11	61.2	64.6
21	12.84	14.885	16.19	71	57.03	62.1	65.6
22	13.65	15.78	17.13	72	57.96	63.1	66.6
23	14.47	16.675	18.08	73	58.88	64.1	67.6
24	15.29	17.58	19.03	74	59.80	65.1	68.6
25	16.12	18.48	19.99	75	60.73	66.0	69.6
26	16.96	19.39	20.94	76	61.65	67.0	70.7
27	17.80	20.305	21.90	77	62.58	68.0	71.7
28	18.64	21.22	22.87	78	63.51	69.0	72.7
29	19.49	22.14	23.83	79	64.43	70.0	73.7
30	20.34	23.06	24.80	80	65.36	70.9	74.7
31	21.19	23.99	25.77	81	66.29	71.9	75.8
32	22.05	24.91	26.75	82	67.22	72.9	76.8
33	22.91	25.84	27.72	83	68.15	73.9	77.8
34	23.77	26.78	28.70	84	69.08	74.9	78.8
35	24.64	27.71	29.68	85	70.02	75.9	79.9
36	25.51	28.65	30.66	86	70.95	76.9	80.9
37	26.38	29.59	31.64	87	71.88	77.8	81.9
38	27.25	30.53	32.63	88	72.81	78.8	82.9
39	28.13	31.47	33.61	89	73.75	79.8	84.0
40	29.01	32.41	34.60	90	74.68	80.8	85.0
41	29.89	33.36	35.59	91	75.62	81.8	86.0
42	30.77	34.30	36.58	92	76.56	82.8	87.0
43	31.66	35.25	37.57	93	77.49	83.8	88.1
44	32.54	36.20	38.56	94	78.43	84.8	89.1
45	33.43	37.15	39.55	95	79.37	85.7	90.1
46	34.32	38.11	40.54	96	80.31	86.7	91.1
47	35.21	39.06	41.54	97	81.24	87.7	92.2
48	36.11	40.02	42.54	98	82.18	88.7	93.2
49	37.00	40.97	43.54	99	83.12	89.7	94.2
50	37.90	41.93	44.53	100	84.06	90.7	95.2

* *CCITT Red Book*, vol. II bis, Geneva, 1961.

established with considerable accuracy. The major considerations are:

1. Separate the traffic operations and the technical operations (which may or may not include maintenance), and list the major tasks in each operating class.

2. Establish the capabilities and skills of the personnel available for the various operating tasks.

3. Determine the likelihood of obtaining personnel with the necessary skills over the projected life of the system. If difficulties are foreseen, estimate the probable additional cost of training, which can be equated with the cost of simplifying the operating requirements by automating the manual functions and maintenance routines.

4. Utilize, if possible, any existing procedures, formats, and operating recommendations whenever they exist to reduce personnel training and simplify interconnection with existing services and other networks.

5. Anticipate any changes in required operating skills which are likely to occur as the service expands in scope and traffic volume.

6. Establish a "service schedule" graduated in terms of when the service is desirable, necessary, essential, and critical. This will serve as the basic guide when establishing the priority, alternate routing, and fallback philosophies.

7. Determine if traffic of one or more priorities can be stored at the source or at some intermediate transit point for a period during peak-hour conditions. Provision of storage facilities for less important traffic can make the difference between a highly acceptable grade of service and a totally unsatisfactory one for higher traffic priorities, often without affecting the timeliness or usefulness of the deferred traffic.

8. Establish overall operating procedures and personnel-utilization plans for light, average, and heavy traffic periods. Evaluation of these plans and procedures may indicate the desirability of automatic equipment and system changes to follow the changes in load.

11. Initial Survey—Geography of the System

The geography of the system or physical extent of the desired communication services will, in many cases, decide whether existing public or private communication facilities can be used to provide all or a major portion of the transmission and reception channels. The initial survey involves the following steps:

1. Lay out a network which includes the initial origin and destination points as well as origins and destinations which may be added in the future.

2. Establish the most desirable primary and alternate circuit routing between all points considering:

    a. Available public communications networks as shown on national and international facilities maps. United States domestic facilities are available from the Federal Communications Commission, the American Telephone & Telegraph Co. and affiliated private telephone companies, and the Western Union Telegraph Company. International facilities can be determined through one of the United States international common carriers; ITT World Communications, Inc.; RCA Communications, Inc.; and Western Union International, Inc. The existing and planned international submarine telephone cable facilities are shown in Tables 5 and 6. Extensive maps of the world's HF point-to-point radio and microwave facilities as well as existing and future traffic volumes can be obtained from the *General Plan for the Development of the International Network*, prepared by the planning committee of the CCITT.[7] See Tables 5 and 6 and Figs. 7 and 8.

    b. Private networks having available capacity for lease (examine governmental regulations and tariff restrictions concerning sharing of communications facilities).

3. Select several primary and alternate routes considering:

    a. Quality of leased private channels or, if the public switched network is to be used, the grade of service that may be expected during the critical time schedule of the operation.

    b. Reliability of the primary and alternate routes considering the possibilities of facilities failure due to weather, political factors, or national emergencies. Political considerations may be necessary if channel routing is to cross national borders using land-based facilities or when using an intermediate country as a transit center.

    4. Compare the tariff rates of the first- and second-choice primary networks, and weigh against the alternate routing capabilities of each network.

12. Facilities Survey—Available Services, Facilities, and Transmission Mediums

    The "initial survey" will establish the geography of the system and determine, in general, if public or private communications networks can be considered (see Figs. 7 and 8 and Ref. 7). If the existing facilities adequately cover the geography of the

Table 5. Existing Submarine Telephone Cables

Number	Name of cable	Terminals	Date of service	Voice circuit capacity	
1	. . . .	Miami (Fla.)–Havana	1950	24	
2	TAT–1	London–(New York/Montreal)	1956	48	+37*
3	ALGER 1	Marseilles–Algiers	1958	80	
4	TAT–2	(Paris/Frankfurt)–New York	1959	48	+37*
5	. . . .	Jacksonville (Fla.)–Puerto Rico	1960	48	+37*
6	CANTAT	London–Montreal	1961	80	
7	. . . .	New York–Bermuda	1962	80	
8	SCOTICE	London–Faroes	1962	24	
9	SCOTICE	Faroes–Reykjavik	1962	24	
10	ICECAN	Reykjavik–Greenland	1963	24	
11	ICECAN	Greenland–Montreal	1963	24	
12	. . . .	Perpignan–(Mers-el-Kebir/Oran)	1962	80	
13	. . . .	Miami (Fla.)–Jamaica	1963	128	
14	. . . .	Jamaica–Panama Canal Zone	1963	128	
15	TAT–3	London–New York	1963	128	
16	ALASKA	Seattle–Ketchikan	1956	48	
17	HAWAII #1	Oakland–Honolulu	1957	50	+11*
18	COMPAC	Sydney–Auckland	1962	80	
19	COMPAC	Auckland–Fiji	1962	80	
20	COMPAC	Fifi–Honolulu	1963	80	
21	COMPAC	Honolulu–Vancouver	1963	80	
22	TRANSPAC	Tokyo–Guam	1964	138	+28*
23	TRANSPAC	Guam–Wake	1964	138	+28*
24	TRANSPAC	Guam–Manila	1964	128	
25	TRANSPAC	Wake–Midway	1964	138	+28*
26	TRANSPAC	Midway–Honolulu	1964	138	+28*
27	HAWAII #2	Honolulu–Oakland	1964	138	+28*
28	TAT–4	(Paris/Frankfurt)–New York	1965	128	
29	. . . .	Florida–Virgin Islands (U.S.)	1964	128	
30	. . . .	Virgin Islands (U.S.)–Antigua	1965	80	
31	. . . .	Perpignan–Tetouan	1965	80	
32	PENCAN	Cadiz–Canary Islands	1965	160	
33	. . . .	Tunisia–Trapani (Sicily)	1964	60	
34	. . . .	Virgin Islands–Caracas	1965	128	
35	. . . .	Panama Canal Zone–Cartagena, Colombia	1965	80	
36	SEACOM	Jesselton–Singapore	1965	80	
37	SEACOM	Jesselton–Hong Kong	1965	80	

* TASI equipments provide additional voice circuits as shown.

Table 6. Submarine Telephone Cables Agreed or Under Negotiation For Service Before 1968

Number	Name of cable	Terminals	Date of service	Voice circuit capacity
1	SEACOM	Hong Kong–Guam	1966	80
2	SEACOM	Guam–Madang	1966	160
3	SEACOM	Madang–Cairns	1966	160
4	. . . .	Marseilles–Bizerte	1966	80
5	SAFEC	Tokyo–Taipei	Before 1968	128
6	SAFEC	Kaohsiung–Hong Kong	Before 1968	128
7	SAFEC	Kaohsiung–Manila	Before 1968	128
8	SAFEC	Manila–Saigon	Before 1968	128
9	SAFEC	Saigon–Bangkok	Before 1968	128
10	SAFEC	Bangkok–Singapore	Before 1968	128
11	SAFEC	Singapore–Djarkarta	Before 1968	128
12	JASC	Nagasaki–Vladivostok	Before 1968	
13	Commonwealth Cable	Penang–Colombo	1967	80
14	Cable	Penang–Bombay	1967	80
15	Cable	Panang–Karachi	1967	80
16	Cable	Karachi ⎫ Bombay ⎬ –London (via South Africa) Colombo ⎭	1968	80
17	. . . .	Calcutta–Rangoon	1968	48
18	. . . .	Chittagong–Penang	1968	60
19	ALGER #2	Marseilles–Algiers	1966	128
20	CARAVELAS	Rio de Janeiro–Dakar (?)	Before 1968	60 min
21	VERA CRUZ	Belem–Trinidad	Before 1968	60 min
22	. . . .	Canary Islands–Cape Verde Island	?	?
23	. . . .	Dakar–Leopoldville (?)	?	?

desired communication service, as well as technically satisfying the transmission requirements for:

Bandwidth
Amplitude and frequency stability
Impulse noise (and overall signal-to-noise ratio)
Phase delay distortion
Continuity of service
Grade of service (if using the switched network)

then the "facilities survey" can be considered complete.

If this initial survey of existing public and private facilities indicates a lack of adequate facilities between the terminal points of the desired service, then the construction of new facilities must be considered. Since landline, coaxial cables, and submarine cable facilities are usually reserved to the licensed common carriers or government Post, Telephone, and Telegraph departments, consideration will be given to factors involved in site selection for point-to-point radio systems. Using the HF spectrum between 4 and 30 Mc for long-haul, point-to-point communications or tropospheric scatter and microwave networks for short-haul communications, the following factors should be considered when selecting suitable sites for transmitting and receiving stations: (The letters "T" for transmitting station and "R" for receiving station after each factor indicate the importance of the factor to one or both types of HF stations with "M" for microwave and "TS" for tropospheric scatter stations.)

1. Adequate area of fairly level well-drained land with allowance for expansion at a reasonable cost. (T, R, M, TS)

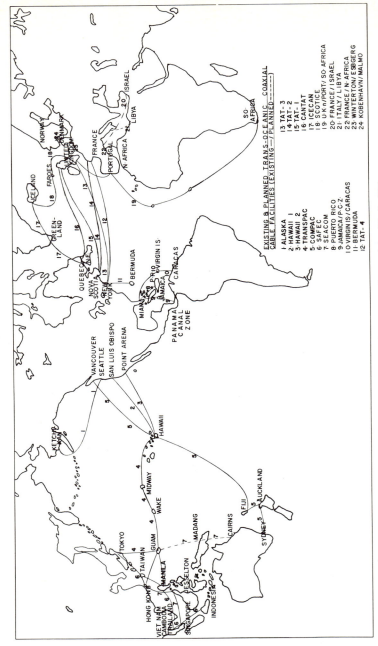

Fig. 7. Existing and planned transoceanic coaxial cable facilities.

EXISTING & PLANNED TRANS-OCEANIC COAXIAL
CABLE FACILITIES (EXISTING—/PLANNED-----)

1· ALASKA	13·TAT- 3
2· HAWAII 1	14·TAT- 2
3· HAWAII 2	15·TAT- I
4· TRANSPAC	16· CANTAT
5· COMPAC	17· ICECAN
6· SAFEC	18· SCOTICE
7· SEACOM	19· U·K n/PORT/ SO·AFRICA
8· PUERTO RICO	20· FRANCE / ISRAEL
9· JAMAICA/ P·C·Z·	21· ITALY / LIBYA
10· VIRGIN IS / CARACAS	22· FRANCE / N·AFRICA
11· BERMUDA	23· WINTERTON/ ESBERG
12· TAT- 4	24· KOBENHAVN / MALMO

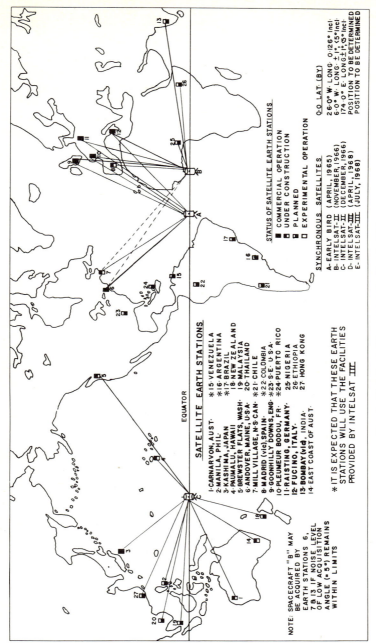

FIG. 8. Global satellite communication facilities.

2. Horizon not over 6° elevation in directions of interest.   (T, R, M)

3. Adequate available public power.   (T, R, M, TS)

4. Available economical signal and control channels between station and operating origin/destination points.   (T, R, M, TS)

5. Reasonably close to a good road, access to an adequate water supply and sewage disposal facilities (desirable but not essential for M and TS).   (T, R)

6. A substantial distance (10 to 30 miles) from existing or planned radio transmitters or sources of substantial electrical interference.   (R)

7. A substantial distance from a radio receiving station.   (T)

8. A substantial distance from a highly developed residential or industrial area which may be a source of radio or television interference complaints.   (T)

9. Convenient residential area for staff.   (R, T, M, TS)

10. A substantial distance from all airports and airway marker beacons, homing stations, and aircraft holding pattern areas.   (R, T, M, TS)

11. If station will use intermediate frequencies (500 kc), it should be close to the shore to provide an all-over-water propagation path, avoiding the attenuation over land.   (R, T)

12. Reasonably good soil conductivity throughout the year to facilitate adequate electrical ground system.   (R, T)

A complete set of standards for radio communication including HF radio, microwave, tropospheric scatter, satellite communications and aircraft and marine communications will be found in Refs. 8 to 10.   (See also Chaps. 15, 16, 17, 19, and 23.)

13. Survey of Communication Techniques and Equipment

With the basic survey parameters in hand, the total evaluation of the available communication facilities, services, and transmission mediums will lead to a conclusion that the available facilities are either adequate and available or not available and/or not adequate as indicated in Fig. 4a.   If they are adequate and available, the next step is to survey the communication techniques and equipment to determine the following:

1. Determine the most desirable form of modulation to render the quality of service desired over the available transmission facilities.   If the transmission facilities are substantially high-quality coaxial-cable channels, the inherent stability and low noise level will permit complete freedom in the selection of the type of modulation.   On the other hand, if the communication path includes a high-frequency radio link, double-sideband amplitude modulation and pulse time modulation are usually excluded from consideration because of the excessive amplitude changes caused by radio path fading and pulse length variations caused by propagation multipath.[11–14]

2. Determine the desirability of multiplexing to conserve bandwidth and to obtain the inherent regeneration advantages of converting to a synchronous system.   Time-division systems, while more costly to construct, conserve bandwidth most efficiently, while frequency-division systems require a guard band between successive channel filters, which is unusable bandwidth, but are extremely simple to operate and economical to build and maintain.

3. Consider the desirability of adding special equipment to detect and/or correct errors in transmission by either self-correction or retransmission techniques.   By the use of totally redundant codes with self-correction such as described by Hamming and Boise Chaduri, a channel in the reverse direction is not necessary.   Less redundant but more efficient codes such as simple parity or the Moore constant ratio codes can detect errors but require a reverse channel to request retransmission to correct the error.   High-quality landlines or coaxial submarine cables can frequently provide error rates of 1×10^6 by simply adding a parity character in each block of information. Long-haul point-to-point HF radio systems require use of error-detecting codes and character-by-character detection and correction to reach a 1×10^5 undetected character error rate.[13,15,16,18,21,23,24]

4. If the available bandwidth is not sufficient to pass the intelligence, consider bandwidth compression and expansion techniques, analogue-to-digital conversion, and redundancy elimination schemes such as slowed-down video over voice-grade telephone lines. These sophisticated techniques for manipulating information are often very expensive but may more than offset their high first cost by the reduced tariffs for narrower bandwidth transmission facilities.[11,14,20]

5. When a voice-communication service requires a group of channels connecting a number of end points through a central office or transit switching center, the switching system to interconnect facilities must be of the "line-switching" or direct line-to-line connection type. If the information is record communication such as message traffic, a second interconnection philosophy should be considered, particularly when expensive long-haul connections are required. This second interconnection method, called "message switching" or "store-and-forward," has many advantages when the traffic volumes are large and a great many destinations are connected by costly long-haul transmission channels (see Chap. 8). Message switching can concentrate traffic on the costly long-distance trunks with maximum efficiency, holding the low-priority traffic until the high priority has been cleared. Line switching, on the other hand, establishes a direct connection by interconnecting the input lines to the output trunks but requires a substantially larger group of trunks to provide the same grade of service as message switching. If large numbers of channels are available at low cost, it is difficult to justify the high first cost of automatic message-switching systems with their large storage facilities and sophisticated format recognition, queueing logic, and message accounting methods. There are, however, a great many advantages when the traffic requires a number of manual processes at the origin and destination and must traverse a number of transit offices. Automatic message-switching systems using large-scale stored-program digital computers are rapidly becoming more widely used than "line-switching" systems for handling large private and public message volumes more efficiently and with higher speeds over minimal trunking facilities.[17,19,20,22,23,25,26]

NOISE AND DISTORTION TYPES

The following definitions of the major sources and types of noise and distortion have been taken from the American Standards Association,[6] while the comments and methods of correction are those of the author. The use of the symbols $N_1 - N_5$ and $D_1 - D_5$ refers to the symbols of Fig. 2.

14. Noise

Noise, as the unwanted disturbance within a useful frequency band, arises from the following sources and exhibits the corresponding characteristics shown.

Ambient noise is acoustic noise existing in a room and usually measured with a sound level meter. It is introduced into the system at the source transducer (N_1) and may be random or periodic, according to its origin. Ambient noise is also present at the information user's location and must be considered as part of the final discrimination process in voice communication systems.

Random noise comprises transient disturbances occurring at random with a Gaussian spectral distribution equivalent to thermal noise. Thermal noise and shot noise are special cases of random noise. Random noise is present in all portions of the system and is part of each N factor in Fig. 2. It is significant only at points of extremely low signal power within the system.

Impulse noise is noise characterized by nonoverlapping transient disturbances. It is usually introduced directly or by induction into terminal equipment or lines by external or directly connected devices such as selector switches or relays. Impulse noise frequently has peak amplitudes comparable to the received signal level and tends to follow the traffic load in a switched network. Impulse noise, dropouts, and circuit interruptions account for a substantial portion of the errors occurring in binary signaling through a switched network.

Radio noise and static: Radio noise is a measure of the field intensity of interfering waves at a receiving point which may be random or periodic in occurrence, while static is interference caused by natural electrical disturbances in the atmosphere.   Both factors combined are represented by N_3 in Fig. 2.

Background noise is the total system noise independent of the presence or absence of a signal.   This is the term for the summation of all noise factors, N_1 to N_5, and represents the noise power in the final signal-to-noise ratio.

15. Distortion

The second category of degrading influences on a communication system are the equipment and path distortions.   The principal sources of distortion are:

1. A nonlinear relation between input and output of a given frequency
2. Nonuniform transmission at different frequencies
3. Phase shift not proportional to frequency

Distortion may be further categorized as follows:

Attenuation distortion (frequency distortion) is either a departure in a circuit or system from uniform amplification or attenuation over the frequency range required for transmission.   This distortion is usually fixed, is easily measured, and therefore is readily controlled or eliminated by appropriate equalizers or compensating nonlinear amplifiers.

Nonlinear distortion is that form of distortion which occurs in a circuit or system when the ratio of instantaneous voltage to current therein (or analogous quantities in other fields) is a function of the magnitude of either.   Voltage- or current-sensitive components and/or circuits in the transducers, amplifiers, passive and active networks, or filters contribute to nonlinear distortion.   The effects can be measured, are usually fixed, and can be corrected or compensated for by such techniques as predistorting the signal to balance the nonlinear distortion or compensating by the addition of networks prior to a linear detection or discrimination device.

Phase distortion is either (1) lack of direct proportionality of phase shift to frequency over the frequency range required for transmission or (2) the effect of such departure on a transmitted signal.

Delay distortion is either (1) phase delay distortion (phase distortion) which is a departure from flatness in the phase delay of a circuit or system over the frequency range required for transmission or (2) envelope delay distortion.   Phase or delay distortion can be measured from end to end when all sections of the communications system have been interconnected and compensated for by the addition of delay equalizers at one or both ends of the system.   If portions of the system involve sections which are part of a switched network, the required delay equalization may vary when one or more sections are switched.   This distortion is of maximum significance in high-speed data-transmission systems and synchronous time-division multiplex systems.

Amplitude distortion is distortion occurring in an amplifier or other device when the amplitude of the output is not a linear function of the input amplitude.   This distortion is readily measured, isolated, and corrected by proper circuit design or by the addition of a compensating nonlinear amplifier or attenuation network.

Intermodulation distortion is distortion which results from cross coupling or leakage of carrier or signal energy from an adjacent channel.   It usually results from an excessive signal level in an adjacent channel or insufficient attenuation of the sidebands or higher-order harmonics by the channel separation filters.

REFERENCES

1. *From Semaphore to Satellite*, International Telecommunication Union, Geneva, 1965 (5 Rue de Varembe, Geneva, Switzerland).
2. Shannon, Claude E., A Mathematical Theory of Communications, *Bell System Tech. J.*, vol. 27, part I, pp. 379–423, July, 1948; part II, pp. 623–656, October, 1948.
3. Shannon, Claude E., Communications in the Presence of Noise, *Proc. IRE*, vol. 37, no. 1, pp. 10–21, January, 1949.

4. Von Bekesy, Georg, *Experiments in Hearing*, McGraw-Hill Book Company, New York, 1960.
5. The International Telegraph and Telephone Consultative Committee (CCITT), 2nd Plenary Assembly, *Red Book*, Volumes II bis, III, IV, V, VI and VII, International Telecommunication Union, Geneva, 1961–1962. (Note: The next set of standards known as *Blue Books* from the 3rd Plenary Assembly, Geneva, 1964, are in process of publication.)
6. American Standards Association, *Definitions of Electrical Terms*, Group 65 Communication, ASA, C42.65, American Institute of Electrical Engineers, New York, 1957.
7. *General Plan for the Development Of The International Network*, CCITT/CCIR Joint Committee for the General Plan for the Development of Telecommunication Networks, Rome, 1963, International Telecommunications Union, Geneva, 1964.
8. The International Radio Consultative Committee, (CCIR), 10th Plenary Assembly, *Green Book*, vols. I, II, III, IV, and V, International Telecommunication Union, Geneva, 1963.
9. O'Grady, F. P., *Artificial Satellites for Telecommunications—Internal and External*, IREE (Australia) Radio and Electronics Engineering Convention, Canberra, 1965.
10. Mueller, George E., and Eugene R. Splanger, *Communications Satellites*, John Wiley & Sons, Inc., New York, 1964.
11. Bell Telephone Laboratories, *Transmission Systems for Communications*, Bell Telephone Laboratories, Inc., Western Electric Publications, Inc., Winston-Salem, 1964.
12. Schwartz, Mischa, *Information Transmission, Modulation, and Noise*, McGraw-Hill Book Company, New York, 1959.
13. Fano, Robert M., *Transmission of Information*, The MIT Press and John Wiley & Sons, Inc., New York, 1961.
14. Everitt, W. L., and G. E. Anner, *Communication Engineering*, McGraw-Hill Book Company, New York, 1956.
15. Davenport, Jr., Wilbur B., and William L. Root, *An Introduction to Random Signals and Noise*, Lincoln Laboratory Publications, McGraw-Hill Book Company, New York, 1958.
16. International Symposium on Data Transmission, Delft, Institute of Technology, Delft, Netherlands, Sept. 19–21, 1960, *IRE Trans. Commun. Systems*, vol. CS-9, no. 1, March, 1961 (entire issue).
17. Parry, C. A., Planning Aspects of a Globe-circling Communications System of High Capacity, Reliability and Performance Quality, *IRE Trans. Commun. Systems*, vol. CS-9, no. 3, pp. 298–311, September, 1961.
18. Strassman, A. J., and A. C. Chapman, A Long Range Digital Communication System, *IRE Trans. Commun. Systems*, vol. CS-9, no. 4, December, pp. 383–389, December, 1961.
19. Montgomery, A. W., Long Line Systems for Global Communication, *IRE Trans. Comm. Systems*, vol. CS-9, no. 4, pp. 396–398, December, 1961.
20. Chipp, R. D., and T. Cosgrove, Economic Analysis of Communication Systems, *IRE Trans. Commun. Systems*, vol. CS-10, no. 4, pp. 416–421, December, 1962.
21. Wildhagen, G. A., Some Results of Data Transmission Tests over Leased Telephone Circuits, *IRE Trans. Commun. Systems*, vol. CS-9, no. 3, pp. 271–275, September, 1961.
22. Halina, J. W., The Telephone Channel in a Global Communications System, *IRE Trans. Commun. Systems*, vol. CS-9, no. 3, pp. 247–252, September, 1961.
23. Unk, J. M., Communication Networks for Digital Information, *IRE Trans. Commun. Systems*, vol. CS-8, no. 5, pp. 207–214, December, 1960.
24. Wilson, F. E., and W. A. Runge, Data Transmission Tests on Tropospheric Beyond-the-horizon Radio System, *IRE Trans. Commun. Systems*, vol. CS-8, no. 1, pp. 40–43, March, 1960.

Chapter 2

USER EQUIPMENTS AND SERVICES

HENRY F. BURKHARD, *U.S. Army Electronics Command*

JOSEPH E. CLARE,* *New Jersey Bell Telephone Company*

User equipments represent a wide variety of types. Typical ones in the fields of telegraphy, data input and output, telephony, and facsimile, will be considered in this chapter. Generally any input-output device forming an interface between the human and the electronic circuit will be considered as user equipment. There are varying degrees of interpretation of this; for example, a device which automatically reads typing, thereby accomplishing what the teletype operator does, is considered a user equipment although all the operator does is load and unload copy into the device.

The usual communications problem is to convey intelligence from one location to one or more other locations with little change in the information content. However, there are other systems, which accept information from an operator employing an input device, perform operations on the information such as in computer systems, and present the result by means of an output device. In the communications system, the output is essentially equal to the input usually delayed and somewhat modified because of input-output device limitations and transmission-circuit distortions and noises. Nevertheless, the output device usually presents the information with little basic difference from that which entered the system. There may be a size variation such as one gets with different-sized television receivers; this is not considered a basic difference between input and output. In communications systems speech is transmitted with reasonable likeness, sketches are transmitted via facsimile and are received as similar sketches, and keyboard operation of a teletypewriter results in a remote printed page resembling the local monitor copy.

The computational-type systems are coming to the forefront, and in many cases the logical processes applied to one or many inputs make the relation between the output

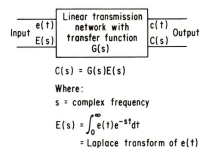

$$C(s) = G(s)E(s)$$

Where:

s = complex frequency

$$E(s) = \int_0^\infty e(t)e^{-st}dt$$

= Laplace transform of e(t)

FIG. 1. Communications system. (*By permission of John G. Truxal, Automatic Feedback Control System Synthesis, chap. 1, McGraw-Hill Book Company, New York, 1955.*)

* Contributed section on Telephone Equipment.

and input far from obvious, preventing a simple mathematical expression as for the communications case above.   In general a digital computer has its storage and logic circuits each in a specified state resulting from previous history, and the introduction of one or more input signals causes changes in the states and in the output circuits in accord with the prescribed program.

For the communications systems, input devices are usually matched to output devices, i.e., telephone to telephone, keyboard to printer, facsimile transmitter to facsimile receiver, etc., and as the transmission circuit approaches being perfect, it becomes more and more similar to a direct connection between the sending and receiving user equipment.

Computer systems on the other hand may have inputs of one type and totally unmatched outputs.   For example, many keyboard inputs may feed a computer and the output may be a display device.   No matter how perfect the computer becomes, its presence is necessary between the input and output, since the signals from devices like keyboards would be meaningless to other devices like displays without the format generation and logical operations of the computer.

TELEGRAPH AND DATA USER EQUIPMENT[2-4]

1. Information Concept and Codes[5]

The input information may be a printed character, a picture, a thought in a person's mind, or some configuration capable of stimulating a response from a human being. Attempts to define and measure information have linked the particular information with the number of other possible information presentations which could have occurred and the relative probability of each.   For example, if the two conditions are rain or no rain and the probabilities are equal, the information contained in one statement of either condition is $H = \log_2 2 = 1$ bit.   This choice of one of two equally probable conditions is the unit of measurement of information.   If there had been eight possible states with equal probability, the existence of one gives information:

$$H = \log_2 8 = 3 \text{ bits}$$

$$H_{\text{avg}} = -\sum_{K=0}^{M-1} P(K) \log_2 P(K) \qquad (1)^*$$

where H_{avg} = average information per Mth-order selection
$P(K)$ = probability of kth choice of M

$$C = Kf_c \log_2 \left(1 + \frac{S}{N}\right) \qquad \text{bits/sec} \qquad (2)$$

where C = channel capacity
K = a factor depending to a degree on system reliability.   It is often almost equal to unity.
S = average signal power
N = average random noise power
f_c = bandwidth of transmission link

This leads to some basis for codes utilized in input and output devices for transforming the information into a form suitable for transmission.   The example above of rain or no rain, representing 1 bit of information, can be coded into a simple binary digital form.   For example, a 3-volt level on a line could represent rain and a 0-volt level could represent no rain.   The input device would be a battery and an on-off switch; the output device could be a lamp.

* Keith Henney, *Radio Engineering Handbook*, 5th ed., pp. 1–34, McGraw-Hill Book Company, 1959.   Based on R. M. Fano, The Transmission of Information, *MIT Res. Lab. Electrons. Tech. Rept.* 65.

The same information could be sent by writing whole sentences and sending them with a keyboard. For example, "It is raining, the area is getting drenched, and continuation of this precipitation for a month will lead to floods." This contains much redundancy. It requires a more complex input device than the on-off switch and more time for a transmission. But the message is capable of a considerable amount of distortion without loss of information. It is the technique of adding more bits of information than the minimum which is employed in error correcting codes.

2. Telegraph and Data Codes

While codes could be derived for multilevel signaling, it is general practice to utilize binary digital signals. The number of bits n used for a given character determines the total number of code combinations in the code: total, $T = 2^n$. These code combinations may or may not agree with the total number of possible characters, since in some codes not all combinations represent a character and in other codes certain combinations are utilized as "escape" combinations, permitting drastic expansion of the total number of characters. If we consider, for example, the Baudot code, which is used for practically all teletypewriter transmissions, and ignore the start and stop pulses required for synchronization, there are 5 information bits. These permit $2^5 = 32$ possible code combinations to be transmitted. One of these, the Figures combination, can be considered an escape combination. After Figures is transmitted, each combination has a symbol assignment as shown in Table 1. (Example: Combination 10100 causes the bell of a receiving equipment to ring in a weather system.) In teletypewriter terminology the binary conditions are called mark and space, corresponding to "1" and "0" in data systems. In the Baudot system the reception of the Figures symbol applies to all following characters as does the Letters combination. In some data codes such as Fieldata the escape symbol, Special, applies to only the code combination immediately following it.

Data codes usually have 6, 7, or 8 bits per character, permitting a larger number of code combinations than for the Baudot code. The military Fieldata code, shown in Table 2, consists of 8 bits per character. The eighth bit is a parity bit, while the remaining seven are information bits. An important system aspect is to recognize that there are two forms of the Fieldata code, namely, the standard form and the paper-tape form. Odd parity is used for the Fieldata standard form; i.e., the parity bit is chosen to be a 0 or a 1 so that the character has an odd number of ones (1's). The Fieldata paper-tape code uses even parity. The parity bit is used for error control but permits detection only of an odd number of errors in sensing or transmitting the bits of a character.

The seventh bit of the Fieldata code (Standard Form), counting from right to left, is the control bit C. For the Standard Form when $C = 1$, the 6 information bits represent alphanumeric characters as shown. When $C = 0$, the 6 information bits represent control characters which are used as instructions for computers and input-output devices. In the paper-tape code form, it is not the control bit alone which determines whether the 6 information bits are alphanumeric or control characters, but it is a comparison of the control bit and the sixth bit I_2. When these are similar, the character is a control, and when dissimilar, alphanumeric. The existence of the two code forms requires that punched paper-tape readers and punches contain logic circuitry such as:

$$C_s = \bar{C}_p I_p + C_p \bar{I}_p \tag{3}$$
$$P_s = P_p I_p + \bar{P}_p \bar{I}_p \tag{4}$$
$$C_p = \bar{C}_s I_s + C_s \bar{I}_s \tag{5}$$
$$P_p = P_s I_s + \bar{P}_s \bar{I}_s \tag{6}$$

where the bar means NOT in logic terminology. These are based on Fieldata character $P\,C\,I_2\,I_1\,D_4\,D_3\,D_2\,D_1$. Subscripts s and p refer to standard and paper-tape code forms.

These equations may be stated as follows: In converting from one form of Fieldata code to the other, change the seventh bit (control) but not the parity bit if the sixth

Table 1. Teletypewriter Codes

Characters[1] Letters Shift	Figures shift Comm	Weather	Code signals Start 1 2 3 4 5 Stop	Intnl[2] Alph #2 Fig shift	ARQ[3] 7-unit Moore code 1234567
A	–	↑		–	0011010
B	?	⊕		?	0011001
C	:	○		:	1001100
D	$	✓		Wru	0011100
E	3	3		3	0111000
F		→		Unasnd	0010011
G	&	↘		Unasnd	1100001
H	#	↓		Unasnd	1010010
I	8	8		8	1110000
J	Bell(△)	✓		Bell (♫)	0100011
K	(	←		(	0001011
L	)	↘		)	1100010
M	.	.		.	1010001
N	,	⊕		,	1010100
O	9	9		9	1000110
P	Ø	Ø		Ø (zero)	1001010
Q	1	1		1 (one)	0001101
R	4	4		4	1100100
S	'	Bell		'(apost)	0101010
T	5	5		5	1000101
U	7	7		7	0110010
V	;	⊕		=	1001001
W	2	2		2	0100101
X	/	/		/	0010110
Y	6	6		6	0010101
Z	"	+		+	0110001
Blank (⌀)		–		Unasnd	0000111
Space (■)				Space	1101000
Car ret (<)				Car ret	1000011
Line feed (≡)				Line feed	1011000
Figures (↑)				Figures	0100110
Letters (↓)				Letters	0001110
				Idle alpha	0101001
				Idle beta	0101100
				RQ signal (Request)	0110100

■ Marking pulse

Note: Upper case H (COMM) may be STOP or #.
1. Some, including military, use Fig. H = motor stop, Fig. J = apostrophe, Fig. F = !, and Fig. S = bell.
2. Letters shift of International No. 2 is the same as the left-hand column of the table.
3. J. B. Moore and R. E. Mathes, Printing Telegraph Systems, U.S. Patent 2183147 and R. E. 23028.
4. Figures shown in parenthesis after BLANK, SPACE, etc. are the symbols printed on tape by a typing reperforator.

Table 2. Fieldata Code with Baudot Code Equivalents

Fieldata standard code	Paper tape code	Data character	Baudot character	Baudot code	
				Case	Code
*7654 3210					*54321
0100 0000	1100 0000	Master space	Blank	E	00000
1100 0001	0100 0001	Upper case	Figures	E	11011
1100 0010	0100 0010	Lower case	Letters	E	11111
0100 0011	1100 0011	Line feed	Line feed	E	00010
1100 0100	0100 0100	Carriage return	Carriage return	E	01000
0100 0101	1100 0101	Space	Space	E	00100
0100 0110	1100 0110	A	A	LET	00011
1100 0111	0100 0111	B	B	LET	11001
1100 1000	0100 1000	C	C	LET	01110
0100 1001	1100 1001	D	D	LET	01001
0100 1010	1100 1010	E	E	LET	00001
1100 1011	0100 1011	F	F	LET	01101
0100 1100	1100 1100	G	G	LET	11010
1100 1101	0100 1101	H	H	LET	10100
1100 1110	0100 1110	I	I	LET	00110
0100 1111	1100 1111	J	J	LET	01011
1101 0000	0101 0000	K	K	LET	01111
0101 0001	1101 0001	L	L	LET	10010
0101 0010	1101 0010	M	M	LET	11100
1101 0011	0101 0011	N	N	LET	01100
0101 0100	1101 0100	O	O	LET	11000
1101 0101	0101 0101	P	P	LET	10110
1101 0110	0101 0110	Q	Q	LET	10111
0101 0111	1101 0111	R	R	LET	01010
0101 1000	1101 1000	S	S	LET	00101
1101 1001	0101 1001	T	T	LET	10000
1101 1010	0101 1010	U	U	LET	00111
0101 1011	1101 1011	V	V	LET	11110
1101 1100	0101 1100	W	W	LET	10011
0101 1101	1101 1101	X	X	LET	11101
0101 1110	1101 1110	Y	Y	LET	10101
1101 1111	0101 1111	Z	Z	LET	10001
1110 0000	1010 0000	)	)	FIG	10010
0110 0001	0010 0001	−	−	FIG	00011
0110 0010	0010 0010	+	&	FIG	11010
1110 0011	1010 0011	<	No equiv.		
0110 0100	0010 0100	=	No equiv.		
1110 0101	1010 0101	>	No equiv.		
1110 0110	1010 0110	—	BREAK		OPEN CIRC
0110 0111	0010 0111	$	$	FIG	01001
0110 1000	0010 1000	(*)	Bell	FIG	00101
1110 1001	1010 1001	(	(	FIG	01111
1110 1010	1010 1010	''	''	FIG	10001
0110 1011	0010 1011	:	:	FIG	01110
1110 1100	1010 1100	?	?	FIG	11001
0110 1101	0010 1101	!	!	FIG	01101
0110 1110	0010 1110	,	,	FIG	01100
1110 1111	1010 1111	⊕ (Stop)	Stop	FIG	10100
0111 0000	0011 0000	0	0	FIG	10110
1111 0001	1011 0001	1	1	FIG	10111
1111 0010	1011 0010	2	2	FIG	10011
0111 0011	0011 0011	3	3	FIG	00001
1111 0100	1011 0100	4	4	FIG	01010
0111 0101	0011 0101	5	5	FIG	10000

Table 2. Fieldata Code with Baudot Code Equivalents (*Continued*)

Fieldata standard code	Paper tape code	Data character	Baudot character	Baudot code	
				Case	Code
0111 0110	0011 0110	6	6	FIG	10101
1111 0111	1011 0111	7	7	FIG	00111
1111 1000	1011 1000	8	8	FIG	00110
0111 1001	0011 1001	9	9	FIG	11000
0111 1010	0011 1010	,	,	FIG	01011
1111 1011	1011 1011	;	;	FIG	11110
0111 1100	0011 1100	/	/	FIG	11101
1111 1101	1011 1101	.	.	FIG	11100
1111 1110	1011 1110	□ Special	No equiv.		
0111 1111	0011 1111	Idle	Idle	Line idle	
0001 1010†	1001 1010	WRU	WRU	FIG	01001
1011 1111†	1111 1111	Delete	Letters		11111

* Indicates the order of transmission in serial form, i.e., ascending order.
† Control characters.

bit (I) is a 1. If the sixth bit is a 0, change the parity bit but not the seventh bit.
Example: $H = 11001101$ in standard form and 01001101 in paper-tape form.
Although these are logic equations, they are solved simply by substituting $I_p = 1$ or 0
or $I_s = 1$ or 0 as appropriate. $\bar{I}$ is always the opposite of I; for example, if $I_p = 1$,
$\bar{I}_p = 0$, etc.

For punched cards, the Hollerith* code is often used. The card has 80 columns,
each representing a character as shown in Fig. 2a. For each column there are 10
basic positions numbered from 0 through 9, and a hole punched in a column represents
the number corresponding to the row punched. In order to include alphabet and
punctuation characters, eleventh and twelfth rows are placed above the other rows
and two holes are punched in each column. A is shown by punching both the twelfth
row and the first row and B by the twelfth row and the second row, etc. The Univac†
card and code are shown in Fig. 2b.

Cards are usually read and punched a row at a time in order to realize higher speeds,
since by this technique 80 bits are handled simultaneously. The information for
the total card is stored and utilized a character at a time. The storage required is in
the order of 960 bits and is expensive, leading some users to punch the information
on the cards as computer words along the rows. This reduces the amount of buffer
storage required, increases the information capacity for each card, and achieves high
operating speed but has the disadvantage of making the punched card less rugged
owing to the large number of holes. Others circumvent the large buffer storage by
reading the card a column at a time at a lower speed. For operation within or between
data systems, codes such as those in Table 3 are used.

Recently the American Standards Association (ASA) has studied a wide variety of
codes and has approved a code known as American Standard Code for Information
Interchange (ASCII) as shown in Table 4. This code has several advantages over
previous codes, such as the arrangement of the alphabetic portion to permit expansion
to alphabets containing more than 26 letters. The ASA code is rapidly approaching
commercial and military adoption for modern teletypewriter transmissions and shows
promise of widespread usage in communications and data systems.

Systems engineering dictates that code compatibility be maintained from transmitter to receiver. This is most easily accomplished by using the same code throughout the system. With code translation in the network it is feasible to have one code at

* IBM Corporation.
† Sperry Rand Corporation

the transmitter and another at the receiver, although there can be much complication when some of the meanings in one code do not exist in the other. For example, the present ASCII does not have underscore, upper case, or lower case while the Fieldata code does. Translation from Fieldata 11100110 (underscore) to ASCII therefore required special consideration. Translation problems of this type may be resolved through limitations placed upon operators to prevent generation of the characters not existing in the other code.

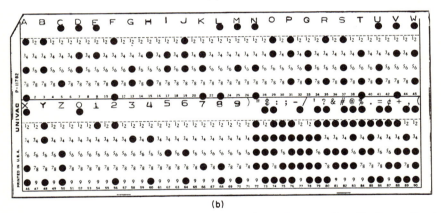

(a)

(b)

Fig. 2. Punched cards. (a) Hollerith coding (IBM 80-column card). (b) Sperry Rand 90-column card.

Although the ASCII has been adopted recently, consideration is being given to making some changes in the left two columns, modifying the lower portions of columns 6 and 8, and assigning lower-case letters in columns 7 and 8. These probable changes are shown in Table 4A.

A special systems problem of interest exists. Newer teletypewriter and data user equipments are being manufactured for which existing transmission circuits are unsuitable owing to the use of an 8-bit code for the user equipments and sensitivity of the transmission circuits to the code for which they are designed (Baudot code). One solution is the establishment of special new transmission circuits. However, there are multimillion dollar long- and short-range networks which have been established for the Baudot transmissions. These are elaborate, employ error correction, and are of

Table 3. UNIVAC and IBM BCD Codes

(a) UNIVAC Code

FIELDS: 2 3 / 1 4

	00	01	10	11
0000	i ˣ / 5	r ′ (E N)	t ᵛ / v	Σ
0001	Δ / 6	, ‼ (o W)		β (Y)
0010	− / −	. (P)	!	:
0011	) / 0	; (:)	)	+ (@)
0100	‡ / 1	A	J	? /
0101	" / 2	B	K	S
0110	# / 3	C	L	T
0111	$ / 4	D	M	U
1000	⅘ / 5	E	N	V
1001	* / 6	F	O	W
1010	& / 7	G	P	X
1011	' / 8	H	Q	Y
1100	(/ 9	I	R	Z
1101	↑ / # (z)	#	$ ᶻ	%
1110	& (c)	¢	*	= (T)
1111	((D :)	@	? (M) ⁸	NOT USED

HIGH-SPEED PRINTER
Normal: prints 1 except for characters with an entry in 4 which are non-printing.
Computer Digit: prints 1 for printing characters; 4 for non-printing characters. The digit 8 prints in place of suppressed zeros.

SUPERVISORY CONTROL PRINTER
Normal: prints 1 except for characters with an entry in 3 which are non-printing. If 3 is a ; then a ; is printed; if not, typewriter action occurs. 2 is upper case where different from lower.
Computer Digit: prints 1 for printing characters; 3 for non-printing.

UNITYPER
Code impulses designated by 1.

HIGH SPEED PRINTER ACTION
@ Fast Feed Symbol 1
↑ Fast Feed Symbol 2
↑ Fast Feed Symbol 3
= Fast Feed Symbol 4
Γ Multi-Line Symbol

SUPERVISORY CONTROL PRINTER ACTION
$ Shiftlock
⅄ Single Shift
↑ Unshift
Γ Carriage Return
t Tabulator

BOTH PRINTERS
i Printer Ignore
Δ Printer Space
β Breakpoint Stop
Σ Printer Stop

(b) Functions

FUNCTIONS	8	4	2	1	B	A	C
					ODD PARITY BCD CODE		
BKSP	8	4	2		B		C
EOT	8	4	2	1			C
DELETE	8	4	2	1	B	A	C
DOWN–SHIFT	8	4	2		B	A	
CR AND LF	8	4		1	B		C
PREFIX	8	4	2	1		A	
IDLE	8	4	2	1	B		
RDR STOP	8	4		1			
SPACE							C
EOB	8	4	2			A	C
UP–SHIFT	8	4	2				
LF	8	4		1		A	C
TAB	8	4		1	B	A	
RESTORE	8	4			B		
BYPASS	8	4				A	
EOA	8		2	1			

(b) Characters

UPPER SHIFT	NORMAL SHIFT	8	4	2	1	B	A	C
					ODD PARITY BCD CODE			
=	1				1			
¢	2			2				
;	3			2	1			C
:	4		4					
%	5		4		1			C
'	6		4	2				C
"	7		4	2	1			
*	8	8						
(	9	8			1			C
)	0	8		2				C
A	a				1	B	A	
B	b			2		B	A	
C	c			2	1	B	A	C
D	d		4			B	A	
E	e		4		1	B	A	C
F	f		4	2		B	A	C
G	g		4	2	1	B	A	
H	h	8				B	A	
I	i	8			1	B	A	C
J	j				1		B	C
K	k			2			B	C
L	l			2	1		B	
M	m		4				B	C
N	n		4		1		B	
O	o		4	2			B	
P	p		4	2	1		B	C
Q	q	8					B	C
R	r	8			1		B	
S	s			2			A	C
T	t			2	1		A	
U	u		4				A	
V	v		4		1		A	
W	w		4	2			A	
X	x		4	2	1		A	C
Y	y	8					A	C
Z	z	8			1		A	
.	.	8		2	1	B	A	
!	$	8		2	1	B		C
?	/				1		A	C
±	#	8		2	1		A	
+	&					B	A	C
—	-					B		
°	@						A	

(a) Sperry Rand, UNIVAC Code   (b) IBM 1050 System to 1050 System Code.

(b) To print this code set, the printer must contain a print element consisting of 88 printable characters (upper and lower case) for 1050-to-1050 use.   Reprinted by permission from *IBM 1050 Data Communication System,* Form A24-3020.   Copyright 1963 by International Business Machines Corporation.

Table 4. American Standard Code for Information Interchange (ASCII)*

1. Scope

This coded character set is to be used for the general interchange of information among information processing systems, communication systems, and associated equipment.

2. Standard Code

$b_4\,b_3\,b_2\,b_1$	b_7=0 b_6=0 b_5=0	0 0 I	0 I 0	0 I I	I 0 0	I 0 I	I I 0	I I I
0 0 0 0	NULL	DC$_0$	ƀ	0	@	P		
0 0 0 I	SOM	DC$_1$	!	I	A	Q		
0 0 I 0	EOA	DC$_2$	"	2	B	R		U
0 0 I I	EOM	DC$_3$	#	3	C	S		N
0 I 0 0	EOT	DC$_4$ (STOP)	$	4	D	T	U	A S
0 I 0 I	WRU	ERR	%	5	E	U	N	S I
0 I I 0	RU	SYNC	&	6	F	V	A S	G
0 I I I	BELL	LEM	(APOS)	7	G	W	S I	N
I 0 0 0	FE$_0$	S$_0$	(	8	H	X	G N	E D
I 0 0 I	HT SK	S$_1$	)	9	I	Y	N E	
I 0 I 0	LF	S$_2$	*	:	J	Z	E D	
I 0 I I	V$_{TAB}$	S$_3$	+	;	K	[	D	
I I 0 0	FF	S$_4$ (COMMA)	<	L	\			ACK
I I 0 I	CR	S$_5$	—	=	M	]		①
I I I 0	SO	S$_6$	.	>	N	↑		ESC
I I I I	SI	S$_7$	/	?	O	←		DEL

3. Positional Order and Notation

Standard 7-bit set code positional order and notation are shown below with b_7 the high-order, and b_1 the low-order, bit position.

EXAMPLE: The code for "R" is:

$$b_7\quad b_6\quad b_5\quad b_4\quad b_3\quad b_2\quad b_1$$
$$1\quad 0\quad 1\quad 0\quad 0\quad 1\quad 0$$

4. Legend

NULL	Null/Idle	DC$_1$-DC$_3$	Device control
SOM	Start of message	DC$_4$(Stop)	Device control (stop)
EOA	End of address	ERR	Error
EOM	End of message	SYNC	Synchronous idle
EOT	End of transmission	LEM	Logical end of media
WRU	"Who are you?"	S$_0$-S$_7$	Separator (information)
RU	"Are you...?"	ƀ	Word separator (space, normally non-printing)
BELL	Audible signal		
FE$_0$	Format effector	<	Less than
HT	Horizontal tabulation	>	Greater than
SK	Skip (punched card)	↑	Up arrow (Exponentiation)
LF	Line feed	←	Left arrow (Implies/ Replaced by)
V$_{TAB}$	Vertical tabulation		
FF	Form feed	\	Reverse slant
CR	Carriage return	ACK	Acknowledge
SO	Shift out	①	Unassigned control
SI	Shift in	ESC	Escape
DC$_0$	Device control reserved for data link escape	DEL	Delete/Idle

* Courtesy of American Standards Association.

Note: For further information see American Standards Association Standard X3.4-1963, UDC 681.3 (Tentative), *American Standard Code for Information Interchange.*

Table 4A. Probable Modification to ASCII Code

7654321					
0000000	NULL	All zeros (time or media fill)	0010000	DLE	Data link escape
0000001	SOH	Start of heading	0010001	DC₁	Device control (one)
0000010	STX	Start of text	0010010	DC₂	Device control (two)
0000011	ETX	End of text	0010011	DC₃	Device control (three)
0000100	EOT	End of transmission	0010100	DC₄	Device control (four)
0000101	ENQ	Enquiry	0010101	NACK	Negative acknowledgment
0000110	ACK	Acknowledge	0010110	SYNC	Synchronous idle
0000111	BELL	Alarm or attention audible signal	0010111	ETB	End of transmission block
0001000	BS	Back space	0011000	CNCL	Cancel (associated data in error)
0001001	HT	Horizontal tabulation	0011001	EM	End of medium
0001010	LF	Line feed	0011010	SS	Start of special sequence
0001011	VT	Vertical tabulation	0011011	ESC	Escape
0001100	FF	Form feed	0011100	FS	Information file separator
0001101	CR	Carriage return	0011101	GS	Information group separator
0001110	SO	Shift out	0011110	RS	Information record separator
0001111	SI	Shift in	0011111	US	Information unit separator

7654321														
0100000	SP	0110000	0	1000000	`	1010000	P	1100000	@	1110000	p			
0100001	!	0110001	1	1000001	A	1010001	Q	1100001	a	1110001	q			
0100010	"	0110010	2	1000010	B	1010010	R	1100010	b	1110010	r			
0100011	#	0110011	3	1000011	C	1010011	S	1100011	c	1110011	s			
0100100	$	0110100	4	1000100	D	1010100	T	1100100	d	1110100	t			
0100101	%	0110101	5	1000101	E	1010101	U	1100101	e	1110101	u			
0100110	&	0110110	6	1000110	F	1010110	V	1100110	f	1110110	v			
0100111	'	0110111	7	1000111	G	1010111	W	1100111	g	1110111	w			
0101000	(	0111000	8	1001000	H	1011000	X	1101000	h	1111000	x			
0101001	)	0111001	9	1001001	I	1011001	Y	1101001	i	1111001	y			
0101010	*	0111010	:	1001010	J	1011010	Z	1101010	j	1111010	z			
0101011	+	0111011	;	1001011	K	1011011	[	1101011	k	1111011	{			
0101100	,	0111100	<	1001100	L	1011100	~	1101100	l	1111100	⌐			
0101101	−	0111101	=	1001101	M	1011101	]	1101101	m	1111101	}			
0101110	.	0111110	>	1001110	N	1011110	^	1101110	n	1111110					
0101111	/	0111111	?	1001111	O	1011111	—	1101111	o	1111111	DEL			

Note:
` = accent grave | = vertical line (absolute)
^ = circumflex   ⌐ = overline (logical negation)
~ = tilde   — = underscore (1011111)

time-proved reliability. A method of utilizing these circuits has been developed as described below.

An 8-bit code can be transmitted over these 5-bit circuits* even though the 8-bit code has 128 combinations (1 bit is parity) and the 5-bit Baudot code has only 58 combinations (allowing for FIGS, CR, LF, space, blank, and LET characters). The 8-bit code is handled by forming two characters with the 8 bits. The first 4 bits have a 0 bit added to form a pseudo "Baudot" 5-bit character, while the second 4 bits have a

* Courtesy of Andrew C. DeRosa, U.S. Army Electronics Research and Development Laboratories, Fort Monmouth, N.J.

1 bit added to form a second pseudo "Baudot" 5-bit character. These may be transmitted through Baudot transmission circuits. What is lost by this technique is speed, since a 10-character-per-second teletypewriter channel will be able to transmit only 5 of the 8-bit characters per second. In many cases this is well compensated for by the ability to use existing reliable teletypewriter circuits for data characters.

3. Signal Types

Signal generation in teletypewriter and data user equipment is generally direct current in nature, while other devices such as facsimile employ a-c signals. For many decades, the bits of the teletypewriter codes have been and still are transmitted in one of three forms: 60 ma neutral, 20 ma neutral, or 30 ma polar.[6] Figure 3a shows a neutral circuit, and Fig. 3b a polar one. The solid wire or ground returns shown are used for both neutral and polar circuits. For the neutral circuits current flowing

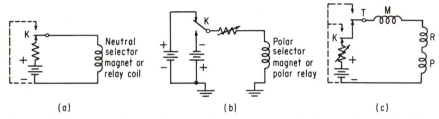

FIG. 3. D-c teletypewriter circuits. (a) Neutral circuit. (b) Polar with ground return. (c) Neutral circuit with monitor M, reperforator R, printer P, keyboard contact K, and TD contact T.

represents mark (1) condition while no current is space (0) condition. For polar circuits mark and space are represented by opposite directions of current flow. Additional teletypewriter equipments are added by placing them in series as shown in Fig. 3c. K is the keyboard contact, T a tape reader contact, M a monitor selector magnet, R a reperforator selector, and P a printer selector. In these circuits the current follows the usual transient equations:

$$i = I_f(1 - e^{-Rt/L}) + I_0 e^{-Rt/L} \tag{7}$$

where I_f = final current ($t = \infty$)
                 I_0 = initial current ($t = 0$)
                 R = total series resistance
                 L = total series inductance

Time constant $= \dfrac{L}{R}$

As shown in Fig. 3c a number of receiving devices may be placed in series, and in order to keep the time constant low and to permit operation over long transmission circuits, the voltage is in the order of 110 volts. With the use of modems* for a-c transmission there is a trend, especially in the military,[7] toward the use of a polar signal of 6 volts (± 1 volt) with a source impedance not exceeding 100 ohms and a

---

* A modem is a modulator and a demodulator set which accepts d-c signals from the sending device, sends the a-c modulated signal, demodulates at the receiving end of the transmission circuit, and delivers a d-c signal to the printer or reperforator. The modem also generates control signals and sometimes timing signals and may contain switching or dialing devices. It is also called a line unit or a data set. These will be discussed separately.

receiver sensitivity of 0.0001 amp (max), 0.5 volt and an impedance of at least 5,000 ohms resistive.   A positive voltage with respect to ground is mark, and negative voltage is space.   Figure 4 shows a typical Baudot code signal, the upper trace being the applied emf, either polar or neutral, and the lower trace the current per Eq. (7). The information bits 1 through 5 are shown in Table 1, and these are preceded by a space signal, used as a start signal, and followed by a stop signal.   A page printer or reperforator receiving a steady mark signal remains in a quiescent state until the

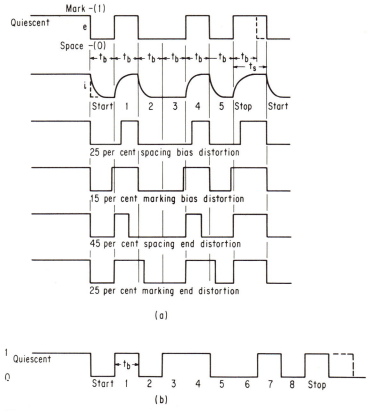

FIG. 4. Teletypewriter and data start-stop signals.   (a) Baudot signals.   (b) Serial data signal.

current changes from mark to space condition, indicating the initiation of a character. This is the reference time which indicates to the receiver that at times corresponding to the five information bit positions sampling should take place to determine mark (1) or space (0) condition of the signal.   The receiver returns to the quiescent state during the stop pulse but initiates the sampling process for receipt of another character as soon as the current drops to space condition again.   A very significant factor is that the receiving device operates fast enough to print or punch a character at the transmission rate and then stops and waits for the next character.   Each individual character is transmitted and received at the transmission character rate even though the number of characters per unit time may be low.   For example, in a 100-word-per-minute (wpm) system a keyboard operator may push only one key.   The character is transmitted with a duration of $\frac{1}{10}$ sec. The receiver goes through its signal sampling cycle in about $6\frac{1}{2}$ bits, about 88 msec.

For certain applications, teletypewriter signals are sent in parallel; that is, all five information bits are sent simultaneously on five different circuits. An additional circuit is used for a strobe signal to indicate to a receiving device when the information may be sampled. A machine for transmitting in this parallel fashion from a paper-tape source is usually called a reader, while the equivalent device containing a distributor and transmitting a serial signal such as in Fig. 3a is called a transmitter-distributor or TD.

Data input-output (I/O) equipments[8] generally operate with parallel signals and ready and strobe channels. This arrangement permits the slowest speed device in the system to control the transmission rate. For example, a reader samples the holes in the tape and, when a signal is received on the ready line indicating that the receiving device is ready, sets each channel to a 1 or 0 corresponding to the information in the tape and then sends a strobe signal to the receiver, which indicates to the receiver that the data channels are ready for sampling. Fieldata devices have the following overall characteristics: ten circuits, eight for information, one ready, and one strobe. Signals are 0 volts (± 0.15 volt) for 0 and -3 volts (-2.85 to -4.50 volts) for a 1 measured across an impedance of 150 ohms. The strobe pulses have a duration of between 1 and 2 μsec. Usually a 19-wire cable connects equipments, the additional wires being used for ground and for control signals such as Write, Forward, Reverse, Rewind, End of storage, Received error, Alarm, Corrected error, Beginning of storage, etc. These controls differ depending upon direction of signal flow and type of user device.

A ready-strobe mode of operation permits operation at the nominal rate which is maximum for the system or at a lower rate determined by either transmitting or receiving device, the slowest essentially being the controlling element for the character transmission or transfer rate. The character information bits may be sent in series, but it is more common in a data center to transfer information to or from the computer over parallel channels. When the ready-strobe method is extended to a communications system, several factors must be considered. Most important is a return path to permit sending of the ready signal from receiving to sending end of the circuit. This can be accomplished over a single (two-wire) bidirectional circuit or a full-duplex (four-wire) circuit. Another factor is the time required for signal transmission.[9] The phase velocity, which is equal to ω/β, ranges about 179,000 miles/sec for open wire, about 58,000 miles/sec for a loaded line, and down to 10,000 miles/sec for some cables. The envelope or group velocity is equal to $d\omega/d\beta$ and is somewhat higher than the phase velocity.[10] ω is $2\pi f$, and β is the wavelength constant in radians per mile. This means that for some circuits the time required for a character to travel from the sending to receiving end of a circuit may exceed the nominal time required for character generation. For a start-stop signal this means that one or more characters may be traversing the line simultaneously. The transmission time will represent a constant time delay, but the nominal transmission rate will be unaffected by the delay. For a ready-strobe signal, however, the nominal transmission rate is seriously affected by the delay in the transmission line, since each character in a sequence requires at least twice the transmission-time delay. For example, a 750-mile open-wire circuit will delay each character of a 240 character-per-second start-stop or synchronous system about 1-character duration. After the initial 1-character delay, the start-stop transmission rate will be 240 characters per second. A ready-strobe operation requires 1-character duration for the ready and 1-character duration for the actual character transmission, reducing the rate to 120 characters per second.

Another consideration is ready-strobe operation in a network where there may be more than one receiving device. The ready signal must be related to the slowest device, and therefore the ready signal of each receiving device is conveyed to the transmitting device through an AND gate, requiring all ready signals to have occurred although not necessarily simultaneously. A reset triggered by the strobe signal for the gate is also needed.

Typical teletypewriter and data speeds are shown in Table 5. These are nominal

Table 5. Typical Teletypewriter and Data Signal Characteristics

Nominal words per minute	Actual characters per sec	Bits per character	Bauds, bits/sec	Oper-ations per minute	Start and infor-mation bit duration, msec	Stop bit duration, msec	Fund. freq. (dot cycles)
60	6.13	7.42	45.5	368	21.97	31.20	22.8
60	6.50	7.00	45.5	390	21.97	21.97	22.8
66	6.74	7.42	50	404	20	28.4	25
66*	6.67	7.50	50	400	20	30.0	25
68	6.82	11	75	409	13.33	26.66	37.5
75	7.67	7.42	56.9	460	17.58	24.96	28.4
75	7.5	10	75	450	13.33	13.33	37.5
100	10.6	7.0	74.2	636	13.48	13.48	37.1
100	10	7.42	74.2	600	13.48	19.14	37.1
100	10	7.50	75	600	13.33	19.99	37.5
150	15	10	150	900	6.67	6.67	75
600	60	10	600	3,600	1.67	1.67	300
1,200	120	10	1200	7,200	0.83	0.83	600
2,400	240	10	2400	14,400	0.417	0.417	1,200

* European.

values and will vary owing to variation in motors and gears among manufacturers. Attempts are being made to standardize teletypewriters employing data codes at speeds of 75×2^m, where m is zero or a positive integer. Eight-level data codes sent as 10- or 11-bit-per-character start-stop signals as in Fig. 4b are being used for communications and data. The average teletypewriter word is considered to be five characters long plus a space and is typified by the word Paris. A simple relationship between word and character rates is the following: Characters per second times 10 equals words per minute:

$$\text{Characters per second} \times 10 = \text{wpm} \tag{8}$$

another relationship is

$$\text{wpm} = \frac{\text{modulation rate} \times 10}{\text{units per character}} \tag{9}$$

$$\text{Example wpm} = \frac{75 \text{ bauds} \times 10}{7.5 \text{ bits per character}} = 100 \text{ wpm}$$

$$\text{Modulation rate} = \frac{1}{t_b} \quad \text{bauds} \tag{10}$$

$$\text{Dot cycles} = \frac{1}{2t_b} \tag{11}$$

where t_b is the bit duration.

Frequently there is need to slave a start-stop transmitter to the remainder of a system. There are two prevalent methods each using a separate control circuit. One is to provide a control signal at the character rate. In mechanical tape distributor transmitters this is often a clutch control, and when the pulse is received, the clutch is released for a revolution sending one character. The second technique is the so-called bit-synchronous mode. An external source sends a stream of clock pulses to the tape transmitter at the bit rate. For each clock pulse received a signal pulse is sent by the transmitter. If there is no tape to be read or the device is a keyboard which has not been operated, a steady mark signal is sent.

4. Keyboards

Message and data signal generation is accomplished by keyboards, paper-tape readers, card readers, and automatic character readers. Keyboards are quite varied, and some are shown in Figs. 5 to 7. Numeric keyboards, not shown, contain keys for 0 through 9 and keys for instructions such as add, subtract, and clear, comprising a

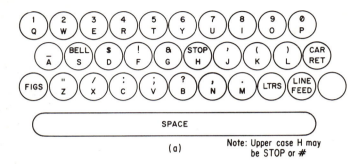

(a)

Note: Upper case H may
be STOP or #

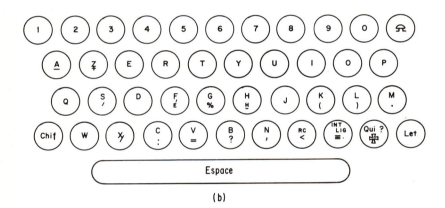

(b)

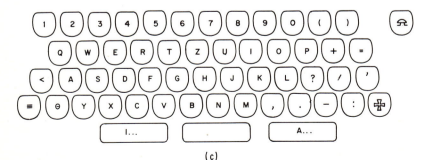

(c)

FIG. 5. Baudot keyboards. (a) American. (b) French. (*Courtesy of Société d'Applications Général d'Electricité et de Mécanique.*) (c) German. (*Courtesy of Siemens & Halske.*)

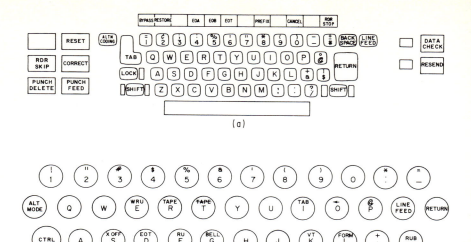

(a)

(b)

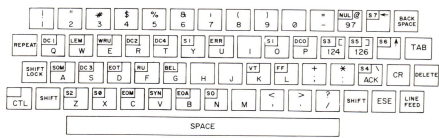

This keyboard is not intended as a standard and is included for information purposes only

Assumption: lower case alphabet will be assigned by ASA to the unassigned portion of ASCII

Key	Bits 765	Bits 765	
SHIFT	110	100 (16 characters)	
	111	101 (16 characters)	
	011	010 (first 12 characters)	
	010	011 (last 4 characters)	
CONTROL	110	000 (16 characters)	
	111	001 (16 characters)	

1. Shift produces upper character on key top (not within small square) or upper case of a letter

2. Unshift produces lower character on key top or the lower case of a letter

3. Control (CTR) produces the character within the small square on the key top

Note: 97, 124, and 126 on the @ [,] keys respectively are the numeric numbers of the ASCII characters that are still unassigned after placing the lower case alphabet in the unassigned position

(c)

FIG. 6. Communications and data keyboards. (a) IBM 1052 keyboard. (*Courtesy of International Business Machines Corporation.*) (b) Eight-level data interchange code keyboard. (*Courtesy of Teletype Corp.*) (c) Proposed U.S. Army ASCII teletypewriter keyboard.

total in the order of 15 keys. The alphanumeric keyboard on the other hand must cover the alphabet, numbers, punctuation, and special symbols.

Common are the so-called QWERTY keyboards with alphabet arrangement similar to that of American office typewriters. The normal French keyboard (Fig. 5b) is referred to as an AZERTY keyboard. The alphabet differs from the QWERTY in the interchange of A & Q and W & Z positions and relocation of M and period. The German keyboard (Fig. 5c) is QWERTZ arrangement. It differs from QWERTY in the interchange of Y and Z location. Punctuation and special symbol assignments

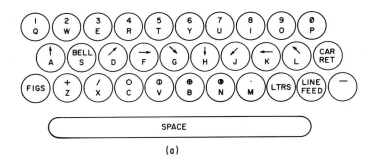

(a)

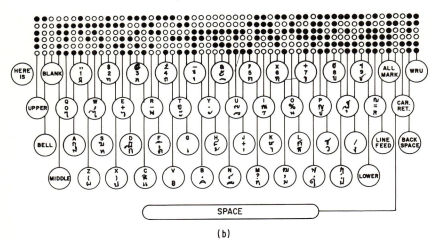

(b)

Fig. 7. Special keyboards. (a) Weather keyboard. (b) Thai keyboard.

vary much more than alphabetic assignments even within a country. When a keyboard such as in Fig. 5a is used for the Western Union Telex Service, FIGS D is "Who are you?", designated as WRU or as a Maltese Cross as on the German keyboard. Sending this combination on a Teletype Corporation Model 32 automatically triggers the answer-back mechanism at the called station. Another key "Here is" triggers the answer-back mechanism of the local station. The Telex keyboard has bell on FIGS J instead of FIGS S, apostrophe on FIGS S, and $ on FIGS F.

A keyboard used for off-line message preparation on punched tape usually has a character counter and an alarm such as a bell or light to let the typist know that the end of a line has been reached. The Mite Corporation Teletypewriter Set AN/TGC-15 has a control on the keyboard which varies the number of characters before the end of line indicator operates.   The counter does not count LTRS, FIGS, LINE FEED, and BLANK.   The keyboard has a break button for opening the series circuit for control purposes and a send receive-receive switch which shunts the keyboard contacts when in receive position to prevent inadvertent transmission during reception.   The keyboard is readily removable and retracts into the set when not in use.

Figure 6a shows the keyboard used for the IBM 1052 Data Communications System.   Although enclosed in the same cover as the primary printer, the 1052 alphanumeric keyboard is similar in appearance and operation to the standard IBM typewriter keyboard.   The 53 keys place at the finger tips of the operator not only 88 different characters and symbols but also the codes and controls necessary for the operation of the system.   Seven of the 53 keys are function keys: space, backspace, line feed, carriage return and line feed (CR/LF), tab, shift, and shift lock.   By pressing one of the numerical keys while holding down the alternate coding (ALTN CODING) key, the operator originates a single-character code such as restore, bypass, reader stop, end of block (EOB), end of address (EOA), prefix, end of transmission (EOT), or cancel.   The output from the keyboard is a 6-bit (plus a parity bit) BCD code (Table 3) that is used throughout the system. When using the keyboard, the operator can enter data manually to be transmitted to a central terminal or to be recorded locally on the printer, paper-tape punch, or the card punch.

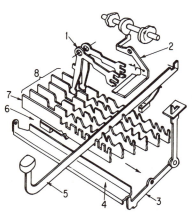

FIG. 8. Keyboard code selection. (*Courtesy of Siemens & Halske.*)

Figure 6b is an ASCII keyboard presently being used for message equipment.   This keyboard is adequate for message requirements. For military data and message needs the keyboard of Fig. 6c is being considered.   It has the advantage that through the use of the shift and control keys in conjunction with the other keys, all 128 code combinations of ASCII can be generated whereas many other keyboards are abbreviated, generating only that portion of the code required for the particular application.

There are many special-purpose keyboards such as for weather and for figurative languages.   The weather keyboard of Fig. 7a is essentially the same as the Baudot of Fig. 5a except that the keytop identifications differ and correspond to the characters printed by the printer as in Table 1.   Pressing corresponding keys on the two keyboards generates the same code combination.

In certain parts of the world, languages differ greatly within short distances, even within one country.   It is hoped that teleprinters will help communications, since it is usually easier to recognize and translate written than verbal messages.   The keyboard of Fig. 7b shows one such arrangement for the Thai language; there are many others such as Burmese and Hindi.

The operation of a typical keyboard may be explained by reference to Fig. 8.   The setting of the five code bars 8 determines the code be transmitted.   When one of the keys is depressed, the key lever 5 presses on the inclined slopes on the upper edge of those code bars which should be moved to create the desired code combination.   Code bars already in correct position are so notched that no code bar motion results when the key is pressed.   The inclined V notches will cause the code bars to move to the left (space pulse) or to the right (mark pulse).

The code bars 8 move bell cranks 1 about the pivot shown, causing them to assume settings corresponding to those of the code bars.   A cam-controlled locking bail 2 locks the bell cranks 1 and therefore the code bars 8 in fixed position during transmission. This prevents other keys from being depressed (key lock) during transmission.

A sixth code bar 7 is shifted to the left when function key "A———" is operated.

This locks all keys assigned to "figures." When function key "1——" is operated the locking bar 7 moves to the right blocking all keys bearing letters. Detent 6 keeps locking bar 7 in position.

Any key which is pressed will cause the common start bail 4 to move lever 3. The vertical arm of lever 3 engages the release lever of the transmitter, releasing a clutch for one revolution. By means of cams and levers, the position of each one of the bell cranks 1 is sampled in turn and results in an open or closed condition of a switch. Start and stop pulses are added, completing the transmit signal.

Of importance for a keyboard are the pressure, interlocking characteristics, and distortion. The last will be discussed under signal distortion. Generally attempts are made to get the touch as light as possible, but owing to the movement of code bars in a mechanical keyboard, appreciable force is required in some teleprinters. This is in the range of 8 oz. The operation of the code bars affords a natural interlock, preventing two keys from being depressed simultaneously. Electronic keyboards have been made or proposed which utilize one of a number of principles such as the making of contacts in a form to generate the code directly; a change of inductance, capacitance, or resistance in the code circuits; or the interruption of light beams to generate the code corresponding to the key pressed. For most of these keyboards, the touch can be very light; however, it is still necessary to operate an interlock mechanism to prevent the simultaneous pressing of two keys and to avoid exceeding the line speed of the teletypewriter circuit.

An additional requirement for some keyboards is to incorporate a slaving mechanism for interoperation in a synchronous system. This usually functions on a character basis wherein a pulse from the synchronous system is a demand for a character. If a character is not available, steady mark condition is sent. The slaving could also be on a bit synchronous basis. For such systems and to keep smooth operation even with bursts of speed, one or two characters of storage are sometimes employed.

Another way to smooth operator's speed irregularities is first to store the message and then transmit the completed message. Means of storage are punched paper tape or small amounts of magnetic tape. For example, a metallic magnetic tape has been used for the storage of up to 500 characters.* When the message is stored, the tape is rapidly rewound by a spring mechanism in order to place the beginning of the message at the read head. This rewinding is unnecessary in one system† which uses a magnetic drum. The drum has a capacity of 3,600 characters.

5. Paper-tape Readers and Distributors

A paper-tape reader is a device which reads the punched holes or code marks on a paper tape and generates corresponding parallel output signals. A tape transmitter-distributor (TD) on the other hand reads similar tapes but generates a single serial stream of bits corresponding to the code on the tape. Sensing is usually accomplished by having mechanical fingers push up against the tape in order to sense whether a hole is present. There are two general forms of punched tape—the fully perforated and the chadless. The holes of chadless tape are partially perforated, leaving the flaps of chad still attached to the tape, hence the term chadless, since there is no need for chad disposal. Finger sensing is equally effective on these two types of tape, whereas light sensing, which has become more prevalent in recent years, is more simply accomplished with fully punched holes. Light sensing is also employed for reading marked tape.

A novel tape sensing technique uses fluid-flow sensing.‡ Filtered but unregulated air is blown through air jets positioned to correspond to the hole pattern of the punched tape to be sensed. The tape being read acts as a valve allowing air flow only through jets where holes exist in the tape. Air flow from each jet expands in a venturi throat drawing air through an associated additional duct in the side of the venturi. This contains a hot wire sensing element (about 600°F) which is one arm of a balanced

* Teletypewriter Set AN/AGC-1 made by Kleinschmidt Division, Smith Corona Marchant Corporation.
† Teleprinter Set AN/AGC-3(XA-2) made by Motorola Inc.
‡ Courtesy of Soroban Engineering, Inc.

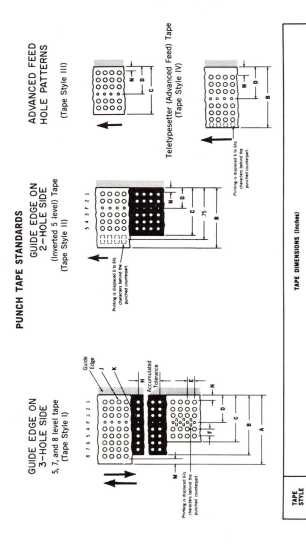

PUNCH TAPE STANDARDS

GUIDE EDGE ON 3-HOLE SIDE — 5, 7, and 8 level tape (Tape Style I)

GUIDE EDGE ON 2-HOLE SIDE (Inverted 5 level) Tape (Tape Style II)

ADVANCED FEED HOLE PATTERNS (Tape Style III)

Teletypesetter (Advanced Feed) Tape (Tape Style IV)

TAPE DIMENSIONS (Inches)

TAPE STYLE	A	B	C	D	E	F	H*	J	K	M	N
I**	1.000 ± .003	.875 ± .003	.687 ± .003	.392 ± .003	.100 ± .003	.100 ± .002	6.000 ± .009	.072 ± .002/.001	.046 ± .002/.001	.072–.047–.059	.056
II	------	.875	.686	.297 ± .002	.100	.100 ± .001	6.000 ± .014	.072 ± .0005	.047	.242–.053	.061
III***	------	------	.687	.387	.100	.100 ± .001	------	.072 ± .0002/.0000	.0475 ± .0002/.0000	.064	.051
IV***	------	.875 ± .003	------	.434 ± .002	.100	.100 ± .001	5.000 ± .007	.072 ± .001	.0476 ± .001	.105	.098

* Soroban Tape Punches maintain feed tolerance of 6.000±.005 inches for 60 punched characters.

** Dimensions presented correspond to EIA Proposed Standard RS-227 for 1 inch 8 level tape.

*** In tape with advanced feed, the leading edge of the sprocket hole is in line with the leading edges of the code holes.

In other tapes, the center line of the sprocket hole is on the center line of the code holes.

PUNCH CARD STANDARDS

Dimensions correspond to EIA Proposed Standard,
Engineering Committee TR 27.6.1 March 1962.

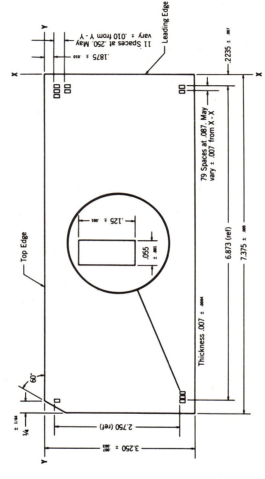

Fig. 9. Punched paper tape and cards. (*Courtesy of Soroban Engineering, Inc.*)

electrical bridge. Output is 4 volts for no-hole and 8 volts for hole. High speeds of 1,000 characters per second continuously or 300 characters per second on a character-for-character basis are claimed with high accuracy while tapes which are transparent or opaque are being read.

Tape readers and TD's are made to read from 5- to 16-hole tape, the most common being 5-hole Baudot. This is usually on $1\frac{1}{16}$ or $\frac{7}{8}$-in.-wide tape. Six- to 8-hole code is usually put on $\frac{7}{8}$ to 1-in.-wide tape. The tape material may be oiled paper, nonoiled paper, opaque paper, Mylar, Mylar and aluminum laminated, or thin metal. Hole separations are usually $\frac{1}{10}$ in. Hole diameters vary somewhat among manufacturers, but attempts are being made to standardize on diameters of 0.072 and 0.046 in. for code and feed holes and a cumulative hole spacing error of no more than 0.015 in. in 6 in. of tape. See Fig. 9.

When code holes or markings are sensed in readers, switches or circuits associated with the holes are operated and a strobe pulse is generated, indicating that the outputs are available to the external circuits. For a TD, the output of these circuits or switches is sampled in sequence for transmission. For some mechanical Baudot TD's, there are the five contacts which are opened or closed, depending on the sensing finger positions and an additional master contact which opens and closes the external circuit. In other Baudot TD's there is only one contact and mechanical linkages to the sensing fingers which determine the opens and closures of it. In one form of electronic TD the mechanical distributor and camshaft are replaced by a ring counter which counts completely around once for each character. The condition of each flip-flop in the ring counter is determined by a light or dark condition on photodiodes, and each stage of the ring counter is presented to the output circuit sequentially. For Fieldata operation conversion circuitry must be included to convert from paper-tape form to the standard code form (see Telegraph and Data Codes, paragraph 2).

The usual TD has three control positions: operate, stop, and tape release. The last position causes retraction of the feed-hole pins and permits tape to be inserted or withdrawn. In addition there is usually a tape lid which holds the tape down for sensing. This lid can be snapped open in order that a continuous tape loop can be inserted for repeated messages or for testing. Another tape sensing finger or photodiode is provided to stop transmission when there is no tape in the sensing position. A tight tape arm is often provided which will stop the transmitter when there is abnormal tension on the tape or when there is a tangle. This is used when a reperforator and TD are mounted together so that the TD stops when a tape loop between it and the reperforator is consumed by transmission and the tape becomes taut. For synchronous modes of operation, there are means for externally controlling the sending of characters. This requires that the reader or TD be capable of stopping on each character being read and starting again under the external control. For data systems extremely high free running speeds are feasible, since reading is electronic but lead and lag times must be used for starting and stopping. For example, a 2,000-character-per-second* reader may require 3 msec start time and less than 1 msec stop time. Because of mechanical limitations starting and stopping times are limited to between $\frac{1}{2}$ and 1 msec.† Some tape readers are designed to read code marks made on paper tape as well as punched holes. These also are high-speed devices operating at speeds up to 2,000 characters‡ per second with start and stop times of less than $2\frac{1}{4}$ and $\frac{3}{4}$ msec, respectively.

Another form of reader is the block tape reader.§ These devices sense from 5 to 20

* Digitronics Model 3500.

† Ferranti Electric Inc. Model 196E reaches speed in a little less than 1 msec and stops in about $\frac{1}{2}$ msec, permitting synchronized stop-start operation up to 380 characters per second.

National Cash Register 472-1 Punched Paper Tape Reader free-runs at 1000 characters per second, stops on a single character, and accelerates in $\frac{1}{2}$ msec.

‡ Omnitronics, Inc. 1,000 characters per second for 10-character-per-inch packing and 2,000 characters per second for 20-character-per-inch Omni-Data recorded tape.

§ Electronic Engineering Company.

Anelex Corporation.

characters simultaneously and offer several advantages such as ability to assign significance to position of characters in the block (formatting), performing logic functions on different portions of the data, parallel feed of characters to an output device, one tape movement for a number of characters, and application of error control on a block basis.   Information from the sensing contacts, pins, or photodiodes is sometimes transferred to a storage matrix, so the tape is allowed to move while transmission takes place.

Special features are included in some readers such as certain code recognition with associated performance of control functions, bidirectional operation, and power-driven feed and take-up reels.[11]   Cassette or tape magazine* storage is available for holding up to 50 ft of 0.004-in. punched tape in the form of a continuous loop.   This features small size because of feed from the inside of a roll of tape and simultaneous rewind on the outside with drive supplied by the reader.   Another auxiliary device, Teletype Model 28 Multiple Wire Distributor, converts five level, parallel intelligence, such as the output of a reader into serial, 7.42 unit, 20 or 60 ma, start-stop signals at 60, 75, and 100 words per minute.   A corresponding Receiving Selector performs the inverse conversion at the receiving end of a circuit.

For a torn tape message center TD's are often ganged in groups of two or three with a common motor.   Automatic message-numbering generators are also integrated into the TD assembly, which has tape holders for messages to be transmitted and bins for holding punched tapes.   Extensive precautions are taken to prevent loss of torn tapes; bins have no slots or openings large enough to lose tapes.   Paper-tape readers are made for a variety of mountings, including table top, rack, panel, or console, and are sometimes mounted on page printers or tape reperforators to form sets.

6. Reperforators and Tape Punches

The term teletypewriter reperforator signifies a device which receives teletypewriter signals and punches corresponding code holes in a tape.   The holes may be fully punched or partially punched (chadless).   A teletypewriter perforator usually denotes a device which is not capable of accepting external signals but punches code holes in tape with information derived from a local keyboard.   The term punch is commonly used in data systems to indicate devices which can punch code holes in tape, cards, or other media from input signals or locally operated keyboards, picks, etc.   The main differences between these communications and data devices are speed, interface, and code employed, and therefore the terms "reperforator" and "punch" will be used synonymously.   Common speeds for reperforators and punches are 60, 66, 75, 100, 200, 300, 600, 750, 1,000, 1,200, 1,500, 2,400, and 3,000 words per minute (wpm).   The usual teletypewriter equipments operate at the lower speeds up to about 100 wpm, whereas most of the data equipments operate at the higher speeds.   More recently there has been a tendency toward combined teletypewriter and data capabilities within single equipments.   This trend has become prevalent owing to the use of voice bandwidth circuits for higher transmission speeds for message traffic and to the use of a common code for communication and data systems.   Tapes are $11\frac{1}{16}$, $\frac{7}{8}$, and 1 in. wide and usually are of oiled-paper stock to reduce punch block and pin wear. The paper is of special but inexpensive type with short fibers to simplify the punching operation and the elimination of the chad.   For the tape widths mentioned, 5 to 8 holes are punched, but reperforators are available for punching up to 16 holes on wider tape.   A chart for determining length of tape on a reel is shown in Fig. 10. This is for unperforated tape.   Perforated tape does not pack so tightly, and the take-up reel of a reperforator should accommodate a slightly larger diameter of tape than the tape supply reel.

Owing to the use of newer codes and to frequent personnel replacement it is becoming more important to use typing reperforators which print characters on the tape as well as punch holes.   This typing may be in the vicinity of the feed holes or may be along the lower edge of the tape.   Chadless tape permits greater flexibility in the location of the printing.   Printed characters are usually displaced from the corresponding

* Ferranti-Packard Type 9169 Tape Magazine.

punched holes because of the difficulty in placing punching and printing mechanisms in the same location and because of the desire to have both operations take place simultaneously. Printing 6, 6½, or 7½ characters behind the punching is common practice. Typical punching arrangements on tape are shown in Fig. 9. Direction of tape motion is very significant; systems exist with opposite directions of tape feed. For interoperation of these seemingly incompatible systems, the tape from the punch of one system can be inserted upside down in the reader of the other system. In some systems the leading edge of the feed holes are in line with the leading edges of the information holes (advanced feed holes*). This helps insert tape which has no printing into a TD properly, since with the usual in-line systems it is possible to have a piece of torn tape upside down and turned end for end without the operator realizing it. The characters would be read correctly but in reversed sequence. The selection

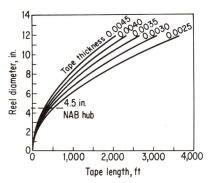

FIG. 10. Paper-tape length vs. diameter. Curves based on 2-oz winding tension. To determine length of tape on any reel, locate curve corresponding to tape thickness. Subtract tape length corresponding to inside diameter from tape length corresponding to outside diameter. (*Courtesy of Omnitronics, Inc.*)

mechanisms, the input circuits, the synchronization requirements, and range control of low-speed teletypewriter reperforators closely resemble those for teletypewriter printers, which will be described subsequently.

High-speed typing reperforators use special mechanisms to achieve the typing speed usually with only one printing head. Soroban Model PT-1 contains a punch and a separate and detachable print head capable of printing at 100 characters per second. The type font consists of raised letters on an octagon drum about ⅜ in. in diameter and 1 in. long. Eight characters on each of the eight faces permit printing of 64 symbols. Simple exchange of type fonts permits rapid change of codes or type styles. For character selection the drum is rotated and positioned axially to align the desired symbol with the print hammer. An inked ribbon is used for printing.

The magnets used for control of the punching, tape feed, printer positioning, and print hammer require well-designed driving circuits. These are external to the punch and printer mechanisms. Figure 11 shows one such control circuit. There is one code interposer circuit for each code hole and sprocket hole which might be punched per character. For example, an eight-level code would require nine circuits. Synchronization is generated within the punch, resulting in timing signals which control the time interval during which the selected code magnets must be energized. For this particular system two printer coils A and B are required for each significant bit of the printing code. Either A or B is energized, depending on the phase position of the printer, either to change an eccentric position (0 to 1 or 1 to 0) or leave the eccentric in the 0 or 1 position it was in after the previous character. The PT-1 printer requires 6 bits for determination of one of 64 characters. Three bits determine one of eight rotational positions, and 3 bits one of eight translation positions. Therefore there are six pairs of A and B coils with corresponding driving circuits. While tape punching is practical at speeds up to 3,000 wpm, typing on the tape is limited to about 1,000 wpm.

Two modes of systems operation are prevalent: start-stop for communications and ready-strobe for data. The significance of the start-stop is that the device must be adjusted for the nominal transmission rate (for example, 100 wpm, 74.2 bits per second). It is usual for a start-stop reperforator operating on 7.0, 7.42, or 7.5 unit codes to cycle in 6.5 units in order to be ready for the start signal and to allow for

* Friden, Inc. Models 5, 6, 7A, and 8B and Soroban LP-2 and GP-2 can be equipped for tape with advanced or centered, in-line feed holes.

speed variation. On a mechanical reperforator speed selection requires use of the proper gears, while on an electronic reperforator speed change requires only switching to another oscillator frequency.

Reperforator applications include off-line tape preparation, torn-tape relay, monitoring, milling and other machine control, and computer operations. Off-line tape preparation was mentioned under Keyboards. It provides a means of saving line

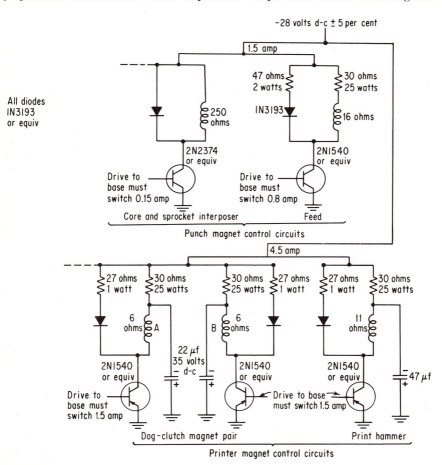

FIG. 11. Typical typing reperforator control circuit. (*Courtesy of Soroban Engineering, Inc.*)

time, simplifying synchronization, and providing message composition and verification prior to transmission. The erratic keyboard rates are eliminated because transmission takes place by reading the tape at a uniform rate. In general a nonprinting reperforator does not need code recognition, since it punches holes corresponding to the incoming signal. A typing reperforator must be cognizant of the code. Functions such as "backspace" are not performed by the reperforator, but back-tape capability is almost always provided by a manual control on the reperforator or by a separate control wire so that errors in tape preparation may be corrected by locally backspacing and punching all holes (LETTERS for Baudot, IDLE for Fieldata, or Delete for ASCII) over the incorrect character. If there is typing, it is overprinted or ignored. Page printers usually ignore the delete characters except for the Baudot case, where some precaution must be taken. For example, if numbers are being punched in the tape and an error is made, the tape is backspaced and all holes (LETTERS) punched. It is then necessary to generate the FIGURES character before continuing to punch numbers.

Characteristics of some available reperforators are shown in Table 6.

High-speed punches operate in a variety of ways. Some have reciprocating drive pins, and when control magnets are operated, interposing pins permit the drive pins to push punch pins. Some restrain a motion which would prevent punch-pin operation. For example, a buckling knee in the punch linkage may prevent punching. When the marking magnets are operated, the knees are prevented from buckling and the punch pins are forced through the paper and punch block hole. This technique is shown in Fig. 12.

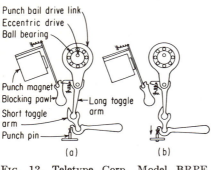

Fig. 12. Teletype Corp. Model BRPE punching principle. (*a*) Punch magnet not energized (space). (*b*) Punch magnet energized (mark).

Figure 13 shows another form of interposing. The punch is a high-speed device designed primarily for use as a computer output equipment. It is applicable in instrumentation and high-speed data transmission. A parallel-type input is fed to the punch magnets H, Fig. 13, causing pins J to be positioned in selected or unselected conditions as shown. When the magnet H is energized, the punch pin J is pushed against the paper. At the appropriate time cam $A3$ will force the die plate C up

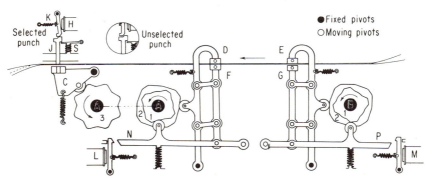

Fig. 13. Creed Model 3000 punch and feed mechanisms. (*Copyright* 1964 *by Creed & Company Limited.*)

against the pin J and punching will take place. D and E oscillate forward and back, out of phase with each other under the influence of cams $A2$ and $B2$. With current flow in feed magnets L and M, levers N and P are free to follow the motion generated by cams $A1$ and $B1$, causing F and G alternately to grip the tape and move it to the left, step by step, $\frac{1}{10}$ in. at a time.

Another concept for a high-speed tape punch uses a tuned reed, punch pin, and magnet for each hole to be punched.* The advantages are high speed up to 2,000 wpm and the quiescent state of the mechanisms when signals are not being received.

* Teletype Corporation Model DRPE.

Table 6. Reperforator Characteristics

Reperforator	Speed without printing	Speed with printing	Code levels	Take-up reel	Tape bin	Input	Tape width, in.	Serial bits per character	Separation code holes and printing	Comments
Teletype Corp. Model 28RT	200 wpm	100 wpm	5	1000'*	100'	Parallel or serial	11/16	......	......	Can receive serial and generate parallel output signals. Manual gear shift available. Multimagnet. External magnet release
Teletype Corp. Model 28LARP	20 char/sec	......	5, 6, 7, 8	Yes	No	Parallel	11/16, 7/8, 1	......	......	
Teletype Corp. Model 35ROTR	100 wpm	......	8	Yes	No	Serial	......	11.0	......	External control
Teletype Corp. Model BRPE	110 char/sec	......	5, 6, 7, 8	No	No	Parallel†	11/16 or 1	......	......	Tuned reed principle. No moving parts when waiting on signal
Teletype Corp. Model DRPE	200 char/sec	......	5, 6, 7, 8	No	No	Parallel	11/16, 7/8, 1	......	......	
Teletype Corp. Model 28LPR	......	100 wpm	5	No	No	Serial	......	......	6 ch	Manual gear shift. Both partial or fully punched tape. Code reading contacts for parallel wire output
Military TT-107‡	100 wpm	100 wpm	5, 6, 7, 8	No	No	Serial	11/16, 7/8	7.42	6 ch	Paper, foil, or Mylar tape
Tally Model 420	60 char/sec	......	5, 6, 7, 8	6, 7½ or 10' diam	No	Parallel	11/16, 7/8, 1	......	......	
Tally Model P150	150 char/sec	......	5, 6, 7, 8	Yes	No	Parallel	11/16, 7/8, 1	......	......	Remote backspace. Paper, foil, or Mylar tape, checks parity
Tally Model 575	75 char/sec	......	5, 6, 7, 8	Yes	No	Parallel	11/16, 7/8, 1	......	......	Bidirectional
Tally Model 820	40 char/sec	......	16	6" diam	No	Parallel	2	......	......	Contains code conversion from standard to paper tape Fieldata code. Has tight-tape and tape-out switches
Kleinschmidt Model 310	40 char/sec	......	8	250'	No	Parallel	1	......	......	
Olivetti T2PN and T2PR	60–75 wpm	......	5	No	No	Serial	11/16	7.5	......	T2PN has keyboard transmit red printing, receive black printing. T2PR receives only
Siemens & Halske T loch 22b	30 or 60 char/sec	......	5	Aux	No	Parallel	11/16	......	......	30 char/sec is for external control; 60 char/sec is for free ability to pick characters from storage
Siemens & Halske T loch 15d	100 wpm	100 wpm	5	Yes	No	Serial	11/16	7.5	7.5 ch	

* On associated TD.
† Auxiliary unit can convert serial to parallel.
‡ Manufactured by Kleinschmidt Laboratories.

Table 6. Reperforator Characteristics (*Continued*)

Reperforator	Speed without printing	Speed with printing	Code levels	Take-up reel	Tape bin	Input	Tape width, in.	Serial bits per character	Separation code holes and printing	Comments
Creed Model 3000	300 char/sec	....	5, 6, 7, 8	Yes	....	Parallel	1⅟16, ⅞, 1	....	....	Optional code check-back facility inspects code photoelectrically three characters after punching point. Tape-low warning device. Step-by-step punching up to 300 char/sec
Soroban Models: PT-1 LP-2	.... 150 char/sec	100 char/sec	5, 6, 7, 8	10½″ diam	No	Parallel	1⅟16, ⅞, 1	....	Printing 6½ ch to right of code holes	Maximum accumulated feed error ±0.005 in. in 6 in. of punched tape. Punching mechanism uses oil splash lubrication
Soroban Model GP-2	300 char/sec	....	5, 6, 7, 8	10½″ diam	No	Parallel	1⅟16, ⅞, 1	....	....	Same comments as for PT-1. Synchronizing is by means of a pulse from a magnetic pickup from a disk
Omnitronics ETR-7	1,500 char/sec	600 char/sec	5, 6, 7, 8	8″ and 10″ diam	No	Serial or parallel	1⅟16, ⅞, 1, 1¼	Baudot 7.0 bits (ASCII) 10 bit 9 bit	In line	Electrostatic printing on tape instead of punching holes. Edge interpretation or printing in sprocket hole area
Omnitronics ETR-8	60 char/sec	60 char/sec	5, 6, 7, 8	8″ and 10″ diam	No	Serial or parallel	1⅟16, ⅞, 1, 1¼	Baudot 7.0 bits (ASCII) 10 bit 9 bit	In line	Electrostatic printing on tape instead of punching holes. Edge interpretation or printing in sprocket hole area

Reperforators for torn-tape relay centers have some special features such as large tape bins with special consideration to prevent possible loss of a tape message through cracks or spaces in the bins. They also have large chad bins and sometimes a large supply reel of tape. Alarms are also provided such as power off or low tape supply; the latter is also provided for many reperforators not used in tape relay centers. Ganging of several reperforators is also resorted to for economy. In some applications a certain amount of automatic blank tape feed-out after the end of message signal is programmed into the reperforator operation. For automatic message centers employing paper tape, reperforators are often used for incoming lines, cross-office storage, and intercept storage and as such are associated with tape readers and TD's through tape loops and tight tape controls.

They are also used as monitors on outgoing lines and have take-up reels because of the large amount of traffic and the value of storing the reels for future reference. When radio circuits are involved, special TD's are sometimes located at the monitor positions in such a manner that reruns of a number of messages on the take-up reels may be accomplished without tearing the tapes. This is essential with multiplexing since 16 and sometimes as many as 40 teletypewriter channels are lost during a fading condition, and if the condition is not immediately detected at the transmitting station, a large number of messages are sent but not received. These must be retransmitted, and it is most simply accomplished from the many monitor positions.

Reperforators or punches used for computer systems have parity checking capabilities and are made to stop the transmission, ring an alarm, or print and punch an error symbol when parity does not check. For computer operation reader and punch need not have the same operating speed. For example, a reader may operate at 300 characters per second while a punch in the same data set may be capable of operating only at 30. Usually the holes punched correspond directly to the incoming coded signals; however, for Fieldata operation there is a paper-tape form of the code, and when this is used, conversion circuitry is required (see Telegraph and Data Code, paragraph 2).

Higher speed paper-tape equipments employ tape marking rather than punching. Instead of code holes being punched, electrostatic charges are formed on the tape corresponding to a hole or a 1 signal condition. Black developing powder makes the charges visible. With this technique high-speed printing of characters is also practical. The tape is coated with a surface capable of accepting and retaining an electrical charge. One technique is to charge the surface and expose selective areas to light, causing bleeding of the charge.* Another technique does not involve light; the charge is produced by electrodes near the surface of the paper.† The latent images are developed by passing the tape through a dry, extremely fine black powder which adheres to the charged surface. Heat is applied to melt a resinous component of the powder, which then adheres firmly to the paper. The time for the charging process is in the order of 1 or 2 μsec. One of the limiting factors on speed is the mechanics of a character-by-character operation. Starting and stopping a tape each require in the order of $\frac{1}{2}$ to 1 msec or more. Electronic or magnetic core buffering or synchronous operation permits speeds in excess of a thousand characters per second. Reading of such marked tapes is effected by measuring reflected light. High degrees of reliability and long "reperforator" and reader life are claimed. A variation of the technique is used for the printing of five columns of characters on a 1-in. tape at a rate of 300 lines (1,500 characters) per second.‡

Tape perforators for off-line tape preparation are made with and without typing. They are used for off-line preparation of punched tape and feature low cost, simplicity, and reliability, since the input is only from a local keyboard. Teletypewriter perforators have existed for many years and have a full alphanumeric Baudot keyboard. A typical data-oriented numeric perforator§ has a keyboard through which 20 different

---

* David Sarnoff Research Center, RCA.
† Omnitronics, Inc., Electrostatic Recorder.
Tele-Dynamics Division, American Bosch Arma Corp.
‡ Omnitronics, Inc., Electrostatic Instrumentation Printer.
§ Tally Corporation Model 430.

codes may be punched in the tape with the corresponding character printed to the left of the code between feed holes.   Punching rate is up to eight characters per second.   The customer has flexibility in choosing the codes and printed symbols, although it is common for 10 of these to be the decimal digits.   One key is a *delete* which punches all holes.   Each character can be an eight-unit code or less.   Tapes are used for computer programming, numeric control, accounting, etc.

Another tape-preparation device* has a full alphanumeric capability permitting a choice of up to 48 codes and symbols consisting of 26 alphabetic, 10 numeric, 5 punctuation, and 7 symbols.   Code changes are made by matrix card change.   Operating speed is up to 10 characters per second.   It differs from the usual perforator in being able to operate to or from remote devices.

7. Page Printers

Until recent years page printers were developed almost exclusively for teletypewriter use and were almost entirely mechanical.   With a trend toward higher speeds and toward data applications many of the mechanisms are being replaced by electronic circuits.[12]   Other printers, still of mechanical design, feature smaller size, longer life, simplicity, or economy.   Significant factors in printer design depend upon code used, speed of operation, type of signal, width of page, and special functions.

Most teletypewriters use a form of the Baudot code shown in Table 1.   The letters shift characters are quite standard.   The Figures shift characters have variations among different countries.   The Figures Comm (Communications) column is fairly representative of American assignments, whereas the French and Germans comply with the International No. 2 Code.   As noted, major differences exist for Figures D, F, G, H, J, S, V, and Z.   For F, G, and H the French assign É, %, and H̲.   To effect interoperation operators are usually instructed not to use these figures shift characters except bell, which is a desirable function.   This may be used in a system of intermixed equipments by having the operator send both Figures J and S.   Some American equipment does use WRU for Figures D.

During the past few years many page printers have also been made for data use. Some of these use commercial codes, and a number were designed to utilize the military Fieldata code of Table 2.   More recently there has been a trend toward the use of the American Standards Association code, Table 4, for both data and teletypewriter message transmission.

a. Printer Response to Code.   Code combinations or characters may be broadly divided into two categories; those that cause printing and those corresponding to functions.   In specifying or designing a printer or system, each character of the code must be examined and a determination made as to the desired printer response.

For the Baudot code, the differences between American and European codes already mentioned require special interoperation consideration.   In addition, the motors of some printers can be stopped by sending Figure H to the printer.   More recently this character is used for the number sign, #, and the printer incorporates a motor time out which turns off the motor if signals are not received for a specified time, usually between 30 sec and 10 min.   Receipt of any character or pressing the break button at the transmitter will turn the motor on, but it may take several character's duration to reach correct speed.   Another symbol varying in response is Figure D.   Receipt of this character may cause printing of $, or this may be recognized as WRU (Who are you), in which case the printer is programmed to automatically transmit its identification.   One technique for the latter is to have a small cylinder with metal code bars projecting radially.   For each position of the cylinder, the bars act as a set of five cams, closing five switches.   By breaking off appropriate code bars, any character code can be formed.   There often are as many as 40 sets of code bars around the cylinder, and upon receipt of WRU the cylinder rotates once, transmitting the characters in sequence.   The code bars or the complete cylinder are replaceable, since choice of characters is made by actually breaking off the code bar at the cylinder.

* Navigation Computer Corporation, Model 1010.

The underscore symbol often poses a problem, since there is no backspace. It becomes necessary to use a carriage return and space the printing mechanism to the desired position. This assumes that the carriage return code does not also cause line feed.

Figure S rings the bell on most printers; this can be continuous ringing or just a gong. On other printers, it may also light an alarm to get the operator's attention.

Most printers ignore the Blank character, although it may also be used to space the printer mechanism one character position or it may cause printing of $\approx$ as on a perforated tape. Line feed is usually adjustable for one-, two-, or three-line spacing for each line feed character received.

The American Standard (ASA) code poses other decisions, since it contains many function (control) characters which must be defined and incorporated by the designer. The user should define to the teletypewriter supplier which characters the receiving equipment should respond to and in what manner, as well as which characters should be ignored. Foremost among these are the characters of colums B_7, B_6, B_5 = 000, 001, and 111 in Table 4. The ERR code is often associated with check of the eighth bit of the code, the parity bit. It may have system implications, since it may involve automatic transmission of the ERR code from a receiving device to the transmitter requiring a two-way circuit.

It is sometimes desirable to print symbols for all code combinations to cover the whole code comprehensively as well as indicating when the parity does not check. These requirements cause such complexity that they are usually compromised. For example, the printer may only respond to 6 of the 7 information bits of the code and print up to 64 characters. It may be satisfactory to use an error symbol when the parity does not check, thus covering a large number of nonacceptable code combinations with one symbol. Receipt of a control character may be indicated by printing the corresponding alphanumeric character and overprinting some mark to show that it is a control character if printer expense and complexity can be tolerated. Some users resort to a reperforator which can punch almost any code combination received.

b. Speed of Operation of Printers. Some of the common speeds are shown in Tables 5 and 7. Up until about the last two decades the maximum teletypewriter speed used was 60 words per minute (wpm). Since then most teletypewriters have been built with capability of operating up to 100 wpm. With the innovation of data systems, page printers have been built with speeds as high as 500,000 wpm. The higher speed printers are expensive, large, complex equipments used solely for computer output and have not had extended applications because of the great amount of traffic handling necessary to justify expense. Printers operating in the range of 12,000 to 15,000 wpm are more commonly used as computer output. These are referred to as line printers and generally have 120 to 150 characters per line. Common speeds are 600 to 1,000 lines per minute. Other printers have been built bridging the gap between the 100-wpm teletypewriter equipments and the 600-line-per-minute (12,000 wpm) line printers of computer systems.

One equipment made by Kleinschmidt Laboratories Inc. is capable of operating at 200 wpm and capable of operating with the Baudot code or the Fieldata code. The conversion from one code to the other is accomplished through exchange of several electrical circuit cards, the device used for printing and the keyboard. The Baudot code uses a three-row keyboard, while the Fieldata and ASA codes use a keyboard with at least 4 rows, as shown in Fig. 6.

Next higher speeds proposed for teletype and data transmission are 400, 600, 750, 1,200, 1,500, 2,400, and 3,000 wpm. The speeds below 200 wpm had been selected in order to use narrowband, essentially d-c circuits, permitting in some cases, teletype and speech multiplexing on a single voice circuit. It is also possible to have from 16 to 40 teletype circuits in a 3-kc band. In addition, lower speeds were adhered to because of the complex mechanical nature of teleprinters and the inability of the mechanical components to have sufficient reliability and ruggedness at higher speeds. The speeds above 100 wpm were selected owing to improved printer design employing more electronics and the use of the entire voice bandwidth circuits for the transmission of the message or data.

Table 7. Typical Printers

Model	Speed	Characters per line	Code	Remarks
1. Teletype Corp. Models 32 and 33	Up to 100 wpm	74	32-5 unit, 33-8 unit	32 uses 7.5 code; 33 uses 11.0 code. Inexpensive. 32 uses Baudot code and three-row keyboard; 33 uses 8 unit data-communications code and four-row keyboard
2. Teletype Corp. Model 28	60, 75	76	5 unit	With or without keyboard. Tabulation, page feed-out, paper winder, stunt box, word counters available
3. Siemens Tele-printer 68	66 wpm	Tape printer 69, 72, 104	5 unit	Has keyboard and answer-back
4. Siemens Tele-printer 100	60, 66, 75, 100 wpm		5 unit	Sprocket feed platen. Paper feed trouble sensing. Tabulation available. Automatic CR and LF
5. Mite Corp. AN/TGC-14A(V)	60, 75, 100 wpm	72, 76	5 unit	AN/TGC-14(V) uses a 7.42 signal, while AN/TGC-14A(V) uses 7.0. This does not affect interoperation
6. Kleinschmidt Model 150	60, 66, 75, 100 wpm	72	5 unit	Sprocket or friction paper feed
7. Kleinschmidt Model 311	200–400 wpm	72	5 unit, 8 unit	Mostly electronic. Keyboard and power supply separate package. Impact printing with drum. Tab, backspace, parity detection
8. IBM Selectric I/O Writer	15.5 char/sec	80	7 bits	Uses ball with raised characters; permits operator's choice of type font
9. Western Union 2-B Tape Tele-printer	66 wpm	N/A	5 unit	No CR and LF necessary. Uses ⅜-in. gummed tape
10. Motorola Mobile	100–400 wpm	72	25–35 bits	Printer turned on and off remotely
11. Radiation Inc. High Speed Strip Printer	40 char/sec	N/A	6 bit	Uses styli and electrothermal paper. Asynchronous. Two printed outputs on ⅜-in. tapes; one is rewound
12. Motorola TP-3000	750, 1,500, 3,000 wpm	72	6, 7, 8 unit	Electrothermal paper. Rapid page feed-out
13. Burroughs Corp. High Speed Tele-printer S203	3,000 wpm	72	5 or 6 unit	Electrographic recording. Pin matrix for charging paper surface.
14. General Dynamics SC-3070	5,000 wpm	72	5, 6, or 8 unit	Uses Charactron ® tube and sensitized paper. Output may be used as offset master for duplication. Parallel input. Optional ready line, serial TTY input, parity check, 120 column, tab, paper take-up reel
15. Data Products Corp. dp/p 3300	300 lpm	12–132	6 unit	Drum type-impact hammers without pivot mechanisms. Paper feed without clutches, brakes, dogs, or springs. Has paper skip speed of 20 in./sec. Prints up to six copies
16. IBM 1403.......	600 and 1,100 lpm	132	6 unit	Used for IBM 1410 Data Processing System, impact printing. Skipping at 75 in./sec
17. Addressograph-Multigraph Corp. Models 960 and 960-A	667 or 1,000 lpm	132–160	6 bit	Drums with carbon paper or ribbon recording. 960-A responds to computer commands on tape channel using BCD code

Table 7. Typical Printers (*Continued*)

Model	Speed	Characters per line	Code	Remarks
18. Potter LP 1200..	1,000 lpm	72, 80, 120, 132, 160	6 bit and 1 parity	Numeric copy 1,800 lpm. Buffered input 100,000 char/sec asynchronous; higher on special order drum and impact printing. Uses magnetostriction buffer storage. Logic $1 = -6$ volts $\pm$ 0.5 volt at 10 ma; logic $0 = 0$ volt $\pm$ 0.5 volt at 5 ma
19. Control Data 1612 Line Printer	1,000 lpm	120	6 bit	Drum type. Paper advance for formatting can be controlled by paper or Mylar tape
20. A. B. Dick Model 9041	3,600 lpm	80	6 bit parallel plus parity	3,600 lpm is typical speed. Character rate is 15,625 char/sec. Printing is electrostatic
21. Dean Printer Model 515	6,000 char/sec			Uses CR tube with fiber optics window and Eastman Kodak (Type 1595) photosensitive paper
22. Briggs Assoc. High Speed Printer and Plotter (with Hogan Facsimile Corp. Recorder)	4,000 lpm	100	See Remarks	Uses single row of styli and marks on moist electrosensitive paper. Can be used as printer, accepting BCD code or binary code for plotting graphics using 36-bit words divided into three 12-bit groups. 10 bits of the 12 specify a data point
23. A. B. Dick Models 910 and 915	13,500 lpm	24 to 30	7 channel	Address printer. Operates off line from magnetic tape transport. Electrostatic printing. Uses 2¾-in. paper width
24. Litton Industries Model 850	3 sheets (8½ × 11) per sec	See Remarks	See Remarks	Uses CRT with a line of conducting elements in face. Records electrostatically on 8½-in.-wide page with resolution up to 200 lines per inch. Can print in excess of 80 characters per line. Can be used for facsimile. Code is for alphanumerics or facsimile
25. General Dynamics SC-4020	17,000 char/sec	Over 72	6 bit	Records on 16- or 35-mm film or with special camera on photo-recording paper. Records alphanumerics, curves, and vectors
26. Radiation Inc. High Speed Printer	30,000 char/sec	107 (or 120)	6 bit plus parity	Records on electrothermal 12-in.-wide paper. Output may be rerolled or discharged
27. Radiation Inc. Super Speed Printer	65,000 char/sec	120	6 bit plus parity	Works with magnetic-tape units. Fanfold paper. Standard sheet 12 by 12 in. Electrothermal paper

c. Recording Media. The most common recording medium is paper which is fed from a roll and the use of an inked ribbon for making ink marks on the paper. Ribbons are of cotton or, for longer life, of nylon. Reinking devices have been built, but most are unsatisfactory owing to clogging, ink leakage, or poor inking capability. Some machines* use a two-color ribbon, printing incoming messages in black and outgoing messages in red. For multiple copies, a sandwich of three sheets of paper with two interspersed carbons is employed. Alarms on most printers will indicate

* Olivetti and Siemens Halske.

when the paper supply is running low, and it is usually necessary to halt operations to replenish the supply. For large traffic loads or high-speed operation, paper is often furnished in fanfold form in boxes which are placed beneath the teletype table. When the supply of fanfold paper is low, it is possible simply to attach the remaining portion to a fresh box of paper without interrupting operation. The serrations of the fanfold simplify the tearing of the output copy into pages.

Two methods of feeding the paper through the teleprinters are prevalent, namely, friction and sprocket drives. When multicopies are bent around the rollers and platens, the difference in radii between the inner and outer layers of paper tends to cause buckling and wrinkling of paper. Staples are sometimes placed near each fold of fanfold paper to prevent this and to maintain registration of sprocket pins which engage in holes in the paper to ensure proper paper feed. Fanfold paper usually permits a larger number of carbon copies than roll form. Other types of recording papers will be described in conjunction with printing techniques.

Normal paper width is 8½ in., and the normal line length is 72 characters, with 10 characters to the inch along the line and six lines per vertical inch. For low-speed data printers there is a tendency toward 80 characters per line to cooperate with card readers. Strip printers are available which present the message on a paper tape about ¼ in. wide; these feature small, lightweight equipment. Other printers have been made using 2-, 2¾-, 4-, and even up to 18-in. wide papers with proportionate printing line length. The last is usually found on high-speed line printers for computer output. For automatic addressing machines 2¾-in. strips with 24 to 30 characters per line are used.* Some narrow page printers are designed to operate in a 72-character-per-line system. For example, the page may be only capable of accepting a line 36 characters long but automatic carriage return and line feed are incorporated, since carriage-return (CR) and line-feed (LF) signals would not occur in the middle of the line of the normal system. Either the automatic carriage return and line feed operate within a fraction of a character duration, or a small amount of storage is employed. This automatic CR and LF and those for a normal machine are sometimes designed to operate on the first space after a predetermined character position along the line.

Some teleprinters print on narrow tape. For instance, Western Union Type 2-B uses a ⅜-in. gummed tape which can be torn into one-line lengths after reception and pasted on a sheet of paper or can be used in tape form, especially when the messages are short and do not have to be filed. Advantages are small equipment size, simplification due to elimination of line-feed and carriage-return functions, and timesavings when small amounts of information must be put into a few widely dispersed boxes of forms.

d. Printing Techniques. Various modes of printing are shown in Fig. 14, and several generalizations can be made. Inked ribbons, carbon paper, pressure-sensitive papers, or NCR (no carbon required) paper may be used in impact printers such as Fig. 14a through *i*. Nonimpact printers[13] such as Fig. 14*j*, *k*, and *l* usually require special papers and are not so adaptable to simultaneous preparation of multiple copies. Therefore, off-line copying devices are sometimes used. Many recorders use preformed characters, while others generate the character shape by a matrix formation, a sweeping process like Fig. 14*k*, or utilizing preformed character strokes. Brief descriptions of the methods follow:

Fig. 14a. Similar to office typewriter basket but much more rugged for long heavy duty. Used for many American and European teletypewriters.

Fig. 14b. Printing block. The block is a mount for sliding rods with type pallets on one end. A hammer strikes the opposite end of the rod, forcing the pallet against the inked ribbon and paper. Character selection is made by moving the block with respect to the paper and hammer for each character. This technique is used at speeds to 100 wpm by Teletype Corporation on Models 28 and 35 teletypewriters. Block movement is sinusoidal, and the whole printing assembly is stepped to the right as characters are printed.

Fig. 14c. Horizontal cylinder. The cylinder contains raised characters per developed surface as shown. The cylinder located behind the paper is rotated and posi-

* A. B. Dick Co. Models 910 and 915 address printers.

tioned along a shaft for character selection. A hammer in front of the paper forces the paper and inked ribbon against the character on the cylinder. Mite Corporation employs this technique in Model 104, AN/TGC-14 and AN/TGC-15, teletypewriters at speeds to 100 wpm.

Fig. 14d. Ball. Has raised characters distributed on surface. The ball rotates and tilts to the correct column and row and impacts against the ribbon and paper to print a character. This technique is employed by IBM for the SELECTRIC® Typewriter. Speed is 150 wpm.

Fig. 14e. Cylinder—moving hammer. The cylinder is made of numerous disks each with raised characters arranged in order of binary count for the code used. An electronic counter indicates the rotary position of the cylinder. This is compared with the incoming code combination, and the hammer is triggered at the proper time. The cylinder does not stop rotating. The hammer moves along the line and must be returned for the CR symbol or at the end of the line. This technique is employed by Kleinschmidt (Division of Smith Corona Marchant) for Model 311. Speed is about 200 wpm for single hammer and 400 wpm for double hammer.

Fig. 14f. Cylinder hammers in fixed location. Similar to Fig. 14e but has a hammer for each character position along the line. Carbon paper or inked ribbon is used. The latter is often wide and at an angle to the printed line to spread wear. Has been used for high-speed "line printers" for data print-out by many companies such as Anelex, Shepard, Potter, IBM, National Cash Register, etc. Speed 100 to 200 wpm per print position. For example, a 120-character-per-line printer may operate at 12,000 wpm. Speed of line printers is usually designated in lines per minute (lpm); the above example would be a 600-lpm printer. The upper limit is about 1,200 lpm for alphanumerics and about 2,200 lpm for numerics.

Fig. 14g. Chain. Raised characters are distributed along a continuous chain loop. Hammers are located at positions to be printed, and when each correct character on the chain is in position, the appropriate hammer is energized. This technique is utilized by IBM for a line printer operating at 600 lpm.

Fig. 14h. Wire matrix. The printing head for each print position consists of a matrix, usually 5 by 7, of fine wires, each of which extends from the printing position through a metal tube to a code plate. The code plate has raised projections, and when it is moved vertically and horizontally to a selected position, projections are aligned with those wires required to form the character. For printing, the code plate is moved forward and forces the selected wires to move longitudinally in the tube, causing the ends to impact against the paper through carbon paper or ribbon. Selection of the correct wires forms the character to be printed. The technique shown is used by IBM for the 26 Printing Card Punch.

Fig. 14i. Vertical cylinder. The cylinder has raised characters. Selection is accomplished by raising or lowering and rotating the cylinder to the correct position. The cylinder impacts against inked ribbon and paper for printing. Used by Teletype Corporation in economical Models 32 and 33. Speed is 100 wpm.[14]

Fig. 14j. Electrostatic. Special paper is used with surface that can accept an electrostatic charge. Each print position along the line has a matrix of wires usually 5 by 7. The ends are close to the paper surface, and when a voltage is applied to selected wires, the paper surface is charged in the shape of the character. The latent image is developed by passing the sheet through a bin containing microscopic-sized, dry black resin particles which adhere to the charged areas of the paper surface. These are fixed (firmly bonded to) the paper through heat or pressure. The technique is used by Burroughs Corporation for data printout. The electrostatic technique is also employed by the Omnitronics Inc.* and Teleregister Corporation for recording code or characters on tape. Since the charging process requires only a fraction of a microsecond, speed is usually limited by transmission or transfer circuits, operational requirements, paper handling, or the fixing speeds. The Burroughs Printer operates at 300 characters per second (3,000 wpm). Other companies such as RCA and A. B. Dick also use electrostatic techniques in conjunction with cathode-ray tubes.

* Subsidiary of Borg-Warner Corp.

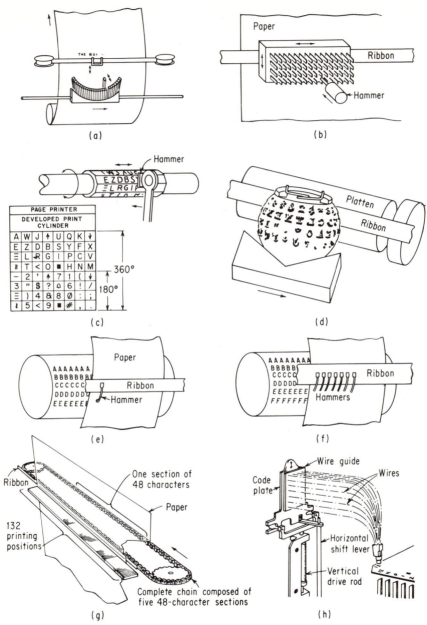

FIG. 14. Printing methods. (a) Type basket. (b) Printing block. (Courtesy of Teletype Corp.) (c) Horizontal cylinder. (Courtesy of Mite Corp.) (d) Ball printer. (Courtesy of IBM Corp.) (e) Cylinder—moving hammer. (Courtesy of Kleinschmidt Div., SCM Corp.) (f) Cylinder—multiple hammer. (g) Chain printer. (Courtesy of IBM Corp.) (h) Mechanical wire matrix. (Courtesy of IBM Corp. Copyright by International Business Machines Corporation.)

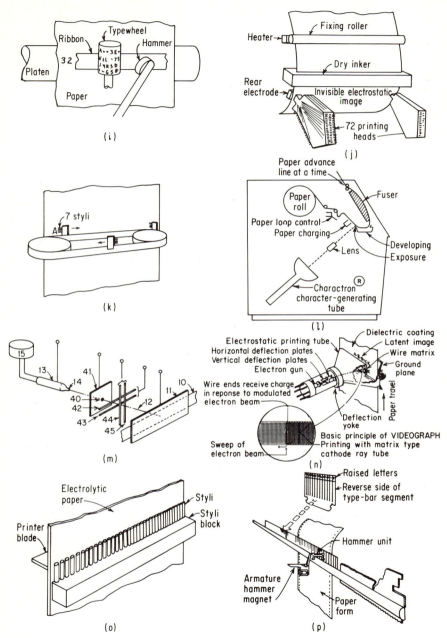

Fig. 14. Printing methods (*continued*). (*i*) Vertical cylinder. (*Courtesy of Teletype Corp.*) (*j*) Electrostatic. (*Courtesy of Burroughs Corporation.*) (*k*) Electrothermal. (*Courtesy of Motorola Inc.*) (*l*) Cathode-ray character matrix. (*Courtesy of General Dynamics Corp.*) (*m*) Ink recording. (*Courtesy of Teletype Corp.*) (*n*) Cathode-ray character generation. (*Courtesy of A. B. Dick Co.*) (*o*) Multistylus. (*p*) Type-bar printer. (*Courtesy of IBM Corp.*)

Fig. 14k. Electrothermal printing. The paper is a semiconductor which changes from white to black with passage of current. Recording heads are mounted on a continuous belt so that, when one reaches the right end of the printing line, another moves into position for starting the next line. Each recording head follows a skewed line moving upward one line height as it proceeds from left to right. This coincides with a simultaneous paper feed. For example, the paper advance is not stepped but is gradual and continuous while the belt moves. Each recording head has seven styli arranged in a vertical line. The styli are in contact with the paper during scanning, and as a voltage is developed between the stylus and a sublayer of the paper, an arc discharge destroys the white paper surface, exposing a black lower layer. The technique is used by Motorola Inc. in their Teleprinter System TP-3000, which operates at 750, 1,500, and 3,000 wpm.

Fig. 14l. Cathode-ray character matrix. A cathode-ray beam is deflected in accordance with the incoming code to a particular position on a character matrix.* The matrix is an array of characters, for example, 8 by 8, each character acting as a mask. A wide variety of matrix masks are available for different codes and symbols. The electron bundle passes through the selected character and is then further deflected to the desired position along the line to be printed. The beam is converted to a light image by the phosphor at the face of the tube, and the image is focused on a charged selenium surface, causing selective leakage of electric charge. The usual electrostatic development is employed, consisting of dusting the selenium surface with black powder and subsequent transfer of the powder image to plain paper. In a variation† of the system the selenium surface is eliminated and the charge is placed directly on a treated paper surface. The G3061 Charactron® tube eliminates the lens system through use of a fiber optics faceplate.

Fig. 14m. Ink recording. Several methods have been conceived employing principles of controlled spray, jet stream with movable baffle, stream with electrostatic deflection, etc. The method depicted in Fig. 14m‡ has a reservoir of ink 15 and nozzle 14. Slight pressure on the ink forms a convex meniscus at 14 but is not sufficient to cause flow. Droplets of ink are electrostatically attracted by the potential difference between the nozzle and platen (anode) 11. 41 is a valving plate for turning the jet on and off. Ink particles are in a charged state and move with a velocity caused by anode attraction. Particles are deflected vertically and horizontally by charges on deflection electrodes 42, 43, 44, and 45. Horizontal deflection is sufficient to cover two adjacent letters of printing, and therefore the number of printing heads required is half the number of characters per line. Speeds are extremely high, since printing time is in the order of microseconds.

Fig. 14n. Cathode-ray character generation. A focused cathode-ray beam is used to generate the character shape. The image at the face of the tube is utilized in one of several ways: (1) by an array of feed through metallic pins§ permitting generation of a charge and subsequent electrostatic printing, (2) by use of a window permitting electron passage followed by use of the electrostatic developing process, and (3) by use of a thin window or a fiber optics bundle which permits greater light-collection efficiency from the phospor and the subsequent use of a photoelectrostatic exposure and developing process, as in Fig. 14l above, or use of a photographic process. For these processes a character generator is required. In one system a monoscope cathode-ray tube scans a character selected from 64 on an aluminum target as shown in Fig. 15. The character selected depends on the input code; the output provides video for the printer cathode-ray tube.

In another version‖ of printing, using a cathode-ray tube with feed-through conductors, the vertical component of character formation is generated by paper feed.

Fig. 14o. Multistylus. The process is line-at-a-time printing. Printing is accomplished by a single row of styli approximately 0.01 in. apart across a page of electro-

* General Dynamics Corp. Charactron®.
† General Dynamics Corp. Model SC-3070.
‡ U.S. Patent 3060429, Charles R. Winston, assignor to Teletype Corp.
§ A. B. Dick Co. Model 9041.
‖ Litton Industries, Model 850.

sensitive paper such as electrolytic paper* or electrothermal paper† used for facsimile. During printing, the paper moves upward; the styli first print the tops of all the letters on the line and, as the paper moves, subsequently the lower portions of the letters. Since all letters on the line are printed simultaneously, speed is high, in the order of 4,000 lpm, about 67,000 wpm. Devices are also usable for plotting graphical data output.

Fig. 14p. Type bar. The printing bar consists of a line of type fonts. Each font is at the top of a thin metallic strip which can flex in cantilever fashion. Hammers which are magnet activated press the fonts against the ribbon and paper, causing printing. The complete bar contains several of the character sets (alphabets) for the machine, and as the bar moves from side to side, each character of the sets is positioned in turn to all positions of the printing line on the paper. Since there is a hammer for each line position, the selection of print characters is a matter of timing the firing of each hammer. The technique is employed for a line printer.‡

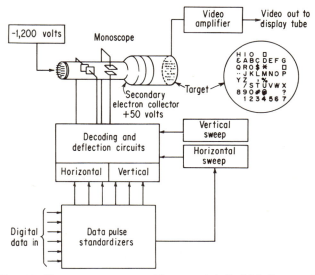

Fig. 15. Character generator. (*Courtesy of A. B. Dick Company.*)

e. Code-controlled Switches (Stunt Box, etc.). This is a class of switches and function devices which operate in response to particular code combinations of received signals. One example is the Teletype Corporation Model 28 Stunt Box. Within a printer the stunt box can close or open electrical circuits to control functions internally or to control external devices. It can also initiate internal functions by mechanical control. The device is housed in a volume about 9¾ by 4½ by 2¾ in. and is designed for much versatility. Up to 42 code slots are available. Function bars for each code slot have tines which engage the code bars. The tines may be selectively removed to make a function bar operate on a particular code. Pawl and lever linkages transfer the function bar movement to an opening or closing of a switch or to the mechanical initiation of a printer function. Thus the receiving of a particular code may cause a printer to start printing or to stop printing, or it may ring an external alarm or operate any other device controlled by a switch. For selective calling of certain stations in a network the stunt box can also operate with a sequence of letters.

* Briggs Associates, Inc., High Speed Printer and Plotter utilizing Hogan Facsimile Corp. Model HPP-110 Multiple Stylus Recorder. See Refs. 15 and 27.
† Radiation Inc., High Speed Printer.
‡ IBM Model 1443 Printer.

For example, receiving ZBT may cause a certain printer to start printing, but receiving any other combination of letters, even inversions of the above combination such as TBZ, will cause no response. Some other functions performed are on-line backspace, motor stop, unshift on space, keyboard lock, vertical or horizontal tabulation, on-line reverse line feed, and form feed-out.

Other companies make devices that perform some of these functions. For example, Kleinschmidt has a code-actuated switch (CASK) which provides a switch closure on one of 32 contacts corresponding to the particular code received.

8. Teletypewriter Operating Procedures

The operation of a teletypewriter bears some relationship to an office typewriter, since the alphabetic portion of the keyboards are similar. One important difference is the speed limitation of 60-wpm machines, since it almost dictates a cadence in typing speed without surpassing the 60-wpm limitation which is built into the keyboards. This factor is often upsetting to an office typist who is used to momentary or sustained bursts of speed. Some teletypewriters have incorporated storage of one or a few characters to realize a faster keyboard, while others achieve this through increased overall operating speeds such as 100, 150, or 200 wpm. A keyboard lock is sometimes used* when the carriage reaches the end of a line to prevent overprinting; it is unlocked when the CR key is pressed.

Most Baudot teletypewriters have a three-row keyboard, making it necessary for numbers and punctuations to have a different location from on an office typewriter. In recent years with the introduction of the eight-level data input-output equipments and teletypewriters, four-row keyboards very similar to typewriters are being used (see Fig. 6).

When initiating a transmission, it is common practice to transmit LET, CR, and LF in order to make sure the receiving machine is in proper state for message reception. This also starts the motor of a quiescent distant machine. For systems equipped for automatic answer back, WRU is sent and as described under Printer Response to Code (paragraph 7a) the receiver automatically transmits a number of characters containing identification. This ensures that the right receiver is reached. This is essential when some degree of privacy is desired.

It is not unusual to prepare the entire message off line in punched-tape form and transmit using a tape reader. This sometimes provides a relatively free keyboard, since a higher speed local reperforator of 100 to 125 wpm may be used. The use of a tape prepared off line also simplifies achieving a synchronous mode of transmission.

With the introduction of TELEX and TWX systems, teletypewriters have been placed on a dialing basis with the ability readily to reach any other subscriber in a network. These systems use circuit switching, whereas the usual teletypewriter system is on a message-switching basis which requires that the operator include sufficient address information in the preamble of the message. In some large systems such as the military, the latter involves considerable operator attention because the originator of the message may refer to an addressee in any one of a number of ways, e.g., Vice-president X Corp., Manager X Corp., X Corp., VP X Corp., President X Corp., etc. There may only be one teletypewriter circuit to the company, and these different means of addressing the corporation must all be converted by the operator at a message center through use of a look-up address file to a proper routing indicator which is put into the address heading so operators at a torn-tape message relay center or the director of an automatic message switching center can perform the routing. An example of an operating procedure for a 20-station switching system is described in Ref. 16.

9. Synchronous vs. Asynchronous Operation

Both synchronous and asynchronous equipments require speed control between sending and receiving circuits. The serial asynchronous teletypewriter mode is com-

* Olivetti.

monly called start-stop, since the receiver operates on a character basis, starting at the beginning of a character, stopping at the end, and starting again upon receipt of the next character. It is insensitive to time elapsed between characters except that some machines have an automatic circuit or device which stops the motor if no characters are received for a certain period of time. The characters can occur at random times such as from a keyboard. There are synchronization requirements within the character, since both transmitter and receiver correspondingly start at the beginning of a character and independently proceed through the character duration with the transmitter generating bauds while the receiver samples the bauds and stops during the stop pulse. The lack of precise synchronous control will in general have more effect in the last bit of the character or in the stop pulse. If a 7.0 code is used and it is

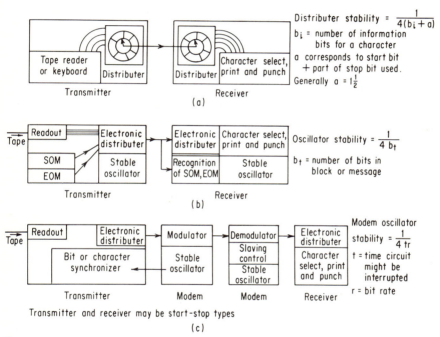

Fig. 16. Asynchronous and synchronous systems. (*a*) Start-stop. (*b*) Synchronous by block or message. (*c*) Synchronous with modems.

desired that the receiver stop within the stop pulse, theoretical required stability would be 1 part in 13 if the transmitter speed was exact. Since this is not usually the case, the accuracy has to be at least 1 part in 26 on a system basis (see Fig. 16). This accuracy provides no allowance for bias and end distortions caused by the transmission circuit or internal printer mechanisms and circuits. It is therefore much more desirable to apply a factor of at least 10, requiring an accuracy of 1 part in 260. With electronic approaches, it is not difficult to obtain economical stabilities of 1 part in a few thousand. This makes it feasible also to send eight-level data codes with start and stop pulses each of one unit duration (10.0 bits per character). Mechanical equipments, however, are made with 1-unit start pulse and a 2-unit stop pulse (11-unit character). A coding efficiency for transmission of information for start-stop could be determined by $E = b_i/b_t$, where b_i are the information bits in a character and b_t are the total bits transmitted for each character. For a Baudot code, efficiency is affected by the need to send Letters and Figures shifts a percentage of the time determined by the statistics of the information.

Synchronous operation may be on message or block basis or on a long-time basis. In the former, a reliable start of message (SOM) signal is sent at the beginning of the message; transmitter and receiver independently generate and sample the information bits of the code which are sent without extraneous bits. Since there are no start or stop signals on a character basis, there must also be some signal (EOM) to cause the receiver to stop. Stability requirements for the user equipment at each end of the circuit may be estimated from:

$$s = \frac{1}{4b_t} \qquad (12)$$

where b_t is the total number of bits in the message or block.

Depending upon the message length selected, the required stability might range from 1 part in 10,000 to 1 part in 100,000.

A coding efficiency would be found from

$$E = \frac{Tb_i}{b_t} \qquad (13)$$

where b_i = information bits per character
$\quad b_t$ = total bits transmitted per message or block
$\quad\quad$ = SOM bits + Tb_i + EOM bits
$\quad T$ = total characters per message or block

Synchronous operation on a long-time basis is usually associated with a modem or overall system with the stability-determining component external to the user equipments. The latter receives control clock signals which essentially are a request to send a character (character synchronous) or to send a bit (bit synchronous). If there is no character to be sent, this must be made known to the modem. This is automatically accomplished in bit synchronous systems employing start-stop subscriber equipment. Since steady mark or 1 condition indicates no character available, the modem may transmit this signal unchanged. The modem, however, may strip the start-stop bits from the signal and send only the information bits. In this case, it is necessary to convey between the two modems by special signaling the condition of no character available for sending.

Stability requirements of the modem oscillator under normal transmission conditions are dictated by the ability to slave the receive modem oscillator to the incoming signals. These systems, however, have major application where the transmission circuit may be broken for short periods of time because of operational or natural conditions.[17] Stability is that necessary to keep the transmit and receive oscillators close enough in frequency during the interrupt period so that, when the circuit is reestablished, there is lock-in of receive oscillator, creating the same bit and phase relationship between the modems that would have existed if the interruption had not taken place.

$$s = \frac{1}{4tr} \qquad (14)$$

where t = time of circuit interrupt
$\quad r$ = bit rate

10. Teletypewriter and Data Sets[18]

These are assemblies of tape readers, keyboards, reperforators, page printers, and auxiliary devices conveniently packaged for the desired application. An RT set is a combination tape reperforator and transmitter. Teletype Corporation Model 28 RT, for example, is built into a cabinet with 3,000-ft tape supply and 1,000-ft tape winder reels located below the punch and reader. Each cabinet is 28 in. wide and houses two sets each 8½ in. wide. A 100-ft-capacity tape storage bin is provided between punch and reader, and the reading head is designed with a pivot so it moves along a taut tape to the reperforator and reads the last character punched. A special design accepts and translates parallel-wire signals to serial for sequential transmission and vice versa for receiving. It is a five-level code device punching at speeds up to 200 wpm; this is limited to 100 wpm when a typing unit is attached. Change of speed is made by gear change, but an optional accessory gear-shift box is available for simple manual gear shifting from 60 to 75 or 100 wpm. RT sets are available at different speeds with various features. Tally Reader-Perforator Model 1477 operates up to 60 characters per second and requires 12 in. of tape between punch head and readout point.

Friden Teledata* Model 8C is an automatic tape transmitter-receiver designed specially for operation in conjunction with a Data-Phone† data set which is designed and leased by the Bell Telephone Company. It transmits eight-channel tape with a special parity checking technique. The transmission cycle is 10.42 units per character.

U.S. Army Teletypewriter Set AN/GGC-3 is a portable, tactical, ruggedized set consisting of a TD, typing reperforator, and keyboard. The transmitted signal has a distortion tolerance of ±5 per cent at 368.1 opm (operations per minute). Received signal tolerances are bias distortion ±40 per cent at 368.1 opm and ±35 per cent at 600 opm and end distortion 35 per cent marking or spacing at 368.1 opm and 30 per cent at 600 opm. The range adjustment has a scale from 0 to 120; 100 scale units equal the width of 1 unit, or one signal pulse.

In many cases a printer is designed so that a tape reader and tape punch can be attached.‡ Usually the printer motor supplies the power, and for simplicity the reperforator with or without typing may use the same input circuitry or mechanism as the printer.

An assembly consisting of keyboard, page printer, tape transmitter, and tape reperforator is called an ASR (automatic send-receive) set. It is common for such a set to include also a switching unit which permits any one or more of the individual components to be placed on line, off line but connected to other units locally or disconnected from other components. Power supplies are included in ASR sets and are also switched. ASR's are available for teletypewriter use employing five-level code§ or for data employing eight-level code.‖

Torn-tape message centers record incoming information on punched tape. The operators tear tape messages, read the addresses, and place the messages in the holding racks of the TD's corresponding to the addresses. Teletypewriter manufacturers cater to this application by providing special packaging for incoming line reperforators, TD's, monitor reperforators, and page printers. For example, Teletype Corporation's Universal Torn Tape System consists of six major modules operating at 60, 75, or 100 wpm as follows: The receiving group is a cabinet of six typing reperforators powered by two motors. Up to 18 in. of blank tape is automatically fed out after each message to prevent tearing tape in the middle of a message. TD's are also ganged in the Transmitting Group. Six TD's are powered by two motors. An optional feature allows two TD's to alternate automatically on a circuit, permitting the operator to load the TD's in pairs. All outgoing messages are automatically numbered. The third module is the Monitor Group, which contains two three-gang printing tape reperforators (total six). Each reperforator is connected to a TD of the transmitter group and is arranged for convenience in retransmitting messages with the automatic numbering generator temporarily disabled, since messages in the 1,000-ft-capacity monitor reel are already numbered.

An optional Transmitter-Monitor Group can be used in place of the two preceding separate Transmitting and Monitoring Groups if only one to three stations are to be

---

* "Teledata" is the registered trade-mark of Friden, Inc.
† "Data-Phone" is the registered trade-mark of Bell Telephone Co.
‡ Siemens 100, Sagem, Teletype Corp. Models 32 (five level) and 33 (eight level), Kleinschmidt Model 311.
§ Teletype Model 28, Kleinschmidt Model 402, Mite AN/TGC-15, Military AN/FGC-25.
‖ Teletype Corp. Model 35.

serviced. The cabinet contains a three-gang TD, three-gang typing reperforator and a three-gang tape winder. The two remaining modules are the Mobile Receiver Group and the Mobile Transmitter Group. The first consists of a typing reperforator on a table, and the second a TD on a table. Each table in turn is mounted on a dolly, permitting each group to be moved to a position where it is substituted for a reperforator or TD which is incapacitated for maintenance.

The Kleinschmidt Division of the SCM (Smith-Corona-Marchant) Corporation also manufactures consoles of typing reperforators and TD's for incoming line, cross-office, intercept, monitor, multiple-call, and various other applications in automatic teletypewriter central offices and for torn-tape centers. Model 179, for example, is a semiautomatic set which houses two typing reperforators and power supply in one console and two TD's in another. The AN/FGC-70 is a complete semiautomatic tape relay set capable of handling six outgoing and six incoming message lines in a full-duplex mode of operation. The Transmit Console contains from six to twelve TRMTD's (tape reader markable tape transmitters) which provide sequential or simultaneous signaling and permit automatic message numbering. The Monitor Console and the Receive Console each have six typing reperforators. Two rerun TD's are included in the Monitor Consoles, and these can be expanded to six. The Receive Console also contains alarm lamp and bells to indicate receipt of a military priority message.

Another type of set offered by several manufacturers is a high-speed communications set operating at speeds in the range of 1,000 wpm over voice bandwidth circuits usually on a circuit switched basis. The Hi-Speed Communication Set of Teletype Corporation consists of separate Transmitter and Receiver Consoles. Each console contains tape equipment, a telephone, dial, and switching so voice or data can be sent. The messages are accumulated on tape, and at an appropriate time the addressee is called by dialing. He may answer by telephone, or the receiving equipment may send back a tone indicating that the tape reperforator is on line. The five-, six-, seven-, or eight-level tape at the transmitting station is read, and the information sent at 1,050 wpm. For the transmission, suitable modulating and demodulating equipment is provided by the telephone company.

To assure that information to be sent is correct, Siemens Halske supply a Tape Verifier Position. It is used for tape preparation from keyboard, verification, comparison, and duplication of tapes. The message to be sent is stored in tape form. Either the message is again stored on another tape, which is subsequently compared with the first, or the keyboard is used. Each keyboard signal is compared with that of the tape first prepared, and if the comparison is favorable, a new tape is punched. In case of disagreement, the system is blocked until the operator determines the correct character to be punched in the new tape.

Data communications sets provide means for recording, transmitting, and receiving data on paper tape, page copy, punched cards, and edge-punched documents. Essentials may be conveyed through description of the IBM 1050 Data Communications System.

The system is a versatile modularized set. While the set is an assembly of eight major components, only those needed for a given application are used. The components are as follows:

1. 1051 Control Unit. This provides electrical interconnection of the system components, performs control functions from the signals, converts incoming serial line signals to parallel and vice versa for transmission, and checks odd-bit parity and longitudinal redundancy. It provides the working-table area for the operator.

2. 1052 Printer-Keyboard. The keyboard is a four-bank version, and in two shifts of 44 characters, it provides 88 characters. It is automatically locked during inquiry or under error condition. It can be used independently of the printer, which is similar to the 1053 printer.

3. 1053 Printer. This is a ball-type printer permitting the operator to change type font in a few moments. It prints a 13-in. line on 15-in. wide paper using pin or friction feed. There is a choice of six or eight lines per inch.

4. 1054 Paper Tape Reader. This accepts the strip of tape, tape loop, or tape being fed from the 1055 Paper Tape Punch. It is also available with edge-punched docu-

Table 8. Card Transceivers

	Channel	Rate	Remarks
IBM 65–66 Data Transceiver	With IBM 67 Telegraph Signal Unit— 60-, 75-, or 100-wpm telegraph channels	3, 4, or 5 cards/min (80-col. cards)	Can multiplex four transceivers over leased telephone circuits, giving 44 cards/min (80-column cards). Higher card rate if less columns used per card. Error cards offset at each end
	With IBM 68 Telephone Signal Unit— two- or four-wire telephone channels	65 card unit 11 cards/min	
		66 printing card unit 10 cards/min (80-col. cards)	
IBM 1013 Card Transmission Terminal	2,400-baud full duplex	50 to 300 cards/min. Speed depends on number of columns per card. Examples: Send, 272 cards/min (50-columns punched); Receive, 77 cards/min (46-columns punched)	Transmits or receives card data. Interoperates with a 7701 or 7702 Magnetic Tape Transmission Terminal, a 1009 Data Transmission Unit (intermediary between line and computer core storage), or another 1013. Includes buffer up to seven cards. Signals are modulated for transmission over dial or private circuits
UNIVAC®* 1004 Card Processor	Telephone (toll)	285 char/sec	Used with UNIVAC Data Line Terminal (DLT) which operates with Bell System's DATA-PHONE† Data Sets 201A or 201B. Automatic character and longitudinal parity check. Communicates with UNIVAC 490 Real-Time Computing System, UNIVAC 1107 Thin-Film Memory, or another UNIVAC 1004 Card Processor
	Telephone (private)	342 char/sec	
		1004-DLT reads 300–400 cards/min. Punches 200 cards/min	

* UNIVAC®—Registered trademark of the Sperry Rand Corporation.
† DATA-PHONE is a trademark of AT&T.

ment feeding. A center feed supply and a take-up reel are available. It retransmits in case of error.

5. 1055 Paper Tape Punch. This punches paper tape at manual or system rates. It has a feed key and delete key. Automatic error correction and edge punching are available. It feeds in tandem to the 1054 Paper Tape Reader.

6. 1056 Card Reader. This accepts single cards or feed from 300-card hopper. It recognizes a cut corner, assuring proper card orientation, and reads 51- or 80-column cards. Facilities are available for card reread in case of error and for automatic error correction. It recognizes control functions such as component selection, end of line, carrier return, turn tape on, etc.

7. 1057/8 Card Punch. This can operate remotely from the keyboard of the 1052 Printer-Keyboard, punching information in the card. 1058 simultaneously prints along the top edge of the card. It provides all normal IBM card punch facilities.

The 1050 system transmits data over ordinary communications company circuits using a telephone company data set at 14.8 characters per second. The system can communicate with other 1050 systems or directly with a data-processing system. It can be used as an independent local data recording system, and any input or output devices not in use can be used for local operations. For circuits other than communications lines an IBM data set is used.

Other data communications sets contain components somewhat similar to the above but operating at different speeds and with variation in features. Usually the speeds are limited by telegraph channels (75 to 150 bauds), telephone channels (up to 2,400 bauds), or multiplex combinations utilizing the telephone channel. However, high transmission rates are feasible. Some card transceiver characteristics are included in Table 8.

Another example of data communications is found in the Univac Standard Communications Subsystem. This serves as a link between the various communications facilities and Univac Data Processing Systems. In this way a central processor can exchange data simultaneously with a number of remote locations using common carrier communications circuits. The Subsystem consists of Communications Line Terminals (CLT), which convert serial signals to parallel and vice versa, and Communications Multiplexers, which interoperate between the central processor and up to 64 CLT's. Three basic CLT speeds are available: low speed up to 300 bits per second, medium speed up to 1,600 bits per second, and high speed 2,000 to 4,800 bits per second. Interfaces are available for AT&T Data Sets 103A, 202A, 202B, 402A, 402B, 402C, 402D, 201A, 201B, and 801A and for IBM 068 Telephone Signal Unit. By means of the Subsystem, communications service is provided for Private Line Teletypewriter (PLT), Teletypewriter Exchange Service (TWX), Wide Area Telephone Service (WATS), private-line telephone, and Direct Distance Dialing (DDD).

Typical input-output devices used with the Subsystem include the Univac 1004 and DLT Card Processor (see Table 8), a Teletype Corporation Model 28 ASR, a Teletype Corporation Model 35 Keyboard, Send-Receive Teletypewriter (KSR), or a high-speed paper-tape set.

11. Effects of Interfering Noise

Signals for teletypewriters and data systems are digital, but effects of electrical noise will depend upon where in the system the noise occurs, whether the system is synchronous or start-stop, and types of noise, modulation, and detection.

For a start-stop system one of the most serious effects of noise is to cause the receiver to start falsely just before a transmission. Usually this causes a number of characters to be garbled, since the teletypewriter receiver will in all probability stop its cycle during a transmitted character and will start again on the next mark to space transition. Because of bit differences from character to character, the receiver will eventually lock into proper phase and, unless noise is severe, will remain locked in. In this condition noise may cause inversion of some of the bits, which will result in receipt of a wrong character. Inversion of the start or stop pulses may trigger off a series of wrong characters as indicated above.

If character integrity is maintained throughout a synchronous system, once the proper start is achieved, short bursts of noise will have no effect on the synchronization of the receiving user equipment, although individual inverted bits may cause individual character errors; these errors will be confined to the one character. For example, a momentary circuit fade will cause loss of some characters but the receiver will continue to look for the bits of each character at the proper time and, when the circuit becomes operable, the first usable character will be received and detected without need for resynchronizing. In such a system the receiving oscillator is often slaved to the transmitted signal, and during a circuit disconnect or fade, the incoming noise in practice may tend to cause frequency drift, although a truly random noise would have little resultant effect.

For a synchronous system which maintains character integrity, the proper start of transmission is extremely important. Several methods may be used to prevent false starts. A simple method is to send a repeated character with a recognizable bit formation such as 00001. This may be locked in during a synchronizing cycle automatically or manually by means of a bit shifting or dropping technique. Another technique is to send some known message repeatedly such as THE QUICK BROWN FOX JUMPS OVER THE LAZY DOG'S BACK 123456789. The receiving operator manually shifts the phase relation of the detection circuit or mechanism until he sees readable copy.

These methods do not indicate the transmission of the first bit in a synchronous stream of information. Where this is necessary, it may be accomplished on a good noise-free circuit in a similar manner to the start-stop operation; i.e., a single transient change from a steady state indicates start. Where circuit conditions are not suitable, one approach is to introduce redundancy into the start signal. For example, the numbers 9876543210 may be transmitted and both transmitter and receiver operate on the basis that the first bit after the 0 is the first bit of the synchronous stream. In this method many components of the count-down sequence may be lost and the proper start still accomplished. In the extreme, all but one of the ten digits may be scrambled and synchronizing still be realized by making the receiver continue the count from the received and recognized number down to 0.

12. Card Readers, Sorters, and Collators

Most card readers are located in the vicinity of a computer and feed signals into the computer or magnetic-tape recorder through a buffer storage. To achieve higher speed the reading is performed on a row-by-row basis (see Fig. 2). Since each character is stored in a column, it is necessary to utilize a buffer capable of storing all 80 characters on the card and sometimes an extra 4 special characters. The buffer storage may be part of the card reader or the computer center. Practical speeds are in the range of 100 to 800 cards per minute. Contact fingers sense the cards, and where a hole exists, the fingers make contact with a metallic subplate or roller. Other sensing means include mechanically moving fingers with switches coupled to the fingers and photoelectric types. A complete matrix of one photodiode for each possible code hole position has been used with proper sensing sequence accomplished by electronic switching of photodiodes. Card reader speeds are usually limited by the mechanics of card handling. The cards are placed in a stack and must be automatically fed one at a time to the sensing positions. Frequently the card must also be guided to different bins depending upon whether the card is successfully read and rejected or a sorting or collating function is being performed.

Card sorters and collators are readers with feed mechanisms, sensing heads, internal logic, and sub-bins to perform these functions usually without external logic operations. While computers perform these functions rapidly, card sorters and collators are economical where a computer is not available. A typical sorter has one read brush which may be moved to any one of the columns of a card. For Hollerith code this consists of one sensing finger; for other codes such as ASCII a multisensing head is more appropriate. The sorter also has 13 bins. Some sorting operations can be best explained by examples. Suppose a stack of cards contains numbers from 0 to 9999 in scrambled order which is to be sorted into numerically increasing order. It is accomplished in four operations as shown in Table 9; this is a very abbreviated stack to simplify the illustration. Stacks may include 1,000 cards. In the table the cards in the stacks are face up. The stacks are turned over for reading, and cards in the bins are face down. The stack is first read with the reading head in the fourth position, i.e., units position (starting with the first, i.e., thousandths, digit would involve a lengthy process). The result is distribution of the cards in 10 bins ranging in regular order with the last numbers from 0 to 9. The eleventh bin will have rejects. The stacks in the 10 bins are carefully placed upon each other, forming a new combined stack with the last digits in numerical order. This is read with the reading head in the third digit position, i.e., tens position. Since the cards in the stack with 0 in the fourth position are read first, these are first to go into the bins on this second sort. The cards are face down. After the sort, the stacks from each bin are carefully

Table 9. Sorting of Numbers

Original stack of cards with numbers shown

```
2147
9999
0000
2222
7436
6215
1290
6310
1549
6216
```

Bin	0	1	2	3	4	5	6	7	8	9	New stack
After first sort	6310 1290 0000		2222			6215	6216 7436	2147		1549 9999	0000 1290 6310 2222 6215 7436 6216 2147 9999 1549
After second sort	0000	6216 6215 6310	2222	7436	1549 2147					9999 1290	0000 6310 6215 6216 2222 7436 2147 1549 1290 9999
After third sort	0000	2147	1290 2222 6216 6215	6310	7436	1549				9999	0000 2147 6215 6216 2222 1290 6310 7436 1549 9999
After fourth sort	0000	1549 1290	2222 2147				6310 6216 6215	7436		9999	

Final stack

```
0000
1290
1549
2147
2222
6215
6216
6310
7436
9999
```

assembled into a new combined stack in the numerical order of the third digit. The process is repeated, reading the second (hundreds) digit and the first (thousandths) digits in turn.

Since only 13 bins are available, sorting names into alphabetical order involves more steps. It can be accomplished with two sorts for each character position alphabetized. The same procedure is used, namely, starting with the rightmost letter to be included in the alphabetizing. Assuming the cards are in Hollerith* code, the first sort is made based on the holes 1 through 9 and reject bin for "no letter" in the position of the word being read (see Fig. 2a). A's and J's will fall in bin 1; B's, K's, and S's in bin 2; C's, L's, and T's in bin 3; etc. A combined stack is prepared by taking the cards in the reject bin and without turning them over placing them in the input bin so the bottom card will be read first. Likewise cards in bin 1 are placed on these, cards from bin 2 on these, etc. Reading is now performed with holes 0, 11, 12, and reject (no holes). "No letter" will be in the reject bin, A to I will be in bin 12, J to R in bin 11, and S to Z in bin 0. This will have sorted a particular letter position of the words, and procedure is continued, as with the numbers, starting at the position farthest to the right in the words and proceeding to the left.

Where many users supply information in a data system, attempts are made to keep the large number of input devices simple. One of the simplest forms is a soft pencil and a stack of cards. Each column represents a numeric or alphabetic character, and the entry is made by marking vertical positions in the column. Reading is accomplished by measuring resistance between sensing fingers which slide on the surface of the card. Another simple entry device† is a manual punching arrangement. A surface is formed with slit resilient material. A card with deeply embossed or scored hole positions is placed on the surface, and the desired holes are punched with a manually held pick. The marked and the manual punching devices have the disadvantage of accepting about one-half the normal number of characters per card. The manually punched card can be processed in machines equipped with standard 80-column feed, but modifications are required on certain machines to ensure proper stacking of the cards. Other small card-preparation devices are available with a keyboard. These usually accept blank cards singly rather than from stacks.

Collators perform the functions of merging two stacks of cards which differ or substituting cards from a primary stack for corresponding cards of a secondary stack. For example, a stack of customer listings in alphabetical order is to be updated. Newer cards comprise another (primary) stack which is also alphabetized. The collator can merge the two stacks. A card from one stack is compared with one from the other; the lowest card is fed through to the collecting bin, while the other card is retained in the reading position and compared with the next card from the opposite stack. This process continues until the two stacks are merged into one properly ordered stack.

The sorter can substitute cards for ones already in a stack. For example, suppose customer information is updated and a new card prepared. This is placed in a stack of other new cards, and these are alphabetized by customer's name. When the customer's old card is reached in the basic, older stack of cards, it is recognized and diverted to a reject bin while the corresponding new card for the customer is sent to the collecting bin. The sorter may also be called upon to check a stack of cards for proper sequence from card to card.

Card readers are often part of a combined reader and punch used in the data center. They may also be part of a card transceiver. A card transceiver is used to transmit information contained in a stack of cards at one location to a blank stack of cards at a remote area. Voice circuits are generally used for the transmission, and this limits the reading speed to between 100 and 200 cards per minute. If a column at a time is read, buffering is kept small.

An economical device‡ for card transmission handles one card at a time. It is used in conjunction with a dial telephone and a modem for conversion of d-c signals

* IBM Corporation.
† IBM Port-a-Punch.
‡ IBM 1001 Data Transmission Terminal.

to a-c.   By means of the telephone dial a distant location is reached where there is a card punch which may be unattended.   The card with up to 36 punched characters is placed on a tray which is manually pushed into the card transmitter.   This stresses a spring which upon release feeds the tray back out at a controlled rate of 12 numeric characters per second, while the characters on the card are read and transmitted. By means of a numeric keyboard up to 44 additional numbers may be sent.   A special symbol indicates to the remote punch that the transmission for a specific card is finished and that the card may be stored.   Another symbol is used in the event that the transmitting operator makes a mistake and detects it.   This will cause reject of the card at the punch.   After data transmissions are finished, the operation may be converted to verbal if an operator is present at the receiving end.

13. Character Recognition[19]

In the last decade the ability of machines to read printed characters has been proved feasible and devices have been marketed which perform this function in a reliable manner.   Reading devices have been put into use in banks for reading printed numbers on checks and in oil company accounting operations for reading numbers which are impressed on sales slips by raised numerals on customers' credit cards. Reliable reading is enhanced by special character formation, efforts to obtain uniform character density, and significant differences between the character density and its background.   Other character readers have been made which read typed or printed alphanumeric characters.   Usually a scanning process is used to examine unit areas of each character, and these are stored in a matrix with incorporated logic such as the equivalent of mask matching or other decision criteria that separate the characters. Since perfect characters can never be assumed, the logic must allow for partially printed characters, characters slightly out of position, and characters with a slight amount of blur.   Sometimes a threshhold action is employed which makes it unnecessary to have exact pattern matching but permits the decision to be based on a "best-match" criterion.   Particular problems with reading typing on a page are the touching of adjacent letters making it difficult for the reader to isolate characters, erasures with the retyped character improperly positioned, partial erasures with resulting overprinting, lines at the bottom of a page not parallel to those at the top, requirement to accept both 10 and 12 characters per inch, and need to read a wide variety of type fonts.   Nevertheless successful readers have been built and are in use, reading thousands of documents daily.   These are greatly enhanced by control of the input characters to eliminate the above problem areas.

Investigational effort in recent years in the field of automatic hand-print readers has shown that a practical, relatively small automatic reading device can be built provided hand printing is controlled in a manner to simplify the decision logic.   For example, the character area is premarked lightly with a grid or pattern upon which the letters are hand printed.   For numbers, two dots vertically displaced indicate how and where to print.   For alphanumerics a special crisscross pattern has been employed with instructions on how to place the character.   With these methods the lines of the characters cross certain areas or lines which are sensed in a binary fashion.   The decision logic becomes a matrix with inputs from the sensing elements and outputs indicating the character selected or the code for the character selected.   The advantage of this device is similar to that for the marked cards; namely, many users can generate input information to a data system or messages for communications with nothing more than a pencil and preprinted forms or cards.   These cards are physically forwarded to centralized transmission points for manual or automatic reading of the hand printing.   A military card form exists for the manual system.   It is of standard size used for punched cards and has room for 25 predetermined typed characters and 55 boxes for entry of the hand-printed message.

14. Audible Noise

Many subscriber equipments are quite noisy, since little attempt is made to reduce noise.   For example, the requirement for simultaneous multiple copies from a printer is most easily satisfied by impact printing, which is noisy. Noise reduction is best applied at the source; the less noise generated, the easier it is to suppress the remainder to a given level. Power applied to printing mechanisms should be only that required for the printing, since excessive power generates more noise. Other noise-reduction techniques are:

Avoid resonances of components in the audio spectrum. Balance rotating components.

Fasten parts firmly to avoid rattles. Even looseness of choke and transformer laminations can cause hums and singing.

Use isolation of noise generators. For example, rubber suspensions can decrease conduction of sound.

Do not depend entirely on thin baffles, since the air pressures of sound may cause the baffle to vibrate, transmitting the sound as though the baffle were transparent.

Use sound-absorbing coatings on the inside surfaces of housings. There are many commercially available materials. A lamination of sheet metal, felt or rubber, and sheet metal has been used for soundproofing.

Where ports and vents must exist, use quarter or three-quarter wavelength passages if feasible. This may only be possible for higher audio or supersonic frequencies.

Supersonic radiation should be detected and suppressed, since it may cause irritability of operators and a subconscious dislike for an equipment for no "apparent" reason.

Special conditions must be detected and eliminated. For example, a device using electrostatic recording at relatively high speed was quite noisy. The cause was found to be the paper acting as a sounding board when it was snapped forward by the line-feed mechanism. A similar amplification of sound by the sounding-board effect of paper was also experienced in an impact-type printer. A cure is isolation of the large, flat area of the paper from the noise source by means of pressure rollers, extra wrap-around angle of the paper on an existing roller, or elimination of large, unsupported areas of paper.

Noise measurements are made through use of precision microphones, accelerometers, amplifiers, output meters, and frequency analyzers.* Appropriate weighting is applied to conform to the intended application of the equipment which is under test.

15. Displays[20]

A display device for messages frequently consists of a teletypewriter with a projection lens system and a rear-view screen. Data displays are much more complex and exist in wide variety, ranging from small numeric displays to extremely large screen group displays. Typical types of information displayed consist of alphanumerics, vectors, special symbols, and graphics, and these may move or be static. The computer output consists of a series of words each indicating the character or symbol to be displayed, the XY coordinates, and other significant data. For example, if a circle is to be displayed, it is identified as a circle, coordinates of the center are given, and the radius is given, or for a vector the angle, location, and length are given.

Figure 17 shows one display device, the 19-in. Charactron® Display Tube which is used in the S-C 1090 Direct View Display Console.† Information from the computer in 6- or 36-bit form enters a tape channel adapter or a core buffer which sends 36-bit data to an input register and to position and selection decoders.

The electron gun (Fig. 17) emits an electron beam which is deflected by the coplanar selection plates under control of signals from the selection decoders to the desired character shaped aperture of the matrix. The electron bundle, in the shape of the character, is converged and deflected to the desired location on the face of the tube. The character size is 0.100 in., and normally 64 characters are provided based on 6 input bits per character. Brightness is 20 ft-L measured at a 30-cycle frame rate with unblanking time of 16 μsec per character. Character positioning results from $9X$ and $9Y$ binary input bits. Full screen position settling time is 24 μsec after application of

* See Leo L. Beranek, *Noise Reduction*, McGraw-Hill Book Company, New York, 1960.
† General Dynamics/Electronics.

binary address. Vectors can vary in length from the spot size up to 2 in. with a writing time of 48 μsec and a brightness approximately equal to the characters. The starting point is based on 18 bits, and the direction and length X and Y components each specified by 6 bits plus sign (14 bits total). Many matrices of characters and symbols are available; one is shown in Fig. 17*b*.

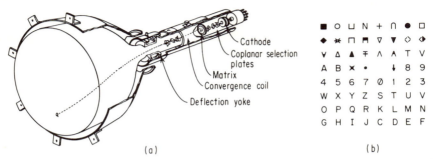

(a)   (b)

Fig. 17. 19-in. display tube and typical character matrix. (*Courtesy of General Dynamics/ Electronics.*)

16. Definitions[21]

Baud. A baud is a unit of modulation rate. One baud corresponds to a rate of one unit interval per second. The modulation rate is expressed as the reciprocal of the duration in seconds of the unit interval. Example: If the duration of the unit interval is 25 msec, the modulation rate is 40 bauds. This corresponds to 40 bits per second.

Break. To break in a communications circuit is for the receiving user to interrupt the sending user and take control of the circuit used, especially in connection with half-duplex telegraph circuits and two-way telephone circuits equipped with voice-operated devices.

Bias Distortion (of start-stop teletypewriter signals). The shifting of the start of all marking pulses from their proper positions in relation to the beginning of the start pulse (see Fig. 4*a*).

Spacing bias distortion is the start of a marking pulse after the proper time, while marking bias distortion is the start before the proper time. Magnitude of the distortion is expressed in per cent of a perfect unit pulse length.

Fortuitous Distortion of Telegraph Signals. The distortion resulting from causes generally subject to random laws, for example, accidental irregularities in the operating of the apparatus and moving parts, disturbances affecting the transmission channel, etc.

Distortion, Teletypewriter Signal (of start-stop teletypewriter signals). The shifting of the transition points of the signal pulses from their proper positions relative to the beginning of the start pulse. The magnitude of the distortion is expressed in per cent of a perfect unit pulse length.

Distortion, End (of start-stop teletypewriter signals). The shifting of the ends of all marking pulses except the stop pulse from their proper positions in relation to the beginning of the start pulse. Shifting of the end of the stop pulse would constitute a deviation in character time and rate rather than being an end distortion.

Spacing end distortion is the termination of marking pulses before the proper time, while marking end distortion is the continuation of the pulse past the proper time. Magnitude of the distortion is expressed in per cent of a perfect unit pulse length (see Fig. 4*a*).

Internal Bias (Teletypewriter). That bias, either marking or spacing, that may occur within a start-stop teletypewriter receiving mechanism and which will have the same effect on the margins of operation as bias external to the receiver.

Modem. Acronym for modulator-demodulator.
Significant Condition of a Modulation.

Passive	*Active*
Frequency high	Frequency low
A	Z
0	1
Current off	Current on
Tone off	Tone on
Space	Mark
−	+
No hole (paper tape, card)	Hole (paper tape, card)

Synchronous System. A system in which the sending and receiving instruments are operating continuously at substantially the same frequency and are maintained by means of correction, if necessary, in a desired phase relationship.

Word (Telegraph). By definition, a telegraph word shall consist of six-character intervals when computing traffic capacity in words per minute.

$$\text{wpm} = \frac{\text{modulation rate} \times 10}{\text{units per character interval}}$$

TELEPHONE EQUIPMENT*

Telephone equipment required at user locations must be capable of providing reliable signaling and supervision to and from each location in addition to furnishing high-grade transmission. In the design of a user communication system, careful consideration must be given to the type of central office serving the customer, since its capability may place restrictions on the type of substation equipment that can be used. This is particularly true if combined voice and nonvoice or record transmission is involved.

Noise generated by switching or connector equipment in central offices may affect the quality of voice transmission and make data transmission impossible. The effect of noise, while important to this subject, is not covered in this section (see Chap. 3).

17. Central-office Systems

Automatic or dial exchanges furnish telephone service to a large percentage of telephone users in the United States, Europe, and most developed countries. However, many magneto and manual systems are still in operation.

a. Magneto Systems. In a magneto exchange, the subscriber signals the operator by cranking a hand generator, producing a low-frequency alternating current that energizes a *signal drop.* The operator signals the subscriber by inserting the plug of an idle cord in the desired line jack and depressing a ringing key. Ringing current from a continuous central-office source is applied to the line for ringing the called subscriber. The bell will continue to ring as long as the operator depresses the key. Transmitter battery for the station is supplied by dry cells located at the subscriber's premises. Magneto systems are also referred to as *local battery* systems.

b. Common Battery Systems. Common battery systems encompass all types of "Manual" and "Dial" central offices. Manual offices are those where all interconnections among local subscribers, and local subscribers and incoming or outgoing trunks are made by operators. In common battery systems, the transmitting power of the telephone is supplied by the battery at the central office.

18. Dial Systems

Dial exchanges provide the customers with the facility for automatically calling numbers without the aid of an operator. Dial exchanges are classified as *step-by-step, panel, crossbar,* or *electronic* as determined by the method of switching employed.

* The Telephony portion of this chapter was prepared by Joseph E. Clare.

a. Four-wire Systems. Four-wire crossbar switching systems developed for large communication systems combine circuit switching and message switching and provide low-error-rate data transmission.

b. Telephone Lines. Central Office Lines terminate at subscriber premises on telephones, switchboards, or key equipment. Switchboards at subscribers premises are classed as Private Branch Exchange (PBX), Private Automatic Exchange (PAX) or Private Automatic Branch Exchange (PABX). When the line terminates at a switchboard, it is usually referred to as "trunk," while the line terminating at a station or telephone is referred to as "station line" or "extension line." Manual switchboards are those where interconnections are made by an operator using cords. Automatic systems enable the user to dial his own connections.

19. Telephone Terminology

The following list of definitions will be used in this section to describe the most common types of telephone services.

a. Tie Line or Tie Trunk. A private communication channel between two PBX's.

b. Circuit. A communication channel between two instrumentalities such as PBX's, key and station equipment, etc.

c. Facility. A means of providing a communication channel. A facility may be derived by means of a carrier channel or a voice frequency channel on cable, open wire, or radio beam. Several different facilities may be interconnected to provide a circuit.

d. Loop. A circuit between a switching equipment and a telephone station.

e. Trunk. A circuit between two switching equipments.

f. Subscriber. A customer supplied with telephone service by a telephone company.

g. Station Equipment. Telephone company instruments and associated equipment furnished to a subscriber. Large switchboards and switching equipment at subscriber locations are classified as PBX equipment.

h. Local Exchange Service. A telephone line furnished to a subscriber from the exchange in his local central-office area.

i. Foreign Exchange Service. A line furnished by the telephone company if the customer requires a telephone line number in an office other than his local central office.

j. An Extension Station. A telephone station at another location in the same general area of the main station. It may be permanently connected to the user's central office line or connected as needed by means of a switch.

k. An Off-premise Extension. One that is located some distance from the main station and may require auxiliary equipment at one or more central offices connected by local or interexchange channels.

l. The Switched Network. This refers to the complicated system of central offices and facilities that enable a telephone subscriber to communicate with any other subscriber.

20. Private-line Services

A *private line* provides service between two or more telephones located at the same address, within the same central office area, or in other than the same central-office area when auxiliary central-office equipment is furnished. When only two stations are involved, it is known as a two-point private line. Three or more stations on the same circuit comprise a multistation private line.

a. A *two-way automatic private line* is equipped with relays at each end and operates from d-c power. When either station lifts the receiver, the other station will be signaled automatically.

b. Private-line circuits which may be interconnected with telephone company networks or other private-line systems are known as *switched private-line circuits.*

c. A *push-button automatic private line* is a line used to communicate between two or more specific telephone stations. A key or button must be operated at the push-button station to signal the other station or stations. Lifting the receiver at the automatic station provides automatic signaling.

d. A push-button–push-button private line is one between two or more telephone stations where either has to operate a key or button to signal the other station.

e. Private-line or full-period service is provided where continuous operation between two or more locations is required. If service is required for a portion of the day, short-period private-line service can be supplied for the number of hours the customer wants service.

Use of either switched network or private-line facilities can be provided on an alternate-use basis for voice communication or data transmission. Telephones with a switching arrangement should be used for this type of service.

f. Intercommunicating lines called *intercoms* provide communication channels between users on the same system. Intercoms vary in size from a simple two-station arrangement, such as between an office manager and his secretary, to one that may encompass an entire plant. To function, each station must be furnished a transmitter battery, a source of signaling power, and an a-c transmission path to all other stations. Signaling is usually done by push button or by dialing.

21. Station Equipment

The choice of equipment is primarily one of satisfying the user's need for attractive, practical instruments that provide the features required for speedy and efficient handling of telephone services.

Fig. 18. Desk telephone, 1927.

Some of the most widely used instrumentalities are as follows:

a. Desk telephones are of two types: the *desk stand* or *handset type* and the *hand combined* set. Desk stands or handsets consist of an instrument that contains a receiver and transmitter and a switch hook or plunger. This is connected by various length cords to a bell box or subset mounted in the knee well of the desk or on a wall or other flat surface and containing the ringer, induction coil, and capacitor (see Fig. 18).

b. Hand combined sets have the ringer and other components enclosed in a mounting with the handset resting in a cradle on top. The entire set is placed on a desk or table top.

c. Wall-type hand combined sets and *hang-up telephones* with separate subscriber sets are available for use where desks are not available or desired.

d. Panel telephones fitting into wall recesses or cabinets have a built-in dial, induction coil, ringer, and cord reel. The cord automatically reels back upon "hang up." Loudspeakers and microphones may be provided in panel sets for use on intercom systems (see Fig. 19).

e. Message waiting telephones are equipped with a small red lamp activated by the message-center attendant to indicate a waiting message or other special information. The lamp flashes to notify the telephone user to call the operator and will continue to flash until turned off by the attendant.

f. Telephones for noisy locations are designed to reduce the amount of noise picked up by the transmitter during a telephone conversation. The handset contains a push-to-talk nonlocking button for cutting the transmitter in or out. When one is

FIG. 19. Panel-type telephone set.

listening to the distant party, the transmitter on the telephone in the noisy location is cut out, eliminating noise on the line.

g. Head telephone sets are generally associated with attendants' telephones at PBX cord-type switchboards and are also available on telephone sets if the telephone user requires use of both hands or is required to stay on the line for extended periods (see Fig. 20). Service dispatching, order taking, telephone dictation or stenographic transcription of telephone conversations, air-to-ground communications, and satellite tracking are some of the uses of head telephone sets.

h. Outdoor Use. Telephones mounted in weatherproof housings, called *outdoor telephones,* provide service for exposed or damp areas and may be used on piers and bridges, at pole locations, in yards, and in tunnels.

i. Portable telephones equipped with plug-ended cords can be moved to locations where a permanent telephone is not required. Flush and nonflush indoor and outdoor jacks are available for use with the portable telephone.

j. Conference Use. It is often desirable that more than two parties participate in a telephone conversation. This can be done by means of a conference circuit whereby several telephones are electrically bridged to one line, either by push buttons or by dialing. While most conference systems limit the number of simultaneous connections to five or six, systems can be designed to handle any number required. Conference calls can also be established by the central-office operator.

k. Hands-free Use. The hands-free telephone is another method of providing conference calls. A separately mounted microphone and a loudspeaker unit are connected to the telephone. When a button is operated, the distant party can be heard on the loudspeaker and any voices directed toward the microphone will be transmitted over the line. A small group gathered in one location can participate in the telephone conversation.

l. Key Telephone Use. When more than one telephone line is required, push-button or key telephone systems offer flexibility and a wide variety of uses, i.e., pickup of several central-office lines, foreign exchange lines, PBX station lines, private lines, and intercommunicating lines. Features of the system include pickup and holding, intercommunication, visual and audible signals, cutoff, exclusion, and signaling. Push

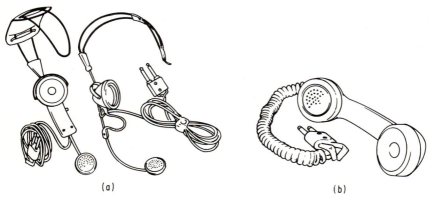

(a)   (b)

FIG. 20. Plug-in sets. (a) Head telephone sets. (b) Plug-in handset.

buttons or keys may be built in or separately mounted and may be illuminated to give a visual indication of incoming calls, held lines, or busy lines. Each terminating line is connected to a specific button on the set. Remotely located indicator lamps can also be provided to display the condition of any line.

Locking keys are provided for the pickup keys, and nonlocking keys for the hold or signaling keys. The nonlocking *hold* key enables the user to hold any line picked up by his set. With a pickup key operated and the handset off hook, momentarily depressing the hold key connects a coil across the line, preventing that line from disconnecting. This action releases the operated line key. The telephone set may now be used for communicating on another line by depressing the pickup key of the desired line. The hold feature is released when any station user picks up the line that has been held.

An incoming call will energize a bell or ringer associated with the line and flash a lamp at the push button of each telephone where that line appears. When the line is answered, a steady light appears. A slow flashing light called *wink-hold* appears when the line is being held.

Cutoff and exclusion features provide secrecy by operating keys or buttons to enable the user to disconnect other telephones from the line. Relay equipment for automatically cutting off other stations operates by simply lifting the handset.

m. TOUCH-TONE* Push-button Telephone Set. Figure 21 shows a type of telephone that provides talking and TOUCH-TONE® calling circuits together with pickup

* A registered service mark of the Bell System.

keys for connecting a maximum of five lines.   Lines may be two- or four-wire common battery or PBX lines, arranged for TOUCH-TONE® push-button calling, or nondial private lines.   This set also provides for holding one or more lines, signal lamps which illuminate the pickup buttons, a ringer which may be associated with one of the lines

FIG. 21. Multibutton telephone set equipped with TOUCH-TONE® dial.

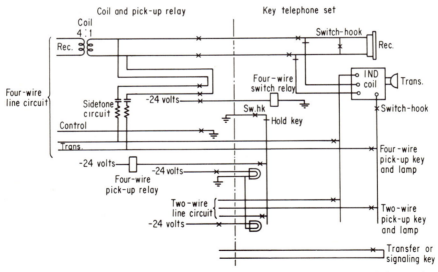

FIG. 22. Simplified schematic two- and four-wire key telephone set transmission paths.

or may be used as a common signal, keys to signal private or intercommunicating lines, and an exclusion key.   A schematic drawing of the circuits in this set is illustrated in Fig. 22.

n. Push-button systems with a capacity up to 120 buttons are available for systems requiring an unusually large number of lines (see Fig. 23).

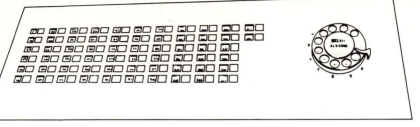

DIMENSIONS OF LARGE PUSH-BUTTON-SYSTEM FLUSH MOUNTINGS

Line capacity	Flash and release	Dial	Size of opening, in.			Size of faceplate, in.	
			Length	Width	Depth	Length	Width
30 button	Yes	None	9⅞	8½	4¾	10½	10
30 button	Yes	Yes	16½	8½	4¾	17⅛	10
60 button	Yes	None	16½	8½	4¾	17⅛	10
60 button	Yes	Yes	23⅛	8½	4¾	23⅝	10
90 button	Yes	None	23⅛	8½	4¾	23⅝	10
90 button	Yes	Yes	29⅝	8½	4¾	30¼	10
120 button	Yes	None	29⅝	8½	4¾	30¼	10

Face mat colors: blue, gold, green, silver, and white.

Fig. 23. 60-button flush mounting.

Fig. 24. 30-button CALL DIRECTOR® telephone.

An 18- or a 30-button keyset called a CALL DIRECTOR* desk-top telephone for large or expanding businesses can be used to provide a common answering point for all lines terminating in the system (see Fig. 24).

Large push-button sets are also used as attendant consoles at dial PBX's on calls requiring operator assistance.

* A registered trade-mark of the Bell System.

o. Selection. Many factors affect the final decisions involved in the development of an efficient communications system to meet a given requirement. One of the most important considerations is the location of the individual stations:

Same building
Adjacent buildings
Same continuous property
Same municipality, state, or country

Complex systems require multiconductor cables containing 50 or more wires. Most telephone companies install these cables without charge for distances up to approximately 150 ft. Beyond this a recurring mileage charge might apply. Long runs of cable, therefore, can involve considerable expense to the customer whether this service is furnished by a common carrier or outside supplier. Where long distances are involved, it is not economical or practical to use multiconductors, so converters must be employed to accomplish transmission on two- or four-wire lines.

22. Power

The power required to operate telephone station equipment varies with the type involved and the size of the installation required. The Telephone Company ordinarily

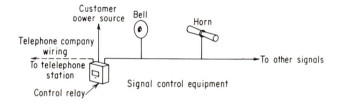

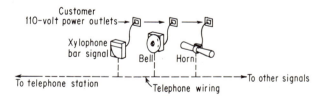

Fɪɢ. 25. Combined signals and control relays.

provides power equipment or a power supply as required. Where the signaling devices are dependent on commercial power, the user is responsible for providing the primary power to energize the signals and operate the rectifiers and frequency converters (see Fig. 25).

Since power sources may vary widely in each locality and power requirements for certain types of telephone equipment and sizes of installations may also vary, the local Telephone Company should determine specific requirements.

When engineering involves customer supplied power, the following provisions should be considered:

1. A steady source of power of the required voltage and, where alternating current is involved, proper frequency.

2. To avoid power fluctuations or noise, use a separately fused branch for the

telephone equipment and not one which would also supply power for lights, fans, or other apparatus.

3. The power circuit or circuits should be permanently installed in accordance with the National Electrical Code and any local regulations which may apply.

4. In order to avoid unintentional interruptions of power supplies for telephone equipment, the associated outlet or receptable should not be controlled by a hand switch.

23. Wiring Considerations

The Telephone Company will, in general, furnish, install, and maintain the wiring between a station location and any instrumentality which operates on current of a type and magnitude not exceeding that normally furnished by the Telephone Company, providing that surface-type wiring will suffice.

Where it is necessary to provide ducts by cutting into the floor, wall, or any part of a building or other structure for wiring of any type, all material and labor are provided, installed, and maintained by the customer. Locating, routing, construction, and

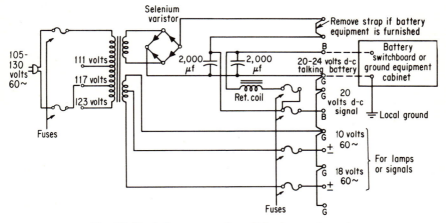

Fig. 26. Typical power-supply unit for key equipment.

type of duct to be provided should be mutually agreed to between the Telephone Company and the customer before any work is started.

Where equipment especially designed for use in explosive atmospheres is provided by the Telephone Company, the commercial power wiring together with the conduit in which it must be run is provided by the user. The engineer must also arrange for the customer to provide approved conduit for telephone company wiring within the hazardous area.

Common battery telephone systems ordinarily supply the battery for the transmitters at individual telephones or key telephone systems and switchboards only when the telephone is connected to a central-office line. The talking and signaling battery and current for visual and audible signals must be supplied locally for these types of systems when not connected to a central-office line.

Power supplies are engineered to meet specific requirements of a communication system. Dry cells of 1½ to 4½ volts are used for transmitter supply for local battery talking sets, bells, and buzzer circuits.

Rechargeable wet cells can be combined to form storage batteries of 14 to 52 volts to supply the current for PBX's, large key telephone systems, and general communication.system use. The batteries are kept charged by permanently connected rectifiers.

Rectifiers are devices for the conversion of alternating current to a current with a unidirectional component. Rectifiers may be tube type or some type of diode capable of handling appreciable amounts of power. Regulated rectifiers are required when variations in the load are such that the voltage fluctuations may exceed the permissible limits for proper circuit operations.

Unregulated rectifiers are satisfactory if the load remains fairly constant. When the power supply is used only for signaling or relay operation, an unfiltered supply may be used. If talking battery is required, the supply must be filtered to remove any a-c ripple, making it electrically quiet.

Rectifiers with good regulation and filtered outputs may be used without batteries to power communication systems, but they are dependent upon the reliability of the a-c power source. In locations where the power lines are susceptible to failures due to storms or other causes and communication must be maintained, reserve power in the form of storage batteries should be furnished (see Fig. 26).

24. Signaling

One of the most economical means of signaling between stations on private lines is to use 20-cycle ringing current. Smaller systems can be equipped with subscriber sets containing hand generators, while larger systems use static frequency generators

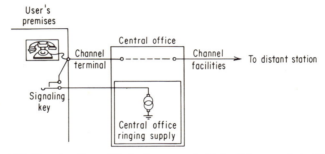

Fig. 27. Key signaling with ringing current obtained from central office.

or subcycle converters that will deliver enough current to operate almost any size system.

When key signaling is required where no local supply is available, ringing current may be obtained from a central-office source (see Fig. 27). Automatic private lines are arranged for signaling by lifting the handset. Battery flowing through a relay in series with the transmitter energizes the relay and supplies battery through its actuated contacts to the indicating device.

Ringers for use on telephone lines are designed to have a high impedance to voice currents but an impedance low enough at ringing frequencies to respond to ringing currents. The Bell System rates ringing bridges as either low impedance or high impedance and limits the number that can be bridged on a line.

A high-impedance ringer has a total d-c coil resistance of 2,500 ohms or more, while a low-impedance ringer has a total d-c coil resistance of less than 2,500 ohms.

A high-impedance ringer in series with a 0.45- to a 0.65-μf capacitor constitutes a high-impedance bridge.

A low-impedance ringing bridge is a low-impedance ringer in series with a 1- or 2-μf capacitor. A neon-type indicator used as a visual signal in lieu of a ringer is classified as a high-impedance bridge.

Polarized ringing bridges using cold-cathode tubes and low-impedance ringers of 1,400 or 1,500 ohms can be used for code ringing. This combination is considered equivalent to a high-impedance bridge.

25. Dialing

The expanding employment of Direct Distance Dialing (DDD) has stimulated the development of push-button sets to provide the user with a more convenient and faster method of dialing* calls.

Automatic dialers of many types, such as card dialers for rotary dial or TOUCH-TONE® telephones, are also available. The card dialer is built into the telephone set and has a card slot, a start bar, and on rotary dial sets a release bar (see Fig. 28).

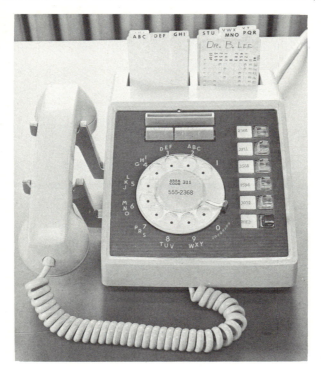

FIG. 28. Multibutton hand combined telephone set with card dialer.

The card dialer for a TOUCH-TONE® dial does not have a release bar, since the card moves fast enough to eliminate its need. Plastic cards with punch-outs are stored in a compartment at the top of the telephone set (see Fig. 29).

To dial a call, a prepunched card is inserted in the slot and after the dial tone is received, the start bar is depressed. The card dialer is arranged to pulse one digit and step to the next. When the last digit is dialed, the dialing mechanism stops. The card is released by pressing the release bar on the dial pulse dialer.

Telephone manufacturers supply dialers that can store and dial from 38 to 1,000 fourteen-digit telephone numbers on magnetic tape. These are known by various names, i.e., Magicall,† Rapidial,‡ and others. The method of operation is essentially

---

* The use of the terms *dial* and *dialing* when explaining push-button calling, while technically incorrect, is a carry-over from the original concept and will probably remain in general usage for some time.

† Registered trade-mark of DASA Corp.

‡ Registered trade-mark of McGraw-Edison Co.

the same for all types. A writing tape is provided for recording the names and telephone numbers. The telephone numbers are then recorded on the magnetic tape by using the dial associated with the dialer. To dial a number, a knob or pointer is moved to the name of the telephone subscriber selected and then, depending upon the type set used, the start bar or call button is depressed. The magnetic memory tape controls

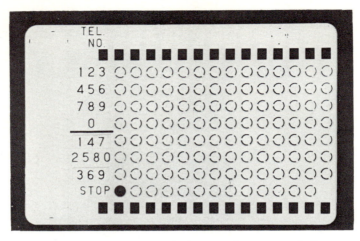

Fig. 29. Card used to dial telephone numbers automatically.

a dialing relay, the contacts of which perform the pulsing of the number including access codes and area codes.

TOUCH-TONE® calling introduced by the Bell System in 1963 for subscriber use employs a push-button-type dial in the telephone set to operate the associated central-office equipment. This dial includes a number of transistor-oscillators capable of generating two frequencies simultaneously. The user lifts the handset, and when the dial tone is received, the number is dialed by pushing down the appropriate buttons. When a button is depressed, two tuned circuits are selected corresponding to the two frequencies used to indicate the selected digit. The two frequencies generated by the oscillator are transmitted over the line to the central-office receiving equipment, which registers the corresponding digit (see Fig. 30).

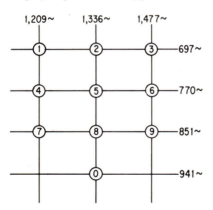

Fig. 30. Basic arrangement of push-buttons used for TOUCH-TONE® dialing.

A TOUCH-TONE® *card dialer* set is a combination of a card dialer and a standard TOUCH-TONE® push-button dial that can be used to place calls both automatically and manually by means of TOUCH-TONE® calling signals. To dial a call automatically, the user selects a prepunched card and inserts it in a card slot, similar to that used for d-c card dialing. The handset is removed, and when dial tone is heard, the start bar is depressed. The start bar moves a mechanical link to hold a common switch in its operated position for the entire dialing sequence. This switch disconnects the handset transmitter to guard against voice-simulated dialing errors and connects the TOUCH-TONE® calling circuit to the telephone-set network. A rotary switch, synchronized with the action of a hole sensing mechanism, opens and closes the excitation path as each line of the card is read.

TOUCH-TONE® dialing provides a means of dialing at a relatively high rate and can be used on private intercom systems.

26. Transmission Considerations

A communication transmission system consists of three essential parts: a source of energy, the medium over which it is desired to transmit energy to a receiving device, and the receiving device itself, which usually converts the electrical energy into some useful form.

Telephone circuits are generally designed to use d-c signaling from the user's location to the central office and 20-cycle signaling from the central office to the user. Direct current is also used to furnish transmitter battery, which is converted to alternating current for transmissions of intelligence.

Telephone facilities are provided by a pair of wires between telephone company premises and the user. Circuit bandwidth is approximately 200 to 3,000 cycles.

While most circuits are of the two-wire type, four-wire circuits provide a much higher grade of service and are almost a necessity for data transmission.

a. Circuit Losses. The amount of direct current available for operating relays, signaling the operator, dialing, or energizing the transmitter is controlled by the resistance of the line conductors. If resistance of the line exceeds maximum limits of the central office, auxiliary line circuits furnishing a higher voltage can be inserted in the line. Voice frequencies generated by action of the transmitter are a-c signals attenuated by the length of the circuit, by energy transferred at the two ends of the circuit, and by energy losses due to apparatus that may be associated with the circuit. Many of these losses can be recovered by use of voice frequency amplifiers. Telephone sets with built-in amplifiers are available for most types of service.

b. Working Limits. Telephone lines should be designed so that they fall within the transmission loss objectives of the central office or switching center serving them. The maximum ranges established are known as the *working limits* and include the maximum permissible d-c resistance, the transmission loss expressed in decibels, and the number of permissible ringing bridges.

c. Conductor Resistance. This term expresses the supervisory or tripping limits of a circuit in terms of the resistance of the wire or cable facilities. This resistance is twice the resistance of the single-wire resistance of the length of wire or cable used.

d. Transmission Loss. Transmission loss of a circuit is the ratio of the input to output power in decibels and is usually measured at 1,000 cycles. The circuit net loss is obtained by adding the losses of all individual portions of the loop to the equipment losses.

e. Ringing Bridges. The number of ringing bridges has a direct bearing on the design of a telephone circuit, since each ringer adds impedance to the line. An excessive number of ringers can result in pretripping or failure to operate ringers or ring up relays.

f. Loop Resistance. Maximum conductor loop resistance for central-office subscriber lines depends upon the type of office. Since telephone companies are continually modifying offices to increase their range, exchanges of the same types may not operate at the same loss. Typical subscriber loop ratings of the most common types of offices are shown in Table 10.

Table 10. Typical Maximum Subscriber Loop Resistance, in Ohms

No. 1 crossbar	800, 1,200, or 1,350
No. 5 crossbar	1,200 or 1,350
Panel system with ground cutoff	700
Panel system with battery cutoff	600 or 1,200
Step-by-step	700, 800, 900, 1,100 or 1,200
Manual	600 or 1,200

g. Loss Objectives. The maximum allowable conductor resistance for a foreign exchange line is the same as for a local central-office line. Transmission-loss objectives for central offices have been established, and by reference to a typical table, the overall end-to-end objectives of a system can be established (refer to Table 11).

h. Private-line Limits. The design of private-line circuits is similar to that of central-office subscriber lines and must conform to specific limitations governing transmission, signaling, supervision, and ring tripping.

Transmission and a-c signaling are affected by the attenuation of the line facilities, whereas d-c signaling, supervision, and tripping are controlled by the loop resistance.

Table 11. Subscriber Line Transmission Loss Objectives

	Local trunk, db	Foreign exchange trunk, db	PBX OP ext., db	FX OP ext., db
Maximum allowable conductor loss..........	4.7	7.7	5.9	7.7
Maximum allowable circuit net loss*.........	5.0	8.0	6.0	8.0
Maximum conductor resistance..............	Varies with Central Office (Refer to Table 10)			
Maximum allowable dial ringing bridges......	4	4	2	4
Maximum allowable manual ringing bridges...	4	4	4	4

* Net loss includes apparatus losses.

Table 12. Two- and Three-point Station Connections Sum of Circuit Net Losses

Decibels

Off-premise extension–tie line–off-premise extension..........	16.0
Off-premise extension–tie line–tie line......................	14.0
Tie line–tie line–tie line..................................	12.0
Off-premise extension–off-premise extension.................	12.0
Off-premise extension–tie line.............................	10.0
Tie line–tie line...	8.0

Off-premise extensions are restricted to 10 pulses per second dials.

Table 13. Allowable Ringing Bridges on Private Lines

Automatic signaling—four ringers at each end
Push-button signaling:
  With auxiliary equipment designed to repeat ringing current—four ringers operated by one push button
  Without auxiliary equipment:
    Two-point private line—two ringers or bells at each end
    Multipoint private line—one ringer or bell at each end
Hand generator limitations—four ringers or bells

The transmission and signaling capabilities of the circuits should be comparable in quality to those normally provided by telephone companies.

Private-line Losses. The maximum allowable conductor resistance for private lines may vary from 750 to 4,000 ohms depending upon the type of equipment used. The maximum allowable net circuit transmission losses vary from 6.0 to 16.0 db.

Table 12 lists the circuit net losses for private-line connections, and Table 13 lists signaling limits.

i. Mobile Radiotelephone Private Line. This is a circuit used to interconnect the console of a privately owned radio system with a remotely located transmitter and receiver. The same channel can be used for two-way voice transmission and for controlling the switching from the receiving to the transmitting equipment.

FACSIMILE*

27. General

Facsimile, in engineering, is the process of scanning graphical information including pictures, converting the information into signal waves, and using these signals locally or remotely to produce a likeness, in record form, of the subject copy. A facsimile system consists of a facsimile transmitter sometimes referred to as a scanner, transmission means, and a facsimile receiver. In some systems a converter is used to change the type of modulation to one more suited to the circuit used. Some prefer to use a facsimile transceiver rather than a separate transmitter and receiver, although the trend in recent years seems to favor separate units because the mechanisms for scanning and recording are usually drastically different and not much is gained by a combination set. In addition, in many cases the transmissions are predominantly in one direction.

The nature of facsimile is such that it poses the most stringent of requirements on the overall system. Special consideration must be given to level change, minute impulse-noise interference, low-level 60-cps pickup, small variations in transmission delay, echoes, multipath, envelope delay distortion, synchronization, and undesirable compression techniques. In the transmission of the million elements comprising a picture, one pulse received considerably different from its correct level, position, or sharpness becomes outstandingly apparent in the received picture, whereas with voice transmissions, considerably more distortion can be tolerated.

Facsimile and television have been compared, and the use of electronic scanning and high speeds for facsimile have necessitated careful examination to determine basic differences. Facsimile results in a final *hard* copy. When a camera is used to take a picture of a received television image, the system essentially becomes a high-speed facsimile system. Although most television systems utilize a 6-Mc spectrum allocation and most facsimile equipments use a voice channel, some facsimile sets have been built to operate at television frequencies. So speed alone is not a basic difference. Television systems are designed so images, with motion, at one location are conveyed to another location, and often there is little difference between succeeding frames, whereas facsimile systems are usually designed for little or no relationship between frames; each is an entity of its own. Television employs interlaced scanning at a sufficient frame rate to eliminate flicker, while facsimile usually is not concerned with these two factors but is required to transmit a stationary image faithfully with a single frame.

28. Components

Figure 31 shows block diagrams of the major components of facsimile sets. Starting with the facsimile transmitter the copy area is the location where the original copy to be transmitted is placed. This may be a drum which the copy is wrapped around either automatically or manually, or it may be a scanning slit which the copy moves past. In some cases a number of copies are stacked and automatically fed to the scanner one at a time. Facsimile machines have been built where the entire copy is placed in a fixed location and scanned without movement of the copy. This is generally accomplished with cathode-ray tube scanning but is not in general use owing to the availability of simpler techniques with better resolution and linearity of scan.

Scanning devices are shown in Fig. 32 and will be described separately.

The synchronizer controls the rate of motion of the scanning device. It sometimes consists of a crystal or tuning fork controlled oscillator which drives a synchronous motor at a constant speed. Accuracies are generally in the order of 1 part in 250,000. When the oscillator is turned on, its frequency must stabilize within the specified

* See Ref. 2, chap. 23; *Facsimile* by Austin G. Cooley; Ref. 3, sec. 19; *Facsimile Transmission and Reception* by Maurice Artzt; and Ref. 22.

tolerance in a few minutes. Sometimes synchronization is accomplished by sending synchronizing signals from the transmitter to the receiver. In systems of this type it is usual to provide a separate channel for sending a 60-cycle signal through the communications channel as a signal separate from the picture signals. For example, this may be frequency shift or amplitude modulation of a carrier in the range of 540 cycles; both transmitting and receiving motors are locked to the 60-cycle modulating frequency. Assuming that both scanning and recording devices are moving at the same velocity owing to the effectiveness of the synchronizer, it is still necessary to phase the two devices. Phasing is the process of assuring that both scanning and recording devices are at the same location along the scanning lines of the original copy and recording medium at a given time. Phasing signals are usually sent prior to scanning the original copy. These consist of pulses of a given duration which are repeated once per scanning line for a specified duration of time, for example, 15 sec of phasing pulses each consisting of 12 msec of black signal occurring once per scanning line.

The modulator of Fig. 31 accepts the photocell signals and causes amplitude or frequency shifts of a carrier. The composite signal is amplified to the appropriate line level. Many variations are in use, such as the application of single-sideband

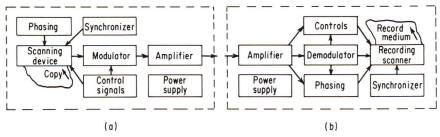

(a)   (b)

FIG. 31. Facsimile block diagrams. (a) Facsimile transmitter. (b) Facsimile receiver.

filtering at the output of the amplifier or amplification prior to modulation (for example in a photomultiplier tube) followed by a buffer output amplifier. In some systems the output signal is a varying d-c signal corresponding to the photocell signals but of appropriate level for transmission to a data set external to the facsimile set. Modulation takes place in the data set. Control signals are also sent in order to change the recorder from standby to active status, to start the actual recording of the picture, and to stop the recording. These are not necessary if manual operation of the facsimile receiver is employed.

At the facsimile receiver, signal level is often low owing to transmission circuit losses, and an amplifier is therefore necessary. This is usually followed by a demodulator and a recording device. Sometimes demodulation is incorporated in the recording device because of its characteristics. For example, a gaseous recording lamp used for photographic recording may act as a halfway rectifier. Input signals from a data set may also be of a pulsating d-c type, again making a demodulator unnecessary. Controls for phasing and synchronizing are functions corresponding to those just described for the facsimile transmitter. Some recording devices are shown in Fig. 33 and will be described later. A wide variety of recording media have been utilized for facsimile. These employ electrothermal, electrochemical, electrostatic, photographic, thermal, pressure, and ink transfer techniques. Only the more common media will be described. Electrothermal consists of a process whereby a stylus is in contact with a special paper which acts as a semiconductor. When voltage is applied to the stylus, the heat generated by current flow causes a dissipation of a top white surface layer of the paper, exposing a black or gray sublayer. In some systems the white layer forms a surface on a carbon-impregnated vinyl material. The passage of current through the material not only dissipates the white surface material but creates small holes in the vinyl material, thereby forming a stencil which can be placed upon an ordinary office duplicator. It is tough and sufficiently durable to permit thousands of copies to be prepared.

Electrochemical processes usually involve use of a wet or damp electrolytic paper. In some, the passage of current through the paper causes formation of a dye. In others there is an actual transfer of metal protons which combine with the electrolyte to form a mark. Recordings may be sepia in tone, magenta, or black. For news photographs and for the transmission of other high picture quality information such as Tiros cloud pictures, photographic processes are used. Exposure is commonly accomplished through the use of a gas discharge lamp such as the Sylvania R-1130. Until recently photographic processing in a darkroom was necessary. However, the use of Polaroid films and paper have made a darkroom unnecessary for many applications.

Pressure-type recording may be accomplished with special papers having surfaces that are pressure sensitive or by the use of ordinary paper with a carbon paper between the recording device and the paper. The latter has the advantage of being very economical, since it permits the use of almost any type of paper with low-cost carbon paper. It is a dry process, utilizes low voltages, and results in acceptable facsimile copy. Ditto-type carbon may be used to prepare a master for the preparation of many copies. Other types of masters may be prepared for use on other office duplicating machines.

Several processes, including the carbon-paper approach, permit the recording to be made on a transparent film. The result may then be projected, and the overall system becomes a remote display system.

29. Facsimile Transmitters

For the scanning systems shown in Fig. 32a and b the original copy is manually or automatically wrapped around the cylinder. As the cylinder rotates, either it progresses axially or the light and photocell system moves, causing the scanning spot to describe a helix on the cylinder. Cylinder speed is accurately controlled by means of a tuning fork or crystal oscillator and synchronous motor. Rotational speeds may range from 45 to 3,000 rpm. Cylinder scanning is adaptable for transceiver operation by replacement of photocell and some of the optics by a glow discharge lamp such as Sylvania R-1130 for photographic recording or by a stylus for direct recording.

The mirror-type scanners in Fig. 32c and d permit the copy to be scanned on a continuous basis. While a flat mirror is shown, a spherical mirror is used in some designs.[23] With a simple lens system it is desirable that the copy conform to a cylindrical surface as shown in Fig. 32c. The copy may be placed in a plane surface if an appropriately corrected lens system is used to give sufficient resolution throughout the scan and satisfactory uniformity of scanning-spot velocity. Much experimental effort is aimed at field flattening and simplification of scanning mechanism, and one of the most promising approaches is through use of fiber optics (see Fig. 32l). In one fiber optic configuration a straight scanning line is transformed to a circle, permitting the linear scanning to be accomplished with simple rotary motion.

Figure 32d involves the use of a multisided prism which must be ground with extreme precision to avoid jitter in adjacent scanning lines. Simpler mechanisms are involved than for Fig. 32c, since the prism is rotated under precise oscillator control. It is also relatively simple to calculate scanning-line repetition rate and scanning-spot velocity. In Fig. 32c the mirror usually oscillates under cam control in a manner to provide linear scanning[24] during the available scanning line and rapid return to an initial position for each scanning line. Accuracy of scanning velocity depends upon control of the cam rotational speed, upon the accuracy of the cam surfaces, and upon tolerances of the system. Scanning-line repetition rate may be calculated from the cam rotational speed, while scanning-spot velocity will additionally depend upon the cam slope.

A number of other systems have been devised for flat scanning. Some of these are shown in Fig. 32e, f, i, k, and l. In general in these systems, the copy is moved past a scanning-line area. It is feasible, especially when using a cathode-ray flying spot

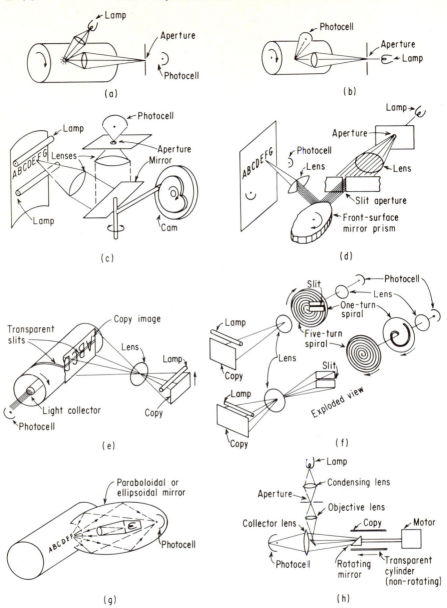

Fig. 32. Facsimile scanning methods. (a) Cylinder—flood lighting. (b) Cylinder—spot lighting. (c) Reciprocating mirror. (*Used by Adler/Westrex, Division of Litton Systems, Inc.*) (d) Flying spot-rotating polysided prism. (e) Cylinder with spiral. (f) Spiral—dual disk. (*Hogan Laboratories, Inc.*) (g) Paraboloid or ellipsoid. (h) Rotating mirror. (*Western Union Ticketfax.*)

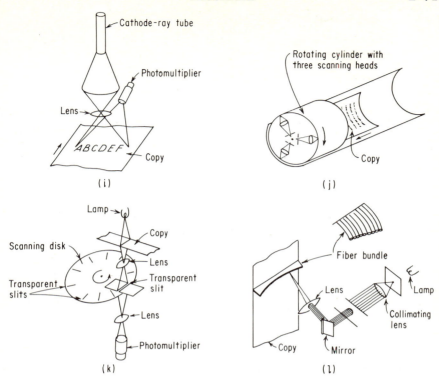

FIG. 32. Facsimile scanning methods (*continued*). (*i*) Cathode-ray tube—flying spot. (*Used by Xerox Corporation.*) (*j*) Rotating optics. (*David Sarnoff Research Center, RCA.*) (*k*) Scanning disk. (*Hogan Faximile Corporation. See Ref. 27.*) (*l*) Optical field flattener.

scanner, to place the whole copy in a scanning area where it remains stationary while horizontal and vertical scanning are accomplished by the motion of the cathode-ray tube spot. This is not usually done because of complexity.

The system of Fig. 32e uses the intersection of a transparent helical slit on a rotating cylinder and a transparent straight-line slit to dissect the image of the original copy into elemental areas as shown. The scanning-line area on the original copy is illuminated by floodlighting, and an optical light-collecting system within the cylinder reflects the light from the elemental area of the copy to the phototube. Scanning-line repetition rate and scanning-spot velocity are proportional to cylinder rotational speed.

In Fig. 32f, dissection of the image of the copy is accomplished by means of the intersection of a transparent multiturn helix and a transparent straight-line slit.[25] Since this combination results in a multiplicity of spots, one for each turn of the helix, an additional single-turn helix is used to limit the light to that from only one elemental scanning area at a time. The multiturn helix rotates at a higher speed than the single-turn helix by a factor equal to the number of turns in the multiturn helix. Another version of this type of scanner uses a single-turn involute instead of the multiturn helix.[26]

The cathode-ray flying spot scanner is shown in Fig. 32i. Although techniques such as those used for TV can be applied for facsimile scanning, it is more common in an electronic scanning approach to use a cathode-ray tube as a small moving light source.   By means of a lens system the spot is focused on the surface of the copy, and one or more photomultiplier tubes are employed to convert reflected light to electrical signals.   One chief advantage of this mode of scanning is the capability of operating at television rates or up to about 30 pages a second (one million words per minute).

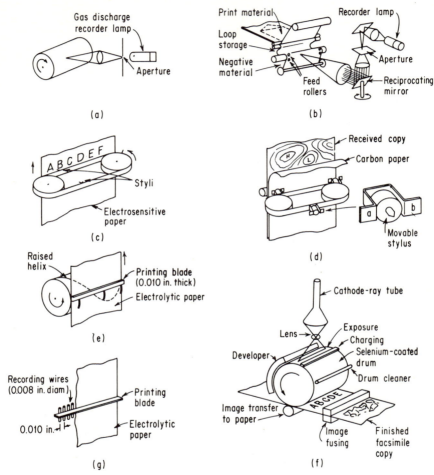

Fig. 33. Facsimile recording.   (*a*) Cylinder (or drum)—photographic.   (*b*) Oscillating mirror—photographic.   (*Army Facsimile Set AN/GXC-4.*)   (*c*) Styli on belt.   (*d*) Carbon-paper recorder.   (*Army-Marine Corps Facsimile Set AN/GXC-5.*)   (*e*) Helix—electrolytic.   (*f*) Xerographic recording.   (*Used by Xerox Corporation.*)   (*g*) Multistylus recorder.   (*Used by Hogan Faximile Corp., Subsidiary of Telautograph Corp.*)

For the parabolic scanner of Fig. 32*g* the copy is mounted on the cylinder, which rotates during scanning.   The paraboloid reflecting surface is an efficient light-gathering system.   The scanning head (phototube and light system) moves axially with respect to the cylinder.   The rotating mirror scanner (Fig. 32*h*) is mechanically simple, since scanning is accomplished by the mirror rotation and line feed, which poses little problem.   The rotating optics system (Fig. 32*j*) exists with a number of variations such as two optical systems 180° apart or four optical systems 90° apart, and either the scanning head progresses axially for line feed, or the scanning head rotates in a fixed position and copy moves beneath it. The latter permits continuous copy feed. With four or eight optical systems and a semicylindrical copy surface a multiplexed mode of transmission can be employed for higher speed transmission and reduced vulnerability to envelope delay distortion and noise. The outputs of more than one phototube are transmitted simultaneously on different carriers. A multispot recorder is used for this system.

30. Facsimile Recording

There are many techniques used for facsimile recording. Some will be described in conjunction with Fig. 33. In Fig. 33a, unexposed photographic film or paper is wrapped around the cylinder and is exposed by the light from a gaseous discharge glow lamp through use of an aperture and a focusing lens. Motion between the glow lamp assembly and the cylinder causes the pencil of light to describe a helix on the surface of the drum. Figure 33b shows another variation of a photographic recording technique: Light from a gas discharge lamp which passes through an aperture is collimated and is reflected from a mirror to a second lens which focuses an image of the aperture upon the recording photographic material. As the mirror reciprocates angularly, the spot of light on the photographic material is made to traverse the page with uniform motion and with rapid return; often polaroid photographic material is used. To maintain uniform motion of the spot and proper focus the page should lie on the surface of the cylinder or a special lens should be designed, if required. In one variation the reciprocating mirror is replaced by a constantly rotating multisided prism. Figure 33c shows styli mounted on a belt. Recording is accomplished by a voltage discharge through an electrothermal sensitive paper. Three or more styli are mounted at equal distances on the belt. Figure 33d shows a carbon-paper recording technique in which three electromagnets are mounted equidistant on a belt.* The electromagnets each have a movable center plunger converging to a point which does the recording. Carbon paper or other pressure-sensitive material is used for the recording. The technique takes advantage of a magnetic path from the flanges a and b to the platen behind the paper and return path through the recording stylus. Figure 33e shows a cylinder with a raised helix and an electrolytic paper and a printing bar. As the cylinder rotates, the intersection of the helix and the printing bar provides a traveling recording spot. Figure 33f shows a xerographic recording technique with a cathode-ray tube being used as a flying spot scanner. The surface of the drum is charged and then subsequently discharged selectively by the light from the cathode-ray tube. Black powder adheres to the charged drum in the form of the recorded image and is transferred to paper where it is fixed by means of heat to form the final copy. The recorder of Fig. 33g has multistyli. The recorder scanning can be accomplished by electronic switching from stylus to stylus rather than by a mechanical process. The technique is used for a microfilm transmission system.[27]

31. Available Facsimile Facilities and Applications

One of the prime uses for facsimile is newspaper work. Reporters proceed to a location where news is breaking, take a number of photographs, process these, and send them back by facsimile means. Small, portable transmitters are available, but often the reporter utilizes the nearest fixed station transmitter rather than carry a portable transmitter. At the main newspaper office a receiver is usually located in a darkroom and is loaded with unexposed film ready for receipt of a picture upon a moment's notice. Wherever feasible Telephoto Schedule II circuits are utilized. However, it is not always practical to obtain as good a circuit expeditiously, and therefore speed of transmission is limited by the use of inferior circuits. Until recently received pictures were developed and prints rushed to the editor's office in order that

* Facsimile Set AN/GXC-5.

he might determine which pictures would be used in the next edition of the paper. In recent years an additional parallel circuit to the editor's office has been set up and direct recordings of incoming pictures are made available to the editor as they are received; simultaneously the negative in the machine in the darkroom is exposed and developed. The direct recordings in the editor's office are made using electrolytic paper with good black-and-white contrast and fairly good rendition of gray tones. The editor selects the pictures to be used, and printing is then effected from the

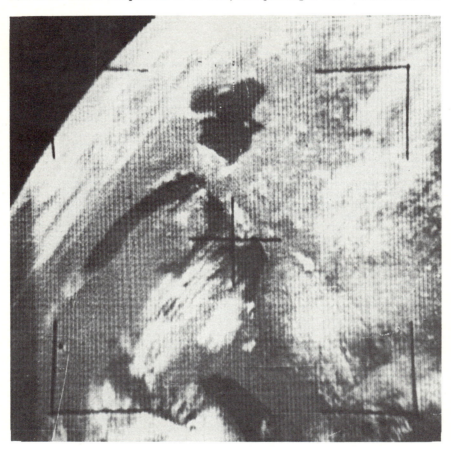

FIG. 34. AN/GXC-4 facsimile transmission of Tiros IV picture of Great Lakes region. (*Courtesy of U.S. Army and Weather Bureau.*)

photograph received in the darkroom or from the electrolytic recording. Most of these pictures are sent over wire circuits in this continent. Overseas pictures are transmitted by radio, using transoceanic circuits. These pictures are then relayed throughout the country by wire. Communication satellites offer the possibility of transmitting from almost anywhere in the world directly to this country.

Another use for facsimile is for the dissemination of weather information. The information is usually initiated as data fed by teletype or other data-transmission means from many weather stations to central locations. Weather maps are then plotted, and various charts prepared. These are sent by facsimile over a world-wide weather network. Standards are 18⅝ by 12 in., 96 scanning lines per inch, 60 and 120 lines per minute, and index of cooperation 576. Drum scanners* with 6 in. diameter and lengths up to 44 in. and flat-bed† scanners are used. Recorders are of the electrothermal,‡ electrolytic§ and carbon paper types.‖ In addition the Tiros satellite provides a meteorological service by taking pictures of the cloud coverage of

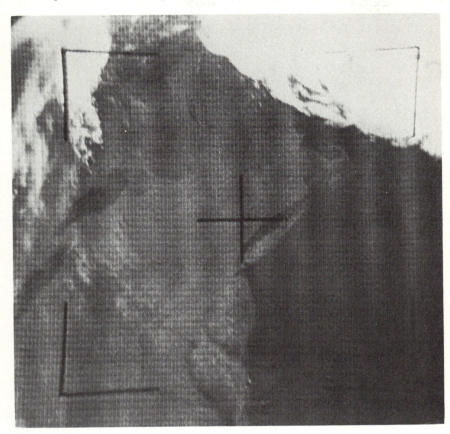

Fig. 35. AN/GXC-4 facsimile transmission of Tiros V picture of East Coast, Cape Cod to Cape Charles. (*Courtesy of U.S. Army and Weather Bureau.*)

vast areas of the earth. Two typical pictures are shown in Figs. 34 and 35. Throughout the Tiros orbits these are sent to several receiving or tracking stations. Approximately 3¼- by 4¼-in. prints (or less) are prepared and sent by high-quality photographic facsimile equipment¶ to Weather Bureau and National Aeronautics and

    * Muirhead and Co. Limited, D-990 Weather Chart Transmitter.
    † Alden Electronic & Impulse Recording Equipment Co., Inc., Alden 18-in. Flat Copy Scanner.
    ‡ Westrex Corporation, a Division of Litton Industries, Weatherfax RJ-3.
    § Alden Electronic & Impulse Recording Equipment Co., Inc., Alden 18-in. Automatic Recorder.
    Muirhead and Co. Limited.
    ‖ Alder/Westrex, Division of Litton Systems, Inc., Weatherfax RJ-4.
    ¶ Facsimile Set AN/GXC-4.

2–76   USER EQUIPMENTS AND SERVICES

Space Administration (NASA) centrals, where photographs are assembled to represent the cloud coverage over a very large area of the earth. This information is used for prognostications and preparation of weather charts. Figure 36 shows a weather map received on a U.S. Army carbon-paper recorder R-1063. As another variation, weather pictures from automatic picture transmission (APT) subsystems on board NASA satellites such as Nimbus are transmitted on a real time basis to small, inexpensive ground stations throughout the world. These are reproduced by facsimile in direct visual form as 8 by 8 pictures using Fairchild electrolytic paper* or as 3¼ by 3¼ photographs using an AN/GXC-4 modified for APT readout. Cloud pictures cover about a 1,000- by 1,000-mile area through the use of a 108° field of view. Operation is automatic and requires 208 sec per picture. Scanning rate is 240 lines per minute.

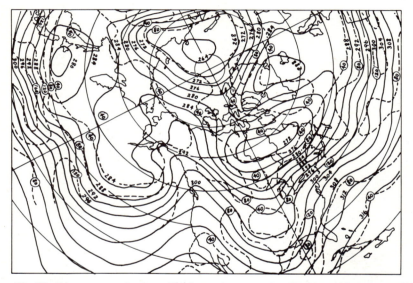

Fig. 36. Weather map received on 8½-in. carbon recorder R-1063. (*Courtesy of U.S. Army.*)

A starting signal is sent consisting of 300 cps for 3 and 5 sec of phasing pulses comprised of white signal interrupted by 12½ msec of black signal once per scanning line. The facsimile recorders start automatically but may also be operated manually. The signal is double-sideband amplitude modulation with black 32 db below white signal. The received carrier is 2,400 cps, and the maximum modulating frequency 1,800 cps.

The electrolytic recorder operates at 100 lines per inch, each with a total length of 8.45 in. on 9⅜-in. wide electrolytic paper supplied in 50- to 400-ft rolls. A useful line is 8 in. long. The index of cooperation is 269 (IEEE 845), and the rate is 2.4 in. of copy per minute. The recorded contrast ratio is 40 to 1. Impedance is 600 ohms.

Large industries, public utilities, and large nation-wide projects are utilizing facsimile as a means of transmitting orders, sketches, and messages among plants and to and from central control points. For example, orders in the Chicago office of a steel company are transmitted at the rate of one every 3 min to the plant miles away. One facsimile corporation† has about 2,000 sets in various uses of this type. These are divided into two major systems as follows: The first employs an analogue Bell

* SCAN-A-FAX, Type F-24A5 recording paper used on a Fairchild Facsimile Recorder Model P392-9.
† Stewart-Warner Corporation.

System DATA-PHONE* 602A Data Set which permits the user to dial other subscribers. Transmission time for an 8½- by 11-in. letterhead is 6 min, with a definition of 96 lines per inch and a speed of 180 lines per minute. Reception is on an electrolytic paper of the catechol type which gives good gray and black rendition. Catechol turns black in the presence of iron ions supplied by the printer blade. The blades are inexpensive and are changed with each 400-ft roll of paper. Synchronization is achieved by sending a reference 60-cps signal through a separate channel of the system. A power amplifier at the receiver amplifies this for driving the synchronous motor. Reception can be automatic and unattended; the receiver starts when the 60-cps signal is received. A 1,000-cps automatic answer-back is employed for unattended operation. Phasing is also automatic, and during the phasing period the receiving paper is advanced from a moisture-retaining chamber to ensure that the paper at the printer blade is damp. The signal from the facsimile transmitter and to the facsimile receiver is a pulsed direct current with 0 volts for white and +7 volts for black.

The second system, also for transmitting messages and sketches, utilizes a leased line with a DATA-PHONE† Data Set. Transmission times for an 8½- by 11-in. letterhead are 6, 3, and 1 min corresponding to 180, 360, and 900 lines per minute, respectively. The 6- and 3-min transmissions are over Schedule II Telephoto Circuits, while the 1 min is over a 15-kc program channel. The recording paper is electrolytic as described above. The signal is amplitude modulated with 2,400- or 4,500-cps carriers. The receiver starts when the carrier appears and phases automatically. Synchronization can be maintained through use of the same power source or by use of stable crystal-controlled oscillators at each end of the circuit. An auxiliary feature is an automatic document loader which will hold up to 100 original copies and automatically feed these to the transmitter one at a time.

Another manufacturer of document-transmission equipment‡ also features an automatic stack feed as an attachment to the facsimile transmitter. A stack of papers up to 1 in. high may be placed in the stack feed. Without the stack feed the facsimile transmitter is a flat-bed scanner and will accept any thickness from tissue to ⅓₂-in. cardboard. Recording is electrolytic.

An outstanding characteristic of another manufacturer's product§ is that transmitter and recorder both have tuning-fork motor control contained within the equipments. This is most significant for noisy circuits and for greater flexibility in being able to operate over a wide variety of power sources. Recording is on dry electrothermal materials requiring no processing. Timefax NDA provides black recording on light background, Timefax A provides a master from which duplicate copies can be made by the hectograph process, and Planofax N provides a paper plate master from which hundreds of copies can be duplicated by offset.

Other systems have been established specifically for message handling. The Western Union Telegraph Co. has a number of such systems. It has been found that small facsimile sets can be installed and operated in small businesses more economically than teletypewriters. By means of a small facsimile transceiver, the Desk-Fax,‖ messages are transmitted to the central office where they are received on facsimile machines and manually transcribed by keyboard to teletypewriter code. The messages are sent through the normal teletypewriter system. Depending upon the equipment at the receiving location the message may be sent from the final central office as a facsimile or a teletypewriter transmission. The blank size is 6½ by 4½ in. with a message area of 5¹¹⁄₁₆ by 3 in. The drum rotates at 180 rpm with a line feed of approximately 130 lines per inch. Transmission time of 2 min and 20 sec is required to handle the full message size. Recording is electrothermal, using Teledeltos.¶ In the order of 40,000 are in use.

* Trade-mark of AT&T.
† Trade-mark of AT&T.
‡ QUIK-FAX, Telautograph Corporation.
§ FAXWRITER, Westrex Company, a Division of Litton Systems Inc.
‖ G. H. Ridings and R. J. Wise, An Improved Desk-Fax Transceiver, *Western Union Tech. Rev.*, vol. 6, no. 3, p. 92, July, 1952.
¶ J. H. Hackenberg, and F. L. O'Brien, Recording on Teledeltos, *Western Union Tech. Rev.*, April, 1962.

Table 14. Typical Facsimile Equipment Characteristics

Copy size,† in.	Total scanning line length, in.	Index of co-oper-ation*	Transmission time or rate	Lines per inch	Scanning lines per minute	Trans-mission circuit	Application
6½ × 4½	....	...	2 min. 20 sec.	130	180	Voice	Messages
8½ × 11	2.75 × π	264	1.88 in. of copy/min	96	180	Voice	Letters and business documents
8½ × 13 8½ × cont.	....	...	3.96 in. of copy/min	....	360		documents
8⅜ × cont.	9.13	379	6 or 3 min/ 8⅜ × 11	96 hor. 100 vert.	180, 360	Voice	Letters and business documents
8 × 11	2.63 × π	263	....	100	90, 180, 360	Voice	Letters and business documents
8½ × 7	2.75 × π	264	0.94 in. of copy/min 0.47 in. of copy/min	96	90 45	Voice	News
187 × 130 mm¶	207 mm	352	11½ min for 130 mm of copy	5⅓ Lines/mm	60 (also 45, 90 120, 150)	Voice	Photographs
8½ × 7	2.75 × π	380	0.725 in./min	138	100	Voice	Photographs
8 × 10	2.75 × π	289	1.43 in./min	105	150	Voice	Photographs
8 × 11.75	3.80 × π	380	1 in./min	100	100	Voice	Photographs
8.5 × 12	2.91 × π	291	12 in./min	100	100	Voice	Photographs
8.5 × 12	2.91 × π	434	0.8 in./min	150	120	Voice	Photographs
18 × 12	18.85	576	0.625, 1.250 in./min	96	60, 120	Voice	Weather maps and charts
18 × 22	18.85	576	0.625, 1.250 in./min	96	60, 120	Voice	Weather maps and charts
8⅜ × cont.	9.13	379	18 × 11 1 min	96	900	Broadcast (15 kc)	Letters and business documents
15 × 17	15.71	264 528	5 min 20 min	52.8 105.6	180 90	Voice Voice	X rays
8½ × 15	8.64	330	8 × 11 45 sec	120	1,800	48 kc	High-speed message transmission
9½ × 42	2.75 × π	275	3.3 pages/min 16.4 pages/min	100 100 190 190	3,600 18,000	48 kc‡ 240 kc§ 48 kc‡ 240 kc§	Requisitions, orders, drawings

* CCIR value given; IRE value is CCIR value multiplied by π.
† Area on original copy is generally less than shown.
‡ Includes TELPAK A (AT&T).
§ Includes TELPAK C (AT&T).
¶ The International Telegraph and Telephone Consultative Committee (CCITT) normal photographic facsimile recommendation. Alternates involve 70- and 88-mm drum diameters and indices of 264 and 352. Transmitting drum factor at least 2.4.

For cases where a small network of these transceivers is to be established, a switching arrangement, designated a concentrator, is available. It permits selection of one station at a time from a patron station or intercommunications among up to five stations.

The Western Union Telegraph Co. provides a facsimile equipment for private systems applications* with or for wide-bandwidth circuits.† The former accepts copy

* Archie S. Hill, Letterfax Equipments for Service Flexibility, *Western Union Tech. Rev.*, 1958.
† C. R. Deibert, F. T. Turner, and R. H. Snider, A High Speed Direct-scanning Facsimile System, *Western Union Tech. Rev.* 1952.
D. M. Zabriskie, A High-speed Telefax Recorder, *Western Union Tech. Rev.*, 1952.
L. G. Pollard, A High-speed Facsimile Transmitter, *Western Union Tech. Rev.*, 1952.

8½ by 11 in. in size and transmits it between customers' offices at 180 or 360 scanning lines per minute requiring transmission times of about 6 or 3 min. The latter system transmits copy up to 8½ by 15 in. at 1,800 scanning lines per minute using a bandwidth from 9.5 to 40.5 kc. Letters 8½ by 11 can be sent in 45 sec. The copy to be sent is simply inserted in a transparent cylinder with printing toward the outside. Scanning takes place through the cylinder.

Facsimile systems are also available for transmission of railroad, airline, hotel, and bus reservation information. The system consists of a network of substations and a reservation control center. The requested reservation is sent out by facsimile to the reservation control center, where it is accepted if the space is available. Information is sent back and recorded directly on the ticket, part of which is given to the customer and part retained for accounting.

X-ray transmission† is also accomplished by means of facsimile. In emergencies an expert radiologist is not always near by. Through use of facsimile, a distant radiologist receives a copy of the X-ray within 15 min, interprets it, and immediately discusses the findings by telephone with the attending physician. Some characteristics of the scanner are maximum copy size 15 by 17 in., 5-in. drum diameter, 90 and 180 rpm, 105.6 (high) or 52.8 (normal) scanning lines per inch, signal voltage linear with density, and index of cooperation 528 (high definition), 264 (normal definition). Receiving is on a recorder similar to that made by Westrex for newsphoto service.

Perhaps the most advanced combination of high definition and high speed is incorporated in a large page-sized press facsimile system.‡ This equipment is capable of transmitting a full newspaper page with a definition of 400 lines per inch in 21 min over 48-kc group circuits. The drum speed is between 300 and 400 rpm. At a higher definition of 800 lines per inch the time is only 4.2 min using a video circuit and 3,000 rpm. Another variation has a definition of 1,000 lines per inch. The significance is impressive when it is noted that for other systems a definition of 200 lines per inch is considered high, page size greater than 8 by 11 in. is considered large, and a speed of 360 rpm is fast. The press facsimile is being used to transmit newspapers from a central editorial location to distant cities where the papers are published without need for typesetting or editing. Reception is on photographic material, which is then processed for large-scale printing using normal printing techniques. In addition to American applications the equipment is essential to the Japanese to replace a very cumbersome hand-printed press communications system.

Typical characteristics of facsimile equipment are shown in Table 14.

32. Facsimile Definitions and Formulas[28]

Converter, Facsimile.*,§   A device which changes the type of modulation.

Usually the change is from amplitude modulation to frequency-shift modulation or to pulsating d-c pulses (sending converter) or vice versa (receiving converter). Typical shift frequencies are 1,500 cps for white and 2,300 cps for black with grays represented by frequencies between these two values. The shift in frequency may vary linearly with the amplitude of the modulating voltage or with the density of the elemented area on the original copy. The former has been proposed as a standard.

Definition.*   Distinctness or clarity of detail or outline in a record sheet or other reproduction.[29]

Also referred to by some as "resolution" or "detail." Often expressed as lines per inch or elements per inch.

Density (in Facsimile).*   A measure of the light-transmitting or reflecting properties of an area. It is expressed by the common logarithm of the ratio of incident to transmitted or reflected light flux.

Direct Recording.*   That type of recording in which a visible record is produced without subsequent processing in response to the received signals.

---

† Westrex Corp., Division of Litton Ind., Pressfax (trade-mark).
‡ Westrex Corp., Division of Litton Industries, Pressfax (trade-mark).
§ Note: The definitions with an asterisk are from *IRE Standards on Facsimile*, Definition of Terms, 1956, courtesy of IEEE. Comments if any are by the author.

Drum Factor.   The drum factor is the ratio of drum length used to drum diameter. Where drums are not used, it is the ratio of the equivalent dimensions.

Echo.[*,30]   A wave which has been reflected at one or more points with sufficient magnitude and time difference to be perceived in some manner as a wave distinct from that of the main transmission.

Common causes of echo are mismatch at some point in a long wire circuit or reflections of radio signals from objects such as metallic structures.   Since scanning speed is usually known and displacement of the images can be measured, the difference in distance can be found from

$$X = \frac{V_W S}{V} \qquad \text{miles} \tag{15}$$

where V_W = velocity of propagation, mps (186,284 mps for radio, about 178,000 mps for open wire, 7,000 to 63,000 mps for telephone cable)
     S = displacement of echo on record medium, in.
     V = recording spot velocity, in./sec

Elemental Area.[*]   Any segment of a scanning line of the subject copy the dimension of which along the line is exactly equal to the nominal line width.

Envelope Delay.[*]   The time of propagation, between two points, of the envelope of a wave.

Note: The envelope delay is measured by the slope of the phase shift in cycles plotted against the frequency in cycles per second.   If the system distorts the envelope, the envelope delay at a specified frequency is defined with reference to a modulated wave which occupies a frequency bandwidth approaching zero.

Envelope Delay Distortion.[*]   That form of distortion which occurs when the rate of change of phase shift with frequency of a circuit or system is not constant over the frequency range required for transmission.

Note: Envelope delay distortion is usually expressed as one-half the difference in microseconds between the maximum and minimum envelope delays existing between the two extremes of frequency defining the channel used.

Envelope delay distortion is one of the most significant distortions affecting facsimile and data transmission.   A precise statement of acceptable amount of envelope delay distortion cannot be made, since it will vary with the modulation system used and type of copy being recorded.   An empirical value of ± 250 μsec per 1,000 cycles of maximum keying frequency is often used.[†]   This is based on an envelope delay distortion of plus or minus one-half the elemental area duration, or bit duration.   Allowable envelope delay distortion:

$$e = \pm \frac{1,000,000}{4f_m} \qquad \mu\text{sec} \tag{16}$$

where f_m = maximum keying frequency
     $= \dfrac{\text{elements/sec}}{2}$

Framing.[*]   The adjustment of the picture to a desired position in the direction of line progression.

Grouping.[*]   Periodic error in the spacing of recorded lines.

This occurs more often in recorders which have multiple recording heads.

Half-tone Characteristics.[*]   A relation between the density of the recorded copy and the density of the subject copy.

Note: The term may also be used to relate the amplitude of the facsimile signal to the density of the subject copy or the record copy when only a portion of the system is under consideration.   In a frequency-modulation system an appropriate parameter is to be used instead of the amplitude.

It is to be noted that the phototube system will integrate the reflections of light within the scanning aperture and that even a black-and-white original copy may cause

---

† See definition of maximum keying frequency.

a scanner to generate signals corresponding to gray when the aperture is located on the sharp edge between black and white or when lines narrower than the aperture are scanned. Therefore, even black-and-white copy is transmitted with better fidelity and quality if the overall system is analogue or reproduces several shades of gray rather than being binary in nature.

Index of Cooperation. For drum (cylinder) scanning, the product of the drum diameter and the number of scanning or recording lines per unit length.†

$$I = DR \qquad (17)$$

where R = scanning or recording lines per inch
$\quad\quad D$ = drum or cylinder diameter
For noncylindrical scanning:†

$$I = \frac{LR}{\pi} \qquad (18)$$

where L = is the scanning-line length (total).

*Jitter (in Facsimile).** Raggedness in the received copy caused by erroneous displacement of recorded spots in the direction of scanning.

Jitter is usually caused by scanning displacements internal to the facsimile recorder. It is possible, however, for the transmission circuit or for a recording device within the system to introduce variable delays in the transmission of pulses.

*Kendall Effect.** A spurious pattern or other distortion in a facsimile record caused by unwanted modulation products arising from the transmission of a carrier signal and appearing in the form of a rectified baseband that interferes with the lower sideband of the carrier.

Note: This occurs principally when the single-sideband width is greater than half the facsimile carrier frequency.

*Maximum Keying Frequency** (Fundamental Scanning Frequency). The frequency in cycles per second numerically equal to the spot speed divided by twice the scanning spot X dimension.

$$f_m = \frac{dv}{2} \quad \text{cps} \qquad (19)$$

where v = spot speed, in./sec
$\quad\quad d$ = No. of elements per/inch

or
$$f_m = \frac{dL(\text{rpm})}{120} \qquad (20)$$

where L = total scanning-line length (or equivalent)
$\quad\quad$ rpm = stroke speed

This is a fictitious maximum, since higher harmonic frequencies will occur for repetitious patterns being scanned and the frequency components of isolated scanned lines in the copy will exceed the maximum keying frequency. The maximum keying frequency, however, is an extremely useful figure, since it is indicative of modulation rate and is frequently used to determine bandwidth requirements of the transmission circuit.‡

In designing a system the maximum keying frequency is usually selected to be compatible with the usable bandwidth of the transmission system. The Kendall effect prevents use of lower frequencies of an available channel, and envelope delay distortion or radio multipath may cause further restrictions on channel usage. A typical example for a voice bandwidth circuit is a maximum keying frequency of 1,200 cps with a carrier of 2,400 cps and suppressed vestigial sideband. For 8½-in. copy

† This is the International Index of Cooperation. The IEEE definition uses total scanning line length instead of the drum diameter: $I_1 = LR$.

‡ See H. F. Burkhard, Considerations on Facsimile Transmission Speed, *AIEE Tech. Paper* 49-108, December, 1948.

the stroke speed is 180 rpm, and the transmission time for an 8½- by 11-in. copy is 5.85 min.

*Maximum Modulating Frequency.** The highest picture frequency required for the facsimile transmission system.

Note: The maximum modulating frequency and the maximum keying frequency are not necessarily equal.

*Multipath Transmission (Multipath).** The propagation phenomenon which results in signals reaching the radio receiving antenna by two or more paths.

Note: In facsimile, multipath causes jitter.

*Phase-frequency Distortion.** Distortion due to lack of direct proportionality of phase shift to frequency over the frequency range required for transmission. Intercept distortion usually does not affect amplitude-modulated facsimile transmission.

Note 1: Delay distortion is a special case.

Note 2: This definition includes the case of a linear phase-frequency relation with the zero frequency intercept differing from an integral multiple of π.

*Phasing.** The adjustment of picture position along the scanning line.

*Picture Frequencies.** The frequencies which result solely from scanning subject copy.

Note: This does not include frequencies which are part of a modulated carrier signal.

*Record Medium.** The physical medium on which the facsimile recorder forms an image of the subject copy.

*Record Sheet.** The medium which is used to produce a visible image of the subject copy in record form. The record medium and the record sheet may be identical.

*Recorded Spot X Dimension.** The effective recorded spot dimension measured in the direction of the recorded line.

Note 1: By effective dimension is meant the largest center-to-center spacing between recorded spots which gives minimum peak-to-peak variation of density of the recorded line.

Note 2: This term applies to that type of equipment which responds to a constant density in the subject copy by a succession of discrete recorded spots.

*Reproduction Speed.** The area of copy recorded per unit time. See Transmission Rate.

*Scanning-line Length.** The total length of scanning line is equal to the spot speed divided by the scanning-line frequency. See Stroke Speed.

Note: This is generally greater than the length of the available line.

*Signal Contrast (in Facsimile).** The ratio expressed in decibels between white signal and black signal.

*Skew (in Facsimile).** The deviation of the received frame from rectangularity due to asynchronism between scanner and recorder. Skew is expressed numerically as the tangent of the angle of this deviation.

*Spot Speed.** The speed of the scanning or recording spot within the available line.

Note: This is generally measured on the subject copy or on the record sheet.

Stroke Speed (Scanning or Recording-line Frequency). The number of times per minute, unless otherwise stated, that a fixed line perpendicular to the direction of scanning is crossed in one direction by a scanning or recording spot.

Note: In most conventional mechanical systems this is equivalent to cylinder speed. This is often stated as rpm of an actual or a fictitious cylinder.

*Synchronizing (in Facsimile).** The maintenance of predetermined speed relations between the scanning spot and the recording spot within each scanning line.

Frequency stability required for synchronization:

$$\text{Stability} = \pm \frac{x}{2LRH} \tag{21}$$

where x is the allowable amount of skew for the total copy (inches of deviation from the nonskewed condition).

Instantaneous frequency stability must be correspondingly high. For example, if max skew is $\frac{1}{16}$ in., total line scanned is 8.66 in., with 96 lines/in. and copy length 11 in.

then, $\quad\quad$ Stability $= \pm \dfrac{\frac{1}{16}}{2 \times 8.66 \times 96 \times 11} = \pm 1$ part in 292,639

Tailing (Hangover).* The excessive prolongation of the decay of the signal.
Transmission Rate. The area transmitted per unit of time.
Total area transmitted per unit time for a drum scanner:

$$A = \frac{L(\text{rpm})}{R} \quad \text{sq in./min} \tag{22}$$

where R = scanning lines per inch
$\quad\quad L$ = total scanning-line length, in.
$\quad \text{rpm}$ = strobe speed
$\quad\quad T$ = time for copy of height H and width W
$\quad\quad W$ = copy width scanned usually less than L
Copy area scanned per unit time:

$$A_1 = \frac{W(\text{rpm})}{R}$$
$$A_1 = \frac{HW}{T} \quad \text{sq in./min} \tag{23}$$

Note that $A = A_1$ if $W = L$.
If measurements are taken on the record medium, T is the reproduction time.
$\quad$ *Transmission Time.* The time required for the transmission of an area or a copy of stated size.

$$T = \frac{HR}{\text{rpm}} \quad \text{min/page} \tag{24}$$

or

$$T = \frac{HLRd}{120f_m} \tag{25}$$

where H = height or length of copy, in.
$\quad\quad R$ = scanning or recording lines per inch
$\quad \text{rpm}$ = stroke speed
$\quad\quad d$ = definition, elements per inch
and where $R = d$

$$T = \frac{HLd^2}{120f_m} \tag{26}$$

A significant point is that for a given transmission channel (f_m = constant) the transmission time increases as the square of the definition!
$\quad$ *Symbols.* These are the symbols used for the facsimile equations of this chapter.
$\quad\quad d$ = definition (number of elements per inch)
$\quad\quad D$ = drum (or cylinder) diameter, in.
$\quad\quad f_m$ = maximum keying frequency, cps
$\quad\quad H$ = height or length of copy, in.
$\quad\quad L$ = total scanning-line length (or equivalent), in.

$$L = \frac{V \times 60}{\text{rpm}}$$

R = resolution, scanning or recording lines per inch.

rpm = stroke speed, scanning-line frequency, or the rotational speed for drum scanning

S = displacement of echo or multipath elements from main image on record copy, in.

V = spot speed, in./sec

V_w = velocity of propogation of signal

W = width of copy scanned (available line), in.

33. Requirements for Interoperation of Facsimile Equipment

In order for facsimile scanners to interoperate properly with facsimile receivers, the following characteristics must be considered:

The scanning line frequencies (rpm) of the scanner and recorder must be essentially equal. The tolerance is usually held between 1 part in 150,000 and 1 part in 300,000. This tolerance results in very slight skew in the received copy. See definition of synchronizing.

Index of Cooperation. This ensures the proper aspect ratio between width and length of copy.

Direction of Scanning. The direction of scanning is the manner in which rectilinear scanning takes place. A normal scanning direction implies scanning in the same manner as a person reading., i.e., the first scanning line starts from the upper left corner of the page and proceeds to the right. Subsequent lines follow from top to bottom. A wrong direction of scanning results in a mirror image and is sometimes purposely used where a negative recording is desired. The international standard for photographic transmission is the opposite direction of scanning from that described above, but for a direct recording system the above and the international standard agree.

Synchronization. For synchronization three techniques are prevalent: tuning fork or crystal control of scanning at each end of the circuit, transmission of a reference synchronizing signal (sometimes a 60-cps modulation of a carrier), and the control of the scanning rate by common commercial power-line frequency. The synchronization of scanner and recorder must be compatible types.

Type of Modulation. The type of modulation such as AM or FM and the direction of modulation such as white as maximum signal must be compatible. In addition the contrast which is the spread in voltage or frequency in going from white to black and the linearity should also be matched.

Control Signals. These are often used to start, phase, and stop a facsimile receiver automatically.

The transmitter drum factor must not exceed the receiver drum factor.

LINE MODULATORS AND DEMODULATORS

Most subscriber devices such as teletypewriters and data input and output devices transmit or receive unmodulated signals which are binary in nature, being pulsed-d-c or polar square-wave signals. These are usually unsuited for transmission over voice bandwidth or wideband circuits, and modulating equipment is used for conversion from the binary digital signal to a modulated signal for transmission or vice versa for reception. These modulation devices have a variety of names, including modems, line units, line terminal equipment, and data sets.

Many types of modulation are employed such as amplitude modulation (AM), frequency shift keying (FSK), phase modulation, etc. Some are essentially transparent, passing signals from one end of the circuit to the other with no concern for code, while other sets require elaborate slaving and timing techniques and stable oscillators. Additional variables are speeds of operation, kind of subscriber device, analogue or digital, and separate channel requirements for controls.

34. Services

The Telephone Company is able to provide services in three ways: (1) as a complete service including Bell Systems communications terminal equipment at the stations, (2) as a data communications facility over voice bandwidth circuits with the customer providing the data terminal equipment, (3) as a group of telephone channels between two points, available with voice channels as follows:

TELPAK A, 12 channels
TELPAK B, 24 channels
TELPAK C, 60 channels
TELPAK D, 240 channels

Private-line data services are set up between two or more points. Digital data speeds are available up to the megabit per second range; analogue facilities are also available at many speeds.

The alternate to private-line services are the exchange-type services ("switched") which include telephone and TWX (Teletypewriter Exchange Service). To permit the use of the exchange telephone system for data, the telephone companies make available DATA-PHONE* service. Nationwide connections are made through normal dialing, and the circuit may then be used alternately for voice and data. The following features and other features available with normal telephone service are usable with DATA-PHONE* service:

PBX services, Private Branch Exchange
DDD, Direct Distance Dialing
FX service, Foreign Exchange
WATS, Wide Area Telephone Service

The data communications facilities are divided into three bandwidths:
Narrowband: up to 200 bits per second includes TWX, which operates up to 150 bits per second.
Voice band: up to 2,400 bits per second, includes DATA-PHONE service and various private-line data channels.
Wideband: above 2,400 bits per second.

35. Sets and Interfaces

Some available data sets are shown in Table 15. Interface characteristics are not described in the table owing to the variety of types and the significance of the control, timing, and alarm circuits, which are as important for proper operation as are the actual data circuits.

As an example, the interfaces for Bell System Data Sets 202A and 202B are shown in Figs. 37 and 38. All leads from the data set to the business machine except ring indicators 1 and 2 and the interlock lead have +8 volts to signal ground (SG) to indicate either an ON or spacing signal and −8 volts to indicate an OFF or a marking signal. Tolerance is ±2 volts. The voltages will be lower if the business machine has an impedance of less than 3,000 ohms. Shorting will not damage the data set circuits. Ring indicators 1 and 2 are shorted by the data set to signify an ON condition. An OFF condition is evidenced by a low-resistance ground signal on the interlock lead. The business machine should limit current on ring indicators 1 and 2 to 500 ma and the open-circuit voltage to 50 volts.

Regarding signals from the business machine to the 202 data sets, signals between +3 and +50 volts on all but ready (RY), remote control (RC), and remote release (RR) correspond to an ON or a spacing signal and either an OFF or a marking condition is indicated by a signal between −3 and −50 volts. Voltages must not exceed 50 volts and are referred to signal ground. The input impedance of the data sets is

* Trade-mark of AT&T.

Table 15. Modulator-Demodulator Sets (Modems)

Type	Analogue/ digital	Baseband characteristic	Line characteristic	Baseband rate	Xmit rec	Primary use and remarks
Data Set 103A[a]	Digital	Serial asynch. Mark = −5 to −25 volts; space = +5 to 25 volts, into 3,000 ohms resistance	Voice-grade circuit	300 baud (150 baud)	T/R	DATA-PHONE[b] Service TWX network
Data Set 103F[a]	Digital	Serial asynch.	Voice-grade circuit	300 baud	T/R	Private-line data service
Data Set 201A[a]	Digital	Serial synch.	Voice-grade circuit	2,000 bps	T/R	DATA-PHONE[b] service. One version has internal transmitter clock; other version can accept external timing. Receiver recovers bit synchronism
Data Set 201B[a]	Digital	Serial synch.	Conditioned voice-grade circuit	2400 bps	T/R	Private-line service. One version has internal transmitter clock; other version can accept external timing. Receiver recovers bit synchronism
Data Sets 202A, B, C, and D[a]	Digital	Serial asynch.	Voice-grade circuit	1800 baud, limited to 1,200 baud in DATA-PHONE service	T/R	For DATA-PHONE[b] service and private-line data channels. A. Integrated housing. B. Separate telephone set. C. Integrated housing, compatible with ACU's, optional reverse channel and conformity with Interface RS-232A. D. Separate housing (requires 804A in alternate voice or switched services), compatible with ACU's, optional reverse channels, conforms with Interface RS-232A and standard arrangement for alternate private-line DATA-PHONE[b] service
Data Set 301B[a]	Digital	Serial synch.	TELPAK A	40,800 bps	T/R	For private-line data channels. Transmitter can be changed from internal to external timing. Receiver operates in bit synchro-

						nization. Complete terminal consists of data set, loop facility, and frequency translating equipment (called wideband modems). Data set is part of TELPAK A2 terminal
DATA Sets 401A, B, E, F, H, J	Digital	Parallel	Voice-grade circuit	20 chs	A, E, H, and H trans. B, F, and J rec.	For DATA-PHONE[b] service. A and B—16 characters. E and F—99 characters. 401—used primarily for telemetry-automatic answer. J—Supersedes 410 B and F and also has voice answer-back capability
Data Sets 402A, B, C, D[a]	Digital	Parallel	Voice-grade circuit	75 chs	A and C trans. B and D rec.	For DATA-PHONE[b] service and private-line data channels. Up to 8 levels. A and B answer-back at 20 chs for two levels. A and C integrated housing; B and D separate housing, contact-type interface, A and B can be combined on one line. C and D same as A and B, respectively, with optional reverse channel, compatibility with ACU's, remote test, and unattended answering
Data Set 602A[a]	Analogue		Voice-grade circuit	Up to approx 900 cps	T/R	For DATA-PHONE[b] service and private-line data channels. Used primarily for 180 line/min facsimile service
Data Sets X603A(M-10) and X603B(M-10)	Analogue		Voice-grade circuit	Up to 120 cps	A trans. B rec.	For DATA-PHONE[b] service and private-line data channels. Used for electrocardiogram transmission, has electrical reverse signaling channel
Data Set X603C(M-10)[a]	Analogue		Voice-grade circuit	Up to 120 cps	T	For DATA-PHONE[a] service and private-line data channels. Used for electrocardiogram transmission, portable, uses X805A reverse channel unit for receiving reverse channel signals and is coupled acoustically

Table 15. Modulator-Demodulator Sets (Modems) (Continued)

Type	Analogue/digital	Baseband characteristic	Line characteristic	Baseband rate	Xmit rec	Primary use and remarks
Data Auxiliary Sets 801A,[a] X801B(M-10), and X801C(M-10) (automatic calling unit)				A and B dial pulses, 10 pulses per sec, C touch tone MF pulsing, 10 digits per sec		A and C any-number automatic calling unit. B one-number automatic calling unit. For DATA-PHONE[b] service, TWX, and private line
804A[a]						Telephone and line control. For DATA-PHONE[b] service and private-line data channels. For use with Data Sets 202D and 402D in alternate voice or switched line service
DATACOM[c] conversion equipment consisting of Data Set AE-101A and Subset AE-691A	Digital	Serial asynch.	Part of voice circuit	75 bauds	T/R	For DDD application. Four wire; transmit tones mark—1,270 cps, space—1,070 cps; receive-mark—2225, space—2025
Modulator Model 612,[d] Demodulator Model 635	Digital	Serial asynch.	Voice circuit	Up to 1,000 bps	T/R	Vestigial sideband, 2-kc carrier transmitted over any schedule four-wire line. Input ±20 volts. Output from 635 is +20 volts (on mark) −20 volts (on space)
Datatel® 25A Data Transmission System	Digital	Serial asynch.	340 cps of voice channel	200 bps 200 wpm 11 cards/min	T/R	Used for teletypewriter, including data code versions; punched tape; card transfer (for example, IBM 1050 or 065/066-068); Data-acquisition (example IBM 7750). Speech plus data. Unipolar or bipolar inputs. Output bipolar (15 volts) or unipolar (20 or 48 volts). Carriers

Equipment	Type	Mode	Channel	Speed	T/R	Remarks
						can be at 1,190, 1,530, 1,870, 2,210, 2,550, 2,890, or 3,230 cps permitting multiplexing when used with Lenkurt 23A Datatel®
TE-210D-2[f]	Digital	Serial synch.	Voice bandwidth schedule-4A	600, 1,200, 2,400 bps	T/R	Accepts serial or parallel data from teletypewriter, telemetry, and other data source. Sends data on four tones and uses KINEPLEX® technique. Optional diversity adapter
T-1200(A) and (B)[g]	Digital	Serial synch.	Voice frequency	300 600 1,200 bps	T/R	T-1200(A) incorporates stable (5 parts in 10^8 per day) internal timing source. T-1200(B)—Similar to (A) but with timing source deleted
Duobinary—Datatel® 26B Data Transmission System[31]	Digital	Serial synch. up to 1,200 bps Serial synch. at 2,400 bps	Voice circuit (see remarks)	600, 1,200, 2,400 bps	T/R	Operates with paper tape, magnetic tape, punched cards, and computers. Unique coding requires ½ bandwidth and permits error detection without introducing redundancy. Can be used for facsimile analogue signal up to 600 cps. Voice circuit should be equalized to within 0.3 msec or optional equalizer Shelf 1 may be used

bps = bits per second
T = transmitter
R = receiver
RS-232A = Electronic Industries Associated Data Interface
ACU = automatic calling unit.

[a] Bell System.
[b] DATA-PHONE is a trade-mark of AT&T.
[c] Automatic Electric subsidiary of General Telephone & Electronics.
[d] Rixon Electronics, Inc.
[e] Lenkurt Electric Co., Inc.
[f] Collins Radio Company.
[g] Stelma Inc.

10,000 ohms. The business machine indicates an on condition by placing a short between RC and either RY or RR; current due to the data set will not exceed 50 ma. Voltages from the data set will not be above 50 volts.

A Cinch or Cannon DB-19604-433 receptacle or equivalent with threaded retaining spacers is used for the data set. The business machine should have a cable not longer than 50 ft terminated in a Cinch or Cannon DB-19604-432 plug with a Cinch DB-51226-1 hood or equivalent with retaining screws to prevent disengagement.

Some of the data sets use a data interface standard described by the Electronic Industries Association in their Document RS-232A, *Interface between Data Processing Terminal Equipment and Data Communication Equipment.*

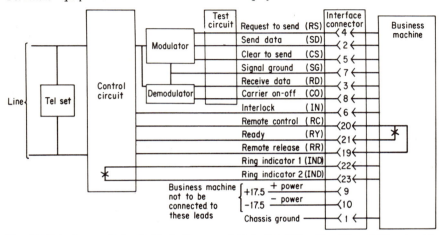

FIG. 37. Interface and simple block diagram of Data Set 202. Pins 24 and 25 not used, pins 11 through 18 reserved. (*Courtesy of AT&T Company. Copyright by American Telephone and Telegraph Company,* 1963.)

A negative voltage between 3 and 25 volts with respect to signal ground is used for data circuit mark (1) condition or control circuit off condition. A positive voltage between 3 and 25 volts is used for data circuit space (0) condition or control circuit on condition. The receiving device presents an essentially resistive load of at least 3,000 ohms with capacitive component not exceeding 2,500 picofarads. Additional details are covered in the document.

Bell System data set 602A is used for facsimile transmissions. Interface circuits terminate in a 25-pin Cinch or Cannon DB-19604-433 Connector-Receptacle with connections as follows:

1. Frame ground (FG)
2. Send data (SD)
3. Receive data (RD)
4. Request to send (RS)
5. is reserved.
6. Interlock (IT)
7. Signal ground (SG)
8. Carrier on-off (CO)
9. Positive power (+18 volts)
10. Minus power (−18 volts)
17. Sync in (SI)
18. Sync out (SO)
19. Remote release (RR)
20. Remote control (RC)
21. Ready (RY)
22. Ring indicator 1 (RI-1)
23. Ring indicator 2 (RI-2)

11 through 16 are reserved. 24 and 25 are not used.

All control voltages conform to RS-232 described above. Control leads RY, RR, and RC require contact closures which should be rated at least 50 ma, 20 volts. Signal and frame grounds are common. Send data uses an analogue signal between

0 and 7 volts into a data set impedance above 30 kilohms; the RS signal must be placed in the ON condition by the facsimile machine in order to send. Likewise receive data signals are analogue 0 to 7 volts into 40 kilohms or greater, and the RS lead must be in the OFF condition.

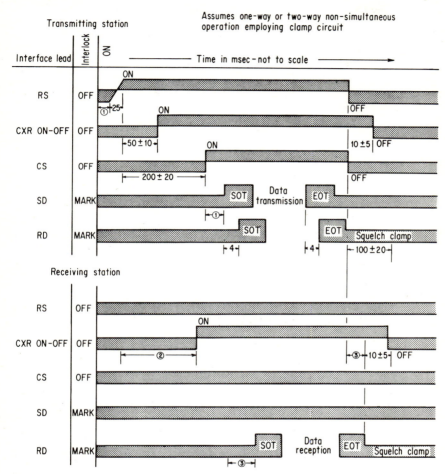

FIG. 38. Timing diagram for proper operation of data systems employing Data Sets 202 Notes: 1. Business machine reaction time. 2. Line delay, echo suppressor turn-around time and receiver clamp removal (50 to 260 msec). 3. Line delay (up to 50 msec). (*Courtesy of AT&T Company. Copyright by American Telephone and Telegraph Company, 1963.*)

Sync in accepts 6.3 volts 60 cps from the sender when RS is ON. Sync out will deliver essentially the same frequency to the customer at 2.0 ± .5 volts; it is OFF when the data set is not ready owing to primary power being off. When carrier (energy above 1,300-cps band) is not detected at the receiving data set, the CO circuit will deliver an OFF to the customer. Two hundred milliseconds of carrier makes CO ON and absence for 5 to 20 msec turns CO to OFF. RC is the common point for RY and RR leads. The data set will not lock in the data mode unless a closure is provided between RR and RC, so this closure must be established when ready to receive or transmit. If this closure is not provided, the telephones will still ring but the call will not be automatically answered. With the data set in the unattended mode an incoming call will cause a contact closure between RI-1 and RI-2 for a duration of about 150 msec whenever ringing occurs. The + and −18 volts power is only for telephone company test purposes.

36. Telex

The Western Union Telegraph Company also has a vast switched teletypewriter network called TELEX. One subscriber equipment consists of a Model 32 (KSR) page teleprinter, a remote-control dial unit, and an automatic answer-back. Another subscriber set is the TELEX ASR containing a Siemens Model T 100 Page Teleprinter with built-in reperforator and tape transmitter. Off-line tape preparation is provided with simultaneous page monitor copy. Connections are established through use of the dials on these sets.[32]

REFERENCES

1. Truxal, John G., *Automatic Feedback Control System Synthesis*, chapter 1, McGraw-Hill Book Company, New York, 1955.
2. Henney, Keith, *Radio Engineering Handbook*, McGraw-Hill Book Company, New York, 1959.
3. Pender, McIlwain, *Electrical Engineer's Handbook*, John Wiley & Sons, Inc., New York, 1950.
4. Freebody, J. W., *Telegraphy*, Sir Isaac Pitman & Sons, Ltd. London, 1959.
   Faugeras, D., *La Télégraphie et le télex*, Publisher-Éditions Eyrolles 61, Boulevard Saint-Germain, Paris V, 1962.
5. Elias, P., A. Gill, R. Price, N. Abramson, P. Swerling, and L. Zadeh, Progress in Information Theory in the USA, 1957–1960, *IRE Trans. Inform. Theory*, July, 1961.
6. DCA CIR 175-2A, *Defense Communications System Engineering Installation Standards Manual*, Superintendent of Documents, Government Printing Office, Washington, D.C.
   Rexroad, William D., *Teletypewriter Fundamentals Handbook*, Computer Design Publishing Corp. Baker Ave., West Concord, Mass. Also see *Computer Design*, November, 1963.
7. MIL-STD-188B, *Military Communication System Technical Standards*.
8. Kaufman, Herbert, and Herbert Ullman, Problems of Mating In-Out Devices to Solid-state Digital Computers, *Sylvania Technologist*, vol. XII, no. 3, July, 1959.
9. Everitt, W. L., *Communication Engineering*, chap. IV, McGraw-Hill Book Company, New York, 1937.
   Reference Data for Radio Engineers, 2d ed., p. 180, Federal Telephone and Radio Corp., 1946.
10. Guilleman, *Communications Networks*, vol. II, pp. 106 and 492, John Wiley & Sons, New York, Inc., 1935.
11. King, George, Tension-control Methods for Winding Problems, *Electromechanical Design*, September, 1963.
12. Sourgens, R., Télégraphie arythmique électronique, *L'Onde Elec.*, No. 403, October, 1960.
13. Webster, Edward, The Impact of Non-impact Printing, *Datamation*, September, 1963.
14. Jacobs, N. A., A New Line of Low-Cost Light Duty Teletypewriter Equipment, *IEEE Conf. Paper* 63-462, Teletype Corporation.
15. Innes, Frank T., High Speed Printer and Plotter, *Proc. 1960 Eastern Joint Computer Conf.*
16. Jorgensen, H., Plan 115 Switching System, *Western Union Tech. Rev.*, October, 1961.
17. Data Timing For HF Radio Transmission, *The Lenkurt Demodulator*, February, 1964.
18. Vincent, George Q., Private Wire Services in Data Processing by Western Union, Reprinted from *Western Union Tech. Rev.* (Summation of data and switching systems).
19. Rosenblatt, F., Perceptron Simulation Experiments, *Proc. IRE*, vol. 48, pp. 301–309, March, 1960.
    Highleyman, W. H., Linear Decision Functions, with Application to Pattern Recognition, *Proc. IRE*, vol. 50, no. 6, June, 1962.

Highleyman, W. H., The Application of Decision Theory to Pattern Recognition, *AIEE-IRE Spring Lecture Series*, New York, May, 1962.

Fischer, G. L., D. K. Pollock, B. Radack, and M. E. Stevens, *Optical Character Recognition*, Spartan Books, 1962.

Falk, Howard, Optical Character-Recognition Systems, Electro-Technol., July, 1964.

Deckert, William W., The Recognition of Typed Characters, *Elec. Mfg.*, June, 1960.

Comeretto, Alan, Advanced Concepts Studied for Pattern Recognition, *Electron. Design*, Mar. 1, 1961.

Akers, Jr., S. B., and R. H. Rutter, The Use of Threshold Logic in Character Recognition, *Proc. IEEE*, August, 1964.

20. King, George, Digital Readout Devices, *Electromechanical Design*, April, 1964.

21. Glossary of Computer Terms, *Sylvania Technologist*, vol. XII, no. 3, July, 1959. *Standards on Electronic Computers: Definition of Terms*, 1956, IRE.

22. Goldsmith, Dr. A. N., Van Dyck, Horn, Morris, and Galvin, *Radio Facsimile*, RCA Institutes Technical Press, 1938.

23. Snider, R. H., Flying Spot Flat-bed Facsimile Transmitter with Automatic Message Loading, *Western Union Tech. Rev.*, vol. 13, no. 1, January, 1959.

24. Turner, F. T., Derivation of a Cam Contour for Minimum Stress, *Western Union Tech. Rev.*, vol. 13, no. 1, 14 January, 1959.

25. Hogan, J. V. L., and G. M. Stamps, A New Continuous-feed Facsimile Scanner, *AIEE Conf. Paper*, Winter Meeting, New York, Jan. 18–22, 1954.

26. Gross, J. F., and A. Portnoy, Telefax Transmitter with Involute Scanner, *Western Union Tech. Rev.*, vol. 13, no. 1, 2 January, 1959.

27. Stamps, G. M., and H. C. Ressler, A Very High-speed Facsimile Recorder, *IRE Trans. Commun. Systems*, vol. CS-7, no. 4, December, 1959.

Shaler, D., An Ultra-high Speed Microfilm Facsimile System, *AIEE Conf. Paper* CP-62-249, Winter General Meeting, January, 1962.

28. Benewicz, T. F., Measuring Line Level on Telephoto Systems, *Bell Lab. Record*, March, 1960.

Temporary Test Standards 1943, *IRE Standards on Facsimile*.

29. Roetling, Paul G., Image Evaluation Techniques, *Research Trends*, Cornell Aeronautical Laboratory, Inc., of Cornell University, Winter 1963–1964.

30. Benewicz, T. F., Carrier Operated Echo Suppressor and Control Device, *IRE Trans. Commun. Systems*, pp. 208–214, June, 1962.

31. Bramhall, F. B., Duobinary Coding Equipment, *Autom. Elec. Tech. J.*, July, 1963. Lender, Adam, The Duobinary Technique for High Speed Data Transmission, *IEEE Conf. Paper* CP-63-283, Winter General Meeting, New York, Jan. 27–Feb. 1, 1963. (Lenkurt Monogram No. 181.)

32. Easterlin, Philip R., Telex in the U.S.A., *Western Union Tech. Rev.*, vol. 16, no. 1, January, 1962. Deloraine, E. M., Evolution of Telephone, Telegraph, and Telex Traffic, *Elec. Commun.*, vol. 39, no. 2, 1964.

Chapter 3

SPEECH CHARACTERISTICS AND ACOUSTIC EFFECTS

WILLARD F. MEEKER, *Radio Corporation of America*

GENERAL SPEECH CHARACTERISTICS

1. The Nature of Speech

Speech sounds are very complex. The primary purpose of speech is the transmission of thoughts from talker to listener. For that reason, in speech communication systems, it is necessary to consider both the production of speech by the talker and the perception of speech by the listener. The physical mechanism of speech production is reasonably well known; that is, the movements of the vocal cords and the various parts of the vocal tract have been studied in considerable detail, and their functions are largely understood.[1-4] However, their control mechanisms, restraints, and interrelationships are not understood. Nor are the processes involved in the perception of speech known. For these reasons, the features of speech which transmit intelligence are not precisely known. Nevertheless, a great deal of information has been accumulated regarding the characteristics of speech signals and the effect of various distortions occurring in its transmission. While such information permits a system designer to estimate the performance of a communication system under many conditions or to design a system to provide desired performance, he may encounter conditions for which experimental data are lacking. In such cases, experiments to evaluate the effect of such conditions may be required.

Speech sounds may be divided into two principal categories having markedly different spectra. *Voiced sounds* result when the vocal cords vibrate; *unvoiced* sounds result from the forcing of air through a constricted portion of the vocal tract to produce noise. A few speech sounds combine the two types of excitation. The various parts of the vocal tract (throat, nose, tongue, teeth, lips) operate to alter the spectrum of the sound produced and impart intelligence, or meaning, to it.

2. Frequency Range and Spectrum

The lowest frequency in speech is that of the fundamental frequency of vibration of the vocal cords. This fundamental frequency (often referred to as "pitch" or "fundamental") ranges from about 90 cps for a deep-voiced man to about 300 cps for a woman with a high-pitched voice. The highest frequencies produced by the various air-stream turbulences extend to very high audio frequencies. However, not all of this frequency range is necessary for the transmission of intelligence. No appreciable loss in intelligibility is observed if the frequency range is restricted to a band of frequencies from 200 to 6,000 cps.[5,6]

When the vocal cords vibrate during the production of voiced sounds, pressure pulses are produced. The vocal cords are then the source of excitation of the vocal

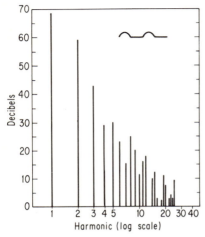

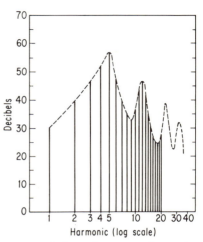

Fig. 1. Typical vocal cord wave (vocal tract excitation pulse) and its harmonic analysis showing the amplitude of the first 25 harmonics. (*From Miller,*[7] *J. Acoust. Soc. Am. Used by permission.*)

Fig. 2. Typical speech spectrum for a voiced sound. The dotted line indicates the envelope of the line spectrum and shows the vocal tract resonances.

tract. The pulses are roughly triangular[7] as shown in Fig. 1 and produce a line spectrum as indicated. Both odd and even harmonics are present. The amplitudes of the higher harmonics may be zero in certain regions, and the harmonics may not be clearly defined at high frequencies because of slight variations in the fundamental period. The amplitude of the harmonics decreases with increasing frequency at the rate of about 12 db/octave.

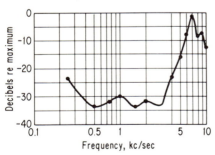

Fig. 3. Energy density spectrum of an unvoiced sound, the "s" in "sect." (*From Hughes and Halle,*[20] *J. Acoust. Soc. Am. Used by permission.*)

The vocal tract resonances serve to accentuate certain regions of the spectrum, producing an output spectrum of the type indicated in Fig. 2. Much of the information in speech resides in the frequency of the first two or three resonances, commonly called *formants*, and in their variation with time.[8,9] The fundamental frequency itself is relatively unimportant to intelligibility but affects such factors as quality and talker identification, as well as providing inflection.

For unvoiced sounds, the vocal cords do not vibrate and the vocal tract is excited not by a line spectrum but by a random-noise spectrum produced by forcing air through a constriction or by directing it against the teeth or lips. Because the vocal tract is now relatively open at the larynx, it reduces the amplitude of the low-frequency components of this random-noise spectrum and may introduce an antiresonance in the mid-region of the frequency range.   Generally broader resonances in the higher-frequency regions are produced by the portion of the vocal tract forward of the constriction.   Again, the frequencies of the resonances and antiresonances carry a major part of the intelligence.   Figure 3 is an example of the spectrum of an unvoiced sound.

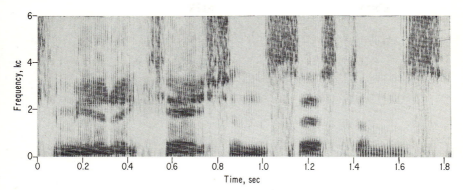

FIG. 4. Sound spectrogram of the phrase "communication systems."   Intensity is represented by the density of the black portions.   The analyzing filter was 300 cps wide.

The spectrum of continuous speech is a constantly changing one.   This is illustrated in Fig. 4, which is a sound spectogram[8] of the phrase "communication systems."   Here the abscissa is time, the ordinate is frequency, and the power density or intensity is represented by density of the black portions.   The analyzing filter was 300 cps wide; consequently, the individual harmonics are not identifiable.   Voiced sounds are, however, identifiable by the vertical striations corresponding to the fundamental pitch period.   The dark, more or less horizontal bars show the movement of the vocal tract resonances (formants) during the course of speech.

The long-term spectrum of speech idealized from the average for a number of male and female talkers[10] is shown in Fig. 5.   It is apparent from Figs. 2 to 5 that the spectrum measured over a short interval may differ markedly from the long-term spectrum.   The long-term spectrum varies considerably with different talkers.   Long-term spectral differences arise from different dimensions of the vocal tract, and these differences no doubt provide some of the cues for talker identification.

FIG. 5. Idealized long-term average speech spectrum at 1 meter from the lips.   The corresponding overall level is 65 db re 0.0002 dyne/sq cm.   (*From French and Steinberg,*[5] *J. Acoust. Soc. Am.   Used by permission.*)

3. Dynamic Range

Speech is such a variable signal that its measurement is difficult.   In normal continuous speech, there is no signal at all about 20 per cent of the time;[6] in addition to pauses for breath or pauses between phrases and sentences, there are short periods of silence which actually form a part of certain speech sounds.   Interestingly, there generally are not gaps between individual words, but one word blends into the next throughout complete phrases or sentences.

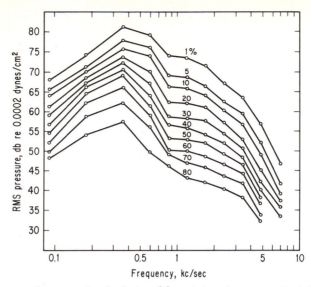

FIG. 6. Rms sound-pressure levels during ⅛-sec intervals measured at 1-ft distance. Composite data for six men. Each curve shows the percentage of intervals in which the level exceeded the ordinate at different frequencies. (*From French and Steinberg,*[5] *J. Acoust. Soc. Am. Used by permission.*)

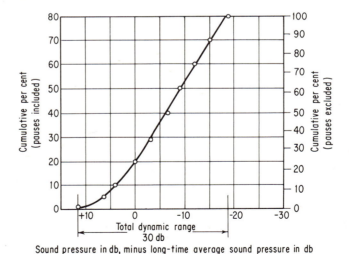

FIG. 7. Cumulative level distribution of speech in ⅛-sec intervals. (*From French and Steinberg,*[5] *J. Acoust. Soc. Am., Beranek,*[6] *Proc. IRE. Used by permission.*)

The particular characteristics of speech chosen for measurement and their method of measurement are generally influenced by known characteristics of hearing. Since the ear responds to sound pressure, pressure microphones are used to pick up the speech signal and the speech signal itself is specified by the sound pressure measured at some standard distance from the mouth. Also, the ear and hearing mechanism appear to integrate over a short period. Figure 6 is one representation of speech. Here measurements were made of the rms sound pressure in $\frac{1}{8}$-sec intervals in one-half octave bands. The $\frac{1}{8}$-sec interval approximates the time constant of integration of the ear. Figure 7 is based upon the 1,000- to 1,400-cps band and shows the cumulative distribution of rms pressures. It is taken to be typical of speech throughout most of its range. Referring to the right-hand ordinate, the conclusion is reached that the total dynamic range of speech is about 30 db and extends from 18 db below its long-term rms average to 12 db above. The peak amplitude of the speech signal encompasses an even larger range, since in $\frac{1}{8}$-sec intervals the speech peaks exceed the rms value by approximately 10 db.

In addition to the amplitude range of normal speech, there are substantial individual variations which extend the range of speech amplitudes which may be encountered. In a given environment one person may talk more loudly or softly than another. Also, other factors, such as ambient noise and sidetone, affect the speech level. These factors are discussed under Input Considerations.

SPEECH INTELLIGIBILITY AND QUALITY

4. Measures of Intelligibility

The primary purpose of speech is to convey thoughts. While some attempts have been made to measure the extent to which listeners correctly perceive the intended thought, a more common procedure is to measure the percentage of words or individual speech sounds uttered by a talker which are perceived correctly by listeners. A number of different tests have been designed for this purpose; the most widely used makes use of a set of phonetically balanced word lists sometimes known as the Harvard PB-50 word lists.[11,12] In the testing procedure, talkers read these word lists over the system under test to a number of listeners. The listeners write down the words heard. The responses are then examined; the percentage of words heard correctly is determined and is termed *per cent word articulation*.

By means of such tests, the effects of a number of system parameters have been explored. From the results, together with the knowledge of the physical nature of speech and certain information about the sense of hearing, methods have been developed for computing speech intelligibility from the system characteristics.[5,6,13,14] The methods are somewhat complex and do not yet provide for all system possibilities but nevertheless give quite good results for most communication systems. In an interpretation of the results of the computation, consideration must be given to such factors as the intended use of the system (normal conversation or code words), the users (highly trained or inexperienced), and the importance of immediate understanding.

The measure of intelligibility which is computed is the *articulation index* (AI). The articulation-index concept is a very useful one because in addition to providing an estimate of the overall system intelligibility, it also permits an understanding of the limiting elements in the system.

The articulation-index concept is based upon a number of observations about speech.

1. Speech obviously must be above the threshold of audibility to be perceived.
2. Noise which exceeds the threshold of audibility masks speech, effectively raising the threshold of audibility.
3. There is an upper limit to the sound pressure which the ear can utilize for the perception of speech.
4. Frequencies from 200 to 6,100 cps are required for substantially perfect intelligibility.
5. Speech has a 30-db dynamic range.
6. Different frequency regions contribute unequally to intelligibility; conversely, the frequency range from 200 to 6,100 cps can be divided into bands of equal contribution to speech intelligibility. The speech range may be divided into 20 such bands, shown in Table 1.

There are additional factors which permit refinements in the calculation.

Table 1. Twenty Frequency Bands of Equal Contribution to Articulation Index*

Band No.	Limits, cps	Mid frequency, cps	Band No.	Limits, cps	Mid frequency, cps
1	200–330	270	11	1,660–1,830	1,740
2	330–430	380	12	1,830–2,020	1,920
3	430–560	490	13	2,020–2,240	2,130
4	560–700	630	14	2,240–2,500	2,370
5	700–840	770	15	2,500–2,820	2,660
6	840–1,000	920	16	2,820–3,200	3,000
7	1,000–1,150	1,070	17	3,200–3,650	3,400
8	1,150–1,310	1,230	18	3,650–4,250	3,950
9	1,310–1,480	1,400	19	4,250–5,050	4,650
10	1,480–1,660	1,570	20	5,050–6,100	5,600

*From Beranek.[6]

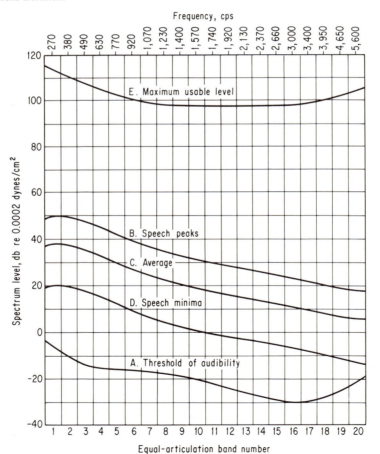

Fig. 8. Diagram illustrating the articulation-index concept. Here the entire dynamic range of speech can be perceived by the listener.

The articulation-index concept is based also upon an analysis of the sound pressures (signal and noise) ultimately produced at the listener's ear. It is convenient to reduce all statements of sound pressure at the listener's ear to *spectrum level,* that is, the sound pressure level in decibels re 0.0002 dyne/sq cm for a 1-cycle bandwidth. The concept is illustrated in Figs. 8 to 10.

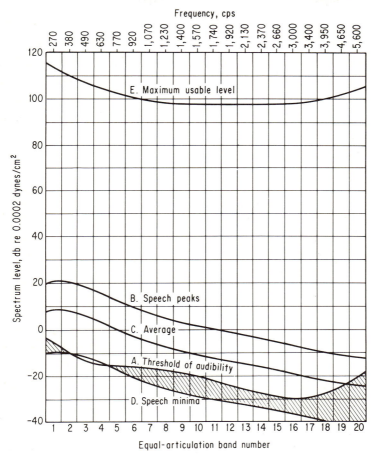

FIG. 9. Diagram illustrating the articulation-index computation. Here a part of the dynamic range of speech (shaded area) falls below the threshold of audibility. The articulation index is the ratio of the unshaded area between curves *B* and *D* to the total area enclosed by curves *B* and *D*.

In Fig. 8, the abscissa is marked in terms of both band number (lower scale) and center frequency of the band (upper scale). In this case, the speech range is divided into 20 bands in accordance with Table 1. The ordinate is the spectrum level of sound pressure at the listener's ear. Curve *A* is the threshold of audibility for continuous spectra. Curve *B* represents the spectrum of speech peaks; curve *C,* the long-term rms spectrum of speech; and curve *D,* the spectrum of speech minima. The spectrum levels shown for speech represent an overall level of 65 db. Curve *E* represents the limit above which no contribution to intelligibility results; the ear can be considered to overload at higher levels.

The area enclosed by curves B and D in Fig. 8 is the base for the articulation-index calculation.

If the speech level were lowered, as shown in Fig. 9, part of the speech (indicated by the shaded area) would not be heard, since it falls below the threshold of audibility. The ratio of the area between curves B and D and lying above curve A to the total area between curves B and D is the articulation index.

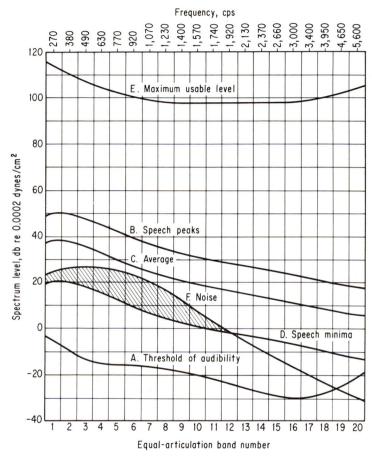

Fig. 10. Diagram illustrating the articulation-index computation. Here noise intrudes into the dynamic range of the speech and masks a part of it (indicated by the shaded area), thus reducing intelligibility.

If the listener is in a noisy location, noise can mask the speech and decrease intelligibility. This may be accounted for by including the noise spectrum level as shown in Fig. 10. Here curve F is the spectrum level of the ambient noise at the listener's ear. Only the area between curves B and D which is above the noise spectrum and the threshold of audibility contributes to intelligibility; the articulation index is now computed as the ratio of this area to the total area between curves B and D.

When the articulation index has been computed, the word articulation (or sentence articulation) can be estimated by means of relationships such as those shown in

Fig. 11. These relationships are determined experimentally and vary somewhat depending upon the experience and proficiency of the listeners and upon the nature of the test material.

In the case illustrated in Fig. 10, the articulation index could be restored to unity by amplifying the speech as indicated in Fig. 12. Here the speech level has been raised 15 db, and none of the area between curves *B* and *D* falls below the noise spectrum or the threshold of audibility. This process could be repeated for higher noise levels until curve *B* begins to exceed curve *E*. Any area above curve *E* is excluded from the articulation-index calculation.

In Figs. 8 to 10, the frequency scale is distorted to give equal weight to each of the 20 frequency bands of equal contribution to intelligibility. The ordinate is in decibels. This permits computation of articulation index by actual measurement of area. It is often more convenient to plot the same data with a logarithmic frequency scale, retaining the decibel scale for the ordinate. The center of each of the equal articulation bands may then be marked on the frequency scale, and the area can be calculated by summing the ordinates at the center of the bands intercepted between the noise-masking or audibility-threshold curve and the speech-peak curve, subject to the limitations mentioned previously.

The above discussion is necessarily somewhat oversimplified; additional details are found in Ref. 14.

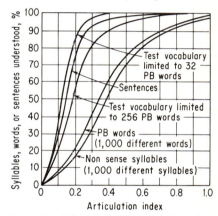

Fig. 11. Relation between articulation index and monosyllable (PB words) or sentence intelligibility. *Note:* These relations are approximate. They depend upon type of material and skill of talkers and listeners. (*From Kryter,*[14] *J. Acoust. Soc. Am. used by permission.*)

5. Articulation Testing

Although the intelligibility provided by a communication system can be estimated by articulation-index computations as indicated above, it is often necessary or desirable to measure directly the performance of a given system. Generally the primary interest is in intelligibility. Other characteristics are important but not readily measurable. Intelligibility is customarily measured by means of "articulation tests."[11,15,16] In such tests a talker reads test material (syllables, words, or sentences) over the system under conditions simulating normal use of the equipment. A group of listeners write down or otherwise record each item heard. The percentage of the items read that are correctly heard by the listeners is called the articulation score.

Articulation tests are psychoacoustic tests and as such require care in their design and conduct typical of good experimental psychology practice. Because of uncontrollable variables, either in the testing environment or in the talker and listeners, the test must be carefully designed in advance to permit a statistical test for significance.[16] Furthermore, because talker and listener proficiencies vary and because there is as yet no means for normalizing these variables, articulation tests do not yield absolute measures of intelligibility.

6. Test Material

The principal types of material used in articulation tests are:

1. Syllables
2. Monosyllabic words (real words or nonsense syllables)
3. Spondees
4. Sentences

The material used may be selected to have approximately the same distribution of speech sounds as occurs normally in the language,[15] or it may contain predominately sounds which are not heard well over the system under test. Rather than record the whole word, the listeners may be asked to identify the word heard as one of several rhyming words differing only in the initial consonant (Fairbanks Rhyme Test[17]).

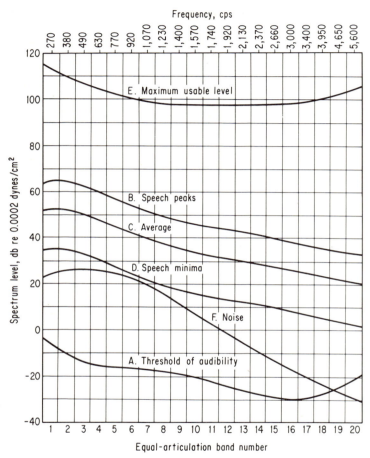

FIG. 12. Diagram illustrating the articulation-index computation. Here the speech level was increased 15 db to raise it above the masking noise level, restoring the articulation index to unity.

7. Test Personnel

There are large individual differences in talker and listener proficiency. Talkers are generally screened for noticeable speech defects or marked accent. Listeners are screened for hearing defects. Listeners require several days of training, for their performance with most test materials improves rapidly during initial tests. Training is continued until the improvement over several trials is not statistically significant. However, after reaching relative stability, the listeners' performance continues to improve at a very slow rate over long periods.[18] Talker proficiency may also change markedly, particularly when the talker is in an unusual environment; talkers can often adapt their talking technique to a particular situation or equipment.

Because of substantial individual variations from trial to trial, six or eight listeners are generally used and their scores averaged. Typically, one to three talkers are used, the number usually being limited by economic or time considerations. Talker-system interactions are sometimes encountered; in this case, the rank order of systems on the basis of intelligibility may differ for different talkers. Such a finding limits the generality of the conclusions.

A reference system may be included among the systems or conditions to be measured and can be useful in evaluating apparent variations in crew performance.

Standardized procedures and test material (phonetically balanced monosyllabic words) are found in Ref. 19.

8. Talker Identification

In most communication situations the listener wants to be able to recognize the person talking. Sometimes the identification of the talker's voice is considered to be essential.

There are many factors which are thought to contribute to talker identification. Some of these are (1) the voice spectrum (affected by both the vocal tract and the glottal excitation spectrum), (2) the average pitch, (3) variations in pitch, (4) diphthongization of vowels, (5) speech rhythm, (6) pauses, (7) speech rate, (8) variations in pronunciation, and (9) insertion of extraneous sounds. Studies to determine the importance of these and other possible factors have been quite limited.

Identification of a talker by voice only over the telephone is a common experience. Pollack, Pickett, and Sumby[21] report that some listeners were able to identify as

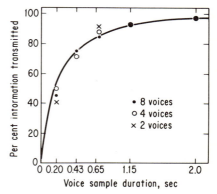

Fig. 13. Informational transmission associated with voice identification as a function of the duration of a voiced speech sample. The parameter is the number of possible speakers. (*From Pollack, Pickett, and Sumby,*[21] *J. Acoust. Soc. Am. Used by permission.*)

many as 65 different recorded speakers. They also found the duration of the sample of the speaker's voice to be an important variable in voice identification. Figure 13 shows the time dependence of the per cent information transmitted, where the information considered is that associated with the identification of a speaker by voice. It is seen that, within the range of two to eight speakers, the percentage of information transmitted by voice identification is independent of the number of voices and that a voice sample of 1 to 2 sec duration is necessary for a high degree of accuracy in identification.

Communication systems can be compared on the basis of accuracy of identification, by listeners, of talkers using the systems.[22] Listeners are first trained to recognize a number of talkers to a high degree of accuracy using a high-quality system. They then listen to sentences or other suitable material over the systems to be compared, with care taken to ensure properly balanced test conditions, and attempt to identify the talker. Systems are then compared on the basis of the percentage of correct identifications. There is at present no standardized test to measure this system characteristic.

9. Speech Quality

Of all the characteristics of a speech communication system, the most difficult to assess is the quality of the reproduced, or transmitted, speech. The term *quality* is not well defined. It is probably not independent of talker identification and speech intelligibility.   It is a subjective characteristic, and judgments of quality may differ among individuals.

The most common method of judging quality is the method of paired comparisons.[23] A system of known characteristics and acceptability is compared with the system under test by listeners who are asked to state which they prefer.

This method can be extended to permit rank ordering of a number of test systems by making paired comparisons of all possible pairs.   Rank ordering is then done on the basis of the number of times each system is judged to be the preferred one.

It must be realized that preference judgments are not necessarily made on the basis of a single characteristic, such as quality, but may be made on the basis of other factors, such as noise, intelligibility, and loudness.   It is entirely possible that some listeners will change their criterion for preference as the tests progress and as they become conscious of certain components not noticed initially.

An attempt to provide a more general method of rating of transmission circuits is the aim of the "isopreference" method[24] for evaluating speech-transmission circuits. This method employs simple *A-B* preference judgments (paired comparisons) in a systematic exploration of a wide range of circuit conditions.   Experience with this method has been limited, and its utility for general use is not yet known.

INPUT CONSIDERATIONS

10. Spectrum of Talkers

While there are both long-term and short-term differences in talkers' spectra,[25] the relation of these differences to system performance is not understood.   That these spectrum differences have a bearing upon talker identification is indicated by experiments in talker identification by computer[26] and from judgments of sound spectrograms.[27]   It is probable that spectral differences contribute also to differences in talker intelligibility and to system-talker interactions.

11. Speech Level of Talkers

Normal conversational speech in a quiet environment produces an average sound-pressure level of about 65 db re 0.0002 dyne/sq cm at 1 meter.[10]   However, there is wide variation among individuals.   This can be seen from Table 2, which shows the

Table 2. Distribution of Talker Levels for Persons Using the Telephone

Per cent of talkers	Talker level range, db*
7	Below 54
9	54–57
14	57–60
18	60–63
22	63–66
17	66–69
9	69–72
4	72–75
0	Above 75

* From Fletcher.[1]

results of measurements of the talker levels of a large number of persons, each one talking into a telephone in a conversational manner.

The average total sound-pressure level at 1 meter for seven young male talkers, when instructed to speak at one-half maximum effort, was 68 db at 1 meter.[6]   The range of speech levels which can be produced has been reported as 46 to 86 db[1] and 39 to 90 db[25] at 1 meter, with 90 db being the highest level that can be maintained without producing painful voice fatigue.   Maximum shout levels are 10 to 15 db higher.

When the voice level is raised substantially above normal, the intelligibility decreases.[28] This effect is shown in Fig. 14, which was obtained by equating the decrease in intelligibility observed at both higher and lower levels than normal to an actual decrease in level, thus providing an effective speech level for intelligibility calculations. The maximum level attainable also varies with individuals.

Most measurements of speech levels are made at some moderate distance from the lips, such as 1 meter or 1 ft. Most communication systems, however, use close-talking microphones, that is, microphones which are held close to the mouth. It is then necessary to determine what sound-pressure level is produced at or near the lips. The apparent source of speech in the mouth varies with the sound produced and is, on the average, about ½ in. behind the lips.[29] Assuming a close-talking microphone to be held about ¼ in. from the lips, and assuming that the pressure varies inversely as the distance, the speech-pressure level at the microphone is about 35 db higher than the level at 1 meter. Thus, a normal conversational speech level of 65 db at 1 meter produces a level of about 100 db at a microphone held close to the lips. This is a sound pressure of 20 dynes/sq cm.

The peak sound-pressure level at a close-talking microphone can be surprisingly high. Taking the 90-db level at 1 meter as the loudest speech level that can be maintained, we find the average sound-pressure level near the lips to be 35 db higher, or 125 db. Since actual peak pressures may be 20 db higher, they may reach 145 db. (For intelligibility calculations, peak pressures are taken to be 12 db higher than the average, or 137 db in this case.) It is thus evident that care must be taken in the design of the microphone amplifier input stage to ensure that it will not be overloaded by these high sound pressures if the talker is in an environment that requires him to speak as loudly as possible.

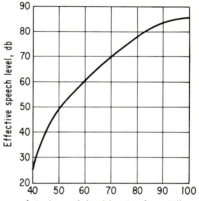

FIG. 14. Effective speech level in decibels as a function of the actual level used by a talker. (*From Kryter*[14] and *Pickett*,[28] *J. Acoust. Soc. Am. Used by permission.*)

12. Sidetone

Most speech communication systems provide sidetone to the talker; that is, the talker's own speech signal is fed back to his ear as well as being transmitted through the system. When the talker is in quiet, he hears his own voice directly. In the case of the telephone, he hears his own voice as sidetone through the telephone receiver at one ear; through the other ear he hears his own voice directly.

Sidetone serves several functions. It gives the talker some assurance that the system is functioning properly. It enables him to adjust his voice somewhat to the system. It can influence his voice level and talking rate.[30] If the sidetone is delayed approximately ¼ sec, it may cause the talker to stutter and disturb drastically his ability to speak normally.[31] When the talker's noise environment is variable, it is advantageous to present his sidetone at relatively constant level by employing automatic volume control (rather than having constant gain around the sidetone loop); this influences the talker to talk louder in the higher noise levels. The optimum sidetone level depends upon a number of system parameters and is usually determined experimentally; however, a sidetone level 0 to 6 db below the normal receiving level is often satisfactory.

13. Background Noise

The talker's acoustical noise environment is a critical factor in the performance of a speech communication system. Its effect upon intelligibility can be accounted for by articulation-index calculations in which the noise spectrum picked up by the microphone is included in the determination of the noise-masking threshold.

The background acoustic noise level also affects the talker's voice level, since he will raise his voice in order to be heard over the noise, within the limits of his ability. The speech level actually used must be determined also in making articulation-index calculations.

14. Noise-canceling Microphones and Noise Shields

Because the source of speech is relatively small, the acoustical wavefront near the mouth is essentially spherical and there is a substantial pressure gradient. By contrast, the wavefront from moderately distant noise sources is essentially plane and has a negligible pressure gradient. By employing a pressure-gradient or pressure-differential microphone[32-34] close to the lips, it is possible to discriminate against the background noise. Physical limitations on microphone size and positioning with

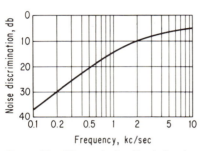

respect to the source of speech, as well as the finite size of the source, reduce the effectiveness of a gradient microphone at the higher frequencies. Figure 15 shows a theoretical noise-discrimination characteristic for a first-order pressure-gradient microphone. Additional discrimination is possible by means of higher-order pressure-gradient microphones.

Fig. 15. Theoretical discrimination against random noise originating at a distance for a first-order gradient microphone $\frac{3}{4}$ in. from a point source. *(From Olson,[33] J. Acoust. Soc. Am. Used by permission.)*

Noise reduction can also be attained by a noise shield over the microphone, enclosing the microphone and the talker's lips. Such a shield is best employed with a pressure-gradient microphone, since it has little effect upon the microphone response and noise-discrimination properties. If used on a pressure microphone, it radically affects response.

The walls of the noise shield must be heavy enough to attenuate noise but flexible enough to allow freedom of movement of the talker's lips and jaw. An opening must be provided for direct air flow; the requirement for such an opening limits the attenuation attainable at low frequencies. At frequencies above 1,000 cps, 20 to 25 db of noise attenuation can be provided by the use of a noise shield.

15. Reverberation

If the talker is in a highly reverberant room and at some distance from the microphone, intelligibility will be reduced because of the reverberation.[35] This is seldom a problem, however, for if the talker is within 3 ft of the microphone, even in a highly reverberant room, the effect on intelligibility is small.[36]

SYSTEM CHARACTERISTICS AND DISTORTIONS

16. Response-frequency Characteristic

The design of a speech communication system should take into account all aspects of the system from the talker to the listener. Thus, the overall response-frequency characteristic from the sound pressure near the talker's lips to the sound pressure produced at the listener's ear should be controlled. While the most natural reproduction would require that this characteristic be approximately uniform throughout the entire speech range, it is generally more economical and under some conditions of noise and distortion provides better intelligibility if the low-frequency content of the input signal is reduced. Often the microphone response is designed to fall off toward the low frequencies at about 6 db/octave starting about 700 cps. In general, a smooth response-frequency characteristic is desirable, but the presence of resonance peaks is often tolerated and does not affect intelligibility markedly, although such peaks degrade speech quality. The effect of the response-frequency characteristic upon intelligibility can be determined in a given situation by articulation-index calculations.

While a frequency band from 200 to 6,100 cps permits transmission of speech without appreciable reduction in intelligibility, this range is seldom employed in practical communication systems. Inspection of Fig. 11 shows that sentence intelligibility remains about 95 per cent even when the articulation index drops to 0.4. Assuming a 30-db signal-to-noise ratio, an articulation index of 0.4 could be provided by the frequency range from 430 to 1,660 cps. Such a system would be usable for many purposes but would require some repeats and possibly the spelling of critical words.

Carrier telephone systems have nominal 4,000-cps channels, but to prevent crosstalk and to provide for signaling, the actual band provided for speech is substantially less. For most practical systems, a 300- to 3,000-cps bandwidth is provided; this gives a maximum articulation index of about 0.7 with a corresponding word articulation of 85 to 90 per cent.

17. Peak Clipping and Overload Distortion

Sharp, symmetrical peak clipping (amplitude limiting) illustrated in Fig. 16, while affecting the quality of speech markedly, does not greatly reduce intelligibility when the reproduced speech is heard in quiet at a comfortable listening level.[37] If the speech is differentiated before clipping, the loss of intelligibility even after "infinite" peak clipping is quite small.[38] Peak clipping can be employed to improve intelligibility under certain conditions.[39] If the

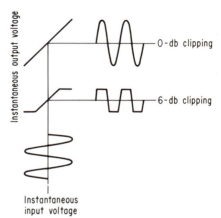

FIG. 16. Waveforms and transfer characteristics illustrating peak clipping. (*From Licklider,*[37] *J. Acoust. Soc. Am. Used by permission.*)

input speech signal is relatively noise-free, it may be clipped moderately, then reamplified so that the clipped peaks are equal to the level of the peaks before clipping. As a result, the low-level portions of the signal are amplified a corresponding amount. If these low-level portions were previously masked by noise but may now be heard, it is apparent that this clipping and reamplification process can result in an improvement in intelligibility. The process is illustrated in Fig. 17. In the figure, (*a*) represents a speech sound having an initial low-amplitude portion and a high-amplitude portion. Noise is assumed small compared with either portion. In (*b*) the signal on the left has been clipped at the level indicated by the dotted lines in (*a*), while on the right it has been amplified so that the peaks are at the same level as the peaks in (*a*). In (*c*) the signal on the left has been clipped at about the peak level of the low-level portion of the speech sound and at the right again amplified to the same peak level as in (*a*). It is seen that the peak amplitude of the low-level portion is now comparable to that of the high-level portion and would be less likely to be masked by noise encountered subsequent to the clipping.

In some cases, peak clipping is accomplished by overloading some stage in the system, usually an output stage. If this produces symmetrical, sharp peak clipping, the results may be satisfactory; if unsymmetrical, intelligibility may be considerably reduced.[37] If the signal-to-noise ratio before clipping is low, clipping has a somewhat more adverse affect on intelligibility, thus limiting its effectiveness. In any case, peak clipping has some adverse affect upon quality, but where intelligibility is of prime importance or where the background noise is severe, the loss in quality may not be objectionable.

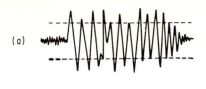

(a)

(b)

(c)

FIG. 17. Schematic representation of the waveform of a speech sound. (a) Undistorted. (b) After 6 db of peak clipping. (c) After 20 db of peak clipping. At the right the waves of (b) and (c) are reamplified until their peak-to-peak amplitudes are equal to the peak-to-peak amplitude of (a). Amplification of the low-level portions is apparent. (*From Licklider, Bindra, and Pollack,*[41] *Am. J. Psychol. Used by permission.*)

18. Center Clipping

In contrast to peak clipping, center clipping illustrated in Fig. 18, has a very deleterious effect upon intelligibility.[37] Figure 19 compares the two types of clipping. The quality of speech is also very sensitive to center clipping.

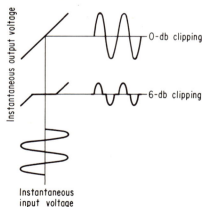

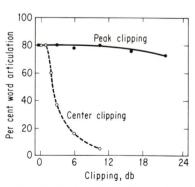

FIG. 18. Waveforms and transfer characteristics illustrating center clipping. (*From Licklider,*[37] *J. Acoust. Soc. Am. Used by permission.*)

FIG. 19. Effects of peak clipping and center clipping. Talkers and listeners were in 110-db noise field; however, the signal-to-noise ratio before clipping was reasonably high. (*From Licklider,*[37] *J. Acoust. Soc. Am. Used by permission.*)

19. Frequency Shift

Frequency shift which results in multiplication of all frequencies (and time) by a common factor, for example, the playback of recorded speech at a speed different from that at which it was recorded, affects intelligibility as shown in Fig. 20.

Frequency shift which results in the addition or subtraction of a constant number of cycles to all frequencies affects intelligibility as shown in Fig. 21. This type of frequency shift occurs in a carrier system when the frequency of the carrier at the receiver differs from that at the transmitter.

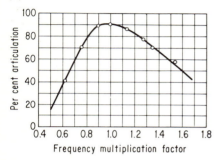

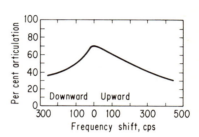

FIG. 20. Effect upon intelligibility of recording speech and playing it back faster or slower, resulting in a multiplication of the frequency by a variable factor. (*By permission, from Fletcher, Speech and Hearing in Communication, D. Van Nostrand Company, Inc., Princeton, N.J., 1953.*)

FIG. 21. Effect upon intelligibility of shifting all frequencies in speech upward or downward a given number of cycles per second. (*By permission, from Fletcher, Speech and Hearing in Communication, D. Van Nostrand Company, Inc., Princeton, N.J., 1953.*)

20. Delay in Voice-operated Relay Operation

The use of a voice-operated relay results in the omission of the initial portion of a message. The effect of removal of a portion of the initial consonant of a word is shown in Fig. 22. It is seen that the relationship is essentially linear. The average duration of the initial consonants in the test from which these data were obtained was about 0.16 sec.

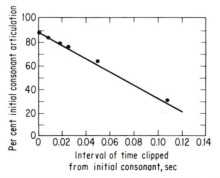

FIG. 22. Effect upon intelligibility of removal of the initial portion of a word. (*From Steinberg,*[36] *reprinted by permission of the copyright owner, American Telephone and Telegraph Company.*)

21. Phase Shift and Transmission Delay

Speech intelligibility is generally very resistant to phase distortion. Phase distortion will probably affect quality quite noticeably before it affects intelligibility. Care must be taken, however, where certain distortions are introduced. For example, if a speech wave is clipped prior to and again after phase distortion, loss of intelligibility may occur.

In two-way transmission, delay in transmission can interfere with the normal interactions in conversation. A talker may pause, expecting a reply, and hearing none because of the delay may start to talk again only to be interrupted by the delayed transmission from the other party. Delays of the order of 100 msec are tolerable. Techniques for accommodating larger delays associated with transmissions via orbiting communication satellites are currently being studied.[40] These delays encountered in satellite communication may be several hundred milliseconds.

22. Echo

A problem which arises in telephone circuits is the echo resulting from electrical reflections at the line terminals which the talker hears via his sidetone. A delay of

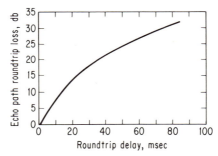

FIG. 23. Average listener judgment of minimum echo attenuation necessary for just tolerable echo conditions on modern telephone sets. (*From Emling and Mitchell,*[42] *reprinted by permission of the copyright owner, American Telephone and Telegraph Company.*)

65 msec produces a distinct echo,[35] and lesser delays tend to blur the sound. Echo suppressors are provided on long-distance circuits to control this effect. The amount of suppression required for acceptability is a function of the delay as shown in Fig. 23.

23. Feedback

Consideration of possible acoustic feedback paths is important in communication system design, especially in complex intercommunication systems where loudspeakers are used and where there may be several communication positions in the same room.

OUTPUT CONSIDERATIONS

24. Output Level

The output level required depends primarily upon the noise environment at the listener. Articulation-index computations which include the effect of the noise at the listener permit determination of the required level for a given intelligibility. Most output considerations can be resolved by considering the signal sound pressure and the noise sound pressure developed at the listener's ear from all sources in carrying out the articulation-index computations.

25. Noise Exclusion

When a headset, earphone, or handset is used by the listener, the ear cap serves to exclude a portion of the ambient noise. When the noise level is high, the noise-exclusion characteristic of the earphone is an important element in determining the intelligibility attainable. If the earphone provides a high degree of noise exclusion, noise transmitted to the listener's ear via his own sidetone loop may limit performance.

REFERENCES

1. Fletcher, H. F., *Speech and Hearing in Communication*, D. Van Nostrand Company, Inc., Princeton, N.J., 1953.
2. Dunn, H. K., The Calculation of Vowel Resonances and an Electrical Vocal Tract, *J. Acoust. Soc. Am.*, vol. 22, pp. 740–753, 1950.
3. Stevens, K. N., S. Kasowski, and C. G. M. Fant, An Electrical Analog of the Vocal Tract, *J. Acoust. Soc. Am.*, vol. 25, pp. 734–742, 1953.
4. Fant, C. G. M., *Acoustic Theory of Speech Production*, Mouton and Company, 's-Gravenhage, The Netherlands, 1960.
5. French, N. R., and J. C. Steinberg, Factors Governing the Intelligibility of Speech Sounds, *J. Acoust. Soc. Am.*, vol. 19, pp. 90–119, 1947.
6. Beranek, L., The Design of Speech Communication Systems, *Proc. IRE*, vol. 35, pp. 880–890, 1947.
7. Miller, R. L., The Nature of the Vocal Cord Wave, *J. Acoust. Soc. Am.*, vol. 31, pp. 667–677, 1959.
8. Potter, R. K., Kopp, and Green, *Visible Speech*, D. Van Nostrand Company, Inc., Princeton, N.J., 1947.
9. Cooper, F. S., P. C. Delattre, A. M. Liberman, J. M. Borst, and L. J. Gerstman, Some Experiments on the Perception of Synthetic Speech Sounds, *J. Acoust. Soc. Am.*, vol. 24, pp. 597–606, 1952.
10. Dunn, H. K., and S. D. White, Statistical Measurements on Conversational Speech, *J. Acoust. Soc. Am.*, vol. 11, pp. 278–288, 1940.
11. *OSRD Rept.* 3802, November, 1944.
12. *American Standard Method for Measurement of Monosyllabic Word Intelligibility*, *S3.2-1960*, American Standards Association, New York, N.Y.
13. Fletcher, H., and R. G. Galt, The Perception of Speech and Its Relation to Telephony, *J. Acoust. Soc. Am.*, vol. 22, pp. 89–151, 1950.
14. Kryter, K. D., Methods for the Calculation and Use of the Articulation Index, *J. Acoust. Soc. Am.*, vol. 34, pp. 1689–1697, 1962.
15. Goffard, S. J., and J. P. Egan, Procedures for Measuring the Intelligibility of Speech; Sound-powered Telephone Systems, *Harvard Univ.*, Psycho-Acoustic Lab., *Rept. PNR-33*, Feb. 1, 1947.
16. Beranek, L., *Acoustic Measurements*, John Wiley & Sons, Inc., New York, 1949.
17. Fairbanks, G., Test of Phonemic Differentiation: the Rhyme Test, *J. Acoust. Soc. Am.*, vol. 30, pp. 596–601, 1958.
18. Radio Corporation of America: unpublished data.
19. *American Standard Method for Measurement of Monosyllabic Word Intelligibility*, *S3.2-1960*, American Standards Association, New York, N.Y.
20. Hughes, G. W., and M. Halle, Spectral Properties of Fricative Consonants, *J. Acoust. Soc. Am.*, vol. 28, pp. 303–310, 1956.
21. Pollack, I., J. M. Pickett, and W. H. Sumby, On the Identification of Speakers by Voice, *J. Acoust. Soc. Am.*, vol. 26, pp. 403–406, 1954.
22. Meeker, W. F., and A. L. Nelson, *Vocoder Evaluation Studies*, Radio Corporation of America, Scientific Report 3 to Air Force Cambridge Research Laboratories (AFCRL 547).
23. Stevens, S. S., *Handbook of Experimental Psychology*, John Wiley & Sons, Inc., New York, 1951.
24. Munson, W. A., and J. E. Karlin, Isopreference Method for Evaluating Speech-transmitting Circuits, *J. Acoust. Soc. Am.*, vol. 34, pp. 762–774, 1962.
25. Rudmose, H. W., K. C. Clark, F. D. Carlson, J. C. Eisenstein, and R. A. Walker, Voice Measurements with an Audio Spectrometer, *J. Acoust. Soc. Am.*, vol. 20, pp. 503–512, 1948.
26. Pruzansky, S., Pattern-matching Procedure for Automatic Talker Recognition, *J. Acoust. Soc. Am.*, vol. 35, pp. 354–358, 1963.

27. Kersta, L. G., Voice Spectrograms for Unique Personal Identifications, *Bell Lab. Record*, vol. 40, pp. 214–215, 1962.
28. Pickett, J. M., Effects of Vocal Force on the Intelligibility of Speech, *J. Acoust. Soc. Am.*, vol. 28, pp. 902–905, 1956.
29. Hawley, M. E., and A. H. Kettler, The Apparent Source of Speech in the Mouth, *J. Acoust. Soc. Am.*, vol. 22, pp. 365–369, 1950.
30. Black, J. W., The Effect of Room Characteristics upon Vocal Intensity and Rate, *J. Acoust. Soc. Am.*, vol. 22, pp. 174–176, 1950.
31. Lee, B. S., Effects of Delayed Speech Feedback, *J. Acoust. Soc. Am.*, vol. 22, pp. 824–826, 1950.
32. Olney, B., F. H. Slaymaker, and W. F. Meeker, The Dipole Microphone, *J. Acoust. Soc. Am.*, vol. 16, pp. 172–177, 1945.
33. Olson, H. F., Gradient Microphones, *J. Acoust. Soc. Am.*, vol. 17, pp. 192–195, 1945.
34. Olson, H. F., *Acoustical Engineering*, D. Van Nostrand Company, Inc., Princeton, N.J., 1957.
35. Knudsen, V. O., *Architectural Acoustics*, John Wiley & Sons, Inc., New York, 1942.
36. Licklider, J. C. R., Effects of Phase Distortion on Telephone Quality, *Bell System Tech. J.*, vol. 9, pp. 550–566, 1930.
37. Licklider, J. C. R., Effects of Amplitude Distortion upon the Intelligibility of Speech, *J. Acoust. Soc. Am.*, vol. 18, pp. 429–434, 1946.
38. Licklider, J. C. R., and I. Pollack, Effects of Differentiation, Integration, and Infinite Peak Clipping upon the Intelligibility of Speech, *J. Acoust. Soc. Am.*, vol. 20, pp. 42–51, 1948.
39. Steinberg, J. C., Effects of Distortion on the Recognition of Speech Sounds, *J. Acoust. Soc. Am.*, vol. 1, p. 121, 1929.
40. Gardner, M. B., and J. R. Nelson, Combating Echo in Speech Circuits with Long Delay, *J. Acoust. Soc. Am.*, vol. 35, pp. 1762–1772, 1963.
41. Licklider, J. C. R., D. Bindra, and I. Pollack, The Intelligibility of Rectangular Speech Waves, *Am. J. Psychol.*, vol. 61, pp. 1–20, 1948.
42. Emling, J. W., and D. Mitchell, The Effects of Time Delay and Echoes on Telephone Conversation, *Bell System Tech. J.*, vol. 42, pp. 2869–2891, 1963.

Chapter 4

PROGRAMMING, PROBABILITY, AND INFORMATION THEORY

LUIS F. NANNI, *Rutgers, the State University*

INTRODUCTION

Mathematical models that have been found very useful in the solution of engineering systems will be developed in this chapter. The first section will include a review of linear programming problems. These kinds of problems are examples of the so-called deterministic models, which require the determination of certain parameters that optimize linear equations subject to linear restrictions. After a review of the basic concepts of probability and statistics, a presentation will be made of non-deterministic models as illustrated by the Markoff-type process of queueing theory. The chapter will close with a review of the basic concepts of information theory.

LINEAR PROGRAMMING

A very large class of problems in industry can be analyzed in terms of a mathematical model called linear programming.

The central problem of linear programming is the minimization (or maximization) of a linear function, the *objective function*,

$$P = c_1 x_1 + c_2 x_2 + \cdots + c_n x_n \tag{1}$$

subject to the linear restrictions

$$\begin{aligned}
a_{11}x_1 + a_{12}x_2 + \cdots + a_{1n}x_n &= b_1 \\
a_{21}x_1 + a_{22}x_2 + \cdots + a_{2n}x_n &= b_2 \\
a_{m1}x_1 + a_{m2}x_2 + \cdots + a_{mn}x_n &= b_m
\end{aligned} \tag{2}$$

$$x_j \geq 0 \qquad j = 1, \ldots, n \tag{3}$$

The coefficients c_i in Eq. (1) are the costs (or profits) associated with the variables x_i.

The linear restrictions can be inequalities or a mixture of equalities and inequalities. Thus, if among the restrictions there are k inequalities of the type

$$\sum_{j=1}^{n} a_{ij}x_j \leq b_i \qquad i = 1, \ldots, k \tag{4}$$

then positive *slack variables* x_{n+i} are introduced so as to have

$$\sum_{j=1}^{n} a_{ij}x_j + x_{n+i} = b_i \qquad i = 1, \ldots, k \tag{5}$$

The objective function becomes

$$P = \sum_{i=1}^{n} c_i x_i + c_{n+1}x_{n+1} + \cdots + c_{n+k}x_{n+k} \tag{6}$$

where
$$c_{n+1} = c_{n+2} = \cdots = c_{n+k} = 0$$

In the same manner, if among the restrictions there are k inequalities of the type

$$\sum_{j=1}^{n} a_{ij}x_j \geq b_i \qquad i = 1, \ldots, k \tag{7}$$

then one can write

$$\sum_{j=1}^{n} a_{ij}x_j - x_{n+i} = b_i \qquad i = 1, \ldots, k \tag{8}$$

where x_{n+i} are positive slack variables. The new objective function will be the same as Eq. (6).

If some of the variables x_j are not restricted to be positive, one can write them in the form

$$x_j = x_j{}' - x_j{}'' \tag{9}$$

where $x_j{}'$ and $x_j{}''$ are both positive.

If some of the restrictions are equalities of the type

$$\sum_{j=1}^{m} a_{ij}x_i = b_i \qquad i = 1, \ldots, k \tag{10}$$

one can also introduce positive slack variables x_{m+i} and change Eqs. (10) to

$$\sum_{j=1}^{m} a_{ij}x_i + x_{m+i} = b_i \tag{11}$$

Since these slack variables must be zero in order for Eq. (10) to be satisfied, very high costs c_i (or very low profits in the case of maximization) are associated with the new slack variables, so that these variables will never be part of the final solution. In general, these "penalty" coefficients will be a large positive value for minimization problems or a small negative value for maximization problems.

Using the Simplex method for the solution of a linear programming problem, the slack variables will constitute the initial solution to the problem so that it becomes necessary to have as many slack variables as there are restrictions of the form (2).

Theorems related to the existence of a solution of the system given by Eqs. (1) to (3) are based on the properties of convex sets. We shall mention here only that the points which are simultaneous solutions of a system of inequalities of the type ≤ form a convex set and that a linear function defined on a convex set attains its minimum (or maximum) value at the extreme points of the convex set.

The methods of solution consist in finding the extreme points of the convex set, substituting the coordinates of each of the extreme points in Eq. (1), and by an iteration process determining the point (or set of points) for which Eq. (1) is a minimum (or maximum).

1. Illustration of Linear Programming

To illustrate the general nature of a linear-programming problem let us consider the following example:
Maximize

$$P = 3x_1 + 2x_2 \tag{12}$$

subject to the restrictions

$$2x_1 + x_2 \leq 10$$
$$x_1 + x_2 \leq 6 \tag{13}$$
$$x_1 + 4x_2 \leq 18$$
$$x_i \geq 0 \qquad i = 1, 2 \tag{14}$$

Figure 1 shows the graphical representation of the three equations (13). Owing to the nature of the inequalities (13) and (14), it can be seen that the solution must be in the convex set bounded by the segments OA, AB, BC, CD, and DO. The extreme points of the convex set are O: (0,0), A: (0,9⁄2), B: (2,4), C: (4,2) and D: (5,0).

Since the maximum of P should be attained at one of the extreme points, substitution of the coordinates of these points in Eq. (6) should yield the optimum value of P. This value is attained at the point C with coordinates $x = 4$, $y = 2$, which produces a value of $P = 16$.

If the linear function to be maximized had been

$$P = x_1 + x_2$$

then it is easy to see that any point between B and C would have been a solution to the problem.

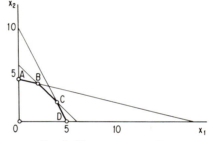

FIG. 1. Linear programming.

2. The Simplex Method

Several numerical methods had been investigated for the solution of the system of Eqs. (1), (2), and (3). The most widely used method at the present time is the Simplex method developed by Dantzig.[1]

For this general discussion, let us assume that the objective function is to be maximized. Using the Simplex method, one generally starts with an initial solution given by the set of slack variables that are introduced for each of the linear equations (2). The initial value of the objective function is thus equal to zero or negative if some negative penalty coefficient was introduced as a result of a restriction given by an equality. This initial solution will generally correspond to an extreme point of the convex set (the point O in the example illustrated in Fig. 1). By moving to an adjacent extreme point of the convex set, the Simplex algorithm will produce a new solution with a value of P larger than the initial value. Since the number of extreme points of a convex set is finite, the process will eventually produce a unique solution or, in the multiple-solution case, a set of solutions to the problem. It is not necessary to examine all possible extreme points of a convex set. The optimum solution will be found by an examination of only a fraction of the extreme points of the set.

The Simplex method will be illustrated by using the same example mentioned above. A Simplex table is formed by using Eqs. (13) and (12) with the help of the slack variables necessary to change the inequalities (13) into equalities. The new equations are

$$2x_1 + x_2 + x_3 = 10$$
$$x_1 + x_2 + x_4 = 6 \tag{15}$$
$$x_1 + 4x_2 + x_5 = 18$$
$$P = 3x_1 + 2x_2 + 0x_3 + 0x_4 + 0x_5 \tag{16}$$
$$x_j \geq 0 \qquad j = 1, \ldots, 5$$

One thus has a system of three equations with five unknowns. These equations are set up in the Simplex table illustrated below.

	Basis				TABLE A_0		
c	Var.	Value	x_1	x_2	x_3	x_4	x_5
0	x_3	10	2	1	1	0	0
0	x_4	6	1	1	0	1	0
0	x_5	18	1	4	0	0	1
	$-P =$	0	3	2	0	0	0

The right-hand side of Eqs. (15) is written in the third column of Table A_0. Equation (16) is written in the last row of Table A_0. Column 2 of Table A_0 gives the names of the basic variables and the negatives of P. Column 3 gives the values of these variables. Column 1 gives the values of c associated with the basic variables.

Using the slack variables as the first solution of the system of Eqs. (15) and (16) one has $x_3 = 10$, $x_4 = 6$, $x_5 = 18$, $x_1 = x_2 = 0$ with $P = 0$. This solution corresponds to the point O in Fig. 1 and is represented in columns 2 and 3 of Table A_0. The initial solution $x_3 = 10$, $x_4 = 6$, $x_5 = 18$ constitutes the first basis of the system. Since all the c's are zero for these variables, the value of $P = 0$.

To improve the solution, let us examine the last row of Table A_0 and select the column that has the largest value in this row, in this case the column under x_1. This selection indicates which variable should be brought into the solution so as to improve the initial value of P of the first approximation. To decide which variable should be eliminated from the basis, the values under the x_1 column are divided into the values in the "value" column and the ratio with the smallest value is then selected. In this case, the smallest ratio is equal to 5, indicating that the variable x_3 should be eliminated from the basis. This elimination is performed by dividing the first row of Table A_0 by the pivot number 2 located at the intersection of column x_1, the variable that comes in into the solution, and row x_3, the variable that goes out. By subtracting the values of this new row from the other rows of Table A_0 as many times as required so as to obtain a new column for x_1 consisting of the number 1 at the pivot location and zeros elsewhere in the column, one gets the second Simplex table A_1, replacing the eliminated x_3 by x_1 in the variable column.

	Basis				TABLE A_1		
c	Var.	Value	x_1	x_2	x_3	x_4	x_5
3	x_1	5	1	$\frac{1}{2}$	$\frac{1}{2}$	0	0
0	x_4	1	0	$\frac{1}{2}$	$-\frac{1}{2}$	1	0
0	x_5	13	0	$\frac{7}{2}$	$-\frac{1}{2}$	0	1
	$-P =$	-15	0	$\frac{1}{2}$	$-\frac{3}{2}$	0	0

The new solution is $x_1 = 5$, $x_4 = 1$, $x_5 = 13$. The new value of P is 15 with a negative sign in the table. To improve this solution let us examine the last row of Table A_1 and select the column with the largest positive value, in this case the column under x_2. To decide which variable should be eliminated from the basis, one proceeds as before; that is, one divides the values of the column under x_2 into the values of the column under "value" and then selects the result with the smallest positive ratio.

The smallest ratio is given by the x_4 row. The pivot number is situated at the intersection of the x_2 column and x_4 row. Dividing the elements of this row by the pivot number and performing the same substraction as indicated above, one obtains the new Simplex table A_2.

	Basis		TABLE A_2				
c	Var.	Value	x_1	x_2	x_3	x_4	x_5
3	x_1	4	1	0	1	−1	0
2	x_2	2	0	1	−1	2	0
0	x_5	6	0	0	3	−7	1
	$-P =$	−16	0	0	−1	−1	0

Examination of the last row of Table A_2 shows that all the values are either zero or negative. No new variable can be brought into the basis that will improve the solution already achieved. The final solution is thus

$$x_1 = 4 \qquad x_2 = 2 \qquad x_5 = 6 \qquad P = 16$$

For the solution of a minimization problem, the final solution will be achieved when the values of the last row are either zero or positive.

a. Alternative Solutions. In a linear-programming problem there might be more than one solution. In fact, the number of solutions can be infinite as in the case mentioned above when the objective function coincides with one of the boundaries of the convex set.

To recognize all the alternative solutions one examines the last row of the final Simplex table. All the variables in the last Simplex table that have a value of zero in the last row are the variables that can be brought into the solution without changing the value of the objective function.

b. Case of Degeneracy. If in trying to remove a variable form the basis two or more ratios have the same minimum value, one would not know which variable to remove. This situation is sometimes referred to as a degenerate case. The rule that will often resolve the degeneracy and lead to a nondegenerate final solution goes as follows. Starting with the column under the first variable in the Simplex table, divide successively each column coefficient of the two rows that produced the same minimum ratio by the two pivoting numbers under consideration. In the first column where the ratios are not equal, select the row that has the smallest ratio.†

If in trying to determine which variable should be removed from the basis the ratios are either negative, indeterminate or infinite, there is no solution to the problem. In this case, either an arithmetical mistake was made or the original statement of the problem did not define a meaningful linear-programming problem.

c. Sensitivity Analysis. After an optimal solution has been found, it is possible to determine the effect of certain changes in the system without having to formulate and solve a new linear-programming problem. The changes in the system can be divided into the following categories:

Changes in the right-hand side (changes in b_i)
Changes in the objective function (changes in c_i)
Changes in the coefficients (changes in a_{ij})
Addition of new variables
Addition of linear restrictions

The effect of these changes can be discussed in terms of the techniques of matrix algebra and will not be developed here. Basically, if the system of Eqs. (1), to (3)

† See Ref. 2, Chap. 4, for a complete discussion of degeneracy.

has a solution, there will exist a system of m equations $(m < n)$ with m unknowns with an inverse A^{-1} in terms of which the modifications to the system can be found. The reader is referred to the literature on linear programming for a thorough development of this topic.[†]

d. Parametric Linear Programming. One of the changes in the system discussed in paragraph 2c was a change in the right-hand side of Eq. (2) from a value of b_i to a value of $b_i + \Delta b_i$, with Δ being a fixed quantity. If one is to examine the change of b_i to a new value given by a linear function

$$b_i(t) = b_i + \alpha_i(t) \qquad t \geq 0 \tag{17}$$

where the α_i are known positive or negative constants, one is led to the solution of a problem in parametric linear programming.

These kinds of changes are of most importance in engineering systems, since the restrictions described by Eq. (2) are never static but of a dynamic nature. For instance, b_i may be related to the availability of a raw material that could possibly go to zero with time.

The reader is referred to Ref. 2 for a further discussion of parametric linear programming.

e. Duality. As has been stated previously, the maximum of the objective function is found as soon as the values in the last row of the Simplex table are either zero or negative. Let us refer to these values as c_j'. For instance, the value of c_3' is equal to -1. The c_j' can be interpreted as showing the effect on the objective function of introducing one unit of the corresponding variable x_j into the solution. For instance, if one unit of x_3 is introduced into the solution, the value of P would decrease by \$1. In the same manner, if one unit of x_4 is introduced into the solution P would also decrease by \$1. At each stage in the computation, the c_j' represent the worth of a marginal unit of the corresponding variable x_j, which is not in the solution. These values are referred to as implicit values or shadow prices.

The total "worth" of variables x_3 and x_4, which are not in the final solution, would be $1 \times 10 = \$10$ and $1 \times 6 = \$6$, respectively. [The numbers 10 and 6 represent the maximum values that x_3 and x_4 can have according to the first two of Eqs. (15).] If we add these values, we have $\$10 + \$6 = \$16$, which is exactly the value of the objective function P in the final solution of the problem. This relation introduces the concept of duality in linear programming.

For every linear-programming problem there exists another linear-programming problem which is called the dual of the original problem. Solving the original problem also solves the dual and conversely.

For Eqs. (12) to (14) the dual problem can be stated as follows: Solve the system

$$2y_1 + y_2 + y_3 \geq 3$$
$$y_1 + y_2 + 4y_2 \geq 2$$
$$y_j \geq 0$$

so as to minimize $10y_1 + 6y_2 + 18y_3$. The coefficients of the first inequality are obtained from the coefficients of x_1 in Eq. (15), of the second inequality from the coefficients of x_2, etc. The right side of the inequalities are derived from the coefficients in Eq. (16), and the coefficients of the minimizing problem are the right side of Eq. (15).

If the dual problem had been solved using the Simplex algorithm the value of x_1 and x_2 would have been determined from the last row of the Simplex table.

In many cases, it is computationally easier to use the Simplex algorithm for the solution of the dual problem than to use the same algorithm for the original problem.

3. The Transportation Problem

The transportation technique was developed for the solution of problems involving the shipping of supplies from m sources to n destinations. However, it can also

[†] See Ref. 2, Chap. 4.

be used in the solution of other types of problems such as scheduling of production, machine and job assignment, etc.

Using the terminology related to the transportation problem, let x_{ij} be the amount to be transported from source i to destination j, c_{ij} be the cost related to x_{ij}, a_i be the production limitation at source i, and b_j the demand at destination j. We can then write

$$\sum_{j=1}^{n} x_{ij} = a_i \qquad i = 1, \ldots, m \tag{18}$$

$$\sum_{i=1}^{m} x_{ij} = b_j \qquad j = 1, \ldots, n \tag{19}$$

$$\sum_{i,j} c_{ij} x_{ij} = \min \tag{20}$$

$$x_{ij} \geq 0 \qquad \text{all } i, j \tag{21}$$

Equations (18) state that the amount to be transported from source i to destinations 1, 2, . . . , n should equal the capacity limitation of source i. Equations (19) state that the amount to be received by destination j should equal its demand b_j. Finally, Eq. (20) states that the cost of transportation should be a minimum and Eq. (21) indicates that the quantities x_{ij} should be positive or zero.

The $m + n$ equations (18) and (19) can be set up in a Simplex table, Table 1, similar to the linear-programming problem described above. $s_1, s_2, \ldots, s_{m+n}$ are the $m + n$ slack variables with associated penalty coefficients C_i that will be needed for an initial solution of system of Eqs. (18) to (20) by the Simplex method. In this simplex formulation of the transportation problem, all the coefficients associated with the variables x_{ij} are either 1 or zero. For some linear-programming problems where the coefficients associated with any particular row are either zero or a constant, not necessarily the same constant for all the rows, by dividing all the rows by the proper constant, all the coefficients will be reduced to zero or 1 and the problem can then be formulated as a transportation problem.

Equations (18) and (19) represent a system of $m + n$ equations with $m + n$ unknown. These equations are not independent, since by adding Eq. (18) with respect to i and Eq. (19) with respect to j we have

$$\sum_{i} \sum_{j} x_{ij} = \sum_{i} a_i$$

$$\sum_{j} \sum_{i} x_{ij} = \sum_{j} b_j$$

Hence

$$\sum_{i} a_i = \sum_{j} b_j = \sum_{ij} x_{ij} \tag{22}$$

Equation (22) states that the total production should equal the total demand. It also indicates that one of the $m + n$ equations (18) and (19) is dependent on the rest, leaving thus a system of $m + n - 1$ independent equations with $m + n$ unknowns.

In general, the total production is never equal to the total demand. Instead of equalities in the system of Eqs. (18) and (19) one has inequalities, and slack variables with associated zero costs must be introduced.

The system of Eqs. (18) and (19) was originally solved by using a technique similar to the Simplex method.[†] Because of the special nature of the matrix of its coefficients, it is possible to use simplified methods that are not directly related to the Simplex algorithm. In what follows the method developed by Dantzig[‡] will be explained.

[†] F. L. Hitchcock, The Distribution of a Product from Several Sources to Numerous Localities, *J. Math. Phys.*, vol. 20, 1941.

[‡] G. B. Dantzig, Application of the Simplex Method to a Transportation Problem, chap. XXIII in *Activity Analysis of Production and Allocation*, T. C. Koopmans (ed.), John Wiley & Sons, Inc., New York, 1951.

Table 1. Simplex Formulation of the Transportation Problem

| | Basis | | | | | | | | | | | | | | | | | | |
C	Var.	Value	x_{11}	x_{12}	$\cdots$	x_{1n}	x_{21}	x_{22}	$\cdots$	x_{2n}	$\cdots$	x_{m1}	x_{m2}	$\cdots$	x_{mn}	s_1	s_2	$\cdots$	s_{m+n}
C_1	s_1	a_1	1	1	$\cdots$	1										1			
C_2	s_2	a_2					1	1	$\cdots$	1							1		
		$\vdots$																	
		a_m										1	1	$\cdots$	1				
		b_1	1				1					1							
		b_2		1				1					1						
		$\vdots$																	
		b_n				1				1					1				1
C_{m+n}	s_{m+n}		c_{11}	c_{12}		c_{1n}	c_{21}			c_{2n}		c_{m1}			c_{mn}	C_1	C_1		C_{m+n}

Equations (18) to (20) are set up in an array as shown below.

Sources \ Destinations	1	2	. . .	n	Capacity
1	c_{11}	c_{12}		c_{1n}	a_1
2	c_{21}				a_2
. . .					
m	c_{m1}			c_{mn}	a_m
Demands	b_1	b_2		b_n	

The quantities c_{ij} represent the cost of transporting one unit from source i to destination j. The system of Eqs. (18) to (20) will be solved when quantities x_{ij} are found so that the capacity and demand restrictions are satisfied and Eq. (20) is minimized. First, an initial solution consisting of $n + m - 1$ assignments is found, and by an iteration process that will be described below, this solution is improved until Eq. (20) is minimized.

Consider the following problem with three sources S_1, S_2, S_3 with capacity restrictions of 13, 7, and 10, respectively, and three destinations D_1, D_2, D_3 with demands equal to 9, 12, and 9, respectively.

TABLE B_0

	D_1	D_2	D_3	
S_1	1	2	2	13
S_2	3	4	1	7
S_3	2	3	5	10
	9	12	9	30

The costs (assumed to be in dollars) associated with each source and destination are indicated in the upper left-hand corner of each cell. Since the system has $n + m - 1 = 5$ independent equations, one must first find an initial solution consisting of five assignments, the x_{ij} of Eqs. (18) to (21), that satisfy the demands and capacities restrictions indicated by Table B_0.

An easy way to find an initial solution is the following: Starting with the first cell S_1, D_1, one makes an assignment that satisfies either the row or column total. The first assignment of 9 satisfies the first column total. The remainder of the row total is then satisfied by assigning 4 to the cell S_1, D_2. The next assignment of 7 satisfies the second row total. To satisfy the second column total an assignment of 1 is made in the cell S_3, D_2, and finally the assignment of 9 to cell S_3, D_3 completes the five initial assignments that the system should have. These assignments are shown in Table B_1 below the diagonal lines of the cells that were selected. The total cost associated with this initial solution is

$$9 \times 1 + 4 \times 2 + 7 \times 4 + 1 \times 3 + 9 \times 5 = \$93$$

Table B_1

	D_1	D_2	D_3	
(cost)	1	2	2	13
(alloc)	9	4		
(cost)	3	4	1	7
(alloc)		7		
(cost)	2	3	5	10
(alloc)		1	9	
	9	12	9	

Initial Solution

Another way to find an initial assignment that might be closer to the optimum value is to use the cells with the minimum costs and assign the maximum values to these cells. The rest of the five initial assignments should then be made according to the costs associated with the cells and the corresponding marginal totals. However, the first method, sometimes called the northwest-corner method, is straightforward and can be readily used in electronic machine computations.

To improve on the initial solution, one should examine the cells that were not used in the first assignment. For instance, if one unit is added in cell S_2, D_3, one unit should be subtracted from S_2, D_2 so as to satisfy the total of 7 for that row. Since one unit was subtracted from S_2, D_2, one unit should be added to S_3, D_2, and finally one unit should be subtracted from S_3, D_3. The effect of these changes in the cost function is to add \$1, subtract \$4, add \$3, and subtract \$5, i.e., a net reduction of \$5. As can be seen by Table B_0, a maximum of seven units can be added to cell S_2, D_3, since no more than seven units can be subtracted from either row S_2 or S_3 owing to the limitation imposed by row S_2. Thus the total net effect of the changes is to reduce the cost function by $7 \times 5 = \$35$. To complete this analysis, all the unused cells should be evaluated in the same manner. If the number of initial assignments is equal to $n + m - 1$ and the row and column totals are satisfied, it will always be possible to evaluate the effect of using all the cells that had no assignments with the initial solution.

The evaluation of all the unused cells can be readily accomplished with the use of the so-called simplex multipliers. First, an extra row of v values and an extra column of u values are added to Table B_1. These u and v values are selected so that

$$u_i + v_j + c_{ij} = 0 \tag{23}$$

for all the cells with initial assignments. Since these u and v values are arbitrary, a row or column with a maximum number of initial assignments is given a value zero and the other values are found according to Eq. (23).

Table B_2

	D_1	D_2	D_3		u
(cost)	1	2	2	13	-2
(alloc)	9	4	(-2)		
(cost)	3	4	1	7	-4
(alloc)	(0)	7	(-5)		
(cost)	2	3	5	10	-3
(alloc)	(0)	1	9		
	9	12	9		
v	1	0	-2		

Selecting $v_2 = 0$ for the second column, one readily finds that $u_1 = -2$, $u_2 = -4$, and $u_3 = -3$ for the first, second, and third row, respectively. With $u_1 = -2$ and $c_{11} = 1$, then $v_1 = 1$. Also, since $u_3 = -3$ and $c_{33} = 5$, then $v_3 = -2$.

Upon determination of all the u and v values, new costs c_{ij}' are estimated by means of the equation

$$c_{ij}' = c_{ij} + u_i + v_j \tag{24}$$

For instance, for cell S_2, D_3 we have

$$c_{23}' = 1 - 4 - 2 = -5$$

The numbers in parentheses in the cells of Table B_2 are the results of this evaluation.

These new costs† estimate the effect of introducing the unused cells in the solution. For instance, if one unit is added to cell S_1, D_3, the total cost would be reduced by \$2; if one unit is added to cell S_2, D_1, the cost would remain the same; and so on. It can be seen that the addition of one unit to cell S_2, D_3 will result in the largest reduction in the total cost. As stated previously, a maximum of seven units can be added to the cell at a total cost reduction of \$35. The new solution is shown in Table B_3.

TABLE B_3

	D_1	D_2	D_3	
S_1	9	4		13
S_2			7	7
S_3		8	2	10
	9	12	9	

Iteration 1

Total Cost: $9 \times 1 + 4 \times 2 + 7 \times 1 + 8 \times 3 + 2 \times 5 = \$58$.

Introducing a new set of u and v values and performing computations similar as above, we have c_{13}' for cell S_1, D_3 is equal to -2, indicating that the addition of one

TABLE B_4

					u
	9	4	(-2)	13	1
	(5)	(5)	7	7	4
	(0)	8	0	10	0
	9	12	9		
v	-2	-3	-5		

unit to this cell would reduce the cost by \$2. Adding one unit to any of the other unused cells would either increase the cost by \$5 or leave the cost unchanged. Since the maximum number of units that can be added to cell S_1, D_3 is equal to 2, we have as

† These new costs c_{ij}' correspond to the shadow prices c_j' described in paragraph 2e of the Simplex algorithm.

TABLE B_5

9	2	2	13
		7	7
	10	·	10
9	12	9	

Iteration 2

new solution the results shown in Table B_5. The new total cost is now $9 \times 1 + 2 \times 2 + 2 \times 2 + 7 \times 1 + 10 \times 3 = \$54$. Introducing a new set of u and v values we have the results shown in Table B_6.

TABLE B_6

				u
9	2	2	13	0
(3)	(3)	7	7	1
(2)	10	(2)	10	−1
9	12	9		
v −1	−2	−2		

The new c'_{ij}'s show that the final solution has been achieved, since they are all positive.

Thus, the transportation problem shown in Table B_0 has the optimum solution indicated in Table B_5 at the minimum cost of $54.

a. Alternative Solutions. If in the final set of computations some of the c'_{ij}'s are zero, the cells with these zero values can be used as alternative routes of transportation at no change in the optimum cost. For instance, if c_{21}' would have been equal to zero, two units could have been introduced in cell S_2, D_1; two units subtracted from cell S_2, D_3; two units added to cell S_3, D_3; and two subtracted from cell S_3, D_1.

b. Degeneracy. If in finding an initial solution the number of assignments is less than $n + m - 1$, a case of degeneracy occurs. To solve this case, additional assignments equal to zero should be used so as to bring the total number of initial solutions to $n + m - 1$. The zero assignment cells are used in the same manner as above for the determination of the u and v values and for the shifting of assignments toward improving the cost function. In some instances this shifting may result in moving a zero assignment from one cell to another with no improvement in the cost function, but eventually a new assignment will be found that will result in net improvement in the total cost.

Degeneracy can also occur in some of the iterations of the transportation technique. This will occur when two or more cells have the same value of x_{ij} and are simultaneously removed when the assignments are shifted. In this case, zero assignments are given to some of the cells removed in order to keep the number of assignments equal to $n + m - 1$, and then one proceeds in the same manner as described above.

c. Unbalanced Production and Demand. If the total production differs from the total demand, then an additional slack row or an additional slack column must be added to the transportation setup in order to take care of the unbalance.

Assume, for instance, that the production from source S_1 equals 18 units instead of 13 units, making the total production equal to 35 units with the same total demand of 30 units. Then the initial transportation table with an initial solution obtained according to the northwest-corner rule will be as follows:

	D_1	D_2	D_3	Slack	
S_1	1 /9	2 /9	2	0	18
S_2	3	4 /3	1 /4	0	7
S_3	2	3	5 /5	0 /5	10
	9	12	9	5	35

The number of initial assignments is six, which is equal to

$$n + m - 1 = 3 + 4 - 1 = 6$$

with the slack column considered as an extra fictitious demand. The solution will be found following the same rules as described above.

The costs given to the slack demand column were assumed to be zero. However, they might be given some other values if such a thing as storage costs or other penalty costs are associated with the slack column.

If, on the other hand, the demand exceeds the production, an extra row must be added to balance the totals. For instance, assume that the demand from D_1 is equal to 14 instead of 9 units. The initial transportation table will then be as follows:

	D_1	D_2	D_3	
S_1	1 /13	2	2	13
S_2	3 /1	4 /6	1	7
S_3	2	3 /6	5 /4	10
Slack	0	0	0 /5	5
	14	12	9	35

The extra row can be thought of as a fictitious source with a total production of five units. The costs associated with the slack row can be quantities different from zero if some penalties are associated with the shortage of units.

Situations can occur where an extra row and an extra column must be added in order to balance demand and production. After an initial solution is determined by the northwest-corner method or any other method, the problem will be solved following the same transportation algorithm described before.

d. Transportation Problem with Transshipment. In the formulation of the transportation problem it was assumed that a source only sends units and the destinations only receive units. It is possible, however, to have situations where units are sent to destinations via other sources or other destinations, in which case a source and a destination both send and receive units. For the formulation of this kind of transportation problem see Ref. 2, Chap. 7.

e. The Generalized Transportation Problem. As stated previously, the transportation problem is a special case of the linear programming problem where the matrix of the coefficients are all unity and have the special configuration shown in Table I. Also, if coefficients associated with Eqs. (18) and (19) are constants for the same value of i in Eq. (18) and the same value of j in Eq. (19), when these equations are divided by the corresponding constants, the linear-programming problem can be solved by use of the transportation technique.

The more general transportation problem can be described as follows:
Find values of x_{ij} so that

$$\sum_j x_{ij} = a_i \qquad i = 1, \ldots, m$$

$$\sum_i e_{ij} x_{ij} = b_j \qquad j = 1, \ldots, n \tag{25}$$

$$\sum_i \sum_j c_{ij} x_{ij} = \min$$

$$x_{ij} \geq 0 \qquad \text{all } i \text{ and } j$$

The constants a_i, b_j, and c_{ij} have the same meaning as in Eqs. (18) to (20). The new elements added to the problem are the constants e_{ij}, which in general are not the same for all the values of x_{ij}. The transportation algorithm can be modified to handle this new situation, but the computations are more involved. For a discussion of this problem see Ref. 2, Chap. 10.

f. The Transportation Problem with Upper Bounds. When some of the cells in the transportation array are bounded by some given numbers, then the transportation algorithm must be modified in order to account for the extra restriction. The problem can be formulated by Eqs. (18) to (21) with the following additional restrictions:

$$x_{ij} \leq U_{ij} \qquad \text{for some or all } i, j$$

If the upper bounds are very restrictive, the problem may not have a solution, since it might not be possible to find an initial solution that satisfies the demand and production requirements. For a discussion of this problem and for an example when $x_{ij} \leq k$ for all i, j, see Ref. 2, Chap. 11.

If some of the upper bounds are equal to zero, which will be the case when some of the cells should not be used in the solution of the problem, then, by introducing a very high cost for these cells, one could apply the standard transportation algorithm for determining initial solutions and solving all the iterations. The scheduling of production problems is an example of this situation.

g. The Transportation Algorithm for Scheduling Problems. The transportation technique is ideally suited to solving problems of production schedule.

Consider the following problem: Two plants A and B can be scheduled to make two products x and y over two time periods 1 and 2. The requirements for the two time periods are:

$$\begin{array}{lll} \text{Period 1:} & x = 75 & y = 150 \\ \text{Period 2:} & x = 100 & y = 200 \end{array}$$

The capacities of plants A and B over the two time periods are

Plant A. Period 1: $x = 50$ $y = 75$ Period 2: $x = 100$ $y = 100$
Plant B. Period 1: $x = 50$ $y = 100$ Period 2: $x = 100$ $y = 100$

Assume for simplicity that the cost of production for both products over the two periods and both plants is equal to \$1.

The following table gives the initial transportation table for this problem with an initial solution found according to the northwest-corner method.

			1 x	1 y	2 x	2 y	Slack	
1	A	x	1 / 1 — 50	1	2	2		50
		y	1 / 25	1 / 50	2	2		75
	B	x	1	1 / 50	2	2		50
		y	1	1 / 50	2 / 50	2		100
2	A	x	100	100	1 / 50	1 / 50		100
		y	100	100	1	1 / 100		100
	B	x	100	100	1	1 / 50	50	100
		y	100	100	1	1	100	100
			75	150	100	200	150	675

Each production possibility over the two time periods can be taken as a source, and each requirement over the two time periods can be taken as a destination. Since the production from time period 2 cannot be used to meet the requirements for time period 1, high costs (in this case costs equal to \$100) were assigned to the cells corresponding to "source" 2, "destination" 1.

This problem can, of course, be extended to take into account several periods, products, and plants with regular and overtime production.†

h. The Assignment Problem. Machine- and job-assignment problems in which n people can be assigned to n machines or jobs can also be solved by the use of the transportation algorithm.

Consider, for instance, the case of four operators that can run each of four machines at a cost indicated in the table below.

	M_1	M_2	M_3	M_4
O_1	2	6	4	3
O_2	3	3	1	4
O_3	5	4	3	2
O_4	4	2	3	5

† See Edward W. Bowman and Robert B. Fetter, *Analysis for Production Management*, Richard D. Irwin, Inc., Homewood, Ill., 1961.

The problem consists in finding the optimum assignment that will minimize the cost of running the four machines. An initial solution should consist of seven assignments, but since the totals for both the rows and the columns should be equal to 1, then additional assignments equal to zero should be inserted in some cells in order to use the transportation algorithm. Table C below illustrates this kind of initial assignment, in which assignments were given to the cells with the smallest cost.

TABLE C

2 / 1	6 / 0	4	3	1
3	3	1 / 1	4	1
5 / 0	4	3	2 / 1	1
4	2 / 1	3 / 0	5	1
1	1	1	1	

At first, only the zeros will be moved from cell to cell, but eventually some of the assignments equal to 1 will be moved and the final solution will be obtained. If not, the initial assignment is optimum.

Other techniques are available for the solution of the assignment problem. Based on a principle that states that, if a constant is added to every element of a row or a column of the cost matrix, the assignment that minimizes the original matrix also minimizes the modified matrix, one would proceed to subtract the smallest element of every row and then subtract the smallest element of every column in order to abtain a modified matrix in which some of the costs will be zero.

For the above problem, we shall have

	M_1	M_2	M_3	M_4
O_1	0	4	2	1
O_2	2	2	0	2
O_3	3	2	1	0
O_4	2	0	1	3

Now, since the costs are either zero or positive numbers, one would then select the cells with zero costs for minimum cost. The cells thus selected would be (O_1, M_1), (O_2, M_3), (O_3, M_4) and (O_4, M_2).

These assignments are the same as the initial assignments used in Table C. Thus, this initial assignment is optimum, giving a minimum cost of $2 + 1 + 2 + 2 = \$7$.

Other algorithms are available for the solution of more complicated matrices for which the method described above does not produce a unique answer. See, for instance, Ref. 3, Chap. 12, or Ref. 4, Chap. 8.

BASIC CONCEPTS IN PROBABILITY AND STATISTICS

4. Random Experiments—Probability of an Event

In the great majority of scientific problems, cases occur when a given experiment can be repeated any number of times under the same uniform conditions. If, in addition, the result of an individual outcome cannot be accurately predicted, the experiment will be referred to as a *random experiment*.

If, in repeating the random experiment n times, one observes the number of times m that a certain event E has occurred, one could test that, as n increases, the ratio n/m appears to oscillate very closely around a constant value P. This value is called the probability of the event E and will be assumed to exist whenever the experiment is defined as a random experiment.

Thus, the probability $P(E)$ of an event E means that, when the random experiment is repeated a large number of times, the relative frequency of occurrence of E will be approximately equal to $P(E)$.

P satisfies the condition

$$0 \leq P \leq 1 \tag{26}$$

If the event E is impossible (or certain to occur), then P will be zero (or 1). However, if, for some event E, $P = 0$ (or $P = 1$), then it is not necessarily true that the event E is impossible (or certain to occur).

A *random variable* is a variable that describes (or measures) all the possible outcomes of the random experiment. Thus, if the random experiment consists of examining manufactured parts, the random variable may take on two values, $x = 0$ for a non-defective part and $x = 1$ for a defective part. The probability of the event "defective," that is, the probability that $x = 1$ $[P(x = 1)]$, will be estimated by the relative frequency of the occurrence of this event in a large number of trials.

5. Sample Space and Events

A sample space will be defined as the set of points describing all the possible outcomes of a random experiment. An event E_1 will be defined by a subset of points in the sample space.

Complement Event. The complement of an event E_1, denoted by $\bar{E}_1$, in a sample space E is defined as the set of points of E not belonging to E_1. The complement is represented by the shaded area of Fig. 2.

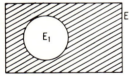

FIG. 2. Complement event.

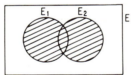

FIG. 3. Union.

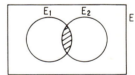

FIG. 4. Intersection.

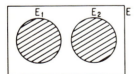

FIG. 5. Mutually exclusive.

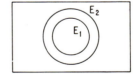

FIG. 6. Inclusion.

Union (or Logical Sum) of Two Events. The union of two events E_1 and E_2, denoted by $E_1 + E_2$, is defined as the set of points of E belonging to either E_1 or E_2 or both. The union is represented by the shaded area of Fig. 3.

Intersection (or Logical Product) of Two Events. The intersection of two events E_1 and E_2, denoted by E_1E_2, is defined as the set of points of E belonging to both E_1 and E_2. The intersection is represented by the shaded area of Fig. 4.

Mutually Exclusive Events. Two events E_1 and E_2 are said to be mutually exclusive if the corresponding sets of points do not have any point in common. (The intersection of E_1 and E_2 is the empty set.) See Fig. 5.

Inclusion of Events. An event E_1 is said to be included in an event E_2, $E_1 \subset E_2$, if the occurrence of E_1 implies the occurrence of E_2. In this case, $E_1 + E_2 = E_2$ and $E_1E_2 = E_1$ (see Fig. 6).

6. Rules of Probability

a. Addition. Let E_1 and E_2 denote two events which may or may not occur when a random experiment E is performed. We shall denote by $E_1 + E_2$ the occurrence of either E_1 or E_2 or both. Or, in other words, $E_1 + E_2$ denotes the occurrence of at least one event.

We shall also denote by E_1E_2, the product of E_1 and E_2, the simultaneous occurrence of both events E_1 and E_2.

The addition rule of probability theory states that

$$P(E_1 + E_2) = P(E_1) + P(E_2) - P(E_1E_2) \tag{27}$$

The third term on the right-hand side is needed so as not to count twice the set of points related to the simultaneous occurrence of E_1 and E_2 (shaded region in Fig. 4).

If the events E_1 and E_2 are mutually exclusive, $P(E_1E_2) = 0$ and

$$P(E_1 + E_2) = P(E_1) + P(E_2) \tag{28}$$

The addition rule can be extended to any number of events. Thus, if $E_1, E_2, \ldots, E_n$ denote n events, not necessarily mutually exclusive, then

$$P(\Sigma E_i) = \sum_{i=1}^{n} P(E_i) - \sum_{i \neq j} P(E_iE_j) + \sum_{i \neq j \neq k} P(E_iE_jE_k) \cdots$$
$$(-1)^{n-1}P(E_1E_2 \cdots E_n) \tag{29}$$

If the events are mutually exclusive, then

$$P(\Sigma E_i) = \Sigma P(E_i) \tag{30}$$

If $E_1, E_2, \ldots, E_n$ account for all possible occurrences of the random experiment E, then $\Sigma P(E_i) = 1$.

b. Multiplication. Conditional Probability. Independence. If two events E_1 and E_2 are independent, that is, if the occurrence (or not occurrence) of one does not affect the occurrence of the other, then the probability of the simultaneous occurrence of E_1 and E_2 is

$$P(E_1E_2) = P(E_1)P(E_2) \tag{31}$$

If the events are dependent, then the occurrence of E_1 will affect the occurrence of E_2, in which case the multiplication rule changes to

$$P(E_1E_2) = P(E_1)P(E_2/E_1) \tag{32}$$

where $P(E_2/E_1)$ denotes the conditional probability of E_2 relative to the hypothesis that E_1 has occurred.

We also have

$$P(E_1E_2) = P(E_2)P(E_1/E_2) \tag{33}$$

The multiplication rule can be extended to any number of events. Thus we have

$$P(E_1E_2E_3 \cdots E_n) = P(E_1)P(E_2/E_1)P(E_3/E_1E_2) \cdots P(E_n/E_1 \cdots E_{n-1}) \tag{34}$$

If all the events are independent of each other, then

$$P(E_1E_2 \ldots E_n) = \prod_{i=1}^{n} P(E_i) \tag{35}$$

7. Bayes' Theorem

For any two events A and B out of a random experiment E one has, from the multiplication rule,

$$P(A/B) = \frac{P(AB)}{P(B)} = \frac{P(A)P(B/A)}{P(B)} \tag{36}$$

Suppose the event B can occur only in combination with one of the m events B_1, B_2, B_3, . . . , B_m, any two of which are mutually exclusive.

Since B_i is included in B, then $B_i = B_iB$ and one can write

$$B = B_1 + B_2 + \cdots + B_m = B_1B + B_2B + \cdots + B_mB$$

For example, suppose there are two boxes each containing 10 elements. Let box 1 have five defective and five nondefective elements and box 2 have two defective and eight nondefective elements. Assume that two elements were extracted from one of the boxes and that they were both defectives. One can think of the event B as being the extraction of the two defective elements. Obviously, B occurred only when either the two elements were extracted from box 1 or they were extracted from box 2.

Using the addition and multiplication rules of probability one has

$$P(B) = \sum_{i=1}^{m} P(B_i/B) = \sum_{i=1}^{m} P(B_i)P(B/B_i) \tag{37}$$

Applying Eq. (36) to a particular event B_i, that is, substituting B_i for A, one has, after using Eq. (37),

$$P(B_i/B) = \frac{P(B_i)P(B/B_i)}{\sum\limits_{i=1}^{m} P(B_i)P(B/B_i)} \tag{38}$$

which is Bayes' theorem.

To summarize: A certain event has been observed, and it is known that it has occurred as a consequence of the occurrence of either B_1 or B_2 or . . . B_m, which exclude each other two by two. The events B_i are called the "causes" or "hypothesis" of B, and Eq. (38) is said to give the probability of one of the causes B_i after the event B has occurred.

The probabilities $P(B_i)$ are called the *a priori* probability of B_i and $P(B_i/B)$ the *a posteriori* probability of B_i calculated under the hypothesis that one has observed the occurrence of B. In most statistical applications, the *a priori* probabilities are unknown and Bayes' theorem does not strictly apply.

As an illustration of a legitimate application of Bayes' theorem, assume that one asks, in the example cited above, for the probability that the two defective elements were extracted from box 2. If it is known that the probability of manufacturing boxes of type 1 is 1/3 and boxes of type 2 is 2/3, then

$$P = \frac{(1/3)[(5 \times 4)/(10 \times 9)]}{(1/3)[(5 \times 4)/(10 \times 9)] + (2/3)[(2 \times 1)/(10 \times 9)]} = \frac{5}{6}$$

8. Probability Distributions

a. Discrete Probability Distributions. Let a random experiment be described by a random variable x that can assume a finite or infinite sequence of values x_i with associated probabilities p_i. We can write

$$p(x) = P(x = x_i) = p_i$$

The set of values p_i completely defines the discrete probability function $p(x)$ shown in Fig. 7.

The probability that x takes on values less than or equal to any particular value x_i will be given by

$$P(x \leq x_i) = \sum_{x \leq x_i} p_i \tag{39}$$

Expression (39) defines the probability distribution of x (sometimes called the cumulative probability function) which will be denoted by $P(x)$ [see Fig. 8 for a graphical representation of Eq. (39)].

If the random experiment is described by two random variables x and y, then the values

$$P(x = x_i, y = y_j) = p_{ij}$$

will completely define the discrete probability function $p(x,y)$.

A two-dimensional table of values of p_{ij} together with a three-dimensional graph will be necessary to describe all the possible occurrences of the random experiment.

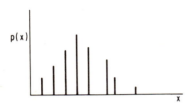

FIG. 7. Discrete probability function.

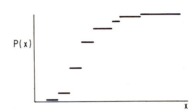

FIG. 8. Probability distribution.

The probability distribution $P(x,y)$ will now be given by

$$P(x,y) = P(x \leq x_i, y \leq y_j) = \sum_{\substack{x \leq x_i \\ y \leq y_j}} p_{ij} \tag{40}$$

The *marginal probability* functions of x and y, denoted by $p_1(x)$ and $p_2(x)$, respectively, are given by

$$p_1(x) = P(x = x_i) = \sum_{\text{all } j} p_{ij} \tag{41}$$

$$p_2(y) = P(y = y_j) = \sum_{\text{all } i} p_{ij} \tag{42}$$

These marginal probability functions can be considered as projections in the x (or y) axis of all the probabilities in the xy plane.

The *conditional probabilities* function of x given y (or of y given x) are defined by

$$p_3(x/y) = P(x = x_i/y = y_j) = p(x,y)/p_2(y) \tag{43}$$
$$p_4(y/x) = P(y = y_j/x = x_i) = p(x,y)/p_1(x) \tag{44}$$

If the conditional probability function is equal to its marginal, then the random variables will be independent. In fact, it can be shown that the necessary and sufficient condition for two random variables x and y to be independent is that the joint probability function of x and y is equal to the product of their marginal probability functions; that is,

$$p(x,y) = p_1(x)p_2(y) \qquad \text{for all } x \text{ and } y \tag{45}$$

b. Continuous Probability Distributions. If the random experiment is described by a random variable x with an associated probability function $f(x)$ which is everywhere continuous with exception of a finite number of discontinuity points, then the random variable x will be said to belong to the *continuous* type. The probability density function $f(x)$ will be defined as

$$P(x \leq x \leq x + dx) = f(x)\, dx$$

$f(x)\, dx$ is known as the probability element of the distribution and is shown in the shaded area of Fig. 9.

The probability distribution is given by

$$F(x) = \int_{-\infty}^{x} f(x)\, dx \tag{46}$$

a graphical representation of which is given in Fig. 10.

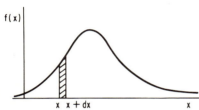

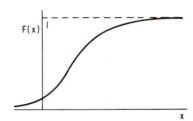

FIG. 9. Continuous probability function.   FIG. 10. Continuous probability distribution.

The generalizations to two or more continuous random variables are the same as for the discrete random variable case. Thus, the joint probability density function of x and y is given by

$$f(x,y)\, dx\, dy = P(x \leq x \leq x + dx,\, y \leq y \leq y + dy)\, dx\, dy$$

The marginal probability function of x is given by

$$f_1(x) = \int_{-\infty}^{\infty} f(x,y)\, dy \tag{47}$$

The conditional probability function of y given x is given by

$$f_4(y/x) = \frac{f(x,y)}{f_1(x)} \tag{48}$$

The two random variables x and y will be indépendent if and only if

$$f(x,y) = f_1(x)f_2(y) \tag{49}$$

9. Moments. Expected Values. Moment-generating Function

a. Moments, Expected Values. For a complete characterization of a random variable one must know the mathematical expression of its probability function, which will assign probability values to all the outcomes of the random variable. Some examples of typical probability functions will be given in paragraphs 12 and 13.

The main characteristics of these distributions are given by two parameters called the mean and the standard deviation.

The *mean value* μ of a random value x is defined by†

$$E(x) = \int_{-\infty}^{\infty} xf(x)\,dx = \mu \tag{50}$$

where the notation $E(x)$ stands for "expected value of x."

The mean value is also referred as the first moment of the variable x. In general, the kth moment of x is given by

$$E(x^k) = \int_{-\infty}^{\infty} x^k f(x)\,dx = \alpha_k \tag{51}$$

The variance of x, denoted by σ^2, is given by the expected mean square deviation of x from its mean value:

$$\sigma^2 = E[(x - \mu)^2] = \int_{-\infty}^{\infty} (x - \mu)^2 f(x)\,dx \tag{52}$$

The *standard deviation* of x is defined as the square root of the variance of x.

The variance is a special case of the kth centralized moment of the random variable x, which is given by

$$E[(x - \mu)^k] = \int_{-\infty}^{\infty} (x - \mu)^k f(x)\,dx = \mu_k \tag{53}$$

b. Properties of the Operator E.   The expected value of a function $g(x)$ of a random variable x is defined by

$$E[g(x)] = \int_{-\infty}^{\infty} g(x)f(x)\,dx \tag{54}$$

It is easy to verify that the operator E is linear with the following properties (see Ref. 5, Chap. 15):

a. $E(k) = k$ k a constant   (55)
b. $E[\Sigma k_i g_i(x)] = \Sigma k_i E[g_i(x)]$ (56)
c. $E[g(x)h(y)] = E[g(x)]E[h(y)]$ if x and y are independent random variables   (57)

By expanding $(x - \mu)^2$ in Eq. (52) one can easily show that

$$\sigma^2 = E[(x - \mu)^2] = E(x^2) - [E(x)]^2 \tag{58}$$

c. Moment-generating Function.   The expected value of the function $e^{\theta x}$ where x is a random variable and θ is another variable, not random, has a very important role in probability and statistics.

The moment-generating function of (MGF) of x is defined by

$$\psi(\theta) = E(e^{\theta x}) = \int_{-\infty}^{\infty} e^{\theta x} f(x)\,dx \tag{59}$$

Assuming the necessary conditions of convergence, one can differentiate Eq. (59) under the integral sign and obtain

$$\frac{d^n \psi(\theta)}{d\theta^n} = \psi^{(n)}(\theta) = \int_{-\infty}^{\infty} x^n e^{\theta x} f(x)\,dx \tag{60}$$

Setting $\theta = 0$ in Eq. (60) one has

$$\psi^{(n)}(0) = \int_{-\infty}^{\infty} x^n f(x)\,dx = \alpha_n \tag{61}$$

† It will be assumed that the random variable x has a distribution of a continuous type, with $-\infty < x < \infty$.

That is, the nth moment of x is obtained by setting $\theta = 0$ in the nth derivative of $\psi(\theta)$. Thus, in particular

$$\psi'(0) = \mu$$
$$\psi''(0) = \alpha_2$$

and so on.

Knowing $f(x)$, one can obtain $\psi(\theta)$ by a Laplace-type integral of the form (59). Conversely, under some general conditions, the knowledge of $\psi(\theta)$ uniquely determines $f(x)$.

10. Bounds of Variation—The Tschebyscheff Inequality

It is possible to give bounds to the probability of deviations of a random variable from its mean without a specific knowledge of the probability distribution involved. The theorem to be quoted here will give these bounds for any probability function with a finite mean and standard deviation.

Theorem.[†] Let x be a random variable having a mean μ and a standard deviation σ, and let k be an arbitrary positive number. Then the probability that an absolute deviation of x from its mean value is larger then $k\sigma$ will be smaller than $1/k^2$.

$$P[|x - \mu| \geq k\sigma] \leq \frac{1}{k^2} \tag{62}$$

The inequality can also be written in the form

$$P[|x - \mu| \leq k\sigma] \geq 1 - \frac{1}{k^2} \tag{63}$$

Substituting $k = 1, 2, 3$ in Eq. (63) one has

$$\begin{array}{ll} k = 1 & P \geq 0 \\ k = 2 & P \geq 0.75 \\ k = 3 & P \geq 0.89 \end{array}$$

Thus, interpreting the results for $k = 3$ one can say that, for any distribution with finite mean and standard deviation, about 89 per cent of the area will be included in the interval $\mu \pm 3\sigma$. Notice that for $k = 1$ the best one can say is that there will be a positive (or zero) area in the interval $\mu \pm \sigma$.

The bounds given above can be improved if, in addition to the μ and σ, one has more knowledge about the type of probability distribution involved.

If the distribution is unimodal and the mode[‡] is equal to the mean, then the Camp-Meidell inequality states that the value of P in Eq. (63) is

$$P \geq 1 - \frac{1}{2.25k^2}$$

The values of P for $k = 1, 2, 3$ are now

$$\begin{array}{ll} k = 1 & P \geq 0.56 \\ k = 2 & P \geq 0.89 \\ k = 3 & P \geq 0.95 \end{array}$$

If the probability distribution is known, the values of P can be computed by an integration process. Thus, for the case of the Gaussian distribution (see paragraph 13c), the values of P for $k = 1, 2, 3$ are 0.683, 0.954 and 0.997, respectively.

11. Propagation of Errors

Errors can be divided into three categories: (1) mistakes that are made in reading or recording observed magnitudes, (2) systematic errors that are the result of personal

---

[†] See Ref. 5, Chap. 15.
[‡] The mode is the value of the random variable with the highest probability.

or instrumental biases, (3) statistical or random errors that are the results of a great variety of causes that are impossible or impracticable to identify.

Mistakes can be eliminated by initial care or independent checking. Automatically recording devices are in many cases the solution for the elimination of this kind of error. Instrumental bias can be eliminated by a proper adjustment or calibration of the instruments being used. This topic belongs to the field of instrumentation and experimental techniques.

Random errors are all other errors that do not show any regularity or the regularity of which is unknown. When a statistical parameter is being estimated, it will be observed that the values obtained are scattered about the actual value of this parameter in a characteristic random fashion. The spread or dispersion of statistical parameter is often referred to as the standard error of this parameter.

a. Linear Propagation of Errors. Let a random variable X be given by a linear combination of k random variables x_i which are independent of each other.

$$X = \sum_1^k a_i x_i \tag{64}$$

Let $$E(x_i) = \mu_i \qquad \sigma_{x_i} = \sigma_i$$

be the mean and standard errors of x_i. The assumption is made that the error made in measuring x_i does not affect the error made in the measurement of the other variables.

Using the properties of the linear operator E (see paragraph 9b), it is easy to show that

$$E(X) = \sum_{i=1}^k a_i \mu_i \tag{65}$$

$$\sigma_x{}^2 = E[X - (EX)]^2 = E[\Sigma a_i(x_i - \mu_i)]^2 = \Sigma a_i{}^2 E(x_i - \mu_i)^2$$
$$= \Sigma a_i{}^2 \sigma_i{}^2 \tag{66}$$

Thus the variance of a linear combination of independent random variables is equal to the sum of the square of the constants times the variance of each variable.

As a special case of Eq. (64), let

$$\bar{x} = \sum_{i=1}^k \frac{x_i}{n} \tag{67}$$

with $$E(x_i) = \mu \quad \text{and} \quad \sigma_{x_i} = \sigma$$

X of Eq. (64) in this case is equal to the mean of k values of x, each value of x having the same mean and standard deviation.

Then, from Eqs. (65) and (66)

$$E(\bar{x}) = \mu$$
$$\sigma_{\bar{x}}{}^2 = \frac{\sigma^2}{n}$$
$$\sigma_{\bar{x}} = \frac{\sigma}{\sqrt{n}} \tag{68}$$

Thus, the expected value of $\bar{x}$ equals the "population" mean and the standard error of the mean equals the standard deviation of x divided by the square root of the number of observations.

b. Nonlinear Propagation of Errors. Assume that X is given by a nonlinear function of independent random variables x_i with means μ_i and standard errors σ_i.

$$X = f(x_1, x_2, \ldots, x_k) \tag{69}$$

Then, by using Taylor's expansion for X about the point $x_i = \mu_i$ and by assuming the deviations $x_i - \mu_i$ to be small, it can be shown that

$$\sigma_X{}^2 = \sum_{i=1}^{k} \left(\frac{\partial f}{\partial x_i}\right)^2 \sigma_i{}^2 \tag{70}$$

12. Some Discrete Probability Distributions

a. The Binomial Distribution. Let there be a random experiment E^* which results in either the occurrence or the lack of occurrence of an event E.

Let
$$P(E) = p$$
$$P \text{ (not } E) = q = 1 - p$$

Let the occurrence of E be measured by a random variable x which will be assigned the value $x = 1$ if E occurs and $x = 0$ if E does not occur.

Let us repeat n times the random experiment E^* and assume that the probability of the occurrence of E at any time is independent of any previous occurrence or non-occurrence of E, and let us ask for the probability of exactly k occurrences of E.

In n times the event E will occur, say, k times and "not E" will occur $n - k$ times. Assume that the event E occurred k times in succession and "not E" the $n - k$ remaining times. By the multiplication rule, the probability of this compound event will be

$$pp \ \cdots \ pqq \ \cdots \ q = p^k q^{n-k}$$

There are exactly

$$\binom{n}{k} = \frac{n!}{k!(n-k)!}$$

ways of obtaining k elements of one kind and $(n - k)$ elements of a different kind out of n elements.

Since, the probability of k occurrences of E should be immaterial to the order in which E occurs, one has by the addition rule

$$b(k) = P(x = k) = \binom{n}{k} p^k q^{n-k} \tag{71}$$

which is the binomial probability function.

Using the properties of the binomial expansion, it can be seen that

$$\sum_{k=0}^{n} b(k) = \sum_{k=0}^{n} \binom{n}{k} p^k q^{n-k} = (p + q)^n = 1$$

The moment-generating function† of $p(x)$ is

$$\psi(\theta) = E(e^{\theta x}) = \sum_x e^{\theta x} \binom{n}{x} p^x q^{n-x}$$

$$= \sum_x \binom{n}{x} (e^\theta p)^x q^{n-x} = (e^\theta p + q)^n \tag{72}$$

Using Eq. (72) and the properties described in paragraphs 9b and 9c, one has

$$\mu = E(x) = np$$
$$\sigma^2 = E(x^2) - \mu^2 = npq \tag{73}$$

† One often says: the moment-generating function of $p(x)$ instead of the more correct expression: the moment-generating function of the random variable x.

b. Poisson Distribution. The Poisson distribution applies to cases where in a single performance of the random experiment E^* the probability of E is very small, but when the random experiment is repeated a very large number of times, the event E will occur a finite, but small as compared with n, number of times.

In fact, it can be shown that, as the number n of trials goes to infinity and the probability p of the occurrence of a single event goes to zero while the average number of occurrences np approaches a constant m, then

$$\lim b(x) = p(x) = \frac{e^{-m}m^x}{x!} \qquad x = 0, 1, \ldots \qquad (74)$$

That is, the limit of the binomial distribution as $n \to \infty$, $p \to 0$ so that $np \to m$ is the probability function given by Eq. (74), which is called the Poisson probability function.

The Poisson distribution is an example of a discrete random variable with an infinite number of values. Adding over all possible values of x, one has

$$\sum_0^\infty p(x) = e^{-m} \sum_0^\infty \frac{m^x}{x!} = e^{-m}e^m = 1$$

Also, it can be shown that the moment-generating function of x is

$$\psi(\theta) = e^{-m}e^{me^t}$$

Differentiating $\psi(\theta)$ and using the properties (61) and (58), one has

$$\begin{aligned} E(x) &= m \\ \sigma^2 &= m \end{aligned} \qquad (75)$$

The Poisson distribution can also be derived by the consideration of events that occur during a period of length t, where the occurrences of these events are governed by the following principles:

1. The probability of a single occurrence of the event in a time dt is proportional to dt (say it is equal to $\lambda\, dt$).
2. The probability that the event will not occur is $1 - \lambda\, dt$.
3. The probabilities are independent of the position of this period of length dt in the time interval $(0, t + dt)$ and of the number of times the event has previously occurred.

Under these assumptions, the probability that the event will occur x times in the time interval $(0, t + dt)$, which will be denoted by $p_x(t + dt)$, can be partitioned into the following mutually exclusive categories:

1. The event will occur exactly x times in $(0,t)$ and not in $(t, t + dt)$.
2. The event will occur exactly $x - 1$ times in $(0,t)$ and once in $(t, t + dt)$.
3. The event will occur $x - r$ times in $(0,t)$ and r times in $(t, t + dt)$.

Then, according to the multiplication rule of independent events and the addition rule of mutually exclusive events, the value of $p_x(t + dt)$ can be written by the following equation:

$$p_x(t + dt) = p_x(t)(1 - \lambda\, dt) + p_{x-1}(t)\lambda\, dt + p_{x-2}(t)(\lambda\, dt)^2 + \cdots \qquad (76)$$

If differentials of order higher than dt are neglected, the above expression can be written in the form

$$\frac{dp_x(t)}{dt} = \lambda[p_{x-1}(t) - p_x(t)] \qquad (77)$$

Expression (70) is a difference-differential equation which by standard methods can be shown to have as a solution

$$P_x(t) = \frac{e^{-\lambda t}(\lambda t)^x}{x!} \qquad x = 0, 1, \ldots \qquad (78)$$

Since λ is taken to be the average number of occurrences per unit of time, λt will be the average number of occurrences during the interval of time of length t and will correspond to the value m used in expression (74).

As examples of the applications of the Poisson distribution from this point of view one can mention the probability of the occurrence of N impulses in a Poisson noise process, the probability of finding x radioactive atoms decaying in the time interval t, the probability that x cosmic-ray particles trip a Geiger counter in the time interval t, the probability of x accidents in the time interval t, the probability of finding x particles (blood, bacteria, grains of cement, etc.) within a field of area t in a microscope, and so on.

c. Bernouilli's Theorem—Law of Large Numbers. Let x be a binomial random variable with mean np and standard deviation $\sqrt{npq}$. Define

$$y = \frac{x}{n}$$

Applying the properties of the operator E one has

$$E(y) = p$$
$$\sigma_y = \sqrt{\frac{pq}{n}} \tag{79}$$

Expressions (79) state that the expected value of the sample proportion y is equal to the "population," or actual, proportion p and its standard deviation is $\sqrt{pq/n}$.

Using Tschebyscheff's inequality (see paragraph 10), then

$$P\left[|y - p| \geq k\sqrt{\frac{pq}{n}}\right] \leq \frac{1}{k^2}$$

Letting

$$\epsilon = k\sqrt{\frac{pq}{n}}$$

then

$$P[|y - p| \geq \epsilon] \leq \frac{pq}{n\epsilon^2}$$

Then for any fixed ϵ, no matter how small,

$$P[|y - p| \geq \epsilon] \to 0 \qquad n \to \infty \tag{80}$$

which is *Bernouilli's* theorem: For a very large sample, the probability that the sample proportion deviates from the actual proportion by more than any preassigned small number ϵ tends to zero as n goes to infinity.

This theorem is one form of the more general *law of large numbers*.

Let x_i be n independent random variables with mean μ and standard deviation σ. As shown previously, $E(\bar{x}) = \mu$ and $\sigma_{\bar{x}} = \sigma/\sqrt{n}$.

Applying Tschebyscheff's inequality to the random variable $\bar{x}$, one has

$$P\left(|\bar{x} - \mu| \geq k\frac{\sigma}{\sqrt{n}}\right) \leq \frac{1}{k^2} \tag{81}$$

Substituting

$$\epsilon = \frac{k\sigma}{\sqrt{n}} \tag{82}$$

then

$$P[|\bar{x} - \mu| \geq \epsilon] \to 0 \qquad n \to \infty$$

Thus, the average $\bar{x}$ of n independent observations of a random variable x with finite mean and standard deviation tends with probability 1 as $n \to \infty$ to the actual mean.

13. Some Continuous Probability Distributions

a. The Rectangular or Uniform Distribution. The random variable x is said to be distributed according to the uniform probability function (see Fig. 11) between two values a and b if

$$
\begin{aligned}
f(x) &= 0 && \text{when } x < a \\
f(x) &= \frac{1}{b-a} && \text{when } a < x < b \\
f(x) &= 0 && \text{when } x > b
\end{aligned}
\tag{83}
$$

The mean of x is $(a+b)/2$, and the standard deviation is $(b-a)/\sqrt{12}$.

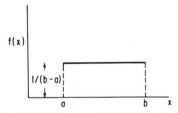

FIG. 11. Uniform probability function.

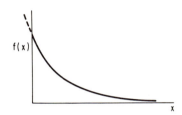

FIG. 12. Exponential probability function.

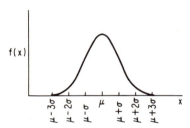

FIG. 13. Normal probability function.

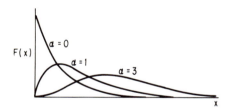

FIG. 14. Gamma probability function.

The error introduced in a numerically calculated quantity by the "rounding off" may be considered uniformly distributed between $-\frac{1}{2}$ and $\frac{1}{2}$ in units of the last figure.

b. The Exponential Distribution. The exponential probability function (see Fig. 12) is defined by

$$
\begin{aligned}
f(x) &= 0 && x < 0 \\
&= \lambda e^{-\lambda x} && 0 < x < \infty
\end{aligned}
\tag{84}
$$

The mean and the standard deviation of x are both equal to $1/\lambda$.

The life of a radioactive atom is distributed according to Eq. (84). Also, the time between two occurrences of a random variable described by the Poisson process given in paragraph 12b is distributed according to Eq. (84).

c. The Normal or Gaussian Distribution. The normal probability function (see Fig. 13) is given by

$$
f(x) = \frac{1}{\sqrt{2\pi}\,\sigma} \, e^{-(x-\mu)^2/2\sigma^2}
\tag{85}
$$

The mean of x is μ, and the standard deviation of x is σ. If the variable x is standardized by the transformation

$$\frac{x - \mu}{\sigma} = z \tag{86}$$

then

$$f(z) = \frac{1}{\sqrt{2\pi}} e^{-z^2/2} \tag{87}$$

The mean of z is zero, and its standard deviation is 1. The moment-generating functions of x and z are

$$\psi(\theta) = e^{\mu\theta + (\theta^2\sigma^2/2)} \tag{88}$$
$$\psi(z) = e^{\theta^2/2} \tag{89}$$

The areas included in the intervals $\mu \pm k\sigma$, for $k = 1, 2, 3$ were given in paragraph 10.

d. The Gamma Distribution. The gamma probability function (see Fig. 14) is given by

$$f(x) = \frac{1}{\beta^{\alpha+1}\Gamma(\alpha + 1)} x^\alpha e^{-x/\beta} \qquad \alpha > -1, \qquad \beta > 0 \qquad 0 \leq x < \infty \tag{90}$$

The moment-generating function of x is

$$\psi(\theta) = \frac{1}{(1 - \beta\theta)^{\alpha+1}} \qquad \theta < \frac{1}{\beta} \tag{91}$$

The mean of x is $\beta(\alpha + 1)$, and the standard deviation of x is $\beta \sqrt{\alpha + 1}$.

This function is skewed to the right for all the values of α and β. As α increases, the skewness becomes less pronounced. Extensive tables of this distribution were published in 1922 by the office of *Biometrika* under the editorship of Karl Pearson.

The numerical value v of the velocity of a molecule with mass m in a gas at absolute temperature T is distributed according to a probability function of the type (90). In fact

$$f(v) = \alpha v^2 e^{-\beta v^2}$$

with $\beta = m/2kT$, in which k is the physical constant called Boltzmann's constant.

14. The Central-limit Theorem

Consider the sum

$$x = \sum_1^n x_i$$

of n independent random variables all having the same probability distribution with mean μ and standard deviation σ.

As indicated previously (see paragraph 11a)

$$E(x) = n\mu$$
$$\sigma_x = \sqrt{n}\, \sigma$$

Form the standardized variable

$$z = \frac{x - n\mu}{\sqrt{n}\, \sigma} \tag{92}$$

The *central-limit* theorem states that, if σ is finite, as n goes to infinity, the random variable z will be distributed according to the normal function (87) with mean 0 and standard deviation 1. The sum $x = \Sigma x_i$ is said to be asymptotically normal with mean $n\mu$ and standard deviation $\sqrt{n}\, \sigma$.

As a corollary of the central-limit theorem, it follows that

$$\bar{x} = \frac{\Sigma x_i}{n}$$

is distributed in the limit according to the normal distribution with mean μ and standard deviation $\sigma/\sqrt{n}$.

If all the variables x_i have different distributions with means μ_i and standard deviations σ_i, it is no longer sufficient to assume that each x_i has a finite second-order moment in order to show the validity of the central-limit theorem.   An additional condition is then necessary in order to reduce the probability that an individual x_i will yield a relatively large contribution to the total value of the sum x.

It can be shown[†] that, if the sum of the third absolute moment of x_i about its mean is small in comparison with the sum of the variances of the x_i, then the central-limit theorem will still be valid for the case of a sum of independent random variables x_i that are not supposed to be equally distributed.

If all the variables x_i are known to be normally distributed, then the sum will also be normally distributed with mean $n\mu$ and standard deviation $\sqrt{n}\,\sigma$.

   a. Shot Noise.   As an application of the central-limit theorem one can illustrate the distribution of the current $I(t)$ in shot noise.

Suppose that the arrival of an electron at the anode of a vacuum tube at time $t = 0$ produces an effect $F(t)$ at some point in the output circuit.   If the output circuit is such that the effects of the various electrons add linearly, the total effect at time t due to all electrons is

$$I(t) = \sum_{-\infty}^{\infty} F(t - t_K)$$

Assume that the probability of an electron arriving at the anode in the interval $(t, t + \Delta t)$ is αt, where Δt is such that $\alpha\,\Delta t \ll 1$, and that this probability is independent of what has happened before time t or will happen after time $t + \Delta t$. α is the average number of electrons per second.

As shown in paragraph 12b, the number of electrons in time t will be distributed according to the Poisson distribution (74).

It can be shown[‡] that

$$\begin{aligned} E[I(t)] &= \alpha \int_{-\infty}^{\infty} F(u)\,du \\ \sigma_I{}^2 &= \alpha \int_{-\infty}^{\infty} F^2(u)\,du \end{aligned} \tag{93}$$

Since $I(t)$ is given by a sum of independently distributed random variables with finite standard deviation, then, according to the central-limit theorem, $I(t)$ will be distributed according to the normal distribution with mean and standard deviation given by Eqs. (93).

15. The Principle of Least Squares

Let there be a set of values of x and y such that

$$y_i = \alpha + \beta x_i + \epsilon_i \tag{94}$$

where α and β are unknown constants and ϵ_i are random errors with mean equal to zero and standard deviation σ.

[†] See Ref. 5, Chap. 17.
[‡] S. O. Rice, Mathematical Analysis of Random Noise, *Bell System Tech. J.*, vol. 24, p. 46, 1945.

Form the quadratic error function

$$S = \sum_{i=1}^{u} (y_i - \alpha - \beta x_i)^2 = \sum_{i=1}^{u} \epsilon_i^2 \tag{95}$$

The principle of *least squares* states that the "best" estimates of α and β are those values that minimize the sum of the square of the deviations S.

These estimates are

1. Linear in the observations y
2. Unbiased
3. Of minimum variance

An estimate $\hat{\theta}$ of a parameter θ is unbiased if $E(\hat{\theta}) = \theta$. Property 3 underlies the meaning of the word "best" in above statement. It means that any other linear and unbiased estimates of α and β will have a larger variance than those estimates given by the principle of least squares.

It can be shown that by minimizing S one obtains

$$\hat{\alpha} = \bar{y} - \hat{b}\bar{x}$$
$$\hat{\beta} = \frac{\Sigma y_i (x_i - \bar{x})}{\Sigma (x_i - \bar{x})^2} \tag{96}$$

The standard error of estimate is given by

$$\sigma^2_{y/x} = \frac{\min S}{n-2} = \frac{\Sigma (y_i - \hat{\alpha} - \hat{\beta}x_i)^2}{n-2} \tag{97}$$

The standard error of the slope $\hat{\beta}$ is given by

$$\sigma\hat{\beta} = \frac{\sigma_{y/x}}{\sqrt{\Sigma (x_i - \bar{x})^2}} \tag{98}$$

and the standard error of $\hat{\alpha}$ is given by

$$\sigma\hat{\alpha} = \sigma_{y/x} \sqrt{\frac{1}{n} + \frac{\bar{x}^2}{\Sigma (x_i - \bar{x})^2}} \tag{99}$$

If, in addition, one knows that the variables y_i are normally distributed, then $\hat{\alpha}$ and $\hat{\beta}$ will be jointly normally distributed with means α and β, respectively, and standard deviations given by Eqs. (98) and (99).

WAITING LINES AND SIMULATION

16. Introduction

Waiting-lines problems belong to a more general kind of random processes referred to as "stochastic processes." A process is stochastic if the random variable that describes it depends on a nonrandom parameter such as time. If the laws governing the arrival of units, their service time, and the order in which the units are taken into service are known, then the characteristics of this stochastic process can be analyzed mathematically.

In the development of a mathematical model for a waiting-line problem, the following characteristics must be known:

a. Arrival Rate and Probability Distribution of Arrivals. *Poisson Arrivals.* If the probability of an arrival during the time t, $t + dt$ is proportional to the length of the interval, say $\lambda\, dt$, and arrivals are statistically independent, then the probability of x arrivals in time 0, t is given by the Poisson distribution described in paragraph

12b. The average number of arrivals will then be equal to λt see Eq. (75), paragraph 12b where λ is the number of arrivals per unit of time, or the *mean arrival rate.* If arrivals are distributed according to the Poisson distribution, then the time between arrivals will be described by the exponential distribution (84), paragraph 13b. The expected time between two consecutive arrivals is then equal to $1/\lambda$.

Non-Poisson Arrivals. Arrivals can be described by distributions other than the Poisson. For instance, when the number of units in the population of arrivals is finite, then the arrival rate λ will not be constant and will, of course, depend on the number of units left in the calling population. This situation develops in waiting-line problems of machine breakdowns where the number of machines in operation or being serviced in the repair shop is finite. If λ represents in this case one breakdown per machine per unit of time, then $N\lambda$ will be arrival rate when no units are in service or waiting in a population of N machines and $(N - n)\lambda$ the arrival rate when n units are in service or waiting.

b. Service Rate and Probability of Services. *Poisson Services.* If the probability of a unit being served in time $(t, t + dt)$ is also proportional to dt, that is, equal to $\mu\, dt$, and services are independent, then the probability of x units being served in time t will also be described by a Poisson distribution, with μ being the mean service rate. The time between services will then be exponentially distributed, and $1/\mu$ will be mean time between services.

Non-Poisson Services. *The Erlang Family of Service Time Distribution.* The Erlang distribution can be considered as a generalization of the exponential distribution and is actually a special case of the gamma distribution described in Eq. (96), paragraph 13d. If k is an integer and μ a real positive constant, the Erlang distribution is given by

$$f(t;\mu,k) = \frac{(\mu k)^k}{(k - 1)!}\, t^{k-1}e^{-k\mu t} \qquad t \geq 0 \tag{100}$$

It can be seen that the mean of t is $1/\mu$ for all values of k and the standard deviation is $1/(\sqrt{k}\mu)$. The mode, or maximum of t, is located at

$$t = \frac{k - 1}{\mu k}$$

When $k = 1$, expression (100) becomes the exponential law with its maximum value at the origin. As k increases, the mode moves to the right, near the mean value $1/\mu$. In fact, as $k \to \infty$, the mode is equal to $1/\mu$. In the meantime, the standard deviation becomes smaller as k increases. In the limit, as $k \to \infty$, the standard deviation is equal to zero. In this case, the distribution is concentrated at $t = 1/\mu$, with a variability equal to zero. This limiting case can be considered as a constant service time with an average service time equal to $1/\mu$.

c. Waiting-line Discipline. The waiting-line model will depend on the manner in which the elements are served. Units may be served in order of arrival, entirely at random, or according to a given set of priorities. Models can also be set up to account for situations where the units leave the line if the waiting time is larger than a certain value $(w \geq w_0)$ or if units decide not to stay in line if there is any waiting at all $(w > 0)$.

d. Number of Servers. The number of servicing stations can be one or some number k. Situations may arise where the number k is large enough that the waiting time is negligible and only the level of server occupancy is of analytical interest.

e. The Number of Stages of Servicing. In many systems the units are processed through several stages before they leave the line. The departure patterns of a given stage become the arrival pattern of the next stage.

f. Size of the Population. If the units come from a very large population, then the arrivals and service times will have a certain average value. If the population is finite, then the arrival rate is not constant and the average waiting time or the average number of units in the line will depend on the size of the population.

g. Measures of System Performance. The characteristics of the waiting-line problem can be expressed in several ways. Examples of these are the average length of the waiting line, the average length of nonempty lines, the average number of units in the system (waiting or being served), the average waiting time of an arrival, the average waiting time of an arrival who waits, and the average time an arrival spends in the system. If the probability distribution of waiting time is derived from the basic assumptions governing the system, then one can also determine the proportion of arrivals that will wait longer than some specified critical time.

Economic considerations may lead to the determination of the optimum number of servicing stations that will minimize cost. Besides the cost of operation of the system, one must, of course, know the cost of waiting, which sometimes is not readily available and is difficult to evaluate.

h. Transients and Steady-state Conditions. A general assumption will be made that the levels of arrival and service are maintained in a steady state so that the probability of n elements in line will be independent of time. In many problems, either the arrival rate or the service rate may change from period to period. One can then break up the total period into a set of steady-state conditions, with a set of transitional states of system loading or unloading. In this chapter, a review will be made of waiting-line models in a steady state, disregarding the transient state, when the probabilities have not yet settled down to their steady-state values, or the explosive state, when delays tend toward infinity.

17. Basic Equations Governing a Waiting Line—Single Service Station

Assuming Poisson arrivals and services, the waiting-line model for a single service station can be formulated as follows:

Consider an interval of length $t + dt$. The probability of zero elements in the line in this interval, denoted by $P_0(t + dt)$, can be written as a function of the probability of zero and one element in time t, $P_0(t)$ and $P_1(t)$, respectively, as follows:

$$P_0(t + dt) = P_0(t)(1 - \lambda\,dt) + P_1(t)(\mu\,dt) \qquad (101)$$

Expression (101) states that, in order to have zero elements in the interval $(0, t + dt)$, one must have zero elements in the interval $(0,t)$ and no arrivals in the interval of length dt or one element in the interval $(0,t)$ and one unit being served in time dt. The rule of multiplication of independent probabilities and addition of mutually exclusive cases (see paragraph 6) are used to formulate expression (101).

In the same manner, the probability of n elements in line during the time $(0, t + dt)$ can be written as a function of the probabilities of n, $n - 1$, and $n + 1$ elements in line, $P_n(t)$, $P_{n-1}(t)$, and $P_{n+1}(t)$, respectively, as follows:

$$P_n(t + dt) = P_n(t)(1 - \lambda\,dt)(1 - \mu\,dt) + P_{n-1}(t)(\lambda\,dt)(1 - \mu\,dt)$$
$$+ P_{n+1}(t)(1 - \lambda\,dt)\mu\,dt \qquad (102)$$

That is, there will be n elements in line during the time $0, t + dt$ if there are already n elements in time $(0,t)$, none arrives, and none is served in time dt or there are $n - 1$ elements in time $(0,t)$, one arrives, and none is served in time dt or there are $n + 1$ elements in time $(0,t)$, none arrives, and one is served in time dt. Expressions (101) and (102) are within infinitesimals of order $(dt)^2$ or higher.

Expressions (101) and (102) can be written in the form

$$\frac{P_0(t + dt) - P_0(t)}{dt} = -P_0(t)\lambda + P_1(t)\mu$$

$$\frac{P_n(t + dt) - P_n(t)}{dt} = -P_n(t)(\lambda + \mu) + P_{n-1}(t)\lambda + P_{n+1}(t)\mu$$

As $dt \to 0$, the left-hand side of the above expressions tends to $P_0'(t)$ and $P_n'(t)$, respectively. Since it has been assumed that the system is in a steady state, both derivates are equal to zero and $P_n(t)$, for any value of n, will be independent of the time t. The above equations then become

$$P_1 = \frac{\lambda}{\mu} P_0$$

$$P_{n+1} = \frac{\lambda + \mu}{\mu} P_n - \frac{\lambda}{\mu} P_{n-1} \qquad n \geq 1 \tag{103}$$

Using induction and the condition $\sum_0^\infty P_n = 1$, the system of equations (103) yields, for the probability of n elements in line, the expression

$$P_n = \left(1 - \frac{\lambda}{\mu}\right)\left(\frac{\lambda}{\mu}\right)^n \qquad n \geq 0 \tag{104}$$

The ratio λ/μ must be smaller than 1 in order for Eq. (104) to give meaningful results.

The probability distribution of the waiting time of an arrival before taken into service can be obtained by using expressions (104) and (78). In fact, the probability of waiting in the interval $(w, w + dw)$ is made up of three events: n elements in the system just before the arrival [with a probability given by Eq. (104)], $(n - 1)$ elements served in the time $(0,w)$ [with a probability given by Eq. (78) with μw as a parameter], and one element being served in time $w, w + dw$ (with a probability given by $\mu\, dw$). Since this is true for all values of n from 1 to infinity, the probability distribution of the waiting time $P(w)$ will be given by

$$P(w)\, dw = \sum_{n=1}^\infty \left(1 - \frac{\lambda}{\mu}\right) \left(\frac{\lambda}{\mu}\right)^n \frac{e^{-\mu w}(\mu w)^{n-1}}{(n-1)!}\, w\, dw$$

$$= \lambda\left(1 - \frac{\lambda}{\mu}\right) e^{-(\mu-\lambda)w}\, dw \qquad w > 0$$

$P(w)$ is discontinuous at the origin, since the probability of no waiting, $w = 0$, is the same as the probability of no elements in line, $P(n = 0)$. Hence we have

$$P(w) = 1 - \frac{\lambda}{\mu} \qquad\qquad w = 0$$

$$= \lambda\left(1 - \frac{\lambda}{\mu}\right) e^{-(\mu-\lambda)w} \qquad w > 0 \tag{105}$$

The average number of units in the system $E(u)$ is obtained from Eq. (104).

$$E(n) = \frac{\lambda}{\mu - \lambda} \tag{106}$$

The average waiting time of an arrival is obtained from Eq. (105).

$$E(w) = \frac{\lambda}{\mu(\mu - \lambda)} \tag{107}$$

The average time in the system $E(v)$ will be the sum of the average waiting time [Eq. (107)] and the average servicing time, which is equal to $1/\mu$ (the average interval between services).   Hence

$$E(v) = \frac{\lambda}{\mu(\mu - \lambda)} + \frac{1}{\mu} = \frac{1}{\mu - \lambda} \tag{108}$$

The average length of the waiting line $E(l)$ can be estimated from

$$E(l) = \lambda E(w) = \frac{\lambda^2}{\mu(\mu - \lambda)} \tag{109}$$

$$\left[E(l) \text{ can also be found from } \sum_{n=2}^{\infty} (n - 1)P_n, \text{ where } P_n \text{ is given by Eq. (104).} \right]$$

The average length of nonempty lines will be given by the conditional expectation of l given that l is greater than zero:

$$E(l/l > 0) = \frac{\mu}{\mu - \lambda} \tag{110}$$

The average waiting time of an arrival who waits is given by

$$E(w/w > 0) = \frac{1}{\mu - \lambda} \tag{111}$$

18. Multiple Service Stations.   Poisson Arrivals and Services

Considerations similar to the ones described in paragraph 17 can be used to determine the probability distributions of waiting lines for k servicing stations.   Expression (102) must be considered separately for the case of $n < k$, when the service rate will be $n\mu$, and $n > k$, when the service rate is $k\mu$.

The probability of n elements in line will then be given by

$$
\begin{aligned}
P_n &= \frac{1}{n!} \left(\frac{\lambda}{\mu} \right)^n P_0 & n < k \\
&= \frac{1}{k! k^{n-k}} \left(\frac{\lambda}{\mu} \right)^n P_0 & n \geq k
\end{aligned}
\tag{112}
$$

P_0 is evaluated using the condition $\sum_{0}^{\infty} P_n = 1$.

The probability that an arrival has to wait can be computed from

$$P(n > k) = \sum_{n=k}^{\infty} P_n$$

We have

$$P(n > k) = \left(\frac{\lambda}{\mu} \right)^k \frac{P_0}{k!(1 - \lambda/\mu k)} \tag{113}$$

Above expressions are valid if $\mu k > \lambda$, that is, if the service rate for the k stations is larger than the arrival rate.

The probability that the total time spent in the system v is larger than a time t is given by

$$P(v > t) = e^{-\mu t} \left(1 + \frac{P(n \geq k)}{k} \left\{ \frac{1 - e^{-\mu k[1-(\lambda/\mu)k - (1/k)]}}{1 - (\lambda/\mu k) - 1/k} \right\} \right) \tag{114}$$

Formulas for the average length of the waiting line, the average number of units in the system, the average waiting time, and the average time spent in the system are given below.

$$E(l) = \frac{\lambda\mu(\lambda/\mu)^k}{(k-1)!(k\mu-\lambda)^2}P_0 \tag{115}$$

$$E(n) = E(l) + \frac{\lambda}{\mu} \tag{116}$$

$$E(w) = \frac{\mu(\lambda/k)^k}{(k-1)!(k\mu-\lambda)^2}P_0 \tag{117}$$

$$E(v) = E(w) + \frac{1}{\mu} \tag{118}$$

Nomographs are available for the computation of these averages. Figure 15† shows a graph of λ/μ (A/S in the graph) for various n's, the number of service stations.

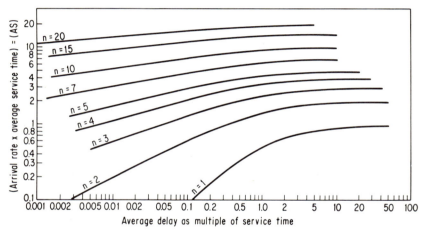

FIG. 15. Average delay for service time exponentially distributed. (*By permission of Richard D. Irwin and Bell System Technical Journal.*)

The abscissa gives the average delay as a multiple of the average service time [$w/(1/\mu) = w\mu$]. Thus, for $\lambda/\mu = A/S = 1$ and $n = 3$, $w\mu = 0.045$ and $w = 0.045/\mu$

19. Waiting Lines with Poisson Arrivals and Constant Service

a. Single Service Station. Using the Erlang family for the limiting case when $k \to \infty$, it is possible to obtain a probability distribution of waiting times and number of elements in line when the arrivals are described by a Poisson distribution and the service is constant.

The average waiting time, the average waiting time of an arrival that waits, and the

† T. A. Mangelsdorf, Waiting Line Theory Applied to Manufacturing Problems in E. H. Bowman and R. B. Fetter (eds.), *Analysis of Industrial Operations*, Richard D. Irwin, Inc., Homewood, Ill., 1959. Information for this chart was obtained from curves published by Edward C. Molina, Application of the Theory of Probability to Telephone Trunking Problems, *Bell System Tech. J.*, vol. VI, 1927, and from work at MIT.

average number of elements in line are given by[†]

$$E(w) = \frac{\lambda}{2\mu(\mu - \lambda)} \tag{119}$$

$$E(w/w > 0) = \frac{1}{2(\mu - \lambda)} \tag{120}$$

$$E(n) = \frac{\mu}{2(\mu - \lambda)} \tag{121}$$

b. Several Service Stations. Formulas are also available for the probability of waiting, $E(w)$, $E(w/w > 0)$, and $E(n)$. They are lengthy and will not be reproduced here. Graphs are available for the computation of these values, and the reader is referred to literature mentioned previously.

Solutions are available for other models with various priority systems in handling the arrivals, waiting-line behavior, and different arrivals and service time probability distributions.[‡]

20. Waiting Lines with Finite Calling Populations

When the number of elements in the population is finite, expression (101) must be modified to account for the fact that the arrival rate is not constant. Fortunately, tables[§] are available for the computation of the average waiting time as a function of the number of elements in the population, for the case of Poisson arrivals and constant and exponential service times.

An interesting illustration of the use of waiting lines with finite calling population is the operator assignment problem.[‖]

Let
$\quad s$ = service time per time unit $(1/\mu)$
$\quad m$ = machine running time per time unit $(1/\lambda$, that is, average time between breakdowns, or arrivals to waiting line)
$\quad w$ = waiting time per time unit
$\quad N$ = number of machines assigned to an operator
$\quad C_r$ = cost of operator per time unit
C_s, C_m, C_w = costs associated with s, m, w, respectively, per machine
$\quad C$ = total cost per time unit per machine

Then the average value of C will be given by

$$C = mC_m + sC_s + wC_w + \frac{C_r}{N}$$

Dividing the above expression by m, we have, with $C_0 = C/m$,

$$C_0 = C_m + \frac{s}{m} C_s + \frac{WC_w + C_r/N}{m} \tag{122}$$

Given $k = s/m = \lambda/\mu$ and the various costs, one can obtain C_0 as a function of N. The tables referred to above give the values of m and W for different ratios k and

---

† See J. R. Shelton, Solution Methods for Waiting Line Problems, *J. Ind. Eng.*, vol. XI, no. 4, 1960. Information for these formulas were obtained from Arne Jansen, The Life and Worries of A. K. Erlang, *Trans. Danish Acad. Sci.*, no. 2, 1948, and other papers referred in the above article.

‡ See Philip M. Morse, *Queues, Inventories and Maintenance*, John Wiley & Sons, Inc., New York, 1958, and T. L. Saaty, Résumé of Queuing Theory, *Operations Res.*, vol. V, pp. 161–200, 1957.

§ Edward W. Bowman and Robert B. Fetter, *Analysis for Production Management*, Chap. 9, Richard D. Irwin, Inc., Homewood, Ill., 1961.

‖ Robert B. Fetter, The Assignment of Operators to Service Automatic Machines, *J. Ind. Eng.*, pp. 22–29, September–October, 1955.

different values of N. By computing C_0 as a function of N one can obtain the optimum value of k for minimum cost.

21. Simulation of Random Processes—The Monte Carlo

The behavior of engineering systems can be readily analyzed once a mathematical model is established showing the relation among dependent and independent variables. However, in a great variety of problems, mathematical models are not readily available for the description of the relationship among the factors or variables that are assumed to have an effect in the solution of the problem. Even if the problem is well defined, the variables that are assumed to influence the final solutions are known, and the measure of the effectiveness of the system (measure of optimization) is also well defined, the mathematical nature of the various relations or interactions among the variables may be unknown or extremely difficult to establish.

Trial-and-error methods can sometimes be used for the solution of these kinds of problems. The solutions thus obtained are not always optimum, and it might be necessary to repeat the computations many times before a solution "near" the optimum may be achieved. For situations such as these, simulation methods have been found to provide an extremely useful and flexible way, not only to define the problem properly but also to give the optimum solution being sought.

According to Webster, to simulate is to "assume the mere appearence of, without the reality." Once the problem is defined, that is, the variables that are assumed to have an effect are established, the general nature of the relation among these variables is described, and the objective function is defined, simulation will attempt to duplicate the system by means of paper, pencil, desk calculators, computers, etc. In some instances, human beings, following certain ground rules, are also used to simulate a system. Certain kinds of management games are now being used where people, following a set of policies and procedures, make decisions which are integrated with other aspects of simulation in an attempt to provide a solution to problems that are not subject to mathematical treatment.

Simulation methods involving a combination of sampling theory and numerical analysis have been used to solve a great variety of mathematical problems.[†] When probability distributions of random variable are explicitly included in the analysis, the simulation is generally referred to as Monte Carlo simulation.

a. Nature of the Monte Carlo. Let there be a random process described by a random variable x whose probability density function is $f(x)$. We shall assume x to be continuous, although the argument carries through for discrete distributions as well.

Let $F(x)$ be the cumulative probability distribution

$$F(x) = \int_{-\infty}^{x} f(t)\ dt$$

The steps in the Monte Carlo simulation of a random process are as follows:

1. Generate a random number between zero and 1, with as many decimals as desired.

2. Plot this random number in the vertical axis of $F(x)$; project it horizontally until the projection intersects the graph of $F(x)$ (see Fig. 10).

3. Obtain the value of x corresponding to this intersection. This value of x is taken to be a random selection from the random process described by $f(x)$.

It should be noticed that, since $F(x)$ is a monotone nondecreasing function, there will be a one-to-one correspondence between the random digit plotted in the y axis and the random value of x obtained in the x axis. If the graph of $F(x)$ were a straight line, which will be the case when x is uniformly distributed, then the frequency of the occurrence of the generated random numbers will agree with the frequency of the occurrence of the x values. This should, of course, be the case, since the distribution of both random numbers is the same. However, when $F(x)$ is not a straight line,

† George W. Brown, Monte Carlo Methods in *Modern Mathematics for the Engineer*, E. F. Beckenbach (ed.), McGraw-Hill Book Company, New York, 1956.

the frequency in which the x values occur will agree with the frequency given by the probability density $f(x)$, with the order of occurrence being completely randomized.

Generation of Random Numbers. Several methods are available for the generation of random numbers. Tables of random digits[†] are available which are convenient to use when the simulation process is performed "by hand." For large-scale problems, random numbers are generated in computers using various schemes described in the literature.[‡] A method that has been found to be very efficient, the "multiplicative congruential method," is one where numbers are generated according to the rule

$$r_1 = 1$$
$$r_{i+1} = \alpha r_i \ (\text{mod } M)$$

where α and M are numbers that depend on the computer configuration. With α and M written as powers of 2, the numbers generated consist of zeros and 1's, which turn out to be binary fractions uniformly distributed between 0 and 1.

Description of $F(x)$. The random variable x may be described by distributions such as the exponential, binomial, Poisson, normal, etc. If the simulation is done in computers, then the cumulative distribution function is described mathematically and the relation between the generated random number and x is also bound mathematically. For instance, if $f(x) = ae^{-ax}$, the exponential distribution, then

$$F(x) = 1 - e^{-ax}$$

and if y represents a random number between 0 and 1 then

$$x = \frac{1}{a} \ln \frac{1}{1-y} \tag{123}$$

will represent a random selection from $f(x)$. Similar relations can be found for other commonly used probability distributions.[§]

In many problems, an empirical cumulative distribution is obtained from actual data and is subsequently used to find the value of x that corresponds to the randomly generated digit y.

When several probability distributions must be used for the analysis of the system, each one should be generated according to the general rules described above.

In some instances, only the expected value of x is used for the analysis. In these cases, the Monte Carlo technique is not necessary, since the usefulness of this method is based on the fact that the sampling is over the complete distribution, which brings not only the expected value of x but also the variability of x around this expected value.

Reliability of Results. The reliability of the results of simulation will depend on the success of introducing all the important and relevant characteristic of the actual problem into the model used to describe the system. Use of high-speed computers makes it possible to simulate several years of actual operations of a system into a few minutes or hours of computer time. Thus it is possible to introduce changes in some of the parameters of the system and observe their effect in the final solution in a very short period of time. In general, the question of accuracy provided by the Monte Carlo is not simple, and in each case the problem should be studied carefully to discover how apparent estimates may be strengthened.[‖]

Reduction of variability of the results is of paramount importance in the simulation techniques. By improving in the method of sampling, it is possible to achieve this reduction of variability. If the variability of the results obtained (the standard

---

[†] Rand Corporation, *A Million Random Digits*, The Free Press of Glencoe, New York, 1955. Fisher and Yates, *Statistical Tables for Biological, Agricultural and Medical Research*, Oliver & Boyd, Ltd., London, 1943.

[‡] IBM Random Number Generation and Testing, *Symposium in Monte Carlo Methods*, John Wiley & Sons, Inc., New York, 1956.

[§] Notes on Operations Research, chap. II in H. P. Galliher, *Simulation of Random Processes*, The M.I.T. Press, Cambridge, Mass., 1959.

[‖] See Ref. 4 pp. 149–151.

deviation) is computed, then confidence interval estimates of parameters evaluated by the Monte Carlo should be given.

b. Monte Carlo Solution of Waiting-line Problems.   Monte Carlo methods are extremely useful for solving waiting-line problems when the relation among arrivals, services, and queue discipline are very complex or perhaps impossible to analyze mathematically.   Examples of these situations are empirical distributions of arrivals and services that do not fit probability models, complicated priority rules for serving the queue, multistage queueing where the departure from one waiting line from the arrivals for the next waiting line, etc.

The following example[†] of a three-stage queueing will illustrate the use of the Monte Carlo method.

Arrivals are Poisson distributed, with an average interval between arrivals of 20 min ($\lambda = 1/20$).   Services for the three stages are distributed as follows:

Stage 1: Normal mean 10 min and $\sigma = 5$ min
Stage 2: Exponential, with $\mu = 1/15$ services per minute
Stage 3: Constant, at 15 min

The arrivals and service at stage 2 are simulated using random digits and a relation such as Eq. (120).   The normal service of stage 1 is simulated using a table of random normal numbers.[‡]   A random normal number is changed to service time using the relation

$$z = \frac{x - 10}{5}$$

where z is a sample from the normal distribution with mean zero and standard deviation 1.

Simulating a sufficiently large number of arrivals, one estimates the time that each unit spends in the system by following the rule that the unit proceeds immediately into a service stage if it is unoccupied and that otherwise it waits until the service station is free.   By dividing the total time that all the units spend in the system, one then calculates the average total time in the system, and by subtracting from this average total time the average service time for each stage (10, 15, and 15, respectively), one estimates the average waiting time.   For a complete discussion of this problem the reader is referred to Ref. 4.

COMMUNICATION AND INFORMATION THEORY

Communication is basically any form of transmission of information between two sources.   Communication problems are of a statistical nature, since the signal transmitted is known only through the probability distribution of its amplitude, which in turn is determined by the statistical nature of the source of transmission.   This fact was fully realized by Norbert Wiener,[6] who stated that for the existence of a message it is essential that variable information be transmitted and that one should have a repertory of possible messages, with a measure for determining the probability distribution of these messages.   Claude Shannon[7] in his pioneering work in the theory of communication formalized the various aspects of information theory, with especial emphasis on electrical engineering problems.   R. A. Fisher[§] introduced the concept of "amount of information" contained in a sample regarding a certain statistical parameter.[‖]   Some basic ideas of the concept of information can also be found in the early work of Hartley.[¶]

[†] See Ref. 4, pp. 149–151.
[‡] Rand Corporation, *loc. cit.*
[§] R. A. Fisher, On the Mathematical Foundations of Theoretical Statistics, *Phil. Trans. Roy. Soc. London*, ser. A, vol. 222, 1922.
[‖] It can be shown that this "amount of information" is related to the second derivative of the entropy function. See S. S. Wilks, *Mathematical Statistics*, p. 409, John Wiley & Sons, Inc., New York, 1962.
[¶] R. V. L. Hartley, Transmission of Information, *Bell System Tech. J.*, July, 1928.

As pointed out by Shannon, the semantic aspects of the messages to be transmitted are irrelevant to the concept of information. Information, as communicated by means of a certain language, symbols, electrical waves, etc., is related to the difficulty of selecting a message from a possible ensemble of messages produced by the information source.

22. Measure of Information

A *unit of information* will be defined as a selection between two equally likely choices. This unit will be called a "bit."

Symbol	Probability	Bits
A_1	1/2	0
A_2	1/2	1

Thus, with 1 bit (zero or 1) one could select either A_1 or A_2. The number of bits per symbol is then 1. Introducing H as a symbol for the amount of information, or number of bits, per symbol, we have

$$H = \log_2 2 = 1 \text{ bit} \qquad (124)$$

N Equally Likely Choices. Assume a source with $N = 8$ symbols all equally likely to occur. The amount of information of this source can be determined by applying the unit of information as many times as it is necessary to identify uniquely any symbol of the source. If the eight symbols of the source are successively subdivided into equally likely subgroups, it can be seen that 3 bits will be necessary for this identification.†

Symbol	Probability	Bits
A_1	1/8	000
A_2	1/8	001
A_3	1/8	010
A_4	1/8	011
A_5	1/8	100
A_6	1/8	101
A_7	1/8	110
A_8	1/8	111

Thus, in the first subdivision, there will be four symbols in each subgroup, the elements in the first subgroup being identified with zero and those of the second subgroup with a 1. In the second subdivision, the elements of the first subgroup are divided into two equally likely subgroups which can again be identified with a zero or a 1. Continuing in this manner, it is seen that three bits uniquely identify any symbol from A_1 to A_8. In this case, we have

$$H = \log_2 8 = 3 \text{ bits}$$

In general, for N being any power of 2, then

$$H = \log_2 N \qquad (125)$$

It can be shown that the above relation also holds true, in the limit, for any value of N.

† This argument was used by Robert M. Fano in "The Transmission of Information," *MIT Res. Lab. Electron. Tech. Rept.* 65, March, 1949.

Not Equally Likely Choices. Assume a source of four symbols with the probabilities $p(i)$ given in the following table:

Symbol	$p(i)$	Bits	Number of bits (b)	Expected number
A_1	1/2	0	1	½
A_2	1/4	10	2	½
A_3	1/8	110	3	⅜
A_4	1/8	111	3	⅜
				$\Sigma = 1\frac{3}{4}$

Proceeding as before, the symbols are divided into groups of equal probability and bits are assigned to each subgroup following the same procedure described above. Thus, since A_1 has the same probability as the group A_2, A_3, A_4, a zero will be assigned to A_1 and a 1 to the symbols in the other group. In the second subdivision A_2 has the same probability as the subgroup A_3, A_4 and consequently a zero will be assigned to A_2 and a 1 to the subgroup A_3, A_4. Finally, A_3 and A_4 are identified by an additional bit.

Now the average number of bits per symbol H is

$$H = E(b) = \sum_{i=1}^{4} b_i p(i) = 1 \times \tfrac{1}{2} + 2 \times \tfrac{1}{4} + 3 \times \tfrac{1}{8} + 3 \times \tfrac{1}{8} = 1\tfrac{3}{4}$$

In this particular case, the number of bits associated with a symbol is in direct relation with the probability of that symbol. One can then write

$$b_i = - \log_2 p(i)$$

Substituting $- \log_2 p(i)$ for b_i, the value of $E(b)$ can be written in the form

$$H = - \sum_{i=1}^{4} p(i) \log_2 p(i) \tag{126}$$

A similar expression for H can be found for sources described by symbols whose probabilities of occurrences are of the form $1/2^k$.

For a source described by a general type of discrete probability distribution it is possible to show† that the expression of H for the average number of bits per symbol is of the same form as Eq. (126); i.e.,

$$H = - \sum_{i=1}^{k} p(i) \log p(i) \tag{127}$$

That proposition is proved under the assumption that H satisfies the following requirements:

1. H should be continuous in the $p(i)$.
2. If $p(i) = 1/k$ for all i, then H should be a monotonic increasing function of k.
3. If a choice is broken down into successive choices, the original H should be the weighted sum of the individual values of H.

Not Equally Likely Messages. Assume a source described by messages of length n that are made up of symbols $A_1, A_2, \ldots, A_k$ whose probabilities of occurrence are $p(1), p(2), \ldots, p(k)$, respectively. For instance, using the symbols A_1, A_2, A_3, A_4

† See Shannon, Ref. 7, pp. 18, 19.

described above, one could make up 16 messages of length two: (A_1A_1), (A_1A_2), . . . , (A_4A_3), (A_4A_4). The probabilities of these messages will be

$$P(A_iA_j) = p(A_i)p(A_j)$$

assuming independent selections.[†]

By ranking all these messages according to their probabilities and by following the above method of subdivision, it is possible to show[‡] that

$$H = -\sum_1^k p(i) \log_2 p(i) = \lim_{n \to \infty} -\frac{1}{n} \sum_1^n P(i) \log_2 P(i) \qquad (128)$$

where $P(i)$ is the probability of a message of length n made up of k symbols with probabilities $p(i)$.

In general, for messages of length n, when n is sufficiently large, Shannon[7] shows that in a typical message there will be, with a high probability of occurrence, p_1n occurrences of the first symbol, p_2n occurrences of the second symbol, etc., so that the probability associated with a typical message is, approximately,

$$P = p_1{}^{np_1} p_2{}^{np_2} \cdots p_n{}^{np_n} \qquad (129)$$

Messages whose probabilities are not given by Eq. (125) form a set of very small total probability.

Thus, for large n, most of the messages have the same probability P as given by Eq. (128). The total number of messages of length n is thus roughly $1/P$ (the sum of the probabilities of all messages of length n must add up to 1). Then the value of H for messages of length n is $H_n = \log_2 1/P$ [see Eq. (125)]. Hence, the average information per symbol is

$$H = \frac{1}{n} H_n = \frac{1}{n} \log_2 \frac{1}{P}$$

Substituting the value of P from Eq. (129) one will have[§]

$$H = -\Sigma p_i \log p_i$$

23. Entropy of a Source

The amount of information in bits per symbol of a source described by symbols with probabilities $p(i)$ is given by Eq. (127)

$$H = -\sum_{i-1}^k p(i) \log p(i)$$

H is known as the entropy of the source. The word entropy was attached to H because of the identity of the expression used to measure information and the expression used in thermodynamics to measure the uncertainty or the disorder of a system. $p(i)$ in this case refers to the probability of one of the possible states of the system.

H is thus referred to as the measure of information, or the uncertainty or randomness of a system.

When logarithms in the base 2 are used, then the unit of information is called a "bit." Because of mathematical convenience, log in the base e is very often used. In this case, the unit of information is called a "nat" (contraction for natural unit). When logarithms in the base 10 are used, then the unit is often referred as a "Hartley."

[†] See Fano, *loc. cit.*, for several examples.
[‡] See Fano, *loc. cit.*, and Shannon, Ref. 7, Theorem 5.
[§] See Shannon, Ref. 7, Theorem 3.

With the use of the notion of expected values (see paragraph 9), expression (129) can be written in the form

$$H = \sum \left(\log \frac{1}{p(i)} \right) p(i) = E \left(\log \frac{1}{p} \right) \tag{130}$$

Thus one can state that the amount of information per symbol is, on the average, equivalent to the logarithm of the reciprocal of the probability of that symbol. Thus symbols with large probability of occurrence provide less information than symbols with small probability of occurrence.

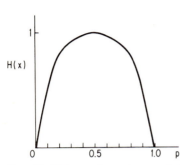

a. Properties of H. 1. H is equal to zero if and only if all but one of the $p(i)$ are zero.

2. H is maximum when $p(i) = 1/k$, in which case $H = \log k$. Figure 16 shows the behavior of $H(x) = -[p \log p + (1 - p) \log (1 - p)]$ for the case of a source with two symbols with probabilities p and $1 - p$. It is seen that H is maximum when $p = 1/2$ and zero when either $p = 0$ or $p = 1$.

FIG. 16. Entropy for a two-symbol source.

3. When a source is described by two random variables x, y (see paragraph 8a) with joint probabilities $p(i,j)$, then one can define the joint entropy

$$H(x,y) = -\sum_i \sum_j p(i,j) \log p(i,j) \tag{131}$$

which is the average information per symbol for the joint occurrences of x and y. The marginal entropies for x and y will be given by

$$H(x) = -\sum_i p(i) \log p(i)$$

$$H(y) = -\sum_j p(j) \log p(j) \tag{132}$$

which are the average information per symbol for x and y, respectively. $p(i) = \sum_j p(i,j)$ and $p(j) = \sum_i p(i,j)$ are the marginal probabilities associated with the random variables x and y, respectively.

The conditional entropy $H_x(y)$, of y given x, will be defined as

$$H_x(y) = -\sum_{ij} p(i)p_i(j) \log p_i(j)$$

where $p_i(j)$ is the conditional probability [see Eqs. (43) and (44)].

$$p_i(j) = \frac{p(i,j)}{p(i)}$$

Thus $H_x(y)$ measures the average information of y for each value of x, weighted according to the probability of the occurrence of x. Since $p(i)p_i(j) = p(i,j)$, then one has

$$H_x(y) = -\sum_{ij} p(i,j) \log p_i(j) \tag{133}$$

It can be shown that

$$H(x,y) \leq H(x) + H(y) \tag{134}$$
$$H(x,y) = H(x) + H_x(y) = H(y) + H_y(x) \tag{135}$$
$$H(y) \geq H_x(y) \tag{136}$$

Expression (136) states that the uncertainty of y when nothing is known about x is larger than the uncertainty of y when x is given. If the two random variables x and y were independent [see Eq. (45)], then $H(y) = H_x(y)$ and also $H(x,y) = H(x) + H(y)$.

4. The relative entropy of a source is defined as

$$\text{Relative entropy} = \frac{\text{entropy of a source}}{\text{maximum entropy}} = \frac{H}{H_{\max}} \tag{137}$$

For a source of k symbols with probabilities $p(i)$, the relative entropy is

$$\text{Relative entropy} = -\frac{\Sigma p(i) \log p(i)}{\log k}$$

The redundancy R is defined as follows:

$$R = 1 - \frac{H}{H_{\max}} \tag{138}$$

$R = 1$ when $H = 0$, which occurs when only one outcome is possible. In this case, no information is provided, since the outcome is certain to occur. On the other hand, $R = 0$ when $H = H_{\max}$, which occurs when all the symbols have the same probability. This is the state of maximum disorder, since there is complete uncertainty about the possible outcomes. Values of R between 0 and 1 give an indication of the various degree of uncertainty (or risks) in the outcomes.

5. The concept of entropy has been used very extensively in the analysis of the structure of languages. The entropy of the English language can be estimated by taking successive approximations to the construction of words and sentences according to the probability of occurrence of simple letters, two letters, etc. The maximum value of H is $\log 27 = 4.76$ bits per symbol, assuming 27 symbols (26 letters and a space). The value of H for the first-order approximation is $H_1 = 4$ bits per symbol if it is assumed that words can be formed using the frequency of letters and space as in English text. Successive values of H, as estimated by Shannon, are $H_2 = 3.56$ bits per symbol, $H_3 = 3.33$ bits per symbol, and $H_n = 1.3$ bits per symbol. With $H_n = 1.3$, the redundancy is about 70 per cent. This redundancy is of importance for combating the effect of noise, since, according to these estimates, it appears to be possible to reconstruct a sentence even if every other letter of the sentence is missing or in error.

24. Communication Channel—Encoding and Decoding Operations

Messages from a given source are encoded into suitable symbols for transmission over a channel. At the destination, decoding operations are necessary to reproduce the information received.

Figure 17 shows a schematic diagram of the components in a communication channel. The information source, with the proper coding operation, may be a sequence of letters as in telegraph or teletype, a function of time representing an amplitude as in radio or telephony, a function of time and x,y coordinates as in television, etc. The transmitter encodes this input into signals suitable for transmission on a channel. This channel may be a coaxial cable, a radio wave, etc. At the destination, the channel and source decoder perform the inverse operation of the channel and source encoder in order to reconstruct the message from the signal received. The input of the

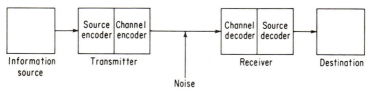

Fig. 17. Communication system.

channel encoder and the output of the decoder will often be in binary representation, regardless of the characteristics of the symbols sent by the source.

a. A Basic Coding Theorem. Assume that the message probabilities are known and that the messages are statistically independent. Let the source x consist of a set of messages x_i, $i = 1, N$ with probabilities $P(i)$. Let K be the number of different symbols used in the construction of the messages x_i. If n_k represents the number of symbols used in messages x_k, then

$$\bar{n} = E(n) = \sum_1^N n_k P(x_k)$$

is the average number of symbols per message. As shown above, log K represents the maximum value of the entropy of the symbols used.

The following theorem is due to Fano.[†]

"Given an ensemble of N messages with entropy $H(x)$, and an alphabet consisting of K symbols, it is possible to encode the messages of the ensemble by means of sequences of symbols belonging to the prescribed alphabet in such a way that the average number of symbols per message $\bar{n}$ satisfies the inequality."

$$\frac{H(X)}{\log K} \leq \bar{n} \leq \frac{H(X)}{\log K} + 1 \tag{139}$$

As stated by Fano, this theorem shows that, for a binary alphabet, the average number of binary digits per message is at least equal to the entropy $H(X)$ and may exceed it by almost one digit.

The above theorem refers to the encoding of individual messages. For theorems related to the encoding of sequences of messages, whether or not the messages are statistically independent, the reader is referred to Ref. 8, Chap. 4.

25. The Discrete Channel with Noise

Let $H(X)$ be the entropy of the source (input to the channel), $H(y)$ be the entropy of the received signal (output of the channel), and $H(x,y)$ the joint entropy of input and output. The conditional entropy $H_y(x)$, called the equivocation, can be considered as a measure of the uncertainty of the signal sent after the signal was received. Let us also assume that these entropies are given in bits per second.[‡]

In a noiseless channel $H(x) = H(y)$. In a channel with noise, following Shannon, the rate of transmission of information will be defined as

$$\text{Rate} = H(x) - H_y(x) \tag{140}$$

Thus, if a source sends 0 and 1 with probabilities 1/2 and 1/2, respectively, and, on the average, one out of 100 symbols are in error, then the conditional probabilities $p_i(j)$ of a zero or a 1, given that either a zero or a 1 were sent, are given by

		j: destination	
$p_i(j)$		0	1
i 0		0.99	0.01
Source 1		0.01	0.99

Since at the destination the probabilities of zero and 1 are also 1/2 and 1/2, respectively (as can be readily verified), then

$$H_y(x) = -1/2 \ (0.99 \log 0.99 + 0.01 \log 0.01) - 1/2 \ (0.01 \log 0.01 +$$
$$0.99 \log 0.99) = 0.081 \text{ bit per symbol}$$

[†] See Ref. 8, Chap. 3.

[‡] Entropy per second H' will be given by $H' = MH$, where M is the number of symbols per second.

and $H(x) = 1$ bit per symbol. Assuming 1,000 symbols per second, then rate = $1,000 - 81 = 919$ bits per second.

a. Channel Capacity. The capacity of the channel will be defined as

$$C = \max_{p(x)} (H(x) - H_y(x)) \text{ bits per second} \tag{141}$$

where the maximum is over all the information sources that can be used as input to the channel, assuming a fixed $H_y(x)$.

In the noiseless channel, $H_y(x) = 0$ and

$$C = \max_x H(x) = \log N \tag{142}$$

If the signals are of different duration, then for the noiseless channel

$$C = \lim_{t \to \infty} \frac{\log N(t)}{t} \tag{143}$$

If $H_y(x)$ is not only fixed but also independent of x, then

$$C = \max H(y) - H_x(y) = \log M - H_x(y)$$

where M is the number of received signals.

Shannon's second fundamental theorem gives full significance to the definition (141). C represents the maximum rate of transmission of information through a noisy channel. The problem, however, is not to receive as much information as possible but to receive as much of the transmitted information as possible. Shannon's theorem shows that it is possible to transmit at the rate C and have a very small probability of an incorrect reception. This theorem can be stated as follows:

Shannon's Theorem for a Noisy Channel. "Let C be the capacity of a discrete noisy channel which is fed by a source with an entropy $H_0(x)$ in bits per second.

"1. If $H_0(x) \leq C$ then there exists a coding system such that the output of the source can be sent through the channel with an arbitrarily small frequency of error (equivocation).

"2. If $H_0(x) > C$, the minimum equivocation is $H_0(x) - C$. It is then possible to encode the source so that the equivocation is less than $H_0(x) - C + \epsilon$, for an arbitrarily small ϵ.

"3. There is no method of encoding that will make the equivocation less than $H_0(x) - C$."

Shannon's theorem is an existence theorem and consequently does not provide a rule for finding an ideal method of matching, or encoding, the source to the channel. Many error-correcting codes have been devised, and the reader is referred to the literature in the subject.†

This theorem shows that it is not necessary to increase the redundancy of the messages transmitted in order to reduce the probability of error. The price that one pays is a long delay, since it might be necessary to wait a long time before the intelligence of a message is completely specified and the encoding procedure can start.

26. Measure of Continuous Information

a. Definition of Entropy for a Continuous Source. When the information source is described by a set of discrete random variables x_i with associated probabilities $p(x_i)$, then the entropy of the source is defined as

$$H(x) = - \sum_{i=1}^{n} p(x_i) \log p(x_i) \tag{144}$$

† See Refs. 7 and 8. Also Ref. 9, Chap. 6, and additional bibliography quoted.

When the source is described by a continuous random variable with a probability density function $f(x)$ (see paragraph 8b), then instead of a discrete probability $p(x_i)$ one has the probability density $f(x) \, \Delta x$ given by

$$p_1(x \leq x \leq x + \Delta x) = f(x) \, \Delta x$$

where Δx is the width of the cell associated with the value of $f(x)$ at the point x.

The natural extension of Eq. (144) for a continuous random variable will then be

$$H(x) = - \sum_{[x]} f(x) \log [f(x) \, \Delta x] \, \Delta x \tag{145}$$

where $[x]$ indicates the region where $f(x)$ is defined.†

When the cell widths are made sufficiently small, then [Eq. (145)] the entropy of a source described by a continuous random variable will be defined as

$$H(x) = - \int_{-\infty}^{\infty} f(x) \log [f(x) \, \Delta x] \, dx \tag{146}$$

Joint and conditional entropies are defined in the same manner. For the case of two random variables, we have

$$H(x,y) = - \iint f(x,y) \log [f(x,y) \, \Delta x \, \Delta y] \, dx \, dy \tag{147}$$

$$H_y(x) = - \iint f(x,y) \log [f(x/y) \, \Delta x] \, dx \, dy \tag{148}$$

$$H_x(y) = - \iint f(x,y) \log [f(y/x) \, \Delta y] \, dx \, dy \tag{149}$$

$$H(x) = - \int f_1(x) \log [f_1(x) \, \Delta x] \, dx$$

$$= - \iint f(x,y) \log [f_1(x) \, \Delta x] \, dx \, dy \tag{150}$$

where $f_1(x)$ is the marginal probability density function of x [see Eq. (47)].

The effect of the cell width in the definition of H will no longer exist when differences of two entropies are considered. For the rate of transmission [Eq. (140)], one has

$$\text{Rate} = H(x) - H_y(x) = - \iint f(x,y) \log [f_1(x) \, \Delta x] \, dx$$

$$+ \iint f(x,y) \log [f(x/y) \, \Delta x] \, dx \, dy$$

$$= \iint f(x,y) \log \frac{f(x/y)}{f_1(x)} \, dx \, dy \tag{151}$$

b. Properties of H. 1. Properties (129) to (131) are also true for a continuous random variable.

2. If x is defined within a finite region A of the x axis, then $H(x)$ is maximum when $f(x)$ is constant. The probability density will then be equal to $(1/A) \, \Delta x$, which is the uniform distribution [see Eq. (83)].

3. If the random variable x is defined in the whole x axis, $-\infty < x < \infty$, and if the standard deviation σ of x is fixed, then $H(x)$ is maximum when the probability function is the Gaussian or normal distribution (see paragraph 13c). The entropy will then be given by

$$H(x) = \log \sqrt{2\pi e} \, \sigma^* \tag{152}$$

where σ^* is the normalized σ in terms of the cell width Δx ($\sigma^* = \sigma/\Delta$).

† See Ref. 9, Sec. 6.4.

4. If the random variable x is defined in the half line, $0 < x < \infty$, and the expected value of x is fixed, say equal to a, then $H(x)$ is maximum when the probability function is the exponential distribution (see paragraph 13b)

$$f(x) = \frac{1}{a} e^{-x/a}$$

and the entropy is

$$H(x) = \log a^*e \tag{153}$$

where a^* is normalized in terms of Δ, that is, $a^* = a/\Delta$.

5. For a n-dimensional random variable defined in the whole space and fixed second moments,

$$A_{ij} = \int \cdots \int x_i x_j f(x_1, \ldots, x_n) \, dx_1 \cdots dx_n$$

then H is maximum when $f(x_1, \ldots, x_n)$ is the multinormal distribution

$$f(x_1, \ldots, x_n) = \frac{|a_{ij}|^{\frac{1}{2}}}{(2\pi)^{n/2}} \exp\left(-\frac{1}{2} \sum a_{ij} x_i x_j\right) \tag{154}$$

where the matrix of a_{ij} is the inverse of the matrix of A_{ij}; that is,

$$\|a_{ij}\| = \|A_{ij}\|^{-1}$$

The maximum value of H is

$$H(x_1, x_2, \ldots, x_n) = \log\left[(2\pi e)^{n/2}|A_{ij}^*|^{\frac{1}{2}}\right] \tag{155}$$

where A_{ij}^* are normalized values in terms of the square of the cell widths.

6. The entropy is a relative measure that changes with the coordinate system used. Thus, H depends not only on the cell widths but also on the coordinate system in which they are measured.

The reason for this dependence is that the probability density $f(x) \, dx$ changes into a probability density $g(y) \, dy$ if the random variable x is changed to a random variable y by means of a function† $y = h(x)$. We thus have

$$g(y) \, dy = f(x) \, dx$$

and

$$g(y) = f[h^{-1}(y)] \left|\frac{dx}{dy}\right| \tag{156}$$

where dx/dy is taken in absolute value.

The relation between $H(y)$ and $H(x)$ is

$$H(y) = H(x) - E_x\left(\log\left|\frac{dx}{dy}\right|\right) \tag{157}$$

where E_x stands for the expected value with respect to x. In this case $\Delta x = \Delta y$, using the same cell widths in both axes.

For the multivariate case, the relation is analogous. We have

$$g(y_1, \ldots, h_n) = f(x_1, \ldots, x_n) \left| J\begin{matrix} x_1, \ldots, x_n \\ y_1, \ldots, y_n \end{matrix} \right| \tag{158}$$

where J stands for the Jacobian of the transformation. The x_i values in f should be changed to the corresponding y values using the relations between the two sets of random variables.

The relation between the entropies is

$$H(y_1, \ldots, y_n) = H(x_1, \ldots, x_n) - E\left(\log J \frac{x_1, \ldots, x_n}{y_1, \ldots, y_n}\right) \tag{159}$$

† $y = h(x)$ will be assumed to be monotonic. Equation (156) is simple to generalize for nonmonotonic functions.

If the relation between y and x is linear

$$y_i = \Sigma a_{ij} x_j$$

then $J = |a_{ij}|$ and

$$H(y_1, \ldots, y_n) = H(x_1, \ldots, x_n) - \log |a_{ij}| \tag{160}$$

For an orthogonal transformation, $|a_{ij}| = 1$ and

$$H(y_1, \ldots, y_n) = H(x_1, \ldots, x_n) \tag{161}$$

27. Sampling Theorem.   Ensemble of Functions

Continuous random functions are common in communication systems.   For band-limited functions, it is possible to quantize the information by a suitable method of sampling.   Sampling of a continuous function has been considered from the mathematical point of view by many authors.   From the viewpoint of communication theory, Shannon[†] showed the following theorem.

Theorem.   "If $f(t)$ contains no frequencies higher than W cps it is completely determined by giving its ordinates at a series of points spaced $1/2W$ seconds apart, so that

$$f(t) = \Sigma f\left(\frac{n}{2W}\right) \frac{\sin 2\pi W[t - (n/2W)]}{2\pi W[t - (n/2W)]} \tag{162}$$

If the source operates during time T, we may consider that

$$|f(t)| < \epsilon \qquad \text{when } |t| > \frac{T}{2}$$

If $f(t)$ is band-limited to W cps, and if it is sampled $1/2W$ sec apart, then $n/2W = T$. The function $f(t)$ will then be represented by a point in a space of $2TW$ dimensions.

For a television program with a band of 5 Mec lasting half an hour one would need $2TW = 2 \times 30 \times 60 \times 5 \times 10^6 = 1.8 \; 10^{10}$ points.

For a single line, with $T \approx 1/15{,}000$ sec, then $n \approx 667$ points.

The size $n = 2TW$ of the space may be reduced by the following considerations:

1. Phase elimination.   If the ear were insensitive to phase, the n of dimensions may be reduced by half, since we would need only $\sqrt{a_n{}^2 + b_n{}^2}$ for each frequency.

2. Restrictions in the information source.

3. Probability considerations.   For instance, messages that are very improbable could be eliminated.   In television, where successive frames are likely to be identical, a reduction in dimensionality may result.

a. Entropy of the Ensemble.   The entropy of the n values x_n is

$$H_n = - \int \cdots \int p(x_1, x_2, \ldots, x_n) \log [p(x_1, \ldots, x_n) \Delta x^n] \, dx_1 \cdots dx_n$$

H_n will not converge as n increases.   We define

$$H = \lim_{n \to \infty} \frac{H_n}{n} \qquad \text{bits per symbol} \tag{163}$$

as the entropy per "degree of freedom."

Since there are $n = 2TW$ points, each point has a duration of $1/2W$ sec.   Hence, the entropy per second is

$$H' = \frac{H}{1/2W} = 2WH \qquad \text{bits per second} \tag{164}$$

b. Power of a Signal Function—White Noise—Entropy Power.   The average

[†] C. E. Shannon, Communication in the Presence of Noise, *Proc. IRE*, vol. 37, no. 1, pp. 10–21, January, 1949.

power is given by

$$P = \frac{1}{T} \int_{-T/2}^{T/2} f^2(t) \ dt$$

Substituting Eq. (162), it can be shown that

$$P = \frac{\Sigma x_n^2}{n} = \frac{\Sigma x_n^2}{2TW} \qquad \text{where } x_n = f\left(\frac{n}{2W}\right)$$

Σx_n^2 is the square of the distance to a point in a space of $2TW$ dimensions. Thus

$$n^2 = 2TWP \tag{165}$$

If $f(t)$ represent the amplitudes of white thermal noise, then the sample values taken at intervals $1/2W$ are uncorrelated and the distribution function of $x_1, x_2, \ldots, x_n$ becomes

$$f(x_1, x_2, \ldots, x_n) = \frac{1}{(2\pi N)^{n/2}} \exp\left(-\frac{1}{2N} \Sigma x_i^2\right) \tag{166}$$

where $N = \sigma^2$ is the average power of the noise. In this case, A_{ij}, $i \neq j$, the covariances are equal to zero.

The entropy per degree of freedom for Eq. (166) will be, using Eqs. (152) and (163),

$$H = \frac{1}{n} \log \left[(2\pi e)^{n/2} N^{n/2}\right] = \log (2\pi e N)^{1/2} \tag{167}$$

The entropy rate H' is

$$H' = W \log (2\pi e N) \tag{168}$$

Entropy Power. For an arbitrary random process described by a function $f(t)$, it is convenient to measure its uncertainty or randomness by comparing it with a white noise with the same entropy. The average power of this white noise N_1 is called the entropy power of the signal $f(t)$ and is given by

$$N_1 = \frac{1}{2\pi e} e^{2H} \tag{169}$$

c. Entropy Loss in Linear Filters. Expression (156) shows that there is a change in the entropy when a transformation of the coordinates is performed. Since the operation of a filter is essentially a linear transformation, we expect that the filtered process will show a change in the original entropy. The following theorem, due to Shannon, establishes this relation:

Theorem. If an ensemble having an entropy rate H per degree of freedom in band W is passed through a filter with characteristic $Y(f)$, the output of the ensemble has an entropy

$$H_2 = H_1 + \frac{1}{W} \int_W \log |Y(f)|^2 \ df \tag{170}$$

As expressed by Eq. (170) it can be seen that the change in entropy does not depend on the properties of the input process and is only a function of the linear filter used. Since $|Y|_{max}$ is generally less than unity, then the change of entropy represented by the integral in (170) will indicate a decrease in the entropy rate H_1. Thus, although the total information of the computed message increases because of the additional information from the noise, the information about the message will tend to decrease, becoming more corrupt with noise.

28. The Capacity in the Continuous Channel

The rate of transmission is expressed by Eq. (141), but since the encoding is now performed using continuous functions, then expression (151) should be used. This equation can also be written in the equivalent form

$$\text{Rate} = \iint f(x,y) \log \frac{f(x,y)}{f_1(x)f_2(y)} \, dx \, dy = E \log \frac{f(x,y)}{f_1(x)f_2(y)} \tag{171}$$

For a source represented by an ensemble of functions, one can define the average rate of transmission by performing a time average over a particular function of the ensemble. We have, for the sequences x, y becoming indefinitely long,

$$R = \lim_{T \to \infty} \frac{1}{T} \iint f(x,y) \log \frac{f(x,y)}{f_1(x)f_2(y)} \, dx \, dy \tag{172}$$

This definition is true if the ensemble of functions in the source is *ergodic*. A source is said to be ergodic if the ensemble of functions is ergodic, that is, if the ensemble is stationary and there is no subset of the functions in the set with a probability different from 0 and 1 which is stationary. For an ergodic ensemble, the average of a function of a random variable taken over the ensemble is the same as the time average taken over all the possible translations of a particular function of the ensemble except for a subset of measure zero.

The capacity of the continuous channel is defined as

$$C = \lim_{T \to \infty} \max_{f(x)} \frac{1}{T} E \left[\log \frac{f(x,y)}{f_1(x)f_2(y)} \right] \tag{173}$$

where the maximum is taken with respect to all the possible inputs to the channel.

Shannon's fundamental theorem for the transmission of information over noisy channel (paragraph 25a) still applies with the capacity defined by Eq. (173). That is, C represents the maximum number of bits that can be sent over the channel with an arbitrarily small probability of error.

As stated by Shannon,[†] if u represents the message, v the recovered message, x is the signal used in the channel, y the received signal perturbed by noise, then

$$H(u) - H_v(u) \leq H(x) - H_y(x)$$

regardless of what type of operations are performed in u to obtain x and in y to obtain v. Thus, no matter what method of encoding is used, the discrete rate for the binary bits cannot exceed $H(x) - H_y(x)$ and, consequently, cannot exceed the channel capacity which is the maximum of $H(x) - H_y(x)$.

Capacity for Additive, Statistically Independent Noise. When the noise is added linearly to the signal, i.e.,

$$y(t) = x(t) + n(t) \tag{174}$$

and $n(t)$ is statistically independent of $x(t)$, then the conditional distribution of y for any x depends only on the difference between y and x; that is,

$$f(y/x) = f(y - x)$$

then from Eq. (140), using property (130) and the fact that

$$H(x,y) = H(x,n) = H(x) + H(u)$$

since x and n are independent, we have

$$\text{Rate} = H(y) - H(u) \tag{175}$$

and

$$C = \max_{f(y)} H(y) - H(u) \tag{176}$$

† See Ref. 9, p. 65.

where the maximization is taken over the ensemble of received signals, subject to the restrictions imposed by the constraints imposed by the transmitted signals.

a. Capacity with an Average Power Limitation.   If the noise is additive, band limited, with an average power N and is also white and Gaussian and the average power of the band-limited signal $x(t)$ is P, then the maximum of $H(y)$ will be attained when $f(y)$ is also Gaussian, with an average power of $P + N$.   This result follows from property 3 of paragraph 26*b*, using $\sigma = \sqrt{P + N}$ for the standard deviation of y.

Then

$$C = \max H(y) - H(u) = \frac{1}{2} \log \frac{P + N}{N} \qquad \text{bits per symbol} \qquad (177)$$

The capacity, in bits per second, for a band-limited signal and noise is,[†] from Eq. (164),

$$C = W \log \frac{P + N}{N} \qquad \text{bits per second} \qquad (178)$$

According to this result, the statistical structure of the signal set needed for an optimum matching of the source to the channel should be the structure of the white Gaussian noise.   Then, by using an appropriate method of coding, it would be possible to transmit binary digits at a rate given by Eq. (178), with a very small probability of error.

This capacity C may not be theoretically realizable for systems with sources described by signals that do not have the characteristics of a Gaussian noise.   Also, the characteristic of the noise may not be Gaussian but of a different nature (static noise, etc.).

For an arbitrary perturbing noise, with a Gaussian signal, the channel capacity is given by[‡]

$$W \log \frac{P + N_1}{N_1} \leq C \leq W \log \frac{P + N}{N_1}$$

where P = average power of the signal
N = average power of the noise
N_1 = entropy power of the non-Gaussian noise

29. The Capacity in the Continuous Noisy Channel with a Continuous Source

So far it has been assumed that the source of information was discrete with a finite rate of generating information (finite entropy).   If the source is continuous, it would require an infinite number of digits for its specification, since a continuous variable can assume an infinite set of values.   This would require a channel with infinite capacity, and since channels are limited by noise, thus giving a finite capacity, transmission of information would be impossible.

However, in a practical situation, exact reproduction is not needed and only reproduction within a certain tolerance is required.   This tolerance is stated in terms of a certain fidelity of recovery, in a such way that a capacity can be defined subject to the fidelity criterion selected.

The fidelity requirement is evaluated in terms of an evaluation function v which can be written in the form $v[f(x,y)]$ to indicate its dependence on the joint probability density function of the random message x and the recovered message y.   Examples of these evaluation functions are

(a) $$v = E[x(t) - y(t)]^2 = \iint [x(t) - y(t)]^2 f(x,y) \, dx \, dy \qquad (179)$$

[†] See Ref. 7, Theorem 17, p. 67.
[‡] See Ref. 9, Theorem 18, p. 68.

which is the rms criterion or the expected mean square difference between the source x and the recovered message y. As a time average, Eq. (179) can be written in the form

$$\lim \frac{1}{T} \int_0^T [x(t) - y(t)]^2 \, dt$$

(b) Absolute value criterion:

$$\rho(x,y) = \frac{1}{T} \int_0^T |x(t) - y(t)| \, dt$$

In general, one can write for the fidelity evaluation function v an expression of the form

$$v = E[\rho(x,y)] = \iint \rho(x,y)f(x,y) \, dx \, dy \tag{180}$$

The rate R_1 of generating information for a continuous source with a fidelity requirement fixed at $v = v_1$ can then be defined as the minimum of the expected value of log $[f(x,y)/f_1(x)f_2(y)]$[see Eq. (171)] for all possible values of the conditional probability density $f_x(y)$. We have

$$R_1 = \min_{f_x(y)} \iint f(x,y) \log \frac{f(x,y)}{f_1(x)f_2(y)} \, dx \, dy \tag{181}$$

To evaluate Eq. (181) one would consider all the communication systems that could transmit the information with the required fidelity (180). The minimum rate obtained will be defined as the rate of transmission subject to the given fidelity requirements.

The transmission of information for a continuous source over a continuous noisy channel is then stated by the following theorem due to Shannon.†

Theorem. If a source has a rate R_1 for a fidelity evaluation v_1, it is possible to encode the output of the continuous source and transmit it over a channel of capacity C with fidelity as near v_1 as desired provided $R_1 \leq C$. This is not possible if $R_1 > C$.

REFERENCES

1. Dantzig, G. B., *Maximization of a Linear Function of Variables Subject to Linear Inequalities*, chap. XXI, The Cowles Commission for Research in Economics, John Wiley & Sons, Inc., New York, 1951.
2. Garvin, Walter W., *Introduction to Linear Programming*, McGraw-Hill Book Company, New York, 1960.
3. Churchmen, C. W., R. L. Ackoff, and E. L. Arnoff, *Introduction to Operations Research*, John Wiley & Sons, Inc., New York, 1961.
4. Sasieni, M., A. Yaspan, and L. Friedman, *Operations Research, Methods and Problems*, John Wiley & Sons, Inc., New York, 1959.
5. Cramer, Harold, *Mathematical Methods of Statistics*, Princeton University Press, Princeton, N.J., 1946.
6. Wiener, Norbert, *Extrapolation, Interpolation and Smoothing of Stationary Time Series*, The Technology Press of the Massachusetts Institute of Technology, Cambridge, Mass., and John Wiley & Sons, Inc., New York, 1950.
7. Shannon, Claude, and Warren Weaver, *The Mathematical Theory of Communication*, The University of Illinois Press, Urbana, Ill., 1949.
8. Fano, Robert M., *Transmission of Information*, The M.I.T. Press, Cambridge, Mass., and John Wiley & Sons, Inc., New York, 1961.
9. Middleton, David, *An Introduction to Statistical Communication Theory*, McGraw-Hill Book Company, New York, 1960.

† See Ref. 7, Theorem 21, p. 78.

Chapter 5

COMMUNICATION SYSTEM MODELING

ENRIQUE FURTH, *Radio Corporation of America**

INTRODUCTION

1. What Is a Model?

A model is basically a representation of some properties of an object, device, or system. The question of what properties are incorporated into a particular model depends on the type, purpose, and complexity of the model. Thus, for example, a map is a model of a particular geographical region; the characteristics of the region which are preserved in a map depend on the purpose of the map. Most maps attempt to preserve certain geometric properties of the represented region; general-purpose maps (conformal maps) preserve distances and shapes in the small, that is, they preserve angles; and some navigation maps preserve straight lines. However, some maps (many city subway maps, for example) preserve no geometric properties; they are "distorted" and preserve only connectivity or incidence; that is, only line intersections and common boundaries of regions can be recognized on such topological maps. Communications systems are also often represented in this fashion, as we shall see later.

Of course, many other types of models exist. Among the more important categories of models are mathematical, logical, and physical models. In a mathematical model the representation consists of mathematical equations or procedures (algorithms). For example, many physical phenomena are represented by models consisting of differential equations. The above-mentioned maps, including the topological kind, are mathematical models.

2. Why Modeling?

Building and operating of communications systems require previous analysis. All analysis must be preceded by simplification and abstraction of the essential characteristics of the real situation; analysis requires an explicit model. The first task is to determine which features are of most interest and to define these quantitatively.

After the interesting features of the system have been identified and described numerically, a model of the system can be devised. By this is meant a set of elements, such as nodes, links, messages, etc., and rules of interaction of the elements. Ideally a model should incorporate all the important features of the system represented (it should be realistic) and at the same time it should be tractable to mathematical or

* Now with Dunlap and Associates, Inc.

physical analysis or simulation.   Actual models of communication systems are generally compromises between these two goals, some leaning more toward tractability while others tend more toward realism.

Since the present discussion cannot describe in detail every model thought to be applicable to communication systems, we shall examine some of them in some detail, briefly describe many others, and supply references where the interested reader can find further information concerning each one.

The criterion applied for choosing those models described in more detail has been threefold: They are easy to use (simple algorithms exist, and computations are easy to perform), they provide insight into the systems they represent, and they are useful in system design, especially for communication system engineering.

COST NETWORK MODEL

In one of the most useful models of communications systems, the various parts of a communication system are represented by cost, delay, or length values.   These cost values can be applied to links and to nodes; however, the costs corresponding to a particular node can be allocated among the links incident upon that node.   Therefore, in this section we shall consider communication systems to be represented by networks in which a cost value is associated only with each link.   These numbers represent the cost of the link, the cost or delay incurred in traversing the corresponding link, the length of the link, or some similar additive link property.

The main applications of this model are in the determination of minimum cost and minimum delay paths and networks and the evaluation of different routing schemes and philosophies.   Certain deleterious phenomena occurring in transmission circuits, such as attenuation, some types of distortion, error rates, etc., are also functions of the magnitudes of the corresponding properties of the component links of the circuit.   As will be seen, the circuit minimizing (or maximizing) such a property can be found by shortest-path techniques, thereby further extending the usefulness of this model. Most-reliable and most-secure routes through networks can also be determined by use of this model.

3. Minimum-cost Paths

The most significant application of cost networks is that of finding minimum-cost paths, or shortest paths or shortest routes as they are also commonly called, through them.   The basic problem consists of finding the path entailing the least cost between two given nodes, sometimes called origin and destination or source and sink, in a cost network.

Important applications of this model include the routing of calls and messages for optimum efficiency and best utilization; routing in networks of variable connectivity (the connectivity can vary in military networks or as a consequence of link failures or totally busy links); testing proposed network configurations, etc.; and the establishment of service charges and tariffs.   A more detailed discussion of specific applications is given in paragraph 10 following the description of the various methods for finding shortest paths.

Basic properties of this model, as is indeed the case with most network models, are defined by the rules of series and parallel combination of links (it should be mentioned that some authors call links branches or arcs; nodes are sometimes called vertices).

4. Basic Properties of Minimum-cost-path Model

Series Rule.   The cost of a series combination of costs is the sum of the individual component costs.

Symbolically, $C_s = a + b + \cdots + m$ (see Fig. 1a), where C_s is the series cost and $a, b, \ldots, m$ are the costs of the individual series components (links or subpaths).

Parallel Rule.   The cost of a parallel combination of costs is the cost of the smallest one of the individual component costs.

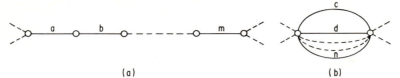

FIG. 1. (a) Series combination of links.   (b) Parallel combination of links.

Symbolically, $C_p = \min [c,d, \ldots ,n]$ (see Fig. 1b), where C_p is the cost of the parallel combination and c, d, $\ldots$, n are the costs of the individual parallel components (links or subpaths).

5. General Methods for the Determination of Shortest Paths

A variety of shortest-path methods has been described.   A review of those methods published up to 1959 exists.[1]   One of the earliest is Shimbel's matrix method.[2]   By this time its interest is mainly historical, since it is rather unwieldy, especially for hand computation.   It does, however, have the virtue of yielding simultaneously the lengths of the shortest paths from every node to every other node in the network.

The most useful and the simplest general shortest-path method, both conceptually and from the viewpoint of computation, is due to Dantzig.[3,4]   It is simple to implement for both hand and machine computation.   If it is desired to find the shortest path and the distance (length of the shortest path) from a given node, hereafter called origin, the following steps comprise Dantzig's shortest-path algorithm:

1. Label the origin zero.

2. Examine all unlabeled nodes j adjacent to any of the already labeled nodes i, and among the nodes examined in this step, find the node j for which $D_i + d_{ij}$ is minimum among all i and j, where D_i is the label of node i and d_{ij} is the length of link ij.

3. Label the node j found in step 2 with $D_j = \min [D_i + d_{ij}]$.   This is its distance from the origin, and the corresponding link ij is part of the shortest path from the origin to the newly labeled node.

4. Repeat steps 2 and 3 until one of the following occurs:

   $a.$ The destination node is labeled if the shortest path to only one node is desired.

   $b.$ All nodes in the network are labeled if shortest paths and distances to all nodes are desired.

   $c.$ No further nodes can be labeled, although not all nodes have been labeled; in this case no path exists from the origin to the unlabeled nodes, and they are essentially infinitely far away from the origin.

An example will best illustrate the procedure.   Figure 2 shows an example of a network; link labels are interpreted as lengths or costs.   If it is desired to obtain the shortest paths from node B, the first step of the shortest-path algorithm indicates that node B (the origin) be labeled zero.   In step 2 we examine nodes A, P, and C adjacent to B, the only node labeled so far.   We have to find the minimum among $D_B + d_{BA}$, $D_B + d_{BP}$, and $D_B + d_{BC}$.   Since the last two are equal, either P or C can be chosen as the node to be labeled in the present step 3.   If P is chosen first, it is labeled with 3. The next node to be labeled will be C, and its label will also be 3.   With nodes B, P, and C labeled, nodes D, E, Q, H, and A must be examined.   The minimum among the following has to be found: $D_C + d_{CD}$, $D_P + d_{PD}$, $D_P + d_{PE}$, $D_P + d_{PQ}$, $D_P + d_{PH}$,

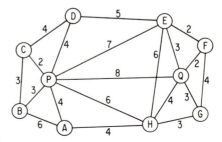

FIG. 2. Example of network.

$D_P + d_{PA}$, $D_B + d_{BA}$. The respective values are 7, 7, 10, 11, 9, 7, and 6. Therefore, the node to be labeled is A, and its label will be 6.

Figure 3 shows the labels (distances) and shortest paths from node B to all nodes. If a node with a label lower than 12 had been the only destination, the procedure could have been terminated upon labeling that node.

As can be seen from the example of Fig. 3, the algorithm generates a tree of all the shortest paths from the origin (node B). It should be noted that, if there is more than one shortest path, as in the case of node D, only one is chosen in order to generate a shortest-path tree.

If only one origin and one destination are of interest, the course of the algorithm can be shortened by starting at both the origin and the destination. In this way distance labels from both these points are progressively obtained. The shortest path between the origin and the destination passes through the node first labeled from both ends; the length of the path is the sum of both labels of the first common node.

The network in the example of Dantzig's shortest-path method has two-way links. It should be noted, however, that the method applies equally well to networks with one-way links and mixed networks, with the obvious precaution of observing the proper traversal senses in the links. Indeed, all the properties mentioned for minimum-cost-path networks, including those of paragraphs 23 to 30, apply regardless of the type of links; the only exceptions are the wye-delta transformations, which apply only to two-way-link networks. The tree of shortest paths *to* a node in a network of one-way links can also easily be constructed with the algorithm described merely by reversing the directions of all links and proceeding in the regular fashion.

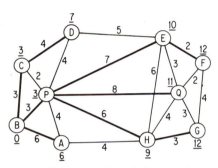

FIG. 3. Shortest paths and distances from node B.

From the nature and simplicity of the method, it can be seen that it is useful for hand computation of even fairly large networks. It is also simple to program and use with electronic computers; several computer-user program libraries have on file shortest-path routines which employ this method. Some of these programs can accommodate networks of up to several thousand nodes.

There are other "labeling" procedures for obtaining shortest paths. Some of these are mentioned in Refs. 1, 3, and 4. They are similar to the one described here; Dantzig seems to have the most compact statement.

6. Network Reduction

In some cases, especially if not all nodes are of interest as origins or destinations, it is possible to reduce or simplify a given network. To this purpose, in addition to the already mentioned rules for the reduction of series and parallel combinations, there exist rules for reducing other configurations. Some of these, including delta-to-wye and wye-to-delta transformation formulas, are given in paragraph 26. Another rule, the absorption rule, which applies specifically to minimum-cost networks, can be derived from the triangle inequalities which appear in the delta-to-wye transformation of this type of network model (see Table 2).

7. Absorption Rule for Minimum-cost Networks

The absorption rule states that, if a link in a minimum-cost network is not the (only) shortest path between its two end points, the link can be removed from the network without affecting the length (or the routing) of any shortest path in the network.

Referring to Fig. 4, the rule states that, if

$$p > a + \cdots + n \qquad \text{(strict inequality)}$$

then the link AB can be removed without affecting *any* shortest paths or their lengths. If

$$p \geq a + \cdots + n \qquad \text{(inequality in the broad sense)}$$

than the removal of the link will not affect *the lengths* of any shortest paths. However, more than one shortest path may exist for some node pairs in this case, and some of those shortest paths utilizing link AB will not appear if that link is removed.

The significance of this rule is twofold: First, when it applies, it serves to eliminate links but not nodes, and second, if applied exhaustively to a network, it transforms it into a network composed only of "essential" links. Such networks are generally somewhat easier to deal with. In the shortest-path algorithm, for example, the nodes adjacent to the origin can be labeled directly by inspection.

The shortest-path problem is a special case of the general linear-programming problem[4,5] and can thus be solved by means of any one of the general linear programming algorithms. Apart from being wasteful, such a procedure does not provide the degree of insight into the solution which the described procedure gives.

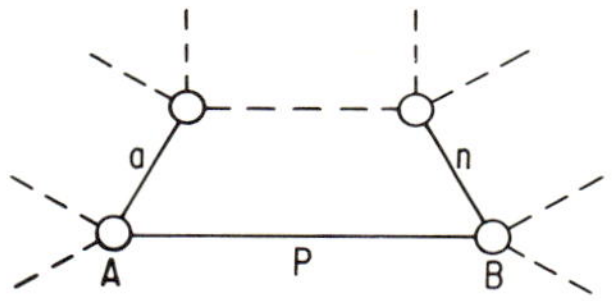

8. Physical Models

Fig. 4. Absorption rule in minimum-cost networks.

There are several physical models (mechanical and electrical devices) of cost networks which have been proposed and in some cases used for solving the shortest-path problem. They are all of interest and will be described briefly.

By far the simplest is Minty's model.[1,6] In it, links are represented by lengths of string proportional to the link costs and tied together at the nodes. In order to find the shortest path between two nodes, the knots representing each one of these are pulled apart. Those links which become taut are in a shortest path, and the "distance" can be measured with a ruler. The shortest-path tree from any origin can be determined by holding the model by the "origin" and attaching weights to the knots; the taut links are in a shortest-path tree. This model applies only to two-way-link networks.

A similar model for one-way-link networks is described in Ref. 7. However it is not quite so simple, since it involves cutting of string links.

Costs (distances) are represented by elapsed times in models described in Ref. 8. Each link is represented by a timer and a light, and each node by digital circuitry. All timers connected to the origin are started simultaneously, and each timer, set to an interval proportional to the link cost, excites the node at the other end of the link at the end of the set interval and lights up the link if that node had not been previously excited. A node so excited starts the timers on the links incident upon it, and so on. The lighted links trace out the shortest-path tree, and the time elapsed between start and the excitation of a node is the distance to it from the origin. This type of model can accommodate all types of links.

Two strictly electrical models have been described. One uses neon bulbs for link costs; the number of bulbs in the link is proportional to the link cost.[9] A slowly increasing voltage is applied between the origin and the destination nodes; when a chain of neon bulbs fires, the applied voltage and the number of bulbs both represent the distance between origin and destination. The lighted links constitute the shortest path. The shortest-path tree cannot be generated by this model. Only two-way links are considered in this model. Reference 9 also describes a similar model employing thyratrons.

In the other electrical model[5] each one-way link is represented by a voltage generator and a diode. The voltage of the generator represents the link cost. Here, too, an increasing voltage is applied between origin and destination; current will flow first over the shortest path between those two points. The voltage applied at that time, equal in magnitude to the sums of the generator voltages of the links in the shortest path, represents the path length. Only the shortest path between the two nodes is traced out. Two-way links can be represented by two equal opposing one-way links.

9. The kth Shortest Path

In many cases it is of interest to find not only the shortest path and its length but also the k next shortest paths. These can serve as alternates to the shortest path; it may be desired to know the identity of all paths differing from the shortest by less than a given percentage (two paths are considered different if they differ in at least one link). Several procedures for obtaining the k shortest paths are described in Ref. 9a. Most of these are generalizations of various shortest-path procedures.

10. Applications

The uses of shortest paths and shortest-path trees in communication systems are many and varied. Among other areas, applications exist in the fields of communications routing, comparative evaluations of existing and proposed communications networks and routing philosophies, evaluation of proposed network additions, establishment of prime and alternate fixed routes, and establishing service tariffs and charges.

Several communications systems based on shortest-path routing have been proposed and extensively studied with a view toward implementation. North Electric Co.'s "saturation signaling" system[10] is based on finding the shortest available path between two nodes having users who wish to communicate. A similar system, "trace routing," has been studied and proposed by RCA for military communication networks. A system based on shortest routes has also been announced by ITT. All these systems utilize the physical propagation delays along the network links as "costs." If desired, these relative costs or delays can be modified by adding artificial delays on the more "expensive" links. These systems all generate the shortest-path tree from origin or destination, depending on the specific implementation. A link which is inoperative or totally busy has infinite delay and will not be seized. These systems can be used with a variety of transmission media, including wire, cable, microwave relays, and troposcatter.

The great merit of these systems lies in the fact that, regardless of the network connectivity which any particular demand is faced with, a route for the demand will be found if one exists. Furthermore, under the shortest-path criterion, this route will be the most economical in terms of the available or surviving facilities. For this reason, this routing philosophy is particularly well adapted to communication systems in which the network connectivity, the offered load, or both, are subject to rapid and sudden variations. This is obviously the situation in many military communication systems. Many of these systems must be prepared to deal not only with suddenly changing demands and varying connectivities due to busyness and outages but also with the addition, deletion, and relocation of mobile nodes and users.

Additional advantages available to shortest-route systems, depending on their exact implementation, include the absence of system-wide status reporting and greatly reduced user directory problem, since each node is concerned only with the identification of subscribers connected to it directly. Also, the channel required for the transmission of routing signals can be designed to be relinquished as the last communication channel on a link, since if a link is totally busy, no signaling is required. Thus service signaling may be made to occupy virtually no link bandwidth. This becomes important in systems where the cost of transmission facilities predominates over that of switching facilities.

In any system in which meaningful relative link costs can be established, computation of least-cost paths and their costs is profitable for establishing prime and preferred alternate routes and routing schemes, for comparing existing and proposed routes with the optimum ones, and for determining the price paid for the use of nonoptimum doctrines. The existence of simple least-cost-path procedures and their straight-forward physical interpretation increases the importance of defining relative link costs in communication systems.

In an evaluation of proposed additional links in a network, shortest-route calculations serve to determine the cost of carrying the added traffic made possible by the new facilities as a function of their relative costs if installed between different nodes in the network.

Another important practical application of this model is in the area of determination of public service tariffs and charges. Known instances of this application have been reported for freight-rate determination in the railroad and trucking industries. Rates are determined based on the shortest existing route between origin and destination, regardless of the length of the route actually used. As railroad and highway networks change owing to the addition and elimination of links, rates are recomputed for all origin-destination pairs. This method should find application in the communications industry also. Time-varying link costs can easily be incorporated in such a procedure.

The applications discussed so far concern path properties which are additive; that is, the numerical value of the property for the path is the sum of corresponding link values. This is the case for costs, delays, and lengths. It is possible to apply the shortest-path model to other properties which, although not additive, can be made additive by a change of variable. For example, in some types of network models path values are obtained by multiplying link values. If instead of these multiplicative link values their logarithms are used, these become additive for the corresponding paths, and the logarithm of the path value is obtained.

This transformation yields some very useful practical applications of shortest-path techniques; only a few will be mentioned here. If loss or attenuation is expressed logarithmically (in decibels, for example) for each link, the path of lowest attenuation between two nodes can be found by the shortest-path procedure.

The probability of occurrence of several independent events is also the product of the probabilities of the individual events. Therefore the probability of a certain path in a network can be considered to be the product of the link probabilities. With the above-mentioned transformation, the most-reliable path through a reliability network can be determined (this is a different problem from that of finding the probability of connection between two nodes, given the probabilities of the links; this problem is briefly treated in paragraphs 23 to 30). Similarly, the most-secure, the most-survivable, and the most-invulnerable routes can be established given the corresponding link values.

11. Minimum-cost Networks

An important application of cost networks is the determination of least-cost networks connecting a given set of nodes. In this case the problem is to select from among all the possible network links a set of links which will result in a network connecting all the desired nodes and such that the sum of the selected link costs is minimum.

This model is of interest in systems which handle communications involving more than two parties. Examples of these would be broadcast networks, establishment of conference call connections, and the routing of multiple-address messages among others. A more detailed account of applications of this model will be given in paragraph 14.

The fundamental problem treated by this model involves determining the set of links, among all those permissible, such that a path exists from every node to every other node and such that the total cost or length of the selected links is minimum. The solution to this problem is strikingly simple for networks of two-way links.[11,12] The same procedure can be adapted to one-way-link networks in some cases, as will be seen. In what follows, two-way links are assumed except where specifically stated otherwise.

12. Shortest Connection Network Procedure

An attractive feature of this model is the simplicity of the various procedures, all similar, for constructing shortest networks. One reason for this simplicity is that the choice of each link is final. Once it has been selected for inclusion in the shortest connection network, it remains in it. Another reason is that there are exactly $n - 1$ links in the shortest network, where n is the number of nodes. Obviously the shortest network is a tree, in fact the shortest tree, since if there were any loop in the shortest network, any link in the loop could be removed without disconnecting the network, yielding an even shorter network, contrary to the assumption. A third reason is that the order of link selection is not fixed; many possible orders of selection exist, as will be seen. The three above-mentioned properties mean that the algorithm is exactly $n - 1$ steps long, and the order of link selection can usually be suited to obtain a further simplification of the procedure or to serve some other criterion or requirement.

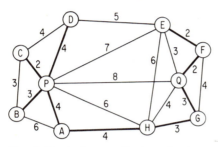

FIG. 5. Shortest connecting tree of a given network.

The simplest algorithm for finding the shortest connection network, hereafter called shortest connecting tree, consists of repeating the following step $n - 1$ times:[11]

Select the shortest link in the given network which has not yet been chosen and which does not form a loop with those links already selected.

The $n - 1$ links thus selected constitute the shortest connecting tree.

Another procedure, suitable when the number of possible links is not much greater than $n - 1$, is the following:[11]

Delete the longest link whose deletion will not cause the remaining network to become disconnected.

This step is repeated as often as possible; that is, until only $n - 1$ links remain undeleted. These constitute the shortest connecting tree.

It should be noted that, if all links have different lengths, the shortest connecting tree is unique; otherwise it may not be, although its length is still unique. A perturbation technique, due to Ford and Fulkerson,[13] can be used to force a unique solution if desired. The technique consists of numbering the links from 1 to m and increasing the length of link i by ϵ^i, where ϵ is an arbitrarily small quantity and $1 \le i \le m$. This causes all links to be of different lengths and therefore gives a unique shortest connecting tree.

The two algorithms given will be illustrated by applying them to the network of Fig. 2. Using the first algorithm, we see there are several shortest links, all of length 2. They can therefore be selected in any order, and since they do not form loops, all will eventually be selected. The first three links selected are therefore links CP, EF, and FQ. All links of length 3 are examined next. Of these, links GH and GQ are selected in any case, link EQ is *not* selected because it would form loop $EQFE$, and *either* link BC or link BP is selected. We select BP. Among links of length 4, links AH and AP are selected, links FG and HQ are not selected, and either CD or DP is selected. We choose DP. Figure 5 shows the original network with the links of the shortest connecting tree marked in heavy lines. The total length of the shortest connecting network is 27.

The second algorithm, performed on the sample network, would proceed thus: Link PQ is deleted first, then PE and then AB, EH, and HP and after that DE. This eliminates all links of length 5 or higher. Among links of length 4, FG and HQ and either CD or DP are deleted. Finally BC and EQ are deleted, leaving the shortest connecting tree (see Fig. 5).

The first algorithm can be modified with respect to the order of link selection. A general principle allows us to select a link from among only a few of the original links rather than having to examine all the prospective links. This principle can be stated in the following way:

The shortest link across any cut of the given network is also a link in the shortest connecting network.

If the nodes in a network are separated into two subsets in such a way that each node is in one and only one of the subsets, the links joining pairs of nodes such that one node is in one subset and the other node is in the other subset are said to form a cut of the network. Obviously a cut in the given network corresponds to a cut in the shortest connecting tree. The above-stated principle therefore says that the shortest link among all the links joining nodes on opposite sides of a cut belongs to the shortest connecting tree. In particular, the principle can be applied to any cut separating a single node from the rest of the network or to a cut separating a group of nodes already connected among one another by previously selected links of the shortest connecting tree.[12]

Applying the principle to node D of the sample network, link DP (or CD) could have been the first link selected (see Fig. 5), or applying it to subset $ABCDP$, link AH could have been selected first. Note that these links are the longest in the shortest connecting tree and were among the last selected by the first algorithm. The principle in effect makes it possible to select the links in any order. This is important if we wish to examine specific sections of the given network or if we want to examine only a small number of links for each step.

The following corollary of the general principle is easily seen:

For each link in the shortest connecting tree there is a cut in which that link is the only one in the shortest connecting tree.

From this corollary it is easy to see how the links of the shortest connecting tree can be chosen in any desired order: it is sufficient to examine the appropriate cuts in just that order.

13. One-way Link Networks

For networks composed of one-way links there is at present no simple algorithm known which will solve the problem of the shortest connecting network, such that there exists a path from every node to every other node. One reason for this is the fact that even the number of links required for such a shortest connecting network is not known in advance for a one-way-link network; it can vary between n and $2(n - 1)$ links for n nodes (therefore it is also obviously not a tree). For the two-way-link case it is always exactly $n - 1$ links. Even the shortest link in the network is not necessarily in the shortest connecting one-way-link network. Some answers are still possible, however, as will be seen in what follows.

If the shortest network having a path from a specified node (the origin) to all other nodes is desired, this can be obtained by means of a variant of the general algorithm, where certain network cuts are examined in succession. First the cut separating the origin from the rest of the network is examined and the shortest outgoing link is selected. Then the cut separating the two nodes already connected in the first step is examined and the shortest outgoing link across it is selected. This process, repeated until all nodes are connected ($n - 1$ steps), yields the shortest network so that there is a path from the origin to every node (this is generally different from and shorter than the shortest-path tree, described in paragraph 5; however, it yields generally longer paths).

When the directions of all links are reversed and this same procedure is carried out, the shortest network having paths from each node to the origin is obtained. The union of these two networks contains a path from every node to every other node, although there is no guarantee that it is the shortest such network. The process can be tried using different nodes as origins, and the shortest of these networks can be chosen.

14. Applications

As already mentioned, the shortest-connecting-tree model is useful in the evaluation and establishment of connections involving several parties. Routing of multiple-address messages and the setting up of conference calls are important examples of this.

Multiple-address messages are billed and usually transmitted as separate single-address messages, except for copies with the same destination. The most efficient use of facilities would consist of transmitting along the shortest tree connecting the origin to the destination points (if the objective should be to minimize transmission or delivery delay to all addressees, transmission should proceed along the shortest-path tree; see paragraph 5).

Similarly, conference calls are often set up as separate calls from the conference call originator and joined together at the originator's location. Not only does this result in duplication, but the amount of facilities used depends on which one of the conferees originates the call. Use of the shortest connecting tree minimizes the usage of facilities and makes the connection independent of the call originator.

The leasing of private networks connecting a number of subscribers is another possible application of this model, since here again it is a matter of finding an efficient way of connecting several points together.

A special case of multiple-user connections is constituted by broadcast networks, in which one station transmits the same information to several receiving points. Many such networks exist, among them the entertainment television and audio networks. The application may be particularly important for these, since the connectivity of such a network changes as often as every 15 min as subscribing stations connect to and drop from the network.

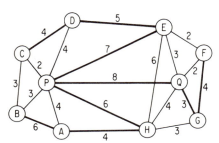

FIG. 6. Longest connecting tree of sample network.

This model can also profitably apply to the connection of what might be called "round-robin" or sequential broadcasting networks, where each station in turn broadcasts for all other stations in the network. The shortest connecting tree will minimize the expenditure for connection links among all the member stations.

Before other applications of the shortest connecting network model are described, it should be pointed out that, in addition to minimizing total connection cost or length, the procedure described can be used to optimize other properties also. Reference 12 describes in more detail the basis for the various generalizations. The first and simplest of these refers to finding the longest connecting tree of a network. This can be done by simply replacing the term "shortest" by the term "longest" each time it appears in the algorithm. Figure 6 shows the longest connecting tree for the sample network of Fig. 2. Its total length is 47.

Another obvious generalization concerns multiplicative parameters, which can be made additive by taking logarithms as described for the shortest-path model in paragraph 10. However, not only does this transformation turn out to be unnecessary, but more general functions of the link parameters than sum or product can be maximized or minimized. All that is required is that the function be monotonic and symmetric in all the arguments. (A symmetric function is one whose value is independent of the order of its variables or arguments. Examples, besides sum and product, would be sum and product of squares, cubes, or exponentials; rms sum; plus many others.[14])

From these considerations it is seen that the shortest-connecting-tree algorithm can be applied to networks involving probabilities, where these represent reliabilities, survivabilities, degrees of link security, etc., and the network connecting all nodes with the minimum number of links and maximum reliability (or similar property) is desired. Furthermore, in view of the property of the algorithm concerning monotonic symmetric functions, it is possible to work with the complements of these probabilities. If the "unreliabilities" are given, the most "reliable" connecting tree can be found by the use of the original minimizing algorithm applied to these "unreliability" link labels. The connecting network thus found has the following interpretation: A message routed along it has maximum probability of being received by *all* nodes when redundant or alternate paths are not permitted. Similar interpretations exist for the case of survivabilities and securities.

As already mentioned in paragraph 13, optimum "all-to-all" connecting networks cannot be found by this method in the case of one-way-link networks. However, the method still applies even in this latter case if there is only one transmitting station. In this case, however, the links of the optimum connecting tree must be generated in a specified sequence, starting from the transmitting node and observing the link directions, as described in paragraph 13. This is not a serious restriction. Thus, if more than one transmitting node exists in a network of one-way links, the optimum connecting tree will in general be different for each transmitting node.

FLOW NETWORK MODEL

Flow networks constitute some of the most important theoretical models of communication systems. The significance of this type of model lies in its applicability and usefulness in determining information-carrying capacities of communication systems. Flow networks are also known as transportation networks or capacitated networks.

In a flow network, the nodes act as sources, relaying points, and/or sinks for one or more commodities. Each link is characterized by a certain numerical capacity. This means that the sum of flows of the various commodities through a link cannot exceed the value of the corresponding link capacity. Commodity ij represents the communications traffic from node i to node j. There can thus exist as many as $n(n-1)$ different commodities. The various commodities are generally not exchangeable for one another; they interact only as they contribute to the saturation of links which they traverse. Commodity flows and link capacities are expressed in the same units, which could be number of wire pairs, bandwidth, transmission rate, or some similar magnitude.

Flow networks are useful in answering questions of the following type: What is the maximum information rate between two given nodes? Can a network with given link capacities simultaneously satisfy a set of communication demands among various nodes? How is the traffic to be routed in order to be able to achieve that maximum rate or to satisfy the given demands? These questions lead to problems of flow-network analysis. Problems of synthesis of flow networks arise in connection with construction of nets which must be able to meet given requirements at minimum total capacity or cost. Even more difficult flow-network problems arise in connection with the augmentation or strengthening of communications systems for the purpose of meeting increased demands.

Flow networks are described from a mathematical point of view by Ford and Fulkerson.[13] A fairly comprehensive account, more oriented toward the communications point of view, is given by Kim and Chien.[15] Some analogies between flow and resistive nets are also pointed out in this reference.

15. Basic Properties of Flow Networks

The series and parallel rules for combining elements in flow networks are closely related to those of minimum-cost-path networks (see paragraph 4). As will be seen later in paragraph 17, the two models are duals of each other.

Series Rule. The capacity of a series combination of capacities is the smallest one of the individual component capacities. Symbolically, $C_s = \min [a, b, \ldots, m]$ (see Fig. 1a), where C_s is the series capacity and $a, b, \ldots, m$ are the capacities of the individual series components (links or subnets).

Parallel Rule. The capacity of a parallel combination of capacities is the sum of the individual component capacities.

Symbolically, $C_p = c + d + \cdots + n$ (see Fig. 1b), where C_p is the capacity of the parallel combination and $c, d, \ldots, n$ are the capacities of the individual parallel components (links or subnetworks).

16. Maximum-flow Theorem

The fundamental problem in flow-network theory is the determination of the maximum amount of flow which can be transmitted through the network from a given source node to a given sink node. This problem amounts to finding the capacity of a single link between source and sink which would be equivalent to the network; it can also be considered as the study of two-terminal flow networks.

In networks consisting entirely of series-parallel combinations this can be done with the rules given above. For networks having a more general structure, the answer is given by the *max-flow min-cut theorem:*

The maximum flow from node i to node j is equal to the capacity of the minimum-capacity cut (min-cut) separating node i from node j.

The capacity of a cut is the sum of the capacities of the links in it; the concept of cut has already been defined in paragraph 11. In the case of networks containing one-way links, only capacities in the appropriate direction across the cut are considered.

It is immediately obvious that the maximum flow cannot exceed the capacity of the minimum cut; various formal proofs show that it is also always possible actually to

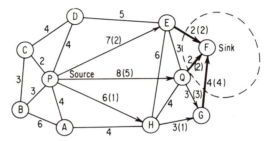

Fig. 7. Illustration of max-flow min-cut theorem.

achieve precisely that level.[13] Reference 16 gives a proof more closely related to communications concepts. Although it is a constructive proof, the procedure it yields for constructing the maximum network flow is not convenient for practical applications. For these, the easiest to use is the one given in Ref. 17. It is a procedure for determining paths from source to sink and maximizing the flow through each until the appropriate min-cut becomes saturated with flows in the forward direction only. Regardless of the algorithm used, the test that the maximum flow has been established consists in finding that a cut separating source from sink has been saturated in the forward direction by virtue of the max-flow min-cut theorem. The saturated cut is obviously a min-cut.

As an example, the max-flow theorem is applied to the network of Fig. 2, which is here interpreted as a flow network. The link labels now represent capacities. It is desired to find the maximum possible flow from node P to node F (in the case of networks composed of two-way links only, source and sink can be interchanged without altering the problem). The minimum-capacity cut which separates nodes P and F contains links EF, GF, and QF, and its capacity is 8 (see Fig. 7). A flow distribution which realizes the maximum flow is also given. The numbers in brackets and the arrows indicate the amount of flow and its direction in each link.

17. Network Duality

A close relation exists between the max-flow problem and the shortest-path problem in the case of some planar networks.[18] A planar network is one which can be mapped onto a plane (or a sphere) in such a fashion that links touch only at nodes. The network of Fig. 2 is planar; link EH can be drawn in such a way that it does not cross any other link.

A network is planar relative to nodes I, J (or IJ-planar) if it is planar after the addition of link IJ. Thus the network of Fig. 2 is BE-planar but not PF-planar.

The dual of a planar network has a node (mesh) corresponding to each mesh (node) of the primal and a link corresponding to each link of the primal. Figure 8 shows the network of Fig. 2 redrawn so that its planar nature becomes evident; superimposed on it is its dual network. The nodes of the dual are drawn square, and its links dotted.

The dual of a planar flow network can be drawn as a cost network, with dual link costs numerically equal to the link capacities of the primal net. Then, to each cut in the flow network corresponds a closed circuit in the cost net. The duality is also very apparent in the series and parallel combination rules of both models. The series rule in one is identical with the parallel rule in the other. In a flow network which is planar relative to source and sink and which has a source-sink link, to every

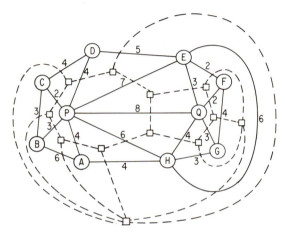

Fig. 8. Planar network and its dual.

cut separating source from sink in the flow network corresponds a path in the cost network if the dual of the source-sink link has been deleted. To the min-cut of the flow net corresponds the shortest path between the nodes at the two ends of the deleted dual link.

It is thus seen that a max-flow problem between source I and sink J in an IJ-planar network corresponds to a shortest-path problem in its dual net. In order to find the max-flow by this method, draw a "dummy" link from I to J and give it infinite capacity; then find the shortest path between the two dual nodes adjacent to the dummy link. The length of this shortest path is the min-cut capacity, and the min-cut links are the duals of the shortest path links. If a link of finite capacity exists originally between I and J, it is removed before the dummy link is inserted. The original IJ link and its capacity must be added to the solution at the end.

The duality between flow and cost networks works both ways, and historically it was first used to determine shortest paths by finding max-flows, since efficient solutions to this latter problem were developed first. At present, however, finding shortest paths is simpler than the determination of max-flows, and the duality property is often used in this latter fashion when it applies.

Although implicitly described here for two-way-link networks, duality and the method just described also apply in the case of one-way-link and mixed networks. Care must be exercised with respect to link directions, however. Reference 1 gives the rules for assignment of link directions in the dual links.

18. Multiterminal Networks

The maximum flow through a network for a pair of nodes having been established, it would seem that, in order to determine the max-flows for all node pairs, $n(n-1)/2$ max-flow problems would have to be solved, since there are that many different node pairs. However, a theorem due to Gomory and Hu[13,19] shows that it is sufficient to solve $n-1$ simple max-flow problems. Furthermore, the $n-1$ min-cuts obtained define a cut tree of the network. This cut tree is equivalent to the original network and shows the structure of the $n-1$ min-cuts (as a result of ties, more cuts may appear; these ties can be broken by means of the perturbation technique described in paragraph 11, leaving $n-1$ "essential" min-cuts).

The sequence of $n-1$ max-flow problems to be solved is easily seen: It is only necessary at each stage to pick two nodes not yet separated by a min-cut and solve the corresponding max-flow problem. The cut tree is also found very easily: A link is drawn between every two nodes separated by *exactly* one min-cut; the capacity of this link is set equal to that of the corresponding min-cut.

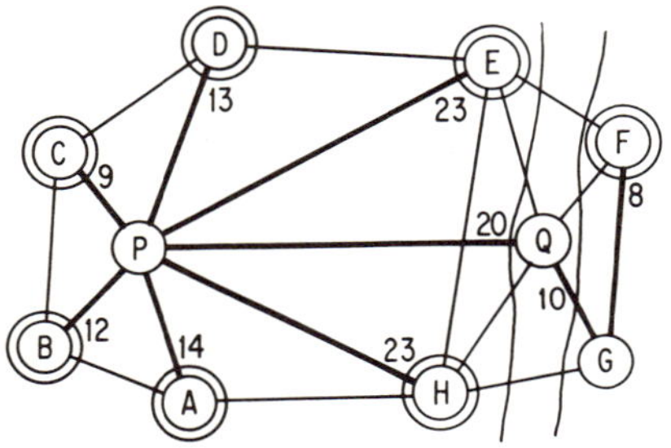

FIG. 9. Essential min-cuts and cut tree of flow network.

Figure 9 shows the $n-1$ min-cuts (after appropriate tie-breaking) and the cut tree (in heavy lines) of the sample network. The cuts and tree links are labeled with their capacities.

It is easy to determine the max-flow between any two nodes by looking at the cut tree: it is the capacity of the path along the cut tree joining the two nodes.

The synthesis of multiterminal flow networks, given the max-flows for all node pairs, is also possible. Solutions of such problems are not unique. The solution using minimum total link capacity is given in Refs. 13 and 20. A more general minimum total cost solution, when the cost per unit capacity for each link is given, is described in Ref. 21.

19. The Multisource Multisink Problem

Although single-source single-sink problems only have been considered so far, these are easily extended to multiple sources and sinks as long as only one commodity is involved, that is, that flow can terminate at any sink regardless of which source it originated at. This is done by adding a dummy master source and a dummy master sink node and appropriate links from these to the given sources and sinks.[13]

20. Multicommodity Problems

Multicommodity flow problems deal with flows which are identified with given node pairs. Thus commodity ij originates only at node i, terminates only at node j, and cannot be interchanged with any other commodity. Since this is the case in many communications systems, these problems are of great importance in communications system engineering. Unfortunately, although several problems have been stated, they do not have simple and elegant solutions like those described so far.

The two-commodity problem (maximization of the sum of the flows of two commodities) is treated and solved by Hu.[22] He also states that the method does not extend to more commodities. What amounts to a general multicommodity problem [maximizing the sum of the flows of all commodities, which could be as many as $n(n-1)$ given a set of commodity demands] has been stated and solved, but only in terms of linear programming.[23] Because of rapid increase in the amount of computation with the size of the problem, hand calculations are practical only for relatively small problems, although machine computations extend the range somewhat. Optimum solutions of this problem do not lead to uniform quality of service for all commodities. Zero flow may result for some of them, leading to no service to those users. However, this may be an acceptable criterion in many cases.

The problem of optimum multicommodity flow under the condition of uniform grade of service has been stated,[24] together with some necessary conditions and possible avenues of solution. Necessary and sufficient conditions for the unrestricted maximum multicommodity flow problem have been conjectured.[25,26] Some multicommodity synthesis problems have also been stated.

21. Applications of Flow-network Model

Single-commodity max-flow and multiterminal problems have application in communication systems in which only traffic between two selected points is transmitted at any one time. This is the case in some command and control nets and in some time-shared networks. The single-commodity multisource multisink model should find application in power-transmission networks and other transportation-type problems. Conceptually, single-commodity problems are important in that they are a first step contributing to the solution of multicommodity problems, which are fundamental in communications.

Multicommodity network models have been used to optimize traffic routing in some networks of small and medium size with the aid of automatic computation.[23] Another multicommodity network model has been applied to obtaining minimum required link capacities in both local and global military communication systems (network synthesis and network augmentation or strengthening) with the use of hand calculation only.[24]

As solution procedures to multicommodity flow problems are found and successively improved, they will play a role of increasing importance in the design and management of communication systems.

22. Physical Models of Flow Networks

Conceptually it is easy to visualize the hydraulic model of a flow network. The links are flow-limited pipes, and the nodes are simply pipe junctions. One-way links would incorporate appropriate valves. Source nodes would be connected to fluid reservoirs under appropriately high pressure. Conceptually such a model is attractive; in practice, however, such details as precise flow measurement and limitation in links and reconnection flexibility, combined with probable limitations on network size, detract from the desirability of constructing such a model, especially in view of the relative simplicity of the max-flow algorithm. This model might still be attractive for dynamic problems; however, no such model has ever been proposed for construction, to this author's knowledge.

The electrical analogue has been described, however.[5] Each one-way link contains a current source whose value represents the link capacity and a diode which ensures current flow in the proper direction. Two-way links are represented by two identical one-way links connected in opposite directions. A master voltage source, connected between the source and sink nodes, forces current through the network. The amount of current through the net measures the maximum flow, the saturated links compose the minimum capacity cut, and the current distribution represents the flow distribution through the various links, saturated and unsaturated.

The difficulty of building flexible and inexpensive current sources has prevented this electrical analogue from being practically implemented; it remains a useful conceptual tool, however, in a world which has become accustomed to thinking in electrical terms.

An electrical analogue of a multicommodity flow model can also be envisaged by representing each commodity by a different frequency. The practical difficulties of construction of such a model would be formidable even compared with the single-commodity model just described.

GENERAL THEORY OF NETWORK MODELS

The previously described models of communications systems can be considered as special cases of the general theory of networks. This theory is concerned with the common properties and similarities of the various types of networks and the application of techniques useful with one type of network to other kinds of networks.

23. General Characteristics of the Various Types of Networks

The links of a network can be either undirected (two-way) or directed (one-way). Based on this classification, we can have undirected networks, directed networks, and mixed networks. The undirected networks (those composed only of two-way links) are generally the most fruitful to study. Most network properties apply to these. All the network properties discussed so far apply to undirected networks. In addition, it is often possible to replace a two-way link by two one-way links if this is convenient. Directed networks (composed only of one-way links) are generally harder to deal with, and their links cannot generally be replaced by two-way links. Mixed networks (containing both one-way and two-way links) have not been extensively studied. From both a theoretical and a practical viewpoint they appear to be less important than the "pure" types previously mentioned.

Networks can also be classified according to the number of parameters associated with each link. Most of the models considered so far are single-parameter networks: Each link is characterized by one number (cost, capacity, etc.). Other important single-parameter networks can be cited. There are probabilistic switch networks where each link contains a randomly operated switch; the parameter indicates the probability of the switch being closed. Switching networks are another particularly simple type of single-parameter network: The link parameters are Boolean variables and can take only the values zero or 1. Resistance networks are another very important instance of one-parameter networks.

Examples of multiple-parameter networks are the capacitated cost networks discussed in paragraph 31 and general RLC electrical networks. In these each link is characterized by more than one parameter.

Networks can also be classified according to whether the link parameters of one link are independent of those of other network links or not. Most useful network types are of the uncorrelated kind (link parameters are independent of those of other links); these include all those previously discussed in detail. Examples of correlated-link networks are multiple-contact relay switching networks, where the same Boolean variable appears in different branches; correlated probabilistic networks, where the closure probability of the switch in one branch is correlated with that in another branch; and electrical networks containing transformers (mutual inductances) whose windings are in different network branches. Generally, correlated networks are much more difficult to analyze than uncorrelated ones and, partly for this reason, tend to be of less practical importance.

24. Single-parameter Networks

The most useful network models are those composed of one-parameter, two-way, uncorrelated links. Examples of these have already been given. We proceed to discuss some properties common to several different models.[9]

The interest in single-parameter networks is usually related to the determination of some characteristic of a particular type of network. The most common problem is that of finding the "driving-point" value of a network as seen from a pair of nodes or terminals. This is essentially the problem of determining the value of a link between the two terminals equivalent to the given network. Reduction to one equivalent link is generally possible in single-parameter networks, as opposed to multiparameter networks, where this is often not possible; this is an additional strong reason for the interest in single-parameter networks.

Properties useful in this equivalence determination and common in the better known types of networks are briefly described in what follows, together with the application of these properties to such networks (see Table 1). The types of networks to be considered are cost, capacity, electrical resistance, uncorrelated probabilistic, and (Boolean) switching networks. Applications of cost and capacity networks have already been discussed. Electrical resistance networks are well known and do not require any further explanation, except possibly to say that they can be generalized to impedance networks (at a single frequency) if complex values and operations are admitted for the single-parameter links. Uncorrelated probabilistic switch networks have been used in reliability and communication system vulnerability (survivability)

Table 1. Properties of Various Types of Networks

	Cost	Capacity	Resistance	Uncorrelated probabilistic	Boolean switching
Series......................	Yes	Yes	Yes	Yes	Yes
Parallel....................	Yes	Yes	Yes	Yes	Yes
Y-to-Δ.....................	Yes	Yes	Yes	No*	Yes
Δ-to-Y.....................	Yes	Yes	Yes	No*	Yes
Star-to-mesh...............	No	Yes	Yes	No	Yes
Mesh-to-star...............	No	No	No	No	No
Quasi linearity.............	Yes	Yes	No	Yes	Yes
Weak reciprocity...........	No	No	Yes	Yes	Yes

* Holds approximately.

studies.[9] Switching networks are also well known. Those considered here consist only of uncorrelated variables (single-contact, no controlled or feedback variables).

25. Series and Parallel Rules

These have already been described for cost and capacity networks (see paraphraphs 4 and 15), and are well known for resistor and switching networks (Boolean product and sum for the latter). For the probabilistic switch network the series rule is that the combination is the product of the individual probabilities; the parallel rule is that the complement of the product is the product of the complements of the individual parallel links. Symbolically, $p_p = 1 - (1 - p_1)(1 - p_2) \ldots (1 - p_n)$, where p_p is the probability of the parallel combination and p_i is the probability of the ith link in parallel.

All generally known types of networks have rules for series and parallel connections, and these are generally sufficient for reducing individual networks consisting only of such series-parallel link combinations. As is well known, however, these are not enough for more general network connectivities. The following properties, where they apply, are helpful in a variety of situations.

26. Wye-to-Delta and Delta-to-Wye Transformations

These are very useful devices for network analysis and reduction. They apply to all the models mentioned except the probabilistic switch network; for the latter approximate transformations apply.[9] These transformations allow the substitution of the three-link configuration of Fig. 10a (wye) for that of Fig. 10b (delta) and vice versa when they are embedded in a network. Table 2 gives the conversion formulas for those networks for which the transformation is possible.

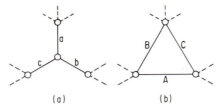

FIG. 10. (a) Wye configuration. (b) Delta configuration.

It should be noted that the wye-to-delta transformation eliminates the center node and therefore simplifies the network; however, if access to this center node is needed, the transformation cannot be used.

Table 2. Wye-to-Delta and Delta-to-Wye Transformations for Different Types of Networks

	$Y \rightarrow \Delta$	$\Delta \rightarrow Y$
Cost.....	$A = b + c$	$a = \frac{1}{2}(B + C - A)$ for $B + C \geq A \geq \lvert B - C \rvert$ $a = 0$ for $A > B + C$ $a = \min [B,C]$ for $A < \lvert B - C \rvert$
Capacity	$A = \dfrac{b + c - a}{2}$, for $b + c \geq a \geq \lvert b - c \rvert$ $A = 0$, for $a > b + c$ $A = \min [b,c]$, for $a < \lvert b - c \rvert$	$a = B + C$
Resistance...	$A = \dfrac{1}{a}(ab + bc + ca)$	$a = \dfrac{BC}{A + B + C}$
Switching....	$A = bc$	$a = B + C$

27. Star-Mesh Transformations

These are generalizations of the wye-delta transformations. For those network types for which the general star-to-mesh transformation is possible (see Table 1), the conversion formulas generalize in the obvious way from the corresponding wye-to-delta formulas of Table 2. Mesh-to-star transformations are not generally possible.[9, 27]

28. Quasi Linearity

This property refers to the characteristics of "black-box" two-port networks. A specific type of network is quasi-linear if, given the driving-point values measured at port 1 for the open-circuit and short-circuit conditions at port 2, it is possible to determine the driving-point value at port 1 *for any arbitrary value* of the load connected across port 2 (see Fig. 11). It should be noted that electrical resistance networks do not have this property, as can be seen from the counterexample of Fig. 12. The two networks have the same open-circuit and short-circuit resistances across port 1, but for

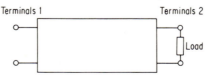

FIG. 11. General two-port network.

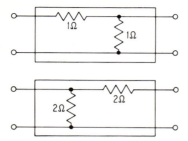

Fig. 12. Quasi linearity does not apply to resistance networks.

an arbitrary resistance load connected across terminals 2, the values seen across terminals 1 will differ.

29. Weak Reciprocity

This property also refers to black-box two-port network configurations. A given type of network is weakly reciprocal if an algebraic operation (denoted by the symbol $\oplus$) exists such that

$$V_{1o} \oplus V_{2s} = V_{2o} \oplus V_{1s}$$

where V_{1o} is the driving-point value measured at terminals 1 with terminals 2 open, V_{2s} is the driving-point value measured across terminals 2 with terminals 1 shorted, with similar meanings for V_{2o} and V_{1s}.

For resistance networks the operation is ordinary multiplication and the property is ordinary reciprocity and amounts to the relation $R_{1o}R_{2s} = R_{2o}R_{1s}$. For probabilistic switch networks the operation is ordinary addition. Boolean switching networks turn out to be weakly reciprocal with respect to both logical addition and logical multiplication.

30. General Conclusions

Examination of Table 1 shows a fairly close relationship between the number of the properties listed which each model has and the extent of its practical application. The electrical and switching network models are undoubtedly the most widely used. The popularity of electrical networks is, of course, due to many other reasons besides their having the properties listed in the table. These reasons include the relative ease with which such networks can be physically built and the wide variety of problems and situations which can be represented by such networks. Cost and capacity networks are also finding progressively wider application. This is due in large measure to the fact that they are easily understood, and although physical models of these types of networks have not been used very much, the simplicity of some of the algorithms and computational procedures allows quick hand calculation of networks of reasonable size. In addition, this same simplicity allows the treatment of very large networks (of up to thousands of elements in some cases) by present-day computers. The probabilistic switch model, although potentially useful, has the drawback of not only the lack of some important properties but also the relative complexity of the formulas for computation even of those properties that it does have.

The theory of network models in general and that of single-parameter networks in particular allow us to examine the different models of communication system models for various properties and to choose the model most suitable to a particular need, based on the properties of the various models.

In the foregoing discussion, some representative properties of the more widely used types of network models have been presented. Other models may be found useful for specific applications; if this should be the case, the general theory and the insight it brings will help us to understand the power of these models and also their limitations. The best known and most widely used model is obviously the electrical network. The theory presented here brings forth clearly which properties of electrical networks carry over to other types of networks.

The field of communication system models is developing very rapidly. New types of models, properties, and applications will continue to appear.

MULTIPLE-PARAMETER NETWORK MODELS

In all the models discussed in detail so far, links have been characterized by a single parameter: cost, capacity, etc. Frequently, however, it is desired to examine more than just one characteristic of a communication system. For these cases, multiple-parameter network models are required. Among the simplest of these are capacitated cost networks.

31. Capacitated Cost Network Models

These constitute generalizations of pure cost and pure flow network models. Each link can be characterized by a capacity, by a figure of cost per unit capacity, or by both. The cost figure represents the cost of building or using a unit of capacity in the link, as the case may be.

Here again we recognize single- and multiple-commodity problems. Some single-commodity problems of this type have become well known in other fields, although they have not found too wide application in the study of communication systems. Among these are the transportation problem and the transshipment problem. In the transportation problem[4] a set of sources and a set of sinks is given, each with its capacity; that is, a source cannot originate more flow than its numerical capacity and similarly for the sinks. The unit cost of transportation from each source to each sink is given, and it is required to supply the required amount of flow at minimum total cost. The transshipment problem[4] is a slightly generalized version of the transportation problem in that sources and sinks are in a cost network with intermediate nodes, rather than being connected only by direct links. The transshipment problem can be reduced to the transportation problem by substituting shortest path lengths (or costs) from each source to each sink for the original network. Simple algorithms exist for solving both problems directly.[4]

The above-mentioned problems are of interest because they serve as starting points for generalizations to multiple-parameter multicommodity problems of direct concern to communication systems. Some of these will be mentioned briefly; others can be imagined readily.

In a cost-flow network, given a set of commodity demands, what is the minimum cost at which the maximum amount of total demands can be served? Another problem might be to assign link costs to a flow network in such a manner that, when each commodity seeks its minimum-cost, or shortest, routes, the maximum amount of flow results. A third problem can be stated if in addition to link capacities and costs, cost charges for each commodity are given from which commodity transportation costs are to be subtracted. Under these conditions, what is the maximum revenue obtainable?

A cost-capacity synthesis problem very much applicable to communication systems and related to those just mentioned exists and has a simple solution. Given a set of nodes, a set of unit costs for the various node pairs, and a set of node pair commodity demands, the problem is to construct the network of lowest total cost capable of satisfying all commodity demands simultaneously. The solution is obtained by repeated application of the shortest-path algorithm to the cost network defined by the set of unit costs. The shortest path is found for each node pair for which a demand is given; links which are used by these paths are marked, and the capacity required by each commodity in each link is noted; finally, the required capacity for each link used is totaled. The network composed of these links is the least-cost net which satisfies all the demands simultaneously, since it has no spare capacity and each commodity flows along the shortest path available to it. If links for all node pairs are allowable and all link costs are equal, the problem reduces to constructing the corresponding network of minimum total capacity.   This problem is trivial; the minimum capacity network is identical with the network composed of all the demands, since in this net all paths are exactly one link long, which is the minimum possible length.   This same answer applies in the case where all links are permitted and unit costs, although not identical, obey the triangle inequalities.   This is due to the fact that in this case also the shortest path between every set of two nodes is composed of only one link.

So far all commodities have been considered conservative; that is, no flow is lost in transit.   Reference 28 considers "lossy" flow networks; each link has both a capacity and a loss coefficient.   Conditions of maximum flow under these circumstances are given in the reference.   These commodity losses can be considered as error rates or unreliabilities; with these interpretations this model is of interest in the study of communication systems.

Many other models involving two or more parameters have been proposed in connection with communication and transportation systems.   Some of the problems which they suggest yield to special procedures.   Others require more general approaches, involving linear and nonlinear programming, variational calculus, or other techniques.   Still others cannot be treated by any of these techniques, and full-fledged simulation is required in such cases.

OTHER MODELS OF COMMUNICATION SYSTEMS*

General model theory is a very vast field which cannot be treated here.   Even when restricted to communications, the subject cannot be covered comprehensively in a few pages, since almost every aspect of any communication system can be modeled.   So far, we have discussed in more or less detail a few models which are useful to the communications system engineer.   In all these models, communication systems are always represented by networks composed of nodes and links, where the nodes are the points where communications originate, terminate, or are retransmitted and the links represent the transmission medium.

Many other types of models of communication systems have been studied, both more general and more specialized than those presented so far.   Reference 9 constitutes a catalogue of virtually all modeling techniques of significance to communication systems.   Each model is explained, many with examples, and pertinent references are included.

32. Typical Examples of Other Models

In what follows, some of the more important modeling techniques applicable to communications systems are briefly mentioned.

In increasing order of generality, starting from network models, the theory of linear graphs can be mentioned.[29,30]   This deals with the basic structure of networks (linear graphs).   Because of its generality, graph theory is important not only in communication systems but also in connection with electrical networks, where its significance has lately been increasingly recognized.   A field of even greater generality is constituted by topology, of which the theory of graphs is a branch.   Topology studies those properties which remain invariant under deformation.

Linear-programming models constitute generalization in a different direction from some of the models described, which are special cases of the general linear-programming model.[4]   The special characteristics of the models treated make it possible to solve meaningful problems by hand computation, while quite large problems can be solved by machine.   The general linear-programming algorithm, however, requires considerably more effort, and the sizes of problems which can be treated must be scaled down accordingly.

In turn, many optimization problems have been artificially "linearized" in order to take advantage of the existence of linear-programming algorithms.   In those cases where this simplification cannot be justified, other optimization techniques which

* See also Chap. 4.

involve more computation may apply. Again in increasing order of generality, they are quadratic programming and convex programming.[4] Beyond this, it becomes increasingly difficult to find globally optimal solutions, and it often becomes necessary to settle for suboptimal or approximately optimal answers. Some results have also been reported with optimization techniques under uncertainty (random variations in the parameters).[4]

Service quality in a communication system, comprising such topics as delays in obtaining service, loss of service due to overloaded facilities, etc., can be studied with the aid of queueing theory,[31] which actually had its origin in the communications industry with Erlang's work. Models of communication systems have been developed recently which allow the study of the effects of various types of traffic priorities, service policies, etc., on service times and delays.[32]

33. Simulation

The line between modeling and simulation is difficult to draw, just as it is difficult to define simulation. On the one hand, almost every instance of simulation can be considered as a special type of model; on the other hand, mathematical and physical models can be considered to be special-purpose simulators. Generally, simulation is considered to involve the use of high-speed digital computers, although there are many exceptions to this statement. Special-purpose and analogue machines have been used, and man-machine combinations and other configurations have also been used.

Generally, simulation is a useful tool in cases where no adequate analytical tools exist or when time considerations do not allow their development and application. It is also often used to confirm analytical results before a system is built.

The advantages of simulating are those that go with the use of all models: Many configurations can be tested under many conditions in a completely controlled environment. With respect to the use of other models, the advantages are that more factors can be considered (more "realistic"). On the other hand, results of simulation runs are less general than corresponding analytical results and apply only under the conditions of the run.

Early instances of what is now called simulation are constituted by the use of "throw-down machines" by the Bell System. These incorporated human switchboard operators and "real" switchboards modified to simulate real traffic. They are very realistic but had the disadvantage of proceeding in real time; that is, the passage of time could not be compressed, and it took very long to gather the relevant statistics under a variety of conditions. Military field exercises, whether directed specifically at testing communications or of a more general nature, also fall into the classification of real-time simulation; they have similar advantages and drawbacks as the previous example.

An early instance of computer simulation of a communication system is that of Ref. 33. Many references concerning communication system simulation will be found in the issues of the *Transactions of the Professional Group on Communication Systems of the IEEE* and in the proceedings of the various communications and computer conferences.

REFERENCES

1. Pollack, M., and W. Wiebenson, Solutions of the Shortest-route Problem—a Review, *Operations Res.*, March–April, 1960.
2. Shimbel, A., Structure in Communications Nets, *Proc. Symposium Inform. Networks*, Polytechnic Institute of Brooklyn, Apr. 12–14, 1954.
3. Dantzig, G. B., On the Shortest Route through a Network, *Management Science*, vol. 6, no. 2, p. 187, January, 1960.
4. Dantzig, G. B., *Linear Programming and Extensions*, Princeton University Press, Princeton, N.J., 1963.
5. Dennis, J. B., *Mathematical Programming and Electrical Networks*, The Technology Press of the Massachusetts Institute of Technology, Cambridge, Mass., 1959.
6. Minty, G. J., A Comment on the Shortest-route Problem, *Operations Res.*, vol. 5, no. 5, p. 724, October, 1957.

7. Klee, V., A "String Algorithm" for Shortest Path in Directed Networks, *Operations Res.*, vol. 12, no. 3, p. 428, May–June, 1964.

8. Rapaport, H., and P. Abramson, An Analog Computer for Finding an Optimum Route through a Communications Network, *IRE Trans. Commun. Systems*, vol. CS-7, no. 1, p. 37, May, 1959.

9. U.S. Army Electronic Proving Ground, *Investigation of Model Techniques*, Final Report, Fort Huachuca, Ariz., July, 1961; Defense Documentation Center document AD275549 (unclassified).

9a. Pollack, M., Solutions of the kth Best Route through a Network—A Review, *J. Math. Analysis Appl.*, vol. 3, no. 3, p. 547, December, 1961.

10. Hambrock, H. E., and C. G. Svala, Saturation Signaling: The Optimum Alternate Routing Scheme, *Ninth Natl. Commun. Symp.*, Utica, N.Y., October, 1963.

11. Kruskal, Jr., J. B., On the Shortest Spanning Subtree of a Graph and the Traveling Salesman Problem, *Proc. Am. Math. Soc.*, vol. 7, no. 1, p. 48, February, 1956.

12. Prim, R. C., Shortest Connection Networks and Some Generalizations, *Bell System Tech. J.*, vol. 36, p. 1389, November, 1957.

13. Ford, L. R., Jr., and D. R. Fulkerson, *Flows in Networks*, Princeton University Press, Princeton, N.J., 1962.

14. Korn, G. A., and T. M. Korn, *Mathematical Handbook for Scientists and Engineers*, McGraw-Hill Book Company, New York, 1961.

15. Kim, W., and R. T. Chien, *Topological Analysis and Synthesis of Communication Networks*, Columbia University Press, New York, 1962.

16. Elias, P., A. Feinstein, and C. E. Shannon, Note on Maximum Flow through a Network, *IRE Trans. Inform. Theory*, vol. IT-2, no. 4, p. 117, December, 1956.

17. Hoffman, A. J., and R. E. Gomory, Finding Optimum Combinations, *Intern. Sci. Technol.*, July, 1962.

18. Ford, L. R., Jr., and D. R. Fulkerson, Maximal Flow through a Network, *Can. J. Math.*, vol. 8, p. 399, 1956.

19. Gomory, R. E., and T. C. Hu, Multi-terminal Network Flows, *J. Soc. Ind. Appl. Math.*, vol. 9, no. 4, p. 551, December, 1961.

20. Chien, R. T., Synthesis of a Communication Net, *IBM J. Appl. Res.*, vol. 4, p. 311, July, 1960.

21. Gomory, R. E., and T. C. Hu, An Application of Generalized Linear Programming to Network Flows, *J. Soc. Ind. Appl. Math.*, vol. 10, no. 2, p. 260, June, 1962.

22. Hu, T. C., Multi-commodity Network Flows, *IBM Res. Rept.* RC-865, T. J. Watson Research Center, Yorktown Heights, N.Y., January, 1963.

23. Kalaba, R. E., and M. L. Juncosa, Optimal Utilization and Extension of Interoffice Trunking Facilities, *AIEE Commun. Electron.*, no. 40, p. 998, January, 1959.

24. Furth, E., Optimum Utilization of Communication Systems, presented at the *7th National Communications Symposium (NATCOM)*, Utica, N.Y., October, 1961.

25. Hu, T. C., On the Feasibility of Simultaneous Flows in a Network, *Operations Res.*, vol. 12, no. 2, p. 359, March–April, 1964.

26. Tang, D. T., Comments on Feasibility Conditions of Simultaneous Flows in a Network, *Operations Res.*, vol. 13, no. 1, p. 143, January–February, 1965.

27. Guillemin, E. A., *Introductory Circuit Theory*, John Wiley & Sons, Inc., New York, 1953.

28. Mayeda, W., and M. E. Van Valkenburg, Properties of Lossy Communication Nets, *Univ. Illinois Coordinated Sci. Lab.*, *Rept.* R-200, Urbana, Ill., April, 1964; Defense Documentation Center document AD 439437 (unclassified).

29. König, D., *Theorie der Endlichen und Unendlichen Graphen*, Chelsea Publishing Company, New York, 1950.

30. Berge, C., *The Theory of Graphs and Its Applications*, John Wiley & Sons, Inc., New York, 1962.

31. Cox, D. R., and W. L. Smith, *Queues*, Methuen & Co., Ltd., London, 1961.

32. Kleinrock, L., *Communication Nets*, McGraw-Hill Book Company, New York, 1964.

33. Brotman, L., and J. Minker, Digital Simulation of Complex Traffic Problems in Communication Systems, *Operations Res.*, vol. 5, no. 5, p. 670, October, 1957.

Chapter 6

TRANSMISSION ENGINEERING OF SWITCHED SYSTEMS

ARDEN C. JOHNSON, *American Telephone and Telegraph Company*

INTRODUCTION

1. General

The early developers of the telephone realized that it would be technically and economically impossible to connect every telephone to every other telephone via direct circuits. Wire centers were established, and the transmission design was arranged to make it possible to provide voice communication between any two points that could be physically interconnected. The same philosophy applies to any switched system. When a private-line user has a large number of locations, a switched system may provide greater flexibility and economy than direct circuits among the many locations. Special consideration must be given to transmission engineering, however, to ensure that the system will perform as required.

As an example, let us assume that two data sets are designed to be connected by a circuit with a 1,000-cycle insertion loss of 8 db. This is not a difficult requirement to meet on a direct circuit. If the data sets are connected to switching centers interconnected by trunks, the insertion loss must be divided between all the circuits in tandem in order to meet the requirement for transmission between the two sets. The problem can be further complicated by alternate routing between the switching centers, involving still other circuits and switching centers. It can be seen that transmission design of each individual circuit must be established so any possible combination will produce the required insertion loss. The station-to-station requirements could therefore affect the type, quantity, and location of switching centers; routing plans; choice of transmission facilities; etc. This chapter will discuss some of these problems.

The discussion in this chapter covers systems which can be developed by the adaption of voice frequency transmission facilities. There are special switching arrangements for television, program, and other broadband and special transmission systems. These systems have to be patched manually, or special means have to be used to make the switches effectively transparent. Special problems arise concerning regulation, regeneration, synchronization, etc. The principles developed here will generally cover these special systems, although the details will be different.

2. Preparation for Transmission Studies

A switched system may be designed for only one form of transmission, or it may be planned for several types of signals. Among these are voice; teletype; low-, medium-

and high-speed data; facsimile; and other signals. Before proceeding with transmission engineering, it is necessary to know what forms of transmission are to be used, locations from which each form of signal will be transmitted, and the locations at which each form of signal will be received. For example, some locations may plan to use voice communications only, some will use voice and teletype, others voice and data, etc. These and other factors are discussed in Chap. 1.

The user equipments and services must be known and analyzed in detail. This is covered in Chap. 2. From this information, the engineer must obtain the requirements for such things as sending levels, received levels, attenuation-frequency characteristics, impedances, noise limits, envelope delay distortion limits, maintenance variations, and other pertinent transmission characteristics. It is also important to determine what types of signaling and supervisory systems are required, since some of the equipment provided for signaling and supervision will have an effect on the transmission design.

Information on user equipment and services should be reviewed carefully, and requirements checked against some of the design parameters stated in other parts of this chapter. Where the requirements appear to be extreme, it may be well to consider changes in the station equipment before freezing design requirements. It may be cheaper to modify, improve, or replace the station equipment than to design and build a switched system to meet stringent requirements.

Acoustic effects should be reviewed as discussed in Chap. 3 if the switched system is to handle voice communications. While other forms of station equipment are not adversely affected by poor acoustic conditions, the telephone or its equivalents may become unusable. Acoustic noise picked up by the telephone transmitter can have the same bad effect on voice transmission as circuit noise.

Before proceeding further, it might be well to review communication system modeling as discussed in Chap. 5. This may suggest a more logical procedure for development of the transmission design.

Switching engineering is discussed in Chap. 7. Switching design is very closely interrelated with transmission design. The types of switching centers, their locations, and the routing arrangements all have a bearing on the control of overall transmission characteristics of circuits connected in tandem. The transmission engineer must expect that switching plans may be developed based on traffic studies and economic selection studies. These plans should be reviewed as they develop to ensure that they will permit satisfactory transmission. The transmission engineer should also expect to have a great deal to say about types of switches, switching features, routing arrangements, etc. It will be shown that all these things have an effect on transmission. Therefore it appears that transmission engineering and switching engineering must go forward concurrently and the transmission and switching designs must be made compatible in all respects. Neither can be considered complete until the other is satisfied.

The transmission engineer will encounter many specialized problems in the process of doing his job. Questions concerning transmission characteristics of facilities, switching methods, data, multiplex, radio, etc., are covered in following chapters of this book. Other sources may also be consulted.[1-3]

3. Other Considerations

The transmission engineer should review the transmission plan finally from two other points of view. First, can the plan be administered? For example:

1. Has standardization been provided to the maximum extent?
2. Is it possible to make the required measurements?
3. Are the specified limits within the state of the art?
4. Can routines be made regularly and easily?
5. Is there a system of control?

Second, is the plan economically sound? What will it cost? Has the user been made aware of the cost of unusual or special transmission features? Chapters 24 and 25 have been prepared to assist the engineer with these types of questions.

4. Terminology

Some of the terms that commonly appear throughout this chapter are defined here to clarify their use.

Switching Center—A location where equipment is assembled to provide for automatic connection of any combination of trunks, access lines, or station lines. It includes the terminal equipment.

Trunk—A transmission path between two switching centers. It includes a transmission facility and the terminal equipment at each end.

Access Line—A transmission path between a switching center and a PBX or similar concentrating point. It consists of a transmission facility with terminal equipment at both ends.

Station Line—A transmission path between a switching center and a station. It consists of a transmission facility and may include terminal equipment at a switching center (or PBX) to connect a transmission facility to switches. It may consist of a trunk circuit, pads, amplifiers, equalizers, echo suppressors, and signaling equipment. Access lines may require terminal equipment similar to trunks. Station lines may require only a line relay, or they may require the use of complex line circuits and conditioning equipment.

Trunk Circuit—An arrangement of relay equipment to connect a trunk to the switches. It performs many functions which are discussed later.

Transmission Facility—A transmission channel with known transmission characteristics. It may be provided with open wire or cable or may consist of a radio or cable multiplex channel.

Drop—The circuit through the terminal equipment which connects the facility to a switch.

EFFECT OF SWITCHING PLANS ON TRANSMISSION DESIGN

5. General

The transmission engineer is concerned with establishing requirements for loss, attenuation-frequency characteristics, envelope delay distortion, noise, etc. This is not difficult when two stations are directly connected because the transmitting and receiving requirements of the circuit are automatically established by the station equipment requirements. When switching is introduced, it becomes impossible to predict the exact combination of lines and trunks that will be connected together. There are so many combinations that adjustment of transmission characteristics can no longer be done entirely at the station. It is evident that to accomplish the transmission plan it may be necessary to specify the types of switching centers, their locations, and the features necessary in the switching equipment if every connection is to provide a uniformly good quality of transmission. We shall therefore review some of the switching arrangements.

6. Switched Connections

The simplest switched connection is shown in Fig. 1. Both station locations are terminated through their station lines to a common switching center and may be connected together through the switches, or they may be connected to other station locations. The transmission requirements for the station lines from *A* to the switch and from the switch to *B* must be sufficiently stringent that the overall requirement from *A* to *B* is not exceeded.

A more complex arrangement is shown in Fig. 2. In this case, the transmission requirements of the station lines between location *A* and its switch and between location *B* and its switch must be still more stringent in order to allow for the transmission effects of the trunk between the switching centers.

The problem may be still further complicated by provision of an alternate route as shown in Fig. 3. The direct trunk route may be designed with a limited number of circuits in the path to develop high usage and therefore greater efficiency. Overflow traffic is switched via the alternate route which has sufficient circuits to handle the traffic. The transmission design must be such that traffic switched via A-C-B is offered as good a transmission path as the traffic switched directly via switches A-B.

The transmission engineer may have several choices. Let us refer again to our first example—a requirement for 8 db of 1,000-cycle loss between locations A and B. For the direct connection, the circuit requirement becomes 8 db. For the arrangement of Fig. 1, we might specify that each connecting circuit have 4-db loss, so any two connected stations would always encounter an overall loss of 8 db. For the arrangement of Fig. 2, we might specify 3-db loss in each station line and 2-db loss in each trunk or any similar arrangement which produces the required overall loss between two locations. For the arrangement of Fig. 3, we might choose to put 2-db loss in each station circuit, 4-db loss in direct trunks, and 2-db loss in each of the tandem paths. We might also choose to put 4-db loss in each station circuit and oper-

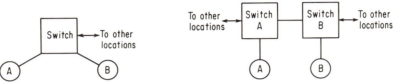

FIG. 1. A switched connection through a common switching center.

FIG. 2. A switched connection through interconnected switching centers.

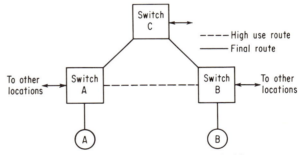

FIG. 3. A switched connection through interconnected switching centers with alternate routing.

ate all trunk circuits at zero loss provided they have sufficient stability. The choice to be made depends on several factors to be discussed further.

7. The Hierarchy Plan

From a transmission point of view, there are two basic arrangements of switching centers. One of these is known as the hierarchy plan. The general arrangement is shown in Fig. 4.

The hierarchy plan is similar in many respects to the national public-message network. A great deal of background information[2] is available which may be of help in designing this type of system. The plan may differ, however, in such respects as the degree of control of facilities, the application of equalization, and the provision of special service features.

The basic features of the hierarchy plan are:

1. The upper and middle level switching centers are four-wire. Lower level switching centers may sometimes be two-wire, although four-wire is preferable from a transmission standpoint.

2. Alternate routing can be provided within the capabilities of the switches. All trunks on the direct route would normally be tested first. If all are busy, the alternate routes would be tested until an idle circuit is found. A definite and fixed routing plan is usually applied for greatest economy in the use of the trunk plant.

3. Trunks and access lines normally use four-wire multiplex facilities.

4. Standard voice grade facilities are used where voice communications only are provided. If more sophisticated transmission schemes are to be provided, each trunk or line can be treated with attenuation and delay equalizers as needed to meet requirements.

5. Trunk groups may be split, and some trunks in the group may have different transmission characteristics from the others. The switches must be capable of selecting the proper grade of trunk on command from the originating station.

6. Two- or four-wire station equipment may be used. Of course, the full advantages of four-wire sets cannot be used unless the switches are all four-wire.

The hierarchy plan can handle voice communications, teletype, and data systems that are normally furnished on telephone-grade facilities. Secure voice and high-speed data services may be limited because of the ability of the network to connect an excessive number of trunks in tandem on final routes, thus making performance somewhat variable.

When alternate routing is provided in the hierarchy plan, each additional choice increases the total number of links in a connection by one. As the number of links increase, the overall transmission characteristics change. In a system with alternate routing, therefore, the transmission characteristics must be flexible enough to include the worst possible connection or the user must be willing to accept degraded transmission on the very small percentage of traffic handled via the final route. The number of levels of offices should be kept to a minimum to help reduce transmission variations unless requirements for circuit assurance or the need for circuit economies are more important.

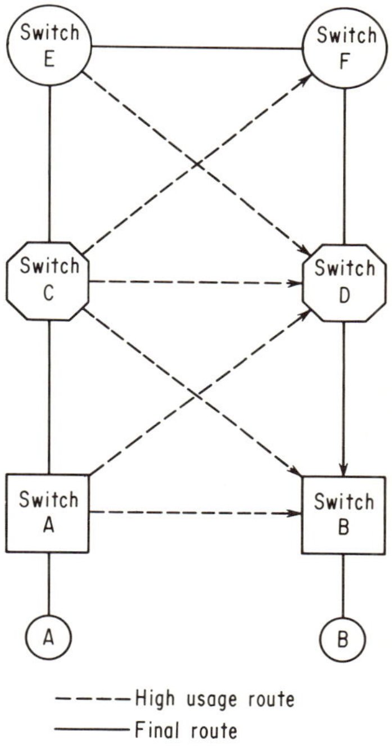

---- High usage route

—— Final route

FIG. 4. A hierarchy switching system.

8. The Hub Plan

Another arrangement with somewhat different features is the hub plan. A typical arrangement is shown in Fig. 5. The hub plan may be used when a high-quality survivable system is desired and economy is not the controlling factor.

In Fig. 5, switch F is an inner ring center and the other switches are middle-ring centers. PBX's connected to the system may be considered outer-ring offices. Inner-ring centers are connected to all adjacent inner-ring centers and middle-ring centers. Also, middle-ring centers are interconnected with all adjacent middle- and inner-ring centers. PBX's normally are homed on a middle- or inner-ring office, but they may have lines to more than one office for survivability purposes.

The chief advantage of the hub plan is that it can continue to function if a switching center fails completely. If a user location has dual access, service can be continued even if one of its serving switching centers fails or if one group of access lines fails.

Alternate routing in the hub plan presents a special problem. Because of the very high number of possibilities, a call may be switched back and forth in the system and accumulate a large number of trunks in tandem before finally reaching the desired office.   It would be impossible to control overall transmission characteristics.   Such a plan must necessarily have some limitations on routing arrangements.   The switches may have to be arranged to advance a routing indicator to control the choices of the next switch.   For example, if transmission requirements dictate that not more than three trunks be connected in tandem, the serving switch must be able to advance a signal, when it selects an alternate route, to indicate that one trunk has already been selected.   The second switch, if it chooses an alternate route, must be able to advance a signal to indicate that two trunks have been connected, and the third switch must be able to interpret the signal to mean that it can select only the final route.   In addition to the above limitations, the switch must be able to examine the destination address and limit alternate routing choices to those which are in a forward or lateral direction.

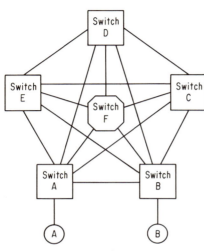

FIG. 5. A hub switching system.

When routing arrangements are provided as discussed above, the facilities can be selected and treated to meet transmission design requirements for a specified number of trunks in tandem with the knowledge that the requirements will never be exceeded under the most extreme switching arrangement.   The switching system can therefore be conditioned to handle high-speed data at a minimum error rate.

9. Effect of Switches on Types of Facilities

Several types of conditioning can be employed for direct connection of user equipment.   Of course, special requirements may also be established.   The introduction of switching affects both the standard grades of service and the special grades in the same manner.   The translation of the overall requirements into requirements for each trunk and access or station line is a major part of the development of the system transmission design.   In this chapter we shall discuss and illustrate the treatment of normal requirements.   Special requirements can be treated in the same general manner.

Some of the characteristics that must be considered when normal requirements are translated into switched circuit requirements are:

1. Transmission levels
2. 1,000-cycle losses
3. Attenuation-frequency characteristics
4. Envelope delay distortion
5. Noise
6. Echo and return loss

10. Types of Transmission Facilities

Treatment of these characteristics is illustrated by considering the following typical types of circuits:

Type A—A basic voice bandwidth channel for data service

Type B—An improved voice bandwidth channel for data services

Type C—A third and more strictly controlled grade of voice bandwidth channel for data services

Other requirements may be established, but they must be treated the same way in switched systems. For the purposes of this chapter, we shall use as examples the three types mentioned above and assume that voice communication is also provided.

When voice communication circuits are transferred to data sets, they may be operated either half or full duplex. Full-duplex circuits must be four-wire, but half-duplex-type circuits may be either two-wire or four-wire. The circuits are usually designed for 600-ohm resistive balanced terminations at both ends. The recommended nominal circuit losses measured between 600-ohm impedances at line-up are 8 db ± 1 db at 1,000 cps for the type of data circuits under discussion. Expected maximum variations should be on the order of ±3 db short term and ±4 db long term.

Attenuation-frequency characteristics vary with the type of circuit. They are shown in Table 1.

Circuit delay requirements are not specified, since absolute delay is not particularly a problem with half- or full-duplex operation. Envelope delay distortion requirements vary, however, with the type of circuit. They are shown in Table 2.

Table 1. Attenuation Frequency Characteristics for Typical Data-type Circuits

Type of circuit	Frequency range, cps	Variation, db*
Type A.........	350–2,000	−2 to +6
	2,001–2,500	−3 to +8
Type B.........	300–1,000	−2 to +6
	1,001–2,400	−1 to +3
	2,401–2,700	−2 to +6
Type C.........	300–500	−2 to +6
	501–2,800	−1 to +3
	2,801–3,000	−2 to +6

* (−) is less loss; (+) is more loss.

Table 2. Envelope Delay Distortion Characteristics for Typical Data-type Circuits

Type of circuit	Frequency range, cps	Delay distortion limits, μsec
Type A.........	1,000–2,400	1,000
Type B.........	1,000–2,400	1,000
Type C.........	1,000–2,600	500
	600–2,600	1,500
	500–2,800	3,000

Table 3. Noise Characteristics for Typical Data-type Circuits*

Type of circuit	Maximum circuit noise	Maximum impulse noise
Type A or B.......	26 db S/N	72 dbrn 0
	54 dbrnc 0	
Type C...........	26 db S/N	68 dbrn 0
	54 dbrnc 0	

* See paragraph 27 for definition of terms.

Noise characteristics also vary somewhat with the type of circuit. These are shown in Table 3.

We shall now proceed to develop the method for application of these requirements to a switched system.

REFERENCE SYSTEMS

11. The Use of Test Level Points

A reference system makes use of test level points. A test level point (TLP) is a location in a circuit or system at which a specified test tone level is expected during line-up. For example, 0 TLP is a point at which the test tone level should be 0 dbm (i.e., a power level of 1 mw). This arrangement serves a number of purposes. It permits establishment of specified losses along a circuit. With 0 dbm sent at the 0 TLP, the circuit can be adjusted to obtain the proper levels at other TLP's, and the losses are thereby adjusted. For example, the inserted loss between a 0 TLP and a −4 TLP must be 4 db. To produce a proper level at a +2 TLP, 2 db of gain must be inserted. By the use of TLP's, trunks and lines can be individually designed so that each will operate at the proper levels when they are connected together. System transmission levels may be referred to TLP's. For example; when a station equipment transmits a signal measured at −8 dbm at the 0 TLP, the transmit level is −8 dbm 0. Noise levels may also be established by reference to a TLP. A transmission system requiring, for example, a 30-db signal-to-noise ratio requires that noise at the 0 TLP not exceed −30 dbm. With a noise-measuring set indicating in dbrn, the noise measurement at 0 TLP must not exceed 60 dbrn to meet the signal-to-noise requirement. The noise requirement would be specified as 60 dbrn 0. The reference system permits noise measurements at other locations in the circuit. If the noise limit is 60 dbrn 0, as above, the noise measurement at a −4 TLP cannot exceed 56 dbrn.

12. Reference System for Direct Circuits

On direct circuits, the transmitter is considered to be the 0 TLP. The TLP at the receiver is established by the overall 1,000-cycle loss of the circuit. For instance, on the data circuits we are using as examples, the receiving terminal is a −8 TLP.

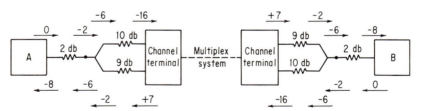

Fig. 6. Typical levels in a direct circuit.

When the circuit consists of a multiplex channel, other TLP's are involved. Multiplex channels are usually lined up to provide 23 db of gain in each direction. The transmitting point is a −16 TLP, and the receiving point is a +7 TLP. Other parts of the circuit must be arranged to accommodate these levels.

Figure 6 illustrates a typical direct circuit utilizing a carrier channel. In this illustration, the end facilities each have 2-db loss and the hybrids have 4-db loss in each direction. Note that any multiplex channel can be patched in when properly lined up, since the transmission level requirements are automatically met at each TLP.

Figure 7 illustrates a typical arrangement of a direct circuit using two multiplex channels in tandem. In this case, the circuit is four-wire and can therefore be oper-

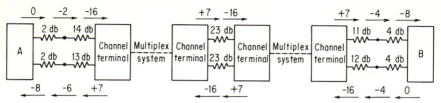

Fig. 7. Typical levels in a built-up direct circuit.

ated full duplex. Note that, even though the end facilities have different losses, requirements can be met at each TLP by the adjustment of pads. Multiplex channels and end facilities can thus be patched out when in trouble without upsetting transmission levels.

13. Test Level Points in Switched Systems

All classes of switching centers are considered to be −2 TLP's. The point of test is theoretically at the center of the switch. This is not usually a practical point at which to make measurements, and in practice they are not made directly at the switch. This will be discussed in further detail in paragraph 54 of this chapter. For design purposes, however, trunks and lines are designed from the switch, and this is a −2 TLP.

Inner- and middle-ring switching centers in hub plans are also considered to be −2 TLP's. PBX's connected to switching centers in any arrangement are considered to be 0 TLP's, except when they also serve as outlets for tributary PBX's. In these cases, because of the need for pad control, which will be discussed further in paragraph 44, the main PBX is a −2 TLP. The tributary PBX is always a 0 TLP.

The transmitting station was considered to be a 0 TLP when connected to a direct circuit. In a switched system, the test level is established so as to provide proper test tone level at the switch (−2 TLP) when the line loss is adjusted. For example, for a station line to have 4-db loss transmitting, the station is a +2 TLP. For 4-db loss receiving, it is a −6 TLP, since a −2-dbm test tone is sent at the switch (−2 TLP).

Figure 8 illustrates the application of the reference system to a typical four-wire zero-loss trunk. Note that the requirements of the −2, −16, and +7 TLP's are met by inserting pads. In actual practice, the loss may consist of trunk circuits, signaling equipment, equalizers, and other terminal equipment. If the terminal-equipment losses are not sufficient, the drop loss is adjusted by varying the pads. If the terminal-equipment losses are too great, the pads are replaced with amplifiers which are adjusted to reduce the total drop loss to the proper value. When the reference system is used, the multiplex channel can be replaced when in trouble by another multiplex channel without upsetting the levels. Also, troubles in the terminal equipment can be cleared by patching in an identical set of terminal equipment. When the circuit is being put together, the terminal equipment can be connected and adjusted without the multiplex channel being available, and the converse is also true. The reference system therefore permits great flexibility in the installation and use of equipment.

Figure 9 illustrates a typical two-wire trunk. In this case, the hybrids are assumed

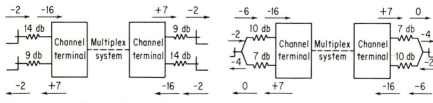

Fig. 8. A typical four-wire zero-loss trunk.   Fig. 9. A typical two-wire trunk with loss.

to have 4-db insertion loss in each direction and make up part of the drop loss. The loss of the trunk is set at 2 db in each direction. It can never be reduced to zero because the hybrids present a singing path which would make the circuit unstable. Note that the loss in each direction is controlled by varying the drop loss at the receiving end.

The typical design shown in Fig. 9 could also be applied to access lines when the multiplex system is terminated at a PBX. In many cases it is terminated at an

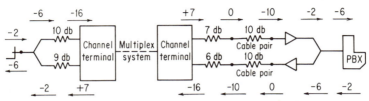

FIG. 10. A typical two-wire line with cable extension.

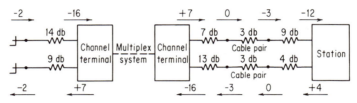

FIG. 11. A typical four-wire station line.

intermediate point, and the line is extended to the PBX on cable pairs. A typical arrangement is shown in Fig. 10. In this example, the cable extensions are assumed to have 10-db insertion loss. Gain is therefore required in the receiving path at the PBX to control the overall loss. Gain is also used in the transmitting path from the PBX. While it could be eliminated by removing the 6-db pad at the multiplex terminal, the use of this amplifier provides a 6-db improvement in signal-to-noise ratio in the cable pair. In this example, the PBX is a −2 TLP.

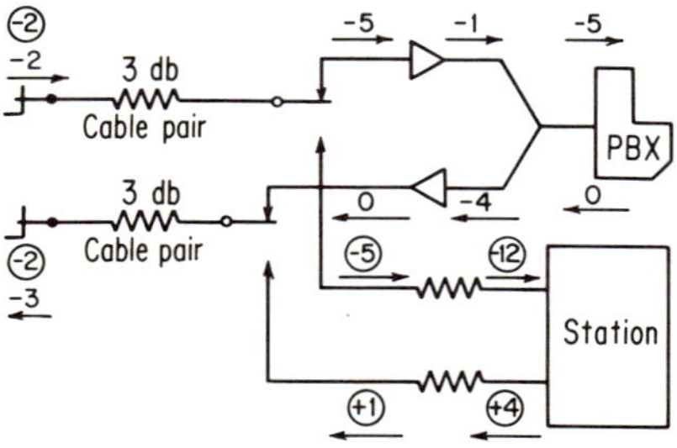

Figure 11 illustrates a typical four-wire station line served by a four-wire switching center. In this case the line is made up of a multiplex channel extended on cable pairs which have 3-db insertion loss. Note that the switching center is a −2 TLP both transmitting and receiving while the station is a +4 TLP transmitting and a −12 TLP receiving. The station line losses are thus established as 6 db transmitting and 10 db receiving.

It is sometimes desirable to terminate a line in a PBX as an access line and provide for alternate use as a station line. Figure

FIG. 12. A typical arrangement of a dual-use line.

12 illustrates a typical arrangement. Note that the PBX is a 0 TLP transmitting in this case but the station is a +4 TLP transmitting. The line can be made to serve both purposes as long as pads are used and are applied on the station side of the transfer.

The previous illustrations showed a few of the many combinations of facilities that might be found in a switched system. Any combination of facilities can be integrated into a system if TLP's are established and the necessary adjustments are designed into the circuits.

1,000-CYCLE LOSS REQUIREMENTS

14. Control of 1,000-cycle Losses

The reference system just discussed provides the means for controlling losses in a system. It presumes that a 1,000-cycle test tone at 0 dbm is used as the standard of comparison. By its use, each line or trunk can be engineered and built separately, and each will be compatible with the rest of the system when interconnected through switches. Having established a reference system, the engineer is now prepared to determine the 1,000-cycle losses for each circuit.

In this section, consideration is given to the control of losses in two-wire voice communications, four-wire voice communications, and data transmission. Other types of transmission require controls of 1,000-cycle losses which, with minor variations, are approached in the same way.

15. Control of Losses in Two-wire Voice Communications

The allowable losses in a two-wire voice communications switched system are fixed by the transmitting levels and required receiving levels. The problem seems rather simple at this point. Consider, however, that transmitting levels are affected by the characteristics of the talker's voice, the acoustic signal-to-noise ratio, the efficiency of the transmitter, the current in the transmitter, sidetone level, etc. The required receiving level is affected by the listener's hearing, acoustic noise, circuit signal-to-noise ratios, etc. It would now seem as if the resolution of all these variables into a transmission plan is almost impossible. Certainly the statistical studies necessary to resolve all these problems are beyond the scope of this chapter. We can use some average values that have been developed from other experience. For example, the modern-type telephone transmitter has an average output of about −15 vu.* Telephone operator headsets use transmitters about 6 db less efficient. Older type telephone sets have about the same efficiency as modern-type sets on short loops, but they are up to 5 db poorer on long loops. For this discussion it is assumed that modern-type sets are used and the transmitter current is maintained at an adequate level. When other types of equipment are used, modification of the transmission design may be required.

Required receiving levels depend on the listener's hearing acuity, acoustic background noise, and circuit signal-to-noise ratios. Received levels lower than −40 vu may sometimes be acceptable. It is generally agreed that −27 vu always provides excellent service. We may therefore conclude that overall losses need not be less than 12 db and should not exceed about 30 db. Experiments have been conducted on long circuits in connection with echo problems. These tests are discussed in more detail later. They proved that there are limitations on the minimum loss acceptable on long circuits. It is also true that some loss must be provided in all circuits terminated on a two-wire basis for stability reasons. We may therefore conclude that a two-wire voice connection will never be much less than 12 db.

As a result of the above considerations and after long experience, voice communications systems are now designed with the objectives of making the overall loss VNL +4 db. Via net loss (VNL) will be discussed in detail in connection with echo. At this point, it is sufficiently defined to say that it is the lowest loss at which a circuit can be operated and remain stable.

In a switched system, all circuits are designed to VNL. The additional 4 db is obtained by adding 2 db in each station line or access line. Station lines are therefore designed to VNL +2 db, not to exceed about 8 db. The average should be 3 to 4 db. Access lines are designed to VNL +2 db, not to exceed about 4.5 db. This allows for some loss in the PBX and extensions. All trunks are designed to VNL.

---

* Volume units. They are measured with a transmission measuring set having a meter specially damped to integrate the complex waveforms of a voice signal. A vu meter reads 0 vu when measuring a single-frequency 1,000-cycle tone of 0 dbm. It has a response time of about 300 msec. See Ref. 4.

Trunks should consist of high-velocity facilities such as radio or cable multiplex. As a result, trunk losses should be relatively low. Direct trunks between lower level offices should not exceed about 2.5 db. Trunks between lower level and middle offices should not exceed about 2.0 db. All other trunks should be operated four-wire and at zero loss. Trunks in a hub plan should also be designed with high-velocity, low-loss facilities. Under these arrangements, trunk losses on any connection should not exceed 10 db.

16. Control of Losses in Four-wire Voice Communications

Four-wire telephones are served by four-wire switching centers. Connections between two four-wire stations therefore may operate on a full-duplex basis. If all trunks are operated at zero loss, alternate routing can have no effect on the overall loss of the connection. Four-wire telephone service is usually required as an accessory to data or enciphered voice equipment, generally requiring four-wire full-duplex operation. Therefore, there is no penalty incurred in requiring four-wire zero loss trunking for four-wire telephony.

Under the zero-loss trunking concept, overall losses on each connection are controlled by controlling the station line losses. It is therefore possible to provide a uniformly high quality of transmission. If the average talker signal level is the same as for two-wire communications, it is not difficult to arrive at the required losses. It was assumed that the average output of the transmitter would be -15 vu and the input to the receiver would not have to be higher than -29 to -27 vu for excellent transmission. The four-wire telephone instrument is basically a two-wire instrument with the receiving path removed from the built-in hybrid. It is therefore effectively about 4 db more efficient. We may therefore conclude that the overall loss should be about 16 to 18 db.

It seems reasonable to provide half the required loss in each station line. This would result in uniformly good transmission regardless of the combination of stations on a connection. In practice it has been found desirable to proceed in a slightly different manner. Four-wire station lines are designed with 6 db of loss in the transmitting path and 10 to 12 db in the receiving path. This results in a higher signal level on multiplex channels involved in each connection and tends to produce a better signal-to-noise ratio.

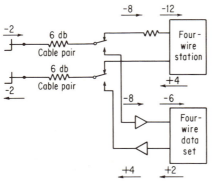

FIG. 13. A typical arrangement of an alternate-use station line.

17. Control of Losses in Data Transmission

Let us consider the provision of data transmission over switched connections which are the equivalent of one of the A, B, or C type of circuits previously mentioned. This means that each overall connection loss should be nominally 8 db. It will be necessary from the standpoints of stability, echo, equalization, etc., to provide four-wire switching. The user equipment may also require full-duplex operation. We shall therefore assume that four-wire switching is provided for this type of service and all four-wire trunks are operated at zero loss. Here again, the overall connection loss can be controlled and made uniform by controlling the loss in the station lines. Station lines designed for 4 db of loss in both the transmitting and receiving paths should give 8-db overall loss consistently on all connections.

Figure 13 illustrates a typical arrangement of a station line providing for data service with an alternate four-wire telephone station for coordination and control. Note that the requirements of each service can be satisfied by means of pads or amplifiers on the station side of the transfer, even though the common line loss does not happen to be the correct value for either case.

TRANSMISSION LEVELS

18. Transmission Levels in Telephony

For two-wire stations with an average output of −15 vu, the signal level at −2 TLP with an average line loss of 4 db is −19 vu. Thus the input to a multiplex channel will be 17 db below the line-up level. For short loops and talkers with strong voices, the level might be up to 10 db higher. Multiplex systems are usually designed to handle the busy-hour load with voice signals averaging about 12 db below line-up levels.

When four-wire stations are arranged with 6 db loss in the transmitting path and have an average output equivalent to about −11 vu, the transmitting level averages about 15 db below the line-up level. For stronger talkers the level will, of course, be somewhat higher but not sufficient to cause serious intermodulation noise in the multiplex system.

19. Transmission Levels in Data Systems

Switched systems providing the equivalent of Types *A*, *B*, or *C* or similar transmission characteristics generally anticipate that the data transmitter will provide composite signals not exceeding −8 dbm. This is equivalent to −10 dbm at 0 TLP, −12 dbm at −2 TLP, and −26 dbm at −16 TLP. Data sets capable of higher output levels should be equipped with an attenuator, or additional pads should be provided at the interface to adjust the line input to −8 dbm. With the transmitting and receiving paths each adjusted to 4-db loss and the trunks operated at zero loss, this arrangement results in a uniform receiving level of −16 dbm at every data set.

The transmission level of −10 dbm 0 assumes, however, that a rather small percentage of channels in a multiplex system is used for data transmission. In a switched system which will have a high percentage of data traffic, the −26-dbm level at the −16 TLP may result in overloading the multiplex system during the busy hour. When this occurs, the intermodulation noise of the system rises. A data channel may effectively crosstalk into another channel and cause a high noise level. Under these conditions, signal-to-noise ratios may actually be improved by lowering the signal levels. Transmission levels for systems heavily loaded with data may require a reduction to about −13 dbm 0 for best performance. This could be accomplished by making the losses in the transmitting and receiving paths of the station lines 7 db and 1 db, respectively, instead of 4 db in each direction.

The arrangement for data transmission could be employed, with modification, for other services such as telephoto or facsimile. For these services the line losses are arranged to maintain the required overall losses while providing the proper signal level at the various TLP's as is done with data transmission.

LOSS VARIATIONS

20. General

After the overall loss design is established and loss requirements have been set up for all the parts of a switched system, two major problems are encountered in actually obtaining the design values. One of these is the actual ability to line up the circuits to the exact values specified, and the other is the ability to hold them near the established values. Each problem is affected by the stability of the facilities, the accuracy of the test equipment, the amount of labor and effort that can be expended, human errors, and other factors. We shall deal with each of these problems in further detail.

21. Line-up Variations

Circuits are usually lined up by installing and adjusting the individual parts. They are then patched together, and overall transmission tests are applied. Consider, for example, the construction of a trunk. The multiplex channel is lined up by men making tests at each end. Each has a different set of test equipment. In addition, the channel line-up may be affected by the quality of installation and adjustment of the high-frequency portion of the system. The terminal equipment at each end is assembled and lined up individually. Each may require the work of two men with different sets of equipment. When the circuit is finally assembled, the overall tests require the work of two men with different sets of test equipment. The actual loss in each part of the circuit will be affected by the accuracy of the test equipment and the amount of effort of the testmen in meeting the requirements exactly. When the overall tests are made, they are again affected by the accuracy of the test equipment. They are also affected by variations in the parts. The individual variations may add, subtract, or cancel out.

The practical approach is to establish some reasonable limits for circuits. If tests show that circuits are within these limits, adjustments should be made in pads or amplifiers to eliminate the small variation. If tests show that the circuits are out of limits, they should be sectionalized and the various parts rechecked. If it appears that the error is due to the accumulation of small variations, then the final adjustment should be made. If the error is due to trouble, setting errors, or design errors, then these errors should be corrected and the circuit retested before making the mop-up adjustments.

The limits or tolerances to be used in this part of the design will depend on the types of facilities involved. Some facilities will tend to vary more from calculated design values as the circuit length increases. Some typical limits based on experience are shown in Table 4.

Table 4. Typical Tolerances in 1,000-cycle Loss Settings of Circuits

Tests	Permissible deviations from specified circuit loss	
	Circuit length, miles	
	Under 500	Over 500
Initial.........	±1.0 db	±1.5 db
Routine.......	±1.5 db	±2.0 db

22. Maintenance Variations

When circuits are properly assembled and adjusted to their designed loss values, the control of losses is not complete. Losses of circuits tend to vary because of temperature changes in facilities, temperature changes in active elements, aging of parts, instability, etc. It is necessary to know something about the probability that the loss of a circuit will stay within some range. This is usually expressed as the maintenance sigma (σ). σ is equal to 1 db if, for example, it is determined from experience that a certain type of channel will be found to be within ±1.0 db of its design loss on about 68 per cent of all measurements assuming a normal distribution curve. A range of 2σ, or ±2.0 db, would include about 97 per cent of all measurements. We must therefore compare the variations in the circuit with the tolerance of the station equipment before deciding the design is satisfactory.

When circuits are switched together, the σ for the overall connection is

$$\sigma = \sqrt{\sigma_1{}^2 + \sigma_2{}^2 + \cdots + \sigma_n{}^2} \qquad (1)$$

This is true in practice because circuits are not all too long or too short at the same time. If we assume a σ of 1.0 for station lines and a σ of 0.5 for all trunks, the σ for various overall connections can be computed from Table 5. From examination of the table, it can be seen that maintenance sigmas increase as the degree of assurance increases and as the number of links in tandem increase. If data sets were involved and were expected to perform 99 per cent of the time when the overall loss does not exceed the nominal value by more than 5.0 db, then the switching system would have to be arranged to connect not more than three trunks in tandem. If the reliability could be relaxed somewhat, the restrictions on the trunking arrangements could be relaxed. The restrictions could also be relaxed if a lower quality of performance is acceptable.

Table 5. Maintenance Sigmas for Various Combinations

Facilities in connection		Maintenance sigmas*		
Station lines	Trunks	Sigma	2 sigma	2.33 sigma
2	0	1.4	2.8	3.3
2	1	1.5	3.0	3.5
2	2	1.6	3.2	3.7
2	3	1.7	3.3	3.9
2	4	1.7	3.5	4.1
2	5	1.8	3.6	4.2

* Sigma represents the 68 per cent probability, 2 sigma the 97 per cent probability, and 2.33 sigma the 99 per cent probability that the maintenance variation will not exceed the value shown.

Control of loss variations is enhanced by proper selection of facilities. Regulated multiplex channels are better than nonregulated channels, and underground cables are more stable than aerial cables, but the better facilities are generally more expensive. The transmission engineer is faced here with reconciliation of transmission requirements with economic aspects.

Control of loss variations can also be improved by more frequent tests and adjustments, but here again the costs must be considered. Doubling the testing frequency will generally not cut the average loss variations in half. Where close control is necessary, the transmission engineer may have to provide automatic test facilities to reduce the labor cost of locating and adjusting circuits that are out of limits.

ATTENUATION-FREQUENCY CHARACTERISTICS

23. The General Treatment

In addition to control of losses at 1,000 cycles, the transmission engineer is usually required to consider losses at all other frequencies. In the design of circuits using voice frequencies, the design requirements are usually stated by reference to the 1,000-cycle loss of the circuit. For example, a requirement in the range 1,000 to 2,400 cps may be stated as -1 to $+2$ db, where $(+)$ is more loss and $(-)$ is less loss. This means that, if the 1,000-cycle loss of a circuit is 5 db, the loss anywhere in the range 1,000 to 2,400 cps cannot be more than 7 db or less than 4 db.

Attenuation-frequency characteristics of switched circuits must meet requirements which are derived from the requirements for an overall connection. Let us assume an overall requirement on a switched connection of 5-db maximum slope in the range 500 to 2,500 cps; i.e., the losses in this range cannot vary more than 5 db. This may also be stated as -2 to $+3$ db or -1 to $+4$ db, etc., depending on where the excursions tend to occur. The process for translating this requirement into the requirement for each circuit in a connection is the same as for the 1,000-cycle losses. This is true because some facilities will have upward slopes while others have downward slopes. Also the excursions may occur at different frequencies. They can be computed from the formula

$$X = \sqrt{X_1{}^2 + X_2{}^2 + \cdots + X_n{}^2} \tag{2}$$

If we assume that the allowable slope of each station line is 3 db, then the allowable slope for each trunk can be computed for various numbers of trunks in tandem. The requirements can be established to yield a 5-db slope and other attenuation-frequency characteristics as shown in Table 6.

Table 6. Slope Requirements per Trunk for Several Combinations of Facilities in Tandem

Facilities in tandem		Overall slope requirement					
Station lines	Trunks	5	6	7	8	9	10
2	1	2.6	4.2	5.6	6.8	7.9	9.0
2	2	1.9	3.0	3.9	4.8	5.6	6.4
2	3	1.5	2.5	3.2	3.9	4.6	5.2
2	4	1.3	2.1	2.8	3.4	4.0	4.5
2	5	1.2	1.9	2.5	3.0	3.5	4.0
2	6	1.1	1.7	2.3	2.8	3.2	3.7
2	7	1.0	1.6	2.1	2.6	3.0	3.4

Analysis of the figures shows that increasing the number of trunks in tandem rapidly decreases the allowable slope per trunk. Tightening the overall slope requirement also rapidly tightens the slope requirement on each trunk. A further analysis would show that tighter requirements on station lines would relax the requirements on the trunks and, conversely, tighter requirements on trunks would permit greater tolerances on the station lines. The limits that can be reached are obviously controlled by the accuracy with which each line and trunk can be equalized.

The transmission engineer must analyze all attenuation-frequency requirements as discussed above. There may be more than one. For instance there may be a requirement in the range 700 to 2,300 cps and another requirement in the range 300 to 3,000 cps. Attenuation equalizers may be required to bring the facilities within limits. The equalizers must be balanced and present the proper impedance to the line. Four-wire circuits must be equalized in both directions. Circuits are usually equalized at the receiving end (postequalization). The engineer may choose to equalize all trunks and attempt to eliminate the need for equalizers on the station lines. Since there are normally fewer trunks, this would result in the least number of equalizers and the greatest economy. If the requirements are difficult to meet, the engineer must remember that trunk circuits will usually use high-quality multiplex facilities while station lines will be constructed from a great variety of facilities. The best solution may be to equalize all trunk circuits as closely as possible, since they must all be equalized anyway. The station lines could then be individually equalized as needed to meet the overall requirements. If the degree of equalization required approaches the limits of the state of the art, the transmission engineer may need to consider restriction of the maximum number of trunks allowed in tandem. It may be necessary to provide large direct trunk groups and accept reduced performance on the small percentage of traffic routed via the maximum connection.

24. Attenuation-equalization Treatment of Data-type Circuits

As an application of the general treatment discussed above, let us consider the treatment for Type *A*, *B*, and *C* circuits. The overall attenuation-frequency requirements are shown in Table 1.

Table 7 shows the requirements for access and station lines with one, three, or five trunks in tandem. Of course, the figures could also be calculated for two or four

Table 7. Attenuation-frequency Requirements for Access and Station Lines to Provide Data-type Services

Type of service	Attenuation-frequency slope requirements (db) with:		
	1 trunk	3 trunks	5 trunks
Type *A*:			
350–2,000 cps.	−1.0 to +3.5	−1.0 to +3.5	−0.8 to +2.5
2,001–2,500 cps.	−1.0 to +4.5	−1.0 to +4.0	−1.0 to +3.5
Type *B*:			
300–999 cps.	−1.0 to +3.5	−1.0 to +3.5	−0.8 to +2.5
1,000–2,400 cps.	−0.5 to +1.5	−0.5 to +1.5	−0.5 to +1.5
2,401–2,700 cps.	−1.0 to +3.5	−1.0 to +3.5	−0.8 to +2.5
Type *C*:			
300–499 cps.	−1.0 to +3.5	−1.0 to +3.5	
500–2,800 cps.	−0.5 to +1.5	−0.5 to +1.5	
2,801–3,000 cps.	−1.0 to +3.5	−1.0 to +3.5	

Note 1: Attenuation-frequency characteristics are with respect to the 1,000-cycle loss.
Note 2: (+) is more loss; (−) is less loss.
Note 3: When applied to access lines, extension-line and tie-line facilities must be included.

Table 8. Attenuation-frequency Requirements for Trunks to Provide Data-type Services

Type of service	Attenuation-frequency slope requirement (db) with:		
	1 trunk	3 trunks	5 trunks
Type *A*:			
350–2,000 cps.	−1.0 to +3.5	−0.8 to +2.0	−0.7 to +2.0
2,001–2,500 cps.	−2.0 to +5.0	−1.0 to +3.5	−1.0 to +3.0
Type *B*:			
300–999 cps.	−1.0 to +3.5	−0.8 to +2.0	−0.7 to +2.0
1,000–2,400 cps.	−0.7 to +1.5	−0.5 to +1.0	−0.5 to +1.0
2,401–2,700 cps.	−1.0 to +3.5	−0.8 to +2.0	−0.7 to +2.0
Type *C*:			
300–499 cps.	−1.0 to +3.5	−0.8 to +2.0	
500–2,800 cps.	−0.7 to +1.5	−0.5 to +1.0	
2,801–3,000 cps.	−1.0 to +3.5	−0.8 to +2.0	

Note 1: Attenuation-frequency characteristics are with respect to 1000 cycle losses.
Note 2: (+) is more loss; (−) is less loss.

trunks in tandem. Note that no figures are provided for Type *C* treatment with more than three trunks in tandem because the requirements on the trunks become extremely strict. The treatment of individual lines is no longer a Type *A*, *B*, or *C* equalization but something much more strict. When applied to access lines, any facilities behind the PBX must also be included in the calculations. When tie lines or off-premise stations are involved, it will sometimes be easier to connect the station directly to the switching center on a station line.

Table 8 shows the attenuation-frequency requirements for trunks when up to five trunks are connected in tandem. Comparison with Table 1 shows that the requirements per trunk are extremely difficult compared with the overall requirements. When all lines and trunks are treated in accordance with the proper requirements of Tables 7 and 8, data-type services can be provided. Note that alternate routing must be limited to the number of trunks in tandem for which equalization is provided. Also, if the number of trunks in tandem is exceeded, performance will be deteriorated. Since data sets tend to be sensitive to excess slope, error rates will increase rapidly if the calculated number of trunks in tandem is exceeded. Other attenuation-equalization requirements may be used. Limits will have to be developed in the same manner as above for each case.

ENVELOPE DELAY DISTORTION

25. The General Treatment

Envelope delay distortion is an effect caused by differences in the velocity of propagation at the various frequencies in the spectrum of the transmission facility. A complex signal generated at one end of a circuit may be composed of signals at several frequencies, all having a definite phase relationship. As the signal is transmitted along the circuit, the portions of the signal at each frequency travel at the velocity of propagation characteristic of the facility at that frequency. The phase relationship gradually changes until the complex signal arrives at the distant end in a distorted condition. A square wave, for example, may no longer be square.

Envelope delay distortion is generally not a problem in voice communication, since the human ear tends to integrate received signals and extract intelligibility by some method relatively insensible to this type of distortion. Purely electronic equipment is not so tolerant. A square pulse, such as may be used in digital data transmission, may have its envelope distorted sufficiently that it cannot be detected correctly by a data receiver. The result is an error. Data sets are usually designed to have a certain tolerance for envelope delay distortion. Once the limit is exceeded, the error rate rapidly increases. This is generally true of all forms of digital and analogue transmission systems. The tolerance for delay distortion tends to decrease as the speed and complexity of the data system increase.

The requirements for delay distortion in a switched system are the same as the requirements of the station equipment. Delay distortion is not a problem in voice communications, and low-speed data and teletype can generally be handled on good voice facilities. The medium-speed systems may require some delay equalization, while high-speed systems require very careful equalization. If the system is extremely sophisticated, it may be necessary to use broadband, direct circuits or specially designed switched systems not capable of any other use.

Envelope delay distortion is measured in microseconds. Absolute delay is not a consideration at this point, but delay distortion at any frequency is basically the absolute delay at that frequency measured from the minimum. In voice frequency multiplex channels, the minimum tends to occur in the range near 2,000 cps. Delay distortion therefore occurs at the high and low ends of the frequency range. Both high- and low-frequency correction must therefore be considered in the design and application of delay equalizers. Other types of facilities may tend to distort at the high or low end of the voice band. Equalizers must therefore be designed and applied accordingly.

Equalizers can be applied at either end of a direct connection to correct the delay distortion of the entire circuit. This cannot be done in advance in a switched system, since the combination of circuits can never be known in advance. Each circuit in the system must be equalized, and the equalization must be good enough that any possible combination of circuits will meet the overall requirements.

Delay-distortion measurements and delay equalization are expensive and time consuming. In a system designed to provide voice communications, teletype, high-speed data, etc., it is customary to equalize only those station lines serving station equipment with delay-distortion requirements. In addition it may be more economical to equalize only a part of the trunks in each route. The switching center equipment must then be able to recognize the class of service required by each station line and select only specially treated trunks on demand of stations with delay-distortion requirements. In systems that are primarily data, it may be more economical to equalize all circuits and thus simplify the switching equipment.

Envelope delay distortion is generally accumulated on an arithmetical basis. For example, assume that there is a requirement of not more than 1,000-μsec delay distortion in the frequency range 800 to 2,800 cps. Dividing the distortion limit arithmetically the maximum permissible delay distortion per circuit for three circuits in tandem would be 330 μsec. For five circuits in tandem, the maximum delay per link would be 200 μsec if divided equally among the circuits. In practice, the distortion does not accumulate on a completely arithmetic basis. For this reason, the requirements per link may be relaxed about 10 per cent.

The transmission engineer can profit from a knowledge of the delay distortion inherent in various kinds of facilities. Delay distortion tends to be inversely proportional to the velocity of propagation. A loaded cable pair has more distortion than a nonloaded cable pair. A channel on a radio system 1,000 miles long may have no more distortion than 10 miles of loaded cable. Radio and cable multiplex systems are relatively free of distortions, and only the channel terminal equipment has substantial delay distortion. Thus, a 1,000-mile multiplex channel may have about the same delay distortion as a 100-mile multiplex channel. The transmission engineer may sometimes find it possible to control delay distortion sufficiently by patching multiplex systems together at the group and supergroup frequencies, coming down to baseband frequencies only at the ends of the system, if the requirements are not too stringent.

A group consists generally of 12 channels, and a supergroup consists of several groups. The bandwidths required to accommodate either a group or supergroup are so broad that the associated filters and accessories insert insignificant amounts of delay distortion as far as an individual channel is concerned. If channelizing equipment is kept out of the middle of a multiplex system, it is possible to confine delay distortion mainly to the channel terminals at each end of the system.

In a switched system the trunks are nearly always high-velocity, uniformly good quality facilities. The station lines tend to be more variable as to make-up. It is therefore practical to divide up the overall requirement in such a way as to make the delay-distortion requirements on station lines less restrictive than those for trunks. Thus, in a system having an overall requirement of 1,200 μsec on a connection consisting of three trunks and two station lines in tandem, we might establish a requirement of 300 μsec on each line and 200 μsec on each trunk.

26. Envelope Delay Treatment for Data-type Services

The discussion above applies to any service having an overall envelope-delay-distortion requirement. As an example, we shall consider its application to data-type services.

Table 9 illustrates the delay-distortion requirements for access and station lines for various numbers of trunks in tandem on a connection to provide Type A, B, or C grade of service. Requirements could also have been developed for two or four trunks in tandem on the connection. Here again no figures are provided for Type C service with more than three trunks in tandem. When the requirements are applied to access lines, extensions and off-premise station lines must also be included as part of the circuit. In some cases it may be simpler to connect the data station directly to the switching center with a station line.

Table 10 illustrates the delay-distortion requirements for trunks to provide Type A, B, or C data services. Note that the requirements for Type C service with three trunks in tandem is at about the limit for reliable measurement. Also note that, when the requirements of Tables 9 and 10 are added, they equal or slightly exceed the overall requirements of Table 2. Owing to small variations in the circuits and the fact that the excursions do not always occur at the same frequencies, there is some tendency for the total delay distortion to be less than the arithmetic sum of all the

Table 9. Delay-distortion Requirements for Access and Station Lines to Provide Data-type Services

Type of service	Maximum delay distortion (μsec) with:		
	1 trunk	3 trunks	5 trunks
Types A and B:			
1,000–2,400 cps......	400	250	175
Type C:			
1000–2,600 cps.......	200	120	
600–2,600 cps........	600	375	
500–2,800 cps........	1,200	750	

Note: When applied to access lines, extension-line and tie-line facilities must be included.

Table 10. Delay-distortion Requirements for Trunks to Provide Data-type Services

Type of service	Maximum delay distortion (μsec) with:		
	1 trunk	3 trunks	5 trunks
Types A and B:			
1,000–2,400 cps......	300	200	150
Type C:			
1,000–2,600 cps......	150	100	
600–2,600 cps........	450	300	
500–2,800 cps........	900	600	

individual distortions. As in the case of attenuation-frequency characteristics, error rates tend to increase rapidly when the distortion limits are exceeded.

CIRCUIT NOISE

27. Definitions

There are two kinds of noise to be considered. One of these is circuit noise, also called steady noise or white noise. It consists of thermal noise in the circuit as well as intermodulation noise and crosstalk. It is of relatively continuous duration and can be measured by a properly weighted meter. The other type of noise is impulse noise. It occurs at random intervals, is of high amplitude, and may be of short duration. It cannot be measured by a normal meter. It is measured by using a meter circuit

Table 11. Relationships in the Transmission and Noise-measuring Systems

	Use	$0\ dbm =$	Unit
Power	Exponential units	10^{-3}	watts
Voltage	Transmission measuring set	0	dbm
Voltage	Effective (rms) voltage across 600 ohms	0.775	volts
Volume	Volume indicator	0	vu
Noise	F I A — 1,000-cycle tone	85	dba
Noise	Weighting — 3KC-band limited thermal noise	82	dba
Noise	C MSG — 1,000-cycle tone	90	dbrn
Noise	Weighting — 3KC-band limited noise	88	dbrn
Noise	Pso-phometric — 1,000-cycle tone	90	dbrn
Sound	Sound level referred to 10^{-16} watts (0.0002 dynes/cm²)	130	db

Scale annotations:

- Volume indicator reference: 0.001 watt
- Voltage at each division = 1.122 × next lower reading

Note: For psophometrically weighted noise in picowatts, dbrnc = 10 log (picowatts), or 90 dbrnc = 10^9 picowatts.

arranged to operate a counter whenever any signal exceeds a predetermined level. The impulse noise level is defined as that level which will not be exceeded more than 90 times per half hour.

It is important to understand the relationships between noise and signal levels in the various scales of measurement. Some of these relationships are shown in Table 11.

Transmission measurements are stated almost invariably by reference to a 1,000-cps tone of 1 mw or 0 dbm (decibels referred to 1 mw). Weighting networks are also shaped with their characteristic curves centered on 1,000 cps. This permits calibration of noise meters with reference to 0 dbm. One type of noise meter, of which the Western Electric Co.'s 2B set is the prototype, uses F1A weighting. This weighting is typical of the acoustic characteristics of Western Electric Co.'s 300-type telephone sets. These sets are becoming obsolete and are being replaced by another type of noise-measuring set typified by the Western Electric Co.'s 3A set. This set provides readings in dbrn (decibels above reference noise). 0 dbrn is equivalent to −90 dbm with C message weighting. C message weighting is typical of the acoustic properties of Western Electric Co.'s 500-type telephone sets. Because of differences in the weighting networks and scale references, noise measurements with the old-type sets can be corrected to dbrn by adding 6 db.

In this chapter, noise values are given in dbrn. When C message weighting is used, the measurements are specified as dbrnc. When noise measurements are corrected to 0 TLP, they are specified as dbrnc 0.

A note on the psophometric scale has been added to Table 11, since it is commonly used in other countries. A more detailed description of noise-measuring systems and scales may be found elsewhere.[4]

Meters for measurement of circuit noise usually have a response time of about 200 msec. They may also have additional damping as an option. The detector associated with the meter adds noise voltages on a root-sum-square basis.

28. Circuit Noise in Switched Systems

Noise requirements in switched systems are stated by reference to the zero test level point, which was originally defined in reference to a manual toll telephone network. In automatically switched systems, the 0 TLP is the milliwatt supply in the switching center, since the test tone level at 0 TLP is 0 dbm by definition. When the test tone is connected to a circuit through a 2-db test pad, the point of connection becomes a −2 TLP.

If a circuit had 4 db of loss in the transmitting direction, the far end would be a −8 TLP when measured through 2-db test pads. If a noise measurement made at the far end of such a circuit gave a meter reading of 36 dbrnc, 8 db would be added to the meter reading. The noise level referred to the 0 TLP would therefore be 44 dbrnc 0.

The arrangement for defining and measuring noise can be extended throughout a switched system. It has the advantage that noise measurements made at any point in the system can be referred to a common point. Since the levels of transmitted signals can also be referred to the same point, signal-to-noise ratios can be readily determined. Conversely, signal-to-noise ratios can be converted readily into noise requirements.

As an illustration, assume a data system designed to transmit at −10 dbm 0, i.e., −10 dbm at the 0 TLP. The system requires a 30-db signal-to-noise ratio for satisfactory operation. The noise requirement is thus 50 dbrnc 0. Assume that the data receiver is designed to receive signals at a nominal level of −16 dbm. What noise measurement is required at the receiver for satisfactory operation? In this case, the data receiver is a −6 TLP. Therefore, a noise measurement of 44 dbrnc or less at the receiver will meet requirements, and the signal-to-noise ratio will be satisfactory at all points in the circuit, even through several switches.

Circuit noise tends to be higher on longer circuits. This seems logical, since the number of exposures to noise sources are greater with length. The transmission engineer can provide some control of noise levels by proper spacing of radio and multiplex repeaters, proper choice of transmission levels, good circuit balance, proper shielding and grounding, etc.

Practical noise objectives for voice frequency circuits range from 31 dbrnc 0 for the shortest circuits to about 48 dbrnc 0 for circuits 6,000 miles long. They are gradually being lowered as the state of the art progresses.

A switched connection 6,000 miles long should have the same noise requirements as a direct circuit of the same length. If a practical limit for 6,000-mile circuits is 48 dbrnc 0, two 3,000-mile circuits connected together would each have a requirement of 45 dbrnc 0 if the contributions of the two circuits are not to exceed the overall requirements when added on a power basis. Similarly, four 1,500-mile circuits connected in tandem would each have a requirement of 42 dbrnc 0, etc.

Circuits in switched systems are of various lengths. The transmission engineer can establish noise requirements for random lengths by setting ranges. A typical arrangement is shown in Table 12. Note that a set of maintenance limits is specified in

Table 12. Typical Circuit Noise Requirements

Circuit noise requirements, dbrnc 0

Circuit length, miles	Design objective	Maintenance limit
0–50	31	44
51–100	34	44
101–400	37	44
401–1,000	41	50
1,000–1,500	43	50
1,500–2,500	45	50
2,500–4,000	47	50
4,001–6,000	48	54

addition to design objectives. Properly engineered and installed circuits should meet the design objectives, but noise levels tend to increase because of aging of parts, etc. In addition to objectives, some values must therefore be established beyond which deterioration of the circuit will not be permitted. These are the maintenance limits, and they should be strict enough to meet the requirements for good service but not so strict as to cause excessive testing and repair work. When the requirements for good service are strict, the engineer must choose between more sophisticated design and higher testing and repair expense.

Noise requirements in a switched system are stated for busy-hour traffic conditions. At this time, radio and cable multiplex systems are most likely to overload and generate intermodulation and crosstalk noise. The chances of crosstalk in cable pairs are also increased. Switching centers also contribute the maximum noise because of switching contacts, signaling and supervision tones, etc., during peak load conditions. Measurements made at any other time usually do not reveal the worst noise conditions.

Some carrier circuits are equipped with compressor-expanders (compandors). Compandored circuits measured with the compandor active should meet requirements about 5 db more strict than noncompandored circuits. If these requirements are not met, the expander should be disabled. The regular requirements then apply.

IMPULSE NOISE

29. The Effect of Impulse Noise

Random pulses of noise of high amplitude, while not seriously affecting voice communications, can cause trouble with other forms of transmission. A noise pulse can obliterate bits of information in a data system or create apparent bits where none should exist. An impulse-noise requirement is generally stated as a level which should not be exceeded by noise pulses more than 90 times per half hour in the voice band.

Impulse noise generally occurs on a random basis. Its rate of occurrence tends to reach a maximum during the busy hour for the facilities which include the circuit being tested. The circuits in a switched connection each tend to encounter their maximum impulse-noise levels at different times of the day. The impulse-noise levels are therefore not added on a power basis when circuits are switched in tandem. The requirement for each circuit is individually established to ensure that adequate signal-to-impulse-noise levels are maintained within the circuit.

30. Impulse-noise Requirements

Switched circuits are designed individually to provide satisfactory signal-to-impulse-noise ratios. Most data systems function satisfactorily if the signal-to-impulse-noise ratio is 9 db or better. For example, if the signal level in a data circuit is −18 dbm 0, the equivalent impulse-noise level is 72 dbrn 0. Therefore, the worst impulse-noise level acceptable for satisfactory operation is 63 dbrn 0. Well-designed circuits, when properly maintained, can generally provide better performance. The transmission engineer may therefore choose to establish a more strict requirement to ensure proper design and maintenance of circuits, but the limit should not be so restrictive as to cause excessive design and maintenance expense. As long as the signal-to-impulse-noise ratios are adequate throughout a switched connection, a further increase in the signal-to-impulse-noise ratio of any link will have little effect on the error rate.

When excessive impulse noise occurs on a switched connection, it is difficult to locate the source of noise. The circuit cannot be sectionalized and tested in parts as is done when circuit noise becomes objectionable. By the time the offending link is isolated and tested, the busy-hour condition which caused the trouble may have disappeared. The control of impulse noise on switched connections must therefore be accomplished by controlling the noise on each individual link.

When high impulse-noise levels are found, the source can sometimes be identified by monitoring the circuit with a receiver. Noises can frequently be identified by their sound. The sources may sometimes be found by looking for simultaneous functions in associated equipment. Common sources of impulse noise are dirty switch contacts, defective filters in power supplies, crosstalk from signaling circuits, improper shielding and grounding, unbalanced circuits, loose connections, etc.

Switching centers are a major source of impulse noise. If the switching center is regarded as an additional short link in a switched connection, an impulse-noise requirement can be established for it. Tests may be made through the switches from the incoming appearance of any circuit to the outgoing appearance of any other circuit during the busy hour for the center. The impulse-noise requirement should be set to provide an adequate signal-to-impulse-noise ratio for any type of signal passing through the switches.

ECHO

31. The Definition of Echo

When a signal is transmitted on a circuit and meets an impedance mismatch, a portion of the signal is reflected back toward the source. The reflected portion of the signal is called echo. The difference in decibels between the transmitted signal level and the level of the echo at any point in the circuit is the echo return loss at that point. Note that echo return loss can vary with the point of measurement, since it also includes any gains or losses between the echo point and the point of measurement. A high echo return loss is always desirable, since it indicates small echoes. Common sources of echo are circuit terminations, line irregularities, and junctions between dissimilar facilities. In a well-constructed line, the echoes occurring at the circuit terminations are generally controlling. Another common source of echo is the junction between a two-wire circuit and a four-wire circuit. This junction is usually accomplished by using a hybrid circuit. In a perfect hybrid circuit, all the energy received from the transmitting side of the four-wire circuit is dissipated equally in the two-wire circuit and the balancing network and no energy returns through the receiving path of the four-wire circuit. If the two-wire circuit and the balancing network are not perfectly matched, some of the transmitted energy is returned through the receive path and appears at the source as echo. Echo problems must always be considered in direct circuits, but there may be additional problems in a switched system, because the switching centers can generate a very large number of circuit combinations. Also, there tends to be more links connected in tandem than would occur with a built-up direct circuit. Echo paths also tend to be longer, creating larger delays.

Figure 14 shows some echo paths in a typical switched connection consisting of two two-wire station lines and two four-wire trunks all in tandem. The switching centers are all two-wire. Two kinds of echo are designated on the figure—talker

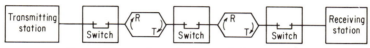

Fɪɢ. 14. Some echo paths in a typical connection.

echo and listener echo. Talker echoes are those that are reflected back toward the transmitting station or talker. Listener echoes are those that are reflected toward the receiving station or listener. Note that both types of echo can occur in a four-wire section. If the sum of all the gains in the four-wire section exceeds the sum of all the losses at any frequency, the circuit will begin to oscillate or sing because of its unstable condition. This suggests that the achievement of high echo return losses will permit reduction of circuit insertion losses to low values. Conversely, insertion losses must be high when the return losses are poor if the circuit is to be stable.

32. The Effect of Echo

Talker echo seriously affects voice communication. If it occurs with very little delay, it tends to increase the sidetone in the talker's receiver and has the result of causing the speaker to talk at a lower level. If the echo occurs with substantial delay, the human ear separates it from the normal sidetone. It sounds somewhat like acoustic echo and is disconcerting to the speaker. Talker echo is of little concern when the transmitting station is a device such as a data set. When the set is in the transmitting mode, its receiver is muted with half-duplex operation and it cannot be influenced by talker echo.

Listener echo has little effect in voice communication if the delay is very short. It tends to reinforce the transmitted signal, and the ear of the listener is not selective enough to detect the echo. If the delay time is substantial, however, the listener is able to detect it. It then becomes confusing for the listener in the same manner as when several people speak at once. Listener echo seriously affects receiving equipment such as data sets. The transmitted signal may echo at the receiving end and again at the transmitting end. On the second trip through the circuit, the echo is out of phase with the original signal. The echo may then appear as a bit of information in a sequence when none should appear, or it may be out of phase with a bit of information and cause it to disappear. The result may be a high error rate.

Figure 15 shows the typical connection of Fig. 14, but the connection is now completely four-wire. For electronic transmitting and receiving equipment, the echo paths may be completely removed and the equipment may be operated on a full-duplex basis. The disadvantage is that the switching centers and circuits are more expensive. Voice communications may still encounter an echo problem. The longest echo path still remains, since a sidetone circuit is required at each end to give the talker the sensation of talking on a live circuit. Sidetone circuits are usually designed to provide a sidetone level about the same as received signal levels. The sidetone circuit may therefore be an echo path, with insufficient echo return loss, unless it has directional characteristics. It is evident that echo must be carefully considered in a switched system. We shall therefore consider the nature of echo in somewhat greater detail.

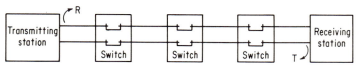

FIG. 15. Echo paths in the typical connection using four-wire circuits.

33. Echo Requirements in Voice Communications

Subjective tests made with large numbers of observers have established the amount of loss required in the talker echo path for various round-trip delay times. The loss values required to satisfy the average user are shown in Table 13.

Table 13. User Tolerance to Talker Echo

Loss required in the round-trip
echo path to satisfy the average user
with respect to round-trip delay

Round-trip delay, msec	Average tolerance to loss in echo path, db
0	1.4
20	11.1
40	17.7
60	22.7
80	27.2
100	30.9

The well-constructed station line terminated in a telephone instrument (off-hook) usually has a return loss on the order of 11 db with a sigma of about 3 db. It can be seen from Table 13 that additional loss must be added as round-trip delay times increase if the circuit is to be acceptable. In extremely long circuits, the required loss may be so great as to result in low signal levels. This situation may be remedied by the use of echo suppressors (paragraph 34).

The conclusions about losses required for satisfactory performance assume that the echo occurring at the station is controlling. To ensure this condition, requirements are usually established for various levels of switching centers if they are two-wire switches. A requirement of 15 db average with a minimum of 9 db is usually specified for the connections to station lines. A requirement of 27 db average with a minimum of 24 db is customarily specified for interconnections between trunks and access lines. Of course, in a system using only four-wire switches, the terminal echo will always be controlling. When main PBX's are designed to interconnect tributary or satellite PBX's, similar requirements are established for the tandem switches at the main PBX.

The losses that must be included in a connection are related to the velocity of propagation of the transmission facilities. A high-velocity facility has a relatively low round-trip delay. Therefore, its losses can be relatively low. Radio and cable multiplex systems have high velocities compared with cable pairs and are therefore used almost exclusively on long circuits.

A reasonable criterion of echo performance is that 99 per cent of users find talker echo to be within acceptable limits. This is 2.33 sigma in terms of a normal distribution curve.

Let L = one way circuit net loss requirement, db
  E = average echo tolerance for a stated round-trip echo delay, db
  R = average return loss (taken as 11 db) for well-constructed station lines
  D_0 = standard deviation of distribution of echo tolerance between observers (2.5 db)
  D_t = standard deviation of terminal return loss distribution (3.0 db)
  D_v = standard deviation of distribution of round-trip circuit loss variation (2 db)
  N = number of trunks in tandem

Then

$$L = \frac{E - R + 2.33(D_0{}^2 + D_t{}^2 + ND_v{}^2)^{1/2}}{2} \quad \text{db} \tag{3}$$

For example, assuming that a single link has 40 msec round-trip delay, the one-way loss requirement becomes

$$L = \frac{17.7 - 11 + 2.33(2.5^2 + 3^2 + 2^2)^{1/2}}{2} = 8.3 \text{ db} \tag{4}$$

For three links having a total of 60 msec of round-trip delay

$$L = \frac{22.7 - 11 + 2.33(2.5^2 + 3^2 + 3 \times 2^2)^{1/2}}{2} = 11.9 \text{ db} \tag{5}$$

If the values of L are obtained for the various values of round-trip delay for various numbers of trunks in tandem, a family of curves may be obtained as shown in Fig. 16.

For single-link calculations the curve is relatively straight for delays greater than about 45 msec and is approximately tangent to a line drawn through the points corresponding to 4.4 db at 0 msec and 9.2 db at 45 msec. The slope is equal to about 0.1 db/msec. This slope is a value commonly used as an approximation in computing overall circuit losses. The required one-way loss is commonly computed as 0.1 db times the round-trip delay in milliseconds plus 4.4 db plus 0.4 db for each additional trunk in tandem. The value obtained by multiplying the round-trip delay by 0.1 db is called via net loss (VNL).

A via net loss factor (VNLF) in decibels per mile can be developed for each type of facility from

$$\text{VNLF} = \frac{2 \times 0.1 \text{ db/msec}}{V \text{ (miles/msec)}} \tag{6}$$

where V is the velocity of propagation.

Formula (6) can be used to determine VNLF's for all kinds of facilities. Some typical values are multiplex facilities, about 0.0015 db/mile; open wire, 0.01 db/mile;

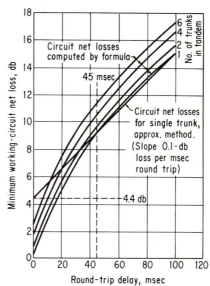

Fig. 16. Approximate minimum working net loss resulting in satisfactory working condition in 99 per cent of cases.

cable facilities, 0.03 to 0.04 db/mile. Let us assume a 4,000-mile connection composed of three trunks in tandem. All facilities, including the station lines, are multiplex with a VNLF of 0.0015 db/mile. The one-way loss for satisfactory operation is

$$L = (4,000 \times 0.0015) + 4.4 + 0.8 = 11.2 \text{ db} \qquad (7)$$

The velocity of propagation from Eq. (7) is 133 miles/msec, and the round-trip delay is 30 msec. From Fig. 16, it appears that the rule of thumb provides adequate performance.

Now assume that the station lines each consist of 100 miles of cable with a VNLF of 0.04 db/mile. Recalculation of the required one-way loss now results in a value of 18.9 db. The values obtained by this procedure may be excessive and result in insufficient signal at the receiver. The engineer has two choices in solving this problem. One is to increase the return loss at the terminations and thus lower the net loss requirement. This is not always practical. The other method is to operate the circuits at arbitrary loss values in the transmitting direction to give the desired signal. Echo control is provided by switching loss in the echo path. This is done with echo suppressors.

34. Application of Echo Suppressors

The echo suppressor is a device which detects signal levels in the transmitting and receiving paths and compares them to determine how to operate switchable pads. It must be installed at a four-wire point in the circuit. In its nonoperated state, it may insert loss in the transmitting path and provide no loss in the receiving path.

This represents the receiving condition. When a voice signal is detected in the transmitting path that is sufficiently stronger than any received signals, as in the transmitting condition, the suppressor removes the pad from the transmit path and inserts a pad in the receive path to maintain the required echo loss. Devices which transfer loss in this manner are called full echo suppressors. Devices which insert loss only in the receive path are called split echo suppressors. When split echo suppressors are used, there must be one installed at both ends of the connection.

Echo suppressors create some new transmission problems. Their operating time must be fast enough to prevent clipping of the first syllable but not fast enough to chatter on impulse noise. The release time (hang-over) must be short enough to prevent clipping the first syllable of received signals, but it must be long enough to prevent chatter between syllables. The release time must always be longer than any maximum round-trip delay that may be encountered in order to reduce the last echo.

Echo suppressors are normally installed at a switching center and associated with trunks. In a switched system with alternate routing, several suppressors could be connected in tandem. The accumulated operate and release times would result in poor service. One approach to this problem is to install split echo suppressors at each end of each trunk. The switching centers are provided with logic to disable the suppressors when connection is made between two trunks but leave them operative when connection is made between a trunk and a station line.

Some forms of transmission will not tolerate echo suppressors. For example, a data system running full duplex or providing synchronizing signals in the receive path would not function with echo suppressors on the circuit. In a shared system, the station equipment must therefore be able to signal its switching center that suppressor disabling is required and the switching center must be able to advance the signal to the next switch. The echo suppressors are usually arranged to be inoperative on receipt of disabling tone signals transmitted by the data set.

The transmission engineer may find it simpler to design two grades of trunk facilities—one equipped with echo suppressors and with compatible transmission parameters and another nonequipped and built for a different set of transmission parameters. With this arrangement, the station equipment must be able to signal the switching centers to indicate the type of circuit needed and the switching center must be able to advance the signal to successive switches in the connection. This results effectively in two switched systems with common station lines and switching centers. Echo suppressors are usually considered when round-trip delay reaches about 45 msec. The engineer can usually estimate the delay for each facility from its velocity of propagation. Other items with substantial delay, such as delay equalizers, must also be included.

IMPEDANCES

35. Switching Centers

Switches in four-wire switching centers are designed to connect 600-ohm impedances. This means that station lines as well as trunks should present a nominal impedance of 600 ohms. When the circuit impedance is substantially different, repeat coils are installed with turns ratios necessary to provide the proper impedance transformation. Multiplex channels are designed to present 600-ohm terminations and can therefore be connected directly to the switches. Two-wire offices are designed to switch at 900 ohms. PBX's are also considered to be 900 ohms. When carrier channels are terminated in these centers, the hybrid must have coils with 1.5:1 turns ratios to match the 600-ohm multiplex channel to the 900-ohm switch. When impedance-matching coils are used, their insertion loss must be included in the calculation of overall losses.

36. Other Locations

Two-wire telephone sets vary considerably as to impedance. They are considered to average about 900 ohms for design purposes. Four-wire telephone sets, data sets, and similar types of equipment are generally designed to work into 600-ohm circuits. When line impedances do not match the station equipment, repeat coils should be installed with proper turns ratios to connect the station equipment to the line. The coil insertion loss must be included as part of the line loss.

SWITCHING CENTER EQUIPMENT

37. Effect of Transmission Design on Equipment

It is obvious from the preceding discussion that the transmission design has considerable effect on equipment design. The major equipment affected by transmission design is the switching center. The switching center consists of electromechanical or electronic switches, control equipment, power supplies, trunk circuits, signaling equipment, equalizers, echo suppressors, multiplex equipment, terminating frames, etc. They must be arranged and adjusted in a way that will permit transmission requirements to be met on the overall connection.

The switches should have no effect on transmission if properly designed. Once a connection is established, there should be nothing to consider except the characteristics of the wiring making up the connection. Control circuits, power plants, and other common equipment should not affect a connection once it is completed. However, all other equipment in the center must be carefully considered.

For transmission design purposes, a circuit is considered to begin and end in a switching center at the center of the switches. The switches are customarily considered to be a -2 test level point (TLP) with one exception. First-level switching centers serving only two-wire lines are considered to be a 0 TLP. A typical arrangement of a four-wire zero loss trunk is shown in Fig. 17.

Multiplex channels are customarily lined up using a -16-dbm test tone level in the transmitting direction, and they are adjusted to deliver the test tone at a level of $+7$ dbm at the receiving end. For a zero-loss circuit, the terminal equipment must insert 14-db loss in the transmitting path and 9 db in the receiving path at each end. These losses may consist of pads or the insertion loss of terminal equipment. If the circuit is to be operated at a gain or loss, it is accomplished by adjusting the loss in the receiving path of the terminal equipment.   This arrangement provides flexibility, because any channel can be used with any terminal equipment.

Figure 18 shows a typical two-wire trunk circuit arranged for 4-db loss in both directions.   Such a circuit must always operate at a loss, since the hybrids in the terminal equipment at each end form echo paths which can produce singing.   The loss of the terminal equipment in the transmitting path is called the transmitting drop loss, and the loss in the receiving path is called the receiving drop loss.   Figure 18 is also typical of the arrangements for access lines.

Terminal equipment for station lines may vary from a line relay associated with a cable pair to an arrangement as complex as Fig. 18.   In the more complex arrange-

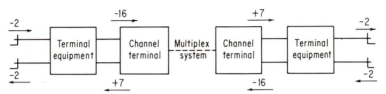

FIG. 17. A typical four-wire zero-loss trunk.

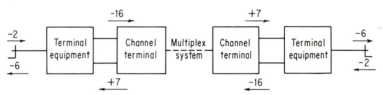

FIG. 18. A typical two-wire trunk with 4-db loss.

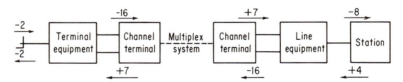

FIG. 19. A typical arrangement of a station line with 6-db loss.

ment, an auxiliary line circuit would be found in place of the trunk circuit, but it would serve somewhat similar functions.

Station lines are treated differently in one other respect.   Figure 19 shows a typical arrangement for a station line with 6-db loss.   Note that the switching center terminal is a −2 TLP for both directions of transmission.

38. Terminal-equipment Arrangements

Terminal-equipment arrangements are important in transmission design.   When lines or trunks are two-wire, the terminal equipment must be two-wire.   This presents some limitations, since each direction of transmission cannot be treated separately.   Four-wire facilities are often used even though the switch is two-wire to permit more flexibility in the control of transmission characteristics.   When four-wire switches are involved, the terminal equipment must, of course, be four-wire.   When four-wire facilities are used with two-wire switches, the hybrid circuit is placed as close to the switch as possible and transmission equipments are installed on the four-wire side of the hybrid. When two-wire facilities are terminated on four-wire switches, a hybrid must be provided and four-wire terminal equipment is provided between the hybrid and the switch.

Figure 20 shows a typical arrangement of terminal equipment. While all the items might not be required, they would be placed in the relationship shown when they are used. The trunk circuit provides idle circuit terminations, makes the trunk busy to other calls when the trunk is in use, translates switch signals to trunk signals, and vice versa. It usually provides for bridging a test jack. It must be either two-wire or four-wire, depending on the type of switch, and it should have no effect on the transmission characteristics. Two-wire trunk circuits may contain a switchable 2-db pad. It is inserted between the switch and the terminal equipment when connected to station lines and removed when connected to other trunks. It thus provides half of the 4-db factor to be added to VNL on over-all connections. When pad control is required, the switching center control equipment must be designed to include the pad control feature.

The signal equipment may be designed for zero insertion loss or may have a fixed loss which must always be included in the transmission design. Hybrids and echo suppressors usually have a fixed insertion loss. Control of the drop loss is usually accomplished next to the hybrid on the four-wire side. When no equalizers are

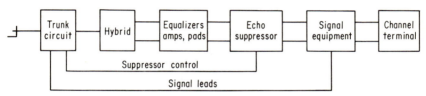

Fig. 20. A typical arrangement of terminal equipment.

required, fixed pads may be all that is required. When equalizers are used, their insertion loss may be too great. Amplifiers must then be added to reduce the loss to the desired value.

The signal equipment indicated in Fig. 20 generally uses a single-frequency in-band tone and is intended for supervisory purposes only. Multifrequency tone equipment provides for transmission of dial pulses. This equipment is part of the common equipment in the switching center. The MF tones are injected or detected by the common equipment through the operation of relays in the trunk circuit.

When the facility consists of cable pairs, the signal equipment is replaced with well-balanced repeating coils. The center taps of the coils are used to apply loop or DX (duplex) signaling to the cable pairs.

FACILITY SELECTION

39. Trunk Facilities

High-velocity systems should always be used, if possible, for trunks in order to meet the low loss requirements. Since trunks constitute the major part of any long connection, the use of high-velocity facilities will reduce round-trip delay and the associated echo effects. Cable or radio multiplex systems are generally necessary. The transmission engineer will, therefore, be greatly interested in the types of transmission facilities available on each trunk route. It may be necessary to specify installation of new facilities or improvement of existing facilities to meet the strict requirements for loss, attenuation-frequency, delay, and noise characteristics that are essential to satisfactory operation of a switched system.

When multiplex-type channels are used, they should be selected to attain the minimum number of channel banks or channel terminals per trunk. Preferably, there should be no more than one pair of channel terminals per trunk. This will reduce the amount of equalization required. This is particularly true for delay equalization, since nearly all of the delay distortion in a multiplex system is contributed by the channel terminals. The end channels of a channel bank may be avoided when delay equalization is necessary. They will tend to require more equalization because of the effect of the group and supergroup connectors and filters.

Multiplex systems with automatic group and supergroup regulation are preferable. They provide better control of loss variations. Systems having a frequency shift of more than ±5 cycles should be avoided, since errors in data are apt to occur. Synchronized systems and systems which transmit carriers are generally acceptable.

Multiplex channels equipped with compandors may be used for trunks not requiring conditioning, providing the channels meet all other transmission requirements. They can also be used for data transmission if the data are at a constant energy level and there is no multiplexing of data signals on the same trunk.

40. Access-line Facilities

Facilities for access lines require basically the same quality facilities as trunks. However, high-quality facilities are not always available. For this reason, transmission requirements are made somewhat less strict than those for trunks (cf. Tables 7 to 10). Access lines will sometimes be short enough to permit the use of cable pairs. Longer access lines may consist of multiplex channels with cable extensions. Cable pairs used for these services should be nonloaded or should use high-quality loading plans. Heavy loading plans such as H135 and H174 should be avoided because they have a low cutoff frequency and excessive delay distortion.

41. Station-line Facilities

Station lines connect directly to trunks. Transmission can, therefore, be no better than the quality of the station lines. The rules for selection of facilities for trunks and access line should, therefore, be applied to station lines so far as possible. If adequate facilities are not available, they may have to be constructed. In those cases where the equalization requirements are stiff, it may be necessary to extend the multiplex channels directly to the user location.

BALANCE IN TWO-WIRE SWITCHING CENTERS

42. Balance Theory

In order to meet overall transmission objectives, it is necessary to adjust the equipment in two-wire switching centers to ensure that echoes are not created because of impedance mismatches in the switching equipment. The process is known as balance. This term probably stems from the adjustments of the hybrid circuits which are essentially Wheatstone bridges.

Figure 21 illustrates the fundamental theory of a hybrid circuit. Sketch A is the typical Wheatstone-bridge circuit. When $R_a = R_b$ and $R_l = R_n$, the bridge is balanced and there is no current flow through the galvanometer. Sketch B shows the a-c version of the bridge. For this circuit, when $Z_a = Z_b$ and $Z_l = Z_n$, the bridge is balanced and no current flows through the measuring set. Sketch C shows the a-c bridge rearranged so that Z_l has become the impedance of the line and Z_n the impedance of the balancing network.

Figure 22 illustrates a typical coil hybrid circuit. Power entering the hybrid over the receive leg divides between the two-wire facility and the network. If the network and two-wire facility balance each other perfectly, the power divides equally between them and no power enters the transmit leg. If the network and two-wire line do not match each other perfectly, then part of the power enters the transmit leg and returns to the originating end. The amount returned depends on the degree of unbalance. The reflected signal is called echo, as discussed before, and the ratio in decibels between received and reflected power is the return loss of the hybrid.

When a signal is transmitted toward the hybrid from the two-wire line, the signal is divided equally between the transmit and receive legs. The portion in the receive leg is dissipated, and the portion in the transmit leg is sent on over the line facility. Note that the transmission loss in either direction can never be less than 3 db plus the coil losses. The capacitor in the two-wire line termination serves no transmission purpose. It is normally provided as a d-c block to permit application of signals through the switches. If the two-wire line is terminated so as to match the network perfectly, the return loss is theoretically infinite. With an open or short termination,

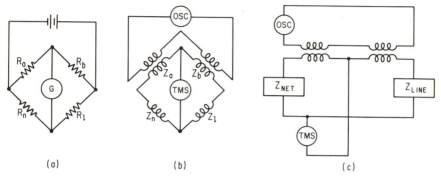

Fig. 21. Fundamentals of the hybrid.

the return loss is theoretically zero. In actual practice, perfect balance is never achieved but return losses higher than 40 db are possible. With an open or short, the return loss is equal to the transhybrid loss, which may be in the order of 7 db. If the transhybrid loss is measured, it can be subtracted from any return loss measurement to get the true balance of the line against the net. When the open or short is at some distance from the hybrid, the return loss will be equal to twice the loss of the intervening wire. This can be verified from transmission-line formulas.[1]

The impedance of the two-wire line and the network will vary with frequency. Also, since the network and the two-wire line are not exactly identical, their impedance will vary independently of each other for a given frequency change. Therefore, the degree of balance between the network and two-wire line will depend on the frequency of the applied signal and will change as the frequency is varied. The theoretical development can be found in other publications.[5]

From amplifier (Receiving)

NET

To amplifier (Transmitting)

Fig. 22. The typical coil hybrid circuit.

It is customary to measure return loss throughout the frequency range 500 to 2,500 cps. The average value is called the echo return loss (ERL). The frequency at which return loss is poorest is usually the frequency at which singing occurs as gain is added. The return loss at this frequency, or the gain required to start singing, is called the singing point. It tends to occur in the frequency ranges 200 to 500 and 2,500 to 3,000 cps. Beyond these frequency ranges, facility characteristics or filters insert higher losses and prevent singing.

Since the network in the hybrid is a fixed value, the degree of balance is contingent on the impedance terminating the two-wire side of the hybrid. In a switching center, the termination is likely to be another hybrid circuit and the intervening wiring. The networks in a 900-ohm switching center consist of a 900-ohm resistor and a 2-μf capacitor in series. The network resistor matches the resistor in any opposing hybrid which may be connected through the switches. The 2-μf capacitor matches the capacitors installed in the hybrids for signaling purposes. The only remaining factor to be considered is the variable length of wire connecting the hybrids through the switches.

Figure 23 illustrates the principle of balance. The network associated with each hybrid is designed to match the impedance seen at the two-wire side of the hybrid. In this case the impedance consists of the network in the opposite hybrid, the resistance of the wiring, and a capacitor in the hybrid. The network therefore consists of a resistor and capacitor in series. The resistance of the wiring is usually limited to 45 ohms in 600-ohm centers and 65 ohms in 900-ohm centers. The network is therefore fairly simple, and the balance is not greatly affected by variations in resistance of the wiring. The network does not, however, allow for shunt capacitance in the wiring. This adjustment is accomplished with an adjustable network build-out (NBO) capacitor connected across the network.

In practice, sample measurements are made of shunt capacitance on all the possible paths through the switches. When the longest possible path has been found, the

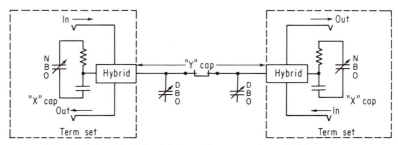

Fig. 23. The principle of balance in a two-wire switching center.

NBO capacitors are adjusted on this connection for the highest return loss and all other NBO capacitors in the switching center are adjusted to the same value. Short wiring paths are then built out to look like the longest path by installing small drop build-out (DBO) capacitors across each pair. The DBO capacitors are equal to the incremental value of shunt capacitance needed to make the total shunt capacitance of each wiring path equal to that of the longest path. When this work is carefully done, it should be possible to connect any circuit to any other circuit through the switches and obtain uniformly high return loss.

The transmission engineer responsible for balance in two-wire switching centers will be very interested in the arrangement of terminal equipment. One group of equipment remotely located from the rest can cause an excessive amount of capacitance build-out. Cabling should also be arranged to minimize path lengths. An excessive amount of capacitance can be accumulated in cross connections and the bridged wiring of test jacks. It is therefore important that the switching center be compact and carefully cabled. NBO capacitors should not exceed about 0.08μf, and in most cases, they should be much smaller. When the NBO and DBO capacitors are large, the balance requirements can be met satisfactorily but there will be an adverse effect on the attenuation-frequency characteristic of every connection through the switching center. This effect tends to accumulate on a switched connection. Of course, this is also true of four-wire switching centers when the cabling is not carefully planned.

43. Balance Requirements in Two-wire Switching Centers

Several steps must be taken to meet balance requirements in two-wire switching centers. They are:

1. Measure the capacitive length of all switching paths.
2. Select the longest path, and determine the setting of the NBO capacitors.
3. Adjust all NBO capacitors to the same value.
4. Adjust all DBO capacitors for best return loss.
5. Measure and record return losses and singing points on all circuits.
6. Compare the results with requirements, and repair all circuits that are unsatisfactory.

Before the work is started, careful study of the office is most helpful. The tests should be listed in order, the equipment locations and cable runs known, and the procedure thoroughly understood.

Figure 24 shows the test arrangement to determine those switching paths which are capacitively longest. Sample circuits are selected from those switching paths which appear from cable layouts to be the longest. The connection is set up through the switches, and the distant hybrid is terminated in characteristic impedances on the four-wire side. All NBO and DBO capacitors are unstrapped for this test. The oscillator is adjusted usually to send at 2,000 cps, and the near-end NBO capacitor is adjusted for maximum return loss as indicated by the lowest reading of the transmis-

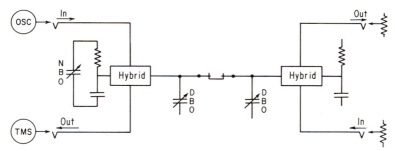

FIG. 24. Test arrangement to determine capacitive lengths.

sion measuring set. The tests are made, and NBOC values recorded for all switched connections which may possibly be the longest.

The path yielding the highest NBOC value is the longest one. If it is much greater than all the rest, consideration should be given to relocation of the equipment. It may be desirable to add a small increment to the NBOC value to allow for expansion of the switching center. When the NBOC value has been decided, all NBO capacitors should be adjusted to this value.

A balance test line should be available at this point. It should consist of a network with a DBO capacitor adjusted to look electrically like half of the longest path through the center. It will generally be much easier to connect every other circuit through the switches to the test line for the remaining tests.

Each circuit is connected in turn to the longest circuit or its equivalent test line. The test sets are connected as shown in Fig. 24, and the DBO capacitor on each circuit is adjusted for maximum return loss at 2,000 cycles.

When all DBO capacitors are adjusted, the tests are all repeated. This time an average return loss is obtained across the 500- to 2,500-cps range. The singing point is also measured. The return loss tests may be made using a noise generator. The transmission measuring set (TMS) automatically does the averaging. In either case, the oscillator and measuring set should both be connected through message weighting networks to simulate in-service conditions.

Table 14 shows typical requirements to be met at first-level two-wire switching centers. The requirements for tests to two-wire station lines are of necessity less than for the other tests. This is because the impedance presented by the station and line is much more variable than the impedance of the balancing networks in the terminating

Table 14. Typical Balance Requirements for First-level Two-wire Switching Centers

Connection		Echo return loss and singing point		Measured
		Average, db	No measurement less than, db	
Trunk to access line.........	ERL	27	25	
	SP	20	16	Term set to term set or equivalent test line
Access line to access line....	ERL	27	25	
	SP	20	16	
Trunk or access line to four-wire station line	ERL	27	25	
	SP	20	16	
Trunk or access line to two-wire station line	ERL	15	9	To station off-hook .
	SP	10	6	

sets. When a circuit does not meet the minimum requirements, a search should be made for such things as wiring errors, bad capacitors, capacitors improperly strapped, bad cable pairs, etc.

44. Balance Requirements at PBX's

The transmission design of a switched system does not necessarily end at a PBX. It may have tributary PBX's and the main PBX serves as a through switching point for traffic between the switched system and the tributary PBX's. When this is the case, pad switching is required at the main PBX. This reduces the access-line loss to VNL design, and the tie trunk to the tributary PBX requires VNL +2 db design.

Table 15. Typical Balance Requirements for a Main PBX

Access line switched to	2-db pad	Echo return loss and singing point			Measured to
			Average, db	No measurement less than, db	
Stations off main PBX	In	ERL	24	20	900 ohm + 2μf termination
		SP	18	14	
		ERL	15	9	On-premises station off hook
		SP	8	6	
		ERL	12	9	Off-premises station off hook
		SP	8	6	
Four-wire tie trunks	Out*	ERL	22	16	900 ohm + 2μf at distant PBX
		SP	15	11	
Two-wire tie trunks	Out*	ERL	18	13	900 ohm + 2 μf at distant PBX
		SP	10	6	

* Pads are left in if the tie trunk loss is less than 2 db.

Balance is also required, and the procedure at the PBX is fundamentally the same as for first-level two-wire switching centers.

Table 15 illustrates typical balance requirements for a main PBX. Note that balance tests on tie trunks are made with the trunk terminated at the far end for convenience. When the tie trunk is four-wire, the terminating impedance is the associated terminating set network. There is a substantial difference in the requirements for tests to extension stations as compared with results with a termination. This is because of the wide variations in length and characteristics of two-wire extension lines and variations in the sidetone path which make a high average balance impossible. This follows from our previous comment in the discussion of echo that the echo path (or return loss) at the end of the connection generally controls the quality of transmission. Balance in the PBX and at first-level switching centers ensures that all other parts of a switched connection are at least as good or better than the station terminations.

TRANSMISSION TESTING

45. Purpose of Testing

Transmission tests ensure that the designer's transmission plan is actually accomplished. The tests verify that the detailed engineering is correct and that no errors crept in during construction. Once the switched system meets all requirements and goes into service, the testing is not finished. In-service tests are also required to verify that the transmission plan meets service requirements. The tests must be repeated at intervals to ensure that transmission does not deteriorate.

46. Types of Tests

The transmission engineer must be concerned with transmission tests. The tests must be adequate to prove the transmission designs, but they should not be more complicated or expensive than necessary. For discussion, the tests are grouped as follows:

1. *Preliminary Tests*—Measurements to be made on individual facilities and equipments before they are interconnected in a trunk or line
2. *Overall Transmission Tests*—Measurements made on trunks and lines from end to end to make final adjustments and meet overall requirements
3. *Routine Tests*—Repetitive measurements to determine whether circuits are remaining within transmission maintenance limits
4. *Trouble-location Tests*—Procedures to bring faulty circuits within limits

47. Preliminary Tests

Preliminary tests are necessarily transmission tests, but they should be performed in advance of overall transmission tests to save time in finding the more obvious troubles, to prevent large offsetting errors, and to prevent covering up maladjustment of circuit components when the final overall adjustments are made.

Some of the tests that should be completed in advance of final testing are:

1. Calibration of test equipment
2. Direct-current tests on cable pairs
3. Multiplex system line-up and transmission tests
4. Multiplex channel line-up and transmission tests
5. Repeater tests and gain settings
6. Signaling-equipment tests
7. Echo-suppressor tests
8. Balance tests
9. Drop loss tests

The first order of business must be the calibration of all test equipment, including test tone supplies. Transmission tests can never be any better than the accuracy of the test facilities.

Tests on transmission facilities should be scheduled. Cable pairs should be checked for opens, shorts, and grounds. Loaded pairs should be checked for proper loading. Tests should be made on cable or radio multiplex systems as soon as the high-frequency lines are available for service. When this work is done, the individual channels can be tested and adjusted.

Repeaters should be tested and adjusted to initial gain settings. Tests should be made on other terminal equipment as required to get it adjusted and connected into the circuit. Sensitivity and timing tests and adjustments should be made on echo suppressors when installed. Signaling tests need to be made, since it is essential to be able to get the circuit properly terminated and conditioned for transmission tests.

When all individual units of terminal equipment are installed and connected into the circuit, it is advisable to make loss tests between the multiplex channel terminals and the switches to verify that the terminal equipment is connected and functioning properly.

48. Overall Transmission Tests

When trunks or lines are installed without equalization, loss measurements should be made at 1,000 cps in both directions and any final adjustments or corrections made to bring the 1,000-cps loss to the design value. Tests should also be made at perhaps four other frequencies to ensure that the circuit does not have unusual attenuation-frequency characteristics. Typical requirements might be -1 to $+3$ db at 700 and 2,300 cps and -3 to $+8$ db at 300 and 3,000 cps.

When circuits are installed with attenuation equalization, the equalizer should be given a nominal initial setting. Tests should then be made on several frequencies in each range of interest to ensure that the requirements are met. This work can be expedited if sweep-type measuring sets are available. After the attenuation-frequency requirements are met, the 1,000-cps loss is finally adjusted to the design value.

When delay equalizers are provided, measurements should be made in both directions of transmission. When fixed equalizers are used and several are connected in tandem to meet requirements, it may be necessary to readjust the 1,000-cps loss.

Circuit noise and impulse noise should be measured on all circuits. When the measurements are within the objectives, the circuits may be considered satisfactory. If they are within the maintenance limits, the circuits may be used for service, but further work should be done to locate the noise sources and improve the noise levels. If noise measurements are not within the maintenance limits, the circuits should not be put in service until the troubles are found. Noise measurements require that the distant end of the circuit be terminated in its characteristic impedance.

49. Routine Tests

Routine test intervals should be established for 1,000-cps loss and circuit noise measurements. These characteristics are most subject to change and usually most easily measured. Tests should be made every 3 months or oftener until experience shows that they may be more profitably made at other intervals. As records are developed, the test intervals can be related statistically to the degree of reliability required.

Impulse-noise tests may be made at less frequent intervals. When impulse noise becomes a problem, it generally indicates a problem common to many circuits. Location and reduction of impulse noise may require a series of tests at various locations in the equipment. The causes are frequently due to design errors, switching-equipment troubles, etc., and are not so likely to be due to deterioration of transmission facilities as circuit noise.

Equalization tests need to be made only at infrequent intervals. They should be made occasionally to detect unreported equipment changes and human errors.

50. Trouble-location Tests

Trouble investigations are made on trouble reports, routine test failures, and alarms. Frequently, the trouble report or routine test may reveal the source of trouble. When no indication is given, the initial test procedure should be followed. When the circuit fails any test, the circuit should be sectionalized and the various parts tested until the trouble is located.

TESTING FACILITIES

51. Facilities at Switching Centers

Switching centers should be equipped with automatic test lines for incoming circuits. These circuits permit making initial and routine transmission measurements without the assistance of a testman at the distant end. Thus more testing can be accomplished for a given amount of effort. Several kinds of test lines are available.

Balance test lines may be provided in four-wire switching centers as a termination for noise measurements. They are also used for noise measurements in two-wire switching centers. In two-wire centers, they are built out with a DBO capacitor to permit their use in balance tests.

Milliwatt supply lines provide the standard 1,000-cps test tone for line-up tests, routine tests, and trouble location. The milliwatt supply line must be calibrated to provide exactly -2 dbm at the -2 TLP. The test line is usually connected to a tap from the milliwatt source. The milliwatt source may also be multipled to other locations in the center. Each outlet must be equipped with an idle circuit termination to eliminate level changes when several outlets are in simultaneous use.

Loop-around test lines may also be installed when four-wire circuits are to be tested. The circuits are first tested in the far-to-near direction, using the milliwatt test line. Two circuits are then connected to the loop-around test line. The loss around the loop is then measured, and the far-to-near loss subtracted to obtain the near-to-far loss. The tests may be repeated until an entire trunk group is measured in both directions from one end.

Jack-ended test lines may also be provided at a convenient location, such as a test board or other test center. They are usually equipped with signals and are used to answer incoming requests from other switching centers for transmission tests on a two-man basis.

Test lines are excellent for the simple transmission tests. They are generally not suitable for the more complex equalization tests. It is also difficult to test a circuit for its data capabilities because the circuit will normally be used in tandem with other facilities and a test of the individual circuit is not significant.

A test jack is usually provided on each circuit in the switching center. The jacks are usually centralized in a test center which also contains the necessary facilities to busy out circuits, provide transmission and signaling tests, and perform miscellaneous other functions. The test board provides a centralized testing and reporting location and is an important part of the transmission testing program.

Transmission testing equipment must also be specified. Rack-mounted high-quality equipment may be provided at the test board. In addition, portable test sets are required for tests on individual components.

52. Facilities at PBX's

PBX's may also be equipped with automatic test lines where the number of circuits to be tested justify their use. They serve the same purpose as in the switching centers and are particularly useful at unattended locations where craftsmen with suitable test equipment must be dispatched. In addition to test lines, portable test equipment must be specified and made available for testing.

53. Facilities at Stations

Portable equipment must be specified and made available for transmission tests at station terminals. At the more complex installations, it may be desirable to provide test jacks for quick access to the circuit.

TESTING METHODS

54. Administration of Test Program

While circuits are designed from the center of the switch, they cannot be tested that way. Tests on trunks are customarily made from outgoing switch to outgoing switch. Only one set of switches is included in a measurement. A carefully planned test program is required so that all switches are included in a measurement at least once. None should be left out, and none should be measured twice.

Circuits normally are controlled by the highest ranking center, and the controlling center originates tests from its outgoing switches. Thus tests are scheduled from the top down in a hierarchy switching plan and from the inner-ring centers out in a hub plan.

Test schedules should be arranged to provide a steady work load. Such a schedule ensures uniform performance of the system. If all circuits were routined at once, transmission performance would be extremely good immediately after the tests. It would then deteriorate steadily until the next test round. Distributed testing keeps the system within the maintenance limits used for design purposes at all times while employing the minimum number of personnel.

55. Use of Test Pads

Test pads are usually provided at switching centers. These test pads serve a number of purposes. They arrange the circuit for testing on a VNL + 4 db basis. This ensures that the circuit remains stable and is tested under in-service conditions. For this reason the pads are standardized at 2 db. When 0-dbm test tones are applied to the circuit through the test pads, the test tones become −2 dbm at the −2 TLP, as required. The test pads provide sufficient termination, when test sets are momentarily disconnected, to prevent singing in the circuit between two-wire switching centers. They also act as a buffer and help to make the terminations provided by test sets appear more like standard terminations.

The test pads are usually associated with the measuring system at a test board. One set of test pads can be connected to any circuit through its test jack.

Because of variations in equipment locations and differences in cable runs, a variable error is introduced by this arrangement. This error is due to losses in the wiring between each circuit and its test jack. The losses of the wiring make the circuit losses appear to be too high. The actual losses of the circuits are then reduced to correct the apparent error. As a result, circuits tend to be slightly short when in service and are slightly less stable.

A sampling should be made of all circuits to be measured. The actual loss of the test-jack wiring is carefully measured for each circuit in the sample, and the average wiring loss obtained. The test pads are then reduced in value by the average amount of the wiring loss. When this adjustment is correctly made, there will be small errors in individual measurements but the average measuring error is zero.

56. Test Analysis

The engineer should make use of the test program in maintaining the system transmission performance close to the design standards. A system should be established for regular reports on circuits tested and measurements obtained. The reports should always show the measurements before adjustment. The measurements may be summarized by centers, by circuit groups, and for the system as a whole. The system averages indicate how well the overall job is being done in keeping performance up to the design standards. They also provide a reference for comparison of the results for switching centers and circuit groups. Such a system sometimes points up areas where poor maintenance work is being done. But much more often they indicate low-quality equipment or facilities or circuit engineering errors.

Analyses are usually based on deviations from the design values. They are summarized to produce distribution grade (sigma) and bias. Objectives are generally established. Typical values might be 1.0 db for distribution grade and 0.25 db for bias. Note that bias standards are in the range of value of the adjustments that are commonly applied to the test pads. This points up the necessity for high-quality test equipment and accurate work.

Meters should have scales with readability of about ±0.2 db or better and should inherently have an even greater accuracy.

A large distribution grade may indicate unsatisfactory regulation, unsatisfactory facilities, careless work, etc. A large bias may indicate improper test methods (errors all in one direction), too long routine intervals, etc. More frequent tests tend to reduce bias but not necessarily distribution grade. More accurate testing may reduce distribution grade but not necessarily bias. The transmission engineer should be vitally interested in these test analyses, since it may be a greater challenge to operate and maintain a switched system satisfactorily than to design it.

REFERENCES

1. Weinbach, M. P., *Principles of Transmission in Telephony*, The Macmillan Company, New York, 1924.
2. *Notes on Distance Dialing*, American Telephone and Telegraph Company, 1961.
3. Albert, Arthur L., *Fundamentals of Telephony*, McGraw-Hill Book Company, New York, 1943.
4. Chappius, C. K., Bell System and International Telecommunications Noise Measuring Systems. *IEEE Paper* 63-4, 1963.
5. *Reference Data for Radio Engineers*, 4th ed., chap. 28, International Telephone and Telegraph Corporation, 1957.

Chapter 7

SWITCHING ENGINEERING OF
SWITCHED SYSTEMS

PHILIP SCHNEIDER, *Radio Corporation of America*

Contributing Authors:

Norman Hovagimyan, *RCA*

Theodor Frankel, *RCA*

Walter Levy, * *RCA*

Leon Stambler, *RCA*

BASIC PRINCIPLES

Switching engineering takes place at two levels: establishing and designing the overall switching plan and selecting, specifying, and designing the individual switching centers. Both will be treated in this chapter, since in practice the design of the one heavily influences the other. For example, the need for tandem switching influences the design of the switching center, and conversely the switching-center design may influence such things as the numbering plan and routing doctrine. All this in turn must be closely coordinated with the transmission design as indicated in Chap. 6 so that the overall information path meets the user requirements.

1. Overall System

Switching engineering begins with the determination of user locations and requirements. Where are the users located geographically, whom must they communicate with, how often, what types of information must be transmitted, what are the statistics of the message lengths, the special features or services desired, and the grade of service required? In general, the switching centers are established at the "center of mass" of the population to be served. However, such considerations as availability and reliability of transmission systems and facilities, real estate, power, skilled personnel, and security may also influence the switch locations. The size of the center will be depend-

---

* Now with Pennsylvania Research Associates, Inc.

ent on the cost of switching as well as the cost of transmission and the "community of interest." Figure 1 illustrates two possible configurations. To make the choice requires a cost trade of the *Plan a*, subscriber loop cost and switching center cost vs. *Plan b*, subscriber loop cost and switching center cost and trunk cost.

To perform this cost trade presumes a prior knowledge of the traffic patterns to allow proper engineering of trunks between switching centers. A further refinement at this stage of planning can be achieved by considering the use of tandem or trunk

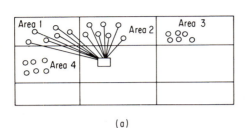
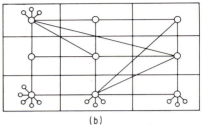

(a)   (b)

Fig. 1. (a) All subscribers connected to one central office. (b) One switching center provided in each area.

switching as shown in Fig. 2, again reducing the cost of transmission at the cost of switching. Variations and combinations of these plans are possible to minimize cost and maximize function as determined by the particular parameters of the problem. The general system topology having been established, the system configuration is more precisely defined in terms of numbers and types of switching centers, the functions and features, the numbering plan, and the trunking plan.

a. Trunking Plan. The trunking plan determines the manner in which calls are routed through the system. For each combination of user-to-user call the primary and alternate routes must be established. The switching centers will then be programmed to select trunks in accordance with this plan. The transmission requirements of the trunks will be governed by this plan. This plan must consider traffic, economics of providing the necessary transmission facilities, and reliability of operation in the face of failures. Trunks are provided between switching offices on the basis of traffic offered. Generally trunks are provided so that the probability of finding all trunks (to the next office) busy during the busy hour is 0.01. On particularly high-cost routes a grade of service as poor as 0.03 may be used. Telephone traffic tables are used to determine the required number of trunks for a given grade of service and traffic offered. The traffic offered is measured in erlangs. The erlang is equal to 3,600 call-seconds or the amount of traffic one trunk can handle in 1 hr if it were occupied 100 per cent of the time. The number of erlangs is determined from the following formula:

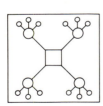

Fig. 2. Tandem switching.

$$\text{Erlang} = \frac{\text{No. of calls per hour} \times \text{average holding time per call (in seconds)}}{3,600}$$

The number of trunks required is then found in a table, such as that appearing in Ref. 10. The per cent occupancy of each trunk group is calculated as follows.

$$\% \text{ occupancy} = \frac{\text{erlangs} \times 100}{\text{No. of trunks in the group}}$$

For a given grade of service (i.e., prob. of blocking = 0.01), the larger the trunk group, the more efficient in terms of occupancy it will be. Therefore, the trunking plan should be laid out to maximize trunk group size and minimize the number of trunk groups. This can be accomplished by use of tandem switching points and overflow or alternate routes.

These same considerations will be applied elsewhere in this chapter to the specification of quantities of time-shared elements of switching centers such as register senders, links, and crosspoints.

Alternate routing provides a means for maximizing trunking efficiency and improving reliability. The direct route between offices can be engineered on the basis of a nonbusy-hour load so that its occupancy when averaged over a 24-hr period is increased. The peak traffic load is handled by routing via an intermediate switching office or a "backbone route." Since this alternate can act as the secondary route for many different switching connections whose traffic may not all peak simultaneously, it can be engineered more efficiently. It is generally a large group, and its peak-to-average traffic load is a smaller number. Reliability is improved by making a bypass route available in the event of circuit or switch failures on the primary route.

Alternate routing introduces complexity in the switching centers, more stringent requirements on the transmission characteristics of the trunks, and more detailed system planning to prevent "ring-around" problems.

2. Central-office Engineering

To provide a group of subscribers with a switching system, a choice must be made which best serves their particular needs. The choice of the system, its adaptation to user needs, and its coordination of control, traffic demands, and capability of switching devices to achieve efficient and economic use of equipment is defined by the "central-office engineering."

The selection of the type of switching system is influenced by a variety of factors, each one having a different degree of impact. Progressively controlled switching systems (step-by-step) tend to be more economical for small sizes of user groups than common control systems. Considerations for expansion or initially larger sizes of switching centers and the need to provide a variety of special features may dictate the use of a common control switching system.

When the subscribers to be served are distributed over a relatively large area, they may be connected to one central office or they may be divided into suitable groups to be served by a number of interconnected switching centers. The organization of such a localized central-office network is one of the most critical considerations in terms of central-office engineering. The trunking arrangement both internally and with regard to related systems must be evaluated from many viewpoints to arrive at the most suitable and efficient implementation. The economics of outside plant, initial cost, installation, and maintenance must be compared with central-office equipment cost and efficiency of use under a variety of configurations.

A configuration in which each switching office, serving a multioffice area, is a completely self-contained central office, of either the step-by-step or common control variety, requires more switching equipment but can be implemented with a less extensive and more efficient trunking arrangement than offices which provide concentration and must home on a master center.

Another major consideration in the selection of the type of switching center, which falls under the heading of central-office engineering, is the adaptation of subscriber requirements to efficient and economic use of switching equipments. Into this category fall such items as subscriber habits, calling rates, holding times, daily and seasonal traffic variations, future demands, ratio of local to trunk traffic, PBX operations, manual and automatic intercept, call accounting procedures, special services and restrictions, register-sender operation, etc.

The design of the switching network, its configuration, and control are affected by most of these above-mentioned characteristics and particularly by the need to anticipate expansion at a future date. Here again advantages and disadvantages between step-by-step and common control systems must be carefully evaluated before a selection, having the most suitable characteristics for a specific application, can be made.

The configuration (the crosspoint matrix) of a step-by-step system is defined by its particular switching element, in general 100- or 200-outlet two-motion selector switches, and can, therefore, be expanded only in terms of fixed steps, e.g., 100, 1000, 10,000. Matrix configurations of common control systems have greater inherent flexibility with regard to units or steps of expansion, but at the cost of a more complex network control.

The size of the matrix and the provisions for internal interconnecting paths are proportional to the number of terminations (lines and trunks) to be served and the grade of service to be provided or specified. Subscriber habits, calling rates, holding times, and ratio of local to trunk traffic are the raw data used in "traffic engineering" to provide the necessary outputs for the design of efficient and economical switching arrays.

PBX operations, intercept and accounting procedures must be viewed primarily from an economic standpoint to justify manual, semiautomatic, or fully automatic implementation. The number of subscribers per PBX, the number of PBX's, and the particular needs of the majority of subscribers served by PBX's must be taken into consideration and evaluated as part of the overall central-office requirements before a decision can be reached for the most suitable approach.

The implementation of intercept features are greatly influenced by the number of subscribers served, their specific needs, and the provisions of manual operator positions for special features. When such manual interfaces are to be held to a minimum, use is made of prerecorded tape announcements.

The selection of a suitable automatic call accounting system—toll ticketing—for a particular central office requires a careful evaluation of the characteristics of those factors which have a major economic impact. In general, common control systems are more suitable for the implementation of fully automatic accounting procedures than step-by-step systems. Subscriber habits (business and residential), the ratio of local to trunk, station-to-station, person-to-person and collect calls, the number and implementation of multiparty lines, on a terminal-per-station or terminal-per-line basis, are some of those characteristics which must be considered. The availability and features of off-line data-processing equipments are other items which will influence the choice of an accounting system.

Central-office engineering is, of course, vitally concerned with all peripheral equipments directly and indirectly connected with the switching equipments. Into this field fall the provisions for test and maintenance, both internal (switching equipments) and external (line and trunk transmission facilities); terminating and distribution frames (MDF, IDF); and grading panels, for switching network access and distribution, to meet a specified grade of service. Power sources, distribution, fusing, and alarming as well as adequate emergency power standby arrangements are an essential part of any central office.

Floor plans for initial installation, with due considerations for future expansion, adequate aisle space for maintenance personnel, provisions for environmental controls (heating, air conditioning), permissible floor loading, and available ceiling heights are some of the factors which must be considered from a selection of equipment point of view and also from a purely economic and availability point of view (see also Chap. 22).

If because of the limited office size and flexibility required, a step-by-step office is selected, the major engineering problems relate to the size and grouping of the selector switches. The allocation of lines and trunks to line and trunk finder switches made on the basis of traffic offered will determine the dial-tone delay for originating calls and the degree of blocking of terminating calls. Tables are available which provide the number of selector and finder switches for a given degree of blocking and delay.*

Similarly in a common control office the network or matrix configuration will determine the probability of success in completing a call during the busy hour. In a small office a simple one- or two-stage network with sufficient paths for all users to call simultaneously (nonblocking) can be provided with small penalty in number of cross points. As the size increases, blocking systems are used to limit the size of the matrix to reasonable numbers. On the other hand, the efficiency of large groups allows

* *Bell System Engineering Practices*, Div. D—Sec. 1, July, 1940.

offices to be engineered with small numbers of paths through the matrix and provide a very low probability of blocking.

The number of registers or other time shared units provided in a common control office will affect the ability to set up calls.   The quantities provided must be sufficient to limit the delay in accepting or setting up a call to some arbitrarily small number. Dial-tone delays greater than 1 sec should be minimized (less than 1 in 100 calls). Excessive delays in dial tone tend to cumulate, since they increase the probability that subscribers will dial before getting dial tone, thus mutilating their calls.   This mutilation adds traffic to the system which further delays dial tone to other users.   If the common equipment cannot complete calls rapidly, similar clogging can result during peak periods.   Typical call setup times in electronic common control offices vary from 50 to 500 msec.

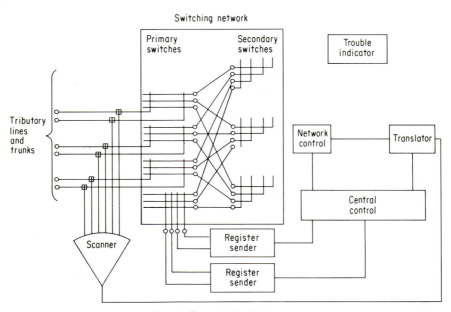

Fɪɢ. 3. General block diagram.

Figure 3 is a block diagram of a typical common control switching center.   The following sections will discuss the switching matrix design and the calculation of the number of required register-senders.

SWITCHING MATRIX DESIGN

3. General[1–10]

A switching matrix is an orderly array of relays.   Operation of one or more relays will provide a transmission path between two terminals.   The simplest of these matrices will be a patch panel.   All terminals are brought to a jack panel, and connection is established by connecting two terminals to each other, manually, with a cord. The number of cords available at each switchboard, regardless of number of terminals, is the maximum capacity of the switchboard; i.e., if a 100-terminal switchboard has 25 cords, this switchboard may serve only 50 terminals at one time.

In order to make this switchboard nonblocking, it is necessary that 50 cords be provided so that no calls will be refused owing to the lack of cords. The cords on this example are comparable to the links of switching matrices. In order to provide non-blocking service to all subscribers, a matrix should have enough links and crosspoint relays so that all the terminals may be connected to each other. It should be kept in mind that the function of a circuit switching unit is to provide a transmission path between two terminals.

a. State of the Art. There are three major techniques used to implement a circuit switch. These techniques will be discussed briefly to give background information. The three major techniques are:

1. Time division[6]
2. Frequency division
3. Space division

b. Time Division. Time-division switching is a technique wherein the intelligence to be switched is sampled at a rate high enough to ensure that no information is lost; these samples are made exceedingly short and are interleaved on a common bus or highway.

For nonsynchronous information, such as analogue signals, the sampling frequency (Nyquist frequency) must be somewhat more than twice the highest sampled frequency. For example, a 4-kc signal would have to be sampled at, say, 9 kc. This sample will carry information in their amplitude, that is, the height of the sample at the instant of sampling. This is known as pulse amplitude modulation (PAM). The samples for each signal appear regularly at the sampling frequency. The multiplexing is achieved by assigning the samples for each connection specific positions in the sampling interval. These positions are known as time slots.

The transfer of information through the system is accomplished by providing all terminals with a gate which is under the control of a control circuit. The gates associated with called and calling terminal are opened simultaneously for a short time. These constitute samples.* The number of lines or time slots which can be accommodated on a single highway or bus is directly related to the sampling rate and the time slot width. If T is the sampling interval and W is the time slot width, the number of time slots n is $n = T/W$.

In order to recover the original information from the multiplexed pulse train, an integration of some sort is required. For analogue transmission this can take the form of a lowpass filter; for digital data a multivibrator may be employed to regenerate the signals. When time division is used for analogue transmission, the input signal must be band-limited to less than one-half the sampling frequency to avoid the generation of ambiguous modulation products which will appear as noise.

A time-division switch seems to have a fixed upper bound on information transfer capacity, a sort of terminal-bit rate product which is analogous to the gain-bandwidth product of an amplifier.

c. Frequency Division. The principle of frequency division is very similar to time division. Different modulation techniques are employed, but the end result is the same. The basic difference between the two techniques is that frequency division uses different carrier frequencies which are selected by the control circuitry to connect two lines together. That is, the two members of a connection are modulated by the same carrier frequency, which is different from the rest of the connections established at that time. A frequency-division switching system would have operational characteristics very similar to time division, i.e., a sharp frequency ceiling, low power level, and some distortion.

d. Space Division. Space division is a name recently given to a method of switching which has been standard for some 70 years. In space division a physical connection is established between terminals which is separate in space from the other

---

* It would appear that, since the two gates are open for such a short time, very little energy will be transferred between terminals. By use of a technique known as "resonant transfer" this transfer of energy can be accomplished with very little loss.

connections in the system. Most commercial telephone switching systems employ space division. Although somewhat bulkier than the previously mentioned two systems, there are no limits on the number of terminals or the bit rate. Today with special precautions there are systems that are capable of transmitting signals up to 10 Mc.

The design of the space-division network involves two areas; the topology of the network and the choice of switch or crosspoint.

Historically there have been numerous types of switches developed; rotary,[2] panel, x-y[®], etc. Modern effort is in the area of coordinate switch structures which lend themselves to crosspoint arrays. The Western Electric and North Electric crossbar switches are examples of the highest development of the "standard art" switch.

In recent years there has been a great effort to replace the electromechanical switch with an electronic equivalent. The hoped-for advantages of such a development are small size, high speed, low cost, low power consumption, and high reliability. Efforts in this area have taken the form of applications of transistors, thyristors, four-layer diodes, gas tubes, tunnel diodes, etc. To date, none of these have been successful enough to leave the experimental stage.

Recently, a device has come into prominence as an approach to the ideal switching element. This is the dry reed relay.[7] This device combines some of the advantages of electronic and electromechanical devices. It is small, cheap, rugged, moderately fast, long-lived, and insensitive to temperature and radiation and has infinite on-off ratio. A network constructed of dry reed relays will contribute nearly zero insertion loss and is easy to control. The physical size of such a network will be greater than a comparable time-division network, and the weight may be greater, but when compared with the other techniques it could be shown that ease of control, bandwidth, and insertion loss characteristics of the space-division technique are superior to the other techniques for large (over 100-line) offices and in some circumstances superior in the smaller sizes.

Since the large part of an automatic circuit switchboard consists of the crosspoint matrix, this section discusses methods of minimization of such a matrix. The network configuration is a function of the number of lines, the desired probability of blocking, the traffic offered per line. For any combination of values each network type discussed below will yield a range of total crosspoints in accordance with the selection of switch size in each stage. The optimum selection results from iterative solutions to equations describing each configuration and thereby determining the minimum number of crosspoints per configuration. The minimum of all configurations is then the optimum. This optimum may not yield the most economical solution, since in general more complicated network arrays require greater complexity in control. Computer programs exist to allow rapid calculation of optimum crosspoint arrays.

4. Types of Nonblocking Switching Networks

A network is an array of relays which, when properly addressed, will furnish a transmission path between any two terminals. The term nonblocking refers to the fact that a network may not lose or fail to complete a connection for any calls because of internal congestion of the links.

These networks may be divided into two general groups: single-stage networks and multistage networks.

a. Single-stage Networks. The single-stage networks provide a transmission path between two terminals through only one operated crosspoint relay. Such arrays, of course, are very simple to control, since only the calling and called terminals need be selected.

These networks (although uneconomical when used for large number of terminals) are used in small, special-purpose switching centers because of the simple control features of these switches. There are two basic configurations of the single-stage networks: (1) square array and (2) triangular array.

b. The Square Array. The simplest type of nonblocking network is shown in Fig. 4. This is a rectangular array having n inlets and m outlets. There are $n \times m$

crosspoints or intersections in such a network.   Clearly this is the most uneconomical arrangement for a given number of terminals.

A square array may be used in a switchboard when only unidirectional connections are required.   They are not generally used as independent matrices but as a building block of higher stage networks.   The inlets are connected to the subscribers, and the outlets are connected to another square array which in turn is serving other subscribers.

The terms inlet and outlet are used to describe the direction of supervision.   Although the matrix itself is inherently bidirectional, in the transmission path, the only subscribers that may initiate a call are the subscribers connected to the inlets.

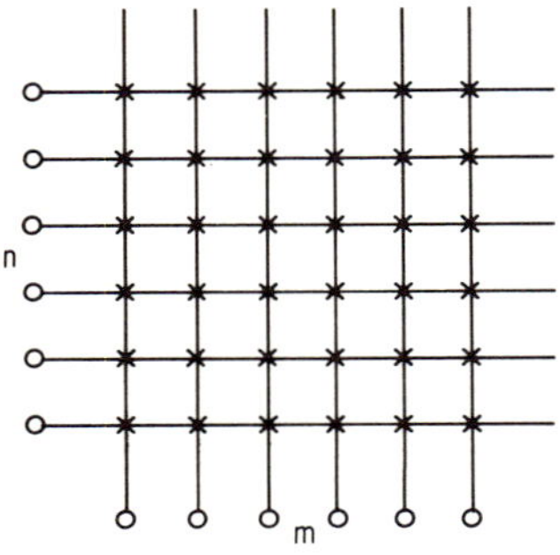

Fig. 4. Square network.

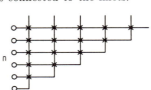

Fig. 5. Triangular network.

c. The Triangular Array.   The square array described in the previous paragraph is used for unidirectional service only.   If all terminals are two-way, i.e., any terminal may be either the originating or terminating end of a connection, a triangular array as shown in Fig. 5 may be used.   This requires $n(n-1)/2$ crosspoints or about half the number required by the square array.   It should be noted that a triangular array is a square array folded over diagonally.

d. Multistage Switching Networks. Considerable economy in crosspoints may be realized in multistage networks, particularly in larger systems.   The larger the number of intermediate stages, the smaller the number of crosspoints becomes, but there are other considerations; e.g., the larger the number of intermediate stages, the more complicated the control becomes.   After about three stages the control circuitry becomes so complicated that a large memory is required to keep track of each call; thus the method becomes cumbersome.   There are many different types of three-stage networks, but all of them conform to the basic network of Fig. 6 recommended by C. Clos of the Bell Laboratories.[1]   This network is particularly suitable for two-way terminal networks.   The network of Fig. 6 will be analyzed to illustrate minimal configurations.

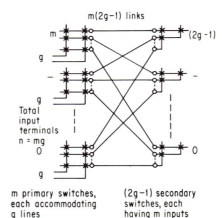

m primary switches, each accommodating g lines

(2g–1) secondary switches, each having m inputs

Fig. 6. Three-stage nonblocking array.

5. Three-stage Network

The network shown in Fig. 6 is a three-stage network, since in general any connection will require three crosspoints to be closed.   An exception to this general statement is the case where a connection is established between two terminals on the same primary switch. The network consists of m primary switches. Primary switches are of square-array type shown in Fig. 4. The primary switches have g inlets to be connected to the terminal equipment and $2g - 1$ outlets to be connected to the switches. The secondary switches are of triangular-array type shown in Fig. 5.

a. The Development of Three-stage Network. Let us assume such a three-stage network will have the following characteristics:

$$T = \text{total number of lines}$$
$$g = \text{number of lines in each primary switch}$$
$$m = T/g = \text{number of primary switches}$$

Every line will terminate on a vertical of a primary. The horizontal of each primary will terminate on different secondary switches. Since there are m primary switches, the secondary switches will have m terminals. The worst case condition will occur when every terminal is connected to another terminal on a different primary switch. In this case, since there will be two links used for each call, the total number of links required for not blocking any of the calls is $m(2g - 1)$. There are m inputs in each secondary switch; therefore $(2g - 1)$ secondary switches are required.

b. Calculation of Total Number of Crosspoints. The total number of crosspoints may be calculated by breaking down the total matrix into integral parts and calculating the number of crosspoints of each component part.

Number of crosspoints per primary switch

$$X_p = g(2g - 1)$$

Number of crosspoints (total) primary switches

$$X_{pt} = mg(2g - 1) \tag{1}$$

Number of crosspoints per secondary switch

$$X_s = \frac{m(m - 1)}{2} \tag{2}$$

Total number of crosspoints in secondary switches

$$X_{st} = \frac{(2g - 1)m(m - 1)}{2} \tag{3}$$

Total number of crosspoints

$$X = \frac{2mg(2g - 1) + (2g - 1)m(m - 1)}{2} \tag{4}$$

$$= (2g - 1)\left(mg + \frac{m^2}{2} - \frac{m}{2}\right)$$

$$m = \frac{T}{g} \tag{5}$$

$$X = (2g - 1)\left(T - \frac{T^2}{2g^2} - \frac{T}{2g}\right) \tag{6}$$

c. Minimum Number of Crosspoints. Equation (6) will yield the total number of crosspoints. In order to find the minimum number of crosspoints, the total number of lines will be assumed constant and Eq. (6) will be differentiated with respect to g.

$$X = 2Tg + \frac{T^2}{g} - 2T - \frac{T^2}{2g^2} + \frac{T}{2g}$$

$$\frac{dX}{dg} = 2T - \frac{T^2}{g^2} + \frac{T^2}{g^3} - \frac{T}{2g^2} \tag{7}$$

$$\frac{d^2X}{dg^2} = \frac{2T^2}{g^3} - \frac{3T^2}{g^4} + \frac{T}{g^3} = \frac{2T^2 + T}{g^3} - \frac{3T^2}{g^4} \geq 0 \tag{8}$$

$$g \geq \frac{3T}{2T+1} \text{ in order for } \frac{dX}{dg} = 0 \text{ to be minimum}$$

$$\frac{dX}{dg} = 0 = 2Tg^3 - T^2g + T^2 - \frac{Tg}{2} = 0$$

$$g^3 - \frac{Tg}{2} - \frac{g}{4} + \frac{T}{2} = 0 \tag{9}$$

$$T = \frac{2(g^2 - \tfrac{1}{4})g}{g - 1} \tag{10}$$

   d. Construction of Tables.   By use of Eq. (10) a table may be calculated. Table 1 is calculated by using the exact formula by giving different values to g and

**Table 1. Number of Terminals per Primary Switch vs.
Total Number of Terminals**

g	T	g	T
5	62		
6	86	26	1,406
7	114	27	1,514
8	146	28	1,626
9	182	29	1,742
10	222	30	1,862
11	266	31	1,986
12	314	32	2,114
13	366	33	2,246
14	422	34	2,382
15	482	35	2,522
16	546	36	
17	614	37	2,814
18	686	38	
19	762	39	3,122
20	840	40	3,282
21	926	41	
22	1,014	42	3,614
23	1,106	43	
24	1,202	44	
25	1,302	45	4,142
		46	
		47	4,514
		48	
		49	
		50	5,102

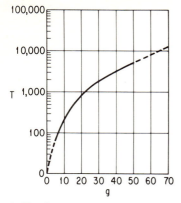

Fig. 7. Total number of lines vs. number of terminals per primary switch.

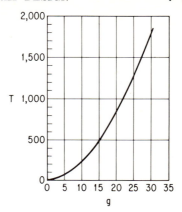

Fig. 8. Total number of lines vs. number of terminals per primary switch (expanded scale).

calculating total number of lines T. Curves of Figs. 7 and 8 are drawn from this table. Figure 7 uses logarithmic scale to contain all the values of $g = 5$ to $g = 50$.

Figure 8 is drawn on decimal scale and contains only values of $g = 5$ to $g = 30$. If the total number of terminals is given, the number of terminals per primary switch for a minimal nonblocking network may be found directly.

e. Crosspoints per Line. An important figure of merit for any network arrangement is the number of crosspoints per line. If the expression for the total number of crosspoints is divided by the total number of lines T, this figure of merit is obtained.

$$\frac{X}{T} = (2g - 1)\left(1 + \frac{T}{2g^2} - \frac{1}{2g}\right) \quad (11)$$

From this expression it is easily seen that the number of crosspoints per line varies as the number of lines on a primary switch and as the total number of lines. Figure 9 and Table 2 show that certain number of lines will yield a minimal network for either of two values of g; i.e., 125-line switchboards may use either $g = 5$ or $g = 10$. In cases like this, the control and addressing of the switch will dictate the value of g. Although this table is calculated for a narrow range of 50 to 200 lines and 5 to 15 terminals per primary switch, a close investigation of 200 terminal values will show that, when $g = 10$, crosspoints per line become minimum.

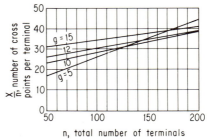

Fig. 9. Three-stage nonblocking array crosspoints per terminal for various sizes of primary switch.

Table 2. Crosspoints Per Line For Different Values of T and g

T	$g = 5$	$g = 10$	$g = 12$	$g = 15$
50	17	23	26	31
100	26	28	30	34.5
150	35	33	34	38
200	44	37.5	38	41

f. Approximations.   Solution of Eq. (10) is very long if T was known.   To solve the equation for g in terms of T is practically impossible, but by manipulating this equation and making some approximations we may arrive at an easier solution.

For
$$g > 1 \qquad g^2 \gg \tfrac{1}{4}$$

Equation (10) becomes
$$T \approx \frac{2g^3}{g - 1}$$

Let $g \gg 1$; then
$$T \approx \frac{2g^3}{g} = 2g^2 \therefore g^2 = \frac{T}{2} \tag{12}$$

Therefore $g = \sqrt{T/2}$.   This is a first approximation which will give us a slightly higher figure of g than is necessary for minimum number of crosspoints.   For a test case we shall assume an average size switchboard of 200 lines: By the approximate formula we find g.

$$g = \sqrt{\frac{T}{2}} = \sqrt{\frac{200}{2}} = 10$$

This figure will be slightly higher than the actual figure for minimum crosspoint. From Fig. 5 a closer approximate value of the g is 9.5.   When T is calculated from the exact formula, $g = 9$ will yield $T = 182$ lines and $g = 10$ will yield 222 lines (Table 3).

Table 3. Comparative Values of g for the same Number of T
(g is calculated from the exact formula; g' is calculated from the approximate formula.)

T	g	g'
62	5	5.54
86	6	6.55
114	7	7.52
146	8	8.5
182	9	9.5
222	10	10.5
482	15	15.5
840	20	20.5
1,302	25	25.5
1,862	30	30.48
2,522	35	35.45
3,282	40	40.42
4,142	45	45.4
5,102	50	50.4

We will use $g = 10$ for the basis of our crosspoint calculations.   If we use a network configuration of 20 primary switches with 10 lines per primary switch and 19 secondary switches, the total number of crosspoints will be

$$X = (2g - 1)\left(T + \frac{T^2}{2g^2} - \frac{T}{2g}\right)$$
$$= (19)\left(200 + \frac{40,000}{200} - \frac{200}{20}\right)$$
$$= 7,410 \text{ crosspoints}$$
$$\frac{X}{T} = \frac{7,410}{200} = 37.5 \text{ crosspoints/line}$$

g. Crosspoint Calculations for Blocking Matrices. A nonblocking matrix can connect any terminal to any other terminal regardless of traffic density and lost-call probability. On present-day commercial telephone systems use of nonblocking switching networks is rare. This is due to the large number of crosspoints required. With the design methods described in previous paragraphs a minimal network may be found. Additional reduction of crosspoints may be calculated by omission of certain internal paths. This unfortunately is a trial-and-error method and very cumbersome. Some computer programs are written to calculate the minimum number of crosspoints given the lost-call probability and traffic density.

In order to demonstrate the extent of calculations, the following lost-call probability formulas are given:

$$P_t = 2P_b + P_{b-1}(a_{11}P_{b-1} + 2a_{12}P_{b-2} + 2a_{13}P_{b-3} + \cdots + 2a_1,\, b_1 P_1)$$
$$+ P_{b-2}(a_{22}P_{b-2} + 2a_{23}P_{b-3} + 2a_{24}P_{b-4} + \cdots + 2a_{2,b-2}P_2)$$
$$+ P_{b-3}(a_{33}P_{b-3} + 2a_{34}P_{b-4} + 2a_{35}P_{b-5} + \cdots + 2a_{3,b-3}P_3)$$

$$+ \frac{P_{b+1}}{2}\left(\frac{a_{b-1},\, b-1}{2}\frac{P_{b-1}}{2} + \frac{a_{b-1},\, b+1}{2}\frac{P_{b-1}}{2}\right) \tag{13}$$

if b is odd. If b is even, the last row term becomes

$$+ \frac{P_b}{2} a_{b/2,b/2}$$

The second equation is

$$P_{b-x} = \frac{A^{b-x}/(b-x)!}{1 + A + A^2/2! + A^3/3! + \cdots + A^b/b!} \tag{14}$$

where A = traffic density, erlangs
$\quad P$ = lost-call probability
$\quad a_{xy}$ = probability that, if there are x idle ends at one switch and y idle ends at another switch, they will not coincide
$\quad P_b$ = probability of b or more calls existing at primary switch
$\quad b$ = number of outlets or links between secondary and primary switches
$\quad A$ = originating traffic offered to the individual primary

However, the reader need not go through all these calculations. There are traffic tables[9],[10] (see Fig. 10) which give the number of secondaries for a given lost-call probability and traffic density.

The lines per primary switch (verticals) are calculated from Eq. (15)

$$g = \sqrt{\frac{T}{2}} \tag{15}$$

and when the traffic density and lost-call probability are known, the number of secondary switches are found from the traffic tables.

The total number of crosspoints are then easily calculated by using

$$X = \frac{2mg(2g-1) + (2g-1)m(m-1)}{2} \tag{4}$$

where m = number of secondary switches obtained from traffic tables
$\quad g$ = number of verticals obtained from Eq. (15)
$\quad X$ = total number of crosspoints

This method of computation will not yield the minimum number of crosspoints, but it will be very close to a minimum.

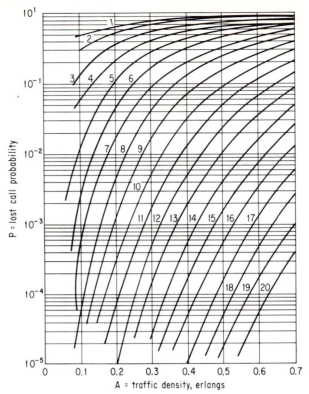

FIG. 10. Number of secondaries for given traffic density and lost-call probability.

6. Control of the Matrix[8]

In a single-stage matrix in order to have a path between two terminals, it is sufficient to operate only one crosspoint. This may be accomplished by addressing the matrix with X and Y coordinates. To operate a crosspoint, one of the two techniques may be used, namely, coincident current or coincident voltage.

In a three-stage network, there are three crosspoints to be operated in order to obtain a path between two terminals (see Fig. 11). In this network, the primary switches have two types of terminals: inlets which are called link-select leads or horizontals and outlets which are called verticals or line-select leads. The terminals appear on verticals, and the links or internal trunks which connect primary switches to the secondary switches appear on the horizontals. In order to connect two terminals together, it is necessary to know the equipment locations (the addresses of the terminals) to be connected together. Second, an idle link must be found to connect the two primary switches. To select an idle link which is not in use for another call may be done in many ways. The most common method is to use an *allotter* and the number of the next idle link ready to be used. When all links are used, the allotter will recycle and start from the top, passing over the links that are still busy. The allotter may be visualized as a ground-seeking electromechanical rotary switch which will be started to seek the next ground when the idle link is made busy.

Another method is to scan the links in sequence until one pair of matching number of links is found to be idle. It should be noted that, in order to find a path between two different primaries through a secondary switch, the same numbered horizontals have to be chosen on both primaries (see Fig. 11). Once both the horizontals and the verticals are known, the crosspoints may be operated according to either of the following two methods.

a. Coincident-current Control. In order to control a matrix in this manner, each relay has to have three coils. The first two coils are wound in such a manner that, when only one coil is energized, the relay contacts will not close; it is necessary to energize both coils to operate the relay. The third coil is connected in series with a contact and has enough ampere-turns to hold the relay in operated condition after the contacts are closed by the two operating coils.

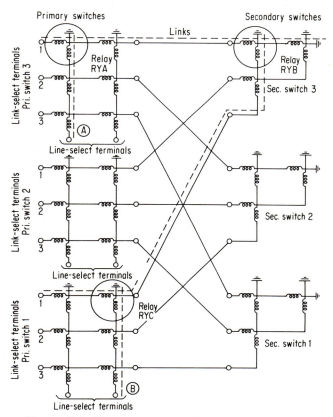

FIG. 11. Coincident current control of a reed relay matrix.

The matrix of Fig. 11 is formed by placing one coil of each crosspoint in horizontal array and the other in the vertical array. In the matrix shown in Fig. 11 one of the coils of each relay belonging to the same horizontal array is connected in series and joined by a link to a secondary switch. The second coil of each relay of the same vertical is connected in series.

In order to describe clearly the operation in Fig. 11 let us assume that terminals marked A and B will connect together, and furthermore it was found that link 1 is idle on both primary switches. Through a pair of contacts not shown in the figure, the current will be applied to vertical A and vertical B. The current will flow through all three coils in series. As previously described, this current is not enough to operate the relay.   When the link is selected and energized in the same manner, whichever relay has both of its coils energized will be the only one that will operate.   In Fig. 11 the dotted lines show the path of current and the circles designate the operated relays. In order to simplify the figure the transmission path is left out.   The transmission path will follow the same configuration.   This path for a four-wire transmission facility is shown in Fig. 12.

The coincident current technique is by far the cheapest way to control a matrix. The two operating coils may be wound using two wires, thus winding two windings together in one operation.   Unfortunately the coincident current control relay itself by nature is a marginal device.   It requires careful consideration of mechanical packaging.   The proximity of two relays will affect each other's operation owing to flux leakage.   The coils should be shielded to minimize this effect.   Tight tolerances should be held to achieve a good margin between operating and nonoperating flux.

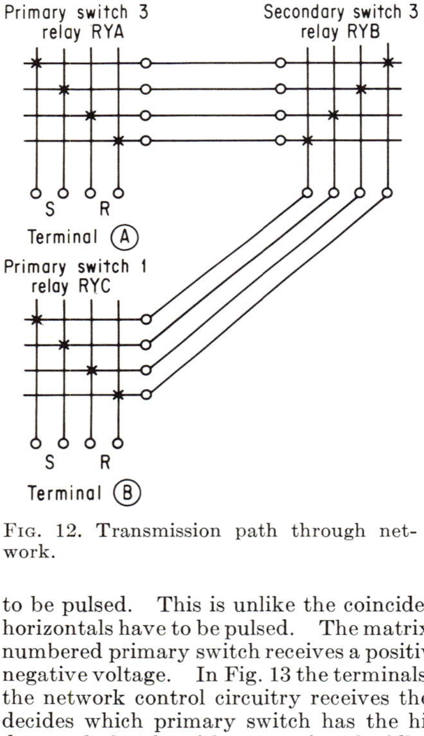

b. Coincident-voltage   Control. This technique allows simpler relay construction.   All that is necessary is for one coil to operate the relay and another coil to hold it operated.   As is shown in Fig. 13 all the relays on the same horizontal array are connected in parallel and all relays on the same vertical are connected in parallel.   This method of matrix operation lends itself to good nonmarginal circuitry with positive action.   The relays are easy to manufacture, and tolerances are not strict.

There are two drawbacks with coincident-voltage control.   The first one is as follows: In order to prevent feedback through the network, a diode should be inserted in series with each coil.   The second shortcoming is that, when a matrix is to be operated, the network control circuitry should know or translate from the matrix address the higher numbered primary switch.   In the matrix configuration, both verticals and only the link on the higher numbered primary switch have to be pulsed.

Fig. 12. Transmission path through network.

This is unlike the coincident-current matrix, where both verticals and horizontals have to be pulsed.   The matrix is operated as follows: The link on the higher numbered primary switch receives a positive voltage, whereas both the verticals receive negative voltage.   In Fig. 13 the terminals A and B will be connected together.   When the network control circuitry receives the equipment locations of both terminals, it decides which primary switch has the higher number.   The link selection may be done as before by either scanning the idle links or allotting them before hand.   After the voltage is applied to verticals of A and B, only one link need be energized, the link on the higher-numbered primary switch.   Upon application of both voltages, RYA relay will close. RYB and RYC relays will be also energized, since these relays are connected in series.   The dotted lines designate the paths of the currents.   It should be noted that RYA will receive twice as much current as RYB or RYC, but this will be for a short time only, after which the voltages will be removed and the hold coils will take over to hold the connection.   In both coincident-current and coincident-voltage controlled matrices, the hold coils of the relays are released from the line circuits.   The links are left operated until the next time when they are picked up for use; then they are released and reused again.

c. Addressing the Matrix.[3–5]   In the previous paragraphs the method of operation of the matrix was described.   In both coincident-voltage and -current methods, in order to operate the matrix, the number of the terminals has to be known. These locations are called equipment locations. Although decimal codes or binary codes might be used to address the matrix, they shall require additional translation. The call number of the subscriber and the equipment location have no relation whatsoever to each other, to the extent that the first vertical of the primary switch number (1) may be connected to a subscriber which has a call number (157). This information is stored in a translator, and every time that the call number 157 is presented to the translator, the translator in turn will come out with the equipment location of the

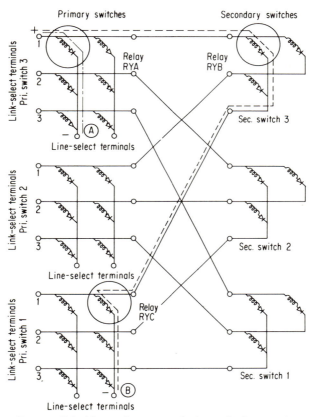

FIG. 13. Coincident voltage control of a reed relay matrix.

first vertical of the first primary switch. This equipment location as mentioned before will be represented by some form of code. Numbering terminals, say for a 200-line switch in decimal form, will require 200 bits of information per line. This definitely is the most expensive way of doing things. A binary code for 200 lines will require 8 bits of information per equipment location which will be the least expensive way, but in turn it will require additional translation to find the number of the primary switch and the number of the vertical. Especially in the case of coincident-voltage control, the higher numbered primary switch should be known. These requirements bring out a fact that a hybrid code will be more suitable than a binary code. For a 200-line switchboard (matrix) we shall have 20 primaries and 10 verticals per primary switch. A hybrid code of 1 out of 20 and 1 out of 10 will be most adequate but still requires 30 bits of information to be stored.   If we divide the primaries into two groups and the verticals into two groups, we end up with a hybrid code of $\frac{1}{2}$, $\frac{1}{10}$, $\frac{1}{2}$, $\frac{1}{5}$.   Total number of bits involved will be 19 bits, which is readily identifiable as primary switch group, primary switch number, vertical group, and vertical number. With these four groups of wires and two tree networks, we may find a path to each individual vertical.   In the previous paragraphs it has been shown that the addressing of the matrix has not yet reached the stage of being a science; it depends mainly on the judgment of the design engineer and amount of hardware involved.

SIGNALING AND SUPERVISION

7. Definition

Signaling and supervision contain that information flow between line terminals and the switching center or between switching centers over trunks which is necessary for call origination, processing, and control.

Supervision is usually associated with the information used to indicate idle-busy conditions and

1. Initiate a request for service
2. Provide for operator recall
3. Alert the subscriber
4. Provide other control features in more advanced systems

Signaling is associated with

1. Call addressing
2. Call precedence information transfer
3. Call security information transfer
4. Other call characteristic information

Other signals such as preempt, call waiting, and call book may be judged as being either signaling or supervision.

8. Forms of Signaling and Supervision

Signaling and supervision can both be accomplished by either

1. Analogue signals
2. Digital signals

a. Analogue Supervision.   Analogue supervision can use either d-c or a-c signals. In the former the supervision is sent as a low frequency (d-c), while for the latter higher audio frequencies, either in-band or out-of-band, are used.   In-band supervision is more prevalent for voice and data communication systems, while out-of-band supervision has greater utility in control systems.

b. Direct-current supervision.   There are many variations of d-c supervision. The most basic is the d-c loop in which the near and far ends of a communications path serve as a complete loop, the state of which might be high or low (open or closed) according to the supervision signal on the communications path.

Usually the d-c loop appears open when a line or trunk is in the disconnect state (ON-HOOK).   It appears as a closed d-c path when the line or trunk is in an occupied state (OFF-HOOK).   By alternating states so as to achieve a specific train of pulses, other supervisory signals such as line busy, trunk busy, etc., can be superimposed on the hook state.   Care must be exercised that the alternating state does not remain in a hook state opposite to its actual supervisory state for a period sufficient to cause false identification of the hook state.

Principal methods used for changing the appearance of the d-c loop are:

1. Opening and closing the d-c path through the loop
2. Reversing the polarity of the loop voltage
3. Changing value of a bridged loop resistance

The appearance of the loop in a line circuit is usually controlled directly by the communicator's instrument.   For instance, lifting of a telephone handset from the cradle results in a sustained d-c level being transmitted toward the switching center.   The sustained d-c level indicates initially a request for service.   After the request is detected, the level serves to maintain the connection.   Return to ON-HOOK after disconnect is noted by the switching center when the d-c loop becomes open for at least a predetermined minimum interval.   The long integration interval necessary for detection of the supervisory signals serves to make d-c supervision a rather slow form of signaling.

Supervision on most trunks and also on some line circuits use E and M lead supervision.   In this method communication between the trunk or line terminating unit and the signaling unit is done over two leads, an E lead that transmits supervisory information from the signaling system to the line terminating unit and an M lead that transmits supervision from the line terminating unit to the signaling system.   This allows the subscriber subset to be simple, since it operates on a d-c basis.   Auxiliary

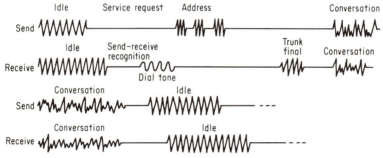

FIG. 14. Alternating-current four-wire trunk supervision.

line signaling units which detect the state of the E and M leads are used to provide conversion of the d-c to a form required by the transmission facility.

In a typical case of E and M supervision on a line circuit, closure of the transmit loop, resulting from lifting of the handset, causes a mark (battery) to be placed on the M lead, which then indicates to the switch a request for service.   At the called end, an interrupted ground is placed by the switch on the E lead to indicate that the line is being rung.   Sensing of the ground by control equipment results in a d-c voltage being supplied to the receiver loop.   A ringer then responds to the interrupted loop voltage.

The obvious advantages of d-c supervision are its simplicity in operation and its low cost of mechanization.   Direct-current supervision, however, does not meet all the requirements of modern-day switching systems.   Its disadvantages are:

1. It needs conversion equipment for use on longer lines and most transmission media.

2. It is not a versatile form of supervision, since it does not allow for many different types of control signals.

3. It is a slow method for supervising.

4. It is not compatible with many modulation techniques.

c. Alternating-current Supervision.   Alternating-current supervision uses two state a-c frequencies.   These frequencies can be within the voice band of the channel or might use a separate out-of-band channel.   The in-band system is more prevalent, since it conserves bandwidth.

Of single-frequency systems on four-wire lines, 2,600 cps is the most popular frequency, although 1,600 cps is also used.   In two-wire lines, different frequencies must be used for send and receive.   In that case, 2,400 cps is used in one direction when 2,600 cps is used in the other direction.   A typical sequence in sending supervision on a four-wire trunk is shown in Fig. 14.   During the idle period 2,600 cps is conveyed in both directions down the trunk. Removal of the 2,600 cps on the transmit side initiates the service request. Recognition of the request results in 2,600 cps being removed on the receive side. Another in-band supervisory burst (1,000 cps) is sometimes used to indicate to the sending switch that it can transmit the address. After sending the address, the calling trunk waits for supervision which indicates either that connection is being set up or that the called party is busy. For the former, a 2,600-cps burst is sent, while the latter might be signified by a 1-kc burst.

The major problem encountered in voice-frequency (in-band) supervision is the possibility of falsely detecting a speech signal to be that of supervision. Since speech is mostly concentrated in the audio range of 400 to 800 cps, a higher frequency, such as 2,600 cps, is more acceptable for in-band supervision. Moreover, this is still not sufficient for voice immunity, since the detectors are still subject to spurious responses owing to harmonics of the stronger lower-frequency signals. To ensure voice immunity, the channel usually is divided into a narrow band centered about the supervision frequency and a guard band which contains all other voice frequencies. The characteristics of these bands is shown in Fig. 15. Since speech currents produce voltages in both guard and supervision bands while supervision currents enter only the supervision band, supervision can be detected when the output of the supervisory band is much greater than that of the guard band.

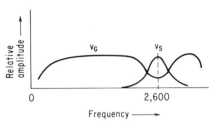

FIG. 15. Guard circuit characteristics. v_s = voltage developed by supervision circuit. v_g = voltage developed by guard circuit.

Sensitivities of the two bands are also changed during the call process. Supervision band sensitivity is raised so as to assist in detection of supervision during call setup. Guard band sensitivity is raised and supervision sensitivity decreased to protect against spurious responses from speech signals during conversation.

Alternating-current supervision is more feasible than d-c supervision for long-haul circuits, since it can be amplified more easily. It is also more compatible with the modulating devices needed for use on various transmission facilities such as HF propagation or carrier systems. It is generally a fast method of supervision, its speed limited only by the bandwidth of the line or trunk circuit. It operates well under moderate noise conditions, especially if multifrequency supervision is used.

d. Digital Supervision. Digital supervision has proved to be the most versatile and adaptable of supervision systems. It is also the most costly, most complex and requires greater bandwidth.

In digital schemes, supervision can be

1. Interleaved (time-division multiplexed) together with an information channel in a continuous bit stream
2. Sent as characters in a fixed position in a prescribed format
3. Sent as distinct codes which are digital equivalents to two-state a-c supervision.

An example of the method described in (1) is the advanced test model of the UNICOM time-division switch. UNICOM uses all digital transmission with supervision and signaling multiplexed in a separate time-division channel (at a slower rate) than the message information. For establishment of this channel a data frame is used. This frame consists of 136 bits, with each seventeenth bit reserved for either framing or supervisory information. The construction of the frame is shown below.

16 Message bits								
Framing F_1	S_1	Encryption check bit	S_2	Framing F_0	S_3	Parity bit	S_4	

The state of the line (idle or busy) is sent in the S channel (bits S_1, S_2, S_3, S_4). Since this channel is used for other than line supervision, it becomes temporarily interrupted for transmission of other supervisory or signaling information; otherwise the hook signal is continuously sent.

Framing bits (F_1 and F_0) are needed to ascertain location of the S bits within the frame. Use is also made of other bits which are in the multiplexed supervision and framing channels. One bit is used for detecting failure of synchronization between encryption equipment at line and switch terminations. Another bit is reserved as a parity check over 136 bits (not all part of one frame).

Advantages of this method are the following:

1. Auxiliary signals can be sent without interfering with message flow.

2. Time-division multiplexing of supervision can be readily used on trunks. This allows the number and types of channels used on the trunks to be varied in accord with trunk characteristics.

3. Common control can be time shared.

4. All supervision and message information can be sent in real time, thereby making the system valid for both voice and data transmission.

The major limitation of the multiplexed scheme is that the information rate is continuously lowered by the ratio of the multiplexed supervision and framing to the total frame rate. In UNICOM this ratio was $8/136$, or about 6 per cent. Another detrimental factor is the large amount of equipment needed in the subscriber terminal for encoding and multiplexing.

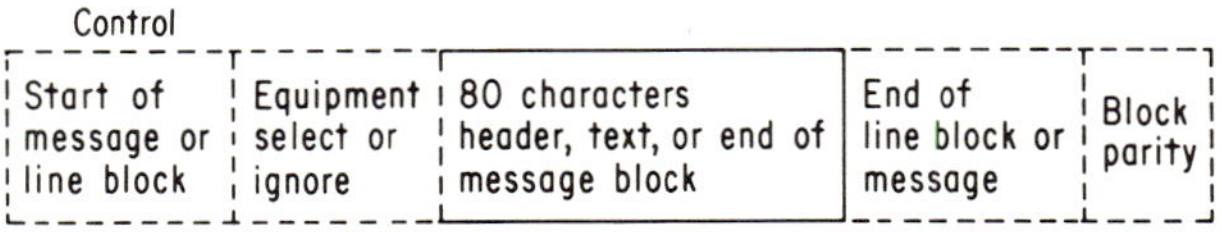

Control

Start of message or line block	Equipment select or ignore	80 characters header, text, or end of message block	End of line block or message	Block parity

FIG. 16. AUTODIN line block format.

A system which uses supervision as characters in a fixed position in a prescribed format is CONUS AUTODIN MESSAGE SWITCH. In mode I transmission, which allows for error control and ultimate correction, idle line characters are continuously sent in order to establish and maintain synchronization and character framing between the terminals and the switch. When a call is initiated the idle line characters are interrupted and a header line block, shown in Fig. 16, is sent by the subscriber to the switch. Sensing of the start-of-message character by the switch enables the switch to guide itself for the valid detection of subsequent header characters, which are inserted in the line block in a prescribed order. When the line block has been fully sent an end-of-line-block (or end of message if the line block is the last of the message) supervisory character is sent. The subsequent character is a block parity character which can be used to detect double errors in any single character of the line block.

Single errors are detectable, since each character is 7 bits plus 1 parity bit. Channel coordination procedures are then used to control the block-by-block transmission. The line block is retransmitted if control ERROR was generated by the switch owing to parity error, or the message might be entirely rejected if format error was detected by the switch. In any case a closed loop of message control is utilized so that a new line block cannot be transmitted until the previous line block was approved as valid by the switch. During line block transmission, when the terminal waits for acknowledgments, idle line characters are continuously transmitted. Message transmission ceases when the line block which contains the end-of-message character receives its acknowledgment.

By utilizing a prescribed line block format and by sending messages on a line block basis, control procedures for error correction and channel coordination become more easily regulated. The code used in AUTODIN uses even parity for control characters while message characters have odd parity. Since the control characters (which are transmitted twice successively) occur at a prescribed place in relation to the line block, they are detectable as control characters only, although the control character code (less parity) is the same as some message character. This permits the utilization of a possible 128 control characters in addition to 128 message characters from a 7 bit plus 1 parity bit code.

Disadvantages of block-by-block transmission are the following:

1. Since it requires buffering of line blocks, it is not feasible for real-time transmission such as voice.

2. Data-transmission rate is degraded by the ratio of control characters to total characters. In the case of Autodin this was $\frac{4}{84}$ = 4.5 per cent.

3. Data-transmission rate is further lowered by the transmission pause intervals necessary for proper channel coordination procedures.

4. Subscriber terminals are quite complex and large, since they must on the receive side perform many of the control actions performed by the switch on transmission.

The third form of digital supervision, which uses digital codes as equivalents to analogue supervision, is shown in Fig. 17. In this case a steady mark level indicates a "free state" (ON-HOOK). Service request is initiated by successive transmission of a specific signal such as idle line characters. This supervisory signal also serves to synchronize on a character basis the sending and receiving clocks. The service

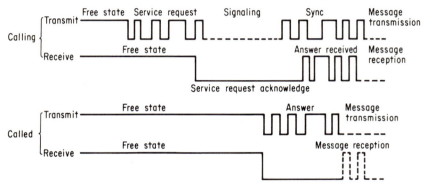

FIG. 17. Digital supervision.

request acknowledge (dial-tone equivalent) appears as a steady space level. After signaling the address, the called party is alerted by the SYNC signal of idle line characters. Answering, results in the SYNC signal being returned through the connection. This informs the calling subscriber that the connection has been established. If the called party is busy, the calling party is informed of this by his reception of alternate "mark" and "space" conditions of unit bit length each. Disconnect is signified by a steady "mark" appearing for a prescribed interval (at least $\frac{1}{2}$ sec) on either the calling or called lines. Although this type of supervision is much simpler than the previously described digital supervision, it still requires a substantial amount of equipment at the subscriber terminal for synchronization and detection. It has the normal advantages of a digital system such as ease of encryption and high signal-to-noise ratio but does not have the capability of features which can be derived from a "framed" system such as UNICOM or a prescribed format system such as AUTODIN. However, unframed serial type supervision, which is faster than the framed or formal system, can be used in a hybrid system in which unframed serial transmission is used for a call setup, while the remaining portion of the call can be on either a framed or format basis. The AUTODIN circuit switch actually uses the latter method.

e. Analogue Signaling. As for analogue supervision, analogue signaling can use either d-c or a-c means.

f. Direct-current Signaling. Dial pulsing, which momentarily changes the level of the hook state, is the most prevalent form of d-c signaling. The character of the signal is usually determined by the number of dial pulses within a finite interval. Care must be exercised in selection of the pulse durations so that the pulse does not serve for misinterpretation of the hook state. Since baud rates of 9.5 to 10.5 pulses per second are most common for d-c signaling, it can be looked upon as a slow form of digital signaling. This method is quite simple in both operation and mechanization. It is, however, quite slow and requires transmission facilities which can pass direct current.

g. *Alternating-current Signaling.* Multifrequency alternating-current has become extremely popular within recent years as a reliable form of a-c signaling. Intensive investigations have shown that from the viewpoint of flexibility, speed of transmission, and reliability, it is best to represent a signal by a controlled multifrequency code rather than by dial pulses. This can be grasped from the simple observation that the loss of one dial pulse results in an uncorrectable error in detection while the loss of one frequency (due to distortion or noise), for a period of time, only delays the final detection, since absolute detection requires multifrequencies. There are many combinational forms that multifrequency signaling can assume. Combinations based upon selection of two-out-of-five frequencies and selections from a 4 × 4 matrix of frequencies appear to be the most popular types in use today. The direct distance dialing network which covers both the United States and Canada makes use of six frequencies, each spaced 200 cycles apart from 700 to 1,700 cps inclusive. These six frequencies provide 15 possible two-frequency combinations. Ten combinations are used for digits 0 to 9 inclusive. Other combinations are used to indicate start of signaling (KP) and end of signaling (ST). Three combinations are reserved for future requirements. Each digit, except zero, can be translated as the sum of two-signal frequencies, each of which represents digits whose sum equals the translated digit. This type of additive plan assists in readily identifying the combinations.

Over trunks, multifrequency signaling is usually slowed somewhat, since some form of control is needed to ensure that the tone combinations are being satisfactorily received. In a typical case the originating sender starts the signaling by sending the first digit. The transmission of the two frequencies, which make up this digit, persists until an acknowledge signal is received from the receiving register. Receipt of the acknowledge signal by the originating sender then terminates the digit and enables transmission of the next digit. This procedure is repeated until all signaling has been performed.

The principal advantages of multifrequency signaling over dial pulsing are speed, accuracy, and capability of operation with different types of transmission facilities. The speed advantage results in less holding time per call, which then results in requiring a smaller number of register senders for common equipment.

h. *Digital Signaling.* Digital signaling can be multiplexed with message transmission such as in UNICOM, or it can be sent as a block of a fixed number of characters in a specific header format. Both CONUS AUTODIN message and circuit switches use the latter approach. In some systems, the entire address is keyed and stored at the subscriber terminal. Push-button keys are used. Pressing of a SEND button enables the header to be serially clocked to the register. If the header is in error, it can be automatically retransmitted under control of the switch. If the header proves to be in error three consecutive times, the subscriber is automatically disconnected and informed of this by an alarm. Ease of storage for control of redundant transmission of signal information is a distinct advantage of digital signaling.

Digital signaling has disadvantages in sensitivity to variable delay and in delay distortion. Because of these factors usually an increased bandwidth is necessary or else limitations are placed on the transmission facility to be used. Digital signaling provides the best interface for the call processors of near-future switching systems. Since these systems will in effect be stored program computers, digital information will require the least conversion equipment for working with these systems.

TRAFFIC CONSIDERATIONS[9,37–42]

The problem of evaluating the performance of a switching system based on traffic consideration will be discussed in this section. This problem is similar in nature to those encountered in connection with logistics, air and highway traffic control, and phases of military strategy. All these require the accumulation of large quantities of data, processing the data according to procedures which are often empirical and producing final information from which the performance of the system can be judged.

9. Throwdown Studies

Since this general problem involves large quantities of input data which are statistical in nature, all possible variations cannot be studied. A sufficient number of typical situations must be tried to obtain statistically reliable results.

These methods have been extensively used in switching-system traffic studies and are called "throwdown" studies. The name stems from the use of dice in the early study of telephone traffic problems. Each die is designated to represent a particular independent event, and the faces of the die are designated according to the probability of the event taking place. When a number of such dice are repeatedly "thrown down" and the results observed, the probability of a particular combination of events taking place can be estimated.

Other similar methods based on selections from lists of random numbers have been used on switching traffic studies for a number of years. Mathematicians, using digital computers, have employed similar statistical methods on problems relating to diffusion of gases, electron ballistics, and the solution to certain types of differential equations. They have called this the "Monte Carlo" method.

The throwdown study involves a method of artificially generating traffic, in which an attempt is made to simulate the traffic which would actually appear to the live operational system. This involves processing a large number of calls systematically and determining what happens to each of the calls, i.e., whether they are blocked (lost) or delayed (for how long), as a function of the switching system parameters such as number of trunks, number of registers, processing (or holding) time of common equipment, type of switching network, transmission characteristics, and variability of human or machine inputs.

A single throwdown test will indicate the performance of the system under a specific set of conditions. In a typical traffic study, a given traffic load is first assumed and a simulated system model to handle this load would be devised. The test run will then show the performance of the system under these particular conditions and indicate the adequacy of the initial model and possible improvements. To obtain a proper balance between equipment quantities and traffic load may require several additional runs.

10. Characteristics of Subscribers Affecting Traffic Data

Traffic presented to a switching system by subscribers is only moderately influenced by the type of system serving their telephones. Subscribers, although they are individuals, exhibit many "group characteristics" dictated not by the requirements of the switching system but by their mode of life. This fact allows statistical treatment of many observed action distributions without introduction of significant error. However, these group actions also present problems of congestion in switching systems which require detailed throwdown study for solution.

As an example of group characteristic, subscribers do not originate a steady barrage of calls over the 24 hr of the day. During midmorning and midafternoon hours traffic is built to a peak value, whereas during certain of the remaining hours it is reduced to a minimum. In some residential areas peak traffic may also occur during the early evening. Throwdown evaluations of simulated switching systems, however, are primarily concerned with the busy hour, the hour in which the greatest number of calls are originated, regardless of its actual time-of-day occurrence.

Useful datum obtained from busy-hour field observations is the calling rate per subscriber (calls per hour), which can be used to set up traffic load conditions on the simulated switching system. The calling-rate characteristics can be measured as average calls per hour placed by subscribers in a number of group classifications.

Weather conditions, preholiday periods, or special events have been found to raise substantially the average calling rate in affected classifications. Values adjusted for these conditions are useful in projecting percentage of overload that can be offered to systems engineered for average daily loads.

Subscribers, however, in originating calls, act independently within their classified group in maintaining the average calling rate. Originating times of calls, therefore, occur at random within the hour. Observations made in the field[37] have shown that subscribers, upon receiving dial tone, do not always follow through to dial a full code. Among possible causes are failure to hang up after completion of a call, answering the wrong telephone where two or more are adjacent, dialing before dial tone, and forgetting the number. Such actions produce waste usage of equipment within the switching system, and their study is pertinent to producing throwdown data.

Since "no dials" and "partial dials" are largely due to subscriber errors in originating calls, many of these calls will be originated again upon discovery of the error. False starts, on the other hand, are attributed to accidental origination with no intent to place a call.

Individually, the subscribers will hold equipment on abandoned no dials, abandoned partial dials, and false starts for varying amounts of time. These varying individual holding times can be quantized into several average values which are equally likely to occur or may be averaged to one value depending upon the throwdown study requirements. The holding times on calls receiving tone are usually assumed to cease a few seconds after tone is received.

Subscribers, as individuals placing ultimate good calls, spend varying amounts of time after receipt of dial tone, before start of dialing, and in dialing a full code. This behavior affects the holding time of registers receiving the dialed digits and must be considered in throwdown studies. Data collected on combined waiting and dialing time characteristics show a frequency and time distribution that can readily be quantized into a number of values, each equally likely to occur. When the number of values for a particular throwdown study are determined, each quantized dialing time is represented by a number. Each ultimate good call is then assigned a number from a random list of the representative numbers to establish the dialing time of the call.

Ultimate good calls will develop one of three terminating conditions attributable to subscriber behavior: (1) DA, called subscriber does not answer, (2) busy tone because of called subscriber line busy, or (3) answer by called subscriber. It is assumed from analysis of "don't answer" (DA) studies that, for certain throwdown evaluations of the switching system, approximately 10 per cent of the ultimate good calls meet the DA condition. The number of busy-tone terminations, of course, will develop during the throwdown study as a result of the average originating and terminating calling rate per subscriber served by the system.

Most subscribers upon encountering a line-busy condition make subsequent attempts to reach the called line. The number and frequency of attempts made depend upon the individual characteristics. A detailed analysis of this characteristic, suitable for use in throwdown studies, has appeared in a paper by Charles Clos.[38]

When calls are answered by called subscribers, the connections will be held for varying amounts of time. It has been determined from field observations that the frequency distribution of these holding times is closely approximated by an exponential distribution.

11. Procedure for a Throwdown Study

The procedure in a throwdown study is first to obtain data representative of the traffic to be handled by the system. Throwdown input data for subscriber traffic is concerned with expected busy-hour traffic (or how many calls should originate in each hour), at what time during each hour will each call be originated, and how long will each call last (holding time). Specific calls using these factors may be generated artificially with the use of random numbers. Since subscribers act independently, originating calls will occur at random within each hour.

Throwdown input data representing subscriber originating time can be produced by assigning to each call, of the total number of calls within a specific hour to be studied, a six-digit number from a list of random numbers. The hour is then divided into one million parts, and the assigned random numbers determine the millionth part of the hour during which the call will originate. Random numbers can be obtained from a subscriber telephone directory, omitting all numbers that could not be considered of a chance nature, or from the use of a table such as Tippett's Table of Random Numbers.

For throwdown purposes, a simplifying assumption can be made that the holding time of a call (length of the call) is not a continuous variable but is quantized, so that a particular holding time will have several values. To determine these values an exponential distribution having the proper average is plotted as shown in Fig. 18. The area under the curve is then divided into equal subareas (10 in this case). The mean value of holding time of each subarea is then used to represent each subarea. Ten holding times are thus produced, which are weighted according to the exponential distribution and which are equally likely to occur. These holding times can be designated with numbers 0 to 9 and then assigned to random calls by choosing single-digit random numbers from a random-number table.

As a simple example of how the throwdown is used, suppose that it is desired to determine how often on the average an "all trunks busy" condition will occur in a particular group of trunks handling interoffice calls. With the data of call origination times and holding times prepared, the throwdown run can be started. The calls are listed in the order of their originating times. The first call is assigned to the first idle trunk. A record that this trunk is busy is made, and the time at which it will become idle determined by adding the assigned holding time to the time of origination. This is also recorded. The call which follows in time of origination is then assigned to the next idle trunk, and the process continued for succeeding calls. Before each call is established, the release times of all busy trunks are scanned to determine whether any busy trunk should be made idle. When each call is set up, idle trunks are chosen from the group in the same order of preference that would be used in the system being simulated.

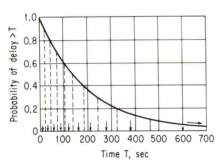

FIG. 18. Distribution of holding times.

Thus, the performance of an actual system is reproduced with considerable accuracy, and detailed records of this performance can be made. From a study of these records the desired information can be determined. The probability of encountering an "all trunks busy" condition can be found, the average number of trunks busy can be determined, or a frequency distribution chart showing the percentage of the time the number of busy trunks is above any given number can be constructed. If proper records are kept, such information as the average number of trunks searched over in locating an idle trunk can be determined. If the trunks were reached through a graded multiple, or if they were in subgroups with a common overflow group, simple extensions of the above procedures would be followed. This particular problem can also be solved by analytical methods and is presented here only to illustrate the application of the throwdown technique.

12. Calls Served on a Delayed Basis

If a telephone subscriber does not receive dial tone from a register immediately after lifting his handset, the call is not considered to be lost but is delayed, because if he waits, he will get the register and receive dial tone. Naturally, the object is to make this delay as short as possible, consistent with providing an adequate grade of service. Calls which are delayed can be either served in a random order (random theory)[39] or stored in a queue and served on a first-come first-served basis (in order of arrival-queue theory). The problem of traffic engineering on a delay basis has been studied by A. K. Erlang,[40] E. C. Molina,[41] F. Pollaczek,[42] and C. D. Crommelin,[44] among others.

The usual way of describing delay distribution is to indicate the probability that a call will be delayed in excess of a certain time (generally expressed in multiples of the average holding time). It has been shown that calls with exponential holding times served in random order yield greater delays than calls served in order of arrival.[43] For calls with constant holding time an analysis has been given by Pollaczek[42] in 1930 which provides the formula which gives directly the probability of a delay greater than t seconds.

The exponential holding time analysis given by E. C. Molina[41] in 1927 for calls served in order of arrival gives the probability of a delay greater than zero and also the probability of a delay greater than t.

$$P(>0) = \frac{(a^c e^{-a}/c!)(c/c - a)}{1 - P(c,a) + (a^c e^{-a}/c!)(c/c - a)} \tag{16}$$

where a = erlangs of traffic offered
c = number of registers (or outlets)

$$P(c,a) = \text{Poisson summation} = \sum_{x=c}^{\infty} \frac{a^x e^{-a}}{x!} \tag{17}$$

$P(c,a)$ is the blocking probability for the case of an infinite number of subscriber sources and where lost calls are held.

The probability of a delay greater than t sec is

$$P(>T) = P(>0)e^{-(c-a)\frac{t}{\bar{t}}} \tag{18}$$

where $\bar{t}$ is the average holding time.

The average delay on calls is given by

$$\bar{d} = P(>0)\frac{\bar{t}}{c - a} \tag{19}$$

and the average delay on calls delayed is given by

$$\bar{d} = \frac{\bar{d}}{P(>0)} = \frac{\bar{t}}{c - a} \tag{20}$$

The following illustrative examples indicate the application of these formulas.

Example 1. Determine the number of registers required so that no more than 1 per cent of the busy-hour calls receive dial-tone delays in excess of 1 sec.

During the busy hour the switching center will be required to handle 1,200 newly originated calls. The subscribers use the registers with an exponentially distributed holding time with an average of 6 sec.

The probability of waiting more than 1 sec is given as

$$P(1) = 0.01 = (1\%)$$

The busy-hour traffic offered in erlangs can be computed as

$$a = \frac{1,200 \text{ calls} \times 6 \text{ sec avg hold time}}{3,600 \text{ sec (busy hour)}}$$
$$= 2.0 \text{ erlangs}$$

The number of registers required = c.
For $c = 6$

The value of $P(c,a) = 0.016$ (from table of cumulative terms of Poisson formula).

and $\dfrac{a^c e^{-a}}{c!} = 0.012$ (from table of individual terms of Poisson Formula)

$$\frac{c}{c-a} = 1.5$$

substituting in Eq. (16) gives

$$P(>0) = \frac{(0.012)(1.5)}{1 - .016 + (0.012)(1.5)} = 0.018$$

substituting in Eq. (18) gives

$$P(>1) = (0.018)(e^{-(6-2)(\frac{1}{6})}) = (0.018)(e^{-0.667})$$
$$= 0.009 = 0.9\%$$

hence $c = 6$ is the correct number of registers.

The percentage of calls delayed by 2 sec is obtained by substituting into Eq. (18) as follows:

$$P(>2) = (0.018)\left(e^{-(6-2)\frac{2}{6}}\right) = \left(e^{-\frac{4}{3}}\right)(0.018)$$
$$= 0.005 = 0.5\%$$

The average waiting time for a register is obtained with the use of Eq. (19):

$$\bar{d} = (0.018)\frac{6}{(6-2)} = (0.018)(1.5) = 0.027 \text{ sec}$$

Example 2.   A 10-trunk route carrying exponentially distributed holding time calls with an average length of 3 min is loaded with 100 calls during the busy hour, and any calls delayed will be served in order of arrival.

The busy-hour traffic in erlangs is

$$a = \frac{100 \times 3}{60} = 5 \text{ erlangs}$$

$$c = 10 \text{ trunks}$$
$$P(c,a) = 0.032$$
$$\frac{a^c e^{-a}}{c!} = 0.018$$
$$\frac{c}{c-a} = \frac{10}{10-5} = 2$$

hence

$$P(>0) = \frac{(0.018)(2)}{1 - 0.032 + 0.036} = 0.035$$

hence, 3.5 per cent of the busy-hour calls will be delayed.

$$P(t > 5) = (0.035)\left(e^{-(10-5)\frac{6}{3}}\right) = 1.5 \times 10^{-6}$$

and 1.5×10^{-4} per cent will be delayed more than 6 min.

13. Lost Calls

The problem of traffic engineering on a lost-call basis has been studied by A. K. Erlang.   The usual way of describing the blocking characteristics of a switching system or subunits of the system, such as trunks, switching matrices, and common control equipments, is to indicate the probability that a call will be lost when a specified amount of traffic is submitted to the system.   If there are no idle paths or equipments in the system, at the instant of time that the system tries to establish a path, then the call is considered to be lost. The lost-call probability is generally referred to as the "grade of service." Typical figures for switching-system grade of service range from 0.01 to 0.001, i.e., from 1 lost call in 100 to 1 lost call in 1,000.

Erlang's lost-call analysis provides the formula which gives directly the lost-call probability when a erlangs of traffic are submitted to a system containing c outlets (on a lost-call cleared basis).

$$P_B = \frac{a^c e^{-a}}{1 - P(c + 1, a)} \qquad \text{(Erlang } B \text{ formula)}$$

where $P(c + 1, a) =$ Poisson summation $\displaystyle\sum_{x=c+1}^{\infty} \frac{a^x e^{-a}}{x!}$

The use of this formula can best be illustrated by the following example.

One thousand subscriber lines originate 300 calls in the busy hour, the average holding time per call being 120 sec.

1. How many trunks would be required to handle this traffic provided all trunks are in a single group and assuming that a grade of service of 0.001 is required?

The busy-hour traffic in erlangs is

$$a = \frac{300 \times 120}{3,600} = 10 \text{ erlangs}$$

$P_B = 0.001$ and $c =$ number of trunks required is unknown. Assume that $c = 21$ trunks. Then

$P(c + 1, a) = P(22,10) = 0.0007$ (from Table of Cumulative Terms of Poisson Formula)

and $\dfrac{a^c e^{-a}}{c!} = 0.0009$ (from Table of Individual Terms of Poisson Formula)

substituting into the formula for P_B we have

$$P_B = \frac{0.0009}{1 - 0.0007} = 0.0009$$

Hence a single group of 21 trunks will handle this traffic with a lost-call probability of 0.0009.

2. If the number of trunks serving a subgroup of these subscribers is 5, how many trunks are required to handle this traffic?

Next it is necessary to determine how much traffic (a) a group of 5 trunks can handle with grade of service of 0.001.

$$c = 5 \text{ trunks}$$

Assume that $a = 0.7$ erlangs

$$P(c + 1, a) = P(6,0.7) = 0.0009$$
$$\frac{a^c e^{-a}}{c!} = 0.0007$$
$$P_B = \frac{0.0007}{1 - 0.0009} = 0.0007$$

Hence five trunks can handle only 0.7 erlang, since the total traffic offered is 10 erlangs; then the number of trunk groups required (each trunk group having five trunks per group) is at least $10/0.7 = 13$ trunk groups so that

$$a = 0.7 \times 14 + 0.2 = 10 \text{ erlangs}$$

Therefore, 14 groups of 5 trunks each are required and a fifteenth group is required to carry 0.2 erlang with $P_B = 0.001$.

$$\frac{a^c e^{-a}}{c!} = 0.0010 \text{ for } c = 3 \text{ and } a = 0.2$$

$P(c + 1, a) = P(4, 0.2) = 0.00005$ (from Table of Cumulative Terms of Poisson Formula)

$$P_B = 1 - \frac{0.0010}{0.00005} = 0.001$$

Hence the fifteenth group will require 3 trunks and a total of 73 trunks will be required.

This example serves to demonstrate clearly that large trunk groups are more efficient than small ones, the difference of $73 - 21 = 51$ trunks being required solely on account of limiting the size of the trunk group to 5. Such limitations are frequently found in switching systems because of the structure of switching mechanisms and economic considerations of switching equipments. However, the economy afforded on reduced cost of switching equipment should be balanced against the additional cost of trunking requirements.

14. Switching-center Call-handling Capability

Proper design of a switching center requires adequate knowledge of expected traffic that the center must handle. Traffic data should be available in terms of calling rates, call holding times, grade of service required (lost-call probability), and tolerable delays (or speed of service) for the subscribers using the system. When these data are known, the proper balance between time-shared (common) equipment and per-line equipment can be established, giving rise to a system which is optimized for service, size, and cost.

For a traffic analysis of the call-handling capability of the switching center, the subsystems which make up the center should be analyzed on a traffic-delay or lost-call basis, using the applicable formulas presented in the preceding sections.

Functional subsystems such as the registers, line-trunk scanner, dialed number translator, and operators positions are usually analyzed on a delay basis. Other subsystems such as switching matrices and trunk circuits are usually analyzed on a lost-call basis. If ringing and busy tones are provided on a per-line basis, there is no waiting with respect to their use and no other common equipment is held.

For each subsystem, a knowledge of the traffic submitted, the holding or processing time, and the grade of service should be known or estimated. Since these subsystems usually perform their functions sequentially, the grade of service of the overall system should be considered to determine whether the holding or processing times of the various functional units are adequate.

NUMBERING PLAN

15. General

A numbering plan is an essential element in a switched communications system. It identifies all users of the system and provides necessary information to the switching equipment to obtain routing information. In general, a numbering plan can be divided into three basic groups of numbers or "codes" which identify (1) a specific area (geographic location), (2) a specific switching center, and (3) an individual user served by this switching center. The number of digits used in each group is determined by the total number of users to be reached, the maximum number of users that can be served by a switching center, and the desired division into geographical or otherwise identifiable areas.

a. Switching-equipment Considerations. Switching equipments are divided into two types in accordance with the method employed to process a received address. *Step-by-step* switching centers process each digit as received, whereas *common control* switching centers store part or the entire address prior to processing. Although step-by-step switching systems inherently have greater flexibility to adapt to varying number of digits in a numbering plan, they present serious limitations with regard to alternate routing implementation and integration into a uniform numbering plan where each destination is reached by the same number of digits regardless of its geographical location. Common control systems which store part or the entire address before processing the call provide greater efficiency in the usage of switching equipments and are indispensable for the implementation of alternate routing. The design of common control switching systems must give careful consideration to the provision of sufficient digit storage capacity, since it sets a limitation to the size of numbering plan that can be accommodated. To minimize the size of the "number store" it can be implemented to recognize, verify, and process that part of the address which is essential for routing, then erase and reuse this portion for the storage of the remainder of the address. However, this method presents other problems when the entire address has to be retransmitted as in the case of tandem calls within a switching system or in calls between related systems. Storage of the entire address has another distinct advantage when it is necessary to translate a received address either in format or in the numerical values of its digits.

With the advent of ever increasing user-to-user dialing encompassing larger and larger areas and the need for a uniform numbering plan to satisfy ease of operation by the users of the system, it has become general practice to adapt step-by-step switching centers to a common control mode of operation as far as address processing is concerned. This is accomplished by the addition of separate register-senders having the same characteristics as employed in pure common control switching centers.

b. Transmission Facility Considerations. Transmission facility characteristics have no direct impact upon the choice of a numbering plan or its implementation provided they satisfy the performance criteria for the specific types of signals used to transmit the numerical values of the individual digits. However, certain types of transmission facilities, such as unloaded cables or open wires, impose a restriction to the use of the numerical value 1 as the first digit when d-c pulses are used. This is primarily a precautionary matter as to avoid the frequent occurrence of wrong numbers. When single or multifrequency types of signaling are used, this restriction no longer exists.

c. User's Instrument Considerations. Basically there are three devices used at the user's instrument to convert the digits of an address into transmitted signals for processing at the switching center. The *rotary dial* is a simple mechanical device which translates the 10 numerical digits 0 through 9 into corresponding serial pulse streams.

The *keyboard*, in a variety of arrangements, allows the inclusion of additional digits (in excess of the numerical value of 10) or symbols which may be used in conjunction with the numbering plan.

The *automatic dialer* may be used as a substitute for either rotary dial or keyboard or may be arranged to perform additional functions in connection with the numbering plan.

Whichever device is used, careful consideration must be given in the choice of a numbering plan to avoid confusion and ambiguity to the user. For example, the similarity between the letter "O" and the figure "0" on the rotary dial has led to certain restrictions in their application to portions of a specific numbering plan.

16. The "Open" Numbering Plan

The "open" numbering plan has no fixed number of digits. In general that portion of the numbering plan which identifies the actual user's instrument is fixed, usually three or four digits, and is dependent only on the configuration and capacity of the final switching center. All digits preceding the user's number may vary and are assigned as routing digits on the basis of the overall system configuration. Efficiency in use of such a numbering plan presupposes knowledge of the various routing patterns by each user and a built-in protection against abuse and misuse. Commercial applications of the open numbering plan are rapidly disappearing and are superseded by the "closed" numbering plan. In military systems, however, particularly for highly mobile tactical applications, the open numbering plan offers a degree of flexibility which would be difficult to duplicate with a closed numbering plan on an economically equal basis.

17. The Closed Numbering Plan

The closed numbering plan contains a fixed number of digits for all users of a system regardless of their geographical location within this system.   This principle of unique destination codes provides each user with a network address made up of a uniform number of digits.   A typical example of such a numbering plan is the one employed in the commercial telephone network of the United States and Canada.   It consists of an "area" code, an "office" code, and a station number comprising a total of 10 digits. For calls between users located in different "areas" it is necessary to dial the entire 10 digits, but for calls between users having the same "area" code only the "office" code and station number need be dialed.   To provide this feature the switching center must be capable of recognizing and distinguishing between the area and office codes.

   a. *The Area Code.*   The area code consists of three digits giving a theoretical number of 999 possible areas.   Because of limitations in the use of digit 1 in the first position, as stated earlier, and first digit 0 being reserved for operator calls, the total available area codes are reduced to 800 combinations.   In order to differentiate between area and office codes the second digit of the area code is restricted to the numerical values 1 and 0.   Furthermore to provide a number of "service" codes, such as information, repair, etc., the third digit of the area code is restricted to numerical values between 2 and 0.*   There are, therefore, 152 possible "area" code combinations available at present.

   Ultimately area codes will be expanded to the 800 possible combinations by restricting only the first digit of the area code to numerical values between 2 and 9.   Differentiation among area, office, and service codes will be accomplished by timing techniques.

   b. *Office Codes.*   The office code also consists of three digits.   At present it is generally represented by two letters and one numerical digit, where the two letters represent the first and second letters of the office name and the third digit is a prefix to the station number.   This code is often referred to as the 2L-5N code.

   Because of restrictions in the English language a certain number of letter combinations cannot be used as office codes.   In order to utilize the maximum possible combinations, the trend is toward "all number calling" (ANC).   The theoretical number of office codes of the 2L-5N form is limited to 640, since letters are assigned only to dial positions 2 through 9 and it is difficult or impossible to find suitable names for certain 2L combinations.   Furthermore because of possible confusion between the letter "O" and the figure "0" office codes ending in zero are avoided.   In contrast to this, ANC office codes allow the use of 792 combinations where only digits 1 and 0 are restricted in the first position and digit 1 in the third position.*

   The expansion in use of office code combinations brings with it a conflict by having identical area and office codes.   Differentiation between the two must be made on the total number of digits dialed; e.g., if only seven digits are dialed, the prefix is an office code, and if ten digits are dialed, it is an area code.   It is quite evident that such a scheme could not be implemented unless *all digits are received* and stored before processing of the address is initiated.

   c. *Station Number.*   The number of digits in the station number, usually four, is determined by the organization and capacity of the switching center.   In many cases the four-digit station number is divided and uses the first three digits as the station identity and the last (fourth) digit to select a specific ringing frequency.   This arrangement is wasteful in the use of office codes.

18. Access Codes

The integration of separate switching networks into a universal numbering plan must take into consideration the specific requirements and switching equipments of
_____
* The restriction on the third digit applies only in the case where the second digit is the number 1.

these systems in order to accomplish efficient and unambiguous procedures in the assignment of station numbers and routing codes.

In general it should be possible to retain a local numbering plan for intrasystem connections without compromising the ability to extend calls into and through related switching networks.   Special access codes which may be different and unique to each network offer the most promising solution to this problem.

The numbering plan discussed previously makes extensive use of such access codes, which vary from one to four digits, to weld the numerous individual switching networks into one homogeneous system.   These access codes are used to overcome certain limitations inherent in some types of switching equipments or may be used to indicate the requirement for special services.

19. Future Requirements

In switching networks of the fixed type it is difficult to anticipate new services or changes in operating procedures which could not be accommodated in the framework of numbering plan considerations as outlined above.   The methods and techniques of implementing such requirements are basically a question of equipment design and are not reflected in a particular numbering plan.   Mobile systems, on the other hand, present problems which are not normally found in fixed installations.   This is the case in many military communications systems where both switching centers and personnel may never occupy the same geographical location for any extensive period of time. This problem is even further complicated when it becomes essential to assign a specific address to an individual regardless of his location in a particular "area" or termination at a particular switching center.   These problems may well require a different approach to the application of a numbering plan but most likely will find their solution in more sophisticated switching equipments.

MESSAGE SWITCHING ENGINEERING[11–36]

20. Introduction

The designer of a communications system is often faced with the necessity of providing store-and-forward message switching centers at one or more locations.   For many years this problem has been solved by use of punched-tape transmitter-receivers at the line terminations combined with manual, semiautomatic, or fully automatic implementation of the cross-office function.   In the last two instances, the function has been provided through use of relay or solid-state logic.   Recently, general-purpose digital computers have begun to be applied to this problem.   While the digital-computer systems are far more complex than the older paper-tape-relay-logic equipment, they possess many advantages which have led to increased acceptance.

In this section, the general characteristics of a digital-computer switching center will be described and material will be presented to illustrate how the load of the communication system influences the design of the center.   Designing a digital computer switching center is a very complex task, involving selection of components such as computers, memories, disk files, tape stations, printers, communications line terminations, etc., from a wide variety of available types and then selecting and implementing a programming system with a multitude of trade-offs (time vs. memory, flexibility vs. efficiency, etc.).

Certain aspects of the communications-system load are easily quantized, while others, not necessarily unimportant, are very difficult.   It is not intended that this chapter serve as a comprehensive detailed treatment of every aspect of the communication-system analysis and switching-center design process.   The intention is to discuss the problem in a way which brings out the principles involved, with quantitative analysis where possible.

a. Functions of the Digital-computer Switching Center.   The basic purpose of a digital-computer switching center is to receive messages, store them, and forward them to the appropriate destination as soon as possible.   This function is basic regardless of the application, but there are, quite naturally, important ways in which the application will influence the design of the system. In certain applications, the only requirement is that the system handle general administrative message traffic. In others, the basic purpose of the system may be to switch messages between outposts and a central data-processing facility in order that a type of centralized control be practiced over a geographically dispersed organization. A simple example of this type of application is an air-lines reservation system. Clearly, there are systems that do both. While each type of application has particular problems, especially in cases where there are complex relations between the message-switching and data-processing functions, the important basic problems are generally common to all applications. For this reason, no specific further references will be made to particular types of applications.

The most important basic function is the "store-and-forward" processing of messages. Other functions commonly found include (1) interpretation of complex addresses and handling instructions, including multiple addresses, group addresses, and narrative addresses; (2) transmission of messages by priority, including preemption; (3) translation of code, speed, format between sender and receiver; (4) record keeping for protection, accounting, and retrieval purposes; (5) data processing based on content of messages. The essential element of a store-and-forward switching center is operation of the center independent of message originator and destination. The center accepts messages from each subscriber when offered and without regard for the immediate condition of the destination. It holds all messages until the destination can accept them and then forwards the message without immediate concern for the condition of the originator. During the interval between receipt and transmission of each message, the center must store it and, furthermore, accept responsibility for the security and integrity of the message. In the older systems, messages were generally stored in a punched-paper-tape loop, one for each outgoing line. In a digital-computer center, the messages are held in a magnetic storage device, generally a disk or drum file but sometimes magnetic tape or high-speed core memory.

The center interprets that portion of the message containing routing and handling instructions, generally called the "header," in order to give the message proper treatment. Depending on the complexity of the system, the message header will contain one or more addresses, a priority indicator, an identification number, date-time code, message format identification code, security code (in military applications), classification or transaction codes for data-processing applications, etc. That a digital-computer switching center can interpret all this information properly and without substantial delay to the message is because of the great speed and flexibility of the general-purpose digital computer usually at the heart of the center. This ability to process messages efficiently in accordance with very complex header information is the most noteworthy advantage which the digital-computer switching center has over the older tape repeater types.

Addresses in messages generally fall into three categories: single, multiple, and group, or "broadcast." A single-address code is the simplest type, requiring transmission as implied in the name. A multiple-address message contains in its header a separate address code for each destination. A group, or broadcast, address is a single code which signifies a predetermined group of destinations. In most applications, addresses are denoted by relatively compact codes, one to seven characters long. There are, however, important cases where addresses (as well as other types of header information) are in narrative form rather than in a compact, easily interpreted code. The ability to interpret addresses in narrative, sometimes ambiguous, form is a great advantage of the digital-computer switching center.

Most communications systems have a requirement for priority treatment of messages. Generally this means that the center must identify the priority class of each message as it arrives and take whatever steps are necessary to deliver the message to its destination ahead of all messages of a lower priority classification. In the simplest case, the system will review the current backlog of messages by priority class for a given destination only in between transmissions of complete messages. In more complex systems, messages of a certain priority class, usually only the very highest, are given preemption rights, which means that even messages already in transmission will be interrupted, if necessary, so that the message with preemption rights can be forwarded without delay.

Priority considerations usually enter the design of a communications system because the designer recognizes the need to minimize delays to important classes of messages. Delays are generally attributable to either line loading or switching-center loading, the latter delay commonly called the "cross-office" delay. In the older paper-tape repeater-type centers, the cross-office delay had to be counted as a significant element in the total message delay. Complex priority and preemption procedures generally require manual handling of messages, which, in itself, tends to be self-defeating, although, perhaps, useful in reducing cross-office delays for important messages from hours to minutes. With a digital-computer switching center, it is a matter of ordinary design to provide an analysis of the entire message backlog as frequently as once every second, with instant action for those messages in the system requiring it. With magnetic drum or disk message storage, messages are accessible in milliseconds; hence the cross-office delay of a properly designed digital-computer switching center is 1 or 2 sec, at most, for all classes of messages.

Another very important function of the digital-computer switching center is the translation of messages from one language to another. Generally this involves differences in code, format, and line speed. There are a number of reasons for the existence of a system where language differences are not eliminated in favor of standardized operation. The communications system under design may be a modernization of an existing system rather than a completely new facility, in which case it is usually necessary to accept existing practices even though they are conflicting. As the needs of individual users of the system vary, there may, necessarily, be a variety of terminal devices each with a different speed, character set, and control procedures. While in either situation it is technically feasible to standardize all terminal devices, it is usually not economical. While language transmission is generally not practical in the older paper-tape repeater-type center, it is easily done in the digital-computer center, thus making unnecessary the forcing of all terminal devices into a standard language.

The switching center usually keeps a record of all messages passing through it. In a digital-computer system this record is normally kept on magnetic tape. The foremost reason for doing this is message protection. As a message may remain in a store-and-forward switching center for quite some time before final delivery (perhaps hours) it must be protected against loss due to failures at the terminals or within the switching center. Records of message traffic have clear value for accounting and statistical purposes. Since the switching center is an element of a communications system itself an element of a larger system serving some particular end, there is often a reason for persons, not necessarily participants in the transmission of a particular message, to examine the message. In such connection, an important switching-center function is implementation of the retrieval of messages which are currently in active storage in the center or which recently passed through.

There are applications where the communication system carries messages between a centralized data-processing facility and the outlying terminals. In such cases, the digital-computer switching center, in addition to implementing the previously described communications functions, provides a very efficient interface to the data-processing facility. In the ideal situation, not restricted by evolutionary factors, it should be possible to handle both the switching-center and data-processing loads in a single digital-computer system with maximum efficiency. Where this is done, the total communications system becomes an "on-line data-processing system," some examples being found in air-lines reservation systems, demand deposit banking systems, industrial inventory control systems, stock brokerage transaction handling or ticker systems.

b. Communications Load Parameters That Affect Switching-center Design.
There are a number of aspects of the communications load in a system which affect the design of a switching center. In this section it is assumed that the need for switching centers at certain nodes in the overall network has been established, and accordingly, there will be no further discussion of justification or placement of the center.

The system designer, in order to determine the need for and placement of a switching center, has had to make estimates of the traffic volume and to make, at least, a preliminary selection of the number and capacity of the communication lines in the system. The quantitative data used to make such a decision, which normally include average message length, message rates and their variation over the day, average line loading, acceptable delays, etc., are also used to determine internal characteristics of the switching center. The design of the center is, however, also influenced by other, perhaps less easily quantitized, factors, such as the type of processing functions required, as their complexity may have a major effect on the power required in the digital computer.

Table 4 contains a listing of some important aspects of the switching-center load which influence its design. It should be noted that each particular system must be analyzed on its own merits, but the listing in Table 4 does cover aspects which are generally important in all systems. The entries in Table 4 are, generally, in order according to ease of quantitative statement and use.

Table 4. Communications Load Parameters

Item	*Important characteristics*
Message length	1. Average
	2. Variance
	3. Complete distribution, if known
	4. Maximum acceptable length
	5. Block size within message
Message rates	1. Peak-hour arrivals
	2. Peak-hour departures
	3. Long-term arrivals
	4. Long-term departures
	5. Maximum input surges (period of seconds)
Line capacity	1. Number of inputs
	2. Number of outputs
	3. Speed of lines, characters/sec
	4. Average peak-hour line loading
	5. Loading on particular lines which departs from average
	6. Expected uncorrected error rate on lines
	7. Error detection, correction, and general control procedures on communications lines
Message delay	1. Allowable in switching center on messages not blocked by line utilization or terminal unavailability
	2. Expected owing to line utilization limits, terminal unavailability, day-night transmission patterns, etc.
Message storage requirements	1. For record keeping $\times$ days of traffic
	2. For retrieval within x min of messages transmitted within most recent y hr
	3. For accounting and statistical data
Reliability	1. Acceptable down time for partial or complete system
	2. Acceptable delays in recovering traffic disrupted when system fails
Addressing and routing	1. Admissable types of addresses in messages: single, multiple, group, narrative form
	2. Average and maximum number of addresses in multiple-address message
	3. Average number of destinations in a group address
	4. Number of destinations in system
	5. Alternate routing procedures: automatic, on-line by supervisory control, off-line
	6. Security checks applicable to routing (military only)
Message formats	1. Specifications for all formats to be recognized and handled
Traffic handling procedures	1. Number of priority classes and functions for each (preemption)
	2. Number of different languages in system (code, speed, format) and exchange requirements
	3. Supervisory handling of traffic problems
	4. Interception of messages for terminals out of service
	5. Interchange of traffic with other networks, not necessarily store-and-forward

21. General Characteristics of Computer Switching-center Equipment

A digital-computer switching center is a system composed of digital-computer equipment commonly used in many applications and certain items relatively specialized for communications purposes. In the first category are found computers, high-speed memories, drums, disk files, tape stations, printers, card readers, etc. In the second category are found principally different types of equipment to integrate the digital computer with the communications lines but also supervisory or inquiry consoles, equipment status monitoring devices, and special displays.

In order to show how the communication load influences the design of the switching center, it is first necessary to describe the types of equipment configurations which may reasonably result from the design process. A typical switching center will be presented, and its various elements and features described. This center is offered as a generic example and is not necessarily optimum for any particular application.

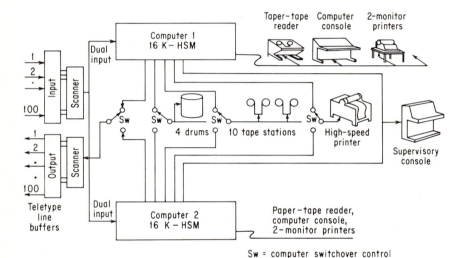

Fig. 19. System block diagram of a computer switching center.

Figure 19 is a pictorial system block diagram showing all major equipment elements. At the heart of the system is a pair of general-purpose digital computers, one on-line, the other a stand-by. Data are exchanged between the communications lines and the core memory of the (on-line) computer through line buffers and a scanner. The line buffers convert incoming serial-bit data into characters and outgoing characters into serial-bit data. The scanner samples all the line buffers at an appropriate rate and generates interrupts in the digital computer whenever an input line buffer has a character formed or an output line buffer can accept a character. The computer responds to the interrupt signal by executing an input/output service request (or program subroutine) which causes the appropriate character to be exchanged between the line buffer and the computer core memory.

While there is a small computer memory zone for each communication line, the principal message-storage device is a group of random-access magnetic drums. Records of all messages are kept on magnetic tape. A high-speed line printer is used for copies of messages and lengthy statistical reports.

A supervisory console contains equipment status indicators and switches and a low-speed electric typewriter to provide a human operator with on-line supervisory control

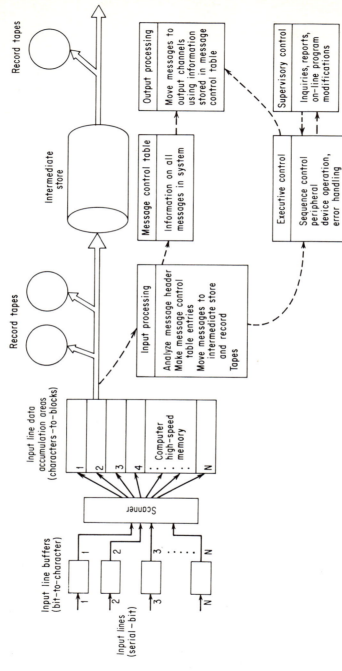

Fig. 20. Functional flow diagram of a computer switching center.

over the system.   There is a subsystem, not explicitly shown, which is continuously monitoring all equipment.   Should a failure develop, a signal goes to the supervisor and the system responds automatically where possible and as programmed.   If the failure appears to be in the on-line control computer, the monitor subsystem will direct the stand-by computer to go on-line and switch over all the auxiliary equipment.   In order to make such a switch-over possible without disrupting traffic, the stand-by computer always obtains duplicate input data on all lines along with a periodic traffic status report from the on-line computer.   When the stand-by computer goes on-line, it uses these data to reconstruct the recent history of processing and then resumes normal traffic.

Figure 20 is a simplified functional flow diagram of the system.   Data enter the system through the input line buffers.   Under control of the scanner, the data are accumulated into individual areas for each input line.   As it is not generally efficient to process messages a character at a time, the data are allowed to accumulate into a block of some practical size before being subject to any significant program control.

The computer program which processes the messages can generally be considered in the following functional divisions: (1) input processing, (2) output processing, (3) supervisory control, (4) executive control.   It is convenient to discuss the data flow through the computer switching center in terms of the functions performed in each of these four divisions.

The basic operation of input processing consists of scanning the input line data accumulation areas for full blocks, analyzing the content of these blocks, modifying processing records as required, and moving the data to the intermediate storage device and the record tapes.

The input subprogram (to give input processing a title) must scan the input line data accumulation areas periodically as these areas fill up in a random fashion based on line activity and not, generally, in a way that is predictable.   This is, of course, due to the fact that arrival of data in the system is at the convenience of the outlying terminals, not at the demand of the switching center.   As an example, suppose the system has 100 input lines operating at 10 characters per second and each input data accumulation area holds 100 characters.   If the input subprogram scans all 100 accumulation areas on an average of once per second, then each time this happens an average of 1,000 characters will have entered the system on 100 different lines and an average of 10 of the 100 accumulation areas will have filled up and require processing. As the messages arriving on the 10 lines whose accumulation areas happen to fill up at this particular time are independent of each other, the input subprogram must maintain individual control tables on each channel and use the information therein to control its processing on an individual channel basis.

The input subprogram operates in two distinct modes depending on whether or not the header of a particular message has been completely analyzed.   Header analysis involves checking all the indicators in the message for validity and functional requirements and then setting up an appropriate message control record, usually in a table in the core memory.   As the most important control information in the header is the priority and routing data, the message control records are generally organized into lists associated with the output lines.   These lists are usually ordered first-in-first-out by priority.   Each block of the message is transferred to the intermediate store and the record tape by the input subprogram.   As the intermediate store is usually a random-access device, the location of the message (or perhaps the locations of its blocks, if scattered) must be noted in the message control record.   For messages still entering the system whose header analysis is complete, the input subprogram merely transfers the blocks to the intermediate store and the record tapes and searches for the end-of-message identification.   Generally the message is not considered eligible for transmission until properly terminated.   Figure 21 is a flow chart of these typical input processing functions.

Output processing is similar to input processing, although simpler and with the data flow inverted.   The computer has output line data distribution areas which are accessible to the electronic scanner.   The scanner transfers data from these areas to the output line buffers a character at a time.   The buffer converts the characters to

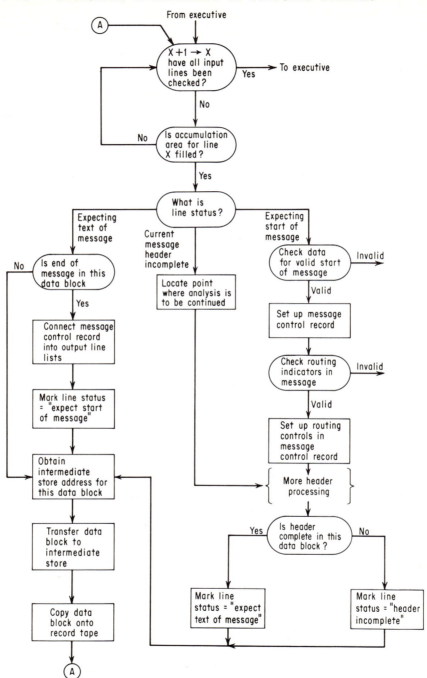

Fig. 21. Input processing flow chart.

serial-bit stream. The output subprogram scans these distribution areas and transfers new blocks of data to them from the intermediate store. Information in the message control record indicates to the output subprogram where in the intermediate store the next block of data is located and whether or not it is the last block of the

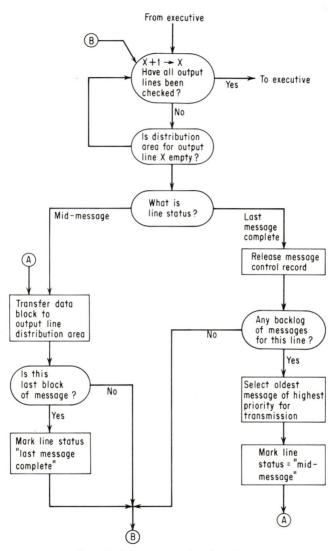

Fig. 22. Output processing flow chart.

message. In the latter case, the output subprogram will complete transmission of the message and move on to the next message in the list for the particular output line. Figure 22 is a flow chart of these typical output processing functions.

The supervisory control subprogram provides an interface between the automatic message-processing aspects of the system, as represented by the input and output subprograms, and the people who supervise the overall operation of the system. This subprogram generates various system status reports for the supervisor either automatically or on specific request. It also accepts commands from the supervisor which cause system operating procedures to be modified.

The executive control subprogram coordinates all other subprograms, controls peripheral device operation, and executes error procedures associated with equipment failures. The executive determines the amount of computer time to be spent in the various programs and the basic sequence of operations. It senses various program backlog indicators and adjusts the program sequence accordingly. When program interrupts due to external stimuli occur, it determines whether to continue with the interrupted program or go to a new one, identified implicitly by the interrupt.

In this illustrative system, the executive follows a simple sequence control approach. As there are no means to throttle input data, the executive guards against overflow by giving the input and subprogram highest priority. From Figs. 21 and 22, it can be seen that both the input and output subprograms are designed to inspect cyclically all communications lines for activity and perform necessary data processing. Under normal circumstances, the executive would allow each of these programs to make one sweep of the communications lines in an alternating sequence: input-output input-output, etc. As the frequency of use of the supervisory control subprogram is relatively low, it being operator initiated, the executive will schedule this function when called for on a low duty-cycle basis. In order to avoid input data overflow, each time the input program completes a sweep, the executive checks a backlog indicator of full input accumulation areas. Should that indicator still show a backlog due to accumulation areas filling up just after they were tested by the input program, the executive will reinitiate another input cycle rather than continue the normal input/output sequence.

It is also possible that, even with this presentation, the system might spend sufficient time in the output and supervisory subprograms to cause an input overflow. The basic approach to avoid this situation involves use of a timer which limits the amount of time that can be spent in the output and supervisory subprograms before the executive returns control to the input subprogram.

This description of a typical digital-computer switching center illustrates the functional relationships between the communications system traffic load and the various items of equipment in the center. It should be clear that the major elements in the center are the general-purpose digital computer, the intermediate storage device, and the equipment that connects the computer with the communications lines.

The other elements, which are secondary in so far as data flow is concerned, are quite important. The remaining sections of this chapter will cover each of these elements of the center as follows:

1. Intermediate storage devices (paragraph 22)
2. Record and retrieval storage (paragraph 23)
3. The central computer (paragraph 24)
4. The communications interface (paragraph 25)
5. Reliability (paragraph 26)

22. Intermediate Storage Devices

The intermediate storage device accepts inbound messages and holds them until released on the appropriate outgoing communications lines. It enables the system to smooth out variations in traffic load due to general traffic randomness, differences in line speed, and long-term unavailability of terminals. The system requirements for the intermediate storage device are directly influenced by the quantitative characteristics of the traffic load. The expected traffic volume through the center can be used to compute the size and access rate of the intermediate storage device.

Message traffic can be described in various units, such as bits per second, character per second, data blocks per minute, messages per minute, etc. It is further possible to describe the load in messages per minute switched through the center or arriving or departing, etc.   In considering the influence of a communication system based on the design of a computer switching center, there appear occasions when it is necessary to describe the load in a bit, character, data block, or message rate.   Furthermore, there is a clear partition between the input and output functions, which means that regardless of what the units are for measuring load, the effect of input and output is additive. This is particularly important in an application where a large portion of the messages are multiple-address, as the number of outbound messages will obviously exceed the number of inbound messages.

   a. Intermediate Storage Devices Available.   In digital-computer switching centers there are four types of devices generally considered for this function: (1) magnetic drum, (2) magnetic disk file, (3) magnetic tape, (4) magnetic core memory.   Magnetic drums or disk files are most commonly used, tape or core memory being suitable only for very low traffic applications.

   Table 5 illustrates the general characteristics of the different types of intermediate storage devices.   From this table it can be seen that the performance and cost characteristics of these devices cover a range of several orders of magnitude.

Table 5. Intermediate Storage Devices

	Random-access time	Max. practical storage, bits	Commercial selling price per bit, cents
Magnetic drum....	1–50 msec	25×10^6	1.0
Magnetic disk.....	25–125 msec	10^9	0.1
Tape.............	Sec–min	4×10^7 per reel	0.05
Core memory......	1.0–10.0 μsec	10^6	50

Note: Figures in this table are based on currently available commercial equipment. They are intended to be broadly representative of the given class of device and are not taken from any particular equipment.

   Random-access time refers to the average time for the first bit of a data block stored anywhere in the device to be brought to some common register.   In the case of the magnetic drum or disk, this time is based on the rotational speed of the device.   The drum and disk are similar devices, differing functionally because of details such as use of fixed or movable heads.   There is no theoretical/practical reason why drums or disk files must differ in general performance characteristics, but current common usage for disk files requires a design which minimizes storage cost per bit at some moderate sacrifice in access time.   The access time to high-speed core memories is in the microsecond region indicated.   Access time to data stored on a tape is largely a function of the amount of tape actually used for storage.

   Consider a tape with a storage density of approximately 333 characters per inch and a recording speed of 100 in./sec.   If storage is allowed over a full reel of 22 ft, then the time to search a full reel is 264 sec, and even if active search is restricted to a small length of tape, the average access time to a data block is on the order of seconds.

   The figures for maximum practical storage capacity are intended to be generally representative of the limits on single devices of each class usually imposed by practical considerations such as cost and commercial availability.   They do not represent an absolute maximum based on either an exhaustive cataloguing of the EDP market or physical limits of the device.

   The figures for price per bit are based on the commercial sales price for typical devices of maximum capacity.   In the case of tape storage, the price is for the tape station, not the reel of tape, since replaceability of tape reels is of no value when considering tape for intermediate storage rather than record storage.   It is important to note that the prices indicated are typical for manufacturers of complete computer systems who furnish substantial system analytic and programming support to individual customers and who maintain large program libraries and other services for general use of all customers. These devices are also sold at much lower prices by firms who specialize in device manufacture, but the purchaser must assume the major engineering, construction, and programming responsibilities generally required to integrate these devices with the computer system.

b. Quantitative Analysis of Intermediate Storage Requirements. *General.* Consider Fig. 20, functional flow diagram of a computer switching center. Associated with each input line is a data-accumulation area which is filled character by character to a specified size by the scanner. At any given instant of time, these areas are filled to various levels in accordance with current inbound traffic. The input program scans these areas periodically and schedules further processing for any input channel whose area has filled past some specified threshold. Such processing includes, ultimately transfer of the data block (up to the threshold) to the intermediate storage device. Once the data block is safely in intermediate storage, its memory area is free to be reassigned to accumulate more input data. While the input program is processing the data block, data continue to accumulate (character by character) beyond the threshold point. Presumably, enough memory is assigned to each input line so that incoming data do not overflow the allocated area.

There are, clearly, deadlines in this process. If the input program cannot keep up with the arrival of data because of momentary surges, one of several events will occur:

1. The system will have to allocate additional core memory to all lines requiring it until the surge subsides.

2. If item 1 is not possible, then the system will have to be equipped with a means for informing the remote terminals to halt transmission in such a way that it assures against loss of data.

3. If 1 and/or 2 are not possible, then the system will lose data on some number of input lines. Recovery from such a loss usually involves supervisory action.

The first step in quantitative analysis of intermediate storage is to chose the amount of memory required for input data accumulation so that, full consideration being given to requirements of the users of the system, a satisfactory grade of service is provided. A similar requirement exists for allocation of memory for the output data distribution areas except that the penalty for overflow is not so severe, as it involves only interruptions of transmission, not loss of traffic. The quantity of core memory required for input and output areas is directly controlled by the traffic load and the access rate of the intermediate store.

The second step is to estimate amount of intermediate storage required. This is a function of the relative loading of the output channels and can be calculated separately. Data accumulate in intermediate store owing to two distinct fundamental causes:

1. Congestion on the output lines due to statistical variations in message length, backlog per line, etc.

2. Messages being undeliverable because of relatively long-term outages of either lines or terminals.

An analysis of intermediate storage requirement due to the first cause (natural congestion) can be made using queueing theory. Storage requirements due to long-term outages can be estimated if the extent of the outages themselves is quantitatively defined. However, it is generally preferable to provide for such situations by transferring messages destined for out-of-service stations from the intermediate store to a less expensive "intercept" store such as magnetic or paper tape. Further discussion of this topic will accordingly be limited to the effects of natural congestion.

Step 1. Choice of Core Memory for Line Buffering and Intermediate Storage Access Rate. Each message passes through a number of sequential steps during input processing (and their inverse during output processing). These are:

1. Accumulation of characters into data blocks
2. Input processing of the data blocks including header analysis
3. Transfer of the data blocks to the intermediate store

Each of these steps requires a certain average amount of time per data character plus another amount of time per data block. In other words, given a constant specified traffic load (in messages or characters per unit time), a large data block tends to increase the traffic capacity of the system. Core memory is quite expensive, and in order to minimize requirements for its use, it is essential to minimize the size of the data blocks. Consequently, it is necessary to give attention to choice of the proper size data block.

The principal limit of system traffic capacity is imposed by the intermediate storage device because of its access time and data-transfer rate. As access time is chargeable per data block, regardless of size, while data-transfer time is chargeable per character, it is possible to compute an *upper* limit on the traffic capacity of the center as a function of:

1. The mechanical characteristics of the intermediate storage device
2. The data block size

The following equation expresses this limit.

$R_{\max}$ = maximum achievable data rate, input plus output, characters/sec
T_a = average access time to a randomly located data block in the intermediate store, msec
T_i = data-transfer time to or from the intermediate store, msec/character
B = data block size, characters/block
$R_{\max} = 10^3 B/(BT_i + T_a)$, characters/sec

This equation is plotted in Fig. 23 for various values of T_a with $T_i = 0.04$ msec (corresponding to a data-transfer rate of 25,000 characters per sec). In order to apply these curves properly, several important points must be considered. The average access time of the intermediate storage device must be determined by evaluating not only mechanical characteristics of the device but the possibility of applying a programming approach which minimizes waiting time due to disk or drum rotation. Once this is done, the block size can be determined for any given traffic load provided, however, that access to the intermediate store is the only limit to data flow. As was previously mentioned, messages must pass through program steps as well as the intermediate store. While it is generally possible to transfer data to and from a mechanical storage element simultaneously with computing, it is still

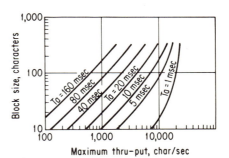

FIG. 23. Maximum through-put, characters per second.

necessary to determine whether computing or data transfer is the limiting factor. In the latter case, Fig. 23 applies. In the former case, an average computing time per data block must be used instead. The possibility that computing time, rather than intermediate storage access, is the limiting factor is very difficult to assess, as it involves a complete analysis of the functional requirements of the application in terms of a particular computer. While the curves of Fig. 23 indicate the maximum traffic capacity for a given data block size (or the minimum data block size for a specific capacity), the decision also involves considering the functional content of the block. There may be format rules which tend to dictate block size in the interest of program simplification.

This analysis appears to assume that there is little choice of the intermediate storage device. This is not completely false, as the number of different devices available with the proper characteristics is relatively limited, particularly when the choice is restricted to the product line of a single computer-system manufacturer. It is, of course, essential that a device be chosen which is at least roughly capable of handling the specific load (as indicated in Fig. 23).   With the device chosen, Fig. 23 then indicates a data block size compatible with the traffic requirements.   It is also possible to consider the possibility of a cost trade-off in the form of less (more) core memory vs. a more (less) expensive, faster (slower) access device.

The curves of Fig. 23 indicate an absolute maximum traffic capacity.   The designer should not, however, minimize block size to the extent that he attempts to use 100 per cent of this capacity.   Because of random variations in input data rate and access time to the intermediate store, a queue of inbound data blocks will develop.   The length of this queue is a function of the relative loading of the intermediate store, and the queue will grow without limit as the loading approaches 100 per cent.   This problem will be explored mathematically below, but it is mentioned here because of interaction among intermediate store service time (access time plus data transfer time), block size, and number of blocks required available in memory.

   c. Memory Allocated for Blocks.   Memory may be allocated to input (output) lines on either of two bases: fixed allocation per line or common pool of data blocks available to all lines on demand.   In an application where data may enter the center at the will of the remote subscriber, it is necessary to provide either a fixed amount of memory for all input lines or some type of signaling arrangement requiring subscribers to bid for entry to the center until receiving an acknowledgment and only then sending the message.   In applications where the subscribers transmit only when polled by the center, memory is required only when a positive response is received to the polling signal.   Considering all the possibilities for coordination systems between center and subscribers, it is generally necessary to allocate at least one data block per input line plus some number of blocks in reserve against the possibility of input data surges and/or delays in the input program's servicing of the line.   Whether the reserve data blocks should be permanently assigned to each input line or in a common pool is a question of computer mechanics.   The data blocks normally have to be addressable by the electronic scanner (see Fig. 20), and in certain designs, this tends to restrict the system to fixed allocation of memory.   In most computer systems, however, it is possible to assign data blocks to input lines at random from a common pool of blocks, and in accordance with fundamental statistical principles, this is usually a far more efficient method.

   The memory requirement (in terms of data blocks) is susceptible to analysis as a queueing problem.   Given estimates of average access time to the intermediate store (or average processing time per data block, if longer), average input data rate, some assumptions on their statistical properties, plus certain other system parameters, it is possible to compute the mean and variance of the number of data blocks in use by input lines.   Under certain conditions, this quantity will be normally distributed; hence, knowing the mean and variance, an upper "3-sigma" limit is easily calculated. If the system were then designed to provide an amount of memory equivalent to this limit (3-sigma) for input data accumulation, it would operate so that, with 99.8 per cent confidence, memory would always be available for inbound data.   Use of the 3-sigma limit is, naturally, illustrative.   The confidence level should be a primary system parameter based upon analysis of requirements, not an arbitrary standard.

   A queueing theory model will be presented which can be used to implement such an analysis of input data block requirements.   A different type of model must be used to estimate the memory requirement for output data blocks.   The total memory requirement is obtained by summing the separate requirements for input and output data blocks, each arrived at by application of an appropriate confidence level.

   While such models are useful in estimating high-speed memory requirements for input and output data buffering, a number of factors tend to limit their usefulness in practical situations.   Although the analytical models are quite sophisticated in themselves, they do not consider all aspects of the physical situation whose behavior is to be predicted.   The analyst must, accordingly, be very familiar with the derivation of the model and adjust the results, perhaps intuitively, to reconcile significant differences between the assumed characteristics of the model and his estimate of the true characteristics of the system.   Digital-computer simulations can overcome this difficulty, but they are very costly to develop.   More significant, perhaps, is the fact that the memory of the central computer must be chosen to hold both programs and the input-output buffer whose requirements are being discussed. No matter how precisely the input-output buffer can be estimated, it must share memory with a program whose size cannot generally be predicted better than plus or minus 10 per cent. Furthermore, computer memories normally come in a restricted range of sizes, e.g., 40,000, 80,000, 120,000, 160,000 characters in the RCA 3301. The best the designer can usually do is to estimate the major memory requirements (input-output buffer, programs, tables, etc.) and choose the size memory which accommodates these requirements with a safety factor.

d. Queueing Model For Input Data Block Estimate.* Assume a system having N input lines whose traffic characteristics are identical. Suppose that the program is so designed that it will allow a queue of full data blocks to accumulate for each input line, assuming blocks available, until serviced. Assume that the program inspects the queues of each input line cyclically and transfers all data blocks found in each queue to the intermediate store. The number of data blocks found in each queue is a random variable whose mean and variance can be computed by use of equations which follow. If we sum this random variable over all the input channels, we obtain a new random variable representing the total number of data blocks found to be occupied by the program during a complete cycle of the input lines. For a sufficiently large number of input lines, this latter random variable tends to be normally distributed; hence its upper "3-sigma" limit is known, etc. (This model is not completely realistic, but the discrepancies tend to compensate.)

Parameters and Variables:

N = number of input lines

μ = average rate at which program (or intermediate store, whichever is limiting) can accept data blocks, blocks/sec

λ = average rate at which data blocks arrive on each individual input line, blocks/sec

R = utilization factor of input system

T = average length of input cycle, sec (average time between successive inspections of queue for any input line)

s = service time per data blocks, a random variable

$\bar{s} = 1/\mu$ = average service time per data block, sec

$\delta^2(s)$ = variance in service time per data block, sec²

$\overline{s^2} = E(s^2) = \delta^2(s) + (\bar{s})^2$

n = number of data blocks queued, a random variable

$\bar{n}$ = average number of data blocks queued, a random variable

$\delta^2(n)$ = variance in number of data blocks queued, a random variable

w = walking time per input line: the time required to inspect the queue of each line regardless of how large it is, a random variable

$\bar{w}$ = average walking time per input line, sec

$\delta^2(w)$ = variance in walking time per input line, sec²

Equations:

$$R = N \frac{\lambda}{\mu} \tag{21}$$

$$\bar{n} = \frac{N\lambda\bar{w}}{1 - R} \quad (22a) \qquad T = N \left(\frac{\bar{n}}{\mu} + \bar{w} \right) \tag{22b}$$

$$\delta^2(n) = \frac{N\lambda^2(\bar{s}^2\bar{n} + \delta^2(w))}{1 - N(\lambda/\mu)} + \bar{n} \tag{23}$$

$$\text{Memory required} = N\bar{n} + K\sqrt{N\delta^2(n)} \qquad \text{data blocks} \tag{24}$$

corresponding to the mean plus k-sigma limit of a normally distributed random variable.

* Based on a model constructed by M. A. Leibowitz, *IBM J. Res.*, July, 1961, and October, 1962. See Refs. 23 and 24.

The model will be explained by an example.   Suppose the following characteristics of the system are assumed:

N = 100 input line
λ = 1.5 blocks/sec average arrival rate per input line
μ = 200 blocks/sec average data transfer rate to the intermediate store

$\bar{s} = \dfrac{1}{\mu} = 5$ ms/block * $\delta^2(s) = 42.7$ ms² *

$\bar{s}^2 = 67.7$ ms² * $\bar{w} = 2.5$ ms/input line *

$\delta^2(w) = \delta^2(s) = 42.7$ ms (assumed arbitrarily)

The utilization ratio is [Eq. (21)]

$$R = 100\, \frac{1.5}{200} = 0.75$$

This is a check.   If $R \to 1$, Eq. (22a) shows the queue length increasing without bounds.   The system will not work under such conditions.   If this were the case, a faster process must be selected or the load reduced.

The average queue size can be calculated from Eq. (22a).   It is

$$\bar{n} = \frac{(100)(1.5)(2.5 \times 10^{-3})}{1 - 0.75} = 1.5 \text{ blocks/line}$$

The variance is [Eq. (23)]:

$$\delta^2(n) = \frac{(100)(1.5)^2[(67.7 \times 10^{-6})(1.5) + 67.7 \times 10^{-6}]}{1 - 0.75(1.5/200)} + 1.5$$

The memory requirement for 3-sigma confidence is [Eq. (24)]

$$M = 100(1.5) + 3\sqrt{100(1.54)} = 187 \text{ blocks}$$

The cycle time [Eq. (22b)] is

$$T = 100\left(\frac{1.5}{200} + 2.5 \times 10^{-3}\right) = 1.0 \text{ sec}$$

e. Probabilistic Model for Output Data Block Estimate.   Suppose we consider the memory requirements for output data blocks for the same system whose input was estimated just preceding:

N = 100 lines
L = 1.5 blocks/sec, average output load per line (must equal input for steady-state system)

Suppose each output line has a transmission capability of

$$r = 2.0 \text{ blocks/sec transmission capability when in use}$$

The program attempts to keep the output lines occupied whenever blacklogs are present.   To do this it attempts to maintain a small queue of data blocks in memory linked to each output line through the electronic scanner.   The program scans these queues periodically, fetching new data blocks from the intermediate store and returning the memory occupied by completely transmitted blocks to a common pool.   If this scan occurs roughly every second, then assuming that a backlog exists, the program must keep at least two data blocks available for each output line to assure continuous use of the line.   A simple program doctrine might be to count the number of completely untransmitted data blocks in the queue for each line and add more blocks in accordance with Table 6:

* These are characteristics of the following process: Average message length = 20 blocks; processing time for first block = 33.3 ms; processing time for second to eighteenth block = 3.33 ms; processing time for last to nineteenth block = 6.66 ms.

Table 6

No. of completely untransmitted blocks in queue	Action required, assuming backlog permits it
0	Add 2 blocks
1	Add 1 block
2	Add 0 blocks
3	Add 0 blocks

Because the scan occurs randomly with respect to transmission of each data block, for the cases where 0, 1, or 2 completely untransmitted blocks are observed, there is almost certainly an additional data block in memory, partially untransmitted.

Neglecting effects at the beginning and end of transmissions, which are small for reasonably long messages (20 blocks average) and trivial for continuous strings of messages, the conditional probability that an output line requires three data blocks is 1. Since the probability that a line is in use is 1.5/2.0 = 0.75, the random variable N = number of blocks of memory occupied by an output line has the following characteristics:

$$P(N = 3) = 0.75$$
$$P(N = 0) = 0.25$$
$$\text{Mean value of } N = 2.25$$
$$\text{Variance of } N = 1.75$$

For 100 lines, with 3-sigma confidence the memory requirement is

$$M = 100(2.25) + 3\sqrt{100(1.75)} = 265 \text{ blocks}$$

f. Step 2: Estimating Size of Intermediate Store. The amount of intermediate storage required can be estimated through use of a queueing model* which will be presented below. In this model, an individual queue of messages is assumed for each output line and its mean and variance computed. The total storage requirement is computed by summing mean and variance over all the output lines. If the number of lines is large and the load per line reasonably uniform, the total storage requirement will be normally distributed and, knowing its mean and variance, an upper limit can be computed based upon a specified confidence level.

This model assumes that no part of the memory required to store a message can be released until the message is completely transmitted. The model is very realistic for a system having only single-address messages. For a system with multiple-address message traffic, the model is slightly conservative, as it counts multiple-transmitted messages as multiple-stored when in fact they are stored but once. As devices used for intermediate storage are normally furnished in discrete sizes, the estimate must only be good enough to permit the proper size unit to be chosen. In applications where the intermediate storage device serves other purposes (program storage, tables, files), the total requirement must, of course, be estimated.

g. Parameters and Variables.

$E(t)$ = mean time between arrivals of messages to be transmitted over a particular output line, sec
$E(s)$ = mean duration of a message transmitted over a particular output line, sec
u = message length in data blocks, a random variable
$E(u)$ = average message length, data blocks
$E(u^2)$ = second moment of message length, (data blocks)^2
$E(u^3)$ = third moment of message length, (data blocks)^3
r = transmission rate of particular output line, data blocks per second
N = number of output lines
Z = number of data blocks of storage required, a random variable
$E(z)$ = mean number of data blocks of storage required
$\delta^2(z)$ = variance in number of data blocks of storage required
Note: $E(u) = rE(s)$.

* Based on work by Plotkin, Einhorn, and Weisgold.[30]

Equations:

$$R = \frac{E(s)}{E(t)} \tag{25}$$

$$E(z) = \sum_{j=1}^{N} \left\{ E(u) + \frac{E(u^2)}{2\left[\frac{E(t}{E(s))} - 1\right]E(u)} + R\left[1 + \frac{E(u^2)}{E(u)}\right] \right\}_j \tag{26}$$

$$\delta^2(z) = \sum_{j=1}^{n} \left(\frac{E(u^3)}{3E(u)\left[\frac{E(t)}{E(s)} - 1\right]} + \frac{1}{4}\left\{ \frac{E(u^2)}{E(u)\left[\frac{E(t)}{E(s)} - 1\right]} \right\}^2 \right.$$

$$+ E(u^2) - [E(u)]^2 + \frac{R}{3E(u)}[2E(u^3) + 3E(u^2) + E(u)]$$

$$\left. - 2\left\{ \frac{R}{E(u)}[E(u^2) + E(u)] \right\}^2 \right)_j \tag{27}$$

Memory required $= E(z) + k\sqrt{\delta^2(z)}$, blocks for K-sigma confidence level   (28)

Use of the model will be illustrated by an example involving the same system discussed in paragraph 22*b*.

N = 100 lines, identical in characteristics and load
r = line capacity, two data blocks per second
L = average line load, 1.5 data blocks per second
$E(u)$ = average message length, 20 data blocks

Assume message length normally distributed with $\mu = 20$, $\sigma = 6$; whence

$$E(u^2) = 436$$
$$E(u^3) = 10{,}160$$

From the line capacity and load, we obtain [Eq. (25)] the service ratio

$$R = \frac{1.5}{2} = 0.75$$

This is a check. R must be less than 1 for a stable system.   As all lines are identical, from (26),

$$E(z) = 100\left[20 + \frac{436}{2(20)(1.33 - 1)} + 0.75(1 + {}^{436}\!/_{20}) \right]$$
$$= 7{,}000 \text{ blocks}$$

From (27):

$$\delta^2(z) = 100\left\{ \frac{10160}{3(20)(1.33 - 1)} + \frac{1}{4}\left[\frac{436}{20(1.33 - 1)} \right]^2 + 436 - (20)^2 \right.$$

$$+ \frac{0.75}{3(20)}[2(10160) + 3(436) + 20]$$

$$\left. - 2\left[\frac{0.75}{20}(436 + 20) \right]^2 \right\}$$

$$= 1{,}372{,}000$$

For 3 = sigma confidence, memory requirements are [Eq. (28)]

$$M = 7{,}000 + 3\sqrt{1{,}372{,}000} = 10{,}500 \text{ blocks}$$

23. Record and Retrieval Storage

In most applications there is generally a requirement that a semipermanent copy be kept of every message passing through the center. Some typical reasons follow:

1. To maintain historical record of traffic
2. To provide data for accounting purposes
3. To protect messages against loss due to switching-center failure
4. To provide data for retrieval or demand

In a digital-computer switching center, record storage is generally accomplished on magnetic tape or in a disk file. (Another type of random-access device, utilizing magnetic cards, is useful in this application. Its primary advantage over the disk is that the storage element is inexpensive and easily removable. It has, however, a much larger access time.) Of the two types of devices, magnetic tape is generally favored for reasons of economy and simplicity. In applications where retrieval of messages in record store must be accomplished in a few seconds, the disk file is required.

Since the record store must, generally, hold a copy of all data passing through the switching center, the data-transfer capability of the record storage device has a simple relation to the traffic load of the center. In most switching-center applications the traffic load, summed over all communications lines, falls in the range of 100 to 10,000 characters per second input plus an equal or higher output load. (The output load generally exceeds the input load owing to multiple addressing, editing, translation, addition of message-identification data, etc.) Magnetic-tape stations with recording rate of 10,000 to 170,000 characters per second and disk files with rates of 25,000 to 75,000 characters per second are commercially available; hence there is no fundamental system limit on traffic load induced by the data rate of the record storage device.

a. Magnetic-tape Record Storage. In choosing a configuration of magnetic-tape stations for a record store, an analysis of data rates is first made to assure device traffic capability, and then the proper number of tape stations is selected to handle the required data rates. A number of factors should be evaluated including the following:

1. Input data rate (peak hourly average plus a safety factor)
2. Output data rate (peak hourly average plus a safety factor)
3. Need for input, output, or both types of traffic to be recorded (copy of each message, properly tagged)
4. Need for separate record other than tagged copy of message
5. Start-stop time of tape station (generally 5 msec nominal)
6. Average size data block to be recorded on tape. (This block may consist of portions of several different messages, each properly tagged.)
7. Central computer capability for simultaneous operation of peripheral devices. (As data transfers must occur between computer core memory and both intermediate storage and record storage, it is important to know whether these transfers can occur in parallel or sequentially.)

The following expression summarizes all of these factors:

R_{min} = minimum acceptable data rate for a record tape, characters per second
D_i = input data load (messages plus tags), characters per second
D_o = output data load (messages plus tags), characters per second
B = average block size on tape, characters
S = start-stop time per tape block, msec
T = allowable elapsed time to write record tape, msec/sec

where $T = 1,000 - T_i - T_x$
T = time, msec/sec, for use of intermediate store, if record store cannot be accessed simultaneously

Otherwise
$$T_i = 0$$

T_x = similar to T_i, but due to other causes, if any

$$R_{\min} \geqq \frac{1,000(D_i + D_o)B}{BT - (D_i + D_o)S} \qquad \text{characters per second}$$

(Note that the significance of the denominator going to zero is that all the available time T is occupied in starting and stopping the tape; hence an infinite data rate is required.)

The quantity of tape stations required depends on several factors. Considering only the pure record-keeping function, there is a minimum requirement for two stations: one in operations, the other standing by with a fresh tape. There may, however, be other functions closely associated with record keeping which require additional tape stations. If there is any sort of inquiry and/or retrieval functional requirement, the system must have a separate group of tape stations, usually two to four in number, to allow tape searching, sorting, reassembly of messages which may have been scattered on the tape, etc. Furthermore, the fact that a system has a requirement for retrieval from record store means that the record tapes will usually be short and, accordingly, the tape-handling problem increases, requiring additional tape stations for security. (A typical 2,400-ft tape reel may hold an hour of traffic if used to full capacity. If, however, a retrieval request is made involving the record tape currently in use, then the tape must be immediately terminated and rewound, no matter how much capacity is yet unused, and the standby tape called into service. Thus, the tape-handling rate is determined by the capacity of the tape.)

As tape stations require substantial periodic maintenance, the system complement must include a certain number of spares. Tape stations may also be required for functions not directly associated with record stores. Examples include:

1. Overflow storage for the primary intermediate storage device
2. Intercept storage for messages destined for terminals known to be out of service

Analysis of tape-handling problems and spares requirements must, naturally, include consideration of all tape usage.

b. Disk File Record Storage. In choosing a disk file for record storage, a different set of problems must be solved. While disk files have roughly the same data-transfer rates as tape stations, there is a major difference in access time. When tape stations are used for record storage, no attempt is made to use the tape as a random-access device. Messages are simply recorded sequentially and usually interlaced with other messages. Consequently, the only access time is the transport start-stop delay, generally about 5 msec. With such an approach, of course, messages can be retrieved only by a tape searching operation which usually requires manual tape handling and, accordingly, takes on the order of seconds to minutes to accomplish. When a disk file is to be used, it is usually to provide random access to any one of a large collection of messages with a delay of less than 1 sec. In order to permit such retrieval, the messages must be copied into the file, individual data blocks being stored in specific locations reserved for each message, as they pass through the system. This means that the time for head motion and rotational latency of the disk file, generally 75 to 200 msec, must be charged to every individual message data block. Considering that message data blocks typically range from 30 to 200 characters, it is obvious that the data-transfer time, being on the order of 1 to 10 msec, is negligible for estimating purposes when compared with the access time. Furthermore, in applications where a permanent copy of each message must be kept for a period of several days, it may be necessary to make a magnetic-tape record of the message in addition to the disk file record.

There are many programming techniques available to minimize disk file access time, but their applicability must be investigated in each particular case. (A system using a disk file because of a retrieval requirement would probably use the same file for intermediate storage.) Disk files have a broader use in real-time systems than as a message store. They are used, quite importantly, in real-time inventory control applications, a prime example being air-line reservation systems. In such an application the messages may not move to and from the disk file, but as they each cause one or more file accesses, the timing problem remains much the same. Thus, the important limit on system capability posed by disk files is the average access time after giving due consideration to applicable techniques for minimizing this time.

24. The Central Computer

a. General: Problems of Choice. The central computer executes the program which operates the switching center, controls the peripheral devices, and transfers message data between the communications lines and the various storage devices in the center. Depending on the requirements of the application, it may range in size from a small single data processor to a complex configuration of large-scale machines. There is no easy way to relate the load of the switching center to its requirements for central computing equipment. This can be done only by a relatively detailed analysis of the problem followed by either a program synthesis from which running time and memory estimates may be made or else comparison with another presumably operating system whose requirements are such as to permit reasonable projections to be made. The end result of such an analysis is usually selection of one computer from a limited choice rather than specification of a machine to meet requirements exactly. This tends to minimize the need for a very accurate program synthesis.

While selection of a computer which is reasonably well fitted to the task of handling a given switching-center load is far from easy, it does submit to a systematic approach. The first, and simplest, step is to choose a computer either hypothetically or from the product line of a manufacturer and determine if it has appropriate basic mechanical capability. Assuming that the chosen machine passes this test, a program running-time analysis must be made to a depth sufficient to determine whether the system will be peripheral device limited or computing time limited. In the former case the analysis is ended. In the latter case, it must continue until a reasonably good estimate can be made of the machine's throughput. The third step is to estimate memory requirements at least to the point of being able to select the proper number of memory modules. The final step is an evaluation of time vs. memory trade-offs with a view toward reducing the cost of the system. This cycle may be repeated on several different computers (or possibly different configurations of computers), and the results compared as to cost, amount of hardware (affecting reliability), reserve capacity, and confidence levels.

b. Mechanical Capability of Computers. Most, if not all, modern stored-program computers have the basic capability to perform reasonably well in a switching center. This is, however, a strictly qualitative statement. Unless the center load is trivial, analysis will reveal that a computer with certain features is inherently more efficient than another which lacks them. However, it is unreasonable to assume that such an analysis would show that two general-purpose computers of roughly the same cost and general capability are so different in performance that one of them is acceptable while the other is useless.

While an analysis of basic machine features is unlikely to permit an obviously best choice to be made among computers of roughly the same cost and capability, it will probably be helpful in rejecting machines that are clearly inadequate and in determining whether one of two machines, otherwise equivalent, has a marginal advantage. What follows is a brief description of certain computer features which are of particular interest in switching-center applications.

1. *Memory Speed and Size.* The cycle time of the memory is a primary restraint on the speed of the computer. As memory requirements for real-time applications tend to be very large when compared with offline scientific or business applications, the ability of the computer to address enough memory must be checked, at least against a first-order memory requirements estimate. Such an evaluation must, however, con-

sider true program storage capability, not just the maximum addressible number of bits or words.

2. *Peripheral Device Handling Ability.*   The peripheral device requirements (tapes, disks, drums, printers) for a switching center can be specified from a traffic load estimate as indicated in Paragraphs 22 and 23.   It is usually necessary to plan that many of these devices will be operating simultaneously, and it is, therefore, essential to determine whether or not the computer can do this.

3. *Program-interrupt Ability.*   Many situations occur in operation of a center where it is desirable to allow a signal, external to the computer, to interrupt a program currently being executed so that the computer can jump to an appropriate routine.   If any one feature is essential for efficient real-time computation, it is this one.

4: *Communications Interface.*   The interface between the computer and the communications lines is somewhat different from between the computer and its conventional peripheral devices.   In low traffic applications this is not very important, but in large ones it is a major factor.   This aspect of computer design will be discussed in detail in Paragraph 25.

5. *Character-handling Ability.*   A certain portion of processing time may be spent in analysis of narrative data in a message and perhaps in editing, translation, etc. While this type of problem can be handled by conventional fixed-word binary computers, a machine which can address characters directly will have an advantage which should be evaluated.

c. Analysis of Program Running Time and Memory Requirements.   There is no easy substitute for a thorough problem analysis followed by a program synthesis if the designer wishes to place any degree of confidence in the ability of a computer to handle a given switching-center load.   Such an analysis is unfortunately a large job, requiring perhaps as much as a man-year of effort by senior programming analysts and system engineers in typical situations.   While such an effort, with its cost and time implications, might be reasonable if it led directly into the detailed development stage of a switching center, there are many situations where it is not practical to do such a complete analysis.

As an example, consider the job of evaluating proposals to furnish a system from several manufacturers, each offering a complete system, particularly if the proposed configurations differ greatly.   It is, therefore, desirable to find methods for analyzing program running time and memory requirements which, while less accurate, are quick and safe.   What follows in this section is a set of guidelines which may be useful toward this end.   They are based on actual experience with digital-computer switching centers handling primarily narrative message traffic.   Figures presented are necessarily rough and should be used only in situations where a direct analysis has at least established their appropriateness.

The usual problem with regard to program running time is to estimate the throughput of the system in messages per unit time.   A first step toward solving this problem is to study the communications interface and peripheral device configuration and establish what percentage of the *central computer's time* is occupied with the purely mechanical job of moving the data through the system at some given message rate. By "purely mechanical" is meant the actual execution of data transfers between computer memory and tape stations, disk files, etc.   Unless the switching-center functions are very simple, it is probably best to assume that a computer cannot give up more than one-third of its time for mechanical input-output tasks and still do the necessary data processing.   The ability of the intermediate and record storage systems to handle a given message rate can be tested by the methods outlined in Paragraphs 22 and 23.   A good rule to follow is that the utilization ratio of these stores should not exceed 0.75 under the peak hour load.

Beyond the question of input-output capability remains the question of computing time limitations.   Generally a message requires a great deal computing time for header analysis and very little for the remainder unless editing translation, or data processing based on message content is required.   If we consider a message to be composed of a header block, a number of text blocks, and and end-message block, Table 7 indicates rough order-of-magnitude indications of the computing time requried per

Table 7. Computing Time Requirements

Portion of message	Input processing, msec	Output processing, msec	Notes
Header blocks.........	25–100	5–10	For all
Text blocks...........	2–5	2–5	Each
End message block.....	5–15	5–10	Only one

block for a center handling narrative message traffic. These figures are reasonable for computers with character-handling ability and instruction fetch plus execution time in the range of 25 to 50 μsec. In particular, the time for text blocks is short because no extensive editing or analysis is assumed. Blocks are assumed 80 characters long.

From these figures it is obvious that the average time per block for relatively long messages (say 20 blocks or more) is close to the time per text blocks. Considering practical limits on accessibility to disk files and drums, it is quite likely that, in a system handling predominantly long messages, the average computing time per block will be less than the average access time per block; hence the system quite likely is not computer limited.

Memory requirements can be similarly treated. The requirement for input-output buffering can be established with reasonably good accuracy as indicated in Paragraph 22. The principal remaining memory requirements, namely program coding, tables, work spaces, and peripheral device buffers, can be established either by direct synthesis or by rough rules based on experience. Even if it is considered acceptable to use a very rough estimate of memory requirements, the designer will still have at least to outline the general layout of memory and prepare basic program flow charts. Under these conditions, it is usually a simple job to specify the major tables and buffers directly. The only major requirement that actually needs rules in order to be estimated roughly is the program coding.

Real-time operating programs tend to be very large when compared with programs used in off-line file maintenance, etc. The principal reason for this circumstance is that the program must handle all problems that arise immediately in a way that does not halt the system when exceptional conditions occur. (It is generally not practical to reject messages to an exception tape and let them accumulate there until the end of the day, etc.) In message switching systems, coding requirements tend to fall in the range of 1,000 to 100,000 instructions (assumed two-address instruction in computer with character-handling ability).

Some use can generally be made of the intermediate storage device to hold segments of the program that are not required very often. However, this does not permit saving all the memory requirement for such programs, as a read-in area in the main memory is still needed. This is an important time-vs.-memory trade-off usually made in the second round of system analysis.

25. The Communications Interface

a. General Discussion. The communications interface defines a combination of hardware and program functions which occupy a position between the communications lines and the central computer of the switching center. The exact form in which this interface is mechanized varies widely, depending on system requirements and available equipment. The value of equipment to implement the communications interface can equal the value of the central processor in practical situations; hence it is important to understand this portion of the switching-center requirements sufficiently to make a judicious choice of equipment.

A more detailed description of the communications interface is best begun by placing it within proper bounds.   On the line side, the division between common carrier and switching center is generally a d-c connection through which data may pass serial by bit, plus a few control lines.   Synchronization may be furnished by either carrier or center, depending on particular circumstances.   The carrier terminal equipment furnishes no storage in excess of the current data bit (or group of bits if the method of signalling employs parallel transmission of bits).   On the central computer side of the interface the data are usually well organized in fixed-length blocks.   They may be converted from the line code to a code more suitable for data processing.   They may have been subject to accuracy tests, and some exchange of control signals may have occurred between center and outlying terminal.

Within these bounds, the communications interface usually includes some, if not all, of the following functions:

1. *Buffering.*   Exchanging bits of information between the computer memory and the communications lines; commonly done in a hierarchy of bits, characters, memory words, data blocks, and messages.

2. *Code Conversion.*   Conversion between line code and internal computer code, normally character by character.

3. *Accuracy Control.*   Checking inbound transmissions for error; normally done on a character and data block basis; on outbound transmissions, generating bits on a character and/or data block basis for use by the outlying terminal.

4. *Channel Control Procedures.*   Acting in response to errors detected in inbound transmissions; generating control signals and exchanging them with the outlying terminal to accomplish coordination, synchronization, rejection of data blocks containing errors; establishing and maintaining synchronization with the outlying terminal.

5. *Computer Memory Allocation.*   Allocation of data blocks to each line as needed and release of these blocks to a common pool after processing.

There are three basic methods for implementing the communication interface in a digital-computer switching center.   The simplest is essentially as shown in the illustrative system of Fig. 20.   There is a minimum amount of hardware to collect bits from each inbound communications line into characters and deposit these characters in the computer memory.   Any analysis of the data required for accuracy control, coordination, code conversion, or signaling procedures is done by a program operating in the central computer.

Naturally such a program must split the time of the computer with the main operating program, and the characteristics of the switching-center load must permit this combination.   If the load is such that the simple time-sharing scheme just described cannot work, then there are two basic alternatives.   The simple approach is use of two separate computers to share the load of the communications interface and the main operating program.   This scheme has certain virtues when the reliability aspects of the center are considered.   It is often possible to connect the two computers so that the center can continue operation at a reduced rate if either machine fails.   The other alternative is use of special-purpose hardware for the communications interface.

Certain requirements of the communications interface, such as control character recognition and code conversion, are likely to place a substantial load on a computer. As an alternative to doubling the number of computers, it is usually possible to employ special hardware in the control logic for interconnecting the communications lines to the memory to perform code conversion, control character recognition, and the like. Such additions, usually less costly than a second computer, accomplish these tasks without taking time from the central computer.

26. Reliability

A digital-computer switching center is an element of a real-time system.   A failure which shuts down the center causes a loss of service to the user of the system.   To the extent that such losses are intolerable, the center must be provided with duplication of equipment and the system must have procedures which secure traffic and provide continuity of service during periods when equipment is down for repair.

Down time may occur as a result of preventive maintenance or catastrophic failures. In certain applications where it is not essential that the system furnish maximum service 24 hr/day, preventive maintenance can be scheduled during the off-peak period of the day without causing a real reduction in system service. In cases where this is not possible, enough extra equipment must be furnished to allow one of each type of unit to be down for preventive maintenance without shutting down the center. In either case, down time due to catastrophic failures is the major problem.

Given estimates of mean time between failures (MTBF) and mean repair time (MRT) for each unit of equipment in the system, it is possible to calculate various aspects of the reliability of the system. While such calculations can offer a guide to selection of an equipment configuration with redundancy, there remains a major problem of selecting a functional design that can utilize the equipment redundancy effectively. If a major element of the switching center fails, not only is service halted but traffic stored in the system may be lost. Switching in a replacement element will restore service, but this alone does nothing toward recovery of the lost traffic. If the application is such that lost traffic can be recovered at leisure, then this can be usually done by retrieval from the record tapes provided there is a practical accounting system. In cases where this approach is not satisfactory, a scheme must be provided for duplex processing and message storage so that a switching-center failure has virtually no effect on the traffic flow.

REFERENCES

1. Clos, C., A Study of Non-blocking Switching Network, *Bell System Tech. J.*, vol. 32 pp. 406–424, March, 1953.
2. *Dial Telephone Switching Circuits*, University of Pennsylvania, Moore School of Engineering, Philadelphia, Pennsylvania. ASTIA AD No. 54515.
3. Davidow, W. H., The Selection Problems for Minimal-state Sequential Circuits, *Stanford Electron. Labs. Tech. Rept.* 1901–2, Stanford University, Stanford, Calif., ASTIA AD No. 260782.
4. Keister, W., A. E. Ritchie, and S. H. Washburn, *The Design of Switching Circuits*, D. Van Nostrand Company, Inc., Princeton, N.J., 1951.
5. Caldwell, Samuel H., *Switching Circuits and Logical Design*, John Wiley & Sons, Inc., New York, 1958.
6. *Electronic Time Division Switching*, Electronetics Division, North Electric Company, Galion, Ohio, 1960.
7. Joel, Jr., A. E., An Experimental Switching System Using New Electronic Techniques, *Bell System Tech. J.*, vol. 37, no. 5, pp. 1091–1102, September, 1958.
8. Joel, Jr., A. E., Electronics in Telephone Switching Systems, *Bell System Tech. J.*, vol. 35, no. 5, pp. 991–1018, September, 1956.
9. *Switching Systems*, American Telephone and Telegraph Company, 195 Broadway, New York, 1961.
10. Wilkinson, R. I., The Interconnection of Telephone Systems, American Telephone and Telegraph Co., *Bell System Tech. J.*, vol. 10, pp. 531–564, October, 1931.
11. Andres, R. K., A Computer Switching Center for International Telegraph Communications, *RCA Engr.*, vol. 8, no. 6, April–May, 1963.
12. Beierle, J. D., Communications Switching and Buffering Networks, *Datamation*, vol. 9, no. 7, July, 1963.
13. Bemer, R. W., The American Standard Code for Information Exchange, *Datamation* (2 parts), vol. 9, nos. 8–9, August–September, 1963.
14. Buegler, R. J., Random Access File System Design, *Datamation*, vol. 9, no. 12, December, 1963.
15. Caplan, D. I., Data Processing for Communication Network Monitoring and Control *Proc. AFIPS Conf.*, vol. 22, 1962, FJCC.
16. Edwards, J. D., An Automatic Data Acquisition and Inquiry System Using Disc Files, *Commun. ACM*, vol. 6, no. 10, October, 1963.
17. Frank, W. L., W. H. Gardner, and G. L. Stock, Programming On-line Systems, *Datamation* (2 parts), vol. 9, nos. 5–6, May–June, 1963.
18. Genetta, T. L., J. F. Page, and J. L. Owings, Autodata, RCA's Automatic Message Switching System, *RCA Engr.*, vol. 5, no. 5, February–March, 1960.
19. Harrison, G., *Message Buffering in a Computer Switching Center*, EIEE Winter General Meeting, 1963, CP 63-276.

20. Head, R. V., Real-time Programming Specifications, *Commun. ACM*, vol. 6, no. 7, July, 1963.
21. Helman, D. R., E. E. Barrett, R. Hayum, and F. O. Williams, Design of ITT 525 "VADE" Real-time Processor, *Proc. AFIPS Conf.*, vol. 22, 1962, FJCC.
22. Kersey, J. R., *The Programmed Transmission Control*, IEEE Winter General Meeting, 1963, CP 63-287.
23. Leibowitz, M. A., An Approximate Method for Treating a Class of Multiqueue Problems, *IBM J. Res.*, July, 1961.
24. Leibowitz, M. A., A Note on Some Fundamental Parameters of Multiqueue Systems, *IBM J. Res.*, October, 1962.
25. Levy, W. A., and A. E. DiMond, Design of a Data Communications System, *RCA Engr.*, vol. 9, no. 4, December, 1963–January, 1964.
26. Lindley, D. V., The Theory of Queues with a Single Server, *Proc. Cambridge Phil. Soc.*, vol. 48, no. 2, 1952.
27. Morse, Philip M., *Queues, Inventories, and Maintenance*, John Wiley & Sons, Inc., New York, 1958.
28. Department of the Navy, *Data Communications Terminal Equipment Guide*, Navy Management Office, Data Processing Systems Division, Navevos P-2414, June, 1962.
29. Perlman, J. A., Data Collection for Business Information Processing, *Datamation*, vol. 9, no. 2, February, 1963.
30. Plotkin, Morris, *Computational Methods for Estimating Storage Capacity Requirements at Autodata Message Centers*, Auerbach Electronics Corp. unpublished report, Dec. 21, 1959.
31. Riordan, John, *Stochastic Service Systems*, John Wiley & Sons, Inc., New York, 1962.
32. Sanders, M., Telefile—A Case Study of an Online Savings Bank Application, *Commun. ACM*, vol. 6, no. 12, December, 1963.
33. Segal, R. J., and H. P. Guerber, Four Advanced Computers—Key to Air Force Digital Data Communication System, *Proc. 1961 Eastern Joint Computer Conf.*
34. Statland, N., and J. R. Hillegass, Random Access Storage Devices, *Datamation*, vol. 9, no. 12, December, 1963.
35. Wier, J. M., Digital Data Communications Techniques, *Proc. IRE*, January, 1961.
36. Wright, E. P. G., Basic Considerations in Calculating Storage for an Electronic Telegraph Switching Center, *Elec. Commun.*, Nov. 3, 1958.
37. Clos, Charles, and R. I. Wilkinson, Dialing Habits of Telephone Customers, *Bell System Tech. J.*, vol. 31, pp. 32–67, January, 1952.
38. Clos, Charles, An Aspect of Dialing Behavior of Subscribers and Its Effect on the Trunk Plan, *Bell System Tech. J.*, vol. 27, pp. 424–445, July, 1948.
39. Riordan, John, Delayed Calls Served in Random Order, *Bell SystemTech. J.*, January, 1953, pp. 100–119.
40. Erlang, A. K., Solution of Some Problems in the Theory of Probabilities of Significance in Automatic Telephone Exchanges, *P.O.E.E. J.*, vol. 10, p. 189, 1917; Calcul des probabilités et conversations téléphoniques, *Rev. Gen. Elec.*, vol. XVIII, Aug. 22, 1925, and vol. XX, Aug. 21, 1926; Application du calcul des probabilités en téléphone, *Ann. Postes, Télégraphes Téléphones*, vol. 7, July, 1925.
41. Molina, E. C., Application of the Theory of Probability to Telephone Trunking Problems, *Bell System Tech. J.*, vol. 6, July, 1927.
42. Pollaczek, F., Various articles including those found in *Math. Z.*, vol. 32, pp. 64–100, 729–750, 1930; *Elec. Nachr. tech.*, June and July, 1931; *Telegraphen Fernsprech-Tech.*, 1930, pp. 71–78.
43. Wilkinson, R. I., Working Curves for Delayed Exponential Calls Served in Random Order, *Bell System Tech. J.*, March, 1953, pp. 360–383.
44. Crommelin, C. D., Delay Probability Formulae, When the Holding Times Are Constant, *P.O.E.E. J.* vol. 25, p. 41, 1932; Delay Probability Formulae, *P.O.E.E. J.* vol. 26, p. 266, 1933.

Chapter 8

TELEGRAPH SWITCHING METHODS AND EQUIPMENT

J. Z. MILLAR, *The Western Union Telegraph Company*

ELEMENTS OF TELEGRAPH SWITCHING

1. General

Telegraph switching has for its principal purpose either the more timely arrival of telegraph information at destination or the more efficient utilization of equipment and facilities to effect a reasonably satisfactory arrival of transmitted information while at the same time achieving economy of manpower and facilities and hence lower costs of operation.   For example, if only one transmission path is available between two points, the arrival time of one bit of information is a function only of the electrical delay of the transmission medium; whereas, if a second path is available via a third point and this second path is not fully loaded, then by the application of switching techniques it becomes possible to deliver two bits of information per unit of time (for a portion of the period) and also to deliver these two bits of information at less total cost than if the three points could only transmit to one another.   Thus it is established that *system loading* and transmission *channel availability* are the two factors that have the greatest influence on system design.

2. Historical

The word telegraph derives from the Greek *tele*, meaning "at a distance," and *graph*, which implies that graphical or written forms of information are to be recorded in contrast to the production of audible sounds as in the telephone.   In recent years data transmission (Chap. 12) by digital methods has still further increased the scope of telegraphy.

Unlike telephonic communication, telegraphic communication permits the introduction of delay in transmission, which is used to advantage.   Availability of the calling person and called person simultaneously is not required.   Furthermore, delayed transmission is frequently used, even of short duration, to reduce the number of circuits required to handle a given load and to provide greater flexibility in handling traffic.

3. System Classification

Telegraph switching systems are classified in several different ways, which also describe characteristics of the systems.   A system is *manual* or *automatic* depending on whether the circuit connections are made by a person; or by equipment under control of switching characters in the message heading or by dial pulses preceding the message. A system is semiautomatic if a person takes an action, such as operating a pushbutton, and the equipment then switches according to a prearranged instruction or program.

Another classification of telegraph switching is *circuit switching*. In this type of switching the telegraph circuits are connected together in tandem, section by section, until a through path is established from the user at the point of origin to the user at the point of destination. Hence, user-to-user service is provided without delay once the connection is established.

Still another classification is *message switching*. When each telegraph message is switched, either in whole or in blocks of characters or bits, while being received or immediately after receipt, to an on-going circuit toward the point of destination, the process is known as message switching. In this form of transmission *store-and-forward* service is provided. While it is necessary to set up circuits at each switching center to accomplish this result, it is not necessary to hold circuits idle and in readiness to accept messages. The storage capacity is sufficient to permit holding messages for reasonable time intervals until circuits are clear, and this process makes for very high circuit loading factors, or circuit efficiency. It would not be unusual to find a loading factor of 85 per cent on the trunks of a message switching system having several centers, whereas to handle the same number of messages with the same average time of transit on a circuit switching network serving the same places, as many as four or five times the number of trunks would be needed and the loading factor might drop to as low as 20 per cent.

Also, operators receive messages by telegraph equipment, not voice, and therefore start and end of message, all destination information, and other necessary information are added to the message (Fig. 3). When the switching is automatic, the recognition of this information as well as the operations to be performed in the switching center are programmed in the equipment.

Frequently, a combination of circuit switching and message switching, or a hybrid arrangement, is employed to meet requirements and thus reduce the total amount of switching equipment needed. Also, a combination of automatic switching through the first center and manual switching through the center nearest to destination is a useful arrangement, since it avoids having large-size directories at all terminal stations. Only a local directory is required, and it can be kept up to date without the necessity of changing all the other directories in the system.

It should be noted that the form of switching employed is independent of the form of transmission or code.

4. Chronology of Telegraph Switching

The electrical telegraph was conceived by Samuel F. B. Morse in 1832, it was demonstrated publicly in 1838, and the first telegraph message was transmitted on May 24, 1844. Alexander Graham Bell conceived telephony in 1874, and the first sentence of speech was transmitted on March 10, 1876.

The Western Union Telegraph Company was incorporated in 1851, and while its principal business was to accept, transmit, and deliver public messages, the furnishing of private-wire service became an important part of telegraphy in 1869. The American Telephone and Telegraph Company in 1887 constructed a private-line telegraph circuit between New York and Philadelphia. The equipment on this circuit was arranged so that the customer, a broker, could switch from telegraph to voice, and thus was established the first *alternate* record-voice switching service.

Furnishing of both telephone and telegraph private-line services soon became an important part of public service, and switching of circuits was employed to meet the varied needs of customers. Thus switching of telegraph and telephone facilities became a standard practice, and since a close technical relationship has existed between these two forms of communication, it is only natural that a switching principle developed for one is soon applied to the other.

METHODS

5. Manual Switching

An early form of telegraph switching was accomplished at telegraph test and service boards or at testing and regulating switchboards. When only local circuits are terminated, the switchboard is known as a "loop" board. Attendants manually establish the connections to provide a particular arrangement of trunks, repeaters if required, and locals. This form of switching is as old as telegraphy, and it should be noted that circuit switching is the functional method used.

Line-terminal switchboards vary in size depending upon the size of the central office and the number of outlying service points, but telegraph switchboards are not

Table 1. Test Procedures

D-c-operated Lines

Function	*Method*
To measure line current and to locate faults such as open line, foreign ground, cross with other lines, etc.	Use of combination volt-milliammeter to measure line current and in conjunction with records describing circuit constants (resistance-capacitance, etc.) to determine location of fault. Also, use of Wheatstone bridge to determine location of resistive type faults
To measure bias, characteristic, end, and fortuitous distortion	Use of bias and distortion meter on line signals transmitted from near or distant end. Use of cathode-ray-type oscilloscopes to examine waveforms. Use of monitor printers to detect presence of impulse noise and improperly formed control signal sequences

Carrier-operated Circuits

To measure transmission on sending and receiving local leg circuits	Same methods as used for d-c-operated lines
To measure carrier telegraph terminal-equipment performance	Use of standard frequency generator and db meter to adjust transmitted marking and spacing channel frequencies and level (or frequency and level on AM systems) and to adjust receiver for proper operation
To measure trunk channel characteristics	Use of db meter and standard frequency generator to determine signal levels at various frequencies in band and to measure rms noise level. Use of delay distortion meter to determine envelope delay of band
To measure overall system performance	Use of telegraph signal pattern generator to transmit pulses into each channel and use of an oscilloscope and/or bias and distortion meter to check margin on each channel receiver

usually multiple boards; that is, one telegraph line appears only at one position, and therefore transfers between operators are made by cord sets and transfer circuit arrangements. Some switchboards are arranged so that the normal assignments of circuits do not require the use of a plug and cord set. These are called *cordless* switchboards, but abnormal or emergency connections are made with cord sets.

Technical attendants who are trained in testing and regulating of circuits are utilized for this work, and the switchboard is therefore equipped with meters and other test equipment in order to provide a quick means of testing lines and apparatus and to ensure that high standards of operation are maintained (see Table 1).

Technical attendants communicate with one another by means of American or International Morse code, by telegraph printers, or by voice in telephone offices. However, in telegraphy only technical operators are employed in the interconnecting of trunk and local circuits, since a need never developed to separate the operating function from the testing function as in the telephone industry, which has employed large numbers of local and long-distance telephone operators.

Manual circuit switching fills a vital role in the handling of press for sporting events, financial reports, and general news; for government reports; and for private-wire customers. The system load dispatcher controls the setting up of connections according to "bids" submitted by users. Today a major special event such as a prize fight, a World Series game, or a political convention requires a large number of manually switched circuits to be put up and taken down. Also, many multipoint networks are needed for certain special events involving scores of drops where reception of a single transmission takes place.

Manual switching is also used to set up "round-robin" networks in which a multiple number of drops may both transmit and receive; that is, when one point is transmitting, all other points receive, and then another point will transmit, again to all the others.

An early form of telegraph switchboard for the manual switching of circuits is the *concentrator*. A concentrator is an arrangement of apparatus by which a minimum number of central-office sets and telegraph printer operators can serve a maximum number of local teletypewriter circuits. The most usual size of concentrator serves 100 lines. If these lines are branch-office lines, as many as 50 operating positions are required. If more lightly loaded lines are connected, for example lines to customers, then as few as 8 operating positions are used. Each operating position is equipped with a teletypewriter and a switchboard turret where all the lines appear, this being the only example of multiple-switchboard use in telegraph plant. Alternatively, some installations use perforators and tape transmitters for heavy outgoing traffic or receive-only printers for heavy incoming traffic. A lamp cabinet is provided in which signal lights show the incoming calls and the operator assignment status. In this way operators are assigned or relieved depending upon load conditions. Also, if an outgoing message is received, it can be handed to the operator connected to that circuit if it is sending to the central office at the time. An arrangement is provided for "double-decking" teletypewriters (two per operator) to accommodate circuits having a heavy incoming load.

Manual switching is employed in military field operations where specially designed telegraph switchboards are employed. The Signal Corps BD-100 telegraph switchboard, perfected in World War II, has a 10-line capacity.

6. Impact of Printing Telegraphy

The first telegraph printer was invented by Royal E. House in 1846 and placed in service three years later, but this was not a start-stop printer.[3]

Between 1855 and 1863 the Hughes Printing Telegraph was developed and subsequently widely used in Europe. This machine used a unison pulse for synchronization.[3,4]

In the early part of the twentieth century, J. C. Barclay, chief engineer of Western Union, invented a page printing system. These printers were first used in 1904, and by 1910 Western Union had manufactured and utilized 63 of these printers on 31,171 miles of circuit.

In 1869 Ludovic d'Arlincourt designed a facsimile system in which the receiving mechanism was driven at a slightly higher speed than the transmitter. At the end of each line the receiver was brought to a stop, following which it was restarted by a pulse received from the transmitter after completion of the cycle. This later became known as start-stop operation.[3]

Morton, Krum, and Kleinschmidt were early American inventors who perfected a line of improved printing telegraph apparatus, and the formation of the Morkrum Kleinschmidt Corporation (now the Teletype Corporation, a subsidiary of the Western Electric Co.) concentrated the talents of these inventors with the result that beginning about 1914 rapid growth in the use of the products of this firm was seen.[4]

One early system still used extensively commercially consists of a keyboard perforator which is a device to convert the alphanumeric-symbol characters arranged in upper and lower case on the keyboard to five rows of perforated holes in paper tape. The five-level Baudot code is employed, and therefore there are 32 combinations. Even though it was necessary to assign a few of these combinations to machine functions, enough remained to satisfy communication needs for the next 45 years. The perforated tape is fed through a *transmitter* where reading pins sense any holes found in the tape and set up contacts which are connected to the *distributor*. Because many of the early systems were synchronous systems, the function of the distributor was to convert the parallel Baudot pulses to serial pulses for line transmission. Time-division techniques were utilized in multichannel systems; hence these were called time-division *multiplex* systems. By substituting the segmented face plates of distributors, it was possible to operate one-, two-, three-, or four-channel systems over one circuit. Reception was accomplished on page *teleprinters*.

7. Teletypewriter Developments

While early printing telegraph equipments were suitable for use by trained commercial operators working constantly on heavily loaded channels, the growth of private-wire or private-line telegraph service brought about requirements that could be satisfied only by the development of new forms of equipment. Since such equipment would be used in customer offices and operated by employees of these firms, it became necessary to design specifically for such use. The new requirements included simplicity of operation, approaching that of a typewriter, and this led to the coining of the word *teletypewriter*.[2] Briefly, other new requirements were the ability to produce page copy, carbon copies, and stencils; positive feeding by sprocket holes and tabulating for form production; quiet operation; good appearance and small size; the ability to transmit from either keyboard or preprepared perforated tape; dependable operation with low maintenance; reasonably high speed; and the transmission of an accurately timed signal sequence for each character of such form that this sequence could be received accurately by receivers with large tolerance to signal distortion. Transmission requirements over circuits of varying length, multistation networks, and variable interference conditions led to the adoption of the start-stop self-phasing code. This consisted of 1 bit for the starting of the receiver selector, 5 bits for the five intelligence pulses, and 1.42 bits for the rest time to permit the selector to deliver the stored character to the printing mechanism and reset itself before the arrival of the next character. Thus the principle of character synchronism was introduced, circa 1917.

The availability of the start-stop teletypewriter gave impetus to the switching of teletypewriter channels, since for the first time all stations of a network could receive accurately under variable line conditions, all stations could either transmit or receive with minimum turnabout time, and a receiving station had the ability to "break" a sender (i.e., to interrupt sending) by means of an added relay for this purpose.

Another new technique was introduced in this period. The *tape teletypewriter* records the characters of an incoming message on a narrow strip of paper tape. The back of the tape is coated with a water-soluble cement, known as gum, and this permits the tape to be affixed to a message blank with a "hand gummer." Another version was a *tape-printer* for receiving-only service, but since these equipments were generally used on synchronous systems, the early models were arranged to accept the five intelligence pulses in parallel, and they were known as five-wire printers.

The first start-stop teletypewriters were ordered by Western Union in 1924 when 50 tape teletypewriters were purchased for experimental use. By 1927 Western Union had 6,500 tape teletypewriters in service. Teletypewriters were soon after adopted by the Bell System, Postal Telegraph, and RCA Communications, many of which were page-printing models.

TELETYPEWRITER SWITCHING

8. Teletypewriter Exchange Service (TWX)

In 1931, The American Telephone and Telegraph Co. introduced a new *manual circuit switching* telegraph service intended for short-period use by customers in the same manner that a leased private-line service would be used by a large-volume customer but with the important distinction that the customer can reach any other TWX subscriber, and therefore the service is essentially a public teletypewriter service.

The Teletypewriter Exchange System includes various forms of teletypewriter equipment at convenient locations on subscribers' premises, switchboards at telephone company offices, and interconnecting circuits both local and intercity trunk. A nationwide directory is issued periodically, which includes a classified section and lists all the subscribers with their respective TWX numbers. Service is available at any time of day or week, and the calling period for any connection is timed by means of clocks with a 3-min initial period the same as in telephone practice. Rates are based on distance zones and time of usage.

After a TWX connection is established, customers type (or send by tape) back and forth, sending messages and receiving replies. Under these circumstances a sender is certain that his communication has been received, since direct acknowledgment from a previously identified station is obtained. Unattended operation (without immediate acknowledgment) is available on calls for stations equipped for such service

Table 2. Growth of Teletypewriter Exchange Service

Year	TWX stations*	Year	TWX stations
1932	2,521	1948	23,423
1933	3,575	1949	25,607
1934	5,773	1950	28,393
1935	7,893	1951	30,815
1936	10,645	1952	33,338
1937	12,499	1953	35,272
1938	13,201	1954	36,672
1939	14,325	1955	38,954
1940	14,842	1956	41,628
1941	16,118	1957	44,715
1942	16,595	1958	47,485
1943	16,013	1959	51,631
1944	15,979	1960	54,734
1945	16,215	1961	57,943
1946	18,462	1962	58,530
1947	20,867		

* Bell System including southern New England and Cincinnati and Suburban.   (Information source American Telephone and Telegraph Co.)

and so listed in the directory.   Conference service is provided among several stations on request.

The TWX system has grown steadily and rapidly.   This growth is shown on Table 2 and illustrates the need for system design to consider the growth factor.

In November, 1962, TWX was converted to an automatic circuit switching system. Each customer's station is equipped with a dial unit, and switching plant at central offices is arranged for direct dialing connections. Simultaneously, 100-word-per-minute teletypewriter equipment was made available for those customers who desire this speed, but a few years will elapse before this becomes the universal speed. Therefore, translators at the switching centers have been provided to permit interoperation with stations equipped with the older 60-word-per-minute machines.

In August, 1963, TWX service was extended to Canada via the facilities of Canadian Bell.

Also, in 1963, the Type 33 teletypewriter and the Type 35 ASR (automatic send-receive) teletypewriter set became available, these models featuring four-row keyboards and utilizing the proposed seven-level data interchange international standard code (see Chap. 2).

ADDITIONAL EQUIPMENT DEVELOPMENTS

9. Reperforation Equipment

In addition to the keyboard teletypewriter, other start-stop machines have great usefulness with resulting impact on telegraph switching design. One of these is the *reperforator*, which automatically perforates paper tape as each character is received. Another machine performing the same function of perforating also types the characters on the tape. This is the *typing reperforator*. There are two models, one which types the characters immediately above the punched holes and which therefore requires a wider tape and the other which only partially perforates the tape, leaving each chad attached on one side so that typing can be located over the top row. This latter model is known as a "chadless" reperforator, and it can be obtained both with and without the typing feature. Chadless tape can be read successfully either mechanically or optically, it is less expensive because it is not so wide, and its use eliminates disposal of the chads. For these reasons, chadless tape equipment has become standard in most switching systems.

With the advent of reperforation, the message could be relayed by passing the perforated tape through a tape *transmitter*, or reader as it is sometimes now called. In early arrangements the transmitter was mounted alongside the reperforator, but this had the disadvantage that the message was delayed by whatever time was required for the tape to pass from the reperforator to the transmitter. Also, it was necessary to insert a large number of "blank" characters after the last message so that it could be read completely. To eliminate this situation, a new form of transmitter was developed, one that could be mounted on an arm located inside the reperforator housing and arranged so that it could read the last character perforated. The transmitter utilizes the feed holes in the tape to pull itself forward as far as the auto-stop position, it can be operated at a speed different from that of the reperforator, and it is equipped with a group of contacts to provide for reading certain characters for switching control purposes. The combination of "creeping" transmitter and perforator is known as a *reperforator-transmitter*. Notwithstanding its advantages this machine does not have widespread use.

In a teletypewriter a *distributor* is necessary to generate accurately timed signal sequences. It is either mechanical in nature, having contacts driven by a cam which in turn is clutch driven from a shaft geared to the motor, or it consists of a face plate and brush arm arranged so that the signal sequence is obtained from the commutator electrically, both types being mounted inside the teletypewriter mechanism. In switching systems it is useful to have the distributor mounted in the same housing as the transmitter, and this mechanism is known as a *transmitter-distributor*. This mechanism has been used extensively in telegraph switching systems for reasons to be developed later.

10. Automatic Send-Receive Sets

A development that improved outstation operation in telegraph switching is the combination of a keyboard teletypewriter and associated perforator and a transmitter-distributor. This combination is known as an automatic send-receive set, or ASR. A Teletype Model 19 ASR set enables an operator to prepare messages in the form of page copy alone, page copy and tape simultaneously, or tape alone. Tape may be prepared off-line at the same time that a message is being received. If it is desired to receive messages in both page and tape form, it is necessary to install an auxiliary reperforator.

On a Siemens-Halske ASR set, messages may be prepared in the form of page copy alone or page copy and tape. Since the tape-perforating mechanism is operated by the typing unit of the set, tape cannot be prepared off-line while a message is being received. The ability to receive incoming messages in both page and tape form is inherent, and no separate reperforator is required.

Additional operating features are available in later models of ASR sets, as, for example, the Teletype Models 29 and 35. When ASR sets are available at outstations, it is possible to automate switching systems to a larger extent.

11. Regenerative Repeaters

The application of the start-stop regenerative telegraph repeater throughout telegraph plant has done away with any practical limitation to the length or complexity of telegraph switching networks. Early forms of start-stop regenerative repeaters were mechanical, consisting of selectors and distributors driven by a motor. Today electronic regenerative repeaters are used almost universally, since they are stable, easy to maintain, and can be set up to operate at different speeds by simple switch settings.

12. Carrier Telegraph Improvements

Briefly, telegraph channels obtained by carrier methods involving frequency- or time-division techniques have been improved continually during the past 30 years. The availability of high-grade telegraph channels has made possible much more dependable switching networks. Frequency modulation of telegraph channel carrier frequencies, sometimes designated as FSK, or frequency shift keying, has been selected as the international standard. This method was developed by Western Union in 1938,[5] and it has been particularly valuable in military radio-teletype transmission, for example, the FGC-29 16-channel Radio-Teletype Terminal.* Phase modulation of telegraph channel carrier frequencies, with coherent detection, has been found to give excellent transmission on long radio circuits.†

TELEGRAPH SWITCHING SYSTEMS

13. Multistation or Way Circuits

The installation in 1917 of a start-stop page teletypewriter system‡ on a circuit from New York to Boston, with a number of intermediate stations, is believed to be the pioneer private-line multistation switching network.[2] While these stations exercised manual control over the use of the circuit, recent systems have been developed utilizing automatic control, or *selection*, of the way stations. Way stations may be arranged to operate duplex; that is, they can send while receiving from the switching center, or in simpler systems the way station can either send or receive. Way-station operation has the big advantage that for lightly loaded stations it is possible to lower the cost of circuits and in some cases it is possible to eliminate one or more switching centers in a system, thereby reducing switching-equipment cost.

14. Manual Reperforator Message Switching

The development of the reperforator brought about the concept of reperforator switching. A simplified version of reperforator switching was the first to be offered to private-wire customers. The establishment of larger leased networks to connect the offices and plants of industrial and government users created the need for switching facilities at central points of these systems to serve the same function as relay points in a manual relay system. Reception from local stations, way stations, or trunks takes place on typing reperforators. The tape transmitter-distributor is connected to a cord circuit, and the operator, after reading the destination of the message, inserts the plug into a jack which is associated with the circuit leading to the destination or to another switching center. Two jacks are provided so that, if the circuit is already in use, the second jack may be employed for making the second connection. Immediately after the connection is made or as soon as the circuit is released from a previously established

* Developed by Western Union.
† Developed by Collins Radio.
‡ American Telephone and Telegraph.

connection, a numbering machine transmits the next sequence number for that circuit, and then transmission starts. The same number is displayed on an indicator, and to complete the record for future reference, the operator "marks off" the received number of the message and writes in opposite to it the outgoing number. Each number, no matter how many times the message is switched, is carried along with the message to final destination, and this makes it possible to trace the transit of a message. The use of numbers has always been considered necessary to avoid lost messages. If an office is closed, or if the circuit is not functioning, the message is torn out and held until it can be transmitted. Otherwise the tape is stored on reels; hence this system is a *continuous-tape system.* It should be noted that the code pulses are completely regenerated each time a message passes through a transmitter-distributor. Therefore, there is no need for regenerative repeaters as in many large circuit switching systems.*

15. Cross-office Reperforator Message Switching

The handling of millions of short telegrams between points scattered throughout the country requires a complicated interchange of messages at all offices where trunk circuits meet other trunk circuits and local plant. In public-message telegraphy the average message is short and takes only a minute to transmit. Also, in a sequence of messages many different destinations are found. The real problem at each central office is to sort out these messages and route them toward their individual destinations without delay and without wasting time on the circuits. Many attempts were made to accomplish this desired goal through switching arrangements of the circuit switching type, but it has never been found economically practical to switch directly from a receiving terminal to an outgoing circuit. The outgoing circuits are so frequently busy that incoming messages would be held up. To avoid congestion on the incoming circuits and to allow distant stations to send at their convenience, it is necessary that incoming messages be moved away as they arrive and held in storage temporarily until they can be transmitted toward destination. Perforated paper tape has been found to be a convenient storage medium, and its use for this purpose has been widespread.

In a cross-office continuous-tape system[1] all receiving terminals are grouped together in one area and all transmitting terminals in another. Each incoming circuit terminates in a typing reperforator and alongside is the tape transmitter. No distributor is necessary because of the use of five-wire intraoffice cabling for the cross-office circuits. These circuits are connected to jacks in a turret associated with each group of receiving positions, and the transmitters are connected to plug and cord sets which have five code wires and four control wires.

As soon as a message is received, a switching clerk notes the destination and immediately connects the plug corresponding to that position to the appropriate jack. Transmission then starts automatically to the unattended sending section of the switching system, where the message is again reperforated and a second continuous tape loop is produced which is the storage for the outgoing circuit. In other words, the switching clerk switches the message, not to an outgoing circuit directly, but to a storage bin which in turn accumulates any messages to be transmitted to that circuit and transmits them as soon as possible. Since the cross-office circuits operate at least twice as fast as the incoming circuit, every incoming message is disposed of long before another message arrives. Without undue fatigue a clerk can switch approximately 200 messages per hour during peak-load periods, but during lightly loaded periods, a single clerk takes care of a whole isle of positions by moving around to switch any messages that appear. Messages are automatically numbered as switched, and incoming numbers are marked off as each message from a distant point is received.†

* A typical example of a private-wire switching system utilizing the technique just described is the Western Union Plan 3A switching system installed for U.S. Steel and Federal Reserve in the early 1940s.

† The first switching center of this type was installed by Western Union at Fort Worth in November, 1934.

The fact that a message was typed manually only once (at the office of origin) brought the error rate down.

A larger center installed at Richmond in October, 1937, included automatic numbering machines on all the outgoing circuits, and the incoming circuits were connected to typing reperforators so that the switching clerks did not need to read the code to determine the proper connection to establish. The benefit of automation was realized by installing a special position for the handling of multiple-address messages. The Richmond system was designated Plan 2. Similar switching centers were installed at Atlanta, St. Louis, Dallas (replacing Ft. Worth), and Oakland, except that at these centers arrangements were made for switching messages to *tie-line* patrons (i.e., patrons connected to the central office by privately operated lines) through a secondary switching unit.

A giant stride in automatic operation was taken on the installations at Philadelphia and Cincinnati which were made in 1947. In these two Plan 20 installations automatic switching from local sending positions into trunk circuits was accomplished by transmitting ahead of each message a two-letter indicator designating the proper outgoing circuit. Connections were controlled by the automatic functioning of relay banks and stepping switches. This was the first use of a director code in public-message telegraphy, and the method was so efficient that telegrams originating at tributary and branch offices, as well as those from local positions, were switched automatically in the Plan 21A centers installed between August, 1948, and January, 1951, at Boston, Detroit, Kansas City, Los Angeles, Minneapolis, New Orleans, Portland, and Syracuse and at Oakland, replacing the Plan 2.

A number of specialized message switching systems have been devised to improve the terminal handling of public telegrams. These are Plans 31, 32, 33, 34, 35, 36, 37, 38, and 40. Another automatic feature which has been installed at many locations is the "burster" page printer. This printer utilizes fanfold sprocket-fed message blanks and an electronic counter which counts the line-feed functions contained in a message, and after reception is complete, the electronic unit adds the correct number of line-feed functions to advance the form just the right amount to bring the printed heading of the next form into exact position. Following this the message form of the last previous message is gripped over a knife edge to burst it apart and the message is then discharged into a hopper or onto a moving belt which carries it to the delivery center. These burster page printers require no operator attention.

16. Torn-tape Reperforator Message Switching

Quite distinct from the Western Union system and yet designed to speed up the handling of telegrams was the system developed by the former Postal Telegraph Co.[1] Both systems utilize typing reperforators for receiving and automatic transmitters for transmission, but instead of the plug and jack arrangement and the cross-office circuit, the Postal method was to transport the received tape manually, message by message, to the sending positions. The tearing apart of the messages for different destinations gave rise to the term "torn-tape switching."

The principal advantage of torn-tape switching is its simplicity. An economic trade-off is made between operator expense and equipment cost. A very high circuit efficiency can be achieved. Also, priority messages can be advanced.

Torn-tape systems were used by Army Communications during World War II, and some of these units were packaged for field use.

Torn-tape message switching is being used in several current commercial private-wire systems. In such service the source of perforated tape does not matter. It can be originated at a keyboard perforator, a teletypewriter ASR set, a card-to-tape converter, or a Flexowriter, but care must be taken to avoid incompatible codes by having similar equipment at opposite ends of the circuit.

The Kleinschmidt Model 179 teletypewriter set is a torn-tape message switching assembly which is available for commercial and military use. It is a prototype of AN/TGC-5. This versatile unit combines two tape transmitters and two typing reperforators, the equipment being housed in two consoles. Either 20- to 60-ma neutral or polar line circuits may be utilized. Messages are automatically numbered. Several unique features are provided which improve operating efficiency.

17. Fully Automatic Switching Systems

Many industrial concerns have requirements for handling a large volume of short messages between their offices in a wide spread organization. Since the distance between various points is large, it is important to use the intercity facilities efficiently, yet it is also important not to have appreciable delay in the transmission of messages. However, these concerns do not find it economically feasible to employ switching operators for duty at irregular hours and in such numbers that peak loads can be accommodated without delay when the average load is likely to be light. For such service fully automatic message switching systems have been designed. An early example of this form of switching was the Type 81-A-1 system* introduced in 1940.[2,6] This system is shown in Fig. 1 and consists essentially of a number of multistation circuits, some of which are operated on a full-duplex basis, that is, in both directions simultaneously, and others on a half-duplex basis, that is, in one direction at a time. Some operate at 60 words per minute, others at 75 words per minute, and in later versions at 100 words per minute.

Each station prepares its messages on perforated tape. Since there are a number of stations on each circuit, arrangements are provided for automatically starting each station in rotation, the messages flowing into the central switching office as "polled." Each message is preceded by certain characters which form a code to control the switching equipment in the control center so that each message will be switched automatically across the office, retransmitted on the proper circuit, and received at the proper distant station without attention by operators at the switching or receiving office. Since it is important to keep the long-haul circuits loaded to capacity, transfer of messages is accomplished by storage in perforated tape, transmitting them across the office at a higher speed than the receiving speed and then, as the sending circuit becomes available, retransmitting them to the proper destination. Storage is provided at both incoming and outgoing positions, primarily to permit operating all cross-office circuits at one common speed (at 100 words per minute in later models of Type 81) and so that intercity circuits can operate at various speeds as is desired. Storage at the incoming circuit termination also permits this circuit to operate continuously without interruption even though cross-office paths are momentarily busy. A relatively high capacity for accepting and storing messages at the outgoing positions tends to speed up cross-office handling and keeps storage at the incoming positions a minimum. It also permits outgoing circuits to be shut down temporarily, as is required when half-duplex circuits are busy with incoming traffic or when there is a circuit interruption. In Fig. 1 the solid lines represent signaling circuits and the dotted lines represent control circuits. The control circuits are mostly automatic in operation, but a control board is provided, and if necessary, one switching clerk can control the center from this position by operating push buttons.

In the cross-office arrangements there are two intercept positions. One, known as the miscellaneous intercept, serves to intercept and record on a typing reperforator all messages having incorrect directing codes, such as codes not assigned. The other, known as willful intercept, is useful in interception of messages requiring special handling. This is a valuable feature when a receiving station is closed for certain periods or when a circuit is out of order for an extended period of time.

Not all stations that originate or receive messages are at distant points. Provision is made for local originating stations and for multiple receiving at common code positions.

The message switching system just described is believed to be the first fully automatic installation for switching messages among a group of telegraph circuits operated in way-station modes and for private-line service.

The Kleinschmidt Type 1145 Switching System is a fully automatic circuit switch-

* American Telephone and Telegraph Company.

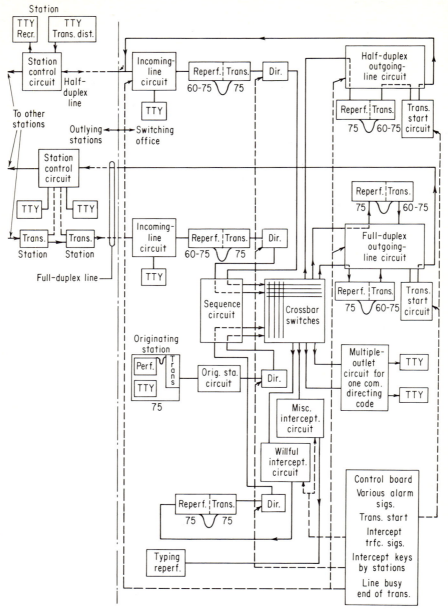

FIG. 1. Functional arrangement of elements in Fully Automatic Message-Switching System Type 81-A-1, American Telephone and Telegraph Company.

ing system designed for light commercial and military message loads. The system consists of a number of teletypewriter stations interconnected and directed by an automatic, unattended switchboard. All switching information is originated from the stations by a line-break switch and the keyboard. When operated, the switch cancels the local teletypewriter motor stop, thus turning on the local teletypewriter equipment, and signals the switchboard that the station wishes to send. The switchboard consists of individual circuit equipment, finder-connector link circuits, and a director-translator control circuit. When the line-break signal is received at the switchboard, the director is activated, which assigns a link circuit to the particular call. The link circuit is then a connection between the calling station and the director-translator. The director signals the calling station that the switchboard is ready for connection information to be transmitted. The calling station then transmits one character from its keyboard to identify the station of destination. On receipt of this identifying character, the connection is established. Release of both stations is accomplished by use of stunt box contacts in the printers. Group-call and all-call are available, the latter being used for emergency messages which preempt busy circuits, and these stations resume transmission when the emergency message has been received unless the station transmitting the emergency message chooses to release all circuits to the idle condition.

18. Military Telegraph Switching Systems

The necessity for faster and more fully automated military teletypewriter communications became apparent soon after World War II, when the armed forces began to establish permanent communications on a world-wide basis. Since the Army, Navy, and Air Force were each to embark on teletypewriter switching network programs, military planners were faced with the realization that there would be a large number of messages transferred from one system to another and that, unless the character code and format used in the message header as well as for the end-of-message sequence were standardized, the electronic directors controlling the passage of messages through the systems would not be able to function. The interservice committee established to act on this problem made recommendations that were to have far-reaching influence on military teletypewriter switching systems. Finalization and adoption of revised joint operating procedures and adoption of a standard message heading format, ACP-127(B), gave system designers a firm basis for arrangements of equipment that would exchange messages automatically. It was also determined by the Communications Panel of the Research and Development Board (DOD) that Joint Military Characteristics for Automatic Teletypewriter Switching Equipment were required, and after serious study of requirements the Joint Communications and Electronic Committee adopted and the Joint Chiefs of Staff approved the document that would also have tremendous impact on the development of equipment for use in joint military operations. The three systems described below fully comply with these requirements, both as to procedures and as to operational requirements.

19. Central Office, Teletypewriter AN/FGC-30*

The FGC-30 is capable of receiving teletypewriter messages on incoming typing reperforators and routing them to outgoing circuits or to temporary storage for subsequent retransmission in a fully automatic manner. The usual intercept facilities are provided to receive traffic which cannot be handled automatically or traffic addressed to switching supervisors. In addition, a special feature separate from the AN/FGC-30 is provided to monitor and scan traffic on marginal radio circuits before it is switched.

* In 1945, the Automatic Electric Co., now a subsidiary of General Telephone and Electronics, initiated the development for the U.S. Army Signal Corps of a fully automatic message switching system specifically designed for military use, Central Office, Teletypewriter AN/FGC-30.[7] Major improvements were possible in this system because new teletypewriter equipment items were specially designed and produced by the Kleinschmidt Division of Smith-Corona Marchant, Inc.

As in other double-storage reperforator switching systems, maximum use is made of incoming and outgoing circuits. Incoming messages may be originated without any delay attributable to the switching center, and similarly, an outgoing circuit not already busy may receive a message from the cross-office circuit immediately. Also, maximum use is made of pooled and common equipment, which results in the highest degree of efficiency in the use of equipment. The extensive use of pooled equipment is a feature of this design, as it provides a unique method of handling precedence messages and is convenient in other respects.

The Kleinschmidt teletypewriter equipment was especially designed for the FGC-30. Several new designs were evolved, among which were a teletypewriter unit consisting of a typing reperforator, a tape reader and transmitting distributor, a tape supply reel, a tape winder, and a tape storage bin, all mounted on a common frame which in turn is mounted on a slide for easy maintenance. The combination tape reader and transmitting distributor, with only 1-in. spacing between reader and transmitting sensing pins, with tape storage between reader and transmitter, has special character sensing at both reader and transmitter and utilizes five-wire output from the reader. A mechanical indicator which registers the amount of tape in storage is available, contacts being closed at 50, 90, and 100 ft of tape. The tape puller associated with the reperforator operates an alarm if the tape is not moving in a normal manner when the reperforator is moving traffic.

Another new device is a special tape reader in the cross-office units, which reads certain characters directly on a single-wire-per-character basis to simplify circuitry and operation of the switching equipment.

A magnetic (solenoid-operated) tape reader at outgoing circuit positions in conjunction with a markable transmitting distributor provides a means of retransmitting a string of messages at the monitor position or of reproducing a message in the monitor tape at a supervisory position.

All teletypewriter tapes in the system are fully perforated and typed in the margin of $7/8$-in.-wide tape.

Modular construction permits expansion of the size of a switching center in 25-line increments up to a maximum of 250 lines, but component equipment items are supplied only as needed.

Messages of any of six degrees of precedence can be recognized, and each message is given access to an outgoing line in strict accordance with whatever cueing order of precedence processing is required. Currently, only four precedence levels are being processed. However, if an outgoing circuit is busy, and if the message being transmitted is one of low precedence, the arrival of a message having high precedence will cause the transmission of the lower precedence message to be interrupted. The portion already sent will be canceled, and the high-precedence message will be forwarded immediately. The canceled message requires manual intervention to pull back the tape, to release the alarm, and to condition the equipment to give the message a new number which will be employed when the circuit is available. This process is known as "can-tran" or "busting" of messages.

Message sequence numbering is employed on incoming circuits and on outgoing circuits, with automatic number comparison being made on incoming messages. Cross-office messages are tagged with a three-letter sequence to indicate which receiving position was employed.

Multiple-address messages are processed as individual outgoing transmissions to appropriate circuits by special equipment which performs the sorting function. Any number of routing indicators may be used, either within the incoming message or within the segregated outgoing messages.

Multipoint circuits may be equipped with as many as 10 stations operating on a full-duplex basis. Transmission to and from these stations is controlled by the switching center, outgoing messages being handled in order of precedence and incoming messages on a three-high over three-low basis. Both outgoing and incoming circuits may be seized to transmit a high-precedence message, that is, by the can-tran technique.

Incoming and outgoing circuit speeds can be 60, 75, or 100 words per minute. Since the use of pooled cross-office circuits normally results in the availability of one of these circuits when needed, cross-office speeds need be only slightly higher than the incoming and outgoing circuit speeds, and these are 75, 100, and 115 words per minute, respectively.

A total of seven characters comprises the routing code that can be recognized by the center. To simplify equipment design, a selection of 4 letters in the first position, 8 letters in the second position, 15 letters in the third position, and all 26 letters in the remaining position has been found to be satisfactory operationally.

Any number of channels may be used to any given destination, and traffic is approximately evenly divided among them.

A number of additional features are provided. An alternate routing patch panel equipped with single-conductor pin-jack patch cords provides flexibility for alternate routing as well as for normal routing. Automatic numbering can be utilized in conjunction with torn-tape operation if this is necessary because of failure of automatic switching apparatus. Most relay and switching equipment is packaged in unit assemblies which are fitted with plug and jack connections. Test messages may be generated with predetermined paths through the center for easy checkout of equipment functions. Alarms and controls are provided for message protection. Any unacceptable message format situation such as failure to receive start-of-message and end-of-message indicators in proper sequence (as might occur in encrypted text) are handled manually in response to an alarm. Return routing can be avoided by intercepting any message that would return to the point that transmitted it. Instead, return routing is used to permit outlying stations to originate self-addressed channel checks. Messages that cannot be routed normally are intercepted. The control supervisor can select any outgoing line for manual forwarding of message tapes. Tapes at the outgoing positions can be rerun in the event of radio or other circuit failure or transmitted to the supervisory position for manual forwarding. The central supervisor's console is equipped with a set of push buttons and lamps which permit an operator to determine which units of the system are temporarily tied together. The supervisor can busy-out sections of the cross-office pool so that all traffic is concentrated in a given area during low-volume periods, thus making supervision easier.

Operation of the switching center can be seen by referring to Fig. 2. All incoming and outgoing circuits pass through the terminal unit where jacks are available for monitoring and patching of circuits to and from the switching equipment. When a message enters the system via one of the incoming circuits, it is received on the typing reperforator in the incoming unit. As the tape advances the associated tape reader registers the *start-of-message* indicator (SOM). This action conditions the channel number comparator to verify the incoming number which follows next on the tape. Nonverification causes an alarm at the supervisor's console and locally at the incoming unit. If the number is valid, the incoming unit demands the services of the director into which the precedence indicator and routing indicator or indicators, if there are more than one, are transmitted by the tape reader. There are normally two directors associated with each 25 incoming line positions. During this operation the tape is stepped through the transmitter-distributor until the first character of the SOM is reached.

The director must determine if the message is a multiple-address message, that is, a message having more than one destination so that there are two or more routing indicators. By reference to the translator the director can ascertain if the routing indicators identify destinations that can all be reached by transmitting the message over one outgoing circuit. If so, the multiple-address message is handled like any other type of message. If not, the multiple-address message is switched to the multiple-call cross-office unit for subsequent processing by one of the sets of multiple-address routing line segregation equipment. This processing includes seizing of the required number of cross-office units and holding them until all the routing indicators have been processed, after which the rest of the message is sent into the cross-office units, and then each resulting message carrying one or more routing indicators is sent to the appropriate circuit. Any number of routing indicators may be included in each multiple-address message.

If the message is a single-address message, the director must find a suitable cross-office unit. It does this by demanding the translator, which is common to all directors. The translator interprets the routing indicator and returns an outgoing-circuit identification to the director. The director now has all the information it needs to make a connection between the transmitter distributor at the incoming circuit position to a cross-office position. If one of the pooled cross-office units is already transmitting to the outgoing circuit, the director examines the precedence that has been

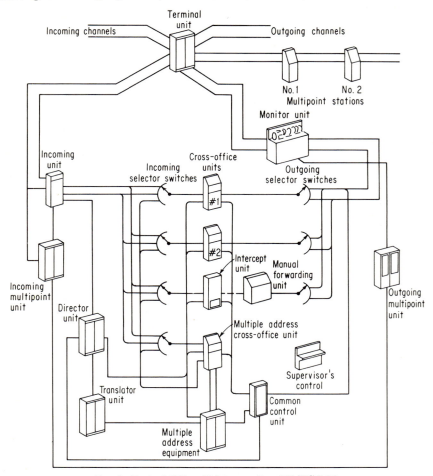

FIG. 2. Central Office, Teletypewriter AN/FGC-30.

set. If it is the same as the new message precedence, and if the earlier message has cleared the incoming side, the director seizes this cross-office unit and causes the second message to be transmitted into it and thus into storage after the other message. If the precedence of the new message is higher than the earlier one, or if the earlier message has not been completely received by the cross-office reperforator, one of the idle cross-office units is seized instead. None of the cross-office units is permanently assigned.

When the director is no longer required, it releases itself, and this takes place after a period of 4 to 6 sec of use.

Outgoing circuits are not actually seized until a message arrives on one of the cross-office units and then in the order of precedence of messages waiting. When a circuit has been seized, a special transmitter temporarily associated with the cross-office unit is used to transmit a three-digit number into the monitor reperforator associated with the particular outgoing circuit, this number not being transmitted to the circuit but being used only to identify the cross-office unit involved. After this, a special transmitter in the monitor unit is used to transmit a new start-of-message indicator (the original SOM being discarded at this point) and the new outgoing channel designation and number to both the monitor and the outgoing circuit. Following this the message, starting with the incoming channel number, is completely transmitted to the monitor and the outgoing circuit. When the message has cleared the incoming unit, the selector switch is released and thus the incoming unit may immediately start the processing of the next message to arrive. Similarly when the message has cleared the transmitter in the cross-office unit, the outgoing selector is reset if this is the last message in storage and the cross-office unit restored to its pool status. At this time the special transmitter in the monitor unit transmits and end-of-message sequence to the outgoing circuit and the monitor unit is restored to normal.

It should be noted that three continuous perforated tapes are generated in the switching center, one at the incoming circuit position, one at the cross-office receiving position, and one at the monitor position. A monitor copy of outgoing transmission is a necessity, justified principally by the kind of communication channels over which the system must operate.

The lower cross-office speed is permitted by starting message processing as soon as the heading has been received from the incoming line unit, this being done by designer's choice to reduce transit delay to an absolute minimum. This, then, requires manual intervention in the event that an incoming message is incomplete for any reason, for example, if the end-of-message combination fails to arrive. If messages are short and the cross-office speed is high, the designer may elect to start processing only after a message is completely received, thus materially reducing the amount of equipment required, but to do this adds delay to the switching of all messages, and the longer they are, the more the delay.

FGC-30 equipment is packaged in cabinets which afford protection for the tele-typewriter and switching equipment. The switching equipment is mounted on jacked-in bases which can be removed quickly and easily for servicing. The tele-typewriter equipment is plug-connected. The equipment is designed for either under-floor or overhead cabling, the former being preferred, and cables from all units terminate in a cable-distributing frame where cross connections are made. This reduces the number of cables and provides test points, easy expansion, and a high degree of flexibility.

20. Automatic Telegraph Switching System, Plan 55-A*

Plan 55-A provides rapid and secure transmission of a very large volume of military traffic over a world-wide network. Messages are switched automatically through any number of switching centers without operating attention. When defects in a

* The 200,000-mile private-wire semiautomatic teletypewriter network of the U.S. Air Force was placed in operation in 1951. The network included five Plan 51 switching centers, and both circuit facilities and switching equipment were leased from Western Union. In 1958, the domestic portion of the network was converted to a fully automatic system utilizing Plan 55-A Automatic Telegraph Switching System[8] equipment developed by Western Union specifically for Air Force use. A prototype was placed in operation in 1956, following which the five domestic switching centers were reequipped to obtain substantially greater message capacity, and the size of the network was increased, particularly the overseas portion, to include five switching centers located outside the United States, making a total of ten switching centers in all. The objectives of replacing the push-button equipment with automatic equipment were to increase the speed and efficiency of message handling, reduce critical manpower needs, gain additional traffic capacity, and make it more convenient to install a system of a certain size and progressively increase its size as required. Equipment and facilities are leased from Western Union.

message heading or circuit conditions prevent automatic switching, means are available for immediate manual (push-button) switching of messages to avoid delay. The message format is such that compatibility is achieved with any other military service without manual retyping. The switching equipment is designed so that it is capable of being operated either automatically or manually for switching of multiple-address messages. Messages of high precedence are given priority over messages of low precedence.

Modular construction is used throughout, and apparatus items are all arranged for plug and jack connections. A minimum number of types of cabinets, each complete in itself, with plug-in connections to the other cabinets, makes it possible to install a center rapidly or to remove and reinstall at another location. All or part of a switching center may be installed as required; for example, a small center may be installed for push-button operation and later converted to automatic operation by adding the necessary cabinets.

Plan 55-A is a double-storage, continuous-tape system. The typing reperforators used on incoming circuits and the reperforators used on the cross-office circuits are chadless models manufactured by the Teletype Corporation, and the tape is $11/16$ in. wide. The cross-office circuits are single wire, electronic, operating at 200 words per minute, which reduces the number of intercabinet cable wires and permits control signals to be sent over the same wire. The transmitter-distributor for sending to outgoing circuits was manufactured by the Teletype Corporation, but the loop-gate transmitter for sending to each cross-office circuit was furnished by Western Union.

The two principal items of equipment required in a Plan 55-A switching center are the incoming circuit cabinet and the outgoing circuit cabinet. Each cabinet houses two operating positions, and these cabinets are connected by means of cables terminated in multiconductor plugs and sockets. All cabinets of each type are identical and require no common equipment to set up an operable manual switching center. These two cabinets contain all the equipment needed for reperforating, transmitting, and push-button switching. Each incoming circuit position can switch messages to a maximum of 400 circuits and 200 destinations, it being necessary only to add the correct number of outgoing circuit positions for the number of destinations to be served.

Each incoming circuit cabinet and each outgoing circuit cabinet contain two positions. Either position can be used to terminate a trunk or a single-station tributary. One position can be used to terminate a multi-station circuit or a trunk having separate high- and low-precedence number sequences.

A third type of equipment cabinet is the director-translator which is required for automatic switching. The cabinet houses two directors and one translator, sufficient to control 25 incoming circuit cabinets (50 incoming circuit positions of which one-half may be multistation circuits). This cabinet is similar to the receiving cabinets and is normally installed at the end of a receiving row. The director receives routing information from the message tape and controls switching operations at the start of each message. The translator receives routing indications from the director and translates them into switch settings. These operations take only a few seconds.

To facilitate maintenance, each of the three cabinets is divided into front and rear sections with a maintenance aisle between. The equipments are mounted on shelves or racks that can be slid or swung outward into the maintenance aisle for easy access. The rear section can be mounted against a wall if desired.

In addition to the three types of cabinets described above, it is convenient and customary to equip each switching center with a supervisory console. This consists of a connection indicator board, a traffic routing board, a close-out indicator board, and supervisory teletypewriter sets.

Numerous safeguards and alarms are provided throughout the system to prevent errors, avoid delays, and guard against lost messages. Each message is numbered in sequence as transmitted, and received numbers are compared automatically. When a message is relayed at the first switching center, the message arriving at a second center contains both the originating number and the number inserted by the first relay center. On the next relay, perhaps to destination, the originating number is deleted, the incoming number to the second relay center is retained, and a new number is added, this being the next serial number expected on the circuit from the center to that station.

Some messages will be transmitted in cipher. If "on-line" or "link" cryptographic equipment is employed in telegraph switching systems, messages are deciphered automatically as received in the switching center, pass through the center in the clear, and are automatically enciphered again as transmitted out of the center. In some cases it is necessary to control the distributor-transmitters that send to outgoing circuits so as to maintain synchronism with the encryption equipment. The distributor-transmitters in Plan 55-A are of this type. With "off-line" cryptographic equipment, messages are enciphered at the point of origin, transmitted through the switching centers in cipher, and deciphered at the point of destination. Such messages must have clear headings for routing purposes and must avoid the chance occurrence of the end-of message sequence in the ciphered message text.

The typing reperforators are Type 28 equipments. At each receiving position there is a message-waiting indicator used for manual switching. It also indicates the number of messages on hand that have not been transmitted, and it controls the blank tape feed-out if incoming reception stops at the end of a message. If start-of-message and end-of-message sequences are not received alternately, an alarm is set which requires manual intervention to determine and remove the cause of the alarm.

Each receiving position has a loop-gate transmitter. This is a tape transmitter which has a slidable latch that holds down the tape and permits sending certain portions of the message twice, once to the director and then repeating into the cross-office circuit. With the loop-gate in its right-hand position the transmitter functions like any conventional magnet-stepped transmitter to step blank tape through without reading. When the start-of-message sequence arrives at the sensing pins of the transmitter, a sequence-number indicator associated with the transmitter checks the call letters and message number in the tape. If these are incorrect, the transmitter stops and an alarm is set. If the message number is correct, the transmitter steps the tape ahead until it detects a line-feed function character, thus deleting any characters that might appear in the tape between the sequence number and that point. The message format prescribes that this line-feed character signifies the beginning of the second line in the message which is the precedence and routing indicator line. Upon detecting the line feed, the loop gate slides to the left, thus holding the tape so that it cannot pass completely through the transmitter but can step past the sensing pins, forming a loop above the transmitter. Characters fed into this loop can be transmitted a second time. For manual switching, when the loop gate slides to the left, a message waiting lamp lights to call the attendant, who looks at the precedence prosign characters on the tape and presses a high- or low-precedence push button which advances the precedence prosign through the transmitter. The attendant then depresses a destination push button for each routing indicator appearing on the tape and a start button which causes the message to be transmitted to all destinations that had been set. Thus single- or multiple-address messages are passed to the appropriate cross-office circuits.

For automatic switching, with the loop gate to the left, a connection is obtained to a director. The precedence prosign characters and the routing indicator or indicators are transmitted to the director. After route information has been acted on by the translator, the director now returns switching information to the receiving position for setting up the cross-office connection to the desired destination. The loop gate then returns to its right-hand position, placing the original character (the first character after the line feed) over the sensing pins. The transmitter is now ready to transmit the message at the proper time.

The electronic cross-office units consist essentially of a pulse generator common to the two operating positions in each cabinet, an electronic transmitter for each receiving position, and an electronic receiver for each sending position. These units generate, transmit, and detect the message and control pulses that are sent across office and, in addition, step the tape transmitter and numbering machine, detect and indicate trouble conditions, read certain message characters as they are being transmitted at high speed, test the idle or busy condition of selected routes, and control the operation of other equipment such as the sequence-number indicators, message-waiting indicators, and the automatic switching directors.

Each outgoing circuit position is equipped with a Model 28 reperforator for reproducing messages received over cross-office circuits and signals from an automatic numbering machine. For multipoint circuits one, two, or three message-numbering machines may send into the reperforator, but no more than three way stations can be operated on one circuit. When a cross-office connection is established, the numbering machine automatically sends into the reperforator the start-of-message characters, a multipoint selection character if required, and the new message number consisting of two characters for station identification, one character for the channel designation, and three numerals for the sequence number. The sequence-number indicator at the receiving position then transmits the incoming message number that it had retained in memory, followed by three format characters. The message is then transmitted across office by the loop-gate transmitter, first the precedence prosign characters and routing indicators for the tape loop, and then the remainder of the message, including the end-of-message characters. The loop-gate transmitter then stops, the numbering machine sends two blanks into the reperforator, and the cross-office circuit is released.

The reperforator is equipped with read-back contacts operated by a set of sensing pins located five characters to the left of the punch pins. These pins read certain character groups in each message to verify the correct operation of all equipment involved in cross-office transmission.

Each outgoing circuit operates at either 60 or 100 words per minute. The Model 28 transmitter-distributor has two camshafts: one a transmitter shaft for feeding the tape and sensing the perforations and the other a distributor shaft for sending character signals in start-stop code. The shafts operate together or separately, so that blank tape or other characters can be idled through the transmitter without sending to the circuit and so that special character signals, controlled by a rotary switch, can be sent by the distributor contacts without stepping the tape through the transmitter.

Two sending positions can be arranged to send to the same outgoing circuit. Normally one position is used for high-precedence traffic and the other for low-precedence traffic. If a high-precedence message has to wait more than 30 sec, an alarm sounds to call the attendant, who can stop a message being transmitted and put the high-precedence message ahead.

To handle heavy traffic loads to a particular destination several outgoing circuits and outgoing positions may be required. All such multichannel groups are addressed as one destination, but the load is controlled by a load-distributor circuit, so that the traffic is evenly divided among the channels. This is the reason for providing twice as many switching circuits as destinations in the incoming circuit cabinets.

An analysis of Air Force multiple-address traffic revealed that such messages were normally addressed to no more than nine destinations. Therefore, to simplify equipment design, no more than nine routing indicators may be included in one multiple-address message. Two or more separate messages can be prepared by a "tape factory" at any station likely to have need for addressing more than nine destinations. Multiple-address messages are automatically processed so as to transmit an individual copy of the message to each ultimate destination. In this process, routing indicators are segregated so that a message to another switching center may contain several routing indicators. As relatively few multiple-address messages require switching to more than four circuits, most multiple-address messages can be switched directly from incoming positions. In rare cases a message may need to be switched to more than four circuits. In such cases, all routing indicators requiring an outlet different from the first three circuits are sent (with a copy of the message) to an extra operating position assigned for this purpose. The message recorded at this position contains at least three less routing indicators than the original message, and this message is then reprocessed as an original. Reprocessing may be required a second time in the event the original message required switching to nine different circuits.

Figure 3 shows the message format utilized in transmitting a message from a tributary station through two switching centers to a destination station.

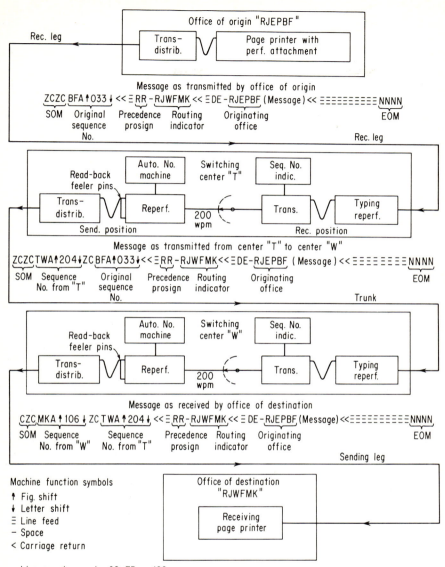

FIG. 3. Plan 55-A, switching single-address message through two switching centers. (*From Western Union Tech. Rev., April,* 1958.)

21. Teletypewriter Switching System Type 82-B-1*

This system, placed in operation in 1959, has greatly increased the speed of message delivery through centralized control and relatively simple operations requiring minimum personnel. It is a flexible, expandable, and reliable network, extending to some 48,000 miles of land-line circuits, linking 236 stations in 103 United States cities. Type 82-B-1 also allows for exchange of messages with a series of semiautomatic systems which stretch another 5,000 miles to reach 85 more cities. From coastal locations, there is a tie-in to the Navy's overseas radio circuits to ships at sea and overseas naval commands and to the other military systems. The system was engineered to handle an initial volume of about 9 million words a day, with ultimate capabilities up to 30 million. Switching centers are at Trenton, N.J., Norfolk, Va., Cheltenham, Md., Stockton, Calif., and San Diego, Calif.

Basic equipment at each center of the Type 82-B-1 system includes cabinets containing electronic devices that switch messages from incoming to outgoing circuits at a speed of 200 words per minute. Terminal equipment at each station includes a control unit and separate incoming and outgoing teletypewriter sets operating at 100 words per minute. The start of transmission is controlled at switching centers where equipment automatically recognizes the destination code on each perforated tape message and sends it on its way to the proper station on the network. Multiple-address messages are also handled automatically. This network is also used for data transmission.

22. Telex

In the postwar period European telecommunications administrations, which are almost entirely government-owned and -controlled, introduced an automatic subscriber-to-subscriber teletypewriter exchange service known as Telex. Rapid growth of this new service was brought about by the operating advantages of the system and its modest cost. Private-wire teletypewriter systems did not compete in effectiveness because traffic volumes were not sufficiently heavy to justify the cost of such systems. Telex was introduced in Canada in 1956 and in other countries of the world as international service became available.

Telex was first utilized in the United States in 1957, when the Western Union Telegraph Co. connected subscribers in New York City with the Canadian system at Montreal.[10] Subsequently domestic service was offered to subscribers in other United States cities. As expansion proceeded, service became available to Mexico from Kansas City, to Europe and South America via direct connections from the New York international gateway and to other points in the world via the San Francisco gateway. Connections can also be established from San Francisco to Vancouver and from Chicago to Winnipeg and Toronto. At the end of 1961, 45 United States cities were equipped, and at the end of 1963, service was provided to 7,000 subscribers in 101 United States cities. While the present United States Telex network has only six junction exchanges, the switching equipment of the type supplied by Siemens-Halske (Munich, West Germany) will ultimately permit junction exchanges in nine cities. Each junction exchange will serve as a parent exchange for 10 district exchanges, and each district exchange will in turn serve up to eight smaller locations having sub-district exchange equipment. Locations having not more than 20 subscribers are served by means of a concentrator operating over five trunks to the nearest exchange. The concentrator does not have a local dial stage because a very high percentage of Telex calls are distant calls. Therefore, calls between subscribers terminated in the same concentrator are double-trunked through the nearest exchange.

Three distinct types of Telex exchange systems are used for the junction, district,

---

* Service for the Navy's fully automatic teletypewriter switching system[9] is furnished by the Long Lines Department of American Telephone and Telegraph Company and local operating companies in the Bell System. Switching center equipment was built and installed by the Western Electric Company.

and subdistrict locations in the United States network.   The Siemens-Halske TWM-2 system, incorporating fundamental design principles for large exchanges, provides the required operating versatility for the junction exchanges.   The subscriber terminations are connected by line finders to centrally controlled dial code translators or registers which can store, evaluate, and interpolate the dial pulses for alternate trunk routing when necessary.   Further concentration of exchange functions is made by having a large group of dial code translators obtain required routing and zoning (for determining charges) information for each call from a centrally controlled electronic coder rack.   Only a register-type system could meet the requirements at junction exchanges that are gateways to the Canadian and Mexican networks or to the facilities of the international overseas carriers.   The ability of the gateway junction exchanges to insert additional digits for routing intersystem calls provides compatible operation with the Canadian and Mexican step-by-step systems and obviates any conflict in the directory code assignments of the three independent networks.   To accomplish this, the gateway exchanges are operated as though they were district exchanges of foreign systems, and such versatility also makes it possible to provide fully automatic service on calls that transit the United States network on a tandem basis.

The Siemens-Halske TW-39 system is used at district and larger subdistrict exchange locations.   This system employs the well-known and simpler step-by-step principle; there the selectors are under the direct control of the subscriber's dial.   In general, the district exchanges are not important trunk routing points, and therefore exchange versatility is not a prime requirement at these exchanges.

The Siemens-Halske TW-56-WU Two-rack Concentrator[11] is used at all small subdistrict exchange locations.   This concentrator was designed to meet the unique needs of Western Union, which required up to 900 tariff points.   Concentrators are frequently replaced with TW-39 equipments as the number of connected customers or the load exceeds the capacity of the TW-56-WU concentrator.

Subscribers have a choice of two basic types of Telex subscriber sets, each of which involves the use of a page teletypewriter equipped with an automatic answer-back device which is coded with the abbreviation of the subscriber's name and city.   Associated with the teletypewriter is a remote-control dial unit with start and stop push buttons for establishing a connection and disconnecting, respectively.   The Telex KSR set is equipped with a Teletype Model 32 keyboard send-receive set and is used where reperforator tape facilities are not required.   A second and more versatile operating arrangement is provided by the Telex ASR set, which includes facilities for transmitting and receiving messages in perforated tape form.   This set is equipped with a Siemens Model T-100 page teleprinter having a built-in reperforator attachment on the left side of the machine and a corresponding transmitter-distributor attachment on the right side for transmitting tape that has been prepared in advance. The remote-control unit of this set has additional controls which permit the teleprinter and associated tape attachments to be used in off-line mode for the preparation of tape simultaneously with a monitor page copy.   For on-line operation, transmission may be either from the keyboard or from the transmitter-distributor.   On either incoming or outgoing transmission the reperforator attachment can be used to produce tape as well as page copy.

When idle, the printer of the subscriber set is turned off and the dial on the remote-control unit is locked.   A call is initiated by depressing the start push button.   This signals the Telex exchange, which immediately recognizes the calling line and transmits a proceed-to-dial signal, thereby unlocking the dial.   The subscriber now dials the number of a distant subscriber obtained from the Telex directory, usually a six- or seven-digit number.   When the circuit connections have been established, as indicated by the fact that the printer motors of both the calling and called subscribers are turned on, and the "operate" lamps are lighted, either subscriber may transmit to the other alternately.   Either subscriber may terminate a call by depressing the stop push button.

Operating procedure permits an exchange of "answer-back" signals to identify subscriber stations.   When the calling subscriber transmits the "who-are-you" signal, the answer-back of the called subscriber is triggered, which is transmitted and recorded at both stations.   The calling subscriber then depresses the "here-is" key, which triggers his own answer-back device, thus recording the identity of the calling station at both stations.   Either subscriber can request answer-back identification at any time to verify circuit continuity.   The answer-back feature is of value in unattended operation.

The operating speed is 50 bauds or approximately 66 words per minute, and with the exception of a few upper-case characters which are assigned on a regional basis and clearly marked on the keyboard, the coded character set corresponds to the international CCITT standards, thus assuring complete compatibility with other Telex systems throughout the world.

There is no minimum time charge for individual Telex calls within the United States. A particular pulse rate based on airline mileage is applied for the actual time the connection is held, and a charge of $2\frac{1}{2}$ cents is made for each pulse.   There are 18 pulse rates ranging from 4 to 72 pulses per minute.   In addition, there is a monthly charge for the subscriber equipment.   There is no charge for any call to Western Union other than the normal rate for a telegram, but subdistrict subscribers may not call Western Union for the purpose of filing a telegram.   The charges for international Telex calls follow a different rate pattern, and subscribers are billed separately by the international carrier for such calls.   These charges include the use of the circuit between the customer's location and the gateway city.   A customer may also dial Western Union in a distant city and file a telegram for local delivery at that point. This service is known as Tel(T)ex.   For such service the customer pays a flat charge for delivery of the telegram plus the normal Telex charges between his location and the distant point.

Telex is growing rapidly in some of the developing regions of the world, for example, in Africa and in Asia.

23. Specialized Private-wire Message Switching Systems

A number of teletypewriter switching systems have been developed for specific customer needs.   Since these systems made use of principles incorporated in larger systems, they need not be described in detail.   However, a few of these switching systems have had such widespread use that they are worthy of mention.

The Western Union Plan 111 Switching System[12] is the most widely used system in service today, over 125 such systems serving almost as many types of industries. It is a single-storage, torn-tape system designed primarily for small- and medium-size networks or as a subcenter of a large system.   The customer may select minimum equipment consistent with his immediate requirements and then add extra equipment or optional features at a later date.   The number of circuits is unlimited.   One receiving console contains up to three receiving positions, and one sending console contains three to six sending positions depending on whether the circuits are single circuits or way circuits.   Either 8 or 16 way stations may be accommodated on a single way circuit by the use of different switching panels.   Master send is available on option.   All outgoing messages are numbered automatically.

The Western Union Plan 51 Switching System[13] is a single-storage, push-button, semiautomatic system having a capacity of 60 different destinations.   Up to eight stations may share a way circuit.   Upon receipt of a message at the switching center, the destination is read from the printed perforated tape by an attendant, who depresses the appropriate push button which sends the message on its way to the destination. At the end of each message the transmitter stops automatically and signals the attendant so that the subsequent message can be switched.   The capacity of Plan 51 may be increased through the use of a technique known as "preselection."   With preselection, the switching center is responsible only for routing the message to the correct way circuit.   Characters in the tape operate the selector of the correct station.   This places a portion of the responsibility for correct routing on the sending operator. Automatic numbering is provided, and this number and the first line of each switched message are recorded on monitor printers so that tracing of messages is facilitated.

The Western Union Plan 54 Switching System[14] is also a single-storage, push-button, semiautomatic system but it has greater capacity (125 circuits) and other improvements.   Plan 51 and Plan 54 systems have their special place in the telegraph switching field.   Certain customers, with unique requirements, prefer to employ a few switching center attendants specially trained and responsible for correct routing rather than to place this responsibility on every transmitting operator as is required in fully automatic operation.

The American Telephone and Telegraph 83-B-1 Teletypewriter Selective Calling System[15] is a half-duplex (or full-duplex in later models) party-line system designed to eliminate operating complications which would arise without station selection. Selective calling eliminates contention for the use of the circuit by the way stations, as all stations are under control of one station called the control station.   The control station gives permission to each of the other stations to send, in a predetermined order, by means of selective signals.   The control station may direct a message to one station or to a selected group of stations, and this avoids the necessity of each station checking all traffic received to select the messages intended for that station.   Selective calling is more efficient where the traffic load is high or where automatic tape sending is employed.

The 83-B-1 system combines the use of a particular type of station-selecting arrangement with an automatic transmitter (or keyboard unlock) start feature.   Station selection is obtained from the response of the stunt box to a two-character code, this device being in the printer, or in a device called a sequence selector, which is a printer with the printing and paper-handling features removed, leaving only the selection features.   Station equipment also includes a station control circuit which functions with contacts in the stunt box to start the tape transmitters or to turn on the printing mechanism of the printer.   Normally an automatic send-receive set is used at each station.   There are usually 10, 20, or 36 stations per circuit.

Stations may be polled by means of a choice of patterns established at the control station, but these patterns may be altered by the use of skip keys.   Stations having priority traffic depress a priority key which causes this traffic to be accepted ahead of other traffic.

Push-button selection may be provided as an optional feature.

The Western Union Plan 115-A[16] (single way system) and the Plan 117-A (duplex way system) are similar selective control systems.

The Kleinschmidt Teleswitcher B System is an automatic selective calling system. It is a half-duplex system and provides facilities for 24 way stations on a single circuit.

24. Electronic Telegraph Switching Systems

In recent years there has been an ever-increasing usage of electronic techniques to achieve both circuit-switching and message-switching designs that would have improved performance.   Circuit-switching systems have followed and are likely to continue to follow developments in the telephone art.   The application of electronic switching principles to message-switching systems began with the electronic director, the electronic translator, and the electronic cross office.   This trend was accelerated by the advent of electronic computers and data transmission.

The most extensive use of electronic switching techniques has been in the data-transmission field.   Several large-scale digital electronic switching and communications data-processing systems have been developed.   These are described in Chap. 12. Mention is made of these systems here because they also switch teletypewriter messages as well as data information.

25. Electronic Switching of Command and Control Communications

Changes in military technology and the world political situation have established the requirement for automated assistance in the exercise of command and control of military forces, both before the outbreak of hostilities and during active warfare. Such automated assistance includes, among other essential requirements, the transmission and utilization of digital communications of a number of different types, such as teleprinters, punched card data, computer inputs and outputs, and information for display purposes.

The U.S. Air Force Strategic Air Command Project 465-L, the SAC Command Control System,* is the most advanced and most sophisticated development of this type that has taken place in the world up to the present time. Basically, the system is capable of receiving inputs; processing, reducing, comparing, and storing information; and, additionally, displaying information in comprehensive display form. To accomplish these tasks the SAC Command Control System has three basic subsystems: Data Transmission, Data Processing, and Data Display. Since this chapter is concerned primarily with telegraph switching, only the Data Transmission Sub-System will be described.

The type of data to be transmitted is largely keyboard-generated message text, although provision is made to transmit information in computer language originated from magnetic tape or punched-card devices.

The Data Transmission Sub-System utilizes a number of equipment arrangements known as centrals. These centrals are the Simplexed Remote Communications Central (SRCC) AN/FYQ-7, the Remote Communications Central (RCC) AN/FYQ-4, the Electronic Data Local Communications Central (EDLCC) AN/FYQ-6, and the Electronic Data Transmission Communications Central (EDTCC) AN/FYQ-5. Assignment and location of these centrals are in accordance with SAC operational requirements.

The SRCC, RCC, and EDLCC equipments include the same equipment assemblies. Messages are inserted by means of a keyboard message composer which is an IBM electric typewriter especially designed for the system. As each character is written, the code elements are transmitted to the concentrator, one of which is included in each central. The character is simultaneously typed on the page copy of the message composer. If an error is made, the typewriter is back-spaced and the correct character is inserted. The operator also enters as part of each message (1) instruction information, that is, information which instructs the EDTCC how to process the message, and (2) address information, that is, where to route the message. One of the functions of the concentrator is to store all characters as they are typed. The characters are accumulated (after correction, if any was necessary) until one complete message has been composed. After the operator has verified each message to make certain that it is correct, a special key is depressed which releases the message out of the concentrator, and it is then transmitted to the EDTCC. The concentrator also inserts format, control, and checking data at appropriate locations. The formated message is encrypted, modulated, and transmitted over a transmission link at the rate of 2,400 bits per second (or at 1,200 or 600 bits per second if the circuits are not operating normally), thus accomplishing speed translation. The EDTCC automatically acknowledges receipt of each message correctly received from the sender.

Another function of the SRCC, RCC, and EDLCC equipment is to receive and reproduce incoming acknowledgments and message arriving from the EDTCC. The receive channel equipment component receives, demodulates, and decrypts this information after which it is stripped of format, control, and checking data and forwarded to a Motorola electrographic printer, which prints the message at the rate of approximately 300 characters per second.

Each SRCC and RCC is connected to the EDTCC by circuits of voice bandwidth, and as these circuits are long-distance transmission links, each circuit is equipped with crypto-modem assemblies. The modems were designed by Collins Radio. The EDLCC is connected by local circuits of voice bandwidth, and these circuits do not require crypto and modem equipments. Each RCC is connected not only to the

---

* Prime contractor: International Telephone and Telegraph Corporation. System manager: International Electric Corporation. Subcontractors: Collins Radio Co., Analex Corporation, Motorola, Inc., System Development Corporation, Dunlap & Associates, Inc., Ark Electronics Corporation, Federal Electric Corporation, International Business Machines Corporation, Bryant Computer Products Division, ITT Federal Laboratories, and ITT Kellogg.

normally assigned EDTCC but to an alternate EDTCC, so that messages originating at or destined to the RCC are transmitted via alternate routes if the normal route is inoperative. Therefore, each RCC has two transmit channels and two receive channels.

Each SRCC is capable of exchanging transmission with Digital Data Transfer System (Sub C), which is part of AF/FYQ-2. In some cases, alternate transmission paths are provided for an SRCC.

The EDTCC, which is duplexed for greater reliability, controls the flow of information to other EDTCC's, and to the Data Processing Central and the Data Display Central with which it is associated. All information enters and leaves the EDTCC through an input-output unit. The heart of the EDTCC is the stored program element which accomplishes all switching and processing functions required. The storage capacity of this element is increased by a drum-memory unit. An automatic channel unit allows the stored program element to communicate with its input-output magnetic tape storage units.

The fault and facility control gives an instant indication of the detailed status of each of the RCC and SRCC systems. It also contains patching and testing facilities to simplify and improve maintenance.

REFERENCES

1. d'Humy, F. E., and P. J. Howe, American Telegraphy after 100 Years, *Suppl. to Trans. AIEE*, vol. 63, pp. 1014–1032, 1944.
2. Duncan, J. A., R. E. Pierce, and R. D. Parker, Telegraphy in the Bell System, *Suppl. to Trans. AIEE*, vol. 63, pp. 1032–1044, 1944.
3. Harrison, H. H., *Printing Telegraph and Mechanisms*, Longmans, Green & Co., Ltd., London, 1923.
4. Williams, Archibald, *Telegraphy and Telephony*, Thomas Nelson & Sons, New York, 1928.
5. Bramhall, F. B., and J. E. Boughtwood, Frequency-modulated Carrier Telegraph System, *Trans. AIEE*, vol. 61, p. 36, 1942.
6. Bacon, W. M., and G. A. Locke, A Full Automatic Private-line Teletypewriter Switching System, *Trans. AIEE*, vol. 70, part 1, pp. 473–480, 1951.
7. Central Office, Teletypewriter, AN/FGC-30, Operating Features, Techniques, Advantages and System Operation, *Automatic Electric Co. Circ.* 1890-A, April, 1960.
8. Vernam, Gilbert S., Automatic Telegraph Switching System Plan 55-A, *Western Union Tech. Rev.*, vol. 12, no. 2, pp. 37–50, April, 1958.
9. Pratt, Capt. R. R., USN, The Role of Landlines in U.S. Sea Power, *AFCEA Signal*, vol. XVII, no. 6, pp. 19–38, February, 1963.
10. Easterlin, Philip R., Telex in the U.S.A., *Western Union Tech. Rev.*, vol. 16, no. 1, pp. 2–15, January, 1962.
11. O'Sullivan, T. J., The TW56-WU Telex Concentrator, *Western Union Tech. Rev.*, vol. 17, no. 3, pp. 110–121, July, 1963.
12. Pessagno, T. S., Switching System Plan 111, *Western Union Tech. Rev.*, vol. 9, no. 2, pp. 55–61, April, 1955.
13. Dirkes, R. F., Switching System Plan 51, *Western Union Tech. Rev.*, vol. 2, no. 4, pp. 138–149, October, 1948.
14. Vincent, G. O., Switching System Plan 54, *Western Union Tech. Rev.*, vol. 10, no. 4, pp. 167–173, October, 1956.
15. Smith, C. W., C. J. Votaw, and A. L. Whitman, The 83-B-1 Teletypewriter Selective Calling System, *AIEE Conf. Paper* 58-443.
16. Jorgensen, H., Switching System Plan 115-A, *Western Union Tech. Rev.*, vol. 15, no. 4, pp. 125–132, October, 1961.

Chapter **9**

TELEPHONE SWITCHING METHODS
AND EQUIPMENTS

L. F. GOELLER, Jr., *Bell Telephone Laboratories*

A telephone switching system makes connections on demand between lines to its customers and/or trunks to other switching systems. The magnitude of the task as well as its innate difficulty presents a challenge to the designer: If a switching system has 10,000 lines or trunks, the almost 50 million potential interconnections must have a high probability of being available when requested.

In addition to establishing interconnections, a telephone switching system must keep itself ready for service with a very high degree of reliability. Furthermore, it must often assist in testing associated transmission facilities.

Many different switching systems are presently in use, and many more have been developed. The classic textbooks in the field[1-3] are out of print; current American texts[4-7] offer a wealth of information in somewhat specialized areas as do those from England.[8,9] Thus the periodical and patent literature must be consulted for detailed information. Although an attempt at generality has been made in this chapter, the material is based primarily on the practices of the Bell System. References cited will provide information on other practices.

1. The Functions of a Telephone Switching System

a. Interconnecting Transmission Facilities. It is convenient to subdivide the interconnecting function as summarized in Table 1. Alternative actions, some of which are shown in the table, are possible at all stages of a call.

b. Maintenance. A telephone switching system must be designed to include a variety of maintenance duties.

Circuits can be designed to check their own operation as they are used, giving a trouble indication on detected failure or excessive time to complete an operation; routine progression testing can be applied to the system; and special diagnostic tests and reporting features can provide maintenance personnel with more specific data when faults occur.

Because a switching system is, among other things, a large automatic connector, it can be used to test customer lines and telephones and also to aid in isolating trunk troubles to the connecting office, the transmission facility, or the originating office.

MECHANICAL AND ELECTROMECHANICAL APPARATUS

All mechanical and electromechanical switching devices have some of the following components: contacts, where two surfaces meet to make an electrical connection;

Table 1. Typical Call-processing Functions

Functions	Typical alternatives*	Par. ref.
Origination		
1. Detect customer's request for service.	. . . .	16a
2. Respond to service request.	. . . .	14a, b
3. Identify calling customer.	. . . .	16c
4. Note class of service.	. . . .	14d
5. Obtain specific request.	Customer does not signal (permanent signal) or does not send enough information (partial dial)	16b
Distribution		
To customer on same switching system		
1. Identify called customer†.	. . . .	14d
2. Test called line for busy.	Called line busy; return busy signal	16g
3. Find path from called to calling customer.	All paths busy; return overflow signal	5, 14a, 15
To customer on another switching system		
1. Identify outgoing trunk group. . . .	No such office. Connect to assistance operator or recorded announcement or other signal	15, 16d, g
2. Test for idle trunk.	All trunks busy; return all-trunks-busy signal or use an alternate route	15, 16d, g
3. Find path from calling customer to outgoing trunk	All paths busy; return overflow signal	5, 14a, b, 16g
3a. Request service from second switching system	Trouble time-out; release circuits	14b, 16d, e, f
3b. Second switching system responds		
4a. Send specific request.	. . . .	14b, 16d, e, f
4b. Second switching system responds	Error detected; return reorder signal	
5. Second switching system identifies outgoing trunk group		
6. Test for idle trunk.	All busy; use alternate route or return all-trunks-busy signal	
7. Find idle path from incoming trunk circuit to outgoing trunk circuit.	All busy; return overflow signal	
8. Request service from third switching system		
9. Process continues until terminating switching system is reached		
Termination		
1. Identify called customer†.	Return number changed or no such number signal	14d
2. Test called line for busy†.	Called line busy; return busy signal	16g
3. Find path from incoming trunk circuit or calling customer to called customer	All paths busy; return overflow signal	5, 14a, 15, 16g
4. Apply alerting signal to called telephone and audible ring signal to calling customer.	. . . .	16g
5. Upon answer, remove alerting signal and establish talking condition. . . .	Called customer does not answer	16g, h, i

Table 1. Typical Call-processing Functions (*Continued*)

Functions	Typical alternatives*	Par. ref.
Talking		
1. Return answer signal to initiate charging....................	Free call, no-answer signal	16*d, e, i*
2. Establish proper transmission circuit..........................	....	17
3. Supervise connection for end of call	Customer may signal for assistance	16*e, i*
4*a*. Detect calling customer hang-up..	Called customer hangs up first	16*e*
4*b*. Detect called customer hang-up and return on-hook signal........	Calling customer hangs up first	
5. Terminate charging.............	....	16*e, i*
6. Take down connection; make circuits available to others.........	....	14*a*, 15

* Calling customer may abandon call at any time until answer.
† Items in Distribution overlap similar items in Termination.

springs, to restore contacts to normal when the activating force is removed; and coils or electromagnets, to permit remote operation of contacts. Often, contacts are mounted on the springs themselves, the springs acting as conductors for the circuits being switched. A number of independent contacts can be used in one device, and several different coils can be provided for operation.

Other members which make electrical contact are wipers and brushes. Both are used in devices which connect one input to one of a number of outputs; both are driven sequentially from one output to the next in order. Wipers differ from brushes in that brushes have to be "tripped" before they make contact, whereas wipers, when driven past an output, always connect to it.

2. Apparatus for Manual Switching*

a. Jacks. Figure 1 shows telephone jacks and a suitable mating plug. Jacks are built with spring-contact members to connect to the plug, but many more contacts can be added if required. These additional contacts can switch independent circuits, all switches being operated by the mechanical force of the inserted plug.

* Reference 3, vol. 2, chap. 6; Ref. 8, vol. 1, chap. 10.

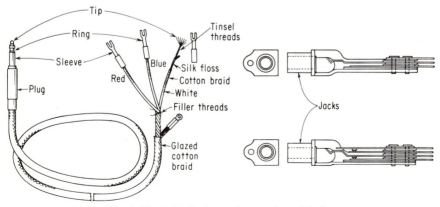

Fɪɢ. 1. Typical telephone plug, cord, and jacks.

b. Plugs and Cords. Conductors in a cord (as well as those connecting to the jack) are named after the parts of the plug: the tip, ring, and sleeve. Battery is associated with the ring lead, and ground with the tip; if the tip touches ground accidentally, no harm will be done. The sleeve lead is usually associated with some form of control, while the tip and ring carry the talking path. In manual telephone switching systems, cords usually have plugs on one end and terminals or lugs on the other. Two of these cords connect to a "cord circuit" which is made up of various pieces of apparatus needed to complete a call.

c. Keys. Keys consist of springs and contacts which make and break electrical circuits. Telephone keys are two- or three-state devices. They have a normal position and one or two separately operated positions. As suggested in Fig. 2, when the switch handle is moved to one "operated" position, only half the springs are affected while the rest remain normal.

d. Lamps. Switchboard lamps mount in sockets which fit into the same panel area as a jack. Only the end of the lamp is visible, and many lamp and jack circuits can be placed within the reach of an operator.

e. Drops and Magnetos. Drops may be used to attract an operator's attention. Certain manual telephone systems, including modern military equipment, which do not supply power to telephones from the central office have small hand-operated a-c generators called magnetos at each telephone which generate an 80-volt signal at 20 cps. The drop at the switchboard responds to a spurt of this signal by lifting a latch which permits an indicator to swing into a visible position. The indicator can be restored manually or automatically when a plug is inserted in the associated jack. Even though drops have nearly vanished from the telephone plant, the "drop side" of a trunk circuit, the side which connects to the central office, remains the opposite of the "line side," which connects to the channel to a distant switching system.

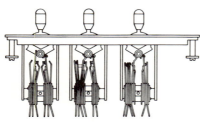

Fig. 2. Typical telephone keys.

3. Stepping Switches*

Stepping switches are in wide use today. Their advantages are ruggedness, well-known technology for both manufacture and use, and the simplicity of circuit design which is possible with multistate devices. In general, mechanical latching holds a stepping switch in the desired position, and no holding power need be expended. On the other hand, such switches are intricate mechanically, require lubrication and maintenance, and are relatively slow.

Stepping switches may be *forward-acting* or *back-acting*. The former move their wipers to the next position upon application of power, while the latter bend a spring with the application of power, permitting the spring to move the wipers to the next position when power is removed. *Homing* switches must be permitted to drop back to an initial position before they can be reused, while *nonhoming* switches can continue indefinitely in the same direction. The wipers on stepping switches may be either *bridging* (making electrical contact with one terminal before disconnecting from the previous one) or *nonbridging* (disconnecting from the first before connecting to the second).

a. Uniselectors.† Small rotary switches are typified by those shown in Fig. 3a and b. The first is called a "minor switch" and is forward-acting. With 10 outputs, it can conveniently be used as a decimal counter. The operate magnet advances the wipers one step for each pulse received up to 10. The release magnet pulls out a

---

* Reference 3, vol. 3, chap. 3; Ref. 8, vol. 2, chap. 3.
† Reference 5, chap. 9; Ref. 10.

detent when energized and allows a spring to return the wipers to the normal position. Switches of this type are available with a number of contact banks to provide a variety of output possibilities.   In addition to the bank contacts, off-normal contacts (which are closed whenever the wipers are not in their home position) are often supplied.   These contacts can be used to establish a path to the reset magnet; when the reset signal is applied by some external switch, the off-normal contacts will remove it automatically when the wipers have returned to normal.

(*a*) Minor switch. Has a normal position and 10 active positions. Must be reset before it can be used again.

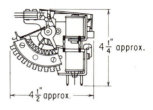

$4\frac{1}{4}$ approx.

$4\frac{1}{2}$ approx.

Typical banks

(*b*) Twenty-two-position rotary selector. Continuous stepping. Switch mechanism can be removed from banks without affecting bank wiring.

Side

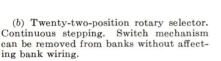

$6\frac{7}{8}$"

2"

$6\frac{1}{4}$"

Top

Typical wipers

Fig. 3. Uniselectors.   (*Courtesy of Bell Telephone Laboratories.*)

The switch shown in Fig. 3*b* has 22 positions over 180 degrees of arc, although various other combinations are possible.   This switch is back-acting and nonhoming but can be made to home with suitable circuitry.   A switch of this sort can be made self-stepping through use of a relay-like contact on the coil which opens when the coil is fully energized.   Breaking the path to the coil permits the spring to step the wipers and reapply drive.

b. Two-motion Switches.[11,12]   When many outputs must be reached, two separate wiper motions can be used.   The Strowger step-by-step (SXS) switch (Fig. 4) connects two wires for the talking path and a third wire for control purposes to any one of 100 outputs; when extra banks of contacts and extra wipers are added, 20 outputs per level can be obtained. Other variations are in use. A train of pulses operates a vertical magnet to step the wipers to a particular level, and another train of pulses operates a rotary magnet to produce stepping to one particular output on that level. The rotary pulses may be generated by control relays mounted on the switch itself.

A "double dog" provides a single mechanical latch for both vertical and rotary motions. Operation of the release magnet disengages the double dog, permitting a spring to restore the rotary motion followed by gravity restoral of the vertical motion. Off-normal springs are provided as in the case of the minor switch to terminate the release operation automatically or perform other functions. One of the newer SXS switches does not use a release magnet; rather, it restores by putting its rotary magnet into a self-stepping condition until it reaches the end of the arc. Vertical restoral occurs first, and a spring then returns the wipers in the rotary direction.

4. Motor-driven Switches*

Most large-output switches are motor-driven. A switch returns a train of

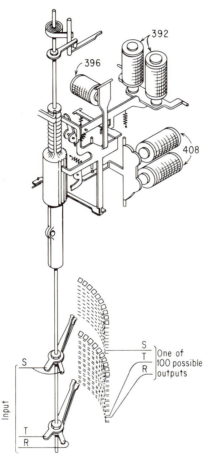

Fig. 4. Early patent drawing of a step-by-step (SXS) switch showing mechanical operation. The vertical magnet is 392, the rotary magnet is 408, and the release magnet is 396.

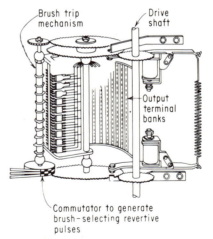

Fig. 5. Rotary switch.

"revertive" pulses to indicate its position to a control mechanism which releases a clutch when the proper position is reached (see paragraph 15a).

Rotary switches (Fig. 5), first with 200 and later with 300 sets of three-wire outputs, have been developed. Ten sets of brushes are provided, and revertive pulses position

* Reference 3, vol. 3, chaps. 4 and 5; Ref. 8, vol. 2, chap. 20.

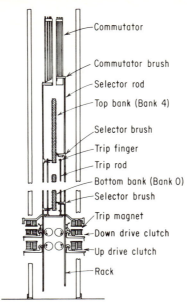

Commutator

Commutator brush

Selector rod

Top bank (Bank 4)

Selector brush

Trip finger

Trip rod

Bottom bank (Bank 0)

Selector brush

Trip magnet

Down drive clutch

Up drive clutch

Rack

FIG. 6a. Panel switch. (From *Telephone Theory and Practice, Vol. 3, by K. B. Miller. Copyright © 1933 by McGraw-Hill, Inc. Used by permission.*)

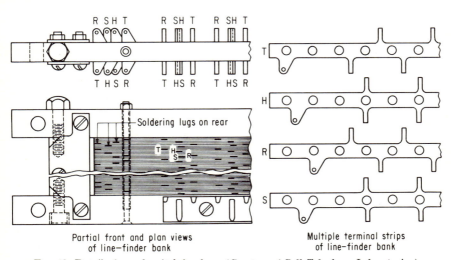

Partial front and plan views
of line-finder bank

Multiple terminal strips
of line-finder bank

FIG. 6b. Detail of panel switch banks. (*Courtesy of Bell Telephone Laboratories.*)

a mechanism to trip exactly one of the brush sets. The brush carriage is then rotated, tripping the selected brush set and sweeping it over the terminals until the desired output is reached. The Rotary System, using switches as described, is extensively used in Europe and South America.

In the United States, the panel switch was developed with an even larger access. This switch, shown in Fig. 6, moves vertically in a straight line over the five panels of output terminals from which it derives its name. In each panel, 100 three-wire outputs are available. Choice between the panels is made by tripping of one of five sets of brushes on a long vertical rod. As in the rotary system, the trip mechanism for one brush is set and then the rod is driven past it to trip the brush and reach the desired terminal. A ratchet mechanism holds the rod at the proper level during use, and a release magnet frees the ratchet to permit restoral. A clutch couples downdrive power to the switch, and the tripped brush is released when the whole mechanism reaches normal.

5. Groupings of Large Motion Switches

Strowger, rotary, and panel switches are usually factory wired in groups with associated apparatus to serve as system building blocks. In large manual systems, a

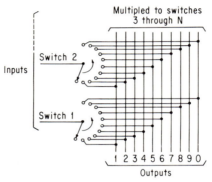

Fig. 7a. Simple multiple. N inputs share 10 outputs. Each switch has 10 output positions.

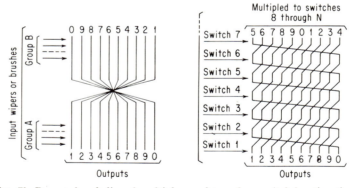

Fig. 7b. Reversed and slipped multiples used to reduce switch hunting time.

customer line will make *multiple* appearances at a number of jacks arranged to permit each operator to terminate calls to any line. Similar multiple appearances are used with switches. Special wiring techniques have evolved which (1) shorten the time required by a progressively hunting switch to find an idle output and (2) allow a group of switches to handle a number of simultaneous calls in excess of the number of terminals on any one switch.

   a. Multiple Wiring to Reduce Hunting Time. A straight multiple is shown schematically in Fig. 7a. All switches start hunting at the same point and hunt in the same direction.   Although the wiring pattern is simple, the first-encountered outputs are used most often while the remainder are used only when the earlier choices are busy.   Thus, in a period of peak traffic, a switch takes longer, on the average, to reach an idle output.

To equalize switch usage and to reduce hunting time, reversed and slipped multiples are used as shown in Fig. 7b.   In the reversed multiple, the wiring is transposed so that half the switches hunt over the outputs in one order and the rest hunt in the opposite order.   In the slipped multiple, the wiring is run so that the hunting order is preserved but the starting point for each switch is different.

The slipped multiple is used chiefly in step-by-step line finders (see paragraph 14a). Panel switches frequently use reversed multiples.   The multiple is simplified by combining wiring and terminals.   Thin metal straps with 30 projections on each side

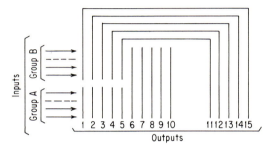

FIG. 7c. Simple graded multiple.   Two groups of 10-output switches are multipled to 15 outputs.   Switches in group A hunt over trunks 1 through 10 in order.   Switches in group B hunt over trunks as follows: 15, 14, 13, 12, 11, 6, 7, 8, 9, 10.

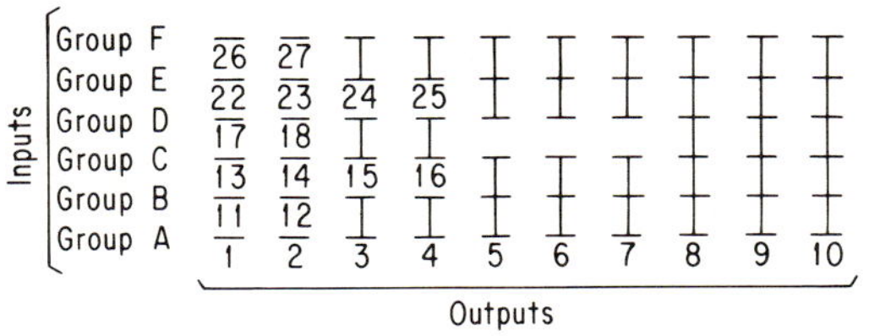

FIG. 7d. More complex graded multiple.   Six groups of 10-output switches are multipled to 27 outputs.

are stacked one above another separated by insulation.   Three hundred straps make up one panel bank, and the five banks in a frame present terminals to the brushes of up to 60 switches.   The use of these straps makes the slipped multiple impractical.

b. Multiple Wiring to Increase Traffic-handling Capacity.   Because the hunting time is usually short compared with the total time a switch is in use, slipped and reversed multiples do not increase the traffic-handling capacity of a group of switches.   Thus when 11 or more 10-output switches in a group may be needed simultaneously during peak traffic periods, the graded multiple is required.   For example, as shown in Fig. 7c, switches can be divided into two groups, each with exclusive access to five outputs.   Both groups of switches then overflow into the remaining five outputs.   The graded multiple idea can be expanded as shown in Fig. 7d (Ref. 8, Vol. 2, Chap. 2).

The graded multiple approach is widely used when small groups of any sort of circuit are provided for high first-choice use while backup groups are available for common overflow.[13]

6. Small Motion Switches

Small motion switches are rugged and inexpensive, need little maintenance, operate quickly, and permit a wide variety of design possibilities. They are usually two-state devices and must be used in relatively complex circuits to carry out even simple functions.

a. Relays.[14–18] Relays are switches operated by electromagnets. The electromagnet can operate many switches at the same time, causing some circuits to open and others to close. Several coils can be used on the electromagnet to permit special control functions. Permanent magnets can be used in association with electromagnets to achieve latching or polar effects, and the electromagnet core can be made of hard magnetic material to produce latching in an alternate way.[19] Figure 8 shows typical relays, and Fig. 9 shows typical relay circuits. Relay circuit design is discussed in Refs. 5, 20, 21, and 22.

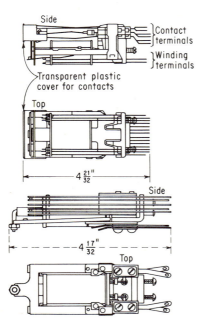

FIG. 8. Typical telephone relays. (*Courtesy of Bell Telephone Laboratories.*)

Relay contact assemblies are available as "makes," "breaks," and "transfers," where make and break refer to the operation taking place when magnetic pull is exerted. A transfer is a make contact and a break contact with one common spring. The path from the common spring through the break contact can be made to open before the make contact closes (early-break-make), or the contacts can be designed to bridge (early-make-break or continuity transfer). More elaborate contact sequences can be provided in special cases.

Relays can be classified into various types according to method of operation. A polar relay is one which operates on current flowing through its coil in one direction but not the other. A sensitive relay is one with a very low operate value; usually 5 ma or less fits this category. A marginal relay is one where the ratio of nonoperate to operate current is very high—approaching 90 per cent. Slow-operate relays are ones which delay closure of their contacts until current has been present for a time on the order of 50 msec or so. Dashpots or thermal delays can be added to provide even slower operation. Slow-release relays can provide release times up to several hundred milliseconds. Differential relays (requiring two or more windings) will operate only when the algebraic sum of current in both windings is above the operate value. Ordinary general-purpose relays will operate in something like 10 to 25 msec, release in a similar period, are insensitive to current direction, and have a nonoperate-to-operate ratio on the order of 50 per cent.

In general, telephone relays operate from the −48-volt central-office battery. Twenty-four-volt batteries are used in local and toll manual switching systems, and many 24-volt relays are also available. In addition, there are special relays for operation in series with lines, trunks, resistors, and other relays.

Energy is stored in the inductance of each relay coil while it is drawing current. If contacts in the circuit interrupt this current, the stored energy causes an abrupt rise in voltage which radiates noise and may cause an arc at the contacts. Such arcs erode contacts and shorten their useful life; contact protection networks consisting of a resistor and a capacitor are connected across relay coils where needed to prevent high-voltage surges.[23] Diodes and varistors (paragraph 9*b*) can also be used.

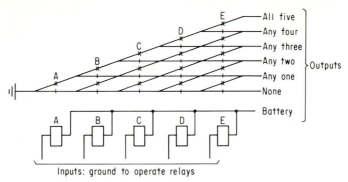

FIG. 9a. Symmetric circuit. Ground appears on an output related to the number of relays operated.

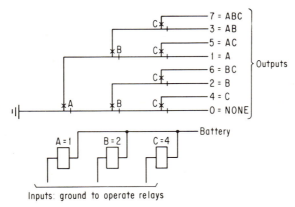

FIG. 9b. Tree circuit. Each combination of operated relays grounds a different output.

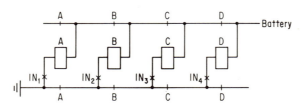

FIG. 9c. Lockout circuit. Only one relay at a time can be operated.

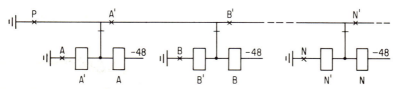

FIG. 9d. Counter circuit. "Prime-pair" counter. P closed operates A. P released operates A'. The next P closure operates B, etc. Twenty relays are needed to count 10 pulses.

In recent years, transistors have come into wide usage in many types of switching circuits formerly dominated by relays. Where speeds 10,000 to 100,000 times greater than a relay can provide are needed, the transistor is necessary; however, where high speed is not required, relays are often a much more desirable choice. Functionally, each contact on a relay is roughly equal to a diode or a transistor and is much less expensive. From a circuit standpoint, current can flow in both directions through a closed relay contact and a relay can switch a number of independent circuits simultaneously, keeping all completely separate and separate from the circuit of the activating coil. Relay contacts and coils can better withstand exposure to voltages and current found on customer lines and trunks than can most semiconductors.

b. Crossbar Switches. A crossbar switch[24–26] has 10 or more specially designed relay-like structures mounted side by side. A typical example is shown in Fig. 10. Fixed contacts in these structures are metal stampings arranged in a vertical position;

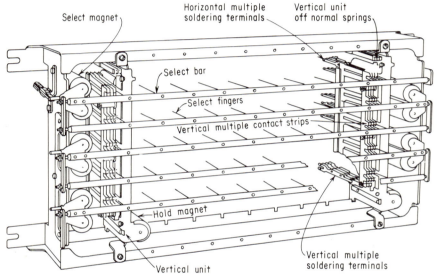

Fig. 10. One hundred-point four-wire crossbar switch, skeleton view. (*Courtesy of Bell Telephone Laboratories.*)

each stamping can make contact with selected members of a large number of moving springs. These moving springs are associated with multipling which runs horizontally across the rear of the array. The frame carries a set of mechanical interlocks which select the contacts to be closed. In this way, one "vertical" (associated with three or more leads) can connect through to a chosen "horizontal."

The sequence of operation is as follows: A select magnet associated with the desired horizontal level is operated first to activate all the mechanical interlocks (called "select fingers" in Fig. 11) on that level. One vertical, or relay-like structure, is activated by operating a hold magnet which pulls its "hold bar" against the select finger. Thus only the contacts at the point where the chosen vertical and horizontal intersect are closed. The select magnet can then be released, and the connection left under control of the hold magnet.

Select fingers on one bar are moved up or down by different select magnets to choose between two adjacent horizontal output levels. Thus the two select magnets associated with one set of select fingers cannot be operated at the same time. It is possible, however, for any two select magnets *not* associated with the same set of select fingers to be operated simultaneously. The hold magnet will then connect the vertical to two horizontals.

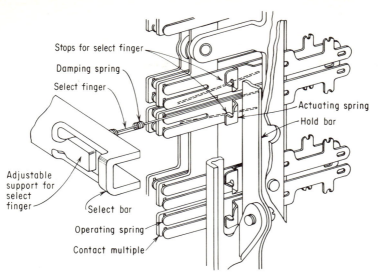

Stops for select finger

Damping spring

Select finger

Actuating spring

Hold bar

Adjustable support for select finger

Select bar

Operating spring

Contact multiple

FIG. 11. Crossbar switch operating principle. (*Courtesy of Bell Telephone Laboratories.*)

In many cases, off-normal springs are used with the hold and select magnets. When a customer line is associated with a vertical, off-normal springs on the hold magnet take the place of the cutoff relay in the line circuit (see paragraph 16a and Fig. 41).

Crossbar switches are generally used in switching networks, but they can also be used for other purposes such as decimal digit registers. Hold magnets can be made to latch magnetically[27] to reduce holding power at the expense of more complex release.

c. Reed Switches and Arrays. Reed switches[28-30] are special types of relays. They consist of contacts controlled by an electromagnet, but as opposed to the relays previously discussed, the springs or reeds are part of the magnetic circuit as well as the electrical circuit being controlled. As shown in Fig. 12, the electromagnet surrounds the reeds and, when a field is applied, the magnetized reeds come together. Restoring force is provided by the reeds themselves as flexed cantilevers. The reeds are sealed in glass tubes to prevent contamination from the atmosphere and air-carried particles. Operate and release times are on the order of 1 to 5 msec. Magnetic latching is possible.

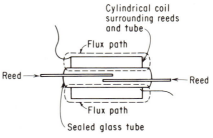

Cylindrical coil surrounding reeds and tube

Flux path

Reed

Reed

Flux path

Sealed glass tube

FIG. 12. Reed relay. Reeds are part of the magnetic circuit as well as the electrical circuit being controlled. When current flows in the coil, magnetic flux inside the tube pulls the reeds together.

In addition to "dry" reeds, mercury-wetted reeds[31] are available. Mercury, by capillary action, works its way back to the contact area from a reservoir to replace any surface erosion. Since minor erosion is thus self-healing, mercury-wetted relays can have very long contact life. Their contacts also have the useful property of not chattering or bouncing on operate or release, and they open with a snap which produces the nearest known approximation to a step function. Thus mercury relays can be used as the outputs from pulse generators where very square wave forms are required as long as 100 operations per second or less are satisfactory.

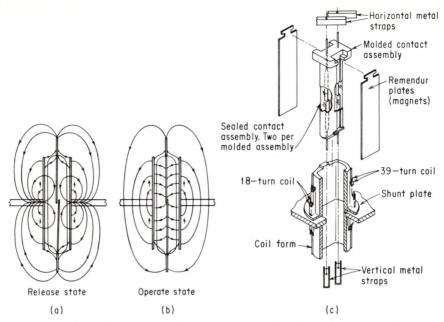

FIG. 13. The ferreed switch. *a* and *b* show the magnetic flux patterns produced by the magnets to permit the reed contacts to open and to cause them to close, respectively. The exploded view (*c*) shows the construction of the switch as well as the method of mounting in a shunt plate of magnetic material. (*Courtesy of Bell Telephone Laboratories.*)

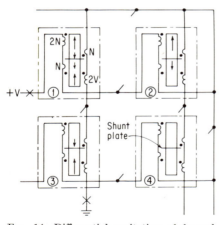

FIG. 14. Differential excitation of ferreed switch magnets. A current pulse flows from $+V$ to ground through the top horizontal and left vertical. Only in switch 1 do both coils with $2N$ turns dominate to set the flux in the same direction (adding) throughout the magnet. All other switch magnets on the same horizontal and vertical have opposite ends set opposing. Switch 4 is unaffected by the operation shown.

Reed switches can be grouped in several ways. One coil can surround several units to get multicontact effects. Much larger arrays of reed switches have found their way into distribution networks.

Of particular interest is the ferreed,[32] a magnetic latching reed switch typically mounted in an 8×8 array and used like a crossbar switch. Since the 8×8 array is only slightly larger than one crossbar switch vertical, considerable space saving is possible. As shown in Fig. 13, the ferreed is made from two magnets of square loop magnetic material, with a flux return path via the reeds. When both halves of each magnet are set in the same direction, the flux is continuous through magnets and reeds to produce closure. When the halves of each magnet are set so that their polarities buck, the net magnetomotive force (mmf) across the reeds is very low and the reeds release. The magnets can be switched with very short pulses, although the reeds may take milliseconds to close.

The ferreeds used in the Bell System's No. 1 ESS (Electronic Switching System) are set by differential excitation. In the simplified array shown in Fig. 14, battery is applied to the top row and ground to the left-hand column. Thus current flows through both sets of windings in switch 1, through only one set of windings in switches 2 and 3, and through no windings in switch 4. In switch 1, the reeds operate. In switches 2 and 3, the reeds release. In switch 4, current flows in neither winding and the switch is left in whatever state it may have been before the pulse was applied.

Use of the same current in both sets of windings of the switch to be set assures tracking of the mmf without any special voltage supply or coil resistance tolerances. Because the act of setting up a path between a given input and output automatically destroys all other connections to that input and output, no special action need be taken to release paths when no longer needed.

7. Inductance, Capacitance, and Resistance

Because of the −48-volt central-office battery and the 10- to 200-ma typical range of currents in relay circuits and customer lines, the ability to dissipate about 5 watts of power under normal operating conditions is implied for most resistive components. Thus inductances (retard coils in telephone terminology) and resistors tend to be relatively large. Transformers (or repeating coils) used in speech paths have very low resistance but must be physically large to prevent magnetic saturation by currents limited by conductor resistance or other components. Capacitors in circuits which extend outside the central-office building must be capable of withstanding lightning surges up to 600 volts and crosses to ordinary power lines.

In addition to power requirements, precision over prolonged periods of time (typically 40 years) is necessary. Thus most resistors are wire-wound, and many early relays also supplied resistors in the form of noninductive windings over their regular coils.

Because speech circuits must be balanced to ground (see paragraph 17), strict balance requirements are imposed on many components. Thus certain pairs of capacitors must be matched with more accuracy than they approach their nominal value. Similarly retard coils, supervisory relay coil windings, and transformers usually require a closer match between the halves of one device than between similar devices.

ELECTRONIC COMPONENTS[33]

Because general information about electronics is readily available, only components or features thereof specifically related to telephone switching will be discussed below.

8. Cold-cathode Gas Tubes

a. Three-element Tubes. When not ionized, these devices present open circuits between their cathode, starter anode, and main anode. For typical tubes, voltages below 70 volts across the starter gap and 150 volts across the main gap will not produce ionization; however, when the starter anode is made more than 70 volts positive relative to the cathode, a glow discharge forms which rapidly transfers to the main anode. The tube breaks down with a sustain voltage of about 70 volts between the main anode and the cathode that is nearly independent of current. A sensitive relay can easily be operated between the tube and the +120-volt power supply normally used for coin control (see paragraph 16i).

Once the main gap is ionized, the starter gap has no further control over the tube. This "latching" property is a major advantage; another is the negligible power required by the starter anode. A high-impedance source can thus trigger the tube without being loaded down. To deionize a gas tube, the cathode circuit is usually opened.

Many circuits for timing, counting, lockout, etc., have been devised with three-element gas tubes (Ref. 5, Chap. 10). Further variety is possible when additional electrodes are added.

b. Two-element Tubes. Two-element gas tubes have a relatively high ratio of breakdown to sustain voltage (190 to 110 volts in one instance). In addition, they exhibit the property of lockout: If two tubes in parallel (sharing a common current-limiting load resistor) break down simultaneously, one robs all the current from the other, which promptly extinguishes. This combination of properties has led to their use in "end-marked" experimental switching networks (see paragraph 15b).

In general, gas tubes have the following disadvantages: They require voltages appreciably higher than the −48 volts normally used in telephone systems. They dissipate power (typically about 1 watt per tube) when conducting. Ionization times are relatively long (up to, say, 75 μsec) even when special activation techniques are used. Finally, deionization times are even longer: on the order of 1 to 10 msec.

9. Semiconductor Devices

a. P-n-p-n Switches.[34] Multijunction semiconductor devices, available with either two or three leads (corresponding to two- or three-element gas tubes), have been used in experimental switching networks as crosspoints and also as control elements for reed switches. They are much faster than gas tubes and operate at voltages well

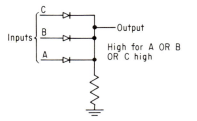

Fig. 15. Diode OR gate for positive-going signals. The output voltage equals the most positive of the voltages at A, B, or C.

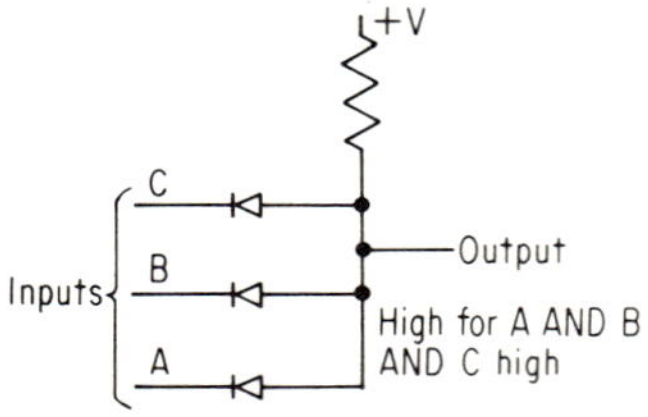

Fig. 16. Diode AND gate for positive-going signals. Output voltage equals the least positive of the voltages at A, B, or C.

within the range of traditional power supplies. As with gas tubes, they present a very high impedance when not broken down and a low impedance when conducting. The sustain voltage is very low (on the order of a volt or two), while the breakdown voltage can be made quite high within fairly wide limits.

b. Diodes and Varistors. Diodes and varistors are two-terminal devices with nonlinear voltage-current characteristics. A diode differs from a varistor in that diode voltage-current characteristics are asymmetrical for opposite polarities.

Varistors[35] are high-resistance devices at low voltages and low-resistance devices at high voltages. The knee in the voltage-current characteristic curve can be designed to occur at voltages as low as 0.2 volt or as high as 60 or 70 volts. The knee between high and low resistance can be quite abrupt or relatively smooth. Varistorlike properties can be created by using two or more diodes in series or parallel; some varistor units are manufactured this way.

Diodes[36] have low resistance when forward-biased and high resistance when back-biased. Above a given reverse voltage, however, they again have a low-resistance region. In many cases, this reverse breakdown is undesirable and is made as high as possible. In other instances it is used as a reference voltage; semiconductor device theory permits the reverse breakdown voltage to be designed quite accurately over a very wide range of voltages. Power dissipated at the junction must, of course, be limited. The forward drop is relatively fixed, usually being approximately 0.2 volt for germanium devices and 0.6 volt for silicon.

Diodes are used extensively in logic circuits. Figure 15 shows an OR gate where the output voltage is the highest of the inputs A OR B OR C, etc. Figure 16 shows an AND gate where the output is high only if the inputs A AND B AND C, etc., are high; that is, the output is clamped to the lowest of the inputs.

Logic design with diodes is inexpensive and effective[37] and is widely used in digital computers and control circuitry for switching systems. However, it can be shown[22] that the logical operations of AND and OR are not sufficient; negation or inversion is also required. Furthermore, as a practical matter, amplification must be added between every few stages of diode logic. Thus transistors enter into design consideration.

c. Transistors.[38,39] Transistors are three-terminal current amplifiers. A small current of the proper polarity flowing in the circuit through the base and emitter terminals will cause a much larger current flow in the circuit including the collector and emitter terminals. In suitably biased circuits, the base current controls the collector-emitter current to produce an enlarged linear replica of itself. In switching circuits, however, either the collector-emitter path is cut off or else it draws enough current through the load resistor and/or the following stages to be driven into saturation. Some typical transistor logic circuits are shown in Fig. 17. It takes only a few standard arrangements of gate circuits to build a wide variety of digital equipment (Ref. 40, p. 2055).

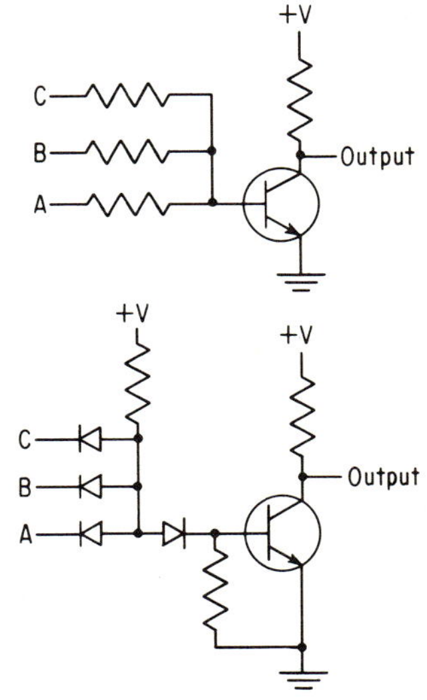

Important parameters are the current gain of the transistor (minimum ratio of collector current to base current), and the difference between the maximum voltage from collector to emitter and the minimum voltage from the base to emitter when the transistor is saturated. Operating speed can be specified in terms of the turn-on time required to switch from cutoff to saturation, storage time to bring the transistor back to the edge of saturation, and turnoff time.[41]

When a transistor is cut off, temperature-sensitive leakage currents flow across back-biased junctions. They pose a considerable problem in germanium transistors, and special design procedures must be taken.[42] Silicon transistors, however, have leakage currents several orders of magnitude smaller. Safe base-emitter back-biasing voltages tend to be relatively small for some transistors. Collector-base junction reverse voltages can be made relatively high, although, to permit the use of small, inexpensive devices, voltages under 30 volts are often specified. A phenomenon known as second breakdown[43] is particularly important where an inductive load such as a relay must be operated by a transistor. The inductive surge when the relay current is cut off can destroy the transistor. A transistor can be protected by a diode clamp, although this increases the release time of the relay.

Fig. 17. Typical transistor logic circuits. Top: Transistor-resistor (TRL) logic. Output is high when A AND B AND C are grounded and low when A OR B OR C are high. Bottom: Low-level logic (LLL). Output is high when A OR B OR C is grounded and low when A AND B AND C are high.

Transistors can be made to operate from both polarities of supply voltages (p-n-p for negative and n-p-n for positive potentials). Unfortunately, silicon transistors can best be fabricated (1965) as n-p-n units, making the use of the existing standard −48-volt supplies inconvenient.

d. Integrated Circuits.[44] Transistors, resistors, capacitors, and diodes can be fabricated in one process, producing entire circuits rather than individual components. If the contents of building blocks are increased, external interconnections are greatly reduced. Because of the small physical sizes involved, however, those interconnections which remain pose some problems. Similarly, reliability and maintainability problems differ from those found in earlier practice.

10. Magnetic Memory Systems

Telephone switching systems need large amounts of memory for storing numbers dialed by customers, translation data for routing between and within systems, and further information concerning the setup of calls and charging. In general, this information breaks down into two categories: that which is unique per call, such as the called number, and that which might be needed on any call, such as routing translations. In electromechanical systems, the former is usually stored in the positions of the switches themselves or in relay registers while the latter is available via special circuits using cross-connection fields to permit occasional changes. Temporary or "scratchpad" memories and semipermanent memories must be carefully organized in electronic switching systems for maximum compatibility with control units. Large capacity and fast access to any stored item become vital; by changing the information stored in memory, it is possible to change the operation of the entire system.

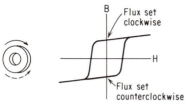

FIG. 18. Magnetic core and *B-H* curve.

Magnetic memory is based on the fact that "hard" magnetic materials will "remember" that they have been in the presence of an external magnetic field by becoming magnets themselves, retaining their magnetism after the external field is removed. The memory field can be sensed by (1) observing its effect on other magnetic circuits or (2) trying to change it and observing the results.

a. Cores, Transfluxers, and Sheets.[45]   Perhaps the simplest magnetic device is the core, a small loop of ferrite or other hard magnetic material. A core and a typical *B-H* curve are shown in Fig. 18. The core can be saturated with its flux going in either direction by simply applying enough magnetomotive force (mmf) of the appropriate polarity. Even when the mmf is removed, the flux in the core remains high. If additional pulses of mmf in the same direction are applied, the flux will not change much, since the core is already saturated, and an output winding of N turns will show only a small voltage $e = -N \, d\phi/dt$, where $d\phi/dt$ is the time rate of change of the flux. If enough mmf of opposite polarity is applied, however, the flux will change from saturation in one direction to saturation in the other, $d\phi/dt$ will be large, and a large output voltage will be available.

Cores are usually organized in rectangular arrays as shown in Fig. 19 to reduce the amount of circuitry required to give access to any particular core. A 1,024-bit (at one core per bit) memory can be made up of 32 rows of 32 cores per row. The selec-

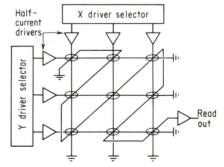

FIG. 19. Coincident current core array.

tion of one row and one column will identify the unique core at the intersection. If something more than half the current needed to saturate a core is applied to the row and to the column, only the core at the intersection will receive enough mmf to be switched. All the others will simply "shuttle" along the saturation curve but will fall short of going around either knee. Whether the selected core switches or not depends on whether it is already saturated in the appropriate direction.

The coincidence of two "half-write" pulses will cause one core to be saturated in the "set" direction. The coincidence of two "half-read" pulses (of polarity opposite to the half-write pulses) will reset the core to produce a voltage on the readout winding.

If the core is already reset, the two half-read pulses will not be able to produce an appreciable output.

The reading of a set core destroys the information stored in it ("destructive readout"). If the information is to be retained, it must be restored with a separate write action. Transfluxors are cores which are read out by attempting to switch only the flux in the vicinity of the small hole as shown in Fig. 20.

As suggested by the transfluxor, the affected region around a hole in a piece of magnetic material can be quite small. A sheet of ferrite pierced with many small holes can be made to behave as so many cores; the holes can be placed quite close together (about $\frac{1}{20}$ in.) for a high bit density. To facilitate overall memory-system fabrication, one access conductor can be deposited or plated on the ferrite sheet directly in the manner of printed wiring. A typical sheet of 256 holes is shown in Fig. 21 (Ref. 40, p. 2147).

Experience has shown that "word-organized" memories are often more practical in telephone systems than "bit-organized" memories in that they permit much larger quantities of information to be withdrawn and returned to memory in a given amount of time. Ferrite sheets lend themselves well to the word-organized approach.

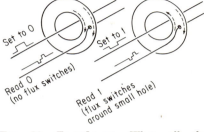

Fig. 20. Transfluxors. When all the magnetic material around the large hole is saturated in the same direction, interrogation of the region around the small hole produces no output. However, when the inner and outer flux paths around the large hole are set in opposite directions, the material around the small hole switches. Bipolar read pulses are used to ensure leaving the interrogated core material around the small hole in its original condition.

A stack of 16 sheets can be placed one above another and threaded with additional address wires required as shown in Fig. 22. With reference to Fig. 22, the Y address puts a half-read pulse through all the holes on one particular sheet. Similarly, an X address puts another half-read pulse through one of the address wires which thread the stack. Where the X "plane" crosses the Y "plane," 16 holes will receive a full readout pulse. Those set will give large outputs, and the remainder will give negligible outputs on the read conductors.

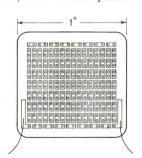

Fig. 21. A ferrite sheet, the equivalent of 256 magnetic cores. The Y address wiring is printed on each side of the sheet and through the holes to produce a continuous conductor.

Once a row of holes has been read out, all the holes in that row are reset. To restore the information in its original form or to insert other information, suitable means must be provided for setting the desired holes. If the X and Y currents are reversed, *all* the holes in the row will be reset. Thus, selective hole setting requires further action which can be applied via the readout leads. If a half-pulse of the inhibit polarity is supplied only to the holes which are not to be set while simultaneous reversed pulses on the X and Y leads are applied, the uninhibited holes will be set to await the next readout.

Ferrite sheet stores can be assembled from modules similar to the one described, providing in one memory over 8,000 words of 24 bits each. The space occupied by ferrite sheets is quite small; most of the space and a large portion of the cost of the overall unit go into the access circuitry. A required memory word is identified by a set of pulses received on a number of parallel buses (see paragraph 14b). In the access circuitry, this address sets a translator which selects and pulses the X and Y leads. The signals on the outputs are detected and transmitted to data-processing equipment over another bus system.   Finally, the access circuits rewrite the information with or without modification, as required.

b. Twistors.[46]   Twistor memories can be organized in a number of ways, but the version used in the Bell System's No. 1 ESS will serve as an illustration (Ref. 40, p. 2097).   Here the basic memory unit is a thin 12- by 7-in. card with space for 64 columns of 45 tiny permanent magnets per column.   These magnets can be demagnetized individually by an external card "writer," and the presence or absence of an active magnet at each location constitutes the information stored.   The cards are arranged in modules of 128 which use common sensing equipment.   A complete memory contains 16 modules for a total of more than 130,000 words of 44 bits each.

The sensing element is a thin magnetic tape twisted around a copper conductor; this is the "twistor" wire.   Figure 23 shows how twistor wires, mounted on a plastic belt for structural strength and proper alignment, pass under the card magnets.   The twistor wire is held saturated by flux from the nearby magnets, but demagnetized magnets have no effect on it.   Thus, when the tape under a card magnet is interrogated, it either switches or does not switch, depending on the state of the nearby

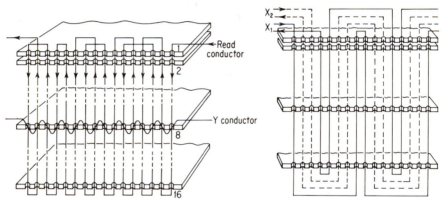

Fig. 22. Address and readout windings threading a stack of ferrite sheets.   (*Courtesy of Bell Telephone Laboratories.*)

magnet.   Because the twistor tape is wrapped around its copper wire some fifteen times in the area dominated by one magnet, a voltage step-up takes place when the tape is switched.   The copper wire is actually a one-turn transformer secondary with the transformer core wrapped around it, the reverse of the usual procedure.   The primary of the transformer is a pulsed "solenoid loop" running at right angles to the twistor wire, on the opposite side from the card magnets.   Thus, the portion of all 44 twistor wires lying under one row of magnets can be pulsed to sense the magnet states. Actually, two sets of twistors are used, one on each side of the solenoid loops; 88 magnet conditions (two words) are sensed at one time, with half of the information selected by external access circuitry.

Each solenoid loop is pulsed by a large magnetic core, selected by coincident currents in a 64 × 64 matrix.   The cores are normally biased "set" with direct current; coincident pulses combine to overpower the bias winding and reset one particular core which switches to produce the interrogate pulse.   When the core-select currents terminate, the bias switches the core back to produce a reset pulse for the twistor wires.

The twistor memory as described is permanent until its magnets are changed externally.   This form of "nonvolatile" memory is highly useful in that power failures and circuit troubles will not cause a loss of information.   Thus, the "program" (or instructions for the data processor) can be safely stored in a twistor memory.   Data, such as line-to-equipment-number translations and trunk group and alternate routing information (see paragraph 14d), are also stored in this manner without fear of loss under trouble conditions.

An electrically alterable twistor memory called the piggy-back twistor[47] uses a second layer of magnetic tape instead of magnetic cards. The inner layer is of soft magnetic material, while the outer layer is hard. The mmf of the hard material magnetizes the soft material under it; the solenoid loop interrogates the soft material without switching the hard material. The hard material can be set or reset by use of coincident pulses in both the solenoid loops and the copper conductors of the twistor wires.

c. Magnetic Drums. Magnetic drums are motor-driven cylinders coated with magnetic material. Record and play-back heads are mounted in a row along a line parallel to the axis of the cylinder. As the drum revolves, these heads can set or reset the magnetic material passing under them or they can read out the set or reset condition without altering it.

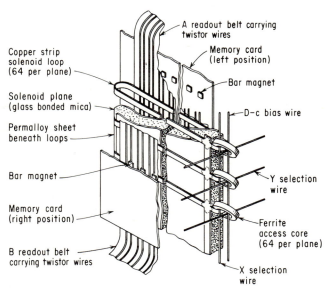

Fig. 23. Cutaway schematic showing the basic elements of a permanent-magnet twistor memory. (*Courtesy of Bell Telephone Laboratories.*)

To identify the location of a particular bit of information, one track (the area which passes under a given head) can have alternate set and reset magnetic states all the way around the drum. As the drum revolves, it reads out pulses for each set area; these pulses step a counter which recycles with every drum revolution. Thus, a particular bit can be found by specifying the counter reading for the angular position and the head for the position along the drum. A drum can be bit organized, or if all the tracks are read out for a given counter reading, it can be word organized. A typical unit has 4,000 words (discrete angular positions) of 120 bits each (120 readout heads) and revolves at 1,800 rpm (Ref. 48, p. 628).

Drums have two disadvantages compared with cores, sheets, and twistors: The drum is a moving part which requires maintenance, lubrication, adjustment of head clearance, etc., and the information stored thereon is available only when it appears periodically under the readout heads. Thus, the drum is not a "random-access" device where any one piece of information is as accessible as any other. However, drums are simple, inexpensive, and particularly well suited to certain tasks such as scanning or translation described in paragraphs 14d and 16a.[49]

d. Magnetic Tapes. Magnetic-tape memories are similar to drum memories where the drum is replaced by a very long ribbon coated with magnetic material. The ribbon is kept on two reels and can be run forward and backward to locate the desired information.   Because tape is relatively narrow, fewer recording tracks can be used.

Magnetic tapes can store vast quantities of information, and if reels can be changed as desired, there is literally no limit to the amount of information which can be made available.   However, access time to a particular piece of information gets longer directly with the length of the tape.

In telephone systems, tape is well suited for the recording of billing information (see paragraph 16*i*).   The large capacity of the tape and the nature of the billing signals to be recorded (continuously progressing in time with no need for reversals of the tape) combine to favor this medium.

11. Delay Lines and Other Memory Systems

Ultrasonic delay lines[50] combine the cyclic access of magnetic drums with the destructive readout of magnetic cores.   Up to 20,000 bits can be stored in a line with slightly more than 11-msec delay, but commercial units with delays of 100 μsec to 2.5 msec which store 1 bit/μsec are more common.

In a delay line, an electrical pulse is converted to a mechanical shock wave by a transducer.   The shock wave travels through a physical medium at a speed slow compared with electrical signals.   When the shock wave reaches the end of the medium, a second transducer converts it back to an electrical pulse.   External circuitry reshapes and times the pulse and reapplies it to the delay line for another trip.   In this way, information can be circulated indefinitely; alternatively, it can be changed between any two trips.   Delay lines have no moving parts, are inexpensive per bit, are small and light, and have a high-level output signal.   They are ideally suited for processing sequential information typical of scanning and other telephone operations.   Stromberg-Carlson's Dynalogic® makes effective use of delay lines[51] in telephone applications.

Various other memory systems have been used, at least experimentally, in telephone systems.   Electrostatic storage tubes, effectively arrays of small capacitors charged as required by the electron beam of a cathode-ray tube, have fast random access, and their contents can be altered electronically (Ref. 52, p. 1195).   For semipermanent memory, photographic plates with exposed or unexposed dots at particular locations can be used to block or pass a light beam.   To change the stored information, an external plate exposure unit is required (Ref. 52, p. 1161).

THE DESCRIPTION OF SWITCHING SYSTEMS

12. Drafting Symbols

Although switching systems can be described broadly in block-diagram form, more complete drawings are usually needed to illustrate and clarify operating principles. The working drawings for an actual system weigh many pounds, and their preparation is a major portion of any development effort.   Drafting standards and practices differ from one telephone administration to another, but there are basic symbols that are useful to all.

a. Early Symbols: Mechanical Operation and Location.*   Early electrical symbols were stylized representations of the actual components and, as such, suggested to even unskilled labor their method of operation.   In telephone drawings where relays, switches, and jacks were the dominant circuit elements, mazes of lines crossed and recrossed one another in an effort to represent the interconnections among contacts on various relays.   Many years of training were required before such drawings could be read easily, and in large circuits, the cost of drafting became prohibitive.

b. Modern Symbols: Circuit Operation.†   To reduce the complexity of drawings containing many relays and switches, several different forms of "detached con-

* Reference 3, vol. 2, chap. 3.
† Reference 8, Vol. 1, chap. 2.

tact" schematics were developed. In such drawings, symbols for contacts and coils are separated from each other and are located in terms of their function rather than their physical construction.

As a further step in reducing complexity, the representation of contacts can be streamlined[53] by showing a make contact as an x and a break contact as a bar. Contacts can thus be superimposed on circuit paths with a saving in drafting time and effort. Although such symbols are convenient and are generally used in the current periodical literature, they do not differentiate among the types of transfers or other special contact arrangements which are available.[16] Figures 24 to 26 show symbols frequently encountered in telephone literature.

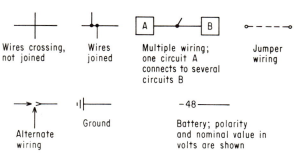

FIG. 24. Frequently encountered wiring symbols.

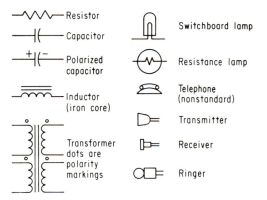

FIG. 25. Commonly used component symbols.

Electronic circuitry usually conforms to the practices of the electronics industry and where digital circuitry is used, the standard gate symbols of the computer field are commonly adopted.

c. Sequence Diagrams. In a circuit containing many relays, it is often difficult to follow circuit operation. Verbal descriptions such as "Relay A operates when the input switch Z is thrown unless relay C is released or relay D is operated" often lead to more confusion than clarity. Sequence diagrams are frequently provided to clarify the situation. A portion of a typical sequence chart, illustrating a useful set of conventions, is shown in Fig. 27.[53]

d. Information Which Working Drawings Convey. The documentation of a system is no simple matter, and different administrations have different procedures. However, in one form or another, the following information is generally required.

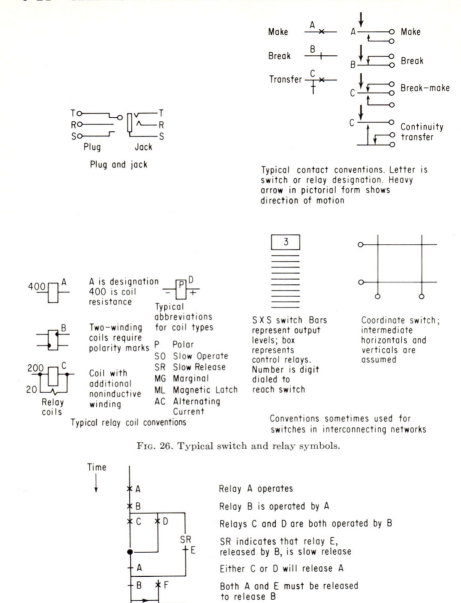

FIG. 26. Typical switch and relay symbols.

FIG. 27. Sequence chart conventions typical of those used with relay circuits.

Administrative Information.   Overall indexes to system drawings are needed, and each drawing is indexed to facilitate the location of critical leads, components, apparatus, options, etc.

Functional Information.   Schematics and sequence diagrams show how a circuit works and how it is related to other circuits.   Written descriptions are usually available to supplement the drawings.

Building and Ordering Information.   This includes options and rules for choosing among them, working limits and critical lead lengths, physical layout, and wiring instructions.

Installation and Maintenance.   Cabling and interconnecting information, transmission test requirements, information for testing and adjusting relays and other apparatus, and timing diagrams for pulse generating circuits are in this category.

13. Classification Terminology

Many types of telephone switching systems have been developed and used successfully.   They have been named after their inventors, as Strowger or Lorimer[54]; after the type of switches used in their interconnecting networks, as rotary, panel, or crossbar; or after other characteristics such as all relay (Ref. 3, Vol. 3, Chap. 6) or electronic. The classification of switching systems must, of necessity, be somewhat arbitrary; Table 2 applies the criteria discussed below to a number of systems.

a. Type of Service.   *PBX, Local, Tandem, and Toll.*   A PBX, or private branch exchange, is a switching system used entirely by one customer such as a store, business, or hotel.   Interconnection between stations is accomplished by switching at the customer's premises, reducing the traffic handled by the central office.   A PBX connects to other PBXs by tie lines or to the public telephone system via trunks.

A local switching system interconnects customer loops within the immediately adjacent geographical area.   It also has trunk connections to other local central offices and to tandem and toll offices.

The boundary between PBX and local switching is often hard to define.   Centrex[91,92] service provides PBX stations with direct inward and outward dialing.   Under such circumstances, it is often more economical to do the PBX switching at the central office (Centrex CO) rather than on the customer's premises (Centrex CU).   This may require longer line loops to the PBX stations, more traffic-handling capacity at the central office, and more elaborate signaling facilities between the central office and the attendant.   However, it centralizes maintenance, simplifies the addition and variation of features, takes less space at customer locations, and greatly reduces installation, removal, and storage costs for PBX equipment.

Tandem and toll[93,94] offices usually interconnect trunks between switching systems rather than lines between customers.   In certain instances, however, large PBX's may connect directly to tandem or toll offices.

Originally, tandem switching was used in metropolitan areas to simplify trunking among large numbers of relatively closely spaced central offices.   With trunks from each office to a tandem switching center, direct trunks between all pairs of offices are unnecessary.   Early tandem switching required little in the way of charge recording. Toll switching, on the other hand, has always required detailed information on customer-originated calls, although the toll switching office was not necessarily involved in the charge recording.   Advances in automatic switching and billing have greatly increased the range of direct customer dialing; as a result, the functions of tandem and toll offices are tending to become similar in many ways.[75]

The Direct Distance Dialing (DDD) Hierarchy.[55,95]   In the general toll switching plan evolved to permit automatic alternate call routing in the United States and Canada, central offices are designated by a class number and a name as shown in Fig. 28.   Class 5 or end offices interconnect customer lines and have direct trunks to other Class 5 offices as traffic requires.   Each Class 5 office, however, has a final trunk group to a Class 4 or higher office in the hierarchy.   The idea is to use a sort of graduated tandem principle, where Class 4 offices home on a Class 3, 2, or 1 office as a tandem point; Class 3 offices home on a Class 2 or 1 office; and Class 2 offices home on a Class 1 office.

Table 2. Classification of Some Automatic Switching Systems

Grouping brackets at left: **Stepping switches** (SXS, Director, Common control, Register-sender, ANI, Automatic ticketing, CAMA) and **Power drive** (Rotary, Panel, 756 A).

Common designation	Originated by or for	References	Principal type of service	Direct	Register	Progressive	Central	Electromechanical	Electronic	Wired logic	Stored program	Translation	Orig. and term.	Combined	Progressive	Central	Hunted path	End marked	Space division	Time division	2-wire	4-wire	Crosspoint	Pass dc, 86-volt ring	Approximate capacity	Comments	Abbreviations
				(Type of control)									(Type of network)														
SXS (Strowger)	AE	1, 2, 3, 8, 56	PBX, local	x		x		x		x			x		x		x		x		x		Wipers	x	50–10,000 lines	Used world-wide: BPO, Bell, Ind. etc.	AE Automatic Electric
Director	AE	57	EAS		x	x		x		x		Cross connect fields														AMA Automatic message accounting
Common control	Bell	59, 60	EAS		note 1	x		x	x	x		Cross connect fields electronic access														Common controls for addition to SXS	ANI Automatic number identification
Register-sender	SC	51	EAS		note 1	x		x		x		Cross connect fields															
ANI	Bell	61	EAS		note 1			x				Cross connect fields														Features for addition of automatic billing	BPO British Post Office
Automatic ticketing	AE	62	EAS			x		x	x	x	x																Bell Bell System
CAMA	Bell	63, 64, 65, 66	EAS																								CAMA Centralized AMA
Rotary	ITT	1, 3, 8	Local, EAS		x	x		x		x		Fixed	x		x		x		x		x		Brushes	x	To 10,000 lines	Used in Europe, South America	EAX® Electronic automatic exchange
Panel	Bell	8, 67, 68	Local, EAS, TDM		x	x		x		x		Out: cross-connect In: fixed	x		x		x		x		x		Brushes	x	To 10,000 lines	For large cities	
756 A	Bell	69	PBX		x		x	x		x				x		x	x				x		XBAR	x	Up to 60 lines, 10 trunks		ESS Electronic switching system

Comparison of switching systems (table rotated 90° on page).

Legend:

- IND — Independent (Non-Bell) companies
- ITT — International Tel & Tel Co.
- PBX — Private branch exchange
- SC — Stromberg Carlson
- SXS — Step by step
- TDM — Tandem
- XBAR — Crossbar
- EAS — Extended area service
- NE — North Electric

Category	System	Co.	No.	Type	Memory (cross connect / store)	Switch device	Capacity	Remarks
Crossbar	No. 1 XBAR	Bell	25, 70	Local	Cross connect	XBAR	To 20,000 lines	Replaces panel in large cities
Crossbar	No. 5 XBAR	Bell	71, 72, 73	Local	Cross connect fields	XBAR	1,000–30,000 lines	Originally suburban; now general
Crossbar	XBAR TDM	Bell	74, 75, 76	TDM Toll	Regular and "Dimond Ring" cross connect fields	XBAR	3,000 incoming 4,000 outgoing trunks	Similar to No. 1 XBAR (outgoing portion)
Crossbar	No. 4 Toll XBAR	Bell	77, 78	Toll	Card translator	XBAR		
Crossbar	AMA	Bell	80, 81		"Dimond Ring" cross connect fields			Adds automatic billing
Crossbar	Pentaconta	ITT	79	Local	Cross connect	XBAR	To 10,000 lines	
Crossbar	Morris	Bell	52	Local	Flying spot store	Gas tubes		Experimental
Reed switch and electronic	No. 1 ESS	Bell	40, 82, 83, 84	Local TDM, Toll	Twistor	Ferreed	To 60,000 lines	New Bell System standard
Reed switch and electronic	No. 101 ESS	Bell	84, 85	PBX	Twistor	Electronic	200 lines per PBX, 3,200 total	Centralized control for several PBX's
Reed switch and electronic	No. 1 EAX®	AE	48	Local	Magnetic drum	Correed	To 10,000 lines	Can be added to SXS
Reed switch and electronic	Broadband	AE	86	Data	Cross connect field	Correed		Customer can select band-width on each call
Reed switch and electronic	HE 60	ITT	87, 88	Local	Ferrite cores, cross connect field	Herkon Reed	To 10,000 lines	Stuttgart field trial
Reed switch and electronic	4-W solid state	NE	89	Military	Ferrite cores	Electronic	Up to 500 lines and trunks	
Reed switch and electronic	2-W EPBX	SC	90	PBX		Electronic	Up to 1,200 lines and trunks	
Reed switch and electronic	800 A	Bell		PBX	Wired, plug-in	Ferreed	To 80 lines	Self-contained

Note 1: Register control is used only on EAS calls.
Note 2: Several markers are used progressively. Each marker sets two or more stages of switches.

There are only nine Class 1 offices in the United States; all are interconnected with direct trunk groups. These regional centers serve large geographical areas; regions are divided into sections and sections in turn are divided into primary areas. Class 2 and 3 offices are sectional and primary centers, respectively; Classes 1, 2, and 3 are called control switching points (CSP's). Class 4 offices are toll centers if operators

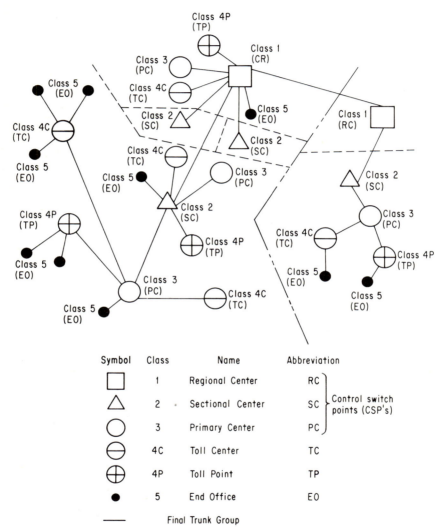

Symbol	Class	Name	Abbreviation	
□	1	Regional Center	RC	
△	2	Sectional Center	SC	Control switch points (CSP's)
○	3	Primary Center	PC	
⊖	4C	Toll Center	TC	
⊕	4P	Toll Point	TP	
●	5	End Office	EO	
——		Final Trunk Group		

Fig. 28. Direct distance dialing (DDD) hierarchy of switching systems.

are available to handle inward traffic; otherwise they are called toll points. A toll point may have operators to handle outward or delayed calls, but these operators, if present, will not handle inward calls.

It is perfectly possible to have switching systems of several classes in one building or even to have one system handling several classes of switching.

b. Type of Control. *Manual vs. Automatic.* Operators are perhaps the most flexible of common controls. From the customer's point of view, nothing is more automatic than speaking a number and getting a connection. Thus, "machine switching" has been suggested as more appropriate than "automatic." "Semiautomatic" systems have used operators to obtain the called number from the customer and key it into machinery which is automatic in all other respects.

Direct vs. Register. Strowger step-by-step switches were originally designed to be controlled directly by customer dials. Each time the dial is wound up and released, a portion of the network path is set up. When the customer dials the last digit, the connection is made to the called customer's (idle) line.

In register control, a circuit is connected to the calling customer's line to register his instructions. Dial pulses (or other signals) are assembled and stored in the form most convenient for the storage medium used. Control of network switches can be carried out by appropriate signals from the register or other circuitry; there is no constraint to use signals generated by a customer dial or even signals based on a decimal numbering system, although there must clearly be a correspondence between customer-generated inputs (usually decimal) and the derived nondecimal signals. Similarly, with register control, station signaling apparatus need not be tied down to the operating characteristics of network switches.

Progressive vs. Centralized. As will be discussed in paragraph 13c below a path through a switching network can be set up progressively, one stage at a time; alternatively, all switches needed for a connection can be set simultaneously. In a progressive network, considerable control equipment is usually associated permanently with each switch to accept sequentially transmitted information from dials or registers, use this information to carry out part of the switch-setting operation, and then generate internally such further control actions as may be needed.

Centrally controlled systems, on the other hand, locate most of the control equipment in some sort of common mechanism shared by many switches. This common equipment takes the dialed number (and sometimes other information) from a register, finds an idle path, and sets up the connection to a trunk or customer all at once. It then disengages itself for use on another call.

Register control of progressive networks is usually referred to as common control because common units control the setting of the switches and then release to serve other calls. There is, however, sufficient difference between a director or a panel sender and a crossbar marker to suggest the value of a clear distinction between register control and centralized common control (see paragraph 15).

Electromechanical vs. Electronic. Until the late 1950's, almost all telephone switching was carried out with mechanical devices operated by electromagnets or motors. Increasingly sophisticated relay circuits were developed, and markers used to control crossbar networks reached an unparalleled degree of complexity with some 1,500 relays. More recently, switching systems have used electronics most effectively in centralized control circuits, replacing relays with devices roughly 10,000 times faster. This greater speed enables one unit rather than a group to carry out all control functions on all calls simultaneously in progress in the office.

Wired-logic vs. Stored Program. It is possible to design both electronic and electromechanical switching systems on a wired-logic basis. In such cases, physical circuits are built for each function and supplied in the quantities required. For example, registers count and store dial pulses, count the number of digits received, and call in control equipment when further action is required. If it becomes necessary to add the ability to count more digits, as when area codes are needed for direct distance dialing, more relays and other equipment must be wired into each register circuit.

When a centralized memory is used in connection with an electronic control unit, all instructions for digit counting (for example) can be stored in the memory in a program similar to that used by a digital computer (Ref. 52, p. 1327). The program tells the control unit what to do with each piece of information it obtains from lines, trunks, or other inputs; rules for counting, the memory locations for digits when counted, the number of digits to expect, etc., can all be stored in memory. To change the rules, only the contents of the memory need be altered. With a new list of instructions, a program-controlled system can receive 10 digits as easily as 7, or a 2-digit abbreviated dialing code which stands for a 10-digit number. The flexibility made possible by stored-program techniques is one of the principal advantages of electronic control of switching systems.

c. Type of Network. *Originating and Terminating vs. Combined.* In metropolitan areas, where the telephone density may approach 100,000 per square mile, many local telephone offices exist in close proximity. The probability of any given call terminating in the office in which it originates is relatively small, and thus it is convenient to divide the switching equipment into two almost independent parts: The first finds a path from the calling customer to an outgoing trunk; then the second finds a path from the incoming end of the trunk to the called customer.

In manual practice, the originating or *A* operator obtained the called number, chose a trunk to the terminating office, and passed the called number to the *B* operator. The *B* operator completed the call by connecting the trunk to the called customer's jack in the multiple. Several automatic systems follow this plan.

In areas where one office serves all customers in the exchange area, most calls terminate in the same office in which they originate. Here, it is sometimes convenient to use the same jacks or switches to connect a customer to a central point for monitoring whether he is calling or being called. This approach does not lend itself conveniently to progressive systems but is widely used with centralized control.

Progressive vs. Centralized Control. The use of progressive control implies at each stage a search for an idle path to the next switch. A Strowger switch, driven to a particular level by the pulses from a customer dial, uses its own control circuitry to hunt over the 10 (or sometimes 20) paths on that level to the next switch group. In a four-digit office, the first digit eliminates nearly nine-tenths of the unwanted lines by the level selection. The switch then has a number of possible paths to hunt over, all of which lead to the same block of 1,000 lines. The second digit eliminates 900 of the 1,000 lines, after which the second switch hunts over a number of paths to the 100 lines served by the final switch. The tens digit eliminates 90 of the possible 100 by choosing one level. In general, last stage hunting is not required; rather, the switch control circuits shift to use the pulses of the units digit to select the particular terminal associated with the called party.

Because of the way in which hunting is carried out, a progressive network can find all paths busy in a second-stage hunt, although, if it had made a different selection in the first-stage hunt, the call might have progressed. Centrally controlled systems avoid this difficulty by making path hunts simultaneously on all stages in parallel rather than sequentially at each stage. The switches are set only if a match is obtained all the way through the network.

Hunted Path vs. End Marked. Whether a network is progressively controlled or centrally operated, most switching systems use some means for hunting over a number of path segments to find a matched set available from input to output. It is possible, however, to use devices or circuits for switches which, owing to their physical characteristics or construction, find the complete idle path automatically when only the terminals at each end of the network are marked with a suitable signal. The path between the marked terminals breaks down, sustains itself, and locks out connections to or from other terminals. Gas tubes and p-n-p-n diodes are devices which have been used for such crosspoints.[52] Electromechanical circuits can also be constructed to use end-marking techniques.[87,88]

Space Division vs. Time Division (Ref. 6, Chap. 5). Most switching systems use different physical paths for simultaneous connections. Space division of this sort is simple and direct but requires a large number of jacks or switches. In recent years, time-division networks have been extensively investigated. Here simultaneous calls use the same transmission path but are interleaved in time. When the two parties to a conversation are connected by suitably designed equipment to a transmission medium for, say, 1 μsec out of every 100 μsec, they can communicate freely (using a bandwidth of something less than 5,000 cps). Furthermore, microsecond samples of many other conversations can be connected via the same transmission medium during the remaining 99 μsec.

Two-wire vs. Four-wire. Telephone systems designed to serve local customers generally use a pair of wires for the transmission path and permit transmission in both directions on that pair. Two wires balanced to ground are used rather than one to reduce noise pickup, and two wires are used rather than four because, when no amplifiers (usually one-way devices) are required, a single pair is entirely sufficient and much more economical for simultaneous transmission in both directions. Local and tandem switching systems in relatively small geographical areas use two-wire switching to match the lines and trunks. Often a third wire is used in the central office for control purposes, but this does not affect the terminology.

Most toll trunks use carrier facilities which are essentially four-wire, the equivalent of one pair for each direction of transmission. This greatly simplifies repeater design and reduces echos. Echos normally occur at hybrid transformers where four-wire circuits are converted to two-wire as described in detail in Chap. 6. When hybrids are used only at the ends of the connection rather than at each repeater and each toll switching point, transmission is improved. Thus, some toll offices switch both pairs and use switches with four contacts for the speech paths (along with others as required for the sleeve or other control leads).

As the cost of carrier systems has decreased with advances in technology, it has become economical to use carrier systems on shorter and shorter trunks. Thus four-wire switching will tend to increase in importance at any office where incoming trunks are connected to outgoing trunks.

Compatible vs. Incompatible Crosspoints. Switching systems have traditionally used their networks to pass a variety of signals differing considerably from speech in both amplitude and frequency. With the metallic crosspoints of step-by-step and crossbar switches, plugs and jacks, etc., this is practical and, for economic reasons, highly desirable. Semiconductor devices, gas tubes, time-division switches, and the like, however, are incapable of transmitting ringing and coin-control potentials (see paragraphs 16*g* and 16*i*) or connecting d-c meters, bridges, or other testing devices to the line and trunk conductors. This incompatibility can be resolved by using special telephone sets, providing more complex circuits for each line at the central office, or both.

SWITCHING-SYSTEM DESIGN

14. Organization of Switching Networks*

The switching network and its control account for a significant portion of the size and expense in a switching system. To interconnect 10,000 lines and trunks in a single office, a network must be capable of finding a path between some 50 million terminal pairs in the presence of perhaps 1,000 already existing connections. A network which *always* permits any input to be connected to any idle output is called *nonblocking*.[96] Economics does not usually permit the use of nonblocking networks. Instead, networks are designed on a probability basis to give the required grade of service as described in Chap. 7.

a. Interconnections. Even during the busiest hour of the day, 80 per cent or more of all telephones on a typical switching system will probably be idle at any given instant. Although a certain amount of per-line equipment (telephone, conductors, line and cutoff relays, etc.) is necessary, switches in networks and trunks between offices need be supplied to handle only the expected number of simultaneous calls with a given probability of blocking. Thus concentration and expansion take place in stages of switching associated with customer lines to reduce the equipment needed for distribution.

Concentration. Early Strowger systems had no concentration; each customer line terminated on its own 100-point selector. There were thus at least ten times as many first selectors idle as busy even in the busy hour. The situation was remedied by the Keith line switch (Ref. 3, Vol. 3, Chap. 3), a device which combined the line and cutoff relay functions (see paragraph 16*a*) with concentration for as many as 100 lines. The 100 inputs had access to 10 outputs. A common drive mechanism preselected an idle

* Ref. 4, Chap. 4.

output, and the first line to come off hook would be connected. The drive mechanism would then choose the next idle output and await another origination. It was possible to adjust the concentration ratio by modifying the drive and multiple arrangements to give 25, 50, or 75 lines access to 10 outputs.

Another solution is to use inexpensive uniselectors as line switches which hunt automatically for idle first selectors when the customer originates a call (Ref. 8, Vol. 2, Chap. 2). Control for such switches can be very simple, and graded multiple techniques make them efficient in handling traffic. They offer very flexible arrangements for concentrating up to, say, 400 lines to 30 first selectors.

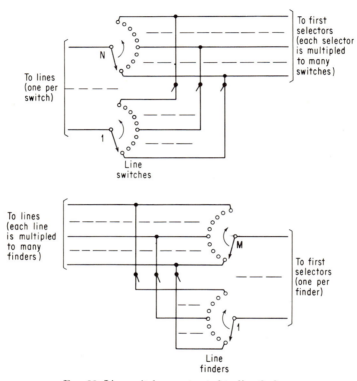

FIG. 29. Line switches contrasted to line finders.

Although British practice makes extensive use of uniselector line switches, line finders are almost universal in the United States.[97] Line finders hunt originating lines rather than idle selectors; only enough line-finder switches are provided to handle the busy-hour traffic at the required grade of service. Line-switch and line-finder action is contrasted in Fig. 29.

The line-finder principle, although occasionally used from about 1905 on, was first applied to the panel and SXS systems in the 1920's. A panel line-finder frame serves 400 customer lines, associating them typically with 56 panel switches, while from 16 to 30 Strowger switches can serve 200 customer lines.

Panel and step-by-step line finders work in a similar manner. When a customer originates a call, one of the switches hunts for, finds, and associates his line with a path to a sender and/or selector. Circuitry common to the unit allots a finder switch to the origination and, in panel, prevents others from working until the finding operation is complete. It is possible for several SXS switches in a unit to hunt simultaneously.

The control circuitry trips the appropriate brush of the panel switch or associates battery with a special commutator terminal on the SXS switch. Both types of switches then start hunting vertically.

When the SXS switch finds the marked commutator segment, it stops its vertical movement and starts rotating across the row of terminals looking for another battery applied to a sleeve lead. When it finds the battery signal, it stops its rotary motion and marks the line busy to terminating calls as described in paragraph 16a. The tip, ring, and sleeve leads are then connected through to the first selector, which returns dial tone.

The panel line finder has two control terminals per line in its multiple: the *sleeve* and the *hunt*. It moves upward until its tripped brush finds battery through a relay winding on the hunt terminal. This battery marks the originating line, and the tip, ring, sleeve, and hunt leads are cut through. Grounding the hunt lead releases the brush tripping mechanism to permit other line-finder switches to answer calls. Ground is also applied to the sleeve lead as a busy signal.

Concentration in crossbar systems resembles the line-switch approach where each vertical is a line switch. Typically, the horizontals can be multipled through several crossbar switches so that 29 to 59 customers or more, each associated with a vertical, can have access to 10 output paths (Ref. 73, Vol. 27, p. 360).

In a ferreed switch, the crosspoints are individually controlled and can have their control leads and their horizontals and verticals factory-wired with greater flexibility than is possible in crossbar switches. Thus two stages of switching using 8 × 8 and 8 × 4 crosspoint groups can be arranged to provide a relatively sophisticated 4-to-1 concentration ratio for 64 lines as shown in Fig. 30 (Ref. 40, p. 2193).

Remote Concentration. From the earliest days of telephony, it has been obvious that location of concentration stages near small groups of customers would result in large savings in wire. A group of step-by-step connectors and a set of Keith line switches could serve 100 remotely located customers over 20 pairs between the switches and a manual office (Ref. 1, Chap. 9). Modern line concentrators[98] are particularly well suited for use in new housing developments in suburban areas.

Expansion. The inverse of concentration is expansion. Any moderately large group of (non-PBX) lines will seldom be receiving calls for more than 10 per cent of its members. Thus, about 10 step-by-step connectors are provided for every 100 such lines. Switches can be added to or omitted from the multiple as required by busy-hour traffic.

In crossbar systems, the same switches can be used for both expansion and concentration. Thus, instead of two sets of switches serving, on the average, about 10 lines per network path, one set of switches serves perhaps 4 lines per network path.

Distribution. Between the concentration and expansion stages, the major portion of the interconnection task must take place. Up to 10 central offices in tandem may be needed to establish a long-distance connection. Within each office, distribution stages are wired to permit the input terminals to reach about the same number of output terminals.

Typical Interconnection Patterns. Figure 31 shows the way a step-by-step system might be built up, starting from the connector and working back. The system shown assumes five-digit telephone numbers with the single digit 0 (10 pulses) to reach the operator.

The basic pattern for No. 5 crossbar is shown in Fig. 32. Two types of switching frames are provided: line-link frames, which act as combination concentration, expansion, and distribution stages, and trunk-link frames, which act primarily as further stages of distribution. A line-link frame is made up of one junctor switch bay and one or more line-switch bays each containing 10- or 20-vertical crossbar switches. The junctor bay has the horizontal multiple on all its switches split so as to be, effectively, twenty 10-vertical switches. Ten verticals in each switch connect to junctors, and the others connect to lines. The "links" between the two halves of the switches fan out so that any line can reach any of the 100 outputs of a line-link frame. The line-switch bays are simply extensions of the line portion of the junctor switch bay, as each set of 10 links is extended to additional lines. With one vertical in the line bay

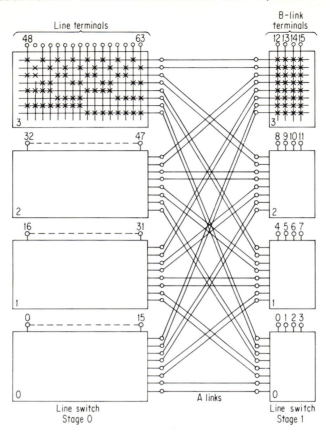

FIG. 30. Two-stage concentrator using ferreed switches.   Sixty-four lines have access to 16 outputs.   (*Courtesy of Bell Telephone Laboratories.*)

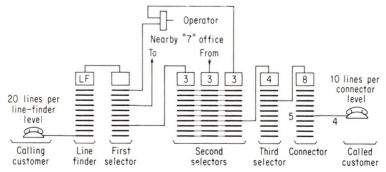

FIG. 31*a*. Typical wiring pattern for a SXS office.   The called customer shown can be reached by dialing 3-4854.   The operator is reached by dialing "O," and customers in a nearby office can be dialed direct by dialing 7 and their four-digit directory numbers. Similarly, customers in the "7" office can dial 3 plus four digits to reach customers in this office.

reserved for no-test use (see paragraph 16g), 19 to 59 lines can reach each 10 links and 190 to 590 lines can reach 100 junctors.

A trunk-link frame in No. 5 crossbar has 200 junctors as inputs and 160 trunks as outputs. Again two stages of switching are used, with the junctor bay switches split

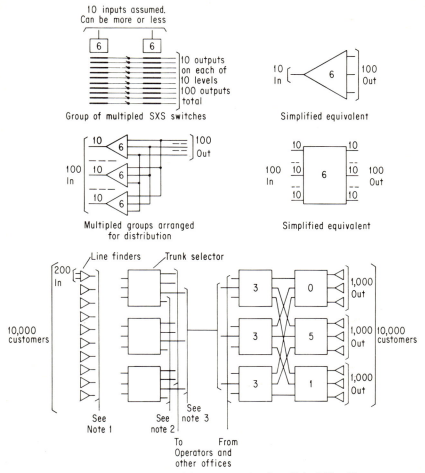

FIG. 31b. More detailed wiring pattern for the office of Fig. 31a.

Note 1. Individual line-finder switch outputs are connected to selectors to distribute as widely as possible the traffic load from each line-finder group.

Note 2. Use of "1" as a first digit in telephone numbers was formerly restricted; early switch hooks caused single false pulses when the receiver came off hook.

Note 3. First selector outputs on level 3, handling local traffic, would probably not be multipled so that enough outputs would be available to handle traffic to the next selector stage.

to act as 20 independent units, each with 10 inputs on the horizontals and 10 outputs on the verticals. The trunk bay, also composed of ten 20-vertical crossbar switches, has 20 inputs per switch (total = 200) to match the $2 \times 10 \times 10$ outputs of the junctor bay. These are six-wire switches, however, and are arranged to provide 16 outputs per switch rather than 10, as shown in Fig. 33. One of the horizontals 2 through 9 is chosen by operating the proper select magnet, and the choice between the two trunks on that horizontal is made by simultaneous operation of the 0 or 1 select magnet. Thus, two select fingers are moved into position where the hold magnet can operate the switches.

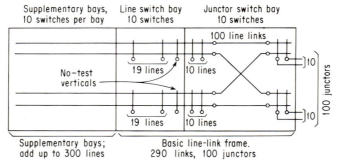

FIG. 32a. No. 5 crossbar line-link frame. A junctor switch bay contains 10 switches with 20 verticals each; the horizontal multiple is split for the link wiring. The line-switch bay has switches with 20 verticals, while supplementary bay switches may have 10 or 20 verticals as required.

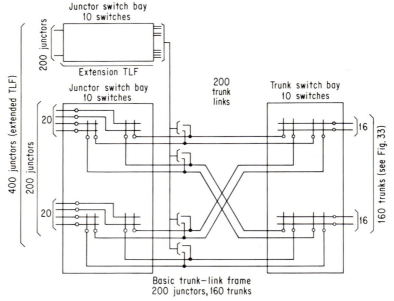

FIG. 32b. Trunk-link frame. Junctor bay switch multiples are split to let each 20-vertical switch work as two 10-vertical switches. Junctor inputs can be increased by paralleling one or two additional junctor bays.

A connection through the No. 5 network proceeds from the originating line through a crosspoint, line link, another crosspoint, junctor, crosspoint, trunk link, and two crosspoints to a trunk circuit. If the call is to another office, the trunk circuit will be outgoing. Incoming calls go from incoming trunk circuits back through the network in reverse order to the called customers. Local calls use intraoffice trunk circuits with ends connected via line- and trunk-link frames to both calling and called customers.

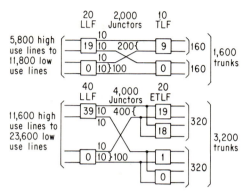

Fig. 32c. Junctor wiring patterns. Trunk-link frames have 200 inputs, while line-link frames have 100 outputs. Thus the ratio of LLF's to TLF's is 2:1, and the number of junctors from each LLF to each TLF = $100/n$, where n is the total number of TLF's. The number of junctors between any two frames must be 10 or greater to keep blocking low. Thus when more than 20 LLF's (or 10 TLF's) are needed, extended trunk-link frames used in pairs are required. Without ETLF's and junctor multipling, 200 junctors have access to 160 trunks (top); with one added junctor bay in each ETLF, 400 junctors can reach 320 trunks (bottom). In very large offices, two junctor bays are added, junctors are multipled to triples of ETLF's, and 600 junctors have access to 480 trunks.

A typical call is set up in the following manner: Upon detecting an origination at the line relay (see paragraph 16a) a marker sets up a path through line- and trunk-link frames from an originating register to the calling customer. The calling customer's switch is identified, and the "equipment number" is stored in relays associated with the register. The register makes a party test (if necessary), returns dial tone, and counts and stores the directory number of the called customer as dialed. When the complete number is received, all information is returned to the marker, which then releases the network connection from customer to register and *calls back* the customer* on the talking connection to a trunk. If the call is local, the called directory number must be converted to an equipment number in a translator called a *number group* so that the marker can find the proper switches.

The panel system and its replacement, No. 1 crossbar, were both developed for metropolitan areas where separate A and B manual switchboards were already in wide use; thus, they are arranged to follow the separate originating and terminating plan. The concentration and expansion stages are somewhat different. A panel office uses separate frames for line finders and final selectors, while the No. 1 crossbar system uses a line-link frame with link wiring and build-out similar to that of No. 5 crossbar. The No. 1 crossbar line-link frame uses 20-vertical crossbar switches for junctor switching with 10 verticals connecting

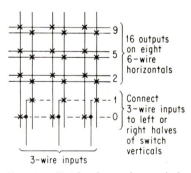

Fig. 33. Ten-level crossbar switch with 16 outputs. For a given connection, two select magnets must be operated simultaneously, followed by the hold magnet associated with the input.

* This is an example of the call-back principle.

to the outgoing distribution network and the other 10 connected to the incoming network.

The panel and No. 1 crossbar outgoing networks are composed of district and office frames; between the line frame and district frame, district circuits are located. These circuits have access via links (see paragraph 14b) to senders which receive the called number from the customer and transmit the called number to the terminating office.   When pulsing is completed, the sender connection is released and the district circuits take over supervision of the calling line and the trunk.

The district and office panel frames are controlled directly by the panel sender using revertive pulsing techniques (see paragraph 15a).   Trunk groups need not use graded multiples or be limited to groups of 10 trunks as in SXS, and here the philosophy of the panel system shows itself most clearly.   Each 100-terminal bank is made up of eight groups of 10 and two groups of 5 outputs, with the remaining 10 terminals used, one to a group, to return an overflow signal when the entire group is busy.   The proper brush is tripped to choose the bank, and the switch is driven to the first output terminal in the desired group.   It then hunts over the group looking for an idle path to the next switch.   If traffic measurements indicate that more than 10 paths are needed for a particular route, the overflow terminal is wired to appear busy.   Thus, a brush finding the first 10 terminals busy goes across the eleventh, the overflow terminal, and into the next group of 10.   Up to 90 outputs can thus be used as a single trunk group if desired.

For district and office frames, No. 1 crossbar uses primary-secondary split switches similar to those in the junctor bays of the No. 5 crossbar trunk-link frame.   The sender, after it has obtained the called number from the customer, calls in a marker; the marker finds an idle outgoing trunk and sets up a path through the district and office frames.   The sender then outpulses to the distant office.   Originally revertive pulsing was used because the No. 1 crossbar system was designed to replace and be compatible with the panel system, but other pulsing methods are available.

At the incoming office, an incoming frame is found in panel as in No. 1 crossbar. Panel uses a separate final selector, while the No. 1 crossbar line-link frame is used for terminating as well as originating traffic, as has been described.   The sender at the originating office controls revertive pulses to set the panel switches directly.   An incoming brush is first selected and tripped, and the switch is driven to the incoming group.   At the group it hunts for an idle path to the final frame independently of the sender.   At the final frame, the final brush is selected followed by decimal selections for units and tens.   The five brushes on the incoming and final frames and the four groups on the incoming frame permit a choice of $5 \times 4 \times 5 = 100$ sets of 100 lines each; thus the 10,000 lines which can be specified by four decimal digits are reached through only two switches (see Fig. 36).

Crossbar switches cannot be set by revertive pulses; incoming registers are associated with the incoming trunk circuits in No. 1 crossbar and are connected via link circuits when needed.   After the number has been registered, a terminating marker is called in to set up the path from the trunk circuit through the incoming and line-link frames to the called customer.   The marker and register are freed for other calls as soon as their work is done.

The No. 1 ESS, designed for use in relatively large offices, can serve as many as 60,000 lines.   Line-link networks and trunk-link networks are used in a manner analogous to line- and trunk-link frames of No. 5 crossbar.   However, each network has 1,024 junctor terminals, and line-link networks can handle up to 4,096 lines each.

A trunk-link network wiring pattern is suggested in Fig. 34.   Four stages of switching are provided in each network, making use of ferreed arrays.   Sixteen 8×8 switches make up a two-stage "grid" with 64 inputs and 64 outputs.   Four grids occupy a 3-ft 3-in. frame 7 ft high.   Four primary frames and four secondary frames make a 1,024 to 1,024 network.

The tip and ring leads are switched in the No. 1 ESS network, but two- and four-wire switches are available for local and toll switching.   The sleeve lead, used in other systems to hold switches operated and to provide a link-busy indication, does not exist. Magnetic latching takes care of the hold function, and a link-busy "map" in a central memory permits high-speed searches for idle network paths by the electronic central control.

Tandem and Toll Switching.[93] Tandem switching can reduce the number of direct trunk groups required between switching systems in an exchange area. Furthermore, toll trunks entering or leaving the area can be used more efficiently when arranged in one large group which completes to local offices via tandem rather than several smaller groups connecting directly to individual local offices.

In SXS areas, it is relatively simple in principle to move a group of second selectors to a central location. All offices in the area will reach these selectors by dialing the same digit into their first selectors. The output levels of the second selectors, driven by the second digit, can choose up to 10 offices; additional digits can then pass through to these offices.

Where several stages of switching are required to reach a large number of offices, a problem is created by the fact that each SXS selector uses up a digit. If a given telephone is to be reached by dialing the same number anywhere in the area, enough

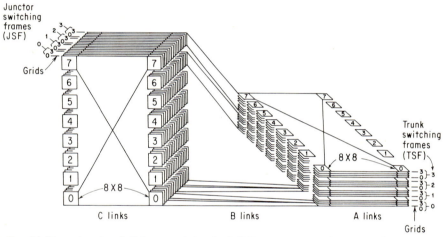

FIG. 34. Four-stage trunk-link network for the Bell System's No. 1 ESS.   1,024 inputs have access to 1,024 outputs.   (*Courtesy of Bell Telephone Laboratories.*)

digits must be provided to get through the largest number of switches in the longest path. Means are also needed for eliminating the excess digits on short paths. *Digit absorbing* selectors which effectively ignore excess digits are widely used, but the director system[57] (Ref. 8, Vol. 2, Chap. 13) offers a more complete solution to the tandem problem in SXS areas. Digit storage and control circuitry with access to a translator is temporarily associated with the connection at a point between the line finders and the selectors. Translation permits the customer-dialed number to be altered as required by the route.

In the panel system, the office frames are not restricted to the same building as the district frames. Thus, office selector tandem was as a natural result of local panel switching. Later, to achieve greater flexibility, panel sender tandem was developed; when pulses were received in a sender which could call for further translation, digits could be added or deleted as required, additional pulsing forms made available, and assistance operators could be located at centralized points.

Just as panel sender tandem resembled the outgoing portion of a panel office (without the line finders), so crossbar tandem, developed to take advantage of the greater flexibility of crossbar switches, markers, translation, etc., resembles No. 1

crossbar without line-link frames.   Another crossbar development, No. 4 type toll crossbar, switches toll trunks on a four-wire basis.   No. 5 crossbar can provide local, tandem, and toll switching.

b. Auxiliary Connections.   In addition to main switching networks used to interconnect customers, certain auxiliary networks are often required to interconnect various functional units used for signaling or control purposes.

Links.   The word "link" is used for the connection between switches within a frame in many common control systems.   It also has another meaning in a somewhat different context: a switching network used to associate a relatively small number of circuits such as senders with a considerably larger number of other circuits such as district junctors or trunk circuits.   This "few to many" function usually comes about when a relatively expensive circuit is needed for only a small portion of a call; a digit receiver or register, needed for less than 5 sec while information is being received, can go on to serve perhaps another 20 calls during the 2 min a given call is in progress.

Both panel and No. 1 crossbar use a sender link (Ref. 70, Vol. 17, p. 321) to connect the district junctor to a sender.   When the sender is connected, the local customer dials into the sender via the link; the link has a separate return path from the sender to the district circuit so that, in the case of panel, the district and office selectors, as well as the selectors in the distant office, can be set.   Number 1 crossbar uses a marker to set up the path through the district and office frames, but the sender pulses to the distant office via the link, outgoing network, and trunk.   In both panel and No. 1 crossbar, digits are being outpulsed even before the customer has finished dialing ("overlap operation").

Panel switches of a special six-wire type are used in the panel sender link.   Each link consists of two switches back to back.   One hunts for and attaches to an idle line-finder-district circuit; when an origination occurs, control circuitry selects a line finder only if it has a link attached.   The line finder then hunts for the line, while the second switch in the link hunts for an idle sender.   When the calling line is connected through the line-finder-district circuit and the sender link to the sender, dial tone is returned and the customer dials into the sender.

A panel sender link frame will have up to 30 links connecting from 14 to 40 district circuits to 100 senders.   The 100 senders are available to many additional district circuits via other sender link frames which are connected in multiple.

Crossbar switches are used for the sender link in No. 1 crossbar; a sender link frame associates 100 district circuits with 100 senders through a two-stage primary-secondary arrangement.   As in panel, the 100 senders are multipled to a number of sender link frames to assure adequate loading of the sender group.   On the average, there are about 10 district junctors per sender and from two to seven lines per junctor.   To connect six wires with regular three-wire crossbar switches, verticals in two adjacent primary switches are used together.   In the secondary switches, two verticals in the same switch are used.

Because a terminating marker is needed to establish a path between an incoming trunk and a customer line (via the incoming and line-link frames), pulsing over the trunk must be converted to a form suitable for marker use.   An incoming register is connected via an incoming link.   This link circuit connects only tip, ring, and sleeve between 100 trunks and 30 registers in a simple primary-secondary crossbar switch array.   The 30 registers are multipled to several link circuits.

Number 5 crossbar uses an even simpler one-stage crossbar switch link which permits about 10 registers on an extended horizontal multiple to reach up to 200 trunks on the verticals.   Links between senders and outgoing trunks differ slightly from those between incoming trunks and registers.   It should be noted that No. 5 crossbar does not use a separate link for its "originating registers," which, like the corresponding senders in panel and No. 1 crossbar, receive customer-dialed digits.   Rather, the regular line- and trunk-link frames are used, and the originating registers appear to the system like trunks.

The crossbar tandem system uses a primary-secondary link to associate senders with incoming trunks.   Since most calls coming into a tandem office go out again (there are usually no local lines on which to terminate), the sender which receives pulsing from the incoming trunk also transmits the proper signals to the outgoing trunk via the link and the switching network. Number 4 type toll crossbar uses links to connect incoming trunks to incoming senders and also to connect some outgoing trunks to outgoing senders when incoming senders cannot outpulse directly in the proper signaling code. In these cases the incoming sender outpulses to the outgoing sender via the two sender links and the switching network; then the outgoing sender outpulses to the distant office.

In the Bell System No. 1 ESS, link switches are omitted and the main switching network itself is used for all connections to senders, register, ringing circuits, etc.

Trees. It sometimes happens that a central control mechanism must select one relay or other device from a large number. Steering circuits for the operate signal can then take the form of relay trees (see Fig. 9) or transistor or diode equivalents (Ref. 5, Chap. 13). For each combination of the n tree-relay magnets operated, a different one of the possible 2^n outputs will be connected to the input. Thus, 10 "address" leads to the tree relays permit the selection of 1,024 different outputs.

In the Bell System No. 1 ESS, such a tree has been arranged to operate magnetic latching relays in trunk circuits, digit transmitters and receivers, etc. (Ref. 40, p. 2255). Called the "signal distributor," it can take an address from the central control and apply a positive or negative current pulse to one of 1,024 relays every 25 msec. In this way, sequences of circuit operation can be stored inexpensively in a central memory; changes in operating sequences can be effected by altering the program.

Buses. In common control systems, it is frequently necessary to transmit large quantities of information from one equipment unit to another. For example, the called number stored in the No. 5 crossbar originating register must be transmitted to a sender associated with the outgoing trunk. Some 10 digits may be involved; because a two-out-of-five self-checking code is used, the status of five leads must be transmitted for each digit. One of the functions of the marker is to act as a path from one circuit to another through marker connector circuits. One marker connector brings all digit leads into the marker from the originating register, and another extends these leads to the sender. All the information is transmitted in parallel and checked, after which the marker connectors and marker release. The marker can also cause the sender to modify the information if necessary.

The marker can act as the originating or terminating point for information. By use of a marker connector to associate one particular frame with a multiwire bus, it can receive the switch position or equipment number of an originating line from line-link frame control circuitry. It can then use other marker connectors to associate itself with test leads to find idle paths through the network, and finally, it can connect to particular line- and trunk-link frames to establish the connection. Because there are several markers and a much larger number of controlled circuits, lockout and allotter circuits are used with the marker connectors to prevent multiple connections (Ref. 73, Vol. 28, pp. 56, 502).

In the No. 1 ESS, high speed made possible by electronic circuitry permits the use of one central control in a time-division process rather than a number of markers working in parallel. This eliminates the need for lockout and allotter circuits on control buses but offers new challenges. The central control works at very high speeds, while relays in the networks and trunk circuits work at about the same rate as in conventional relay circuits. Thus, electronic buffers located near the circuitry to be controlled pick off and store the high-speed digital bus information.

The central control communicates with network frames, signal distributors for trunk circuits, scanners, etc., over a high-speed bus system with $\frac{1}{2}$-μsec pulses (Ref. 40, p. 2021). At these frequencies, balanced pairs are used as transmission lines. Care must be taken to minimize reflections. Delay, too, is important, and lengths of interconnections are limited. Up to 36 bits are required by various units, and the address bus, connecting through the pick-off circuitry of one unit after another, carries these bits in parallel to terminating resistors at the end of the run. Because these words pass through a large number of pick-off circuits, some means must enable the desired unit. Thus, each unit has an individual enable pair which is pulsed to open gates associated with the pick-off circuitry in the unit. The central control uses a separate circuit, called the "central pulse distributor," to send out these enable signals.

For reliability, each unit provides an acknowledge signal via an answer bus back to the central control. Scanners (see paragraph 16a) also use the answer bus to return information about the status of lines, trunks, and other circuits. Scan points are interrogated 16 at a time in response to a word applied to the address bus and return a 16-bit word showing the status of the interrogated scan points on the answer bus.

The central control also has separate bus systems to its memory units. Because of the great amount of information exchanged between the memories and the central control, transit-time delay must be held to a minimum; thus these buses are less than 150 ft in length.

c. Cross Connections. Large bundles of wires must be run between outside plant, frames, and equipment units in a telephone office. These wires are usually cabled and permanently connected to terminal blocks on the equipment. Because

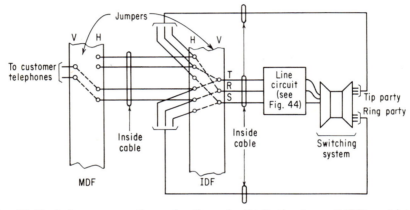

Fig. 35. Typical cross connections using the main distributing frame (MDF) and intermediate distributing frame (IDF). In the situation shown, a "terminal per station" progressive system provides individual and two-party service. All connectors are cabled uniformly, but the IDF jumper wiring is reversed to apply ringing to the tip conductor of the party whose ringer is from the tip side of the line to ground. *V* and *H* stand for the vertical and horizontal arrangements of the frame terminals, respectively. Message register connections are not shown.

connections must be changed from time to time, special distributing frames, grouping frames, grading panels, etc., are used to provide flexibility and simplify growth without disconnecting the cables from their terminals.

For example, cables entering a building terminate on the main distributing frame (MDF), with each individual pair of wires connected to its own terminals among those which are usually arranged in vertical columns (Ref. 3, Vol. 3, Chap. 9; Ref. 40, p. 2428; Ref. 99). Protection devices[100] (see Chap. 21) to guard against lightning surges and foreign potentials may be part of such terminals, although sometimes it is more economical to have a separate protector frame. Switching equipment is cabled to MDF terminals usually arranged in horizontal rows. Jumpers are run between the vertical and horizontal terminals to provide changeable connections between particular switches or jacks and customer lines. In systems where the directory number has a fixed relation to a switch position, a jumper change permits a customer to keep his directory number when he moves to a different location in the same general area. Similarly, intermediate distributing frames (IDF's) are sometimes used to balance the traffic load on line finders without changing a customer's directory number. Figure 35 shows a typical MDF-IDF arrangement.

d. Addresses and Translations. The multiple jacks in a manual switchboard were numbered sequentially to help the operator find the called number. In the SXS system, decimal digits of the directory number are usually made to match exactly with selector and connector levels and with the connector terminals on a given level; thus the directory number has an obvious physical meaning owing to the way the switches are used.

In the early 1900's, the rotary system was developed using a switch with 20 terminals on each of its 10 levels. In selectors this worked well. A decimal number selected the level, and rotation permitted hunting over 20 paths to the next switch. In a connector, however, a 1-out-of-20 selection was impractical with only 10 pulses. At this point E. C. Molina[102] invented "translation." The sender, which received the

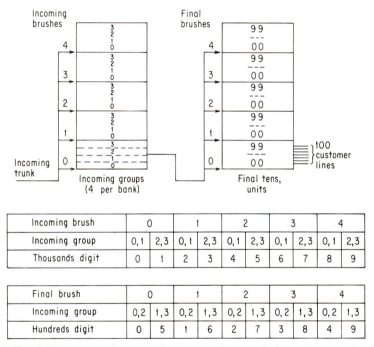

Incoming brush	0		1		2		3		4	
Incoming group	0,1	2,3	0,1	2,3	0,1	2,3	0,1	2,3	0,1	2,3
Thousands digit	0	1	2	3	4	5	6	7	8	9

Final brush	0		1		2		3		4	
Incoming group	0,2	1,3	0,2	1,3	0,2	1,3	0,2	1,3	0,2	1,3
Hundreds digit	0	5	1	6	2	7	3	8	4	9

FIG. 36. Panel directory number to equipment number translation. An example of fixed translation between nondecimal switches and decimal numbers.

dialed number in decimal form, converted it to the form needed for each switch. In the panel system, where 500 point switches are used, the final tens and units are controlled on a decimal basis but one incoming brush of five, one group of paths out of four to the final frames, and one final brush out of five have to be selected on the basis of the thousands and hundreds digits. Figure 36 shows the translation rule. Note that numbering starts with 0 rather than 1.

In an incoming panel office, the relation between a customer's directory number and his switch terminal is as fixed as in the SXS system in spite of translation. In the outgoing office, however, the situation is quite different. Trunk groups must be changed from time to time; for example, direct trunks may replace tandem connections and new systems with different signaling requirements may replace older ones. Thus, flexibility is needed in the choice of outgoing trunks and signaling methods.

To permit the panel sender to select brushes and groups in the district and office frames in whatever way is required, a translator called a decoder is used for the office code. The three digits of the office code, as dialed into the sender by the customer, are a destination code. The signal returned from the decoder is a route code. With flexible translation, the route code can be changed without requiring the customer to change the destination code he is used to dialing.

Figure 37 shows the form of the decoder. Patterns of opens and grounds on several leads transmit the office code to the decoder as soon as the customer has dialed the third digit. The decoder uses this pattern to operate one of a group of route relays; more than one office code can operate the same route relay in some tandem connections. Each route relay causes a unique signal to be returned to the sender. The sender uses this information to select the proper district and office brushes and groups and various

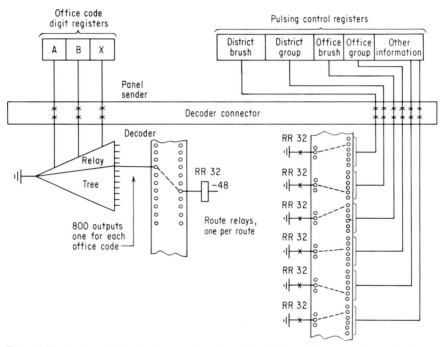

FIG. 37. Route translation in the panel system (simplified). Offices can be reached over different routes by changing the cross connection to the route relay coil. Parameters within a route can be changed by moving jumpers associated with route relay contacts.

other items related to pulsing and charging. Two cross-connection fields in this scheme permit changing the route relay selected by a given office code and, independently, changing the parameters which can vary within a given route such as the type of signaling required.

The whole translation operation takes place in less than half a second. Five decoders, shared by perhaps 300 senders or more, can serve a number of offices. In crossbar systems, a similar scheme is used. A decoder, cross-connection field and all, is built into each of the relatively small number of markers required in an office.

The director system was developed by the Automatic Electric Co. to provide SXS with senders and translators. Freed from a permanent relation between destination and route codes, universal numbering plans are possible and full advantage can be taken of both tandem trunking plans and standard SXS equipment. Because of the wide use of SXS, a number of other companies have developed similar equipment.

Directory number translation went through three stages: decimal directory numbers to nondecimal switch locations, office (destination) codes to route codes, and finally, directory numbers to arbitrary switch locations. Even with sender operation, the last of these would have been impractical in systems where the sender in the originating office controlled the switches in the terminating office directly. Every change in every terminating office would have to be reported and cross-connected in every originating office.

Where incoming numbers can be intercepted and translated, this limitation is removed. Number 1 crossbar offers an illustration (Ref. 70, Vol. 17, pp. 356 and 373). An incoming trunk circuit, when seized, connects the trunk to a register via a link circuit. The register receives the directory number and passes it to the marker via a connector circuit. The marker, in turn, connects to an external translator called a "number group" which converts the directory number to an "equipment number" identifying the line-link frame, the intersection of the horizontal row of switches in that frame (horizontal group) with the column of switch verticals one above another in the frame (vertical group), and the particular vertical within the vertical group. Ringing information is obtained at this time, and the busy test is made.

The number group is typical of most translators. The representation of the digit causes certain relays similar to route relays to operate. The outputs from these relays pass through a cross-connecting frame where changes can be made and return to the marker in a form that corresponds to the equipment number. The number group is at least an order of magnitude larger than the outgoing trunk-group translators. It is divided into sections so that two or more markers, processing widely separated numbers, can use the different sections simultaneously.

A similar scheme is used in No. 5 crossbar, although there are minor differences. For instance, the No. 5 crossbar marker makes the busy test after translation is completed except in the case of such multiple line groups as PBX's and key telephone systems.

In toll offices, the large number of alternate routes makes special translators necessary. The card translator,[103] developed for the No. 4A toll crossbar system, is an example of combined electromechanical and electronic design. A translator can hold slightly more than 1,000 thin metal cards, each pierced by 118 holes. When no translation is being made, the holes in all cards are lined up and an optical system sends light through these "tunnels" to phototransistors which operate amplifiers which ultimately control relays. Each metal card has certain of its holes enlarged; when a translation is to be made, exactly one card is selected mechanically and caused to drop a small distance.

The enlarged holes still permit the light to shine through to the phototransistors, but the remaining holes pass out of alignment, cutting off the light. A particular set of relays is thus operated to return the required information. Cross connection as a means of altering a translation is replaced by changing cards.

In addition to the switch location of the outgoing trunk, the card translator gives alternate route information, digits to be added or deleted, and the type of pulsing to be used to the next switching point, what to do if all trunks are busy, etc.

Random-access memory systems, such as those used in digital computers (see paragraphs 10 and 11), can, by their very nature, be used as translators. The address of a memory location is the input, and the information stored in that location is the output. Rewriting the stored information corresponds to making changes in a cross-connecting field. Cyclic access memories, such as magnetic drums, are also useful as translators.[49] A drum, revolving continuously, steps its angular-position counter and reads out the information passing under its many heads. A match circuit can be set to watch for a given word on either the position counter or the readout and use the related indication on the readout or position counter, respectively, as the translated information (Ref. 48, p. 628).

A telephone system requires a variety of stored information in addition to the translations described above. Much of this information can be lumped under the general heading of "class of service"[104] and refers to specific needs of individual lines or trunks in processing calls. Originally, customer lines had service classes (primarily related to charging) such as flat rate, message rate, and coin. These were indicated to the operator by different-colored answering lamps. In automatic systems with line finders, each class is assigned to separate line-finder groups. Similarly, it is possible to handle party-line ringing by using different connectors, each connector group specializing in returning one of the several ringing signals. The need to make an automatic party test on a given line, receive dial pulses or TOUCH-TONE® signals, etc., must also be recorded. PBX's add many more special classes of service.

In common control systems, the class information can be stored in many ways including switch positions, cross-connecting fields, etc. To be useful, however, it must be associated with the registers or senders on a per-call basis. Typically, stored information causes class relays in senders or registers to be set, and these relays control the functions involved. New classes of service can be added by altering the relatively few common circuits.

New requirements and new services have produced an increasing need for ever-larger numbers of classes of service. The difficulty of increasing these numbers is one of the most serious limitations of electromechanical systems. On the other hand, electronic systems, built around large memory units, are virtually unlimited in this respect (Ref. 40, p. 2533). A class mark can be stored as a number in memory and identified by translating the equipment number of the line or trunk. In theory, at least, each line can have a separate class number if desired.

15. Control of Switching Networks

a. Progressive Networks.* In most progressive networks, each switch is driven to a particular group of terminals and then allowed to hunt over those terminals until it finds one idle.

Forward and Revertive Pulses. When a switch is driven to a particular terminal, pulses may be sent forward to operate a magnet which moves the switch a step at a time or else sent back to report the position of a motor-driven switch. Forward pulsing has achieved wide applicability in SXS, while revertive pulsing is used in the panel and rotary systems.

Forward pulsing requires only brief energization of the appropriate switch magnets. SXS switches mount relays which operate over the customer loop and close a break contact when a dial pulse (momentary loop open) appears. Thus only a simple mechanism to open and close the loop repetitively need be located at the customer's premises. Revertive pulses, on the other hand, must be served by a sender. The sender contains a "down-counter" which is set to the desired number; a start signal is transmitted to the switch by closing a control path called the "fundamental circuit" through a pulse-counting relay. The fundamental circuit usually includes the trunk tip and ring conductors when the switch is in the terminating office, but district and office switches in the originating office can use a ground return. Closure of the fundamental circuit energizes a clutch which applies motor drive to the switch. The switch starts moving, but periodically a commutator short-circuits the fundamental circuit while holding the clutch control operated as shown in Fig. 38. Each shorting of the fundamental circuit is a progress report on the position of the switch, releasing the pulse-counting relay in the sender to step the counter one step nearer the normal state. When the counter reaches normal, it causes the sender to open the fundamental circuit. When the commutator short is removed after the fundamental is opened, the clutch drive is disengaged. The switch stops and is held by a mechanical latch.

End-of-digit Signal vs. Interdigital Timing. When a train of pulses is used as a signal, the end of the pulse train must be detected. Early SXS systems sent dial pulses down one side of the loop (with ground return) and an end-of-digit signal down the other side at the end of the pulse train. Similarly, revertive pulsing uses a positive signal (opens the fundamental circuit as described above) at the end of a pulse train.

Use of a ground return is undesirable in dial pulsing because of earth potential

* Ref. 4, Chap. 5.

differences, noise pickup, electrolysis, etc. Fortunately, the interval between pulse trains is appreciably greater than the interval between pulses within trains, and timing can be used to derive an end-of-digit signal. In the *A-B-C* circuit shown in the upper part of Fig. 39, the *A* relay operates over the customer's loop, operating the *B* relay. The first release of the *A* relay upon receipt of a pulse operates the *C* relay. Both the *B* and the *C* relays are made to release slowly. The *B* relay holds over dial pulses and prevents momentary loop opens from releasing the switch. It differentiates between dial pulses and the abandonment of the call. The *C* relay operates when the loop is opened on the first pulse of a digit train and holds over the momentary loop closures. It releases only when the loop remains closed for a period longer than an interpulse interval and thus indicates end-of-digit. This circuit, using one

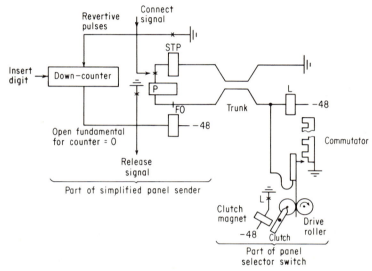

FIG. 38. "Fundamental circuit" for revertive pulsing. The connect signal operates the *L* relay to apply drive. The commutator shorts the fundamental circuit to release the STP relay. Each STP release steps the counter. The counter opens the fundamental circuit when it reaches zero to stop the updrive. The polar relay responds to a reversed battery signal to release the sender.

fast pulse-repeating relay and two slow-release timing relays, is used on all direct-controlled SXS switches as well as in almost all senders which receive dial pulses. In systems where an electronic control scans for dial pulses on a time-division basis, timing internal to the control system can replace the *B* and *C* relays.

Progressive Hunting and All Paths Busy. Once a progressive switch has been driven to a particular group of outputs, it must hunt over these outputs in multiple with many other competing switches to find an idle path (see paragraph 5). Internally generated pulses take over the switch drive upon release of the *C* relay. A third wiper on a SXS switch contacts the sleeve lead of a path in the multiple; upon finding a ground (busy signal) it permits the generation of a pulse to cause a step of the rotary magnet (Fig. 39). When an ungrounded sleeve lead is found, the stepping will stop. The sleeve is immediately grounded to make the path busy to other switches, and control is passed forward to the next switch, which repeats the process.

In revertive-controlled selectors, revertive pulses help set the switch so that the proper brush will be tripped. Reclosing the fundamental circuit causes the switch to reoperate its clutch to drive the switch to the proper group of outputs. When the proper number of revertive pulses has again been received, the fundamental circuit opens to signal the switch to use its tripped brush in a hunt for the first nongrounded sleeve in the selected group.

If a switch hunts over all the terminals available to it and finds no idle path, the call cannot proceed. The calling customer is informed so that he can hang up and try again. Because SXS switches usually have 10 positions on each level for path outputs, a set of contacts associated with the eleventh rotary step provides a tone connection. If less than 10 paths are needed to the following switches, the extra sleeve terminals can be strapped to look permanently busy.

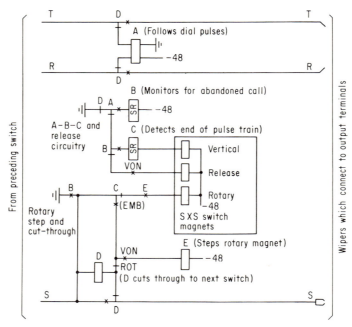

Fig. 39. Typical SXS selector circuit (simplified). The vertical magnet steps each time A releases. The E relay operates through B and C contacts. When C releases, the rotary magnet steps to release E. E then releases the rotary magnet. If the S wiper finds ground (busy), E reoperates to operate the rotary magnet which again releases E. When the S wiper finds no ground, D operates. T and R are cut through to the next switch where an A and B relay operate to ground S. A and B of this switch release to prepare the switch release circuits.

Note 1. VON contacts are operated when the switch has moved vertically off normal.
Note 2. ROT contact opens when the rotary magnet operates.

Both the rotary and panel systems were developed to increase the number of paths over which a switch could hunt. In metropolitan areas, it is not uncommon to have 100 trunks or more from one office to another. Many more than 10 groups of 10 trunks are needed to carry the traffic which a single group of 100 can carry easily (see Chap. 7 and Ref. 4, Sec. II).

Switch Release. Switches in most progressive systems are released by removing the "dog" or latch that holds the switch in place. Upon hang-up, supervisory relays release to activate a release magnet. The release magnet is held operated until the switch returns to normal; Fig. 39 shows the SXS release system commonly used in the United States.

b. Centrally Controlled Networks.* *Separation of Pulsing and Control.* Once the network is removed from direct customer control, the need to maintain compatibility between customer signaling and the switches is eliminated.[105] This permits introduction of such features as TOUCH-TONE® pulsing by customers and, even where dials are retained, at least theoretically permits longer range, faster pulse speeds, or both.

In trunk signaling, the gains are greater. Direct-current pulsing can be accommodated only with difficulty over carrier systems; MF (multifrequency) pulsing is ideally suited for transmission over voice frequency channels and is much faster (see paragraph 16*f*). Digit senders and receivers for trunk circuits optimize signaling, just as localization of control optimizes switch operation.

Centralized Path Hunting, Operating, and Releasing.[106] The crossbar switch is a simple piece of apparatus which offers a variety of possibilities to the designer (see paragraphs 6*b*, 14*a*, and 14*b*). Although sometimes used with progressive control, it has come into its own in centrally controlled systems. A switching system handling more than 30,000 lines is easily designed around crossbar switches, with up to 12 markers sharing the control operations as required.

Connector circuits associate markers with network frames, connecting through large numbers of leads in parallel. These leads carry, among other things, the busy-idle status of the links between switches. In Bell System networks, each switching frame contains two stages of switches with links between them; two-frame paths are set up by markers. Thus three sets of paths must be examined: the links between the primary and secondary stages of each frame (called line links, trunk links, office links, etc., depending on the name of the frame) and the junctors between frames. Three idle links must be "matched" for a complete connection, and then the proper operating sequence of select and hold magnets is carried out. Finally, the path is checked for continuity.

A simplified three-link network of 10 line-link frames and 10 trunk-link frames, each frame made up of 10 by 10 crossbar switches, is skeletonized in Fig. 40 to suggest the type of strategy employed. Paths are set up between trunks and lines. At the start, the marker has identified the trunk terminal and line terminal. In the network there are 100 bundles of junctors between the 10 line-link frames and trunk-line frames. The marker identifies the correct bundle in terms of the trunk- and line-link frame numbers. The bundle required in the illustration is numbered 43 between trunk-link frame 4 and line-link frame 3. As can be seen, all junctors in this bundle are connected to fourth verticals on the junctor switches in the line-link frame and the third horizontals in the junctor switches of the trunk-link frame. The TLF number matches the LLF vertical and the LLF number matches the TLF horizontal.

The vertical associated with the desired customer line is part of a particular crossbar switch in Fig. 40. It can be connected to any of 10 horizontals. Similarly, the trunk can be connected to any of 10 verticals. But a match can be found only among similarly numbered line links, trunk links, and junctors owing to the nature of the wiring pattern. The path through the 0 links and junctor is shown darkened; any of the other nine paths might have been chosen if idle.

To test for a matching idle path, network connectors bring all 30 sleeve leads to the marker, where they are examined; additional leads for magnet control for all 10 paths are also connected. When an idle channel is found, its select magnets are operated followed by the hold magnets, after which the select magnets are released. The leads involved are shown in Fig. 41. The trunk circuit should now be holding a ground on the sleeve lead, which, in turn, locks all the hold magnets operated after the marker releases. The trunk circuit will normally maintain this holding ground until the calling customer hangs up. Usually a slow-release relay is used between the supervisory relay and the holding ground to prevent momentary opens, voltage surges, etc., from prematurely releasing the connection.

An actual switching network as shown in Fig. 32*c* requires somewhat greater sophistication, but the principles are the same. The chief feature of this technique is

* Ref. 4, Chap. 6.

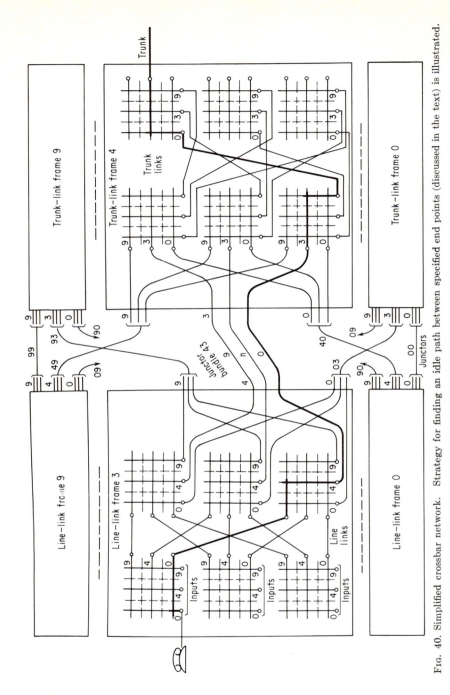

FIG. 40. Simplified crossbar network. Strategy for finding an idle path between specified end points (discussed in the text) is illustrated.

9–50

the way the wiring between switches and frames simplifies identification of the switch verticals and horizontals; also of interest is the way select, hold, and test leads are combined to reduce the total number of leads brought to the marker. Nevertheless, a marker connector requires a very large number of contacts.

The No. 1 ESS, developed for medium to very large offices, uses line- and trunk-link networks of ferreed switches. Each network serves 1,024 junctors and has four stages of switching with *A*, *B*, and *C* links. Each link and junctor have a bit in memory which is marked busy or idle by the central control in the course of operation. The central control, operating entirely as a high-speed digital information processor under program control, hunts in memory for matches between idle links and junctors. When a path is found, it can be operated immediately or stored for future use. For example, the calling customer is connected to a tone circuit on the trunk-link network which returns audible-ring tone until the called customer answers. Similarly, the

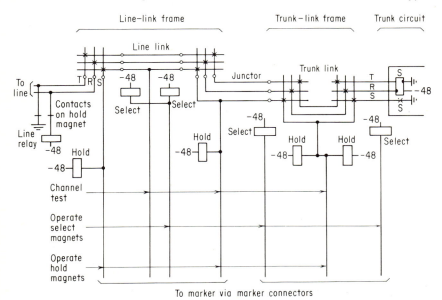

FIG. 41. Test and control leads brought to a marker for one path through a crossbar network. Ten paths are tested simultaneously. Select and hold magnets are operated by the marker; hold magnets are held by the supervisory relay in the trunk circuit.

called customer is connected to a ringing circuit. The path to be used for conversation is found and reserved before ringing and tone connections are made; upon answer the paths to the ringing and tone circuits are marked idle in memory and the reserved talking path is put into use.

Network control is organized so that one address on the peripheral bus (see paragraph 14b) will set a path through two stages of switching. A frame composed of sixteen 8 × 8 switches in a primary-secondary grid requires a 1 of 8 selection for the particular primary switch, a 1 of 8 selection for the desired vertical within the switch, a 1 of 8 selection for the secondary switch, and a final 1 of 8 selection for the vertical within the secondary switch. This requires 12 bits and defines exactly the one path between 1 of 64 inputs and 1 of 64 outputs. Two 12-bit words are needed to set up a path through the four-stage trunk-link network of Fig. 34.

End-marked Networks. In an end-marked network, it is necessary to know only the addresses of the terminals to be connected; nothing need be known about the busy-idle status of the internal links.

In a certain sense, a single crossbar switch can be thought of as an end-marked network. If the select magnet of the output terminal is marked with a suitable operate signal and the hold magnet of the input terminal is similarly marked after a slight delay, the path through the network is established. There is, however, no choice of links to be made. In a two-stage crossbar network, every input can reach every output but only one path from a given input to a given output exists. Thus again there is no choice of links. Only when at least three stages are used, as in Fig. 42, is there a choice of paths from input to output which requires a hunt for idle links.

Breakdown devices (see paragraphs 8 and 9a) are well suited for use in end-marked networks (Ref. 52, p. 1421), although with proper circuitry almost any type of switch

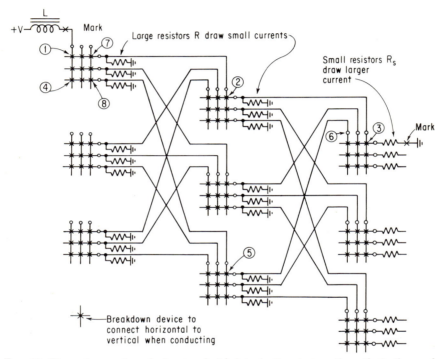

Fig. 42. Three-stage end-marked network (skeletonized). Automatic path finding and lockout are described in the text.

can be used.[87] The properties desired in a breakdown device are lockout and a large ratio of breakdown to sustain voltage. To illustrate the principle, Fig. 42 shows a three-stage network with terminals marked at each end. The crosspoint labeled 1 breaks down between the input mark voltage V and ground through the large resistor R. V minus the sustain voltage of 1 appears at crosspoint 2, which also breaks down. V minus the sustain voltages of 1 and 2 is then applied across 3, which breaks down via R_s, a small resistor, to the output terminal marked with ground. A suitable path has now been selected. However, a path through crosspoints 4, 5, and 6 would also have broken down if all its links had been idle. Assuming that it did, the lockout property described in paragraph 8b will cause one path to be extinguished. Even when several paths in parallel break down, the lockout property (usually) forces the selection of just one.

Assuming that the path through 1, 2, 3 is operated, the voltage across R_s is V minus three small sustain voltages. Now if a second mark is applied to the vertical containing crosspoint 7, only the sustain voltage will appear across 7 and no breakdown will occur. Crosspoint 8 and all others on the vertical which are not associated with a busy horizontal will break down and propagate the second mark signal through the network. Similar biasing will block mark signals out of busy paths in the second stage.

A mark voltage breaks down all idle crosspoints on the first-stage vertical. Each of these fans out to second-stage switches, where again all idle crosspoints break down. Thus in 10×10 switches, each first-stage crosspoint must be prepared to provide enough mark current for 10 second-stage crosspoints, and, if the network continues, 100 in the third stage, etc. When the marked path breaks down and locks out other parallel paths to the same end points, crosspoints fanning out toward other terminals must also be extinguished. The coil in series with the mark voltage helps achieve this. Breakdown of the last stage to the ground mark via R_s attempts to draw a much larger current through the selected path. Since the inductor will tend to hold the current constant, current for the selected path will drain out of the fan-out crosspoints, which promptly extinguish.

It can be seen that fan-out mark currents become excessive if the network contains very many stages. One way around this is to use propagator circuits to take over the supplying of current after a stage has broken down. Another technique uses a so-called progressive search[107] for the marked terminal. If the resistors R are replaced by capacitors and the coil L is replaced by a resistor, lockout permits only one crosspoint at a time to break down in any given stage. After the capacitor associated with the conducting crosspoint charges up to a high enough voltage to break down the following stage, a similar lockout takes place. If there is no mark at the end of a particular hunt, the last stage to break down is choked off and current continues to charge the preceding capacitor until a parallel crosspoint breaks down. Until the marked path is found, capacitors continue to charge and apply breakdown voltages.

A path through the breakdown devices can be used to operate reed-switch crosspoints to provide a metallic speech path,[87,108] or else the breakdown devices themselves can be arranged to carry the conversation.[82,109]

Time-division Networks.[85,90,110,111] A time-division network has two parts: a highway to which regularly spaced voltage samples from each pair of communicating customers can be connected on a time-shared basis and a mechanism to control these highway connections. The highway can be operated on a two- or a four-wire basis, depending on the type of transmission facility (see paragraph 13c). A four-wire network requires either twice as many time slots per highway or two highways, one for each direction of transmission, to permit the same number of connections.

Transmission through the highway can be implemented at low loss (less than 2 db on the average in one system) by means of resonant transfer.[112] In Fig. 43, the customer on line 1 is speaking and the charge on capacitor $C1$ is following the speech voltage through the low-pass filter. When the time slot assigned for this conversation comes up, the switches for line 1 and line 2 are closed and each line is connected to the highway through inductors $L1$ and $L2$. Because $C1$ is charged and $C2$ is not, the charge on $C1$ moves toward $C2$ via the inductors. $C1$ and $C2$ resonate in series with $L1$ and $L2$; if the gate switches are left closed for half a cycle of this resonant frequency, all the charge on $C1$ will be transferred to $C2$. Opening the switches traps the charge on the capacitor of line 2 where it passes to customer 2 through the low-pass filter. $C1$, of course, recharges quickly, since the period during which transfer takes place is much shorter than the period of the highest speech frequency passed by the filter. The direction of transfer is clearly arbitrary for bilateral switches, depending only on which capacitor has the higher voltage. To prevent charge from being trapped on the highway to produce crosstalk, the highway is grounded for brief guard intervals between time slots.

Line-gate control requires a circulating memory with two words for each time slot. Each word must have enough bits to hold a binary address for each input: 9 bits for 512 inputs, 10 bits for 1,024, etc. To establish a connection, the calling and called line addresses must be written in an idle time slot in the memory. To take a connection down, the words are rewritten as all zeros.

Each word pair, upon being read out, triggers a pair of high-speed selectors.   These selectors operate the two electronic line-gate switches simultaneously, every time the memory cycles.   To ensure the transmission of a 4,000-cps bandwidth, a pair of line gates must be sampled 8,000 times a second (actually, a somewhat higher rate is used to allow more component margins).   Allowing 2 μsec for each resonant transfer and $\frac{1}{2}$ μsec for a guard interval, about 50 simultaneous conversations can take place. Thus to handle more than 500 lines, a time-division switch must work rapidly indeed.

Time-division networks, like space-division networks with electronic crosspoints, will not usually pass signals outside the frequency and power range of ordinary speech.

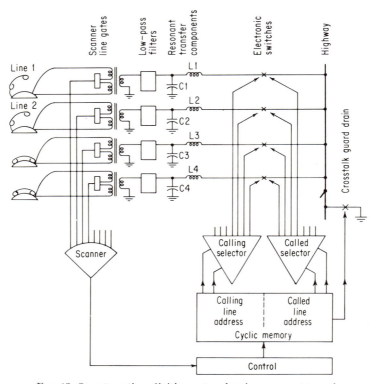

Fig. 43. One-stage time-division network using resonant transfer.

Thus other means must be found for applying ringing and coin control potentials, bringing dial pulses to suitable detectors, and making tests on lines and trunks with such standard maintenance instruments as voltmeters and bridges.   Signals such as ringing can be switched in at the line gate under control of a voice-frequency tone transmitted through the highway.   Dial pulses can sometimes be made to pass through the highway as extremes of maximum and minimum level; although they are not capable of operating a relay directly at the output, they do provide a distinctive signal which can be detected with suitable circuitry (Ref. 85, p. 338).

The simplest time-division networks have only one stage of switching.   When all inputs have access to the same highway, full access is provided, and when the number of time slots is equal to or greater than half the number of inputs, the network is non-blocking.   While the number of time slots is limited by speed considerations, the number of gates is limited only by wiring capacitance, the number of bits per word in the cyclic memory, and the specified probability of the call being blocked because all time slots are in use.

Highways can interconnect other highways to increase system size. Under such circumstances, a space-division network of time-division switches results. Here, synchronization between stages poses the principal problem. Ideally, all stages of switching should be gated at the same instant to permit resonant transfer from input to output. When a prior call is using the desired time slot in one or more switches, this is impossible. Two solutions have been used: momentary storage of the sample and reassignment of the time slots. A storage capacitor associated with a highway gate can receive the sample from an input in one time slot and pass the sample to the next switch in a different time slot; the delay is less than the interval between samples and is not noticeable. Alternatively, moving one or more existing connections to different time slots to free matching time slots in the switches needed by a new connection requires only the alteration of addresses in the cyclic memory, a simple matter for a common control.

Common and Centralized Control. Common control is divided into two parts: the register-translator-sender equipment, which can be used with any form of switching network, and centralized network control concerned with path hunting, operating, and reserving. As has been discussed, common control, with or without centralized network control, permits the introduction of flexibility into a switching system. Senders and registers permit independent optimization of signaling and switch operation, while translators permit alternate routing and free directory numbers from specific switch locations.

With centralized network control, equipment associated with each switch can be minimized. Path hunts can be carried out in parallel, and trunks need not be placed in consecutive positions on switches. Busy-testing of lines and trunks can often be expedited, and *pocketed calls*[113] (where a trunk group with known idle members becomes busy between the time a register makes a group-busy test and the switches to the trunk group are set) can be eliminated within any one office. Furthermore, it takes far less time to set a number of switches simultaneously rather than sequentially.

Common control systems have several disadvantages. The first is the obvious complexity of the control equipment itself. This leads to relatively high start-up cost and the difficulty of designing economical small offices. Common control offices are somewhat more vulnerable to extended traffic overloads than are offices under direct customer control, a result of the small number of control circuits used at high traffic efficiency and also the behavior of customers and operators when calls are blocked.[105] Blocked calls produce increased traffic in the form of retries, and dialing before dial tone produces wrong connections, partial dials, etc., which add to the congestion when the system is least able to handle it. Finally, from the end of dialing until connection is made to the called line, common control is always slower than direct dial control. The SXS connector is on the called terminal when the last pull of the dial has returned to normal.

Registers and senders must remain in contact with calling customers or distant offices for prolonged periods of time, while translators and network controllers usually have holding times of appreciably less than $\frac{1}{2}$ sec. Because of the interdependence of the control functions and the dissimilarity of holding times for various operations, several different approaches to common control design are used.

First, register-translator-senders such as the director have been developed to modernize and add features to SXS. A small number of translators serves an appreciably larger number of registers and senders. Next, network controlllers for small modules of crossbar or other coordinate switches are used with register-translator-senders. Two or three stages of switching serving up to 1,000 lines or so form a module, and connections between such groups are carried out by markers working progressively. Communication between markers uses the speech paths being set up or intermarker data links.[87] Relatively small modules permit considerable flexibility in office size, and modules of newer design can be added to existing groups of older equipment to provide growth without premature retirement of existing apparatus.[48]

When offices are to be installed as complete entities in sizes greater than, say, 1,000 lines, truly centralized network controllers capable of establishing entire connections become practical.   Because a common group of markers now serves a large number of frames, competition for access must be carefully regulated.   However, once concentration is achieved, the two to twelve markers can take on more (and more complicated) functions than can a larger number of simpler controllers.

Perhaps the ultimate in flexibility is stored-program control (Ref. 40, p. 1923; Ref. 52, p. 1327).   Once stimuli from the outside world have been converted into zeros and ones at suitable interfaces, anything that can be done with circuitry can be done with a stored program in the same way that a digital computer can solve a wide variety of problems.[114]   Two advantages are of particular importance: First, only one program segment is required for each function (such as dial pulsing, for example) no matter how many calls require that function simultaneously.   Second, the large-scale memory required for stored-program operation opens the door to electronic memory for per-call storage.   In the first case, large numbers of identical control circuits can be eliminated, and in the second, bulky relay registers can all be replaced by compact locations in memory.   Memory systems themselves are completely general, and their contents are readily alterable to meet varying situations.   Given a suitable interface with customer lines and trunks, they bring a new degree of flexibility to telephone switching.

16. Call-processing Functions

Network operation, while necessary, is not sufficient to carry out the switching task.   A number of other functions must be considered and these may turn out to dominate the overall design.

a. Detecting Originations.   Whenever a customer wishes to use his telephone, he must signal his intent to the switching system.   This is usually done by removing the handset from the cradle, which automatically switches on direct current supplied from the central office for the carbon transmitter.   Such current can be made to operate some sort of d-c-sensing device associated with the line.   In most instances, a simple "line relay"[14] has been used; a contact of this relay closes a path which makes other equipment aware of the service request (Ref. 3, Vol. 2, p. 317).

A line relay is usually disconnected for the duration of a call after the service request has been acted upon.   Thus a "cutoff" relay or its equivalent is provided to remove the line relay until hangup.

A line and cutoff relay make up a typical line circuit, although sometimes one or more message registers or other circuitry may be included.   The cutoff relay is used on both originating and terminating calls because the line relays are disconnected from both the called and calling lines.   The line circuit applies three signals to the sleeve lead: idle—available to receive a terminating call; busy—in the process of setting up or using a connection; and seizure—customer off-hook awaiting service from the central office.   Figure 44 shows a common line circuit (see also Figs. 41 and 46).

Start Signals.   Operation of the line relay upon seizure usually provides some sort of start signal in manual and electromechanical offices.   The start signal initiates a chain of events which prepares the system to receive the customer's detailed instructions.

Scanners.   Some electronic switching systems use scanners (Ref. 40, p. 2255; Ref. 52, p. 1383) to sample the status of each customer line periodically on the order of ten times a second.   This is a continuing process, active night and day.   On the basis of changes in scanner outputs, the control circuitry decides upon the proper course of action.

Scanners are usually composed of "line gates," which are opened or closed depending on whether the associated line is off or on hook, and access circuitry, which interrogates the gates by attempting to pass pulses through them.   This is suggested in Fig. 45. An off-hook line usually blocks the interrogate pulse as a fail-safe measure.   Line gates can be contacts on line relays, diodes biased by loop current, or such special devices as the ferrod sensor.[115]   The ferrod sensor can be thought of as a transformer with a variable coefficient of coupling.   When no current flows through its control

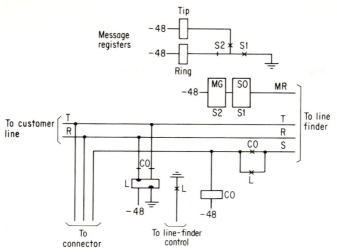

Fig. 44. Typical line circuit for use with line finders and message registers. Operation of the L relay causes the finder to hunt for battery on the S lead through the winding of the CO relay. Grounding the sleeve operates CO to cut off the L relay and mark the line busy.

Direct ground on the MR lead causes the $S2$ relay to operate followed by the $S1$ to score the tip party message register. Resistance ground causes only the ring party register to be scored. Message registers when used are mounted separately from line and cutoff relays.

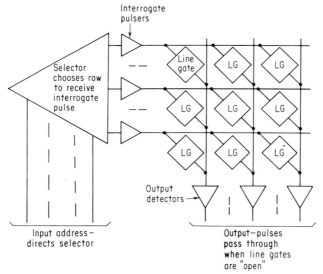

Fig. 45. Scanner line gates in an array to minimize access circuitry. All the line gates on a row are scanned simultaneously.

windings to the customer loop, sampling pulses are coupled from an interrogate winding to a readout winding. When current flows, the transformer is saturated, the coefficient of coupling is greatly reduced, and the pulses are blocked.

b. Determining the Customer's Needs. When a switching system has acted upon a service request, it informs the customer that it is ready for more detailed instructions by returning dial tone or "number please."

Dial-pulse Detection and Counting. Dial pulsing was developed to operate Strowger switches but is used in almost all automatic systems. The mechanism required at the customer's telephone, the dial, is relatively simple, and many years of technological development have reduced its cost to a minimum. The customer rotates the dial a given number of degrees for each digit, and releases it. Upon its return, the dial breaks the direct current flowing in the customer loop. In the United States, the number of breaks is equal to the digit sent, 0 being represented by 10 pulses. In some foreign systems, 1 pulse represents 0, 2 pulses represent the digit 1, etc.

In the SXS system, pulses from customer dials are repeated by relays associated with each switch in the network. The relay provides a local circuit for operating the vertical magnet and, on connector switches, the rotary magnet as well. In register-controlled systems, 10 position minor switches (Fig. 3a), one per digit, can be used to count and store information. Relay counting circuits are also used (Fig. 9d). Because pulse counting requires very critical operation of relays, particularly with regard to timing, it is desirable to have only one set of counting relays per register and use simpler, less critical devices such as reed relays for storing the digit once it is counted. Thus counting relays often have contact networks which translate the digit to a 2-out-of-5 self-checking code for storage and handling.

In electronic offices, dial pulses can be scanned and then counted under program control. Operations such as counting, translating, and storing are typical of the traditional hardware operations which can be carried out with greater economy and convenience by program in a central control.

Voice Frequency Pulsing. A number of attempts have been made to speed up the dialing operation. A dial takes, on the average, $\frac{1}{2}$ sec to transmit a digit. Interdigital times range up to 1 sec or more, and the total time for dialing a number may be from 10 to 15 sec. TOUCH-TONE® pulsing,[116,117] the first serious competitor to the customer's dial in many years, uses logarithmically spaced voice frequency tone pairs generated by a transistor oscillator at the station (see Chap. 2). The total time for a seven-digit number averages around 5 sec. This speeds service for the customer and reduces holding time for the digit detector. Register control must replace direct control on TOUCH-TONE® lines in SXS systems.[60]

Other Possibilities. Systems have been proposed[54,108] in which the calling customer stores the called number in his telephone before lifting the receiver to initiate the call. The central office then interrogates the telephone and does not necessarily require register circuits of its own. Such systems provide the customer with a visual check of the called number but require expensive telephones which can send only a limited number of digits.

Repertory dialers store a group of numbers frequently called by a customer. When the customer selects the desired party and initiates a call, the repertory dialer transmits the information to the central office by conventional (Ref. 8, Vol. 1, p. 171) or TOUCH-TONE[118] pulsing.

c. Calling Customer Identification and Verification (*When Needed*). A switching system, whether manual or automatic, does not necessarily need the number of the calling customer to complete a telephone call. Operators simply plug in at illuminated answering lamps; line finders search for terminals marked with a voltage. Thus until recently the calling customer could be found automatically without being identified.

Direct distance dialing has changed the situation radically. Not only must the long-distance connection be set up automatically, but the charging information must be obtained and recorded automatically as well. This requires, among other things, knowledge of the calling number. Number 5 crossbar, using the "call-back" principle (paragraph 14a), records the equipment number of the calling line in the originating register before dial tone is returned. When dialing is completed, both the calling and called switch locations are used for setting up the talking connection. Because the (equipment) number of the calling line is available as a natural consequence of network operation, automatic message accounting[80,81] was first used with No. 5 crossbar. Automatic number identification (ANI)[61,62] facilities are available to augment earlier systems.

Line Identification.[101] When the calling number must be obtained, the simplest procedure is to route the call via an operator. Before direct distance dialing, this was no hardship because an operator was needed to set up all connections beyond the exchange area. In the early days (Ref. 3, Vol. 3, Chap. 9), the customer would be routed to a "recording" operator who would make out the toll ticket. The customer would hang up, and the ticket would be passed to a "completing" operator who would try to set up the call. When the connection was ready, the completing operator would call back the originating customer; this act automatically verified the calling number.

With the coming of the panel system for large metropolitan exchange areas, assistance operators were reached via selector switches. Rather than drop the connection and call back the customer, a number-checking tone was applied to the sleeve lead at the cord circuit. This tone followed the sleeve back through the switches to the MDF, where it was cross-connected to number-checking jacks at the assistance operators' positions. An operator would touch the tip of her calling cord to the jack corresponding to the calling number. If she heard the tone in her headset, the number was verified.

Automatic number identification (ANI) follows a similar procedure. Some piece of common control circuitry replacing the operator applies a tone or distinctive d-c voltage to the sleeve lead. At the MDF, the sleeve is cross-connected to a detector whose input terminal corresponds to the directory number. Upon locating the marked sleeve lead, the detector transmits its input terminal number in a suitable code to the common control equipment which requested the identification. The whole process is carried out in less than $\frac{1}{2}$ sec, and one identifier is usually sufficient even when there are several switching systems in one building.

Party Identification. Identification of the calling line is sometimes not enough; station (or party) identification must also be provided when required. Two-party identification is relatively simple. One of the two stations has a resistance to ground either when the handset is off-hook (Bell System) or the dial is off-normal (Automatic Electric). The trunk, district junctor, originating register, or similar equipment tests for and records the presence or absence of this ground.

For lines with more than two parties, the Automatic Electric Co. makes use of "spotter" dials[62] which, while sending regular dial pulses, ground the tip side of the line one or more times during the time the loop is closed. Some systems simply require the dialing of an additional identifying digit prior to dialing the called number.

d. Trunk Circuits. The word "trunk," although used in a variety of ways in telephone literature, refers, in the Bell System, to a channel between switching systems used to provide telephone connections between customers. Internal to any one system, the term *path* may be applied. Paths are made up of links and junctors between crosspoints. In foreign practice a path is called a trunk and a trunk is called a junction.

When a customer line is connected to a trunk, circuitry is needed to carry out such necessary functions as supervising the customer and the distant office for answer and/or hang-up, supplying current for the customer's transmitter, applying and removing ringing, etc. In some manual systems, cord-ended trunk circuits were associated with the terminating end of a trunk. The originating end was multipled to a number of positions where originating operators could use the same cord circuits to complete to trunks as would be used to terminate to the local multiple. The supervisory relay in the cord circuit which monitored the local called customer also monitored the trunk (see Fig. 46a).

Panel and No. 1 crossbar follow this plan. At the originating end, the trunk conductors are multipled to terminals which can be reached via district and office frames. The monitoring point is in the district circuit between the line finder and the district selector switches.   Because any district circuit can be used in a call via any trunk, all trunks must return the same supervisory signals from the called or tandem office or else the district circuit must be able to handle several types of supervision.   Number 5 crossbar and No. 1 ESS both use the call-back principle, and thus both require outgoing trunk circuits to carry out the functions of the eliminated district circuit.   With trunk circuits at both ends of a trunk, compatibility between a large variety of switching, transmission, and signaling systems is possible.

Supervisory (on-hook/off-hook) information flows in both directions over trunks between switching systems.   However, trunks are classified as one-way (incoming or outgoing) and two-way.   A one-way trunk permits calls to be initiated only at one end, while a two-way trunk can be seized at either end.   Where trunks are long and expensive or available in very small groups, two-way use provides greater efficiency.

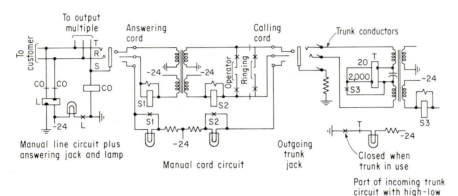

Fɪɢ. 46a. Manual cord circuit and high-low (one way) trunk supervision.   Calling customer operates L relay.   When answering cord is plugged in, CO relay operates from the cord battery. $S1$ operates over the loop to keep the cord lamp off. $S1$ releases to light lamp when calling customer hangs up.

The operator signals the called office over an order wire (not shown), plugs into trunk jack.   The T relay in the incoming trunk circuit operates.   The called customer answering operates $S3$ which shorts the high-resistance winding of T. $S2$ now operates.   When either party hangs up, a cord lamp lights.

At the end of the call, the operator in the originating office pulls down the cords.   This releases the CO relay in the line circuit and the T relay in the trunk circuit.   When the trunk lamp lights, the trunk cord is pulled down by the operator in the terminating office.

However, one-way trunks greatly simplify switching by eliminating "glare," a condition which exists when two switching systems seize the same trunk simultaneously from both ends (see paragraph 16f, Start and Stop Pulsing Signals).

Reference is frequently made in the literature (Ref. 3, Vol. 3, Chap. 11; Ref. 55, Sec. 6) to certain types of trunks which may be defined[119] loosely here:

Intertoll: Trunks interconnecting toll offices (Class 4 and higher in the DDD network).   These trunks are operated at via net loss (VNL) as discussed in paragraph 17.

Toll Connecting (Toll Switching): Trunks connecting toll offices to Class 5 offices. These trunks are operated at VNL +2 db.   There is a toll connecting trunk at each end of a toll connection.

Interoffice: Trunks between local central offices.

Tandem: Trunks to and between tandem offices.

Tandem Completing: Trunks from tandem to local offices.

Recording-Completing: Trunks from a local office to a toll operator who records the billing information and completes the call.   Originally recording trunks were separate from completing trunks (see paragraph 16c, Line Identification).

CLR (Combined Line and Recording): Same as Recording-Completing.

AB Toll (obsolete): Operator interoffice trunk. The operator makes out a toll ticket on each call.

e. Supervision Signals on Trunks.[120] Supervision has two parts: monitoring the switch-hook status of the customers on each end of a connection and transmitting that information to all switching systems in the connection to permit proper holding

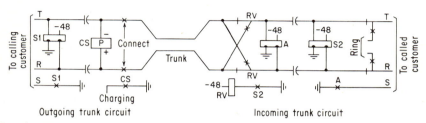

FIG. 46b. Reverse battery (one-way) supervision for automatic offices. Sleeve leads hold network switches operated.

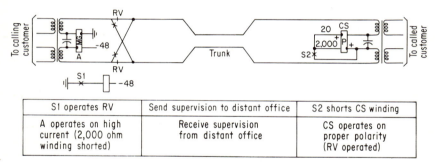

S1 operates RV	Send supervision to distant office	S2 shorts CS winding
A operates on high current (2,000 ohm winding shorted)	Receive supervision from distant office	CS operates on proper polarity (RV operated)

FIG. 46c. High-low/reverse-battery (two-way) supervision.

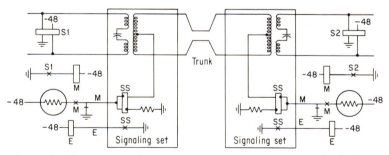

FIG. 46d. E and M leads supervision. Battery on M operates the distant SS relay but not the nearby SS relay.

of the switches and charging. Figure 46 shows common supervision circuits; these and others are described below.

Ring-down Supervision. Magneto switchboards used magnetos and drops for signaling between customer and central office and between one central office and another. The customer would crank his magneto to activate a drop to alert the operator; the operator, after connecting a cord between the calling and called lines, would operate her magneto to ring the called customer's bell. When the call was to another office, the operator would plug her calling cord into a suitable trunk and operate her magneto just as for a local call. This would activate a drop to attract the distant operator, who would receive the called number verbally, make the appropriate connection, and ring. Magnetos will work over 40-mile ranges and can be designed for 400 miles if necessary. This form of "spurt" signaling is very convenient in low traffic systems using pairs of wires as the transmission medium.

High-Low Supervision. When common battery switchboards replaced magneto boards, the line relay and lamp replaced the line drop and the supervisory relays and cord lamps replaced the "clearing-out" drop in the cord circuit. To permit use of the same cord circuit for local and trunked calls, high-low supervision was devised. The trunk circuit at the tandem or terminating office contained a two-winding relay to monitor the originating office. When the A operator plugged into her trunk jack, the distant trunk relay would operate. One of its two windings had a very high resistance and would not permit enough current to flow to operate the supervisory relay in the A operator's cord circuit. When the called customer answered, the high-resistance winding was shorted out, leaving the other winding (of very low resistance) in the circuit. This operated the A operator's supervisory ($S2$) relay which turned off the cord lamp as usual. Called-party hang-up would reinsert the high-resistance winding and turn the lamp on again. If the originating party hung up first, the A operator would pull down her cords, releasing the relay in the distant office. This would light a lamp for the B operator telling her to disconnect. Thus current or no current sent supervision forward, and high current or low current sent supervision back.

Reverse-battery Supervision. Reverse-battery supervision makes use of a polar device at the originating office and a means for transposing tip and ring at the terminating office. Here forward supervision is current or no current as in hi-low (but current is supplied from the called rather than the calling office), and supervision from calling to called office is current polarity rather than magnitude.

High-Low Reverse Battery. High-low and reverse battery can be combined to produce a two-way supervisory system. A marginal relay supplying current is provided at one office, and a polar two-winding relay is inserted in the trunk path at the distant office. Reversing the battery at one office operates the polar relay, and shorting the high-resistance winding of the polar relay at the other office operates the marginal relay.

E and M Lead Supervision. The most common two-way supervisory system is named E and M lead after arbitrarily designated leads in a long-forgotten toll trunk circuit. The M lead sends information to the distant office. Battery through a resistance lamp on M means off-hook, and ground means on-hook. The E lead receives information from the distant office. Ground means off-hook, and open means on-hook.

The E and M leads connect to special signaling sets which provide facilities good for very long distances on either wire or carrier trunks. Where wire conductors are used (pairs or quads), the signaling sets reapply the signals to the tip and ring conductors on a d-c basis. Carrier channels which do not have built-in signaling facilities generally use single-frequency (SF)[121] signaling in which a voice-frequency tone is applied to indicate on-hook and is removed to indicate off-hook. A tone transmitter and a tone receiver are required for each end of a trunk. The receivers must be able to respond properly to the SF tone without being falsely operated by speech or noise. Filters for frequency discrimination, timing for discrimination against momentary simulations by speech, and comparison circuits which reject tone in the presence of other signals but accept it in their absence are the basis of operation.

f. Pulsing. Signaling between central offices is usually divided into pulsing and supervision. Supervision, as discussed in paragraph 16e, monitors the status of the customers and the speech path, while pulsing is concerned with the information needed to set up the proper connection.

Pulsing Channels. Pulsing can be sent over the trunk to be used by the customer for conversation, or it can be sent over another channel specially reserved for signaling.[93] In manual systems, these separate channels were called order wires (Ref. 3,

Vol. 2, p. 298).   Each trunk group had an order wire, and each *A* operator had a key per order wire.   An *A* operator wishing to use a given trunk group operated her key and gave the called number to the *B* operator, whose headset was connected at the far end.   The *B* operator then chose an idle trunk (since all trunks in the group appeared at her position) and returned the trunk number to the *A* operator over the order wire. The *B* operator then plugged into the called customer's jack in her multiple or into a busy jack to return a busy signal if the called line was engaged.   The *A* operator plugged her calling cord into the trunk designated, and the calling customer heard an audible ringing signal or a busy tone.

As the number of terminating offices increased, the number of order-wire keys at each operator's position became unduly large.   Furthermore, direct trunk groups to many offices could not be used efficiently because of the small amount of traffic involved.   Thus tandem trunking was developed, and direct trunking was used only where the traffic was large enough to warrant it.   Order-wire signaling was difficult with tandem trunks because of the large numbers of trunk groups before the tandem operator.   "Straightforward" signaling solved the problem.

In straightforward signaling, the *A* operator made a busy test and plugged into an idle trunk to the tandem or terminating office.   The trunk and control circuits at the tandem or *B* positions were quite complex and associated the operator's headset in order with the trunks requesting service.   When the operator was ready to answer a call, she pushed a key to permit the automatic equipment to step to the next call waiting; at the same time "zip tones" were returned automatically over the trunk to the *A* operator.   These tones were high-pitched "beeps"; three beeps meant that a tandem operator had been reached and the *A* operator should give the office code. Two beeps meant pass the directory number to the *B* operator.   In this way, a call could be routed through several operators.   The many automatic features in trunk and position circuitry meant that operators could handle large quantities of traffic.

Automatic systems generally use straightforward pulsing rather than order wire. When registers are not available at intermediate offices, the sender in the originating office, like the *A* operator, stays on the circuit until the called office is reached; with registers, each sender can transmit the number and drop off, leaving the sender in the following office to retransmit as required.

Information Transmitted.   In American practice, a sender transmits the DDD area code if needed, the office code if needed, and the called directory number.   The general rule is never to send unnecessary information.   If a direct trunk to a given office is selected, there is clearly no need to send the office code.   Similarly, a DDD area code need not be transmitted over a trunk connecting to a toll office in the terminating area.

When SXS offices are part of a tandem or toll switching system, office code and area code must be converted to as many pulse trains as necessary to operate selectors. When a call must be routed via a switching point in another area back to the original area, the area code must usually be reinserted.   Thus switching systems may be required to alter the pulsing information they receive by adding digits, deleting digits, and converting digits from one method of pulsing to another.   In general, only switching systems with some form of central control and translation facilities can carry out such complex functions.

Pulsing Systems.   To manipulate numerical information, switching systems must be able to use various forms of electrical coding.   SXS offices send and receive dial pulses most economically; panel offices prefer revertive and panel call-indicator (PCI) pulsing.   Crossbar and electronic offices, because of the flexibility provided by common controls, senders, and registers, can use any of the above types of pulsing but prefer a faster voice-frequency system called multifrequency pulsing (MF).   Toll operators and test positions adjacent to or part of automatic systems sometimes use d-c key pulsing, which, although limited in range, is as fast as MF and uses less expensive receivers.   Principal characteristics of these systems are summarized in Table 3. Brief descriptions follow.

Dial Pulsing.   Dial pulsing over trunks is similar to dial pulsing from customers: A loop is opened at a rate of approximately 10 pps with the break period approximately

Table 3. Pulsing Signals

Digit or signal	Dial pulses (breaks)	Revertive pulses (shorts)	Panel call indicator (current magnitude and polarity) — Th digit (Quadrants) A	B	C	D	Other digits (Quadrants) A	B	C	D	D-c key pulsing (current magnitude, polarity, and lead) — B board (Lead designations) FT	FR	K	A board KT	KR	Multifrequency pulsing (voice frequency tones) Frequency, cps		TOUCH-TONE® pulsing (voice frequency tones) Frequency, cps	
0	10	1	—	n	—	n	—	n	—	n	—	—	N	p	—	1,300	1,500	941	1,336
1	1	2	—	n	—	N	p	n	—	n	N	—	N	P	—	700	900	697	1,209
2	2	3	p	n	—	n	—	N	—	n	n	—	N	n	—	700	1,100	697	1,336
3	3	4	p	n	—	N	p	n	—	n	—	N	N	N	n	900	1,100	697	1,477
4	4	5	—	N	—	n	p	n	p	n	—	n	n	p	N	700	1,300	770	1,209
5	5	6	—	N	—	N	—	n	—	n	N	—	n	P	P	900	1,300	770	1,336
6	6	7	p	N	—	n	p	n	—	N	n	—	n	P	P	1,100	1,300	770	1,477
7	7	8	p	N	—	N	p	N	—	N	—	—	n	N	P	700	1,500	852	1,209
8	8	9	—	n	p	n	—	N	—	N	N	n	n	n	P	900	1,500	852	1,336
9	9	10	—	n	p	N	p	n	p	N	n	n	n	p	N	1,100	1,500	852	1,477
KP	Time	Open loop	None				None				Open leads			Open leads		1,100	1,700	No tone	
ST																1,500	1,700		
Inter-digital	Time	Open loop	None				None				Open leads			Open leads		No tone		No tone	
Notes	1	2	3				3				3			3, 4					

1. Interdigital time is a minimum of 200 msec of loop closure, but some systems may require up to 1 sec.
2. Pulses are returned from the called office. Only tens and units digits are decimal. See Figs. 36 and 37.
3. n = small negative current. N = large negative current.
   p = small positive current. P = large positive current.
   — = open conductors or loop.
4. KP (key pulse) button associates key set with operator cord circuit.
   ST (start) button releases key set and signals sender to begin setting switches.

60 per cent of the pulsing period. The pulsing rate can vary from 9.5 to 10.5 pps, and the per cent break from 58 to 64 per cent. Pulsing at 20 pps has been used between senderized systems but is tending to give way to more modern methods. In addition to loop pulsing, battery-ground pulsing is sometimes used to extend pulsing range. Here batteries at the sending and receiving offices are connected to the loop series aiding during pulsing. In the interdigital interval, the battery in the transmitting office is removed in favor of a bridge from one conductor to the other.

The interdigital time in dial pulsing ranges between about 200 msec and 1 sec depending on the system and sometimes the digit. With dial pulsing, digits average more than $3/4$ sec to send.

Revertive Pulsing. Revertive pulses are repetitive shorts between the two conductors of the trunk, applied at the receiving office to indicate or simulate the position of a motor-driven switch (see paragraph 15a). When the loop is closed and the receiver is ready, the latter starts a train of revertive pulses at the rate of about 27 pps. Because timing is not used to detect the interval between pulse trains (called the "interselection time"), accuracy of speed and the equivalent of per cent break are not so critical as in dial pulsing as long as maximum brush speed is limited. When revertive pulses are simulated by relay circuits, 25 to 27 pps, with the loop shorted for 45 to 55 per cent of the pulse interval, are usually specified. When the transmitter observes the correct count for the given selection, it must open the fundamental circuit within 10 msec and hold it open for at least 100 msec to prevent the next revertive pulse from being generated.

When the panel sender has completed its task, a reverse battery release signal is returned to a polar relay in the sender. If the called line has been reached, the signal is called "incoming advance," but if all paths between switches are busy, it is called "overflow." If a customer abandons a call before the path is set up or if some malfunction occurs, the counter in the sender does not halt the next train of revertive pulses; selector motion continues to a group busy terminal where the reverse battery signal is initiated. In trouble cases, the sender is released as soon as the customer finishes dialing to permit the reorder tone to be returned; if the customer abandons, the sender and switches are released at a convenient point in the operating cycle of the sender.

Panel Call Indicator (PCI). Panel systems were designed to be directly compatible with manual equipment; to enable a panel sender to give a manual B operator the called number, digit lamps were provided at the B position. These lamps were lit by register relays which monitored both the polarity and strength of the direct current in the loop.

PCI is still used between some automatic systems. A digit is sent in four sections, or "quadrants," each of about 70-msec duration. In the second and fourth quadrants, a regular battery connection is made to the loop, through either a small or large resistance. In the first and third quadrants, either an open circuit or a reversed battery through a large resistance is sent. These signals are usually shown as N (heavy negative), n (light negative), −, and p (light positive). The receiver counts the operations and releases of a polar relay which responds to large or small negative currents, and after every second negative pulse the counter causes the information registered during the preceeding four quadrants to be registered, corresponding to lighting a digit lamp for an operator. No pause is made between digits, and a four-digit number can be sent in slightly more than 1 sec. Because the receiver has only to recognize two levels of current in each of the two polarities, one sensitive, one marginal, and one polar relay serve to detect digits in the inexpensive manner shown in Fig. 47. The PCI code is not the same for all digits; by making the thousands digit's code different, a simpler receiver can be made for small offices.

Direct-current Key Pulsing (DCKP). When an automatic switching system is used with nearby manual switchboards, toll switchboards, and test facilities, it is economical to use d-c key sets, one button per digit, to key information into the automatic system. This reduces operator holding time and speeds service while taking advantage of the low cost of d-c signaling methods. Where toll operators have access to trunks via a No. 5 crossbar system, special incoming registers are used which take DCKP from the operator and, using the marker and the appropriate senders associated with the outgoing trunks, convert the signal to whatever is required by the trunk. This method of operation is called "marker pulse conversion."[72] Two slightly different methods of DCKP have had wide use. One uses tip and ring plus a third wire, with no current, light current, and heavy current in the tip and ring conductors and light and heavy current in the third wire. This system requires marginal but no polar relays.

A second method uses small and large positive and negative current separately on each lead of a pair with ground return. Such signaling requires only a single pair but needs twice as many relays at the detector: two sensitive, two polar, and two marginal.

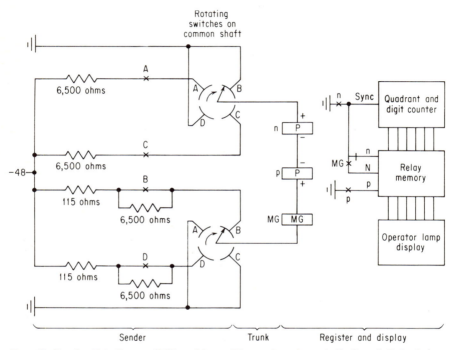

Fig. 47. Panel call indicator (PCI) pulsing. The sender relays *A*, *B*, *C*, and *D*, by being either set or reset, can be arranged in 16 possible combinations of which 10 can be used for digits. The rotating switches transmit one digit for four consecutive "quadrants" at about four digits per second. At the receiver, marginal and polar relays detect current amplitude and direction. The counter directs these inputs to the proper memory relays which, in combination, light the appropriate lamps in the operator display.

This system uses a KP (key pulse) button to connect the keyset to the operator's cord circuit and, when the digits have been keyed into the sender, a ST (start) signal to release the keyset and notify the sender to begin outpulsing.

MF Pulsing.[122] Multifrequency pulsing has all the advantages of d-c key pulsing when an operator wishes to send digits: speed and very simple sending circuits. Tone buses are multipled to all operator positions, and tones are gated out through keysets. MF, using frequencies between 700 and 1,700 cps, can be transmitted over any channel that can carry speech including, in particular, repeatered and carrier trunks. Its disadvantage lies in the relatively complicated digit receiver required. Early models use vacuum tubes, while those developed later are transistorized.

MF pulsing uses a code based on exactly two out of six arithmetically spaced tones to represent digits and additional signals. This permits a simple receiving check using a symmetric circuit (Fig. 9a) on relays following the detector. Key pulse and ST (start) are sent to the receiver; the first is an enabling combination of longer duration that makes the receiver more sensitive. Normally the receiver is insensitive or otherwise arranged to reject noise "talk-offs." The ST signal indicates that all digits have been sent.

TOUCH-TONE® *Pulsing.* Although TOUCH-TONE® is primarily intended for customer signaling (see paragraph 16b), it could be used for interoffice signaling as well.

Start- and Stop-pulsing Signals (Ref. 55, Sec. 4). In automatic systems, pulsing must not begin until the distant office is ready to receive it. Occasionally, when one sender sends signals first to a SXS tandem and then to a "link-type" office beyond, pulsing must be stopped after the SXS switches are set to await operation of the incoming register link in the second office and then started again. In general, supervisory signals carry the start- and stop-pulsing information but not always. Revertive pulsing, for instance, required only a connect signal at the outgoing trunk; the sender seizes the trunk (makes it busy to other originating calls) but does not send the connect signal until it is ready. The terminating office, in turn, does not send revertive pulses until it, also, is ready.

In SXS systems under direct customer control, the only start pulsing signals which can be used between central offices are additional applications of dial tone. PBX's and other small SXS systems frequently use this procedure, and directing codes such as 9 are required to get into the next higher office in the hierarchy. These offices then return the second dial tone to indicate that the area code, office code, and directory number can be dialed. Such directing codes are undesirable in that they prevent the use of a uniform numbering plan and demand that customer dialing behavior differ as a function of telephone used.

To eliminate second dial tone, senderized (link-type) systems provide fast "by-links" to permit registration of the customer-dialed number to begin immediately upon detection of seizure. A customer dialing from SXS into a No. 5 crossbar office will thus not even be aware that some of his digits are used up in the SXS office and the remainder go directly to the No. 5 incoming register.

When a senderized system outpulses to SXS, again no start pulsing signal is required. If the proper polarity of battery and ground is seen on the trunk conductors, the selector is ready to go and pulses can be sent. The reverse polarity is used as a make-busy signal on one-way trunks.

PCI pulsing is started as follows: The sender in the calling office seizes the trunk to a manual or tandem office which accepts PCI pulsing. Battery and ground are seen on the trunk conductors via the trunk supervisory relay. The battery and ground signals are removed when the sensitive, marginal, and polar relays are connected across the conductors; this constitutes the start signal.

Pulsing with dial pulses or MF signals into senderized systems such as Nos. 1 and 5 crossbar requires time to associate a digit receiver. Thus "wink" and delay dial signals have been developed. The general procedure is to have a trunk appear on-hook from both ends when idle. Off-hook is sent forward as a connect signal, and off-hook is returned to indicate awareness of seizure but unreadiness to yet accept digits. The terminating office then returns on-hook when it is ready.

The change from off-hook to on-hook is the start-pulsing signal recognized by wink-start senders, while observation of on-hook alone is enough for delay-dial senders which postpone testing for on-hook until the original (idle period) on-hook has had time to be changed to off-hook at the terminating office. It is sometimes possible for the terminating end to connect a digit receiver to receive pulsing before the originating end has its sender connected. Thus the off-hook signal may be missed altogether at the sender. Usually wink start is used on one-way trunks while delay-dial is used on two-way trunks. A one-way trunk is seized (made busy to other outgoing calls) and associated with a sender. The sender then applies the connect signal. On two-way trunks, the connect signal is sent from the trunk circuit immediately upon seizure so that the far end will not seize the trunk during the interval the sender is being attached.

At the receiving end, the off-hook part of the wink is sent by switching from the trunk to the digit receiver. The change to on-hook is then generated in the receiver after a timed interval. With reverse-battery supervision, this procedure is particularly simple: The trunk supervisory relay applies battery and ground of one polarity to the trunk conductors while the polarity of the digit receiver is reversed. Thus a single rereversal in the digit receiver terminates the wink. If the called office is busy, there may be a delay before a digit receiver can be attached. This delay in sending off-hook makes the transition from off-hook to on-hook necessary as a start pulsing signal.

Complications arise with delay-dial on two-way trunks because the off-hook signal from the terminating end must remain until a digit receiver is available, but a delay-dial off-hook cannot be distinguished from an off-hook signaling a seizure by the terminating end. A seizure will remain on the trunk for a prolonged period of time, but a delay-dial will hardly be longer than, say, $\frac{1}{2}$ sec. Thus timing circuits in senders guard against an indefinite period of "glare." One sender times out before the other, drops the trunk, and chooses another or else sends a reorder signal to the originating operator or customer. The other sender, left in possession of the trunk, should not be tricked by the on-hook sent to it as a result of the far-end abandon. Pulsing is delayed long enough to permit a digit receiver to be connected (off-hook again) and send the start signal (a second on-hook). Thus a sender on a two-way trunk must delay digit transmission after seeing the start signal from the distant office for a long enough period to be sure the start signal is genuine. A minimum interval of 1.9 sec is sometimes specified.

g. Connecting to the Called Customer. As soon as the terminating office knows the directory number of the called customer, steps can be taken to establish the final link of the connection. In progressive controlled systems, the connection is complete upon receipt of the final digit from the originating or intermediate office. In common control systems, the called number is stored in a register and, in general, must be translated to an equipment number to identify the actual switch position. At this time, information can also be obtained from a central memory or translation point concerning the required ringing signal, the existence of special services such as dial transfer, or unusual or trouble conditions to prevent completion of the call (Ref. 40, p. 2533; Ref. 70, Vol. 17, p. 356; Ref. 73, Vol. 28, p. 298).

Busy Test and No Test. Before a path can be completed to the called customer, the switching system must be sure the line is not busy with another call. In manual systems, a busy line has battery applied to its cutoff relay via the sleeve lead from the cord-circuit lamps (see Fig. 46a). This potential is available on all sleeve appearances in the multiple, and busy test is effected by the operator touching the exposed sleeve terminal at the nearest multiple jack with the tip of her calling cord. With her listening key operated, her headset is capacitively coupled to the line; the click produced upon tip contact indicates a busy condition. If the line is idle, no click is heard and the operator inserts the cord all the way into the jack; ringing can then be applied automatically or with the ringing key.

In automatic systems the sleeve lead usually holds switches operated and a busy condition is a grounded sleeve at the line circuit. Thus a test for a busy line looks for this ground. In progressive systems the last pulse train drives the connector to the desired line; in centrally controlled systems the equipment number is known before the path is found. Thus busy test is not always made at the same point in the call. In those electronic systems which use common memory to keep a constant record of busy and idle lines, the busy test can be made quickly on a memory location entirely within the high-speed control circuitry.

There are certain occasions when a connection is made to a line without a busy test. This *no-test* situation occurs when an operator is asked to verify that a line is really busy and also when certain maintenance operations are to be performed. Extra switches, accessible only to operators and maintenance personnel, provide the access. In progressive controlled systems, these switches are usually in the multiple with regular connectors and are driven to the line terminal in question. In crossbar systems a no-test vertical is used to connect to the same line link to which the vertical of the tested line is connected. (See Fig. 32a.)

Ringing, Ring Trip, and Audible Ring.[123]   When the called line is found to be idle and the connection is established, the called customer must be alerted.   This is usually carried out by ringing the bell in the telephone with a relatively large, low-frequency voltage.   The most commonly used signal is 20 cps at about 86 volts rms.   The a-c ringing signal has a d-c battery in series with it; a specially designed slow-operate relay is used to detect answer by responsing to direct but not alternating current.   The on-hook ringing current can be appreciably bigger than the off-hook direct current when the telephone is answered between rings, particularly on party lines where loops are long and many ringers operate at once; thus the design of the ring-trip relay is critical.   The system must "remember" that ringing has been tripped to prevent reapplication of ringing if the called customer hangs up before the calling customer.

For individual lines and party lines where only one customer hears the applied ringing, a 2 sec on, 4 sec off cycle is generally used in the United States.   This permits division of ringing loads into three equal intervals during the 6 sec and triples the effective output power of the ringing generator.   Various codes of ringing such as two shorts, long-short-long, etc., are used for party lines.

Two-party ringing can be fully selective by simply connecting one ringer from each side of the line to ground and applying ringing to the proper conductor with a ground return.   Selective ringing for up to four parties often associates different polarities of d-c battery with a-c ringing signal to operate electrically or mechanically polarized bells.   Different ringing frequencies can be used with tuned ringers.   Frequencies such as $16\frac{2}{3}$, $33\frac{1}{3}$, 50, and $66\frac{2}{3}$ cps can be used at high power levels, while voice-band signals at speech levels can be used with special station apparatus such as tone ringers.[124]   Voice-band ringing has the great advantage of being able to pass through time-division and other electronic networks.

The multiplicity of ringing signals poses problems for the designer of switching systems.   An incoming trunk must be able to terminate to any customer in the office, regardless of the type of ringing he may need; similar requirements exist for intraoffice calls.   Several solutions have been used to select the proper ringing signal.   In manual practice and some automatic systems, notably those progressively controlled, an extra party digit has been used.   This digit is designated by a letter such as M, J, R, or W after the numerical digits of the directory number: 6493-W.   Cord circuits for party-line ringing had four push buttons to select the appropriate type of ringing; the operator depressed the proper button and pushed the ringing key.   In SXS systems, a minor switch can be associated with each party-line connector; dial impulses corresponding to the station letter set the switch to connect the proper signal.   This type of operation, called "terminal-per-line," requires only one output multiple for a given line, no matter how many parties, and thus saves equipment.   "Terminal-per-station" operation is presently standard, however, to eliminate party-line digits in the uniform numbering plan for DDD.   If a uniform numbering plan is not used, a sender in a tandem or toll office has no way of knowing when a customer has finished dialing and must time for a reasonable interval on all calls.   This delays service.

When terminal-per-station operation is used in SXS systems, lines with particular types of ringing are grouped on connectors which provide one kind of ringing only.   Thus party-line differentiation must be made by means of a digit before the connector tens and units.   In a common control system such as No. 5 crossbar, each line has only one appearance on the network; the directory-to-equipment-number translation also provides the information needed to apply ringing.   Trunk circuits used in terminating calls have access to all types of ringing used in the office through a ringing switch composed of one vertical on a crossbar switch (Ref. 73, Vol. 28, p. 168).   A 10-vertical crossbar switch is mounted with each 10 trunk circuits in the trunk frames.   The marker sets the ringing switch as part of its functions in connecting to the called customer.

Number 1 ESS uses a different approach (Ref. 40, p. 2331).   Ringing circuits independent of trunk circuits are provided, and a called line is connected to the appropriate ringing circuit.   When answer is detected, the system takes down the ringing path through the network from called customer to ringing circuit and sets up the talking connection from customer to trunk.   This simplifies the trunk circuits considerably and permits great flexibility in the introduction of new types of ringing as they are developed.

While the called customer's telephone is being rung, a tone is returned to the calling customer to indicate the progress of the call.   In most systems this tone, called audible ring, is mixed with the ringing signal and is returned to the calling customer from the same trunk or connector circuits that apply ringing to the called customer.

Call Progress Tones, Recorded Announcements, and Assistance Operators.   Various tones must be returned to customers when a line is busy, when all paths or all trunks are busy, or when no such number or no such office exists.   Line-busy and path- or trunk-busy tones are low-pitched sounds interrupted usually at 60 and 120 ipm, respectively.   In manual practice, the interruptions are also superimposed on the supervisory signal to flash operator cord lamps; this enables an operator to monitor a number of circuits simultaneously for unusual conditions.   With almost all local calls and more and more toll calls dialed directly by customers, the need for visual operator signals has decreased to the point where the flashing signals have been almost completely removed.

All-paths-busy (or overflow) tone is returned automatically in progressive controlled systems; if all the paths to the next stage of SXS switchs are found busy, overflow tone is connected via springs operated on the eleventh rotary step.   In common control systems, overflow tone is available on separate network appearances.   If the trunk group and its alternates are all busy, the customer is connected to an overflow trunk.   Other call progress tones include dial tone and audible ring.   Table 4 summarizes these tones, their constituent frequencies, and their purpose.

Table 4. Call Progress Tones

Name and purpose	Frequencies (Bell System)		Interruptions
	Older standard*	Present standard†	
Dial tone—inform customer he may dial......	600, 120 (low tone)	350 + 440	Continuous
Busy—called customer's line in use...........	600, 120	480 + 620	60 IPM
Overflow—all paths through office busy.......	600, 120	480 + 620	120 IPM
Audible ring—inform customer called line is being rung	420, 40	440 + 480	Follows ringing
Permanent signal or partial dial—inform customer he has dialed too slowly or not at all	500 (high tone)	....	Continuous
Howler—alert customer to telephone off-hook.	480‡	4 frequencies selected for maximum radiation from receiver	

  * Frequencies modulated to produce harmonics.
  † Frequencies linearly mixed.
  ‡ Starts softly and increases to high level.

Recorded announcements have been used since before 1905 in place of such signals as busy tone.   At present, elaborate recorded announcement facilities are available to provide customers with information about their calls; similar systems can also return time-of-day signals, weather announcements, and other information as desired.[125]   Often, in metropolitan areas, these facilities are located at tandem offices for more efficient usage.

It is a basic principle that no automatic system is completely foolproof; there will probably always be a need for some skilled human being to provide assistance.   Thus it is good practice to make sure that *any* abnormality will be routed to an operator if all other means fail.[126]   Time-outs are provided in almost all common usage circuits in common control systems to assist in such operations.

h. Talking, Rering, and Hang-up. The entire process of setting up a connection through one or more offices exists for the sole purpose of permitting customers (or their machinery) to talk to each other. Upon answer, a signal is returned to the originating office to permit the start of charging; if an intercept operator or other free-call termination is reached, the charge signal is inhibited. The flashing signals formerly used for line, path, or trunk busy consisted of on-hook/off-hook signals; lightning hits and other noise signals can simulate supervision changes. Thus the originating office delays the start of charging briefly to prevent a false charge on flash.

In general, a connection is held by the originating party, but if the called party hangs up while the calling party remains off-hook, it is desirable that the called party's line be freed for other calls. A time-out is often provided, however, to prevent a false call origination at the calling end if the called party hangs up first. This timing also permits a called customer to switch from one extension to another or a PBX operator to switch cords without losing the connection.

Joint-holding trunks are often provided to operators. These circuits require that both the operator and the customer disconnect before the connection is released. This feature is important in emergencies when the customer may hang up before giving the full information needed by the operator.

When toll calls are set up between operators, it is sometimes necessary for the outward toll operator to get the inward operator back on the connection or vice versa. The rering signal, originally a spurt of ringing on the trunk, serves this purpose. At present, many calls are dialed direct, and almost all calls which require operators need only one. Thus the rering signal finds little use on ordinary calls. However, where conference calls are required among groups at two or more distant points, operators are needed to set up the connections. Inward operators are also needed on collect calls to coin stations. For such situations, the rering signal is retained.

In all systems with party lines, "reverting calls"[127] between customers on the same line require special treatment. The called party cannot be rung while the calling party is off-hook; if he hangs up to permit ringing, the calling party cannot hear the called party answer. The general procedure is for the calling customer to hang up after making his request; the system then rings both the called and calling parties. When the called party answers, ringing is tripped. The calling party then lifts his own receiver. If the called party does not answer, the calling party can trip ringing by momentarily lifting his receiver.

The customer may not know that he has dialed another customer on his own line. The switching system in turn may not know which party on a multiparty line has initiated the call and thus may not know the proper ringing signal to apply. It is often simplest to route reverting calls to an operator, who can instruct the calling customer, obtain all necessary information, and supervise the call.

Hang-up at the end of a call releases the connection and ends charging. The final on-hook signal must persist longer than the rering signal if suitable differentiation is to be made, but it must not be so long that the customer is inconvenienced in originating a new call. Alternatively, a false origination must not be produced by the remaining customer when the other hangs up.

i. Charging and Billing. Proper charging for telephone service requires thorough consideration in the design of a switching system. Great care must be taken to avoid false charges; at the same time, the collection of earned revenue must be assured.

Flat Rate. The easiest method of billing is to charge customers a flat rate, regardless of how much they use their telephones for local calls. Flat rates for businesses are higher than those to private individuals. Toll calls are routed through operators for billing.

Message Registers. Metered service provides an inexpensive and accurate way to permit charging in accordance with usage. A message register must be associated with every customer who elects this type of service, and some means of operating the register is required on every successfully completed call. In most systems, either local or tandem, the charge signal is returned when the called customer goes off-hook. In SXS systems, for instance, the connector reverses battery toward the calling customer. This signal is picked off at a circuit usually located between the line finder and first selector and applied to a fourth wire through the line finder to the message register in the line circuit. If two-party service is provided, party identification is made and the message register circuit set so that a marginal relay can operate the proper message register. The panel system and No. 1 crossbar function in nearly the same way. The reverse battery signal from the incoming selector is returned via tip and ring to the district circuit where it is picked off by a polar relay; two leads extend through the line frames to message registers for two-party lines. Number 5 crossbar offices use more elaborate circuitry (Ref. 73, Vol. 28, p. 404), since the outgoing trunk circuit receives the reverse battery signal and extra paths for message registers through the line- and trunk-link frames would require all crossbar switches to have extra contacts. The sleeve lead is used with superimposed a-c signals; in this way crossbar switches need only three contacts per crosspoint but can score message registers for two-party lines.

Manual Toll Recording. In early manual practice, the customer asked for long distance and was connected via a "recording" trunk to a recording operator who took down the calling and called numbers, obtained routing information, etc. The ticket she prepared was forwarded to a toll operator who set up the connection, recalled the originating customer, and supervised the call. The toll operator used an automatic time stamp to put the answer and hang-up time on the ticket.

Automatic Ticketing.[63-65] SXS systems were originally developed for local flat-rate switching; all outside connections went through an operator, who could easily write a toll ticket. In areas such as Los Angeles, however, tandem and toll operators were swamped with large numbers of tickets for calls between nearby offices. Message registers could have been used with provisions for operating them one or more times per call, depending on distance and holding time, but flat-rate service in the direct dial area had been the rule and the addition of per-customer message registers would have been expensive. Thus automatic ticketing was developed to be used in association with tandem and toll trunks, permitting flat-rate local calls to be ignored while the extra-cost calls could be recorded at a concentrated point.

Several automatic toll ticketing systems have been developed. Automatic number-identification equipment[61-62] is used to obtain the calling number (although an operator can be called in to obtain this information verbally and key it into the system), and registers and transmitters are needed to store the called number as dialed and retransmit it when the proper information is available from translation. Although the ANI system is common to all charge trunks, the small sender groups are shared, and other common equipment is used, each trunk circuit contains relay registers or ticket printers and tends to be relatively expensive. In the Bell System, a special form of automatic message accounting has replaced automatic ticketing; in this system the per-trunk ticket printers are no longer required.

Automatic Message Accounting.[80,81] To provide automatic billing for DDD calls on a nation-wide basis, automatic message accounting was developed. In this system, 100 trunks share a recorder which stores a wide variety of per-call information in a form suitable for machine processing. Holes are punched in rows across a 3-in.-wide paper tape. Three kinds of entries may be required for each call. The first contains the calling number, the called number when needed for detailed billing, information concerning the charging rate for the particular customer, and a number identifying the particular trunk. The remaining two entries are the answer time and the disconnect time. Many calls are intermixed on the same tape, and the per-call entries are tied together by the trunk identity number. In the accounting center, the tapes are "played back" to permit assembly of the per-call information. Computers calculate the cost of each call and facilitate the preparation of customer bills.

When an office is too small or too incompatible to permit economical modification to AMA, centralized AMA (CAMA)[66] located at a tandem or toll office can be used. ANI or operator identification of the called number is needed; when used, the ANI system transmits the calling number to the tandem or toll office over the same trunk which has been used to transmit the called number and which will be used for the conversation.

Electronic switching systems, with their large, inexpensive-per-bit memories and elaborate central controls, often find it convenient and economical to store all billing information locally until the call is complete and then make a single entry per call on a magnetic tape.

Coin Telephones. Coin telephones are arranged in several different ways.[128] The most common is to require the receiver to be off-hook and the coin in the slot for the line relay to operate via ground at the subset over the ring conductor alone as shown in Fig. 48. Coins are collected or returned by applying the proper polarity of 130-volt battery to the tip side of the line, a path unaffected by the state of the switch hook. The coin magnet pulls a mechanism one way or the other to collect or return a coin, depending on the polarity of voltage with respect to ground. The signal is sent from a trunk circuit, district circuit, or some other convenient point associated with answer supervision.

Another method of coin operation permits the call to be originated from a pay telephone in the same way as from any other telephone. However, when the call is

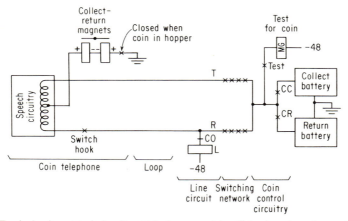

Fig. 48. Typical coin-control circuit. Telephone must be off-hook and a coin must be in the hopper to originate a call. To collect or return a coin, the proper d-c potential must be applied to the collect and return magnets.

answered, the calling customer must deposit his coin to activate his transmitter. On calls to operators, fire departments, etc., the coin is not needed. An advantage of this system is the complete lack of any need for coin-collect or -return equipment. A disadvantage is the inability of the operator to make coin returns.

In British practice, the coin-collect and coin-return functions are provided by "button A" and "button B," respectively, on the coin box, again eliminating any need for coin control equipment at the central office. The customer must deposit his money before he can dial; when he is connected to the called party, he pushes button A to activate his transmitter and tip the money into the coin box. If the call fails, he pushes button B, tipping his money into the coin-return chute. Special provisions are made to permit dialing certain emergency calls without depositing a coin (Ref. 8, Vol. 1, p. 172).

17. The Transmission Properties of a Telephone Switching System[129]

a. The Switching System as a Transmission Circuit.
From the viewpoint of the transmission engineer, a telephone switching system is simply a lumped network inserted into a transmission facility. As such, the ideal switching system should be invisible. It should have no loss, no phase shift, no frequency dependence, and it should generate no noise. For many reasons, this ideal cannot even be approached.

Because a switching system is continually connecting different lines and trunks together, the best strategy for good transmission is not only to minimize loss but also to minimize the difference between any two different paths through the office. Reasonable transmission requirements for a local central office are summarized in Table 5. These requirements are not always met in practice.

Perhaps the closest approach to a direct connection was the cord circuit in local battery magneto switchboards where the wires from the calling customer were extended

Table 5. Typical Transmission Requirements for a Class 5 Switching System as a Whole

I. Insertion loss:

Average at 1 kc, line to trunk connection (MDF to MDF). . . . . 0.5 db

Maximum at 1 kc, allowing for different paths through the network and trunk circuit component tolerances and variations with direct current. . 0.8 db

Variation with frequency in terms of maximum deviation from 1 kc loss:

At 200 cps. . . . . 0.5 db

At 3,200 cps. . . . . 0.3 db

For line-to-line connections, the above values may be doubled

II. Harmonic distortion and overloading:

Maximum compression of a +10-dbm 1-kc signal. 0.2 db

Maximum level of harmonic distortion below a 0-dbm 1-kc signal. . . . . . . 55 db

Maximum level of intermodulation products below the combined level of two equal amplitude frequencies. . . . . 55 db

III. Delay

Absolute delay. . . . . 660 μsec

Envelope Delay Difference

600–3,000 cps. . . . . 40 μsec

500–3,200 cps . . . . 80 μsec

IV. Return loss balance:

Nominal office impedance. . . . . 900 ohms

Standard impedance against which office is measured (office consists of customer lines, switching network, office wiring and trunk circuits as viewed from the trunk transmission facility)
$\quad$ 900 ohms in series with 2.14 μf

	Average over all customer lines	Sigma
Singing return loss.	6 db	2 db
(The smallest value of return loss in the band between 250 and 3,200 cps)		
Echo return loss.	11 db	3 db
(The average value of return loss over the band between 500 and 2,500 cps)		

V. Longitudinal balance:

	Average	Minimum
200 cps.	66 db	61 db
1,000 cps.	60 db	55 db
3,000 cps.	58 db	53 db

VI. Crosstalk and noise:

Crosstalk coupling loss. . . . . 75 db minimum

Maximum noise in a talking circuit (3A noise measuring set, C message weighting). . . . . 16 dbRN

Impulse noise (at 0 TLP). . . . . −52 dbm

directly to the trunk wires or to the called customer's conductors. Even here, however, something had to be added. The clearing-out drop, effectively a high inductance bridged from tip to ring, was necessary to permit the customers to signal the operator at the end of the call (Ref. 3, Vol. 2, Chap. 7; Ref. 8, Vol. 1, Chap. 15).

The common battery system (which eliminated the magneto and local battery at each customer's premises) required inductors in the cord circuit in series with the central-office battery to prevent the low impedance of the battery from shunting speech signals. The resistance of the inductors, in turn, demanded the further addition of series capacitors in the cord circuit to prevent a short low-resistance customer loop from shunting current away from a high-resistance long loop. Transformers were also used, but capacitors were still required at the center tap to bypass supervisory relays.

Sometimes very low resistance noninductive relays were used to eliminate the need for the capacitor (see Fig. 46a).

Transformer battery-feed circuits permit impedance matching where required and attenuate longitudinal signals. Bridged inductor circuits will not do either but have the advantage of low cost when they use the windings of supervisory relays. Bridged inductors insert a high impedance in series with noise which may be superimposed on the battery potential. This noise can be further reduced where the same battery is connected to both the calling and called sides of the circuit if the battery leads are transposed to produce cancellation of the noise signal.

In general, a battery-feed circuit must limit the direct current drawn by a short loop. Approximately 400 ohms, limiting customer loop current to slightly more than 100 ma, is standard for −48-volt batteries. This resistance may be the d-c resistance of a retard coil, but it cannot be that of a transformer because such resistance would be in series with the speech signal and would produce considerable attenuation. Customer telephones use the current flowing in the loop to compensate for level variations due to loop length. A long loop with associated high attenuation has a small direct current; a small current controls the compensator in the telephone to increase transmitter output level and remove loss in the receiver circuit. These compensators are based on the assumption that a central office will have 400 ohms of internal resistance.

A battery-feed supervisory circuit has the basic configuration of a high-pass filter. The low-frequency roll-off is produced by the series impedance of capacitors, the shunt impedance of inductors, or some combination of both. High-frequency roll-off comes from the shunt capacity between conductors and transformer leakage inductance. When pulsing equipment is associated with a speech path, open circuits, short circuits, and other extreme variations are often introduced. The multiplexing of supervision and pulsing signals onto the talking path almost always leads to transmission degradation.

b. Nominal Characteristic Impedance. Local and tandem switching systems have a nominal characteristic impedance of 900 ohms, while toll systems are 600 ohms. Both values are related to the characteristic impedance of open wire lines used in the early days of telephony and have become deeply ingrained in transmission practice. However, 900 ohms in series with a 2.14-μf capacitor is the impedance which best matches the universe of customer loops terminated in subsets,[130] and 600 ohms represents a good compromise between the insertion loss and high-frequency roll-off produced by the series resistance and shunt capacity of the switching network and office wiring. It is not unusual to have 1,000 loop feet of wire (80 ohms and 0.015 μf) between carrier terminals. In broadband switching systems, high-frequency response is of vital importance. In one instance, the series resistance of office wiring is rigidly limited to permit a nominal impedance of 150 ohms; both low- and high-frequency roll-off produced by shunt elements are thus greatly reduced (Ref. 86, Vol. 83, p. 163).

c. Return Loss Balance. The word "balance" is used in two different ways in transmission theory: Return loss balance refers to the impedance match between two circuits or one circuit and a standard, and longitudinal balance refers to the symmetry of both sides of a talking circuit with respect to ground. The latter is discussed in paragraph 17d.

In transmission-line theory, any discontinuity in impedance gives rise to reflections. These reflections subtract from the transmitted power and produce echoes when they return to the source. If the impedance of a transmission line is measured looking in both directions from a given point, a perfect impedance match implies infinite return loss; no energy will be reflected or returned at that point. Alternatively, if the impedance looking one way differs from that looking the other, a reflection will occur and the returned signal will be smaller than the incident signal by the amount of the return loss. The standard method of expressing return loss in decibels is

$$RL_{\mathrm{db}} = 20 \log \left| \frac{Z_1 + Z_2}{Z_1 - Z_2} \right|$$

In modern carrier systems where separate unilateral paths are used for each direction of transmission, the principal reflections come at the hybrid coils where four-wire facilities are converted to two-wire for connection to local customer loops.   Carrier systems are used nearly exclusively for toll transmission today, and even exchange trunks are being converted to carrier at a rapid rate.   Thus four-wire transmission systems can be considered the rule rather than the exception.   Because loops to local customers are likely to remain two-wire for obvious economic reasons, hybrid coils must be closely associated with switching circuits.   Hybrid coils are discussed in Chap. 6.

When four-wire transmission facilities are used, echoes consist of energy that passes across the hybrid from the incoming to the outgoing side.   This transhybrid loss is composed of three parts: the loss from the four-wire incoming side to the two-wire path via the hybrid (usually slightly greater than 3 db because approximately half the incoming energy goes into the balancing network of the hybrid), the return loss at the point where the hybrid connects to the two-wire facility, and the loss from the two-wire path back through the hybrid to the four-wire outgoing side (again slightly more than 3 db).

As can be seen from Table 5, the return loss looking toward any given customer loop is likely to be rather low.   Thus "S" pads can be inserted in the two-wire path to

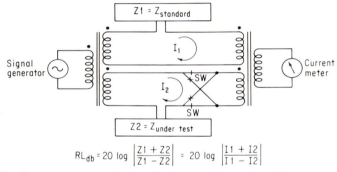

$$RL_{db} = 20 \log \left| \frac{Z1 + Z2}{Z1 - Z2} \right| = 20 \log \left| \frac{I1 + I2}{I1 - I2} \right|$$

Fig. 49. Circuit to measure return loss.   Generator and meter impedances seen through the transformers are very small compared with $Z1$ and $Z2$; thus $I1$ is inversely proportional to $Z1$ and $I2$ is inversely proportional to $Z2$.   Operation of the switch SW permits the addition or subtraction of $I1$ and $I2$.

increase the return loss.   A suitable value is 2 db (although higher values have been used in the past); because the impedance mismatch now occurs on the customer side of the pad, the reflected signal must pass through the pad twice and the return loss is increased by approximately 4 db.   The insertion loss is, of course, only 2 db.   The 2-db loss and S pads will be discussed further in paragraphs 17g and 17h.

Return loss can be measured with the circuit shown in Fig. 49.[131]

d. Longitudinal Balance.   If both sides of a speech path are balanced with respect to ground, equal and opposite voltages will be induced in each wire by stray electric or magnetic fields; both wires will "go up and down" together.   However, if the impedance of one wire to ground is different from that of the other, the induced voltages or currents will not cancel and "metallic" current will flow as noise interfering with the signal.   Thus good longitudinal balance is essential for good transmission.

In the past, longitudinal balance was measured in terms of microamperes per milliamperes, that is, microamperes of noise signal in the metallic path produced by milliamperes of longitudinal current.   Present practice is to rate longitudinal balance in decibels, using the circuit shown in Fig. 50.   The test set consists of a very well balanced transformer (to permit the introduction of a truly longitudinal signal) with a meter on the secondary side.   The meter measures the net effect of the current difference in the primary windings produced by unbalance in the circuit under test. The longitudinal voltage signal is introduced through a 0.5 μf capacitor which simulates the coupling impedance between noise sources such as nearby power wires, cable sheaths, etc. This measuring circuit is obviously more sensitive to high-frequency signals than those in the power range. As measured on this set, the balance of any path through a switching system should average 60 db as shown in Table 5.

It should be noted that good longitudinal balance is not sufficient to assure freedom from the ill effects of longitudinal signals. Where unbalanced circuits, coupled by transformers, are inserted into the speech path, distributed capacity between the transformer windings can couple the longitudinal signal directly into the unbalanced circuit. Longitudinal noise can often be on the order of tens of volts in amplitude, while the metallic signals are usually much smaller than 1 volt. Thus even relatively small interwinding capacitances can inject large amounts of interference. In critical locations, transformers with electrostatic shields or other means for achieving low interwinding capacity should be used.

e. Crosstalk and Noise. A telephone switching system is a natural generator of noises. Every electromagnet stores energy in its coil; this energy must be dissipated between the time the hold path of the relay is opened and the relay releases.[23] Each transformer and bridged inductor which carries the direct supervisory current in the speech paths must discharge its stored energy periodically. These transient voltages are associated with wires cabled in tight bunches where electrostatic coupling is hard to eliminate. Where circuit densities are high, electromagnetic coupling from

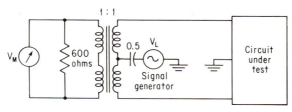

$$LB_{db} = 20 \log (V_L / V_M)$$

Fig. 50. Circuit to measure longitudinal balance.

relay and transformer coils and electrostatic coupling from all components in close proximity add additional noise. Data transmission is particularly susceptible to electrostatically coupled noise pulses from transients, while electromagnetic coupling from one circuit to another is annoying to listeners. Shielding, careful placement of parts both within a circuit and between circuits as mounted on frames, separation of cables to isolate different types of signals, and careful attention to longitudinal balance must be basic to switching-system design.

Contacts in the switching network and trunk circuits produce several kinds of noises. Contacts which bounce or chatter will make and break inductor currents several times in each operation. Base-metal contacts when held together act as noise generators. Contact pressure which is too low leads to high-resistance connections. This resistance can be varied by changes in the direct current flowing through it to produce noise. If direct current does not flow, chemical films can form to increase resistance and attenuation. In general, high-contact-pressure-nonchattering contacts made of precious metals which meet without wiping action are the most satisfactory, and the presence of direct current seems to minimize noise in most instances. Sealed contacts eliminate contamination by the environment.

Electronic systems with central controls using high-speed pulses must be designed to minimize impulse-noise pickup from other circuitry. At high frequencies, "ground" points only a few inches apart may be at different potentials. Unbalanced wiring runs should be as short as possible, and all signal paths between frames should be carefully planned. Filtering of the d-c power supply at each frame is more critical than in electromechanical offices, and the voltage drop in the battery and ground feeders as well as the changes in this drop with circuit operation must be taken into consideration.

f. Delay. It is well known that phase distortion has little effect on the understandability of speech. Thus until the advent of digital-data transmission, voice circuit phase shift has been of little concern to the switching system designer. Indeed, the use of metallic telephone trunks for low-frequency telegraph signals required low-frequency roll-off (with its attendant phase shift) below 300 cps. As has been discussed (paragraph 17a), supervisory circuits tend to be high-pass filters while voice frequency transmission facilities limit high-frequency response.

The delay requirements stated for one broadband switching system are 50 μsec absolute delay for a band between 14 and 55 kc, with delay distortion of 20 μsec. For narrower bandwidths, the requirements are somewhat less stringent. Very careful coil design, short cable runs, and the use of active equalizer circuits where needed solved the problem (Ref. 86, Vol. 83, p. 163).

g. Echos and VNL. With four-wire transmission facilities, the principal sources of reflection in a telephone connection occur at the end points. These reflections can cause echoes in the speech band and oscillations or singing at the high- or low-frequency extremes. The DDD network requires a number of trunks to be connected in tandem with as little difference between built-up and direct connections as possible. This means that intertoll trunks must operate at very low loss depending primarily on delay; such loss is called "via" net loss (abbreviated VNL). If VNL is based on the minimum value required to control echo, it turns out that singing and noise requirements are usually met as well, providing there is at least 2 db of loss in the toll-connecting trunks at each end of the connection. Thus the minimum loss between an originating and terminating central office is VNL + 4 db.[55,132]

h. Pads and Pad Switching. Providing all toll connecting trunks with a minimum of 2-db loss as described above represents a major rearrangement of the toll plant in the United States. Prior to the introduction of direct distance dialing, S pads (paragraph 17c) were placed in all intertoll trunks. When intertoll trunks were connected together, the S pads were switched out; on intertoll to toll-connecting paths, they were inserted to improve return loss. Thus a toll connection had not only the losses of the intertoll and toll-connecting trunks but also the losses of two sets of S pads. By building out all short toll-connecting trunks to a minimum of 2 db, the VNL + 4 objective has been gained, the maximum loss has been reduced by 2 db at each end of the connection, and the S pad switching equipment has been eliminated from all toll-connecting and intertoll trunk circuits. These advantages justify the considerable effort expended.

As is described in Chap. 6, the outgoing switches of a Class 5 (local) central office are defined as the 0 transmission level point (TLP). Because of the 2-db loss in toll-connecting trunks, the outgoing switches in Class 4 and higher offices are considered to be at the −2 TLP.

Carrier system terminals in the Bell System are usually adjusted to have their outputs at the +7 TLP and their inputs at the −16 TLP. This 23-db difference is available to counteract losses. In general, pads must be provided in each trunk to adjust its losses to the exact amounts required. In the No. 4 type (four-wire) toll crossbar system,[77] 6.5-db A pads are switched out (partially or wholly) when necessary to overcome hybrid or trunk losses and P pads are selected to compensate for losses in the office wiring, switching network, and trunk circuits. On the incoming side of a trunk, the P pad is chosen to make the loss between the carrier output terminal and the opposite end of a network connection equal to 9 db + VNL. On the outgoing side, the loss is made 14 db from the switching network (−2 TLP) to the carrier terminal (−16 TLP). This uniformity facilitates carrier line-up and trouble shooting.

In two-wire systems, A pad equivalents are not used although P pad equivalents are necessary. With the increase in the use of exchange carrier systems and negative resistance repeaters, A pad switching is tending to become uneconomical even in its original field of application.

REFERENCES

1. Smith, A. B., and W. L. Campbell, *Automatic Telephony*, McGraw-Hill Book Company, New York, 1914.

2. Mitchell, J. G., *Principles and Practice of Telephony*, McGraw-Hill Book Company, 1924.
3. Miller, K. B., *Telephone Theory and Practice* (3 vols.), McGraw-Hill Book Company, New York, 1933.
4. American Telephone and Telegraph Co., *Switching Systems*, 1961.
5. Keister, W., A. E. Ritchie, and S. H. Washburn, *The Design of Switching Circuits*, D. Van Nostrand Company, Inc., Princeton, N.J., 1959.
6. Pearce, J. G., *Electronic Switching*, Telephony Publishing Co., Chicago, 1965.
7. Crowley, T. H., G. G. Harris, S. E. Miller, J. R. Pierce, and J. P. Runyon, *Modern Communications*, Columbia University Press, New York, 1962.
8. Atkinson, J., *Telephony* (2 vols.), Sir Isaac Pitman & Sons, Ltd., London, 1948, 1950. Reprinted, 1964.
9. Morgan, T. J., *Telecommunication Economics*, Macdonald & Co., Ltd., London, 1958.
10. Huffman, R. L., The Type 45 Rotary Switch, *Autom. Elec. Tech. J.*, vol. 1, p. 41, July, 1948.
11. Salvesen, N., Some Variants of the Versatile Strowger Selector, *Autom. Elec. Tech. J.*, vol. 2, p. 128, January, 1951.
12. Molnar, Imre, Strowger Switch Bank Capacities, *Autom. Elec. Tech. J.*, vol. 1, p. 92, October, 1948.
13. Wilkinson, R. I., Theories for Toll Traffic Engineering in the U.S.A., *Bell System Tech. J.*, vol. 35, p. 421, March, 1956.
14. Shackleton, S. P., and H. W. Purcell, Relays in the Bell System, *Bell System Tech. J.*, vol. 3, p. 1, January, 1924.
15. Keller, A. C., New General Purpose Relay for Telephone Switching Systems, *Bell System Tech. J.*, vol. 31, p. 1023, November, 1952, and *Trans. AIEE*, vol. 71, part 1, p. 413, 1952.
    Keller, A. C., H. N. Wagar, R. L. Peek, Jr., and M. A. Logan, Design of Relays, *Bell System Tech. J.*, vol. 33, p. 1, January, 1954.
16. Leeds, M. B. (ed.), The Electronic Design Relay Directory, 1965, *Electron. Design*, Nov. 29, 1965.
17. Peek, Jr., R. L., and H. N. Wagar, *Switching Relay Design*, D. Van Nostrand Company, Inc., Princeton, N.J., 1955.
18. Knapp, H. M., Design Features of Bell System Wire Spring Relays, *Commun. Electron.* (AIEE), September, 1956, p. 482.
19. Grau, T. G., and A. K. Spiegler, Core Materials for Magnetic Latching Wire Spring Relays, *Bell System Tech. J.*, vol. 44, p. 999, July–August, 1965.
20. *Basic Circuits* and *Relay Magic*, Circulars 1927 and 1012-A, Automatic Electric Co., Northlake, Ill.
21. Keister, W., S. H. Washburn, A. E. Ritchie, G. R. Frost, A. E. Joel, and O. Myers, Some Basic Relay Circuits (Six Papers), *Trans. AIEE*, vol. 67, p. 720, 1948 and vol. 68, p. 571, 1949.
22. Caldwell, S. H., *Switching Circuits and Logical Design*, John Wiley & Sons, Inc., New York, 1958.
23. Gumley, R. H., Relay Contact Protection, *Bell Lab. Record*, vol. 34, p. 350, September, 1956.
24. Reynolds, J. N., The Crossbar Switch, *Bell Lab. Record*, vol. 15, p. 338, July, 1937.
25. Scudder, F. J., and J. N. Reynolds, Crossbar Dial Telephone Switching System, *Bell System Tech. J.*, vol. 18, p. 76, January, 1939.
26. Kruithof, J., and M. Den Herton, Mechanoelectronic Telephone Switching System, *Elec. Commun.*, vol. 31, p. 107, June, 1954.
27. Zupa, F. A., Magnetic Latching Crossbar Switches, *Bell System Tech. J.*, vol. 39, p. 1351, September, 1960.
28. Hovgaard, O. M., and G. E. Perreault, Development of Reed Switches and Relays, *Bell System Tech. J.*, vol. 34, p. 309, March, 1955.
29. Hovgaard, O. M., Capability of Sealed Contact Relays, *Commun. Electron.*, vol. 76, p. 466, September, 1956.
30. Keller, A. C., Recent Developments in Bell System Relays, *IEEE Trans. Commun. Electron.*, vol. 70, p. 79, January, 1964.
31. Husta, P., and G. E. Perreault, Magnetic Latching Relays Using Glass-sealed Contacts, *Bell System Tech. J.*, vol. 39, p. 1553, November, 1960.
32. Feiner, A., The Ferreed, *IEEE Trans. Commun. Electron.*, vol. 70, p. 73, January, 1964, and *Bell System Tech. J.*, vol. 43, p. 1, January, 1964.
33. Joel, A. E., Electronics in Telephone Switching Systems, *Bell System Tech. J.*, vol. 35, p. 991, September, 1956.
34. Moll, J. L., M. Tanenbaum, and J. M. Goldy, P-N-P-N Transistor Switches, *Proc. IRE*, vol. 44, p. 1174, September, 1956.

35. Dienel, H. F., Silicon Carbide Varistors: Properties and Construction, *Bell Lab. Record*, vol. 34, p. 407, November, 1956.

36. Wilson, D. K., Semiconductor Diodes, *Bell Lab. Record*, vol. 33, p. 227, June, 1955.

37. Yokelson, B. J., and W. Ulrich, Engineering Multistage Diode Logic Circuits, *Commun. Electron.*, vol. 74, p. 466, September, 1955.

38. Bell Laboratories Staff, Ten Survey Articles on Transistors, *Bell Lab. Record*, vol. 36, p. 190, June, 1958.

39. Kleimack, J. J., and H. C. Theuerer, Epitaxy and Transistor Fabrication, *Bell Lab. Record*, vol. 40, p. 289, September, 1962.

40. Bell Laboratories Staff, No. 1 ESS—19 papers, *Bell System Tech. J.*, vol. 43, page references cited in text, September, 1964.

41. Ebers, J. J., and J. L. Moll, Large-signal Behavior of Junction Transistors, *Proc. IRE*, vol. 42, p. 1761, December, 1954.

42. Pressman, A. I., *Design of Transistorized Circuits for Digital Computers*, John F. Rider, Publisher, Inc., New York, 1959.

43. Schiff, P., Preventing Second Breakdown in Transistor Circuits, *Electronics*, vol. 37, no. 18, p. 66, June 15, 1964.

44. IEEE Symposium Members, Integrated Circuits, *IEEE Spectrum*, vol. 1, p. 62, June, 1964.

45. Huskey, H. D., and G. A. Korn (eds.), *Computer Handbook*, Sec. 12, McGraw-Hill Book Company, New York, 1962.

46. Barrett, W. A., F. B. Humphrey, J. A. Ruff, and H. L. Stadler, A Card-changeable Permanent Magnet Twistor Memory of Large Capacity, *IRE Trans. Electron. Computers*, vol. EC-10, p. 451, September, 1961.

47. Piggyback Twistor, *Bell Lab. Record*, vol. 42, p. 206, June, 1964.

48. Automatic Electric Company Staff, No. 1 EAX®, 6 papers. System Parameters (p. 621), Register-Sender (p. 232), Translation with Magnetic Drum (p. 628), Switching Units (p. 655), Switching Unit Markers (p. 439), Control Center (p. 453), *IEEE Trans. Commun. Electron.*, vol. 83, pages as above, May, September, and November, 1964.

49. Malthaner, W. A., and H. E. Vaughan, Control Features of a Magnetic Drum Telephone Office, *IRE Trans. Electron. Computers*, vol. EC-4, p. 21, March, 1955.

50. May, Jr., J. E., Ultrasonic Delay Lines, *Bell Lab. Record*, vol. 34, p. 212, June, 1956. Meitzler, A. H., Ultrasonic Delay Lines Used to Store Digital Data, *Bell Lab. Record*, vol. 42, p. 315, October, 1964.

51. Pearce, J. G., An Electronic Register Sender for Step-by-step Automatic Telephone Systems, *IEEE Trans. Commun. Technol.*, vol. COM-13, p. 149, June, 1965.

52. Bell Laboratories Staff, Morris ESS, 10 papers. Introduction (p. 1091), Central Control (p. 1125), Flying Spot Store (p. 1161), Barrier Grid Store (p. 1195), Stored Program (p. 1327), Scanner (p. 1383), Signal Distributor (p. 1405), Breakdown Device Network (p. 1421), *Bell System Tech. J.*, vol. 37, pages as above, September, November, 1958. Experience with Morris ESS (p. 605), Automatic Trouble Diagnosis (p. 575), *Commun. Electron.*, vol. 81, pages as above, January, 1963.

53. Meyer, F. T., An Improved Detached-contact-type of Schematic Circuit Drawing, *Commun. Electron.*, vol. 74, p. 505, September, 1955.

54. Lorimer, J. H., and G. W. Lorimer, Automatic Exchange, U.S. Patent 1,187,634, issued June 20, 1916.

55. Engineering Dept., AT&T Co., *Notes on Distance Dialing*, American Telephone and Telegraph Co., New York, 1961.

56. Robb, W. F., A. M. Millard, and G. M. McPhee, Dial Switching of Connecticut Toll Calls, *Elec. Eng.*, July, 1936.

57. Smith, A. B., The Director in Automatic Telephone Switching Systems, *Autom. Elec. Tech. J.*, vol. 1, p. 53, July, 1948.

58. Young, J. S., Common Control for Step-by-step Offices (Series 100 Director), *Autom. Elec. Tech. J.*, vol. 8, p. 258, October, 1963.

59. Riddell, G., A New Crossbar Link Design for Step-by-step Common Control, *Bell Lab. Record*, vol. 42, p. 285, September, 1964, and *IEEE Trans. Commun. Electron.*, vol. 83, p. 446, September, 1964.

60. Martins, A. S., Common Control Features for the Step-by-step System, *Bell Lab. Record*, vol. 42, p. 167, May, 1964, and *IEEE Trans. Commun. Electron.*, vol. 83, p. 466, September, 1964.

61. Pennoyer, D. H., Automatic Number Identification and Its Application to No. 1 Crossbar, Panel and Step-by-step Offices, *Bell System Tech. J.*, vol. 37, p. 1295, September, 1958.

62. Taugner, J. W., Automatic Number Identification in SATT Systems, *Autom. Elec. Tech. J.*, vol. 8, p. 86, July, 1962.

63. Friend, O. A., Automatic Ticketing of Telephone Calls, *Trans. AIEE*, vol. 63, p. 81, 1944.
64. Naylor, J. T., Application of Strowger Automatic Toll Ticketing and Punch Card Accounting in Metropolitan Telephone Networks, *Autom. Elec. Tech. J.*, vol. 2, p. 45, February, 1950.
65. Ostline, J. E., The Expanding Field of Application for Strowger Automatic Toll Ticketing, *Autom. Elec. Tech. J.*, vol. 3, p. 3, July, 1951.
66. King, G. V., Centralized Automatic Message Accounting System, *Bell System Tech. J.*, vol. 33, p. 1331, November, 1954.
67. Craft, E. B., L. F. Morehouse, and H. P. Charlesworth, Machine Switching Telephone System, *J. AIEE*, vol. 42, April, 1923, and *Bell System Tech. J.*, vol. 2, April, 1923.
68. Bell Laboratories Staff, The Panel System, 18 papers, *Bell Lab. Record*, vols. 6–10, page references cited in text, May, 1928, through December, 1931.
69. Abbott, H. H., A New Small Crossbar Telephone System for Private Branch Exchanges, *Commun. Electron.*, vol. 77, p. 911, January, 1959.
70. Bell Laboratories Staff, No. 1 Crossbar, 15 papers, *Bell Lab. Record*, vols. 17, 18, page references cited in text, February, 1939, through May, 1940.
71. Korn, F. A., and J. G. Ferguson, The No. 5 Crossbar Dial Telephone Switching System, *Trans. AIEE*, vol. 69, part 1, p. 244, 1950.
72. Dehn, J. W., and R. E. Hersey, Recent New Features of the No. 5 Crossbar Switching System, *Commun. Electron.*, vol. 75, p. 457, September, 1956.
73. Bell Laboratories Staff, No. 5 Crossbar, 30 papers, *Bell Lab. Record*, vols. 26–29, page references cited in text, December, 1948, through January, 1951.
74. Collis, R. E., Crossbar Tandem System, *Trans. AIEE*, vol. 69, p. 997, 1950.
75. Adam, A. O., Crossbar Tandem as a Long Distance Switching System, *Bell System Tech. J.*, vol. 35, p. 91, January, 1956.
76. Meszar, J., Full Stature of the Crossbar Tandem Switching System, *Commun. Electron.*, vol. 75, p. 486, September, 1956.
77. Abraham, L. G., A. J. Busch, and F. F. Shipley, Crossbar Toll Switching System, *Trans. AIEE*, vol. 63, p. 302, 1944.
78. Bell Laboratories Staff, Crossbar Toll, 15 papers, *Bell Lab. Record*, vols. 22, 23, page references cited in text, November, 1943, to December, 1945.
79. Gohorel, F., Pentaconta Dial Telephone Switching System, *Elec. Commun.*, vol. 31, p. 75, June, 1954.
80. Meszar, J., Fundamentals of the Automatic Telephone Message Accounting System, *Trans. AIEE*, vol. 69, part 1, p. 255, 1950.
81. Bell Laboratories Staff, Automatic Message Accounting, 23 papers, *Bell Lab. Record*, vols. 29–31, page references cited in text, February, 1951 to March, 1953.
82. Ketchledge, R. W., The No. 1 Electronic Switching System, *IEEE Trans. Commun. Technol.*, vol. COM-13, p. 38, March, 1965.
83. Bell Laboratories Staff, No. 1 ESS, 15 papers, *Bell Lab. Record*, vol. 43, p. 194, June, 1965.
84. Higgins, W. H. C., A Survey of Bell System Progress in Electronic Switching, *Bell System Tech. J.*, vol. 44, p. 937, July–August, 1965.
85. Bell Laboratories Staff, No. 101 ESS, 4 papers. Electronic PBX (p. 329), Memory and Logic (p. 332), Common Control (p. 321), Time-division Switch (p. 338), *IEEE Trans. Commun. Electron.*, vol. 83, pages as above, July, 1964.
86. Automatic Electric and Western Union Staffs, Broadband Switching System, 6 papers. System Philosophy (vol. 82, p. 687), Register-Sender Capabilities (vol. 83, p. 167), Versatility in Translating (vol. 82, p. 682), Matrix Switching Equipment (vol. 82, p. 674), Automatic Ticketing (vol. 82, p. 427), Transmission Reliability (vol. 83, p. 163), *IEEE Trans. Commun. Electron.*, vols. 82, 83, pages as above, July, November, 1963, and March, 1964.
87. Schonemeyer, H., Quasi-electronic Telephone Switching System HE-60, *Elec. Commun.*, vol. 39, p. 244, 1964.
88. Oden, H. S., and H. Schonemeyer, Aspects of, and Experience with, a Reed Switching System Using Wired Program Logic, *IEEE Trans. Commun. Technol.*, vol. COM-13, p. 50, March, 1965.
89. Curran, C., and C. G. Svala, A 4-wire Solid-state Switching System: Application, Concept and Configuration, *IEEE Trans. Commun. Electron.*, vol. 83, p. 662, November, 1964.
90. Brightman, B., An Electronic 2-wire Private Branch Exchange, *IEEE Trans. Commun. Electron.*, vol. 83, p. 680, November, 1964.
91. Shea, P. D., Centrex Service: A New Design for Customer Group Telephone Service in the Modern Business Community, *Commun. Electron.*, vol. 80, p. 474, November, 1961.

92. Young, J. S., A Centrex System for Areas Served by Independent Telephone Companies, *IEEE Trans. Commun. Electron.*, vol. 83, p. 692, November, 1964.
93. Bronson, F. M., Tandem Operation in the Bell System, *Bell System Tech. J.*, vol. 15, p. 380, July, 1936.
94. Davidson, J., Toll Switchboard No. 3, *Bell System Tech. J.*, vol. 6, p. 18, January, 1927.
95. Clark, A. B., H. S. Osborne, J. J. Pilliod, W. H. Nunn, and F. F. Shipley, Automatic Switching for Nation-wide Telephone Service, 4 papers. Automatic Switching (p. 245), Fundamental Plans for Toll Plant (p. 284), Nation-wide Numbering Plan (p. 257), Automatic Toll Switching Systems (p. 261), *Trans. AIEE*, vol. 71, part 1, pages as above, September, 1952.
96. Close, C., A Study of Non-blocking Switching Networks, *Bell System Tech. J.*, vol. 32, p. 406, March, 1953.
97. Dodge, W. L., Development of Step-by-step Line Finders, *Bell Lab. Record*, vol. 7, p. 236, February, 1929.
98. Whitney, W., The New Line Concentrator No. 1A, *Commun. Electron.*, vol. 81, p. 83, May, 1962.
99. Salvesen, N., The Use of Distributing Frames in Strowger Automatic Exchanges, *Autom. Elec. Tech. J.*, vol. 2, p. 97, April, 1950.
100. Koliss, P. P., A New Enclosed Main-frame Connector, *Bell Lab. Record*, vol. 38, p. 347, September, 1960.
101. Schneckloth, H. H., Some Basic Concepts of Translators and Identifiers Used in Telephone Switching Systems, *Bell System Tech. J.*, vol. 30, p. 588, July, 1951.
102. Molina, E. C., Translating and Selecting System, U.S. Patent 1,083,456, issued Jan. 6, 1914.
103. Hampton, L. N., and J. B. Newsom, The Card Translator for Nationwide Dialing, *Bell System Tech. J.*, vol. 32, p. 1037, September, 1953.
104. Brooks, Charles, Class-of-service Markings—What and Why, *Bell Lab. Record*, vol. 40, p. 12, January, 1962.
105. Myers, O., Common Control Telephone Switching Systems, *Bell System Tech. J.*, vol. 31, p. 1086, November, 1952.
106. Holbrook, B. D., Some Logical Requirements for the Control of Switching Networks, *Proc. Intern. Symp. on Theory of Switching*, part II, The Computation Laboratory of Harvard University, Annals, vol. 30, p. 235, 1939.
107. Macrander, M. S., A New "Progressive" Technique for Establishing a Connection through Crosspoint Networks, *Autom. Elec. Tech. J.*, vol. 8, p. 12, January, 1962.
108. Malthaner, W. A., and H. E. Vaughan, An Electronically Controlled Switching System, *Bell System Tech. J.*, vol. 31, p. 443, May, 1962.
109. Automatic Electric Co. Staff, Electronic Crosspoint PABX, 3 papers. Features (p. 471), Transmission Network (p. 491), Logical Control (p. 496), *Commun. Electron.*, vol. 80, pages as above, November, 1961.
110. Vaughan, H. E., Research Model for Time-separation Integrated Communication, *Bell System Tech. J.*, vol. 38, p. 309, July, 1959.
111. Brightman, B., and M. P. Tubinis, Considerations Pertaining to the Design of an Electronic Telephone Switching System. Perkins, Jr., J. C., Transmission Aspects of an Electronic Switchboard Employing Time-division Multiplexing, *Commun. Electron.*, vol. 78, pp. 777, 949, January, 1960.
112. Haard, H. B., and C. G. Svala, Means for Detecting and/or Generating Pulses, U.S. Patent 2,718,621, issued Sept. 20, 1955.
113. Molnar, I., An Appraisal of Pocketed Calls in Common Control Switching, *Commun. Electron.*, vol. 81, p. 556, January, 1963.
114. Lee, C. Y., Representation of Switching Circuits by Binary-decision Programs, *Bell System Tech. J.*, vol. 38, p. 985, July, 1959.
115. Baldwin, Jr., J. A., A Magnetic Device for High-speed Sensing of Small Currents, *Commun. Electron.*, vol. 80, p. 1, March, 1961.
116. Schenker, L., Pushbutton Calling with a Two-group Voice-frequency Code, *Bell System Tech. J.*, vol. 39, p. 235, January, 1960.
117. Bell Laboratories Staff, TOUCH-TONE Calling, 4 papers. Applications (p. 1), Central Office Arrangements (p. 5), Signal System and Receiver (p. 9), Caller for Station Sets (p. 17), *IEEE Trans. Commun. Electron.*, vol. 82, pages as above, March, 1963.
118. Ham, J. H., and J. F. Ritchey, TOUCH-TONE Card Dialer Set, *Bell Lab. Record*, vol. 41, p. 268, July–August, 1963.
119. American Standard Definitions of Electrical Terms—Group 65, *Communications*, 1957.
120. Breen, C., and C. A. Dahlbom, Signaling Systems for Control of Telephone Switching, *Bell System Tech. J.*, vol. 39, p. 1381, November, 1960.

121. Weaver, A., and N. A. Newell, In-band Single-frequency Signaling, *Bell System Tech. J.*, vol. 33, p. 1309, November, 1954.
122. Dahlbom, C. A., A. W. Horton, Jr., and D. Moody, Application of Multifrequency Pulsing in Switching, *Trans. AIEE*, vol. 68, p. 392, 1949.
123. Stacy, L. J., Vacuum Tube Improves Selective Ringing, and Increasing the Range of Tripping Relays, *Bell Lab. Record*, vol. 15, p. 111, December, 1936, and vol. 17, p. 353, July, 1939.
124. Meacham, L. A., J. R. Power, and F. West, Tone Ringing and Pushbutton Calling, *Bell System Tech. J.*, vol. 37, p. 339, March, 1958.
125. Bennett, W., Telephone-system Applications of Recorded Machine Announcements, *Trans. AIEE*, part 1, vol. 72, p. 478, September, 1953.
126. McGuinness, T. P., TSP Circuits, *Bell Lab. Record*, vol. 42, p. 258, July–August, 1964.
127. Glenner, E. J., Reverting-type Calls in an Automatic Telephone System, *Autom. Elec. Tech. J.*, vol. 3, p. 143, March, 1954.
128. Steinhauer, K. F., Pay Stations for Ten-cent Service, and The Prepay Paystation Repeater, *Autom. Elec. Tech. J.*, vol. 3, p. 37, April, 1952, and vol. 3, p. 91, May, 1953.
129. Bell Laboratories Staff, *Transmission Systems for Communication*, 3d ed., Bell Telephone Laboratories, 1964.
130. Hinderliter, R. G., Transmission Characteristics of Bell System Subscriber Loop Plant, *IEEE Trans. Commun. Electron.*, vol. 82, p. 464, September, 1963.
131. Smethurst, J. O., A Repeater Test Set, *Bell Lab. Record*, vol. 34, p. 17, January, 1956.
132. Huntley, H. R., Transmission Design of Intertoll Telephone Trunks, *Bell System Tech. J.*, vol. 32, p. 1019, September, 1953.

Chapter 10

MULTIPLEXING

MAURICE E. COOKSON, * *Lenkurt Electric Co., Inc.*

EWELL M. THOMPSON, *Lenkurt Electric Co., Inc.*

In the early days of telephony, a separate transmission line was required for each voice communication channel. With the rapid development and growth of telephony, the need arose for a more efficient means of utilizing the existing telephone plant. The need was particularly noticeable in large urban areas and in long-distance lines, where enormous numbers of wires and cables had to be installed to provide sufficient voice channels.

These conditions led to the development of a technique whereby a number of voice-frequency signals could be combined into a composite signal and transmitted over a single communications channel. This technique was first called *carrier telephony* but more recently has become known as *multiplexing*.

The first so-called carrier system, with a capacity of four channels, began operating in the United States in 1918 over an open-wire transmission line. Today, modern multiplex systems carry up to 2,700 or more voice channels using coaxial cable and microwave radio transmission facilities.

FUNDAMENTALS

Voice frequencies transmitted over telephone systems range from about 300 to 3,400 cps. To transmit a number of these signals simultaneously over the same transmission medium, the signals must be kept apart so that they do not interfere

*Now with Electro-Mechanical Research, Inc.

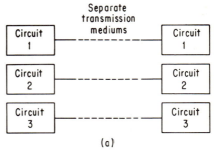

(a)

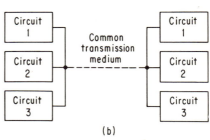

(b)

FIG. 1. Multiplexing permits two or more communications circuits to be combined on a common transmission medium.

with one another and so they can be separated at the receiving end.   This is accomplished by separating the signals either in frequency or in time.   The process of separating the signals in frequency is called *frequency-division multiplexing (FDM)*, whereas the process of separating the signals in time is called *time-division multiplexing (TDM)*. The concept of multiplexing is shown in Fig. 1.

1. Frequency-division Multiplexing (FDM)[2-7]

Frequency division is a method of multiplexing in which two or more voice-frequency signals are translated to separate frequency bands by modulation processes so that they can then be combined and transmitted over a single medium.   Any one of a number of modulation processes may be used for this purpose.   The two types of modulation used in frequency-division multiplexing are amplitude modulation (AM) and frequency modulation (FM).   The concept of frequency-division multiplexing is illustrated in Fig. 2.

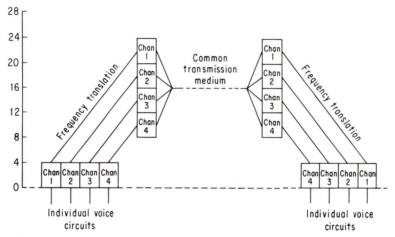

Fig. 2. In frequency-division multiplexing, each circuit is translated to a separate position in the frequency spectrum before being applied to a common transmission medium.

a. Amplitude Modulation.   In amplitude modulation, the amplitude of the carrier is controlled by the modulating wave.   As shown in Fig. 3, the resultant modulated wave has the same frequency as the carrier, but the carrier-wave amplitude varies in direct relation to the modulating wave.   The curves through both the positive and negative peaks of the modulated wave are identical with the modulating wave and are called the wave envelopes.   The modulation factor m is a measure of the degree of modulation.   For a sinusoidal variation as shown in Fig. 3, the modulation factor—normally called the *modulation index* and sometimes the *degree of modulation*—equals the peak amplitude of the envelope minus the amplitude of the unmodulated carrier divided by the amplitude of the unmodulated carrier.   For more complex signals, the modulation index is more difficult to determine, since it varies from one instant to another.   The modulation index is the fractional extent by which the modulation varies the amplitude of the carrier and can be expressed as a percentage if multiplied by 100.

The maximum amount that the carrier amplitude can be varied without loss of signal is equal to the carrier amplitude.   When this occurs, 100 per cent modulation is obtained.

The modulated wave consists essentially of the carrier wave and frequencies above and below the carrier wave.   These side frequencies are separated from the carrier by a frequency equal to that of the modulating wave.   Where a complex modulating wave, such as speech, is used, the side frequencies above and below the carrier both consist of a band (sideband) of frequencies.   A sideband includes all the frequency components of the modulating wave.

Three important factors in the use of amplitude modulation are derived from an analysis of the modulated wave: (1) The sidebands obtained from a complex wave each have the same bandwidth as the original modulating wave, (2) the same intelligence is contained in each sideband, and (3) the frequencies in the upper sideband have the same relative relationship as the modulating wave, but those in the lower sideband have an inverse relationship.   Power distribution in the sidebands is directly related to the distribution of power in the modulating wave.

Double Sideband.   As a result of amplitude modulation, the frequency band of the modulating wave is translated to a different position in the frequency spectrum.   Such translation is used in combining circuits for frequency-division multiplexing.   If both sidebands and the carrier are used, the technique is called double-sideband transmitted-carrier modulation (DSBTC).   If the carrier is not transmitted, the technique is called double-sideband suppressed-carrier modulation (DSBSC).   There is a disadvantage in using DSBTC modulation because of the amount of power in the carrier in relation to the sideband (information) power.   Even with 100 per cent modulation, the power of each sideband is only one-fourth that of the carrier.   Since the power of the sidebands is proportional to the square of the modulation index, for low modulation levels the sideband power becomes only a fraction of the carrier power.

In multiplex systems, this is quite important because of the number of channels that are involved.   The various elements of the system common to more than one channel must be capable of handling a large amount of power that is not useful in the transmission of information.   In addition, the sideband-to-noise power ratio is relatively low.

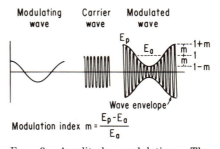

Modulation index $m = \dfrac{E_p - E_a}{E_a}$

Fig. 3.   Amplitude modulation.   The amplitude of the carrier is varied by the modulating wave.   The frequency of the modulated-wave envelope is the same as the modulating wave.

For these reasons, it is more common to suppress the carrier (DSBSC) or to transmit the carrier at a relatively low level.   Where the carrier is suppressed, it is necessary to derive a carrier frequency at the receiving terminal.   This may be done either by separately generating the carrier frequency, deriving the carrier frequency from the transmitted sidebands, or by transmitting a separate tone from which the demodulating frequencies may be derived.

After the carrier is suppressed, it is possible to increase the sideband power and still keep the power-handling capacity of common equipment below that which would have been required if the carrier were transmitted.

Single-sideband Suppressed Carrier.   The most widely used modulation technique in frequency-division multiplexing is single-sideband suppressed carrier.   As previously stated, when a carrier is amplitude-modulated by a signal, two sidebands are produced, both containing the intelligence.   Therefore, both sidebands are not required in order to transmit the message.   A balanced modulator is used to suppress the carrier, leaving only the two sidebands.   This reduces the power of the signal by one-half but does not suppress one of the sidebands.   Sideband suppression is accomplished by applying the signal to a filter which passes only one of the sidebands while effectively attenuating the other.   The bandwidth of the signal is now approximately equal to the original voice-frequency signal, that is, about 3,100 cps.   Further bandwidth reduction, therefore, is not possible without lowering the quality of the signal.   Figure

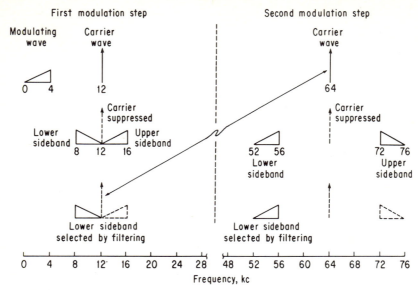

FIG. 4. In SSBSC modulation, the carrier and one sideband are removed during each modulation step.

4 illustrates two typical SSBSC modulation steps used to shift one v-f signal to a carrier-frequency channel.

From a practical standpoint, there is still a slight amount of wasted bandwidth because of the imperfect characteristics of the channel filters (see Fig. 5). A perfect filter would have vertical sides in its frequency-amplitude characteristic, thus passing all frequencies within its passband perfectly while completely attenuating all frequencies outside the passband. Interference that might be caused by the imperfect performance of filters is avoided by spacing the voice channels slightly apart so that signal energy from adjacent channels is attenuated 60 db or more. Since the effective bandwidth of voice signals is approximately 3.1 kc, a standard spacing of SSBSC channels has been set at 4 kc.

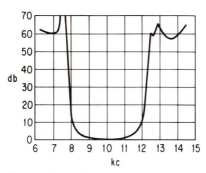

FIG. 5. Typical attenuation-frequency characteristic of a voice channel in a multiplex system.

Because SSBSC multiplexing is so widely used in high-density radio and coaxial-cable transmission systems, a standard frequency allocation and modulation plan has been adopted which has received world-wide acceptance. In this plan, up to 12 voice channels are assembled into a *group band* with a frequency range of 60 to 108 kc. The signals processed in each of these channels appear as the lower sidebands of 12 channel carrier frequencies which are spaced 4 kc apart. Five such group bands can be combined to form a 60-channel system by two additional modulation steps. In the first step, each of the five group bands modulates separate group carriers to form a *supergroup band* with a frequency range of 312 to 552 kc. In the final modulation step, the signals in the supergroup band modulate a suitable carrier to form a *line frequency band*.

Using the basic 12-channel groups, the present standard modulation scheme can be expanded to multiplex 2,700 or more voice channels. A typical frequency-allocation and modulation plan for a 960-channel frequency-division multiplex system is shown in Fig. 34 (see System Operating Characteristics).

Although SSBSC modulation permits economies in bandwidth and transmitted power, it complicates the transmission problem. An ordinary amplitude-modulated signal in which a carrier and two sidebands are present requires the simplest transmitters and receivers. The carrier and two sidebands establish a definite harmonic relationship between themselves which is maintained through subsequent modulation or demodulation processes. This harmonic relationship is essential to avoid distortion.

If a balanced modulator is employed to remove the carrier, this missing carrier must be reinserted before the message can be recovered. If the substitute carrier is as much as 90° out of phase with the original carrier and both sidebands are still present, the resulting distortion completely destroys the message. In order to use this method, there must be absolute phase and frequency synchronization between all terminals.

The requirement for synchronization is greatly reduced in the case of SSBSC modulation. The necessary phase relationship that must exist between the single information-bearing sideband and the carrier is much simpler than in the case where the carrier and *two* sidebands must maintain an exact relationship.

In SSBSC, the reinserted carrier should be as close to the frequency of the original carrier as possible. If the frequency is not identical, slight distortion results. Carrier frequency errors add or subtract a fixed number of cycles from all frequencies in the channel, thus changing their harmonic relationship. This has the effect of garbling the fidelity of the transmission and giving a strange quality to speech. If excessive, it adds to the chance of error in telegraph or data transmission.

Unlike DSBSC, where errors of more than 2° or 3° of phase shift, which is less than $\frac{1}{100}$ cycle, cannot be tolerated, errors as great as 10 to 20 cycles are tolerable for speech transmission in SSBSC systems. In the case of low-speed pulse transmission such as telephone dialing or manual telegraphy, 10-cps error may be tolerated before transmission errors increase appreciably.

In the case of frequency-shift teletype (60 words per minute), frequency errors ranging from 5 to 15 cycles may be tolerated. At 100 words per minute, the error limit ranges from 3 to 10 cycles. The error tolerance is stated here as a range, instead of a fixed limit, to allow for systems having a number of transmission sections. The upper figure is the maximum tolerable error in any system, even where several transmission sections or repeaters are used. The lower figure is the frequency error that should not be exceeded in one section, since such frequency error may be additive in going from one section to the other.

The frequency tolerance for higher speed data transmission becomes more lax as the transmission rate increases. As the data-transmission rate increases, the bandwidth required by the data channel increases similarly, and the absolute amount of error tolerance will be a percentage of this bandwidth rather than being a certain number of cycles. The different tolerances stated above for two different speeds of frequency-shift teletype were based on the fact that teletype channels are more or less standardized, and the error rate given was for identical bandwidth occupancy.

Synchronization. The proper frequency relationship between the transmitter and the receiver depends upon the carrier which is modulated by the message signal. In ordinary amplitude modulation, the carrier is transmitted with the sidebands and no problem occurs. When the carrier is eliminated, however, some frequency reference is required between transmitter and receiver to make sure that the reinserted carrier frequencies are kept well within the error limits stated above.

Two basic methods are commonly used to maintain this frequency relationship. Both require a pilot frequency to be transmitted with the outgoing signal. This pilot frequency undergoes whatever frequency translations or phase shifts are imposed on the various information channels.

In one method of frequency control, this transmitted pilot is used to modulate a local oscillator in the receiver to obtain the carrier frequencies used in demodulating the information channels. If the transmitted signals, including the pilot, have undergone additional modulation steps or other frequency changes during transmission, this method cancels these errors out, leaving only the possibility of a fixed error between the transmitter and receiver local oscillators. This error, of course, can be eliminated by simply adjusting the frequency of the local oscillator.

The second method of frequency control again requires the use of a transmitted pilot frequency. This pilot is used as an absolute frequency and phase reference. At each receiver, a single "master" oscillator may be used, and all channel carriers and a local pilot frequency are derived from it. Both the transmitted pilot and the locally generated pilot frequency are compared in a phase discriminator circuit. Any phase or frequency difference between the two results in an error voltage being generated in the output of the discriminator. This error voltage is then applied to the master oscillator to "pull" the oscillator frequency, thus correcting the local pilot frequency. Since all local carriers are derived from this oscillator, changes in its frequency which correct differences between the incoming and the local pilot also remove any frequency or phase errors between the transmitted and local carrier frequencies.

Loading. An important engineering consideration in the design of a multiplex system, particularly one to be used over great distances, is that of loading. Where a large number of channels are multiplexed, total average signal power may be considerable. If the channels are used to transmit speech conversations, the *range* of power may be extremely large—as much as 70 db.

In order to obtain the best signal-to-noise performance from both the multiplex system and the transmission medium, it is desirable to operate the system at the highest signal level that can be handled by the equipment without excessive distortion. If the range of signal power is too great because too many loud talkers are using the system simultaneously, for instance, the system will overload during periods of peak use, with the result that intermodulation distortion and noise will be excessive. If modulation levels are reduced to prevent this, there will be many periods when the signal level will be far too low and background noise becomes predominant.

One way of overcoming this difficulty is to restrict the range of signal levels applied to the system. This may be accomplished by some sort of peak limiting or by automatic gain control at the transmitter. This permits the transmitted signal level to be relatively constant regardless of the range of input signal levels.

b. Frequency Modulation. Although frequency-division multiplexing is most often accomplished by amplitude modulation, various types of angle modulation may also be used. Angle modulation is the general term used to describe any form of modulation in which the frequency or phase of a sinusoidal carrier wave is controlled by the modulating wave. Frequency and phase modulation are the two types of angle modulation most commonly used. Although these types of angle modulation are somewhat different, they are closely interrelated and both are often used in a single modulation system. Because the basic considerations are similar, the following discussion is restricted to frequency modulation.

Frequency modulation (FM) is the process in which amplitude changes of the modulating wave are used to vary the instantaneous frequency of the carrier wave from its unmodulated value. An example of the action of the modulating wave on the carrier wave to produce a frequency-modulated output wave is shown in Fig. 6.

The magnitude of frequency change for a given amplitude of the modulating signal is called frequency shift, frequency swing, or frequency deviation. Frequency swing normally denotes the maximum frequency shift that occurs when a sinusoidal modulating wave is employed. Frequency deviation is the maximum value of frequency shift permitted by equipment design and is also called peak deviation. The term frequency deviation is also often used to denote the instantaneous difference between the instantaneous frequency of the modulated wave and the carrier frequency.

While frequency shift is controlled by the amplitude of the modulating wave, the rate at which the carrier frequency is shifted, called the deviation rate, is controlled by the frequency of the modulating wave. If a 1-Mc carrier is modulated by a 1,000-cps signal, the frequency swing depends upon the amplitude of the 1,000-cps signal. If a signal of a different frequency but the same amplitude is used, the frequency swing will be the same but the deviation rate will be equal to the frequency of the modulating wave.   This is shown in Fig. 6.

The individual cycles of the modulated wave are not sinusoidal because of the instantaneous variations in frequency which occur during modulation.   This complex wave contains a large number of sidebands rather than the two normally associated with amplitude modulation.

Where a single sinusoidal modulating wave is used, the spectrum of the modulated wave is symmetrical with respect to the carrier frequency.   In this case, the sideband frequencies are displaced from the carrier by integral multiples of the modulating frequency.

If a more complex modulating wave, such as speech, is used, the frequency spectrum of the modulated wave becomes very complicated.   The sideband frequencies present

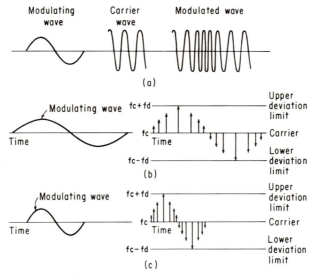

FIG. 6. Frequency modulation.   Diagrams (*a*) and (*b*) show how the carrier frequency varies with the amplitude of the modulating wave.   A comparison of (*b*) and (*c*) shows how the deviation rate is controlled by the frequency of the modulating wave.

include not only those that would be obtained with each modulation frequency acting separately but also various combination frequencies.

Although the total bandwidth of a frequency-modulated wave is quite large, the higher-order sidebands often contain only a small portion of the total wave energy. The actual bandwidth can, therefore, be reduced considerably without introducing undue distortion.   The energy distribution depends upon the amplitude of the different frequency components.   Component amplitudes, in turn, are related to the modulation index and can be calculated with the aid of a table of Bessel functions. The results of a number of such calculations are shown graphically in Fig. 7, from which the bandwidth of a variety of conditions can be determined.   Bandwidths determined for Fig. 7 will contain all but about 1 per cent of the energy in the modulated wave.   The loss of 1 per cent of the energy will cause distortion.   Although the distortion introduced can be tolerated in some applications, in others the distortion requirements are more severe, and a greater proportion of the energy must be included. For these cases, a detailed analysis must be made and Fig. 7 cannot be used.

When the modulation index is greater than 1, the bandwidth is approximately equal to twice the sum of the frequency deviation plus the modulating frequency.   As the modulation index decreases to values of 0.5 and below, the bandwidth becomes essentially equal to twice the modulating frequency. In this case, the bandwidth is the same as for an amplitude-modulated wave.

Because both methods are used in frequency-division multiplexing, the advantages and disadvantages of frequency modulation are normally expressed in terms of similar characteristics in amplitude modulation. On this basis, the chief advantage of FM is in its ability to exchange bandwidth occupancy in the transmission medium for improved noise performance.

The noise-power reduction advantage of FM over AM for random noise is often given as $R = 3f_d{}^2/B^2$, where f_d is the frequency deviation and B is the output bandwidth of the receiver. (B is equal to the highest modulating frequency.) This

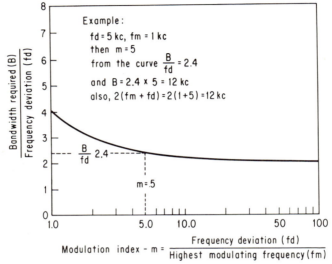

Fig. 7. Required bandwidth for different modulation indexes of an FM system. All sideband components except those having less than 1 per cent of the wave energy are included.

advantage is also expressed in terms of FM advantage and may be written as db $= 10$ log $3f_d{}^2/B^2$.

On a peak power basis, the noise advantage of FM as compared with SSBSC is $R = 3f_d{}^2/2B^2$.

In addition to noise advantage, FM exhibits a characteristic often called *capture effect*. Where two signals in the same frequency band reach the receiver, the one appearing at the higher level is accepted while the weaker signal is almost entirely rejected.

The noise advantage of FM is obtained for normal signal and noise levels at the input to the receiver. This advantage increases as the frequency deviation (modulation index) is increased. However, as the peak signal level is decreased to that of the peak noise level, there is a rather sharp transition between good and poor signal-to-noise ratios. The point at which this transition occurs is called the FM improvement threshold (sometimes shortened to threshold).

Although the noise advantage increases as the modulation index is increased, the corresponding increase in bandwidth increases the noise. For large bandwidths, the threshold becomes more critical and is reached at higher signal levels. The optimum value of modulation index is thus a compromise between signal-level range and noise advantage.

Other features of FM make its application to frequency-division multiplexing attractive where bandwidth is not a critically limiting factor. These are: (1) Separate limiting is not necessary, since in a properly designed FM receiver, the output signal level is insensitive to input signal level variations above threshold level, and (2) synchronization is not a problem because of the detection method.

2. Time-division Multiplexing (TDM)[1,3,4,8,22,24]

Messages are separated in time by briefly sampling each channel in a regular sequence. The receiver then separates the samples and reconstructs the original messages. This process is called time-division multiplexing. The concept of time-division multiplexing is shown in Fig. 8.

The samples of each message are transmitted by a train of pulses. The pulse train for one channel is interleaved with the pulse train representing another channel. The pulse train is then the carrier, and it can be modulated by varying any character-

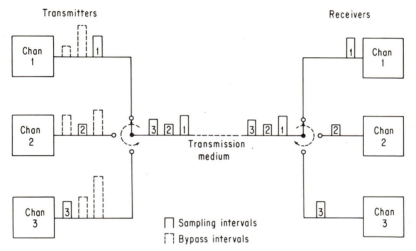

Fig. 8. In time-division multiplexing, a number of circuits share the same transmission medium, but at different times.

istic of the pulses, such as amplitude, width, or position. Regardless of the modulation method used, the minimum sampling rate is determined by the highest frequency to be transmitted. A sine wave must be sampled at least twice during each cycle in order to reconstruct it accurately at the receiving end. If a voice channel passes a highest frequency of 4,000 cps, the minimum sampling rate is 8,000 samples per second. This sampling requires a pulse every 125 μsec. If the pulse is short enough to occupy only a small fraction of this 125-μsec interval, the pulse from another channel may be transmitted in the remainder of the interval, thus permitting multiplex operation.

The four most commonly used time-division modulation techniques are pulse-amplitude modulation (PAM), pulse-duration modulation (PDM), pulse-position modulation (PPM), and pulse-code modulation (PCM). Figure 9 illustrates the pulse trains resulting from these four modulation techniques.

a. Pulse-amplitude Modulation (PAM). In pulse-amplitude modulation the amplitude of each pulse represents the amplitude of the modulating wave at a specific instant in time. A simple commutator will provide PAM at a slow rate. For faster sampling, electronic switching devices are used.

PAM is simple but quite susceptible to noise. Because the signal intelligence is represented by the pulse height, a small amount of noise can change the pulse height enough to render the signal unintelligible.

b. Pulse-duration Modulation (PDM). In pulse-duration modulation (often called pulse-length modulation or pulse-width modulation) the instantaneous value of the modulating wave is represented by the width of the pulse. The leading edge or the trailing edge (or both) of the pulse may be modified by the modulating wave to produce a pulse of a particular length. The performance characteristics of a PDM system can be improved by using extra bandwidth. This is in contrast to PAM, which is not helped by a wider bandwidth.

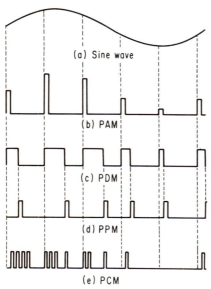

(a) Sine wave

(b) PAM

(c) PDM

(d) PPM

(e) PCM

FIG. 9. Types of modulation employed in time-division multiplexing systems.

Since the pulse shape is used to reconstruct the signal, any distortion of the pulse shape results in distortion of the reconstructed signal. A wider bandwidth produces less distortion in the pulse and hence a more accurate reconstruction at the receiver. Thus, PDM is less sensitive to noise than PAM.

c. Pulse-position Modulation (PPM). In pulse-position modulation all pulses are the same size and the same shape. The instantaneous value of the modulating wave displaces the pulse from its normal position by an amount proportional to the amplitude of the modulating wave.

PDM and PPM are both particular examples of pulse-time modulation. Either of them (or PAM) permits a completely accurate reconstruction of the signal, provided that the sampling rate is at least twice the highest frequency to be transmitted and assuming that there is no distortion of the pulse in route.

d. Pulse-code Modulation (PCM). In pulse-code modulation, the message is sampled periodically and the values observed are represented by a coded arrangement of several pulses. Each separate signal value has a unique arrangement of pulses. Thus, only the presence or absence of pulses—not their shape—determine the received message and its quality. Distortion of the transmitted pulses does not degrade message quality. It merely increases the chances of a mistake in interpreting the code and the printing of a wrong character.

FIG. 10. In PCM systems, repeaters are able to retime and regenerate distorted signals. (*Courtesy of Lenkurt Electric Co., Inc.*)

PCM or related approaches are now the primary methods seriously considered for time-division multiplexing systems. This stems from the resistance of PCM to interference and noise. Even more important, PCM permits the use of regenerative repeaters.

Regenerative repeaters detect the presence or absence of pulses (which may be badly distorted) and replace them with perfect new ones (see Fig. 10). The regenera-

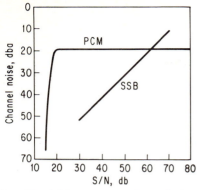

FIG. 11. Arbitrary comparison of signal quality versus transmission signal-to-noise ratio of PCM and SSB modulation processes. (*Courtesy of Lenkurt Electric Co., Inc.*)

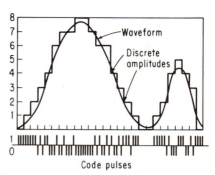

FIG. 12. A waveform can be represented by a number of discrete amplitudes. Each of these discrete amplitudes can then be converted to a specific arrangement of code pulses, as shown at the bottom of the figure.

tive repeaters must be spaced closely enough to identify the incoming distorted pulses correctly and be able to send out new pulses which precisely match the original pulse stream. It is theoretically possible to transmit messages for unlimited distances without degradation, but in practice, small timing errors tend to add up as the number of repeaters increases, eventually placing a limit on system length.

Pulse-code modulation requires more bandwidth than amplitude modulation, but it uses this bandwidth more efficiently in overcoming noise and interference than almost any other modulation method. For instance, frequency modulation (FM) also trades bandwidth for noise improvement, but rather slowly. With FM, the signal-to-noise ratio improvement (in decibels) is proportional only to the logarithm of the increase in bandwidth. The PCM improvement in decibels is in direct proportion to the increase in bandwidth. This signal quality is inherent in the transmitted signal and is not the result of noise encountered in transmission as long as the signal remains above the noise threshold of the system (see Fig. 11).

A continuous waveform with its infinite number of amplitude values must be quantized into discrete steps for representation by a pulse code, as shown in Fig. 12. The random differences between the actual waveform and the quantized approximation result in "quantizing noise." With

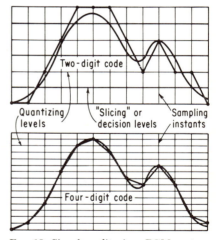

FIG. 13. Signal quality in a PCM system is determined by how closely the quantized signal approximates the original signal. (*Courtesy of Lenkurt Electric Co., Inc.*)

error-free transmission, this is the only source of noise in the system. Quantizing noise may be minimized by using as many steps as possible, but this increases the number of code pulses which must be transmitted (see Fig. 13). The number of steps is determined by the number of binary digits in each code group: Number of steps

$= 2^n$, where n is the number of digits in the code. Thus a five-digit code yields 32 amplitude steps, while a seven-digit code provides 128 steps. The latter reduces the quantizing error considerably. The bandwidth required is directly proportional to the number of digits; a seven-digit code requires seven times as much bandwidth as a binary (single-digit) code but improves the signal-to-noise ratio 36 db. (The signal-to-quantizing-noise ratio in decibels can be calculated from the expression

$$S/N = 10.8 + 6n.)$$

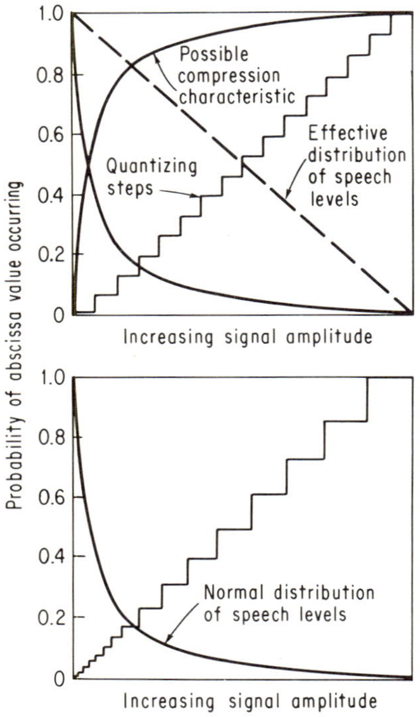

Satisfactory PCM requires that all the code pulses be regenerated with great precision and timing accuracy. The problem of pulse regeneration and retiming determines the major limit on the length and performance of present systems. The number of channels transmitted is limited only by the ability of repeaters to identify and restore pulses. This is controlled by the pulse rate and the transmission characteristics.

Most present systems use a seven-digit code, with an eighth digit for signaling. Since each channel should be sampled 8,000 times per second, 64,000 pulses per second are required for each channel. To transmit 24 channels by this means, more than 1.5 million pulses per second must be transmitted.

The higher the pulse rate, the more the pulses are degraded and the harder they become to identify. The practical significance is that regenerative repeaters must be closer together as more channels are transmitted.

If signal quality is improved by the use of additional code digits, the number of channels must be reduced or more frequent or more expensive regenerative repeaters must be used.

One way of reducing the total quantizing noise present in PCM transmission without increasing the number of code digits is to vary the size of the quantizing steps to take advantage of the nature of speech. If the quantizing characteristic is altered to favor the weak signals at the expense of the high amplitudes, as shown in Fig. 14, more of the speech energy is finely quantized, thus lowering the total amount of noise. High signal amplitudes suffer more degradation than with uniform quantizing however.

FIG. 14. Linear quantizing (a) uses few quantizing steps for a majority of speech amplitudes. By using a compression characteristic which amplifies the low amplitudes more than the high, more uniform quantizing is obtained. More quantizing steps for low amplitudes (b) result in less quantizing noise. (*Courtesy of Lenkurt Electric Co., Inc.*)

This instantaneous companding technique significantly improves signal quality. As long as the compressor at the input and the expandor at the output complement each other, a wide range of compression-expansion characteristics may be used. The optimum characteristic depends on the nature of the talkers, instrument weighting, and similar factors. With the use of a certain typical compression-expansion characteristic that varies logarithmically with signal amplitude, an improvement of better than 26 db in the signal-to-quantizing-noise ratio is obtained. This is the equivalent of about four additional digits in the pulse code. In systems which use a seven-digit code, instantaneous companding provides quality equivalent to an eleven-digit code.

Rigorous control of the timing of the transmitted pulses is necessary in a multiplex system based on time separation. When the system employs PCM, timing-accuracy requirements become even more stringent. The ability of PCM to overcome crosstalk and other interference is based on its ability to recognize the presence or absence of the code pulses with great accuracy. Transmission errors result in *clicks* and *snaps*, since a false amplitude—one that has little or no relation to the correct signal wave-form being reconstructed—is produced by the decoder. Such transmission errors become increasingly likely as the pulse rate increases. Noise and crosstalk contribute considerably to increased transmission errors.

In addition to masking the signal pulses, crosstalk or other interference tends to shift the position of individual pulses, creating "timing jitter" which adds to the uncertainty about the presence of pulses. To avoid an accumulation of timing error from repeater to repeater (which would sharply restrict system length), each repeater must regenerate pulses which have the correct duration and spacing. This requires precise timing information at each repeater. The timing information may be derived from the data stream itself. The periodically recurring pulses of the signal are used as a reference for controlling an oscillator or for exciting a resonant circuit to produce a timing wave. The sine wave obtained from this circuit is amplified and used to derive a narrow sampling pulse. A small amount of timing jitter is unavoidable, and this shows up in small perturbations of the reconstructed pulse train. These irregularities affect pulse retiming at the next repeater, and timing error accumulates, eventually limiting the overall length of the system. If a precision timing generator is included at intervals, this source of error can be greatly reduced.

3. Telegraph Multiplex[7,14,15]

The principles of multiplexing can also be applied to telegraph transmission systems. Information in telegraph systems is transmitted by means of d-c pulses which operate the various telegraph machines. In multiplexing telegraph signals amplitude- and frequency-modulation techniques are used to convert the *mark* and *space* d-c pulses to a-c signals. In each case, the d-c pulses modulate different carrier frequencies in a manner analogous to voice multiplex systems.

In both AM and FM techniques, a tone oscillator provides the necessary carrier frequency. In AM systems, the tone oscillator is keyed on or off. For this reason, this method is sometimes referred to as on-off modulation. The telegraph mark and space conditions are therefore indicated by either the presence or absence of the carrier frequency. In FM systems, the carrier frequency is shifted in one direction to indicate a mark condition and the opposite direction to indicate a space condition. The amount of shift is the same in both directions and varies from about 30 to 85 cps depending upon the requirements of the system. This method is also referred to as frequency-shift keying (FSK) or frequency-shift modulation.

The frequency band available for telegraph multiplex systems has varied somewhat with the evolution of voice systems. Currently, two bands are most commonly used. The first band is referred to as the low-frequency group and ranges from about 340 to 3,400 cps. This band is approximately the same as the voice-frequency band found in telephone multiplex systems. The second band is referred to as the high-frequency group and ranges from about 3,500 to 5,150 cps. Although this band is above the usual voice-frequency band, it is still below the lowest line frequency used in most voice-frequency multiplex systems.

The bandwidth required for a telegraph channel depends on such things as information rate, noise, filter attenuation to adjacent channels, and whether or not both sidebands are transmitted (AM systems). In the low-frequency group, the usual bandwidth for telegraph signals at 100 words per minute for both FM and double-sideband AM is 170 cps as shown in Fig. 15. Systems operating at 75 words per minute require a bandwidth of only 120 cps. The bandwidth in the high-frequency group typically ranges from about 200 to 240 cps.

The number of telegraph multiplex channels available in most systems depends upon the required bandwidth or channel spacing. In the low-frequency group, 18 channels can be obtained with 170-cps spacing while up to 26 channels can be obtained with 120-cps spacing.   The high-frequency group is limited to about 8 channels.

Since the low-frequency group in a telegraph multiplex system is within the same frequency range as a voice channel, one such group may be assigned to one channel of a voice (telephone) multiplex system.   This means that up to 18 or 26 telegraph multiplex signals can be processed in a single voice channel, as shown in Fig. 16.   The characteristics of the voice channel which are important to telegraph depend to a great extent on whether the telegraph signal is amplitude modulated or frequency modulated.

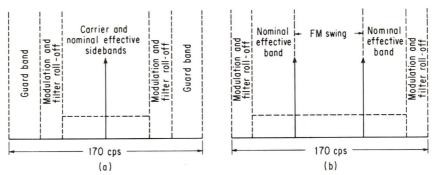

Fig. 15. Bandwidths required for 100 word-per-minute telegraph channel using AM or FM.

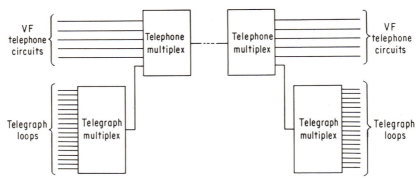

Fig. 16. Up to 18 telegraph channels can be applied to one channel in a VF multiplex system.

Variations in level are much more important to amplitude-modulated systems than to frequency systems.   Where the signal is keyed either on or off to indicate marking and spacing conditions, the received output level may vary over a wider range than is permissible where marking and spacing conditions are represented by different signal amplitudes.

While level variations are relatively unimportant for FM telegraph, the frequency stability of the multiplex system over which the telegraph signals are transmitted is much more critical for FM than for AM.

MULTIPLEX SYSTEMS

Multiplex systems are classified according to the transmission media over which they operate—open wire, multiple-pair cable, radio, and coaxial cable.   Because the transmission characteristics of these media are so different, the system characteristics are also different.

4. Open-wire Systems[2]

a. Typical System Description. A typical open-wire carrier system is the Lenkurt Electric 45A. This is a single-sideband suppressed-carrier system which transmits up to 12 voice channels, with signaling, over a single open-wire pair. The two directions of transmission are frequency stacked to provide "equivalent four-wire" operation on the single pair. West-to-east transmission is in the 40- to 88-kc range, while east-to-west transmission is normally in the 100- to 148-kc range.

Each voice-frequency band enters the channel unit where it modulates one of four carrier frequencies: 12, 16, 20, or 24 kc. Filters select the lower sideband of this modulation process, and four such sidebands are combined to form a pregroup. Each pregroup then modulates one of three carrier frequencies at 64, 80, or 96 kc. The proper sideband (containing the intelligence of four voice channels) is selected by a bandpass filter. The three pregroup sidebands are then combined to form a complete 12-channel base group occupying the frequency range from 40 to 88 kc. This is the line frequency for west-to-east transmission. Pilot frequencies of 40 and 80 kc are added before the base group is filtered and placed on the line. For east-to-west transmission, the base group is modulated to a higher frequency range, usually 100 to 148 kc. Pilot frequencies of 99 and 150 kc are added before the base group is filtered and placed on the line.

At the repeater, the low group (40 to 88 kc) being transmitted from west to east enters the repeater at its west side and passes through the low-pass section of the directional filter and directional-filter equalizer to the regulator. The regulator provides broadband slope and flat correction to offset changes in line attenuation,

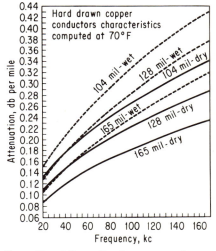

Fig. 17. Attenuation-frequency characteristics of typical open-wire lines used in multiplex systems. (*Courtesy of Lenkurt Electric Co., Inc.*)

so that all channels are retransmitted to the line at approximately the same level. After regulation, the low group is further amplified by the transmitting power amplifier and is passed to the line via the low-pass directional filter. The high group passing from east to west is treated in a similar manner by the other branch of the repeater.

b. Transmission Characteristics. The attenuation vs. frequency curves of Fig. 17 indicate how the loss of open-wire lines increases with frequency, putting a practical limit on the usable bandwidth. Moisture, ice, and sleet significantly increase loss, particularly in the higher frequencies. Because most systems operate below 150 kc to keep loss and crosstalk within tolerable limits, they are usually limited to about 16 voice channels.

c. Crosstalk. Crosstalk is classified by the far-end coupling loss and the near-end coupling loss. The far-end crosstalk coupling loss between two pairs may be determined by transmitting a signal on the first pair and measuring the resulting crosstalk on the other pair at the distant terminal. The far-end crosstalk coupling loss is the difference between the level of the disturbing signal as received on the disturbing pair and the level of the same signal received on the disturbed pair. Near-end crosstalk coupling loss is measured by transmitting a signal over the first pair and then measuring the resulting crosstalk on the other pair at the same station.

In making crosstalk measurements it is important that all wires of the line, including those under test, be terminated in their characteristic impedances. Since noise present in the disturbed pair adds to the crosstalk reading, measurements on the disturbed pair must be taken with the disturbing tone both on and off. The difference between these two readings on a power basis is indicative of the crosstalk level.

Specific values of crosstalk coupling loss vary widely, depending on such factors as line length, the transposition scheme used, the frequencies involved, and the level differences on the particular pairs. The manufacturer of a carrier system normally specifies crosstalk requirements for the lines to which his system is to be applied. For example, if the manufacturer specifies a minimum crosstalk coupling loss of 50 db, this means that the minimum loss must be 50 db in order for his system to meet the performance specifications.

d. Repeaters. Repeaters provide primarily two things: flat gain to overcome the line loss and slope correction to overcome the variation of loss with frequency. Repeater spacing is usually dictated by the loss at the highest frequency to be used. Repeater gain is set to counteract this loss, and the slope-correction network attenuates the lower frequencies more than the upper frequencies.

For example, if the line is 104-mil hard-drawn copper, the attenuation at 150 kc is 0.32 db/mile (from Fig. 17). When the line is wet, this increases to 0.395 db/mile. If the maximum repeater gain is 40 db, the repeater spacing is limited to about 100 miles or less to allow for the wet lines. Ice, snow, and sleet further restrict distances between repeaters. Under extremely severe sleet conditions, attenuation may be as high as 2 db/mile.

Repeaters are designed for one-way operation. In the case of a two-wire circuit, hybrids are used to provide a short section of four-wire line to accommodate two one-way repeaters.

5. Cable Systems[2,5,25]

Multiplex systems for operation on twisted pair cable operate in much the same fashion as do open-wire systems. The chief difference is in overcoming the much higher transmission loss of the cable. This necessitates closer repeater spacing, but the usable frequency range of cable systems is higher, reaching perhaps 500 kc.

Cable is much less affected by weather conditions than is open wire. Cable attenuation does increase with increasing temperature, however. Since aerial cable is subject to greater temperature extremes than buried cable, a larger regulation range is necessary.

Most cable systems operate on a four-wire basis, using a different pair of wires for each direction of transmission. Additional separation is often provided by using different frequency ranges for the two different directions of transmission.

a. System Description. A typical cable system is the Lenkurt 45BN. It is a single-sideband suppressed-carrier system which transmits up to 24 voice and signaling channels over two pairs in a nonloaded cable. It uses the frequency band between 40 and 264 kc. The frequency range from 40 to 140 kc is used for transmission in one direction (low-group transmit), while the frequency range from 164 to 264 kc is used for transmission in the other direction (high-group transmit).

The channel equipment used by the 45BN is identical with that used by the 45A described above. The two systems also have other common characteristics, such as the use of the same carrier frequencies. The group equipment of the 45BN combines two 12-channel base groups into a single 24-channel group occupying the frequency band from 40 to 140 kc. This band is then amplified and preequalized to compensate for approximately one-half of the line slope in the following section. In the preequalizing process, either positive or negative preequalization is introduced into the band. Positive preequalization is the introduction of increasing loss as the frequency is increased. Since line attenuation will affect the high frequencies more than low frequencies, positive preequalization is used if the terminal is designed to transmit a high group (164 to 264 kc). This is done because the 40- to 140-kc band is modulated an additional step, reversing the position of the band in the 164- to 264-kc range. Negative preequalization is used if the terminal is designed to transmit a low group, since no further modulation of the band takes place in the terminal.

After the 40- to 140-kc band is preequalized, a 96-kc synchronizing pilot is introduced into the transmission path. This synchronizing pilot is used both for synchronization at the distant terminal and for regulation of flat line losses at the system repeaters and the distant terminal.

For low-group transmission, the 40- to 140-kc band is filtered to suppress unwanted out-of-band frequencies, amplified, and passed to the transmitting line. If the terminal is arranged to transmit high group, the band is filtered and then modulates a 304-kc carrier frequency. The lower sideband is selected, producing a 164- to 264-kc band which is amplified and passed to the transmitting line facilities.

The repeaters used for this system provide four functions: (1) gain, (2) frequency frogging so the signals will be transmitted inverted and in a different portion of the frequency spectrum in the next section, (3) broadband system regulation to correct for varying flat loss in the preceding cable section, and (4) fixed slope correction to compensate for the attenuation-vs.-frequency characteristics of the cable.

The 45BN system is designed for repeater sections which have the approximate flat loss slope characteristics of 8 miles of 19-gage toll cable. Smaller cable, with higher

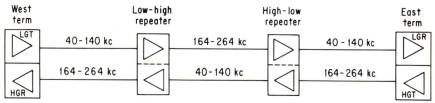

FIG. 18. Carrier-frequency bands are interchanged by frequency frogging at repeater stations.

attenuation, normally requires closer repeater spacing. The repeater provides a nominal gain of 46 db to compensate for the fixed flat loss.

Power for the repeaters is supplied over the same cable pairs used for message transmission, thus eliminating separate repeater power supplies.

Because the cable pairs are fanned and unshielded at the repeater inputs and outputs, a single cabinet repeater rack must be restricted to repeaters of one type, either high-low or low-high, to eliminate the possibility of excessive near-end crosstalk coupling between systems. See Fig. 18.

The Western Electric N2 system uses a double-sideband transmitted-carrier technique. This system provides 12 two-way voice channels on two pairs of toll or exchange cable. It uses different frequency bands for each direction of transmission. The frequency bands are inverted (frogged) at each repeater point to reduce crosstalk and provide self-equalization of the cable attenuation-vs.-frequency characteristics.

The transmitted carriers of the N2 are spaced 8 kc apart, requiring a 96-kc frequency range for each direction of transmission. The low-group frequencies are 36 to 132 kc, and the high-group frequencies are 172 to 268 kc.

Power is fed to the repeaters over the cable pairs. Typical repeater spacing is about 6 miles on 19-gage cable.

Signaling information is transmitted by an in-band tone at 2,600 cps.

The Lenkurt 81A2 is an example of a cable system designed specifically for short-haul applications, such as interlocal, extended-area service, and toll-connecting trunks. It provides 24 two-way channels over two pairs in a cable. The system uses a double-sideband transmitted-carrier modulation process to form two 12-channel groups. Each group is then modulated to the line frequency in a single-sideband suppressed-carrier process.

Channel carrier frequencies are spaced at 14-kc intervals. This permits interleaving of the channel carrier line frequencies for the two directions of transmission to minimize near-end crosstalk. The line frequency spectrum extends from 21.0 to 402.9 kc.

All 81A2 voice circuits share a common signaling channel on a time-division basis. This signaling channel, placed above the line carrier frequencies, is also used to transmit a slope-regulating pilot.

b. Frequency Frogging. The interchanging of the frequency allocations of multiplex channels in cable systems to prevent singing, reduce crosstalk, and correct for line slope is called frequency frogging. This is accomplished by having the modulators in a repeater translate a low-frequency group to a high-frequency group and vice versa. Because of this frequency-inversion process, a particular channel appears in the low group at one repeater section and is translated to the high group at the next section. This results in nearly constant attenuation with frequency of two successive repeater sections and eliminates the need for large slope equalization and adjustment. Singing and crosstalk are also minimized because the high-level output of a repeater is at a different frequency from the low-level input to other repeaters.

6. Radio and Coaxial-cable Systems[2,3,10,11]

Because of the similarities in transmission characteristics of radio and coaxial cable systems designed for operation over these two media are often considered together Often a system designed primarily for one application is easily adapted to the other application. Unlike the attenuation-vs.-frequency characteristics of wire lines, the attenuation characteristics of coaxial cable are comparatively linear; that is, the attenuation increases at a constant rate with increasing frequency. The usable bandwidth then depends upon the practicability of placing the repeaters close enough together to provide sufficient gain at the highest frequency used. The use of radio is restricted by the availability of the radio-frequency spectrum.

Since a 600-channel system normally requires a bandwidth of about $2\frac{1}{2}$ Mc, the use of such a system virtually demands a transmission facility consisting of either microwave radio or coaxial cable. Thus, microwave and coaxial-cable systems are typically used on heavy-density routes, often for long-haul circuits.

a. Typical System Description. A typical single-sideband suppressed-carrier radio multiplex system is the 46A manufactured by Lenkurt Electric Company (see Fig. 19). Although designed for application to microwave radio facilities, this system may also be used on a suitable coaxial cable. The system accommodates up to 600 channels, each with a usable bandwidth of 3.1 kc and a nominal bandwidth of 4 kc. These channels can be used for toll-quality voice transmission, data transmission, telegraph, or any other type of circuit operating in the voice-frequency range of 300 to 3,400 cps.

In a system of this type, many optional arrangements are usually available to fit the particular application. In the case of the 46A, 4-, 12-, and 120-channel versions are available for routes which do not require the full 600-channel capability. Other options such as redundant circuitry to increase reliability are also available.

The 46A is a single-sideband suppressed-carrier system with the channel carrier frequencies spaced at 4 kc. Filters select the lower sideband in each channel-modulation process. Figure 20 shows the three-stage modulation plan. The three stages form group, supergroup, and line frequencies compatible with CCITT recommendations.

In the first stage each voice-frequency input modulates one of 12 channel carrier frequencies to provide a basic 12-channel group of 60 to 108 kc.

In the next stage of modulation five of these 12-channel groups modulate five group carrier frequencies to produce a basic 60-channel supergroup of 312 to 552 kc.

In the final stage of modulation, ten 60-channel supergroups are placed on the line within the frequency range of 60 to 2,540 kc. This is accomplished by placing supergroup 2 on the line without modulation and by direct modulation of the remaining nine supergroups with the appropriate supergroup carrier frequencies.

A group pilot frequency is supplied at 84.08, 92.0, or 104.08 kc for in-service monitoring of system levels and performance. A line pilot is supplied at 32, 60, 64, 308, or

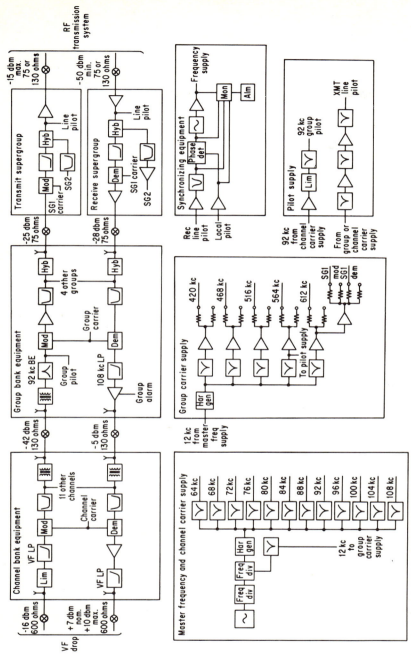

FIG. 19. Simplified block diagram of typical system (Lenkurt 46A, arranged for 120 channels).   (*Courtesy of Lenkurt Electric Co., Inc.*)

564 kc for monitoring and frequency synchronization. Supergroup pilots, a result of modulation of group pilots, appear at 424, 411.92, or 315.92 kc. Group, supergroup, and line pilots coordinate in both level and frequency with CCITT and Western Electric L-type equipment.

The Western Electric L Multiplex system is used by the Bell System for long-haul circuits transmitted over coaxial cable or microwave radio. Voice channel capacity of

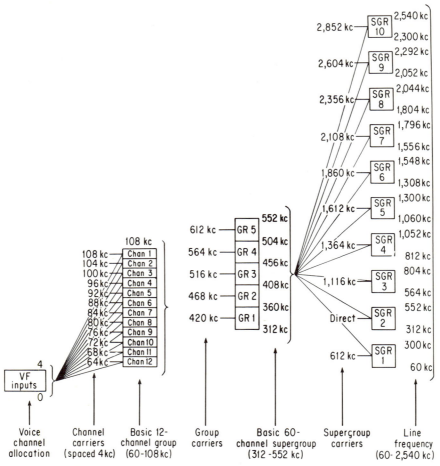

Fig. 20. Three-stage modulation plan for Lenkurt 46A multiplex system. (*Courtesy of Lenkurt Electric Co., Inc.*)

the L3 system ranges from 60 to 1,860, depending on the arrangement. The L system uses the same channel and group frequencies as the 46A system described above. It also uses the same supergroup frequencies through supergroup 8. (Supergroup 9 is 120 kc higher in the L system, and supergroup 10 is 248 kc higher.)

For high-density applications, the L system "frequency-stacks" 10 supergroups to form a 600-channel master group. Three master groups are then stacked to occupy the frequency spectrum from 564 to 8,284 kc. One additional supergroup occupies the band from 312 to 552 kc, giving a total system capacity of 1,860 channels.

7. Coordination between Multiplex Systems[16]

When two or more multiplex systems must share the same cable or open-wire transmission path, they may interfere with each other, causing increased crosstalk and noise in both systems. Although this may be reduced by suitable design of the multiplex equipment, increased crowding of transmission facilities may bring together systems having characteristics which increase interference.

a. Coordination of Levels. Interference between two multiplex systems is directly proportional to the difference in their operating levels. If the signal level in the disturbing circuit is high, the crosstalk will tend to be high, on a decibel-for-decibel

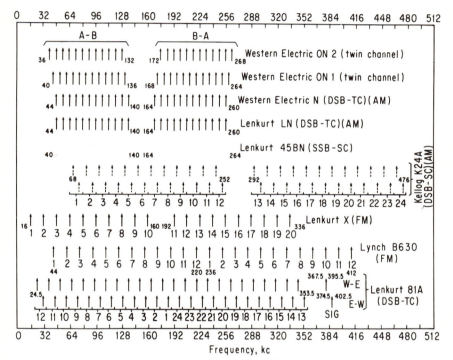

FIG. 21. Frequency allocations of several typical cable multiplex systems. (*Courtesy of Lenkurt Electric Co., Inc.*)

basis. For instance, if circuit A operates at a level 10 db higher than circuit B, crosstalk from A will appear in B 10 db higher than if the operating levels were equal. However, interference from B will be 10 db lower.

Systems operating between the same two points should maintain the same nominal transmission levels, and this level coordination should be maintained at all points along the line. If a system joins another at an intermediate point between terminals, the transmission level of the entering system should be adjusted to correspond to the levels of the other systems in the cable.

Where cables with mixed wire gages are employed, the difference in attenuation characteristics may require a special "compromise" in operating levels in order to minimize level differences. For instance, if one signal is carried on a 22-gage pair and another on a 19-gage pair, the signals transmitted over the 22-gage pair will be attenuated more rapidly than those on the 19-gage pair. In such cases, it is generally necessary to reduce the transmitting level of the system on the 19-gage pair by half the attenuation difference between the two.

b. Frequency Coordination. When multiplex systems share the same cable, interference can be further controlled by proper selection of frequency allocations and type of modulation. Each multiplex channel is vulnerable only to interference which falls within its band; other interference is rejected by the filters (see Fig. 21).

As shown in Fig. 22, weighting characteristics and channel filters effectively "reshape" the frequency distribution of energy entering the system. The desired messages transmitted over the system are also altered, but because they are received at a much higher level than interfering energy, this

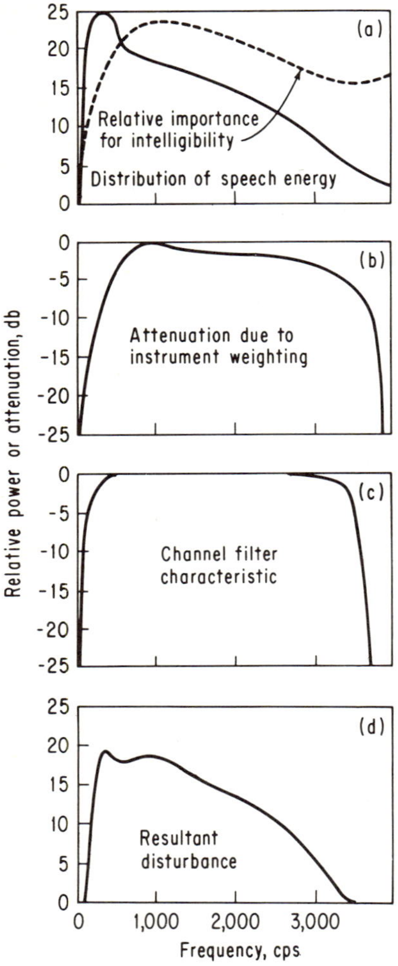

Fig. 22. Effect of filter and instrument characteristics on speech power spectrum. (*Courtesy of Lenkurt Electric Co., Inc.*)

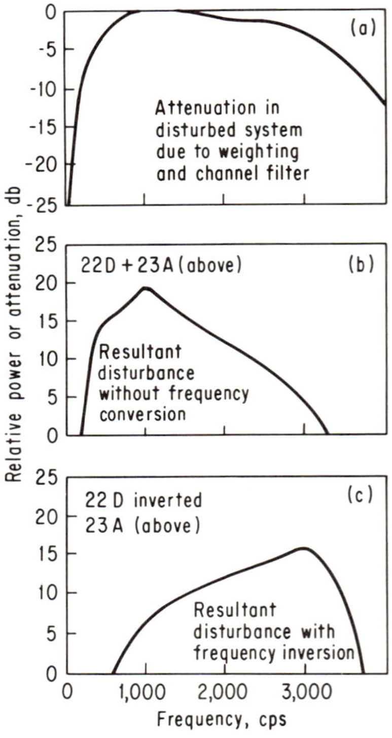

Fig. 23. Characteristics of disturbed channel reduce interference. (*Courtesy of Lenkurt Electric Co., Inc.*)

shaping has less effect on clarity and intelligibility. The most important frequencies for intelligibility lie between 800 and 1500 cps, while most of the energy present in speech is concentrated between about 200 and 500 cps. Part *d* of Fig. 22 represents the equivalent signal that is transferred from the "disturbing" system into the "disturbed" system.

Figure 23*a* illustrates the transmission or frequency-response characteristics of the "disturbed" multiplex channel. If the frequency allocations of the two interfering systems are the same, the disturbing energy appears in the disturbed channel with the relative magnitude shown in Fig. 23b, after having undergone attenuation by channel filters and weighting characteristics of the telephone equipment.

c. Frequency Inversion. Frequency inversion is possible when single-sideband modulation is used if one system transmits the upper sideband of each channel carrier while the other system transmits the lower sideband. In addition to affecting intelligibility of the crosstalk, this reduces the energy coupled into the disturbed system significantly by shifting energy peaks of the interfering signal to new locations on the transmission characteristic where there is more attenuation. Figure 23c shows the resulting energy spectrum when the disturbing signal diagrammed in Fig. 22d is inverted in frequency and passed through the channel having the characteristics of Fig. 23a. Since most of the disturbing energy lies in the vicinity of 1,000 cps, it is sharply reduced by the relatively high attenuation introduced by the transmission characteristic of the disturbed channel. A reduction in interfering energy of about 3 db is realized by frequency inversion.

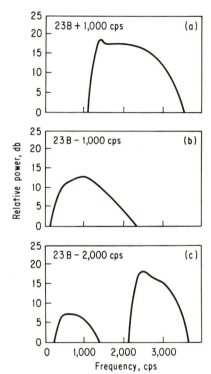

d. Frequency Staggering. The interfering energy may be made to fall outside the passband of the disturbed channels by shifting the carrier frequencies relative to each other. Figure 24a shows the energy distribution of interference when the disturbing channel carrier is shifted 1,000 cps higher in frequency than the disturbed channel. When the disturbing channel frequencies are shifted 1,000 cps lower than the disturbed channel, the interference appears as shown in Fig. 24b. Further shifts would result in even more improvement but for the presence of adjacent channels. As channel carrier frequencies are shifted further, this disturbed channel begins to pick up energy from two channels, and interference increases with further shifting.

When channel frequencies are inverted or shifted in frequency 1,000 cps or more, the crosstalk becomes unintelligible, and the same amount of interfering energy is less disturbing to talkers. Unintelligible crosstalk can be as much as 3 db higher in level than intelligible crosstalk to produce the same disturbing effect. Thus, frequency

Fig. 24. Interference between two SSB channels when (a) disturbing signal is 1,000 cps above disturbed channel; (b) disturbing channel is 1,000 cps below disturbed channel; (c) disturbed channel is 2,000 cps below disturbed channel. (*Courtesy of Lenkurt Electric Co., Inc.*)

inversion can yield a total improvement of about 6 db, while staggering of carrier frequencies can produce even more improvement, depending on the nature of the disturbing signal and the transmission characteristics of the disturbed channels.

In the case of pulse transmission, "intelligibility" is not a factor and interference is strictly a function of the amount and frequency of the energy coupled into the disturbed circuit. If the data are transmitted over a single frequency assignment within a voice channel, the use of frequency inversion or staggering may cause the disturbing tones to fall outside the passband of the receiver channel filter. However, where several data channels are transmitted in a single voice channel, frequency inversion and staggering may merely transfer the interference from one data channel to another.

The frequency allocations of two systems which share a cable should not be staggered in such a fashion that at one terminal the high level transmission of one system coincides in frequency with the much weaker incoming signal of the other system. In such a case, near-end crosstalk is increased by the transmission loss of the path, assuming that both cable pairs are of the same gage and that both systems operate at the same nominal transmitting level.

e. Single-sideband vs. Double-sideband. Most of the power in a double-sideband (DSB) signal is in the transmitted carrier. Even at 100 per cent modulation, the carrier has twice the power of both sidebands together. Carrier power remains constant regardless of modulation, while sideband power varies directly with the degree of modulation.

If such a signal is staggered in frequency from another, so that the carrier frequency falls within the passband of another channel, it creates an interfering tone in the disturbed channel. For this reason, systems that are designed to operate in the same cable with other transmitted-carrier systems almost invariably employ the same carrier frequencies.

In single-sideband (SSB) systems, the transmitted signal power depends entirely on the modulation of each channel. The interference caused by such a system varies with the modulation characteristics and the operating levels used.

When DSB systems are used in the same cable with SSB systems, the DSB systems are inherently less susceptible to interference than the SSB systems, other factors being equal. If there is no frequency staggering or inversion, only one sideband of the DSB transmission is coupled into the SSB channel, so that the resulting crosstalk is directly proportional to the difference in the levels of the two—the same as between two SSB systems.

When a DSB signal is demodulated, the voltage of the two sidebands adds in phase for an effective level increase of 6 db. Thus, an interfering SSB signal produces interference in a DSB channel 6 db weaker than in another SSB channel.

Even when two SSB channels interfere with a single DSB signal, one into each sideband, the two interfering channels have no coherent phase relationship and so add on a power basis rather than on a voltage basis. This results in a 3-db increase in interfering energy, which is still effectively 3 db lower than the DSB signal.

f. Frequency Modulation. Most FM systems restrict the modulation index to little more than unity in order to conserve bandwidth, resulting in noise performance essentially equal to that of a DSB system. Although most noise and interference are amplitude-varying phenomena which the FM receiver should eliminate, practical deficiencies of the receiver limiters allow some interference to pass.

When no modulation is present, the carrier of the FM channel may appear as an interfering tone in other channels which take in the FM carrier frequency. This source of interference is reduced when the carrier is modulated because of the distribution of carrier power into the sidebands.

Interference from FM sidebands is much like that from frequency-inverted SSB channels, unintelligible but proportional to the amount of energy present within the passband of the disturbed channel. Unlike AM sidebands, the energy distribution across the band varies from instant to instant in a way that is not proportional to the energy distribution of the modulating signal but changes as carrier energy is distributed into farther sidebands with increasing modulation level.

g. Pulse-code Modulation. Pulse-code-modulation (PCM) multiplex systems introduce special problems of coordination with other systems. Practical PCM systems transmit pulses at about $1\frac{1}{2}$ million per second. The nearly 1-Mc bandwidth required for transmission at this rate imposes severe transmission requirements on the cable pair. Coupling between pairs becomes much greater than at the lower frequencies which characterize SSB, DSB, or FM transmission. Although the basic nature of PCM usually permits adequate transmission under these conditions, interference with other systems in the same cable becomes intolerable unless they are also of the PCM type. Even so, the transmission of several PCM systems over a single cable increases mutual interference so badly that a separate cable may be required for the return direction.

8. Signaling in Multiplex Systems[19]

Signaling provides the means for managing and supervising a communications system; it establishes connections, helps select the route (when there is a choice), announces the incoming call, reports the fact if a line is busy.

Signaling through multiplex systems is done in many ways. Some communications systems use separate channels to convey information used in controlling the operation of others. In most systems, however, each channel carries its own signaling. In local (physical or metallic) telephone circuits, this is achieved by d-c signals which share the line with the message signals. In multiplex systems, however, different techniques are required. Voice channels customarily occupy a 4-kc bandwidth. About 300 cps of this bandwidth is used for isolation from adjacent channels. The rest, usually about 3,700 cycles, must carry both speech and signaling. In some cases, the 4-kc channel is used for voice and also for one or more telegraph or teletypewriter signals. In such cases, channel filters are used to prevent mutual interference between the speech, telegraph, and supervisory signals which share the channel.

In most multiplex systems, three basic methods are used for signaling: (1) separate channel, (2) out-of-band, and (3) in-band. The two most widely used signaling methods are in-band and out-of-band.

a. Separate-channel Signaling. Separate-channel signaling is often used on very high density "backbone" routes or under special circumstances where signaling cannot be conveniently handled within the voice channels. Separate channel signaling leaves the entire voice channel free for communications, without the possibility of mutual interference between the speech and signaling, but it is not always economical, since certain channels must be set aside to handle only signaling functions. Also, repair and maintenance are more complicated when signaling and speech are sent over separate channels. Reliability may be less, since both channels are subject to failure independently of each other.

b. Out-of-band Signaling. With out-of-band signaling, channel filters are designed with an upper cutoff frequency below the top edge of the channel. This leaves a portion of the spectrum free to transmit signaling tones. Generally, a single tone is used, and this is keyed to convey signaling information.

When signaling is completely separated from the speech portion of the channel, it is possible to maintain relative freedom from mutual interference between the speech and the signaling tones. Signaling tones can be transmitted during the conversation, thus permitting extra functions such as regulation, which might be desirable during the period the speech channel is in use.

In addition to being more flexible, out-of-band signaling is easier and more economical to accomplish, particularly if some sacrifice in channel bandwidth is allowed. In telephone circuits, there is very little speech energy present at the upper end of the channel. Accordingly, filtering requirements may be somewhat relaxed (since telephone instrument weighting also provides a degree of "filtering").

A careful compromise between speech quality and equipment cost is required. If the frequency of the out-of-band signaling tone is too low, the restriction on bandwidth may impair speech quality. If the signaling tone frequency is raised so that it is close to the top edge of the 4-kc band, channel filters must be made more complex.

A case in point is the use of a signaling frequency of 3,825 cps, standard in many countries (CCITT Recommendation). Where channel filters are not sufficiently effective, a 3,825-cps signaling tone appears as a 175-cps tone in the adjacent channel, but at a fairly low level. This results because the 3,825-cps tone is at the same frequency as the "image" or unused sideband of a 175-cps tone in the adjacent channel. Although filters essentially eliminate this sideband, even the best channel filter characteristics are far from ideal. Figure 25 shows typical attenuation characteristics. Note that, although the 175-cps component from the next channel can appear following demodulation, it is attenuated about 45 db. Although attenuated to a low level, the tone may still be audible and disturbing. In such cases, it is necessary to provide additional filtering following the demodulator in order to eliminate the tone.

Conversely, if a high-pass filter is not used to block the 175-cps component before modulation, speech energy at this frequency may appear as 3,825-cps energy in the adjacent channel, thus causing false signaling. In order to attenuate the 175-cps component adequately with conventional filters, the low-frequency channel cutoff is

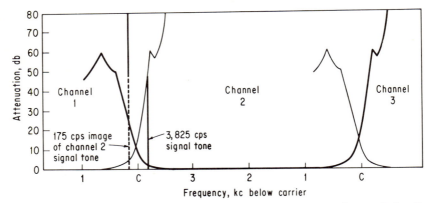

FIG. 25. Typical multiplex channel bandpass characteristic. 175-cps image of signaling tone is attenuated about 45 db. (*Courtesy of Lenkurt Electric Co., Inc.*)

approximately 300 cps. In conventional practice, the high-frequency cutoff remains at approximately 3,400 cps. Although this frequency range provides acceptable performance over most communications circuits existing today, longer circuits would be excessively degraded. The cumulative effect of the repeated filtering that would be required to prevent adjacent channel interference would result in a channel characteristic like that shown in Fig. 26. This type of frequency characteristic reduces intelligibility. Accordingly, to maintain high standards of speech quality, it would be necessary to employ channel filters having much sharper attenuation characteristics, thus raising the cost of equipment.

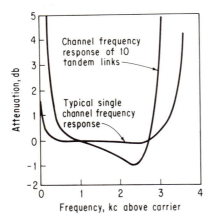

FIG. 26. Frequency-response variations become exaggerated as they are repeated in tandem links. (*Courtesy of Lenkurt Electric Co., Inc.*)

One way of overcoming this problem is to lower the frequency of the out-of-band signaling tone so that it is farther from the edge of the 4-kc band. However, this may lower the highest speech frequency that can be transmitted. Thus, the voice-frequency cutoff might be reduced from 3,400 to about 3,300 cps. This would allow the signal tone to be reduced from 3,825 to, say, 3,700 cps. Although the top frequency would be lower, this permits the lower cutoff frequency also to be reduced, perhaps from 300 to about 200 cps. In this case, the image frequency of the signaling tone goes from 175 to 300 cps, well within the effective rejection capability of the channel bandpass filters. Because of the increased attenuation of the signaling image frequency, supplementary filtering following the demodulator is not required. Subjective tests by the British Post Office and Bell Laboratories have shown that a frequency band from 200 to 3050 cps provides better quality than the band ranging from 300 to 3,150 cps, where channel bandwidth must be restricted, as in submarine cables. Voice intelligibility is more dependent on the low-frequency end of the voice band than on the high-frequency end.

One disadvantage of out-of-band signaling is that it requires some sort of d-c repeater at the end of each link. The signal pulses are detected and made to operate a relay which, in turn, keys the signaling equipment in the succeeding link. Thus, signaling terminals are required at both ends of each link, which increases the complexity and cost of the system and the possibility of signal distortion.

c. In-band Signaling. With in-band signaling, the signaling tones are transmitted at a frequency within the speech band, usually 1,600, 2,400, or 2,600 cps. The principal objection to in-band signaling is that the signaling tones lie right in the speech band. This leads to the possibility that speech energy at the signaling frequency may be able to "talk down" the signaling, that is, cause false signals with voice energy. Conversely, signaling tones are audible and thus cannot be used during conversation.

The biggest advantage of in-band signaling is the extreme flexibility that it provides. The speech and supervisory signals share the same transmission facility, but at different times. The system is arranged so that supervisory signals are on the line only before and after a call. Since the signaling becomes a part of the transmission, it is not necessary to use d-c repeaters when going from one link to another. At branching points, a similar flexibility is obtained. The lack of d-c repeaters eliminates the delay and pulse distortion which characterize out-of-band signals sent through several links.

In-band signaling provides unusual flexibility and economy in large offices. Since the signals are carried over the speech circuit, it is unnecessary to cable the so-called E & M (receive and transmit) signaling leads through the office. Signaling equipment can be associated directly with the switching equipment, thus allowing a trunk circuit to be obtained from any available transmission medium rather than being restricted to certain multiplex systems.

To prevent spurious signaling by voice energy, a "guard" circuit is commonly used to distinguish between speech and signaling tones. Typically, the guard circuit consists of a network which detects the presence of other frequencies. When other frequencies are present, the guard circuit "assumes" that the signaling tone frequency is caused by speech and therefore prevents signal circuit response.

Further protection can be obtained by proper choice of frequency for the in-band signaling tone. In general, it is desirable to use the highest frequency that can be transmitted easily through the worst transmission channel that might be used. Speech energy declines rapidly at the higher frequencies, thus reducing the likelihood of talkdown. Certain older multiplex systems have a cutoff frequency near 2,800 cps. For this reason one of the most commonly used in-band signaling frequencies is 2,600 cps. Speech energy at 2,600 cps is relatively low.

As a further precaution, a brief time delay on the order of 30 msec reduces the likelihood of speech or noise energy causing spurious signals. Normally, most noise frequencies are very transient. When a delay is introduced, the circuit is made relatively insensitive to noise energy at the signaling frequency.

d. Time-division Signaling. In some new multiplex systems supervisory messages are transmitted by means of time-division signaling. In this type of signaling all voice channels share a common signaling channel. Each of the signaling leads is connected to a sampling gate. Each channel is sampled in sequence, and the presence or absence of a signaling tone is transmitted to the receiver.

At the receiver, the incoming pulses are sorted and distributed to the appropriate channels. Although this arrangement lacks the flexibility of in-band signaling, it does provide unusually reliable and economical signaling without encroaching on the bandwidth available for each channel. If most of the signaling functions are associated with the common equipment, cost of the system is substantially reduced without reducing quality or reliability, an important consideration in short-haul systems.

9. Auxiliary Equipment[3,17,21,23]

a. Four-wire Terminating Sets (Hybrids). A four-wire terminating set is a hybrid connection used at the transition between four-wire and two-wire operation.

The most common point of application is where the two-wire voice-frequency drop connects with the four-wire channel in the multiplex system.

A terminating set has four sets of terminals: four-wire transmit, four-wire receive, two-wire line, and balancing network. It has high loss between opposite sets of terminals and low loss between adjacent terminals. The isolation between four-wire transmit and four-wire receive terminals is typically about 60 db, while the loss between either pair of four-wire terminals and the two-wire line varies from about 3 to about 8 db, depending on the type of hybrid used.

The two main types of hybrids, the transformer hybrid and the resistance hybrid, are shown in Figs. 27 and 28. The transformer hybrid normally has a lower loss in the transmission path between the two-wire line and the two sides of the four-wire line. The minimum theoretical loss is 3 db, while the minimum theoretical loss for the resistance hybrid is 6 db. Resistance hybrids are normally smaller, lighter, and cheaper, even though they introduce more loss.

Standard impedances for four-wire terminating sets are 600 ohms for the four-wire line and either 600 or 900 ohms in series with 2.1 μf for the two-wire line.

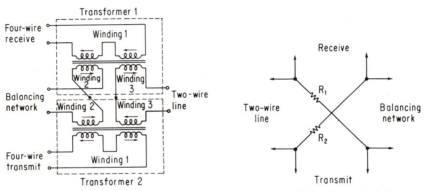

FIG. 27. Transformer hybrid.   FIG. 28. Resistance hybrid.

b. Compandors. A compandor provides an effective improvement in the signal-to-noise ratio in a voice circuit, thus permitting satisfactory operation over noisier circuits than would otherwise be possible. The use of compandors can permit the relaxation of performance standards elsewhere in the system. The compandor can be used to upgrade transmission quality in existing noisy circuits used for speech by improving the apparent signal-to-noise ratio by about 25 db.

A compandor consists of a volume compressor at the transmitting end of a multiplex channel and a volume expandor at the receiving end. By imparting more gain to low-intensity signals than to high-intensity signals, the compressor reduces the dynamic range of signals transmitted between terminals as shown in Fig. 29. The expandor at the receiving end of the circuit restores the signal to its original dynamic range. Thus, the weak signals which are most vulnerable to noise are lifted clear of the noise, while the strong signal components, which are relatively invulnerable to noise, are amplified less or possibly attenuated. This avoids overloading the transmission-path equipment.

Compandors are employed effectively on amplitude-modulated multiplex systems to reduce the effects of crosstalk as well as to improve the apparent signal-to-noise ratio. Repeater spacing of multiplex systems operating over wire lines or cable pairs is often limited by line noise conditions or excessive near-end crosstalk. With the noise advantage offered by compandors, repeater spacing may be extended—limited only by maximum system gain.

Multichannel microwave radio systems can achieve greater fade margins and longer transmission paths because of the additional signal-to-noise advantage of the compandor. Such an advantage can also reduce the radio antenna gain requirements, thus permitting the use of smaller antennas.

When compandors are employed on multiplex systems, it is necessary to consider an average power increase of about 5 db for voice circuit loading. This additional power must be either attenuated or added to the nominal figures used to calculate the loading effect.

Compandors offer little or no noise improvement for data. In addition, compandors tend to introduce intermodulation distortion and can be detrimental to pulse-type data signals with changing power levels. When practicable, therefore, compandors should be removed from voice circuits that are to be used for data. Nevertheless, it is possible to transmit data signals through compandors without too much signal degradation. It is currently done over telephone networks that provide data-transmission service on a dial-up basis and in systems employing in-band signaling.

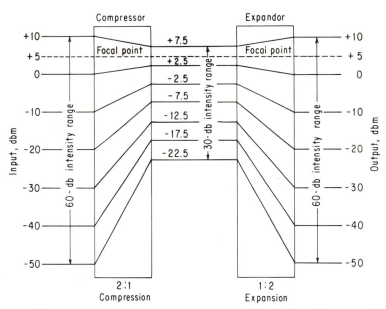

Fig. 29. The effect of a typical compandor on various power levels. (*Courtesy of Lenkurt Electric Co., Inc.*)

c. Equalizers. An equalizer is a voice-frequency line treatment network which produces either attenuation or delay characteristics (or both) opposite to those already inherent in the channel. Most multiplex systems include some measure of built-in attenuation correction, but the relative delay is primarily a function of the filter characteristics.

Equalization is seldom required for voice transmission but is often necessary if the voice channel is to be used for data. The usual data band is 1,000 to 2,600 cps. This is normally the flattest portion of the channel, but it often requires additional correction as shown in Figs. 30 and 31.

An equalizer may consist of either a passive or an active network. The use of a passive equalizer introduces considerable loss, which must be made up by additional amplification. An active equalizer incorporates amplifiers in its design and is usually adjustable over a considerable range.

d. Echo Suppressors. Echo suppressors are used on long circuits to prevent the echo of reflected speech from annoying the talker. The use of echo suppressors is much more important on long lines than on short lines because the annoyance increases as the time delay of the echo increases.

Most echo suppressors in use on commercial telephone circuits permit transmission in only one direction at a time.   This often results in the loss of some speech syllables as the echo suppressor shifts from one direction of operation to the other.

During data transmission an echo suppressor must be removed from the transmission path, or it will not permit a data receiver to ask for the repeat of a message in which an error has been detected.   This disabling of the echo suppressor is often accomplished by a device which is sensitive to a tone of 2,250 cps generated by the data-transmitting equipment.   When the disabler receives this tone for some specified length of time, it holds open both directions of transmission while data are being transmitted.

e. Order Wire.   An order wire is an auxiliary communications circuit for the use of maintenance personnel in the line-up and maintenance of communications equipment.   Some multiplex systems include a built-in order-wire facility, while in other systems this service is provided by accessory equipment.   Order-wire circuits are

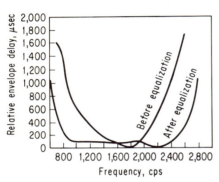

FIG. 30. Typical envelope delay characteristics. (*Courtesy of Lenkurt Electric Co., Inc.*)

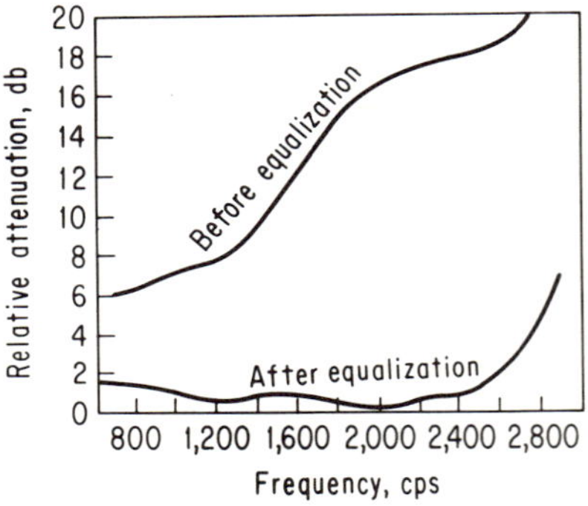

FIG. 31. Typical attenuation characteristics. (*Courtesy Lenkurt Electric Co., Inc.*)

often restricted in bandwidth to conserve frequency spectrum.   In a typical arrangement a 1.8-kc order-wire talking circuit, an order-wire signaling circuit, and up to six additional signaling tones can be transmitted over a standard 3.1-kc voice channel.

The order-wire signaling tone may produce an audible tone from a speaker, or it may be used to operate a buzzer or call lamp.   Some arrangements permit voice calling by the use of a speaker at the receiving end.

f. Alarm and Control Equipment.   Most multiplex systems are equipped (often as an optional accessory) to provide alarm indications for equipment or transmission-path failure.   Typical functions monitored are loss of power supply, loss of carrier, and loss of pilot signals.   Many alarm arrangements include both major and minor alarms, with a minor alarm indicating, for example, the failure of individual channels and a major alarm indicating the failure of one or more groups of channels.

Equipment is also available for remote monitoring and control of unattended stations from a master location.   One such system, for example, reports the status of up to 20 alarm functions from each of 10 separate remote stations.

g. Speech-plus-data Equipment.   One or more telegraph channels or a low-speed data channel can be provided over a nominal 4-kc voice channel without interfering with the speech.   Speech-plus-data equipment multiplexes the telegraph or data channels into the voice channel along with the speech.   A typical arrangement reduces the speech bandwidth by approximately 400 cps to provide a 200-bit-per-second data channel.   This reduces speech intelligibility by about 10 per cent.

SYSTEM OPERATING CHARACTERISTICS[3,9,12,13,20,23]

Operating characteristics for multiplex systems are generally established by telephone system practices and commercial, government, and international standards.

Standard operating characteristics are essential in long-distance transmission systems, where coordination among different types of multiplex equipment is necessary. Telephone industry standards and standards and recommendations issued by the Defense Communications Agency and the International Telegraph and Telephone Consultative Committee (CCITT) generally govern the operating characteristics of multiplex systems. For economic reasons, operating characteristics for exchange or subscriber multiplex systems do not necessarily conform to universal standards and generally vary according to particular needs.

Most high-density or long-distance multiplex systems employ single-sideband suppressed-carrier modulation techniques. Unless otherwise stated, the operating characteristics described herein apply mainly to systems of this type.

10. Channel Bandwidth

As a result of tests performed by the telephone industry, the standard voice channel bandwidth has been established at about 3,100 cps. Typically, the range of useful voice frequencies is from about 300 cps to 3,400 cps. Accordingly, most multiplex channels are designed to process a signal within this frequency range. In SSBSC multiplex systems, voice channels are usually spaced at 4-kc intervals to accommodate the voice frequencies. If it is necessary to pass a wider band of frequencies, such as required for certain video signals, multiplex systems sometimes provide a means of combining groups of voice channels to acquire a single wideband channel. Typical v-f ranges are:

Lower frequency limit (3-db point): 250 to 300 cps
Upper frequency limit (3-db point): 3,100 to 3,400 cps

11. Frequency-allocation and Modulation Plans

Each type of multiplex system employs a particular modulation scheme to shift the v-f signals to various carrier-frequency bands. These schemes are referred to as *frequency-allocation and modulation plans* and usually appear in the form of diagrams. Standard frequency-allocation and modulation plans have been established for SSBSC multiplex systems. One such plan for SSBSC multiplex systems assembles up to twelve 4-kc channels into a basic group frequency band ranging from 60 to 108 kc. In this first modulation stage there is a separate carrier frequency for each of the 12 channels. (A modified version of this plan, called twin-channel modulation, requires only six carrier frequencies.) Additional channels are obtained by expanding the number of 12-channel groups and by using a number of intermediate modulation steps. A final modulation step is required to combine the multiplexed signals into the line-group or baseband group (output) frequency range.

a. Open-wire Systems (SSBSC). The standard frequency-allocation and modulation plan provides up to 12 channels for open-wire systems. The line-group frequencies for open-wire systems are different for each direction of transmission in order to establish an equivalent four-wire system over one wire pair. The two directions are conventionally referred to as the east-west direction and the west-east direction. After the 12 voice channels have been combined into the 60- to 108-kc basic group frequency band, they are then shifted to one of four staggered line-group frequency allocations. A typical frequency-allocation and modulation plan for an open-wire system is shown in Fig. 32. The line-group frequency range for the four staggered groups are as follows:

System (CCITT)	West-east, kc	East-west, kc
SOJ-A-12	36–84	92–140
SOJ-B-12	36–84	95–143
SOJ-C-12	36–84	93–141
SOJ-D-12	36–84	94–142

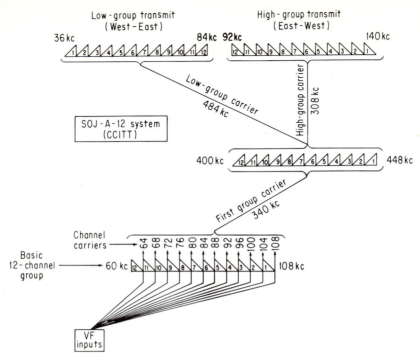

Fig. 32. Frequency allocation and modulation plan, typical open-wire multiplex system.

b. Multipair Cable Systems. The frequency-allocation and modulation plan used in most toll applications provides up to 24 channels for multipair cable systems, using two cable pairs. Two basic 60- to 108-kc 12-channel groups are required. The line-group frequencies are different for each direction of transmission. The line group for one direction is referred to as the low-group transmit (LGT), while the line group for the other direction is referred to as the high-group transmit (HGT). A frequency-allocation and modulation plan for a typical 24-channel multipair cable system is shown in Fig. 33. Typical line frequency ranges are as follows:

	Transmit, kc	Receive, kc
LGT terminal............	36–140	164–268
		(8-kc separation between 12-channel groups)
	40–140	164–264
		(4-kc separation between 12-channel groups)
HGT terminal............	164–268	36–140
	164–264	40–140

c. Microwave Radio Systems. The present CCITT standard frequency-allocation and modulation plan provides up to 1,800 channels for wideband (microwave) radio systems. In this plan, five basic 12-channel groups are assembled by a second modulation step to form a basic 60-channel supergroup with a frequency range of 312 to 552 kc. Larger capacity systems are formed by additional modulation steps using

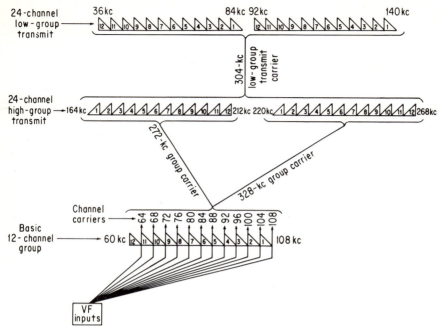

Fig. 33. Frequency allocation and modulation plan, typical multipair-cable multiplex system.

higher-order master groups and supermaster groups. One master group consists of five supergroups. One supermaster group consists of three master groups. A standard frequency-allocation and modulation plan for a system with 960 channels is shown in Fig. 34. Baseband frequency ranges for systems with different channel capacities are as follows:

Channel capacity	Baseband frequency range, kc
12	12–60, or 60–108
24	12–108
60	12–252, or 60–300
120	12–552, or 60–552
300	60–1364
600	60–2,792
900/960	60–4,287
1,800	300–8,248

d. Coaxial-cable Systems. There are five standard coaxial-cable systems recommended by the International Telegraph and Telephone Consultative Committee (CCITT) which use the same basic frequency-allocation and modulation plans specified for microwave radio systems. These standard coaxial-cable systems are based upon five different cable configurations, each designed to transmit a certain line frequency band. Each of the five systems is described as follows:

2.6-Mc System. 600-channel system using 10 supergroups with a line-group frequency band of 60 to 2,540 kc

4-Mc System. 960-channel system using 16 supergroups with a line-group frequency band of 60 to 4,028 kc

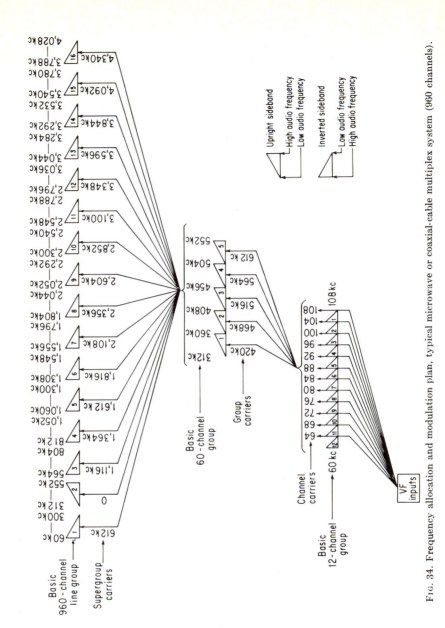

Fig. 34. Frequency allocation and modulation plan, typical microwave or coaxial-cable multiplex system (960 channels).

6-*Mc System.* System specified for transmitting a vestigial sideband television signal with a video-frequency band of 5 Mc

12-*Mc System.* 2,700-channel system using 45 supergroups in 9 master groups or 3 supermaster groups with a line-group frequency range of 308 to 12,435 kc

12. Voice-frequency Transmission Levels

Voice-frequency circuits extend from the switching center of a communications system to the channel units of the multiplex terminal. The amount of v-f power required at the input and output of the channel units is based on the normal amount of power required by the switching center equipment. The standard used to measure the transmission levels of multiplex systems is typically a 1,000-cps test tone at a level of 0 dbm0. Such a test tone is commonly available at telephone switching offices. Typical per channel values are:

1. Input level:
   Four-wire operation, -4, -13, or -16 dbm
   Two-wire operation, 0 dbm
2. Output level:
   Four-wire operation, $+4$ or $+7$ dbm
   Two-wire operation, adjustable from -10 to $+5$ dbm

13. Carrier-frequency Transmission Levels

A number of factors influence the amount of power required in the carrier-frequency signals, such as the transmission medium (metallic lines or radio), noise level of the line, line attenuation, operating frequencies, and crosstalk considerations. In the case of radio transmission, the signal levels are determined by the input requirements of the particular radio. The amount of power transmitted to metallic lines is usually high enough to permit maximum spacing of repeaters while not overloading the amplifiers. It is also necessary that the carrier-frequency power be sufficiently higher than line noise in order to maintain a proper signal-to-noise ratio. Typical per channel test-tone levels at the line are:

1. Open-wire systems: 0 to $+18$ dbm
2. Multipair cable systems: -20 to 0 dbm
3. Coaxial-cable or radio systems: -50 to -6 dbm

14. Channel Capacity

The number of voice-channels that a SSBSC multiplex system can handle are determined primarily by the bandwidth limitations of the transmission medium and the signal-to-noise ratio that must be maintained. The capacity of present-day systems are:

1. Open wire: 16 channels
2. Multipair cable: 24 channels
3. Coaxial cable and radio: 2,700 or more channels

15. Input and Output Impedances and Return Loss

Input and output impedances are specified at the various interface points of a multiplex system. Such impedances are based upon the nominal characteristic impedance of the transmission lines connected at the interface points. A minimum acceptable return loss is normally prescribed at these points.

1. Voice-frequency drops: 600 ohms four-wire and 600 or 900 ohms two-wire, balanced with 26 db minimum return loss
2. Group frequency interface (60 to 108 kc): 135 ohms, balanced, or 75 ohms, unbalanced with 20-db minimum return loss
3. Supergroup frequency interface (312 to 552 kc): 75 ohms, unbalanced, with 20-db minimum return loss

4. Baseband or line frequency interface: 75 ohms, unbalanced, or 135 ohms, balanced with 20-db minimum return loss

16. Carrier Leak

The electrical balance of suppressed carrier modulators used in SSBSC multiplex systems is never perfect.   Because of this, some carrier frequency energy usually leaks into other parts of the system.   This unwanted energy is called *carrier leak*. High carrier leak increases crosstalk and also the loading of line amplifiers.   The amount of tolerable carrier leak is specified in decibels below the reference test-tone level or in some cases as dbm0.   CCITT recommendations specify that the power of the carrier leak transmitted to the line should not exceed −17 dbm0 for any channel and −14.5 dbm0 at any 12-channel group level.   For equivalent four-wire systems, the maximum tolerable carrier leak per channel is specified as −16 dbm0 for the upper frequency group and −17 dbm0 for the lower frequency group.   The maximum allowable carrier leak for all channels in such a system is specified as −14.5 dbm0.

17. Drop Level Stability

The losses in a multiplex system are not constant but vary because of such factors as changing weather conditions and unstable power sources.   In many types of multiplex systems, regulation is used to compensate for much of the variations in power level that may occur.   In practice, it is desirable to keep the levels of v-f signals transmitted from the multiplex terminal within a tolerance of ±0.5 db over short time intervals.   Over a long period, the permissible level variations may be much larger but must still be held within well-defined limits.

18. Frequency Stability

The use of many carrier oscillators in multiplex systems introduces a problem of frequency stability.   Although frequency stability is often associated with the change in frequency of an oscillator over a period of time, the frequency stability of concern in a multiplex system is the net change in voice frequency that occurs between the sending end of the system and the receiving end.   In telephone systems, the amount of frequency change, and indirectly the frequency stability, is related to the change which may occur without the change being discernible to the ear.   Frequency stability (end-to-end) in the order of ±3 to ±5 cps is satisfactory for most telephone circuits. However, frequency stability is more critical in multiplex systems carrying telegraph and high-speed data.   For this reason, most standards for frequency stability in multiplex systems specify that the difference in frequency from one end of the system to the other shall not exceed 2 cps.

19. Pilots

Pilots are auxiliary signals employed in multiplex systems for such functions as level regulation, frequency synchronization, alarm systems, and maintenance monitoring.   The transmit line levels at multiplex terminals and at repeaters must be maintained within close tolerances.   Line noise and crosstalk from adjacent systems increase if the level is too low, while too high a level causes overloading, which can result in intermodulation distortion and crosstalk into other systems.   Regulating pilots are used to operate compensating devices throughout the multiplex system in order to control line levels.

Frequency synchronization pilots are often used to maintain end-to-end frequency stability.   This is especially important in SSBSC systems, where the line carrier frequency must be reinserted at the receiving terminal, and for systems carrying telegraph and high-speed data.   Frequency synchronization is accomplished by phase-locking the carrier frequencies in one terminal with the carrier frequencies at the opposite terminal by means of one or more pilots transmitted in the system.   Thus, if the carrier frequencies at the transmit terminal begin to drift, the respective carrier frequencies at the receive terminal will drift a like amount.

The frequency of the pilots and the number required in each multiplex system depend mainly on the particular frequency-allocation and modulation plan and, of course, any special needs of the system. Pilots are normally transmitted at a level 10 to 20 db below the system test-tone level.

20. Loading Effects

a. Intermodulation Distortion. Intermodulation distortion occurs when traffic is so great as to overload the amplifiers or exceed the design rating of the modulators and associated circuit elements. Present-day multiplex systems carry many voice-frequency channels that are handled by common amplifiers, modulators, and demodulators which are not perfectly linear.

In the case of a single channel, modulation products resulting from nonlinearities are directly related to the power and frequency of the signal applied to the channel and appear as distortion of the signal. In multiplex systems, however, the case is entirely different. Most modulation products appearing in any given channel are unrelated to the signal applied to that channel. At higher multiplex frequencies, modulation products from any one channel may be distributed over many other channel allocations. Thus, instead of distorting only the impressed signal, the entire multiplex system overloads, resulting in background noise and crosstalk that increases as the system load becomes greater. In some cases, crosstalk may even be intelligible. In larger systems, however, most distortion appears only as random noise which reduces the signal-to-noise ratio.

Distortion increases significantly when large signal voltage peaks drive some element in the system beyond its region of linear operation. In the case of amplifiers employing negative feedback, there is only a slight increase in distortion and intermodulation products as signal level increases until the amplifier "break" point is reached. In vacuum tubes this point is reached when grid current begins to flow or plate-current cutoff occurs, while in transistors this occurs when the saturation or cutoff point is exceeded. Accordingly, distortion increases very rapidly with signal voltage after the break point is reached.

Intermodulation products occurring below the amplifier break point consist mostly of second- and third-order distortion products. When the signal voltage exceeds the break point of the amplifier, sufficient higher-order intermodulation products are generated to cause interference in most or all of the channels in the system.

In reducing the effects of interference, it is necessary to maintain the signal level as high as possible without causing overload. A high signal-to-noise ratio greatly enhances the quality of communication. In a multiplex system, even where the transmission level of each channel is fixed, the level of the composite signal varies. Since many different frequencies are transmitted, the phase relation between these various frequencies varies randomly. Sometimes several frequencies will reach a peak together, causing a momentary rise in total voltage. At other moments, the various frequencies may combine to lower the total signal voltage well below average.

Since numerous signals are applied to a common amplifier in a multiplex system, the phase combinations that may occur greatly increase, and there is always the chance that many or all of the signals will achieve a peak value simultaneously. However, as the number of signals increases, the chance they will all reach a peak value simultaneously decreases.

b. Speech Loading. Speech signals are more complicated than telegraph tones and data signals. Each signal consists of a variety of frequencies and a great range of amplitudes.

In telephone multiplex systems, many factors influence the load on the system. Some of the more important include speech habits of the telephone user, hourly variations in system use, the psychological effect on the speaker of the circuit quality, and the technical characteristics of the subscriber's equipment and local telephone plant.

By using statistical methods of computing load, it is possible to provide a multiplex system that will not overload except during the very busiest periods. Telephone company practice has been to design communication systems for a "break" point of 1 per cent of the busiest period. Thus, for the busiest hour, overload may occur during 36 sec (1 per cent of 3,600 sec). Since this total time is distributed throughout the entire period as a number of brief moments of possible audible disturbance, the net effect on an individual conversation is negligible.

Since amplifiers, modulators, and the like are responsive to instantaneous peak signal values rather than average or rms values, multiplex systems must be adjusted to accommodate the possibility of large peak voltages. In normal telephone practice, the total average input voltage (or power level) to the carrier system is set so that the overload or break point is not reached more than 1 per cent of the busiest hour. Since the voltage distribution varies with the number of talkers using a system, the amount the average signal level must be reduced (the peak factor) will vary with the size of the system. Figure 35 shows peak factors for speech (and tones) in terms of the number of active channels.

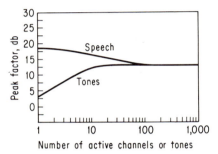

FIG. 35. Multichannel peak factor. Values of signal peaks (above rms) exceeded only 1 per cent of the busiest hour. (These curves are based on material by B. D. Holbrook and J. T. Dixon of Bell Telephone Laboratories, and originally published in *The Bell System Technical Journal*.)

As an example of how the peak factor is used in determining input levels to a multichannel amplifier, take the case of a 24-channel system. From Fig. 35 the peak factor for 10 active channels is approximately 16.5 db. Now, if the maximum tolerable power into the common amplifier or radio transmitter is −17 dbm (a typical value), the power exceeded 1 per cent of the time must be reduced to a value of 16.5 db below −17 dbm, or to −33.5 dbm. For simplicity, assume that the power exceeded 1 per cent of the time is −30 dbm. Then the composite signal must be reduced to −33.5 dbm by 3.5 db of attenuation before being applied to the common amplifier or transmitter.

c. Telegraph and Data Loading. The channel capacity and transmission levels for a multiplex system are stated in terms of voice circuits, referenced to the rms value of a test tone (usually 1,000 cps). Because of their narrow bandwidth, however, up to about 26 ordinary telegraph or data signals can be applied to a single voice channel of a multiplex system. Since these multiple-tone signals have a greater average power than a voice signal, a loading factor must be considered.

In the calculation of the loading effect of a multiple-tone signal, peak power is used. As the number of telegraph or data signals is increased, the potential peak power level that the composite signal may reach also increases, even though the rms level is held constant by lowering the level of the individual tones. To avoid the possibility of peak overloading, a *peak factor* is used when the maximum permissible level of multitone signals applied to a single voice channel is calculated. The *absolute* peak factor is expressed mathematically as

$$\text{Absolute peak factor (in db)} = 3 + 10 \log N$$

where N is the number of single tones. Thus, for a single tone, the absolute peak factor is 3 db. Since the absolute peak factor can become quite high, it is not practical to build high-density multiplex systems that are completely free from the possibility of peak overloading. In practice, it is more common to express the peak factor as the statistical probability that the peak power of a composite signal will exceed the sum of the rms value of the signal and the peak factor a given percentage of the time, usually 1 per cent of the busiest hour (see Fig. 35).

Once the peak factor is known, the theoretical maximum permissible level of the composite signal applied to a single v-f channel can be calculated from the following equation:

$$P_m = P_s + 10 \log N + K \tag{1}$$

where P_m = maximum permissible peak power of the composite signal, dbm
P_s = rms power of single tone, dbm
$10 \log N$ = rms power addition of N number of tones, db
K = peak factor of the composite signal, db

The value of P_m is determined by the input power limitation of the multiplex system. Hence, the value of P_m for a multiplex system with a v-f input level specified as -16 dbm would be -13 dbm, or 3 db above the -16 dbm (rms) level.

It should be noted that Eq. (1) is valid only when all tones are of equal amplitude. Where the tones have different levels, P_m is found by simply converting the power of each tone from dbm to milliwatts, adding the resulting powers, and then converting the sum from milliwatts to dbm. The result is the maximum permissible peak power of a composite signal containing tones of different levels.

The above discussion applies only to the loading of one typical v-f channel unit amplifier in a multiplex system. When loading more than one v-f channel with telegraph and data signals, additional peak loading factors must be considered, which may further limit the total number of such signals that can be applied to the system.

21. Voltage Requirements

Multiplex systems employing electronic tubes conventionally require either 24 or 48 volts for the tube filaments and 130 volts for the tube anodes or plates. These voltages were established originally by the battery plants used for early telephone and telegraph systems. Multiplex systems employing transistors require only a 24- or 48-volt supply. Voltages applied to multiplex systems must be closely regulated to maintain proper signal power levels specified for the system. See Chap. 20.

OPERATING STANDARDS[9,13]

Long-distance communications depend upon the cooperation and coordination of thousands of different communications systems. In the United States alone, there are more than 2,600 independent telephone systems in addition to the Bell System and the many military and private systems. Inherent in the development of multiplex systems, therefore, is the need to comply with existing standard practices to provide a uniform system of communications networks. This need has fostered the growth of agencies that develop and issue written standards, practices, and recommendations that govern the operating characteristics of multiplex systems. Most of these agencies do not have legal status, but because of the benefits derived from standardization, their recommendations are usually accepted by general or implied consent of most communications companies and manufacturers.

The standards or recommendations issued by these agencies provide the basis for preparing the technical specifications that prescribe the performance characteristics for a multiplex system. The particular standards adopted depend, of course, on the type of multiplex system, its intended use, and the performance requirements necessary to interconnect it with other systems. The following is a list of the most prominent agencies and organizations that issue standards, practices, recommendations, and regulations that govern the operating characteristics of multiplex systems in the United States.

1. International (Recommendations)
   a. International Telegraph and Telephone Consultative Committee (CCITT)

    b. International Radio Consultative Committee (CCIR)
2. Commercial and Industrial (Standards and Practices)
    a. American Telephone and Telegraph Company (Bell System Practices)
    b. Electronic Industries Association (EIA)
    c. American Association of Railroads (AAR, Communication and Signal Section)
3. Government (Regulations and Standards)
    a. Defense Communications Agency (DCA)
    b. Federal Communications Commission (FCC)
    c. Rural Electrification Administration (REA)

REFERENCES

1. Aaron, M. R., PCM Transmission in the Exchange Plant, *Bell System Tech. J.* vol. XLI, no. 1, pp. 99–141, January, 1962.
2. Albert, A. L., *Electrical Communication*, 3d ed., John Wiley & Sons, Inc., New York, 1950.
3. Bell Telephone Laboratories, Incorporated, *Transmission Systems for Communications*, 3d ed., Western Electric Co., Inc., Winston-Salem, N.C., 1964.
4. Black, H. S., *Modulation Theory*, D. Van Nostrand Company, Inc., Princeton, N.J., 1953.
5. Boyd, R. C., Redesign of Short-Haul Carrier Systems, *Bell Lab. Record*, vol. 41, no. 7, pp. 275–279, July–August, 1963.
6. Combellick, T., Synchronization of Single-sideband Carrier Systems for High-speed Data Transmission, *IRE Trans. Commun. Systems*, vol. CS-7, no. 1, pp. 110–114, June, 1959.
7. Cowan, F. A., Modulation in Communication, *Trans. AIEE*, vol. 66, pp. 792–796, 1947.
8. Davis, C. G.: An Experimental Pulse Code Modulation System for Short-haul Trunks, *Bell System Tech. J.*, vol. XLI, no. 1, pp. 1–24, January, 1962.
9. Defense Communications Agency, *DCS Engineering-installation Standards Manual*, DCA Circular 175-2A, Government Printing Office, Washington, D.C., April, 1963.
10. Graham, R. S., W. E. Adams, R. E. Powers, and F. R. Bies, New Group and Supergroup Terminals for L Multiplex, *Bell System Tech. J.*, vol. XLII, no. 2, pp. 223–278, March, 1963.
11. Hallenbeck, F. J., and J. J. Mahoney, Jr., The New L Multiplex—System Description and Design Objectives, *Bell System Tech. J.*, vol. XLII, no. 2, pp. 207–221, March, 1963.
12. Holbrook, D. B., and J. T. Dixson, Load Rating Theory for Multi-channel Amplifiers, *Bell System Tech. J.*, vol. XVIII, pp. 624–644, October, 1939.
13. International Telegraph and Telephone Consultative Committee, *Red Book*, vol. III, IInd Plenary Assembly, New Delhi, 1960.
14. International Telephone and Telegraph Corporation, *Reference Data for Radio Engineers*, 4th ed., American Book—Stratford Press Incorporated, New York, 1956.
15. Lenkurt Electric Co., Inc., Carrier Channels for Telegraph and Data, Parts 1 and 2, *Lenkurt Demodulator*, vol. 7, nos. 10 and 11, October and November, 1958.
16. Lenkurt Electric Co., Inc., Coordination between Carrier Systems, *Lenkurt Demodulator*, vol. 11, no. 8, September, 1962.
17. Lenkurt Electric Co., Inc., Hybrids, *Lenkurt Demodulator*, vol. 13, no. 1, January, 1964.
18. Lenkurt Electric Co., Inc., Multiplexing and Modulation in Carrier Telephone Systems, Parts 1 and 2, *Lenkurt Demodulator*, vol. 7, no. 12, and vol. 8, no. 1, December, 1958, and January, 1959.
19. Lenkurt Electric Co., Inc., Signaling in Carrier Channels, *Lenkurt Demodulator*, vol. 12, no. 3, March, 1963.
20. Lenkurt Electric Co., Inc., Speech Loading of Carrier Systems, *Lenkurt Demodulator*, vol. 8, no. 8, August, 1959.
21. Lenkurt Electric Co., Inc., The Theory and Use of Compandors in Voice Transmission Systems, *Lenkurt Demodulator*, vol. 13, no. 10, October, 1964.
22. Lenkurt Electric Co., Inc., The Transmission of PCM over Cable, *Lenkurt Demodulator*, vol. 12, no. 1, January, 1963.
23. Lenkurt Electric Co., Inc., The Universal Voice Channel, *Lenkurt Demodulator*, vol. 13, no. 3, March, 1964.
24. Lenkurt Electric Co., Inc., Time-division Multiplex—New Promise for Old Technique? *Lenkurt Demodulator*, vol. 11, no. 11, November, 1962.
25. Wilson, L. H., Exchange-trunk Carrier for Plant Expansion, *Autom. Elec. Tech. J.*, vol. 9, no. 5, pp. 180–187, January, 1965.

Chapter 11

WIRE AND CABLE TRANSMISSION CHARACTERISTICS

JACK H. WOLFF, *Radio Corporation of America*

1. Introduction

Wire and cable transmission lines are utilized in communication networks to transmit frequencies from near zero cycles for telegraph, low-speed data, and signaling to several megacycles for high-capacity carrier systems and television. The principal types of available wire lines may be classified as follows:

1. Open-wire
2. Multipair cable:
    a. Aerial
    b. Buried
    c. Underground (in conduit)
    d. Submarine (river and shallow-water crossings)
3. Special facilities:
    a. Coaxial cable including submarine
    b. Spiral-four (disk-insulated)
    c. Video cable

These facilities might also be categorized on the basis of their electrical or physical characteristics or their intended application. The choice of the most suitable wire line is dependent on the transmission criteria and economic considerations. In planning a new wire-line installation, consideration should be given to future usage, i.e., circuit growth and changing transmission requirements, in addition to meeting the basic transmission objectives. Meeting these objectives requires a knowledge of the attenuation, phase shift, and characteristic impedance, as well as the attenuation and phase distortion characteristics of these facilities. It requires knowing their primary parameters of resistance, inductance, capacitance, and leakage conductance. Cross-talk couplings, loss variations, and line stability must be determined. The latter involves engineering of the line facilities for the worst expected weather conditions. An analysis of both natural and man-made interference expected to be induced into the transmission line is required if acceptable S/N ratios are to be obtained. Finally, considerations must be given to the nature of the terrain and geographical location of the proposed installation, as this will affect the type of wire line to be used, the interference sources, and the economics of the installation.

WIRE TYPES

2. Open-wire

"Open-wire" refers to bare line conductors supported on insulator-equipped cross-arms which in turn are mounted on poles spaced approximately 130 ft apart (40 poles per mile). Compared with cable, open-wire has the one outstanding advantage of having substantially lower transmission loss. However, it is much more susceptible to mechanical damage from storms and man action and to electrical interference from other communication lines, power lines, and all other sources of natural or man-made electrical noise.

Wet weather will increase the attenuation of open-wire lines, as will snow and ice conditions. Economically, open-wire should be considered only where the circuit requirements are very small (less than 10 circuits) and comparable cable costs would be prohibitive. They are primarily of importance in this chapter because there is currently a substantial mileage of open-wire plant in existence.

a. Types of Open-wire. Three types of open-wire facilities are in general use: hard-drawn copper, high-tensile-strength steel wire, and copper-clad steel wire (Copper-weld).

Hard-drawn copper wire is used where its advantages of high conductivity and corrosion resistance outweigh its high relative cost. Typical applications are on long,

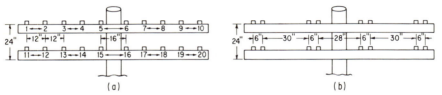

(a)   (b)

Fig. 1. Pole line arrangements.

high-quality toll circuits and in highly industrialized areas where atmospheric conditions would lead to corrosion of other wire types.

The most commonly used sizes of hard-drawn copper wire are 104, 128, and 165 mils in diameter.

High-tensile-strength steel wire is the most durable mechanically of the three types in general use. This allows the use of longer pole spans, thus reducing installation and pole line costs. Because of its relatively low conductivity, it is normally utilized only on relatively short circuits. Steel wire is generally galvanized with one of three weights of zinc per square foot of wire surface, i.e., grade A, 0.8 oz; grade B, 1.6 oz; and grade C, 2.4 oz. This in turn results in the various conductivities and breaking strengths, depending on the grade used. The nominal diameter of steel wire is generally 109 mils.

Copper-clad steel wire is high-strength or extra-high-strength steel wire surrounded by copper. It has either a 30 or 40 per cent conductivity ratio compared with pure copper wire (annealed copper standard). The physical and electrical characteristics of copper-clad steel are a compromise between the superior strength characteristics of steel and better electrical characteristics of copper. The attenuation characteristics are superior to steel wire, particularly at the higher frequencies, where the skin effects become pronounced. This type of open wire is available in nominal 104, 128, and 165 mil diameters.

b. Pole Arrangements. A typical pole arrangement for an open-wire voice-frequency system is shown in Fig. 1a, while Fig. 1b shows a typical pole arrangement for a carrier system with a maximum frequency of approximately 150 kc. The standard side circuits are connected to insulators 1–2, 3–4, 5–6, etc., in Fig. 1a. The phantom circuits, which are derived from the side circuits, would be associated with the wires connecting to insulator groups 1–4, 7–10, 11–14, 17–20. It should be noted that the spacing between insulators 5–6 and 15–16 is larger than the others, resulting in different transmission characteristics for these *pole pair side* circuits. Similarly, the *pole pair phantom* circuit associated with the wires on insulators 5–6 and 15–16 also has different transmission characteristics from the nonpole pair phantoms.

Loading is not used on open-wire circuits principally because of the added transmission instability which may be introduced by loading during periods of "wet weather."

As indicated in the typical pole line arrangement for a carrier system in Fig. 1*b*, there is closer spacing between the wires comprising a side circuit and greater separation between individual circuits than for the voice-frequency arrangement shown in Fig. 1*a*. This is necessary, since at carrier frequencies the crosstalk coupling becomes greater. For this reason phantom circuits are not generally derived at frequencies above the voice range.

3. Multipair Cables

The four basic types of multipair cables are (1) aerial cable placed on poles or other aboveground structures, (2) cable buried directly in the ground, (3) underground cable placed in conduit, and (4) submarine cable* for shallow stream, river, and lake crossings or for use in swampy areas. All these cables normally utilize either paper (pulp or tape) or polyethylene for the conductor insulation.

Aerial cable installations are cheapest on a per-mile basis for mountainous terrain. When additions or changes are required to existing plant, aerial cables are more flexible and fault location is easier, faster, and less costly. They have the disadvantage of high vulnerability to severe damage from storms, hurricanes, tornadoes, floods, earthquakes, and man-caused action. Since aerial cables are subject to greater temperature variations than buried or underground cables, this must be considered in the transmission engineering of these facilities.

Buried cable is relatively free from the catastrophic effects to which aerial cable is subject. It can be installed faster than aerial cable and has lower annual operating and maintenance costs, as well as reduced right-of-way problems.

Underground cable in conduit is expensive in first cost compared with direct-burial cable, but it offers maximum protection to the cable, longer life, and more flexibility if additions are required.

4. Special Facilities

This category includes the following wire and cable transmission facilities:

Coaxial cables including submarine
Spiral-four (disk-insulated cable)
Video cable

These facilities, although widely used, are considered for purposes of this chapter to be too specialized in application to be extensively discussed.

Coaxial cables are basically used for wideband communication circuits and are usually economically justified only when a very high number of channels (over 600) and/or television programs are to be transmitted.

Spiral-four cable consists of 16-gage disk-insulated individually shielded quads. Each quad is composed of four wires placed at the corners of a square. This cable is used primarily as an intermediate and entrance cable for high-frequency open-wire carrier systems because of its low-attenuation characteristics.

Video pairs consist of two 16-gage copper wires each insulated with expanded polyethylene. The conductors are assembled with two glass-fiber-reinforced polyethylene insulators. This pair is covered with an inner longitudinal copper tape and an outer spiral copper tape. The video pairs are used primarily for TV transmission

---

* This should not be confused with submarine cable for ocean and other deep-water crossings, which is generally of a coaxial-type construction.

between exchange area points in the television network and between studios and transmitters.

TRANSMISSION CHARACTERISTICS

5. General

The electrical characteristics of wire and cable transmission lines are generally termed their primary and secondary parameters. The primary parameters are the distributed resistance, inductance, leakage conductance, and capacitance per unit length. The secondary parameters consist of the characteristic impedance Z_0, attenuation α, and phase change β. The propagation constant γ is a complex quantity equal to $\alpha + j\beta$.

The primary parameters of a transmission line can be approximately calculated from the physical line characteristics, computed from the measured secondary parameters, or measured directly. The secondary parameter can be calculated from the primary ones if they are known, measured directly with appropriate instrumentation or indirectly from open- and short-circuit measurements of the line.[1-4,30]

6. Primary Electrical Cable Parameters

The primary electrical cable parameters of resistance, inductance, leakage conductance, and capacitance can be approximately determined from the physical characteristics of the transmission line or by direct measurement.

7. Resistance

In the voice-frequency range the d-c line resistance is approximately the same as its a-c resistance and can be computed readily.[18] The calculated value of d-c loop resistance for copper conductors at 68°F is

$$R_{dc} = \frac{0.1095}{d^2} \tag{1}$$

where R_{dc} = resistance, ohms per loop mile
d = conductor diameter, in.

As the frequency is increased above the voice-frequency range, the "skin" and "proximity"* effects tend to increase the a-c resistance of the wires above the d-c resistance. A third factor which increases the a-c resistance is the eddy-current losses in the surrounding pairs and sheath. The a-c resistance thus becomes

$$R_{ac} = R_{dc} + \Delta R_s + \Delta R_p + \Delta R_i \tag{2}$$

where R_{dc} = d-c resistance
ΔR_s = incremental resistance due to skin effect
ΔR_p = incremental resistance due to proximity effect
ΔR_i = incremental resistance due to eddy currents in other pairs and sheath

Computation of the a-c resistance becomes complex, and various approximations for estimating this resistance have been developed.[4,8-12] Typical resistance values of various line facilities as a function of frequency are shown in Fig. 2.

Conductor resistance also varies with line temperature in accordance with the following relation:

$$R_1 = R_0[1 + K_0(T_1 - T_0)] \tag{3}$$

R_0 and T_0 represent the known resistance and temperature, and R_1 is the unknown

---

* This effect is negligible in open-wire lines.

resistance at temperature T_1. The temperature coefficient of resistance K_0 as well as the conductor resistance is a function of frequency. In the voice-frequency range $K_0 \cong 0.00222$ per degree Fahrenheit, and in the high-frequency range (above approximately 100 kc) $K_0 = 0.00111$ per degree Fahrenheit.

The temperature range for buried cables is normally assumed to vary from 32 to 78°F and 0 to 110°F* for open-wire and aerial cables with an annual mean temperature of 55°F.

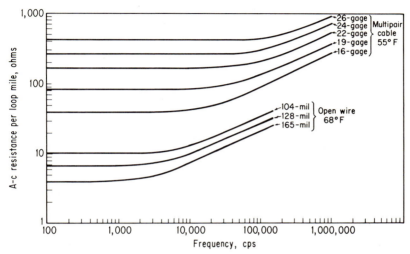

FIG. 2. Typical resistance values of wire and cable facilities.

8. Capacitance

Unlike resistance, the mutual capacitance between the two wires of a pair varies very little with frequency and temperature.

The mutual capacitance of an open-wire pair depends primarily on the wire spacing, number and type of insulators, and the air dielectric, which varies with weather conditions. In cable, the mutual capacitance between two conductors of a cable pair is dependent on the wire diameter, the type of insulation and thickness, the spacing between wires, and the dielectric constants of the insulation and air space between conductors. The mutual capacitance of a cable pair is further influenced by the presence of the surrounding wires which effectively form an electrostatic shield, as well as any other metallic shields which are present as a basic part of the cable construction. Exact computations of the mutual capacitance of a cable are complex; however, various simplifications are possible to allow fairly accurate calculation of this primary cable parameter.[6,8,18]

The nominal capacitance of an open-wire pair is 0.01 μf/mile. This will vary with the spacing between wires and increase under wet-weather conditions. Large increases in capacitance will also be caused by ice deposits. Nominal cable capacitances range from 0.06 to 0.09 μf/mile depending on specific applications. Nonquadded cables intended for exchange area service generally are designed for a nominal 0.083-μf/mile capacitance while for toll quality usage a 0.066-μf/mile capacitance is used as the design objective.

* The maximum temperature of open-wire and aerial cable may be 25°F greater than the ambient-air temperature.

Table 1. Typical Inductance Values of Wire Line Facilities

Transmission facility	Inductance mh/mile	
	1 kc	100 kc
Open wire (copper):		
104 mil:		
6-in. spacing.....................	3.2	3.1
8-in. spacing.....................	3.4	3.3
12-in. spacing.....................	3.7	3.6
128 mil:		
6-in. spacing.....................	3.1	3.0
8-in. spacing.....................	3.3	3.2
12-in. spacing.....................	3.5	3.4
165 mil:		
6-in. spacing.....................	2.9	2.8
8-in. spacing.....................	3.1	3.0
12-in. spacing.....................	3.4	3.2
Cable:		
19 AWG (0.066 μf) paper..............	1.0	0.93
19 AWG (0.066 μf) polyethylene.......	1.2	1.1
19 AWG (0.083 μf) paper..............	0.78	0.73
19 AWG (0.083 μf) polyethylene.......	1.0	0.93
22 AWG (0.083 μf) paper..............	0.80	0.77
22 AWG (0.083 μf) polyethylene.......	1.0	0.98

9. Inductance

Inductance changes very little with frequency and temperature compared with resistance. A typical inductance of 1.2 mh/mile for a 19-AWG low-capacitance cable pair at 1,000 cycles decreases to approximately 1.0 mh/mile at 1 Mc. Table 1 shows typical values of inductance for various line facilities at 1,000 cycles and 100 kc. The inductance variation between different wire gages and types of cable pair insulation is also small.

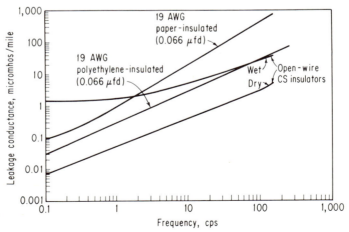

FIG. 3. Typical line leakage conductances.

10. Leakage Conductance

The leakage conductance of a wire pair varies with frequency, moisture, and the type of conductor insulation. Typical leakage conductance values for various line facilities are shown in Fig. 3.

11. Calculation of Primary from Secondary Parameters

When the attenuation, characteristic impedance, and phase change of a transmission line are known or can be obtained by measurement, the primary parameters can be computed from the following relations:[1,2]

$$R + j\omega L = Z_0(\alpha + j\beta) \tag{4}$$

$$G + j\omega C = \frac{\alpha + j\beta}{Z_0} \tag{5}$$

where R = series resistance, ohms per loop mile
  L = series inductance, henrys per loop mile
  G = leakage conductance, mhos/mile
  C = mutual capacitance, farads/mile
  Z_0 = characteristic impedance, ohms $(R_0 + jX_0)$
  α = attenuation, nepers/mile
  β = phase change, radians/mile
Equations (4) and (5) can be rearranged to solve for values of R, L, G, and C directly

$$R = \alpha R_0 - \beta X_0 \tag{6}$$

$$L = \frac{1}{\omega}(\beta R_0 + \alpha X_0) \tag{7}$$

$$G = \frac{\alpha R_0 + \beta X_0}{R_0{}^2 + X_0{}^2} \tag{8}$$

$$C = \frac{1}{\omega}\left(\frac{\beta R_0 - \alpha X_0}{R_0{}^2 + X_0{}^2}\right) \tag{9}$$

12. Calculation of Secondary from Primary Parameters

When the primary electrical transmission characteristics are known from either calculation or measurement, the secondary parameters can readily be calculated from the general transmission-line equations. The parameters used in the equations in this section are defined as follows:

  R = resistance, ohms per loop mile
  C = mutual capacitance, farads/mile
  L = series inductance, henrys per loop mile
  G = leakage conductance, mhos/mile
  ω = $2\pi f$
  f = frequency, cps

The attenuation α in nepers per mile is

$$\alpha = \sqrt{\tfrac{1}{2}[\sqrt{(R^2 + \omega^2 L^2)(G^2 + \omega^2 C^2)} + (RG - \omega^2 LC)]} \tag{10}$$

The phase shift β in radians per mile is

$$\beta = \sqrt{\tfrac{1}{2}[\sqrt{(R^2 + \omega^2 L^2)(G^2 + \omega^2 C^2)} - (RG - \omega^2 LC)]} \tag{11}$$

The resistive and reactive components of the characteristic impedance Z_0 in ohms are

$$R_0 = \sqrt{\frac{\sqrt{(R^2 + \omega^2 L^2)(G^2 + \omega^2 C^2)} + (RG + \omega^2 LC)}{2(G^2 + \omega^2 C^2)}} \qquad (12)$$

and

$$X_0{}^* = \pm \sqrt{\frac{\sqrt{(R^2 + \omega^2 L^2)(G^2 + \omega^2 C^2)} - (RG + \omega^2 LC)}{2(G^2 + \omega^2 C^2)}} \qquad (13)$$

The characteristic impedance Z_0 in ohms can also be obtained directly from the following relation

$$Z_0 = \sqrt{\frac{R + j\omega L}{G + j\omega C}} \qquad (14)$$

The general equations (10) to (13) can be simplified when approximate computations will suffice. At low frequencies where ω is small compared with R/L and G/C, the general equations reduce to

$$\alpha \cong \sqrt{RG} \qquad (15)$$

$$\beta \cong \frac{\omega L}{2R} \sqrt{RG} + \frac{\omega C}{2G} \sqrt{RG} \qquad (16)$$

$$R_0 \cong \sqrt{\frac{R}{G}} \qquad (17)$$

$$X_0 \cong \frac{\omega L}{2R} \sqrt{\frac{R}{G}} - \frac{\omega C}{2G} \sqrt{\frac{R}{G}} \qquad (18)$$

For the ordinary types of wire line facilities where G/C is usually much smaller than R/L, Eqs. (16) and (18) may be further reduced to

$$\beta \cong \frac{\omega C}{2} \sqrt{\frac{R}{G}} \qquad (19)$$

$$X_0 \cong - \frac{\omega C}{2G} \sqrt{\frac{R}{G}} \qquad (20)$$

Equations (15) to (20) are applicable to a frequency range which for most of the line facilities in this chapter is below voice frequencies.

For the middle frequencies where ω is large in relation to G/C but small compared with R/L, which is the general case for cable pairs at voice frequencies, Eqs. (10) to (13) simplify to

$$\alpha \cong \sqrt{\frac{\omega RC}{2}} \qquad (21)$$

$$\beta \cong \sqrt{\frac{\omega RC}{2}} \qquad (22)$$

$$R_0 \cong \sqrt{\frac{R}{2\omega C}} \qquad (23)$$

$$X_0 \cong - \sqrt{\frac{R}{2\omega C}} \qquad (24)$$

The high frequencies where ω is large compared with both R/L and G/C apply to cable pairs at multiplex and video frequencies, coaxial cables, and most of the open-

* X_0 has the same sign as $(GL - RC)$.

wire circuits from the voice to the multiplex range.   The general equations (10) to (13) for this frequency range reduce to

$$\alpha \cong \frac{1}{2} \sqrt{\frac{C}{L}} \left(R + \frac{GL}{C} \right) \tag{25}$$

$$\beta \cong \omega \sqrt{LC} \tag{26}$$

$$R_0 \cong \sqrt{\frac{L}{C}} \tag{27}$$

$$X_0 \cong \frac{1}{2\omega} \sqrt{\frac{L}{C}} \left(\frac{G}{C} - \frac{R}{L} \right) \tag{28}$$

Again since generally G/C will be much smaller than R/L, Eqs. (25) and (28) can be further simplified for the high-frequency range to

$$\alpha \cong \frac{R}{2} \sqrt{\frac{C}{L}} \tag{29}$$

$$X_0 \cong - \frac{R}{2\omega \sqrt{CL}} \tag{30}$$

13. Calculation of Phase Delay, Velocity of Propagation, Wavelength, and Envelope Delay

The phase constant β is the angle in radians that the phase of an unmodulated single-frequency wave is retarded in a unit length of line.   The transmission time of this single-frequency wave is phase delay, which is obtained from the following relation:

$$t_p = \frac{\beta \times 10^6}{\omega}$$

where t_p = phase delay, μsec/mile
β = phase change, radians/mile
ω = frequency, radians/sec
The velocity of propagation or phase velocity of the line is the reciprocal of the phase delay; i.e.,

$$v_p = \frac{\omega}{\beta} \tag{31}$$

where v_p is the velocity of propagation in miles per second.
The wavelength is given by

$$\lambda = \frac{2\pi}{\beta} \tag{32}$$

where λ is the wavelength in miles.   The envelope delay of a modulated wave is

$$t_e = \frac{d\beta}{d\omega} \tag{33}$$

where t_e is the envelope delay in seconds per mile.

SECONDARY PARAMETERS OF TRANSMISSION LINES

14. Attenuation

The attenuation-frequency characteristics of open-wire lines vary with many factors such as wire spacing, method of transposition, type of insulators, physical line length,

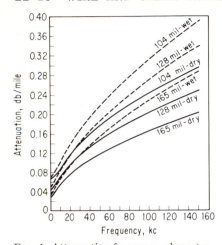

FIG. 4. Attenuation-frequency characteristics of typical open-wire lines. Calculated values at 68°F for hard-drawn copper, 53 pairs CS insulators per mile, 12-in. separation between wires.

condition of the facilities, and the wire size and type. These variables are exclusive of the variations due to the immediate weather conditions in effect. Typical values of attenuation during dry and wet conditions are shown in Fig. 4.*

In areas where ice, snow, and sleet are prevalent, the attenuation will increase beyond that of the wet-weather conditions owing to the lowered dielectric values between conductors. This is especially prevalent at the higher frequencies. For example, at 40 kc, a ⅜-in. radial ice thickness will result in an approximate 0.25-db/mile increase in attenuation over that of dry weather. At 150 kc the increased attenuation would be about 2.5 db/mile.

The smaller gage conductors (16, 19, 22, 24, 26 AWG)† used in cable cores compared with open-wire results in greater attenuation per unit length. The close spacing of the conductors in the cable core increases the mutual capacitance and therefore the attenuation.

Typical attenuation values of paper-insulated conductors through the carrier frequency ranges are shown in Fig. 5. Attenuation values for typical 19- and 22-AWG

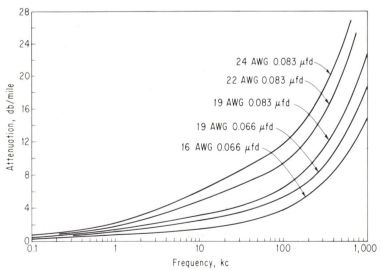

FIG. 5. Typical attenuation-frequency characteristics of paper-insulated cable pairs. Temperature 55°F.

* The dry- and wet-weather values are calculated for lines using 53 pairs of CS (Pyrex glass) insulators per mile. The dry weather values are also approximately correct for DP (double-petticoat) insulators. However, the attenuation difference between dry and wet weather conditions for DP insulators is approximately twice that for CS insulators.

† Although 10- and 13-AWG cable conductors have been used in the past, current utilization of these sizes is almost nil.

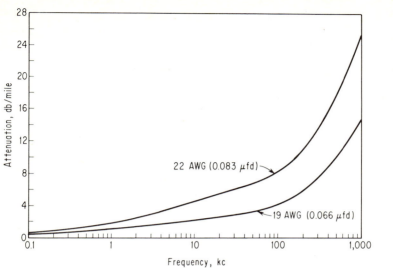

Fig. 6. Typical attenuation-frequency characteristics of polyethylene-insulated cable pairs. Temperature 55°F.

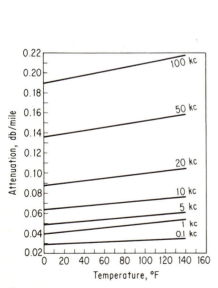

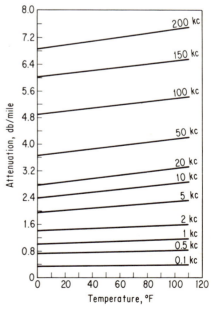

Fig. 7. Typical attenuation-temperature characteristics of 128-mil open-wire hard-drawn copper, 12-in. spacing.

Fig. 8. Typical attenuation-temperature characteristics of 19-gage, 0.066 μf/mile cable pairs.

polyethylene-insulated cables are plotted in Fig. 6. The attenuation characteristics of the paper- and polyethylene-insulated cables are approximately the same through the voice-frequency range to about 10 kc. At 100 kc polyethylene-insulated cable has approximately 1-db/mile lower attenuation, and at 1,000 kc polyethylene-insulated pairs have about 4.5-db/mile less attenuation.

Since all the primary cable parameters, particularly resistance, vary with temperature, the attenuation of wire line facilities will also change with temperature variations. Figure 7 indicates the approximate variation of attenuation of a 128-mil open-wire pair with temperature and Fig. 8 shows the relative variation in attenuation for a 19-AWG cable pair with temperature.

15. Phase Shift

The phase-shift–frequency characteristics of typical open-wire lines and nonloaded cables are shown in Fig. 9. Unlike the attenuation, the phase shift varies only slightly with temperature. For open-wire lines the difference in phase shift between dry and

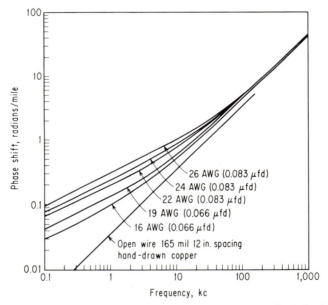

Fig. 9. Typical phase-shift–frequency characteristics of nonloaded cable.

wet weather is small, with a 0.03-radian/mile shift at 100 kc under extreme weather conditions.

16. Characteristic Impedance

Typical characteristic impedance-frequency values are indicated in Table 2 for open-wire and nonloaded cables. The variation of characteristic impedance with temperature is greatest at the lower frequencies. Above 100 kc this variation may usually be neglected. Table 2 also shows that above 100 kc the decrease in characteristic impedance with frequency is very small.

Table 2. Typical Values of Characteristic Impedance

Characteristic impedance Z_0*

Frequency, kc	Open-wire 165 mil‡	Cable†					
		16 AWG	19 AWG	19 AWG	22 AWG	24 AWG	26 AWG
		0.066 µf/mile		0.083 µf/mile			
0.2	714/−22.1	708/−44.1	1,001/−44.5	887/−44.6	1,270/−44.8	1,710/−44.8	2,203/−44.9
0.5	625/−10.7	449/−42.9	635/−43.9	562/−44.1	804/−44.5	1,083/−44.7	1,394/−44.7
1	608/−5.6	319/−40.7	450/−42.8	399/−43.2	569/−44.1	767/−44.3	988/−44.5
2	605/−2.9	230/−36.7	320/−40.8	283/−41.6	403/−43.2	542/−43.7	698/−44.1
5	600/−1.5	160/−27.2	208/−34.8	184/−37.0	257/−40.6	344/−41.7	442/−42.9
10	597/−1.1	136/−18.0	161/−27.0	137/−30.3	185/−36.4	246/−38.7	315/−40.9
20	594/−0.7	126/−11.4	138/−18.0	113/−21.5	139/−29.7	181/−33.3	226/−37.0
50	591/−0.4	122/−6.1	126/−9.6	100/−12.1	109/−17.6	132/−21.7	155/−27.1
100	590/−0.3	117/−4.4	122/−6.1	96/−7.8	102/−11.3	120/−14.0	132/−18.2
200	....	113/−3.0	117/−4.4	92/−5.6	97/−7.8	114/−9.1	122/−11.3
500	....	110/−2.1	113/−2.5	88/−3.3	92/−4.4	108/−5.3	116/−6.4
1,000	....	107/−1.6	110/−1.6	86/−2.0	90/−2.5	105/−2.7	112/−3.6

* Impedance in ohms, angle in degrees.
† Paper insulated, quadded, temperature 55°F.
‡ 12-in. spacing, 53 pairs of insulators per mile, dry weather, temperature 68°F.

LOADING

17. General

Loading is used on voice-frequency circuits to reduce the transmission loss of a transmission line by inserting series inductance (loading coils) into the line at regular intervals.* In addition to reducing the overall attenuation, loading also improves the attenuation and phase distortion characteristics. The addition of inductance to the line results in a higher line impedance and a lower velocity of propagation. Loading is no longer used on open wire, since it tends to decrease the transmission stability of this type of circuit.

When entrance cables are used to extend open-wire carrier circuits into urban areas, loading is used primarily for matching the impedance of the entrance cables to the open wire at the carrier frequencies.

On new installations economic justification for loading for a given transmission objective is dependent on cost trade-offs between the decreased wire gage made possible by its use and the reduction or elimination of repeaters vs. the cost of loading (material, installation, etc.). When decreased attenuation and or lower attenuation and phase delay distortion are required on existing circuits, loading should also be considered. In this case the cost trade-off would be between the cost of adding loading vs. the cost of repeater additions and the required attenuation and delay distortion-correction networks.

18. Load Coil Spacing

Loaded cables are designated by standard codes such as 19H-44 or 22B-88, where the first term denotes the cable conductor gage; i.e., 19 AWG, 22 AWG, etc., the letter designates the spacing between load coils (see Table 3); and the last term denotes the loading coil inductance in millihenrys, i.e., 44 or 88 mh.

Table 3. Spacing Code for Loading Coils

Code designation	Load coil spacing, ft
A	700
B	3,000
C	929
D	4,500
E	5,575
F	2,787
H	6,000
J†	640
X	680
Y	2,130

† Spiral-four cable.

The most commonly used cable loading spacings are the B, D, and H systems, i.e., 3,000-, 4,500- and 6,000-ft separations between loading coils. In practice these theoretical spacings cannot normally be adhered to because of geographical and man-made interferences (highway crossings, river crossings, highly developed urban areas, lakes, etc.). As a result load coil spacing limits have been established which allow deviations from the standard separations but will still maintain minimum transmission impairment.

For cables having the same mutual capacity per unit length and when repeaters

* This is generally referred to as "lumped" loading. "Distributed" loading is theoretically more desirable, but very limited attempts have been made because of practical considerations.[14,15]

are used or may be added at a future time, the following spacing limits* should generally be adhered to:

1. The deviation of the average spacing of a loading system shall not deviate from the standard spacing by more than ±2 per cent.

2. The deviation of any individual loading section shall not deviate from the average spacing by more than ±2 per cent.

3. Disregarding the signs of the differences between all loading sections and the average spacing, the average deviation shall not exceed ±½ of 1 per cent.

If no repeaters are used or contemplated for future use, the above limits may be relaxed.

When a spacing irregularity results in an objectionable transmission characteristic, various remedial measures are available. When the distance between load coils is too short, the section may be electrically lengthened by the use of building-out capacitors or the use of a section of cable (stub-cable) shunted across the main cable or by the series insertion of a cable section. The use of fractional loading coils should also be considered, but only if the transmission objectives cannot be met by alternate means. In the case of excess distance between two load coils, the correction consists of dividing it into two sections and building out one or both sections to the normal separation of the loading system used.

19. Loaded-line Terminations

Loaded lines are normally terminated either at the center of a loading coil (mid-coil termination) or at the center of the line between two load coils (mid-section or end-section termination), since only at these two points is the impedance in either direction the same. The mid-section termination for an H spacing (6,000 ft) would thus be 3,000 ft, which would be the length objective for the end sections. Various corrective practices are applied when the end-section objective cannot be met. Normally, building-out capacitors are used to lengthen the end section electrically to equal one-half of the normal section.

20. Cutoff Frequency

The addition of inductance to a cable results in its assuming the characteristics of a low-pass filter. The approximate theoretical cutoff frequency of a loaded cable can be determined from the following relation:[17]

$$f_c = \frac{10}{\sqrt{LC}} \tag{34}$$

where f_c = cutoff frequency, kc
  L = inductance of one loading coil plus the line inductance of one loading section, mh
  C = capacitance of one loading section, μf

Nominal values of f_c for the common types of loading are shown in Table 4. If the spacing between the loading points and the inductance of the loading coils is varied, the cutoff frequency may be varied. Practically, a cutoff frequency above 30 kc† is not feasible, thus limiting the use of loading generally to voice-frequency circuits. The "effective" or useful frequency range of a loaded line is generally considered to be restricted to frequencies below approximately 0.8 of the cutoff frequency.

---

* CCITT load coil spacing recommendations[5] are that the nominal load coil spacing in a repeater section should be equal to the theoretical value within ±2 per cent. The actual load coil spacing may differ by 10 meters (approximately 33 ft) from the nominal spacing.

† The cutoff frequency of spiral-four cable is approximately 28 kc.

Table 4. Typical Values of Cutoff Frequency for Various Loading Systems

Loading system	Mutual capacitance, μf/mile	Nominal cutoff frequency, cps
B-88	0.066	5,500
B-88	0.083	4,900
H-88	0.066	3,900
H-88	0.083	3,500
H-44	0.066	5,500
H-44	0.083	4,900
D-88	0.083	4,000

21. Transmission Parameters of Loaded Cables

The exact computations of the attenuation, phase shift, and characteristic impedance as a function of frequency of loaded cable facilities are complex procedures.

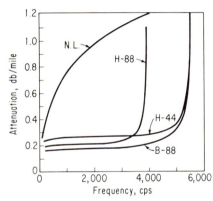

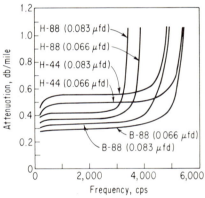

FIG. 10. Typical attenuation-frequency characteristics of 16-gage, 0.066 μf/mile, polyethylene-insulated loaded cable. Temperature 68°F. (*Courtesy of Anaconda Wire and Cable Company.*)

FIG. 11. Typical attenuation-frequency characteristics of 19-gage polyethylene-insulated loaded cable. Temperature 68°F. (*Courtesy of Anaconda Wire and Cable Company.*)

However, for frequencies well below the cutoff frequency, such as 1,000 cycles, these
transmission parameters may be approximated by the following simplified relations:

$$\alpha \cong \frac{R}{2} \sqrt{\frac{C}{L}} \tag{35}$$

$$\beta \cong \omega \sqrt{LC} \tag{36}$$

$$Z_0 \cong \sqrt{\frac{L}{C}} \Big/ -\tan^{-1} \frac{R}{2\omega L} \tag{37}$$

where α = attenuation, nepers/mile
β = phase shift, rad/mile
Z_0 = characteristic impedance, ohms
R = cable loop resistance plus load coil resistance, ohms/mile
L = cable loop inductance plus load coil inductance, henrys/mile
C = mutual capacitance, farads/mile

The approximate equations (35) to (37) are based on the assumption that the resistance and inductance of the load coils are uniformly distributed. The load coil constants must therefore be converted to per-mile values. Thus for the H spacing

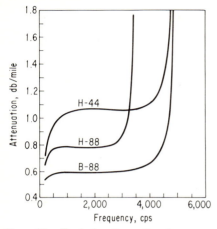

FIG. 12. Typical attenuation-frequency characteristics of 22-gage (0.083 μf/mile) polyethylene-insulated loaded cable. Temperature 68°F. (*Courtesy of Anaconda Wire and Cable Company.*)

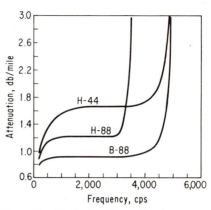

FIG. 13. Typical attenuation-frequency characteristics of 24-gage (0.083 μf/mile) polyethylene-insulated loaded cable. Temperature 68°F. (*Courtesy of Anaconda Wire and Cable Company.*)

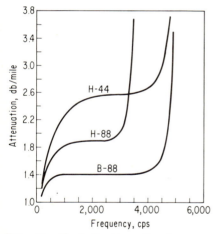

FIG. 14. Typical attenuation-frequency characteristics of 26-gage (0.083 μf/mile) polyethylene-insulated loaded cable. Temperature 68°F. (*Courtesy of Anaconda Wire and Cable Company.*)

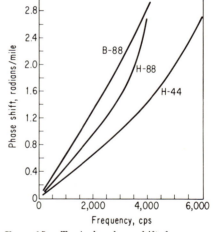

FIG. 15. Typical phase-shift–frequency characteristics of 16-gage (0.066 μf/mile) polyethylene-insulated loaded cable. Temperature 68°F. (*Courtesy of Anaconda Wire and Cable Company.*)

(6,000 ft) the coil values of resistance and inductance must be adjusted by a factor of 1.136 for use in the above formulas.

Typical values of attenuation and phase shift are plotted in Figs. 10 to 19 over the voice-frequency range for polyethylene-insulated cables using the most common loadings, i.e., H-44, H-88, and B-88. Table 5 shows typical characteristic impedance

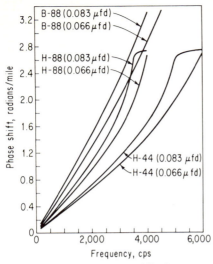

Fig. 16. Typical phase-shift–frequency characteristics of 19-gage polyethylene-insulated loaded cable. Temperature 68°F. (*Courtesy of Anaconda Wire and Cable Company.*)

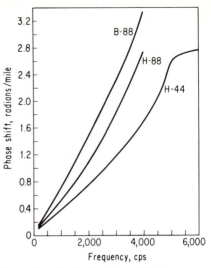

Fig. 17. Typical phase-shift–frequency characteristics of 22-AWG (0.083 μf/mile) polyethylene-insulated loaded cable. Temperature 68°F. (*Courtesy of Anaconda Wire and Cable Company.*)

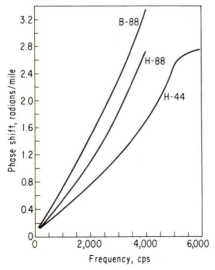

Fig. 18. Typical phase-shift–frequency characteristics of 24-AWG (0.083 μf/mile) polyethylene-insulated loaded cable. Temperature 68°F. (*Courtesy of Anaconda Wire and Cable Company.*)

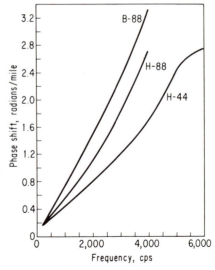

Fig. 19. Typical phase-shift–frequency characteristics of 26-AWG (0.083 μf/mile) polyethylene-insulated loaded cable. Temperature 68°F. (*Courtesy of Anaconda Wire and Cable Company.*)

Table 5. Typical Impedance-frequency Characteristics of Polyethylene Loaded Cables*

Characteristic impedance (68°F) †

Loading and wire gage	Mutual capacity, μf/mile	Frequency, cps							
		200	500	1,000	2,000	3,000	4,000	5,000	6,000
16H-44	0.066	906/−21.3	804/−10.1	795/−5.2	833/−2.6	922/−1.8	1,121/−1.6	1,819/−2.9	1,793/−87.2
16B-88	0.066	1,569/−8.1	1,548/−3.3	1,563/−1.6	1,648/−0.8	1,827/−0.6	2,220/−0.6	3,574/−1.3	3,661/−88.4
16H-88	0.066	1,155/−13.4	1,110/−5.7	1,130/−2.8	1,267/−1.5	1,695/−1.3	4,914/−81.8		
19H-44	0.066	1,109/−30.2	864/−17.5	814/−9.6	874/−3.9	926/−3.3	1,124/−2.8	1,821/−4.9	1,779/−85.5
19B-88	0.066	1,630/−13.4	1,560/−5.7	1,567/−2.8	1,649/−1.4	1,828/−1.0	2,222/−0.9	3,577/−1.9	3,654/−87.8
19H-88	0.066	1,278/−21.5	1,137/−10.2	1,138/−5.2	1,271/−2.6	1,698/−2.2	4,745/−77.8		
19H-44	0.083	986/−30.5	764/−17.9	719/−9.8	753/−5.0	863/−3.4	1,165/−3.4	3,451/−65.2	989/−87.7
19B-88	0.083	1,449/−13.5	1,387/−5.7	1,398/−2.9	1,495/−1.4	1,720/−1.0	2,327/−1.2	8,288/−76.6	1,981/−89.0
19H-88	0.083	1,136/−21.7	1,009/−10.3	1,017/−5.2	1,180/−2.7	1,884/−3.1	1,729/−87.7		
22H-44	0.083	1,320/−37.0	911/−27.1	778/−17.3	776/−9.3	879/−6.4	1,182/−6.4	2,722/−61.4	977/−86.0
22B-88	0.083	1,610/−21.7	1,424/−10.3	1,411/−5.3	1,501/−2.6	1,725/−1.8	2,335/−1.9	7,625/−72.5	1,975/−88.5
22H-88	0.083	1,399/−30.5	1,090/−17.8	1,044/−9.7	1,192/−5.0	1,898/−5.6	1,712/−86.1		
24H-44	0.083	1,645/−39.8	1,084/−32.8	860/−23.8	803/−14.1	891/−9.9	1,184/−9.9	2,308/−57.1	976/−84.0
24B-88	0.083	1,830/−27.9	1,484/−15.2	1,428/−8.1	1,505/−4.0	1,727/−2.8	2,335/−2.8	7,119/−67.9	1,975/−88.0
24H-88	0.083	1,692/−35.3	1,208/−24.2	1,085/−14.6	1,203/−7.7	1,893/−8.4	1,707/−84.2		
26H-44	0.083	2,070/−41.7	1,334/−37.1	1,003/−30.2	864/−20.3	922/−15.1	1,192/−15.1	1,923/−54.6	974/−81.1
26B-88	0.083	2,180/−33.3	1,608/−21.4	1,468/−12.3	1,516/−6.3	1,732/−4.3	2,337/−4.3	6,483/−63.7	1,974/−87.4
26H-88	0.083	2,098/−38.7	1,411/−30.5	1,172/−20.8	1,230/−11.8	1,886/−12.8	1,693/−81.5		

* By courtesy of Anaconda Wire and Cable Company.
† Magnitude in ohms, angle in degrees.

values for these facilities. Figure 11 indicates the attenuation-frequency character-istics for both low- and high-capacitance (0.066 and 0.083 μf/mile) 19-AWG cable, and Fig. 16 shows corresponding values of the phase-shift–frequency characteristics.

CABLE AND WIRE TRANSMISSION CONSIDERATIONS

22. Two-wire Circuits

Two-wire voice-frequency circuits utilize a single pair of wires for both directions of transmission. Their primary advantage is the obvious savings in line facilities. Consideration is also given to their use when a limited number of circuits on existing open-wire or cable plant are required and new plant additions are not economically justified. The maximum distance for these circuits on open-wire is generally 300 to 700 miles, depending on the conductor size, and 150 miles for 19-gage loaded cable when two-wire repeaters are utilized. These distance limitations are due primarily to line balancing considerations which directly affect the singing margin and echo loss of these circuits and the instability resulting from additional repeaters.

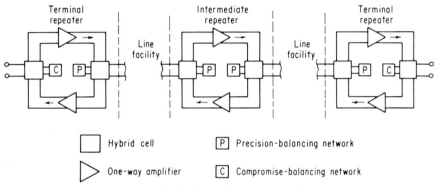

FIG. 20. Typical two-wire repeatered line.

Figure 20 shows a repeatered two-wire line consisting of two terminal repeaters and one intermediate repeater. Since the impedance of the balancing network should match that of the line, the transmitted power through the repeater is nominally 6 db less than the gain of the amplifier itself. Depending on the type of intermediate two-wire repeater used, the maximum transmit level is generally no greater than 10 db* referred to the zero transmission level and the gain of the repeater is normally limited to a maximum of 25 db. The gain of terminal repeaters is lower because of the gener-ally higher impedance mismatch between their compromise networks and the terminal equipment to which they connect.

Unlike four-wire circuits, which have only one complete path for echo currents, two-wire circuits have a large number of return paths, which increase with the number of repeaters and limit the practical use of two-wire circuits to relatively short distances.

Near-end crosstalk, which affects the gains at which the repeaters may operate and thus the net loss of the circuits, is also a limiting factor on two-wire repeatered circuits. Near-end crosstalk is generally the controlling crosstalk factor on these circuits.

23. Return Loss and Singing Margin

Since, in practice, perfect balance cannot be obtained between a line and the balanc-ing network designed to simulate the line, there is a finite loss between the output and

* An output of +6 db for V-type repeaters and +3 db for 22-type repeaters is generally used if allowed by other transmission considerations.

input repeater circuits.   There is thus a closed transmission path through the amplifiers and hybrids of a two-wire repeater.   If the gains in this path exceed the losses, sustained oscillation, or "singing," may result.

The return loss of a two-wire repeater is a measure of the degree of unbalance that exists at a frequency between the balancing network and line impedance.   The design and adjustment objectives are to keep the return loss (RL) as high as possible. The return loss in decibels, is

$$RL = 20 \log_{10} \frac{Z_N + Z_L}{Z_N - Z_L} \tag{38}$$

where Z_N = magnitude of the balance network impedance
$\quad\quad Z_L$ = magnitude of the line impedance
If in a repeater the sum of the gains of the two amplifiers exceeds the return losses at the two ends of that repeater, singing will occur at any frequency.   To guard against this, the repeater gains are generally reduced enough below the level where singing will occur.   This singing margin is generally 10 db.

This limits the repeater gain (sum of the two amplifiers) to 10 db less than the return losses at the two ends of the repeater.

The gain of the amplifiers, as well as Z_N and Z_L in Eq. (38), varies with frequency. Thus, singing margin computations must be made at various frequencies in the voice-frequency range to determine the point in the frequency range at which the poorest balance conditions occur.

24. Negative-impedance Repeaters

The negative-impedance or E-type repeater[19,20,32] was developed basically for use as a two-wire repeater in exchange or local telephone cable plant.   This repeater functions on the concept that, when a negative impedance is introduced into a transmission

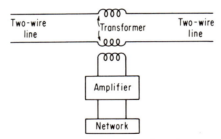

Fig. 21. Block diagram of negative impedance repeater.

line, it increases the current in that line by reducing the overall impedance.   This results in a transmission gain, thus effectively producing the same result as a normal amplifier, with the added advantage of providing amplification in both directions.

Since this type of repeater is bridged across the line, as shown in Fig. 21, failure of the amplifier will not result in complete loss of the transmission, although it will be substantially reduced.   A further advantage of this repeater is that it requires no bypass or other arrangement if d-c or low-frequency signaling is transmitted on the same line.

The "gain" or maximum magnitude of negative impedance that can be inserted into a transmission line is limited by noise, crosstalk, singing margins, repeater overload, and echo loss considerations.   Generally, singing is the controlling limitation on gain. Usually the E-type repeater can be operated at a gain sufficient to reduce the loss of a loaded line to one-half of its nonrepeatered value and the loss of a nonloaded line to one-third of its nonrepeatered value.   The maximum gain of the repeaters is generally limited to 8 to 10 db.

25. Four-wire Circuits

Unlike two-wire circuits, four-wire repeatered lines have only one singing path, near-end crosstalk can be practically eliminated through the physical separation of the two directions of transmission, and the overall circuit attenuation can be easily regulated on each separate transmission path. These factors allow longer transmission distances for four-wire circuits. For voice transmission over open-wire, repeater spacings of 150 to 350 miles are generally used, while for cable, repeater spacings varying from about 40 to 90 miles are used depending on the specific line facilities available.

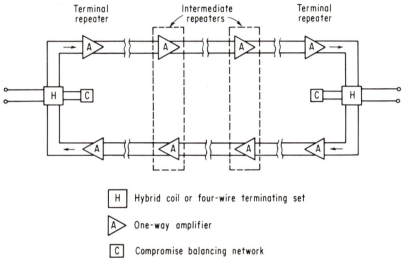

FIG. 22. Typical four-wire repeatered line.

Figure 22 shows a typical four-wire repeatered line. Repeaters in this type of circuit can operate at a maximum output of 10 db above zero reference level and a nominal minimum input of 25 db below zero reference level. This allows a maximum gain of 35 db. The spacing of four-wire repeaters is based primarily on crosstalk and noise considerations.

INTERFERENCE WITH WIRE AND CABLE TRANSMISSION

26. General

The prime interference sources to wire communication systems may be divided into natural and man-made origins. The principal sources of natural interference are lightning, atmospheric static, and thermal noise. Man-made interference may originate within the communication plant through crosstalk from other systems or direct coupling from terminal station equipment such as power supplies, repeaters, carrier equipment, relays, etc. Man-made interference also originates from sources external to the communication plant, primarily inductive interference from adjacent and nearby power sources and from external sources of RF energy.

27. Lightning

Although relatively infrequent, lightning interference is the most severe from the point of view of actual physical damage and consequent interruption to wire line facilities and from induced electrical interference into the wire line.   Direct lightning strokes to a cable produce severe damage but are relatively infrequent.   Induced current in cables due to lightning occur more frequently than direct hits, but their effect is generally much smaller.

Physical damage in the case of open wire can result in insulation breakdown of wooden poles and insulators.   In the case of cable, a direct stroke of lightning to earth may result in holes being burned in the cable sheath by the extremely high currents,* breakdown and burning of the dielectric, and actual fusing of the conductors.   Holes in the cable sheaths will allow the subsequent entry of water and its resultant insulation and conductor damage, either directly or by eventual corrosion.

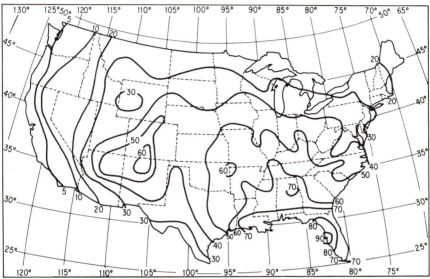

Fig. 23. Average annual thunderstorm days.   (*Weather Bureau, Office of Hydrologic Director, Thunderstorm, Rainfall, Hydrometeorological Rept. 5, Pt. 2, Aug. 22, 1945.   U.S. Weather Bureau Library No. 77.37/73, U 587 H, No. 5, Pt. 2.*)

The characteristics of lightning itself, i.e., the current amplitude, wave shape of the stroke, and frequency of occurrence, determine the possibility of damage and severity. In the case of cable the resistance of the cable sheath and the impulse breakdown voltage of the core-to-sheath insulation also affect the frequency and extent of physical damage.   For buried-cable plant the earth resistivity in the vicinity of the cable also affects the frequency of trouble occurrence.   For example, a cable with 0.5-ohm/mile sheath resistance will have theoretical trouble expectancy of approximately five failures per 100 miles for 10 thunderstorm days when the earth resistivity is 5,000 meter-ohms compared with approximately 0.5 failure for an area with an earth resistivity of 1,000 meter-ohms.

Methods for estimating the number of cable faults per year have been developed[7,21,22] based on data as shown in the map in Fig. 23,[23] which indicates the frequency of thunderstorm occurrence in various areas.

As it is impracticable to prevent lightning surges from reaching wire lines, one of the basic principles of protection is to provide a low-impedance ground path of adequate current-carrying capacity.   In cables, increased core-to-shield dielectric strength and

* About one-third of the lightning strokes to ground have crest currents in excess of 30,000 amp.

increased shield thickness, thus lowered shield resistance, are effective in reducing the trouble expectancy. For buried cables, external shield wires are used to reduce the currents reaching the cable.

28. Atmospheric Static

Atmospheric static, which is impulsive in character, is present at all times and affects wire lines from the very low frequencies to many megacycles, particularly during periods of severe and nearby thunderstorms. The static field spectrum has its maximum intensity at 10 kc. A generally accepted design objective for the static noise induced into wire lines is that it should not exceed approximately 26 dba at zero level more than 1 per cent of the time. This is equivalent to about 36 hr during the 5-month static season. The static field strength is a function of the thunderstorm incidence (TSI).

29. Thermal Noise

Thermal noise, resulting from the agitation of electrical charges in wire conductors, is usually a negligible contributing factor to the total noise existing on a transmission line at voice frequencies. It can become significant in long coaxial- or submarine-cable carrier systems. Over a 3,000-cycle band thermal noise is in the order of −135 dbm.

30. Power-line Induction

The inductive fields generated by power lines can induce relatively large signals in wire line facilities[2,7] through inductive and capacitive coupling. Open-wire lines are particularly susceptible to this type of interference, and induced power noise levels of 28 dba are common and 40 dba not too infrequent when there is joint usage of poles for power and communication circuits. Methods for reducing the induced power-line interference on open-wire lines include the use of open-wire transpositions, balancing of the two sides of the circuit, coordinated transpositions of the power-line circuits, and physical separation of the communication and power circuits wherever feasible.

In cables, in addition to the reduction in power-line induction provided by the twisting of the conductors comprising the pairs, decreased power-line interference is due to the metallic shields in the cable.[4,24]

31. Crosstalk

Crosstalk in wire and cable lines is the result of inductive and capacitive coupling between the lines, with the inductive coupling resulting from the electromagnetic field and capacitive coupling from the electrostatic field generated by the "disturbing" circuit or circuits.

At voice frequencies the major crosstalk effect in wire lines is directly related to the capacitance unbalance between lines, with crosstalk due to inductive coupling negligible. At carrier frequencies both inductive and capacitance coupling are important with inductive coupling controlling.[27−29]

On open-wire lines[25] the principal method of reducing crosstalk is through the use of transpositions. Many transposition systems have been developed[31] to meet the requirements of the various frequency bands utilized in open-wire transmission. In cable, the close physical spacing of the pairs aggravates the problem of crosstalk even further. Cable pairs are effectively transposed by twisting the conductors of each pair together and varying the rate of twist between pairs. Other methods, including the use of compandors, circuit balancing, random splicing etc., are used to reduce the effects of crosstalk.

Typical rms values of crosstalk coupling[8,26] per mile for 19- and 22-gage polyethylene cables in the voice-frequency range are 90 db for near-end and 94 db for far-end. For paper-insulated cable pairs the near-end rms crosstalk coupling is approximately 87 db with far-end 90 db. At 150 kc near-end and far-end crosstalk coupling values are approximately 75 and 71 db, respectively, for polyethylene- and paper-insulated cable. At 750 kc the near-end and far-end rms crosstalk coupling values approximate 68 and 61 db, respectively, for polyethylene-insulated cables and 68 and 58 db, respectively, for paper-insulated cables.

REFERENCES

1. Ryder, John D., *Networks Lines and Fields*, Prentice-Hall, Inc., Englewood Cliffs, N.J., 1949.
2. Everitt, William Littel, *Communication Engineering*, 2d ed., pp. 323–343, McGraw-Hill Book Company, New York, 1937.
3. Karakash, John J., *Transmission Lines and Filter Networks*, The Macmillan Company, New York, 1950.
4. Johnson, Walter C., *Transmission Lines and Networks*, McGraw-Hill Book Company, New York, 1950.
5. CCITT Second Plenary Assembly, New Delhi, *Red Book*, vol. III, pp. 8–16, December, 1960.
6. Maupin, J. T., Measurements in Multipair Cables, *Bell System Tech. J.*, vol. 30, pp. 652–667, July, 1951.
7. Sunde, Erling D., *Earth Conduction Effects in Transmission Systems*, D. Van Nostrand Company, Inc., Princeton, N.J., 1949.
8. Eager, G. S., Jr., L. Jachimowicz, I. Kolodny, and D. E. Robinson, Transmission Properties of Polyethylene Insulated Telephone Cables at Voice and Carrier Frequencies, *Commun. Electronics, IEE*, November, 1959.
9. *Radio Instruments and Measurements*, 2d ed., p. 309, National Bureau of Standards, Government Printing Office, Washington, D.C., 1937.
10. Arnold, A. H. M., Proximity Effect in Solid and Hollow Round Conductors, *J. IEE London*, vol. 88, pt. II, pp. 349–359, August, 1941.
11. Carson, John B., Wave Propagation over Parallel Wires: The Proximity Effect, *Phil. Mag.*, vol. 41, pp. 607–633, April, 1921.
12. Dwight, H. B., Proximity Effect in Wires and Thin Cables, *Trans. AIEE*, vol. 42, pp. 961–970, September, 1923.
13. Meyerhoff, L., Pipe Losses in Nonmagnetic Pipe, *Trans. AIEE*, pt. III (Power Apparatus and Systems), vol. 72, pp. 1260–1275, December, 1953.
14. Raisbeck, G., Attenuation in Continuously Loaded Coaxial Cable, *Bell System Tech. J.*, vol. 37, no. 2, pp. 361–374, March, 1956.
15. Buckley, O., The Loaded Submarine Telegraph Cable, *Bell System Tech. J.*, vol. 4, no. 3, pp. 355–374, July, 1925.
16. Harden, W. H., Practices in Telephone Transmission Maintenance, *Bell System Tech. J.*, vol. 4, no. 1, pp. 26–51, January, 1925.
17. Shaw, T., and W. Fondiller, Development and Application of Loading for Telephone Circuits, *Bell System Tech. J.*, vol. 5, no. 2, pp. 221–281, April, 1926.
18. Windeler, A. S., Design of Polyethylene Insulated Multipair Telephone Cable, *Trans. AIEE*, pt. I (Communications and Electronics), no. 46, pp. 736–739, January, 1960.
19. Merrill, J. L., Jr., A Negative Impedance Telephone Repeater, *Trans. AIEE*, vol. 69, pt. 2, pp. 1461–1466, 1950.
20. Merrill, J. L., Jr., A. F. Rose, and J. O. Smethurst, Negative Impedance Telephone Repeaters, *Bell System Tech. J.*, vol. 33, no. 5, pp 1055–1092, September, 1954.
21. Horn, F. W., and R. B. Ramsey, Cable Sheath Problems and Design, *Elec. Eng.*, December, 1951, pp. 1070–1075.
22. Sunde, E. D., Lightning Protection of Buried Toll Cable, *Bell System Tech. J.*, vol. 24, April, 1945.
23. Weather Bureau, Office of Hydrologic Director, *Thunderstorm, Rainfall, Hydrometeorological Rept.* 5, pt. 2, Aug. 22, 1945 (U.S. Weather Bureau Library No. 77. 37/73, U 857H, no. 5, pt. 2).
24. Gooding, F. H., and H. B. Slade, Shielding of Communication Cables, *Trans. AIEE*, vol. 74, pt. I (Communications and Electronics), pp. 378–387, July, 1955.
25. Chapman, A. G., Open-wire Crosstalk, *Bell System Tech. J.*, vol. 13, pp. 19–37, 195–238, January and April, 1934.
26. Henneberger, T. C., and M. D. Fagen, Comparative Transmission Characteristics of Polyethylene Insulated and Paper Insulated Communication Cables, *Trans. AIEE* (Communications and Electronics), vol. 59, pp. 27–33, March, 1962.
27. Weaver, M. A., R. T. Tucker, and P. S. Darnell, Crosstalk and Noise of Cable Carrier Telephone System, *Bell System Tech. J.*, vol. 17, January, 1938.

28. Dr. G. A. Campbell's Memoranda of 1907 and 1912, *Bell System Tech. J.*, vol. 14, pp. 558–572, October, 1935.
29. Booth, R. P., and R. N. Hunter, Cable Crosstalk—Effect of Non-uniform Current Distribution in the Wires, *Bell System Tech. J.*, vol. 14, pp. 179–194, April, 1935.
30. Hebbert, C. M., The Transmission Characteristics of Toll Telephone Cables at Carrier Frequencies, *Bell System Tech. J.*, vol. 20, pp. 293–330, July, 1941.
31. Chapman, A. G., J. V. Budcemi, and J. M. Flannigan, The REA-1 Transposition System, *Trans. AIEE* (Communications and Electronics), vol. 50, pp. 440–448, September, 1960.
32. Merrill, J. L., Jr., Theory of E. Type Repeaters, *Trans. AIEE* (Communications and Electronics), vol. 15, pp. 443–447, November, 1954.

Chapter 12

DATA-TRANSMISSION EQUIPMENT

J. Z. MILLAR, *The Western Union Telegraph Company*

1. General

Telegraphy is characterized generally as the transmission of language messages (or other information in graphic form) destined to and read by human recipients. The expanding use of electronic digital computers for business data processing beginning in about 1954 created a demand for communication systems that could transmit data in a form suitable for automatic introduction into computing systems and, after processing, return data to the same or other locations where the information would be acted upon. Thus, data transmission is understood generally to be the transmission of machine language. This form of information is most easily understood by the machine, and therefore processing time is reduced to extremely small time values, even for complicated processing. The distinction between telegraph transmission and data transmission lies not in how the signals are transmitted, since in a digital system there is no distinction. Rather, the distinction lies in how the information is utilized. Also, data may include analogue type signals, particularly if the computer is an analogue computer. On the other hand, telegraphy may also be analogue in nature as in facsimile transmission.

Manufacturers of business machines and computers categorize computers as "on-line" and "off-line." *On-line* computers process information at the time of receipt. Consequently, transmission is unscheduled. Examples of on-line data processing include reservation systems, telemetering systems, etc. These systems are sometimes called real-time systems, although, of course, there are always time delay in propagation on transmission circuits and time delay in processing. An *off-line* computer is operated on a scheduled basis. Typically, it prepares reports, payrolls, orders and billing, statistical summaries, etc., on a daily or weekly basis, sometimes on a monthly basis. Bulk data transmission is frequently required in the operation of off-line computers.

It should be noted that early data-transmission systems were merely point-to-point communication channels connecting input and output devices to a remote location where computation or some other process took place. The evolution of such data links into complex processing systems in which the computer not only processes information but also controls the communication system took place only a few years later, starting in about 1956. These systems are known as communication-oriented data processors, and they will be discussed more fully in subsequent paragraphs.

TERMINAL EQUIPMENT

2. Perforated-paper and Magnetic-tape Devices

The availability of telegraph channels that can transmit digital information reliably and conveniently makes these channels of great use in data transmission. Input and output signals are frequently in the form of d-c pulses. Thus, interface connections to most data-processing machines are convenient. However, it was soon found that better overall efficiency could be achieved if signal storage methods were incorporated in the transmission system. As in telegraph switching systems, perforated paper tape became quite useful for storage of data in data communication systems. As a recording medium at the source of information, perforated paper tape is an inexpensive mechanism for storing and transmitting multicard data.

Two types of paper tape are commonly used for recording business data: the five-track telegraphic (Baudot) coded tape and the eight-track business machine coded tape. Five-track tape requires auxiliary error-control devices, but eight-track tape contains a vertical parity bit which materially reduces but does not completely eliminate errors.

Perforated paper tape may be prepared by a telegraph perforator or by a Flexowriter, which can be coded to punch five-, six-, seven-, or eight-level tape.

Another common recording medium is magnetic tape. The most usual code is the seven-track computer code. However, one track is used for the vertical parity check, and consequently, the adoption as a universal standard of the new ASA Code for Information Interchange (which has information bits in seven levels, with the eighth reserved for vertical parity) will require equipment changes. Magnetic tape is usually spaced 200, 556, or 800 characters per inch.

3. Converters

Business-machine manufacturing companies market converters of various types, and sometimes these converters are integral parts of transmission systems. An example is the IBM 63 Card-Controlled Tape Punch. This unit converts data from punched cards to five-track paper tape at the rate of 10 columns per second (100 words per minute). Accuracy of each card prepared in tape form is provided by the Checkable Code Generating Unit used with the IBM 63. This unit counts the holes punched in each tape record and then punches the units position of this count at the end of the record. The tape is not verified by reading and comparison to the card at the transmitting end, but the check information is utilized at the receiving end to establish accuracy for each card block received.

The IBM 46 Tape-to-Card Punch and the IBM 47 Tape-to-Card Printing Punch are converters which convert five- or eight-track paper tape into punched cards at a maximum rate of 20 and 18 characters per second, respectively (200 and 140 words per minute). When the converter is equipped with the Telegraph Checkable Code Comparing Unit, the count of the holes appearing in the tape is accumulated and the units position is compared against the original count established by the IBM 63 Punch. If the two disagree, a code digit is punched in the card being created to identify it as an errored card. This card must then be retransmitted.

The IBM 7765 Paper-Tape to Magnetic-Tape Converter is a solid-state device which operates at 150 characters per second (1,500 words per minute). Input may be eight- or five-track, chad or chadless, paper tape. As the paper-tape characters are read, they are converted to seven-track (200-character-per-inch density) magnetic tape which can be read directly into a computer.

Systems considerations may necessitate direct entry of paper tape into a computer. Most IBM computers permit this. The device used with the IBM 1401 or the IBM 1410 Data Processing Systems, for example, is the IBM 1011 Paper Tape Reader. Five- or eight-track, chad or chadless, paper tape can be read into either system at rates of up to 500 characters per second (5,000 words per minute). Operation of the

IBM 1011 is under control of the stored program in the computer. The 2671 Paper Tape Reader provides 1,000-character-per-second input to the IBM System 360.

A companion unit for the IBM 1011 Paper Tape Reader is the IBM 1012 Tape Punch. This unit produces both punched paper tape for telegraphic transmission and punched Mylar (Du Pont registered trade-mark) tape or paper tape for the automatic control of machine tools. A unique feature of the IBM 1012 is its ability to punch tape and check for errors in a single operation. All characters punched are checked by a reading station. If an error is sensed, the tape is backspaced and the erroneous character or characters are deleted by overpunching of a delete code. The correct characters are then punched automatically. The speed of operation is at the rate of 150 characters a second in five-, six-, seven-, or eight-track tape code, and the tape may be either 1 or $1\frac{1}{16}$ in. wide.

4. Card-transmission Terminals

Beginning in 1954, telephone channels began to be utilized for transmission of data. While telephone channels were used for transmission of digital data for military purposes somewhat earlier than this, the advent of the IBM 65 and the IBM 66 Data Transceivers made possible the transmission of punched-card data over telephone channels as well as standard speed or high-speed telegraph circuits. In such an arrangement one transceiver reads punched cards at the transmitting point and another transceiver punches the data into cards at the receiving location. Since either transceiver is capable of transmitting or receiving data, one machine at each location will provide transmission in either direction, alternately. If card interpretation is required, the IBM 66 Printing Card Punch is used; if not, the IBM 65 Card Punch is used. With the use of standard transceivers, the card-transmission rates for telegraph circuits are three, four, or five cards per minute, these rates corresponding to the standard telegraph circuit speeds of 65, 75, and 100 words per minute. The code employed in this form of transmission is a start-stop, serial-by-character code composed of 1 start bit, 8 information bits (in IBM 4 out of 8 code), and 2 stop bits. When telephone channels are employed, the card-transmission rates are 11 cards per minute for the IBM 65 and 10 cards per minute for the IBM 66 (which employs a longer stop interval to allow time for the print function). Amplitude-keyed audio-frequency carrier transmission is employed for transmission on telephone channels.

Also, IBM Data Transceivers can be equipped with the Digital Subset Feature which is arranged to be compatible with the type of transmission facility to be utilized. This feature allows use of a transceiver over toll (dial) or leased telephone channels or 180-bit-per-second leased telegraph circuits. When telephone channels are used, either toll or leased service, the transmission speed is 10 fully punched cards per minute for the IBM 65 and 9 cards per minute for the IBM 66, while with telegraph service, the speed is 11 cards per minute for either machine. Modulating and demodulating devices (modems, see paragraph 10), available from telephone and telegraph companies, are required when telephone-type channels are employed, but when high-speed telegraph circuits are employed, these circuits are already digital in nature. Another feature, successive card checking, increases productivity by continuing transmission when an occasional error occurs. Errored cards are offset in the stacker of both the sending and receiving units, and these cards are then retransmitted at the end of the batch. Up to four transmissions can be made simultaneously over the same telephone circuit by using four pairs of transceivers, but these transmissions must all be made in the same direction, as operation in the duplex mode is not provided. When four transmissions take place over a single telephone channel, the carrier frequencies are 1,000, 1,500, 2,000, and 2,500 cps.

The Sperry Rand UNIVAC 1004 Data Line Terminal (Type 1) is another data card equipment which may be operated over either leased or toll-type telephone channels. This terminal makes use of the UNIVAC 1004 Card Processor as a basic component. The UNIVAC 1004 Data Line Terminal is designed to operate with the Bell System 201A or 201B Data Sets or similar devices provided by other carriers. While the UNIVAC 1004 Data Line Terminal normally transmits to or receives from another identical unit, it is also suitable for intercommunication with a UNIVAC 490 Real-Time System or a UNIVAC 1107 System.   The system will also accept perforated-paper-tape input.   The operating and processing abilities of the UNIVAC 1004 Card Processor are not impaired when it is being used for data communication.   Immediate error detection and correction as well as voice communication for verification, if necessary, assure accurate results.

After a telephone connection has been established by use of the dial or push buttons of the telephone set provided with the Data Set, the operators switch the telephone channel to Data.   Synchronous transmission is employed, which requires that clocking pulses at the rate standard for the type of facility (2,400 bits per second for leased service or 2,000 bits per second for exchange switched service) be supplied by the Data Sets.   As soon as bit synchronization is established, either terminal may initiate transmission, in accordance with the programs previously established in both terminals.

The Univac 1004 Data Line Terminal at the transmitting end receives characters in parallel bit form from the magnetic core storage and delivers the characters to the Data Set in serial bit form.   The stored characters may be in either XS-3, 90-column, Univac card code or any other 6-bit code form.   The Data Line Terminal also adds a character parity check bit, where necessary, to make each data character odd parity, except that the end-of-message character is made even parity.   Thus the line code is 7 bits per character.

Character synchronization is established by the transmission of four synchronizing characters at the beginning of each transmission.   Start of message is established by the transmission of one space code (the null or all 0 character).   Next follows the data characters of one message which is normally one individual card in length (except that the first data character may be reserved for program control purposes to indicate the message status or transmission code to follow).   Each message is completed by the transmission of one end-of-message code character, one longitudinal parity character, and one space or null character.   One transmission can consist of any number of messages (cards), but a four-card message format is considered optimum with respect to the capacity of the core storage.   Request to transmit and end of transmission are control signals provided by the Data Set.

The longitudinal parity character generated at the transmitting end is essentially a count for even parity of all the bits in each of the bit levels for all data characters in the message, including the start-of-message but not the end-of-message codes.   The receiving equipment compares this character with a locally generated longitudinal parity character and, if correct, removes the interlock, thus allowing the receiving program to proceed, and the data message is then punched into a card or printed out. If the two longitudinal parity characters do not agree, or if the vertical parity of any received character is incorrect, the message is retransmitted.

All messages are acknowledged on receipt; that is, following the transmission of each message, both terminals turn around and the receiving terminal transmits an acknowledgment sequence which has a format similar to that of a transmitted message.   The acknowledgment consists of four synchronizing characters: a start-of-message character (space), an acknowledgment character (Code 1, NG; Code 2, OK; Code 3, end of transmission is acknowledged), an end-of-message character, a longitudinal parity character, and a final space character.   This acknowledgment is checked for accuracy in the same manner as any message transmission.

In addition to the parity checking operations, the following additional errors in data communication are detected: short message, long message, no end of message, operate without request to transmit, and request to transmit without operate.   However, it should be noted that only transmission errors are deleted, and therefore errors in punching or reading of cards and certain other equipment malfunctions will produce errors that will not be detected.

The operating speed of the UNIVAC Data Line Terminal depends on the bit rate of the transmission facility and whether printing is required, but up to approximately 150 cards per minute can be achieved.

The IBM 1013 Card Transmission Terminal reads and transmits data from punched input cards.   It receives and punches data into output cards.   (Transmitting and receiving are not simultaneous.) It is capable of handling a 56-character set of alphabetic, numeric, or special characters. The IBM 1013 can communicate with IBM synchronous transmission units such as another IBM 1013, an IBM 7701 or IBM 7702 Magnetic Tape Transmission Terminal, an IBM 1009 Data Transmission Unit with associated data-processing system, an IBM 7711 Data Communication Unit, or System 360 via the 2701 Data Adapter Unit.

Such a combination provides not only versatility in data transmission but also convenient conversion of data from card to magnetic tape, tape to card, card to card, and tape to tape. The IBM 1013 transmits, under stored-program control, at speeds of 150, 250, or 300 characters per second (selected by the operator) over leased, toll, or private telephone facilities.

Cards feed from the 1,200-card hopper of the serial reader-punch (face down, 9-edge first) to the reading station, then to the punching station and the 1,300-card stacker. As the card moves through the read station (at 700 columns per second), a solar sensing unit reads the card. Reading starts in column 1, with any number of columns from 1 to 80 being read and entered into storage. Any number of columns can be punched at 80 columns per second. When punching is completed, the card is ejected and stacked.

The number of card transmissions per minute (50 to more than 400) depends on the number of characters per card and the number of cards per record.

The operator selects either a fixed or variable program to control the functions of the IBM 1013. The fixed program, which requires no program-card entry, does not vary the card format. The variable program makes it possible to alter the card record being transmitted or received. The variable data are stored and later used to control the selection of data, column by column.

The card record (1 to 80 characters) can be transmitted as 1 character or 127 characters (80 columns of card plus 47 columns from duplication buffer storage). Grouping as many as 71 card records into a single transmittal record is possible if the 329-character capacity of the main buffer storage is not exceeded.

5. Mark Sense Data Cards

While the use of punched cards has been accorded widespread and almost universal acceptance as a means of introducing data into a system, it has been found that errors on the part of key-punch operators have been troublesome. Also, while the Hollerith card code is convenient to use in card reading and punching equipment, it is not a code that is suitable for transmission over serial binary data-transmission channels, and code converters are required except when parallel transmission channels are used. The Mark Sense Data Card and the Mark Sense Data Card Reader-Transmitter recently introduced by the Western Union Telegraph Company have attracted much interest. The Mark Sense Data Card is available in two versions, the numeric version and the alphanumeric version. These cards are printed in an ink that is conductive and are custom-arranged to meet the requirements of the application. A portion or portions of the card may be printed to contain fixed format information, while the remaining or alternate portions are printed in variable format, that is, printed in such a way that the user merely marks with a pencil the character to be transmitted, one mark for each character. This marking is done by the person most familiar with the information to be transmitted, and consequently fewer errors are made. If an error is made, the responsibility is that of the originator only. The cards are then stacked and placed in the hopper of the Mark Sense Data Card Reader-Transmitter. The reader consists of a mechanical drive unit that passes each card under reading contacts at the proper speed to agree with the data channel. The reading contacts are arranged to be error free; that is, the contacts must find one mark in each column. If no mark is found, or if two or more marks are found, the machine stops and the operator must clear the deficiency. The transmitter portion of the equipment is an all-solid-state electronic assembly that generates the output code, which may be five-level teletype code, seven-level magnetic-tape code, eight-level ASCII code, or any similar code. A choice of speeds is also available, depending on the rapidity with which the cards are driven through the reader. However, the timing of each code group is generated by the electronic circuitry and must be slightly faster than the reading speed. This Mark Sense Data Card technique is applicable to many different data-transmission requirements. It has been found to reduce errors and to be a less expensive method of introducing data into a system.

Two peripheral components of IBM computers are the IBM 1418 Optical Character Reader and the IBM 1428 Alphanumeric Optical Reader. These equipments can make use of data cards on which the user places pen or pencil marks in rectangular areas which are opposite printed designations and which are also opposite the black printed lines at the edge of the card. As the card passes through the reader, any pencil marks present are read photoelectrically, the black line at the edge of the card acting as a trigger. An example of the utility of this process is the counting of ballot cards individually marked opposite the names of candidates by the voter. After collecting of the cards, the vote is counted by the computer, which is programmed to ensure accurate results.

6. Direct Keyboard Data Entry

The Model 33 Self Contained Keyboard which provides direct parallel-wire entry of variable data into computers and associated business machines has been developed by the Teletype Corporation. This unit may be used to replace punched-card systems for data collection and transmission. Key arrangement is similar to an ordinary four-row alphanumeric office typewriter.

TRANSMISSION CHANNELS AND EQUIPMENT

7. Characteristics of Low- and Medium-speed Communication Channels Suitable for Data Transmission

Data-transmission channels, such as telegraph channels and channels of voice bandwidth, are available from common carriers of communications services. Wideband (up to 8 Mc) channels are currently in use, as are some privately owned channels. Data channels are provided by common carriers under two general classes of service, one designated *leased facility service* (private-wire or private-line service) and the other *public service* (switched or exchange service).

Leased facility service includes private-wire or private-line circuits and switching equipment suitable for data transmission and operated over various types of data-transmission channels, the service being utilized for the private use of a customer or a customer group which may legally subscribe to such service. TELEX systems are also being furnished for private use, plant in such service being devoted to the exclusive use of one customer.

Exchange public services include DATA-PHONE Service, TWX (Teletypewriter Exchange Service), Broadband Exchange Service, and TELEX (an international and domestic teletypewriter service).

Many applications of switched data-communication services require an attendant during the call setup process. Such calls may be originated automatically by an Automatic Calling Unit. An ACU differs from a device that only dials numbers automatically in that it can perform all the functions usually performed by an attendant in originating a data call. This includes putting the station in the "off-hook" condition, detecting a dial go-ahead signal, dialing, determining whether the call has been established properly, and placing the data set in the data mode. Also, the ACU transfers the telephone line between itself and the data set in order to exclude the data set during the time that the ACU is performing its functions. Three types of ACU have been developed: any number, repertory, and one-number dialing. This range will probably be further broadened by touch-tone calling techniques, introduction of faster switching and standard call progress signals into the telephone network, and the use of one ACU to originate calls on more than one telephone connection.

Services of communications common carriers may be furnished in two different ways, as a complete service in which the common carrier furnishes both communications equipment and facilities and data terminal sets, or the common carrier may furnish only the communications equipment and facilities for the transmission of data between units of data terminal equipment provided by the customer. Hybrid arrangements involving full common-carrier services at one location and partial common-carrier service at another location are permitted. Connections between customer-provided data-processing equipments and data terminal equipment provided by common carriers are made at an interface.* These interface connections are arranged in a multiconductor cable terminated in a plug, the jack being located on the common-carrier equipment.

While standard telegraph channels are being used satisfactorily for data transmission, requirements special to digital data handling, for example, the provision of signaling arrangements for the control of data-processing equipments or special interface arrangements, have been accommodated by the offering of data-transmission channels† by common carriers.

a. Telegraph Speeds. Data-transmission channels operating at standard telegraph speeds are supplied by both AT&T and Western Union. These are:

60 speed: Schedule 1, 45 bits per second (AT&T); Class A, 50 bauds (WU)
75 speed: Schedule 2, 55 bits per second (AT&T); Class B, 57 bauds (WU)
100 speed: Schedule 3, 75 bits per second (AT&T); Class C, 75 bauds (WU)

b. Medium Speed. Data-transmission channels operating at higher speeds are tariffed as follows: 180 baud, Class D (WU). These are duplex (two-path) channels suitable for the transmission of binary signals at nonsynchronous rates up to 180 bauds. Service is available only on a 24-hr-per-day 7-day-per-week basis. Multipoint service is available. A terminal subset (modem, see paragraph 10) is furnished by the Western Union Telegraph Company to convert signals generated by apparatus furnished by the customer to signals suitable for transmission over facilities furnished by Western Union.

c. Schedule 4 (AT&T). Four types of channels of a grade similar to those furnished for private-line telephone service are furnished for the transmission of data signals. These channels are furnished on a two-point or a multipoint basis for transmission in one direction or for simultaneous or nonsimultaneous transmission in both directions. Channels furnished under this schedule may be used in connection with terminal equipment provided by the customer or the Telephone Company. The transmission rate, the number of stations that may be connected, and the distance over which satisfactory transmission is possible are limited by operating and transmission factors. These channels meet the following specifications:

Type 4: Similar in transmission characteristics to those channels furnished for private-line telephone service.

Type 4A: Envelope delay between 1,000 and 2,400 cps shall not exceed a maximum difference of 1,000 μsec. The loss deviation with frequency (from a 1,000-cps reference) shall not exceed −1 to +3 db from 1,000 to 2,400 cps and −2 to +6 db from 300 to 2,700 cps.

Type 4B: Envelope delay shall not exceed the maximum difference as follows:
Between 1,000 and 2,600 cps a maximum difference of 500 μsec.
Between 600 and 2,600 cps a maximum difference of 1,500 μsec.
Between 500 and 2,800 cps a maximum difference of 3,000 μsec.

* See Draft Recommendation V-24, CCITT, Standards of the type and form of signals to be exchanged at an interface between data-processing terminal equipment and data communication equipment, *Spec. Study Group A Contrib.* 92-E, Data Transmission, Oct. 18, 1963.

† See Tariff FCC No. 237—Channels for Data Transmission, filed by American Telephone and Telegraph Company, Long Lines Department; and Tariff FCC No. 237, Data Transmission Service (Sec. 12), filed by the Western Union Telegraph Company.

The loss deviation with frequency (from a 1,000-cps reference) shall not exceed -1 to $+3$ db from 500 to 2,800 cps and -2 to $+6$ db from 300 to 3,000 cps.

Type 4C: Envelope delay shall not exceed the maximum difference as follows:

Between 1,000 and 2,600 cps a maximum difference of 300 μsec.

Between 800 and 2,800 cps a maximum difference of 500 μsec.

Between 600 and 3,000 cps a maximum difference of 1,500 μsec.

Between 500 and 3,000 cps a maximum difference of 3,000 μsec.

The loss deviation with frequency (from a 1,000-cps reference) shall not exceed -2 to $+3$ db from 500 to 3,000 cps and -2 to $+6$ db from 300 to 3,200 cps.

d. Class E (*WU*). Class E channels are suitable for the transmission of binary signals and are capable of meeting the following channel specifications:

1. Envelope delay between 1,000 and 2,400 cps shall not exceed a maximum difference of 1,000 μsec.

2. The loss deviation with frequency (from a 1,000-cps reference) shall not exceed -1 to $+3$ db from 1,000 to 2,400 cps and -2 to $+6$ db from 300 to 2,700 cps.

Multipoint service is available. A terminal subset furnished by either the Telegraph Company or the customer is required. When the terminal subset is provided by the Telegraph Company, binary signals up to 1,200 bauds may be transmitted. When the terminal subset is provided by the customer, the Telegraph Company does not state what maximum speed can be obtained or whether a particular speed will be satisfactory to the customer. This is a matter of experiment on a particular channel with assistance by the Telegraph Company as required.

e. Class F (*WU*). Class F channels are furnished on a two-point basis only, with no more than one station located at each point. These channels are suitable for transmission of binary signals and are capable of meeting the following channel specifications:

1. Envelope delay between 800 and 2,700 cps shall not exceed a maximum difference of 100 μsec.

2. The loss deviation with frequency (from a 1,600-cps reference) shall not exceed -3 to $+3$ db from 800 to 2,700 cps.

A terminal subset is required. The present tariff specifies that this subset is to be furnished by the customer, but these channels are conditioned to transmit up to 2,400 bauds with a Telegraph Company subset to be tariffed later or at any baud rate by a patron-provided modem set that is found to be satisfactory. The network furnished to the Department of Defense previously designated DATACOM and now designated AUTODIN was equipped with 2,400-baud *synchronous* terminal subsets by Western Union.

f. Schedule 5 (*AT&T*). These are channels specially conditioned and maintained with regard to steady and impulse noise, insertion loss, and envelope delay characteristics to transmit SAGE-type data signals at rates of 1,300 or 1,600 bits per second at a normal error rate of no more than 1 in 100,000 bits.

It should be noted that, whereas data-transmission channels operating at 50, 57, 75, and 180 bauds are tariffed to include modems furnished by the common carriers, data channels operating at higher rates require additional terminal equipment (data sets or modems) which if furnished by common carriers are tariffed at monthly rental rates in addition to the line charges. Therefore a prospective user of common-carrier data channels should secure copies of appropriate tariffs and study them carefully to obtain a detailed knowledge of the services to be rendered and equipment to be furnished.

Also, a prospective user desiring services of an intrastate nature or service in areas not served by AT&T and Western Union should consult tariffs filed by Bell System associated companies or independent telephone companies with the corresponding state public utility commissions.

8. Probable Error Rates On Low- and Medium-speed Data-transmission Channels

CCITT Study Group A (Data Transmission) has analyzed transmission tests on 50-baud leased telegraph channels and on the International Telex network.* The observed error rate was from 1- to 2-bit errors per 100,000 bits transmitted. On leased circuits the error rate was from one to eight character errors per 100,000 characters transmitted, and on Telex circuits the error rate was from four to five character errors per 100,000 characters transmitted. On circuits of these types, these error rates correspond to approximately one error per half hour to one error per 4 hr of transmission. The most probable time interval between any kind of error was 1 hr. A "drop-out" condition defined as an interruption exceeding 300 msec of time was observed for approximately 1 per cent of the time. Errors caused by interference (noise) fell into the following pattern:

50 to 60 per cent of the errors were 1-bit errors.
10 to 20 per cent of the errors were 2-bit errors.
3 to 10 per cent of the errors were 3-bit errors.
2 to 6 per cent of the errors were 4-bit errors.

A similar report on the error rate to be anticipated on telephone networks within the United States has been reported to CCITT.† This report indicates that on the average one character error will occur for each 100,000 characters transmitted. Specific curves of error rates are shown for different kinds of facilities and for two representative transmission speeds, namely, 600 and 1,200 bits per second.

The above-anticipated error rates are representative, and while plant improvements are being made to reduce these error rates, it is not anticipated that error-free transmission will be available in the foreseeable future. For this reason, the use of error-control systems is recommended.

9. Wideband Data-transmission Channels

The above-listed data-transmission channels are intended for private-line or private-wire leased service at telegraph speeds or speeds that can be accommodated by a channel of voice bandwidth. When higher speeds are required, channels of larger bandwidth are necessary. Such wideband channels have been tariffed by AT&T as TELPAK‡ channels. The tariff provides that TELPAK channels may be utilized by the customer as a multiplicity of channels of lesser bandwidth, that is, channels of voice bandwidth and/or channels of telegraph bandwidth, for which bulk rates apply. Alternately, the entire broad band may be used (with appropriate modems) as a high-speed data channel. In such service the bandwidths are:

TELPAK A, 48 kc
TELPAK B, 96 kc
TELPAK C, 240 kc
TELPAK D, 1 Mc

However, the FCC is investigating TELPAK tariff matters in Docket 14251, and therefore the reader is cautioned to obtain the latest available information about this service when contemplating its use.

The Western Union Telegraph Company has provided private-wire leased channels of 8-Mc bandwidth for use in missile-tracking systems, utilizing a microwave radio relay system.

* See CCITT, *Spec. Study Group* A (Data Transmission) *Contrib.* 92, Annex XIII, p. 131, Oct. 18, 1963.
† See CCITT *Red Book*, Telegraph Technique and Data Transmission, vol. VII, p. 439.
‡ See Tariff FCC No. 250, TELPAK Service and Channels, AT&T.

As data transmission is expanding rapidly, it is likely that other types of wideband data channels will be provided by communication common carriers.

10. Use of Modems or Data Sets

As has been previously indicated, data may flow on metallic circuits by means of d-c pulses, that is, data are introduced or received, or may flow over local cable pairs in serial or parallel binary digital pulse form.   When telegraph channels of longer length are employed, or when channels of voice bandwidth or broadband channels are used, a modem or data set must be inserted between the transmission channel and the data terminal equipment.   A modem is a modulation-demodulator set which converts the binary digital pulses at the interface to signals that are suitable for transmission over the data channel (usually a modulated carrier).   The use of modems not only improves the quality of transmission but provides auxiliary circuits for signaling and for control of the data sets.   Modems arranged for the transmission of analogue information are also available.   This type of modem would be suitable for use with facsimile, direct writer (telautograph), medical devices, or meter reading (telemeter) terminal apparatus.

Functions pertaining to transmission, signaling, or control over the transmission network are usually performed within the data set.   Functions pertaining to acquisition, storage, processing, or display are usually performed within the data terminal equipment supplied by the customer.   Separation of these functions is accomplished at the interface (see paragraph 7).   Many advantages are derived from the clear division of responsibility afforded by the data-set–interface concept.

Transmission frequencies, levels, wave shaping, and other transmission parameters may be optimized with respect to the diverse and variable characteristics of the transmission network.

Incompatibility between the data-transmission system and existing communications equipment and services may be avoided.   Such incompatibility might result in interference either to the particular data service or to other services being provided by the network.

The designer of the data equipment is able to utilize a uniform, reproducible signal at the interface without extensive study of the variability of the communications network.

Both data terminal equipment (viz., business machines) and communications portions of a system may be independently tested by their respective proprietors at a point of uniform signal conditions.

Each party may modify or update his own portion of the data system without repercussion upon the other.

Data Sets (modems) provided by AT&T for low-speed serial data services are classified as follows: Data Sets 101A, 101B, and 101C are used in TWX service.   Data Set 101A was first employed when TWX was mechanized (cut over to dial operation on the telephone network).   It operates with Model 28 and the older Type 15 Teletypewriters at 60 words per minute (wpm) (45.45 bauds).   Data Set 101B is identical with the 101A except that it contains apparatus needed to convert to 100-wpm service (75 bauds).   Data Set 101C is similar to the 101A but is for use with Models 33 and 35 Teletypewriters, which employ the eight-level 11-unit-pattern 110-baud code.   The newest set of this series is Data Set 105A, which is similar to the 101C but which was made with newer construction methods and contains more sophisticated operating features.   The above sets are employed when no voice operation is involved, although toll network dialing features are provided.   When alternate voice-data service is required, Data Set 103A is employed.   This set includes telephone and control apparatus for use with customer-furnished data terminal equipment.   Data Set 103B consists essentially of transmitting and receiving portions of the 103A less dialing, control logic, and voice facility.   This set is intended for use on private lines where no switching is required.   The new Data Set 103E provides data-only operation and includes features not available in other sets, such as disconnect on loss of signal, clearing on uncompleted incoming call, and loudspeaker monitoring of network tone signals.

All the above Data Sets are furnished normally as common-carrier service under appropriate tariffs by the Bell System companies. They are arranged for installation in the base cabinets of TWX station teletypewriter sets, such as the Model 33 KSR TWX station set, except for Data Set 103A, which is supplied in two unit packages suitable for use on tables, desks, etc. These Data Sets utilize a channel of voice band-width normally.

Data Sets (modems) are provided by AT&T for medium-speed serial data services. Data Sets 201A and 201B are in use in computer inquiry systems such as airline reservation systems, on-line banking systems, and stock-quotation networks and in systems with business-machine terminals, using magnetic tape, paper tape, and cards. The 201A operates at 2,000 bauds and is used both on switched telephone network and on private-line systems. The 201B operates at 2,400 bauds and is used on private-line systems only. Both these data sets are four-phase synchronous sets, and both require a channel of voice bandwidth. Both sets are available with internal synchronization, that is, the transmitter derives timing from an internal oscillator, or with external synchronization where the data set derives timing from the bit synchronizing signal provided by the business machine. Interface connections are provided in accordance with EIA RS 232, which is the same as that currently being proposed for CCITT standardization. Data Sets 202A and 202B are medium-speed serial data sets for operation on channels of voice bandwidth in switched network at rates up to 1,200 bauds or in private-line service at rates up to 2,000 bauds. The data pulses at the interface are converted into frequency-modulated carrier signals suitable for trans-mission. Data Set 202A has a built-in telephone and is arranged for two-wire switched network and for two-wire private-line operation only. Data Set 202B employs a separate telephone set if voice communication is desired and is arranged for either two- or four-wire private-line network operation where no telephone is required, for private-line or switched network applications where a separate telephone arrange-ment is required, and for two-way simultaneous operation on suitable four-wire facili-ties. Frequency modulation was selected as a means for transmitting data signals over the telephone network because this method allows data signal recovery which is independent of transmission level changes and which is of special importance when the d-c property of the data signal must be preserved. Other advantages of frequency modulation are present, but to a small degree because of the characteristics of line noise and the high signaling rate employed. Some development work has been done on a vestigial-sideband data set for serial operation on voice band circuits at signaling rates approaching 3,000 bauds, but no data sets of this type have yet been standardized.

Data Sets are provided by AT&T for low-speed parallel data-transmission systems, as exemplified by Data Set 401E (transmitter) and Data Set 401J (receiver), which are used for the transmission of information read from punched cards. This system uses 14 frequencies within the voice band, divided into three groups (5, 5, and 4). The lowest frequency in each group is designated a rest or null frequency, and the higher frequencies are designated as data frequencies. Eight of the frequencies used are the same as those used in Touch-Tone calling (push-button dialing). Thus, it is possible to use a Touch-Tone telephone set to transmit data. There are two coding methods, one of which has two variations. In the first method the all-rest combination is used as a character separator, or space. In one variation, each character is represented by closing one data lead in each of the three frequency groups, thus $4 \times 4 \times 3$, or 48, characters may be transmitted. In the other variation a character is formed by closing a data lead in one, in each of two, or in each of the three frequency groups, and in this system one or two rest frequencies may remain in a character. Thus there are $5 \times 5 \times 4$, or 100, combinations, of which one is used as the intercharacter separator, which leaves 99 combinations to represent characters. The other code method is often used for transmitting information read from IBM Hollerith alphanumeric tabu-lating cards. These cards are punched to represent a letter by two punched holes, one in the upper zone and one in the lower, numeric, zone. A numeral is represented by a hole in the numeric zone. A single hole in one of the three rows of the upper zone is used to represent certain special characters. Each hole in the numeric zone closes two contacts, one in the first frequency group and one in the second. Thus each pair of data frequencies represents a numeral. Each of the zone holes is transmitted by one of the three data frequencies in the third frequency group. An alpha character is thus represented by three data frequencies. However, if a numeral is being transmitted, no zone hole is present, and in this case the null frequency in the third group is transmitted together with the pair of frequencies representing different numerals. Similarly, a zone-only special character is transmitted by the null frequencies of the first and second group, together with the data frequency in the third group. Since the three null frequencies are transmitted for space, and since every character is represented by three frequencies, this system may be called a three-out-of-fourteen system, which coding possesses self-contained error-reducing properties. The power for Data Set 401E is obtained from the telephone line, but the 401J receiver requires commercial power supply. The maximum speed of this family of Data Sets is 20 characters per second (210 words per minute equivalent).

A medium-speed, binary, parallel data-transmission system is also offered by AT&T. Data Set 402 utilizes eight frequencies within the voice band and operates up to 75 characters per second (750 words per minute). Timing signals are also transmitted. This set uses two-out-of-eight coding. Both the 401 and 402 systems have the advantage that parallel-to-series conversion is not required. Also, timing is furnished to the receiving business machine. This simplicity, resultant low cost, and the error-protection features make the system particularly attractive for data-gathering applications. All the signals interchanged at the interface are contact closures. Special added features are also provided, such as an answer-back signal, recall, etc. Also, alternate telephone transmission is provided.

Data Sets are provided by AT&T for analogue data transmission on switched or private-line facilities. Data Set 602A permits two-way nonsimultaneous transmission of analogue data such as facsimile or other analogue data that match the bandwidth, distortion, and signal-to-noise capabilities of the data set. Facsimile requires a baseband transmission as wide as possible, and this set transmits a frequency spectrum from direct current up to 1,000 cycles in the main channel. A secondary channel is provided for the transmission, as a carrier modulation, of 60-cycle control or synchronizing information. Frequency modulation is employed to minimize interaction between the two subchannels and to maximize baseband frequency response within limits imposed by switched telephone facilities, especially those employing companders or those subject to line level variations. An integrated telephone set and a six-button key set are included. Also, a remote testing arrangement is provided. For the transmission of narrowband analogue data, that is, information that requires a frequency band only from direct current to 120 cycles, three Data Sets are available. Data Set 603A is a nonportable transmitting-only set, Data Set 603B is a nonportable receiving-only set, and Data Set 603C is a completely portable transmitting-only set. These sets transmit medical analogue data such as electrocardiograms, etc. Data Set 805A is a completely portable reverse channel unit to be used with the portable transmitter to receive coordination signals from the distant end of the telephone connection.

A High Speed Data Set has been developed by the Bell Telephone Laboratories for use over TELPAK A (48 kc) private-line channels. This set has been assigned the type number 301B. A system equipped with these sets will transmit serial binary digital data at 40.8 kilobits per second, and simultaneous transmission may take place in both directions when the terminals are properly equipped. These are synchronous sets operating under control of a stable timing signal at the transmitter and presenting both recovered data and 40.8-kilobit timing signals (20.4 kc) at the receiver. They will employ N2-carrier facilities for short-distance transmission and L-carrier facilities for long-distance transmission. A voice coordination channel is provided. Four local loops are used to extend the above carrier channels to the customers' premises at each terminal. Four-phase modulation of the 30.6-kc data carrier is employed, and the bandwidth of the data channel extends from 10.2 to 51 kc when it is transmitted over local cable pairs. To permit testing, an 803A Data Auxiliary Set is used to connect the 301B Data Set to the local loop at customers' premises. It provides arrangements for testing both Data Set 301B and the local loop, it has jacks for plugging in of test equipment, and a back-to-back test of the data set may be made. At the central office the carrier band of the data-transmission channel (the 10.2- to 51-kc data band) arriving over the local loop is frequency-shifted. Existing N2- and L-carrier terminals cannot provide access to the carrier frequency space required for data transmission. Therefore, both N2-carrier and L-carrier wideband modems have been developed, which replace voice bandwidth channel equipment normally used. In the case of N2 carrier, the band is translated to the range 203.5 to 244.3 kc, whereas in either an L1-carrier or an L3-carrier system the band is translated to the range 60 to 108 kc. In the case of N2 carrier the voice coordinating channel is placed adjacent to the lower end of the data band, whereas in an L carrier, it is placed at the high end (104 to 108 kc). These wideband modems include equalizers and filters, as well as amplifiers, pads, modulators, and demodulators.

AT&T-developed modems are manufactured by Western Electric and are normally furnished under tariffs as Bell System services, sale of modems normally being only to independent telephone companies.

The Western Union Telegraph Company furnishes modems of the following types under common-carrier tariffs. The Type 60 modem is utilized for low-speed nonsynchronous data transmission, and this set is the standard set for Class A data channels (50-baud), Class B data channels (57-baud), and Class C data channels (75-baud). Frequency modulation (FSK) is employed because of its transmission advantages on multichannel systems, a deviation of ±35 cycles in a channel width of 150 cycles being optimum for this set. The Type 70 modem is a 180-baud nonsynchronous data set utilized to furnish Class D data channels. These channels are suitable for 11-card-per-minute card transmission, stock-quotation service, and similar services. The channel spacing in a voice band of these modems is 300 cycles, and the deviation is ±80 cycles. The Type 74 modem is a 1,200-baud nonsynchronous data set utilized to furnish Class E data channels. These channels are suitable for use with 100-card-per-minute data terminal equipments, high-speed perforated-tape transmission, and similar services. This modem utilizes a channel of voice bandwidth, and a deviation of ±400 cycles of a 1,800-cycle carrier is standard. The Type 76 modem is a 2,400-baud synchronous data set utilized to furnish Class F data channels. These channels are suitable for magnetic-tape transmission or similar medium-speed services. This modem utilizes two carriers (1,100 and 2,500 cycles) which are modulated with a deviation of ±400 cycles, each operating at a 1,200-baud rate. A synchronous combining unit is employed to stagger-key these two subchannels to transmit the 2,400-baud serial data stream. This combiner provides timing for crypto and data terminal sets which are connected to it. Delay and amplitude correction are included in the above modems when channels of voice bandwidth are used. An asynchronous vestigial sideband modem for 2,400-baud operation, known as Type 77, is currently in production and is intended primarily for use in switched network service. Timing can be provided on an optional basis. Modems of a given type may be packaged in different ways depending on the specific application, i.e., for use in the telegraph central office or in the patron's office, for data only, or for alternate data or voice purposes, etc. Rack, table, or wall mountings may be used as conditions require. Western Union has under development a series of high-speed modems which provide serial data nonsynchronous channels of 5 kilobits (8-kc band required), 10 kilobits (16-kc band required), and 30 kilobits (48-kc band required). These data channels will be suitable for switched or private wire services, and like all Western Union modems, full-duplex service is provided.

The Western Union Telegraph Company developed and produced the AN/FGC-29 Radio Teletype Terminal* for the U.S. Signal Corps. Large numbers of this modem are currently in use on long-distance HF radio circuits by the Army, Navy, and Air Force. It provides 16 full-duplex teletype channels operable at 100 words per minute. These are derived by FSK modulation of 16 carriers located between 425 and 2,975 cps. The deviation is ±42.5 cycles. Equipment is arranged for space-diversity combining for all 16 channels or frequency-diversity combining, in which case only 8 channels are

* J. E. Boughtwood, and F. H. Cusack, Telegraph Terminal AN/FGC-29—Circuit Design Aspects and Equipment Features, *Western Union Tech. Rev.*, vol. 9, no. 1, January, 1955.

available. The equivalent data rate for the FGC-29 is 1,200 bits per second (75 bits per second per channel). As start-stop operation is the normal mode of operation, each channel may carry a synchronous data stream, as, for example, in crypto operation, independently of the other channels. The 16 channels may also be utilized for parallel data transmission. Series-parallel and parallel-series converters may be employed to obtain a synchronous series binary 1,200-bit-per-second data channel. Alternately, eight channels can be combined to obtain a synchronous data channel for transmitting 100 data cards per minute, leaving the other eight channels for a second 100-card-per-minute data transmission or utilizing them for eight teletypewriter circuits. Further information on this modem may be found in TM 11-2245.

A number of electronic-equipment manufacturers have developed data-transmission terminal equipment for sale to military and commercial customers who have access to communication channels of various types, such as high-frequency radio channels, very high-frequency forward scatter radio links, and voice bands leased from common carriers.

The Collins Radio Company* has supplied a series of modems of different design incorporating Collins Kineplex® predicted wave signaling. These Kineplex systems are phase-modulated synchronous transmission systems—hence the signal baud arrival times can be predicted accurately, and differentially coherent detection, developed by Collins Radio, can be used. This technique can provide two major improvements over conventional FSK start-stop systems: improvement in circuit performance as measured by error rate vs. signal-to-noise ratio and improvement in spectrum utilization as measured by the signaling rate vs. bandwidth of the transmission medium. Trade-off between these two factors is optimized in equipment designed for different applications. Operation of the system requires a high order of frequency stability and a linear transmission medium to realize maximum capabilities. However, doppler correction circuits and the use of differentially coherent detection make operation over long HF radio circuits practicable.

One of the early Collins Kineplex equipments designed for data transmission was the Collins TE-202 Terminal Equipment, which was made available in two configurations. Both used quaternary phase modulation and differentially coherent predicted wave detection. For the transmission of data by teletype channels, the equipment was arranged to permit operation of 1 to 40 teletype channels at a maximum channel speed of 100 words per minute. Buffering equipment was used on each teletype channel to convert from start-stop teletype code to the synchronous five-element code employed in the Kineplex. Twenty audio-frequency carriers spaced 100 cycles in the range from 700 to 2,700 cps were employed. Each such carrier was modulated with two 75-bits-per-second data channels. A synchronizing tone of 2,900 cycles was also transmitted. The second configuration of the TE-202 was employed for the transmission of serial binary data at the rate of 3,000 bits per second by the use of series-parallel and parallel-series data converters.

The Collins TE-207B-1 Kineplex Airborne Digital Data Communication Modem is a solid-state full-duplex airborne transmitter and receiver capable of a 2,400-bit-per-second data rate over a voice bandwidth HF radio channel. It is designed to provide reliable HF radio data communication to and from aircraft. The modem accepts synchronous serial or parallel digital data from external sources, quaternary phase modulates one or more audio carriers for transmission, and delivers regenerated data signals in either serial or parallel form. Three data rates may be selected automatically by dial: 600, 1,200, or 2,400 bits per second. When the modem is operating in the 600-bit-per-second serial binary mode, the data are separated into two 300-bit-per-second channels for quaternary phase modulation of a 1,375-cps carrier. Alternately, two parallel synchronous channels may be accepted. When the modem is operating in the 1,200-bit-per-second serial binary mode, the data are separated into four 300-bit-per-second channels, two of which are used to modulate a 1,375-cps carrier and the other two to modulate an 1,815-cps carrier. When the modem is operating in the 2,400-bit-per-second serial binary mode, the data are separated into eight 300-bit-per-

* Information Science Center, Newport Beach, Calif.

second channels which are used in pairs to modulate four carriers at 935, 1,375, 1,815, and 2,255 cps. A continuous tone (495 cps) is combined with the carrier tones and transmitted. At the receiver, the 495-cps tone is used for doppler correction or correction of frequency drift in equipment. Transmit and receive data rates may be the same or different during duplex operation. The composite signal is amplitude-modulated for bit synchronization. Diversity reception, signal-presence indication, and remote control are additional features. The receiver supplies data timing (serial operation) or 300- and 600-cps (parallel operation) timing, or timing may be supplied to the modem from an external source of suitable stability.

The Collins TE-210D-2A Digital Data Communication Modem is a solid-state, full-duplex transmitter and receiver arranged for rack or cabinet mounting for use at ground locations. Operating rates and features are similar to the TE-207B-1, and the design is suitable for operation over a voice bandwidth channel obtained from a wire line, cable pair, carrier system, or microwave system as well as over HF radio circuits. An automatic level control maintains relative input level by correcting for long-term level variations. A differential phase delay compensator (equalizer) is included. This modem was employed on the SAC Command Control System (465-L). The TE-210-D-2A communicates digital data among computers, business machines, teletypewriters, and other digital data sources. It is designed for continuous, unattended, automatic, full-duplex operation at 600, 1,200, or 2,400 bits per second.

The Collins TE-211A-1 is a completely solid-state, full-duplex data modem capable of data rates up to 4,800 bits per second over wire line, cable, carrier, or microwave. An important feature of the modem is its ability to accept two 2,400-bit-per-second serial or parallel data inputs simultaneously for transmission over a single voice channel. The modem also accepts synchronous serial or parallel binary data at 600, 1,200, 2,400, 3,600, or 4,800 bits per second.

The Collins AN/GSC-4 is the fastest data modem available for use over a voice bandwidth facility. It is a completely solid-state, full-duplex modem capable of data rates up to 5,400 bits per second over wire line, cable, carrier, or microwave. The modem accepts synchronous serial or parallel binary data at 600, 1,200, 2,400, 3,600, 4,800, or 5,400 bits per second.

The Collins TE-216 Adaptive Kineplex Data Communications Modem Line is comprised of circuit cards and modules which can be assembled into modems for data transmission over nominal 3-kc voice bandwidth facilities. From the family of circuits, tone frequencies, and pulse timing, a wide range of data rates and channel characteristics can be accommodated with the same basic circuits and modules. The various standard configuration modules of operation are electrically compatible with existing Collins data modems such as the TE-207, TE-210, AN/GSC-4, and TE-202. A choice of mounting arrangements are available, as the equipment packages may be mounted in racks, cabinets, drawers, or airborne racks (ATR's). The Collins TE-216A-4 is a four-tone full-duplex transmitter and receiver designed to replace the TE-210. It is capable of operation at 600, 1,200, 2,400, and 3,600 bits per second. The TE-216A-20 is designed to replace the TE-202. It is available in cabinet mounting. Like the TE-202, 20 carriers are quaternary modulated for compatible operation with a TE-202, in which case the data rate is 2,400 bits per second. A third channel per carrier has been added by subjecting the modulated carrier to an amplitude modulation so that a 50 per cent increase in the data rate is obtained. Therefore, when a TE-216A-20 is operating against a similar unit, the data rate is 4,500 bits per second (60 input channels at 75 bits per second on each channel). This arrangement may be used for 100-word-per-minute teletypewriter operation or for serial or parallel binary data transmission for use with computers, business machines, telemetry, supervisory control, or other applications.

The Collins TE-217A Adaptive Kineplex Data Communication Modem is designed for a nominal transmission rate of 92,400 bits per second over a 48-kc bandwidth communication channel. The transmission rate may be altered to provide operating tolerances or to compensate for variations of amplitude or phase, noise, or frequency-stability characteristics of the line. The two binary channels are quaternary phase modulated on each of eight carrier frequencies, and a third binary channel is obtained by simultaneous amplitude modulation of the carrier. This method of phase and amplitude multiplexing provides 3-bit transmission capability on each carrier tone. Thus each channel transmits data at the rate of 3,850 bits per second, which corresponds to a pulse length of 260 μsec. The equipment may be mounted in cabinets, drawers, or airborne racks.

The Collins TE-218A Adaptive Kineplex Data Communication Modem is designed for a nominal transmission rate of 450,800 bits per second over a 240-kc communication channel. The transmission rate may be altered to provide operating tolerances or to compensate for variations of amplitude or phase, noise, or frequency-stability characteristics of the line. The two binary channels are quaternary phase modulated on each of eight carrier frequencies, and a third binary channel is obtained by simultaneous amplitude modulation of the carrier. This method of phase and amplitude multiplexing provides 3-bit transmission capability on each carrier tone. Thus each channel transmits data at the rate of 19,200 bits per second, which corresponds to a pulse length of 52 μsec. The equipment may be mounted in cabinets, drawers, or airborne racks.

A number of other equipment manufacturers also make modems for sale and in some cases for lease. Among these are Automatic Electric, IBM, RCA, Lenkurt, Rixon, Hughes, General Dynamics Electronics, Robertshaw-Fulton, Hallicrafters, Minneapolis Honeywell, and Ericson.

11. Reverse-channel Operation

It is frequently necessary to use the reverse channel of a duplex (four-wire or two-path) data channel for return control signal transmission or for error-control purposes. For example, provision is made in modems of certain types for the transfer of control signals both to and out of the interface for the more complete automation of data terminal equipment operation. Similarly, error-control systems may require the reverse transmission of signals that control the acceptance or rejection of a block of data being transmitted in the forward direction. Also, some system designs require the reverse-channel transmission of everything transmitted in the forward direction. Transmission requirements may be the same, or there may be added requirements with respect to the reverse-channel characteristics.

MAINTENANCE CONSIDERATIONS

12. Test Sets of Particular Value in Data Transmission

As has been noted above, the inclusion of interface (see paragraph 7) makes possible the testing of data-transmission channels independently of tests performed to determine the proper operation of data terminal equipments. Maintenance of low-speed data channels operating at signal rates below 150 bauds and in which the line signals are of the d-c type may be tested by means of relatively simple test sets. Voltmeters and milliammeters are useful in measuring steady-state line voltages and currents.

a. Bias and Distortion Measurement. A Bias and Distortion Meter is the most frequently used instrument for determining the dynamic performance of a low-speed data-transmission channel. Such meters measure bias distortion, characteristic distortion, and total distortion, including fortuitous effects. To measure *bias distortion* the distant transmitter is caused to send alternate marking and spacing signals, i.e., alternate 1 and 0 binary pulses. If the received signals arrive at equal time intervals and with equal amplitudes, the channel is free from bias distortion. If, however, the marking pulses integrate to a larger energy content, either because of longer bit length and/or because of larger amplitude, the channel operates with positive bias distortion and it becomes necessary to adjust circuit and apparatus characteristics until no bias distortion is present. Similar adjustments are made to correct negative bias distortion. *Characteristic distortion* is present when a series of marking bits is followed by a spacing bit that is displaced in timing or that arrives at a lower amplitude than normal when alternate marking and spacing bits are transmitted. The inverse of the above signal train may also occur. To detect the presence of characteristic distortion, it is necessary to transmit a repetitive signal pattern, composed of various combinations of pulses, and observe the displacement of single pulses at the receiver with the Bias and Distortion Meter. Some Bias and Distortion Meters utilize an indicating-type milliammeter to permit observation of distortion, and other distortion meters utilize a small cathode-ray tube. An oscilloscope is often of considerable value in observing characteristic distortion and in making circuit adjustments to eliminate any observed transmission impairment. *Fortuitous distortion* is caused by noise bursts, rapid level changes, cross modulation, or extraneous disturbances from power systems, spurious signals, etc. Testing for the presence of such interfering signals can be done by the use of an oscilloscope, but usually location and removal of the source of the interference are rather complex and time consuming. The procedure found best for analysis of these intermittent distortion effects is to transmit a repetitive test pattern and synchronously receive this coded pattern by means of an identically coded device. The received signals are simultaneously recorded on a magnetic-tape loop. If any one of the bits is errored in transmission, the receiver counts an error and the tape loop is stopped. The incoming signals are transferred to another tape recorder, and the first tape loop is transcribed into a Memoscope or similar oscilloscope so that whatever incoming interference or noise can be displayed and the source determined. The Western Union Synchronous Pattern Generator, No. DT-120, can be used to supply synchronous test patterns of selectable coding ranging from 33 to 40 bits in length. The pattern is programmed by changing plug-in diode holders, and a crystal oscillator is divided digitally to obtain a series of speeds ranging from 75 to 4,800 bauds. The generator can also be programmed to produce 1:1 and 2:2 bit ratios of reversals.

b. Envelope Delay Distortion Measurement. As channel speeds became higher than 150 bauds, it became necessary to utilize channels of voice bandwidth and broadband channels for transmission of data. This in turn made it necessary to amplitude and delay equalize the channel over a relatively large frequency range. This equalization must be accomplished even though the transmitting and receiving ends of the channel are located far apart. For proper equalization, it was necessary to measure the transmission characteristics of the channel accurately (at the point where the modem is connected). The Western Union Telegraph Company has developed an envelope delay distortion-measuring set that accomplishes this purpose. This set (No. 7782-C) is for use with channels of voice bandwidth and covers the frequency range 300 to 10,000 cps. The accuracy at any point of this range is 10 μsec, but readings can be estimated to 5 μsec. The set consists of transmitting and receiving units. The amplitude vs. frequency characteristic can also be measured, the incoming signal level being indicated on one meter and its frequency on another. A standard decibel meter is also useful in measuring levels at various points in a system.

c. Additional Data Test Sets. As bit rates increased, it became necessary to measure bias and distortion on the data equipment side of the modem at speeds higher than 150 bauds. The Western Union High-Speed Bias and Distortion Meter (No. 10266-B) can monitor data transmission at 150, 300, 600, 1,200, and 2,400 bauds. It has provision for six added speeds. As the input impedance is high, the meter is bridged on the circuit while it is operating.

One other type of test set has been found necessary. This is a test set to align the modem being used on a particular data channel. The Western Union On-Line Data Channel Monitor (No. 11210-A) and the Western Union Transceiver Test Set (No. 10499-A) are designed to test Types 70 and 74 Modems at speeds ranging from 75 to 4,800 bauds. These devices can generate test signals to check the receiver portion of the modem and a demodulator-receiver section to monitor incoming transmission. A High-Speed Bias and Distortion Meter may be used to generate signals with which to test the transmitter portion of the modem while the sending channel side is being monitored.

Dot generators (square wave), frequency counters, vacuum-tube voltmeters, and audio oscillators are useful test instruments that are frequently used by maintenance personnel.

In addition to the above-listed test instruments, the Bell System now makes extensive use of relatively simple data test sets which are capable of generating and measuring both the distortion and the error performance of an overall data system at a variety of speeds of transmission. These test sets, coded 901 (Interface Adapter), 902 (Distortion Measuring and Error Checking), 903 (Word Generator), and 904 (Data Test Center), are widely used both by field forces testing on subscribers' premises and by central-office maintenance forces. Developmental work has been continuing at the Bell Telephone Laboratories on simplified methods of determining the data-transmission performance of a circuit utilizing time domain methods for estimating the error performance of data sets in the presence of distortion and impulse noise.

ERROR CONTROL

13. Basic Considerations

Error control in data-transmission systems has been achieved by the incorporation of electronic error-control circuits or by adding error-control devices external to the transmission equipments which perform error-control functions solely. The application of such devices can control only those errors produced by transmission or equipment deficiencies, and therefore other techniques must be utilized to prevent human errors, which often occur between the source of raw data and the transmission system and in the utilization of output data. Assuming that a means has been provided to eliminate human errors, it then becomes the function of the transmission system to transfer the data without errors, additions, or subtractions so that the output of the data link is identical with the input. It is sometimes considered sufficient to detect errors and resort to human intervention to eliminate input of false information in subsequent data processing. These control methods are known as error-detection methods, for example, the offsetting of a received data card known to contain an error and the subsequent retransmission of the errored card. Such methods are wasteful of manpower and time, and the savings in equipment cost can rarely be justified. Error detection and automatic correction methods require more costly equipment, but overall system annual costs are materially reduced. A number of developments of this latter type have been completed and are in service.

a. Redundancy. Basically, all systems performing error detection require transmission redundancy whether automatic correction is employed or not. There are two transmission techniques utilized in present systems, those systems that transmit only in the forward direction and those that transmit in both forward and return directions. Both types of systems require the dividing of the input signals (bits or characters) into word blocks of short enough length to be managed without excessive cost for storage elements. A forward system requires a transmission channel in one direction only, whereas a forward and return type of system requires a transmission channel of the two-path type (a duplex channel). However, to achieve a high degree of error control in a forward system, it is necessary to resort to coding methods that contain a much higher order of redundancy, approaching 100 per cent, which requires a transmission channel of twice the speed. Such codes are known as systematic or Hamming codes. While these codes have been treated extensively in a theoretical manner, equipment utilizing these techniques has not yet been employed extensively. On the other hand, the availability of a duplex channel permits reducing redundancy to small percentages, and since such a system will permit data to be transmitted in both directions simultaneously, the overall traffic-carrying capacity of the system remains high, very little time being required for efficient error control. It therefore becomes a designer's choice as to which method will be used in an operating system. However, in data links that are short or between equipment sections that are connected by short wires, the designer may employ error-control techniques that are quite wasteful of transmission time to lower the overall system cost. For example, in the "loop-check" method, all data bits are transmitted to the distant point and relayed back to the transmitter where the check is made. If an error occurs, the transmitter sends a control signal which causes the last block transmitted to be dumped at the receiver, and the same block is retransmitted. If no error occurs, the transmitter sends a control signal to the receiver which causes it to accept the last transmitted block.

b. EDAC. The Western Union Error Detection-Automatic Correction System (EDAC)* is a highly accurate and low-cost error-control method of error detection and correction for both directions of transmission on a full-duplex channel. The speed of operation can be high on channels of relatively long propagation time, for example, as high as 2,000 bauds on long HF radio circuits. EDAC transmissions are divided into blocks, and since the trunk side is operated synchronously, the same block length is required in each direction. Various block lengths are possible, but one current model operates with blocks of four data characters. These characters are transmitted from nondestructive storage, and 5 check bits derived from a binary summation of the data bits are also included. At the receiving end, each block is stored and checked before delivery. A 3-bit control signal is then returned to the sending end (a corresponding control signal is sent simultaneously on the other side of the channel). When the control signal indicates that an error was made, the preceding four data characters are retransmitted; otherwise these are erased and four new characters are transmitted. Mutilation of the control signals does not destroy EDAC accuracy. Retransmission is carried out until acceptance is indicated. Long-period circuit interruptions also have no effect except the lost time involved. If five-level teletypewriter code is employed, there would be 20 information bits in the block in each direction; and since 5 plus 3 bits are required for error control and with a 2-bit synchronizing signal, the total number of bits is 30, which is the same number required for four characters of start-stop 7.5 unit code. The redundancy in this example is zero. If EDAC is operated against a nonsynchronous input or output system, buffers are required.

c. Beckman ECT. Beckman Instruments, Inc.,† has developed an Error-Control Transceiver (ECT) to transmit and receive digital data over channels of voice bandwidth. Data are redundantly coded in such a way that the probability the equipment will pass erroneous data is 10^{-14}. At an operating speed of 1,200 bits per second this would be one incorrect block of data in about 2,500 years. Transmission is full-duplex, and any type of input and output can be accommodated by suitable buffers.

d. SECO. Sequential coding and decoding equipment for the detection and correction of errors in high-speed digital transmission has been developed and tested by the Lincoln Laboratory, MIT, Lexington, Mass. Data were transmitted at an average rate of 7,500 bits per second on an 800-mile toll-grade telephone circuit with noteworthy results. This was the first demonstration of an adaptive electronic system for error control on communication channels.

DATA-COLLECTION AND DATA-TRANSMISSION SYSTEMS

14. DATASPEED

DATASPEED is a Bell System Service providing transmission of data on perforated paper tape at high speeds. The Teletype Corporation manufactures the tape readers and punches, together with associated electronic and mechanical equipment. Appropriate assemblies are mounted in cabinets, which also provide space for Data Sets that are employed for signal transmission over circuits of voice bandwidth.

Five different models of Data Speed equipment are in service or under development. These are identified in the Bell System as Types 1, 2, 4, 5, and 8. Type 1 transmits data in five-level paper-tape form at 105 characters per second. Type 2, also in service, is an improved model which accepts five-, six-, seven-, or eight-level tape. Types 4 and 8, now under development, will provide error detection and automatic retransmission of errored data. Type 5, also under development, provides eight-level parallel transmission.

Type 2 equipment, like Type 1, operates at 105 characters per second. The digital

* J. J. Durachinski, EDAC, *Western Union Tech. Rev.*, vol. 18, no. 4, pp. 140–148, October, 1964.

† Systems Division, 2400 Harbor Boulevard, Fullerton, Calif.

signal sequence which is transmitted via the Data Set is in a 10-unit start-stop coded format. When a tape for fewer than eight code levels is to be transmitted, the signal is not changed from the 10-bit form. Instead, the unused bit positions in the character format are placed under control of a manual switch which is set for the number of code levels to be used. Thus the unused bit positions are recoded by the operation of the switch. The tape reader and punch are designed so that the unused code levels are not sensed or punched. The tape guides are adjustable to accommodate tape widths from $1\frac{1}{16}$- to 1 in. The signaling speed is constant at 105 characters per second and 1,050 bauds.

The Type 4 DATASPEED provides automatic error detection correcting errors by automatic retransmission of errored portions. Each data block contains 80 characters, each block being followed by a pair of check characters. A switch allows the error-control system to be deactivated, and the equipment then is capable of interoperation with Type 2 stations. The signaling speed is constant at 105 characters per second and 1,050 bauds. Types 1, 2, and 4 DATASPEED Sets utilize the Type 202-A Data Sets.

The Type 8 DATASPEED equipment is basically similar to Type 4 but operates at 2,400 bauds using a Type 201-B Data Set. Improved readers and punches have been developed for operation at this speed.

The Type 5 DATASPEED equipment uses the Type 402 Data Set which provides eight-level parallel transmission at 75 characters per second.

Transmission of perforator paper tape has proved satisfactory when channels are used in the private line or toll telephone plant. DATASPEED equipment may be arranged to transmit through a communication channel into a computer, a magnetic-tape terminal, or a teletypesetter.

15. Use of Telex Network for Data Transmission

It has long been recognized that the Telex network has great potential as a means of transmitting data. (For a description of Telex refer to Chap. 8, Telegraph Switching.) Connections may be quickly established to a distant point, and thus Telex lends itself well to the introduction of data into a computation center from a large number of outlying points and the delivery of read-out information to the same or other points. This use is particularly applicable for lightly loaded terminal stations where the amount of data to be transmitted is not sufficient to justify the leasing of a private-wire channel. The rapid growth of Telex service both internationally and nationally brings this potential closer to universal realization. Telex can also serve as a means of extending a private-wire data network to lightly loaded tributary points. The use of the Telex network without teletype equipment and thus without the answer-back feature (but retaining the dialing units for setting up calls), the use of synchronous codes or codes with more than 5 information bits for data transmission, the use of error-correction techniques, and the use of speeds higher than 50 bauds when in the data-transmission mode are all matters being given further study by Study Group A, Data Transmission, CCITT. Perforated paper tape is also used in Telex data transmissions.

16. Use of Teletypewriter Exchange Service (TWX) for Data Transmission

The use of TWX for data transmission is enhanced by design features discussed in Chap. 8, Telegraph Switching. Coincident with the conversion of the TWX network to automatic operation utilizing the switched telephone network, provision was made for the use of the 7-bit information code (ASCII) which is now being proposed for international standardization. Also, Models 33 and 35 Teletypewriter sets were placed in service at a portion of the TWX stations, and provision was made for translation of codes and speed between old and new equipment. These teletypewriters operate at 100 words per minute and employ a four-row keyboard. The application of these new teletypewriters is expected to be not only for the traditionally man-to-man communications but also in new data services where the teletypewriter is essentially the remotely located input or output device for a computer.   Since either unattended answering or automatic calling units are used for origination, there will be no need for human intervention at the computer location.   Operation with perforated paper tape also permits data to be transferred efficiently and automatically.

17. DATA-PHONE

Bell System DATA-PHONE Service provides for the transmission of data between telephone stations using regular exchange telephone service.   Normal telephone dialing techniques are employed to "address" the calls.   The connections so established may be used alternately for voice and data communications; the voice mode may be used to coordinate the manual operation of the data terminal equipment at each end.   Normal telephone message charges apply, plus a monthly charge for the data-communication equipment.

DATA-PHONE Service currently provides for the transmission of digital data at rates up to 2,000 bits per second.   Analogue data may also be transmitted at commensurate speeds.   Future developments are expected to produce even higher speed capabilities.   Most systems may be equipped to answer calls automatically to permit unattended operation of the distant station.   Adjuncts are available to permit the calling station to place and administer calls automatically under control of the data terminal (i.e., business machine) equipment, allowing fully automated "machine-to-machine" calls.

DATA-PHONE Service provides the data user with most of the features known to telephone customers.   The full capabilities of the nationwide telephone network are at his disposal, permitting data communication between widely dispersed geographic locations.   This includes automatic alternate routing to help calls find a path even during periods of heavy traffic or damage to intercity circuits.

Most features available with normal telephone service are usable with DATA-PHONE Service, among them being Direct Distance Dialing (DDD), Private Branch Exchange (PBX), and Foreign Exchange (FX) Services.

The TWX service is complementary to DATA-PHONE Service, since the former handles digital rates up to 200 bauds while the latter is useful at the higher data speeds.

Telemetering, alarm, and supervisory control systems are a specialized class of data transmission for which DATA-PHONE Service may be used.   Instruments at remote locations can be read over switched telephone circuits from a central monitoring station, where the information may be displayed automatically or recorded for use in data-processing equipment.   Similarly, devices at remote locations may be controlled from a central location or another remote location.   Automatic calling equipment permits the remote locations to call the central station when a monitored condition changes state, thus providing a form of alarm service.   More sophisticated systems provide for transmission of command and indication information in both directions. DATA-PHONE transmission may provide attractive new methods of operating telemetering, alarm, or supervisory control systems.

18. Broadband Switching Service

The Western Union Telegraph Company recently placed in operation a switching network known as the Western Union Broadband Switching Service that permits subscribers to select various types of channels by push-button telephone for the direct, two-way exchange of digital or analogue data and voice communications.   Service is available in 19 major city exchange areas coast to coast, and the network is soon to be expanded.   Toll charges are based on distance, bandwidth of channels selected by the customer, and line time used with a 1-min service minimum.

Broadband service provides for the automatic selection of high-quality circuits for high-speed transmission of various forms of analogue- and digital-data communications required by business, government, and the military.   Such communication includes the exchange of data in the form of punched tape, punched cards, or magnetic tape; facsimile at high and low speeds; and voice on an alternate basis.

A seven-digit number code is used in making Broadband calls. Using a specially designed, 10-button Western Union telephone, the subscriber lifts the handset and depresses, for example, the numbers 322-2-131.

The first three numbers connect the caller with the Western Union exchange center at the destination city; the fourth number selects the desired bandwidth; the last three numbers are those of the called subscriber. The circuit is then ready for the immediate exchange of record and voice communication. All circuits will be four-wire, full-duplex bands. The advantages of a four-wire system are elimination of echo and singing distortion, improved quality of the transmission path, and the availability of the return path for error control, synchronization of data, or computer equipment, and if the user so elects, the system will permit simultaneous transmission of data in both directions.

During the introductory period, Broadband Exchange Service will be offered in two bandwidths, one with 2-kc spacing and the other having 4-kc spacing, equivalent to a half voice band and a full voice band, respectively. Wider bandwidths, nominally of 8, 16, and 48 kc, are intended ultimately to be offered.

19. AUTODIN, Automatic Digital Network

The world's largest and most advanced data-transmission network, AUTODIN,* was originated by the U.S. Air Force in 1958 as Project 423-L, Combat Logistics Network (COMLOGNET). This designation was changed to DATACOM by the Air Force during the development cycle of the system, and in February, 1962, when management of the network was assumed by the Defense Communications Agency, it became AUTODIN. The counterpart of AUTODIN for voice communications is the Automatic Voice Network, AUTOVON. These two major networks comprise the nation's Defense Communications System. AUTODIN was placed in service by the Western Union Telegraph Company for the Department of Defense in February, 1963.

The original configuration of the domestic portion of AUTODIN consisted of five electronic switching centers that interconnected over 350 local and overseas tributary stations. Overseas stations were connected into the system at two of the switching centers which served as gateway stations. An expansion program, designed to accommodate a maximum of 2,700 domestic stations, was commenced immediately after the basic system was completed. This program includes four new switching centers and the expansion and updating of the five existing centers. Facilities and equipment in the continental United States are leased to the Department of Defense by Western Union, utilizing solid-state switching equipment purchased from RCA and data terminal equipment purchased from IBM and RCA. Large-scale use of Types 70, 74, and 77 Western Union modems is included. Western Union technical control equipment is also used.

To improve the overseas data communication service, DCA has implemented installation of an overseas automatic digital network compatible and interconnected with the domestic AUTODIN network. The overseas network provides for the establishment of 10 automatic electronic switching centers and associated tributary stations to accommodate a maximum of 1,600 overseas subscribers. Western Union is providing engineering assistance for the overseas program. This includes program implementation, preparation of engineering specifications, evaluation of proposals submitted by equipment manufacturers, site preparation, and system integration. It is anticipated that the two networks, domestic and overseas, will function as a single AUTODIN complex, providing a high-speed, extremely reliable and accurate data-communications service to all the elements of the Department of Defense. Switching equipment and Technical Control facilities for overseas installations are being fur-

* H. A. Jansson, AUTODIN—System Description, part I, Network and Subscriber Terminals, vol. 18, no. 1, January, 1964, and part II, Circuit and Message Switching Centers, vol. 18, no. 2, April, 1964, *Western Union Tech. Rev.*

nished by the Philco Corporation, and data terminal equipment by IBM and RCA. AUTODIN is operated by the Air Force Communication Service (AFCS).

The system requirements encompass both message switching (store and forward) and circuit switching (direct user-to-user facility). Data originating at local stations are automatically switched to one or several destinations (single or multiple address). A set of transmission control procedures and a family of data terminals were designed to satisfy the data source, speed, and accuracy requirements; message formats and rules for conversion from one format to another were established for the message switching centers; and high-speed modems were developed to meet the transmission requirements. Some of the major functional requirements initially imposed on the system by the Air Force were the following:

Messages accepted by the system must be on IBM cards, on magnetic tape, or on perforated-paper teletype tape, except that other forms of input may be accommodated as required. A message may be a single card or a large number of cards, or it may consist of small or large lengths of perforated paper or magnetic tape, or it may consist of a single teletype message.

Codes must include Air Force card code, AF teleprinter code (Fieldata code), and a variety of magnetic-tape codes characteristic of the computer generating the tape (IBM, RCA, NCR, etc.).

Transmission rates must be adapted to the data source and type of transmission facility and should initially adhere to one of the following standards: 75, 150, 600, 1,200, 2,400, and 4,800 bauds.

Exchange of messages must be permitted between all types of terminal devices, for example, from card input to teletypewriter output.

Special security-protection features must be included to ensure against system compromise of message security (lost messages, messages delivered to wrong address, classified messages addressed to stations not authorized to receive, etc.).

Compatibility with government-furnished link-encryption equipment must be assured.

Routing facilities must include unlimited multiaddress capability, group code addressing, and routing by precedence, security, and message type.

Automatic error-detection and -correction features must be provided for transmission to and from data terminals and internally in switching centers.

Messages must be processed and released independently in the order of their priority and time of arrival at the center. Flash and emergency messages must preempt lower priority transmissions if other direct circuits are not available.

Statistics of various types must be accumulated by message switching centers.

A computer-based message switching center was developed to meet the above diverse requirements. A stored-program controlled message processor was selected for the basic switching functions to provide the flexibility required. Modular design of the center equipment into functionally independent units or subsystems permitted test and maintenance routines to be performed on each unit and also enhanced flexibility in design.

A solid-state four-wire circuit switching center having its own common control facility and using dry reed relays for setting up connections was developed. A unique supervisory signal and control procedure was established for this center which permits the terminals to call the desired destination automatically, using the "header" or address portion of the message.

Special control procedures, designated as Mode I, were specified to attain high accuracy and speed.

Four types of station terminal equipments are utilized. The *Teleprinter Terminal* is a standard five-channel, asynchronous (start-stop), perforated-paper-tape, full-duplex, Model 28 ASR set, operating with no return control procedures (Mode II) and using channel sequence numbers to avoid lost messages. The standard speed of 100 words per minute is employed (75 bauds). This type of terminal is used only at stations which are not equipped for Mode I operation, such as a station in another military network which must have access to AUTODIN or at a station that is scheduled to be equipped with a data terminal at a later date. The *Compound Terminal* transmits or receives teletype messages utilizing a modified Model 28 ASR set and messages composed of IBM cards utilizing two Type 536 IBM Card Printing Key Punches, one for transmitting and one for receiving. Transmission is asynchronous and in eight-bit Fieldata code (the common language code of the system), of which seven levels are used for information and the eighth level for odd parity check. Transmission on the line is at 150 bits per second corresponding to 200-word-per-minute teletype or 12 cards per minute (Western Union class D channel, using Type 70 modems). Data messages, card or teletype, are transmitted in 80-character blocks, preceded and followed by two unique block control characters, called framing characters, thus making the line block 84 characters long. The first three framing characters identify the type of block (first block of message, intermediate block, or last block) and the type of message (card or teletype). The fourth framing character is used solely to verify the accuracy of the transmitted block and is called the block parity character. It corresponds to the even longitudinal parity count of the bits in each of the eight levels in the code, excluding the first framing character. Thus the system includes a combination of vertical and horizontal parity counts to determine accuracy of transmission. Each block must be correctly acknowledged before the next block is transmitted, but if an error is indicated, the block is retransmitted. However, two blocks are stored at both terminals and alternate acknowledgment codes are used to identify uniquely to which of the two consecutively transmitted line blocks the acknowledgment pertains. Also, the receiver may reject a message or halt transmission, and the transmitter may discard a portion of a message already accepted by the receiver so that a new message may be processed. Thus, core storage of two 84-character blocks at both transmit and receive sides of the terminal equipment and switching center equipment permits blocks to be transmitted without error and with maximum utilization of line time. Code conversion is included in the terminal for teletype to and from Fieldata, and Hollerith to and from Fieldata. The Compound Terminal operates as either a message switch or circuit switch terminal. When used as a circuit switch terminal, it is equipped with additional facilities required for supervisory signaling with the center. The *High Speed Card Teletype Terminal* functions in a manner identical with the Compound Terminal except that the asynchronous line rate is 1,200 bauds and additional buffering is provided to operate with an IBM 1931 Card Reader-Punch. The Type 74 Western Union modem is employed which requires a channel of voice bandwidth. The card-transmission rate is 100 cards per minute, but the teletype rate remains 200 words per minute as in the Compound Terminal. The *Magnetic Tape Terminal* serves as the input-output communication device for computer installations. It reads and writes tapes which are compatible with the computer being served. Code and format conversion is performed to permit one terminal to communicate with other terminals serving other computer types, on either a circuit switch or message basis. Transmission is synchronous, full-duplex, at a line rate of 2,400 bauds. The Type 76 Western Union modem is employed. The block length is 1,200 characters, and as in the other data terminals, blocks are transmitted alternately from storage. The card rate equivalent is 200 cards per minute.

AUTODIN operating in Mode II interfaces with existing military teletype networks. Messages so transferred and processed appear in the same format as that employed in these networks. As this is the standard mode of operation of these networks, no modifications need be made at the teletype switching centers to interconnect with AUTODIN.

All tributary and trunk communication channels to and from an AUTODIN switching center are connected to appropriate modems located in the modem area. The lines carrying digital bit streams (d-c pulses) then pass through the Technical Control Area where channels may be tested, routing established, or transmission monitored. The lines then pass to the Crypto Area where government-furnished encryption and decryption equipments are located. From the Crypto Area the lines pass to the Center Operations Room where the Circuit Switching Units and the Message Switching Units are located.

Circuit switching provides direct user-to-user circuit connections through one or more switching centers. Circuit switching is usually found best for transmissions between a large number of like data sources each having a low message volume.

Message switching provides facilities for data transmissions between data sources that are unlike, that have large message loads (particularly long messages), and when the traffic distribution pattern is such that long delays would be expected during "busy" conditions if circuit switching were employed.

By providing a compatible interchange facility between the two switching units, AUTODIN preserves the best features of each method of switching. For example, circuit switching calls encountering busy conditions or those destined for unlike terminals can be routed to the message switching unit for subsequent delivery to the addressee.

The Circuit Switching Unit performs the primary functions normally attributed to a dial switching system for telephony; that is, it monitors the state of each incoming line to determine when a call requires service, processes incoming calls on a selective sequential basis, establishes and holds connections which are determined to be valid for circuit switching, transfers calls which are determined to be proper for message switching, and releases connections which are no longer required. However, no dials are used in AUTODIN terminal equipments. Rather than relying on manual call generation techniques such as the telephone dial, the standard AUTODIN message header is used. This feature, together with a supervisory control procedure, permits calls to be automatically generated and messages to be transmitted immediately upon establishment of the connection. The circuit switch unit employs space-division switching technique. Solid-state circuitry is used to the fullest extent practicable. The switching network and much of the relay logic are composed of dry reed switches. The modular size of this unit is 50 lines.

The Message Switching Unit employs store-and-forward concept and consequently is much larger and more complex than the Circuit Switching Unit. Each switched message is received in its entirety before it is relayed to the next center or to its destination. In addition to the functions normally associated with transmitting and receiving messages, which in the Message Switching Unit are time-shared among the input-output channels, the unit is provided with extensive random-access storage for received messages and with logic and memory necessary to recall and transmit these stored messages to the output channels in the required order, code, and format. These functions are performed by a computer and its peripheral devices under the control of a stored program. The input-output functions are performed by buffers and the Accumulation and Distribution Units. The message-processing functions are performed by the Communication Data Processor, which is a large-scale computer designed primarily for data handling and logical operations. In-transit message store is performed by Drum Storage Units and three Tape Stations designated as intermediate storage. Intercept Tape Storage is provided to accept messages for closed stations or those temporarily inoperable. Reference and Journal information are written on tape units. At some sites, messages for colocated data-processing facilities are received and transmitted by EDPE tapes via a Tape Station Converter. In addition to these equipments there are the usual system control console, monitor facilities, and a high-speed printer.

The reader desiring more detailed information on AUTODIN should study available publications.

20. IBM Data-collection and Data-transmission Systems

The International Business Machines Corporation manufactures for rental or for sale a number of data-collection and data-transmission systems that require the use of communication channels of various types. While these systems make use of some types of terminal equipments already discussed in previous paragraphs of this section, the assembly of these devices together with control equipments and other devices into systems requires a somewhat broader consideration of communication channel requirements.

The IBM 357 Data Collection System is a system for collecting and transferring data from remote in-plant locations to a central in-plant point where it can be used as input to a data-processing system.   The main advantages of the IBM 357 are speed, economy, flexibility, ease of operation, and automatic entry of time information.   As many as 20 IBM 350 Input Stations consisting of Model 4, badge only; Model 5, card only; or Model 6, badge and card readers, can be employed.   An optional feature at each input station is the IBM 372 Manual Entry Unit: Model 1, 6 slides; Model 2, 9 slides; or Model 3, 12 slides.   The recommended maximum transmitting distance is 7,500 ft.   Therefore only local cables or wire pairs are required.   At the central station an IBM 358 Input Control Unit scans all input stations, determines the one ready to transmit, and establishes a path for data flow from the input unit to a modified IBM 24 Nonprinting Card Punch or an IBM 26 Printing Card Punch which serves as an output for the system.   Read-out time is supplied by an IBM 360 Clock Read-Out Control and an IBM 361 Read-Out Clock, which are optional.   Time is entered in four digits which express hours and hundredths of hours.   One clock will serve up to 20 IBM 357 systems, and one Clock Read-Out Control will serve up to 35 output punches.   When desired, the IBM 357 may be operated on-line with an IBM 1440 or IBM 1460 Processor by the use of an IBM 1408 Transmission Unit and an IBM 1448 Transmission Control Unit.

The IBM 1030 Data Collection System uses the same principles as the IBM 357 system but offers greater speed and capacity, increased flexibility (including data-cartridge input), two-way communication, and the ability to operate on-line with a central computer.   There are two models of stations, the IBM 1031A for use with two-wire lines over which up to 10 stations can operate and the IBM 1031B for use with multiwire cabling.   Data may be entered via punched card, plastic badge, or keyboard or from a pocket-size data cartridge on which up to 12 numeric digits of data may be logged at a work point.   When inserted in the cartridge reader at the transmitting station, the data are transmitted to the central processor.   Each IBM 1031A station may be equipped with up to nine IBM 1033 printers for data dissemination.   When operated off-line (like the IBM 357) data are transmitted to an IBM 1034 Card Punch which can accommodate an appropriate combination of up to 24 Models A and B input stations at a speed of 60 characters per second.   When operated on-line, transmission to the IBM 1033 printers is at the rate of 14.8 characters per second, and the printer may be installed as much as 3,000 cable-feet from one of the IBM 1031A stations.   Time entries are made in conjunction with an IBM 1032 Time Unit, which serves an IBM 1448 Transmission Unit for on-line use with a computer, or the IBM 1034 Card Punch for off-line operation.

The IBM 870 Document Writing System is a system which prepares as many as two typewritten documents, two punched cards, and one punched telegraphic, five- or eight-channel paper tape simultaneously.   Information may be entered manually via a keyboard or automatically sensed by a punched-card-reading or telegraphic tape-reading mechanism.   These equipments are usually located at the branches of a multiple-branch data-transmission system.   While orders, bills, requisitions, checks, labels, and the like are being written at the branch point, the same data are simultaneously recorded on punched cards or paper tape.   The cards or tape (or both) can then be sent by companion transmission equipment to the central-office data-processing center.   These transmission devices have been described in previous paragraphs. The basic unit of the IBM 870 system is the IBM 834 or IBM 836 Control Unit, with or without five-channel paper-tape reader.   Data from input tape, cards, or keyboard may be entered.   In addition to program control by means of its wired panel, the control unit can perform key-punch operations.   Both the IBM 834 and IBM 836 have a card-reading and card-punching station and are equipped with a combination alphanumeric keyboard.   An auxilliary drum, as a source of constant data, can also be included.   The IBM 836 has a print unit in addition.   Optional input-output units are the IBM 972 Auxilliary Keyboard for prekeying of fixed and/or variable numeric data while the machine is performing other functions, IBM 866 nontransmitting typewriter (one or two) to permit typewriter output from any input source, IBM 962 Five-Channel Paper Tape Punch for tape punching, IBM 534 Card Punch (Auxiliary), and IBM 536 Printing Card Punch (Auxiliary). Operation is serial in nature, with one sensed or keyed character at a time being directed to a common channel through which it activates any output unit that is programmed ON. Operating speeds range from 9 to 20 card columns (characters) per second. Transmission speed to the central data processor depends on the type of transmission equipment employed and the type of communication channel available.

The IBM 1001 Data Transmission System is a low-cost means of transmitting alphabetical and numerical data from one location to another. It is composed of one or more transmitting terminals from which data may be transmitted over leased channels of voice bandwidth or over switched channels of DATA-PHONE (AT&T) or BROADBAND Switching Service (WU). Each terminal consists of a card reader, a 10-digit keyboard, functional keys, and an audio speaker. Numerical data may be either manually keyed into the system via the keyboard or entered automatically by inserting a prepunched card. Alphabetic information is from punched cards only. Up to 22 columns of information can be read from a single card with the numeric model and 35 columns with the alphanumeric model at a reading rate of 12 columns per second. The speaker emits different sounds to indicate that a connection has been established, that the remotely located card punch is not ready to receive, that a record was correctly transmitted, or that a record was incorrectly transmitted. At the central location an IBM 24 Card Punch or an IBM 26 Printing Card Punch is used. Several automatic checking features ensure validity of all transmitted characters and the correct length of each message. Transmitting and receiving data sets (modems) are supplied by the common carrier furnishing the communication service.

The IBM 1094 Line Entry Keyboard is a device for entering data directly into a system. It stores and transmits numeric-coded data over common-carrier communication facilities to a central processing location. Data to be transmitted are keyed into the keyboard by operation of push buttons arranged in rows on the panel of the keyboard. Operation is similar to that of the IBM 1001 Data Transmission System, and therefore a system may include under certain conditions a mixture of these two data input terminals. Data transmitted by either of these terminals is recorded on an IBM 24 Card Punch, Model 5 or 6, or an IBM 26 Printing Card Punch, Model 5 or 6. The keyboard is scanned and data transmitted at the rate of 12 characters per second. The communication channel may be a switched data channel of voice bandwidth as provided by DATA-PHONE or BROADBAND Switching Service or a leased private-line channel. These facilities must be equipped with an AT&T Model 401A for numeric or Model 401E Data Set for alphanumeric or equivalent at the keyboard location, and with an AT&T Model 401J Data Set or equivalent at the central processing center. The type of data set required at the keyboard location depends on whether special features are installed on the IBM 24 or IBM 26 Card Punches or whether the network is a mixed network. These special features are the alphabetic receive feature and the intermix feature.

The IBM 1009 Data Transmission Unit is a Tele-processing® unit that permits high-speed two-way transmission of alphabetic and numeric data between data-processing systems, including the IBM 1401 (except Model A), IBM 1440, IBM 1460, or any system using an IBM 1414 Input-Output Synchronizer (Model 4, 5, or 6). Transmission speeds range from 75 to 150 characters per second over switched or private leased circuits of voice bandwidth. A complete system terminal consists of an IBM 1009 with a companion data-processing installation and the communications-company facilities (modems, lines, etc.). Such a terminal can communicate with another IBM 1009 terminal with data-process system or other synchronous transmission units such as the IBM 7702 Magnetic Tape Transmission Terminal, an IBM 1013 Card Transmission Terminal, an IBM 7711 Data Communication Unit, or IBM System 360 via the 2701 Data Adapter Unit. The IBM 1009 accepts characters, one at a time, from processor storage. It checks for character validity and changes a valid character from binary-coded decimal form to transmission code. It then serializes the character bits and feeds them to communications-company facilities for transmission. At the same time it accumulates longitudinal check bits to be used by the receiving terminal to check the accuracy of transmission (a block error-control system). A communication channel of voice bandwidth is required. At the receiver, after accuracy of the block characters has been established, the characters are recoded into the proper code required by the output device. The speed of operation of the IBM 1009 is such that it is compatible with transmission requirements for intermediate to large-scale solid-state computers such as the IBM 1410, IBM 7070, IBM 7074, IBM 7080, or IBM 7090, including such computers when equipped with IBM 1301 Disc Storage Units. In such usage the IBM 1014 Remote Inquiry Unit can be used to provide remote access for as many as 20 inquiry stations each as much as 8 miles distant (when connected by metallic wires to the computer). Inquiries typed on the 44-character keyboard of the unit are transmitted to an IBM 1410 Computer, and after processing, the replies are transmitted back to the IBM 1014, where they are typed out on its printer. The IBM 1410 Computer may be equipped with an internal optional feature, the Telegraph Input-Output Feature, which makes telegraph stations capable of sending to or receiving from the core memory of the computer. If processing of an inquiry is requested, it is carried out automatically. Also, the IBM 1009 Data Transmission Unit, originally introduced for use with the IBM 1401 Computer, may be equipped with an IBM 1009 Data Transmission Unit Adapter, which makes it usable with an IBM 1410 Computer, in which type of operation the speed is increased to 250 or 300 characters per second. Thus, computer-to-computer transmission is available between IBM 1401's and IBM 1410's or between an IBM 1410 and an IBM 1401. At the higher speed of the IBM 1009 Adapter, the communication channel should be a leased circuit of voice bandwidth or a BROADBAND Switched Service circuit.

The IBM 1448 Transmission Control Unit directs and regulates the flow of data into a data-processing system (such as the IBM 1440 or IBM 1460) from as many as 40 half-duplex stations on either multipoint or point-to-point circuits. It links the data-processing system with a data-collection system (such as the IBM 357 or IBM 1030) or with data-communication systems (such as the IBM 1050 or IBM 1060) and thus becomes an integral part of the processor. The IBM 1448 is a multiple low-speed communication-line scanning device that polls the outstations or lines under control of the processor program, accepting priority data first (priority interrupt) and subsequently accepting data from other points in sequence. Operation is similar to the AT&T 83-B-1 Teletypewriter Selective Calling System (see Chap. 8, paragraph 23). Speed of operation of the IBM 1448 is 14.8 or 60 characters per second, depending on the type of station terminal. A communication channel of voice bandwidth is normally employed.

The IBM 1050 Data Communications System is a multipurpose office-oriented Tele-processing® system. It operates over leased or switched communication channels that can accommodate 14.8 characters per second (approximately 150 words per minute). Transmission is half-duplex between a central point and a remote terminal or between two remote terminals. This modular system consists of units that can be used in various combinations to meet specific transmission, receiving, and recording needs of business operations. It is specifically applicable for management control systems for firms with more than one location. The system consists of an IBM 1051 Control Unit and six different units for input and output which can be added to tailor the installation to individual requirements. A remotely located IBM 1050 can communicate with a central processing center if its processor is equipped with a suitable transmission control unit or console. Choice of equipments to transmit or receive are the IBM 1052 Printer Keyboard, the IBM 1053 Printer, the IBM 1055 Paper Tape Punch, the IBM 1054 Paper Tape Reader, the IBM 1056 Card Reader, the IBM 1057 Card Punch, or the IBM 1058 Printing Card Punch. Both on-line and off-line functions can be performed, and data transmissions are electrically checked for errors.

The IBM 1060 Data Communications System is intended for use by such institutions as savings banks, savings and loan associations, and commercial banks. It links teller-window locations, either within the main office or in branch offices, with a central data-processing system. The local stations are cable-connected to the IBM 1448 Transmission Control Unit of the processor, and the remote stations are connected by leased local communication-company facilities. Basic units are the IBM

1061 Control Unit and the IBM 1062 Teller Terminal, both specialized devices to handle banking transactions.

The IBM 7710 Data Communication Unit provides data transmission between an IBM 1401 Data Processing System and another IBM 7710 with IBM 1401, an IBM 1009 Data Transmission Unit with data processor, an IBM 7701 or IBM 7702 Magnetic Tape Transmission Terminal, or an IBM 1013 Card Transmission Terminal or other IBM synchronous transmission units. Transmission is at the rate of 150, 250, or 300 characters per second, which requires a communication channel of voice bandwidth, or at the rate of 5,100 characters per second on broadband communication services such as microwave or coaxial-cable channels. The processor program controls the movement of data, a character at a time, in parallel bits, from the computer storage to the IBM 7710. It in turn then controls the transmission to the distant terminal. The IBM 7710 can transmit intermixed binary and BCD alphanumeric records, and at the receiving station each character is checked automatically. This transmission check and record check, plus the input and output checks of the IBM 1401, maintain consistent accuracy of received data.

The IBM 7740 Communication Control System, operating alone or with a computer, provides complete control for a communications network. The IBM 7740 (with optional disk storage) is a center not only for communication control but also for message accounting, logging, traffic reporting, error handling, and other procedures allied to complete communications control. It may also perform operations such as remote data collection, in which batched input records are stored in its disk files for later processing while already updated records are available for immediate retrieval on request from remote terminals. In addition, the IBM 7740, with or without disk storage, may be connected to a data-processing system to provide a Tele-processing® system of large capability, with provision for both present and expected increases in communications traffic. The IBM 7740 is a computer tailored to communications operations. The result is automatic, integrated, rapid handling of the many communications-center operations that previously were manual or semiautomatic, often performed in separate areas. The specific operating characteristics of the system follow: The IBM 7740 supervises the communications network; that is, it controls traffic by polling sending terminals, addressing receiving terminals, checking communications lines, and maintaining queue control. The IBM 7740 Communication Control System with disk storage accepts signals simultaneously from a number of high- and low-speed lines, converts these signals to bits, and then assembles them into characters and, finally, messages. Also, it performs message protection, error checking, code conversion, and editing of the messages. It relays them to disk storage or to a remote terminal on one of the communications lines, or both. With disk storage it is an integrated system encompassing all the functions currently performed by manual, torn-tape, and automatic and semiautomatic switching systems. The speed of the system is designed for peak load traffic. It receives and processes incoming messages and immediately switches them to output queues. As soon as each of the required lines becomes available, the messages are transmitted. With its error-checking features, the IBM 7740 assures not only that messages are speedily forwarded from incoming to outgoing lines but also that they are forwarded accurately. With its code and format conversion and flexibility of acceptable character formats (5 through 8 bits per character), it provides for changes in format and character structure of a message between its incoming and outgoing states, thus allowing different types of terminals to communicate with each other. The system processes for immediate forwarding to the addressees individual messages, group messages (those coded for certain standard distributions), or multiple messages (those containing several addresses in the header). Since programmed control is flexible, the system can handle any header format that has been anticipated by the programmer. The IBM 7740 Communication Control System with disk storage, while operating independently to control a communications network, provides the option of direct entry to and exit from IBM 7000 series or the IBM 1410 data-processing systems. When the IBM 7740, with or without disk storage, is attached to an associated data-processing system, it relays messages to the computer.   Messages originating in the computer or stored in the IBM 1311 Disk Storage Drives are received from the computer system or disk storage, checked, code converted, edited, and transmitted.   IBM 1311 Disk Storage can provide, in addition to the IBM 7741 core storage, more than 13.4 million characters of random-access storage for logging data, for traffic reporting, and for billing information, as well as for tables, programs, and extra space for message buffering.   A disk pack can be changed in less than 2 min and is interchangeable with disk packs produced by other systems using IBM 1311 disk storage.   This provides for later expansion of data-processing operations.   Average access of 150 msec to any of the 13.4 million characters makes it possible to dump files rapidly into (or recover files from) core storage.   Thus, core storage in the IBM 7740 system is freed by using the IBM 1311 Disk Storage Drives for a much greater input-output message buffering potential than would be possible otherwise.   The IBM 7740 without disk storage can connect to as many as 84 half-duplex, low-speed communications lines with transmission rates up to 200 bits per second and as many as two half-duplex, high-speed lines with transmission rates up to 2,400 bits per second or as many as 56 half-duplex, low-speed communications lines and as many as 4 half-duplex, high-speed lines.   In the IBM 7740 system with disk storage, the maximum number of communications lines is 84 low-speed, half-duplex lines and no high-speed lines or 56 low-speed, half-duplex lines and two high-speed, half-duplex lines.   Note that, in planning line configurations, one full-duplex line can be substituted for two half-duplex lines for telegraph and high-speed lines.   The number and kinds of remote terminals that can communicate with the IBM 7740 and the data-processing systems to which it may be locally attached, together with additional detailed information concerning the system, are all readily available in an IBM publication.*

The IBM 7770 Audio Response Unit can be attached to the IBM 1401, 1440, or 1460 Data Processing Systems through the IBM 1311 File Control Channel.   It provides audio response to inquiries made from telephone-type terminals, IBM 1001 Data Transmission Terminals, or other similar terminals.   The composed audio response (speech) comes from a vocabulary prerecorded on a magnetic drum in the IBM 7770. Connection between the inquiry terminals operates in a half-duplex mode over common-carrier switched telephone/data networks, common-carrier leased private-line circuits, or privately owned communication networks of voice bandwidth.   With the IBM 1001 as the inquiry device, an appropriate data-set transmitter (AT&T Data Set 401K or equivalent) must be employed, and at the IBM 7770 location an appropriate data-set receiver (AT&T Data Set 401L or equivalent) is required.   Other types of data sets are required for various types of telephone terminal sets.   Further information is readily available from an IBM publication.†   Similar audio response units have been developed for attachment to IBM 1410 or 7010 Data Processing Systems (7770 Model 2) and the IBM System 360 (7770 Model 3).   In addition, a smaller audio response unit, the IBM 7772, may be connected to the IBM System 360.   Full information on these systems is contained in current IBM publications.‡

The IBM 7711 Data Communication Unit is a Tele-processing® system component that provides medium-speed and high-speed digital-data transmission between an IBM magnetic-tape unit and a remote IBM magnetic-tape unit or other input-output equipment.   It can communicate over communication channels of voice bandwidth as well as on broadband facilities as provided by communications common carriers, or it can be used with equivalent private communication facilities.   The IBM 7711 reads data from a magnetic-tape unit, record by record, into its core storage buffer for speed synchronization with the communications modem and for conversion to a fixed-count serial transmission code (block code).   The basic IBM 7711 will attach to either an IBM 7730 Tape Unit or IBM 729 II, 729 IV, 729 V, or 729 VI Tape Units.   It can

* *IBM 7740 Communication Control System, System Summary*, IBM Reference Library, Publication File T-P-00, Form A22-6752.
† *IBM 7770 Audio Response System*, IBM Reference Library, Publication File 7770-01, Form A22-6800.
‡ *IBM 7770 Audio Response Unit with 1410 and 7010 Systems* (Model 2), IBM Reference Library, Publication File 7770-01, Form A22-6804.

transmit data to or receive data from such remote input-output equipment as the IBM 1009, 1013, 7701, 7702, 7710, 7740, 7750, or 7711 Tele-processing® Terminals or System 360 via the 2701 Data Adapter Unit. Serial transmission is used between local and remote units at signaling speeds of 1,200, 2,000, or 2,400 bits per second on channels of voice bandwidth or at speeds up to 230,000 bits per second on broadband channels (150 to 28,800 characters per second). An EIA Standard RS232A communication interface is usable with voice-type channels, and a special high-speed interface is usable with broadband channels. All characters entering or leaving an IBM 7711 are error-checked by parity or 4-of-8 code, longitudinal parity, fixed bit count, and lost or duplicate records check. Erroneous records detected by the transmitting terminal are corrected before transmission. Erroneous records detected by the receiving terminal are corrected by automatic retransmission. Additional information may be obtained from an IBM publication.*

The IBM System 360 is a new information-handling equipment that combines significant advances in computer organization with a built-in communications capability not previously available. The system can utilize, as required, more than 40 types of peripheral equipment which store information and enter it into and retrieve it from the computer. The IBM System 360 can respond to inquiries and messages from hundreds of simultaneously communicating terminal devices while the computer continues to process information already received. The system has universal code capability, as it provides defined operations for processing either the widely used BCD interchange code or the recently adopted American Standard Code for Information Interchange or any other code containing not more than 8 information bits. An additional advantage of the 8-bit character is the ability to pack two decimal digits in each byte (8 code bits plus 1 parity), thus providing high data-packing efficiency, and this is implemented because decimal data occur in business records more than twice as frequently as alphabetical. In communicating with terminal input-output devices, the code, format, and speed must be identical with values designed into the device; hence a wide range of capabilities were incorporated in the IBM System 360. Communication lines and data devices are attached to a control unit that performs character assembly and transmission control. The control unit may be either an IBM 2701 Data Adapter Unit or the IBM 2702 Transmission Control. Each IBM 2701 Data Adapter Unit may be equipped with any one of 10 different transmission adapters, and there are possible combinations of adapters that can be connected to the same channel or to a second channel, depending on the selected configuration. Operation is either simplex (one-way) or half-duplex (two ways alternately). The IBM 2702 Transmission Control enables IBM System 360 to combine data processing and data communications within the same system configuration. It directs and controls information flow between the system and a variety of remote terminals utilizing various terminal controls, line adapters, and special features. A complete description of the IBM System 360 is available.† A description of the IBM 1070 Process Communications System is also included, which enables the IBM System 360 to control natural gas and petroleum pipelines, utility distribution systems, and other collection and control functions associated with process control.

21. AT&T 1A Data Selective Calling System

This data-transmission system can be considered as two separate systems, one a high-speed network and the other made up of the low-speed extensions of the high-speed network. The low-speed extensions may be five-level Baudot code versions of the AT&T 83-B-1 Teletypewriter Selective Calling System (see paragraph 23, Chap. 8), or the extensions may be eight-level Data Interchange Code (ASCII) operated, in which case the system is known as the 1A-8X System. The high-speed network will operate in the same manner whichever code version is used on the low-speed exten-

* *IBM 7711 Data Communication Unit*, IBM Reference Library, Publication File T-P-01, Form A22-6808.
† *IBM System 360 System Summary*, IBM Reference Library, Publication File S 360-00, Form A22-6810-1.

sions. The speed of operation of the low-speed extensions will be either 75 or 110 bits per second in the half-duplex mode. Loops of the d-c type will be used at 75 bits per second, but loops of the a-c type obtained with modified Data Sets 130 and 103 will be used at the 110-bit-per-second speed between the Teletypewriter Set and the local central office. The network may include hubbing points serving more than one terminal station, and either a standard 143A2 Regenerative Repeater or a 143A2 Regenerative Repeater modified to handle the eight-level code at 110 bits per second may be used as required. Feeder channels of the carrier type are derived from a standard 43A1 System for 75-bit-per-second transmission or a modified 43A1 for 100-bit-per-second transmission. Since both data channel speeds are higher than the standard speed (45 bits per second) of the 11 Type Station Controller (part of the 83-B-1 System), speed converters are required when data stations are connected to it.

The multipoint high-speed network is operated over Schedule 4 Type 4A channels of voice bandwidth utilizing Data Sets 202B at a speed of 1,050 bits per second. Local lines of the four-wire type are utilized for connection (with Data Sets 202B) to either 44-type or 46-type bridges on the four-wire intercity backbone channels of the trunking network. This arrangement permits each station to communicate with each other station on the network. The stations operate under control of a master station, and in the half-duplex mode only one station can transmit data traffic at a given time. However, a full four-wire multipoint data network can be established by using two bridges at each interconnection point, in which case full-duplex transmission is available over the trunking portion of the network. Some of the existing multipoint networks provided to the airlines are designed in this manner.

22. ITT 7300 ADX, Automatic Data Exchange

The ITT Data and Information Systems Division of the International Telephone and Telegraph Corporation designed the system, supplied the equipment, and developed the programming for a number of 7300 ADX installations, the first of which was for Eastern Air Lines in November, 1963. Thus it became the first commercially available electronic switching system designed to meet the record-communications and data-transmission requirements of modern business enterprise. The 7300 ADX is a fully automated message switching and control system utilizing reliable solid-state circuitry that represents a synthesis of communication and computer technologies. As systems for different applications require different configurations, modularity of design for growth in volume and expansion of functions, flexibility through use of internal stored program, and the capability for instantaneous or delayed transmission, random or batched data handling and automatic processing of priorities all become important design features of the ADX system. Additional capabilities are automatic translation of messages from any coded format to any other coded format, automatic control of a large number of transmission devices which operate at different speeds, and the ability to perform message accounting.

The ADX system, under program control, monitors and controls the switching and processing of data messages. A complete system includes all necessary units and hardware to permit interconnection of computers and as many as 400 terminal devices over various transmission channels such as telegraph, telephone, or broadband circuits. Flexibility brought about by the integration of engineering and programming makes possible a number of operating features:

Flexible and automatic routing of multiple-address, multiple-priority, group-address, and broadcast transmissions.

Better line utilization and faster handling of multiple-addressed messages.

Automatic handling of overflow traffic through diversion to magnetic-tape or magnetic-disk storage.

Automatic speed conversion is inherent owing to design (for example, 60-word-per-minute teletypes can be interconnected with 2,400-bit-per-second machines).

Automatic code conversion by programming.

Automatic detection of transmission and internal errors, with error control programmed in accordance with customer requirements.

Insertion of the date and time on messages at input to or output from the communication center.

Insertion of channel-identification numbers at input to or output from the communication center.

Automatic on-line trouble diagnosis routines to detect, bypass, and report faulty lines and equipment.

Automatic message traffic accounting.

Automatic alternate routing.

The 7300 ADX System components generally consist of a Multiplexed Message Processor (MMP), Magnetic Tape Units and their associated controls, and cabinets containing Line Termination Units, Level Converters, and the Master and Dual Line Switches. Optional units include Magnetic Disk Units, Electric Typewriters, Cathode Ray Display Units, as well as additional core and magnetic-tape storage.

The Multiplexed Message Processor serves as the control and nerve center for the ADX System. It includes the Operator Console, Control Circuits, and the core memory, which provides for program and working storage. The storage element is magnetic core, arranged in 1 to 16 banks of 4,096 words each, for a maximum capacity of 65,536 MMP words. The additional storage units require additional cabinets, each of which will accommodate two storage elements and their associated circuitry.

An MMP word consists of 18 binary bits. Assuming three 6-bit characters per word, it is possible to store 12,288 characters per memory bank, or a maximum of 98,304 characters.

Cycle time in the MMP is 5 μsec. For single-access instructions, the system is capable of performing 200,000 operations per second. For double-access instructions, the system can perform 100,000 operations per second.

Tape Controls are capable of controlling a maximum of eight Magnetic Tape Units each. Three Tape Controls can be included in the standard ADX System.

A Line Termination Unit is required for each incoming and outgoing communications line. The cabinets in which the line units are housed also include the Level Converters and the Line Switches. Transmission lines of the same speed may be grouped and terminated as a group. A maximum of eight lines may be terminated in a group. High-speed data lines are terminated individually.

Each ADX System is independently engineered to meet the special requirements of the customer, and system configurations can be easily adapted to changing needs. Many changes can be made without changing hardware. If, for example, the user adds stations or assigns different priorities to the operating stations, complete reprogramming is not necessary; simple modifications will suffice. All programs operate independently, and changes in a program are easy to make.

Programming concepts, information on message processing, a detailed description of system components, and information on applications may be obtained from an ITT publication.*

23. Collins Data Central

The Collins Radio Company† has developed an automatic, electronic message switching and processing system designated the Collins Data Central. This system provides direct, multiple-address, and mixed-speed telegraph input and output. It converts code, speed, and format automatically. Messages are electronically transferred to addressed circuits immediately as they are received. There is no manual handling of telegraph tapes or cross-office delay other than processing time of the computer. If an addressed circuit is busy, the message for that station is temporarily stored and forwarded as soon as the circuit is free. Automatic message accounting, queueing, and checkoff prevents lost messages, and historical files permit easy message retrieval for analysis, logging, or retransmission.

The central processing equipment is the Collins C-8401 Communication Oriented

---

* *ITT 7300 ADX System Features Manual*, ITT Data and Information Systems Division, 320 Park Ave., New York, N.Y.

† Communication and Data Systems Division, Collins Radio Company, Dallas, Tex.

Processor. The processor regulates traffic, acknowledges receipt of messages, and assigns sequence numbers for message accounting and retrieval. Outstations are controlled automatically to achieve maximum system efficiency. The unique "Programmed Logic" of the C-8401 processor has built-in communications flexibility and can be programmed to fit a wide variety of applications or changed to meet new requirements.

A typical outstation is equipped with a Model 28 ASR Teletypewriter equipped with stunt box or equivalent control device. The common language for system control is five-level Baudot teletypewriter code.

Peripheral equipment to complement the Data Central Processor includes data-transmission sets, C-8046 47 Magnetic and C-8811/12 Paper Tape Units, card read-punch units, high-speed printers, C-8076 Magnetic Disc Files, teletypewriters, and other specialized equipment optionally.

Modular construction and a common set of circuit cards ensure system flexibility to increase capacity to a maximum of 256 circuits.

Polling sequences are used on multiple-station circuits for transmission control. Polling tables containing station addresses are held in main core storage for each circuit. Outstation equipment includes selective calling units and answer-back units so that the processor can sequence the messages from each outlying station.

24. CONTROL DATA® 8050 Message and Data Switching System

The Control Data Corporation* is producing for commercial sale a general-purpose data-transmission system for handling information of varying codes and speeds and for collecting and disseminating data to remote points. A typical 8050 Message and Data Switching System can function with communication channels of the telegraph type operating at 60, 75, or 100 words per minute or with medium-speed digital data channels operating as high as 2,400 bits per second. Incoming information is inserted into the magnetic core memory of a stored program computer. Outgoing data are then automatically relayed to remote stations from core memory.

Each 8050 electronic switching system utilizes a standard Control Data computer, such as the 160-A, as the controlling element which performs the functions of inspecting incoming messages and data, temporarily storing this in memory and then sending the data to remote stations. Choice of the message heading and format is flexible and may be tailored to the user's needs. Provision is also made for single and multiple addressing, group addressing, and priority handling. As in other electronic switching systems, the number and type of buffering units required depend on the operational features desired in the system. The 8050 system is generally equipped with 8155-A-01 16-channel Digital Multiplex Units arranged for 5-bit code operation or 8155-B-01 16-channel Digital Multiplex Units arranged for 8-bit code operation. These multiplexer units are appropriately connected to Telegraph Channel Units, Digital Data Transfer Units, or a selection of several different terminal units that are arranged for specific purposes. Also, the system may be operated through Control Data 8061 Digital Communications Terminal, which provides a means of connecting to geographically separated data computing or control systems by means of communication channels of voice bandwidth. These communications terminals may be equipped with Digital Communications Units, of which there are several varieties, that are arranged for interfacing with the modems or data sets provided by the common carriers. Also, high-speed digital communications terminals are available which operate at 40.8 kilobits per second or at 2.5 megabits per second, which, of course, require broadband channels.

The 8050 computer may also be equipped with various peripheral devices, such as magnetic-drum memory units, magnetic-tape units, or high-speed perforated paper-tape equipments.

The Western Union Telegraph Company utilizes Control Data equipment in the Western Union 301 System. While each 301 System is especially engineered for

* Industrial Data Processing Division, Control Data Corporation, 8100 34th Ave. South, Minneapolis 20, Minn.

different applications, a system generally includes a CDC 160-A computer, a CDC Magnetic Drum Memory, and one or more CDC Magnetic Tape Transports. However, the Western Union 301 System uses a Communications Line Multiplexer Unit of Western Union design which also contains specialized buffering devices essential to different applications.

25. UNIVAC® 490 Real-time System

A number of business firms have made use of centralized order processing and inventory control systems. One such firm, Westinghouse, had been utilizing a tele-type switching network to receive data messages as well as administrative messages. The switching center for this network was located at Pittsburgh, and transit messages were handled by reperforator switching of perforated tape, but data messages were first perforated on paper tape, and then punch cards were prepared which were used as input to the computer. The computer program then performed certain checking and editing functions and proceeded to locate the desired numbers in its file inventory records. If the item was out of stock at the warehouse nearest the customer, it was searched for at warehouses progressively nearer the factory and finally at the factory itself in order to minimize transportation costs. The item having been located, the computer provided cards from which a teletype message tape was generated, directing the warehouse to ship the item and preparing the labels, bills of lading, and packing lists for the item on receiving machines at the warehouse. At the same time, price extensions and sales taxes were calculated and the invoice printed ready for mailing. Inventory records were updated in the process and examined for reordering points. Additional cards, used later for the development of sales statistics, were also produced during the processing of an order.

As the above system grew in size and included a larger number of product lines, it became possible to consider large-scale, high-speed computers with high-capacity, random-access memory units and significant advances in data transmission. It was decided to utilize the computer as an integral part of the existing communication system as well as for processing functions. The UNIVAC 490 System was selected to perform the above functions, and installation was completed in the fall of 1962. The initial equipment complement consisted of a UNIVAC 490 Central Processor with 32,000-word core memory, a Type FH-880 Drum Memory with 75,000-word capacity, a disk memory of 80-million-character capacity, and 12 Uniservo II-A Tape Units. Communications control units, which provide the linkage between teletype lines and central processor, make up a substantial portion of the system. The common language is the five-level Baudot code. Approximately 100 offices, warehouses, factories, and other points are served. Additional information may be obtained from The Director, Westinghouse Tele-Computer Center, Westinghouse Electric Corporation, Pittsburgh, Pa.

While the above installation may be considered typical, there have been a substantial number of additional UNIVAC 490 computer installations which serve the specialized needs of different firms.

26. GSA Advanced Record System

In January, 1962, the General Services Administration announced plans for an ultramodern Advanced Record System (ARS). The principal purpose was to develop a common-user record communications system to replace fragmented and duplicative communications among civilian agencies and in so doing to provide an economical and efficient communications system having enhanced transmission capabilities, including data transmission, and one that would meet day-to-day requirements as well as emergency needs. The GSA prepared a general description of the desired system, and this was submitted to industry for bids in August, 1963. On December 31, 1963, The Western Union Telegraph Company was selected as the prime contractor to provide overall system supervision and engineering and to supply the equipment for the system. The UNIVAC Division of Sperry-Rand and the ITT Federal Laboratories

Division of IT&T are subcontractors for computer and switching equipments.

Basic requirements call for a switching system among 1,600 data stations which will link government civilian agencies in 600 cities. The ARS system has to be economical, modular, and simple to manage. All communications are to be completely private, fully protected against disaster, and capable of cryptographic transmission. Printed telegraph, facsimile, and high-speed data transmission is to be the fastest and most accurate that advanced technology can produce. In addition to handling routine communications, the system must provide sufficient information-processing capabilities to serve the data requirements of its many and varied users. Furthermore, the system is to be compatible with the AUTODIN network of the Department of Defense for instant interconnection during any national emergency.

Automatic, direct, dial-up connections between any two stations in the network are being provided by exchange offices and a circuit-switching network, using solid-state common control equipment and dry reed switching matrices. Also, automatic acknowledgment and end-of-message cutoff are provided. The additional requirements of handling of messages in order of priority, communications with noncompatible subscribers, code and format conversion among different agencies, and storage and routing to other networks are met by the marriage of automatic direct-dialing circuit switching with high-speed, solid-state computer-oriented message switching. Computers will be used for automatic processing of multiaddress messages and messages destined to other networks, such as AUTODIN, and for the automatic handling of refile messages to Western Union Telex and TWX and manual handling of refile messages to the Western Union Public Message System. The Advanced Record System will accommodate printing telegraph, facsimile, and high-speed data communications.

Terminal equipment will consist of Model 33 Teletypewriter Sets for low-volume stations and Model 35 Teletypewriter Sets where there is heavy volume and where stations require an automatic dialing unit and/or the ability to simultaneously prepare perforated tape as a message is being received. From the foregoing it will be observed that the network will be a full-duplex network utilizing the ASCII code as common language. Data are introduced by means of preprepared perforated paper tape or by means of card readers associated with UNIVAC 1004 subsystem equipment.

Initially, the circuit switching network will consist of three junction offices and 24 switching exchanges, or district offices. Subscribers to the ARS will be connected to the district offices, and each of the district offices will have two trunking groups to junction offices. Circuit switching equipment will be supplied by ITT Federal Laboratories.

Message switching is provided by a UNIVAC 418 computer and various peripheral equipments located at each of the three junction offices. Peripheral equipment includes the control console, tape control, and a multiplicity of UNISERVOS IIIC, as well as drum control and a UNIVAC magnetic drum. A UNIVAC 1004 subsystem will also be installed at each junction station.

The ARS system is to be a completely private network and will not share its switching equipment or trunking facilities with commercial networks or any other government system. Protection from interruption is enhanced because of the usage of the Western Union microwave radio beam network for a large part of the transmission facilities used for ARS.

27. Additional Data-transmission Systems

A number of other electronic-equipment manufacturers have designed data-transmission systems for various uses. As data transmission is a rapidly growing art, it is impossible to cover all systems because of space limitations. However, there are a few new systems worthy of note. For example, Automatic Electric* has designed a digital telemetering and supervisory control system for pipelines. Another example

* Automatic Electric Sales Corporation, Subsidiary of General Telephone and Electronics Corporation, North Lake, Ill., *Bull.* 1008, April, 1962.

is the installation of an Electronic Telegraph System* by RCA Communications, Inc., which utilizes two RCA high-speed digital computers to receive, examine, store, route, and transmit messages to and from commercial customers all over the world. A third example is the new Electronic Data Transmission System developed by the Kleinschmidt Division of the Smith Corona Marchant Corporation, which combines a high-speed fully transistorized data printer, paper-tape punch, and photoelectric tape reader. A foreign development of interest is the British Post Office DATEL 600 Service, which will begin operation in January, 1965. This system will enable data to be transmitted over regular telephone lines between special equipment at the calling and receiving stations. Transmission will be within the speed range of 600 to 1,200 bits per second. Alternately, DATEL services will include data-transmission facilities on telegraph circuits at speeds up to 100 bits per second and also the intermediary speeds of 200 and 300 bits per second, which are to be introduced in 1966. Also, Standard Telephones & Cables, Ltd., the British subsidiary of IT&T, will market advanced on-line real-time communication message switching systems and other computer-based systems.

28. Glossary

A glossary of communication terms has been compiled by members of American Standards Association Committee X 3.3.2 entitled *Tentative Communications Glossary*. This glossary is to be considered as a working document, as it is being subjected to considerable editing and critical review. However, it is hoped the final edition will be available soon. This document contains many terms applicable to data transmission.

* World's First Computorized Commercial Telegraph System, *Wire Radio Commun.*, February, 1964, pp. 24–26.

Chapter 13

CLOSED-CIRCUIT TELEVISION

PIERRE MERTZ, *Consultant**

FUNDAMENTALS

1. Fundamental Characteristics

Television is a communications system for converting a picture into an electrical signal, which is transmitted over a distance, then converted back into a picture. The pictures are changed so frequently that they appear continuous to the eye.

The picture is a brightness variable in two dimensions, which in addition changes with time. The electrical signal is merely an intensity variable in time. It is therefore necessary to establish a code relation between the picture and the signal.[1,2] This code is the scanning process, as in Fig. 1. A spot area, as at *a*, moves across and down the field of view to cover the entire picture in a scanning pattern. It repeats this so fast that in an inverse process at the receiver the eye of the viewer sees only the picture and cannot trace the spot motion.

This coded scanning pattern has imperfections which lead to picture defects and also to inefficiency in the carrying of the picture information by the electrical signal. Both of these have important consequences to be noted later.

Some of the imperfections of the scanning process are relieved by the use of an interlaced pattern, as in Fig. 2. Here alternate passages through the picture are different. The first and successive odd passages are shown by the solid lines. The second and successive even passages are interlaced between the first and shown by the dashed lines. The interlaced pattern is used universally in broadcast television.

The electrical signal, with either sequential or interlaced scanning, has a frequency spectrum of Fourier components[1] which looks generally like Fig. 3. It consists of clusters of high-energy components separated by low-energy regions. The intervals are equal to the scanning line frequency.

Besides the details in a picture it is generally important to indicate to the receiver its overall or average brightness or darkness. The Fourier component in the signal that carries this information is at zero frequency, near and including which there is usually a cluster of high-intensity components, as shown in Fig. 3. This has serious consequences because most transmission lines do not normally transmit zero frequency, or direct current, satisfactorily if at all. Thus some arrangement to be noted later is necessary to carry this part of the information.

The coding that ties together the picture and the signal in Figs. 1 and 2 needs a very close correlation with the inverse decoding at the receiver if the picture is to be

* Retired from Bell Telephone Laboratories, Inc.

acceptable. Consequently in most television systems (and in all broadcast television) a signal to carry this coding is multiplexed on to the picture signal. This is called the "synchronizing signal." It is illustrated in Fig. 5.

In the first place, in the scanning pattern of Fig. 1, the spot a, after having moved to the extreme right of the picture, must get back to the extreme left to start a fresh scanning line. This return path is shown dotted and is traversed very fast. It is necessary that the electron beam at the receiver be suppressed during this interval so that it will not show in the picture. This is accomplished by a pulse in the signal, inserted at the camera end, as in Fig. 4. The black-and-white signal levels show here, and the "blanking" level is set a little beyond the black ("blacker than black") to give some margin to ensure suppression. The duration of the pulse is a little more than the retrace time, again to give some margin.

Fig. 1. Sequential scanning.

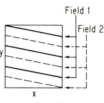

Fig. 2. Interlaced scanning.

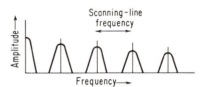

Fig. 3. Television video spectrum.

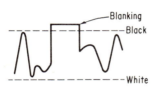

Fig. 4. Blanking pulse in video signal.

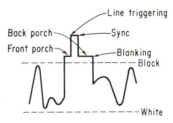

Fig. 5. Blanking and line synchronizing pulses in video signal.

A further correlation is required in the code, namely, the exact instant when the spot a starts to move to the right to begin a new scanning line. This is indicated by a "sync pulse" superposed on the blanking, as in Fig. 5, whose leading edge triggers the line. A so-called "front porch" and a "back porch" remain to indicate blanking level.

It is, of course, similarly necessary to deal with blanking and synchronization when the spot a in Fig. 1 returns from the bottom to the top of the picture. Essentially this is done in exactly the same way as for the line return, but it is complicated by the fact that it must be superposed on and at the same time distinguished from the line return signal, and also it is complicated by the interlace.

The distinction from the line return is secured by making the signal very much longer, since the spot requires, say, on the order of a dozen line durations to go back to the top of the picture. The superposition problem is to continue triggering scanning lines during this interval, and this is solved by serrating the long pulse in a rather complicated pattern. The details of this pattern will not be gone into here, as they have already been amply covered.[2]

The broadcast television signal configuration for the United States has been very carefully standardized by the Federal Communications Commission, with technical help from the industry. European and Japanese signals are generally similar, but there is some difference in detail[2] (particularly marked in the case of the British signal).

2. Distinctive Characteristics of Television Transmission

The immediate distinctive characteristic of television transmission is the wide frequency band which it needs.[1,3] If one assumes 500 scanning lines and 600 picture elements along each, this makes 300,000 total elements in the entire picture. To permit smooth merging of one picture in the next, through persistence of vision, one can assume 30 complete pictures per second. This means 9 million picture elements per second. If each picture element is reproducible by a half cycle of signal wave, it requires a 4.5-Mc frequency band, which is about the width used for the standard television broadcast picture signal.

This is an enormous width as compared with that for a telephone facility (about 3,000 cycles) or for a high-fidelity sound circuit (about 15,000 cycles). Because of this, television has revolutionized communications engineering.

This great bandwidth has generated a constant pressure to devise means for compressing it. In the first place it is obvious from Fig. 3 that a significant part of the spectrum is wasted in the regions of very low signal intensity. Also a normal picture is not completely random and shows broad structure correlations among adjacent and near-adjacent picture elements, which in turn means redundancy in the conventional signal.[4] An enormous amount of effort has gone into the study of ingenious ways to circumvent these inefficiencies.[5] Some of this has been perfectly sound physically and has led to perfectly feasible solutions. However, so far very few if any of these solutions have found real commercial application. The reason in most cases is that they require much apparatus complication, more than economically warranted. This means, of course, that there is always the possibility of a simple and successful device. Sometimes the bandwidth is reduced by a very simple reduction in frame repetition rate or detail or both.

The second distinctive peculiarity of television signals is the need for transmitting direct current. This is not altogether new, as it also has long existed in telegraphy and facsimile. Telegraphy has surmounted it by using special types of circuits and later, carrier systems. Facsimile has also gone to carrier. Television also uses carrier where it can, but the extremely broad band makes this more of a problem. Thus considerable transmission is necessary at baseband (which, in television, is called "video" transmission as an analogy to "audio" transmission for voice signals). For video transmission the d-c requirement imposes special devices, to be noted later.[6]

A third peculiarity of the television signal (which again it shares with telegraphy and facsimile) is that it requires a faithful retention of its wave shape. This means that the transmission medium through which it is propagated must be free of phase distortion.[7] It is in quite marked distinction from telephony, around which by far the greatest part of the communications art has been developed and for which phase-measuring and -correcting equipment has never been considered important. Phase measurement and correction are used in data and facsimile transmission, but the problem is more serious in television because of the wider frequency band and the much tighter tolerances, as measured in microseconds of delay.

For convenience, the phase is not measured directly, but instead a related quantity, the "envelope delay," is used. This is the delay of the envelope of a spurt of signal, or wave, of a given frequency in its passage through the transmission medium. For distortionless transmission it is necessary that this spurt envelope arrive at the same time at the receiver for all frequencies in the utilized range. That is to say, the envelope delay of the transmission medium must be flat over the frequency range within a rather narrow tolerance.[8]

Still another peculiarity of television is its sensitiveness, in the quality of the reproduced picture, to pattern types of interference. This it shares with facsimile. The eye is extremely sensitive to patterns in a picture and detects superposed patterns even when they are very faint. The most common pattern is a simple single frequency superposed on the field of view, where it appears as parallel bars. It can come about either as additive single-frequency noise or as a modulation product. This means that both of these must be kept well down if the use planned for the television system involves artistic or other sensitiveness to the appearance of bar patterns (see paragraphs 13*d* and *e*).

3. Closed-circuit Television Uses

Closed-circuit television can be utilized wherever it is desired to extend the human visual function over a distance or to inaccessible or dangerous or even only uncomfortable places and where this is not used for broadcast to the public. Or it can be used for picking up a television broadcast signal and channeling it to private locations for reception. Its use has obviously, therefore, many varied facets.[9]

One form of this is special visual monitoring, which can have many aspects. It can monitor dangerous locations, such as furnace combustion chambers or the processing of explosive or atomic materials. It can monitor multiple locations, such as the concentrated supervision of a large number of scattered processing operations in remote parts of a plant, the visual supervision of a number of loading or unloading operations or of the aisles of a department store, or the automobile traffic in specific sections of a highway. The visual monitoring can be extended to announcements of arrivals and departures at an airport or railroad terminal from one central to many scattered locations. Or again, visual examination can be made of the interior of sewer or other pipes, underwater construction, or other operations.

The closed-circuit television may also be used for conferences among executives in different buildings or even remote cities. It can permit an address to a convention from a distant city or a viewing of a current remote convention by a large number of employees.

Closed-circuit television is also used in education to permit one especially qualified teacher to instruct classes in different rooms or in different towns, where it is desired that this be in "real time," i.e., not from a stored program (such as a film).

Closed-circuit television is also used for communities that cannot receive regular television broadcast stations because they are in the shadow of a mountain or for some other reason. Here some individual or company sets up an antenna that can receive the signals and then forwards them to subscribers by private facilities. This is called a "community antenna" system, and it is sometimes also used in hotels and apartment houses. It is but one step from this to an enterprise which itself generates programs (or uses leased entertainment films) and circulates them to private subscribers for a fee. Sometimes this is done on a national scale with spectacular sporting events that are circulated to theaters that project the image, in various cities, to paying audiences.

Military television is usually of the closed-circuit type. It takes on various forms. Sometimes it is used for instruction, sometimes for entertainment of personnel in off hours. Television equipment for combat use has been described,[10] but it is so special that it will not be considered here.

In most of the cases mentioned above, the closed-circuit television system is completely apart from broadcast television. Hence it does not need to use the same scanning and other signal standards that are used in broadcasting. In many situations these are used nevertheless, since broadcast equipment is sold in large quantity and apt to be more easily and economically available than specially designed equipment. However, there are cases where the resolving power, frame frequency, and sometimes general signal complexity of the television broadcast standards are higher than really needed for the special application and require more frequency bandwidth than is considered economical. For these cases a reduced version is applicable.

On the other hand a number of the applications could profitably use a higher quality image. This can especially be secured where the number of linked transmission facilities in tandem is considerably less than for a television broadcast. For these purposes commercial equipment is manufactured which, for instance, goes up to 675 scanning lines and a 12-Mc frequency band.

In most cases the sound channel is not multiplexed on to the picture signal as is done in broadcast television, but it is carried separately or omitted altogether.

Television from satellites is also, in a way, "closed-circuit" television.[11] This is in the sense that the receiving equipment to pick up the signals that come from the satellite is far more sophisticated and expensive than can be installed by the general public. This sort of television is very special and will not be covered here.

SYSTEMS

4. Closed-circuit Television Transmission

Perhaps the first question to confront the prospective user is whether to carry on his own transmission or hand it over to a common carrier. If the transmission is of a longer haul nature, it can be expected to require a sophisticated consideration of all the component parts of the system if it is to give satisfactory results. The various common carriers have given all aspects of the problem extensive study, and it would be difficult to compete with them.[12] For the shorter haul transmission (and here the exact dividing line between "long" and "short" haul depends upon the circumstances of use) there are cases where the private ownership of the transmission medium can be advantageous.

The transmission can be carried out over a variety of media—wire or radio. The radio medium has the advantage of flexibility, which is very important if the terminals may need changing of location at times or if they may need to be mobile. The problem of securing a right of way for the wire line is exchanged for that of securing a radio-frequency allocation. The cost depends largely upon the distance and the nature of the intervening terrain. For very short distances and where the right of way is already owned or easily obtained, the wire circuit is probably cheaper. Where the distances are greater and the right of way uncertain and where power supply to repeaters is a problem, the advantage probably lies with the radio link.

The radio link[13] is carried over a microwave band. The bands allocated to the various purposes such as closed-circuit television vary over the years. At present a band is allocated and commercial equipment is being made for the 10.7- to 13.25-kmc region. A common carrier band is allocated from 5.925 to 6.425 kmc, and an operational band for television broadcasters (not, of course, closed circuit) from 6.575 to 6.875 kmc (see Chap. 19).

Here it is assumed that the radio link under consideration is relatively short haul. Long-haul multichannel microwave radio links have been built and described by the common carriers, and they will not be discussed here.

The short-haul microwave link usually involves several pieces of equipment, at both transmitter and receiver. The transmitter proper has a radio output power of some $\frac{1}{10}$ to 1 watt. The antenna is usually a parabolic dish 2 to 4 or 5 ft in diameter. The combination can handle distances of 1 mile up to, in extreme cases, some 20 miles. There are also a control unit and sometimes a separate power-supply unit. The receiver comprises the receiver proper, an antenna (also a parabolic dish), a control unit, and a power-supply unit. These are in addition to the one or more cameras and their control unit at the transmitter and one or more display units at the receiver. These are all interconnected with cables, which involve coaxial or waveguide elements where they carry the picture signal.

Sometimes these relatively short-haul systems are provided with repeater stations that each contain a transmitter and receiver with antennas. The transmitter and receiver are connected directly through the IF strip, without going down to video. These repeater stations extend the range of transmission, but not too many are generally used in tandem, the exact number varying with the circumstances.

5. Community Systems

A somewhat less ambitious system than the long-haul common carrier circuit is sometimes used for community antenna or community television service or again for a school-system closed-circuit network. The engineering involved in the installation and operation of some of these has been described with a fair amount of detail in the literature.[14,15] The design of these is far from standardized, and the few notes here show the general range of possibilities.

The first point to be noted is the general range of frequencies which is transmitted over the system. In descending order these ranges are as follows:

a. VHF Radio Channels as on the Air. This range includes the entire gamut of VHF channels, 2 to 13 inclusive, as they occur on the air, with a band from 54 to 216 Mc. Where it is desired to pick up the entire band from the air, it is convenient to furnish to community subscribers without any frequency transformation whatever. This simplicity is obtained, however, at the cost of high attenuation in the wire circuits (to be noted below). It is most appropriate, therefore, for short-distance transmission, in the extreme case, for example, for apartment house or hotel distribution.

b. Lower Band Only. This range includes only Channels 2 to 6, inclusive, of the VHF band, namely, 54 to 88 Mc. It permits a more limited service if the channels are picked directly from the air. However, the higher-frequency channels may be translated down to lower-frequency channels that are unused in the particular locality. The five channels can be transmitted economically over somewhat longer distances than the full 13.

c. Low-frequency Spectrum. Here the design is to set up paralleling multiple channels on the same cable, spaced in an economical way and regardless of the allocation of broadcast channels. This leads to frequencies generally lower than the broadcast channels, which start at 54 Mc. It also permits, if desired, the use of wider

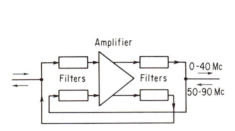

FIG. 6. Two-way repeater with frequency separation of directions.

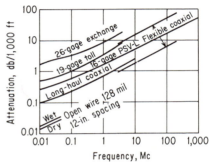

FIG. 7. Attenuation of various conductors for television.

individual bands than allocated for the broadcast channels. Sometimes the entire utilized band runs completely below the broadcast frequencies, and sometimes it overlaps a portion of them. In nearly all such cases it is necessary to provide a modulator and demodulator to transfer the television band used, either up from video or down from the broadcast frequency as picked up. In such arrangements even a single channel can be used, displaced up from video frequency.

d. Two-directional System. In this arrangement the lower-frequency range is used for transmission in one direction while the higher-frequency range is used for transmission in the other. The separation is accomplished by filters as shown in Fig. 6. This represents a simple intermediate repeater station. At a terminal the transmitter would feed into the input of the amplifier in one band while the receiver would be fed by the output in the other band.

There are varying possibilities in the frequency ranges used with these schemes. The allocation outlined in the figure is merely illustrative and shows how a system devised for five channels in one direction is adapted to send two channels in each direction, with a sacrifice of the middle channel. For the allocation, translation would be needed in the frequencies of the channels from their normal positions on the air.

6. Conductors

Generally the lowest attenuation and, initially, the lowest cost conductor system is open wire. A simple example for dry and wet weather is plotted in Fig. 7. A primary difficulty comes from exposure to the weather, as rain increases the attenuation as illustrated and sleet may even be worse. Storms can also cause physical damage. Other difficulties come from the need of the insulator supports, which may cause echoes in the picture from the transmission-line irregularities involved.

Of the enclosed conductors, namely cable, there are two general types. The more common of these is the coaxial cable, which is unbalanced to ground. It consists of a central conductor surrounded by a tubular outer return conductor. Between the two is insulation in the form of beads, twisted cord or tape, washers, plastic foam, or even a solid plastic. The outer conductor may be solid or made of tapes or braided wire, sometimes in two or more layers. It is usually protected, sometimes by iron tapes and finally by a further outer covering of plastic.

There is a large variety of designs of coaxial cables that are commercially available. The choice is a compromise among attenuation, size, cost, flexibility, shielding, and general convenience in handling. A few of these are plotted in Fig. 7. The "long-haul" coaxial is generally not too flexible but cheaper. There is a wide range of flexible coaxials that vary in size and insulation and number of layers of outer conductor braiding. The range shown, in attenuation, is not exhaustive.

The enclosed conductors may be paired and therefore balanced to ground. Because of their somewhat higher attenuation, they are less likely to be used for channels that have been displaced well up from the video band. The PSV (polyethylene-shielded video) pairs are plotted in Fig. 7. Their particular advantages, such as better handling of the region near direct current, are described under video transmission below. Occasionally, plain telephone pairs are used for video transmission, and two of these, 19-gage toll pairs and 26-gage exchange pairs, are plotted in Fig. 7.

In general it is necessary to transmit 60-cycle power to the intermediate repeaters by means of a separate and often paralleling cable. The principal precaution needed is to keep down 60-cycle interference and modulation on the picture signals (see paragraphs 13*d* and *e*).

7. Amplifiers

The one element that has made the multichannel closed-circuit television transmission possible is the development of the amplifier of the distributed type.[16] The effect of the distribution has been to remove the fundamental restriction on bandwidth and gain caused by grid and plate capacitances of the tube in a single stage of gain. The principle is illustrated in simplified form in Fig. 8.

The grid capacitances and similarly the plate capacitances are isolated by inductances, so that they form elements of two separate low-pass transmission lines. The capacitances in the figure are inherent to the tubes, but the inductances are external and added. The lines can be of different impedances but must have exactly the same delay so that the outputs will add in phase.

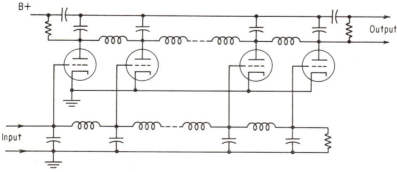

FIG. 8. Schematic distributed repeater. (*After Ginzton, Hewlett, Jasberg, and Noe, in Television Engineering Handbook by D. G. Fink, McGraw-Hill Book Company, New York*, 1957.)

A large variety of such amplifiers, in band location, bandwidth (up to hundreds of megacycles), and gain, is now commercially available. Some are available even in solid-state elements. It is possible to meet the bandwidth requirements generally set for the various systems outlined above, with gains from 20 to 40 db, at modest costs. The noise figure of such amplifiers varies, of course, with the bandwidth and top frequency, but it is generally advertised to run in the order of 10 db.

When the amplification is not strictly linear, intermodulation of input frequencies appears in the output. There, two channels can intermodulate to generate interference into a third channel. This is one of the very important limitations on long-haul transmission because it cumulates at every repeater. It is also of some importance for the short-haul systems which have just been discussed.

"Second-order" modulation, operating on two frequencies a and b, generates frequencies $a + b$ and $a - b$ (or $b - a$). If the channels are allocated at odd multiples of a given frequency, say

$$a = (2m + 1)f \quad \text{and} \quad b = (2n + 1)f \tag{1}$$

then
$$a + b = (2m + 2n + 2)f \tag{2}$$

which is an even multiple of f, and

$$a - b = (2m - 2n)f$$

or
$$b - a = (2n - 2m)f \tag{3}$$

Both of these are even, and therefore the products land between channels.

Most of the television channels are allocated on this basis, though Channels 5 and 6 are a little off, and thus, in any undisplaced community system (as in paragraph 5 above), have some protection against second-order modulation. Also, some amplifiers are designed push-pull to reduce second-order modulation and increase the protection somewhat.

"Third-order" modulation generates frequencies $2a + b$, $2a - b$ (or $b - 2a$), $2b + a$, and $2b - a$ (or $a - 2b$). The locations of these products do not give this protection, but in general third-order modulation is apt to be less intense than second-order.

The modulation products appear in a number of forms in the picture of the channel that they interfere with, but in many cases they appear as bar patterns. As was mentioned earlier, the eye is particularly sensitive to bar patterns in general. The result is that the reduction of such interference to tolerable proportions is often an important factor in making multichannel systems work satisfactorily.

However, the exact engineering of a system against such interference is highly complex, and the measurement of the nonlinearity of the amplifier characteristics rather difficult, so that often the matter is handled merely by experiment and by using a conservative spacing between repeaters. This permits operating amplifiers at a more modest output level, which quickly reduces the modulation effects. Such a solution is permissible where the number of repeaters is not too large. It would not do for a long-haul system.

8. Video-frequency Transmission

Where the distances are not great, the television signal is transmitted directly without modulation, i.e., at "video" frequency. Here again the direct current is not usually transmitted directly, and an auxiliary arrangement is used to carry it indirectly.[6]

A simplified version of the problem and of its solution is indicated in Fig. 9. At a is shown the signal which corresponds to a picture area that is gradually changing from white to black as transmitted through a channel which does not pass the direct current. The synchronizing pulse level is gradually displaced toward a center zero. If this is impressed on a picture tube, the alignment with blacks and whites may start out correctly, as shown at the left. However, to the right it has drifted so that the dark signal has gone only to a middle gray. The synchronizing pulses are no longer "blacker than black" and will appear in the picture.

The remedy consists in "clamping" the synchronizing pulses (either at the tips or at the porch levels) to a fixed voltage, as in Fig. 9b. Then when this is impressed on a picture tube, the signal levels continue to appear at the correct levels; i.e., the direct current has been "restored."

There is, of course, more to the problem than appears in this simple illustration. The actual carrying out of the indirect transmission of the direct current has been the subject of considerable effort and ingenuity.

Video-frequency transmission can be handled over either coaxial or balanced-pair cable conductors. The vacuum-tube or transistor circuits generally used in the amplifiers are of the unbalanced type to ground. Thus an unbalanced wire circuit is the easier one to handle, and the coaxial cable (particularly in a more flexible form that uses a braided type of outer conductor) is very popular for this use where distances are not over the order of 100 ft.

The major difficulty with an unbalanced wire circuit, however, lies in its vulnerability to low-frequency interference, particularly 60-cycle power.

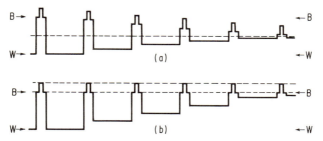

Fig. 9. Direct-current restoration in video signal. (a) Signal without direct current. (b) Direct current restored.

Chiefly because of this it is desirable to transmit the signal over conductors that are balanced to ground. This requires specially designed transformers, because the range of octaves in the video signal frequencies, from near direct current (say 30 to 45 cycles) to 4 or 5 Mc, is very large.

For conductors, plain pairs of the type used for telephony have been used. However, much more desirable paired conductors have been developed, designated as television or PSV pairs. These use some form of polyvinyl insulation and have a substantial shield around, almost like the coaxial outer conductor. This, together with the balanced pair and twist structure of the conductors, gives considerable protection against inductive interference. Details on the engineering of video circuits using this type of cable have been described at some length.[17]

9. Tapping Devices

It is frequently necessary in any closed-circuit television system to branch or tap a circuit, and in a community distribution system this is imperative. The simplest tap on a circuit is illustrated in Fig. 10a. The major problem which this brings up is the discontinuity introduced in the circuit, which generates echoes both in the through and branched circuits. The echoes can considerably degrade the received picture quality. An arrangement which almost eliminates the echoes is shown at b, assuming that the resistances are properly proportioned. Here, however, the loss that is introduced is significant and when repeated, as in a community system, would be too great. By using reactive instead of dissipative elements, as at c and d, it is possible to cut down the loss somewhat.

The best tapping arrangement is, of course, one which is accomplished inside an amplifier to two paralleling outputs (or inputs) or at least by the high-impedance

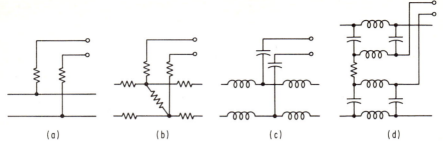

Fig. 10. Passive bridged taps on television circuit. (*a*) and (*b*) Dissipative taps. (*c*) and (*d*) Reactive taps. (*Adapted from Fig.* 3 *of article by Kirk.*[14])

input (or output) of an amplifier onto a line. This, however, is more expensive than the passive tap.

10. Control

Line attenuations in a closed-circuit system will change with temperature or weather, and amplifier gains will also change, even with regulated power supplies. If the signal level falls, it will become noisy at the receiver, and if it rises, intermodulation will increase. Thus some form of regulation or control is usually necessary. In common carrier systems this is generally done automatically by the use, say, of a pilot tone. In the more modest closed-circuit systems it is more likely to be done by hand. Commercially available equipment usually includes a test or control panel with a meter to monitor signal level at a suitable crucial point. The gain can be adjusted to bring the signal level back to a normal value.

11. Economics

The economics of community systems are apt to be rather critical, and it is necessary to consider very closely the balance between quality of service and cost. A rather detailed discussion of this, including the general aspects of design choices for cable, amplifiers, etc., has been given by Kirk.[14] The same conditions, of course, do not necessarily apply exactly to other types of closed-circuit systems, but familiarity with these economics helps to secure sound engineering design in the other systems.

PRACTICAL TRANSMISSION

12. Terminal Equipment

For general purposes the terminal equipment used in closed-circuit television is of much the same character as that used for broadcast television.[18] For direct live pickup there are the high-grade cameras of the studio type and the more portable small cameras that compromise somewhat on quality characteristics to get this compactness. Since the use to which the system is put, however, is often specialized, one can expect more specialized cameras also. For these, the compactness is frequently likely to be a required characteristic, and in many cases, picture sharpness cannot be compromised. Flexibility, representation of motion, or other factors may have to be compromised. For really special purposes, such as examination of processing or combustion in a furnace, an entirely separate design of camera may be called for. For such things as film scanning, however, standard equipment is likely to be suitable. The control unit is usually of a more simplified type than for broadcast television.

The display unit, again, is much of the same kind as used for broadcast reception. In fact, it may economically be a broadcast receiver of which some parts are not used. When picture quality and extra sharpness are desired, specially designed display units are commercially available. Where the display is to be viewed in a large room or theater, projection-type units are also commercially available.

13. General Transmission Requirements

Aside from specialized systems, the transmission requirements for closed-circuit television can be expected to be much the same as for broadcast television. A brief recapitulation of the more immediate of these can be noted.

a. Frequency Band. Essentially this comprises three regions. In the first, which can be called the "passband," fairly close tolerances can be expected on the attenuation distortion with frequency, delay distortion, noise, intermodulation, etc. The second region can be called the "elimination" band, which is not expected to contain any appreciable picture signal, to the extent, in fact, that it can be expected to be devoted to a paralleling independent channel or any other different use. A third region exists between the two and can be called the "roll-off" band or transition region. Its exact width can vary from something like one-tenth up to approximate equality with the passband. Its lines of demarcation are not exact, but it is often considered to start where the passband has dropped by 3 db and to end where the transmission has dropped by 30 or 40 db below the passband.

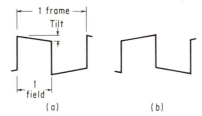

Fig. 11. Low-frequency distortion in video signal, causing tilt.

In carrier systems there are two roll-off bands, one above and one below the passband. Similarly there are two elimination bands.

For a video transmission system the passband is about half the number of picture elements to be transmitted per second.[3] In a double-sideband carrier system, the passband is about equal to the number of picture elements per second. In a "vestigial-sideband" system one of the carrier sidebands is mostly cut off, so that the pass bandwidth lies somewhat between the single- and double-sideband width (usually nearer the former). Frequency-modulation systems are designed in various ways and require a passband which can run from about that needed for a vestigial-sideband system to rather more than needed for a double-sideband system.

b. Attenuation Distortion with Frequency. Various figures have been named for this according to tolerances accepted on image quality, but a typical figure is a tolerance of ±1.5 db from flatness over the passband. This corresponds[8,19] to an echo 15 or 16 db down. This is not for an image of the highest quality. A better figure for the latter would be an echo around 25 db down, which would correspond to limits around ±0.5 db. In either case the figure can be increased in the roll-off band.

c. Delay Distortion. This is also connected with echoes and can be expressed in a variety of ways.[8,19] Probably the simplest is an expression for the tolerance on the envelope delay departure from flatness over the passband. A typical figure is ±0.5 picture-element duration, but for the best image quality this figure is to be reduced. The problem is really more complicated, but in practice the involvements are usually ignored.

Toward very low video frequencies the distortion is usually expressed in terms of the "tilt" which is caused on a square wave of period equal to a full frame duration.[20] The tolerance runs between 1 to a few per cent tilt. This is illustrated in Fig. 11, where negative tilt is shown at *a* and positive at *b*.

d. Noise. Noise has been expressed in terms of its rms value, as a ratio to the peak-to-peak value of the composite television signal.[21] The tolerance varies, of course, with the image quality desired but has run in the order of 45 db.

It has been found, as to be expected, that the composition of the noise also affects its objectionable character. Some theoretical work has yielded a weighting function with video frequency. Experimental work has shown that a single weighting function cannot take care of all types of noise, but a good compromise value has been developed for broadcast television. With this weighting function, illustrated in Fig. 12, random noise that is uniformly distributed over the passbands and roll-off bands reads 11 db lower than the noise when unweighted. Thus if one takes the 45-db tolerance mentioned previously, the tolerance to noise measured in the weighting network would be 56 db below peak-to-peak composite television signal.

The case of 60-cycle noise in the video signal warrants special consideration. If it is not exactly synchronized with the field repetition frequency, the tolerance in a high-

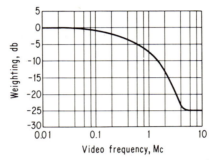

Fig. 12. Noise weighting function for broadcast television.

grade picture runs to some 40 or 50 db below the video signal, each measured peak to peak. However, a good clamper can reduce the effect of 60-cycle noise by a factor up to 30 db, thus placing the tolerance at 10 to 20 db below signal (again peak to peak). This is done at the expense of increasing the impairing effects of video noise by about 6 db at somewhat higher frequencies (2.5 to 100 kc), as discussed by Doba and Rieke.[6]

e. Modulation. Nonlinear modulation is expressed in a variety of ways, one method being to state the intensity of a modulation product (of a new frequency) generated from two components of equal intensity and different frequencies. The ratio of the new product to either of the two basic intensities is expressed in decibels. The tolerance on this ratio varies. In multichannel systems or frequency-displaced systems, it usually runs somewhere between 50 and 60 db for good image quality.

For 60-cycle modulation the tolerance is not too different from that for 60-cycle noise, i.e., where nonsynchronous, 40 to 50 db below the signal. However, a clamper cannot reduce the effect of the modulation as it can the effect of the additive noise. In video monochrome systems the tolerance is significantly less severe.[22]

14. Equalization

The equalization of television circuits differs from that of other circuits principally in that the frequency band is wide and that the phase is involved in the equalization as well as the amplitude. In broadcast television the signal must usually go through a very large number of links between the pickup camera and the final viewing screen. Thus the overall tolerance has to be apportioned among all these links, and the amount allocated to any particular link is very small. For closed-circuit television the total number of transmission links is usually very much smaller and the tolerance for each is correspondingly greater. Thus for equal image quality the equalization is much simpler.

In such a simpler case one can first equalize the attenuation by conventional means, taking care to choose equalizing networks that give a minimum amount of phase disturbance.   Then the envelope delay can be equalized with all-pass delay-correcting networks.[23] These have the property that for truly nondissipative elements within them there is no loss over the entire frequency band.   This condition can be approached closely enough, with practical networks, that the residual attenuation distortion is not significantly increased by the delay correction (remembering here that we are discussing the simpler case of closed-circuit television).

Some correction, both of attenuation and of delay, is obtained over video cable circuits at the lower frequencies merely by the use of a resistance termination as against an exact characteristic impedance match.   This is illustrated roughly in Fig. 13 for a 5-mile length of PSV cable.

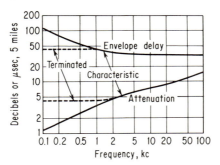

Fig. 13. Partial equalization in short PSV-L video circuit caused by resistance terminations. The attenuation and delay depart from the infinite length characteristics.

The formula which expresses the modified propagation characteristic is

$$\exp G = \cosh \Gamma + \frac{Z^2 + K^2}{2ZK} \sinh \Gamma \tag{4}$$

where $G = a + ib$ = insertion constant for terminated line
$\Gamma = A + iB$ = pure propagation constant over line length
Z = characteristic impedance
K = terminating resistance
Near zero frequency, this becomes

$$G = \frac{R}{2K} - \frac{R^2}{8K^2} + i\omega \left(\frac{CK}{2} + \frac{L}{2K} + \frac{CR}{4} - \frac{RL}{8K^2} \right) \tag{5}$$

where R = resistance in line length
C = capacitance in line length
L = inductance in line length
ω = radian frequency = 2π times cyclic frequency
The coefficient of $i\omega$ is the phase delay, but since it shows no distortion, it is also the envelope delay.

The all-pass lattice network[23] used for delay correcting is shown in Fig. 14.   The equalization procedure was suggested by Nyquist.[24]   The network contains two pairs of members that are reciprocally related with the terminating resistance, as

$$X_1 X_2 = -K^2$$

$$iX_1 = i\frac{Kb}{2}\left(F - \frac{1}{F}\right) \tag{6}$$

$$iX_2 = \frac{-i(2K/b)}{F - (1/F)}$$

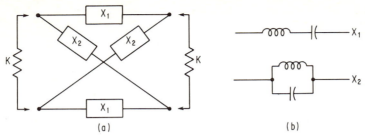

FIG. 14. Lattice network all-pass delay equalizer. (a) Overall network. (b) Network members.

Here b = magnitude parameter on X_1 and X_2
    F = normalized frequency = f/f_0
    f_0 = resonant frequency of X_1 and X_2
  The phase shift of such a network is

$$\phi = 2 \tan^{-1} \frac{X_1}{K} \tag{7}$$

The envelope delay D is

$$D = \frac{d\phi}{d\omega} = \frac{2K}{K^2 + X_1^2} \frac{dX_1}{d\omega} \tag{8}$$

$$D = \frac{b}{\omega_0} \frac{1 + (1/F)^2}{1 + (b^2/4)(F - 1/F)^2} \tag{9}$$

where $\omega_0 = 2\pi f_0$

or $Df_0 = \dfrac{b}{2\pi} \dfrac{1 + (1/F)^2}{1 + (b^2/4)(F - 1/F)^2} \tag{10}$

Plots of Eq. (10) for several values of the parameter b are illustrated in Fig. 15. The process of equalization consists of adding sections of the lattice network, with

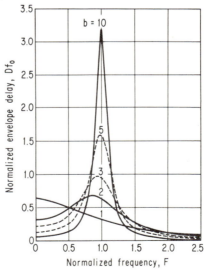

FIG. 15. Normalized envelope delay of lattice network. (*After Nyquist.*[24])

appropriately selected values of b and f_0, to add to and flatten out the measured (or computed) envelope delay of the circuit and its equipment (amplifiers, attenuation correcting network, etc.) until the residual envelope delay distortion is brought down to an acceptable amount.

For the broadcast television video and other circuits it has been found desirable to go to more sophisticated methods.[25]

15. Measurements

A detailed discussion of measurements is beyond the scope of the small space available here. A brief listing is given of the various types of measurements and measuring instruments involved, with references to more complete literature.

Some guidance can be obtained from the various IRE (now IEEE) standards on test procedures.[26] These describe a number of recommended standard tests on the various portions of the television systems, and the standards on definitions of terms make clear the technical terms used. Another excellent and more extensive guide on such measurements has been published by Dillenburger.[27] There are, of course, in addition a number of test handbooks designed for television technicians.[28]

The general nature of these measurements comprises:

1. Basic voltage and current measurements (up to the high frequencies of television)
2. Basic resistance, reactance, capacitance, and inductance measurements (also up to television frequencies)
3. Basic optical picture characteristics, including luminance, contrast gradation and range, sharpness and resolving power, and nature of ambient surroundings, viewed toward measurements on television equipment
4. Measurement of picture impairments, such as noise (or graininess), transient effects (which appear like fringes, or echoes), and geometrical distortions and use of test film
5. Measurements on camera tubes and picture tubes
6. Electronic measurements on amplifiers, modulators, and communication lines
7. Phase angle and delay measurements

Many of the above measurements are not peculiar to television. On the other hand, in many cases special precautions are necessary when they are made for television application because of the special characteristics of the eye or for other reasons. Among the references are a number of papers that seem appropriate, but these cannot really be considered as comprehensive.[29,30]

In particular it must be noted that, of all the measurements listed, those related to delay or phase measurements[30] are generally the least familiar in the communications art. Adequate measuring apparatus for these is least likely to be found commercially available, and the common carriers usually make their own. It is difficult to find apparatus that is sufficiently accurate both in the megacycle range and down toward low frequencies (say 100 to 1,000 cycles). For this reason it is often found that the delay distortion in video circuits at the lowest frequencies is calculated, from primary constants or such other data as are available, rather than measured. Phase measurement per se is not usually sufficiently accurate on anything but completely passive circuits, and even here it can be questionable.

REFERENCES

1. Mertz, P., and F. Gray, A Theory of Scanning, *Bell System Tech. J.*, vol. 13, pp. 464–515, July, 1934.
   Mertz, P., Television—The Scanning Process, *Proc. IRE*, vol. 29, pp. 529–537, October, 1941.
2. Fink, D. G., *Television Engineering Handbook*, pp. 2–3 to 2–29, McGraw-Hill Book Company, New York, 1957.
3. Nyquist, H., Certain Topics in Telegraph Transmission Theory, *Trans. AIEE*, vol. 47, pp. 617–644, April, 1928.

4. Kretzmer, E. R., Statistics of Television Signals, *Bell System Tech. J.*, vol. 31, pp. 751–763, July, 1952.

Harrison, C. W., Experiments with Linear Prediction in Television, *Bell System Tech. J.*, vol. 31, pp. 764–783, 1952.

Averbach, E., and A. S. Coriell, Short-term Memory in Vision, *Bell System Tech. J.*, vol. 40, pp. 309–328, January, 1961.

5. Gicca, F. A., Optimum Television System Design, *Electron. Progr.*, vol. 7 no. 3, pp. 1–6, November–December, 1962.

Graham, R. E., and C. C. Cutler, Predictive Quantizing of Images, *IRE Wescon Conv. Record*, pt. 4, p. 147, 1958.

Schreiber, W. F., C. F. Knapp, and N. D. Kay, Synthetic Highs—An Experimental TV Bandwidth Reduction System, *J. Soc. Motion Picture Television Engrs.*, vol. 68, pp. 525–537, August, 1959.

Deutsch, S., The Possibilities of Reduced Television Bandwidth, *Trans. IRE Broadcast Television Receivers*, October, 1956, pp. 68–82.

Kretzmer, E. R., Reduced Alphabet Representation of TV Signals, *IRE Conv. Record*, pt. 4, pp. 140–147, 1956.

6. Doba, S., and J. W. Rieke, Clampers in Video Transmission, *Trans. AIEE*, vol. 69, pt. 1, pp. 477–487, 1950.

Wendt, K. R., Television DC Component, *RCA Rev.*, vol. 9, pp. 85–111, March, 1948.

7. Nyquist, H., and S. Brand, Measurement of Phase Distortion, *Bell System Tech. J.*, vol. 9, pp. 522–549, July, 1930.

8. Wheeler, H. A., The Interpretation of Amplitude and Phase Distortion in Terms of Paired Echoes, *Proc. IRE*, vol. 27, pp. 359–385, June, 1939.

Mertz, P., Tolérancessur le déphasage et l'affaiblissement des circuits internationaux de télévision, *CCIF*, 3 ème Comm. Rapporteurs Document 21, p. 4, 1947–1948.

Mertz, P., Influence of Echoes on Television Transmission, *J. Soc. Motion Picture Television Engrs.*, vol. 60, pp. 572–596, May 1953.

9. Mayers, M. A., and R. D. Chipp, *Closed Circuit TV Planning*, Hayden Publishing Company, New York, 1957.

Becker, K. F., J. R. Hefele, and W. T. Wintringham, Experimental Visual Communication System, *Bell System Tech. J.*, vol. 38, pp. 141–176, January, 1959.

Lovell, R. E., coordinator, Television as a Means of Instructional Communication, *Symp. Soc. Motion Picture Television Engrs.*, Los Angeles, 1962; *J. Soc. Motion Picture Television Engrs.*, vol. 72, pp. 145–176, March, 1963.

10. Marshall, C. J., and L. Katz, Television Equipment for Guided Missiles, *Proc. IRE*, vol. 34, pp. 375–401, June, 1946.

11. Project Telstar, Communications Experiment, *J. Soc. Motion Picture Television Engrs.*, vol. 72, pp. 91–96, February, 1963.

Dickieson, A. C., The Telstar Experiment, *Bell System Tech. J.*, vol. 42, pp. 739–1908, July, 1963.

Courtney-Pratt, J. S., J. H. Hett, and J. W. McLaughlin, Optical Measurements on Telstar to Determine the Orientation of the Spin Axis, and the Spin Rate, *J. Soc. Motion Picture Television Engrs.*, vol. 72, pp. 462–484, June, 1963.

12. The L-3 Coaxial System—Symposium, *Bell System Tech. J.*, vol. 32, pp. 779–1005, July, 1953.

Ford, G. T., and E. J. Walsh, Tubes for a New Coaxial Transmission System, *Bell System Tech. J.*, vol. 30, pp. 1103–1128, October, 1951.

13. Reference 2, pp. 18–48 to 18–49.

14. Kirk, D., Economic Considerations in Closed-circuit Television System Design, *J. Soc. Motion Picture Television Engrs.*, vol. 66, pp. 661–671, November, 1957.

15. Zorbaugh, H., Television—Technological Revolution in Education? *J. Soc. Motion Picture Television Engrs.*, vol. 66, pp. 671–676, November, 1957.

Warman, W. C., Washington County Closed-circuit Television Network, 1956–1957, *J. Soc. Motion Picture Television Engrs.*, vol. 66, pp. 677–679, November, 1957.

Pawlowski, A., Design of Wide-band RF Systems for Closed Circuit Television, *Military Systems Design*, vol. 6 no. 6, pp. 22–25, November–December, 1962.

Brugger, J. R., Television in Washington County Schools, Hagerstown, Maryland, *J. Soc. Motion Picture Television Engrs.*, vol. 66, pp. 680–682, November, 1957.

Zworykin, V. K., E. G. Ramberg, and L. E. Flory, *Television in Science and Industry*, John Wiley & Sons, Inc., New York, 1958.

16. Ginzton, E. L., W. R. Hewlett, J. H. Jasberg, and J. D. Noe, Distributed Amplification, *Proc. IRE*, vol. 36, pp. 956–969, 1948.

Horton, W. H., J. H. Jasberg, and J. D. Noe; Distributed Amplifiers—Practical Considerations and Experimental Results, *Proc. IRE*, vol. 38, pp. 748–753, July, 1950.

Payne, D. V., Distributed Amplifier Theory, *Proc. IRE*, vol. 41, pp. 759–762, June, 1953.

Also Ref. 2, pp. 11-21 to 11-27.

17. Nebel, C. N., Local Wire Video Television Networks, *Trans. AIEE*, vol. 69, pt. II, pp. 1451–1460, 1950.

Doba, S., and A. R. Kolding, A New Local Video Transmission System, *Bell System Tech. J.*, vol. 34, pp. 677–712, July, 1955.

Bodle, D. W., and P. A. Gresh, Lightning Surges in Paired Telephone Cable Facilities, *Bell System Tech. J.*, vol. 40, pp. 547–576, March, 1961.

Also Ref. 2, pp. 18-49 to 18-63.

18. Reference 2, chaps. 15–17.

19. Mertz, P., A. D. Fowler, and H. N. Christopher, Quality Rating of Television Images, *Proc. IRE*, vol. 38, pp. 1269–1283, November, 1950.

20. Reference 2, pp. 11-4 to 11-6.

21. Mertz, P., Perception of Television Random Noise, *J. Soc. Motion Picture Television Engrs.*, vol. 54, pp. 8–34, January, 1950.

Mertz, P., Data on Random Noise Requirements for Theater Television, *J. Soc. Motion Picture Television Engrs.*, vol. 57, pp. 89–107, August, 1951.

Schade, O. H., Electro-optical Characteristics of Television Systems, *RCA Rev.* vol. 9, pp. 5–37, 245–286, 490–530, and 653–686, 1948.

Barstow, J. M., and H. N. Christopher, The Measurement of Monochrome Video Interference, *Trans. AIEE, Communication and Electronics*, vol. 72, no. 10, pp. 735–741, January, 1954.

22. Reference 2, p. 11-6.

23. Everitt, W. L., and G. E. Anner: *Communication Engineering*, 3d ed., pp. 536–538, McGraw-Hill Book Company, New York, 1956.

24. Nyquist, H., Distortion Compensator, U.S. Patent 1,675,460, July 3, 1928.

H. Nyquist, Phase Compensating Networks, U.S. Patent 1,735,052, Nov. 12, 1929.

Nyquist, H., and P. Mertz, Secret Telephone System, U.S. Patent 1,726,578, Sept. 3, 1929.

25. Rounds, P. W., and G. L. Lakin, Equalization of Cables for Local Television Transmission, *Bell System Tech. J.*, vol. 34, pp. 713–738, July, 1955.

Brogle, A. P., The Design of Reactive Equalizers, *Bell System Tech. J.*, vol. 28, pp. 716–735, October, 1949.

Sperry, R. V., and D. Surenian, Transversal Equalizer for Television Circuits, *Bell System Tech. J.*, vol. 39, pp. 405–422, March, 1960.

26. 60 IRE 17.S1 Standards on Television: Methods of Testing Monochrome Television Broadcast Receivers, 1960.

54 IRE 23.S1 Standards on Television: Methods of Measurement of Aspect Ratio and Geometric Distortion, 1954.

55 IRE 23.S1 Standards on Television: Definitions of Television Signal Measurement Terms, 1955.

58 IRE 23.S1 Standards on Television: Measurement of Luminance Signal Levels, 1958.

60 IRE 23.S1 Standards on Television: Measurement of Differential Gain and Differential Phase, 1960.

61 IRE 23.S1 Standards on Video Techniques: Definitions of Terms Relating to Television, 1961.

50 IRE 23.S2 Standards on Video Techniques: Methods of Measurement of Time of Rise, Pulse Width, and Pulse Timing of Video Pulses in Television, 1950.

60 IRE 23.S2 Standards on Video Techniques: Measurement of Resolution of Camera Systems, 1961.

50 IRE 23.S3 Standards on Television: Methods of Measurement of Electronically Regulated Power Supplies, 1950.

27. Dillenburger, W., *Fernseh Messtechnik*, Fachverlag Schiele & Schön GMBH, Berlin, 1960.

28. Middleton, R. G., *TV Troubleshooting and Repair*, Hayden Publishing Company, New York, 1963.

Herrington, D. E., and C. P. Oliphant, *Photofact Guide to TV Troubles*, Howard W. Sams & Co., Inc., Indianapolis, 1962.

29. Wilhelm, H. T., Impedance Bridges for the Megacycle Range, *Bell System Tech. J.*, vol. 31, pp. 999–1012, 1952.

Blackman, R. B., and J. W. Tukey, Measurement of Power Spectra from the Point of View of Communications Engineering, *Bell System Tech. J.*, pp. 185–282, 485–569, January and March, 1958.

Green, E. I., H. J. Fisher, and J. G. Ferguson: Techniques and Facilities for Microwave Radar Testing, *Bell System Tech. J.*, vol. 25, pp. 435–482, July, 1946.

Alsberg, D. A., Principles and Applications of Converters for High-frequency Measurements, *Bell System Tech. J.*, vol. 32, p. 506, March, 1953.

May, A. S., Microwave System Test Equipment, *Bell System Tech., J.*, p. 1510, November, 1953.

30. Alsberg, D. A., and D. Leed, A Precise Direct Reading Phase and Transmission Measuring System for Video Frequencies, *Bell System Tech. J.*, vol. 28, pp. 221–238, April, 1949.

Robertson, S. D., A Method of Measuring Phase at Microwave Frequencies, *Bell System Tech. J.*, vol. 28, pp. 99–103, January, 1949.

Ring, D. H., The Measurement of Delay Distortion in Microwave Repeaters, *Bell System Tech. J.*, vol. 27, pp. 247–264, April, 1948.

Elliott, J. S., High-precision Direct-reading Loss and Phase Measuring Set for Carrier Frequencies, *Bell System Tech. J.*, vol. 41, pp. 1493–1517, September, 1962.

Rudisill, J. A., Simple Phase-angle Measurement Technique, *Bell System Tech. J.*, vol. 32, p. 257, January, 1953.

Chapter 14

POWER-LINE CARRIER SYSTEMS

T. M. SWINGLE, *Tennessee Valley Authority*

H. I. DOBSON, *Tennessee Valley Authority*

INTRODUCTION

1. Electrical Power Transmission Lines

These lines are also used extensively for the transmission of many forms of communication including telephony, telegraphy, telemetry, and various other types of information. This transmission of intelligence takes place simultaneously with the transmission of electrical energy and without mutual interference.

Power-line carrier facilities are used primarily to provide company-owned communication services for power-system operation. On large power systems, it is not possible to provide all communication needs[1] by means of power-line carrier because of the limited frequency spectrum available. However, high-voltage transmission lines provide a very reliable medium for communication, and in general, power-line carrier is used for the *more important* services.

Power transmission lines of any voltage can be used for communication purposes; however, lines operating at 33 kv and higher are most prevalent. Lower voltage lines are usually tapped or looped through a greater number of stations between terminals, which increases the cost of maintaining a suitable path for communication signals.

Communication over power lines is accomplished by superimposing carrier frequencies on the transmission lines in somewhat the same manner as is done on open-wire telephone lines, except for the method of connecting the carrier transmitters and receivers to the line conductors.

Within the United States, power-line carrier frequencies fall between the limits of 30 and 200 kc. While 30 kc is not an absolute lower limit, the bandwidth obtainable at lower frequencies with standard line coupling and tuning equipment is not adequate for most practical needs. The upper limit is due to the extensive use of frequencies between 200 and 415 kc for aircraft navigation radio facilities.

2. Control

In the United States, power-line carrier, along with other means of communication, is under the jurisdiction of the Federal Communications Commission. Unlike some types of communication, however, licensing is not required for its operation. Excessive radiation to an extent which causes interference to licensed communication services is prohibited; otherwise, transmitted power levels are not restricted. If interference among different carrier services occurs, it is the responsibility of the parties involved to cooperate in working out a solution.

In many countries other than the United States, power-line carrier is licensed by some agency of the government and power levels are controlled. The allowable frequency range, however, is considerably higher, extending in some countries to frequencies above 500 kc.

3. System Considerations

In power systems, requirements for additional communication facilities usually accompany the expansion of power facilities. The type of overall communication system and how large a part is played by power-line carrier are the prerogatives of each user.

Some users may establish certain rules as standard policy; e.g., all substations with full-time operators shall be equipped with at least one power-line carrier telephone circuit to the appropriate load-dispatching center. Regardless of whether by policy or by individual decision, the following items must be established for any expansion project: (1) What new services are required, and what improvements or expansions to existing facilities are necessary? (2) Which of the new services are to be provided by power-line carrier? (3) How can the new carrier services best be coordinated with one another and integrated into the existing carrier system?

Assuming that the first two items are determined satisfactorily in preliminary system design, the third may be expanded, beginning with telephone services. There are several types of power-line carrier telephone equipment with a variety of options available. The application engineer must consider what he wishes to accomplish from a system standpoint, i.e., whether a carrier channel is required as a trunking circuit in a switching network, strictly as a point-to-point circuit, or as a multiparty circuit. Additional items for determination include method of signaling—dial, ring-down, or common battery—and whether two-wire or four-wire voice operation is required.

Determination of the most practical method for providing other required services, i.e., protective relaying, telemetering, supervisory control, or other telegraphic functions, might begin by consideration of the terminal locations. If several services are required between the same end stations, a choice must be made between combinations of audio tones on a common facility or the use of individual facilities directly in the carrier-frequency range. Economic factors enter into this choice. Two or three services might be provided as individual facilities, whereas a tone combination may be less expensive if several services are to be provided. A limited number of services suitable for audio tone transmission may share a portion of a telephone voice band provided the carrier circuit is of the duplex type.

Other factors to be considered in choosing the best method for a given application include the relative importance of each service and the availability of unused space in the carrier frequency spectrum.

4. Design Considerations

Carrier equipment of sufficient variety is available today to enable the user to select a type well suited for practically any required service. Published manufacturers' specifications, covering each individual type, provide information concerning its capabilities, including transmitter power output, receiver sensitivity and selectivity, threshold noise level, maximum distortion, and other pertinent technical data. On the basis of this information a carrier system can be designed to provide almost any desired degree of reliability and quality. However, any carrier terminal equipment will provide satisfactory service only when used within the limits of its capability. Therefore, the realization of optimum reliability and quality requires a transmission path which is fully adequate under all switching conditions of the power system and all weather conditions. The transmission path is the responsibility of the user.[2]

Some economy may be realized on long circuits if full-time utilization is not essential. In many systems, short periods may be tolerated during which a carrier telephone circuit is unusable because of severe weather. Such a *fair-weather* circuit may be extremely valuable for handling excess traffic during normal periods.

In order to assure that transmission-path requirements are met, it is necessary that preliminary design work on the carrier channel be done before the terminal equipment is selected. This preliminary work includes estimates of carrier channel loss, noise level, and the effect of weather upon each. Knowing these characteristics, a design engineer can calculate the *minimum* transmitter power required to maintain a satisfactory signal-to-noise ratio at the receiver input.

On long carrier channels, a choice must be made between the use of high-powered transmitters or the use of intermediate repeaters. In recent applications in the United States, most carrier telephone circuits too long for 10-watt transmitters use cascaded circuits connected together on an audio basis.

COUPLING TO THE POWER LINE

5. Objective

A most important consideration in connecting to a high-voltage three-phase power transmission line is the protection of personnel and communication apparatus from the hazards of normal line voltages, switching surges, and lightning strokes. At the same time, an efficient path must be provided for the carrier frequency or band of frequencies to be transmitted. A carrier circuit may be coupled either phase-to-phase or phase-to-ground.[3]

6. High-voltage Coupling Capacitor

A high-voltage coupling capacitor whose reactive impedance is very high at the power frequency but nominal at carrier frequencies is connected directly to the transmission-line conductor. For phase-to-phase coupling, two capacitors are required.

Coupling capacitors are physically composed of several paper-capacitor elements connected in series and immersed in oil. The outer container is a porcelain cylinder with multiple rain-shed skirts which provide a long creepage path. They are equipped with a circular metal fitting on each end which serves both for physical support or mounting and for electrical connection. A gas bubble is usually left inside the cylinder to allow for thermal expansion of the oil, although some European types use an expanding metal diaphragm so that the inside can be completely oil filled.

7. Voltage Rating

The voltage rating of a coupling capacitor is the nominal phase-to-phase voltage of the transmission line on which it is to be used, although the actual potential across the capacitor is the phase-to-ground voltage. In addition to continuously withstanding this voltage, the capacitor insulation must withstand high-voltage impulses caused by lightning and switching surges and is sometimes subjected to considerable overvoltages for extended periods. Tables 1 and 2 list capacitor requirements and values which are standard in the United States.

The capacitance and voltage rating of every capacitor of a particular manufacturer's type are coordinated so that the units may be stacked to provide direct addition of their voltage ratings. For example, a 0.006-μf 115-kv unit stacked with a 0.015-μf 46-kv unit will make a 0.0043-μf 161-kv unit, or three such 115-kv units stacked together will make a 0.002-μf 345-kv unit.

The capacitance of coupling capacitors is determined somewhat as a compromise between the economics involved in their manufacture and the requirements of carrier applications. Particularly at lower carrier frequencies, coupling efficiency is higher with higher capacitance. Also for any chosen tuning method, the bandwidth of the carrier channel will be broader with larger values of coupling capacitance.

Table 1. Coupling Capacitor Voltage Ratings, Operating Voltages, Dielectric Test Voltages, and Minimum Creepage Distance*

Voltage rating, kv	Maximum operating voltage, kv	Dielectric test voltage, kv			Minimum creepage distance, in.
		Low-frequency withstand tests		Impulse withstand tests	
		Dry 1 min	Wet 10 sec	1.5 × 40- μsec full wave	
2.4	1.6	15	13	45	7
4.8	3.2	21	20	60	8.5
8.32	5.5	27	24	75	10.5
14.4	9.0	36	30	95	14.5
14.4	9.0	50	45	110	14.5
23	15	65	60	150	19.5
34.5	22	85	80	200	26.5
46	28	110	100	250	33.5
69	42	165	145	350	45
115	70	265	230	550	76
138	84	320	275	650	88
161	98	370	315	750	103
230	140	525	445	1,050	145
288	174	655	555	1,300	177

* Taken from National Electrical Manufacturers Association Standards Publication SG 11-1955.

Table 2. Capacitance Ratings for Main Capacitance*

Circuit voltage, kv	Capacitance, μf†		NEMA creepage required, in.	BIL, kv‡
	Low	High		
46	. . . . . .	0.015	33.5	250
69	0.0032	0.010	45	350
115	0.002	0.006	76	550
138	0.0016	0.005	88	650
161	0.0013	0.0043	103	750
230	0.001	0.0030	145	1,050
287.5	0.0008	0.00250	177	1,300
345	0.00064	0.002		

* Taken from National Electrical Manufacturers Association Standards Publication SG 11-1955.

† The values of capacitance shall be within a tolerance of minus 10 per cent to plus 15 per cent. Values do not include bottom tap capacitance in potential device.

‡ Basic impulse insulation level.

8. Protection

Some form of protection must be provided at the bottom of the stack or low-potential end of the coupling capacitor. Attitudes regarding the proper quantity and location of protective elements vary among different users. However, at some point, there is usually a drainage coil to provide a low impedance to ground at the power frequency, a grounding switch to facilitate safe testing and maintenance, and an arc gap to limit lightning and other surge voltage peaks. These elements are normally housed in the supporting base of the coupling capacitor. Duplicates of some or even all of these protective elements are also preferred in line-tuning unit cabinets by some users.

9. Coupling Capacitor-potential Devices

These provide a means whereby joint use is made of a capacitor for carrier coupling and for deriving a power-frequency reference potential proportional to the line-to-ground voltage of the transmission line. The potential is obtained from a tap in the bottom unit of the capacitor stack or by the use of an auxiliary capacitor in the supporting base to create such a tap. Other components of the potential device include a choke coil in the tap lead to prevent influence to the carrier circuit, protective elements, a transformer with appropriate ratios (usually with multiple secondary windings), and a phase-angle correcting network. In devices with low capacitance, the corrective network is adjustable so that maximum accuracy may be obtained for any burden within its rating. With high capacitance, field adjustment is unnecessary.

The voltage derived from the potential device has many uses, the most common of which is line synchronization. Other uses include emergency relaying potential and in rare instances a limited source of power in remote locations not provided with a normal source of 115 volts a-c. Some of the more sophisticated devices have accuracies sufficient for billing meter application.

10. Line-tuning Units

These serve to cancel or reduce the effects of coupling-capacitor reactance and to provide matching of impedances for maximum transfer of carrier energy. They are usually located in the switchyard beneath the coupling capacitors with which they are used. Connection to carrier terminal equipment, usually indoors, is made with coaxial cable. Line-tuning components may be located indoors in some instances. However, this procedure is inefficient and should be avoided where losses are critical and where coaxial-cable lengths exceeding approximately 100 ft are involved. Coaxial cables in general use in the United States have nominal characteristic impedances of 50 to 75 ohms. Typical types are RG-8/U, ⅜ in. diameter, 52 ohms, and RG-34/U, ⅝ in. diameter, 71 ohms. In general, the larger size is more suitable for long runs and other applications where excessive physical stress may be encountered.

Line-tuning units may be divided into two general categories: resonant types and broadband types.

11. Resonant Line-tuning Units

These units employ reactive tuning elements to produce series resonance with the reactance of the coupling capacitor at one or more frequencies. In the single-frequency unit, this requires only one series inductance coil for tuning. The complete single-frequency line-tuning assembly also includes a protector unit and an impedance-matching transformer as shown in Fig. 1a. The inductance coil is usually variable by means of taps on the coil and a movable core. The coil is wound with Litz wire to provide an extremely high Q.

Two methods by which the coupling capacitor may be tuned to series resonance at two frequencies are shown in Fig. 1b and c. The circuit in Fig. 1b is used when the two frequencies are to be directed over separate coaxial cables. A parallel-resonant trap unit in each branch is tuned to reject the frequency of the signal which passes through the opposite branch. The series inductance in each branch is adjusted to obtain maximum current of the desired-frequency signal. The circuit of Fig. 1c may be used when it is desired to tune both frequencies into the same coaxial cable. The upper inductance is resonated with the coupling capacitor at the higher of the two frequencies. The branch of the lower portion which contains both a capacitor and an inductance is also series-resonated at the higher frequency. The lone inductance in the opposite branch is finally adjusted to obtain maximum net current of the lower-frequency signal.

It is possible to tune a single coupling capacitor to resonance at three or more frequencies. However, this is seldom done because the cumbersome circuitry required

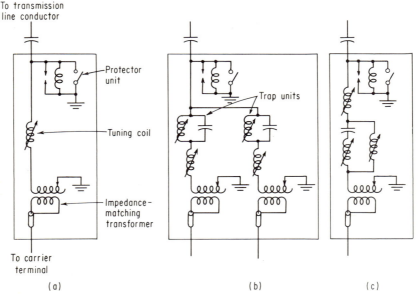

Fig. 1. Resonant line-tuning units. (a) Single frequency. (b) Double frequency, two coaxial cables. (c) Double frequency, one cable.

causes excessive losses and produces undesirable antiresonances at frequencies often very near the desired resonant frequencies.

For many years tuning has been accomplished in resonant line-tuning units by connecting an RF milliammeter in series with the circuit and adjusting for maximum current. In two-frequency tuning units, parallel-resonant trap units are disconnected from the circuit, tuned separately, and reconnected before adjustment for maximum current is made independently in each branch. Impedance matching is a process of minimizing the standing-wave ratio on the coaxial cable. This is accomplished by selecting the impedance-matching transformer setting which causes the input cable current to approach the value obtained into a dummy resistive load of the correct value.

These techniques are still widely used, although many RF milliammeters have a considerable amount of thermal inertia and proper settings are not sharply defined. Power-frequency current, particularly with high-capacitance coupling capacitors, can mask low-level high-frequency current in the nonselective milliammeters. Also, this method for transformer ratio setting does not assure a proper match.   More recent methods make use of an impedance bridge whereby adjustments are made and transformer ratios selected on the basis of obtaining the correct magnitude of input impedance with zero reactive component at center frequency.   Multiple measurements at closely spaced frequencies are made in order that averages of variations due to standing waves on the transmission line may be considered.   Impedance measurements with normal laboratory accuracy are not required for this purpose.   Simplified bridge circuits and substitution methods of impedance measurement can be used successfully.

12. Broadband Line-tuning Units

These use filter networks rather than series-resonant circuits to compensate for the reactance of the coupling capacitor. The coupling capacitor itself serves as one element of the filter, which is usually a constant-K highpass or bandpass arrangement such as the examples shown in Fig. 2.   The objective of broadband tuning is versatility; that is, several frequencies may fall within the passband of the filter without the necessity of individual tuning.   The primary aim in the design of filters for broadband coupling units is to obtain maximum efficiency within the passband.   Complex circuitry such as would be required to obtain high rejection outside the passband is not needed.

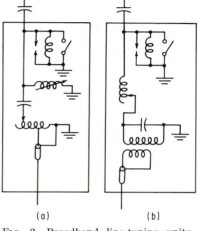

(a)   (b)

Fig. 2. Broadband line-tuning units. (*a*) High-pass T section.   (*b*) Bandpass half section.

The units illustrated in Fig. 2 are capable of coupling the carrier-frequency band from 50 through 200 kc with very good efficiency using coupling capacitances of 0.006 μf or higher.   Capacitances of 0.004 μf or less may be used, but with some degree of impedance compromise which becomes greater as the capacitance becomes lower.   Broadband units of these types may be made quite compact and are usually mounted within the coupling capacitor base, making a self-contained assembly.   Variations of these units and others using different circuitry are sometimes mounted in separate cabinets.

Some broadband tuning units are equipped with test terminals so that adjustments can be made for maximum carrier-frequency voltage at specified frequencies within the passband.   Others are made with single-value elements set for average conditions and require no field adjustments.   Measurements of input impedance are very useful in determining if a broadband coupling unit is functioning properly.

13. Phase-to-phase Tuning

Phase-to-phase tuning requires two coupling capacitors and duplicates of all elements in the line-tuning assembly except the impedance-matching transformer.   Ordinarily only one transformer is used as shown in Fig. 3.   The impedance-matching transformers in some broadband couplers form integral parts of the units.   In such cases phasing leads

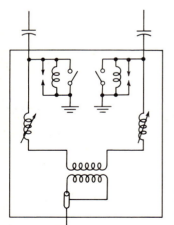

Fig. 3. Phase-to-phase single-frequency line-tuning unit.

or cables connect the two transformers together so that they may be driven 180° out of phase via a single coaxial cable.

14. Line Trap

A line trap is a device for isolating the carrier channel from certain detrimental conditions found in the power system and to isolate carrier channels from one another to prevent interference. To accomplish this isolation, the trap must have a high impedance at carrier frequencies. Line traps are installed in series with the power-line conductors and therefore must carry the power-frequency current that is present.

The main-coil conductor in modern traps may be copper or aluminum and may be large stranded cable or composed of multiple flat layers. The conductor is helically wound, usually in a single layer, to form an air-core inductance coil which is rigidly supported by several columns made of suitable high-strength material. Tensile strength is provided by a steel strain rod through the center of the trap attached to rigid assemblies at each end. American traps are enclosed on the ends by insulating covers. Openings near each end, where the coil winding terminates on external connecting studs, are covered by suitable barriers to prevent small birds from nesting inside the trap.

Line traps may be mounted in substation structures by vertical suspension or placed either vertically or horizontally on insulated pedestals. In isolated locations such as a transmission-line tap point, traps may be suspended horizontally in series with a short span of transmission-line conductor.

15. Line-trap Ratings

A line trap has three power-frequency current ratings: continuous-current rating, thermal short-circuit rating, and mechanical short-circuit rating.

Table 3. Line Trap Thermal and Mechanical Current Rating*

Continuous-current rating, rms symmetrical, amp	2-sec thermal current rating, rms symmetrical, amp	Mechanical current rating, rms symmetrical, amp
400	15,000	15,000
800	20,000	20,000
1,200	36,000	36,000
1,600	44,000	44,000
2,000	63,000	63,000

* Taken from National Electrical Manufacturers Association Standards Publication SG 11-1955.

The continuous-current rating is the current which a line trap is capable of carrying continuously without exceeding a specified temperature. The short-time thermal current rating is the current which a line trap is capable of carrying for 2 sec without exceeding a specified temperature. This 2-sec fault current duration is standard in America, although other test durations ranging from 1 to 4 sec are sometimes used in Europe. By American standards, the mechanical current rating is the current which a line trap is capable of withstanding when the initial peak is completely offset. This rating is given as an rms figure, but because of the offset characteristic of transmission-line fault currents for which this rating is made, tests must be made with a peak current 2.83 times the rms value. European mechanical ratings are given as peak values.

Table 3 lists ratings of standard line traps available in the United States.

It is desirable that the impedance of line traps be negligible at the power frequency. The nominal inductance of most traps currently being manufactured in the United States is 265 μh, which corresponds to a 60-cycle reactance of approximately $\frac{1}{10}$ ohm. European traps are manufactured with inductances ranging from 180 μh to more than 1 mh.   The higher inductance traps have lower short-circuit current capability.

16. Protection

Line traps and their tuning elements are protected from lightning surges by one or more suitable arresters.   Arrester characteristics are coordinated with the trap inductance and with its short-circuit ratings so that line faults within rated limits will not flash the arrester.   In older model traps, arresters were enclosed within tuning pack housings.   Newer styles have a separate arrester connected in parallel with the main coil, sometimes with auxiliary protection in the tuning packs.

17. Tuning

A line trap may be tuned as single-frequency, double-frequency, or broadband. Typical schematics of each are shown in Fig. 4.

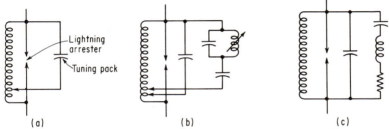

FIG. 4. Schematics of typical line traps.   (a) Single frequency.   (b) Double frequency. (c) Broadband.

For traps made in the United States, single-frequency tuning requires one pack which will provide tuning ranges of either 30 to 90, 50 to 150, or 70 to 200 kc, depending on the capacitance values within the pack.   Each pack contains two capacitors which may be connected individually, in series, or in parallel to select one of four values of capacitance which divide the frequency range into narrow bands.   Fine adjustment within a band is accomplished by locating a tap on the main coil.

Line traps currently available require three packs for double-frequency tuning.   The higher of the two frequencies is tuned in the same manner as a single-frequency trap. The pack containing the inductance is also tuned independently to the higher frequency.   The third pack completes tuning requirements for providing maximum impedance of the entire circuit at the lower frequency, with fine adjustments made by a second tap on the main coil.   The frequency ranges for *second-frequency* tuning packs are the same as for single-frequency traps.   Both frequencies of the double-frequency trap, however, need not fall within the same range.   With appropriate pack combinations, widely spaced frequencies, e.g., 40 and 180 kc, can be tuned.

The tuning components of a broadband trap are arranged as a filter circuit with the main-coil inductance of the trap serving as one element.   The network includes a terminating resistance.   Most broadband line traps are fixed-tuned, either to the appropriate geometric-mean frequency to obtain maximum bandwidth in the upper portion of the carrier-frequency spectrum or to another geometric-mean frequency specified on order.   These traps provide minimum impedances of 400 to 600 ohms over their bandwidths, depending on the inductance of the main coil and the associated design parameters.   One 265-μh trap provides an impedance of 400 ohms or more within the band from 90 to 200 kc. A 1-mh trap provides 600 ohms or more over a bandwidth from approximately 51 to 200 kc. Adjustable auxiliary broadband tuning packs for use with standard single-frequency 265-μh traps are available with minimum impedances as high as 1,000 ohms over usable bandwidths, e.g., approximately 147 to 200 kc.

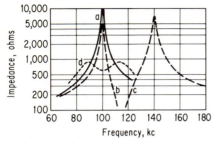

FIG. 5. Impedance characteristics of 265-μh line traps. (*a*) Single frequency. (*b*) and (*c*) Respective resonances of double frequency. (*d*) Broadband.

18. Impedance Characteristics

Impedance characteristics of typical line traps are shown in Fig. 5. In general, both the impedance magnitude and bandwidth of single-frequency and double-frequency line traps are higher with higher resonant frequencies. The magnitude of each impedance peak associated with a double-frequency trap is approximately one-half that which is obtainable by tuning each frequency individually with separate single-frequency traps. The impedance magnitude of the broadband trap is a function of design parameters and is independent of the geometric-mean frequency to which the trap is tuned. The bandwidth, however, is directly proportional to the square of the geometric-mean frequency. Although the impedance magnitude obtainable with a broadband trap is much less than the peak impedance of a single-frequency trap, its bandwidth is greater and the impedance has a more predominant resistive component.

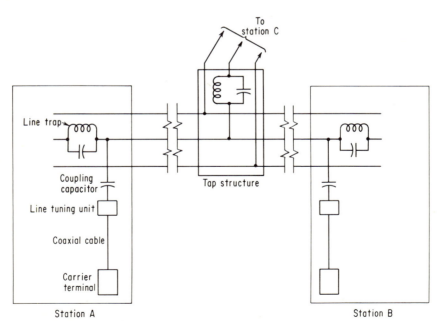

FIG. 6. Typical application of line traps. The relationship of the trap to other coupling equipment in normal usage is illustrated at stations *A* and *B*. The installation at the tap structure is one example of special usage.

19. Line-trap Application

The purpose of line traps is to minimize carrier losses by confining carrier energy to the desired path as nearly as possible. In addition, traps are effective in reducing noise levels at carrier receivers and in minimizing interference to and from other carrier circuits.

Proper and complete trapping of a carrier channel will eliminate or reduce many of the propagation problems encountered in ordinary practice described in paragraph 23. One example is the reduction of resonant conditions and reflections due to stub lines tapped onto transmission lines with through circuits, as in Fig. 6.

TRANSMISSION-LINE CHARACTERISTICS

20. Carrier Propagation[4-7]

The behavior of a carrier signal being propagated along a multiconductor power-transmission line is governed by the same physical laws that have been developed into the classic equations normally applied to isolated two-conductor communication transmission lines. A true analysis of propagation along the multiconductor line, however, is much more difficult because of the multiplicity of self- and mutual impedances which exist.

Several conducting paths exist on a power line, including at least three power phases and a ground path, which usually consists of one or two ground wires and earth. When a carrier voltage is impressed upon an input circuit consisting of any two of these paths, a wave of carrier energy will proceed down the line, apparently beginning the same as on the theoretical two-conductor communication circuit. Except with special cases of symmetry, however, the energy is not confined to the intended route. Mutual coupling between conductors causes a continuing interchange of energy which results in changing proportions of current as the signal progresses down the line.

Various component theories have been applied to obtain better concepts of the complex phenomena associated with wave propagation on the multiconductor lines. The *natural mode* concept is particularly helpful in predicting the distribution of carrier-frequency current among the power conductors and in explaining many apparently strange aspects of carrier behavior.[6] A transmission line with n conductors has n natural modes which, when combined in appropriate proportions, can represent any current or voltage distribution among the n wires. No mutual coupling exists between modes, all conductor impedances are equal for a given mode, and the attenuation to a carrier signal in each mode may be considered independently as a linear function of distance.

On a *transposed* power line the three phases approach electrical symmetry for any basic configuration. A *phase-to-phase* carrier circuit on any two of the three phases, therefore, approaches balanced circuit behavior.[4] On *untransposed* lines, e.g., single-circuit with the three phases in the same horizontal plane, phase-to-phase carrier signals are balanced only when applied to the two outside conductors.[5] The unbalanced wave propagated when a carrier signal is applied on adjacent phases may be perceived as having two components, one of which is the same as the previous balanced arrangement and another which involves all three phases. Despite the lack of symmetry, a carrier signal on adjacent phases has lower attenuation because the added three-phase component has less loss per unit length of line than the pure phase-to-phase component. However, a small additional loss is experienced at the receiving terminal because there is energy on the uncoupled phase which cannot be recovered.

Every carrier signal coupled *phase-to-ground* has, in addition to interphase components, a true phase-to-ground component consisting of carrier-frequency current flowing in the same direction in all three phases and returning through ground wires or earth. The attenuation per unit line length of this component is very high in comparison with those not directly involving ground currents, so that, within a short distance from a transmitter, the phase-to-ground component is essentially lost. This apparently fixed loss at the sending end is very significant if one is to obtain consistent estimates of phase-to-ground carrier attenuation. Energy on uncoupled phases which cannot be recovered at the receiving end of a line is also significant.

Because of electrical symmetry, phase-to-ground carrier on either phase of a transposed line will be propagated in a manner similar to that on the center phase of an untransposed line. Phase-to-ground carrier on an outside conductor of the untransposed line, because of different component distributions, is attenuated more within a given line length than the previous examples.

Further discussion of various carrier arrangements on different transmission-line configurations is beyond the scope of this chapter. Each user must decide, on the basis of the number of applications, the severity of the requirements of each application, and other pertinent factors, whether extensive analysis of such systems is justified. With sufficient mathematical treatment, application engineers may use modal component studies to determine which of several practical methods of coupling will provide lowest loss for a given line configuration.

21. Characteristic Impedance

The value of impedance to which carrier terminals and coupling equipment are adapted in an effort to achieve minimum mismatch attenuation is called the characteristic impedance of a carrier circuit.

Table 4. Range of Characteristic Impedance of Power-line Carrier Circuits

Transmission-line conductor, each phase	Characteristic impedance, ohms	
	Phase-to-ground	Phase-to-phase
Single....................	350–500	650–800
Bundled (two wires).........	250–400	500–600
Bundled (four wires).........	200–350	420–500

On phase-to-ground carrier circuits and to a lesser extent on phase-to-phase circuits, reflections on uncoupled phases cause a return of energy toward the coupled-circuit input. Therefore, reflection-free termination, according to the classic concept of characteristic impedance, is not possible.

One method for determining characteristic impedance is to make several bridge measurements at closely spaced frequencies near the operating frequency. A plot of these values vs. frequency discloses standing-wave patterns. Characteristic impedance is calculated as the geometric mean of maximum and minimum values.

Comparative measurements of this type on phase-to-phase circuits indicate only a slight dependence on the terminating impedance of the uncoupled phase. Significant differences, however, may be observed in similar measurements on phase-to-ground circuits with the power line deenergized, depending upon whether the uncoupled phases are open-circuited or shorted to ground.

There is surprisingly little variation of characteristic impedance among lines of different voltage with single conductors per phase. It is mathematically proportional to the logarithm of the ratio of conductor radius to spacing. In general both these quantities are larger on lines of higher voltage. Lower values of characteristic impedance exist on lines with bundled conductors,* where effective radius is considerably larger than the radius of each individual conductor.

Table 4 shows the range of values that may be expected on a wide variety of lines.[4]

* Two or more subconductors per phase at the same potential serving as one conductor of the line.

22. Attenuation and Carrier Frequency Response

Any predetermination of power-line carrier channel attenuation by calculation is, at best, an estimate. Considerable variation may be found in measured results on completed installations because of individual peculiarities which are likely to exist on any line and because of many minor factors for which consistent predicting methods have not been formulated.

Figure 7 displays limits of attenuation per unit length that may be expected on a wide variety of lines based on published data from field measurements. The following descriptions of various factors affecting attenuation will be found helpful in estimating where, within these limits, the attenuation of a particular carrier circuit might fall.

On balanced two-conductor telephone lines, attenuation is approximately proportional to the square root of the frequency. On multiconductor power lines with much wider proportionate spacing, attenuation increases more steeply at higher frequencies, so that it is more nearly a linear function of frequency than parabolic.

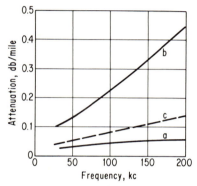

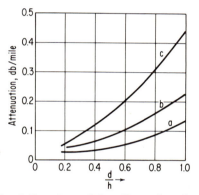

Fig. 7. Approximate range of phase-to-phase attenuation per unit line length. Because attenuation is not necessarily linear vs. distance, unit values are averages dependent on line length. Curve c may be considered typical on lines between 110 and 220 kv less than 100 miles long.

Fig. 8. Experimental data illustrating effect of proportionate line dimensions on carrier attenuation. The symbols d and h represent geometric mean spacing and height, respectively. Curves a, b, and c are for frequencies of 50, 100, and 200 kc, respectively. Data will vary with other line characteristics and earth resistivity.

Lower voltage lines, i.e., 46, 69 kv, etc., ordinarily have smaller conductors with correspondingly higher resistance and higher attenuation per unit length. Transmission lines in the 300- to 500-kv range usually have bundled conductors with lower resistance at carrier frequencies. However, the conductors are usually spaced wider in comparison with their height above ground, which tends to offset the lower attenuation trend of the larger conductor area. The proportionate spacing and height of any power transmission line are such that induced carrier-frequency current flows in the earth even when the carrier signal theoretically consists solely of interphase components. The magnitude of this current and the resulting carrier attenuation are functions of the ratio of conductor spacing to height and also of ground resistivity (see Fig. 8).

As discussed in the preceding paragraphs on carrier propagation, a signal which has energy existing on uncoupled phases is attenuated at a nonuniform rate along the length of a line, particularly near the sending end. This nonuniformity is significant on phase-to-ground circuits and should be considered in attenuation estimates. One method used with fair success on *transposed lines* is to add 3 to 6 db to an estimate of total phase-to-phase attenuation based on uniform attenuation per unit line length. For phase-to-ground circuits which include carrier bypasses around intermediate substations, this fixed loss must be added anew for *each additional line section*.

The use of this method is inadequate for most applications on *untransposed lines*. Here the rate at which a signal is attenuated along the line depends not only on whether the coupling is phase-to-phase or phase-to-ground but also on the line configuration and the choice of conductors. More accurate estimates can be made by the use of graphs such as Fig. 9. Unfortunately, this method requires separate charts for each transmission-line configuration and is further complicated when a section of line contains two or more configurations within its length.

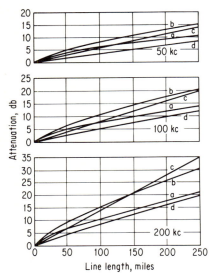

FIG. 9. Calculated attenuation as a function of distance on medium-voltage, untransposed, single-circuit power lines. (*a*) Center phase-to-ground. (*b*) Outer phase-to-ground. (*c*) Phase-to-phase, outer phases. (*d*) Phase-to-phase, adjacent phases.

The curves in Fig. 9, made from calculated data, have only a limited amount of experimental correlation and should be used accordingly. They do not include the effects of reflections on uncoupled phases, which will vary with different applications. An interesting and useful point is that, on *long lines*, a carrier circuit coupled phase-to-ground can be almost as efficient as one coupled phase-to-phase.

Coupling losses at end terminals and bypasses should be included in estimates of total carrier circuit attenuation. The loss through most broadband line-tuning units will fall between 0.5 and 2 db. Resonant tuning units have higher losses at both low and high frequencies.

Figure 10 illustrates typical losses of many resonant tuning units which are now in service. Losses in modern units currently being installed are somewhat less.

Approximate losses in coaxial cables range from 0.3 db per 1,000 ft at 30 kc to 0.7 db per 1,000 ft at 200 kc.

Carrier-frequency hybrid units are frequently used between closely spaced transmitters and receivers connected to the same coaxial cable or to prevent intermodulation between two transmitters operating in parallel at closely spaced frequencies. An allowance of approximately 3.5 db for each hybrid should be included in attenuation estimates. Other miscellaneous items include attenuation due to branching of the carrier circuit and bridging loss. For example, if a carrier bypass is made from one line in a station to two other lines, the equal division of carrier energy represents a 3-db bridging loss, three-way divisions represent 4.8-db loss, etc. Where multiple branches are involved, the low impedance of the paralleled coaxial-cable circuits can cause, in addition, an appreciable mismatch loss that must be calculated.

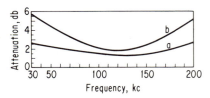

FIG. 10. Attenuation in resonant line-tuning units. (*a*) Single frequency. (*b*) Double frequency.

Cross-coupling of carrier signals through a high-voltage switchyard by *unintended paths* can affect the efficiency of a carrier bypass in either a favorable or unfavorable manner. This depends on the frequency-related phasing of the cross-coupled signal relative to the bypassed signal. Cross-coupling takes place through coupled phases where line trap impedances are less than ideal but, for the most part, occurs on untrapped phases and on parallel sections of transmission lines approaching the station. Of greater significance than the influence on bypass efficiency, cross-coupled carrier energy is dispersed among all lines in a station to create potential interference to other services. In locations where a carrier repeater is installed, cross-coupling provides a feedback circuit which will limit the amount of gain that can be realized. For such installations, the actual value of cross-coupling attenuation should be measured.

Bad weather causes increased attenuation on transmission lines. If a carrier circuit is to function properly at all times, sufficient margin must be allowed in its design that the effects of weather will not disable it.

In areas where insulators accumulate heavy deposits of fly ash, chemical vapors, or other industrial contaminants, attenuation, even in fair weather, will be higher than average. Rain in such contaminated areas causes an appreciable increase in attenuation. In areas remote from industrial centers, rain will usually cause attenuation to increase by 20 to 40 per cent of the decibel value.

In moderate climates, occasional sleet coatings may approximately *double* line attenuation. In more severe climates, heavier sleet and hoarfrost coatings on line conductors may cause attenuation to increase by a factor of 3 to 4. Tests made in extremely severe weather have indicated attenuation as much as fifteen times normal values. It may be assumed, however, that such extreme increases will not occur over the entire route.

In estimating reserve signal margins, it is necessary to apply weather factors only to attenuation which occurs on transmission lines. Losses in line-tuning units, coaxial cables, and other miscellaneous items are not dependent on weather conditions.

To be assured of proper operation, measurements of *carrier frequency response*, i.e., actual attenuation as a function of frequency, should be made on all newly installed carrier channels. A poor frequency response causes distortion to carrier signals and is particularly detrimental to services employing frequency modulation. This characteristic can be as serious a problem as a high magnitude of attenuation. Typical evidences of poor frequency response include a bandwidth too narrow for the required carrier service and periodic or erratic variations of attenuation vs. frequency.

23. Problems Encountered in Practice

Once an adequate path is established, power lines provide very rugged, efficient, and reliable media for carrier communications. Nevertheless, they are constructed primarily to meet the requirements of high-voltage power transmission in the most economical and practical manner. Several problems exist in using these lines for communication, some of which are common to all carrier circuits, while others are due to peculiarities of individual transmission-line arrangements.

Reflections from any impedance discontinuity contribute to poor frequency response. Minor reflections occur continuously along any transmission line because of nonuniformity of line characteristics at towers and variable conductor heights above uneven terrain. Changes in line configuration, e.g., where a double-circuit line separates into two individual lines, cause more severe reflections. However, reflections due to these inherent characteristics of power transmission lines are rarely serious enough to be prohibitive.

Reflections from terminations on coupled phases may be minimized by proper application of line traps and careful matching of impedances. Reflections on uncoupled phases depend on the impedance presented by high-voltage apparatus and other lines entering a station. The severity of their influence on carrier-channel frequency response is largely unpredictable but does depend to some extent on the line configuration and symmetry in the method of coupling. Extreme cases, exhibiting excessive attenuation and poor frequency response, often called *absorption effects*, may require rephasing of the carrier circuit to obtain satisfactory performance. An example is shown in Fig. 11.

Tap lines cause a variety of problems. Unless specifically coupled for a carrier terminal at the end of the tap line, the terminating impedance of all phases is usually high and of reactive nature. The input impedance at the tap point will be low at and near each frequency for which the electrical length of the tap line is an odd multiple of quarter wavelengths. It is customary to insert line traps in series with tap lines of this type as in Fig. 6, where the periodically recurring low impedance causes intolerable attenuation within the bandwidth of any carrier channel on the through circuit. Because of the way in which carrier propagation involves uncoupled phases, minor losses due to reflections on the tap line will still exist at periodic frequencies. In installations where carrier is required at the station on the end of the tap line, the coupled phase at that location is appropriately terminated. An unavoidable 3-db bridging loss and some reflection losses to the through circuit are to be expected at the junction. In practice, however, much more severe attenuation at periodically recurring frequencies might be experienced, particularly if the circuit is coupled phase-to-ground. This may also be attributed to reflections on uncoupled phases of the tap line. Where such conditions are intolerable or severe enough to justify additional expense, the through circuit may be converted to phase-to-phase coupling, the uncoupled phases may be trapped at the tap, or the uncoupled phases may be equipped with coupling capacitors and terminated at the station on the end of the tap line.

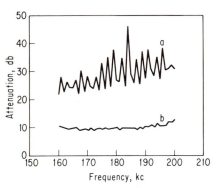

FIG. 11. Absorption effects which apparently result from lack of symmetry. (*a*) Frequency response of carrier channel coupled phase-to-ground on outer phase. (*b*) Frequency response on center phase-to-ground.

Occasionally, an existing power line will be cut and a *double-circuit loop* constructed to provide a station off the through-line route with two independent lines rather than a single tap line. When a carrier circuit terminates in this station, carrier energy will be cross-coupled into the opposite side of the loop. It might be necessary to install traps in the opposite line at the loop-in point to prevent excessive attenuation. A carrier signal which bypasses the station will exhibit the characteristics of multiple-path transmission unless the bypass is installed *at the loop-in point*.

Multiple-path transmission occurs when carrier signals from a single source follow separate paths of different electrical lengths to a common receiving point. Differences in propagation time cause the phase relationship of the recombined voltages to be a function of frequency; i.e., the received voltages are in phase and additive at one frequency and yet approach cancellation at another. The resultant frequency response of such a carrier channel is periodic in nature and may exhibit variations of 20 db or more in attenuation. In most instances, line traps applied in the normal manner prevent serious multiple-path effects.

24. Noise and Interference[2,4]

The noise level at the input to a carrier receiver determines the minimum received signal level necessary for adequate performance. Increasing receiver sensitivity beyond *nominal requirements* is of no benefit because performance depends on the *ratio* of *signal to noise*.

According to various studies, power-line noise consists of discrete *impulses*, occurring either erratically or periodically, superimposed on a *random* background noise at a lower level. The random component has a continuous frequency spectrum and is sometimes called *white* noise. It is recognizable in telephone circuits as a background hissing sound. Impulses cause an irregular frying or crackling sound or a buzz if they occur periodically. These noises are augmented by occasional rather violent bursts caused by circuit-breaker operation, disconnect-switch arcs, or nearby lightning discharges. On higher voltage lines, corona discharges cause a modulation of the noise voltage envelope by superimposing additional noise impulses in cadence with the positive peaks of the power-frequency voltage. Because corona also causes losses to power transmission, appropriate measures are taken, i.e., bundled conductors and special hardware on extra-high-voltage lines to hold corona losses within reasonable economic limits and to keep radio interference fields in power-line vicinities at tolerable levels. Fortunately, these same measures are effective in preventing noise in the carrier-frequency range from continuously rising as a function of transmission-line voltage.

Because of its complex nature, power-line noise, at a given location, has different interference effects on different kinds of carrier receivers. Various methods of measuring and describing noise voltages are used in the attempt to relate these values properly.

Peak value refers to the maximum amplitude of repetitive noise impulses and bears no relationship to the repetition rate of the impulses or to the random level between impulses.

Quasi-peak value is related to the peak value and also to the impulse repetition rate. Quasi-peak instruments have a measuring circuit with time constants allowing fast charge and slow discharge.

Average value represents the area under the amplitude-time curve divided by the appropriate time interval.

The *rms value* is an *effective* voltage representative of noise power.

In modern analytical studies, rms values are convenient in accurately relating a value within one bandwidth to that which may be expected within another bandwidth. Also with rms values, the effective noise level due to a number of known sources can be calculated.

In high-selectivity receivers, such as those used for frequency-shift telemetering, the effective differences in the interfering qualities of various types of noises are diminished. Rms values may be used for performance estimates; however, average values, which are easily measured with available instruments, are also sufficient. Interference to protective relaying, telemetering, or other services, when applied on keyed carrier facilities, is more closely related to average values of noise.

Listening tests show that quasi-peak values are more closely related to true nuisance values of noise in telephone circuits.

Peak noise is particularly detrimental to circuits which involve electronic switching devices, e.g., automatic simplex carrier telephone circuits.

Noise levels depend upon the voltage, dimensions, degree of contamination, and other conditions of a power line. They may vary over a wide range with variable weather conditions. No practical method has been devised whereby an *accurate* prediction of noise on a given voltage and type of line can be made. The determination of actual ranges of noise, to be accurate, must be obtained by a series of measurements or a continuous recording during all weather conditions.

In power-line carrier techniques, a normal assumption is that similar lines in a given area have comparable noise levels. This approach deviates from accepted telephone-line techniques in which every line is measured individually. This distinction is justifiable on the basis that power-line carrier techniques, from many aspects, cannot be so exacting as those in telephone-line practice.

If it is not feasible to make specific measurements and if more accurate data concerning a specific area are unavailable, the values in Fig. 12 may be used as a reasonable approximation of noise levels on typical lines, subject to the qualifications in the following paragraphs.

The ratio of peak to quasi-peak and average values shown are simply representative of the lines upon which measurements for these data were made. While they may be typical, no fixed relationship exists for universal application.

The levels shown in Fig. 12 are based on phase-to-ground measurements. Noise on phase-to-phase circuits may differ depending primarily upon the nature of the noise source.[3] Corona noise on phase-to-phase circuits is usually higher by approximately 2 db, whereas noise due to external sources such as lightning may be as much as 6 db less.

To be realistic, these levels should be reduced by a factor to account for the protective influence of the line trap at the receiving location. The actual improvement is a function of the high-voltage station characteristics and other parameters which may differ with each individual circuit. A probable average is 8 to 10 db.

For application to circuits with receiver bandwidths other than 400 cps, correction factors must be used. These corrections may be based on the assumption that noise *voltage* is proportional to bandwidth raised to a power of x, where x approximates unity for peak and quasi-peak values and is approximately 0.5 for average and rms values. Figure 13 provides decibel values for these corrections.

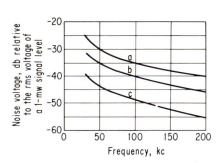

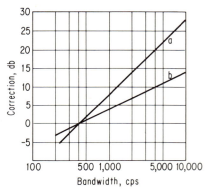

Fig. 12. Noise levels on typical power lines, 220 kv and below, based on measurements within a 400-cps bandwidth. (*a*) Peak. (*b*) Quasi-peak. (*c*) Average. Noise on extra-high-voltage lines may be 5 to 10 db higher.

Fig. 13. Approximate corrections for noise level within bandwidths other than 400 cps. (*a*) Peak and quasi-peak. (*b*) Average and rms.

Design criteria for power-line carrier telephone circuits are directed toward obtaining an adequate signal-to-noise ratio in the *audio* circuit. Such criteria are not normally set so high as those for commercial toll-grade communication circuits. Actual noise levels are higher, and for the purposes intended, fidelity and quality beyond that required for good *intelligibility* cannot in many cases be economically justified.

Audio noise meters in current usage provide a reading related to something between the average and rms values of noise in the audio band. They have selectable frequency characteristics which enable them to measure *flat* or *weighted* noise. Weighted readings are based on frequency characteristics representative of the combined response of a typical telephone handset receiver and an average human ear. These readings are less than flat noise by a factor ranging from about 2 to 10 db, depending upon the audio bandwidth and the nature of the noise.

The minimum signal-to-noise ratio considered acceptable for power-line carrier circuits by various users ranges between 25 and 30 db based on weighted audio measurements. The corresponding lower ratio based on flat noise is more easily related to carrier-frequency measurements.

The method for relating carrier-frequency measurements to audio measurements is different for different modulation systems and also depends upon the type of noise measurements made at carrier frequencies. European practice favors rms noise values where for a single-sideband circuit, the carrier-frequency signal-to-noise ratio based on noise measured in a 3-kc bandwidth is the same as the audio ratio. For amplitude modulation the ratio of unmodulated carrier level to noise measured in a 6-kc bandwidth is considered equal to the audio signal-to-noise ratio with 100 per cent modulation. For frequency modulation the audio signal-to-noise ratio is assumed to be improved over the ratio obtained with AM by a factor dependent on the modulation index. For an index of unity, this factor is approximately 5 db.

The preference by some users for the higher quasi-peak values requires different criteria for carrier-frequency signal-to-noise ratios. For single-sideband application a 10-db ratio of test signal level to quasi-peak noise in a 3-kc bandwidth is considered minimum. For amplitude-modulated circuits a 10-db ratio of unmodulated carrier level to quasi-peak noise in a 6-kc bandwidth may be used. Frequency-modulated circuits are considered to have as much as a 5-db advantage over AM provided the circuit is relatively distortion free.

For services other than telephone, adequate margin for correct performance, rather than audio measurements, serves as the criterion. Signal-to-noise ratios are based on noise values within bandwidths corresponding to that of the carrier receiver. Minimum carrier-frequency ratios considered acceptable by various users include pilot relaying, 20 db, rms signal to average noise, and keyed-carrier telemetering, 15 db on the same basis. Some recommended ratios on an rms basis include 26 db for telemetering with amplitude keying and 20 db for frequency-shift telemetering.

Various attitudes are prevalent concerning satisfactory methods for establishing bad-weather margins. Exact figures for maximum noise levels are difficult to obtain and are too high for practical design consideration. These extremely high levels, however, are usually of short duration and infrequent occurrence. Designs may be based on noise levels which statistically are not exceeded more than 10 per cent of the time. These may not exceed average fair-weather values by more than 5 to 10 db, although for short periods noise levels may be 20 to 25 db above normal.

Full-time satisfactory operation can usually be assured by a margin equal to the estimated bad-weather increase of attenuation plus 10 db for increased noise.

Carrier facilities are vulnerable to *interference* from other carrier signals which exist at specific frequencies rather than being distributed throughout the spectrum as power-line noise. Excessive interference to and from other carrier services may be prevented by careful coordination of frequency assignments, proper use of line traps, and special attention to any situation which might contribute to an interchange of carrier energy between lines, e.g., parallel line sections and switchyard cross-coupling.

25. Measurements

Measurements on a completed carrier channel and on the applied carrier circuit are very desirable to ensure proper operation and to gain detailed experience with the user's own system.[8,9]

Impedance measurements have been discussed briefly in preceding paragraphs. Input impedance of deenergized power lines may be made directly on the power conductors provided suitable protective elements are used to eliminate hazards of electrostatic voltages and induced surges. The protection circuit used should be tested to determine if its influence on the measurement requires correction factors. Measurements on energized lines may be made through normal coupling capacitors or complete line-tuning assemblies. If a coupling capacitor alone is used, suitable accuracy is usually obtained by simply deducting the capacitive reactance computed for each frequency at which measurements are made. Measurements through line-tuning units usually require some form of substitution method to arrive at the true input impedance of the line.

Measurements of *attenuation and frequency response* may be made by the method shown in Fig. 14. Measurements should be made at frequency intervals close enough to establish the complete standing-wave pattern so that results may be interpreted accurately. All carrier channels will exhibit some periodic variation of attenuation vs. frequency at a rate dependent upon line length. The frequency interval between successive peaks is approximately 90 kc divided by the line length in miles. On long lines it will be found impractical to measure complete broadband carrier channels at frequency intervals small enough to disclose every maximum and minimum point. This difficulty can be overcome by *sampling* appropriate narrow bands at very close increments to determine the range of attenuation variation. Unless these variations are severe, the remainder of the band may be measured with more suitable increments.

Noise measurements for general application may be made through ordinary coupling apparatus to obtain voltages directly on the power-line conductors provided suitable corrections are made for attenuation in the coupling equipment. It is usually safe to assume that noise levels on lines of the same type and voltage are closely enough related, over a fairly large area, that every line need not be measured individually.

Noise levels on line conductors are often applied for general use in preference to values on coaxial cables or other points so that the effects of different line-tuning methods and different cable lengths may be considered individually for each application. For design purposes, coupling losses at a receiving location may be deducted from total circuit attenuation to establish a net received signal level on the line itself.

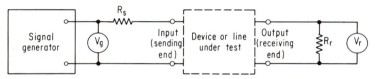

Fig. 14. Method of measuring carrier-frequency attenuation and frequency response. If there is a net impedance transformation in the circuit, the source impedance and receiving termination will not be equal.

$$\text{Loss} = 20 \log \frac{V_g}{2V_r} + 10 \log \frac{R_r}{R_s}$$

Beyond this point, signal and noise are attenuated equally, causing no change in signal-to-noise ratio.

Noise voltage at any point is composed of contributions from local sources and other contributions which are propagated to that point from distributed sources along the line. Because of this propagation, noise-level behavior is similar to carrier-signal behavior in that it will exhibit standing-wave patterns and will be dependent upon the impedance terminating or bridging the line at the measuring point. Attitudes regarding the feasibility of placing a fixed terminating resistance at the measuring point vary among different users. However, if the noise level is to be compared directly with a measured carrier signal, the terminating impedances should be equivalent.

The signal-to-noise ratio of an individual carrier circuit may be obtained by direct comparison of noise level and received signal level measured at a common point, e.g., a terminated coaxial cable.

Carrier-frequency *reserve signal margin* may be determined by inserting a variable attenuator in the output of a carrier transmitter and increasing its attenuation until erratic performance is observed at the receiving end of the circuit; e.g., speech intelligibility becomes impaired or telemeter recorders draw wide bands. Attenuating the input signal at a receiving location normally is not sufficient for this test because the signal-to-noise ratio is not changed by attenuation placed at this point. However, on directional-comparison pilot relay circuits, it is customary to adjust receiver sensitivity for a fixed margin above the minimum setting required to operate the receiver relay with a normal input signal. A margin defined in this manner *can* be checked by attenuating the received signal if an adequate signal-to-noise ratio has already been proved.

CARRIER ON POWER CABLE

26. High-voltage Power Cables

These cables are manufactured in several types. It is possible to operate carrier facilities over these cables if due consideration is given to the fact that both characteristic impedance and attenuation are quite different from that of overhead power lines. Although very little information has been published regarding *detailed* characteristics of power cables at carrier frequencies, the following general statements are applicable.

Each conductor in a power cable is enclosed in a grounded metallic shield. This provides two important advantages. First, the characteristics of each phase are essentially independent, i.e., not vulnerable to the problems caused by uncoupled phases on overhead lines. Second, cable conductors are not exposed to the same sources of noise that plague overhead power lines. In fact, the noise level is usually so low that carrier-receiver gain is not limited beyond that determined by inherent noise in the receiver itself.

Coupling to power cables is usually made on a phase-to-ground basis. The range of characteristic impedance of various types of cable is in the order of 25 *to* 50 *ohms,* and attenuation is approximately *ten times* the decibel value of that on overhead lines. These figures are approximations which may be used as a general guide. Actual measurements preferably should be made before operation of carrier equipment is attempted.

Because of the low impedance, carrier-channel bandwidth obtainable with conventional coupling facilities is highly selective compared with that on overhead lines.

In some power systems, relatively short lengths of power cable are occasionally used to provide underground entrances for overhead power lines. A carrier circuit on the overhead line may be operated in several ways. If space is available, and if the location is a true terminal point, the carrier transmitters and receivers may be located at the transition point and the services extended into the main substation by means of audio cable. A second method is to install coupling facilities on the overhead line and bypass the carrier frequencies into the substation via coaxial cable. A third method is to install coupling facilities on both the overhead line and the cable so that impedance-matching facilities may be installed at the transition point. The fourth method is to permit direct connection of the carrier circuit from the overhead line to the cable. Only rarely, however, will this last method provide enough efficiency for satisfactory operation because of the extreme impedance mismatch which will exist. A choice among these methods will depend upon the length of cable involved, the required bandwidth of the carrier channel, the importance of the carrier services, and the amount of reserve signal that is available.

CARRIER ON OVERHEAD GROUND WIRES

27. General

In many parts of the United States and other parts of the world where lightning activity is high, it is common practice to construct power transmission lines with "ground" or shield wires strung above the power-carrying conductors. These ground wires are normally grounded to each tower or pole ground throughout the length of the line and are positioned to shield the transmission line from lightning.

These ground wires can be used for carrier communication without adversely affecting the lightning protection of the transmission line by insulating them from the towers or poles with relatively low voltage insulators. Communication is accomplished by properly connecting the insulated ground wires to carrier-frequency equipment through suitable line-protection facilities at the terminals.

28. Insulated Ground Wires vs. Phase Conductors

While experience in the use of insulated ground wires for carrier communication is limited at this time, there are already certain apparent advantages in their use compared with using the phase conductors.

The coupling facilities necessary for connecting the carrier terminal equipment to the line conductors are much less expensive. A greater overall bandwidth is realized over the insulated ground wires than can economically be provided over the phase conductors using commercially available coupling equipment. Maintenance costs are lower because practically all maintenance work on both line and terminal equipment can be done without deenergizing the transmission line. No line outages are required for changing frequencies of carrier terminal facilities or adding new facilities to the insulated ground wire channel.

One major disadvantage is that more momentary interruptions are likely to occur to carrier communication when used on insulated ground wires than when used on phase conductors. This is because the ground-wire insulators will flash over on heavy and close-in lightning discharges and possibly from fault current induction, including that from closely paralleling transmission lines. However, experience indicates that only the most exacting communication services such as carrier pilot relaying and transfer trip protection might not be entirely satisfactory. Even this should not be considered conclusive at this time. Other services, such as telephone, telemetering, and supervisory control, are providing entirely satisfactory operation over insulated ground wires.

No experience is available to indicate the extent of attenuation increase due to complete sleet covering of the relatively small insulators.

29. Insulation

Several different types of insulators are in use. For example, a 15-kv guy strain insulator of the fluted type has proved to be suitable. This insulator, along with two strain clevises at each point of suspension, provides a reliable arrangement in that the ground wires cannot fall down onto the phase conductors if an insulator is broken.

The flash-over voltage of the 15-kv insulator is approximately 32 kv. In current practice, arc gaps are ordinarily added to provide a lower potential flash-over and to prevent scoring of the insulator glaze with repeated lightning strokes.

Other satisfactory insulators in use include pedestal and higher voltage types.

30. Conductor

Usually, shielded transmission lines designed for 110 kv or higher use two ground wires, while in some cases lower voltage lines use only one. While one insulated ground wire can be used for carrier communication, the two wires used as a pair have definite advantages in noise and attenuation. Normally the ground wires are not required to have particularly high conductivity when used only for shielding the phase wires from lightning, although there are some conditions, such as extremely high earth resistivity, where higher conductivity ground wires might be helpful for this purpose. The ground wires for many existing transmission lines are galvanized-steel messenger.

If the ground wires are to be used for carrier communication, it may be economical to use higher conductivity wires, depending on the length of line and type of carrier equipment proposed. Several types of wires having varying degrees of conductivity are now available which are suitable for insulated ground-wire application.

31. Transpositions

The insulated ground wires are subject to induction from the phase conductors, and if two ground wires are used, it is highly desirable that they be transposed. Transpositions are made by dead-ending the ground wires on a structure and supporting cross connections on pin-type insulators. The use of span transpositions between structures has not been considered practical.

The primary aim of a transposition scheme is to provide *balance* of power-frequency influence so that induced current in the two wires will be equal and in phase. However, proper transposing will also reduce the *magnitude* of drainage current flowing in the ground lead of the terminal protectors by a proportion depending on the power conductor configuration. Independent calculations have indicated that a substantial saving in power loss results from insulating and transposing the ground wires. The proportionate loss reduction depends on individual line dimensions and other characteristics. In one example, with a 185-mile, 500-kv power line carrying 1,000 Mva, the calculated losses in the ground wires were approximately 500 kw with no transpositions and between 5 and 10 kw with a suitable transposition scheme.

A transposition scheme should take into account any change in the uniformity of the line, such as closely paralleling transmission lines for more than a few hundred feet or a change from single-circuit to double-circuit tower construction. For best results, transpositions should be located so that the power-frequency induction is balanced out independently in each uniform section of transmission line. Thus, the number of transpositions in any particular line will depend upon its uniformity.

Usually, within a relatively long uniform section of line, transpositions located at the one-quarter point and the three-quarter point are sufficient for most carrier communication applications. On very short sections, one transposition located at the midpoint may be adequate.

Some users prefer more frequent transpositions in long sections in order that a minimum of changes will be required in the event an intermediate station is to be added and also to obtain more rejection of crosstalk to and from the phase wires. It is not practical to attempt elimination of crosstalk at carrier frequencies between the insulated ground wires and the phase wires to the extent that identical frequencies may be used on each. However, minimizing cross-coupling in this area is highly beneficial from an interference standpoint.

Careful attention to the matter of transpositions is well worthwhile from the standpoint of reducing cross-coupled interference as well as obtaining balance of power-frequency induction.

32. Line-coupling and Protective Equipment

Since the insulated ground wires will be subjected to rather high power-frequency induction, it is necessary to provide a safe and suitable means for connecting to the carrier transmitter-receiver equipment. A satisfactory arrangement for a two-ground-wire system is shown in Fig. 15.

The drainage coil connected to the line must be capable of carrying the power-frequency drainage current continuously. The magnitude of the drainage current may be anywhere from a fraction of an ampere up to 15 amp or more and will depend upon the configuration of the transmission line, the power-line voltage and loading, and the accuracy of the transposition scheme. The coupling and protective unit shown will tend to block the basic power frequency and lower-order harmonics from the carrier terminal equipment but will offer very little loss at carrier frequencies.

An insulating-transformer type of line-coupling and protective unit such as is used on underbuilt telephone lines may also be used successfully. This type of unit, however, has one disadvantage in that unbalanced power-frequency voltage on the ground wires will be transferred to the drop side of the transformer. If the ground wires are accurately transposed, this will not be a serious problem. A high-pass filter may also be used between the transformer and the carrier terminal equipment to block the power frequency and lower-order harmonics.

During transmission-line faults the insulated ground wires may carry appreciable portions of resulting fault current depending upon the location of the fault with respect to either end of the line. For this reason it is desirable that suitable heavy-duty arc gaps be provided for bypassing the coupling unit during fault conditions. Grounding relays are also used in some versions.

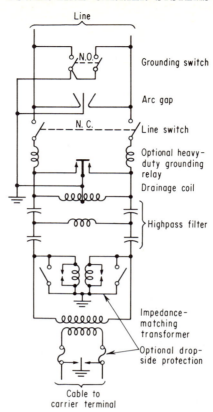

Fig. 15. Coupling unit suitable for terminating insulated ground-wire communication circuit.

Either telephone cable or coaxial cable may be used for connecting the coupling unit to the carrier terminal equipment.   The required impedance transformation will depend upon the type of cable used.

33. Carrier-frequency Characteristics

Carrier-frequency characteristics of insulated ground wires are affected by several factors, such as conductivity of the wires, spacing of the wires when a pair is used, distance between ground wires and phase wires, number of phase wires on the towers, and height of ground wires above earth.

Unlike telephone lines, power transmission lines are not standardized to any appreciable extent.   Ground-wire spacing and height for different lines, even for the same voltage, may vary considerably.   Where two ground wires are used, they may be spaced as much as 65 ft apart on transmission lines designed for 500-kv operation.

Standard telephone lines with conductors normally spaced 8 to 12 in. apart and accurately transposed lend themselves to quite precise determination of the various parameters.   In contrast, designs of insulated ground-wire circuits are still based on averages of available experimental data which serve as general guides for performance estimates.

The characteristic impedance of a balanced ground-wire pair is approximately 900 ohms, nominal.   Only small variations are to be expected in wide ranges of proportionate line dimensions. Quite satisfactory operation has been obtained with standard telephone-line carrier equipment, normally used on 600-ohm circuits, without intermediate impedance-matching facilities. The characteristic impedance of a single-wire with ground return falls within the approximate range from 500 to 600 ohms.

Attenuation on insulated ground-wire circuits varies widely, depending largely on the conductivity of the wires, and to a much lesser extent on line dimensions, proximity to phase conductors, transposition intervals, and line lengths. Figure 16 shows two extremes of attenuation based on actual measurements on lines varying in length from 12 to 100 miles. The lower loss is for Alumoweld conductor, and the other for galvanized steel. For estimating purposes, conductors with different resistances per unit length should cause attenuation of proportionate magnitude. Limited experience indicates losses on a single wire with ground return to be approximately 50 per cent higher than that on a balanced pair.

Noise on insulated ground-wire circuits is maximum at low frequencies. On most circuits the noise level is prohibitive at frequencies below approximately 6 kc, although,

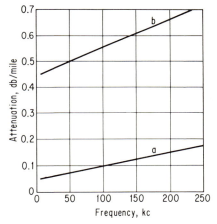

Fig. 16. Typical attenuation on insulated ground wires with wide range of conductor resistivity. (*a*) Three-strand No. 6 Alumoweld. (*b*) $\frac{7}{16}$-in. galvanized steel.

in some cases, more frequent transpositions may allow limited carrier operation down to 3 kc. At carrier frequencies a balanced ground-wire circuit has noise advantages over phase-to-ground power-line carrier. One example based on audio measurements on a single-sideband telephone circuit operation in the 150-kc range indicates this advantage to be about 10 db within a 3-kc bandwidth.

34. Carrier Terminal Equipment

A carefully designed insulated ground-wire pair will have a usable bandwidth from about 6 to well over 200 kc. Standard open-wire telephone-line carrier equipment, including repeaters, can be used quite satisfactorily if attenuation and noise limits are considered in the application. Standard power-line carrier equipment, which normally operates at higher output levels, can also be used successfully.

TELEPHONY

35. Power-line Carrier Equipment

This equipment for telephone service is available for either simplex or duplex operation and may be frequency modulated, amplitude modulated, or single sideband.

The majority of carrier telephone equipment currently available uses vacuum-tube electronics; however, the trend is toward transistor types, which have advantages in space and power requirements. Most transistorized equipment can be operated directly from station batteries, eliminating the need for emergency a-c power sources.

36. Simplex

Only one carrier frequency is used in a simplex carrier circuit, and only one terminal transmits at a time. Control is transferred back and forth during a conversation by means of a coordinated relay scheme which causes the receiver to be blocked during transmission and, in turn, prevents transmitter operation while a signal is being received. In *manual simplex* systems, this control is usually accomplished by means of a push-to-talk button on the telephone handset. In *automatic simplex* systems, voice-operated relays or electronic transfer units are actuated by speech from the telephone transmitter. This makes it possible to route the audio two-wire drop of a circuit through telephone switchboards to various parties who need not be equipped with push-to-talk handsets.

The greatest advantage of simplex carrier telephone equipment is that it can provide party-line service. With only one terminal transmitting, any reasonable number of terminals may be receiving; consequently, multiple-party conversations are possible. An additional advantage is frequency conservation.

37. Duplex

Continuous transmission of carrier from each terminal is made possible by the use of separate frequencies for the two directions. Advantages of this method over simplex systems are as follows: Untrained personnel find it easier to carry on a normal conversation, duplex equipment is more readily adapted for routing the voice circuit through manual or automatic exchanges, and speech-plus-tone operation is possible. There are some limitations in that party-line service is possible only with cascaded individual circuits (except with some single-sideband systems, paragraph 41). The system requires more frequencies, particularly with cascaded circuits.

38. Speech-plus-tone Service

Telemetering and control functions are often transmitted via frequency-shift, amplitude-modulated, or keyed audio tones. Tone transmitters and receivers are available with various bandwidths for spacings of 100 to 510 cps, depending upon the functions to be transmitted.

The audio bandwidth available over power-line carrier telephone circuits is usually in the order of 300 to 3,000 cps. With sufficiently high-quality circuits, speech with good intelligibility can be transmitted over a bandwidth of approximately 300 to 2,000 cps. When appropriate filters are used and a sufficient guard band is allowed, the remainder of the audio bandwidth available in a duplex carrier circuit may be utilized for the continuous transmission of audio tones.

Nominal crossover frequencies range from 1,800 to 2,400 cps with 2,200 probably most widely used. Guardband requirements depend upon the characteristics of the separating filters but in some cases are as low as 100 cps above the crossover frequency.

39. Frequency Modulation (FM)

The frequency modulation used in power-line carrier is of the narrowband type in order to conserve space in the frequency spectrum. The total bandwidth of the transmitted signal is limited to approximately 6 kc by restricting the deviation ratio to unity or less.

Noise encountered on power lines affects the amplitude of a received signal. Therefore FM systems inherently offer an advantage over amplitude-modulated equipment in that FM receivers do not respond so readily to noise voltages. However, the advantage to be gained is far less in narrowband power-line carrier equipment than in broader band commercial radio broadcast systems. With a good-quality carrier channel, an FM system operating with unity deviation ratio may have a signal-to-noise ratio 4 to 5 db higher than an AM system of equal transmitter power. Frequency-modulated receivers are critically dependent on a carrier-frequency channel of high quality. Nonlinear phase shift and erratic frequency response in the carrier path will cause distortion. If the carrier path cannot be made suitable, the deviation ratio must be reduced, and the noise advantage over AM carrier is reduced accordingly.

40. Amplitude Modulation (AM)

The amplitude modulation used in power-line carrier is basically the same as that used in the standard broadcast band of commercial radio. It is the most vulnerable of all systems to atmospheric and other line noises. Many of the older type transmitters were easily overmodulated, causing *sideband splatter* and high harmonic output. On the other hand, the older equipment was of straightforward design, very rugged, and easily maintained. Until the mid 1940s powerline carrier telephone equipment was almost exclusively AM. Very little AM equipment is currently being manufactured in the United States, although there are still a great number of earlier vintage terminals in service.

41. Single Sideband

This is basically a form of amplitude modulation but has several important advantages. It requires only one-half the bandwidth of an AM or FM system transmitting the same audio-frequency range. The entire power output of the transmitter can be concentrated in the intelligence, although in some single-sideband transmitters, the normally suppressed carrier is reinserted at a reduced level to be used for synchronizing or regulating purposes at the receiver.

With transmitters of equal power capability, each 100 per cent modulated single-sideband equipment usually provides a signal-to-noise advantage of 6 db or more over AM equipment. The exact advantage is dependent on the nature of the noise, i.e., how the effective noise level within the 6-kc bandwidth required for the AM signal compares with that in the 3-kc bandwidth required for the single-sideband signal.

A single-sideband receiver does not require the presence of a continuous carrier in the received signal. For this reason it is possible for a single-sideband receiver to receive intelligible signals from more than one source simultaneously by reinserting one carrier to demodulate all signals. Single-sideband equipment is available which makes use of this principle to provide party-line operation of duplex equipment. A repeater is used at intermediate points which transmits the same band of frequencies in each direction and likewise receives the same from each direction. *Through signals* are partially demodulated and translated in frequency before being amplified and retransmitted. This technique of exchanging and re-using the same frequencies is called *frequency frogging*. Although equipment requirements for a *breakout* repeater of this type are only slightly less than with cascaded individual channels, additional frequencies are not required.

42. Regulation

Transmission-line attenuation is subject to variation with weather conditions. Also a receiver in a simplex party-line carrier circuit must operate on signals from transmitters at different distances. Compensation in the receiver is therefore required to maintain a relatively constant output with wide variations in input signal level. In AM and FM receivers, a continuous carrier frequency is contained in the received signal; therefore, regulation is accomplished in the same general method as automatic volume control (avc) in ordinary radio receivers. In AM sets the composite signal is rectified to produce a gain-controlling bias which is applied to a number of amplifier stages.   In FM receivers, limiter grid current is used to produce the d-c bias voltage. This method is not satisfactory for single-sideband systems because of the wide range and rapid rate of change of normal speech levels which comprise the composite signal. A pilot signal of constant level is inserted at a frequency outside the normal voice range.   This signal is retrieved through a filter at the receiver and used to produce gain-controlling bias or to control variable attenuators.   The modulating carrier frequency is often reinserted at a low level and used as the pilot signal.

43. Signaling

Signaling is an essential part of telephony.   Loudspeaker calling is sometimes used for carrier circuits in remote locations; however, in most instances, a carrier circuit must be capable of behaving as a telephone extension or as a trunking circuit between telephone exchanges.

The two positions of a relay are made to represent the presence and absence, respectively, of a magneto ringdown signal or a dial pulse, or on-hook and off-hook status, respectively, for common battery telephones.

Several different methods are used for transmitting a signal in such a way that a relay in the remote receiver will follow the action of the local transmitter relay.   A shift in the carrier frequency, modulation by 60 cycles, or the application of audio tones are some of the methods used.   In single-sideband systems *out-of-band* frequency-shift tones are often used with methods similar to the speech-plus-tone principles described in paragraph 38.   In some types the frequency of the tone used for pilot regulation is shifted for signaling purposes.

CARRIER FUNCTIONS IN PROTECTIVE RELAYING

44. General

The protection of high-voltage transmission lines and apparatus requires high-speed operation of relays which detect the presence and location of faults and cause appropriate circuit breakers to trip, isolating the fault from the power system.   Carrier provides an exchange of information between opposite ends of the line which permits higher speed clearance of faults.   While apparatus protection is usually internal to each station, carrier is sometimes used to provide transferred trip signals to remote circuit breakers for removing power from faulted apparatus at a local station.

45. Directional-comparison Pilot Relaying

This method employs carrier as a means of comparing the direction to a fault from a local station with its direction from a distant station.   This information enables the relays to determine if a circuit breaker should be tripped or kept in service.   Fault-detecting relays can be set to operate for faults which occur within a predetermined distance from their location.   However, because of variable system conditions, these relays cannot distinguish between a fault very near the end of a protected line section and one just beyond.   If a fault occurs beyond the remote station, directional relays at that location will cause carrier to be transmitted to the local station to block tripping.   Part of the security within this scheme lies in the fact that carrier transmission is not required for faults within the protected line section which is also the carrier path.

A fault on one transmission line leading into a multiline station causes carrier to be transmitted from a pilot relay terminal on each of the other lines.   It is highly desirable that all pilot relay carrier channels be trapped so that close-in faults of this type will not cause excessive attenuation on the circuits that are required to operate.   When this is done, receivers may be desensitized to increase their security against false operation caused by noise voltages.

46. Phase-comparison Pilot Relaying

This depends upon carrier signals for comparing the phase angle of fault current at the two ends of a protected line section. A fault anywhere in the vicinity causes the carrier terminal at each end to send pulsed carrier during alternate half cycles of a 60-cycle voltage. This voltage is derived from fault current flowing in the protected line section. An electronic circuit in each terminal compares the phase of the local fault current with the pulses received from the remote terminal. Circuit breakers at each end are blocked from tripping if the comparison indicates a fault outside the protected section but permits tripping for internal faults. As in directional-comparison systems, circuit-breaker tripping will take place correctly without carrier for faults within the protected line section. This gives a fail-safe characteristic to the scheme; i.e., transmission of carrier through a faulted path is not essential.

Applications of phase-comparison relaying include combinations enabling better protection of lines having close mutual coupling with other lines. Such combinations may perform as phase comparison for certain types of ground faults and as directional comparison for phase-to-phase fault protection. Ordinary directional-comparison and phase-comparison systems are referred to as *carrier-blocking* schemes.

47. Permissive Trip Relaying

Unlike directional- and phase-comparison relaying schemes, permissive trip systems require carrier to be propagated through a faulted transmission path. Studies and tests have shown that the arc of a powerline fault does not produce a short circuit but has a finite impedance at carrier frequencies. Nevertheless, considerable additional attenuation to the carrier signal is produced which depends on the type of fault, the distance to the fault, and the method by which the carrier circuit is coupled. Permissive trip circuits are most often applied on phase-to-phase coupling, which is least vulnerable to increased losses due to faults. In addition, the power output of the carrier transmitter is usually boosted during faults to compensate for increased losses.

Equipment for permissive trip circuits is usually a frequency-shift type which continuously transmits a guard frequency during normal operation, then shifts to a trip frequency when a fault occurs in the protected line section. Actual tripping at the remote end, however, occurs only if relays there have also detected a fault in the protected section, hence the term *permissive*. This type of scheme is divided into further categories called permissive underreach and permissive overreach, which indicates whether the distance settings of the fault detector relays extend only to the near side of or beyond the station at the remote end of the protected line section.

Frequency-shift audio tone equipment on speech-plus-tone circuits can be used for permissive trip service and is quite common on single-sideband circuits in Europe and Canada. It is customary to drop the voice circuit and any other tone functions momentarily to devote the total power of the transmitter to the trip signal.

48. Transferred Trip Relaying

This is used to clear a local fault by tripping a breaker at one or more remote locations. For example, a station not equipped with a high-side circuit breaker (or with its breaker bypassed for maintenance) sends a tripping signal to the station or stations supplying power in the event of an internal transformer fault.

Since there is no permissive aspect in ordinary transferred trip functions, utmost precautions are desirable to guard against false trips. Frequency-shift equipment is commonly used which continuously transmits a guard signal except during fault conditions. Most receivers are designed so that in the event the guard signal fails, random noise will tend to hold the discriminator biased toward the guard side. In addition, a lock-out relay may be provided which, upon guard signal failure, will disable the trip circuit after a short time delay and provide a channel-failure alarm.

Another precaution sometimes taken is the use of dual-channel facilities wherein simultaneous trip signals must be received in two receivers to cause tripping.

TELEMETERING AND CONTROL

49. Equipment

The equipment for telemetering and control purposes includes a wide variety of types. Within the United States widespread use is made of both narrowband frequency-shift carrier and audio tones superimposed on telephone-type carrier circuits. There are some examples of narrowband *amplitude-modulated* carriers in service, although this type is becoming obsolete. Audio tones are sometimes applied as secondary functions of telephone circuits (paragraph 38) and sometimes applied in independent tone groups superimposed on unidirectional carrier circuits using equipment of the basic type normally used for telephone service. In Europe, audio tones are by far the most popular media for telemetering and control functions.

In some applications where circuits are required over single line sections and where occasional short-duration interruptions are tolerable, telemetering or control is accomplished as a secondary function of pilot relay circuits. Protective relays take preference and temporarily drop these secondary facilities during fault conditions.

Another secondary facility used occasionally for supervisory control involves keying or shifting the carrier of a duplex FM carrier telephone circuit or keying the carrier of an AM circuit.

50. Telemetering

A telemeter circuit is a facility which permits a measured quantity to be interpreted at a distance from the measuring point. It differs from *remote metering* in that the measured quantity, called the *measurand*, is converted into a different form to facilitate transmission to the distant point.

Typical quantities telemetered in power system operation include generation at power plants, tie-line power interchanged with neighboring companies, and other appropriate items of load information, which are recorded in load dispatching centers. More detailed measurements such as voltage, current, power, hydro plant gate position, etc., are telemetered from supervisory controlled stations to corresponding master stations where they may be recorded or displayed visually.

There are five basic forms of analogue-type telemetering. *Ratio type, voltage type,* and *current type* are normally used where metallic circuits are available and are of no further interest for carrier application. The remaining two which are commonly used on carrier systems are *frequency type* and *pulse type.*

Frequency-type telemetering employs the frequency of a periodically recurring electric signal as a translating means. Systems in common usage represent measurands by variation of frequency between the limits of 6 to 27, 10 to 30, 18 to 30, 15 to 35, 20 to 25, or 80 to 100 cps.

Practically all manufacturers making equipment operating within any of these ranges also make another version with a top frequency of 15 cps to take advantage of special rates on leased telephone facilities. However, this lower-frequency equipment is seldom used in power-line carrier applications.

Pulse-type telemeter employs the characteristics of intermittent electric signals, other than their frequency, as the translating means. The most common system of this type uses the duration of an impulse to represent the measurand.

Digital telemetering uses some form of pulse code to represent numerically the quantity being measured. With the advent of solid-state devices, it has become possible to produce *digital* telemetering systems which have a high degree of reliability. Most of the available pulse code schemes contain self-checking techniques which will provide an indication at the receiving terminal if errors have occurred in transmission.

Digital telemetering is inherently capable of greater accuracy than analogue methods. Data-transmission systems which are compatible with modern digital computer activities are becoming increasingly attractive.

51. Supervisory Control

This is a service which permits control and supervision of a number of functions at a remote station without requiring individual circuits for each function.

In a typical system, a point to be controlled or interrogated is selected at the master station by a push button. Automatic equipment, usually a relay counting chain, emits a series of coded pulses which are transmitted via carrier or other means to the remote station where similar relays are actuated by the pulses. A connection is established to the desired point in a manner similar to automatic switching in dial telephone systems. At the same time the equipment at the remote station causes an identical series of pulses to be returned to the master station where it may be checked for errors in transmission. Upon receipt of the check indication, the operator then must push an appropriate button to cause the desired operation to take place. Control functions include tripping and closing of circuit breakers, insertion and removal of automatic reclosing equipment from circuit-breaker controls, selection of telemeter functions, setting gate positions, and any other required function.

In addition to control operations initiated at the master station, changes in status of any supervised device at the remote station will be reported automatically. This report involves a pulse coding and checking scheme identical with that used for control selection.

Almost every supervisory control system contains a *master check* feature which, upon interrogation, systematically verifies the status of all controlled devices connected to the system.

The carrier requirements for a supervisory control system include a circuit in each direction for transmitting pulses plus, in most cases, one or more telemeter circuits from the remote station to the master station.

Metered indications are an essential part of the supervision of a remote station. Three methods of telemetering may be used depending upon the importance attached to the ability for continuous presentation of metered quantities. In the simplest method, telemeter signals are transmitted from the remote station over the same circuit as the supervisory pulses. The telemeter indication, therefore, must be interrupted for each control operation. In another method, one additional circuit is provided for telemetering. Once connected to a desired point, the telemeter circuit provides continuous indication of the quantity measured while the supervisory control equipment is free for other functions. The third and most sophisticated method makes use of individual telemeter circuits for each desired indication. This method is normally used with supervisory control of generating stations where continuous presentation of all measured quantities is highly desirable.

52. Remote Alarm Systems

Occasionally, unattended stations with relatively simple power switching schemes may be equipped with a remote alarm system rather than complete supervisory control facilities. In a typical system, repeated pulse trains are continuously transmitted from the unattended station to an appropriate attended station. Each pulse in the train represents one point being monitored. An abnormal condition causes the width of the appropriate pulse to be increased, which in turn provides an alarm indication at the receiving location. It is very desirable that such a system have error-detection facilities to prevent false alarms due to lightning and switching disturbances. Only one unidirectional carrier circuit is required.

Typical abnormalities for which alarms are provided include circuit-breaker trip and lock-out, low oil-circuit-breaker air pressure, high transformer temperature, differential relay operation, low battery voltage, loss of station service, and any other condition which requires the attention of maintenance personnel.

53. Load-frequency Control

A modern power system interconnected with neighboring utilities must satisfy its own load requirements, maintain scheduled interchanges, and maintain system frequency within satisfactory limits.

The most commonly used load-frequency control system requires that all *tie-line loads* be telemetered to a central load dispatching office where they are totalized and compared with *established power-interchange schedules.* A frequency bias is added to the difference between actual and scheduled interchange to establish a *net area generation requirement.*

Because of energy storage in the physical mass of rotating synchronous generators, additional energy must be added to restore a low system frequency to normal. The required *generation per unit deviation* is a function of the natural characteristics of the system. This quantity is called frequency bias with units of megawatts per $\frac{1}{10}$ cycle.

In some power systems the net area requirement is transmitted directly to a regulating plant over a carrier circuit of the type used for telemetering of tie-line load information. At the plant, the requirement signal is converted to raise-hold-lower impulses which control governor mechanisms.

In other systems, the raise-hold-lower impulses are derived at the central load dispatching office and routed to the appropriate plant or plants. For this type of service, the carrier circuit must be able to transmit three discrete frequencies sometimes called mark, space, and center frequency.

Because of thermal inertia in steam-operated generators, hydro plants are generally preferred for system regulation. Often a single plant provides regulation for the entire system. However, carrier facilities must be provided to transmit the master load-control signal to a number of plants, so that the load dispatcher may select the most appropriate regulating plant or plants for any particular system condition.

REFERENCES

1. Bibliography of Power System Communication Literature, 1920–1954, *Trans. AIEE*, vol. 75, pt. III, pp. 1–12, 1956.
2. Guide to Application and Treatment of Channels for Power-line Carrier, *Trans. AIEE*, vol. 73, pt. III-A, pp. 417–436, 1954.
3. Jones, D. E., and P. W. Waddington, Coupling Methods for Power Line Carrier Systems, *Trans. AIEE*, vol. 76, pt. III, pp. 1284–1287, 1957.
4. Alsleben, E., Experimental Values of the Characteristics Determining the Behavior of High-voltage Power Networks at the Frequencies Used by Carrier Current Communication Circuits, *CIGRE Rept.* 340, 1958.
5. Adams, G. E., Wave Propagation along Unbalanced High-voltage Transmission Lines, *Trans. AIEE*, vol. 78, pt. III, pp. 639–646, 1959.
6. Perz, M. C., A Method of Analysis of Power Line Carrier Problems on Three-phase Lines, *IEEE Tech. Paper* 63-937.
7. Jones, D. E., and B. Bozoki, Experimental Evaluation of Power-line Carrier Propagation on a 500-kv line, *IEEE Tech. Paper* 63-935.
8. Report of Methods of Measurements for the Application of Power Line Carrier; *Trans. AIEE*, vol. 80, pt. III, pp. 1046–1052, 1961.
9. Alsleben, E., Guiding Values for the Planning of Power Line Carrier Communication Systems and Information on the Determination of Line Characteristics, *CIGRE Rept.* 319, 1962.

Chapter 15

HIGH-FREQUENCY COMMUNICATION CIRCUITS

E. D. BECKEN, *RCA Communications, Inc.*

INTRODUCTION

High-frequency (3 to 30 Mc) communication engineering requires careful attention to system concepts. All portions of a circuit between the locations involved are closely interrelated, and the equipment design not only must perform its functions correctly but must enhance the performance of related equipment often located elsewhere. Such system engineering concepts as the pass bandwidth of the propagation path and the bandwidth of the antennas will influence the design considerations of many of the facilities involved.

As often happens, some of the functions in the communication circuit are controllable and others are uncontrollable in that they exist in nature and the designer must recognize their characteristics and take advantage of the opportunities. In a high-frequency communication circuit it is the propagation path, the ionized layer surrounding the earth, that exists in nature and is uncontrollable. The nature and characteristics of this path dictate what must be done in system engineering such circuits. It is for this reason that attention will first be directed to the propagation path or communications media. The natural physical function of getting into and through the path or, in other words, radiation and transmission through space must be understood. The parameters of this path—reliability, accuracy, bandwidth, and speed capabilities—will be discussed along with short- and long-term variations. In the early years it was astonishing to learn that in some mysterious way it was possible to transmit signaling through space by electromagnetic radiation. However, now that much is known about the physics of electromagnetic transmission through space it remains an astounding phenomenon. Exhaustive experimental and theoretical work of many over the years has developed the present body of knowledge that makes system engineering of a high-frequency communication circuit feasible.

Another important factor in system engineering is a thorough awareness of the nature and speed of the communications required. If the telegraph or data circuit is used for slow speeds, the system design will be different from high-speed requirements. If voice communications are involved, other factors become relevant.

The third major element in design is the characteristics of the physical plant needed to handle most efficiently the type of communication required over the propagation path involved. System engineering will require examination of the nature of the antennas, transmitters, receivers, and terminal devices. It will consider alternative modulation and demodulation techniques, frequency and space diversity, error-protected codes, channeling characteristics, redundancy, signal-to-noise ratios, reliability, and accuracy. It will also give careful attention to the economics involved. Good service commensurate with acceptable costs is essential.

These three major areas for system engineering consideration will be dealt with, and concurrently and very importantly the matters of operation and maintenance will be considered. The system must be engineered so that appropriate levels of trained staff can cope with the operation and maintenance in providing good service in an economic manner.

The following sections develop these four factors, namely, (1) propagation, (2) the nature of the communication service required, (3) the equipment and plant, and (4) concurrently, the operation and maintenance.

PROPAGATION

1. Ionization[6,9,27,28,34,35,48]

High-frequency radio communication is made possible by the existence of a number of ionized layers above the surface of the earth whose ion density varies with height and time. During the daytime the ionization appears in four layers D, E, F1, and F2 at heights of about 50 to 90, 100, 200, and 300 km, respectively. Typical height patterns of the E, F1, and F2 layers are shown in Fig. 1 for both daytime and nighttime. The F2 layer has the greatest ionization density and is therefore able to reflect the higher frequencies. The F1 layer is similar to the F2 layer, but its ionization density is not so great. During daylight hours the E layer is dependable for path lengths up to about 2,000 km, particularly in the temperate zones. Sporadic E, which is clouds of more intensely ionized particles which wander around the E region, mainly in the equatorial belt, is not very dependable and tends to be a nuisance. The D layer is mainly an absorption layer. The D and E layers disappear at night. The D layer disappears very rapidly at sunset followed by the E layer. The F1 and F2 layers merge into one at nighttime at about the height of the F1 daytime layer,

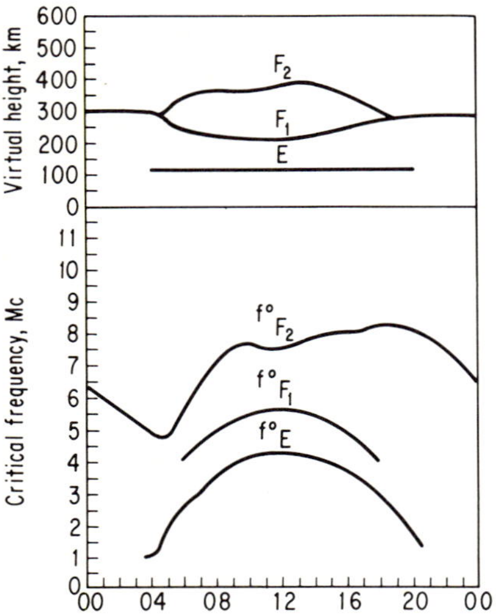

FIG. 1. Typical diurnal curves. (*Ionospheric Radio Propagation, Natl. Bur. Std. Circ. 462, Fig. 3.5, p. 18, June 25, 1948. By permission of the Director, Central Radio Propagation Laboratory.*)

Table 1. Description of Earth's Atmosphere*

Earth's surface.........	Temperature $273 \pm 20°K$
Troposphere...........	Temperature decreases with height. Altitude 0 to 8 miles
Tropopause............	Temperature minimum $210 + 20°K$ and altitude 13 ± 5 km (8 miles)
Stratosphere..........	Temperature increases with height. Altitude 8 to 30 miles
Stratopause...........	Temperature maximum $273 \pm 20°K$ and altitude 50 ± 5 km (30 miles)
Mesosphere...........	Temperature decreases with height. Altitude 30 to 55 miles
Mesopause†...........	Temperature minimum $190 \pm 25°K$ and altitude 85 ± 5 km (55 miles)
Thermosphere.........	Temperature increases
Thermopause..........	Should be the beginning of an isothermal region

* M. Nicolet, The Constitution and Composition of the Upper Atmosphere, *Proc. IRE*, vol. 47, p. 142, February, 1959.

† The earth's atmosphere below the mesopause is referred to as the homosphere and above as the heterosphere.

although it is designated as the F2 layer in the nighttime. The F2 layer thus becomes of greatest importance in long-distance radio communications for the frequency range from about 3 to 30 Mc.

A knowledge of the composition of the upper atmosphere is necessary in order to understand better the mechanics of high-frequency communications and its determinants. The atmosphere can be described as shown in Table 1.

The atmosphere at sea level is made up of about 78 per cent nitrogen (N_2) by volume with oxygen (O_2) 21 per cent and small traces of carbon dioxide (CO_2) and argon (A). The atmospheric data between 50 and 100 km at the height range of the lower ionosphere are tabulated in Table 2.

These variations in temperature and pressure are an indication of the relative changes in ionization and chemical composition at different levels and the different

Table 2. Atmospheric Data between 50 and 100 Km*

Altitude, km	Temperature, °K	Pressure, mm Hg
50.0	274	0.67
52.5	274	0.49
55.0	274	0.36
57.5	263	0.27
60.0	253	0.19
62.5	242	0.14
65.0	232	$9.6 \ 10^{-2}$
67.5	221	$6.6 \ 10^{-2}$
70.0	210	$4.5 \ 10^{-2}$
72.5	207	$3.0 \ 10^{-2}$
75.0	203	$2.0 \ 10^{-2}$
77.5	200	$1.3 \ 10^{-2}$
80.0	197	$8.7 \ 10^{-3}$
82.5	193	$5.7 \ 10^{-3}$
85.0	190	$3.7 \ 10^{-3}$
87.5	193	$2.4 \ 10^{-3}$
90.0	197	$1.6 \ 10^{-3}$
92.5	200	$1.0 \ 10^{-3}$
95.0	203	$6.8 \ 10^{-4}$
97.5	207	$4.5 \ 10^{-4}$
100.0	210	$3.0 \ 10^{-4}$

* M. Nicolet, The Constitution and Composition of the Upper Atmosphere, *Proc. IRE*, vol. 47, p. 144, February, 1959.

rates of chemical recombination and dissociation and diffusion. The ionosphere is subject to great ionization changes with height and time and is therefore subject to wide variations in its high-frequency communication capabilities. It varies with time of day, season, and solar activity. There are also random fluctuations and severe disturbances many of which are not clearly understood.

The sun is the source of ultraviolet radiations causing most of the ionization of the molecules in the earth's atmosphere. In addition, some of the ionization and effect on the earth's magnetic field are caused by corpuscular radiation actually shot out of the sun and arriving at the earth 1½ to 2½ days later, which compares with about 8.4 min for the ultraviolet ray to make the same trip. Sunspots are believed to be the source of much of the sun's radiation affecting radio communications. It has been known for many years that the number of sunspots varies over an 11-year cycle. The last maximum was in 1958, and the next maximum will be in 1969. The present minimum sunspot cycle occurred in 1964. Sunspots are born and increase in size and/or activity or remain rather stable, linger over the 27-day period of rotation of the sun, or die away quickly. Some seem to have very little effect on the earth.

Scientific correlation of sunspot types, size, groupings, polarity, and other characteristics with radio conditions has not yielded many answers, but some facts have become quite clear.　The number of sunspots tends to increase the ionization of the earth's atmosphere and results in an appreciable rise and fall in such ionization over the 11-year sunspot cycle.　In addition, it has been determined that sunspots will not cause disturbances on the earth unless they exist in the certain areas of the sun bounded by* "a semi-circle about 26° in radius centered at the sun's optical center, on the eastern hemisphere, and terminating at the axis of rotation."　It takes a spot

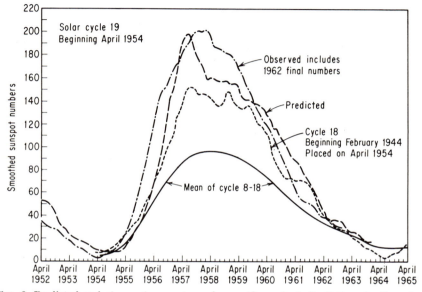

Fig. 2. Predicted and observed sunspot numbers.　(*Ionospheric Predictions for December, 1963, Natl. Bur. Std., Fig. A.　By permission of the Director, Central Radio Propagation Laboratory.*)

about two days maximum to cross this imaginary zone of 26° of latitude or longitude, and during this time the effect on the earth is maximized.　The reasons for this zone effect are not clear.　A spot can exist for 27 days and show up again in the critical zone in the next rotation of the sun and again cause radio disturbances.　The sunspot-activity variations over the current and past 11-year solar cycles are shown in Fig. 2.

2. Frequencies[34,36,42,69]

The effect of these variations in ionization on the radio frequencies will be examined. The greater the ionization, the higher the frequency that can be reflected or refracted back.　This is true whether the transmitted energy strikes the ionized layer at vertical incidence or some oblique incidence.　There is a definite established relationship between the maximum frequency that can be used (i.e., the maximum frequency that will be refracted from the ionized layer) and the angle at which the transmission strikes the ionized layer.　This relationship is given by the following formula:

$$f_0 = f_n \sec i$$

* Reference 66, p. 232.

where f_0 = maximum usable frequency at oblique angle i
   f_n = maximum frequency that will be refracted back at vertical incidence
   i = angle of incidence, that is, angle between direction of propagation and the
      perpendicular to the earth

This means that with a given maximum frequency for vertical-incidence refraction, which is determined by the extent of ionization, a higher frequency can be used for communication as the angle of incidence is increased and the sec i increases. It is therefore desirable to strive for concentration of radiated energy at low angles with respect to the horizon because maximum frequencies can then be used which are less affected by absorption than the lower frequencies. It is therefore apparent that a world-wide knowledge of the maximum frequencies possible at vertical incidence (called critical frequencies) for any time, plus information on the height of the ionized layers, would enable world-wide calculation of the maximum usable frequencies on any international radio circuit. The National Bureau of Standards accumulates critical frequency and ionized layer height data as submitted by many ionospheric measuring stations located throughout the world. These data are used to prepare maximum usable frequency charts which appear monthly in the National Bureau of Standards bulletin *Ionospheric Predictions*, TB11-499-9/TO 31-3-28. These frequency contour charts are prepared for paths of 4,000 km or less (single-hop transmission) and for paths greater than 4,000 km (multihop transmission) for the F2 and E layers (see Fig. 3 for multihop transmission). Transmission is almost always by the F2 layer for distances greater than 4,000 km. As noted above, the ionosphere is subject to considerable variations from the average, and the use of the maximum usable frequency (MUF) will result in considerable periods of time when the frequency is so high that it penetrates through the F2 layer and is lost in space. The term optimum working frequency (OWF) or in French "fréquence optimum de travail" (FOT) is therefore introduced and is equal to 85 per cent of the maximum usable frequency for the F2 layer.

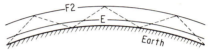

FIG. 3. Propagation by a number of hops. (*Ionospheric Radio Propagation, Natl. Bur. Std. Circ. 462, Fig. 6.8, p. 75, June 25, 1948. By permission of the Director, Central Radio Propagation Laboratory.*)

The E layer, however, is more stable, and the optimum working frequency for the E layer is taken as equal to the maximum usable frequency for the E layer. Frequency contour charts are also prepared covering the sporadic E layer designated Es. This is a layer that appears at sporadic intervals mainly during the day and does account for some communications, although it cannot be depended upon extensively.

It is also possible to calculate the lowest useful high frequency (LUHF). This calculation takes into consideration the absorption that the radio frequencies encounter in passing through the D, E, or F1 layer before refraction from the F2 or lower layers. The lower the frequency below the optimum working frequency, the greater will be the absorption, so it is necessary to operate the radio circuits as near to the optimum working frequency as possible, and this results in a practical compromise between number of diurnal frequency changes and the number of megacycles of the operating frequency below the optimum working frequency at any time. It has been found that the tolerable spread between actual operating frequency and the optimum working frequency is greater for the higher frequencies than for the lower frequency in the 3- to 30-Mc frequency band.

It must be emphasized that these data for MUF, OWF, and lowest usable frequency (LUF) are for average conditions and represent information obtained from a somewhat limited number of recording points. It is found that there is considerable variation from the average conditions, and there are a sufficient number of disturbed days during a year to leave much room for variations. Information on the calculations of MUF and OWF are given in the *National Bureau of Standards Handbook* 90 issued December 21, 1962. Following are extracts from this handbook.*

* Reference 35, pp. 4–13.

"The graphical prediction maps give, for each even hour of Universal Time (GMT or UT), values of predicted F2-Zero-MUF and predicted F2-4000-MUF. [Figures 4 to 7 inclusive] are maps of these characteristics for December 1958 [for Universal Time 00, 06, 12, and 18]. The F2-Zero-MUF and F2-4000-MUF [inclined incidence for 4,000-km hop] maps for the same hour appear on the same page. The values predicted are monthly medians. It should be remembered that there is considerable day-to-day variation about the monthly medians, especially on disturbed days. Areas in or near the auroral zone are particularly affected by ionospheric disturbance.

"The graphical prediction maps are prepared using Universal Time (UT), defined as the local mean time at the Greenwich (zero longitude) meridian. Therefore, if local or standard time is desired, transformation must be made after the prediction calculations are completed.

"As the number of hops in a single path and the number of possible paths increase, the detailed analysis of ionospheric propagation for a given circuit becomes impractical by manual methods. Simplifying assumptions, however, work fairly well for most applications. In the MUF calculation for distances up to 4,000 kilometers, one-hop F2-layer propagation with the point of reflection at the mid-point of the path is assumed. Since the characteristics of the ionosphere at the point of reflection affect the MUF, that point is called a 'control point.'

"For distances beyond 4,000 kilometers, a two control point method is used. This method assumes: (1) that there are F2-layer control points on the great circle 2,000 kilometers from each terminal . . . ; (2) ionospheric conditions at the control points determine the frequencies that can arrive at or leave the respective terminals; (3) the highest frequency which can be propagated over the circuit (the MUF) is approximated by the highest frequency which can arrive at or depart from both terminals.

"Experience indicates that the two control point method, which ignores the details of propagation between the control points, provides a useful approximation for ionospheric effects which tend to increase the MUF (e.g., scatter, ionospheric tilts, and the high angle, or Pedersen ray), since this method tends to make the MUF higher than that predicted in terms of an integral number of hops, each less than or equal to the maximum distance for a single hop.

"[Figure 8][11] is a map of the world on a modified cylindrical projection of the same size and scale as the maps appearing monthly in the CRPL predictions [National Bureau of Standards bulletin *Ionospheric Predictions*]. An area 10° in latitude by 15° in longitude appears as a square with a side of 0.6 cm, which has the same proportion of latitude to longitude as that recommended by the C.C.I.R. The scale recommended by the C.C.I.R. is 1 cm for 10° latitude and 15° longitude. The distortion in this map is greatest at the poles. The earth is assumed to be a sphere of 40,000 kilometers circumference, the error being negligible in most radio propagation problems. [Figure 9] is a chart to the same scale as [Fig. 8] indicating great circles on the surface of the earth (solid lines). Because of the symmetry of the sphere, the set of great circles crossing the equator at two points 180° apart can be used for any great circle. Distances, in thousands of kilometers, are indicated by the numbered dot-dash lines crossing the great circles, while the intermediate dotted lines show 500-kilometer intervals.

"The great circle charts may be used to obtain great circles graphically, measure distances, and determine the location of 'control points,' proceeding by the following steps:

"1. Prepare a transparency by placing a piece of transparent paper or suitable plastic over the map [Fig. 8]. Draw the equatorial line (zero degrees latitude), the Greenwich meridian (zero degrees longitude), and any other reference meridians desired. (It is often useful to mark the reference meridians for the time zones of the path end points.) Place dots over the two terminal locations of the transmission path, marking one end A, and the other end B.

"2. Place the transparency over the great circle chart [Fig. 9]. Keeping the equatorial line of the transparency on the equatorial line of the great circle chart, slide the transparency horizontally until the terminal points, A and B, fall either on the

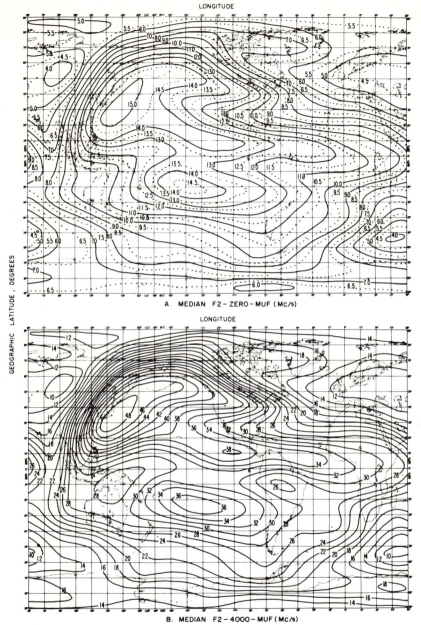

DECEMBER, 1958
UT = 00
LONGITUDE

A. MEDIAN F2 – ZERO – MUF (Mc/s)

LONGITUDE

B. MEDIAN F2 – 4000 – MUF (Mc/s)

GEOGRAPHIC LATITUDE, DEGREES

Fig. 4a. Median F2-zero-MUF (Mc/sec) December, 1958, UT = 00. (*S. M. Ostrow, Handbook for CRPL Ionospheric Predictions Based on Numerical Methods of Mapping, Natl. Bur. Std. Handbook 90, Fig. 1A, Dec. 21, 1962. By permission of the Director, Central Radio Propagation Laboratory.*)
Fig. 4b. Median F2-4,000-MUF (Mc/sec) December, 1958, UT = 00. (*Ostrow, op. cit., Fig. 1B.*)

15–7

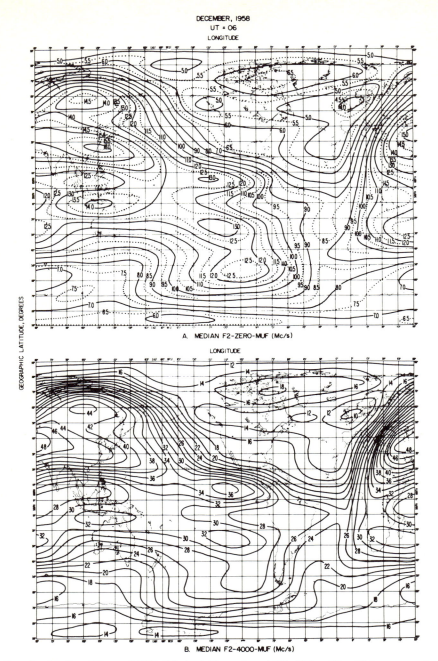

FIG. 5a. Median F2-zero-MUF (Mc/sec) December, 1958, UT = 06. (*Ostrow, op. cit.,* *Fig.* 4*A*.)
FIG. 5b. Median F2-4,000-MUF (Mc/sec) December, 1958, UT = 06. (*Ostrow, op. cit.,* *Fig.* 4*B*.)

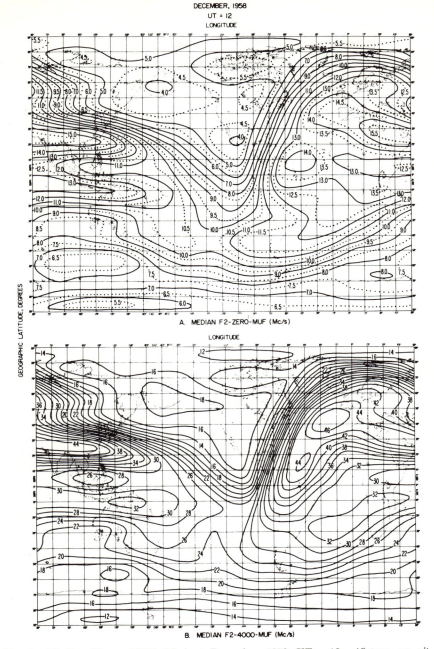

A. MEDIAN F2-ZERO-MUF (Mc/s)

B. MEDIAN F2-4000-MUF (Mc/s)

Fig. 6a. Median F2-zero-MUF (Mc/sec) December, 1958, UT = 12. (*Ostrow, op. cit.,*
Fig. 7A.)
Fig. 6b. Median F2-4,000-MUF (Mc/sec) December, 1958, UT = 12. (*Ostrow, op. cit.,*
Fig. 7B.)

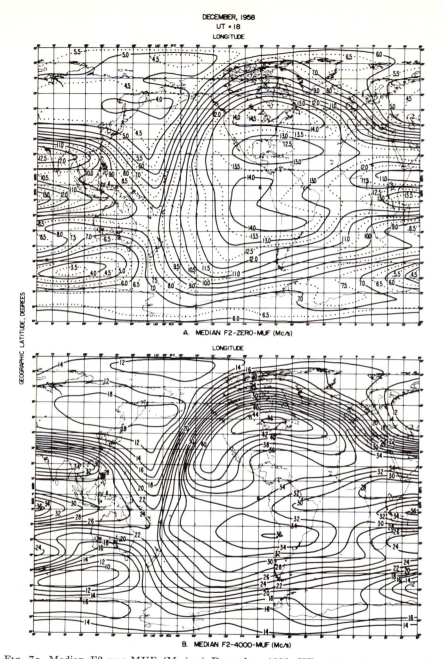

A. MEDIAN F2-ZERO-MUF (Mc/s)

B. MEDIAN F2-4000-MUF (Mc/s)

FIG. 7a. Median F2-zero-MUF (Mc/sec) December, 1958, UT = 18. (*Ostrow, op. cit.,*
Fig. 10A.)

FIG. 7b. Median F2-4,000-MUF (Mc/sec) December, 1958, UT = 18. (*Ostrow, op. cit.,*
Fig. 10B.)

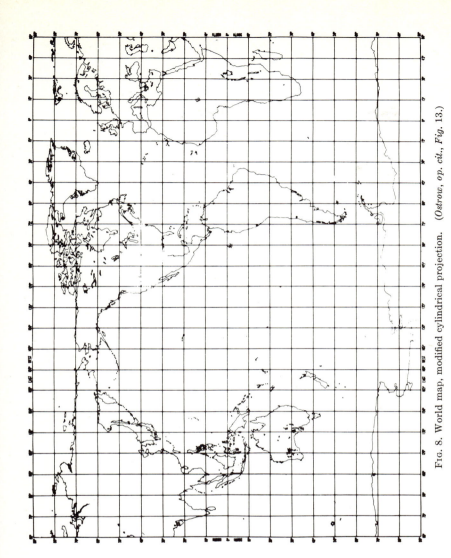

FIG. 8. World map, modified cylindrical projection. (*Ostrow, op. cit., Fig. 13.*)

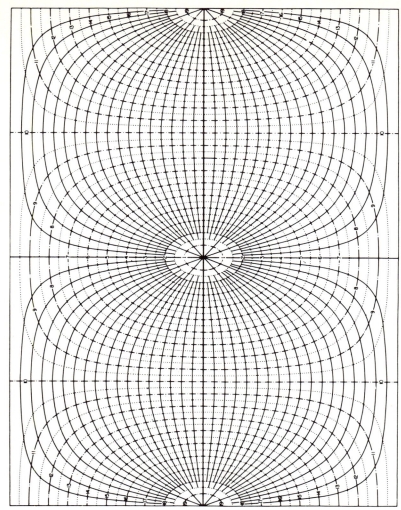

FIG. 9. Great-circle chart centered on equator for modified cylindrical projection world map. (Solid lines represent great circles. Numbered dot-dash lines indicate distance in thousands of kilometers.) (*Ostrow, op. cit., Fig.* 14.)

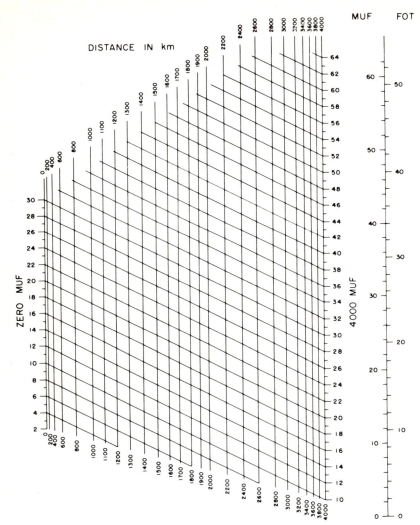

DISTANCE IN km

ZERO MUF

4000 MUF

MUF   FOT

Fig. 10. Nomogram for transforming F2-zero-MUF and F2-4,000-MUF to equivalent maximum usable frequencies at intermediate transmission distances, conversion scale for obtaining optimum working frequency (FOT). (*Ostrow, op. cit., Fig. 39.*)

same great circle or are at the same proportional distance between adjacent great circles.

"3. The length of transmission path is determined from the dot-dash lines, numbered in thousands of kilometers from the center of the system of great circle curves, and the intermediate dotted curves, indicating 500-kilometer intervals. Halfway point and 'control points' may be measured off by the distance scale and their position marked on the transparency.

"4. Place the great circle transparency over each F2-Zero-MUF map [Figs. 4a, 5a, 6a, 7a] taking care to superpose accurately the equator and reference meridian lines on their corresponding lines on the map. Read and tabulate the values of F2-Zero-MUF at the halfway point for each map.

"5. In the same way, read and tabulate F2-4000-MUF at the halfway point for each map [Figs. 4b, 5b, 6b, and 7b].

Table 3. Typical Multiple Frequency Assignments around the World*

(Partial list only)

Frequency	Assignment	Circuit	Type of modulation
14,800	Diggers Rest, Australia	Sydney, New South Wales	12A9b
14,800	Hankow, China	Nanking and Shanghai, China	10A34
14,800	Pointe-Noire, Republic of Congo	Dakar and Senegal	0.1A1
14,800	Koenigswusterh, Germany	South America	2.1A12
14,800	Paris, France	Dakar and Algiers	0.1A1
14,800	Port Stanley, Falkland Islands	Domestic	0.1A1
14,800	Saint Claude, Guadeloupe	Cayenne and Fort de France	0.1A1
14,800	Cayenne, French Guiana	Fort de France and Saint Claude	0.1A1
14,800	Fukuoka, Japan	Sapporo	8A13
14,800	Diego Suarez, Malagasy Republic	Dakar, Paris	0.1A1
14,800	Mexico City, Mexico	Nogales, Sonora	6A3
14,800	Fort de France, Martinique	Cayenne, Saint Claude	0.1A1
14,800	Las Pinas, Philippine Islands	Tokyo, Honolulu	0.1A1, 0.1F1
14,800	Juba, Republic of the Sudan	Khartoum	6A3b
14,800	Dakar, Republic of Senegal	Pointe-Noire, Paris	0.1A1
14,800	Bizerte, Tunisia	Paris, Casablanca	0.1A1
14,800	Moscow, USSR	Not specified	6A134, 4F6
14,800	New York, N.Y.	Stockholm, Frankfurt Cairo, Caracas	10A1234, 8F14 8F14

* International Frequency Registration Board, *International Frequency List*, 2d ed., vol. VIII, p. 387, Geneva, Feb. 1, 1963.

"6. Obtain F2-MUF for the transmission path from the distance interpolation nomogram [Fig. 10] by placing a straight edge between values of F2-Zero-MUF (left hand scale) and F2-4000-MUF (right hand scale) for the same Universal Time (UT). Read and tabulate the MUF at the intersection of the straight edge with the appropriate vertical distance line, interpolating between the oblique lines."

Frequencies are authorized for use by each national government to its own citizens but in accordance with plans cooperatively arrived at with all nations of the world. The international cooperative organization consists of the division of the Telecommunication Union with about 124 member nations scattered throughout the world. The assignment of frequencies in the 3- to 30-Mc band by each nation is reported to the International Frequency Registration Board in Geneva, Switzerland (see Chap.

19). This agency of the Telecommunication Union will record the assignments and advise the filing nation if there are prior filings with which the new registration might interfere. Such findings can be enforced if necessary only by the nation involved. The important thing to remember from a systems engineering aspect is that the same frequency is assigned many times over and over by the different nations around the world and for use at different or the same times. Under normal conditions such geographical and time sharing of multiple assignments is entirely workable, but at times of abnormal or changing propagation conditions interference problems may arise. These instances are in most cases resolved by mutual arrangements between the communications operating administration or organization involved. Even at best frequency management by a communications carrier is a difficult one in order to obtain maximum efficiency and capacity out of the licensed assignments. It requires close and continuing attention to frequency usage, and the change of frequencies during the day can be rapid at times, thus necessitating flexibility of facility arrangements and a keen awareness of momentary effectiveness of the frequency in use.

An example of multiple-frequency assignment around the world is shown in Table 3. If all were operated at the same time even with the geographical separations shown there might be serious cases of interference. Relative authorized powers and bandwidths are also shown in the IFRB assignments, and these all have a bearing on the possibility of potential interference.

3. Noise[26,34,38]

Noise in the frequency range from 3 to 30 Mc varies inversely as the frequency and is due to local and distant thunderstorms. The noise from distant thunderstorms is transmitted to the receiving station via the same ionosphere that transmits the radio waves, so considerable variation of such noise is to be expected. Diurnal and seasonal variations are marked. Extensive measurements indicate that in nearly all cases 50 per cent of the noise values lie within ± 5 db and 90 per cent within ± 10 db of the median value, and these measurements tend to indicate that predominate noise sources are local in character.

The rms and peak value of random or fluctuation noise in a receiver is proportional to the square root of the effective bandwidth of the receiver. The peak value of the receiver response to impulse noise is proportional to the effective receiver bandwidth, and the rms value of the receiver response to impulse noise is proportional to the square root of the bandwidth. Thunderstorm noise, unless in the immediate area, is generally random in character.

The *National Bureau of Standards Circular* 462 gives data on the distribution of noise throughout the world, and it is to be noted that the high-noise zones are in the tropical land areas. This information is shown in Figs. 11 to 14, inclusive. Following are extracts from the U.S. Department of Commerce, *National Bureau of Standards Circular* 557 issued August 25, 1955.*

"This paper deals with external radio noise of three types; namely, atmospheric, galactic, and manmade. Each type has different characteristics and thus will be discussed separately.

"Of the three types, atmospheric noise is the most erratic in character, consisting in general of short pulses with random recurrence—superimposed upon a background of random noise. If these short time variations of instantaneous noise power are averaged over a period of several minutes, this average power level is found to be relatively constant during a given hour, the variations seldom exceeding 2 or 3 db, except during sunrise or sunset periods. The median value of this average noise power within the hour has been taken as the basic unit . . . and will be referred to as its hourly median or hourly value.

"The hourly values of the noise vary with time of day because of changing propagation conditions and frequency of thunderstorms, but fortunately this variation tends to follow a particular pattern. At the lower frequencies, the nighttime noise is

* Reference 24, p. 3.

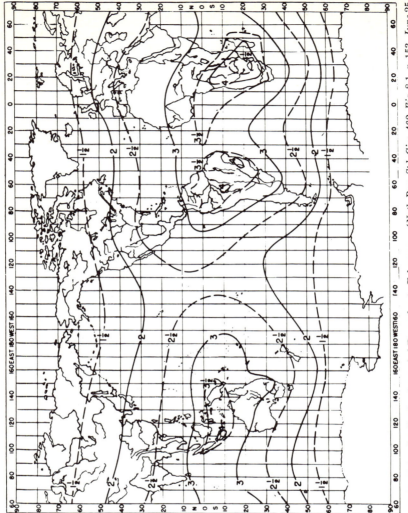

Fig. 11. Noise distribution for period December to February. (*Natl. Bur. Std. Circ.* 462, *Fig.* 8.1, *p.* 153, *June 25*, 1948.)

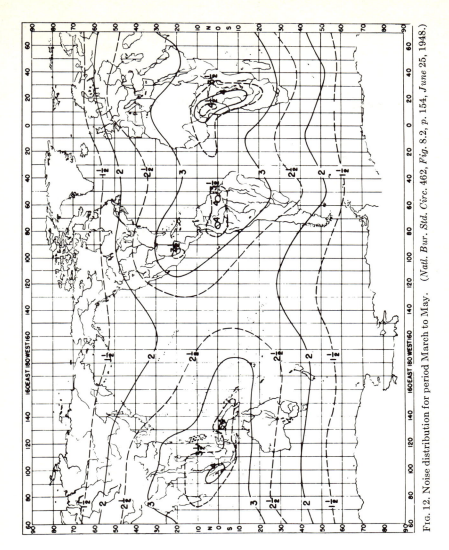

FIG. 12. Noise distribution for period March to May.   (*Natl. Bur. Std. Circ. 462, Fig. 8.2, p. 154, June 25, 1948.*)

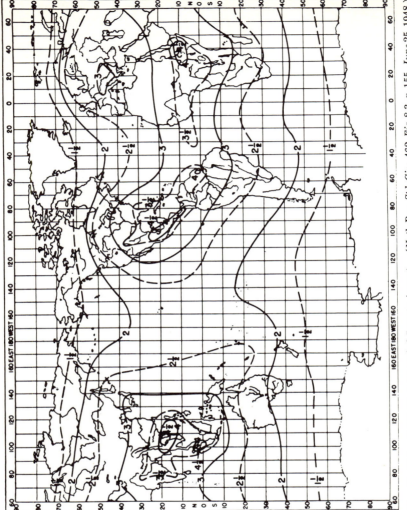

FIG. 13. Noise distribution for period June to August.   (*Natl. Bur. Std. Circ.* 462, *Fig.* 8.3, *p.* 155, *June* 25, 1948.)

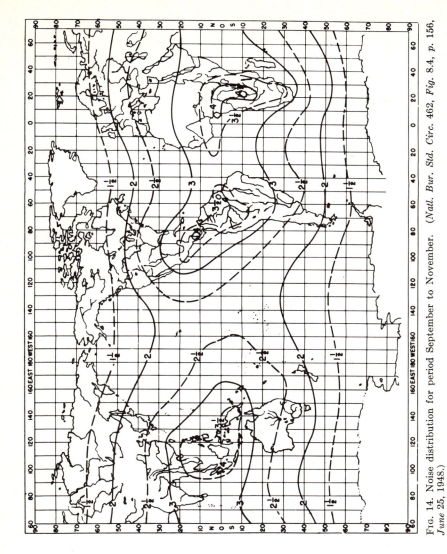

Fig. 14. Noise distribution for period September to November. (*Natl. Bur. Std. Circ.* 462, *Fig.* 8.4, p. 156, *June* 25, 1948.)

high because noise is propagated by means of the ionosphere from storms at large distances. During the daytime, the ionospheric absorption is high, and the noise received from distant storms is reduced, the received noise being principally from local storms. Local storms tend to predominate during the afternoon hours, and therefore the noise level is enhanced somewhat during those hours. Thus we have maximum noise levels at night, minimum during the morning, a moderate increase in the afternoon, and again high levels at night.

"At the higher frequencies, the shape of the diurnal curve tends to reverse itself because the ionosphere will support propagation only during the daylight hours. Actually, at these frequencies the diurnal curve of the received noise becomes relatively flat because of the presence of galactic and manmade noise.

"There is also a regular seasonal trend to the noise that is influenced by ionospheric absorption, as well as the location and number of thunderstorms. The thunderstorm centers tend to shift above and below the equator from summer to winter and in addition the ionospheric absorption is higher in the summer, which tends to offset the increased thunderstorm activity at that time. Nevertheless, the received noise tends to be highest in the summer and lowest in the winter at tropical and temperate latitudes.

"The received noise level varies with frequency because the noise radiated by the thunderstorm and its efficiency of propagation are functions of frequency. In general, the received noise level decreases with increasing frequency.

"There are variations in the noise level with geographic location, the highest levels being encountered in equatorial regions and the lowest levels in the polar regions. The received noise levels are influenced by topography, as well as weather and propagation conditions.

"There are also variations in received noise levels with sunspot activity, but no attempt has been made to take this into account in this paper.

"In the prediction curves that follow [Figs. 11 to 14 inclusive] those variables for which the noise has definite trends, i.e., time of day, season, frequency, and geographic location, have been taken into account directly. However, there are certain unpredictable variations which can be taken into account statistically. For example, at a particular time of day and season, the hourly value will vary from day to day because of random changes in thunderstorm activity and propagation conditions. Because of this random variability, it has been found desirable to designate time blocks for prediction purposes. Each time block is for four consecutive hours within a given season. The median of the hourly values within the time block is referred to as the time-block median. Deviations of the hourly values from the time-block median are expressed in terms of the upper and lower decile values of the cumulative distribution of the hourly values.

"The limits of the time blocks were chosen as 0000 to 0400, 0400 to 0800, 0800 to 1200, etc., so that the sunrise-sunset periods changing through the seasons would fall within only the 0400 to 0800 and 1600 to 2000 time blocks. Also this choice of time-block limits prevents splitting a time block between successive days."

4. Multipath[32]

Radio signals traversing different paths between two points can cause the multipath effect which in telegraphy results in a lengthening or shortening of the received signal and in telephony oftentimes a noticeable distortion of the incoming signal. Most of the times this multipath is caused by different length paths in the same vertical plane, and these paths may differ in the number of hops or the type of hops between the same ionized layer and the earth or via different ionized layers. Sometimes they may be caused by variations in the horizontal direction of arrival. Much of the problem arises because of different number of hops between the communicating points involving the same ionized layer. However, the sporadic E layer is another source of multipath.

If the frequency in service lies below the maximum usable frequency (MUF), there will be increased tendency to develop multipath because the ionosphere will continue to refract or reflect signals of greater incidence, i.e., nearer to perpendicular incidence as the operating frequency becomes separated further below the optimum working frequency (OWF). The longer paths support more multipath because there is more opportunity for increasing the number of hops in the transmission involved up to a maximum limit in distance where there would be sufficient absorption of the longest delayed signal to render its interference potential negligible.

"The MRF [multipath reduction factor] may be defined as the lowest per cent of the MUF for which the range of multipath propagation time differences is less than a specified value.

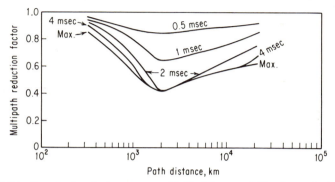

Fig. 15. Multipath reduction factor curves. (*R. K. Salaman, A New Ionospheric Multipath Reduction Factor (MRF), IRE Trans. Profess. Group Commun. Systems, vol. CS-10, no. 2, Fig. 4, p. 221, June, 1962.*)

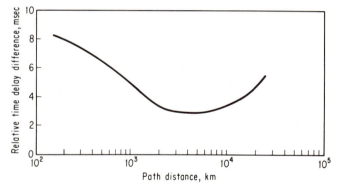

Fig. 16. Maximum expected time-delay difference. (*Salaman, op. cit., Fig. 5, p. 221.*)

"To illustrate the use of the MRF curves, consider a path length of 2500 km. When operating at a frequency between the MUF and 85 per cent of the MUF, the multipath spread, in general, will be less than 500 microseconds. Between 85 per cent and 65 per cent of the MUF, the spread will be less than 1 millisecond. Between 65 per cent and 43 per cent of the MUF, the spread will be between 1 and 2 milliseconds . . . there is a frequency at which the maximum multipath spread occurs. This frequency is given in [Fig. 15] and the value of the maximum multipath spread is given in [Fig. 16]. For the 2500-km path, the maximum spread expected is about 3 milliseconds. For a 1000-km path, the maximum expected spread is about 5 milliseconds, and will occur at about 56 per cent of the maximum regular layer classical MUF."*

It is estimated that these curves are probably less than 15 per cent in error.

* Reference 72, pp. 221, 222.

Multipath is evidenced by a variability of the front or trailing edge of the signal element.   Such a variability in the order of greater than 3-msec duration will effect four-channel ⁔ime-division multiplex keying speeds where the keying element length is approximately 5⅙ msec.   Assuming a 210-mile F2 layer height on the New York/San Francisco radio circuit, for example, it is calculated that time delays over the direct path will be 0.45 msec for one hop, 1.0 msec for two hops, 1.8 msec for three hops, and 2.9 msec for four hops.   This circuit should therefore be operable on four-channel multiplex without any difficulty due to multipath except under unusual conditions. Propagation via large number of hops is not likely owing to two reasons, the first being that the angle of incidence decreases with the number of hops and finally the angle is so small that the radio energy penetrates the ionized layer.   Second, the longer the path, the greater the absorption, so that what energy does get through via multiple hops is usually of considerably smaller amplitude.

Multipath is controlled to some extent by narrowing the vertical directivity pattern of both the transmitting and receiving antennas and by minimizing the high angles of

Table 4

Circuit	Length, miles	Hops	Range of vertical angle over sun-spot cycle, deg
New York/Tangier........	3,720	2	1.2–11.2
		3	7.2–21.4
		4	12.0–32.6
San Francisco/Tokyo......	5,147	3	1.7–10.7
		4	5.7–17.3
		5	9.0–23.3
		6	12.3–27.9

radiation and reception as, for example, the use of two-tier rhombics to minimize the radiation that would otherwise occur on a single-tier rhombic at, for example, 25 to 30°.   The vertical beams cannot be restricted too much, however, since there is a variation from day to day and from season to season of the best vertical angle of transmission.

Typical range of vertical angles of reception over the 11-year sunspot cycle is shown for certain circuits in Table 4.

5. Fading[7,16,17,32,37]

The radio signals arriving at a receiving antenna may have come via two or more different paths, thus resulting in phase differences and often phase cancellation in part or in whole.   These different paths may be the result of differences in the number of hops traversed in traveling from the transmitting to the receiving point.   Variations in the degree of absorption along these paths can cause fading effects as well as variations in the height of the ionospheric layer involved.

Variations in the propagation path resulting in changes in the polarization of the signal can result in fading.   Even changes in the horizontal direction of arrival of signals such that there can be some measure of phase cancellation will produce fading effects.

Fading rates which may be relatively slow measured in seconds or even minutes result from fluctuation in ionospheric height or absorption.   Flutter fading measured in several cycles per second may occur particularly on radio propagation paths in or near the vicinity of the auroral zone.   When signals of about equal amplitude arrive at the receiving points via two paths, flutter-type fading often occurs as a result of the variation in phase relations.

The depth of fading can vary from a few to 20 or 40 db or more.   The depth of fading can vary relatively across the frequency band involved, and this effect, known as selective fading, results in distortion of radio telephone signals.*

"The time distribution of the instantaneous field intensity resulting from the combination of a large number of waves of random phases and of nearly the same amplitude has been studied and found to be represented by the following formula:

$$T = \epsilon^{-0.693(E/E_m)^2}$$

where T is the fraction of the time the instantaneous field intensity exceeds the value E, and E_m is the median value of the field intensity. This is called the 'Rayleigh' distribution. For such a distribution, the median field intensity is equal to $\sqrt{0.693}$ times the rms value, i.e., 0.832 times the intensity of the homogeneous or nonfading field. For such a distribution the lower decile value, or value of field intensity exceeded 90 per cent of the time by the instantaneous values, is 0.39 times the median value. The upper decile value, or value exceeded only 10 per cent of the time by the instantaneous values, is 1.8 times the median value.

"Very bad interference fading is experienced in cases where the sky wave returns to earth at a distance from the transmitter such that the ground wave is of comparable amplitude. The combination of a randomly fluctuating sky wave with the steady ground wave produces much more severe fading than is commonly experienced with the sky wave alone."

In addition to these three types of fading, namely, different paths, polarization, and absorption fading, the length of the path may be such that fading results because of the so-called "skip distance" where the geometry of a single-hop transmission is such that the signal skips over the receiving location. As the ionospheric path fluctuates, this skip distance may move back and forth and cause changes in the received signal strength.

6. Absorption[36]

The degree of ionization of the earth's atmosphere particularly at the lower heights affects the amount of absorption of the high-frequency radio transmission. This ionization resulting from charged corpuscular radiation from the sun seems to have its origin in sunspot activity. The increase in this activity on the surface of the sun often creates large variations in the earth's magnetic field, and consequently this result is referred to as a magnetic storm in the earth's atmosphere.

Extensive magnetic storms are accompanied by increases in the ionization at lower heights and at lower latitudes and consequently result in greater absorption of the radio transmission. Such magnetic storms are visible at night as auroral displays. If the great-circle path passes through these disturbed areas, the radio signals are subjected to large and varying absorption, and the effect is greater at the higher latitudes.

Absorption is greatest in the daytime and occurs most in the D layer of the earth's ionosphere. The absorption index is defined as the logarithm of the ratio of the unabsorbed field intensity to the absorbed field intensity.† The variation of absorption with frequency is shown in Fig. 17 at noon on January, 1942, at Washington, D.C. The frequencies shown are incident and have to be related to the nonincident or oblique-incident radio transmission signals in proportion to the secant of the angle of incidence upon the D region.

For vertical-incidence frequencies above 1 Mc or equivalent for oblique incidence the absorption index α is given by the following formula:‡

$$\alpha = JQ\bar{K}S = AS$$

where J = seasonal variation factor
  Q = solar-cycle variation factor
  $\bar{K}$ = average value of K for transmission path
  S = function of frequency and distance traveled by the radio signal through the absorbing region

The seasonal variation factor J is given in Table 5.

* Reference 34, p. 108.
† Reference 34, p. 111.
‡ Reference 34, p. 113.

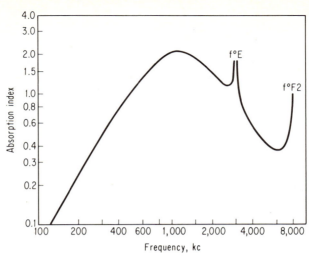

FIG. 17. Variation of absorption of the ordinary wave with frequency, noon, January, 1942, at Washington, D.C. Absorption is given as log (incident field intensity/reflected field intensity). Cusps on curve correspond to indicated critical frequencies. (*Natl. Bur. Std. Circ.* 462, *Fig.* 7.6, *p.* 111, *June* 25, 1948.)

Table 5. Seasonal Variation Factor J*

Month	Both terminals of transmission path		One terminal of transmission path north and the other south of equator
	North of equator	South of equator	
January............	1.3	1.0	1.15
February..........	1.3	1.0	1.15
March.............	1.15	1.15	1.15
April.............	1.15	1.15	1.15
May..............	1.0	1.3	1.15
June..............	1.0	1.3	1.15
July..............	1.0	1.3	1.15
August............	1.0	1.3	1.15
September........	1.15	1.15	1.15
October...........	1.15	1.15	1.15
November.........	1.3	1.0	1.15
December.........	1.3	1.0	1.15

* Ionospheric Radio Propagation, *Natl. Bur. Std. Circ.* 462, Table 7.1, p. 112, June 25, 1948.

The solar-cycle variation factor is given by the following formula.*

$$Q = 1 + 0.005R$$

where R is the sunspot number which is given in the National Bureau of Standards Central Propagation Laboratory's monthly publication, series TB 11-499-9/TO 31-3-28.

  * Reference 34, p. 113.

The diurnal variation factor K is given by the following formula.*

$$K = 0.142 + 0.858 \cos \chi$$

where χ is the sun's zenith angle (equal to zero when the sun is overhead).

The above absorption calculations are for average conditions, and the daily absorption variations are such that 90 per cent are less than about twice the monthly median and 90 per cent are greater than about one-half the monthly median, i.e., ± 6 db. Furthermore, these limits are approximately independent of the frequency, path length, season, and time of the day.†

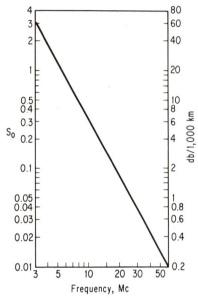

Fig. 18. Long path absorption constant S_0 for distance in thousands of kilometers, F2-layer propagation. (*Natl. Bur. Std. Circ.* 462, Fig. 7.31, p. 133.)

The absorption on long paths where the single hop on the F2 mode is 1,000 km or greater is given in the following formula:‡

$$\alpha = S_0 J Q \bar{K} d$$

in which JQ and $\bar{K}$ are the same as in the formula above, d is the path distance, and S_0 is a function of frequency only. The values for S_0 are given in Fig. 18.

Knowing the formula for absorption, it is therefore possible to calculate the field intensity for long paths over 3,200 km by the following formula:§

$$F = F_0 + \tfrac{1}{2} \log P - S_0 J Q \bar{K} d$$

where F is expressed in log microvolts per meter for 1 kw effective radiated power and F_0, the median incident unabsorbed field intensity for 1 kw effective radiated power, is given in Fig. 19. P is the effective radiated power and can be considered equal to the average value between radiation angles 0 and 30°.‖

* Reference 34, p. 112.
† Reference 34, p. 113.
‡ Reference 34, p. 114.
§ Reference 34, p. 118.
‖ Reference 34, p. 119.

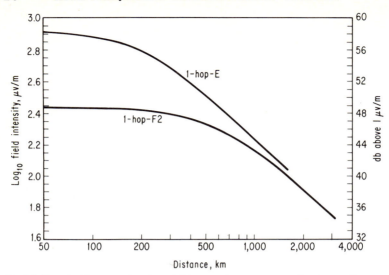

FIG. 19. Median incident unabsorbed field intensity for 1-kw effective radiated power based on virtual reflection heights, E layer, 105 km; F2 layer, 320 km. (*Natl. Bur. Std. Circ.* 462, *Fig.* 7.4, *p.* 109.)

$\bar{K}d$ is determined by calculating the values of K from the above formula and substituting these values of K_1 and K_2 in the following formula.*

$$\bar{K}d = 0.142D' + (K_1 + K_2 - 0.284) \tan D'/2R$$

where D' = distance of path where K is not zero, km
R = radius of earth, km

7. Sudden Ionospheric Disturbance (SID) and Solar Flares[68]

There is a very close relationship to sudden ionospheric disturbances (SID) and solar flare. Almost all SID are found to occur at times when there are solar flares, but for reasons not clearly understood, there are times of solar flares where there are no sudden ionospheric disturbances.†

At the time of these particular short-wave disruptions the radio transmission decreases very rapidly in a minute or so to a much lower level. Recovery, on the other hand, is much slower and may take as much as a half hour or more. A typical overall duration would be 20 min with a range from 5 min to 1 hr. It is believed that these sudden disturbances are caused by bursts of ultraviolet radiation from solar flares which cause heavy ionization and consequently greatly increased absorption in the D region of the earth's atmosphere. These effects are usually world-wide, and they are more prevalent during the 1 year before maximum until 1 or 2 years after maximum.

8. Auroral Zones[34]

The auroral zones encircle the geomagnetic north and south poles and encompass belts about 20° in radius. They are areas of erratic high-frequency radio propagation and are subject to normal ultraviolet radiation from the sun but are also involved with more corpuscular radiation from the sun than belts of lower latitude. Consequently absorption is high in these zones. Figure 20 is an auroral-zone absorption map giving information on the absorption factor K'.

* Reference 34, p. 119.
† Reference 25, p. 1.

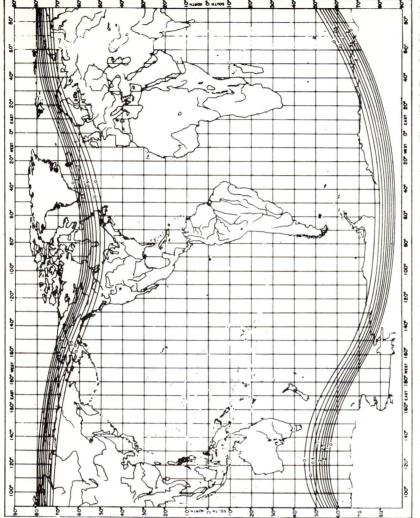

FIG. 20. Auroral zone absorption map.   Numbers on curves are absorption factor K'.   (*Natl. Bur. Std. Circ.* 462, *Fig.* 7.48, *p.* 150.)

Radio signals whose great-circle path passes through these auroral zones are affected by high fading rates (20 cps or more) and phase perturbations.* Observations show that the received carrier level drops 10 to 20 db when fast fading occurs, and these fading rates are generally higher on days with magnetic disturbances.

9. Direction of Arrival[56,57]

The horizontal direction of arrival of high-frequency radio transmission varies from about 3 to 5° over short periods of time on single-hop F-layer transmission.† Short-period variation or single-hop E-layer transmission directional variations are greater. Scatter propagation arrival angles can vary up to 20° or more. Azimuth deviations are also greater for paths traversing the auroral zones.‡ Variations occur both to the right and left of the great-circle path.

10. Signal-to-Noise Ratios

The signal-to-noise ratio at the demodulating point in a high-frequency radio communication system is one of the important elements in determining the ultimate level of the accuracy of the service. The strength of the radio signal at the receiving point is determined by the transmitter power, the gain and directivity of the transmitting and receiving antennas, and the characteristics of the propagation path. Fading ratios in the long-distance high-frequency radio path can vary from about 2 to 100 or from 6 to 40 db. The receiving antenna, even though directive, is subjected to local noise disturbances and atmospheric noise, originating mostly in thunderstorm activity, which is propagated long distances. This atmospheric noise has a random or Gaussian background with bursts due to lightning strokes riding on top. Local man-made noise can be either Gaussian or impulse in nature.

The head end of the receiver, i.e., the first RF amplifier circuit, introduces thermal noise which, under good radio propagation conditions particularly above 16 Mc/sec and in the absence of local noise, may be the controlling noise in the system. This is usually not the case, however, in high-frequency radio circuits. The use of a frequency-modulated (FM) signal or frequency-shift-keyed (FSK) signal will improve the signal-to-noise ratio in the output of the demodulated stage because of the amplitude characteristics of noise. The demodulator is then sensitive to frequency changes and has relatively little sensitivity to amplitude changes. The bandwidth occupied by the signal is also significant, since the greater the signal bandwidth, the greater the amount of noise that is received. These factors are discussed in greater detail under modulation parameters. At this time the minimum signal-to-noise levels at the receiving location are of interest for the various types of services. It is estimated that a practical minimum ratio of signal to noise at the receiver input for high-frequency single-sideband (SSB) or FSK radio teletype telegraphy is 12 db in the channel involved. More information is given on signal-to-noise ratios in the section on Type of Service.

The error rates not only are affected by the signal-to-noise ratio but are somewhat sensitive to the fading rate as well. If the fading rate in telegraphy approaches or exceeds the keying speed, there will be a larger increase in the error rate than otherwise.

11. Distortion[53]

High-frequency radio systems are subject to the same three basic distortion problems affecting all telecommunication systems. The three types of distortion are fortuitous, characteristic, and bias and are defined as follows.§

"Fortuitous distortion is a random distortion of telegraph signals such as that commonly produced by interference.

* Reference 30, pp. 1, 2, 5.
† Reference 32, p. 35.
‡ Reference 32, p. 41.
§ Reference 54, p. 68.

"Characteristic distortion is a displacement of signal transitions resulting from the persistence of transients caused by preceding transitions.

"Bias in telegraph transmission is a uniform displacement of like signal transitions resulting in a uniform lengthening or shortening of all marking signal intervals."

A fourth distortion, namely multipath, has already been covered in some detail and can be classified as a characteristic distortion of the ionosphere. All equipment elements of the high-frequency system tend to introduce varying amounts of distortion, and synchronous and nonsynchronous regeneration of signals is often used.

FREQUENCIES

12. Availability[11]

Of the 24,500 kc of spectrum space between 3 and 27.5 mc, there are 14,820 kc, or about 60 per cent, available in total for all international fixed point-to-point communi-

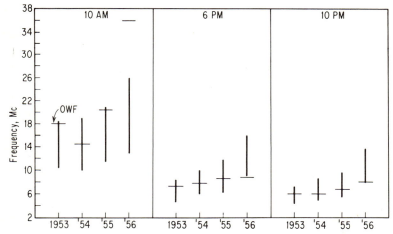

Fig. 21. Average useful frequency spread for December between New York and Central Europe years 1953–1956. Frequency spread data above 20 Mc is incomplete. Horizontal cross line represents optimum working frequency (OWF).

cation carriers under the jurisdiction of the FCC, and the spectrum must be shared geographically and by time with the rest of the world. Frequency percentages are somewhat misleading, however, in that the same frequency is assigned over and over again throughout the world on the basis of both geographical and time sharing (see also Chap. 19).

The long-distance communication channel capabilities of high frequencies vary with diurnal, seasonal, and 11-year sunspot cycle conditions. The ionosphere in the 3- to 30-Mc portion of the radio spectrum resembles a bandpass filter with time-variable bandwidth and insertion loss characteristics. At any given time on any specific long-distance circuit, the ionosphere will be capable of carrying a particular bandwidth of frequencies. Under stable propagation conditions, this bandwidth characteristic may remain relatively constant for as much as several hours. At other times, the frequency spread of the passband varies considerably from hour to hour. It also varies from day to day, seasonally, and over the sunspot cycle. The passband spread for day frequencies is usually greater than for the lower nighttime frequencies. Propagation disturbances generally result in a reduction in the optimum working frequency as well as in the width of the passband of the ionosphere. Generally, the longer the circuit, the narrower the passband.

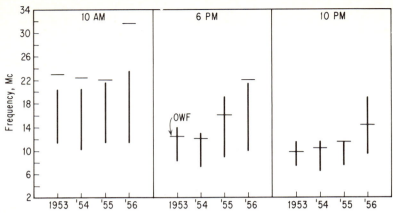

FIG. 22. Average useful frequency spread for December between New York and South America years 1953–1956. Frequency spread data above 20 Mc is incomplete. Horizontal cross line represents optimum working frequency (OWF).

These bandwidth and attenuation characteristics of the ionosphere, which vary with time and among geographical paths, are the major factors which limit the number of communication channels that any available number of frequencies will provide. Generally the lower the optimum working frequency, the narrower the passband, the higher the attenuation, and the poorer the signal-to-noise ratio. Since low frequencies

Table 6. Summary of Frequency Spreads for Several Sunspot Cycle Conditions

Month	Year	Time	Ionosphere spread in frequencies, Mc		Total spread, Mc	Period of sunspot cycle
			Lowest	Highest		
Europe						
September............	1956	10 A.M.	13.5	23.0	9.5	Near high
September............	1956	6 P.M.	12.0	19.5	7.5	Near high
September............	1956	10 P.M.	9.0	15.0	6.0	Near high
December............	1950	10 A.M.	13.2	20.8	7.6	Intermediate
December............	1950	6 P.M.	5.5	10.4	4.9	Intermediate
December............	1950	10 P.M.	4.9	8.8	3.9	Intermediate
December............	1953	10 A.M.	10.3	18.1	7.8	Low
December............	1953	6 P.M.	5.0	8.1	3.1	Low
December............	1953	10 P.M.	4.4	7.0	2.6	Low
South America						
December............	1953	10 A.M.	11.7	20.4	8.7	Low
December............	1953	6 P.M.	8.3	13.5	5.2	Low
December............	1953	10 P.M.	7.3	11.5	4.2	Low

are required during the low part of the sunspot cycle, it can be anticipated that the greatest difficulty will be encountered in carrying the communication load at that time.

Figures 21 and 22 show typical frequency spreads at various times of the day and from day to day during the low part, 1953, and the high part, 1956, of the sunspot cycle on the New York to Central Europe and New York to South America circuits. Optimum working frequencies are also shown, as indicated by the horizontal lines. The possible frequency spread shown is not complete for frequencies above 20 Mc, as can be seen where the OWF is appreciably above 20 Mc, and the reason for this was simply the limitation of the recording equipment at the higher frequencies. The range of daily spreads and the narrower spreads during the low part of the sunspot cycle are clearly indicated. These spreads are the passband characteristics of the ionosphere.

Table 6 summarizes information on frequency spreads experienced on New York to Europe and New York to South America circuits for several sunspot-cycle conditions.

One fortunate condition exists, as can be seen by an examination of Table 6 and Figs. 21 and 22, and that is the extensive opportunity to time-share the frequency spectrum between the European and South American circuits, particularly in the evening and nighttime and during the low portion of the sunspot cycle when operation is most difficult.

An examination of this table for the New York to Europe circuit shows that 2.6 Mc of high-frequency spectrum is available at the low part of the sunspot cycle, December, 1953, at 10 P.M., as against 7.8 Mc available on the same day at 10 A.M. or a ratio of

Table 7. Frequency Spectrum for Fixed Services between 3 and 27.5 Mc at Locations under Jurisdiction of FCC

(Data shown for various blocks of frequency spectrum per the Table of
Frequency Allocations)

Frequency band and range, kc	No. of kilocycles
3,155– 3,200	45
3,200– 3,230	30
3,230– 3,400	170
4,000– 4,063	63
4,438– 4,650	212
4,750– 4,850	100
4,850– 4,995	145
5,005– 5,060	55
5,060– 5,250	190
5,250– 5,480	230
5,730– 5,950	220
6,765– 7,000	235
7,300– 8,195	895
9,040– 9,500	460
9,775– 9,995	220
10,100–11,175	1,075
11,400–11,700	300
11,975–12,330	355
13,360–14,000	640
14,350–14,990	640
15,450–16,460	1,010
17,360–17,770	410
18,030–19,990	1,960
20,010–21,000	990
21,750–21,850	100
22,720–23,200	480
23,350–24,990	1,640
25,010–25,070	60
25,110–25,600	490
26,100–27,500	1,400
Total	14,820

3:1 in the amount of usable spectrum space available. Similarly, for the New York to South America circuit at 10 P.M., December, 1953, 4.2 Mc of space is available as against 8.7 Mc at 10 A.M. on the same day for a 2:1 ratio. The available spectrum space is greater at the high end of the sunspot cycle.

Table 7 shows the frequency spectrum in kilocycles available for the fixed services between 3 and 27.5 Mc at locations under the jurisdiction of the FCC. The data are shown for various blocks of frequency spectrum per the International Table of Frequency Allocations.*

13. Utilization[67]

The efficient and economical operation of high-frequency radio circuits on an hour-by-hour basis requires a proper knowledge of the characteristics of the radio propagation path and the use of the frequencies. It is not sufficient just to engineer the plant

Table 8. SINPO and SINPFEMO Signal Reporting Code*

	S	I	N	P				O
	S	I	N	P	F	E	M	O
Rating scale	Signal strength	Degrading effect of			Fre-quency of fading	Modulation		Overall rating
		Inter-ference	Noise	Propaga-tion dis-turbance		Quality	Depth	
1	Barely audible	Extreme	Extreme	Extreme	Very fast	Very poor	Continuously overmodu-lated	Unusable
2	Poor	Severe	Severe	Severe	Fast	Poor	Poor or nil	Poor
3	Fair	Moderate	Moderate	Moderate	Moderate	Fair	Fair	Fair
4	Good	Slight	Slight	Slight	Slow	Good	Good	Good
5	Excellent	Nil	Nil	Nil	Nil	Excellent	Maximum	Excellent

Special Remarks:
1. A signal report shall consist of the code word SINPO or SINPFEMO followed by a group of five or eight numerals, rating respectively, the five or eight characteristics of the particular signal code.
2. The letter X shall be used instead of a numeral for characteristics not rated.
3. Although the code word SINPFEMO is intended for radiotelephony, it may be used for radiotelegraphy.
4. The over-all rating for radiotelegraphy shall be as indicated in Table 9.
5. The over-all rating for radiotelephony shall be as indicated in Table 10.
   * *Codes and Abbreviations for the Use of The International Telecommunication Services*, 2d ed, p. 105, International Telecommunication Union, Geneva, 1963.

facilities to make the most of the propagation path and frequencies, but it is equally important to change frequencies currently in accordance with the requirements of the propagation path whether these requirements come along in an orderly anticipated manner or as a sudden change.

Supervisory technicians who have the appropriate technical background to cope with changes as necessitated by the ionospheric path must be available if the system is going to perform properly. The assignment of this responsibility to one specific knowledgeable area for each major operating location can make a great deal of difference in the ultimate performance, and the system engineer should see to it that this operating function is adequately staffed with trained personnel.

Since high frequencies are assigned many times around the world in geographic and time-sharing arrangements and because of the vagaries of the ionosphere, interference

* Reference 45, pp. 44–56.

does result at times.   When it occurs, the party suffering the interference will have to identify, if possible, the source of the interference.   Many transmissions today are multiple channel with special codes and speeds, all of which makes standard monitoring and identification extremely difficult during regular operation.   Upon identification, the interfering party is notified and is expected to take rapid remedial action.   There are cases where, after investigation and under domestic and international recognition of priority claims, it may be found that the complaining party is actually at fault and will then have to take the necessary steps to alter transmissions.

If it is possible for the party suffering the interference to identify the source, it is in order to ask other operating administrations and the national and international regulatory agencies to help identify the source.   Under these circumstances it is sometimes

Table 9. Radiotelegraphy Operation*

Signal strength	Mechanized operation	Morse operation
1. Unusable........	Unreadable	Unreadable
2. Poor............	BK's, XQ's, and call signs readable	BK's, XQ's and call signs readable
3. Fair............	Marginal single start-stop printer	50 words per minute
4. Good...........	2-channel time-division multiplex	100 words per minute
5. Excellent.......	4-channel time-division multiplex	High speed

* *Codes and Abbreviations for the Use of the International Telecommunication Services*, 2d ed., p. 106, International Telecommunication Union, Geneva, 1963.

Table 10. Radiotelephony Operation*

Signal strength	Operating conditions	Quality
1. Unusable........	Channel unusable by operators ⎫	Not
2. Poor............	Channel just usable by operators ⎭	commercial
3. Fair............	Signal quality seriously affected. Channel usable by ⎱ operators or by experienced subscribers ⎰	Marginally commercial
4. Good...........	Signal quality slightly affected ⎱	Commercial
5. Excellent........	Signal quality unaffected ⎰	

* *Codes and Abbreviations for the Use of the International Telecommunication Services*, 2d ed., p. 106, International Telecommunication Union, Geneva, 1963.

necessary for the damaged party temporarily to shift frequency assignment and use in order to clear the channel for traffic.

Again, because of the multiple-frequency assignments, it is inadvisable to allow a frequency to go unused for any long period of time because others will have found the inactive portion of the frequency spectrum and the practical future use of the then idle frequency may be jeopardized.

Radio transmissions must be identified regularly, particularly at start-up and shut-down, under domestic and international regulations.   Identification by giving the call sign of the station involved is sufficient, and this is permitted by Morse manual or machine code, by voice, or by teleprinter keying.   The radio carrier or subcarrier can be keyed manually or automatically, or more complicated identification techniques involving phase modulation on FSK telegraph carrier or amplitude modulation of the reduced carrier in a single-sideband system can be used.   These latter offer the advantage that normal keying of the transmission is possible simultaneously and traffic movement therefore need not stop during these identification transmissions.

Another important adjunct to proper operation and testing of the systems is the use of standard reporting codes. The "SINPO" and "SINPFEMO" reporting codes were adopted by the International Telecommunication Union for this purpose. Tables 8 to 10 give the details of these reporting codes which have been used internationally for radiotelegraph and radiotelephony for many years.

14. Radio Propagation Forecasts[36,55,66]

There are several organizations in the world that prepare forecasts of ionospheric disturbances from a few hours to as much as several weeks in advance. These forecasts, some of which are generally available to the public and others of a somewhat private nature, attempt to predict with varying degrees of success the advent of disturbed conditions in the future.

This public service in the United States is available as listed below. Forecasts are regularly transmitted on radio via the National Bureau of Standards stations WWV and WWVH.

1. Central Radio Propagation Laboratory, National Bureau of Standards, Boulder, Colo.
2. North Atlantic Radio Warning Service, Box 178, Fort Belvoir, Va.
3. North Pacific Radio Warning Service, Box 1119, Anchorage, Alaska

Radio propagation forecasts are not an exact science and involve assembly of a considerable amount of ionospheric information from locations throughout the world. This information, together with data on traffic conditions, is studied by experts who are thoroughly knowledgeable on sunspots and ionospheric behavior and is used in the development of periodic short-term and long-term radio propagation forecasts. These forecasts, some of which originate from private sources, have many times proved very useful to the communication carriers in arranging for plant facilities, circuit utilization, and diversionary routes. Experience in the use of the information contained in these forecasts is also an essential ingredient in the success of the service.

Radio forecasting is possible for all areas of the globe, but those areas that currently accumulate the most information on ionospheric conditions and trends are the areas where the greatest accuracy of the forecasting is possible.

ANTENNAS

15. General[18,21]

The characteristics of the ionosphere at high frequencies have a large bearing on antenna design. The transmitting antenna must favor the operating frequencies required and the vertical and horizontal directions involved. Similarly, the receiving antenna must favor the frequency and the direction of arrival of the signal both vertically and horizontally and must discriminate against signals arriving from unwanted directions. The required overall gain of the antenna systems is very important and is dictated by signal power, attenuation in transmission, and noise.

The variation in the height of the ionosphere as well as the physical separation of the points between which communication is desired determines the range of optimum vertical angles. The variation in horizontal direction of arrival must be considered. The transmission beam must be concentrated to obtain maximum advantage of the higher gain realized with such concentration, but the beamwidth should not be so small that ionospheric changes carry it beyond the limits of communication for the propagation path involved.

Since the optimum working frequency of the propagation path varies, the antenna used must be capable of working efficiently over the band of frequencies required. Antennas having a narrow high-frequency bandwidth can be used efficiently only at more limited times of the day, season, or sunspot cycle, while those having a wide bandwidth are more versatile operationally. High gains in the order of 8 to 16 db are desirable, but in view of the wide variation in signal strength at the receiving point, antenna gain at the expense of too limited an operating frequency versatility or horizontal or vertical antenna directivity is not generally warranted.

Interference caused by other high-frequency signals restricts the successful operation on the limited available frequencies, and it is therefore important to reduce the amount of power transmitted in undesired directions, since this can cause interference to other transmissions and severe congestion in the frequency spectrum. Similarly, at the receiving end the antenna should discriminate against signals from unwanted directions and thus develop greater frequency utilization efficiencies.

16. Antenna Sites[71]

Vertical angle clearance between the transmitted beam and natural obstructions such as mountains should be sufficient for both transmitting and receiving antennas. It is preferable that the vertical angle subtended by such obstruction not exceed 2 or 3°, but up to 6° is acceptable. For long-distance circuits the optimum angle of transmission can range from about 9 to 18°.

On shorter circuits where the angle of maximum transmission is higher, approximately 20 to 40°, the angle subtended by the obstruction may be about twice this, or

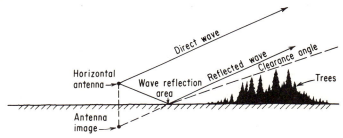

Fig. 23. Direct and reflected waves and clearance angle. (*E. A. Laport, "Radio Antenna Engineering," Fig. 313, p. 228, McGraw-Hill Book Company, New York, 1952.*)

from 5 to 12°. This would permit use of a rather rugged site, which might otherwise not be feasible.

When the circuits are long, i.e., 1,500 miles and over, and the antenna field surroundings very rugged, it may be necessary to design the system for three hops rather than a two-hop transmission. The direct and reflected transmission from the antennas should have clearance from trees and buildings as shown in Fig. 23.

If the land should slope away in front of the antenna, this must be taken into consideration, and by proper sloping of the antenna the vertical directivity can be controlled. This principle holds for both transmitting and receiving antennas, and it is generally best to design antennas for both ends of the circuit for the same vertical and horizontal directivities.

The conductivity of the soil in the area immediately underneath and ahead of high-frequency antennas is not so important as for the lower frequencies, but for those frequencies from 3 Mc to a few megacycles above, good conductivity of the soil is preferable particularly if the path length is so short that the ground wave is of some importance. The receiving site should be as free as possible from man-made noise sources, and a survey should be made to ensure relative freedom from such noise around the clock and on typical days of the week (see Chap. 23).

In some areas "precipitation static" caused by snow, fog, or wind-blown dust coming into contact with the metallic portions of the receiving antenna and feeder is troublesome and can be minimized by insulating the wires, thus reducing the electrical static.

17. Rhombic Antennas[22,50,71]

Rhombic antennas are used extensively for high-frequency communication, both transmitting and receiving. These antennas have relatively broad frequency band-width capabilities and thus can span much of the required frequency range as propagation conditions vary daily and over longer periods. They provide substantial gains, and their relatively simple construction leads to moderate cost. They do require substantial land area and involve the loss of transmitting power in the terminating load.

A typical horizontal rhombic antenna and directive radiation patterns are shown in Figs. 24 and 25. These typical radiation patterns are for three different frequencies

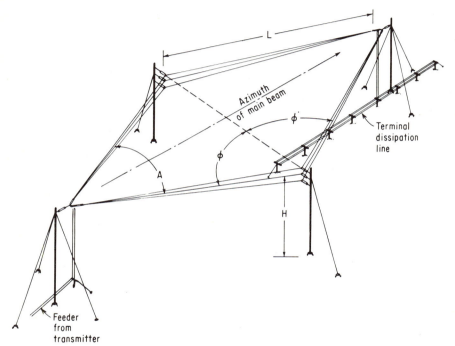

Fig. 24. Horizontal rhombic antenna (common three-wire form). (*Laport, op. cit., Fig.* 3.77, p. 316.)

within a band of 2.25 to 1, as will be noted by the different values for the length of the sides L as measured in the three different wavelengths involved. The center is the zenith, and the outer periphery is the flat horizon. The radiation lobes are shown by 3-db contour lines. Similar radiation patterns are shown in Fig. 26 for the same rhombic but with an apex angle of 18° instead of the 22° apex for the previous figure. The three-wire side curtains improve the frequency response by making the input impedance flatter over the frequency range normally used.

Radiation is the result of an attenuated traveling wave along the side length, and the radiation along the main axis is horizontally polarized for horizontal rhombics. The vertical angle of radiation can be changed by constructing the rhombic antenna with a tilt, i.e., with its front support at a higher level than the rear or feed support. Such tilting increases the vertical beamwidth of the major lobe, which can be of value on long circuits where there is a varying vertical angle of optimum propagation. The input impedance is usually about 600 ohms. Static drains are usually provided between

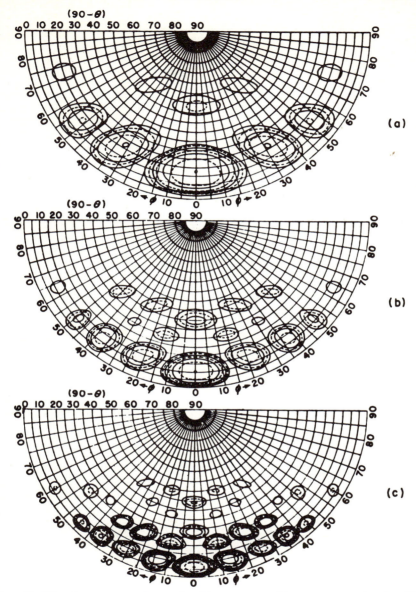

FIG. 25. Directive patterns for a rhombic antenna with an apex angle $A = 22°$. (a) $l = 3.33\lambda$; $h = 0.8\lambda$. (b) $l = 5.0\lambda$; $h = 1.2\lambda$. (c) $l = 7.5\lambda$; $h = 1.8\lambda$. (After Christiansen in "Radio Antenna Engineering" by E. A. Laport, Fig. 3.78, p. 317, McGraw-Hill Book Company, New York, 1952. Used by permission.)

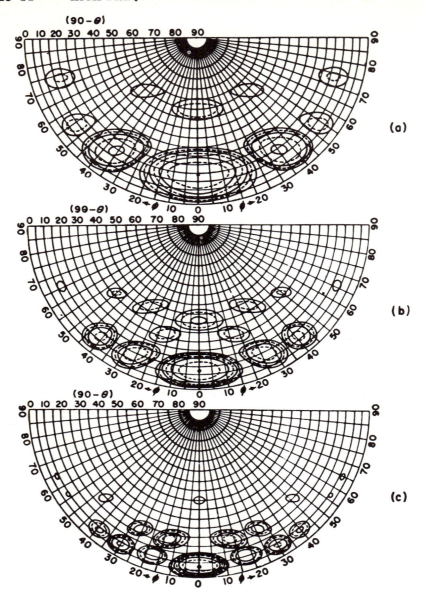

Fig. 26. Directive patterns for a rhombic antenna. Same as Fig. 25 except that apex angle $A = 18°$. (*After Christiansen in "Radio Antenna Engineering" by E. A. Laport, Fig. 3.79, p. 318, McGraw-Hill Book Company, New York, 1952. Used by permission.*)

Table 11. RCA Fishbone-antenna Dimensions*

Length dipoles, ft	Optimum frequency, Mc	Useful range, Mc	Width (two bays), ft	Length total, ft	Pole height, ft	Useful angle (azimuth), deg
34	18	13–22	120	312	60	10
48	14	10–19	148	312	90	10
66	9	3–13	200	312	120	10

* E. A. Laport, *Radio Antenna Engineering*, Table 3.4, p. 340, McGraw-Hill Book Company, New York, 1952.

each side and ground. The main diagonal of rhombic antennas generally varies from 400 to 1,000 ft, the lesser diagonal from 200 to 400 ft, and the height from 60 to 120 ft.

There has been some trend to consideration of large rhombic antennas. Extensive investigation on circuits into London indicates that there is a possibility of at least a 6-db gain from doubling the height of a rhombic antenna.* Longer side lengths also result in greater gain.

18. Fishbone Receiving Antenna[71]

Another somewhat less popular receiving antenna is called the fishbone antenna because its active elements when viewed from above have a similarity to the configuration of the backbone of a fish. A typical fishbone antenna is shown in Fig. 27. Horizontal dipoles with a length of about one-half wavelength in series with high-reactance capacitors are attached to the central feeder as shown. The capacitors are used to adjust the velocity of propagation along the antenna giving about 90 per cent of free-space velocity, which results in a characteristic impedance of about 400 ohms. The maximum directivity is along the line of the feeder toward the receiver.

Typical horizontal and vertical radiation patterns for a fishbone antenna are shown in Fig. 28, and the absence of large secondary lobes is to be noted. Typical fishbone antenna dimensions are shown in Table 11.

Construction details are shown in Fig. 29. These antennas are often used in two bays, i.e., with two systems fed in parallel from the same transmission line for a further improvement in directivity gain.

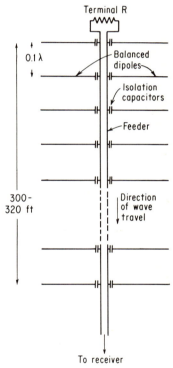

Fig. 27. RCA fishbone-antenna circuitry. (*Laport, op. cit., Fig. 3.92, p. 339.*)

19. Horizontal Half-wave-dipole Antenna System

The simplest antenna for high-frequency circuits is the horizontal half-wave dipole, but its radiation pattern and gain leave much to be desired. However, a knowledge

* D. E. Watt-Carter and S. G. Young, Survey of Aerials and Aerial Distribution Techniques in the H.F. Fixed Service, *Proc. IEE, London*, vol. 110, p. 1530, sec. 2.1, September, 1963.

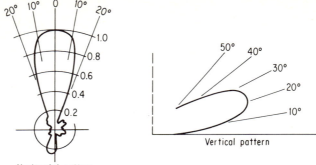

Horizontal pattern

Vertical pattern

FIG. 28. Horizontal and vertical patterns for RCA fishbone unit. (*After Beverage and Peterson in "Radio Antenna Engineering" by E. A. Laport, Fig. 3.93, p. 340, McGraw-Hill Book Company, New York, 1952. Used by permission.*)

of its radiation pattern and characteristics is useful because many varieties of antennas are made up of groups of dipoles stacked in different manners with improved gains and directivities over the single dipole.

The length of a half-wave high-frequency antenna in feet is given in the following formula:

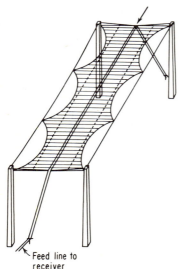

$$\text{Length of half-wave antenna (ft)} = \frac{468}{\text{freq. (Mc)}}$$

This is shown in Fig. 30.

The formula for the radiation pattern in the vertical plane perpendicular to a horizontal dipole over perfectly conducting ground is*

$$F(\alpha) = \sin (h \sin \alpha)$$

where α = angle to the horizon
h = height above ground, electrical degrees

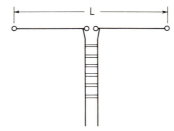

FIG. 29. Construction details of fishbone antenna.

FIG. 30. Horizontal half-wave doublet antenna (dipole antenna).

The radiation pattern for such a horizontal dipole is shown in Fig. 31. A typical array of horizontal dipoles is shown in Fig. 32, and the radiation pattern is shown in Fig. 33.

20. Miscellaneous Antennas[31,39,49]

It is not practical to describe here in detail the variety of additional types of high-frequency antennas, but it is worthwhile listing a few of the more prominent ones with a brief statement as to their areas of usefulness.

* Reference 71, sec. 3.6.1, p. 233.

Figure 34 shows the general outline of a V antenna which is similar in appearance to one-half of a rhombic antenna. It is a simple antenna to construct, does not require so much land area as a rhombic antenna, and has a relatively high gain.

Figure 35 shows the general outline of the lazy H antenna, which is a simple broadside type.

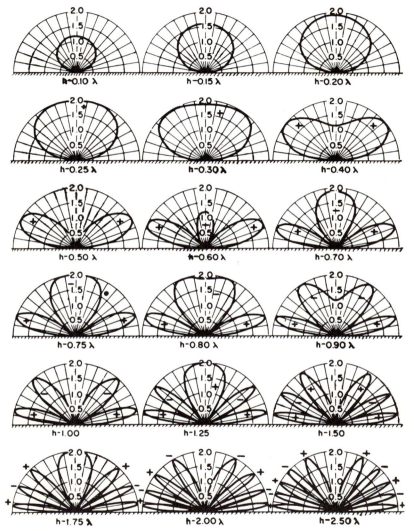

FIG. 31. Vertical polar radiation diagram in the plane normal to a horizontal dipole antenna. (*By permission of the Controller of Her Majesty's Stationery Office, London, British Crown copyright, RAF Signals Manual.*)

Logarithmic antennas have radiating elements whose resonant frequencies are in geometric progression, and this results in appreciable effectiveness at a wide range of frequencies. They are of some interest for special purposes in the high-frequency field. The general construction is as shown in Fig. 36. The radiating elements are folded

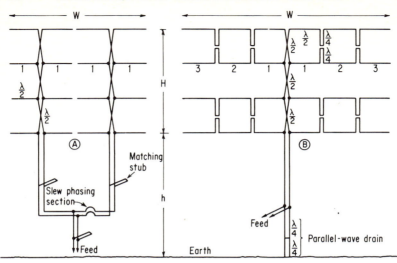

FIG. 32. Arrays of horizontal half-wave dipoles and typical feeder connections. (*Laport, op. cit., Fig.* 3.49, *p.* 276.)

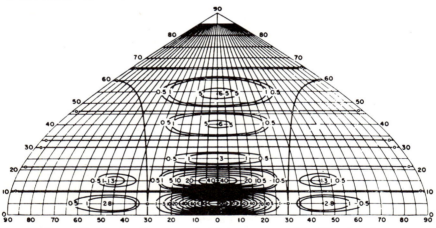

FIG. 33. Power-distribution diagram. Aerial type same as Fig. 32a when *h* equals two wavelengths. (*After Hayes and McLarty in Laport, op. cit., Fig.* 3.50, *p.* 277.)

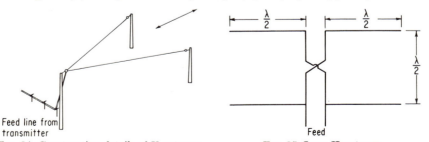

FIG. 34. Construction details of V antenna.

FIG. 35. Lazy H antenna.

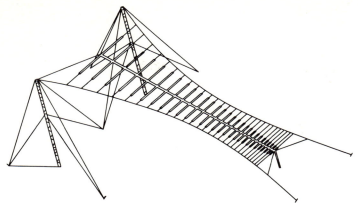

FIG. 36. Logarithmic antenna with folded dipoles array. (*M. F. Radford and E. W. Woloszczuk, New H. F. Logarithmic Aerials, Fig. 4, p. 143, Convent. HF Commun., The Electronics Division of the Institution of Electrical Engineers, Savoy Place, London, March, 1963.*)

dipoles. Only those elements that are sufficiently near to resonance will contribute to the radiation, and therefore the aperture efficiency is rather low. An antenna constructed in accordance with this sketch with 29 dipoles ranging in length from 6 ft 2 in. to 31 ft and with an overall length of 96 ft and a height of 50 ft gave a gain of 15 to 20 db in the 20- to 50-Mc range.* The radiation angle was 50° horizontal and 7 to 25° vertical (all 3-db points).

Another type using broad dipole arrays as the radiating element and shown in Fig. 37 with the same general dimensions as given for the logarithmic antenna gave a 13- to 25-db gain over a 13.5- to 41-Mc range.† The radiation angle was 58° horizontal (down to 3 db points) and 11 to 38° vertical (3 db points).

21. Diversity Receiving Antennas[7]

The signal arriving at a point at one moment may have considerable differences in amplitude from that arriving at a nearby location. This characteristic of the propagation path therefore led to the suggestion that two or more antennas separated by some suitable distance be used to feed different radio-frequency chains in a receiver incorporating a combining unit which would automatically select the stronger of the two signals at any time. Two-set antenna diversity with the appropriate receiver used in 50-baud teleprinter service gives an improvement of at least 12 db over a single receiver-antenna arrangement.‡ When

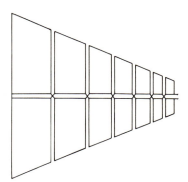

FIG. 37. Logarithmic antenna with broad dipole array. (*Radford and Woloszczuk, op. cit., Fig. 3, p. 143.*)

the rate of generation of errors on the single antenna receiver system is running as high as 20 errors per 1,000 characters, the use of two-set diversity over a single receiver gives a gain of about 12 db, and with an error rate of 1 error per 1,000 characters this same improvement may run as much as 30 db.

* M. F. Radford and E. W. Woloszczuk, New HF Logarithmic Aerials, *IEE Conv. on HF Commun.*, p. 142, Savoy Place, London, March, 1963.
† *Ibid.*
‡ Reference 71, sec. 3.3.4, p. 223.

A third receiver-antenna system might also be used for three-set diversity, but with frequency-shift-keyed (FSK) telegraphy such diversity would show only a slight gain, perhaps up to 4 db, over two-set diversity. If amplitude-modulated telegraphy is used, the gain for the third-set diversity is well worthwhile. The limiting used in the demodulation system with FSK provides an improvement which makes two-set space

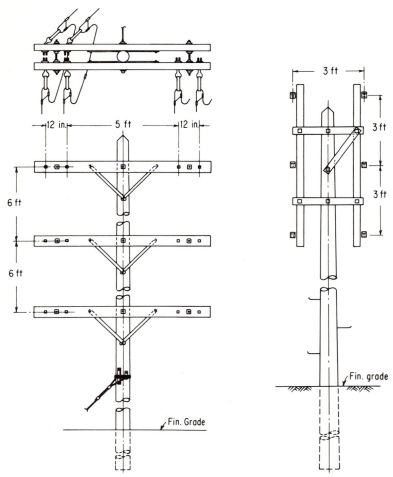

FIG. 38. Typical arrangement for several two-wire transmission lines between antenna field and transmitters. (*Courtesy of RCA.*)

FIG. 39. Typical arrangement for several four-wire transmission lines between antenna field and receivers. (*Courtesy of RCA.*)

diversity FSK equivalent to three-set space diversity with amplitude-modulated telegraphy.

Comparable space diversity gains are not possible with telephony, but some improvement in signal performance is noted.

Antennas for space diversity reception are usually placed about 1,000 ft apart or at the apexes of an isosceles triangle with two of the antennas situated on a perpendicular to the line of transmission or reception.

It is also possible to use two transmitters with frequencies 90 cycles apart and associated antennas spaced about 1,500 ft to obtain transmitter space diversity gain for telegraphy somewhat comparable to that for receiving antenna space diversity. The transmitter frequencies are spaced 90 cycles apart to avoid momentary phase cancellation at the receiving end and consequent loss of a telegraph signal element.

22. Transmission Lines—Transmitting

Open-wire transmission lines with 600-ohm characteristic impedance are used extensively for connecting the antennas to the high-frequency radio transmitters. These lines are relatively easy to construct and maintain, and if care is taken to see that they are uniformly maintained and that the separation between lines is at least about six times their own wire spacing, then the level of crosstalk is held to an acceptable level.

A typical arrangement for several two-wire transmission lines is shown in Fig. 38. The standing-wave ratio on these systems ranges from about 1½:1 to 2:1. Care

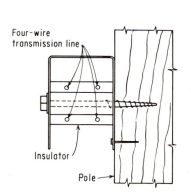

FIG. 40. Details of balanced four-wire receiving transmission-line supports. (*Courtesy of RCA.*)

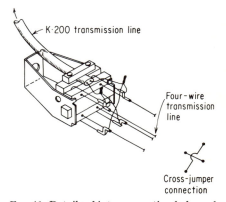

FIG. 41. Details of interconnection balanced four-wire receiving transmission line and two-wire coaxial transmission line. (*Courtesy of RCA.*)

must be taken to avoid sharp bends at corners or other sources of impedance discontinuities. The attenuation is 0.03 db per 100 ft for No. 4 copper wire at 16 Mc/sec.

When space is limited, particularly near the transmitter buildings, it is sometimes necessary to use coaxial transmitter transmission lines. Their use has the advantages of flexibility and freedom from crosstalk and impedance discontinuities, and they are less hazardous to personnel. Unbalanced coaxial feeders with about 50 ohms impedance together with wideband transformers are employed to some extent. The attenuation is about 0.075 db per 100 ft for ⅞-in.-diameter rigid or semirigid line at 16 Mc/sec.

23. Transmission Lines—Receiving

At the receiving station greater attention must be given to ensure a minimum amount of crosstalk and extraneous noise pickup. Consequently, balanced four-wire transmission lines with characteristic impedance of 200 ohms are used and considerable care is taken to ensure uniformity of spacing of the individual wires and appropriate distance of at least 3 ft between adjacent transmission lines. Also in order to maintain proper balance to ground and adjacent exposures, each pair of the four wires located diametrically across from each other are connected together to ensure maximum balance. Details of this four-wire-line construction and a view of a typical receiving-antenna trunk-line system are shown in Figs. 39 and 40.

Again, because of space limitations in the receiving building, two-wire transmission lines or coaxial transmission lines are used. The physical details of the 200-ohm balanced four-wire transmission line to the 200-ohm two-wire line are shown in Fig. 41.

24. Multicouplers—Receiving

Multicouplers provide sufficient amplification and isolation so that several receivers can be connected to the same antenna on different frequencies without any harmful degradation of the signal-to-noise ratio. This permits substantial cost savings, particularly at the larger radio stations.

TYPE OF SERVICE

25. Telegraphy

This is considered first partly because it was the initial radio communication service and because, at least in its early stages, it required a minimum of bandwidth. Both continuous carrier frequency-shift-keying transmitter/receiver systems and single-sideband transmitter/receiver systems with subcarriers are used for telegraphy. Telegraph or record communications encompass Morse, teletype, data, and similar signaling and require efficient and accurate service. The mutilation of a single signal element will produce a character mutilation and could change the meaning of a word or message. Therefore, radiotelegraph circuits require very careful system engineering to minimize the possibility of introducing mutilations.

Care must be taken to ensure that there are no signal drop-outs or fill-ins or dangerously high levels of distortion. Telegraphy lends itself to the efficient introduction of regeneration of the signal elements and to the restriction of the bandwidth required to that needed to pass the rate of keying involved. Automatic error-correction equipment is used extensively on high-frequency radio circuits operating at the international standard of 50 bauds. A circuit with an error of 1 character in 10^4 or better is considered a good commercial circuit, and if the errors rise to 1 in 10^3 characters, the circuit is very marginal.

26. Data Communications[13]

The amount of multipath limits the keying rate in telegraphy to a signal element length of 5-msec or more duration, which is equivalent to 200 bauds, but for the best accuracy it is preferable to key the individual frequency-shift-keyed circuits at about 100 bauds or less, which gives an individual signal element length of 10 msec or greater. Therefore, if the telegraph keying or data-communication rate requirement exceeds approximately 100 bauds, the signal stream must be broken down into several parallel paths each of the same speed. An example of this would be a 600-baud data signal which would be broken down into six 100-baud channels or as an alternate into eight 75-baud channels each spaced 170 cycles apart at midband. Each of these data channels would then have to be recombined at the receiving end into the serial form for transmission on the local wire line circuits to the customer. Such conversion should preferably incorporate regeneration of the signals. Error-detecting and self-correcting codes and automatic error-correction facilities are now coming into use for higher speed data communications on high-frequency radio.

In each subcarrier of the independent sideband transmission system the energy is contained in the subcarrier and the two keying sidebands which are spaced from the subcarrier by an amount equal to the keying frequency. Where long marks or spaces occur, the energy appears at the mark or space frequency near the edge of the channel. Some of this energy appears at low levels in the adjacent channels, where it is attenuated by the adjacent channel filter. Each subcarrier may fade in these high-frequency operations, and this can result in differences in levels in adjacent channels of 10 to 15 db. The 170-cycle midband spacing used for these subcarriers provides the opportunity to obtain the necessary attenuation so that crosstalk is minimized.

The system should be designed so that there will be no more errors than 1 in 10^4 bits or preferably 1 per 10^5 bits. It has been estimated that to obtain no more than 1 in 10^4 bit errors 90 per cent or more of the time on 90 per cent or more of the days requires on these high-frequency radio circuits 10 watts per cycle of bandwidth on the night frequencies. Less power is required for these data-communication circuits in the daytime.

27. Telephony

Independent or single-sideband radio transmitter/receiver systems are used for telephony today. This voice bandwidth service is provided on the basis of one to four channels per system. With the latter a 12-kc frequency bandwidth is required, which enables the operation of two voice channels each above and below the reduced carrier. Voice-channel shifters and restorers are used. These systems also incorporate privacy facilities of either a complex split-band type or in some cases a simple inversion privacy whereby the frequency order of the voice bandwidth is inverted so that the high frequencies are changed to the low frequencies at the transmitting end and reversed to their original frequency position at the receiving end.

It is not possible to regenerate telephone signals as is done with telegraphy, and consequently the signal-to-noise ratio of each section is very important. The desired signal must be sufficiently above the noise so that with appropriate thresholding techniques the signal can be amplified and passed along without amplifying the noise level at that point.

Since telephony is a two-way service and is operated on a half-duplex basis on landline systems, special arrangements are required in interconnecting such a system with the full-duplex system represented by the radio paths where the outbound and inbound radio systems are independent of each other. Echo-suppressing and automatic gain-controlled amplifiers are essential elements in these systems.

Space diversity has not shown the spectacular advantages for radio telephony as it has for radio telegraphy, and consequently its use in radio telephony has been very limited.

28. Facsimile and Radiophoto[1, 62−65, 70]

The end product of the transmission of facsimile and radiophoto is visually observable, and problems with multipath, frequency instability, and interference are immediately evident. Since the greater portion of a voice bandwidth is required, the opportunity for degradation of the signal is greater than for the narrower band telegraphy. The engineering of a system for this service can therefore be very demanding.

The two principal systems for transmission of facsimile and radiophoto on high-frequency radio are by an audio-frequency shift (AFS) of an independent sideband subcarrier, which is also referred to as subcarrier frequency modulation, and by a radio-frequency shift (RFS) of a continuous-wave (CW) transmission.

The audio-frequency shift of a single-sideband subcarrier as shown in Fig. 42 covers the frequency range from 1,500 to 2,300 cps with the keying sidebands spilling out beyond these two frequency positions. The receiver bandwidth is about a standard voice channel wide, since it must accept the reduced carrier at its lower end and the top keying sidebands at the other end. More than one of these modulated voice bandwidths can be carried on one independent sideband system.

The radio-frequency shift system is shown in Fig. 43 with a CW transmitter, and this system has been found to have a performance advantage when noise and multipath are prevalent.

Double-sideband amplitude modulation is also used, but since it requires twice the transmitting bandwidth of an independent sideband transmission and consequently has twice the noise, its use is very limited now. In addition when the two sidebands in double-sideband amplitude transmission are subjected to selective frequency fading, then the demodulated signal at the receiving end will be distorted upon recombination.

International and national regulations are designed to eliminate its use sometime in the future.

In the design of the overall system for radiophoto and radiofacsimile service, care must be taken to ensure that there is a high order of freedom from noise, distortion, and interference and that the phase characteristics are reasonably flat across at least the working part of the voice bandwidth. It is generally desirable and necessary to provide phase equalization for this service.

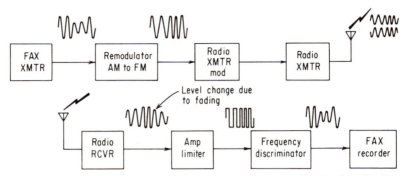

FIG. 42. Audio-frequency shift system using a voice frequency radio circuit. (*K. Henney, "Radio Engineering Handbook,"* 5th ed., *sec.* 23-34, *Fig.* 37, *McGraw-Hill Book Company, New York,* 1959.)

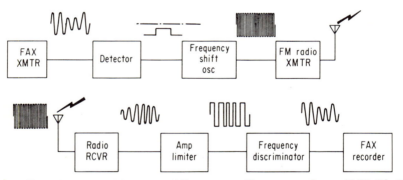

FIG. 43. General features of frequency-shift system.   (*Henney, op. cit., sec.* 23-36, *Fig.* 39.)

It is usually very desirable to provide alternate voice control so that, when the terminal facsimile units are being lined up at both ends of the circuit, it can be done by voice coordination over the voice bandwidth available for this service.

Normal bandwidth and keying speed limitations result in several minutes time required for sending the facsimile or photo transmission. Consequently, considerable effort has been devoted to the investigation of means for reducing the bandwidth or keying speed requirements. Variable speed and skip scanning techniques can result in some timesavings in facsimile, but these techniques have not yet been placed in service on high-frequency radio. Similar timesavings for radiophoto have not been possible.

In radiophoto transmissions there are often gradual changes from black to white as the picture passes through shades of gray. Consequently analogue modulation is preferable. Analogue modulation is also used for facsimile, but an overall improvement can be made in clarity and resolution if digital-modulation techniques are used instead.   The reason for this is that there is generally a very sharp change from black to white and vice versa with facsimile material and the digital modulation will make such a sharp distinction on the received copy.

Space diversity reception will provide some gain for facsimile and radiophoto, and the actual gain for two-receiver space diversity over a single-receiver operation is probably in the order of 3 db.

Table 12 gives the important specifications as established by the International Telegraph and Telephone Consultative Committee (CCITT) for various facsimile and radiophoto terminal equipment as used on high-frequency circuits and on the interconnecting landline systems.

Table 12. Facsimile-systems Standards*

The International Telegraph Consultative Committee (CCIT) has set up the following specifications for facsimile machines as standards to be used in international communications:

For photographs:

Cylinder circumference ($D \times \pi$)	8.66 in.
Min useful circumference	8.00 in.
Max skew	0.15 in.
Useful cylinder length	11.81 in.
Max picture size	8.07 × 11.65 in.
Drum factor	2.5 min
Line advance (scanning density)	96 and 127 lines per in.
Indexes of cooperation	264 and 350, respectively (830 and 1,100 IRE)
Stroke speed (rpm for drums)	90 and 60, respectively, and multiples and submultiples

For meteorological charts and large documents:

Index of cooperation (proposed)	576
Drum speed (recommended)	60 rpm

Proposed for business documents:

Index of cooperation	264
Drum diam†	2.6–2.95 in.
Min usable drum length of transmitters	11.8 in.

Drum speeds of 90, 120, 150, 180, 240, and 360 rpm have been proposed

* K. Henney, *Radio Engineering Handbook*, Sec. 23-38 on Facsimile Standards by A. G. Cooley, 5th ed., McGraw-Hill Book Company, New York, 1959.

† U.S. State Department proposes standard of 2.75 in.

29. Signal-to-Noise Ratios and Reliability[23, 29, 39, 68]

Full and accurate information is not available on the required signal-to-noise ratios for given degrees of reliability for the various communication services.   However, the CCIR has dealt with this in Recommendation 161, Volume I, of the IX Plenary Assembly in Los Angeles, 1959, and these data are shown in Table 13.

The required ratio of peak RF signal noise in the 6-kc/sec band of −2 db for a 50-baud teleprinter channel with F1 frequency shift telegraphy and a ratio of 26 db for a one-voice-channel single-sideband telephone system gives an indication of the relative signal-to-noise ratio comparison for the two services.   The teleprinter circuit with its much smaller receiver audio bandwidth of 100 cps requires considerably less signal than telephony with its 3,000-cps bandwidth.

Great care must be taken in interpreting any of these data, since many factors such as bandwidth at several points in the system, type of modulation, type of recording, etc., are all involved.   In high-frequency radio communications there is a great variation in circuit attenuation and atmospheric noise levels, and care must be taken to design systems with sufficient safety factors so that the degree of circuit reliability remains at a high order over appreciable periods of time.   At times specific tests using calibrated transmitters, antennas, and receivers are the best way to determine what

Table 13. Signal-to-Noise Ratios Required*
(Stable conditions, note 6)

Type of service	Receiver audio bandwidth, kc/sec	Audio signal-to-noise ratio, db	Receiver bandwidth, kc/sec	Ratio of peak RF signal-noise in 6 kc/sec band (Note 1), db
A1 telegraphy:				
8-baud low grade	1.5	−4	3	−7
24-baud	1.5	11	3	8
120-baud recorder	0.6	10	0.6	0
50-baud printer	0.25	16	0.25	2
A2 telegraphy:				
8-baud low grade	1.5	−4	3	−3 (Note 2)
24-baud	1.5	11	3	12 (Note 2)
F1 frequency-shift telegraphy:				
120-baud recorder	0.25	4	1.5	2
50-baud printer	0.10	10	1.5	−2
Phototelegraphy F4:				
Subcarrier frequency-modulation single-sideband emission	3	15	3	12
Hellschreiber, frequency-shift	1.5	6	3	3
A3 telephony:		(Note 3)		
Double-sideband, just usable quality, operator to operator (Note 4)	3	6	6	18
Double-standard, marginally commercial (Note 5)	3	15	6	27
Double-sideband, good commercial quality (Note 5)	3	33	6	35†
Single-sideband:				
1 channel	3	33	3	26†
2 channels	3	33	3‡	28†
3 channels	3	33	3‡	29†
4 channels	3	33	3‡	30†
Broadcasting	5	33	10	47

* *CCIR Document of the IXth Plenary Assembly Los Angeles* 1959, vol. VI, Recommendation 161, p. 120, Geneva, 1959.

† Assuming 10-db improvement due to the use of noise reducers.

‡ Per channel.

Note 1: Measured as the ratio of r.m.s. signal corresponding to peak output of the transmitter and r.m.s. noise in a 6 kc/s band, assuming stable conditions.

Note 2: Carrier keyed. Beat-frequency oscillator used.

Note 3: For A3 telephony the figures in this column represent the ratio of the audio signal as measured on a standard vu meter, to rms noise for a bandwidth of 3 kc/sec. (The corresponding peak signal power, i.e., when the transmitter is 100 per cent tone-modulated, is assumed to be 6 db higher.)

Note 4: For 90 per cent intelligibility of unrelated words.

Note 5: When connected to the public service network.

Note 6: These values are based on Washington Doc. No. 138, Doc. No. 112 of Geneva, 1951, and The Hague Doc. No. 11.

might be expected on a circuit, and these data together with the general information in Table 13 can give the designer some guidance for the particular circuit under consideration.

Typical commercial telegraphy reliabilities are shown in Table 14 for several circuits at various periods of the sunspot cycle. These data should not be used for guidance for design of a circuit without having all details as to transmitter power, antenna gains, receiver types and gain, transmission-line attenuation, type of modulation, bandwidths

Table 14*

[The following table shows the *averaged* reliability of four New York–Central European circuits (Paris, Berne, Brussels, and Frankfurt), the reliability of the New York–Tangier circuit, and the reliability of the New York–Buenos Aires circuit for the intermediate (downward), minimum, and maximum part of the sunspot cycle for March, June, September, and December of 1952, 1954, 1957, 1958 along with the monthly sunspot number.]

	Central Europe	Tangier	Baires	Sunspot Number	
1952:					
March......	58 %†	80 %	95 %	39	
June........	94 %	99 %	96 %	15	Intermediate
September...	76 %	92 %	98 %	27	
December...	85 %	94 %	99 %	26	
1954:					
March......	82 %	93 %	97 %	10	
June........	97 %	99 %	96 %	0	Low
September...	84 %	95 %	96 %	2	
December...	94 %	97 %	98 %	7	
1957:					
March......	96 %	95 %	94 %	167	
June........	94 %	98 %	90 %	205	Maximum
September‡..	84 %	87 %	85 %	244	
December...	96 %	96 %	95 %	233	
1958:					
March......	91 %	94 %	96 %	191	
June........	91 %	94 %	95 %	171	Maximum
September...	91 %	93 %	98 %	201	
December...	92 %	96 %	95 %	187	

 * J. H. Nelson, Circuit Reliability, Frequency Utilization, and Forecasting in the H.F. Communication Band, in G. J. Gassmann (ed.), *The Effect of Disturbances of Solar Origin on Communications*, Symposium of the Ionospheric Research Committee AGARD AVIONICS Panel, Naples, Italy, Pergamon Press, New York, 1963.

 † March, 1952, was a severely disturbed month. The bulk of night-hour traffic was relayed via Tangier, which increased effective reliability considerably.

 ‡ September, 1957, had three very severe disturbances.

used, etc. It does, however, give some indication of the range of variation on typical commercial circuits with given characteristics over a period of time.

CHANNELING

30. Frequency Division[2,39]

The world-wide assignment of high frequencies carries a stipulation as to the allowable bandwidth. These bandwidths can vary up to about 12 kc/sec. Assignments below the 3-kc bandwidth can be used for telegraphy and data communications, while assignments from the 3- to 12-kc bandwidth can be used for telephony or with further frequency division for telegraphy and data communications as well.

A 3-, 6- or 12-kc bandwidth can be used for one-channel, two-channel, or four-channel telephony on an independent sideband system. A two-channel system can be carried with one voice channel on each side of the reduced carrier on an independent sideband system or on one side of a single-sideband system. The four-channel voice system would be operated with two channels on each side of the reduced carrier.

The voice channels can be further divided by frequency division techniques into several telegraph channels in the manner shown in Fig. 44. Each one of these single-sideband subcarriers can carry a keying speed of about 200 bauds. Another common frequency division technique is shown in Fig. 45, where the keying speed for each subcarrier is about 100 bauds.   In both these cases the frequency-shift-keying speed ratio or modulation index is about 2.

Each of the subcarriers in Fig. 44 can be used to carry a time-division multiplex system with automatic error correction capable of carrying four 50-baud teleprinter channels.   The time-division system is operated at about a 2 per cent slower speed or at an aggregate keying speed of 192 bauds, which is in accordance with CCITT standards.   Similarly, each of the subcarriers in Fig. 45 can be used to carry a time-division system with automatic error correction capable of carrying two 50-baud teleprinter channels. The aggregate keying speed in this case is 96 bauds.   These same subcarriers will handle data-communication signals of about the aggregate speeds indicated.

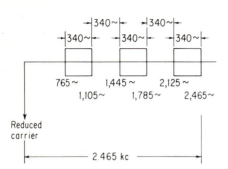

Fig. 44. Subcarrier arrangements with 192-baud keying on independent sideband transmitter.

The bandpass filter for the 200-baud system is shown in Table 15, and the bandpass filter for the 100-baud system is about one-fourth as wide as those shown in the table.

Since automatic frequency control is used, the bandwidth at the receiver can and should be narrower in order to reduce the possibility of harmful interference.   In a typical case the filter characteristics for 200 bauds and 340-cycle total shift are about as shown in Table 16.

A somewhat common two-channel system, twinplex or four frequency diplex, is provided by using an FSK signal with four positions of frequencies where each of the mark and space signaling on the two end channels is connected to one or the other of the four frequency positions available on the high-frequency circuit.   The assignment of these mark space signals on the two channels to the four radio-frequency positions is shown in Table 17.

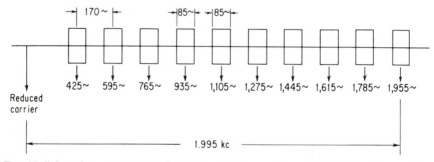

Fig. 45. Subcarrier arrangement with 96-baud keying on independent sideband transmitter.

These systems are generally operated with the two end channels synchronized so that the radio signal element length or keying speed is always the same as that for each individual channel.   If the single-channel speed is 50 bauds, then the radio-frequency aggregate keying speed for two channels is also 50 bauds.   This gives the advantage of minimizing the effect of multipath degradation but adds the complication of having to synchronize the individual channels.   It is also possible to place two separate two-channel time-division multiplex with automatic error correction (ARQ) or two four-channel time-division systems on this two-channel frequency shift system, thus providing four or eight 50-baud telegraph channels over one radio system.   However, it is essential that these multiplexes be synchronized with each other for satisfactory per-

Table 15. Designed for Approximately 200-baud Keying*

Frequency, either side of center frequency, cps	Attenuation above that of center frequency attenuation, db
170	1.5 + or −0.5
340	6.0 + or −0.5
510	13.5 + or −1.0
680	24.0 + 2.0 − 4
Above 680	Maximum practicable

* E. D. Becken, Single-sideband Operation for International Telegraph, *Proc. IRE*, vol. 44, Table I, p. 1785, December, 1956.

Table 16. Designed for Approximately 200-baud Keying and Total Frequency Shift of 340 Cps*

Attenuation relative to midband, db	Approximate frequency either side of center frequency, cps
40	485
30	455
20	385
10	345
5	305
2.5	265
0	95

* E. D. Becken, Single-sideband Operation for International Telegraph, *Proc. IRE*, vol. 44, Table II, p. 1785, December, 1956.

Table 17. Four-frequency Diplex System*

Frequency of emission†	Channel 1		Channel 2	
	Teleprinter	Morse	Teleprinter	Morse
f_4 (highest frequency)......	A‡	Mark	A	Mark
f_3........................	A	Mark	Z	Space
f_2........................	Z§	Space	A	Mark
f_1 (lowest frequency).......	Z	Space	Z	Space

* C.C.I.R. *Documents of the IXth Plenary Assembly, Los Angeles*, 1959, vol. 1, p. 141, Rec. 247, Geneva, 1959.

† f_1, f_2, f_3, f_4 designate the frequencies of the emissions, the spacings between adjacent frequencies $(f_4 − f_3)$, $(f_3 − f_2)$, $(f_2 − f_1)$ being equal.

‡ A represents the start signal of the teleprinter.

§ Z represents the stop signal of the teleprinter.

formance on the radio path; otherwise some radio signal element lengths can be very short and susceptible to multipath interference.

31. Time Division[12]

High-frequency radio telegraphy makes extensive use of time-division multiplex. Both two-channel and four-channel systems are in use for the 50-baud standard telegraph keying used in the international communication networks. Most of these multiplexes operate with a slightly lower aggregate speed than the number of channels indicates for reasons of phasing in of the local or end channels with the synchronous operation of the multiplex. Consequently, the aggregate keying speed of a two-channel multiplex is 96 bauds and a four-channel MUX is 192 bauds. For those systems, particularly in the United States, that are still operated with a terminal printer speed of 45.5 bauds per channel, the two-channel aggregate speed is 85⅚ bauds and the four-channel aggregate speed is 171¾ bauds. The vast majority of these time-division multiplexes in use on high-frequency radio circuits incorporates ARQ. These systems are carried either on individual standard frequency-shift-keying transmitters or on subcarriers of independent sideband transmitters. With the latter there is a strong trend to use two-channel subcarrier systems, and it is not uncommon to have 15 of these two-channel systems on one independent sideband system for a total capacity of thirty 50-baud telegraph channels.

This ARQ equipment employs the seven-unit Moore constant ratio code. Since it is a synchronous system, the start-stop elements are eliminated and the five remaining signal elements of the standard teleprinter or Baudot code are converted automatically into a seven-element code having always three marking and four spacing elements. Any disruption of this 3:4 constant ratio will cause the ARQ equipment at the receiving

Table 18. Improvement Ratio and Net Channel Speed with Automatic Error Correction (ARQ)*

Net channel speed, words/min	Signal mutilation rate, characters	Divide by improvement factor of	Probability of transpositions, characters†
60 nom	1/1,000	10,000	1/10,000,000
55	25/1,000	400	1/16,000
50	50/1,000	200	1/4,000
40	150/1,000	70	2/1,000
30	240/1,000	40	6/1,000
20	350/1,000	30	12/1,000

* J. B. Moore, Constant-ratio Code and Automatic-RQ on Transoceanic HF Radio Services, *IRE Trans. Commun. Systems*, vol. CS-8, no. 1, Table II, p. 75, March, 1960.
† Tabulated values are indicative only.

end to request a repetition of the signaling automatically until a proper ratio signal is received. In this manner the vagaries of the propagation path may slow down the overall rate of communications, but a very high percentage of only good signals is accepted and passed along. The rate of improvement of overall performance of an ARQ protected circuit is shown in Table 18. The possibility of a double mutilation, called a transposition, resulting in an incorrect but still valid 3:4 ratio signal is slight and is shown in the final column of this table. The second column gives the signal-mutilation rate in characters on a radio circuit without ARQ, and the figures in the first column show the corresponding net channel speed in words per minute with the ARQ functioning. The third column shows the improvement factor with ARQ when the signal-mutilation rate without ARQ is as shown in the second column.

The time-division multiplex with ARQ also lends itself readily to the provision of synchronous subdivided channels at one-half or one-fourth speeds, and auxiliary terminal devices called subdividers are used extensively for this purpose. Again, these subchannels are protected by ARQ from the mutilations of the high-frequency radio circuit.

32. Phase Division

Channeling of the high-frequency radio path can also be provided by using terminal equipment sensitive to the phase of the keyed signal. The most common of this equipment makes use of four different phase positions, phase quadrature, for providing four channels. This principle can also be used on a two-phase basis. All are of the synchronous type.

CODING [12,14,15,33,51,52,58,61]

A few of the world's high-frequency point-to-point telecommunication circuits and the majority of the marine ship-to-shore operations still employ the Morse code for telegraphy. This code shown in Fig. 46 is characterized by signal elements of unequal length. Furthermore, the keying rate is very apt to be variable, depending on the requirements of the manual transmitting or receiving operator. This variable keying rate does not lend itself to the use of regeneration.

Character	Code	Character	Code	Character	Code
A	· —	1	· — — — —	Ü	· · — —
B	— · · ·	2	· · — — —	(OR)	— · — — · —
C	— · — ·	3	· · · — —	"	· — · · — ·
D	— · ·	4	· · · · —	—	· · — — · —
E	·	5	· · · · ·	=	— · · · —
F	· · — ·	6	— · · · ·	S O S	· · · — — — · · ·
G	— — ·	7	— — · · ·	Attention	— · — · —
H	· · · ·	8	— — — · ·	CQ	— · — · — — · —
I	· ·	9	— — — — ·	DE	— · · ·
J	· — — —	0	— — — — —	Go ahead	— · —
K	— · —			Wait	· — · · ·
L	· — · ·	·	· · · · · ·	Break	— · · · —
M	— —	;	— · — · — ·	Understand	· · · — ·
N	— ·	,	· — · — · —	Error	· · · · · · · ·
O	— — —	:	— — — · · ·	OK	· — ·
P	· — — ·	?	· · — — · ·	End message	· — · — ·
Q	— — · —	!	— — · · — —	End of work	· · · — · —
R	· — ·	'	· — — — — ·		
S	· · ·	-	— · · · · —		
T	—	/	— · · — ·		
U	· · —	Ā	· — · —		
V	· · · —	Á or À	· — — · —		
W	· — —	É	· · — · ·		
X	— · · —	CH	— — — —		
Y	— · — —	Ñ	— — · — —		
Z	— — · ·	Ö	— — — ·		

FIG. 46. International Morse code.

Most of the point-to-point high-frequency radio circuits now use the five-unit Baudot code which is converted to the Moore seven-unit constant ratio code as described in the section on time-division channeling. The five-unit code and its seven-unit counterpart on the radio circuit operated with automatic error correction are shown in Fig. 47. The five-unit Baudot code employs a start element and five-signal elements of equal length and a stop element generally of one or one and one-half

Five-unit Baudot code

Signal Nr.	1	2	3	4	5	6	7	8	9	10	11	12	13	14	15	16	17	18	19	20	21	22	23	24	25	26	27	28	29	30	31	32
FIGURES	−	?	:	✛	3	▢	⊟	⊘	8	BELL	(	)	.	,	9	0	1	4	'	5	7	=	2	/	6	+	CARRIAGE RETURN	LINE FEED	LETTERS	FIGURES	SPACE	BLANK
LETTERS	A	B	C	D	E	F	G	H	I	J	K	L	M	N	O	P	Q	R	S	T	U	V	W	X	Y	Z						
1	1	1		1	1	1	1			1	1			1		1		1		1		1	1	1	1				1	1		
2	2		2			2			2	2	2	2			2	2	2			2	2	2							2	2	2	
FEED HOLE	o	o	o	o	o	o	o	o	o	o	o	o	o	o	o	o	o	o	o	o	o	o	o	o	o	o	o	o	o	o	o	o
3			3			3		3	3		3		3	3		3	3		3		3	3		3	3				3		3	
4		4	4	4		4	4			4	4		4	4	4		4			4		4			4			4	4			
5		5					5	5			5	5		5	5	5			5		5	5	5	5	5				5	5		

some D keytops read "WHO ARE YOU" in U.C.; others are blank. F,G,H keytops are blank in U C

Seven-unit Moore constant-ratio code

FIGURES	−	?	:	✛	3	▢	⊟	⊘	8	BELL	(	)	.	,	9	0	1	4	'	5	7	=	2	/	6	+	CARRIAGE RETURN	LINE FEED	LETTERS	FIGURES	SPACE	BLANK	SIGNAL ∑	IDLE α	IDLE β
LETTERS	A	B	C	D	E	F	G	H	I	J	K	L	M	N	O	P	Q	R	S	T	U	V	W	X	Y	Z									
1			1				1	1	1		1	1	1	1		1		1		1				1	1		1								
2				2		2		2	2		2				2	2		2		2			2				2	2			2	2	2		
3	3	3		3	3	3		3	3			3	3				3		3	3	3	3					3						3		
FEED HOLE	o	o	o	o	o	o	o	o	o	o	o	o	o	o	o	o	o	o	o	o	o	o	o	o	o	o	o	o	o	o	o	o	o	o	o
4	4	4	4	4	4				4				4	4		4		4					4	4	4		4				4	4			
5		5	5							5	5		5	5		5			5	5	5		5				5	5		5	5		5		
6	6				6		6		6	6	6			6	6			6			6		6				6	6	6						
7		7			7	7			7	7			7			7			7		7	7		7	7	7					7		7		

FIG. 47. Five-unit Baudot and seven-unit Moore constant-ratio code.

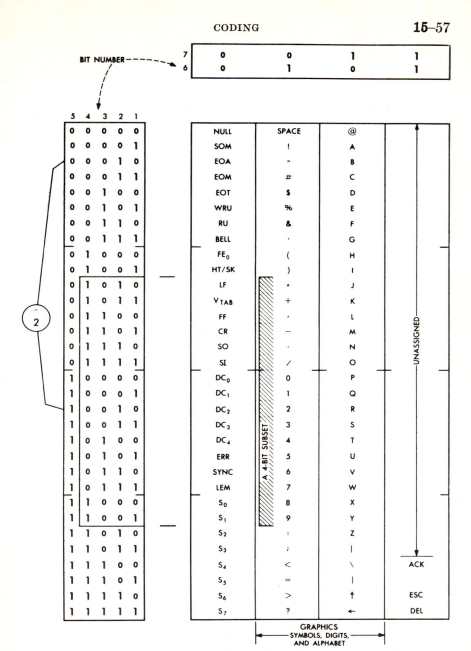

FIG. 48. The new ASCII code shown in binary order. The extreme left of the drawing shows the five lower-order bits. At top are the two higher-order bits. The symbol (″) appears on the same key as 2 by a "Shift" of bit 5. (*Reprinted by permission of the copyright owner, Bell Telephone Laboratories, Inc., and the author, J. F. Auwaerter. This article originally appeared in the Bell Lab. Record, pp. 397–398, November, 1963. Drawing p. 397.*)

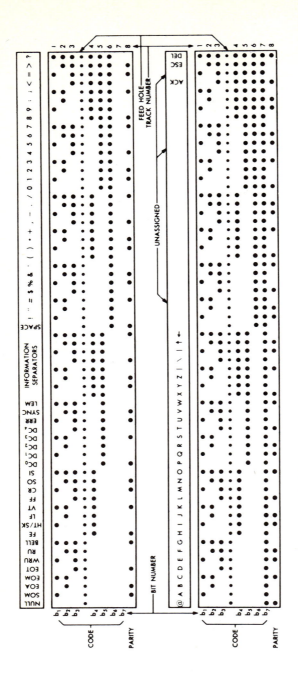

FIG. 49. A rendering of the eight-level teletypewriter tape. Present Bell System usage is to make the eighth track always marking. Optional vertical parity in the eighth track will accord with this drawing. (Reprinted by permission of the copyright owner, Bell Telephone Laboratories, Inc., and the author, J. F. Auwaerter. This article originally appeared in the Bell Lab. Record, pp. 397–398, November, 1963. Drawing p. 398.)

15–58

element length.   Solo radio circuits operated on such a basis often employ five-unit start-stop regeneration for the 20-msec signal elements involved.

The multiplex automatic error-correction equipment provides the conversion from the five-unit signaling on the local side to the seven-unit signaling on the radio side. The multiplex itself provides the synchronous regeneration on these circuits.

A six-unit start-stop code is used for teletypesetting and a few other special purposes but has had limited usefulness even though it offers more combinations (64) than the 32 for the five-unit code and the 35 for the seven-unit balanced ratio code.

The growth of data communications has required the development and use of other codes.   For example, terminal equipment is available for reading punched card information and transmitting the data in eight-unit code form with a unity start element and 1.5 length stop element for a total of 10.5 elements per character.   The eight signal elements are assigned so that there are always four marking and four spacing.   This makes it possible to detect errors at the receiving terminal which then can result in an error indication on the return path which stops the sending position for manual attention.   Start-stop regenerative repeaters are usually essential on these transmissions on high-frequency circuits.   Keying is at the rate of 3 to 5 cards per minute per channel with one type of terminal equipment and 11 cards per channel with another type of terminal equipment.   The former speed is equivalent to about 60 to 100 words per minute, and the latter to about 240 words per minute or 180 bauds.

Another newer code for teletypewriters uses seven elements based upon the American Standard Code for Information Interchange (ASCII).   It is based on the binary logic of computers and is shown in Fig. 48.   This code with an eighth bit for parity is shown in Fig. 49.   This code has a 128-character capacity and with a unity start element and a two-signal element length stop element for 11 signal elements per character.   The eighth bit is made a marking bit and is added only when needed to make the total number of marking bits an even number.   These parity bits can detect a single error in the transmission but not a double one.

Other codes under consideration for use on high-frequency circuits employ multiple parity checks which consist of the addition of more than one check bit.   With these check bits it is possible to detect single or multiple errors.   All these special codes involve varying degrees of redundancy and consequently varying degrees of protection on the radio path.

TRANSMITTERS

33. General[70]

The general design characteristics of a radio transmitter are dictated in part by consideration of other elements in the transmission circuit.   System engineering must consider the transmitting frequency range, attenuation of the radio propagation path and transmission lines, antenna gains and speed, complexity, and accuracy requirements of the keying signal.

A short, easy circuit for daytime operation only suggests the use of a low-power transmitter, and if the capacity requirement is also low, a simple, relatively inexpensive transmitter would be sufficient.   On the other hand, if the radio propagation path is long and difficult and the load requirement is high, then the transmitter will require high power, wide operating frequency range, and multichannel capability.

Telegraphy alone or with telephony will have further bearing on the nature of the transmitter.   If there is sufficient site room for large high-gain antennas, the power gains of the transmitter can be more moderate.   Similarly, the antenna and transmission-line considerations at the receiving site and the gain and noise factors of the receiver are all factors.

Standard transmitter modulation symbols are shown in Table 19.   Frequency tolerance for high-frequency radiation for power above 500 watts is 15 parts/10^6 for the 4,000 to 29,700 band per CCIR recommendations.

Table 19. Radio Transmitters Modulation Symbols*

Type of modulation of emission	Type of transmission	Supplementary characteristics	Symbol
1. Amplitude	Absence of any modulation	. . . .	A0
	Telegraphy without the use of modulating audio frequency (on-off keying)	. . . .	A1
	Telegraphy by the keying of a modulating audio frequency or frequencies or by the keying of the modulated emission (special case: an unkeyed modulated emission)	. . . .	A2
	Telephony	Double sideband, full carrier	A3
		Single sideband, reduced carrier	A3a
		Two independent sidebands, reduced carrier	A3b
	Facsimile	. . . .	A4
	Television	. . . .	A5
	Composite transmissions and cases not covered by above	. . . .	A9
	Composite transmissions	Reduced carrier	A9c
2. Frequency (or phase) modulated	Absence of any modulation	. . . .	F0
	Telegraphy without the use of modulating audio frequency (frequency-shift keying)	. . . .	F1
	Telegraphy by the keying of a modulating audio frequency or frequencies or by the keying of the modulated emission (special case: an unkeyed emission modulated by audio frequency)	. . . .	F2
	Telephony	. . . .	F3
	Facsimile	. . . .	F4
	Television	. . . .	F5
	Composite transmissions and cases not covered by the above	. . . .	F9
3. Pulsed emissions	Absence of any modulation intended to carry information	. . . .	P0
	Telegraphy without the use of modulating audio frequency	. . . .	P1
	Telegraphy by the keying of a modulating audio frequency or frequencies, or by the keying of the modulated pulse (special case: an unkeyed modulated pulse)	Audio frequency or frequencies modulating the pulse in amplitude	P2d
		Audio frequency or frequencies modulating the width of the pulse	P2e
		Audio frequency or frequencies modulating the phase (or position) of the pulse	P2f
	Telephony	Amplitude-modulated pulse	P3d
		Width-modulated pulse	P3e
		Phase (or position) modulated pulse	P3f
	Composite transmissions and cases not covered by the above	. . . .	P9

* Warren B. Bruene, Radio Transmitters, in K. Henney (ed.), *Radio Engineering Handbook*, 5th ed., Sec. 18-31, Table 1, McGraw-Hill Book Company, New York, 1959.

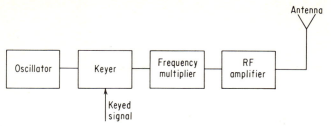

FIG. 50. Frequency-shift-keyed (FSK) transmitter.

34. Frequency-shift-keyed (FSK) Transmitter[59]

The simplest transmitter is the relatively low-capacity frequency-shift-keyed transmitter for telegraphy. With this type the continuous carrier is shifted a total of about 400 cycles in accordance with the marking and spacing elements of the keying signal. Since the radio propagation path is often subjected to multipath, the keying rate is such that the signal elements should not be shorter than 5 or 6 msec. If 50-baud telegraph signals are used to key the transmitter directly, the signal elements are of 20-msec duration, and if four-channel time-division multiplex is used, the keying aggregate is at the CCIR speed of 192 bauds.

The system elements of an FSK transmitter are shown in Fig. 50. The bandwidth of such a system is given by the following approximate formula:

$$fw = 2(f_d + 3f_k)$$

Table 20. Estimating Bandwidth*

$\dfrac{\Delta F}{f}$, radians†	Carrier frequency and successive significant side currents expressed in percentage of the unmodulated carrier level I_m‡	Required band-width	Maximum equivalent phase shift, deg, $\Delta\theta = 57.3\dfrac{\Delta F}{f}$
0.4	96.04, 19.6	$2f$	22.92
0.5	93.85, 24.23, 3.1	$4f$	28.7
1.0	76.52, 44.01, 11.49, 1.96	$6f$	57.3
2.0	22.39, 57.67, 35.28, 12.89, 3.4	$8f$	114.6
3.0	26.01, 33.91, 48.61, 30.91, 13.2, 4.3, 1.14	$12f$	171.9
4.0	39.71, 6.6, 36.41, 43.02, 28.11, 13.21, 4.91, 1.52	$14f$	229.2
5.0	17.76, 32.76, 4.66, 36.48, 39.12, 26.11, 13.1, 5.34, 1.84	$16f$	287
6.0	15.06, 27.67, 24.29, 11.48, 35.76, 36.21, 24.58, 12.96, 5.653, 2.12	$18f$	343.8
7.0	30.01, 0.5, 30.14, 16.76, 15.78, 34.79, 33.92, 23.36, 12.80, 5.9, 2.3, 0.8	$22f$	401.1
8.0	17.17, 23.46, 11.3, 29.11, 10.54, 18.58, 33.76, 32.06, 22.35, 12.63, 6.1, 2.6, 0.96	$24f$	485.4

* A. Hund, *Frequency Modulation*, 1st ed., p. 33, McGraw-Hill Book Company, New York, 1942.
† Holds also for $\beta = \Delta\theta$.
‡ First number denotes the percentage amplitude of carrier frequency F; second number is the percentage amplitude for side frequencies $F - f$ and $F + f$; the third number holds for frequencies $F \mp 2f$, etc.
$\Delta F = \frac{1}{2}$ carrier shift
f = keying frequency, cps

where fw = bandwidth required

f_d = one-half total shift (frequency deviation = one-half total shift)

f_k = fundamental keying speed, cps

Bandwidth can be determined more accurately from Table 20, where ΔF is one-half carrier shift and f is the signal frequency in cycles per second. A 50-baud signal is equivalent to 25 cps, since a cycle corresponds to two signal elements per second or 2 bauds. This table is based on Bessel functions, and it is to be noted that the energy distribution for the higher modulation index, that is, $\Delta F/f$ varies up and down for the discrete frequencies on each side of the carrier frequency.

An example of such a calculation using Table 20 would be for a total shift of 400 cycles and a 50-baud keying signal, $\Delta F = 200$ and $f = 25$ cps,

then
$$\frac{\Delta F}{f} = \frac{200}{25} = 8 \qquad \text{modulation index}$$

and the bandwidth from the table would be $24f$ or 600 cycles.

Frequency-shift-keying transmitters vary in power from about 200 watts to 50 kw, but the general range for commercial telegraphy would be about 1 to 40 kw. The shift should be kept within ± 3 per cent of its nominal value.

35. Double-sideband Transmitter

Continuous-wave transmitters with audio modulation for telephony result in two sidebands each carrying the intelligence. Since these double-sideband transmitters require a bandwidth of at least 6 kc for a 3-kc voice bandwidth, they are inefficient and have been and are being replaced by single-sideband facilities. The peak power is four times the power of the unmodulated carrier.

36. Independent or Single-sideband Transmitter[3–5,19,60]

Independent or single-sideband transmitters with their efficient frequency utilization are used extensively in both telegraphy and telephony. They can be arranged so that either one or both of the sidebands are utilized, each independent of the other.

An independent sideband system lends itself to the use of several subcarriers for telegraphy as described under the frequency division section of channeling. Since each subcarrier is shifted in frequency, its own bandwidth requirements and energy distribution are given by the Bessel function data in Table 20. If a two-channel time-division multiplex with an aggregate keying rate of 96 bauds (48 cps) is carried on a subcarrier with an 80-cycle total shift, the modulation index is less than 1, and since the bandwidth is limited to about 150 cycles, it will pass the carrier and second-order harmonic but these carry most of the energy.

A simple functional block diagram of an independent sideband transmitter is shown in Fig. 51. Independent sideband transmitters are rated in peak envelope power, which for long-distance high-frequency communications generally varies from about 1 to 30 kw. The peak power rises as the square of the number of subcarriers when there is a fixed bandwidth per tone channel. Peak power is defined in the American Standard Definitions Of Electrical Terms as follows.* "In a modulated carrier system, the peak power output is the output power, averaged over a carrier cycle, at the maximum amplitude which can occur with any combination of signals to be transmitted."

Independent sideband transmitters are being used almost exclusively in new installations for carrying telephony largely because of their efficient utilization of the available bandwidth. Systems with two voice channels each on two independent sidebands are in extensive use. These systems require the use of channel shifters and channel restorers in order to place the second voice channel above the first in frequency sequence. This technique is shown in block diagram in Fig. 52.†

* Reference 54, p. 76, definition 65.32.144.

† Reference 39, p. 32.

"The two-tone method of rating the power of single-sideband radiotelephone transmitters consists in setting the level of each of two equal tones applied to the audio-frequency input so that the resulting cross-modulation term $(2f_1 - f_2)$ is 25 db below the level of either tone, measured in the r.f. output of the transmitter; the peak power rating of the transmitter is taken as four times the r.f. power output, after removal of one of the two tones.   Single channel speech is applied at the audio-frequency input at a VU level equal numerically to the mean dbm level of one of the two aforementioned tones.   For multichannel single-sideband transmission, the level of each channel is reduced 0.5 $(N - 1)$ db, where N is the number of channels, up to a total of about four."

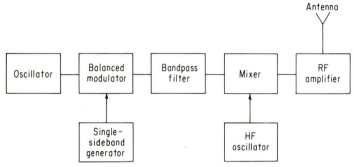

FIG. 51. Independent sideband transmitter.

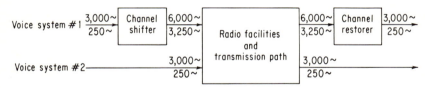

FIG. 52. Channel shifters and channel restorers.

RECEIVERS

37. Superheterodyne Receiver[3]

Superheterodyne receivers are used to receive FSK telegraph transmission.   In these the local oscillator voltage is combined with the signal and the converted new signal has the same modulation as the original keyed signal but at an intermediate frequency.   A functional block diagram of a superheterodyne receiver is shown in Fig. 53.   The audio-frequency bandwidth has to be limited to the necessary bandwidth for the keying involved.   Two-set space diversity and sometimes three-set space diversity are used.

The noise factor of a receiver gives an overall system indication of the efficiency of the receiver.   It is defined in the American Standard Definitions Of Electrical Terms as* "The noise figure of a transducer is the ratio of (1) the output-noise power to (2) the portion thereof attributable to thermal noise in the input termination at standard noise temperature (290°K).   The noise figure is thus the ratio of actual output noise to that which would remain if the transducer itself were made noiseless.

"Note: In heterodyne systems, output noise power (1) includes spurious contributions from image-frequency transformations, but portion (2) includes only that which

* Reference 54, p. 37, definition 65.08.183.

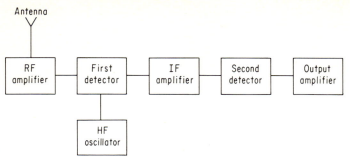

Fig. 53. Superheterodyne radio receiver.

appears in the output via the principal frequency transformation of the system and excludes that which appears via the image frequency transformation.

"The term *noise figure* is sometimes used synonymously with *average noise figure*, and sometimes synonymously with *spot noise figure.*

"Since the noise figure is a function of both the magnitude and phase of the output impedance of the input circuit, and of the frequency, a complete specification requires that these conditions be stated.

"A transducer developing no noise within itself has a noise figure of unity (zero db). Increasing the internally generated noise increases the noise figure."

Typical noise factor levels for high-frequency radio receivers would vary from about 4 to 10 db.

These receivers with suitable IF and AF bandwidth are also used for telephony.

38. Independent or Single-sideband Receivers

Independent sideband receivers are in extensive use in the high-frequency communication service largely again as a result of the high order of bandwidth efficiency. They are used for both telegraphy and telephony. A functional block diagram is shown in Fig. 54 for a dual space diversity system. A typical noise factor for such a system would be 6 db.

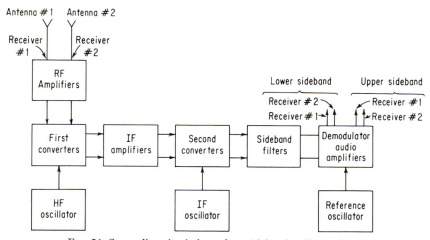

Fig. 54. Space diversity independent sideband radio receiver.

An SSB receiver has a 9-db advantage over an equivalent DSB receiver because with SSB the radiated sideband amplitude is equal to the 100 per cent modulated carrier of the DSB receiver and the AF amplitude is therefore twice that for the double sideband, giving a gain of 6 db. Since the SSB uses half the bandwidth, this gives another 3 db or a possible total gain of 9 db.

MONITORING AND MAINTENANCE[39,43]

The provision and proper utilization of monitoring facilities are very productive of results. They often can mean the difference between a good circuit and a poor or failing circuit even though the circuit itself was carefully engineered. Any long-distance high-frequency network involves many elements, and the failure of any one going unnoticed or unidentified can be disastrous. The availability and staffing of an appropriately designed monitoring system cost only a fraction of the circuit costs, and yet they can make a profound difference in overall performance.

There are a number of basic monitoring requirements which must be met at least in a simple way. In a larger station or facility it is important to plan and arrange these basic facilities for maximum efficiency. If the main equipments such as transmitters are operated on a remote-control basis, additional monitoring facilities must be provided.

Some of the monitoring equipment is related specifically to the equipment, such as the alarms for indicating failure in the high-voltage supply of the transmitter, etc. Others have a direct relationship to the performance and quality of the signal passing through the facility. An example of the latter would be a "carrier-off" alarm which rings when there has been an interruption to the radio-frequency carrier.

The provision of alarms can be overdone, so that the operating staff cannot cope with all the ringing bells and alarms. Rather than have carrier alarms or their equivalent at all points in the system, many of which might alarm at the same time, fewer such indicators strategically located in the system could result in quicker operator attention and maintenance results. Even though automatically activated alarms are not required at all points, it is important that provision be made so that the performance of the system can be checked with ease and speed at all equipment points. Such a multiplicity of check points is essential in identifying the nature and location of the problem. Appropriate portable or rack-mounted meters and test instruments with plug-in jack features are necessary.

Some of the basic monitoring facilities that must be included are as follows:

Monitoring facility	*Location*
1. Transmitter carrier alarm	Transmitting station
2. Transmitter RF signal quality indicator	Transmitting station
3. Transmitting transmission-line standing-wave indicator	Transmitting station
4. Transmitter frequency	Transmitting station
5. Frequency-shift stability	Transmitting station
6. Independent sideband subcarrier frequencies and levels	Transmitting station
7. Connecting-line carrier-off alarm	On lines to transmitting and receiving stations
8. Power-supply failure	All locations
9. Equipment fire alarms	All locations
10. Distortion monitoring	All locations
11. Oscilloscopes	All locations
12. Frequency counters	All locations
13. Signal level meters	All locations

While the ability to monitor and test locally each and every equipment in the circuits is necessary, it is also essential to concentrate signal monitoring at consoles or their equivalent in the major locations. This will be of considerable assistance in locating and identifying the source of the difficulty. The console operator should be able by key selection or plug and jack arrangement to take measurements and observations at certain key points in the circuit.   Once the source of the trouble is identified, the console operator should be able by means of interfacility communication channels to refer the specific trouble to the maintenance staff responsible for the equipment for correction or substitution.

There is much current interest in more automatic and sophisticated monitoring units which would ride the circuit and call the attention of the operating staff when the signals deteriorated below a previously determined threshold.   An example of this is the facility which rings an alarm on a circuit in a central telegraph office when the rate of ARQ cycling reaches a predetermined level and is maintained there beyond a specific time.

Another example is an automatic distortion monitor on the high-frequency radio circuit at the receiving station which rings an alarm for the attention of the receiving-station attendant when the distortion reaches a certain predetermined level.

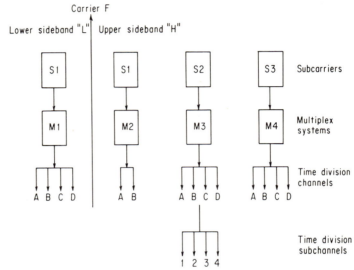

Fig. 55. Telegraph channeling designation on independent sideband transmitter.   (*CCIR Documents of the IXth Plenary Assembly, Los Angeles*, 1959, *Vol.* I, *Recommendations, p.* 143, *Recommendation* 248.)

The complexities of world-wide high-frequency radio communications operation have necessitated organization of various national and international monitoring stations, particularly in the matter of frequency stability.   Many nations of the world under the auspices of the International Telecommunication Union have joined together in systematic scanning and reporting on frequencies in operation.   These world-wide monitoring stations identify many cases of interference, and once identified, it is usually possible to correct or adjust for the frequency discrepancy.   These stations also make systematic field strength measurements and are able to suggest appropriate adjustment in power to alleviate interference which may be somewhat temporary in nature.

It is of considerable importance in both monitoring and operating for both ends of the circuit to identify with certainty the specific channel or channels affected. Consequently, standard channel numbering is recommended by the CCIR as outlined in VI Recommendation of the IXth Plenary Assembly at Los Angeles in 1959, and these are shown in Fig. 55 for telegraphy and Figs. 56 and 57 for four-channel and three-channel telephony.

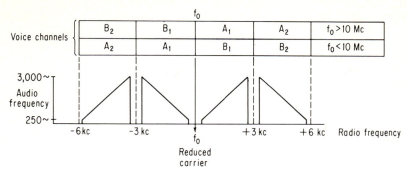

FIG. 56. Telephone channeling designations on independent sideband transmitters for four voice channels. (*CCIR Documents of the IXth Plenary Assembly, Los Angeles,* 1959, *Vol.* I, *Recommendations, p.* 145, *Recommendation* 249, *Fig.* 1.)

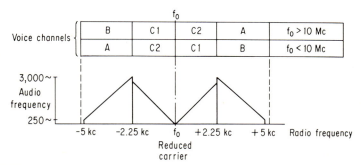

FIG. 57. Telephone channeling designation on independent sideband transmitters for three voice channels. (*CCIR Documents of the IXth Plenary Assembly, Los Angeles,* 1959, *Vol.* I, *Recommendations, p.* 146, *Recommendation* 249, *Fig.* 3.)

INTERNATIONAL ORGANIZATIONS

39. The International Telecommunication Union (ITU)[10,20,44-47]

An understanding of the system engineering problems affecting high-frequency radio communications would not be complete without a review of the international organization concerning itself with much of the engineering standardization, investigation, and recommendations involved. The International Telegraph Union was founded in Paris in 1865, and the initial membership included 20 European states. Since then its name has been changed to the International Telecommunication Union, and it has grown steadily until it now has about 124 member countries throughout the world. The Bureau of the Union is its permanent body, which is located in Switzerland and which takes care of the continuing, varied, and extensive administrative responsibilities. The organization of the ITU is shown in Fig. 58.

40. Consultative Committees of the ITU[39-41]

Most of the technical work of the Union has been handled by its consultative committees, which have advisory capacity only and no authority for executive action. However, the steady flow of recommendations from these consultative committees is generally ratified at the periodic Plenary Assemblies of the ITU. These recommendations are characterized by their thoroughness and by the great breadth and depth of the supporting studies.

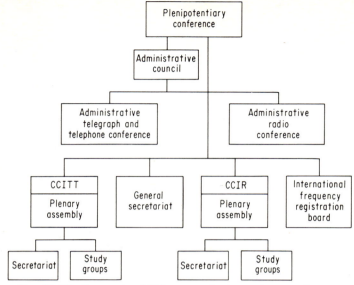

FIG. 58. The International Telecommunication Union organization.

The International Telegraph Consultative Committee, the CCIT, began in 1925 as did the International Telephone Consultative Committee (CCIF). A few years ago the CCIT and the CCIF were merged in the International Consultative Committee on Telegraph and Telephony (CCITT).

The growth of high-frequency radio activity, particularly in the long-distance communication field, led to the creation of another committee, the International Radio Consultative Committee (CCIR) in 1927.

These various consultative committees have the following active study groups at the present:

CCITT

Study group number	Name
I	Telegraph operating and tariffs
II	Telephone operating and tariffs
III	General tariff principles and rental of telecommunication circuits
IV	Maintenance of the general international telecommunication network
V	Protection against dangers and disturbances of electromagnetic origin
VI	Protection and specification of cable sheaths and poles
VII	Definitions, vocabulary, and symbols
VIII	Alphabetic telegraph apparatus
IX	Telegraph transmission performance and telegraph channels
X	Telegraph switching
XI	Telephone switching and signaling
XII	Telephone transmission performance and local telephone networks
XIII	Automatic and semiautomatic telephone networks
XIV	Apparatus and channels for facsimile telegraphy
XV	Transmission systems
XVI	Telephone circuits
SP.A.	Data transmission (special committee)

SP.B.	World-wide automatic and semiautomatic telephone network (special committee)
SP.C.	Noise (special committee, joint CCITT/CCIR committee)
CMTT	Television transmission over long distances (joint CCITT/CCIR(
PLAN	General development plan for the international network (joint CCITT/CCIR committee)
S-COM PLAN (AFRICA)	Plan subcommittee for Africa (joint CCITT/CCIR committee)
S-COM PLAN (ASIA)	Plan subcommittee for Asia (joint CCITT/CCIR committee)
S-COM PLAN (LATIN AMERICA)	Plan subcommittee for Latin America (joint CCITT/CCIR committee)
RIT (temporary)	Committee for the Inter-American telecommunication network (joint CCITT/CCIR committee)

CCIR

Study group number	Description of area of interest
I	Transmitters
II	Receivers
III	Fixed service systems
IV	Space systems
V	Propagation, including the effects of the earth and the troposphere
VI	Ionospheric propagation
VII	Standard frequencies and time signals
VIII	International monitoring
IX	Radio-relay systems
X	Broadcasting
XI	Television
XII	Tropical broadcasting
XIII	Mobile services
XIV	Vocabulary

When the work of the CCITT and the CCIR overlaps, it is closely coordinated and specific questions for study are often referred by one committee to the other.

An examination of the recent documents released by these consultative committees will provide much useful and current information on the state of the art. These study groups of the consultative committees perform an essential work of screening, recording, and reporting on the extensive work that is being done by many operating administrations and research and development laboratories throughout the world.

41. The International Frequency Registration Board (IFRB) of the ITU[69]

The Administrative Radio Conference at Atlantic City in 1947 authorized the establishment of an International Frequency Registration Board (IFRB). The latter is an international administrative body to administer matters pertaining to the notification, registration, and use of frequencies internationally. Its principal purpose is to regularize the recording of frequency assignments and use and to assist in the elimination of harmful radio-frequency interference on the world's radio communications circuits.

42. The International Scientific Radio Union (URSI)[8]

The International Scientific Radio Union was founded in 1913 in Brussels, Belgium, and now has a world-wide membership. It began as a study of means for controlling the frequency of radio transmissions and a study of the fluctuations of the received signals arriving over long paths. The years saw its membership grow and its breadth of activity broaden to take in most aspects of radio. Close cooperation exists between URSI and the consultative committees of the ITU.

REFERENCES

1. IRE Standards on Facsimile: Definition of Terms, *Proc. IRE*, vol. 44, pp. 776–781, June, 1956.
2. Becken, E. D., Single-sideband Operation for International Telegraph, *Proc. IRE*, vol. 44, pp. 1782–1788, December, 1956.
3. Goldstine, H. E., G. E. Hansell, and R. E. Schock, SSB Receiving and Transmitting Equipment for Point-to-Point Service on HF Radio Circuits, *Proc. IRE*, vol. 44, pp. 1789–1794, December, 1956.
4. Buff, C., Application of Single-sideband Technique to Frequency Shift Telegraph, *Proc. IRE*, vol. 44, pp. 1692–1697, December, 1956.
5. Kulinyi, R. A., R. H. Levine, and H. F. Meyer, The Application of SSB to High-frequency Military Tactical Vehicular Radio Sets, *Proc. IRE*, vol. 44, pp. 1810–1823, December, 1956.
6. Nicolet, M., The Constitution and Composition of the Upper Atmosphere, *Proc. IRE*, vol. 47, pp. 142–147, February, 1959.
7. Pierce, J. N., and S. Stein, Multiple Diversity with Nonindependent Fading, *Proc. IRE*, vol. 48, pp. 89–104, January, 1960.
8. Smith-Rose, R. L., URSI: The International Scientific Radio Union, *Proc. IRE*, vol. 51, pp. 1406–1407, November, 1963.
9. Munro, G. H., and L. H. Heisler, Ionospheric Dynamics, *Proc. IRE*, vol. 51, pp. 1475–1481, November, 1963.
10. deWolf, F. C., The ITU and Global Communications, *Trans. IRE*, P. G. on Communication Systems, vol. CS2, pp. 18–21, November, 1954.
11. Miles, P. D., International Radio Frequency Management, *Trans. IRE*, P. G. on Communication Systems, vol. CS2, pp. 22–25, November, 1954.
12. Moore, J. B., Constant-ratio Code and Automatic—RQ on Transoceanic H. F. Radio Services, *IRE Trans. Commun. Systems*, vol. CS-8, pp. 72–75, March, 1960.
13. Goldberg, B., H. F. Radio Data Transmission, *IRE Trans. Commun. Systems*, vol. CS9, pp. 21–28, March, 1961.
14. Croisdale, A. C., Error Rates and Error Detection on Telegraph Circuits, *IRE Trans. Commun. Systems*, vol. CS9, pp. 28–37, March, 1961.
15. Van Duuren, H. C. A., Error Probability and Transmission Speed on Circuits Using Error Detection and Automatic Repetition of Signals, *IRE Trans. Commun. Systems*, vol. CS9, pp. 38–50, March, 1961.
16. Glenn, A. B., and G. Lieberman, Performance of Digital Communication Systems in an Arbitrary Fading Rate and Jamming Environments, *IRE Trans. Commun. Systems*, vol. CS11, pp. 57–68, March, 1963.
17. Barrow, B. B., Diversity Combination of Fading Signals with Unequal Mean Strengths, *IRE Trans. Commun. Systems*, vol. CS11, pp. 73–78, March, 1963.
18. Schelkunoff, S. A., and H. T. Friis, *Antennas—Theory and Practice*, Johy Wiley & Sons, Inc., New York, 1952.
19. Hund, A., *Frequency Modulation*, 1st ed., McGraw-Hill Book Company, New York, 1942.
20. Codding, Jr., G. A., *The International Telecommunication Union*, E. J. Brill, NV, Leiden, Netherlands, 1952.
21. *The Radio Amateur's Handbook*, 40th ed., 1963.
22. Harper, A. E., *Rhombic Antenna Design*, D. Van Nostrand Company, Inc., Princeton, N.J., 1941.
23. Florman, E. F., and J. J. Tary, Required Signal to Noise Ratios, RF Signal Power, and Bandwidth for Multichannel Radio Communications Systems, *Natl. Bur. Std. Tech. Note* 100, January, 1962.
24. Crichlow, W. Q., D. F. Smith, R. N. Morton, and W. R. Corliss, Worldwide Radio Noise Levels Expected in the Frequency Band 10 Kilocycles to 100 Megacycles, *Natl. Bur. Std. Circ.* 557, 1955.
25. Lincoln, J. V., SID Type Phenomena and Solar Flares, *Natl. Bur. Std. Rept.* 5065, 1957.
26. Watt, A. D., R. M. Coon, E. L. Maxwell, and R. W. Plush, Performance of Some Radio Systems in the Presence of Thermal and Atmospheric Noise, *Natl. Bur. Std. Rept.* 5088, 1957.
27. Smith, Jr., E. K., and R. W. Knecht, Some Implications of Slant E_s, *Natl. Bur. Std. Rept.* 5503, 1957.
28. Thomas, J. A., and E. K. Smith, A Survey of the Present Knowledge of Sporadic E Ionization, *Natl. Bur. Std. Rept.* 5564, 1958.
29. Plush, R. W., A. D. Watt, and E. F. Florman, Carrier to Noise Requirements for Tele-

type Communication Via Non-fading or Fading UHF Carriers, *Natl. Bur. Std. Rept.* 5585, 1960.

30. Koch, J. W., W. M. Berry, and H. E. Petrie, Experimental Studies of Fading and Phase Characteristics of High-frequency CW Signals Propagated through Auroral Regions, *Natl. Bur. Std. Rept.* 6701, 1960.

31. Viezbicke, P. P., Measured Performance of an HF Log-periodic Antenna, *Natl. Bur. Std. Rept.* 6705, 1960.

32. Salaman, R. K., W. B. Harding, and G. E. Wasson, Fading, Multipath and Direction of Arrival Studies for High Frequency Communications, *Natl. Bur. Std. Rept.* 7206, 1961.

33. Nesenbergs, M., Practical Trends in Coding Theory, *Natl. Bur. Std. Rept.* 6757, 1961.

34. Ionospheric Radio Propagation, *Natl. Bur. Std. Circ.* 462, 1948.

35. Handbook for CRPL Ionospheric Predictions Based on Numerical Methods of Mapping, *Natl. Bur. Std. Handbook* 90, 1962.

36. Ionospheric Predictions (Monthly), U.S. Dept. of Commerce, National Bureau of Standards, Central Radio Propagation Laboratory, TB11-499-9/TO 31-3-28.

37. Salaman, R. K., Historical Survey of Fading at Medium and High Radio Frequencies, *Natl. Bur. Std. Tech. Note* 133, 1962.

38. Thomas, H. A., and R. E. Burgess, Survey of Existing Information and Data on Radio Noise over the Frequency Range 1-30 MC/S; *Natl. Phys. Lab. Gr. Brit. Radio Res. Spec. Rept.* 15, 1947.

39. International Radio Consultative Committee, *C.C.I.R. Documents of the IXth Plenary Assembly*, vol. 1, Recommendations, International Telecommunication Union, Los Angeles, 1959.

40. International Radio Consultative Committee, *C.C.I.R. Documents of the IXth Plenary Assembly*, vol. III, Reports, International Telecommunication Union, Los Angeles, 1959.

41. International Radio Consultative Committee, *C.C.I.R. Documents of the Xth Plenary Assembly*, vol. I, Emission, Reception, Vocabulary, International Telecommunication Union, Geneva, 1963.

42. International Radio Consultative Committee, *C.C.I.R. Documents of the Xth Plenary Assembly*, vol. II, Propagation, International Telecommunication Union, Geneva, 1963.

43. International Radio Consultative Committee, *C.C.I.R. Documents of the Xth Plenary Assembly*, vol. III, Fixed and Mobile Services, Standard-frequencies and Time-Signals, Monitoring of Emissions, International Telecommunication Union, Geneva, 1963.

44. *Final Acts of the International Telecommunication and Radio Conferences*, International Telecommunication Union, Atlantic City, 1947.

45. *Radio Regulations*, Additional Radio Regulations—Additional Protocol—Resolutions and Recommendations, International Telecommunication Union, Geneva, 1959.

46. *Final Report—Panel of Experts* (Resolution No. 3 of the Administrative Radio Conference Geneva, 1959), International Telecommunication Union, Geneva, 1963.

47. *Point to Point Telecommunications—A Journal for the Telecommunication Engineer*, vol. 7, Marconi's Wireless Telegraph Co. Ltd., Chelmsford, England, June, 1963.

48. *Point to Point Telecommunications—A Journal for the Telecommunication Engineer*, vol. 8, Marconi's Wireless Telegraph Co. Ltd., Chelmsford, England, October, 1963.

49. *Convention on H. F. Communications*, The Electronics Division of the Institution of Electrical Engineers, Savoy Place, London, March, 1963.

50. Electronics—Power—Science and General, *Proc. IEE*, vol. 110, September, 1963.

51. Hayton, T., C. J. Hughes, and R. L. Saunders, Telegraph Codes and Code Converters, *Proc. IEE*, vol. 101, pp. 137–150, London, May, 1954.

52. Carter, R. O., and L. K. Wheeler, Code Converters for the Interconnection of Morse and Teleprinter Systems, *Proc. IEE*, vol. 101, London, May, 1954.

53. McGuire, E. R., Telegraph Signalling and Distortion, *ATE J.*, vol. 13, no. 2, 1957.

54. *American Standard Definitions of Electrical Terms*, Group 65 Communications, American Institute of Electrical Engineers, 1957.

55. Standard Frequencies and Time Signals from NBS Station WWV and WWVH, *Natl. Bur. Std. Misc. Publ.* 236, 1961.

56. Kanaya, S., and H. Kokoi, The Lateral Deviation of Radio Waves Propagated along the Longer Great Circle Path from Europe, *Rept. Ionospheric Space Res. Japan*, vol. XIV, no. 2, 1960.

57. Miya, K., and K. Ueno, Influence of Directivity of the Transmitting Antenna on the Bearing of H. F. Signals, *Rept. Ionosphere Space Res. Japan*, vol. XIV, no. 2, 1960.

58. Moore, J. B., Accuracy and Speed on Short-wave Teleprinter Services, *Proc. Natl. Electron. Conf.*, vol. 9, February, 1954.

59. Lyons, W., Design Considerations for Frequency Shift Keyed Circuits. *RCA Rev.*, vol. XV, no. 2, June, 1954.

60. Lyons, W., Considerations in SSB and ISB Systems for Long-Distance Radiotelegraph Communications, Communications and Electronics, American Institute of Electrical Engineers, January, 1960.
61. Auwaerter, J. F., The New Teletypewriter Code, *Bell Lab. Record*, vol. 41, no. 10, November, 1963.
62. Ridings, G. H., Facsimile Communication—Past, Present, Future, *Signal*, November, 1962, pp. 34–40.
63. Simpson, S. H., and R. E. Hammond, Radiophoto Standards, *RCA Rev.*, vol. III, no. 4, December, 1947.
64. Facsimile Systems, *Dept. Army Tech. Manual* TM 11-489, March, 1958.
65. Bliss, W. H., Advancements in the Facsimile Art during 1961, *AIEE Paper* 62-313, 1962.
66. Nelson, J. H., A. Arzinger, and H. E. Hallborg, Sunspots and Radio Weather, *RCA Rev.*, vol. IX, no. 2, p. 229, June, 1948.
67. *Codes and Abbreviations for the Use of the International Telecommunication Services*, 2d ed., International Telecommunication Union, Geneva, 1963.
68. Gassmann, G. J. (ed.), *The Effect of Disturbances of Solar Origin on Communications*, Papers presented at Symposium of the Ionospheric Research Committee AGARD AVIONICS PANEL, Naples, Italy, Pergamon Press, New York, 1963.
69. International Frequency Registration Board (IFRB), *International Frequency List*, 2d ed., vol. III, February, 1963.
70. Henney, K., *Radio Engineering Handbook*, 5th ed., McGraw-Hill Book Company, New York, 1959.
71. Laport, E. A., *Radio Antenna Engineering*, McGraw-Hill Book Company, New York, 1952.
72. Salaman, R. K., A New Ionospheric Multipath Reduction Factor (MRF), *IRE Trans.* PGCS-10, no. 2, pp. 221, 222, June, 1962.

Chapter 16

RADIO RELAY COMMUNICATION

R. GUENTHER, *Radio Corporation of America*

S. J. MEHLMAN,* *Radio Corporation of America*

B. SHEFFIELD,† *Radio Corporation of America*

INTRODUCTION

Historically, transmission facilities as part of a communication system have used conductive wire and cable circuits (covered in detail by Chap. 11). Radio circuits using propagation through the atmosphere made their first appearance in the communications field around the turn of the century after G. Marconi demonstrated the global propagation capability of high frequencies (HF) between 3 and 30 Mc.

The world-wide coverage by way of the HF mode of propagation is possible through reflection from the ionized upper part of the atmosphere (ionosphere) which is produced by irradiation from extraterrestrial sources. Unfortunately, the process creating the ionosphere fluctuates widely, causing drastic changes in HF propagation (as described in detail in Chap. 15). Consequently, the operating circuits provided by the HF propagation mode are subject to very complex transmission fluctuations reducing their availability to a great extent. Furthermore, atmospheric noise, generated by electric storms and propagated by the same ionospheric mode around the world, presents another degrading factor in HF transmission. Also the total HF spectrum is limited by nature (<30 Mc), and wide use around the globe was possible only through strict international regulation of spectrum utilization.

As technology advanced in the early 1930s, components and circuits for higher frequencies were developed which opened up the frequency band from 30 to 10,000 Mc and higher. Early experience with these higher frequencies indicated distinct advantages over the HF mode of propagation.

BASIC PROPAGATION FACTORS

1. General Characteristics

The outstanding characteristics and potential use of VHF (30 to 300 Mc), UHF (300 to 3,000 Mc), and higher frequencies (microwaves) are the following:

1. Atmospheric noise decreases with increasing operating frequency and falls below the thermal-noise level, putting radio circuits above 100 Mc on a par with wire and cable circuits.

* Collaborated in paragraphs 18, 19, and 20.
† Collaborated in paragraphs 21, 22, 23, 24, and, 25.

2. Propagation is not affected by the ionosphere; instead it takes place in the lower atmosphere (troposphere), which is more stable and affected only by meteorological influences such as pressure, temperature, turbulence, and stratification of the atmosphere.

3. Propagation approaches with increasing frequency an optical straight-line path, so that the same frequency can be used in many sites provided they are not in optical sight of one another.

4. The spectrum available is about 1,000 times wider than that of HF.

5. The modulation bandwidth is also wider by about 1,000 times compared with HF, which makes it possible to transmit broadband signals such as TV and multi-channel traffic.

6. The only restriction is the line-of-sight operation with limited diffraction around the curved earth which limits the range to about 20 to 40 miles per link depending on the topology of the terrain. Since the power needed is relatively low (usually less than 10 watts), relay stations are feasible-equivalent to repeaters used in conjunction with wire line or cable circuits.

2. Free Space

The basic line-of-sight propagation factors are closely related to free-space propagation. A radio source at the point T (Fig. 1), radiating uniformly in all directions (isotropic radiator) with a transmitting power P_t, will produce the radiation density S at the distance d

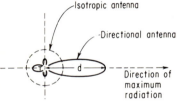

FIG. 1. Gain of directive antenna.

$$S = \frac{P_t}{4\pi d^2} \qquad (1)$$

where S is Poynting's vector equaling radiation density, e.g., in watts per square foot. The vector S, indicating the power flow per unit area, is perpendicular to the surface of a sphere with a diameter d centered around the radiating source. If an antenna is placed anywhere on the surface of this sphere with an effective area (aperture) of A_r, it will produce the available power

$$P_r = P_t \frac{A_r}{4\pi d^2} \qquad (2)$$

Thus the attenuation between an isotropic transmitting antenna and a receiving antenna of aperture A_r or path loss is

$$10 \log_{10} \frac{P_t}{P_r} = 10 \log_{10} \frac{4\pi d^2}{A_r} \qquad (3)$$

The physical explanation of Eq. (3) can be reduced to simple geometrical factors, namely, the ratio between the total surface area of the sphere of radius d divided by the effective area A_r of the receiving antenna.

For a transmitting antenna with a directive radiation pattern, the radiation density S_d will be increased by the gain G_t of the transmitting antenna in the direction of maximum radiation as shown in Fig. 1.

$$S_d = G_t S \qquad (4)$$

Thus the basic loss between two directive antennas in free space will be

$$\text{Free-space path loss} = 10 \log_{10} \frac{4\pi d^2}{G_t A_r} \qquad (5)$$

From electromagnetic theory[1-3] it follows that for an antenna of arbitrary shape or form the ratio between gain G and effective area A is constant.

$$\frac{G}{A} = \frac{4\pi}{\lambda^2} = \text{constant} \tag{6}$$

From this general relationship and Eq. (5), the path loss in free space between an arbitrary antenna with aperture A_t and A_r is

$$\text{Free-space path loss} = 10 \log_{10} \frac{d^2\lambda^2}{A_tA_r} \tag{7}$$

where $\lambda = c/f$ wavelength of radiation
c = 186,000 mps = velocity of propagation in free space
f = frequency of radiation

3. Line of Sight

In propagation over the surface of the earth, neglecting for the moment all atmospheric effects, there are two propagation paths as shown in Fig. 2. One path is

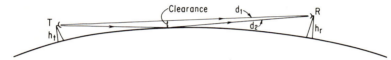

FIG. 2. Clearance for line-of-sight path.

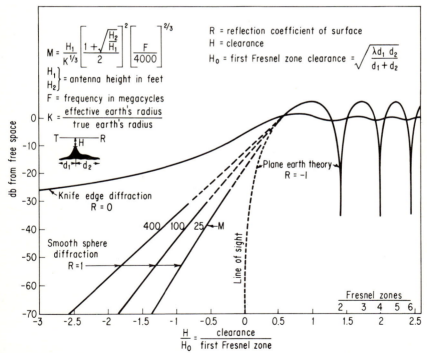

FIG. 3. Height-gain curve for constant distance d_1 = const. and transmitting antenna height h_t = const. (*From "Antenna Engineering Handbook," by Henry Jasik. Copyright* © 1961. *McGraw-Hill Book Company. Used by permission.*)

along a direct line between T and R with distance d_1; the other path is reflected from the surface with distance d_2. At a certain path-length difference between d_1 and d_2 there will be a cancellation and the receiver power will drop to zero or near zero depending on the reflection coefficient. For a fixed distance d_1 between transmitter T and receiver R the variation of antenna height will, therefore, ideally produce a variation in received power according to the height-gain curve as shown in Fig. 3.

The first maximum from the surface corresponds to the first Fresnel zone clearing; that is, in the middle of the path the distance from d_1 to the ground includes one Fresnel zone. Because it is broader than the other maxima, it will be less susceptible to atmospheric fluctuations affecting the effective height of the antenna.

4. Atmospheric Fluctuations

The influence of atmospheric fluctuation on the line-of-sight propagation falls into three major categories:

1. Changes in the average gradient of the refractive index which produce up and down shifts of the maxima and minima shown in Fig. 3 corresponding to changes in antenna height
2. Reflection from horizontal stratification in the troposphere with discontinuities in the refractive index resulting in higher-order multipath propagation effects
3. Absorption in the atmosphere due to raindrops, water vapor, and molecular oxygen (O_2).

The first one usually produces slower variation and depends primarily on humidity, barometric pressure, and temperature distribution with increasing elevation. The second results usually in fast fluctuations of the composite received power in the receiving antenna and is determined primarily by the existence and motion of horizontal strata and atmospheric turbulence producing fluctuations of the receiver signals which can be best described in a statistical way by a Rayleigh distribution.

The median values are not much different from the free-space calculations (1.6 db below free space) and fluctuations which are customarily plotted in terms of distribution curves (cumulative probability), indicating the percentage of time the signal will be below the value on the abscissa as shown in Fig. 4 for the theoretically worst case of a typical link. For a required operating reliability, say 100–0.1 per cent = 99.9 per cent, the fading margin to be allowed is about 30 db below free space. These figures mean that the path loss will be larger than free-space attenuation by 30 db for only 0.1 per cent of the total operating time.

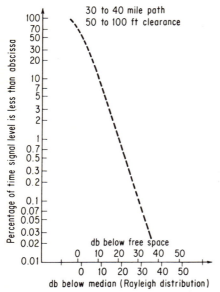

FIG. 4. Fading margin due to multipath for typical path.

Statistical data from many operating circuits are used to obtain representative data for use in planning radio relay circuits. It must be borne in mind, however, that the exact data of a particular link may be different because of differences in the topological profile and peculiarities in the local tropospheric conditions. Statistical data presented later (paragraph 18c) are representative for fairly average conditions in the temperate zone and for typical, gently rolling terrain. Usually propagation data are defined as path loss decibel attenuation normalized for either half-wave length dipoles or isotropic antennas at the transmitter and receiver. The free-space path loss between half-wave dipoles is

$$d_{\lambda/2} = 10 \log_{10} \left(7.68 \frac{d}{\lambda} \right)^2 = 17.7 \text{ db} + 10 \log_{10} \frac{d^2}{\lambda^2} \tag{8}$$

and between isotropic antennas

$$d_{\text{iso}} = 10 \log_{10} \left(4\pi \frac{d}{\lambda} \right)^2 = 22 \text{ db} + 10 \log_{10} \frac{d^2}{\lambda^2} \tag{8a}$$

When directive antennas are used, the gain over an isotropic dipole has to be added to the signal power for each antenna separately. Since the gain of one half-wave dipole over an isotropic antenna is 2.15 db, the difference between Eqs. (8) and (8a) is 4.3 db.

5. Tropospheric Scatter

In the late 1940s, the effect of forward scattering due to turbulence and other tropospheric phenomena, usually summarized by the term tropospheric scatter, showed that it was possible to extend the range of one radio link beyond the optical horizon at the cost of:

1. Appreciably higher transmitter power (over 10 kw)
2. Larger antennas (up to 120 ft diameter)
3. Narrower modulation bandwidth

This mode of propagation is the so-called beyond-the-horizon or "troposcatter" mode. It is possible to trade off the above three factors and cover distances up to 600 miles for modulation bandwidths of about 100 kc/sec or 100 miles distance at about 2 Mc/sec bandwidth.

The line-of-sight and the beyond-the-horizon modes are transmission media which complement each other very well in the VHF through the microwave region. In planning radio relay stations, it will be necessary to make a choice based on conditions of terrain, topology, accessibility, line-of-sight conditions, and cost of construction and maintenance of sites. A summary of the basic propagation is shown in Fig. 5, which demonstrates the gradual transition between the two modes of propagation. Detailed data for each mode will be given later in paragraphs 18c and 20c. In general, operational experience has shown that line-of-sight circuits are more economical than beyond-

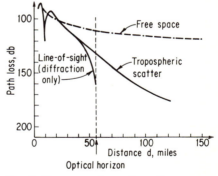

FIG. 5. Summary of smooth earth propagation below and beyond the horizon. Frequency = 150 Mc/antenna height = 400 ft.

the-horizon circuits in well-developed countries with accessible sites. The troposcatter operation is the only feasible one in island hopping. However, it is more economical only in poorly developed countries, where cost of site development requires appreciable expenditures including perhaps substantial road construction.

6. Satellite Repeaters

For the general discussion of propagation factors, it is obvious that a communications satellite orbiting the earth represents a line-of-sight repeater of a specialized kind which permits extension of the line-of-sight operation over very large distances, 5,000 to 8,000 miles, depending on the altitude of the satellite. The design consideration will, for this reason, follow the same procedure except that free-space propagation is almost preserved, with absorption in the atmosphere the only other propagation factor.

7. Power and Noise-level Planning

The planning of a link in terms of power and noise level is best done by means of a level diagram[4] between transmitter and receiver. In addition to an increase in transmitter power and antenna gain, an overall signal-to-noise ratio improvement is also possible through the use of wideband modulation methods, which permit a trade between transmitted bandwidth and signal power such as that offered by frequency

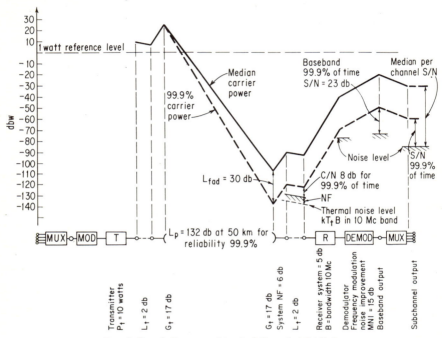

Fig. 6. Level diagram of typical line-of-sight link.

modulation (FM), pulse position modulation (PPM), or pulse code modulation (PCM). Data for determining the design factors for the modulation system will be presented later in paragraph 14. Summarizing the systems planning procedure for noise performance, it may be seen from the level diagram in Fig. 6 that it is best to start at the right from the S/N ratio required by the user and express all noise performance factors of the subsystems in decibels (db) of power ratios.* By adding noise improvement or subtracting loss it is thus possible to arrive at the needed transmitter power for a given distance and modulation system parameters. Note that the noise enters first at the output of the receiving antenna (expressed by systems noise figure F_s or noise temperature T_s) and the received carrier power at this terminal determines the carrier-to-noise ratio (C/N), which will stay the same throughout all linear amplifiers in the receiving system. The modulation noise improvement is realized in the nonlinear part of the receiving system, the demodulator, on the output of which (baseband output) the signal-to-noise ratio (S/N) will be larger by the amount of the noise improvement factor.

* Power ratio in db = 10 $\log_{10}$ (power a/power b).

8. Comparison with Other Transmission Media

In order to make proper choices among the various types of transmission media, the transmission engineer has to compare them from the standpoints of transmission performance, reliability, first cost, and annual cost of operation (including capitalization).

In performance, the line-of-sight radio circuits are very close to the traditional wire and cable circuits. Thermal noise is the limiting factor in both cases, and the attenuation increases with the distance. The radio medium is not quite so stable as the cable circuit and for this reason has to be designed to a certain statistical reliability of service determined by the statistical path loss variation mentioned before.

Fig. 7. Broadband microwave radio relay systems in the Bell System. (*Courtesy of American Telephone and Telegraph Co.*)

Open wires are also subject to fluctuations from changes in weather conditions and produce, depending on distances, sometimes comparable loss fluctuations relative to radio circuits (as shown in more detail in Chap. 11).

In line-of-sight circuits, basebands comparable in width to the best coaxial cables with careful equalization are possible as proved, for example, by the wide use of these circuits throughout the Bell System shown in Fig. 7. By the end of 1963 a total of 63,100 miles of microwave routes were in use in the Bell System.

In summary, the comparison shows the radio relay circuits to be, in performance, fully equivalent to wire and cable circuits for toll quality, long-haul trunks, and industrial-quality special applications. Economically, radio relay circuits usually prove to be more advantageous for distances greater than about 10 to 30 miles and channel capacities larger than 10 to 60 voice channels. The exact prove-in distance and capacity depend on specific conditions, including the environment of each site, and have to be evaluated for each specific application. In addition, radio relay circuits have the capability to provide temporary links mounted in vehicles, a service that cables and wires could not provide so easily. In very congested nodal points of communication networks, the available frequency spectrum may be insufficient, and a complement with cable links is indicated to carry the traffic from the congested point to the periphery by cable.

OPERATIONAL REQUIREMENTS AND PERFORMANCE STANDARDS

Since the radio relay circuits are in performance equivalent to conventional wire and cable circuits, the operational requirements will be the same and are usually defined by the operating systems organization in a set of transmission standards.

9. CCITT and CCIR Standards*

In common-carrier applications, for this reason, existing transmission performance standards are applicable to radio relay circuits as well. The generally accepted international standards are the recommendations developed by the CCITT and CCIR which are up-dated quite frequently. The latest pertinent edition comprises the results of the Tenth Plenary Assembly, published[5] in Geneva in 1963.

European and other foreign postal administrations providing public communication services usually follow these recommendations. They are in very close agreement with Bell System standards, which are usually determined in detail for each transmission system individually. The latest publication of systems noise objectives[6] will be a very useful reference.

10. U.S. Military Standards

The military communications agencies in the United States, particularly the newly established Defense Communications Agency (DCA), largely follow the international recommendations. However, they have developed specific military standards now available in preliminary form and known as MIL-STD-188B. It is expected that these MIL standards will be used by all services and incorporated into a new set of standards to be issued by DCA.[7]

11. U.S. Industrial Standards

In industrial and right-of-way applications such as pipeline installations, power-distribution systems, toll-road networks, railroad systems, etc., requirements are usually somewhat different, since there is no need for interconnection with other systems or public service systems. Following is a list of most applicable standards used in the United States.

1. Electronic Industries Association

TR-141	Microwave Relay Systems for Communications	December, 1955
TR-142	Microwave Housing Facilities	December, 1955
RS-158	Mechanical Consideration for Transmission Lines in Microwave Relay Applications	June, 1956
RS-173	Emergency Stand-by Power Generators and Accessories for Microwave Systems	December, 1956
RS-195	Mechanical Characteristics for Microwave Relay System Antennas and Passive Reflectors	July, 1957
RS-203	Microwave Transmission Systems	January, 1958
RS-210	Terminating and Signaling Equipment for Microwave Communication Systems Part I: Telephone Equipment	August, 1958
RS-222	Structural Standards for Steel Transmitting Antennas Supporting Steel Towers	August, 1959
RS-250	Electrical Performance Standards for Television Relay Facilities	October, 1961
RS-252	Baseband Characteristics of the Microwave Radio and Multiplex Equipment	October, 1961

2. Association of American Railroads Standards
Communications Section 12-23
Revised Edition—October, 1960

* CCITT-Comité Consultatif International des Téléphonique et Télégraphique. CCIR-Comité Consultatif International des Radio. Both CCITT and CCIR are members of the International Telecommunication Union, an agency of the UN.

TRANSLATION OF STANDARDS INTO SYSTEMS AND EQUIPMENT SPECIFICATIONS

12. System Design Procedure

To meet customer requirements it is necessary to translate standards into systems and equipment parameters or specifications. The basic system design procedure, as mentioned before, starts best at the receiving end, where the noise performance standards prescribed by the user have to be met (shown at the right-hand side of Fig. 6). Working toward the left, the noise improvement due to the modulation system, such as FM, PPM, or PCM, has to be subtracted to obtain the carrier-to-noise power ratio (C/N) required at the input of the receiver. The noise power at the input of the receiver can be determined from the receiver noise figure, taking into account signal losses and noise contributions of the transmission line between the antenna and the receiver and the noise picked up in the antenna. The receiving antenna gain G_r will then determine the power required at the receiving location to meet the performance standards. From the path-loss attenuation and statistical fluctuations of the path loss, it is now possible to determine the required radiated transmitter power. Taking into account the transmitter antenna gain G_t and the transmission-line loss, we finally obtain the transmitter power output P_t. The modulation system parameters were introduced already when the modulation noise improvement was calculated at the receiving end.

The most practical way to accomplish the design process described above is to express all gains, losses (power ratios), and signal-to-noise ratios in decibels, so that only simple additions and subtractions are necessary. Referring to the example illustrated in Fig. 6, the whole process can be reduced to the following expression of the S/N ratio at the baseband output.

$$\text{Baseband } \frac{S}{N} = 10 \log_{10} \frac{P_t}{1 \text{ watt}} + 10 \log_{10} G_t - 10 \log_{10} L_t$$

$$- 10 \log_{10} L_{fs} - 10 \log_{10} L_{fad} + 10 \log_{10} G_r - 10 \log_{10} L_r$$

$$- 10 \log_{10} \frac{F_s K T_R B}{1 \text{ watt}} + 10 \log_{10} \text{MNI} \quad (9)$$

where P_t = transmitter output power
L_t = transmitter antenna line loss
G_t = transmitter antenna gain
L_{fs} = free-space loss
L_{fad} = fading loss
G_r = receiver antenna gain
F_s = system noise figure
B = receiver noise bandwidth
T_r = room temperature, 290°K
K = Boltzmann's constant = 1.38×10^{-23} watt-sec/°Kelvin
L_r = receiver transmission line loss
MNI = modulation-noise improvement

$$\frac{S}{N} = \frac{P_t \, G_t \, G_R}{K T B \, (NF)} \frac{(MNI)}{\left(\frac{4 \pi R}{\lambda}\right)^2} L$$

All the items listed above will be covered in greater detail later, including quantitative data and computational aids.

In most cases the system will consist of a number of tandem links, and the system parameters then have to be reduced to equipment parameters on a per-link basis. The most important part of the CCIR standards noted before in paragraph 9 is related to the noise measured at the end of a 2,500-km reference circuit at the zero transmission level point of the system.

13. Noise Units and Objectives

For this noise measurement there exists a number of different units which are defined in Table 1.[8] The noise power measurement without any weighting of the

Table 1. Definitions of Terms

db: Logarithmic power ratio expressed in decibels; db = 10 $\log_{10} (P_2/P_1)$. In the same impedance, decibels can also be expressed in voltage ratio: db = 20 $\log_{10} (V_2/V_1)$.

dbm: Decibels above or below 1 mw = $\pm$dbm.

dbr: Used to refer the signal level at any point in a transmission system to an abitrary point in the system known as the point of zero relative level.

dbw: Decibels above or below 1 watt.

dbv: Decibels above or below a reference voltage of 1 volt. Normally used to define the relative level of a video signal relative to 1 volt peak to peak.

dbc ("db Collins"; used by Collins Radio Co.): Decibels above or below a reference voltage of 0.775 volt rms, that is, the db scale on HP-400 series VTVM's.

dbx: Crosstalk coupling measurement; decibels above or below reference coupling.

dbrn: Decibels above reference noise, where the reference noise power is 10^{-12} watt, or -90 dbm at 1,000 cps. Established by the Bell Telephone Co. for measurements with 144-line weighting for noise-interference measurements. Now largely superseded by measurements in dba and dbrn-Cm.

dba: Decibels above reference noise, adjusted. Similar to dbrn, except used for interference noise measurements with Bell F1A-HA1 telephone sets and with the reference noise power at -85 dbm at 1,000 cps.

dbrn-Cm: Decibels above reference noise, adjusted for C-message circuits. Similar to dbrn except used for interference-noise measurements with Bell 500 telephone sets and with reference noise power at -90 dbm at 1,000 cps. This is the interference-noise measurement term used on the new Bell Type 3A noise meter.

O-TLP: Abbreviation for zero-transmission-level point. A measuring point in the telephone system where the zero reference of 1 mw appears. Also known as the RTLP (reference-transmission-level point).

dba0: Decibels of adjusted noise-interference power referred to 0 dbm at the RTLP. That is, noise of $+20$ dba at OTLP is $+20$ dba0; at a -4 db TLP, it is $+24$ dba0.

dbm0: Decibels of sinusoidal signal, or noise, power referred to 0 dbm at the RTLP. A signal of $+7$ dbm at OTLP is $+7$ dbm0; a signal of -15 dbm at a -15 dbr point is 0 dbm0.

Psophometric voltage: Interfering noise voltage present at a measuring point in a telephone system, measured as recommended by the CCITT using a psophometer (noise voltmeter). This noise voltage is then converted into psophometric emf (equivalent noise generator emf) by considering the circuit source and load impedances referred to 600 ohms resistive. Psophometric emf = 2 $\times$ (psophometric voltage) for 600-ohm resistive circuits.

dbm0p: An interfering sinusoidal or noise power level in a telephone system measured with a CCITT standard telephone psophometer and giving the same reading as an 800-cps tone of equal power level in dbm0.

Note: Interfering noise levels can also be measured psophometrically in picowatts (10^{-12} watt) by relating the psophometric emf to equivalent output power in a 600-ohm matched system.

---

frequency response of the instrument is expressed usually in picowatts (10^{-12} watt = pw) or dbm which represents the ratio, expressed in decibels, below 1 mw. There are various weighting curves in use determined by the average frequency response of operating equipment and facilities. Figure 8 shows the relationship between the different units with the CCITT weightings generally used abroad shown on the right side. On the left side there is the older Bell System 144 and F1A weighting and the new C weighting. According to CCIR documents the subjective effect of noise at the zero transmission level may be roughly rated as follows:

7,500 pw (CCITT weighting).........	Highest permissible mean value for any 1 hr
7,500 pw (CCITT weighting).........	Mean value for 1 min not to be exceeded more than 20 % of any 1 month
47,400 pw (CCITT weighting).......	Mean value of 1 min not to be exceeded 0.1 % of any 1 month
1,000,000 pw (unweighted)...........	Measured with 5-msec integration time not to be exceeded 0.01 % of any 1 month

This may be compared with the Bell System noise objectives illustrated in Fig. 9.
The total noise is generally caused by two mechanisms:

1. Thermal-noise effects
2. Modulation effects

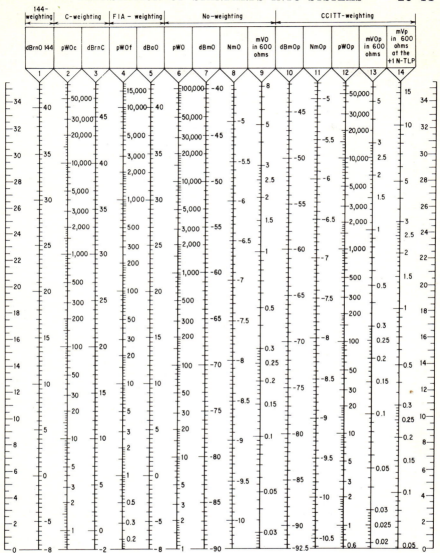

Fig. 8. Conversion chart for the most widely used noise units.   (*From W. Zaiser in Nachr. Tech. Z.*, no. 9, p. 462, 1963.)

Thermal noise is generated by thermal agitation within the receiver, the transmission line, space, and the propagation medium, giving rise to random noise commonly called "Gaussian noise." Modulation noise usually arises from a multitude of causes within the system, depending on the type of modulation used.

14. Modulation-noise Improvement

The noise improvement due to modulation is given by the S/N power ratio at the output of the receiver divided by the C/N ratio at the input of the receiver. This

$\frac{B}{b} = 2m+1$ $\left(\frac{B}{b}\right)^2 \stackrel{\sim}{=} 4m^2$

improvement is always obtained at the price of an increased bandwidth in the receiver and the transmission path.

For system design purposes it is necessary to divide the total permissible noise between thermal and modulation noise, keeping in mind that thermal noise will fluctuate with propagation fluctuation whereas modulation noise is usually a constant

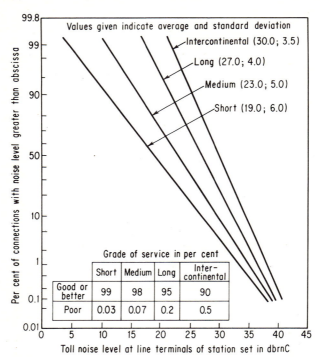

FIG. 9. Noise objectives in the Bell System. (*By permission of American Telephone and Telegraph Co.*)

design factor based on maximum traffic load conditions. CCIR recommendation of partitioning noise allowance in FM/FDM systems is shown in Fig. 10.

a. FM Noise Improvement. The modulation-noise improvement in FM systems[9] may be derived from

$$\left(\frac{S}{N}\right)_b = \frac{C}{N} \times \frac{3}{2} m^2 \times \frac{B}{b} \stackrel{\sim}{=} \frac{C}{N\left(\frac{b}{B}\right)} \cdot \frac{3}{2} m^2 = \left(\frac{S}{N}\right)_i^{\frac{3}{10}} \quad (10)$$

where $(S/N)_b$ = signal-to-noise ratio after demodulation (baseband)

C/N = carrier-to-noise ratio at the receiver input

C = received carrier power before demodulation

$N = KT_rBF$ = equivalent receiver noise power before demodulation

F = receiver noise figure

$K = 1.38 \times 10^{-23}$ watt-sec/°K = Boltzmann's constant

$T_r = 290°K$ = room temperature

B = receiver bandwidth before demodulation

b = bandwidth at receiver output after demodulation (baseband)

$m = \Delta F/b$ = modulation index or deviation ratio — *see Fig. 13*

ΔF = peak frequency deviation for peak modulation by the full baseband signal

$2 \Delta f = B$

for $\frac{B}{b} = 30$, $m = 14$

$\frac{3}{8}\left(\frac{B}{b}\right)^2 = 337$

$\left(\frac{B}{b}\right)^2 = 900$

$m^2 = 294$

$\frac{3}{2} m^2 =$

$\frac{3}{2} m^2 = \frac{3}{8} \left(\frac{B}{b}\right)^2$

Equation (10) represents the signal-to-noise ratio improvement referred to the full baseband output. In the case of multichannel operation, for example z voice channels, each of bandwidth b_1, the S/N ratio for the highest (worst) channel is

$$\frac{S}{N} = \frac{1}{2}\frac{C}{N}(m_{1,p})^2\frac{B}{b_1} \tag{10a}$$

where $m_{1,p} = \Delta F_{1,p}/b$ peak modulation index of the worst (upper) channel
$\quad \Delta F_{1,p}$ = peak deviation for one channel
$\quad b_1$ = bandwidth of one subchannel

$\quad \dfrac{b}{b_1} = z$ = number of channels (when baseband is fully occupied)

For large $z(z > 100)$ Fig. 13 shows peak modulation of z channels to be proportional to $\sqrt{z}$ or $m = \sqrt{z}\, m_1$ and $b_1 = b/z$ which, if inserted in Eq. (10a), shows that it

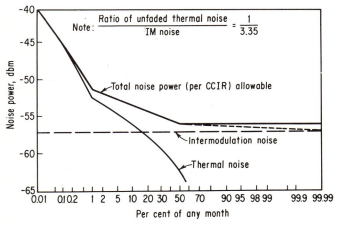

Fig. 10. CCIR recommendation for intermodulation and thermal-noise allocation. (*Courtesy of Comité Consultatif International des Radio Communications CCIR.*)

holds for the worst single channel and the worst part of the baseband signal as well, so that Eq. (10a) can be also written as

$$\frac{S}{N} = \frac{1}{2}\frac{C}{N}m^2\frac{B}{b} \qquad \approx \frac{S}{N}\frac{3}{10}\left(\frac{B}{b}\right)^2 \tag{10b}$$

many channels

Equations (10), (10a), and (10b) are only valid above threshold power[9] of $C_{th} \cong 10N$, where N is the total noise power at the receiver input. This threshold power may be reduced through FM feedback or phase-locked detectors[64] by as much as 8 db and more, depending on the modulation index m (the larger improvements are possible only for $m > 1$). The modulation index m and the bandwidth ratio B/b are related to each other, since the frequency deviation ΔF will determine the bandwidth occupancy of the radiated spectrum, particularly for m larger than 1. For details of the relationship see Fig. 11.

b. Multichannel Loading. In multichannel systems using frequency-division multiplexing, peak modulation as expressed by peak frequency deviation ΔF in FM systems or the corresponding peak modulation voltage has to be split among all subchannels. In multitone applications where each subchannel carries a constant and equal amplitude sinusoidal signal, the peak of z subchannels is z times the peak of one signal. For a small number of z sinusoidal tones we have to use

$$\text{Peak deviation per tone} = \frac{\Delta F}{z} \tag{11}$$

$$\text{rms deviation per tone} = \frac{\Delta F}{\sqrt{2}\, z} \tag{12}$$

where ΔF is the peak deviation of the composite signal. For voice multichannel applications the peaks from all channels add statistically and the composite signal can be expressed only in terms of probability to exceed a certain peak value (overload expectation). In the limiting case of large numbers of channels ($z > 100$) the statistical distribution of the composite signal is independent of the individual channel distribution. From the central-limit theorem, it follows that the composite signal

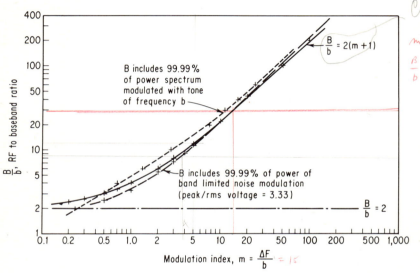

Fig. 11. FM power spectrum bandwidth vs. modulation index (deviation ratio). $\Delta F =$ peak frequency deviation; $b =$ baseband upper cutoff; $B =$ bandwidth of modulated spectrum.

will in this case approach white Gaussian noise. Thus for an overload expectation of 4×10^{-4} (with either positive or negative amplitude) the peak-to-rms voltage ratio can be calculated from the normal (Gaussian) distribution to be 3.52 or a power ratio of about 11 db. Figure 12 shows the probability to exceed a certain "peak-to-rms" ratio. This leads to the rms modulation of z channels:

$$\Delta F_{z,r} = \frac{\Delta F}{3.52} \tag{12a}$$

If all subchannels are carrying equal average power, the rms deviation of one channel is

$$\Delta F_{1,r} = \frac{\Delta F}{3.52\, \sqrt{z}} \tag{12b}$$

In the case of small numbers of channels ($z < 100$) the central-limit theorem cannot be applied and Eq. (12b) will depend on the peak-to-rms ratio in the individual

better when $2 < m < 10$ to use $\frac{B}{b} = 2\left(\frac{m+2}{3.5}\right)$!!

carson's rule $\frac{B}{b} = 2(m+1)$

$\frac{B}{b} = 2(m+2)$

$\frac{B}{b} = 21$

m	B/b	
1	4	
2	6	7
5	12	14
10	22	24

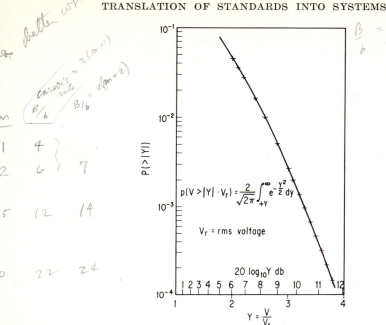

$$p(V > |Y| \cdot V_r) = \frac{2}{\sqrt{2\pi}} \int_{+Y}^{\infty} e^{-\frac{Y^2}{2}} \, dy$$

V_r = rms voltage

$20 \log_{10} Y$ db

$Y = \dfrac{V}{V_r}$

FIG. 12. Probability of exceeding peak voltages ($> |Y| \times v_r$) in white gaussian noise.

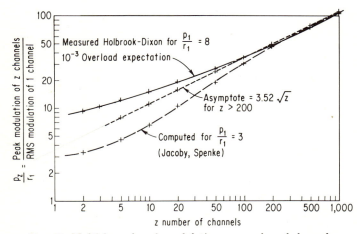

Measured Holbrook-Dixon for $\dfrac{p_1}{r_1} = 8$

10^{-3} Overload expectation

Asymptote = $3.52 \sqrt{z}$ for $z > 200$

Computed for $\dfrac{p_1}{r_1} = 3$ (Jacoby, Spenke)

$\dfrac{p_z}{r_1} =$ Peak modulation of z channels / RMS modulation of 1 channel

z number of channels

FIG. 13. Multichannel peak modulation vs. number of channels.

channel. In this case the relation between peak modulation of z channels to the rms of one channel is a very complex one and is shown in the form of two plots in Fig. 13, as a function of the number of channels z. One of the curves marked Holbrook and Dixon[10] is derived experimentally from an individual channel distribution with the voice peak-to-rms voltage ratio of 8. The other curve marked Jacoby and Spenke[11] was calculated from a single-channel distribution with a peak-to-rms voltage ratio of 3 and the assumption that only one-half of the channels are active at any one time

30% for telephone co.

because of the two-way nature of a telephone conversation (traffic factor 0.5). The ratio p_z/r_1 plotted in Fig. 13 indicates how much smaller the rms deviation of one channel ($\Delta F_{1,r}$) has to be compared with peak deviation ΔF of the composite signal for all z channels. The dotted line indicates the asymptote used for Eq. (12) or (12a) given before for the case of a large number of channels. The average noise power equivalent to the composite signal power of z channels recommended by CCIR[5] for noise loading tests is shown in Fig. 14.

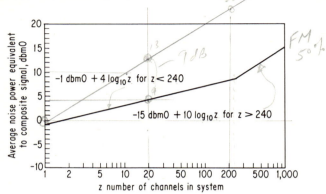

FIG. 14. CCIR recommended noise loading (Geneva 1963, No. 2173-E). (*Courtesy of CCIR.*)

c. PPM Noise Improvement.

The noise improvement in PPM systems can be expressed in a similar way as that for the FM system. The following expressions are based on a sampling rate of each subchannel at the theoretical Nyquist rate (two times the upper cutoff frequency b_1) and modulating the position of each sample from its quiescent position by Δt. In the case of multichannel operation each channel is given a separate time slot, so that, when the receiver is gated, each channel can be separated. The signal-to-noise ratio at the output of each channel can be computed from the same basic relation independent of the number of channels z used, since both b and b_1 have to be multiplied by z.

$$\frac{S}{N} = \left(\frac{S}{N}\right)_b \times 5.5 \times 10^{-2} \left(1 - 4\frac{b_1}{b}\right)^2 \left(\frac{b}{b_1}\right)^3 \tag{13}$$

where $(S/N)_b$ = signal-to-noise ratio in the baseband after RF demodulation, before conversion from PPM to the original signal
$b = 1/t$ = bandwidth of baseband
t = width of rectangular pulse before passing through the baseband filter
$t_1 = 1/2b_1$ = Nyquist interval of samples
$\Delta t = t_1/2 - t$ = peak deviation

In contrast to the expression for the FM improvement, the PPM expression [Eq. (13)] shows the relation to the signal-to-noise ratio in the baseband after the demodulation of the RF carrier. The RF carrier could in the simplest case be amplitude-modulated, using an envelope detector in the receiver, which by itself would not provide any change in S/N ratio, and Eq. (13) will also hold for the C/N ratio at the input of a RF bandwidth $B = 2b$. The same is true for small deviation frequency modulation (frequency-shift-keying) of the baseband signal. This leads to the S/N improvement expressed in terms of carrier-to-noise ratio:

$$\frac{S}{N} = \frac{C}{N} 0.69 \times 10^{-2} \left(1 - 8\frac{b_1}{B}\right)^2 \left(\frac{B}{b_1}\right)^3 \tag{13a}$$

where $B = 2b$ is the receiver bandwidth. Both Eqs. (13) and (13a) hold for signal or carrier power above threshold given by the following expression:

$$C = \geq 200N \frac{b_1}{B} \tag{14}$$

d. PAM Noise Improvement. In PAM systems the signal-to-noise ratio is

$$\frac{S}{N} = \left(\frac{S}{N}\right)_b 0.41 \times \frac{b}{b_1} \tag{15}$$

or

$$\frac{S}{N} = \frac{C}{N} 0.205 \frac{B}{b_1} \tag{15a}$$

PAM has no threshold effect, for $b/b_1 = 2$, PAM would degenerate into conventional AM with no noise improvement. It is obvious from this equation that the trade of systems bandwidth B for signal-to-noise ratio in PAM is not nearly so favorable as in FM and PPM. PAM, however, is sometimes preferable because of the simplicity and low cost of implementation. Because of time-division multiplexing used in conjunction with PAM, Eqs. (15) and (15a) hold for any number of channels as in PPM.

e. PCM Noise Improvement. Another quite different modulation system is pulse code modulation (PCM). It differs from all the other modulation systems mentioned before primarily because of the quantization of the signal to be transmitted. The quantization process converts in general an analogue signal into discrete values of amplitudes at suitable sampling intervals (Nyquist interval $= 1/2b_1$). The discrete amplitude values can then be expressed in a digital code. The signal-to-noise ratio of such a PCM system is determined by the quantization errors introduced through the approximation process of the analogue signal by a finite number of discrete levels. It is given[12] by the following expression:

$$\frac{S}{N} = \frac{3}{2} \times d^{2\kappa} = \frac{3}{2} \times d^{B/b_1} \tag{16}$$

where κ = number of digits in the PCM code ($\kappa = b/b_1$)
$\quad d$ = base of each digit, for example ($d = 2$ for binary)
$\quad d^\kappa$ = number of quantizing levels
$\quad B = 2b_1\kappa$ = optimum RF bandwidth
The signal-to-noise ratio after decoding given by Eq. (16) represents modulation noise and is independent of the input carrier-to-noise ratio as long as the carrier power is large enough.

In most practical applications the base $d = 2$ or binary code is used. Since there are $2b_1$ samples with κ digits each, the required base bandwidth is

$$b = b_1\kappa$$

and rate of transmission is $2b_1\kappa$ for a single-channel system. For good speech quality with PCM, at least $\kappa = 6$ digits are needed and a sampling rate of $2b_1 = 8$ kilobits per second for a total of $2b_1\kappa = 48$ kilobits per second per channel. In multichannel systems with time-division multiplexing the composite signal for z channels would have the transmission rate of $2b_1\kappa z = z \times 48$ kilobits per second. Figure 15 shows how quantizing S/N varies with the selected code as determined by Eq. (16).

An approximate expression for the carrier-to-noise ratio needed for bases $d \geq 2$ was shown[12] to be related to d. For an error rate of $\approx 10^{-6}$:

$$\frac{C}{N} \cong 8(d^2 - 1) = K^2 \frac{d^2 - 1}{12} \tag{17}$$

where K is a constant depending on permissible error rate (for $d = 2$ and 10^{-6} error

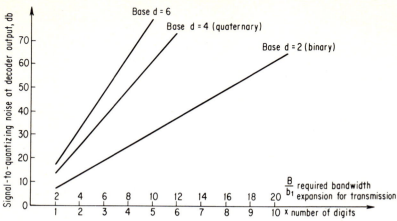

FIG. 15. PCM signal-to-quantizing noise ratio for different codes.

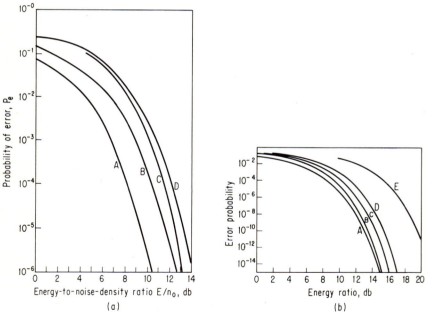

FIG. 16a. Probability of error in binary FM and PM transmissions. $E = C/B$; $n_0 = F \times K \times T_r$. A, optimum binary system, coherent detection of PCM-PM; B, coherent detection of PCM-FM; C, discriminator detection of PCM-FM; D, noncoherent detection of PCM-FM. (*Courtesy of IEEE Transactions of PTGSET.*)

FIG. 16b. Comparison of binary systems with different shaping networks. A, ideal; B, differentially coherent (PSK); C, discriminator (all IF shaping); D, discriminator (equal shaping); E, fixed narrow filter. (*Reprinted from RCA Rev., vol XXII, no. 4, p. 699, December, 1961.*)

rate $K = 10$). The minimum C/N ratio for satisfactory performance in such a system is usually determined by the specified error rate with which the digital decoder is required to operate. This relationship will be also determined by the receiver bandwidth B and the type of carrier modulation and detection used, such as FM and PM and different pulse shaping and detection methods.[13,14] Figure 16a and b shows the error rate plotted for the almost optimum receiver bandwidth of $B = 2b_1\kappa = 2b$ as a function of the energy ratio at the input of the receiver. By energy ratio it is meant the carrier energy per signal element divided by the noise power per unit bandwidth. The design engineer will have to pick the C/N ratio from this plot to meet a given specified error rate of the received PCM code after carrier demodulation. Note that pulse shaping in the IF amplifier only, gives best results, and so does coherent detection. The noise improvement for the different modulation systems enables the designer to relate the required noise performance standards to the C/N ratio needed at the receiver input.

15. Comparison of Modulation Systems

A comparison of all modulation methods[15,16] discussed before is shown in Fig. 17, which allows one to pick the required C/N ratio for a chosen bandwidth ratio B/b_1 and output $S/N = 55$ db.

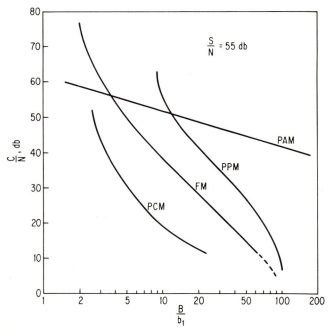

FIG. 17. Comparison of modulation systems for trade of bandwidth against carrier power.

Now it is necessary to determine the required noise level at the receiver input from the parameters of the modulation system.

16. Noise Figure and Temperature

The actual thermal-noise power effective at the input of a receiver can be computed from the bandwidth B and the noise figure F of the receiver. The definition[17–19] of the noise figure is

$$F = \frac{(C/N_r)_{\text{in}}}{(C/N)_{\text{out}}}$$ (18)

where $N_r = KT_r \times B$ = reference-noise power at the receiver input
$T_r = 290°K$ = room temperature,°K

This results into a total equivalent noise power at the receiver input of:

$$N = F \times N_r = F \times K \times T_r \times B$$ (19)

For a receiver consisting of a number of cascaded stages each of noise figure F_1 to F_n and optimum gain of each stage G_1 to G_n the overall receiver noise figure is

$$F = F_1 + \frac{F_2 - 1}{G_1} + \frac{F_3 - 1}{G_1 G_2} + \frac{F_n - 1}{G_1 G_2 G_{n-1}}$$ (20)

where G is the optimum gain as computed for maximum available power, that is perfect match of the load. (For examples of computation of noise figures F_1 to F_n for various types of circuits refer to the literature.[20–23])

The signal source at the receiver input, usually the antenna, is receiving, besides the desired signal, also noise power which can be expressed by

$$N_a = K \times T_a \times B$$ (21)

In this case it may be said that the antenna noise temperature is equal to T_a, which

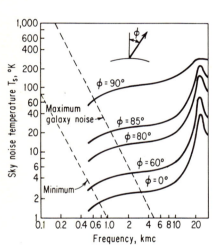

FIG. 18. Antenna temperature vs. antenna angle. (*Courtesy of Proc. Natl. Electron. Conf.*, 1959.)

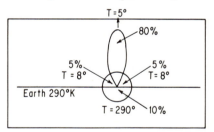

FIG. 19. Example of the evaluation of antenna temperature. (*Courtesy of Proc. Natl. Electron. Conf.*, 1959.)

may be higher than room temperature. This is particularly true if the antenna operates in the frequency range below 100 to 500 Mc, where the cosmic noise radiated by the Milky Way can be fairly large[24] as shown in Fig. 18. The minimum and maximum indicated by the two dashed lines represent the extreme conditions for an ideal antenna pointing either in the direction perpendicular to the Milky Way (low noise) or in the plane of the Milky Way (high noise). An antenna pointing horizontally on the surface of the earth will also be affected by the temperature of the surface of the earth, which could be computed from the actual pattern of the antenna including side lobes and back lobes. A typical antenna may have the power distribution over the horizontal angles of direction as indicated in Fig. 19. The percentage indicated can be used as weighting factor for the noise temperature in each sector to obtain the overall equivalent noise temperature T_a. For example,

$$T_a = 0.8 \times 5° + 0.10 \times 8° \times 0.1 \times 290° = 33.8°K$$

The noise figure of an entire system (see Fig. 20), consisting of an antenna, transmission line, and receiver, can be combined into the system noise figure F_s and can be expressed by

$$F_s = \frac{T_a}{T_r} + \frac{T_l}{T_r}(L - 1) + (F - 1)L \tag{22}$$

where $\qquad T_l$ = temperature of transmission line
$(F - 1)KT_r \times B$ = equivalent noise power from the receiver alone
$\qquad \alpha$ = attenuation constant of transmission line, db/ft
$\qquad l$ = length of transmission line, ft

$$\alpha l = 10 \log_{10} L$$

For easy conversion from decibel attenuation into power ratio L see Table 2. The noise figure is conveniently expressed in decibels:

$$NF_s \text{ (db)} = 10 \log_{10}\left[\frac{T_a}{T_r} + \frac{T_l}{T_r}(L - 1) + (F - 1)L\right] \tag{23}$$

When the receiver antenna is pointing toward the sky such as in ground-to-air or ground-to-space applications using satellite repeaters, Fig. 18 shows that the antenna noise temperature can be much lower than room temperature.[24] In such cases it is more practical to use the concept of equivalent "noise temperature" rather than noise figure. The two are related to each other in the following way:

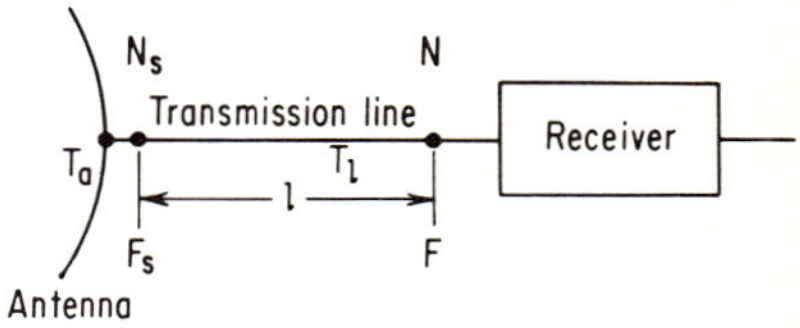

FIG. 20. Receiving system noise figure and noise temperature.

$$T = FT_r$$

Applied to an overall system as shown in Fig. 20 the systems noise temperature computed from Eqs. (22) and (23) is

$$T_s = T_a + T_l(L - 1) + (F - 1)T_r \times 1 \tag{24}$$

Table 2. Conversion from db Attenuation to Power Ratio

Attenuation αl, db	Power loss $L = \dfrac{\text{power in}}{\text{power out}}$
0	1.000
0.1	1.023
0.2	1.047
0.3	1.072
0.4	1.096
0.5	1.122
0.6	1.148
0.7	1.175
0.8	1.202
0.9	1.230
1.0	1.259
2.0	1.585
3.0	1.995
4.0	2.570
5.0	3.162
6.0	3.981
7.0	5.012
8.0	6.310
9.0	7.943
10.0	10.000

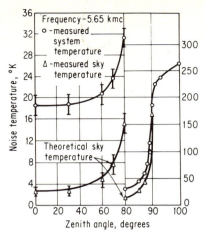

FIG. 21. Measured system noise temperature and theoretical sky noise temperature vs. the antenna angle with respect to the zenith. (*By permission of Proc. Natl. Electron. Conf., 1959.*)

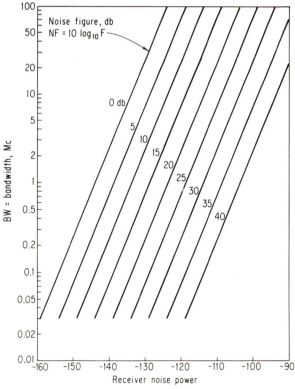

FIG. 22. Equivalent noise power at receiver input. $N = -204$ dbw $+ 10 \log_{10} B$ cps $+ 10 \log_{10} F$.

In applications such as radio astronomy and communications satellite systems the use of amplifiers using MASERS provides systems noise temperatures in the order of 10 to 20°K. Figure 21 represents noise temperatures actually measured in an experimental system. The effective systems noise power at the antenna terminals is thus

$$N_s = F_s \times KT_r \times B = KT_s B \tag{25}$$

For easy computation of system noise power from receiver bandwidth and noise figure use Fig. 22.

17. Interchannel Crosstalk in FM Systems

As mentioned before, thermal noise is only one source of noise. The other is modulation noise. In FM systems the modulation noise affects performance of multichannel systems with frequency-division multiplexing (FDM) because of nonlinear distortion resulting in interchannel crosstalk produced by four causes:

1. Microwave transmission-line echoes (reflections)
2. Nonlinear amplitude response of the modulator and demodulator
3. Group delay distortion in RF and IF selective circuits
4. Nonlinear amplitude response of baseband amplifiers

The first and the third items are different causes producing the same effect, i.e., nonlinear phase-transfer characteristics. In FM systems all amplitude variations and distortions are eliminated by the amplitude limiter in the receiver as far as modulation goes. However, all nonlinear distortions of phase characteristics the FM signal is going through are converted into nonlinear amplitude distortions after detection, thus causing intermodulation.

Reflections in the RF transmission lines will produce phase (φ) distortion, commonly referred to as group delay distortion (group delay $= d\varphi/d\omega$, where φ is phase and ω is angular frequency), depending on the phase of the echo with respect to the signal (modulated carrier). If a discontinuity exists, producing an echo in quadrature, the resultant modulation distortion is shown[25] to be of second order. The ratio of second-order distortion amplitude D_2 over the modulation signal amplitude S for short echo delays can be expressed as

$$\frac{D_2}{S} \approx 0.2 \times r_q \tau^2 \times \Delta F \times f \sqrt{1 - 0.5 \frac{f}{b}} \tag{26}$$

where r_q = magnitude of reflection coefficient (echo amplitude/signal amplitude)
τ = echo delay with respect to signal at end of line
ΔF = frequency deviation, radians/sec
f = modulation frequency within baseband
b = bandwidth of baseband

Similarly the in-phase echo r_p produces third-order distortion products given by the following expression:

$$\frac{D_3}{S} \approx 0.019 \times r_p \times \tau^3 \times (\Delta F)^2 \times f \sqrt{1 - 0.333 \left(\frac{f}{b}\right)^2} \tag{27}$$

Both relations hold for "short delays" when

$$\tau \Delta F < 1$$
$$\tau b < 1$$

Note that the distortion is directly proportional to the reflection coefficient and modulation frequency as shown for D_2/S in Fig. 23. Echoes in quadrature or in phase produce the distortion as illustrated in Fig. 24. For the larger frequency deviations $\Delta F > 1.5$ Mc the approximate equations (26) and (27) do not hold and the

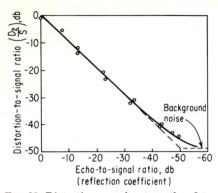

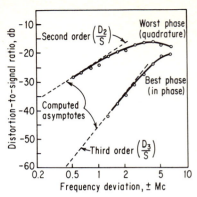

Fig. 23. Distortion vs. echo strength. 0 to 2 Mc noise loading; ±2 Mc frequency deviation; 0.5-μsec echo delay; carrier echo in phase quadrature; 1.95 Mc test frequency. (*Courtesy of Proc. IEEE.*)

Fig. 24. Distortion vs. echo phase. 0 to 2 Mc noise loading; −15 db echo strength; 0.2 μsec delay; 1,950 kc test frequency. (*Courtesy of Proc. IEEE.*)

solid lines represent measured data. All tests were done with flat noise loading in the baseband (2 Mc), −15-db reflection coefficient (10 $\log_{10} r^2$), $\tau = 0.2$-μsec echo delay, and a modulation frequency $f = 1.95$ Mc. Similary for large echo delays Eqs. (26) and (27) break down, and only the limits for very large echo delays may be computed provided that the modulation index (deviation ratio) is large; that is, $m = \Delta F/b \gg 1$

$$\frac{D}{S} \approx 5.8 \times r \times b^{(\Delta F/2 - 3/2)} \times f \times e^{-2.88 f^2/(\Delta F)^2} \tag{28}$$

In Fig. 25 there is an example of the transition from small delay (computed) to larger delay (measured).

In the region of very large echo delays the effect of phase between signal and echo is practically without effect on distortion because the echoes act more like a noncoherent disturbance. The asymptotic values reached for long delays are shown for one example in Fig. 26.

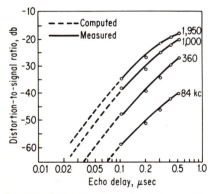

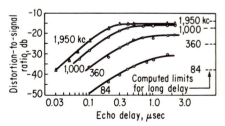

Fig. 25. Distortion vs. echo delay. Echo strength, −15 db; noise input, 0 to 2 Mc; frequency deviation, ±1 Mc. (*Courtesy of Proc. IEEE.*)

Fig. 26. Distortion vs. echo delay. Echo strength, −15 db; noise input, 0 to 2 Mc; frequency deviation, ±4 Mc. (*Courtesy of Proc. IEEE.*)

Because of the very complex relationships there are no simple solutions in closed form which cover all practical cases. The most general presentation of the parameters involved was given by Curtis, Bennett, and Rice[26] and is summarized in Fig. 27. The numbers on the curves indicate the number of decibels by which the interference power in the top channel of a multichannel system is smaller than the relative echo $(10 \log_{10} |r|^2)$. For example, an echo caused by reflection coefficient $r = 0.1$ or $10 \log_{10} |r|^2 = -20$ db will produce 20 db less or a total of -40 db interference ratio along the curve marked -20 db. Note that rms frequency deviation ΔF_r is used rather than peak deviation ΔF used before. In multichannel systems with more than 100 channels the peak-to-rms ratio of the baseband signal is very nearly 3.52 (11 db), as has been shown before. If, for example, the interference power has to be 50 db

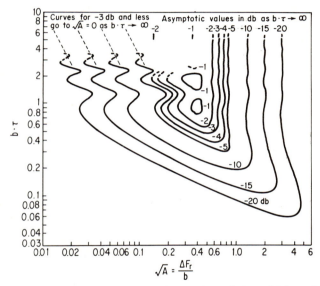

FIG. 27. Contours of constant interference in the top channel of a multichannel FM system. The numbers on the curves indicate the ratio of the interference power to the echo power in decibels; i.e., they are the values of $10 \log_{10} (P_I/r^2 P_S)$. $\Delta F_r =$ rms frequency deviation, cps; $b =$ top frequency of noise signal; cps; $\tau =$ echo delay, sec. *(By permission of the American Telephone and Telegraph Co.)*

below signal in the top channel of an FM frequency-division multiplex channel and the reflection coefficient in the transmission line is 40 db, the system parameters on the -10-db contour line or outside it have to be chosen.

Another example will show the results related to a practical system with following parameters:

Operating carrier frequency	$F_c = 2$ Gc
Baseband	$b = 2$ Mc
Transmission line length (waveguide 4.3 × 2.15 in.)	$l = 200$ ft
Echo delay	0.296 μsec
Peak deviation	4 Mc
RMS deviation (11 db or 3.52 peak-to-rms ratio)	1.13 Mc
Number of channels (4 kc each)	480

If mismatches exist only at each end of the transmission line of equal magnitude $(|r|)$, the voltage standing-wave ratio (VSWR) required for a specified intermodulation-to-signal ratio is shown in Fig. 28. For a single line to contribute not more than −65 db would require a VSWR ≤ 1 to 1.05, which represents a very stringent requirement for all microwave components. The group delay distortion of RF and IF tuned circuits (item 3 listed before) which results from nonlinear phase responses will likewise cause intermodulation noise. An ideal transfer characteristic should have constant delay over the entire passband. If the actual delay of the RF and IF amplifiers is known, its deviation from the ideal can be related to intermodulation distortion power in an FM system using frequency-division multiplex (FDM) terminals.

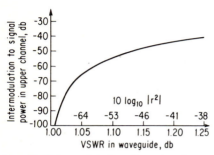

FIG. 28. Intermodulation distortion due to equal mismatches on each end of waveguides 4.3 by 2.15 in. at 2 Gc.

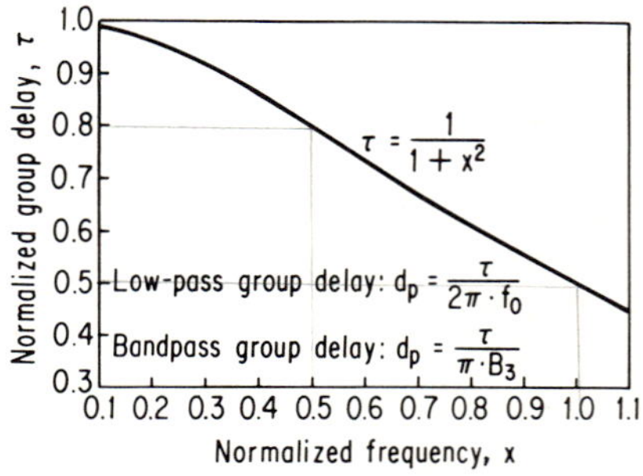

FIG. 29. Normalized group delay for single-tuned (pole) bandpass and low-pass filters.

A group delay response which is linear over the passband (constant slope) of a network will produce second-order harmonic distortion (voltage ratio) according to[27-29]

$$d_2 = \pi f \times \Delta d_s \tag{29}$$

where f = fundamental modulation frequency in baseband ($<b$)

Δd_s = magnitude of group delay change from center frequency (f_c) of filter to frequency of peak deviation ($f_c + \Delta F$)

Frequently the constant slope group delay distortion is given as slope

$$S = \frac{\Delta d_s}{\Delta F} \tag{30}$$

commonly expressed in nanoseconds per megacycle, with ΔF equaling peak deviation.

A parabolic group delay response within the passband of a network will produce third-order harmonic distortion according to

$$d_3 = \frac{\pi}{2} f \times \Delta d_p \tag{31}$$

where Δd_p is the magnitude of group delay change from the center frequency f_c of the filter to frequency of peak deviation ($f_c + \Delta F$).

The parabolic group delay is commonly given as

$$P = \frac{\Delta d_p}{(\Delta F)^2} \tag{32}$$

expressed in nanoseconds per megacycle squared. Most filters produce predominently a parabolic group delay and result consequently in third-order distortion. Filters with one pole (single-tuned circuit) with low-pass as well as bandpass characteristic may be readily computed from the normalized group delay[30] shown in Fig. 29 and the following expression given for each type of filter:

$$d_p = \frac{\tau}{2\pi f_0} \qquad \text{for the low-pass filter} \qquad (33)$$

where f_0 = cutoff frequency
$\tau = 1/(1 + x^2)$ (see Fig. 29)
$x = f/f_0$
f = frequency within passband

group delay

$$d_p = \frac{\tau}{\pi B_3} \qquad \text{for the bandpass filter} \qquad \simeq .3 \text{ nsec or} \qquad (34)$$

(handwritten: 1000 Mh₂)   *(handwritten right margin: .3 µsec / 1 MHz)*

where B_3 = bandwidth between 3-db points
$\tau = 1/(1 + x^2)$
$x \cong 2 |f - f_c|/B_3$, for a narrowband filter* where $B_3/f_c < 10$ per cent
f_c = center frequency
f = frequency within passband

Filters with more than one section or pole designed for maximally flat response have a (parabolic) delay distortion given by the following relation:

$$P = \frac{\Delta d_p}{(\Delta F)^2} = \frac{4C}{\pi} \frac{1}{(B_3)^2} \qquad (35)$$

(handwritten: nsec / (MH₂)²)

where C is a constant, determined by the type of filter (see Table 3).
For the most widely used double-tuned, critically coupled filter

$$\frac{\Delta d_p}{(\Delta F)^2} = 1.8 \left(\frac{10 \text{ Mc}}{B_3}\right)^3 \qquad (35a)$$

All group delay distortion relations hold only for a restricted part of the 3-db bandwidth (up to about 40 per cent) so that the following restriction has to be observed.

$$2 \Delta F < 0.4 \times B_3 \qquad (36)$$

Because of the very stringent linearity required in multichannel systems it is always necessary to operate well within these limits so that Eqs. (33) to (35) will give good results where they are needed most.

Table 3. Constant C in Eq. (35) for Group Delay Distortion

n, number of sections (poles)	C
2	1.414
3	1.000
4	1.083
5	1.236
6	1.414
7	1.604
8	1.799
9	2.000
10	2.202

In many cases it will be necessary to employ phase equalizers, which usually are rather complex all-pass type of network. Very simple phase equalization within limited bandwidth can be obtained by judicious combinations[30] of single- and double-tuned circuits. The two types of circuits have within the center portion of the passband opposite parabolic curvature and may be used to compensate the group delay distortion of each other.

* Exact relations for x in broader filters (>10 per cent) are more complex and depend on the type of filter; for details see Ref. 31.

In most multichannel systems, particularly for voice operation, it is customary to specify intermodulation distortion in noise power produced in an idle channel while all other channels are fully loaded with white Gaussian noise to operate the system at peak modulation.[27] In some cases the ratio of intermodulation noise to normal signal power, the so-called noise-power-ratio (NPR), is specified, which is, of course, related to harmonic distortion although not in a simple way.[28,29] Since the intermodulation noise power spectrum produced within the baseband owing to group delay distortion in filters is increasing with increasing modulation frequency, it will be worst for the top channel.

In Fig. 30 the absolute distortion power (measured with F1A weighting) in pico-watts in the highest channel (at the 0-dbm transmission level) of a 600-channel system loaded with the CCITT recommended (see Fig. 14) noise power at +12.8 dbm0 is shown as a function of delay distortion for the two types of delay characteristics. Curve A (to be used with the S scale on the top) is for the case of linear distortion (constant slope) vs. frequency. From curve A it is possible to read the permissible slope of the delay distortion characteristics for a tolerable power of intermodulation noise. The second curve B (to be used with the P scale on the bottom) shows the same relationship when the delay vs. frequency response is parabolic. The P scale on the bottom shows also the 3-db bandwidth for one double-tuned circuit according to Eq. (35a). The effect of delay distortion thus places another restriction on bandwidth, since the phase characteristic and the resulting delay distortion are related to the bandwidth for a given design of the selective circuits.

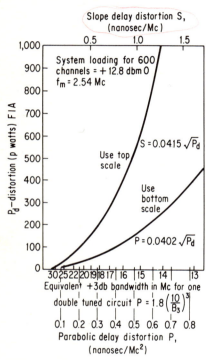

FIG. 30. Intermodulation noise (F1A weighting) in top channel of multichannel FM/FDM system.

The nonlinearities in the baseband amplifiers (item 4) are also a source of intermodulation distortion; however, they are easily controlled by the application of negative feedback as used in linear repeaters for wire and cable multichannel systems (see Chap. 10) and do not represent a limiting factor in radio systems. However, the nonlinearities of the modulation and demodulation circuits are not so easily controlled. Here the deviation from the ideal (linear) modulation characteristic, i.e., the nonlinearity of the dynamic characteristic, can be translated into intermodulation noise. From the analysis of Gaussian noise through square law devices,[32,33] it can be shown that baseband noise with upper cutoff frequency b applied to a dynamic characteristic with second-order nonlinearity

$$V_{out} = \alpha V_{in} + \beta V_{in}^2 \tag{37}$$

will produce an intermodulation spectrum within b:

$$w_2(f) = \frac{\beta^2 N_{in}^2}{b^2} (2b\text{-}f) \tag{38}$$

There will be an additional spectrum extending from $f = b$ to $f = 2b$, which falls, however, outside the baseband. From Eq. (38) it follows that intermodulation noise due to a second-order nonlinear dynamic characteristic is highest in the lowest channel of narrow bandwidth $b_1 \ll b$. The worst (lowest) channel second-order noise power N_{b_1}

$$N_{b_1} = \int_0^{b_1} w_2(f) \, df = 2\beta^2 N_{in}^2 \frac{b_1}{b} \tag{39}$$

and the second-order noise-power-ratio in the worst (lowest) channel

$$(NPR)_{2d} = \frac{1}{2} \frac{1}{\beta^2 N_{in}}$$

For any modulator or demodulator tested with sine-wave modulation of the same peak deviation ΔF as total noise used in the NPR test, we can relate the second harmonic test tone ratio d_2 to $(NPR)_{2d}$:

$$(NPR)_{2d} \cong \frac{p_n^2}{8(d_2)^2} \cong \frac{1.55}{(d_2)^2} \tag{40}$$

where $\quad d_2 = \dfrac{\beta}{2d} \times V_{tone} = \dfrac{\text{2d harmonic}}{\text{fundamental}} \quad$ (2d harmonic voltage ratio)

p_n = noise voltage peak factor = 3.52 ($\sim$11 db) for 2.10^{-4} overload expectation. This means that for one modulator/demodulator, for example, if an NPR = 60 db is required, $d_2 \leq 0.155$ per cent, an extraordinarily high linearity requirement. An equivalent relation has been derived by L. Thompson[28] giving for the second-order distortion practically the same result as Eq. (40) if the same peak ratio $p_n = 4$ is used. For the third-order distortion we have for the worst channel

$$(NPR)_{3d} = \frac{p_n^2}{2.3 d_3^2} \cong \frac{5.40}{d_3^2} \qquad \text{for } p_n = 3.52 \tag{41}$$

The spectrum of intermodulation noise due to third-order nonlinearity in the dynamic characteristic has a fairly flat spectrum with a broad maximum in the center of the baseband.

Because of time-division multiplexing prevalent in PPM, PAM, and PCM systems, nonlinear distortion is not a factor in interchannel modulation noise for these modulation systems. However, there is crosstalk between samples of adjacent channels caused by intersymbol interference due to the shape of filter characteristics. In Eq. (13) it is assumed that b is the upper cutoff frequency (ideal low-pass filter) of the baseband and the sampling pulses are generated by rectangular pulses of duration $t = 1/b$. In most cases there will be also a lower cutoff frequency because of the difficulty in transmitting a d-c component which also contributes to the intersymbol interference.

The following expressions[34] for the channel crosstalk resulting from the upper and lower cutoff frequency are fairly good approximations for the PAM application. If the upper cutoff frequency f_u is defined as the 3-db point of attenuation characteristic with a 6-db/octave slope (such as an RC low-pass filter), the worst case is represented by

PAM worst interchannel crosstalk loss in db (NPR)

$$(NPR)_{f_u} \approx 2\pi \times 8.686(t_g + \delta)f_u \tag{42}$$

where $\quad \delta = (1/2\pi f_u) \log_e 2\pi f_u t_s \approx 3.22/\pi b \approx 1/b$

$\qquad t_g$ = duration of guard slot between two adjacent sampling pulses of width t_s

$\qquad t_s = 1/b$ = duration of sampling pulse in the baseband

$1/(t_s + t_g)$ = resultant repetition rate

$\qquad f_u$ = upper cutoff frequency $\approx b$

$$(NPR)_{f_u} \approx 2\pi \times 8.686(t_g \times b + 1) \tag{42a}$$

If the lower cutoff frequency f_l is defined in the same way as f_u the resulting crosstalk is

$$\text{PAM total interchannel crosstalk in db } (\text{NPR})_{f_l} \approx 20 \log_{10} \frac{f \times b}{2b_1 \times f_l} \qquad (43)$$

where $f(\ll b_1)$ = modulating frequency in the interfering channel
b_1 = bandwidth of one channel

The postdetection noise power in a channel b_1 produced by a statistical noncoherent disturbance at the input to the detector (baseband) is in both PAM and PPM proportional to the mean square of the interfering crosstalk voltage. However, the crosstalk noise power relative to the signal power produced in a single PPM channel is smaller because of the larger PPM noise improvement (13) compared with that of PAM (15).

Thus the crosstalk in PPM systems with more than 10 channels ($b/b_1 \geq 10$) is approximately

PPM worst interchannel crosstalk loss in decibels due to upper cutoff:

$$(\text{NPR})_{f_u} \approx 2\pi \times 8.686\ (t_g \times b + 1) + 10 \log_{10} 0.1 \left(\frac{b}{b_1}\right)^2 \qquad (44)$$

Total interchannel crosstalk in decibels due to lower cutoff:

$$(\text{NPR})_{f_l} \approx 20 \log_{10} \frac{f}{2\sqrt{10 f_l}} \left(\frac{b}{b_1}\right)^2 \qquad (45)$$

Modulation noise in PCM systems is a function of the system parameters which determine the so-called quantizing noise. The quantizing process in PCM is always restricted to a number of the discrete amplitude levels expressed by a digital number of base d and length κ (number of digits) as discussed before.

Figure 15 showed the signal-to-noise ratio due to the quantizing noise in a PCM channel after decoding as derived from Eq. (16) for different values of d (base) and κ (number of digits). Because PCM systems are always used in conjunction with time-division multiplexing, there is no intermodulation noise due to nonlinear characteristics of the modulation system. Intersymbol interference producing crosstalk in PPM and PAM will not affect the quantizing noise in PCM unless it produces a disturbance comparable to the carrier energy-to-noise ratios indicated in Fig. 16a and b required for an expected error rate. As a rule the crosstalk loss given by Eqs. (42) and (43) can be used to check the energy-to-noise ratio produced by intersymbol interference.

18. Line-of-sight Path Loss

The total path loss (L_{tp}) computation for line of sight has to take into account *all* factors in addition to free-space propagation loss. As mentioned before, in all practical cases for operation over the curved earth surrounded by the atmosphere, certain atmospheric factors such as varying temperature, water-vapor content, wind velocity, and air density are at work to cause variations called fadings.

Ray bending is particularly severe if there is a substandard refraction due to an unusual set of meteorological circumstances where the troposphere has a refractive index which increases with altitude instead of the usual decrease. As a consequence of any change from the standard decrease (the so-called standard atmosphere), the transmission rays bend up or down from the earth, creating the same effect as an increase or decrease in antenna height. When this bending is severe, it may happen that a hitherto line-of-sight path degenerates into an obstructed path with a consequent drop in signal level which may last several hours. This corresponds to a shift of the maxima in the height-gain characteristic as shown in Fig. 3. The effects of this kind of fading are slow and can be reduced only by safer site location, which would provide greater path clearance and hence permit inverse bending to take place without the received signal dropping into the first null or an apparent drop below the line of sight.

Multipath fading is due to the simultaneous reception at the receiving antenna of one or more reflecting rays in addition to the direct ray. The vectorial summation of the direct and reflected rays can change quite rapidly with time, and resultant amplitudes can go from well below to well above the free-space value due to phase cancellations and additions.

For line-of-sight operation, multipath fading may result from the summation of two or more significant rays arriving at the receiver. Links over water or desert exhibit a relatively strong earth-surface reflection. However, at any location, a perfect reflection will be approached for rays within 1° from their horizontal. The field strength will then alternate between zero and a value twice the free-space field as the reflective layers change in height. For a reflection coefficient less than perfect, the resulting signal will fall between these extreme values. The receiver antenna is ordinarily located close to the first maximum for the average propagation conditions. With multipath fluctuations of the

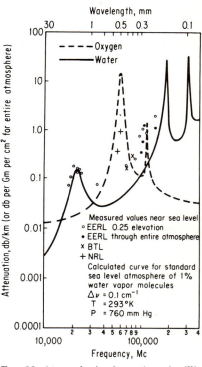

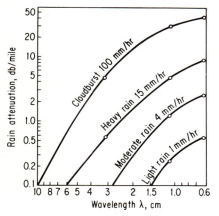

Fig. 31. Rain attenuation L_r vs. wavelength. (*By permission of the American Telephone and Telegraph Co.*)

Fig. 32. Atmospheric absorption of millimetric radio waves. (*Courtesy of Proc. IEEE.*)

transmission medium, the path lengths of the direct and reflected rays vary, and what was hitherto a maximum resultant signal can now change to a minimum signal. This type of multipath effect causes relatively fast fadings. Use of the high-low siting technique will help maintain a relatively steady resultant signal by increasing the ray angle with the predominantly horizontal layers of reflection. Diversity techniques will also reduce the effects of this kind of fading and will be discussed later.

a. Rain Attenuation. Absorptive energy fading is due to the attenuation of signal energy in the transmission medium due to the presence of rain, snow, or fog. At 10,000 Mc, a heavy rain will cause an attenuation[35] of about 0.6 db/mile, falling to 0.1 db/mile at 5,000 Mc. At 2,000 Mc, this effect is quite negligible (see Fig. 31).

b. Water Vapor and Oxygen Absorption. Selective absorption due to water vapor and oxygen[36] in the atmosphere first occurs at 24,000 and 60,000 Mc, respectively, and is hence of no consequence at 2,000 to 10,000 Mc (see Fig. 32).

c. Multipath Fading. The evidence is that on good optical paths, multipath fading is controlling. Multipath fading is generally much worse in the summer months than for any other times of the year, and for any particular day, the fading is greatest in the early morning hours. Bullington[35] has tabulated the following worst months for 11 different sites, 25 to 77 miles in length, at 4,000 Mc, with first Fresnel zone clearance or more: May 1, June 2, July 1, August 1, September 2, October 2, and December 2 as shown in Fig. 33. In the wavelengths from 3 (10 Gc) to 20 cm (1.5 Gc) there is a small reduction in fading as the wavelengths increase. The actual depth of fades as a function of percentage of total time these or deeper fades were reached is given in a number of references. It is not possible to draw generalized conclusions, since the conditions (frequency, site location, etc.) will not always be the same. However, certain results can be used as a basis for designing margins into the equipment to handle fades on a statistical basis:

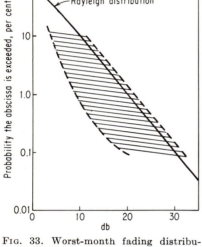

FIG. 33. Worst-month fading distribution. Fading range at 4,000 Mc on 11 typical line-of-sight paths from 25 to 77 miles in length. The crosshatched area shows the range of worst-month fading distribution. (*Bullington, Beyond-the-horizon Transmission, Proc. IRE*, p. 1178, October, 1955.)

1. Almost all the serious fading is due to complex multipath transmission, with at least four important contributing rays.

2. The worst fading is usually the early morning hours during the summer months.

3. Fades as deep as -45 db are occasionally encountered.

4. The fading is frequency selective, and virtually no correlation is shown on frequencies separated by 160 Mc or more.

5. The average maximum rate of change of fade was found to be 10 db/sec with rates as high as 100 db/sec occurring very seldom.

6. The variation of fading with time approaches the Rayleigh distribution on a statistical basis for the "worst month" (usually a summer month). This appears to be a conservative approach to the time variation of fading, and data are available to show close correlation between the Rayleigh distribution and a physical microwave system.

7. It is extremely unlikely that all hops have the same worst month. From the scant evidence on hand, probably no more than 30 per cent of the hops are in simultaneous worst month or Rayleigh fading during any one month. The remaining hops are probably experiencing less-than-Rayleigh fading of varying severity.

d. Multihop Systems. Figure 34 gives the S/N ratio probability distribution[37] for an FM radio relay system made up of n hops when all hops are identical with respect to hop lengths and equipment characteristics such as transmitter power, antenna gains, receiver noise figure, receiver bandwidth, and peak deviation and assuming that the fading in each hop is random and is Rayleigh distributed. The distribution for a dual-diversity 60-hop system is given in Fig. 35. The significance of these curves is that, for probabilities of less than 1 per cent, the noise power from n hops is n times the noise power of one hop while for higher probabilities, the noise power for n hops is considerably more than n times the one-hop noise. For example, at the median probability for a 60-hop system, the overall noise power is 6 db greater than 60 times 0 db.

e. Diversity Considerations. The fact that most of the deep fading is due to the simultaneous reception of a number of rays reaching the receiver antenna by

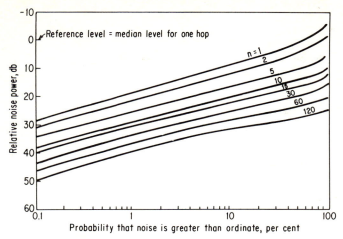

FIG. 34. Noise power probability distribution for an n-hop FM radio relay system assuming Rayleigh-distributed fading with no diversity.

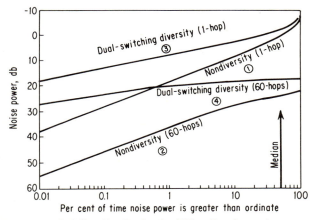

FIG. 35. Noise power probability distribution for FM radio relay systems assuming that all hops are in Rayleigh-distributed fading. 0 db = median level for one nondiversity hop.

separate paths and adding up to bring about a resultant signal of low amplitude is of significance because it permits systems arrangements for mitigating these fades. The method by which this is done is referred to as diversity and takes advantage of the fact that fading is largely independent and uncorrelated when:

1. The signals are transmitted and received at two frequencies spaced 160 Mc or more.[38] This is known as frequency diversity.

2. The single transmitted signal is received by two antennas spaced 100 or more wavelengths apart. This is known as space diversity.

It will be recalled that line-of-sight microwave propagation indicates that, for the worst month in the year, the received signal power will vary in what is known as a Rayleigh distribution. Mathematically, this can be expressed by

$$P_1 = 1 - \exp \frac{-V^2}{v^2} \tag{46}$$

where P_1 = probability that the amplitude is less than some value V
      V = instantaneous measured signal amplitude
      v = unfaded value of V

When $P_1 = 1/2$, the median is about 1.6 db below the unfaded value.

It also can be shown[38] that, if two uncorrelated signals having the same median value are available at a pair of terminals to be selected according to which signal is the stronger (switching diversity), the probability that the stronger signal level is equal to or less than V is

$$P_2 = \left(1 - \exp \frac{-V^2}{v^2}\right)^2 \tag{47}$$

when V/v is small,

$$P_2 = \left(\frac{V}{v}\right)^4 \tag{48}$$

but exactly

$$P_2 = P_1{}^2 \tag{49}$$

If instead of switching to the stronger signal, the signals may be combined in a linear adder,[39]

$$P_2 = 1 - \exp\left(-2\frac{V^2}{v^2}\right) - 2\sqrt{\pi\frac{V}{v}} \exp - \frac{V^2}{v^2} Q \tag{50}$$

where

$$Q = \frac{1}{2\pi} \int_0^{\sqrt{2V/v}} \exp \frac{-X^2}{2} \, dx \tag{51}$$

For small V/v, P_2 approaches

$$P_2 \approx \frac{2}{3}\left(\frac{V}{v}\right)^4 \qquad \text{for } \frac{V}{v} \ll 1 \tag{52}$$

Therefore, the probability that the received signal level is less than a given level V is greater for switching diversity than for linear combining by a factor $\sqrt{2/3}$, or 0.9 db.

Now, if instead of switching we are to employ ratio squarer combining,

$$P_2 = 1 - \left(1 + \frac{V^2}{v^2}\right) \exp \frac{-V^2}{v^2} \tag{53}$$

and for small V/v, P_2 approaches

$$P_2 \approx \frac{1}{2}\left(\frac{V}{v}\right)^4 \qquad \text{for } \frac{V}{v} \ll 1 \tag{54}$$

This results in a 1.5-db advantage for ratio squarer combining over switching. The above-listed probability distributions are plotted in Fig. 36. For higher-order diversity systems, diversity gains are plotted in Fig. 37.

f. Free-space Loss and Median Path Loss. The free-space loss L_{fs} is the propagation loss between two isotropic antennas in a region free of obstructions or atmospheric effects as given by Eqs. (8) and (8a). Free-space loss increases by 6 db with every doubling of distance (see nomogram, Fig. 38).

$$L_{fs} = 4.56 \times 10^3 f^2 d^2$$

or
$$L_{fs} \text{ in db} = 36.6 + 20 \log \frac{f}{\text{Mc}} + 20 \log \frac{d}{\text{mile}} \tag{55}$$

where f = frequency, Mc
      d = distance between isotropic antennas, miles

The median path loss for a Rayleigh faded signal was determined to be 1.6 db less than the unfaded or free-space loss.

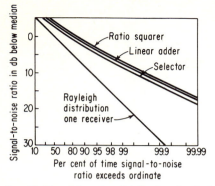

FIG. 36. Signal-to-noise distribution for dual diversity using various combining methods.

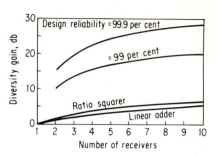

FIG. 37. Diversity gain as a function of the number of receivers. Upper two curves for selector combining; lower two curves for linear and ratio squarer combining.

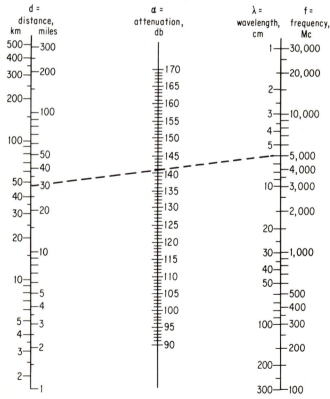

FIG. 38. Nomogram for solution of free-space path attenuation α between isotropic antennas. $\alpha = 37 + 20 \log f + 20 \log d$ decibels. Example shown: distance 30 miles, frequency 5,000 Mc; attenuation = 141 db.

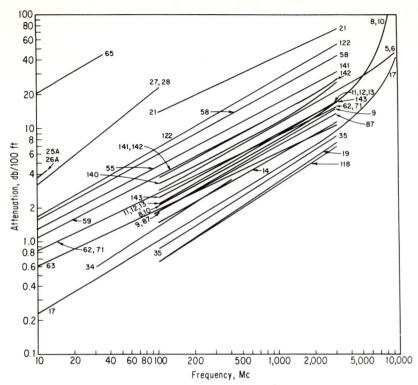

Fig. 39. Attenuation of some flexible coaxial lines. (*From "Antenna Engineering Handbook," by Henry Jasik. Copyright © 1961. McGraw-Hill Book Company. Used by permission.*)

The mean path loss for a Rayleigh faded signal is 10.4 db greater than the free-space loss.

The total path loss L_{tot} is the summation of the fading margin and atmospheric losses added to the free-space loss.

g. Total Path Loss. The total loss L_{tot} of a line-of-sight radio relay path is the sum of all loss factors (in decibels).

$$L_{tot} = L_{fs} + K_1 + L_O + L_r + L_v + L_{fad} \tag{56}$$

where L_{fs} = free-space loss from Eq. (8) or Fig. 38, db
K_1 = correction factor of free-space loss/median loss
= −1.6 db for a Rayleigh-faded path
L_O = oxygen loss from Fig. 32, db
L_r = rain loss from Fig. 31, db
L_v = water-vapor loss from Fig. 32, db
L_{fad} = fading margin, db
For Rayleigh fading, use Fig. 33. For multiple-hop systems, use Fig. 34. For diversity systems, use Fig. 37.

h. Example of Path-loss Calculation. What is the total path loss without diversity for 99 per cent propagation reliability between isotropic antennas for a 20-mile path at 5,000 Mc? Assume heavy rain to persist over 10 miles of the path and Rayleigh fading.

$$L_{fs} = 36.6 + 20 \log f + 20 \log d$$
$$= 36.6 + 74 + 26$$
$$= 136.6 \text{ db}$$
$$K_1 = -1.6 \text{ db}$$
$$L_0 = 0.013 \times 20 = 0.3 \text{ db}$$
$$L_r = 0.1 \times 10 = 1 \text{ db}$$
$$L_v = \text{negligible}$$
$$L_{fad} = 18.4 \text{ db}$$
$$L_{tot} = 136.6 - 1.6 + 0.3 + 1 + 18.4$$
$$= 154.7 \text{ db}$$

19. Transmission Lines and Antennas

A loss of signal energy is incurred in the transmission lines[40] connecting the antenna with the receiver and the transmitter. The choice of line is generally made on the basis of minimizing line loss, consistent with the fulfillment of other requirements including:

1. Economics
2. Power-handling capabilities
3. Transportability (rigid vs. nonrigid, size and weight, bending radius)
4. Temperature, environment
5. Impedance

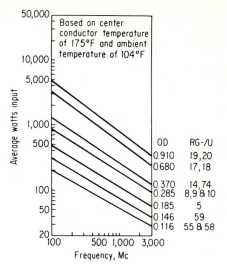

FIG. 40. Power-handling capacity of some flexible coaxial lines. (*From "Antenna Engineering Handbook," by Henry Jasik. Copyright © 1961. McGraw-Hill Book Company. Used by permission.*)

Generally, at lower microwave frequencies (below 2 kMc) and medium powers (50 watts) coaxial cable is used; for the higher microwave frequencies and higher powers, rectangular waveguide is generally the choice. A useful chart giving attenuation of commonly used coaxial cables and waveguides is given in Fig. 39. From this chart, for example, the loss per 100 ft at 1 Gc for RG-8 is 8.9 db per 100 ft.

The power-handling capacity of flexible coaxial lines as a function of frequency is given in Fig. 40. For the general case of a coaxial line,

$$A_e = A_c + A_d$$
$$A_e = 4.34 \frac{Rt}{Z_0} + 2.78f \sqrt{\epsilon} \, F_p \tag{57}$$

where A_e = attenuation, db/100 ft
A_c = attenuation due to conduction losses, db/100 ft
A_d = attenuation due to dielectric losses, db/100 ft
Rt = copper resistance per 100 ft

$$= 0.1 \left(\frac{1}{d} + \frac{1}{D} \right) \sqrt{f}$$

D = inner diameter of outer conductor, in.
d = outer diameter of center conductor, in.
f = frequency, Mc/sec
ϵ = dielectric constant relative to air
F_p = power factor of dielectric at frequency f
Z_0 = characteristic impedance, ohms
$$= (138/\sqrt{\epsilon}) \log_{10} D/d$$

Assuming that the dielectric constant and power factor are independent of frequency,

The conductor losses are proportional to the square root of frequency.

The dielectric losses are directly proportional to frequency.

As the frequency increases, the dielectric losses become increasingly important.

For the general case of a rectangular copper air-filled waveguide operating in the TE_{10} mode, the attenuation is given by

$$A_e = \frac{1.107}{a^{3/2}} \left[\frac{(a/2b)(f/f_c)^{3/2} + (f/f_c)^{-1/2}}{\sqrt{(f/f_c)^2 - 1}} \right] \sqrt{r} \tag{58}$$

A_e = attenuation, db/100 ft
a = larger inner dimension, in.
b = smaller inner dimension, in.
f = operating frequency, Gc
f_c = cutoff frequency, Gc = $5.9/a$
r = metal resistivity relative to copper

Metal	*r*
Copper............	1
Aluminum.........	1.28
Brass.............	2
Chromium.........	1.23
Rhodium..........	1.71
Silver............	0.97

Directive antennas[41] as used at UHF and microwave frequencies for point-to-point service provide a "radiation gain" as discussed before (see Fig. 1). The gain of an

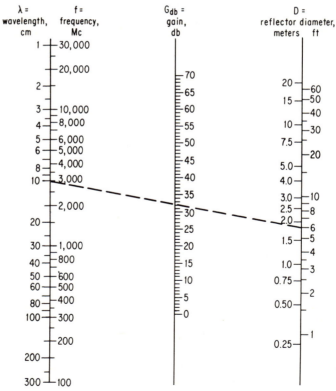

Fig. 41. Nomogram for determination of apparent power gain G_{db} (in decibels) at a parabolic reflector. Example shown: frequency, 3,000 Mc; diameter, 6 ft; gain = 32 db.

actual antenna relative to an isotropic antenna is

$$G = \frac{A}{A_{\text{iso}}} = \frac{A 4\pi}{\lambda^2} \qquad (59)$$

where λ = wavelength of signal

A = effective area of antenna under consideration

$A_{\text{iso}} (\lambda^2/4\pi)$ = effective area of an isotropic antenna

The effective area of the actual antenna is equal to the true area of the antenna multiplied by its efficiency.

The most commonly used directive antenna in UHF and microwave point-to-point applications is the parabolic antenna. This antenna consists of a horn or dipole feed which illuminates a large curved surface whose shape is that of a parabola of revolution. The efficiency of the antenna is less than unity for a number of reasons, primarily nonuniform dish illumination and energy spillover past the edges of the dish. In practice the efficiency can vary from 55 to 65 per cent. A conservative figure to use is 55 per cent.

Thus the gain of a parabolic reflector antenna is

$$G = 0.55 \left(\frac{\pi d}{\lambda}\right)^2 \qquad (60)$$

or $\qquad G_{DB} = 20 \log f + 20 \log d - 52.5 \text{ db}$

where G_{DB} = antenna gain, db

f = frequency, Mc

d = diameter of parabola, ft

A nomogram for this relationship is given in Fig. 41.

The 3-db beamwidth α of the parabolic antenna is given by

$$\alpha = \frac{68,700}{f \times d} \qquad (61)$$

where α = 3-db beamwidth, deg

f = frequency, Mc

d = diameter of parabola, ft

20. Beyond-the-horizon Path Loss

Communication is possible at UHF and microwave frequencies beyond the horizon up to about 600 miles by means of tropospheric forward scatter. The energy arrives at the receiver essentially by "scattering" due to discontinuities in the air scatter volume "seen" by both antennas. The path loss under these conditions is much higher than the free-space loss (Fig. 5) but considerably less than that due to the diffraction mode alone. The large loss must be compensated by high transmitter power, very large and high-gain antennas, low-noise receiver front ends, and diversity reception. However, increasing antennas beyond a certain limit has been found to limit the gain of the antennas that can be realized in operation. The theoretical increase is diminished by a factor related to the ratio of the scatter angle to the antenna beamwidth and sometimes called the "aperture-to-medium coupling loss." The physical mechanism causing this loss is not yet fully understood.

In a calculaton of system performance,[42-44] such as S/N ratio obtained or transmitter power required, the relations developed in the preceding section for line-of-sight systems hold, except for the path-loss calculation. The path loss for operation in the troposcatter mode has to take into account all factors for this mode of propagation:

$$L_{tot} = L_{fs} + L_s + L_{ref} + L_{fad} + L_{cpl} - G_T - G_R$$

where L_{tot} = total path loss, db

L_{fs} = free-space loss from Eq. (8) or Fig. 38, db

L_{cpl} = aperture-to-medium coupling loss, (see Fig. 44), db

L_s = median scatter loss [see Eq. (62)], db

L_{ref} = correction factor due to variation of refractive index n_s of air near the surface

L_{fad} = fading margin (see Fig. 43)

G_R or G_T = receiver or transmitter antenna gain, respectively

a. Median Scatter Loss

$$L_s = 57 + 10(\theta - 1) + 10 \log_{10} \frac{f}{400 \text{ Mc}} \qquad \text{for } \theta > 1° \tag{62}$$

where θ = scattering angle, deg

f = carrier frequency, Mc

The scatter angle is determined from the scatter geometry as shown in Fig. 42.

$$\theta = (\theta_0 - \theta_1 - \theta_2) \qquad \text{deg}$$

$$= \frac{d - d_1 - d_2}{R} \frac{180}{\pi} \qquad \text{deg} \tag{63}$$

$$d_1 = \sqrt{2Rh_1}$$

$$d_2 = \sqrt{2Rh_2}$$

where R is the radius of the earth ($= 3,963$ miles) (R and h are in the same units).

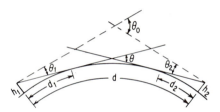

FIG. 42. Tropospheric scatter geometry.

b. Correction for Refraction. This factor corrects for the variation of mean yearly surface refractive index N_s of air for other years and other areas.

$$L_{ref} = -0.2(N_s - 310)$$

where $N_s = (n - 1)10^6$

n = surface refractive index of air[47]

c. Fading Margin (L_{fad}). Fading in tropospheric scatter is a combination of fast fading and slow fading. The fast fading is due to multipath and has a Rayleigh distribution.[46] The slow fading is due to changes of the refractive index of the atmosphere. It has a normal distribution with a standard deviation varying from 2 to 10 db, depending on the scatter angle.

A set of fading margins which statistically combine both fast and slow fading distributions for various scatter angles and orders of selection diversity are shown in Fig. 43. For combiner diversity, the fading margin can be reduced as follows:

Order of diversity	Combiner improvement, db
1	0
2	1.5
3	3.0
4	4.0

d. Aperture-to-Medium Coupling Loss. Theoretical calculations of aperture-to-medium coupling loss L_{cpl} have been made by Booker and deBettencourt, Staras, and others. The Staras[45] results appear to fit limited experimental data most closely. A plot of results is given in Fig. 44.

When propagation takes place over mountain peaks representing a relatively sharp knife edge, diffraction will occur which may result in reduced scatter loss. In such cases the median scatter loss L_s [Eq. (62)] may be replaced by knife-edge diffraction loss shown in a nomogram in Fig. 45.

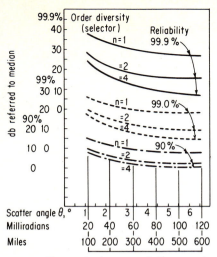

FIG. 43. Tropospheric scatter fading margin with selector diversity. (*Courtesy of Trans. IEEE-PTGCS.*)

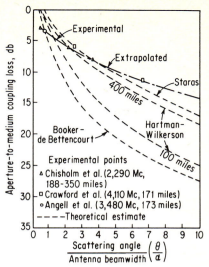

FIG. 44. Aperture-to-medium coupling loss in tropospheric scatter antennas. (*Courtesy of Trans. IEEE-PTGCS.*)

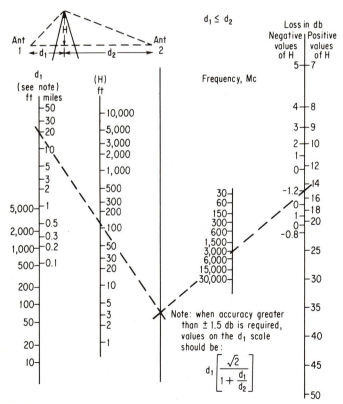

FIG. 45. Knife-edge diffraction loss relative to free space. (*Courtesy of Trans. IEEE-PTGAP.*)

RELIABILITY AND MAINTENANCE

21. General Considerations

A radio relay system must be maintained systematically to minimize losses of income due to traffic outages and to keep the level of cash expenditures for system upkeep to the lowest practical amounts (see also Refs. 48 to 50, 52, 54.) A principal ingredient of maintainability is reliability, some pertinent aspects of which are considered in the following paragraphs.

Responsibility for achieving high operating reliabilities must be shared by the designer and the operator (user) of a system. It is assumed in this section that the equipment designer will have taken adequate precautions to ensure a minimum of equipment failures. For example, all units will have been designed with forethought to easy maintenance, whereby the more failure-prone components such as tubes are readily accessible and easily replaced; an ideal design would group together components having similar failure modes. Modular construction with the least possible number of movable contacts will have been used to reduce maintenance inspections and troubles. The maintenance routines themselves should involve the simplest practicable techniques using a few basic test equipments of moderate accuracy.

Upon the owner-operator, or user, of a system rests the responsibility of high operating reliability starting at the time of ordering a system, long before it is installed. A system must be properly engineered from the start to ensure reliable operation at full channel capacity under all conceivable propagation conditions and environments. No amount of maintenance can compensate for signal loss due to systems underplanning, such as improper station siting or excessive spacing between relay stations. An important feature of modern planning of extensive radio relay systems includes automatic fault detection and reporting as well as automatic switching to standby equipments when faults occur.

The system operator must choose between two categories of maintenance: preventive and corrective. Preventive overmaintenance can be as harmful as undermaintenance; hence a happy medium must be sought in the early stages of a maintenance program. A certain amount of preventive maintenance is desirable in the form of routine inspections and systematic recording and review of selected meter readings. Such procedures help to provide warnings of approaching end-of-life conditions and simultaneously serve as a continuing training aid in the rapid testing and correcting of equipment troubles.

Preventive maintenance consists of locating a fault, diagnosing its cause, and correcting it with the least possible effect on communication services before a breakdown occurs. Of these factors, the trouble correction is often less problematical in radio relay systems than the fault location and diagnosis, especially with remotely located, unattended stations. Many schemes have been devised to simplify the diagnosis and location of remote faults,[55–58] and several typical fault supervision systems are described in paragraph 24.

Efficient maintenance requires the establishment of a maintenance organization. With extensive systems this may involve the designation of several maintenance centers, but their number and location are determined by the reliability demands of the system, availability and capability of technically trained personnel, existence of a training center, and geographic and economic factors. In large systems it is often the practice to perform only minor repair work at the remote sites. Fieldmen are generally equipped only with sufficient test equipment to localize troubles and with spare panel units which are used to replace a defective unit. Faulty units are then taken to a central repair depot for a thorough overhaul. In order to perform these operations efficiently, a maintenance organization must be equipped with suitable vehicles and also with intrasystem communications, including local and express order wire, as well as mobile radio to connect intermediate and base stations with the repairmen's vehicles.

Traffic outages can be attributed broadly to three basic categories of faults: loss of signal, failure of electronic equipment, and power failures. These faults are related, but they can be treated separately within certain limits.

A signal may be lost owing to any of the above three basic causes; i.e., it may be a mechanical fault in antenna or tower structure; it may be an equipment failure; it may result from loss of power. However, antenna circuitry is obviously a vital part of maintenance. Contact surfaces and junctions of cable and waveguides must be kept weather resistant; dielectric cable performance must be checked to prevent signal loss due to aging; antenna and tower structures must be kept rigid to ensure that path losses remain within specified limits, such as 3 db. Complete antenna failures are experienced so rarely that it is uneconomical to provide them in duplicate.

22. Component Failure Rates

Failures of components can contribute seriously to transmission outages, but preventive maintenance can help to minimize these outages. Estimates of average failure rates of typical components given in Table 4 highlight where major troubles may be expected. In particular, vacuum tubes appear to suffer the highest failure rates. Accordingly it is rewarding to stress close observation of vacuum-tube operation, especially of the high-power transmitting-type tubes. Aging, break-in periods, and reliability review programs are essential parts of successful vacuum-tube maintenance programs. The failure rates of tubed equipments can be reduced by a factor or 7 to 8 in many cases by operating continuously instead of on a cyclical basis. For tubes or other components with predictable and reasonably uniform life expectancy, it would be possible to eliminate periodic tube checking by replacing them when they reach 80 to 90 per cent of life expectancy. Another benefit of a reliability program is the product improvement which can result from maintenance reports and failure analyses.

Table 4. Typical Average Failure Rates*

Components	Estimated failure rates % per 1,000 hours of operation
Capacitors (general-purpose)	0.01 –0.6
Capacitors (electrolytic)	0.02 –2.0
Crystal diodes	0.05
RF inductors	0.05
Meters	0.2
Motor/generators (below 1,000 rpm)	0.04
Potentiometers	0.3
Relays	0.001–0.5
Resistors, fixed	0.01 –0.3
Switches	0.01 –0.1
Transformers	0.05 –2.0
Transistors	0.2
Tubes (receiver types)	1.0 –2.0
Tubes (high-power, transmitter types)	1.0 –20.0
Soldered joints (dipped)	0.0001
Wrapped joints	<0.0001

* Ref. 50.

Table 4 shows also the comparative advantage of stressing vacuum-tube performance as distinguished from capacitor or resistor performance. For example, analyses confirm that tubes have the highest failure rates and in many instances are responsible for about half of equipment failures. Resistors and capacitors have considerably lower failure rates, but they are more numerous than tubes and account for over one-third of all failures. Even if tubes could be made perfect, this would only halve the equipment failures. On the other hand, improving reliability of resistors and capacitors would result in only small improvements in overall equipment reliability in view of the low failure rate of these components. These considerations highlight the false economy of purchasing equipment by price alone when traffic reliability is an important factor.

In operating digital equipment in particular, it is desirable to verify the operability of all components, including especially those which are used only occasionally. A dynamic check, as distinguished from a static test, can often be devised to ensure that all components are exercised during a checkout procedure.

It should be noted that components have been developed for special applications with much lower failure rates than those shown in Table 4. However, such components must be specially defined and selected.

23. Power-supply Considerations

Radio relay routes often traverse desolate areas where commercial power may be unreliable or unavailable. An independent reliable power source is therefore invariably required as backup reserve power, if not as a primary source, and the most practical supply is the automatic diesel-electric power plant.

In high-reliability radio relay system installations used to transmit dial telephone, data, or other synchronous transmissions, even momentary power interruptions are intolerable. For these installations the power units generally include a no-break feature using either a floating battery system or flywheel techniques. Triple diesel alternator installations may be specified for isolated cases where sites are completely inaccessible during certain periods.

In installations of lower reliability, single diesel-electric plants may serve as backup for unreliable commercial power; voltage regulators serve to smooth fluctuating supplies.

In all the above installations it is mandatory that periodic maintenance inspections and routines be carried out in order to attain operational continuity. In the case of dual diesel installations, for example, some operators have found it profitable to alternate engines on a 5-to-1 basis and to overhaul the machine with heaviest service at periodic intervals. Fault detection and reporting techniques, customarily applied to dual installations, ensure operational reliability (See Chap. 20).

24. Supervisory and Alarm Systems

The operation of radio relay systems can be supervised by direct inspection or by semiautomatic or automatic means, ranging from very simple systems using on-off tone signals to indicate remote troubles to highly sophisticated systems for reporting faults automatically. A complete automatic fault-reporting system employing computer techniques and components for use at a central control point would provide automatic continuous reports of urgent and nonurgent trouble conditions, audible and visual alarms, and a display of the operational status of all local and remote equipments. Many such systems have been described in the literature, some are referenced in this section and in the bibliography, and several typical remote monitoring systems are described briefly below.

A relatively simple supervisory system[55] for ensuring reliability of uninterrupted service and for meeting the requirements for unattended operation is used in a broadband microwave system of the Bell Telephone Co. Alarms report equipment failures and abnormal conditions at remote stations to an attended control point from which maintenance personnel can be dispatched. A voice frequency facility is also provided to enable communication between radio stations. Six major alarms are provided, including:

Failure of commercial power
Failure of transmission
Low battery and lightning arrestor failure
Failure of air-navigation tower warning lights
Recovery of tower warning lights
"Signal in" by maintenance personnel from one of the radio stations

The fault-reporting system operates in the baseband spectrum below 4,000 cycles and uses a 2,600-cycle tone which originates at the control station and is continuously transmitted to the far terminal and looped back over the return path to the control station. Trouble or abnormality is indicated by office alarms which are activated if the tone either fails to return or is interrupted. When an alarm is registered, the 2,600-cycle tone oscillator is manually returned to one of a series of interrogation frequencies which are assigned on an exclusive basis to each remote station. A bridging filter at each remote station allows the assigned tone to be returned to the control point. A simple analysis thus isolates the trouble location to one or an adjacent pair of stations for most faults. Thus on interrogation the control point receives a tone, no tone, or a pulsed tone. A continuous tone indicates no trouble; no tone indicates a transmission failure; a pulsed tone indicates commercial power failure or low battery voltage. The returned tone can be coded to obtain additional information required by air-navigation authorities regarding status of tower lights, top light flasher, low battery voltage, and lightning-arrestor failure.

Another efficient automatic supervisory system[56] is one in which the basic design has been simplified by application of solid-state components. A simple switch-over logic arrangement eliminates the need for sensing tones and waveguide switching (by means of a power splitter using a tunnel diode amplifier).

Figure 46 shows both the alarm transmitting and decoding equipment used at a supervisory terminal station. Under normal conditions when no alarms are being transmitted, an FSK tone transmitter at each station sends a steady tone on its assigned spacing frequency. In the referenced design 18 channels are provided in the 4- to 8-kc range.

When an alarm condition develops, the alarm encoder causes the tone transmitter to send out a continuous series of frequency-shift signals consisting of short pulses for normal conditions and long pulses for fault conditions. The supervisory equipment at an attended station detects the transmission of alarm data by means of a tone receiver associated with the channel of each remote station. After the alarm has been acknowledged, the operator switches an alarm decoder unit to the receiver which is reporting data and the particular type of alarm is displayed on a lamp panel. This alarm system continuously monitors all stations. No interrogation is required. It does not depend on two-way continuity of the microwave system and is designed for maximum reliability plus freedom from false alarms.

A complete alarm system of this type was designed to monitor up to 14 conditions at each of 18 remote stations. This may be expanded, and any of the 14 alarm inputs may be combined into a single alarm. As shown in Fig. 47, the main and standby alarms are combined in an OR gate, since there is seldom any need to differentiate between the two alarms. In order to avoid alarms that might be caused by momentary fades or by equipment adjustments, a time-delay circuit is included. For alarms such as illegal entry or power failures, this delay may be bypassed. Operation of the waveguide transfer switch is also monitored, and an alarm may be initiated on transfer or on loss of the normal polarizing current supplied to the circulators.

Other sophisticated faults supervision schemes include project FIST (Fault Isolation by Semiautomatic Techniques) and project ADMIRE (Automated Diagnostic Maintenance Information Retrieval), which are described in Refs. 59 and 63.

25. Operating Statistics

In subsequent paragraphs it is shown how maintenance efficiency is improved by keeping detailed performance records, analyzing them periodically, and applying corrective measures where possible. Published information on performance and maintenance experience with radio relay systems is very limited, probably mainly because of the time which must elapse before reliable conclusions can be drawn. Brief excerpts are given below of the available experiences with radio relay systems of a United States power system company and of two communications organizations, one in the Arctic region and the other in Nigeria, Africa.

A microwave system owned and operated by the American Electric Power Service Corp.[53] includes 1,210 RF path miles, 18 terminal stations, 21 drop-channel repeaters, 22 through repeaters, and five passive (billboard) repeaters. The extent of these

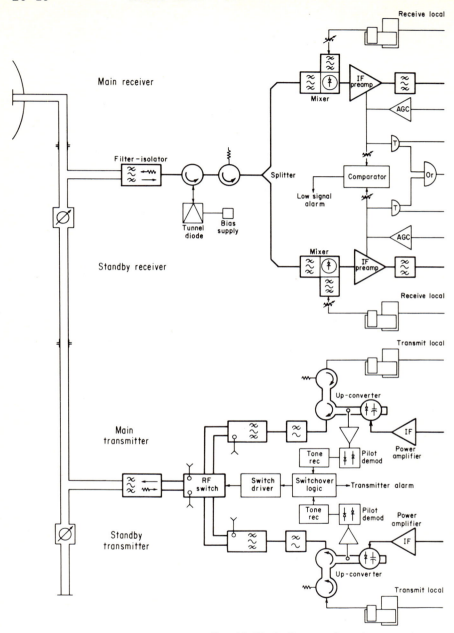

Fig. 46. Block diagram of an alarm system at a

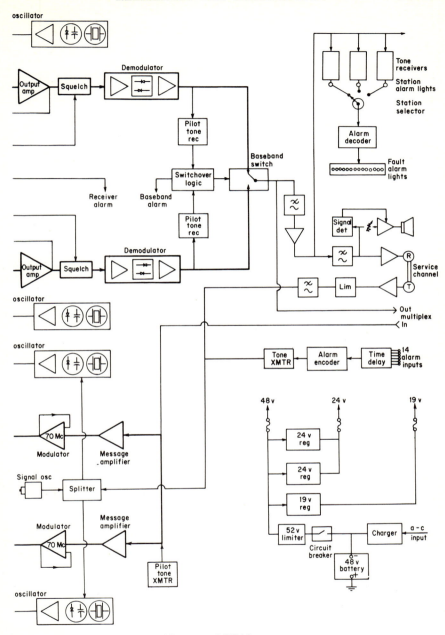

supervisory terminal station. (*Courtesy of RCA.*)

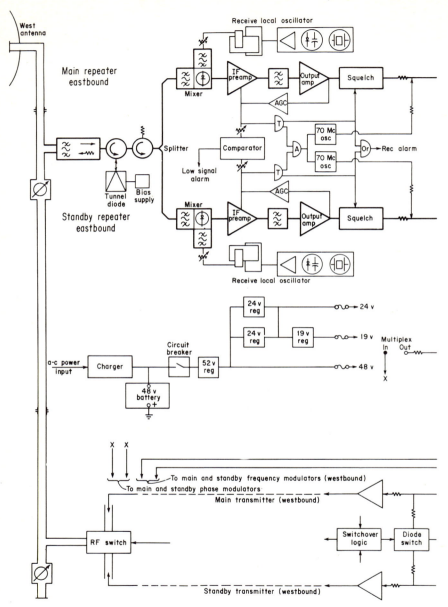

Fig. 47. Block diagram of an alarm system

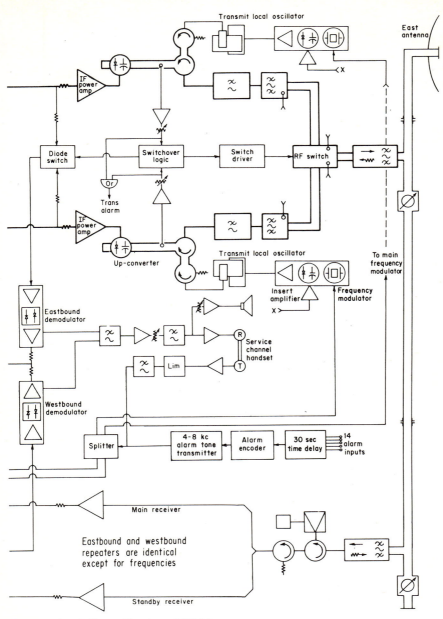

at a repeater station. (*Courtesy of RCA.*)

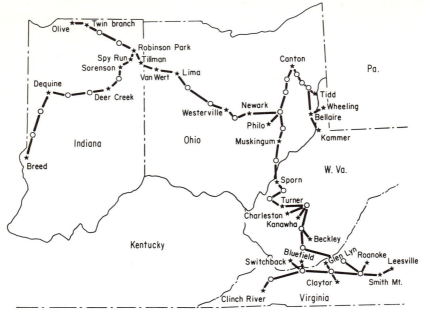

Fig. 48. American Electric Power microwave system. (*Courtesy of American Electric Power Service Corp.*)

facilities is shown on the system map, Fig. 48. The system operates mostly in the 6,700-Mc range, using vacuum-tube transmitters with FDM and SSB modulation to transmit voice and telemetering data channels using mostly FSK. The channel utilization is shown in Table 5. Records of outages have been accumulated and analyzed in considerable detail. Detailed performance records have been kept for several subroutes of the microwave system and are summarized in Table 6.

Table 5. Channel Utilization

Service	Channel, miles	Per cent of total
Voice dispatch................	7,290	43.5
Voice administrative...........	6,170	36.8
Telemeter and deviation........	2,805	16.7
Radio control................	290	1.7
Supervisory...................	205	1.2

The causes of microwave system interruptions shown in Table 7 are listed as percentages of the totals observed. An outage is defined by this company as "any degradation of signal sufficient to cause spiking for one or more minutes on telemeter charts." An interruption due to fading is considered to extend over the entire period during which the telemeter signal is unsuitable for input to automatic control apparatus without repetitive manual attention or alarm resetting. During fade outages of up to about 15 min, voice channels are intermittently operative for all but about one-third of this time.

Experience gained from analysis of past outages has permitted continued improvements in system reliability. For example, outages attributable to lightning during

Table 6. Continuity of Microwave Subsystems

Number of hops	Frequency, Gc	Standby mode	Hours observed	Hours outage	Average per hop reliability, %
4	6	Cold	61,864	28.4	99.988
7	6	Cold	45,422	32.7	99.990
4	2	Cold	39,360	21.4	99.986
3	6	Hot	30,070	51.9	96.942
4	2	Cold	29,928	26.3	99.978
9	6	Hot	25,942	3.6	99.998
6	6	Hot	25,125	37.4	99.975
1	6	Hot	21,120	6.3	99.970
6	2	Hot	6,696	17.2	99.957

Table 7. Causes of Interruptions

Cause	Per cent
Fading	33
Communications equipment failure	45
Power equipment failure	21
Miscellaneous and unknown	1

the summer and autumn of 1955 accounted for 50 per cent of all interruptions to the original Canton-Philo system in the first 2 years of operation. Improvements were made in repeater station tower, building, and equipment grounding. Service entrances were protected with low-voltage pellet-type arresters, and power wiring was revised to remove regulating transformers from the emergency power-generator circuits. This work was effective in that no outages directly attributable to lightning have occurred at these stations in the past 7 years. Outages due to this cause on the system as a whole have been negligible.

Another example, the White Alice System, consisting of 66 radio relay stations, spans the length and breadth of Alaska to form a vital link between the far north and Air Defense Headquarters in Colorado Springs, Colo. It is the first large-scale network of its type established to provide high-quality, reliable communications facilities where, previous to White Alice, communications media were erratic or nonexistent. Many of the radio relay sites are located on mountain tops in some of the remotest, most inaccessible terrain a communications network has ever encountered. A variety of equipment, which includes tropospheric scatter, microwave, VHF, and land lines, relay the signals over distances up to 395 statute miles. The system of 66 sites has 45 attended and 21 unattended relay stations and provides 1,453,300 miles of broad-band voice channels.

A typical reliability report[58] shows system outage of $\frac{3}{100}$ of 1 per cent, during a typical month. The actual total outage hours were 315 of total assigned circuit hours of 1,142,712. Accordingly, the system reliability index for the particular month was 99.97 per cent, which attests to the high reliability made possible by careful design and efficient maintenance management. The comparatively minor outage time was accumulated as shown in Table 8 with the contributing factors arranged in order of decreasing importance.

The experiences gained with the Nigeria VHF network have been described in considerable detail in an excellent article,[50] and some of the maintenance highlights are described in the following paragraphs. The system comprises a main radio network of 1,000 route miles, forming 43,000 channel miles. It includes 18 link paths with 18 terminal stations at sites and 23 repeaters. The system is unique in that instead of standby equipment a twin-path arrangement was provided in which priority traffic is carried on one path and nonpriority in the other. Priority groups are patched to the

Table 8. White Alice System Typical Outage Time

Cause of outage	Circuit outage hours (of 1,142,712)	Per cent of hours (of 315 outage hours)
Power....................	81.85	25.98
Radio....................	73.77	23.42
Terminal equipment........	62.75	19.92
Operating error............	50.58	16.06
Carrier...................	25.25	8.02
Miscellaneous..............	20.8	6.60
Total outage hours........	315.00	100.00

nonpriority path upon failure of the main path, which results in loss of nonpriority traffic. The nonpriority traffic is the basis for quoting system performance, as priority traffic is lost only when both paths fail simultaneously—a rare occurrence. However, the absence of a standby path increases the need for efficient maintenance and fault control.

In the Nigerian system maintenance is performed by a group of 26 radio and transmission engineers who maintain all VHF radio and all telephone channeling and telegraph equipments. Six diesel engineers maintain diesel engines and generators. There are also about 75 locally employed technical officers and technicians with a force of 100 miscellaneous staff such as riggers, clerks, and drivers.

Faults are classified by categories so as to simplify the reporting by operators and the subsequent analyses. A fault analysis for 1960 for 16 links revealed an average outage time of 1.09 per cent, or 1.83 hr per link path, distributed as shown in Table 9.

Table 9. Typical Outage and Fault Record

Fault type	Outage time, %	No. of faults, %
Radio equipment.......................	21.8	15.8
Fading and FNF (fault not found).......	19.9	28.0
Fuses................................	18.5	19.6
Tubes (other than rectifiers).............	14.8	12.2
Components (as resistor, capacitor)......	12.2	4.7
Rectifier tubes........................	6.5	6.3
Power supply.........................	5.5	13.0
Lightning.............................	0.8	0.4

Partly as a result of the fault analyses numerous remedial steps were taken. To reduce outages due to rectifier tubes, they were changed to another type and the new types were replaced semiannually at unattended stations. Grounding was improved at radio stations, and surge diverters were provided to reduce the number of blown fuses. Surge diverters were installed also at tower lighting cables, at overhead power lines, and between radio equipment and dual engine sets. Slow-blow and antisurge fuses were also introduced as a further improvement measure. Fading and Fault Not Found outages were also reduced by thorough overhauling of radio equipment.

TYPICAL COST DATA

Among the most important considerations in planning transmission facilities such as microwave radio relay links are the cost of putting them in operation (first-cost installed) and the cost of maintaining service (annual cost). The first usually includes

Table 10. Cost Comparisons of Some Typical Private Microwave Systems

Illustration No.	0	1	2	3	4	5	6	7	8	9	11	12	13	14	15	18
General features:																
Frequency band, Gc	6	6	...	...	6	...	6	6.7	...	6	...	6	6	6	6	...
No. of stations	30	...	...	...	10	...	...	20	15	...	17	52	...	...	...	7
No. of terminals	3	3	4	3	...	2	...	...	...	3	...	...	2	6	...	...
No. of repeaters	4	4	8	8	...	...	...	...	...	1	...	...	...	9	...	...
Path miles	700	125	195	240	200	6	16	400	400	30	255	1,033	156	594	135	156
Channel capacity	48	48	...	24	25	2	0.6	...	...	20	48	48	12	48	10	...
Active voice channels	28	33	18	...	...	...	...	36	24	...	23	...	...	...	...	3
Investment cost (approx):																
Total thousand $	600	250	800	345	414	13	42	640	500	60	510	1,772	40	180[14]	70[15]	178
Per path-mile, thousand $	0.850	2	4.1	1.44	2.1	2.5	2.7	1.6	1.3	2	2	1.7	0.255	0.3	0.52	1.2
Per channel-mile, dollars	30	60	228	60	83	1,200	110	45	52	100	87	36	21	6	52	400
Annual operating costs (approx.), (thousand $):																
Power	6	...	...	15	...	...	...	8	...	1.2	6	...	...	...	...	5.4
Service and inspection	60	...	...	32	...	...	...	9	...	6	25	...	...	...	...	17.8
Total	13[1]	40	47	...	37[4]	0.25[5]	0.26[6]	17	52	7.2	31	12[12]	...	...	...	23.2

Notes:

[1] Maintenance averaged.
[4] Principally labor, operating, and maintenance.
[5] Tubes.
[6] Tubes and parts.
[12] 10-man maintenance crew.
[14] System used mountain tops, hence very few towers, also existing buildings and staff.
[15] All sites previously owned.

(handwritten marginalia:)
Investment cost
$ 2000/path mile
$ 60/channel/mile
$ 13,000/year ÷ 45 ÷ 125 miles = $2
$ 31,000/year ÷ 48 ÷ 255 = $2.50
Plus
Total =

Table 11. Basic Installed Cost for Various Communications Systems

System	Cost per mile	Cost per channel-mile	
		600 channel	72 channel
Microwave relay:			
Military........................	$12,000	$20	$167
Commercial..................	8,000	13	111
Average......................	10,000	17	139
Submarine cable:			
High..........................	11,500	...	160
Low..........................	8,500	...	118
Average......................	10,000	...	139
Troposcatter:			
High-grade military.........	25,000	...	347
Low-capacity military.......	15,000	...	208
Average......................	20,000	...	278
Coaxial cable (L-3):			
High..........................	45,000	75	625
Low..........................	20,000	33	278
Average......................	32,500	54	451

equipment cost, shipment to the site, installation, and acceptance tests. The latter consists of capitalization of first cost (depreciation and replacement) plus operation and maintenance. It is obvious that methods of keeping account of the various cost items will differ widely; however, the following survey gives a good picture of typical cost data.

A survey[60] of privately owned microwave communications systems is summarized in Table 10, which shows a wide variety of initial costs and channel usage. The table is based on replies to questionnaires sent to electric utilities, airlines, highway departments, and a railroad. Several of these systems include TV circuits, but these costs are not shown because of incomplete data. It must be kept in mind that the cost data represent a wide area of circumstances; in some cases the systems were installed in existing buildings and required neither additional shelter nor personnel. In other cases, the need for towers was obviated by using the terrain advantages of naturally elevated sites.

A useful comparison[61] of circuit costs as a function to trunk cross section in the Bell System is shown in Fig. 49. It is seen that the annual costs drop steadily with increasing numbers of circuits. The curves display clearly the economic superiority of microwave radio relay, particularly type TD-2 for high-capacity applications. The costs are relatively highest for low-capacity open-wire and cable circuits.

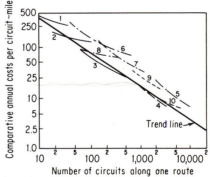

Legend:

Open wire ───────   Cable ─ ·─ ─ ─
1. VF
2. VF and C system
3. VF and C and J system

Microwave relay ── ── ──
4. TD-2 4 kmc
5. TH 6 kmc

6. VF-2 wire
7. K system-SSB
8. N system-DSB

Coaxial cable ─ ─ ─ ─
9. L1 3.5 Mc baseband
10. L3 8.5 Mc baseband

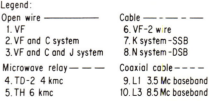

Fig. 49. Circuit costs vs. trunk cross-section. (*Courtesy of the American Telephone and Telegraph Co.*)

Table 11 shows the installed costs for various types of relatively high-capacity military and commercial radio relay and cable systems from another study.[62] These costs include facility costs, power plant, and electronics equipment. The total annual operating costs for such communications systems are shown in Table 12, which includes

Table 12. Total Annual Operating Cost of Various Communication Systems

System	Cost per mile	Cost per channel-mile	
		600 channel	72 channel
Microwave relay............	$2,652	$5	$37
Submarine cable............	1,584	...	22
Tropospheric scatter.........	4,032	...	56
Coaxial cable..............	4,600	8	64

(handwritten: iNiTiAL + MtcE (annual)) *(handwritten: rather low)*

both the capitalization and the yearly maintenance operating costs. How these data were obtained is described in some detail in the study by R. Chipp and T. Cosgrove.[62]

SITE SELECTION AND PATH PROFILING

A good radio relay installation largely depends on proper site selection and accurate profile determination. Although topographic maps which are usually fairly accurate exist in well-developed countries (± 10 ft in height), they are not available in many other parts of the world. For first planning, they will do well; however, more and more field engineers prefer aerial photogrammetry—or measurement of elevations from stereophotography. This highly specialized task is usually best performed by experienced specialists whose services are readily available through consulting firms.

Aerial photogrammetry is very accurate and the method has speed and economy not possible with other means. One organization* rendering this type of service lists the following accuracies at three typical flight altitudes realized from aerial photography and the resulting trade-offs.

Table 13. Accuracy Measurement for a Range of Flight Altitudes

Flight alt. above avg. terrain, ft	Scale of 9″ × 9″ film negatives	Scale of projected photographs (map scale)	Area of coverage of 1 model (60 % overlap), miles	Horizontal accuracy of path, ft	Vertical accuracy of spot elev. along path, ft
6,000	1″ = 1,000′	1″ = 200′	1.19 × 0.68	±15	±2
12,000	1″ = 2,000′	1″ = 400′	2.6 × 1.3	±25	±4
18,000	1″ = 3,000′	1″ = 600′	3.9 × 2.04	±35	±6

"The lower the altitude of the photography the greater the accuracy of the horizontal and vertical measurements that can be extracted from the photography. Low altitude photography would generally be more costly on a per mile basis and would cover smaller areas, a disadvantage in site selection. The engineer must weigh the higher costs of low altitude flights against the construction savings resulting from lower tower heights due to more refined measurements of the critical point. When a great deal of construction detail is present in the photography, low altitude photog-

* Microwave Services International, Inc., Denville, N.J.

raphy is necessary to evaluate potential reflections and to read the elevations of the tops of small structures on the path, such as radio towers and church steeples. The

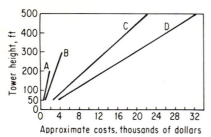

FIG. 50. Approximate costs-thousands of dollars. Maximum allowable bending moment above top guys in foot-pounds: A, 5,000; B, 2,500; C, 150,000; D, 420,000. (*Courtesy of Trylon Wind Turbine Co.*)

following is a tabulation [Table 13] of accuracy of measurement for a range of flight altitudes."

Another organization* using radar techniques quotes accuracies better than ±3 ft over smooth terrain. For a typical survey, over a 69-mile path in Pennsylvania, radar survey gave an error of ±6 ft using 142 random observations and for selected treeless areas ±4 ft against a U.S. Geodetic Survey map 1:24,000 with 20-ft contours.

In connection with site selection or surveying, it is important to consider all variables for possible trade-offs. Increased tower height raises cost depending on top loading from wind and ice; however, greater height permits operation over a longer path.

Between towers of equal height the line-of-sight distance over smooth earth (radius ≈ 4,000 miles) is $d = 2 \sqrt{2rh} = 2.46 \sqrt{h}$, where h is measured in feet and d in miles; i.e., it increases with the square root of tower height. In practical applications, the

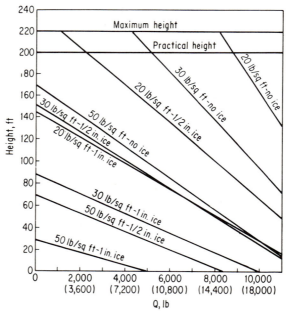

FIG. 51. Loading graph, maximum allowable bending moment of vertical top loading under any wind and icing conditions, 8,000 ft-lb. (*Courtesy of Trylon Wind Turbine Co.*)

height has to be traded against distance on economical grounds. The data in Fig. 50 show the approximate relationship between tower heights and cost for several typical towers. The bending moments shown are the maximum values allowable above the

* Aero Service Corp. (Division of Litton Industries), Philadelphia, Pa.

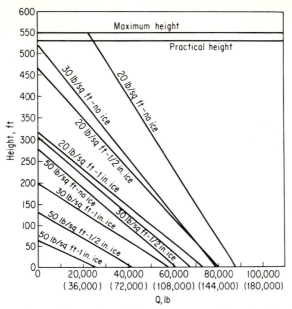

Fig. 52. Loading graph, maximum allowable bending moment of vertical top loading under any wind and icing conditions, 150,000 ft-lb. (*Courtesy of Trylon Wind Turbine Co.*)

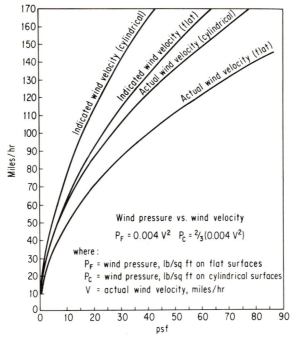

Fig. 53. Wind pressure vs. wind velocity. (*Courtesy of Trylon Wind Turbine Co.*)

top guys. In addition to the bending moment, it is necessary also to specify all effective dead weights of external loads, guying, anchors, weather proofing, painting, and lighting.

Figures 51 and 52 show typical loading graphs for various wind and ice conditions for two different towers. Figure 53 shows a convenient wind-velocity–wind-pressure conversion chart to facilitate computations of bending moments (see Chap. 23).

ACKNOWLEDGMENTS

The authors would like to express their thanks for the cooperation and contributions of technical information made by the following organizations.

Aero Service Corp., Div. of Litton Industries
American Electric Power Service Corp.
American Telephone and Telegraph Co.
Association of American Railroads
Bell Telephone Laboratories, Inc.
Boston Technical Consultants, Inc.
Electronic Industries Association
Microwave Services International, Inc.
Radio Corporation of America
RCA Service Co.
RCA Victor Co., Ltd.
Windturbine Co.

REFERENCES

1. Stratton, J. A., *Electromagnetic Theory*, chap. V, pp. 268–296, McGraw-Hill Book Company, New York, 1941.
2. MIT Radiation Laboratory Series, vol. 13, *Propagation of Short Radio Waves*, chap. 2, pp. 27–41, McGraw-Hill Book Company, New York, 1951.
3. Schelkunoff, S. A., *Electromagnetic Waves*, chap. IX, pp. 331–374, D. Van Nostrand Company, Inc., Princeton, N.J., 1943.
4. Guenther, R., Radio Relay Design Data 60 to 600 Mc, *Proc. IRE*, vol. 39, pp. 1027–1034, September, 1951.
5. CCIR, Document of the X Plenary Assembly, Geneva, 1963, vol. IV, Radio Relay Systems, Space Systems, Radio Astronomy, No. 2173-E; CCITT, Telephone Transmission Quality, Local Lines and Telephone Sets, vol. V, Geneva, 1963.
6. Lewinski, D. A., A New Objective for Message Circuit Noise, *Bell System Tech. J.*, vol. XLIII, pp. 719–740, March, 1964.
7. MIL Spec 188, DCS-Eng., *Installation Standards Manual*, DCA-CIR, 175-2a.
8. Brock, F., private communication of unpublished work at RCA.
9. Crosby, M. G., Frequency Modulation Noise Characteristics, *Proc. IRE*, vol. 25, pp. 472–514, April, 1937.
10. Holbrook, B. D., and J. T. Dixon, Load Rating Theory for Multi-channel Amplifiers, *Bell System Tech. J.*, vol. 18, pp. 624–644, October, 1939.
11. Jacoby, H., and F. Spenke, Ein neuer Beitrag zur Ermittlung der erforderlichen Leistung von Tragerfrequenz—Vielfachverstarken, *Veroeffentl. Gebiet Nachr. Technol.*, vol. 9, no. 1, pp. 135–144, 1939.
12. Oliver, B. M., J. R. Pierce, and C. E. Shannon, The Philosophy of PCM, *Proc. IRE*, vol. 36, pp. 1324–1331, November, 1948.
13. Meyerhoff, A. A., and W. M. Maser, Optimum Binary FM Reception Using Discriminator Detection and IF Shaping, *RCA Rev.*, vol. XXII, pp. 698–728, December, 1961.
14. Shaft, P. D., Error Rate of PCM-FM Using Discriminator Detection, *IEEE Trans.* SET-9, pp. 131–137, December, 1963.
15. Piloty, R., Uber die Beurteilung der Modulationssysteme mit Hilfe des nachrichtentheoretischen Begriffes der Kanal-Kapazitat, *Arch. Elektr. Ubertr.*, vol. 4, pp. 493–508, December, 1950.
16. Jelonek, Z., A Comparison of Transmission Systems (London Symposium, 1952) in W. Jackson (ed.), *Communication Theory*, Butterworth Scientific Publications, London, 1953.

17. North, D. O., The Absolute Sensitivity of Radio Receivers, *RCA Rev.*, vol. 6, pp. 332–343, January, 1942.

18. Friis, H. T., Noise Figures of Radio Receivers, *Proc. IRE*, vol. 32, pp. 419–422, July, 1944.

19. Franz, K., Uber der Empfindlichkeitsgrenze beim Empfang electrischer Wellen, *Elektr. Nachr. Techn.*, vol. 16, pp. 92–96, 1939.

20. Bennett, W. R., *Electrical Noise*, McGraw-Hill Book Company, New York, 1960.

21. Goldberg, H., Some Notes on Noise Figure, *Proc. IRE*, vol. 36, pp. 1205–1214, October, 1948.

22. Nielsen, E. G., Behavior of Noise Figure in Junction Transistors, *Proc. IRE*, vol. 45, pp. 957–963, July, 1957.

23. Uenohara, M., Noise Considerations of the Variable Capacitance Parametric Amplifier, *Proc. IRE*, vol. 48, pp. 169–179, February, 1960.

24. DeGrasse, R. W., D. C. Hagg, E. A. Ohm, and H. E. B. Scovil, Ultra-low Noise Antenna and Receiver Combination for Satellite or Space Communication, *Proc. Natl. Electron. Conf.*, vol. 15, pp. 370–379, 1959.

25. Albersheim, W. J., and J. P. Schafer, Echo Distortion in the FM Transmission of Frequency-division-multiplex, *Proc. IRE*, vol. 40, pp. 316–328, March, 1952.

26. Bennett, W. R., H. E. Curtis, and S. O. Rice, Interchannel Interference in FM and PM Systems under Noise Loading Conditions, *Bell System Tech. J.*, vol. 34, pp. 601–636, May, 1955.

27. Golding, J. F., The White-noise Method of Measuring Crosstalk and Noise Interference in Multi-channel Telephone Links, *Electron. Eng.*, vol. 30, pp. 349–351, May, 1958.

28. Thompson, L. E., Distortion in Multi-channel Frequency Modulation Relay System, *RCA Rev.*, vol. XI, no. 4, pp. 453–464, December, 1950.
Medhurst, R. G., and T. H. Roberts, Multiple Echo Distortion in Frequency Modulation, Frequency Division MUX Trunk Radio Systems, *Trans. IRE Commun. Systems*, CS-10, pp. 61–71, March, 1962.

29. Fedida, S., and D. S. Palmer, Some Design Considerations for Links Carrying Multichannel Telephony, *Marconi Rev.*, part I, vol. 18, pp. 132–155, 4th Quarter 1955, and Part II, vol. 19, pp. 1–46, 1st Quarter, 1956.
Brockbank, R. A., and C. A. A. Wass, Non-linear Distortion in Transmission Systems, *J. IEE*, part III, vol. 92, pp. 45–46, 1945.

30. Kurzrock, R., A Technique for Equalizing Parabolic Group Delay, *Proc. IRE*, vol. 50, pp. 1840, August, 1962.

31. Cohen, S. B., Direct Coupled Resonator Filters, *Proc. IRE*, vol. 45, pp. 187–196, February, 1957.

32. Rice, S. O., Mathematical Analysis of Random Noise, *Bell System Tech. J.*, vol. 23, part I, pp. 282–332, July, 1944, and vol. 24, part II, pp. 46–156, January, 1945.

33. Collings, R. H. P., and J. K. Skwirzynski, The Distortion of FM Signals in Passive Networks, *Marconi Rev.*, vol. 17, pp. 113–136, 4th Quarter, 1954.

34. Straube, H. M., Dependency of Crosstalk on Upper and Lower Cutoff Frequencies in PAM Time Multiplexed Transmission Paths, *IRE Trans. Commun. Systems*, vol. CS-10, no. 5, pp. 268–276, September, 1962.

35. Bullington, K., Radio Propagation Fundamentals, *Bell System Tech. J.*, vol. 36, pp. 593–626, May, 1957, and "Antenna Engineering Handbook," chap. 33, McGraw-Hill Book Company, 1961.

36. Tolbert, C. W., and A. W. Straiton, An Analysis of Recent Measurements of the Atmospheric Absorption of Millimetric Radio Waves, *IRE Proc.*, vol. 49, pp. 644–650, March, 1961.

37. Dutka, J., and S. J. Mehlman, "Noise Power Probability Distribution for Multi-hop FM Radio Relay Systems," Lecture of GLOBECOM II Conference, December, 1958.

38. Webler, I., H. W. Evans, and G. A. Pullis, Protection of Service in the TD-2 Radio Relay System by Automatic Channel Switching, *Bell System Tech. J.*, vol. XXXIV, pp. 473–510, May, 1955.

39. Altman, F. J., and W. Sichak, A Simplified Diversity Communication System for Beyond-the-horizon Links, *Elec. Commun.*, vol. XXXIII, pp. 151–160, June, 1956.

40. Packard, K. S., and R. V. Lowman, Transmission Lines and Waveguides, "Antenna Engineering Handbook," pp. 30-1 to 30-42, McGraw-Hill Book Company, New York, 1961.

41. Schelkunoff, S. A., and H. T. Friis, "Antennas—Theory and Practice," Chapman & Hall, Ltd., London, and John Wiley & Sons, Inc., New York, 1952.

42. Kennett, D. B., Tropospheric Scatter Multi-channel Communications, *Point-to-Point Telecommun.*, vol. 8, pp. 35–54, February, 1964.

43. Yeh, L. P., Simple Methods of Designing Troposcatter Circuitry, *IRE Trans. Commun. Systems*, vol. CS-8, no. 3, pp. 193–198, September, 1960.

44. Parry, C. A., A Formalized Procedure for the Prediction and Analysis of Multi-channel Tropospheric Scatter Circuits, *IRE Trans. Commun. Systems*, pp. 211–221, September, 1959.
45. Staras, H., Antenna to Medium Coupling Loss, *IRE Trans. Antennas Propagation*, vol. AP-5, pp. 228–231, April, 1957.
46. Rice, P. L., A. G. Longley, and K. A. Norton, Prediction of the Cumulative Distribution with Time of Ground Wave and Tropospheric Wave Transmission Loss, *Natl. Bur. Std. Tech. Note* 15, July, 1959.
47. Bean, B. R., and J. D. Horn, Radio-refractive-index Climate Near the Ground, *J. Res. Natl. Bur. Std.*, vol. 63D, no. 3, pp. 259–271, November–December, 1959.
48. Woodman, G. B., New Ideas in Microwave System Maintenance, *Western Union Tech. Rev.*, vol. 17, no. 2, pp. 78–81, April, 1963.
49. Woodman, G. B., Maintenance of a Radio Relay System, *Western Union Tech. Rev.*, vol. 5, no. 4, pp. 141–147, October, 1951.
50. Lewis, C. A., and M. Telford, Reliability Criteria in Radio Relay Systems, *Point-to-Point Telecommun.*, vol. 7, no. 3, pp. 30–43, June, 1963.
51. Summerland, L. A., Performance and Maintenance of the Nigeria VHF Trunk Network, *Point-to-Point Telecommun.*, vol. 5, no. 3, pp. 21–39, June, 1961.
52. Woodman, G. B., Improved Vacuum Tube Reliability through Maintenance, *Western Union Tech. Rev.*, pp. 164–167, October, 1957.
53. Nelson, T. R., "Expansion of Microwave Facilities to Meet Increased Power System Needs," American Electric Power Service Corp., Paper DP63–859. Presented at the Northeastern District Meeting of IEEE, Portsmouth, N.H., May 20–22, 1963.
54. Maintainability Digest for the Electronics Industry, *Maintainability Bull.* 1, Electronic Industries Association Engineering Department, 11 West 42nd Street, New York 36, December, 1960.
55. Hathaway, S. D., D. D. Sagaser, and J. A. Wood, The TL Radio Relay System, *Bell System Tech. J* , vol. XLII, no. 5, p. 2297, September, 1963.
56. CW-60, *Total Solid State 6 kmc Microwave*, published by RCA Microwave Department (Broadcast and Communications Products Div.), Camden 2, N.J.
57. Sheffield, B., Radio System Controls Railroad in Venezuela, *Electronics*, vol. 29, no. 12, pp. 159–163, December, 1956.
58. *System Reliability—White Alice System*, Government Services—RCA Service Company, Box 7014, Anchorage, Alaska, May, 1963.
59. New Ammunition for Trouble Shooters, Staff Article, *Electron. Design*, vol. 11, no. 25, pp. 46–51, Dec. 6, 1963.
60. Broadus, T. H., et al., *Microwave Communications—Commercial Possibilities in the Sixties*, Micom Associates; data quoted by permission of copyright owner Dr. O. V. Mody, Boston Technical Consultants Inc., 5 Bryant Road, Lexington 73, Mass.
61. Abraham, L. G., The Complexity of the Transmission Network, *Bell Lab. Record*, vol. 38, no. 2, p. 47, February, 1960.
62. Chipp, R. D., and T. Cosgrove, "Economic Analysis of Communication Systems," presented at Seventh Communications Symposium of the IRE, Utica, N.Y., October, 1961.
63. Shapiro, G., G. J. Rogers, O. B. Laug, and P. M. Fulcomer, Jr., Project Fist—Fault Isolation by Semiautomatic Techniques Parts I and II, *IEEE Spectrum*, vol. 1, no. 8, pp. 98–111, August, 1964, and no. 9, pp. 130–144, September, 1964.
64. Enloe, L. H., The Synthesis of Frequency Feedback Demodulators, *Proc. Natl. Electron. Conf.*, vol. 18, pp. 477–497, 1962.
    Giger, A. J., and J. G. Chaffee, FM Demodulation with Negative Feedback, *Bell System Tech. J.*, vol. 42, pp. 1109–1135, July, 1963.
    Jaffe, R, and E. Rechtin, Design and Performance of Phase Locked Circuits Capable of Near Optimum Performance Over a Wide Range of Input Signal and Noise Levels, *IRE Trans.-Information Theory*, vol. IT-1, pp. 66–76, March, 1955.

Chapter 17

MOBILE RADIO SERVICES

NEAL H. SHEPHERD, *General Electric Company*

1. Service Grouping

The growth[1] of vehicular mobile radio in the United States between the years of 1946 to 1959 has been at a rate of greater than 20 per cent per year. By 1963, the total number of vehicular mobile transmitters has exceeded the 1,000,000 mark, although the per cent rate of growth has been on a slight decline since 1959. About 41 Mc/sec of the frequency spectrum in the 25 to 50-, 150 to 162-, and 450 to 470-Mc/sec ranges is allocated by the Federal Communication Commission to the various services. The three frequency ranges are commonly referred to as low band, high band, and 450 Mc, respectively. The services are divided into five basic groups by a separate part of the FCC Rules and Regulations.

The number of services have shown a steady increase until now there are more than 20. The use[2] to which radio is placed in each service is slightly different, although the basic equipment meets the needs of all services. The various services and different applications within a given service place an ever-increasing demand on the systems engineer to provide custom-built systems with little or no increase in cost. To meet such demands, it is necessary that each basic base or mobile station be easily modified to accept a variety of options. The number of equipment models formed by various options are in the hundreds.

Each mobile radio service is formed on the basis of similar usage within a group or on the basis of grouping similar industries together. The FCC Rules and Regulations provide for licensing of transmitting equipment according to services in the following parts. (An abbreviated description follows certain more important services. Consult the FCC Rules and Regulations for complete information.)

2. Maritime Mobile Services,[3] Parts 81 and 83

The VHF Maritime Mobile Service provides ship-to-shore communications in coastal waterways, in inland waterways, and on the Great Lakes.

3. Public Safety Radio Services, Part 89

a. Police Radio Service.[4] The oldest application of mobile radio for dispatching and directing the movements of police in cars, on horseback, or on foot. It is also used in directing police movements on a limited local scale or a large coordinated operation for the investigation or eliminating of illegal operations.

b. Forestry Conservation Radio Service. For reporting origin of forest fires, directing the deployment of men and materials to areas of need, and coordinating the hundreds of workers involved in forestry operations.

c. Highway Maintenance Radio Service. For coordinating the movement of trucks, materials, and workmen in the day-to-day work schedules to provide a maximum of efficiency.

d. Special Emergency Radio Service. To provide mobile communication for hospitals, physicians, disaster relief, and ambulances.

e. Fire Radio Service
f. State Guard Radio Service
g. Local Government Radio Service

4. Industrial Radio Services, Part 91

a. Power Radio Service.[5] For directing the movement of the line maintenance crews to areas of construction or maintenance.

b. Petroleum Radio Services.[6] Mobile radio is used to coordinate the complex refining process from the well to the end use. It plays a vital role in every phase of the well-drilling process.

c. Forest Products Radio Service
d. Motion Picture Radio Service
e. Relay Press Radio Service
f. Special[1] *Industrial Radio Service*
g. Business Radio Service
h. Manufacturers Radio Service
i. Telephone Maintenance Radio Service

5. Land Transportation Radio Services, Part 93

a. Motor Carrier Radio Service. For dispatching fleets of trucks to collect and deliver constructed materials and services with minimum of delay or deadheading.

b. Railroad Radio Service
c. Taxicab Radio Service
d. Automobile Emergency Radio Service

6. Domestic Public Radio Services, Part 21

a. Domestic Public Land Mobile Radio Service.[7] For use by the general public to provide an extension on the telephone system of the office and home.

b. Rural Radio Service. To provide telephone facilities in remote areas where landlines cannot be economically provided.

7. Frequency Allocations[8]

In general, frequencies are allocated to services on a block system; thus each given service has only a fraction of the total frequencies allocated to choose from. The blocks for each service are actually several blocks of closely spaced frequencies scattered through the mobile radio-frequency bands. Closely spaced channels within a block coupled with a limited number of available frequency assignments tend to create both intersystem and intrasystem radio-frequency interference problems. To provide the maximum possible utilization of the frequency spectrum, the system engineer must be able to analyze interference problems and provide fixes which can reduce or eliminate the interference.

MOBILE EQUIPMENT

8. Mobile Equipment Packages

Prior to 1958, virtually all mobile radio products used tubes throughout the transmitter and receiver. Plate voltage power was obtained by converting 6- and 12-volt d-c with vibrators and generators. In 1957, transistorized power supplies[9] were introduced as add-on modification and later as integral part of the mobile package. To protect transistors from excessive temperature rise, heat sinks were added to the mobile packages.

Transistorized receivers[10] and partially transistorized transmitters quickly followed the introduction of transistorized power supplies, now required for the transmitter only. The introduction of transistors to mobile equipment resulted in the following more important benefits:

a. Small Size. A reduction in volume of about 50 per cent, facilitating under-the-dash mounting.

b. Lower Standby Drain. Standby drain with the transmitter filaments turned on has been reduced about 60 per cent. With the transmitter filaments turned off, sometimes called battery saving, the current drain is as low as 40 ma from a 12-volt power supply.

c. Higher Reliability. Reliability of transistors has reduced spare parts requirements and maintenance costs.

Some of the more popular mobile packages are as follows:

a. Standard Mobile—Trunk Mount. The main unit consisting of the transmitter, receiver, and power supply is mounted in the trunk, and the control head and speaker are under the dash.

b. Standard Mobile—Front Mount. The main unit, control head, and speaker are mounted as an integrated package under the dash.

c. Motorcycle Mobile—Rear Mount. The main unit is mounted on the rear fender, and the control head with speaker is mounted on the handle bars.

d. Motorcycle Mobile—Front Mount. The entire unit is mounted on the handle bars.

e. Mobile Telephone. A vehicle trunk-mounted equipment with control unit on the dash including headset cradle and optional dial. Units are called from landline telephones connected through a base station. Other mobiles may also call through the base station in a similar manner.

9. EIA Standards[11–13] and Manufacturers' Specifications

EIA Standards are maintained by the Electronics Industries Association, 11 West 42d Street, New York, N.Y. These standards are adopted in the public interest to provide the maximum possible compatibility among units built by different manufacturers and to eliminate misunderstandings between the manufacturer and the purchaser. The standards are not binding on either member or nonmember companies; however, practically all equipment manufactured for mobile communications use meet or exceeds EIA Standards. A comparison of EIA Standards with a typical manufacturing specification is shown in Table 1.

10. Transistorization

The introduction of germanium transistors in mobile communications equipment brought many other changes in equipment design such as:

a. Printed Circuits. Printed circuit boards contain strips of copper on both sides of an insulated board which provide connections to the various components which are directly soldered to the board.

b. Sockets. Most transistors require only three connections through a socket, while other transistors are soldered directly into the circuits, thus eliminating or reducing the number of connections formerly required by tubes.

c. Cooling Requirements. Smaller size of mobile equipment and temperature-sensitive transistors have forced designers to place cooling fins on various sides of the mobile equipment to limit temperature rise. A new generation of silicon transistors will make the limit of temperature rise less critical; however, heat sinks will continue to be required to provide smaller size mobile equipment.

d. Maintenance. Transistors may be destroyed by momentarily applying excessive base or emitter current.

Table 1. EIA Standards and Manufacturers' Specifications
Mobile Receiver

Quantity measured or specified	EIA Standards			Typical manufacturers' specifications		
	25–54 Mc/sec	144–174 Mc/sec	400–470 Mc/sec	25–54 Mc/sec	144–174 Mc/sec	400–470 Mc/sec
Sensitivity, μv	1.0	1.5	2.5	0.3	0.5	1.0
Selectivity, db	70	70	70	80	80	80
Spurious attenuation, db	85	85	80	100	100	90
IM spurious attenuation, db	50	50	50	60	60	60
Audio power output, watts	1	1	1	2	2	2

Mobile Transmitter

Power output	*N/A*	*N/A*	*N/A*	80 watts	80 watts	60 watts
Frequency stability, %	±.002	±.0005	±.0005	±.0005	±.0005	±.0005
Spurious radiation, db	43 + 10 log (power output)			53 + 10 log (power output)		

e. Selectivity. One of the greatest advances in improving the performance of receivers was the introduction of multiple-tuned helical circuits preceding the first RF amplifier.

11. Tubed Equipment

While transistors were being introduced into mobile equipment, tubed equipment was also going through design changes such as:

a. Size Reduction.[14] Multipurpose tubes have allowed reduction in the size of mobile equipment. Tubes designed for conductive cooling, which must be heat-sinked to the equipment case, have facilitated considerable size reduction.

b. Power Reduction. Standby power has been slightly reduced through the use of multipurpose tubes.

12. Power-supply Voltage

EIA Standards list the following standard voltages for mobile equipment as shown in Table 2. Figure 1 shows typical variation in terminal voltage of 12-volt batteries

Table 2. EIA Standard Test Voltages

6 volts d-c		12 volts d-c	
Operating current, amp	Test voltages, volts d-c	Operating current, amp	Test voltages, volts d-c
Below 10	6.6	Below 6	13.8
10–22	6.5	6–16	13.6
22–36	6.4	16–36	13.4
36–54	6.3	36–50	13.2
54–70	6.2	Over 50	13.0
Over 70	6.1		

which was measured on a large number of vehicles under various conditions of battery charging and operation of the radio equipment. The curves show that about 4 per cent of receivers may be operated out of the ±20 per cent variation in battery terminal voltage as allowed by EIA Standards. No EIA performance requirements are specified outside the ±20 per cent range. Applicable EIA Standards allow for a variation of battery terminal voltage of ±10 per cent. Above the range of +10 per cent the life of certain components may be reduced.

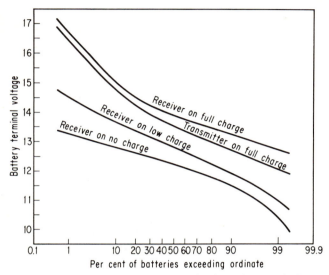

FIG. 1. Typical variation in battery terminal voltage for 12-volt d-c supply.

13. Voltage Regulation

Voltage regulation in 6- and 12-volt d-c mobile is in general limited to lower power stages, such as oscillator and audio stages. Tubed equipment rarely uses voltage regulators, whereas transistorized equipment frequently includes some form of regulation. Mobile equipment operating voltages above 12 volts are subjected to voltage surges and greater than ±20 per cent variation in voltage, thus making a voltage regulator a necessity in most applications. A typical example is found in the railroad industry. The battery power voltage available in diesel locomotives is 72 volts d-c. Typical variation in voltage on the 72-volt d-c bus is from 58 to 87 volts d-c. One type of voltage regulation is described by Gilbert.[15]

14. Vehicular Noise Interference[16]

A major source of interference encountered by mobile radio equipment is the vehicle in which the equipment is installed. Engineering committees of Society of Automotive Engineers (SAE), EIA, and IEEE have for several years been in the process of formulating standards of measurements, procedures, and limits of radiation. The SAE's standard for levels was established at a distance of 50 ft from the vehicle with the maximum levels as follows:

Frequency range, Mc/sec	Levels horizontally polarized
25–78	−2 db above 1 μvolt/meter/kc
78–216	+8 db above 1 μvolt/meter/kc

The frequency range, distance, and polarization are primarily suited to reducing interference to TV receivers and do not necessarily meet the noise-reduction needs of equipment installed in or near the vehicle. The IEEE proposed standard of Subcommittee 27.7 utilizes a vertical antenna near the front and rear of the vehicle, thus giving results more suited to estimating noise interference to an installed receiver. An average of data taken on several vehicles of various makes by the IEEE proposed method is shown on Fig. 2. For comparison, an estimate is also shown of the expected level of a vehicle which meets but does not exceed SAE standards for radiation at 50 ft.

Also the typical levels of front-end noise generated in the mobile receiver is shown. As can be seen from these curves, the peak noise level in a 10-kc/sec bandwidth can be greater than 75 db above the receiver front noise, thus creating a potential serious reduction in receiver sensitivity.

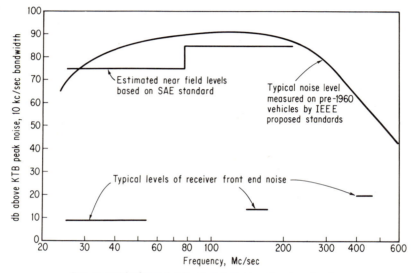

Fig. 2. Vehicular noise level and receiver front-end noise.

Vehicular noise is primarily of the impulsive type; thus the degradation of the receiver's sensitivity is considerably less than the above-mentioned 75 db. The actual degradation of the receiver's sensitivity is somewhat dependent on the design of a particular receiver and the frequency stability of the transmitter and receiver as well as the impulse noise level. Design factors affecting impulse noise susceptibility are discussed in a separate paragraph in this chapter. In many applications, the degree of noise reduction through appropriate receiver design and frequency stability is not sufficient to meet required standards of performance and other "fixes" must be used such as impulse noise blankers.

15. Impulse Noise Blankers[17,18]

Several types of blankers have been applied to mobile receivers with various degrees of results. Each system of blanking has certain advantages and also disadvantages. Figure 3 shows three basic types of blankers.

a. Blanker A. This method of blanking in the audio portion of the receiver has several advantages as well as disadvantages. By detecting a trigger signal beyond the low IF filter it is practical to eliminate any susceptibility due to intermodulation or desensitization from adjacent channels. This type of blanker can be adjusted to blank on only those impulses which would eventually reach the audio, whereas other types of blankers look at a broader noise spectrum and are frequently triggered into blanking by interference, which would not appear at the desired frequency of the receiver. At the low intermediate frequency the impulse noise has been stretched to such an extent that required repetitive blanking pulses must remove an appreciable portion of the audio, thus establishing an upper blanking limit of about 200 pps.

b. Blanker B. In this system the trigger voltage is derived after the RF selectivity but before the main selectivity of the receiver. This method of selecting the trigger voltage opens the blanking channels to intermodulation and desensitization interference. Blanking may occur on strong desired signals when not normally required, thus producing interference rather than eliminating it. Blanking pulses are considerably shorter than those used in blanker *A*, thus providing an advantage of operating at rates of greater than 2,000 pps. The high IF blanker must be considered an in-between type having some of the advantages and disadvantages of both blankers *A* and *C*. Since it operates on the same RF bandwidth as the receiver itself, it has the further disadvantage of being incapable of avoiding interference by a change in the operating frequency as is possible in the case of blanker *C*.

c. Blanker C. Blanker *C* operates on the principle of a completely separate channel for deriving the trigger signal. Here the blanking pulse may be less than 4 μsec in length, thus facilitating blanking at greater than 100,000 pps. While this method certainly is best as far as the ability to remove all undesired impulses, it continues blank when not required unless other means are provided to turn it off. This type of blanker can degrade the receiver's performance because of intermodulation in the blanking channel and through blanking a very strong adjacent channel signal near the desired frequency channel.

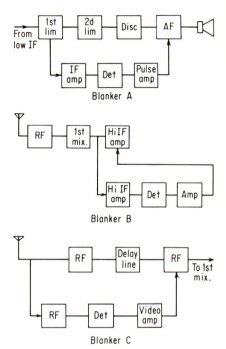

Fig. 3. Impulse blankers used in mobile equipment.

BASE-STATION EQUIPMENT

Base-station equipment is formed by the proper application as required of one or more of the following equipments:

Transmitter and receiver combination
Antenna and transmission line
Control equipment
Cavity filters

The maximum performance of the base station is dependent on the proper application of the individual equipments. An essential part of proper application is a full understanding of the characteristics of the individual equipments and their interface requirements.

16. Transmitter and Receiver Combination

Transmitters and receivers normally operate in the same cabinet but can operate in separate cabinets at the same or widely separated location. Whether operating in the same cabinet or separated, they normally operate independently of each other except for antenna and audio. Standard cabinets provide a limited amount of shielding to radio-frequency radiation both from inside and from outside. When greater shielding is required, special cabinets must be specified to meet the particular needs. The following types of base-station combinations are in common use:

a. Vertical Base Station. This is a standard EIA rack in a cabinet with front and rear access doors. Power, audio, and control leads enter from the bottom or back, while coaxial cable enters from top or back.

b. Pole on Wall-mounted Outdoor Station. This is a small rack cabinet for outdoor installations with front-door access. To facilitate maintenance, the rack can usually be swung outward to provide access to both sides of the equipment. Cables enter from the bottom or rear.

c. Desk Top Station. This is a compact cabinet, usually lower power, for mounting on a desk to provide local control at the same point. Cables enter from the rear.

d. Desk Side Station. This is a compact cabinet for mounting next to a desk, whose height is the same as a desk. Cables enter from the bottom or rear.

17. Antennas and Transmission Line

Mobile radio systems are limited in range by authorized transmitter power and local noise conditions. However, through proper application of the antenna and transmission line, the coverage pattern of a system can usually be tailored to fit the user's needs. A variety of antennas is available to meet system coverage requirements. A later paragraph on antennas will discuss several types and illustrate how they are applied.

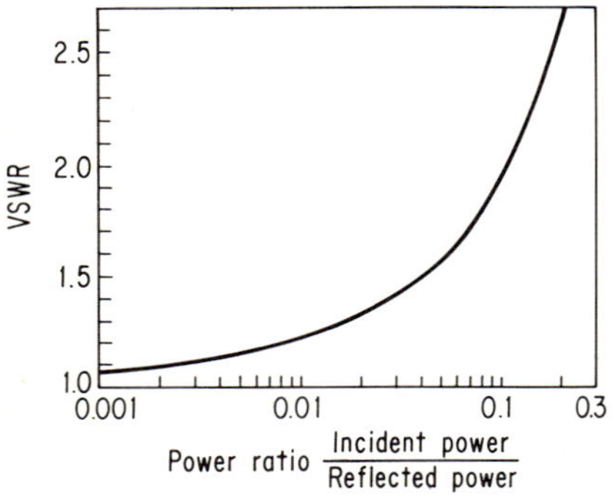

Fig. 4. Power ratio and VSWR relations.

After an antenna is installed, its performance can be partially checked by measuring the voltage standing-wave ratio (VSWR) at the equipment end of the transmission line. VSWR is related to the incident power P_i and reflected power P_r by the following

$$\text{VSWR} = \frac{\sqrt{P_i} + \sqrt{P_r}}{\sqrt{P_i} - \sqrt{P_r}} \tag{1}$$

The incident and reflected power are readily measured with directional wattmeters or calibrated directional couplers. From the ratio of the reflected and incident power, the VSWR can be found with the aid of Fig. 4.

The VSWR at the antenna is usually greater than that measured at the input to the transmission line. The VSWR at the antenna can be read from Fig. 5 when the line loss and the VSWR at the end of the line are known.

Defective transmission lines frequently cause higher VSWR readings at the equipment end of the line than that presented by the antenna. Most systems operate with a VSWR of 1.5 or less. If higher VSWR's are measured, both the transmission line and antennas should be measured separately to aid in locating defects.

Transmission-line loss should be held to a minimum to conserve transmitter power. Power loss in the transmission line will reduce communications range. However, loss in the receive mode may not reduce range if local noise level is greater than the receiver noise figure. Table 3 gives a list of the most frequently used types of transmission lines including the attenuation per 100 ft and the power rating.

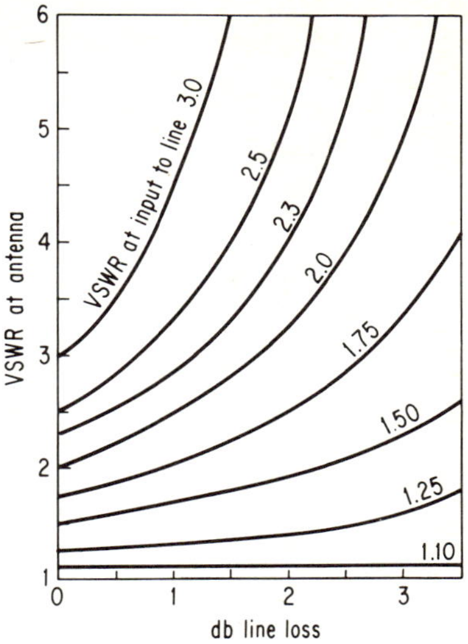

18. Control Equipment

Base-station equipment may be operated locally or remotely. When it is operated remotely, either telephone lines or radio may be used.

Local control of a base station may be at the transmitter and receiver or extended up to several hundred feet by a control unit. A control unit usually consists of an audio amplifier, indicator lights, microphone, speaker, push-to-talk switch, squelch control, level controls, and control voltages.

Level adjustments are provided in a control unit to maintain the optimum modulation level at the transmitter or to adjust the receiver audio at the control unit to meet the needs of the operators. A control unit should always be located in a low sound level area to eliminate background noise as heard by the mobile unit. Push-to-talk switches are located on the floor or desk as required to fit the particular needs of the operator.

Remote control of base stations is provided over one or two telephone pairs. A telephone pair is usually a pair of wires ranging from wire-size AWG 19 to 26, providing a d-c circuitry. However, in recent years, the use of d-c circuit telephone lines has

FIG. 5. Variation of VSWR with transmission-line loss.

Table 3. Line Loss in Decibels per 100 Ft and Maximum Average Power Rating for Transmission Lines

	Frequency, Mc/sec					Velocity of propagation
	30	40	50	150–170	450–470	
RG8/U	1.0 2 kw	1.2 1.65 kw	1.5 1.50 kw	2.65 0.68 kw	5.5 0.30 kw	0.67
RG17/U	0.38 7.8 kw	0.46 6.15 kw	0.65 5.4 kw	1.1 2.35 kw	2.4 1.0 kw	0.67
½-in. foam	0.43 4.8 kw	0.5 4.0 kw	0.56 3.5 kw	1.0 1.8 kw	2 0.95 kw	0.79
⅞-in. foam	0.24 9 kw	0.28 7.5 kw	0.32 6.5 kw	0.60 3.5 kw	1.2 1.7 kw	0.79
⅞-in. pressurized	0.23 7.5 kw	0.25 6.5 kw	0.28 5.8 kw	0.50 3.2 kw	0.9 1.8 kw	0.89
Bal. line RG86/U	0.3 5.0 kw	0.4 4.0 kw	0.5 3.3 kw	1.0 1.5 kw	2.0 0.75 kw	0.72

been on the decline. As a substitute for d-c circuits, the telephone companies have been supplying circuits with only voice frequency capabilities. The impedance of a telephone pair is nominally 600 ohms with a transmission loss proportional to the line length. The frequency response of the line is deemphasized, rendering a bassy quality unless some form of frequency compensation is used.

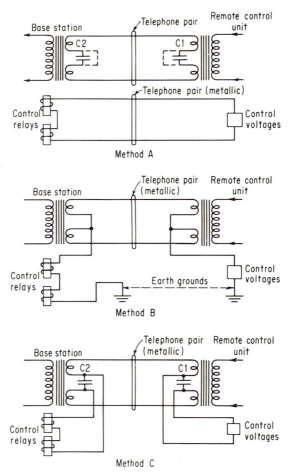

FIG. 6. Telephone-line control circuits.

Where a d-c circuitry is provided, the loop resistance must be less than about 1,500 ohms. Depending on the size of wire used, a 1,500-ohm circuit can be from about 5 to about 15 miles long. Figure 6 shows three methods of providing audio and control circuits which are most frequently used between the control unit and the base station.

Method *A* uses two telephone lines, one pair for audio, one metallic pair for control. Since the control circuits are separate from the audio circuits, parallel control units will be free from key clicks caused by d-c control circuits. Capacitors C1 and C2 are used to compensate for loss in high frequencies. If lines are short, the capacitors are usually shorted.

Method *B* uses one metallic pair for both audio and control. By simplexing the control voltage with ground, key clicks are minimized, since d-c control currents are balanced out in the audio transformer. Ground currents may cause improper relay operation.

Method *C* uses one metallic pair for both audio and control, and the control voltage is simplexed from one line to the other by splitting the output transformer with a capacitor. This method is recommended in locations where earth currents are large enough to affect relay operation. Since control voltage is applied line-to-line, key clicks will be heard when control units are operated in parallel.

To prevent interference among the various users of telephone lines, the following maximum transmission levels are typical of these specified by telephone companies:

D-c voltage line-to-line........................ 270 volts
D-c voltage line-to-ground.................... 135 volts
Line current, a-c or d-c...................... 0.35 amp
A-c voltage 1,000 cps line-to-line............ 2 volts
Speech levels................................. +8 vu

19. Cavity Filters

When two or more transmitters or a transmitter and receiver are operated simultaneously from the same base-station location, it is usually necessary to provide additional selectivity for both the transmitters and receivers to reduce interference to receivers. One effective means of reducing interference is through the use of cavity filters inserted in the transmission lines between the antenna and the transmitters and receivers. Cavity filters are formed from a quarter-wavelength line open-circuited at one end and shorted at the other. Input and output circuits are provided through loop coupling near the shorted end of the line. The unloaded *Q* for quarter-wavelength coaxial cavity shown in Fig. 7 is given by the formula

$$Q_0 = \frac{\lambda}{4 + 7.2(h\delta/b)} \qquad (2)$$

where $b/a = 3.6$ for optimum Q
λ = wavelength, meters
a = outer radius of inner conductor, in.
b = inner radius of outer conductor, in.
$h = 2,950/f$, length of center conductor, in.

$$\delta = \frac{1}{2\pi}\sqrt{\frac{10}{\sigma f}} \qquad \text{skin depth, in meters}$$

where σ = conductivity, mhos/meter
f = frequency, Mc/sec

and $\delta = \dfrac{6.61 \times 10^{-5}}{\sqrt{f}}$ meters for copper

For copper cavities with optimum proportions *Q* reduces to

$$Q_0 = \frac{4.53 \times 10^5}{[4 + 7.2(h/b)]\sqrt{f}} \qquad (3)$$

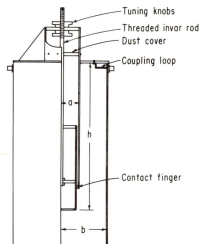

FIG. 7. Coaxial cavity filter showing essential parts in partial cross section.

Curves giving unloaded *Q* for two typical sizes of cavities are given in Fig. 8. The loaded *Q*, which is the *Q* obtained when connected in a circuit, is dependent on the

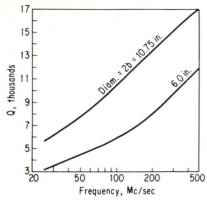

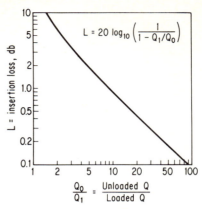

FIG. 8. Unloaded Q of coaxial cavity filters of optimum proportions ($b/a = 3.6$).

FIG. 9. Relation of insertion loss in decibels to the ratio of unloaded to loaded Q.

insertion loss L shown in Fig. 9 as given by the following formula:

$$L = 20 \log_{10} \frac{1}{1 - Q_1/Q_0} \qquad \text{db} \qquad (4)$$

where $Q_1 = $ loaded Q

$Q_0 = $ unloaded Q, typical values about 95 per cent of theoretical maximum given in Fig. 8

To provide greater attenuation with less insertion loss, cavities can be combined by $\lambda/4$ coaxial lines between each cavity to form a multiple circuit filter. Attenuation curves for one, two, and three cavities are shown in Fig. 10. The attenuation curves are for equal loaded Q and do not include insertion loss at the center frequency. The interference and special systems paragraphs of this chapter include specific information on how cavity filters are applied.

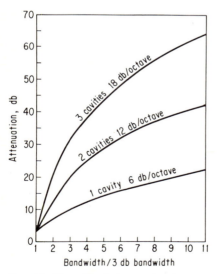

FIG. 10. Normalized attenuation for one, two, or three coupled cavities with equal loaded Q. Insertion loss at center frequency not included.

RECEIVERS

Radio receivers commonly used in the mobile radio services are double or triple conversion superheterodyne designed for the reception of frequency-modulated signals. The design of these receivers is discussed in various articles of the *IRE PGVC Transactions* and will not be reviewed herein; however, overall performance characteristics and specifications required in optimizing system design will be covered. The more important receiver performance characteristics are defined by EIA and IEEE Standards. Measurement standards provide a means of comparing equipments of different designs. The relative importance of each performance characteristic is dependent on the particular system application and the environmental condition in the vicinity of the equipment.

20. Receiver Sensitivity[16]

The relative importance of receiver sensitivity is controlled by the radio-frequency noise condition in the vicinity of its antenna. In a large majority of both base- and mobile-station installations, receiver sensitivity is limited by radio-frequency noise. Receiver sensitivities measured in a system are degraded from 6 to 20 db. This variation in degradation is caused by the highly random nature of the noise. Two methods of measuring receiver sensitivity are in common use. The oldest and probably the most frequently used is the 20-db quieting method. Because of certain limitations in the quieting method, it should not be used to compare receivers of different designs. The quieting method of measuring sensitivity should be limited to

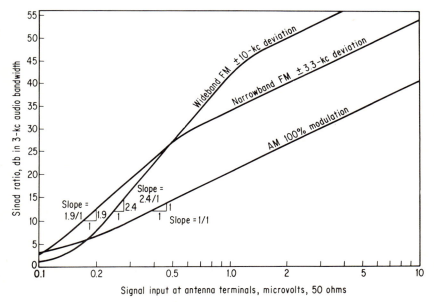

Fig. 11. Comparison of typical sensitivity performance of wideband FM, narrowband FM and AM receivers. All three receivers are compared on the basis of 10-db noise figure.

comparing receivers of the same design because of its inability to distinguish differences in audio response and IF bandwidth.

The second method, 12-db SINAD, is the most widely accepted approach from a system engineering standpoint and is called "usable" sensitivity. SINAD is used as an abbreviation for the quantity measured of Signal + Noise + Distortion to Noise + Distortion and is expressed by a decibel ratio. The 12-db ratio is chosen as an arbitrary point for establishing a reference sensitivity, whereas a complete sensitivity curve can be drawn for a wide range of input signals as shown in Fig. 11. Included in this figure for comparison purposes is a typical amplitude-modulated receiver curve.

Sensitivity by the SINAD ratio method is presented in a slightly different manner by constant SINAD ratio curves illustrated in Fig. 12. The constant SINAD curves give a complete picture of the desired signal characteristics of the receiver, including IF nose bandwidth and audio distortion. The bending of each curve above the 5-kc deviation level is primarily controlled by the IF nose bandwidth. The higher SINAD ratios at 24 db and above are controlled by receiver audio distortion. The closeness

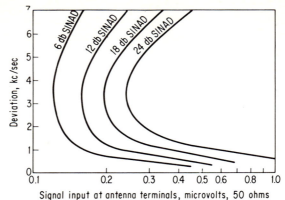

FIG. 12. Typical constant SINAD ratio curves for 40-Mc/sec narrowband FM receivers. Maximum system deviation for narrowband FM is ±5 kc.

of the curves is an inherent characteristic of FM and limiting performance of a particular receiver.

21. Noise Figure

The most universal method of measuring receiver sensitivity or front-end perform-ance is by noise figure expressed in decibels. This quantity is defined by

$$F = 10 \log_{10} \frac{N_o}{N_i G} \tag{5}$$

where F = noise figure, db
$\quad N_o$ = noise power output at point where G is measured
$\quad N_i$ = noise power at input to receiver
$\quad G$ = available gain between points noise power is measured

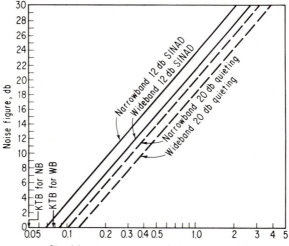

FIG. 13. Conversion of sensitivity to noise figure for typical wideband and narrowband receivers.

Since noise figure must be evaluated in the linear portion of the receiver, it is necessary, in an FM receiver, to measure N_o and G prior to any limiters. Such measurements are difficult to perform accurately in FM receivers. For comparison purposes Fig. 13 gives typical curves comparing noise figure for both SINAD and 20-db quieting sensitivity for wideband and narrowband receivers. Also shown is the equivalent voltage in 50 ohms for KTB at a bandwidth B of 13 kc/sec for narrowband and 30 kc/sec for wideband.

22. Capture Effect

Although not defined in standards, capture effect is sometimes erroneously referred to as a sudden take-over of the stronger signal in FM receivers. Capture by stronger signals is not sudden but is a gradual improvement in signal-to-noise ratio. The rate of improvement is dependent on the bandwidth of the FM system. The slope of the curves in Fig. 11 demonstrates capture improvement of wideband, slope 2.4/1, and narrowband, slope 1.9/1, over white noise. In the case of white noise, this capture improvement continues to the threshold improvement point, which represents a signal input about 13 db above the rms noise level. Above the threshold improvement point the FM curves follow a 1/1 slope as shown for the AM curves for all signal levels.

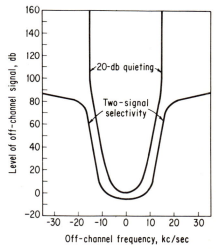

Fig. 14. Typical two-signal selectivity curve of narrowband receiver. Also curve of off-channel signal level to produce 20-db quieting in audio.

23. Selectivity[16]

The selectivity of a receiver as defined by IEEE Standard 52 IRE 17.51 is "that characteristic which determines the extent to which the receiver is capable of differentiating between the desired signal and disturbances at other frequencies." To meet the requirements of this definition, the two signal generators method of measuring selectivity has been standardized by both IEEE and EIA. One signal generator is adjusted as a desired signal with standard test modulation, while the other provides the disturbances at other frequencies. A typical narrowband receiver selectivity curve is shown in Fig. 14. The portion from 0 to about 16 kc/sec off-channel frequency is a function of the low IF filter. The low IF filter provides the greatest attenuation to interfering signals. Beyond the 16-kc/sec point the low IF filter relinquishes its control of receiver performance to circuits near the antenna. The converters, high IF and RF, play their parts in receiver performance, as stronger off-channel signals cause limiting and generate spurious responses due to nonlinearities. The paragraph on radio-frequency interference will cover receiver selectivity in greater detail.

24. Audio Quieting

Quieting is produced in the receiver audio output when an unmodulated signal at or near the center frequency of the receiver is being received. As the signal is raised or lowered in frequency above or below the center frequency of the receiver, its amplitude must be increased to maintain a given degree of quieting. When the level of the off-channel signal is plotted as shown in Fig. 14, a 20-db quieting curve is formed.

Such a curve follows the shape of a portion of the selectivity curve but deviates rapidly above the 80-db level.   At about the 100-db level, the quieting curve takes on a vertical or near-vertical slope.   Obsolete specifications sometimes refer to the 20-db quieting curve as being a measure of selectivity, although no EIA or IEEE Standards includes either a definition or method of measurement.

25. Impulse Response

Above the threshold improvements point for impulse noise, the signal-to-noise ratio at the output of an FM receiver compared with that of an AM receiver is two times the deviation ratio of the FM receiver.   This improvement for wideband receivers is about 28 db, while the improvement for narrowband receivers is about 18 db. For signal levels below the threshold for impulse noise the improvement in S/N ratio decreases.   The degree of improvement below the threshold is a function of the IF bandwidth of the receiver, the impulse response of the receiver, and the frequency of the desired signal.   The degree to which each affects the signal-to-noise ratio is relatively independent of the other; thus they can be examined separately.

Interference produced in an FM receiver as a result of impulse noise increases as the bandwidth decreases.   Filters in narrowband receivers produce increased stretching of an impulse, thereby causing interference over a greater percentage of the time between each impulse as compared with wideband receivers.   The stretched pulse can be considered as a spectrum of frequency vs. amplitude. Typical amplitude and frequency impulse responses of narrowband receivers are shown in Fig. 15.   The amplitude response shown is representative of the

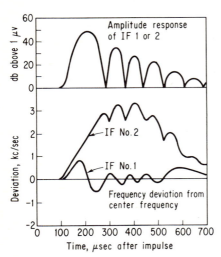

Fig. 15. Typical amplitude and frequency impulse response for two types of narrowband IF receivers.

receiver IF output for an impulse input of 90 db above 1 μV/Mc.   The frequency deviation is representative of two different IF filters with IF filter 1 tuned for minimum frequency deviation while IF filter 2 was not tuned to provide minimum deviation. The interference potential of the two IF filters will be considerably different, although the pulse length and pulse amplitude are the same.

Figure 16 illustrates the degradation of a narrowband receiver due to impulse noise. For receiver 1 the degradation is minimum at the channel frequency, while the degradation of receiver 2 is minimum at about 2 kc/sec higher than the channel frequency. Receiver 1 is said to be "phase tuned."   The difference in the degradation between receivers 1 and 2 of the stretched pulse is as shown in Fig. 15.

26. Multifrequency Operation

Receivers can be equipped to operate two or more channel frequencies either simultaneously or sequentially by switching crystal or separate front-end switching.   The frequency spread between the most widely spaced frequencies is a function of the RF selectivity when the first oscillation is to be changed in frequency.   Dual front-end receivers are generally recommended for frequency separations greater than 0.4 per cent.   See Fig. 17.

Simultaneous monitoring can be provided for certain frequency separations; how-

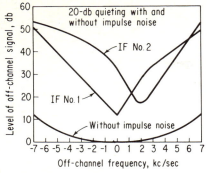

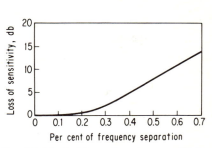

Fig. 16. Typical 20-db quieting curves for two types of narrowband receivers showing degradation due to impulse noise of 70 db above 1 $\mu v/Mc$.

Fig. 17. Typical loss of receiver sensitivity for multifrequency operation.

ever, about 3-db loss in sensitivity will be experienced owing to increased noise bandwidth. Simultaneous monitoring receivers are also subject to increased spurious-response problems.

TRANSMITTERS

Radio transmitters used in the mobile radio services, schematically shown in Fig. 18, consist of a crystal oscillator, audio section, phase modulator, multipliers, and power amplifiers. The performance of the transmitter is primarily determined by the low-

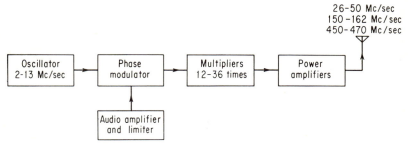

Fig. 18. Typical block diagram for a phase-modulated radio transmitter used in the mobile services.

power-level stages comprising the oscillator, modulator, and audio-limiter circuits. Low-power transmitters are generally transistorized throughout, while intermediate-power transmitters may be partially transistorized, and high-power are usually tubed.

27. Oscillator

Close control of the transmitted frequency is maintained by a crystal controlled oscillator. Maximum frequency change is held at 0.0005 per cent and for some special applications at 0.0001 per cent. Close frequency control is necessary to minimize impulse noise interference as illustrated by Fig. 16. A variation in frequency of 1 kc/sec, or about 0.00065 per cent, at 160 Mc/sec will cause a degradation in receiver sensitivity from 6 to 8 db for typical levels of vehicular impulse noise.

Multifrequency operation of the transmitter can be provided by changing the crystal through a switch or by switching to a second oscillator of a different frequency.   Two, three, and four frequency transmitters are usually provided as standard equipment, while greater than four frequencies are treated as specials.   The maximum frequency spacing between the initial frequency to which the transmitter is tuned and any other frequency is about 0.4 per cent.   Greater frequency separation can be obtained by optimizing the tuning between the two wide-spaced frequencies or by suffering some reduction in power outputs.   For frequency separations greater than 0.6 per cent, it is recommended that a tune-up crystal midway between the extreme frequencies be provided to optimize tuning on each frequency.

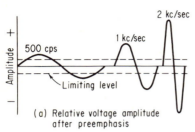

(a) Relative voltage amplitude after preemphasis

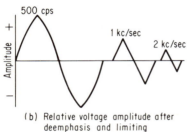

(b) Relative voltage amplitude after deemphasis and limiting

FIG. 19. Relative amplitude of three audio voltages after preemphasis, limiting, and deemphasis.

28. Audio/Limiter

The audio from the microphone in mobile units and from the telephone line of remotely operated base stations must be amplified about 20 to 30 db before passing to the limiter and the phase modulator.   FCC Rules limit the maximum deviation at any audio frequency to 5 or 15 kc/sec.   To meet the requirement some form of instantaneous limiting must be used instead of automatic gain control requiring time-delay action.   In phase-modulated transmitters the most frequently used type of limiter consists of three functional parts: preemphasis, peak limiter, and deemphasis.   The preemphasis prior to limiting forces the higher audio frequencies to be limited first, thus giving the same peak deviation for all audio frequencies.   Simple RC circuits with the time constants of about 750 μsec are used for both preemphasis and deemphasis.   Typical wave shapes for three audio frequencies are shown in Fig. 19.   The limited voltages after deemphasis have the same slope at the zero crossing as shown in Fig. 19b.   Frequency deviation D of the phase modulator is given by

$$D = mf \tag{6}$$

where m = index of modulation, radians
f = modulating frequency
Since the amplitude of the limited voltage decreases at a rate of 6 db/octave of frequency and the deviations, as given by Eq. (6), increases 6 db/octave of frequency, the maximum frequency deviation is limited for frequencies within the audio passband.

In accordance with FCC requirements, an audio filter is installed between the modulation limiter and the phase modulator.   At audio frequencies between 3 and 15 kc/sec, the filter must provide an attenuation greater than the attenuation A in decibels at 1 kc at least

$$A = \log_{10} \frac{f}{3} \tag{7}$$

The filter is intended to reduce adjacent channel interference due to extra sideband radiation created by modulation peaks.   The effectiveness of the filter is dependent on certain characteristics of the modulation limiter which tend to produce harmonics of the limited frequency.   When the modulation limiter restricts the generation of harmonics, the audio filter is found to have little or no effect on adjacent channel, whereas if the modulation limiter generates high-level harmonics, the audio filter will provide considerable reduction in adjacent channel interference.   Adjacent channel interference measurement using VOSIM modulation has been described by Shepherd[19] and Smith.[20] The VOSIM utilizes pulse noise to provide a standard type of simulated noise modulation.

29. Phase Modulators

Typical phase modulators are capable of about 1 radian deviation at an acceptable distortion. The relationship between phase and frequency modulation as given in Eq. (6) requires that the frequency deviation D decreases as the modulating frequency decreases. Thus a transmitter may have full modulation capability for all frequencies down to 400 cps but not below.

The operation of a phase modulator can be illustrated by voltage vectors as shown in Fig. 20. The two out-of-phase voltages A and B are derived from the oscillator. The vector sum of these two voltages which is found at the output of the

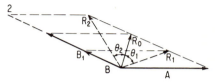

Fig. 20. Vector voltage relationship in a typical phase modulator.

modulation is represented by R_0. The phase modulator is capable of amplitude-modulating the voltage B between the points 1 and 2, which produces new vector sums R_1 and R_2. The new vectors R_1 and R_2 have been changed in phase from R_0 by about 1 radian or 57°.

30. Frequency Multipliers

Frequency multipliers serve the dual purpose of amplifying the voltage from the oscillation and multiplying phase deviation of the modulator. The orders of multiplication most commonly used are, 12, 18, 24, and 36. The phase deviation in radians is increased by the order of multiplication; thus the output of the last multiplier produces peak deviation 12, 18, 24, and 36 radians. Power output of the last multiplier is nominally in the 1- to 3-watt range.

An undesirable product of multipliers is a wideband noise spectrum which is both generated in the multipliers and increased in amplitude in proportion to the increase in phase deviation. The spectral noise generation and amplification are important from a system-interference standpoint, since they limit the interference ratio for adjacent channel operation. Figure 21 shows typical levels of transmitter noise[21] for two different multiplications for a 30-Mc transmitter. The higher level noise is produced by a multiplication of 12, while the lower level noise is for a multiplication of 4. The difference in noise level of the two multipliers is about 18 db near the carrier frequency and 8 db at a frequency spacing of 1 Mc. The transmitter noise output is given in units of decibels per kilocycle below carrier power. Noise level received by a narrowband receiver, with a bandwidth of 13 kc, would be about 11 db above that shown in Fig. 21.

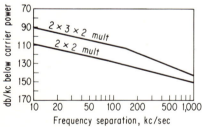

Fig. 21. Comparison of transmitter noise generated by different multipliers.

31. Power Amplifiers

The output of the multipliers is amplified by one or more Class C power amplifiers. Tube power amplification can provide about 13-db gain per stage, while the gain of transistor amplifiers is usually less than 10 db. Low-power base stations up to 30 watts when followed by a second power amplifier can provide up to 330 watts output. Mobile stations are available in various output ratings from 10 to 100 watts. The output of the power amplifier is coupled to the antenna through a low- or bandpass filter to reduce the amplitude of spurious response generated or amplified in the power-amplifier stage. Bandpass filters, when designed with high-Q circuits, have the further advantage of reducing transmitter noise.

ANTENNAS

The antenna system is like a door to a mobile communication system, the design of which can play an important part in the transmission and reception capabilities of the overall system, since the transmitter power output is limited by FCC Rules and receiver sensitivity is limited by local man-made noise. When the power is limited by noise only, the antennas can be used to adjust the coverage area to meet the needs of a given system.

32. Base-station Antennas[22,23]

Of prime importance in the choice of the most efficient base-station antenna is first to determine the area to be covered and then calculate the propagation loss between the base-station location and the extremities of the required coverage area. The coverage areas should be shown on a scale map with possible cases of interfering stations also located. A variety of antenna types is available to provide the best possible signal level to fit the attenuation calculated for each direction from the base station.

a. Omnidirectional Antenna. The most frequently used type of base-station antenna, it gives the best performance when the attenuation in all directions is approximately the same. Omnidirectional antennas should normally be installed on the top of a tower to prevent pattern distortion by the tower. For low band frequencies, the usual type of antennas are ground plane and coaxial. High-band and 450-Mc equipments use some form of stacked array to provide gain by reducing the vertical radiation angle. A simple type of stacked array, in frequent use, is formed by four or more folded dipoles arranged around a support pipe to provide a circular pattern.

Omnidirectional patterns can also be obtained at various levels on a tower where it is not possible or practical to provide mounting at the top, provided the cross section of the tower is less than $\lambda/4$ across. This type of antenna usually consists of two folded half-wave dipoles mounted on opposite or adjacent ledges of a tower and staggered in the vertical plane by $\frac{1}{2}$ to 1 wavelength between centers.

Where the tower is longer than $\lambda/4$, an omnidirectional pattern cannot be easily obtained. The antenna is usually mounted out from one side or leg, which produces an offset. If the tower is very large in cross section and has large openings on each side, it is sometimes possible to obtain a nearly circular pattern by mounting the antenna inside the tower.

b. Shaped Antenna. Patterns can be obtained to cover specific needs by combining unidirectional, bidirectional, and omnidirectional patterns. Figure 22 shows the pattern of a bidirection antenna which provides a quasi-rectangular coverage area with a length-to-width ratio of about 2. The length-to-width ratio can be increased by stacking the same type of antenna or by other types of higher gain bidirectionals.

Figure 23 shows the pattern of a six-element unidirectional antenna with each of its three-element parts horizontally spaced $\lambda/2$ and fed in phase. This type of antenna is also suited to cover a rectangular area where the base station is not located at the center of the required coverage area.

A keyhole-shape pattern can be provided by combining the unidirectional antenna with an omnidirectional antenna. Both antennas are fed in phase and connected through a matching section to a single transmission line. This shape coverage pattern is useful where general coverage is required in all directions except in one particular direction where greater coverage is desired.

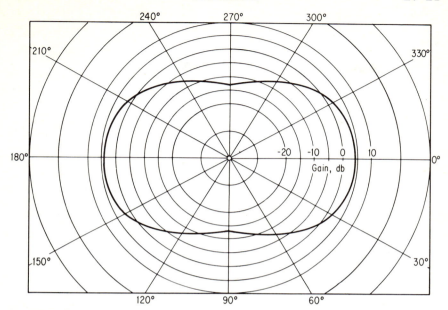

FIG. 22. Bidirectional antenna patterns formed by two half-wave vertical dipoles spaced $\lambda/2$ apart horizontally and fed in phase. (*By permission of Tom J. McMullin, Decibel Products Inc.*)

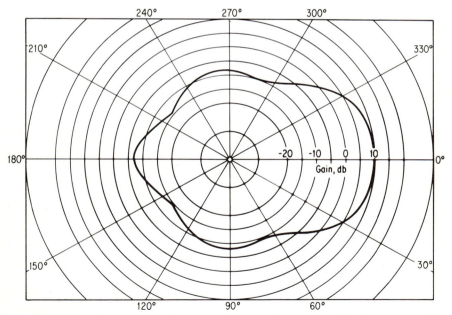

FIG. 23. Unidirection antenna pattern formed by two three-element Yagi antennas vertically polarized, spaced $\lambda/2$, and fed in phase. (*By permission of Tom J. McMullin, Decibel Products Inc.*)

Offset patterns can be provided by arranging folded-dipole, omnidirectional arrays with the dipoles stacked on only one side of the support pipe. Tower-side-mount folded dipoles when placed on only one side of the tower will also provide an offset pattern.

Cardioid-shaped patterns are most frequently obtained by installing a reflector element on a folded-dipole ground-plane antenna. The cardioid pattern is useful in reducing interference in a small angle for a given direction.

33. Mobile Antennas[24]

Mobile antennas are available in two basic types; ground-plane $\lambda/4$ and bumper-mount coaxial $\lambda/2$ antennas.

Ground-plane antennas are commonly mounted in a variety of places on a vehicle, such as rooftop, fenders, trunks, and trunk lid. Considerable pattern distortion can be expected from these antenna locations except for the rooftop mount. High-band and 450-Mc antennas can usually be mounted on the rooftop because of their limited height. However, low-band antennas, owing to lengths from 5 to 8 ft must be mounted on the bumper or on the lower part of the fender. To reduce the physical length of low-band antennas, loading coils have been widely used. With a decrease in length up to 2:1, it is possible to mount the antenna on the trunk lid or sometimes on the rooftop. Loading coils or any other matching device used on shorted antennas reduce the bandwidth of the antenna; thus greater care must be taken in tuning and loading to obtain maximum efficiency for radiation and reception.

Coaxial $\lambda/2$ antennas are well suited to high-band and 450-Mc use when mounted on the rear bumper. The pattern distortion resulting from this type of antenna is not great provided that a considerable percentage of the radiating portion of the antenna is above the rooftop level of the vehicle.

34. Interference Reduction[25]

At base-station locations where transmitters and receivers are operated simultaneously from the same locations, it is usually necessary to obtain the maximum possible attenuation between transmitting and receiving antennas to minimize interference due to transmitter noise and receiver desensitization. Either vertical or horizontal antenna spacing can be used to obtain attenuation. Horizontal spacing follows free-space attenuation, whereas vertical spacing depends on mutual coupling between antennas. Figure 24 gives the attenuation between antennas for vertical spacing. Horizontal spacing for the same attenuation would be up to ten times as great.

Directive antennas designed to provide sharp nulls can be used to reduce or eliminate

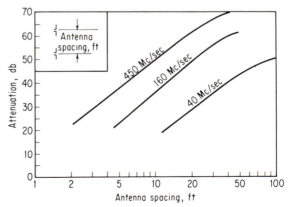

Fig. 24. Attenuation between vertically spaced vertically polarized antennas for various spacings center to center. Curve applies to gain antennas as well as dipoles.

interference originating from some distance from the base station by pointing the null in the direction of the interference. Buildings, hills, and other obstructions can also be used to reduce interference at locations when it is practical to place obstructions between the interference and receiving antenna.

Side-mount antennas reduce noise due to discharge to charged clouds. Wrapping of elements with a good dielectric tape reduces precipitation static build-up from dust and snow.

35. Lightning Protection[26]

Base-station antennas are susceptible to lightning damage unless the proper protective measures are taken. The following protective methods have been successfully employed in various base-station installations:

Apply common grounding to all equipment, metallic structural members, and facilities entering the station.

Provide a discharge gap in the transmission line or use a dc short on the antenna.

Provide heavy-gage conductors in parallel with the coaxial line.

Provide lightning arresters in primary power lines to the equipment.

PROPAGATION

Radio-wave propagation in the mobile frequency bands is subject to several types of losses. The type of losses to consider for a given path is primarily controlled by the distance between transmitting and receiving antennas and secondarily by type of terrain and obstructions. The type of losses to consider are free space, plane earth, smooth spherical earth, shadow, ionospheric scatter, and tropospheric scatter.

36. Free-space Propagation

Free-space transmission loss α in decibels between half-wave dipole antennas is given by the following formula:

$$\alpha = 33 + 20 \log_{10} f + 20 \log_{10} d \qquad (8)$$

where f = frequency, Mc/sec

d = distance between antenna, miles

Except for certain cases the median transmission loss between any two antennas is always equal to or greater than the free-space loss. Figure 25 shows the free-space transmission loss for various frequencies between 30 and 960 Mc/sec.

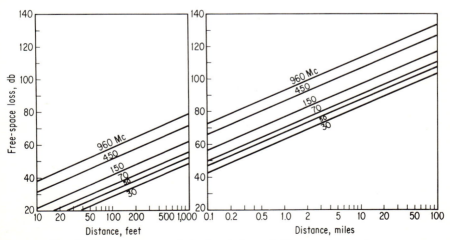

FIG. 25. Free-space transmission loss between half-wave dipole antennas.

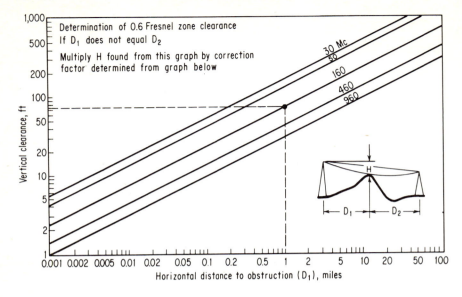

Determination of 0.6 Fresnel zone clearance
If D_1 does not equal D_2

Multiply H found from this graph by correction factor determined from graph below

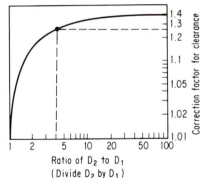

Example (see broken lines)
If: $D_1 = 1$ mile $D_2 = 4$ miles
$F = 160$ Mc $D_1/D_2 = 4$

Then: correction factor = 1.265
$H = 1.265 (74) = 93.5$ ft

Basic formula for calculating 0.6 Fresnel zone clearance

$$H = 1,316 \sqrt{\frac{D_1\, D_2}{f(D_1 + D_2)}}$$

FIG. 26. Required clearance for free-space transmission loss.

Free-space loss between antennas is controlling as long as 0.6 Fresnel zone clearance from obstructions is maintained around the line-of-sight path. The amount of clearance H required is shown on Fig. 26 and given by the following formula:

$$H = 1,316 \sqrt{\frac{D_1 D_2}{f(D_1 + D_2)}} \tag{9}$$

where D_1, D_2 = distance, miles, shown in Fig. 26

    f = frequency, Mc/sec

When the clearance is less than 0.6 Fresnel zone, the attenuation of the path follows the plane earth loss.

37. Plane Earth Propagation

Plane earth transmission loss in decibels differs from free-space loss in that it is independent of frequency but is dependent on the antenna heights as shown by the follow-

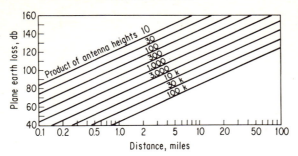

Fig. 27. Plane earth transmission loss between half-wave dipole antennas.

ing formula and Fig. 27:

$$\alpha = 144 - 20 \log_{10} h_1 h_2 + 40 \log_{10} d \qquad (10)$$

where h_1, h_2 = antenna heights, ft
d = distance between antennas, miles

The formula for plane earth loss includes the term $40 \log_{10} d$, which increases the attenuation vs. distance twice as fast as in the free-space case, or 12 db per double distance. As the path between the two antennas falls below the line-of-sight condition, the diffraction losses must be added to plane earth to obtain smooth spherical losses.

38. Smooth Spherical Earth Propagation

Diffraction around the earth's curvature makes it possible for transmission at beyond line of sight with losses greater than for plane earth. The magnitude of this increased loss is dependent on frequency and distance when antenna heights are less than a limiting height as shown in Fig. 28.

Other losses similar to diffraction losses are caused by objects such as hills, buildings, and trees casting a shadow in the transmission path.

39. Shadow Losses

Transmission losses through or around hills, buildings, and trees cannot easily be calculated because of the complex nature of the obstruction. It can in general be shown that the loss increases with frequency, which is to some extent offset by reflection from objects not located in the direct path. Egli[27] has treated shadow losses as terrain factor, giving median deviation based on available FCC and RCA data. Such data are representative of average conditions for specific locations. However, they are useful as a guide to follow in all types of locations. The following median terrain factor losses are typical:

Frequency band	Losses, db
Low band	0
High band	12
450 Mc/sec	22

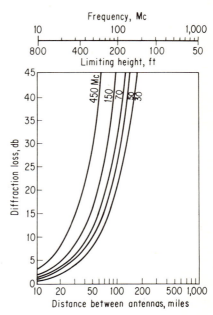

Fig. 28. Diffraction loss caused by bending of radio waves around the earth.

Trees[28] produce considerable loss at 450 Mc, particularly in full foliage. Foliage losses at high band should be considered in heavily wooded forests but can generally be neglected in urban areas.

Shadow losses for rough terrain[29] can be included on a statistical basis when the average height between valley and top of hill is known for the area of interest. Figure 29 shows the estimated shadow loss.

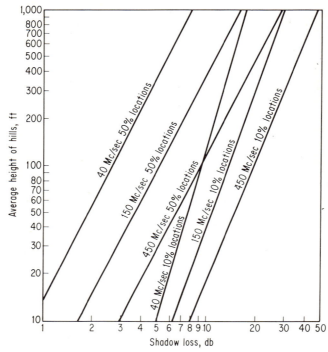

FIG. 29. Estimated shadow loss exceeded at 10 and 50 per cent of locations in areas where the average height of hills is known.

40. Ionosphere Scatter Losses

Ionosphere scatter is important to users of mobile systems because of the potential of creating interference and not as a means of communicating. Probably the most severe interference problem is caused by reflections from the F2 layer during years of high sunspot activity. Propagation loss during high sunspot periods on low-band frequencies can vary from a few decibels less than free space until the loss is controlled by other scatter modes.

41. Tropospheric Scatter

Interference between mobile systems by signals propagated by tropospheric scatter is rare except in those cases where system gain is greater than 165 db and the distance between systems is less than 100 miles. As a means of communicating between base station and mobile stations tropospheric scatter is seldom used; however, some point-to-point circuits do depend on tropospheric scatter.

42. Base-to-base Propagation Losses

The expected propagation loss between base-station antennas is illustrated in Fig. 30. In ranges between 1 and 5 miles, the attenuation is controlled by free-space loss, which

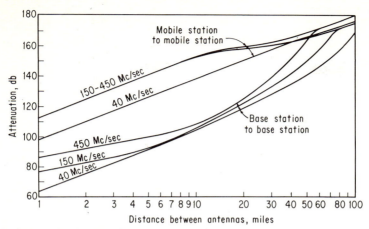

Fig. 30. Attenuation between base-station antennas with antenna heights 100 ft above average terrain. Attenuation between base-station and mobile-station antennas for base-station antenna height of 100 ft.

increases at a rate of 6 db per double distance. Beyond 5 miles and up to 60 miles plane earth loss is controlling, and the attenuation increases at a rate of 12 db per double distance. Diffraction loss, due to curvature of the earth, must be added to plane earth losses beyond the 10-mile range, which gradually increases the slope to about 30 db per double distance. Scatter loss reduces the attenuation slope at attenuations greater than 170 db to about 10 db per double distance. Beyond 100-mile range, the attenuation slope due to scatter gradually increases to 40 db per double distance.

43. Base-to-mobile Propagation Losses

The expected propagation loss between base and mobile stations is shown in Fig. 31 for 100-ft base antennas. Below the 160-db attenuation level, the propagation loss follows plane earth plus diffraction losses. Above 160 db, the losses are gradually

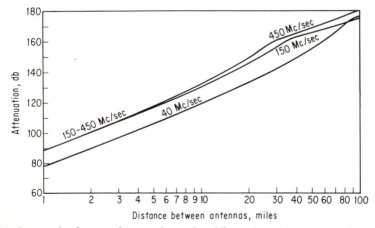

Fig. 31. Attenuation between base station and mobile station with antenna height of base station at 100 ft. Effective antenna height of 40 Mc/sec mobile assumed to be 20 ft.

controlled by scatter. The effective antenna height, assumed to be 20 ft, of the mobile station at 40 Mc/sec accounts for the 10-db difference between 40 and 150 Mc/sec. The effective antenna height of low-load mobile antennas is controlled by the type of soil or water under the vehicle. The range of variation is from 6 ft for poor soil to over 100 ft for sea water. The attenuation between two mobile antennas at 150 Mc/sec is greater than that for two mobile antennas at 40 Mc/sec as shown in Fig. 30. Here again, the effective antenna height is assumed as 20 ft at 40 Mc/sec and 6 ft for 150 Mc/sec, for each mobile antenna.

RADIO-FREQUENCY INTERFERENCE

Radio-frequency interference[30] has gradually become the most important problem to consider in design, operation, and maintenance of mobile communications systems. Interference problems have been increasing at a faster rate than equipment performance improvements can cope with. Except in rare cases, it is no longer possible to operate an interference-free system. However, systems should be designed to minimize interference problems by the use of all available information and equipment. In order to make the best use of available equipment, it is necessary to have a thorough understanding of all types of interference, to maintain accurate spectrum signatures of radio-frequency equipment, and to utilize rapid means of predicting interference potential.

44. Types of Interference[31]

The various types of interference come under the grouping of either co-channel or off-frequency. The first and usually the most difficult problem is to determine whether a particular type of interference is co-channel or off-frequency. Both co-channel and off-frequency interference create the same type of undesired output from a receiver; thus it is not possible to distinguish one from the other without performing specific tests for identification. The main sources of interference are as follows:

Co-channel interference:
    Local co-channel transmitters
    Skip co-channel transmitters
    Restricted radiation devices
    Incidental radiation devices
    Galactic and atmospheric noise
    Transmitter noise
    Transmitter sidebands
    Transmitter spurious and harmonics
    Transmitter intermodulation
Off-frequency interference:
    Receiver desensitization
    Receiver selectivity
    Receiver spurious responses and images
    Receiver intermodulation

For each type of interference created by extra band radiation from a transmitter, there is a corresponding type of interference generated in the receiver. The degree of importance of the transmitter extra band radiation from the standpoint of interference potential is a direct function of the receiver performance. In general, receiver performance is superior to transmitter performance for all types of interference except intermodulation. A closer examination of each type of interference will reveal the superiority of typical receiver performance.

45. Receiver Selectivity and Modulation Splatter[32]

Modulation splatter interference is produced by modulation sidebands of a transmitter entering a receiver IF passband at either the nose or along the skirt. The point of entry of the sidebands is controlled by the attenuation slope of the receiver's IF filter. Figure 32 shows a typical narrowband receiver selectivity curve and two narrowband transmitter spectrums operating in adjacent channels. The attenuation slope of the receiver shown is 10 db/kc, whereas the attenuation slope of the transmitter sidebands is 5.7 db/kc. The optimum slope for a narrowband receiver is between 6 and 8 db/kc. Any greater slope would have very little effect on reducing modulation splatter interference created by an adjacent channel transmitter. The example shows transmitter noise interference entering at the nose of the receiver. Any interference entering the nose of the receiver must be treated as co-channel interference. In this case, the transmitter noise is similar to the white noise generated in the

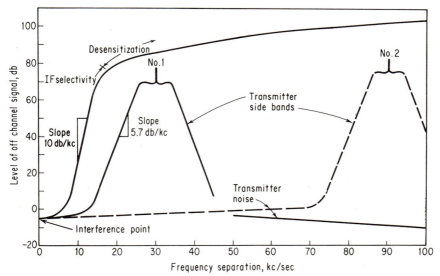

Fig. 32. Typical narrowband receiver selectivity curve of a high-band receiver operating adjacent to a typical modulation sideband of a narrowband transmitter on the next 30-kc channels. Also shown is a similar transmitter operating three channels removed from the receiver.

receiver front end, thus making it extremely difficult to recognize the presence of any interference.

46. Receiver Desensitization and Transmitter Noise[21]

That portion of the receiver selectivity curve adjacent to the IF selectivity portion is commonly referred to as desensitization. It results from interfering signals causing limiting in certain amplifier or mixer stages in the receiver. Recent improvements in receiver performance have considerably reduced interference problems due to receiver desensitization. Receiver specifications are not always adequate to show such improvements unless comparison is made between the complete two-signal selectivity curve of two similar receivers. The main design technique which has brought about this improvement in receiver selectivity is through the use of helical-tuned circuits in the RF portion of the receiver.

Figure 32 illustrates how improved receiver selectivity has practically eliminated interference due to desensitization. Transmitter spectrum 2 is about 20 db below the desensitization point, whereas the transmitter noise spectrum is interfering with the receiver at its nose. In this example transmitter noise is controlling the adjacent channel interference ratio.

47. Receiver and Transmitter Spurious Responses and Radiation

The importance of spurious responses of receivers and spurious radiation of transmitters as a cause of interference has been gradually decreasing. Prior to the year 1948, when mobile systems were being rapidly introduced into the VHF band, spurious response or spurious radiation was responsible for a large percentage of interference. The installation of low-pass filters in transmitters and improvements in receiver selectivity have virtually eliminated interference due to spurious response except in base-station locations where several transmitters and receivers are operating simultaneously. Typical spurious response of receivers and spurious radiation of transmitters are attenuated 85 db or greater. In those cases where transmitters and receivers are operated in close proximity and cause interference, it may be necessary to provide improved cabinet shielding and/or antenna filtering by cavity filters.

48. Receiver[33] and Transmitter[34] Intermodulation Spurious Responses

The most serious interference problem, with the exception of co-channel in certain services, has gradually become intermodulation. Intermodulation interference is

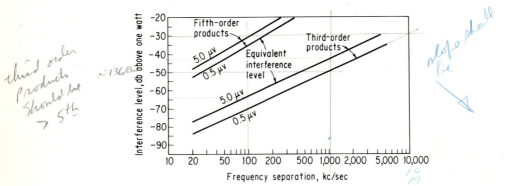

third order products should be > 5th

Fig. 33. Typical intermodulation interference curves for high-band receivers. Each curve represents the equivalent signal level of a third- or fifth-order product generated at the desired signal frequency by two equal-level interfering signals.

generated in any nonlinear circuit by two or more signals whose products land on or near the frequency being interfered with. Transmitter intermodulation is usually produced in the power amplifier of the transmitter, and receiver intermodulation is usually produced in the first converter of the receiver.

Receiver intermodulation is usually controlling when the transmitters causing the interference have their respective antennas separated more than 500 ft. Typical high-band receiver equivalent intermodulation interference level, IM_3 in dbw, for third-order products is given by the following:

$$IM_3 = 7 + 2A + B - 60 \log_{10}(|f_A - f_B|) \tag{11}$$

where f_A, f_B = frequency of interfering transmitters, Mc/sec
A = signal level at receiver's input of nearest frequency transmitter, dbw
B = signal level at receiver input of farthest frequency transmitter, dbw

Similarly, the fifth-order product IM_5 is given by the following:

$$IM_5 = -60 + 3A + 2B - 135 \log_{10}(|f_A - f_B|) \tag{12}$$

where f_A, f_B, A, B are as defined in Eq. (11). These two expressions for IM have been plotted on Fig. 33 for two values of equivalent interference level of -143 and -123 dbw, which are 0.5 and 5.0 volts, respectively.

Transmitter intermodulation is usually controlling when the transmitter antennas are less than 500 ft apart. Transmitter intermodulation products are expressed in decibels below the level of the signal injected as shown in Fig. 34. If, for example, two high-band transmitters have 50-db attenuation between their antennas and are spaced in frequency by 1 Mc/sec, the third-order product radiated from each transmitter would be about 63 db below their respective carrier power outputs.

Cavity filters can be used to reduce or eliminate intermodulation generated in the receiver or in transmitters where the level is high enough to cause objectional interference. The amount of attenuation required by the filter can be calculated by using either Fig. 33 or 34. In some cases where intermodulation is eliminated in the transmitters through the use of filters, it may be generated by loose connections in the

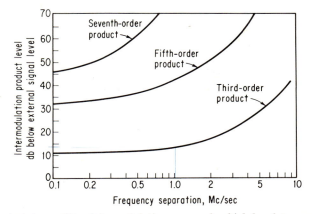

FIG. 34. Typical transmitter intermodulation curves for high-band transmitter. Each curve represents the level of intermodulation product radiated from a transmitter in decibels below the level of external signal injected into the transmitter.

antenna system or in metal objects, not securely bonded together, in the vicinity of either transmitting antenna.

49. Incidental Radiation Devices

The radio-frequency spectrum between 30 and 100 Mc/sec is being gradually polluted[35] by man-made noise. The noise generators causing the spectrum pollution are numerous, with each producing its distinctive type of noise from white to peak levels up to 80 db above the average noise level. No up-to-date data are available for areas of the United States with sufficient accuracy to use in system design. The following estimates can be used as a guide for system installations in urban areas:

Frequency, Mc/sec	Average noise level, db	Peak noise level, db
40	16 above KTB	80 above 1 μv/Mc
150	14 above KTB	80 above 1 μv/Mc
450	8 above KTB	70 above 1 μv/Mc

An article by Ellis[36] from Australia has evaluated man-made noise in terms of site noise factors expressed in decibels above KTB. Typical values of site noise factors at 150 Mc for an average time is shown to be about 20 db above KTB. It is further shown that site noise factor for horizontal polarization has close correlation to vehicle traffic density as follows:

$$\text{Site noise factor} = 7 \log_{10} (D - 5) \qquad \text{db at 150 Mc/sec} \qquad (13)$$
$$\text{Site noise factor} = 7 \log_{10} (D - 10) \qquad \text{db at 420 Mc/sec} \qquad (14)$$

where $D = 10 \log_{10}$ (total traffic within the immediate vicinity).

MOBILE-SYSTEM DESIGN[16,37]

Mobile systems are designed to meet the specific requirements of each application. Although the various system applications are too numerous to describe individually, each can be divided into one or more basic system types. Each basic type is formed by the proper application of standard equipments. The communication range required for a system is usually given first consideration.

50. Communication Range

Communication range can be adjusted to fit the needs of a given system within certain limits. The following parameters have the greatest effect on range:

1. Frequency band used
2. Transmitter power
3. Base-station antenna height
4. Man-made noise levels
5. Terminal variations
6. Receiver sensitivity

Typical base-to-mobile ranges for the three mobile bands are shown in Fig. 35. As can be seen, the range is limited by man-made noise generated near the mobile receiver. Figure 36 shows the mobile-to-base range for the same systems. These curves show the importance of selecting quiet receiver locations, since the range is almost completely controlled by local noise except for the case of 450 Mc/sec.

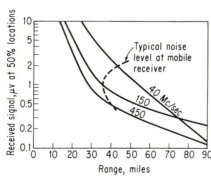

Fig. 35. Typical base-to-mobile range at 50 per cent locations for 250-watt transmitters, 100-ft antenna height, and average hill height of 100 ft. Antenna gains of high band and 450-Mc band are 5.6 and 10 db, respectively.

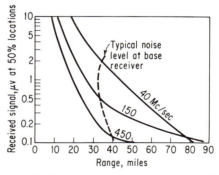

Fig. 36. Typical mobile-to-base range at 50 per cent locations for 100-ft base-station antenna height and average hill height of 100 ft. Power output of mobile transmitter 20 watts for 450 Mc and 50 watts for low and high bands. Base-station antenna gains of high band and 450-Mc band are 5.6 and 10 db, respectively.

Figure 37 shows that mobile-to-mobile range is considerably reduced when compared with mobile-to-base range. In general, mobile-to-mobile range is not sufficient to serve the need of a system except where vehicles are operating in a limited area.

51. Single-frequency Simplex Systems

Both base and mobile stations transmit on a common frequency in a single-frequency simplex system as shown in Fig. 38. Each station must take turns to transmit on a push-to-talk basis. When two or more systems operate in the same area on the same frequency, considerable interference can be expected among base stations. Such interference effectively reduces mobile-to-base range in proportion to the distance between base stations. Mobile-to-mobile operation is seldom used owing primarily to limited range.

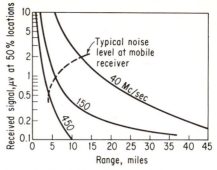

Fig. 37. Typical mobile-to-mobile range at 50 per cent locations for 100-ft average hill height. Power output of transmitters 20 watts for 450 Mc and 50 watts for low and high bands.

52. Two-frequency Simplex Systems

The two-frequency simplex system corrects the problems of mobile stations competing with base stations, thus allowing several systems to operate on the same pair of

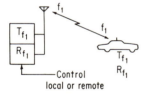

Fig. 38. Single-frequency simplex system. Both base and mobile stations operate push-to-talk on a single frequency.

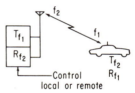

Fig. 39. Two-frequency simplex system. Base and mobile stations operate push-to-talk on different frequencies.

frequencies as shown in Fig. 39. Both base and mobile stations operate on a push-to-talk basis. Systems containing a large number of mobiles can operate on a zone basis by locating a base station near the center of each zone to be covered. Shaped pattern antennas can be used to cover the area of a given zone. Interference created in the zone overlap areas can be minimized by placing zone boundaries along rig lines or through low-populated areas.

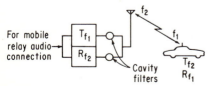

Fig. 40. Two-frequency base duplex systems. Base operates duplex, while mobile operates push-to-talk two-frequency simplex.

53. Two-frequency Base Duplex Systems

When provision is made for the base station to transmit on one frequency and simultaneously receive on a second frequency, it is capable of operating duplex while the mobile operates push-to-talk two-frequency simplex. One type of two-frequency with base duplex operation is shown in Fig. 40. The two cavities shown in the antenna connections are necessary to provide the required attenuation between transmitter and receiver. The amount of attenuation is a function of the frequency separation between the transmitter and receiver as shown in Fig. 41. The required attenuation can also be provided by separating the transmitting and receiving antennas as previously described by Fig. 24. When the duplex frequencies are spaced closer than about 0.5 Mc/sec, it is frequently necessary to use both cavities and antenna spacing. Although Fig. 41 can be considered a typical curve for state-of-the-art equipment, it will be necessary to evaluate the performance of any specific equipment which may not provide equal or better performance.

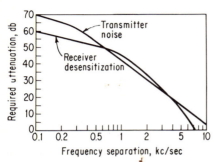

FIG. 41. Typical duplex operation curves for high-band 50-watt transmitters and 0.5-μv receiver. The transmitter noise curve shows required attenuation between transmitter and receiver at the frequency of the receiver, while the receiver desensitization curve shows required attenuation between transmitter and receiver at transmitter frequency vs. frequency separation between transmitter and receiver to prevent greater than 3-db loss of desired signal performance.

54. Mobile Relay Systems[38]

A mobile relay system is a special application of the base duplex system described above, which has the primary purpose of extending the communications range between mobile stations. A mobile relay, as the name implies, retransmits automatically the audio received from mobile stations to other stations. To retransmit automatically, it is necessary that either a carrier-operated relay or coded signal be used to activate the relay transmitter. In order to prevent undesired activation of a repeater, it is frequently necessary to provide both coded signal and carrier-operated relay. As a further safeguard for deactivation a repeater station is also necessary to provide a time delay which must be reset after each 3-min operating period. Duplex operation curves similar to those shown in Fig. 41 are essential in the design of a mobile relay station.

55. Two-frequency Duplex Systems

For two-frequency duplex operation, both the base and mobile stations can transmit on one frequency while simultaneously receiving on a second frequency as shown in Fig. 42. The base-station design is similar to the base duplex system, while the mobile station must be equipped with a second antenna or the necessary filters between the transmitter and receiver to prevent interference. Duplex systems eliminate some confusion when used by persons not accustomed to radio systems, since any push-to-talk requirement is optional. The mobile station must be designed for greater duty cycle,

FIG. 42. Two-frequency duplex system. Both base and mobile stations operate duplex. Both base and mobile stations must have either antenna filters or two antennas each.

resulting in large size and heavier components, whereas the base station is usually designed for continuous duty. Duplex systems have limited application except in extension of a landline telephone system, which is commonly called a mobile telephone system.

56. Backbone Systems[39–41]

The backbone type of system is becoming increasingly important in operation and maintenance of limited-access highways, railroads, and pipelines. Each backbone system covers a narrow strip along the right of way which usually follows one direction for a considerable distance. Backbone systems when operated side by side can also provide wide area coverage. Whether singular or side by side, they generally create the following problems:

1. Interference in the overlap area between two adjacent base stations
2. Uncertainty in selecting the receiver providing the best signal-to-noise ratio from a mobile unit when two or more receivers have detected the signal
3. Design of control and audio system feeding the various transmitters and paralleling audio outputs from receivers

A variety of special equipment and techniques has been developed to alleviate these problems. However, each new system brings slightly different requirements thus increasing the need for special equipment. Interference in the overlap area between two adjacent base stations occurs when the mobile receiver is accepting two carriers of nearly equal amplitude but slightly different in frequency and modulation. A variety of techniques, such as the following, has been used to reduce interference in the overlap area:

1. Installation of manually or automatically switched directional antennas in the mobile unit. Directional antennas allow the mobile unit the possibility of selecting the signal which is slightly stronger while also attenuating the weaker signal.
2. Maintaining accurate frequency control of the base station. When the frequency difference between base stations is less than 300 cps, the beat note created by the two carriers will be below the audio passband of the receiver.
3. Polarize and phase-equalize audio to maintain synchronized deviation on all base stations. Interference is minimum where the instantaneous deviation between each base station as received at the mobile receiver is minimum.
4. Equip the base station with directional antennas providing the greatest possible front-to-back ratio. The forward lobe of each directional antenna will overlap the back lobe of the next antenna when both are pointed in the same direction, thus reducing the width of the overlap area by a considerable amount.

Uncertainty in selecting the base receiver providing the best signal-to-noise ratio is eliminated by voting-unit devices. A voting unit gives precedence to the base receiver receiving the strongest signal by connecting it to the audio line and locking out all other receivers.

Control equipment in backbone systems serve a variety of functions such as the following:

1. *Standby Switching.* Automatic or manual switching to standby equipment requires elaborate signal, control, and testing equipment.
2. *Console Design.* Control consoles are usually a special design for each system, although certain parts may be common.
3. *Backbone Audio and Control Circuits.* The audio and control portion of the backbone system is usually furnished through microwave relay equipment or landlines.

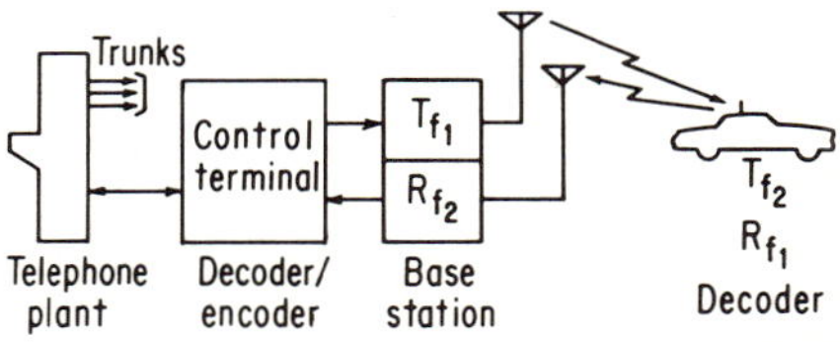

Fig. 43. Mobile telephone system. The base station operates duplex, while most mobile stations operate push-to-talk.

57. Mobile Telephone Systems[42–44]

Mobile telephone systems differ from other mobile radio systems in several ways. They require a greater degree of flexibility, since a mobile telephone is not identified with any one base station. A subscriber is generally free to travel from city to city and in each place may wish to use his mobile telephone as he would his office telephone. The essential elements of a mobile telephone system are shown in Fig. 43.

A practical selective signaling unit which is capable of a large number of calls is an essential part of any mobile telephone system. The majority of systems uses two-tone 600 and 1,500 cycles, which were first applied to the Gill selector consisting of a polar relay activating a stepping mechanism which can latch at predetermined points along its travel. The predetermined points form the numbering systems between which the mechanism returns to its starting point. The calling sequence is complete when a complete series of points has been passed. In recent years, a variety of stepping mechanisms have been designed which follow the basic principle of the Gill selector; however, they are capable of increasing the number of possible codes. Mobile telephone systems have been going through a constant evolution of changing technology as follows:

1. Single-channel mobile push-to-talk, with calls made through a mobile telephone operator by mechanical selectors.

2. Manually selected multiple-channel mobiles, mobile push-to-talk, with dial direct from landlines. Some solid-state selectors are being used.

3. Manually selected multiple-channel mobile with duplex operation and two-way dial.

4. Automatically selected channels with duplex operation and two-way dial. Selector and channel-switching circuits are all solid state.

The introduction of multichannel operation has provided a greater utilization of available radio-frequency channels.

58. Open-wire Transmission Systems[45]

Radio-frequency propagation in tunnels, in buildings, and in underground facilities is subject to constant attenuation vs. distance. The attenuation varies with frequency, size, and configuration of the enclosure. Typical values of attenuation in an automobile tunnel for two lanes of traffic is about 10 db/100 ft. The range of two mobile units within such a tunnel is limited to about 1,000 ft. Attenuation in other tunnels, such as train and subway, is slightly more dependent on their size and construction. Attenuation in buildings constructed of concrete and steel varies from 10 to 20 db/100 ft, with 10 db/100 ft representative of open construction and 20 db/100 ft found where there are one or more partial partitions for each 100 ft. Attenuation through 8-in. concrete walls varies from 8 to 10 db, while attenuation through solid metallic walls varies from 10 to 40 db.

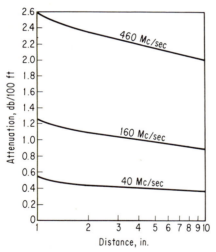

FIG. 44. Attenuation of RG-86/U when placed at various distances from a concrete surface.

To provide adequate received signal level in large enclosed areas, it is necessary to install either distributed or lumped radiator feed from low-loss transmission lines. Lumped radiators in the form of dipole antennas are adequate to meet most needs where the area can be covered with 10 antennas or less. Enclosed areas requiring more than 10 antennas can be more easily covered with unshielded twin-conductor transmission such as RG-86/U. The attenuation of RG-86/U varies with frequency and the distance spaced from dielectrics similar to concrete, as shown in Fig. 44. The radiation from a twin line is about 35 db below the radiation of a dipole antenna operating at the same power level.

A tunnel communication system can be designed for a given frequency band by using an attenuator 55 db between the transmission line and the mobile equipment. If a 50-watt base station were installed in the middle of the line, the power would divide with 25 watts in each direction. Assuming a received signal level of −130 dbw, the total system gain is about 144 db. The allowable attenuation in each section of transmission line is 89 db, giving a 8,900-ft line at 160 Mc/sec for each direction from the base station. The total length of line would be 17,800 ft or a distance of about 3.4 miles each base station can cover at a frequency of 160 Mc/s. Where it is necessary to provide communication in longer tunnels, a series of base stations may be used as required.

To feed the RG-86/U from a base station, it is necessary to use a balun providing a 50- to 200-ohm impedance transformation. A balun can be constructed from RG-8/U cut to $\lambda/2$ electrical length L, in inches, as given by the following expression.

$$L = \frac{3,950}{f} \quad \text{in.} \tag{15}$$

where f is the frequency in megacycles per second.

When the center conductor at each end of the $\lambda/2$ coaxial line is connected to the input of the 200-ohm line and a 50-ohm feed line is connected to either side of the balanced 200-ohm line with its shield connected to the shield of the $\lambda/2$ line, a balun is formed.

PERSONAL/PORTABLE SYSTEMS

Rapid technical advances have pushed mobile communications equipment into smaller and smaller packages, with the ultimate being a personal transmitter/receiver which will fit conveniently into a man's clothes. This continual reduction in size has invariably been accompanied by a significant reduction in communication range. Reduction in range is a direct result of reduced antenna efficiency and available battery power. The reduction in antenna efficiency is applied to both the transmitter and receiver, whereas the reduction in battery power must be applied to the transmitter only. With lower transmitter power output and poor antenna efficiency, the talk-back range has been decreased to a greater degree than the talk-out range. To offset this difference in range, it is necessary to provide more base stations receiving at various locations.

59. Personal/Portable Receivers[46]

The introduction of production quantity transistors made it practical to market a small-size receiver capable of being carried or worn by a person. The performance specifications of personal receivers are somewhat reduced as compared with mobile receivers, as can be seen from typical values in Table 4.

Table 4. EIA Standards and Manufacturers' Specifications for Personal Receivers

Quantity measured or specified	EIA Standards (proposed)		Typical manufacturers' specifications	
	25–54 Mc/sec	144–174 Mc/sec	25–54 Mc/sec	144–174 Mc/sec
Average sensitivity, μv	*	30	12	15
Selectivity, db	30	30	45	45
Spurious attenuation, db	35	35	45	45
IM spurious attenuation, db	30	30	45	45
Audio power output, mv	*	*	100	100

* No standard given.

Sensitivity is the most significant reduction of performance rated in personal receivers. In this case, the sensitivity is the average radiation sensitivity, which includes antennas and the influence of a man on its average efficiency. The reduction of audio power output is not so significant, since audio is usually delivered close to the user's ear through an earpiece inserted in the ear or through a small horn attached to clothing near the ear.

A personal receiver can be operated as a separate unit or combined with a transmitter in a common package. When operated as a separate unit, it usually functions as part of a one-way signaling system for transmitting messages or telephone numbers to call. Personal signaling systems are limited in range to a few miles and much less when operated from inside buildings.

60. Personal/Portable Transmitters

Transistorization of personal/portable transmitters has been much slower than for their companion receivers owing mainly to the lack of production-quantity high-power, high-frequency transistors. Suitable transistors are becoming available in sufficient quantities, although their high price forces the continued use of some tubed transmitters.

The power output of personal/portable transmitters ranges from 0.5 to 20 watts in the 25- to 54-Mc/sec band and from 0.2 to 8 watts in the 144- to 174-Mc/sec bands. Transmitters designed for the 450- to 470-Mc/sec band have not as yet found wide usage. The effective radiated power of personal transmitters is reduced considerably from the above-mentioned power output. The actual reduction is to a great extent dependent more on the mode of operation than on the antenna design. Transmitters are usually provided with telescoping antennas which are most frequently left in the down position. The effective radiated power of a transmitter with the antenna fully extended and operated at head height is usually less than 3 db below its rated power output.

The use of transmitter and receiver equipment for person to person has been widespread where the required range is less than 1 mile. The state of art in battery development is the biggest factor in reduction of weight and size. Typical applications are as follows:

1. All types of construction
2. Harbor pilots to bridges
3. Outside maintenance such as power lines and right of ways

61. Antennas[47]

Because of size limitations, portable equipment and particularly personal equipment must use small antennas for radiation and reception. A small antenna is a radiating device whose largest dimension is less than one-half wavelength. Its effective radiated or received power is theoretically less than that of a half-wavelength dipole. The radiation resistance of small antennas is usually less than their self-reactance. The ratio of radiation resistance to self-reactance is called the radiation power factor, which is a measure of its efficiency as a radiator. A perfect antenna, meaning that all energy delivered to the antennas is radiated, will have a radiation power factor of unity. Small antennas, having a radiation power factor much less than unity, retain an appreciable amount of energy from cycle to cycle as stored energy, thus limiting the amount of radiated energy. When energy is stored from cycle to cycle, the antenna appears as a high-Q circuit, thus also restricting its bandwidth of operation to much less than that of a half-wave dipole.

Small antennas take on a variety of forms several of which are listed with their average loss as compared with a half-wave dipole as follows:

1. High-Q ferrite core resonant coils. Average loss is 20 db.
2. Earpiece cord and short wires. Average loss is 25 db.
3. Resonant loops.

4. Combinations of 1 and 2.   Average loss is 15 db.

5. Quarter-wave whips without adequate ground planes.   Average loss is 10 db.

Typical radiation patterns of three types of small antennas are shown in Fig. 45. The ferrite-core antenna is the most compact antenna for the minimum loss.

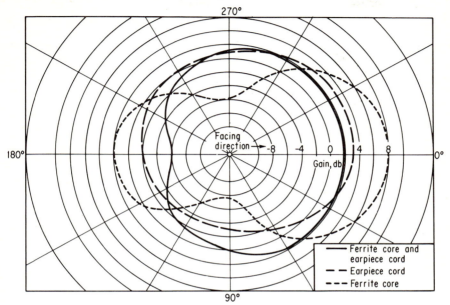

FIG. 45. Field patterns of personal receivers in left-hand shirt pocket.

62. Power Supplies

A variety of battery types is used in personal/portable equipment.   Those most frequently used are the following:

1. Zinc-carbon
2. Alkaline
3. Mercury
4. Nickel-cadmium

Except for the nickel-cadmium, none of the above batteries are suitable for recharging.   The alkaline and mercury batteries provide the longer life before discharge, whereas the nickel-cadmium, after repeated recharges, provides the longest overall life.

The minimum power-supply life, according to EIA Standards, is one day.   One day of life is defined as 8 hr of operation with a duty cycle of 6 sec receive at rated audio power output, 6 sec transmit at rated RF power output, and 48 sec standby. Rechargeable batteries normally provide several days of life, while nonrechargeable batteries provide up to 3 weeks of life.   The most economical type of battery is dependent on both the type of operation and the battery cost per hour.   Required temperature range is an important consideration, as some batteries are not capable of operating below 0°C.

63. Equipment Accessories

Carrying cases, lapel speakers, and external antennas are typical examples of equipment accessories.   Portable cases are weatherproof with protection against rain, rough handling, and dirt. Lapel speakers are necessary for most types of operation, particularly in high-noise-level areas. Considerable extension of range can be obtained through the use of external antennas.

64. Paging Systems[48,49]

Paging systems, as the name implies, are one-way signaling systems which frequently include a voice message to a personal receiver following an alerting signal. They may also be considered as an extension of the telephone bell on a wireless basis. The usual type of signal is transmitted by several tones and received by resonant reeds, thus permitting an individual to be selectively signaled without disturbing others capable of being called on the same system.

Systems in present use can be divided into two basic types, the first being a wide-area general-use type providing subscription service to the public. Such a service is frequently part of a telephone-answering business which provides this service, through the use of radio, by relaying messages. Such messages are transmitted over a high-power base station of usually 250 watts. The outdoor coverage of such a system is from 3 to 5 miles. Beyond 5 miles the coverage varies greatly with receiver sensitivity and terrain conditions. With the 3-mile range, the coverage in buildings is highly variable. Coverage in offices with outside windows is generally adequate, while in halls and below-ground areas it becomes more difficult as the distance from the nearest outside opening increases.

The second type, an in-plant paging system, is a private system limited to the general area of a manufacturing plant or commercial building. The required equipment is a personal receiver usually carried in a pocket or case on the belt and a base station located near the center of the designed coverage area.

Paging is sometimes performed by passing messages through a telephone operator. On receiving a message, the telephone operator will send the proper selective signal over the radio system followed by a message. Recent improvements have removed this added duty from the operator by an automatic dial system which can be initiated from any telephone on the same switchboard, providing direct voice messages to the receiver.

Both these paging systems lack the capability of the called party initiating an answer without calling through a landline telephone. When the calling party receives no answer, he is then not sure whether further effort should be made to get the message through.

65. Personal Two-way Systems

The development of personal two-way systems has been slow because of decreased talk-back range. Wider use of wired communications as discussed in paragraph 58 will result in a more general use of two-way systems.

SELECTIVE SIGNALING AND CONTROL

The use of selective signaling and control as both a primary and secondary application to mobile radio systems has in general exceeded the rapid growth of mobile radio. Selective signaling and control are accomplished by modulating the transmitted carrier with one or more audio-frequency tones in a coded sequence. The coded tones, after being demodulated in a receiver, are used to perform a variety of functions such as the following:

1. Operate an audible or visual device
2. Control traffic lights
3. Operate warning signs along highways
4. Remote control of equipment
5. Remote indication of equipment status
6. Report failure of equipment
7. Remote reporting of a variety of metering devices

A typical selective signaling and control system consists of one or more tone generators, one or more tone receivers, and the necessary logic circuits to decode the tone sequence.

66. Tone Generators

Tone generators are designed to operate in the frequency range form 60 to 3,000 cps. The frequency of operation depends to a great extent on the type of tone signaling system used and the function provided. The various type of tone generators

Table 5. Tone Generators

Use	Frequency range, cps	Type of oscillator
Tone squelch............	60–250	Twin T network
Tone squelch............	60–250	Resonant reed
Selective signal.........	300–1,000	Resonant reeds
Selective signal........	300–3,000	Twin T network
General use............	300–3,000	RC or RL circuits

can be divided into frequency ranges and type of use as shown in Table 5. The performance characteristics of the tone generator in order of importance are as follows:

1. Frequency stability
2. Starting time
3. Harmonic distortion
4. Amplitude stability

The frequency stability requirements depend on the bandwidth of the tone receiver, the operating margin established for the system, and the spacing between adjacent tone channels. Harmonic distortion must be held to a minimum in tone squelch systems and other systems below the margin established for each tone frequency. Starting time is of considerable importance when it is greater than 10 per cent of the response time of the tone receiver. Variations in the amplitude at the output of a tone generator must be maintained to a small percentage of the established margin.

67. Tone Receivers[50]

Tone receivers are generally designed to operate on corresponding frequency to the tone transmitter by using similar selective circuits employed in tone transmitters. The performance characteristics of tone receivers in order of importance are as follows:

1. Frequency stability
2. Response time
3. Amplitude stability

Frequency stability is generally a function of supply voltage and temperature. Frequency variations should be maintained over the voltage range and temperature range at some small percentage of the operating bandwidth of the filter. The response time T of a single tuned filter circuit is described by the following:

$$T = \frac{1}{\pi \, \Delta f} \tag{16}$$

where Δf = bandwidth, cps
$ T$ = time, sec

The time T represents the time required for the output to rise in one time constant which is 4 db below the maximum output. The decibel margin is given by the following expression:

$$M = 20 \log \frac{e^{\pi T \, \Delta f}}{e^{\pi T \, \Delta f} - 1} \tag{17}$$

Margin M is defined as the ratio in decibels between the voltage delivered to the filter and the minimum voltage required to operate the output device. From the required margin and response time, the bandwidth of a filter can be determined from Fig. 46.

After the margin and filter bandwidth have been fixed, the number of tone channels available in the audio band between 300 and 3,000 cps can be determined from Fig. 47.

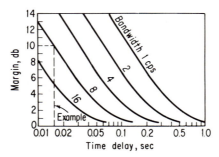

Fig. 46. Margin related to time delay and bandwidth.

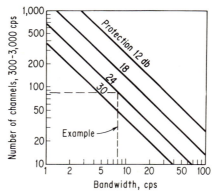

Fig. 47. Number of tone channels available in the audio-frequency band between 300 and 3,000 cps for various bandwidth and protection in decibels.

The protection must include the margin; thus the actual protection between two adjacent tone frequencies decreases as the margin increases. Figures 46 and 47 can be used to determine the number of available tone frequencies by the following example:

   Filter bandwidth.............. 8 cps
   Margin...................... 10 db
   Time delay.................. 0.0156 sec
   Desired protection............ 14 db

When the margin is added to the desired protection, the total number of channels is 85, as found from the 24-db protection line.

It is important that tone receivers be protected from false operation.   False operation can be caused by one or more of the following:

1. Adjacent channel tones
2. Intermodulation in simultaneous tone systems
3. Noise
4. Voice modulation

In general, false operation can be reduced or eliminated by reducing the filter bandwidth to provide sufficient protection from interference listed opposite 1, 3, and 4 above.   Intermodulation, caused by two or more simultaneous tones, creates interference at or near the tone frequencies of the receiver; thus any reduction in bandwidth will have little effect.   The only satisfactory method for eliminating interference due to intermodulation in a simultaneous tone system is to select tone frequencies.

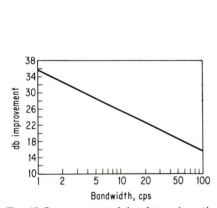

Fig. 48. Improvement of signal-to-noise ratio due to a single tuned circuit.

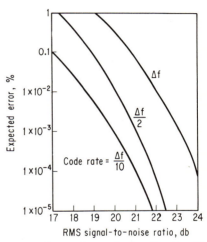

Fig. 49. Expected error in decoding for different code rates and signal-to-noise ratio at output of filter.

The selection of tone frequencies to eliminate all third-order products will cause considerable reduction in the total number of usable frequencies.   The approximate improvement I in signal-to-noise ratio for a flat noise spectrum from 30 to 3,000 cps is given by the following:

$$I = 24 + 10 \log \frac{15}{\Delta f} \quad \text{db} \qquad (18)$$

where Δf is the bandwidth of the filter in cycles per second.

When the signal-to-noise ratio of a tone before filtering is added to the signal-to-noise improvement found from Fig. 48 for a particular filter bandwidth, the total is obtained. The total signal-to-noise ratio obtained at the output of the filter is reduced by the margin in decibels giving the effective signal-to-noise ratio at the detector.   The detector must make a decision based on this effective signal-to-noise ratio.   If the period for the detector to make a decision is long compared with the filter bandwidth, then the probability of the detector making a correct decision is greatest.   As the period is reduced, the expected error at the detector increases, as shown in Fig. 49. The expected error for sequential and simultaneous tone systems may be estimated by taking the product of the error for the individual tones.

68. Single- and Frequency-shift Tone Systems

A variety of selective signaling and control systems performs the required functions by either single-tone or frequency-shift tones. The single tones are transmitted by a short pulse or continuous as in tone squelch systems. Frequency-shift tones are used almost exclusively in mobile telephone systems (MTS). Single-tone systems, with a short pulse, are commonly used in paging systems where the number of required combinations are less than about 30.

Frequency-shift tone systems used in MTS are coded by several groups of frequency-shift tones with each group representing a digit. The sum of the digits is a constant. The total number of available codes is determined by the number of digits and their sum. Frequency-shift tones are also used for transmitting binary codes with the number of codes doubled for each added bit.

69. Simultaneous Tone Code Systems[51]

Simultaneous tone code systems are coded by transmitting two or more tones by a short pulse. The number of possible codes N are given by the following expression:

$$N = \frac{n(n-1)(n-2)(n-8) \cdots (n-a+1)}{a!} \tag{19}$$

where n = total number of tones

      a = number of simultaneous tones

The ability to transmit a large number of codes in the shortest possible time is the main advantage of a simultaneous tone code system. Figure 50 shows the required number of tones and combinations for possible codes between 10 and 1,000.

The main disadvantage to the use of simultaneous tone coding is its inherent problem of false operation due to generation of intermodulation spurious tones. The generation of spurious tones by intermodulation can be minimized by restricting the number of simultaneous tones and maintaining the tone amplitudes below the level required to cause amplitude distortion in the radio-frequency transmitter.

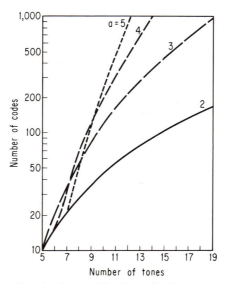

FIG. 50. Number of codes provided by a simultaneous tone system.

70. Sequential Tone Code Systems

Sequential tone code systems are coded by transmitting two or more tones separated by the required time. If one tone is lost or does not follow in a given time sequence, the entire code must be repeated. The number of possible codes N are given by the following expression:

$$N = n^a$$

where n = total number of tones

a = number of tones per code in sequence

The sequential coding system normally requires a longer time to transmit a given code as compared with the simultaneous coding systems; however, the number of possible codes is much greater as shown in Fig. 51. The protection from false operation is

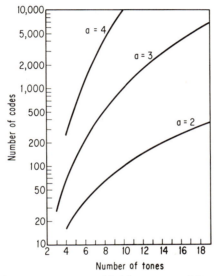

Fig. 51. Number of codes provided by a sequential tone system.

improved, since the problems due to intermodulation spurious response have been eliminated. The probability of missing a code because of signal fades has increased, since the codes are longer. In some systems, it is common practice to repeat a code two or three times to improve the reliability of decoding.

Some coding systems have combined simultaneous and sequential code in order to utilize the advantages of both. One such system is a numeric coding system which utilizes five tones transmitted two at a time. As can be seen from Fig. 50, five tones transmitted two at a time provide ten codes with each code representing a digit between 1 and 10. If the tones are chosen to eliminate all third-order intermodulation combinations, a highly reliable and fast digital coding system is provided.

71. Tone Squelch Systems[52]

The use of tone squelch systems is an established method for the elimination of botherance, although their use tends to increase rather than decrease interference. Botherance is a term which describes the effects of an undesired signal when no desired signal is being received. To eliminate botherance, the tone squelch systems hold a receiver squelched closed for all signals except those coded with the desired tone. If each co-channel user is provided with a separate tone code, only those stations with the same code will have squelch openings. Interference between co-channel systems, may be increased when a station in one of the systems, assuming that it has a clear channel, proceeds to transmit without first monitoring the channel. A means of monitoring the channel has been used in some equipments by providing a combination tone and noise squelch. To provide monitoring capabilities in mobile stations, the hang-up microphone is used to operate a switch which disables the tone squelch when off hook. On base-station microphones, a key level switch in the microphone forces the operator to monitor the channel momentarily before transmitting.

Tones for operating tone squelch systems have been chosen from the frequency band between 67 and 250.3 cps. These tone frequencies, being outside the audio-frequency passband of the communications systems, cause only slight degradation providing their distortion and audio level are controlled as required in EIA Standard RS220.[53]

Most tone squelch systems operate two-way, even though it is generally most important to provide tone squelch at the base station only. The base station is normally receiving about twice as many transmissions as a mobile station. As the number of mobile units per system increases, the advantages of tone squelch systems decreases. A large system with 100 or more mobile units may find very little advantage to tone squelch unless the type of dispatching requires the mobile units be divided into particular groups.

With slight modification tone squelch systems can be used to provide a variety of functions in addition to their intended use. For example, specific mobile relay stations can be selected from a group by initiating the proper tone. By selecting a particular relay station it is possible to eliminate interference due to overlaping coverage of two stations. Another application is in the remote control of one base station by several users. Each user's control equipment would be equipped with a different tone as also would his associated mobile units.

72. Mobile Telephone Systems[42]

The two-tone system, using 600 and 1,500 cps, is most frequently used for selective calling in mobile telephone systems. To start the sequence of tones, 600-cps tone is applied first to modulate the radio-frequency transmitter, followed 500 msec later by 1,500 cps. The purpose of the first shift from 600 to 1,500 cps, which is performed automatically, is to clear all selectors with the digit 1. This procedure has the disadvantage of eliminating the digit 1 from any numbering systems. The usual dialing procedure can then be followed with each digit sent by frequency shift at a rate of 8 to 11 pps. Between digits, the last tone is carried over until the dial starts pulses for the next digit.

The Gill selector has set the pattern for a variety of mechanical selectors found in common use. Improved types of decoders are all solid-state devices which have eliminated some of the undesired features of the mechanical selector. With new solid-state selectors, the digit *one* can be used, thus increasing the number of possible combinations. Other features have been added which include automatic selection of radio-frequency channels and transmission of a mobile billing number.

73. Personal Paging Systems[54]

In general, personal paging systems use the simplest form of selective signaling. Selective receiving devices used are necessarily the smallest possible, since they must operate as part of a personal receiver. A single reed selector is used as the selective receiver when the number of required codes are less than about 30. Two or more simultaneous or sequential tones are used when the number of required codes are greater than 30. In most cases, sequential tone systems are preferred to eliminate false calls due to generation of intermodulation products.

When connected through terminal equipment to an automatic telephone system, personal paging systems have provided an improved in-plant signaling system. Any party who is equipped with a personal receiver can be signaled through the dial telephone by calling a special number. A message may then be given by voice transmission. To provide two-way communication, the called party must go to the nearest telephone and call the calling party.

REFERENCES

1. Bailey, A., Developments in Vehicular Communications, *Proc. IRE*, vol. 50, pp. 1415–1420, May, 1963.
2. Floegel, M. E., Mobile Radio Changes the Pace of the Nation, *IRE Trans.* PGVC, vol. PGVC-6, pp. 31–35, June, 1955.
3. Green, M. E., VHF Marine Mobile Systems in British Columbia, *IRE Trans.* PGVC, vol. PGVC-6, pp. 67-70, July, 1956.
4. Brookes, G. A., Police and Fire Department Communications Centers: A Systems Approach to the Control Console and Related Facilities, *IRE Trans.* PGVC, vol. VC-9, no. 3, pp. 58–71, December, 1960.
5. Humphrey, T. G., Tone Signaling Increases Mobile Radio Efficiency, *IRE Trans.* PGVC, vol. VC-10, no. 2, pp. 45–48, August, 1961.
6. Keller, J. E., Radio Speeds Flow of Oil, *IRE Trans.* PGVC, vol. PGVC-8, pp. 66–73, May, 1957.
7. Talley, D., A Prognosis of Mobile Telephone Communications, *IRE Trans.* PGVC, vol. VC-11, no. 2, pp. 27–39, August, 1962.
8. Plummer, C. B., Reasons for Establishing a "Service," *IRE Trans.* PGVC, vol. PGVC-8, pp. 74–75, May, 1957.
9. Hansen, R. E., Mobilization of Transistors, *IRE Natl. Conv. Record*, vol. 6, part 8, pp. 28–32, March, 1958.
10. Myers, R. T., A Transistorized Receiver for 150-Mc Mobile Service, *IRE Trans.*, vol. VC-9, no. 2, pp. 70–79, August, 1960.
11. *EIA Standards*, Land-mobile Communications System Using FM or PM in the 25–470 Mc Frequency Spectrum, RS-237 August, 1960.
12. *EIA Standards*, Land-mobile Communications FM or PM Transmitters 25–470 Mc, RS-152-A, October, 1959.
13. *EIA Standards*, Land-mobile Communications FM or PM Receivers, RS-204, January, 1958.
14. Gaylord, J. W., The Conductive Cooling of Power Tubes in Vehicular Communications Equipment, *IEEE Trans.* PTGVC, vol. VC-12, pp. 71–80, September, 1963.
15. Gilbert, E. M., Time Ratio Regulation, A New Inverter Design Technique, *IEEE Trans.* PTGVC, vol. VC-12, no. 1, pp. 60–70, September, 1963.
16. Buesing, R. T., and N. H. Shepherd, System Performance, Compatibility, and Standards, *IRE Trans.* PGVC, vol. VC-9, no. 3, pp. 32–43, December, 1960.
17. Smith, J. S., Impulse Noise Reduction in Narrow-band FM Receivers: A Survey of Design Approaches and Compromises, *IRE Trans.* PGVC, vol. VC-11, no. 1, pp. 22–26, August, 1962.
18. Germain, J., Extender Operation—Ignition Noise Suppression Built Right into the Radio Receiver, *IRE Trans.* PGVC, vol. VC-1, no. 1, pp. 40–43, August, 1962.
19. Shepherd, N. H., SINAD Interference Evaluation by VOSIM, *IRE Natl. Conven. Record*, vol. 5, part 8, pp. 23–30, March, 1957.
20. Smith, J. S., Adjacent Channels and The Fourier Curse, *IRE Trans.*, vol. PGVC-9, pp. 3–11, June, 1957.
21. Shepherd, N. H., and J. S. Smith, The Gaussian Curse—Transmitter Noise Limits Spectrum Utilization, *IRE Trans.*, vol. PGVC-10, pp. 22–32, April, 1958.
22. McMullin, T. J., The End of the Line, *IRE Trans.*, vol. PGVC-11, pp. 20–32, July, 1958.
23. Scheldorf, M. W., New Type High Gain Station Antenna, *IRE Trans.*, vol. PGVC-11, pp. 10–16, July, 1958.
24. Harris, E. F., Investigation of Antennas for Two-way Mobile Communications in the VHF and UHF Regions, *IRE Trans.*, vol. PGVC-12, pp. 2–4, April, 1959.
25. Biggerstaff, W. F., Operation of Close-spaced Antennas in Radio Relay Systems, *IRE Trans.*, vol. PGVC-12, pp. 11–15, September, 1959.
26. Bodle, D. W., Lightning Protection for Mobile Radio Fixed Stations, *IRE Trans.*, vol. PGVC-1, pp. 122–131, February, 1952.
27. Egli, J. J., Vehicular Transmission, *IRE Trans.*, vol. PGVC-11, pp. 86–90, July, 1958.
28. Glentzer, K. V., 450 Mc Coverage Tests at Chicago, *IRE Trans.*, vol. PGVC-6, pp. 20–32, July, 1956.
29. Bullington, K., Radio Propagation Variations at VHF and UHF, *Proc. IRE*, January, 1950, pp. 27–32.

30. Shepherd, N. H., and A. C. Giesselman, Interference—A Look at the Ounce of Prevention, *IRE Trans.*, vol. PGVC-13, pp. 4–10, September, 1959.
31. Shepherd, N. H., Analyzing Interference in FM Communications Systems, *Electron. Ind.*, March, 1958, pp. 15–29.
32. Gifford, R. P., The Knee of the Nose, *IRE Trans.*, vol. PGVC-4, pp. 40–51, June, 1954.
33. Schultz, Curt, Spectrum Compression and Its Problems, *IRE Trans.*, vol. PGVC-6, pp. 76–82, July, 1956.
34. Shepherd, N. H., A Report on Interference Caused by Intermodulation Products Generated in or near Land Mobile Transmitters, *IRE Trans.*, vol. PGVC-13, pp. 16–19, September, 1959.
35. Gifford, R. P., Spectrum Pollution, *IRE Natl. Conven. Record*, vol. 10, part 5, pp. 223–225, March, 1952.
36. Ellis, A. G., Site Noise and Correlation with Vehicular Traffic Density, *Proc. IRE Australia*, January, 1963, pp. 45–52.
37. Farmer, R. A., SINAD System Design, *IRE Trans.*, vol. VC-10, no. 1, pp. 103–108, April, 1961.
38. Dodrill, G. E., and J. F. Atkinson, Mobile and Fixed Radio Relay Operation in the Power Radio Service, *IRE Trans.*, vol. PGVC-4, pp. 62–85, June, 1954.
39. Giesselman, A. C., The Meat of the Backbone, *IRE Trans.*, vol. PGVC-11, pp. 64–70, July, 1958.
40. Davids, H. H., Control Consoles—The VHF Integrators, *IRE Trans.*, vol. VC-10, no. 1, pp. 64–74, April, 1961.
41. Dewire, D. S., and E. A. Steere, Precision Carrier Frequency Control and Modulation Phase Equalization of Base Transmitters in a Mobile Radio System, *IRE Trans.*, vol. VC-9, no. 1, pp. 54–68, May, 1960.
42. Chaney, W. G., Selective Signaling in the Bell System—Relay to Transmitter, *IRE Trans.*, vol. PGVC-12, pp. 67–70, April, 1959.
43. Culbertson, A. F., A New Manual Mobile Telephone System, *IRE Trans.*, vol. PGVC-13, pp. 73–82, September, 1959.
44. McDonald, R., "Dial Direct" Automatic Radio Telephone System, *IRE Trans.*, vol. PGVC-11, pp. 80–85, July, 1958.
45. Monk, N., and H. S. Winbigler, Communication with Moving Trains in Tunnels, *IRE Trans.*, vol. PGVC-7, pp. 21–28, December, 1956.
46. Dewire, G. M., Strictly Private, *IRE Trans.*, vol. PGVC-9, pp. 86–87, May, 1960.
47. Shepherd, N. H., and W. G. Chaney, Personal Radio Antennas, *IRE Trans.*, vol. VC-10, no. 1, pp. 23–31, April, 1961.
48. Mitchell, D., and K. G. Van Wynen, 150 Mc Personal Radio Signaling Systems, *IRE Trans.*, vol. VC-10, no. 2, pp. 57–70, August, 1961.
49. Mitchell, J. F., Personal Radio Paging in the VHF Band, *IRE Trans.*, vol. VC-9, no. 3, pp. 48–57, December, 1960.
50. Roualt, C. L., Voice Frequency Tone Signaling, *IRE Trans.*, vol. PGVC-2, pp. 6–16, August, 1952.
51. Hassel, E. W., VHF Radio Coordinated Traffic Light Control System, *IRE Trans.*, vol. PGVC-7, pp. 1–9, December, 1956.
52. Buesing, R. T., Botherance Rejection with Double Barreled Channel Guard, *IRE Trans.*, vol. PGVC-10, pp. 7–15, April, 1958.
53. *EIA Standard*, Continuous Tone-controlled Squelch Systems (CTCSS), RS-220, April, 1959.
54. Bostwick, L. G., A Miniature Tuned Reed Selector of High Sensitivity and Stability, *IRE Intern. Conven. Record*, vol. 10, part 8, pp. 21–27, March, 1962.

Chapter 18

ADDRESS COMMUNICATION SYSTEMS

HENRY MAGNUSKI, *Motorola Inc.*

DESCRIPTION OF THE ADDRESSED
COMMUNICATION SYSTEMS

1. Introduction

Address communication systems are frequently called RADA, an acronym for Random Access Discrete Address.[1] This abbreviation will be used throughout this chapter.

Address communication (RADA) is a relatively new concept in communication system design. It can be defined as a system in which many users can send, independently, different messages over a common wideband frequency channel at the same time and in the same geographical area by utilizing a technique of continuously addressing each communication. In such a system, many messages will coexist and will carry not only different modulating signals but also different addresses, so that each one can be distinguished from all others. A typical RADA system will be a time-sharing system in which each user transmits short RF pulses occupying a small portion of the time. The transmissions are not synchronized, so the pulses from different users may coincide, causing a certain amount of self-interference that is characteristic of the RADA system. A wideband channel is required because short pulses are transmitted.

RADA also requires a different frequency-assignment policy. Instead of the usual division of the available frequency spectrum into many narrowband channels and assigning separate channels to each user or user group as shown in Fig. 1, a common single wideband channel will be used by all users. It is subdivided into a few subchannels as shown on Fig. 2, which are called frequency slots. Each user of the RADA system will transmit short pulses on some of these frequency slots, occupying each frequency slot for only a small fraction of time.

Thus it can be seen that this is a time-sharing system, as contrasted to the narrowband channel assignment in which a channel is occupied continuously for long periods of time. Two such narrowband channels are shown to be occupied by two users on Fig. 1.

In RADA the spacing between received pulses will be different for each user. This spacing constitutes the user's private address. Figure 2 shows two different private addresses being transmitted.

A RADA transmitter has to perform two separate functions, namely, modulation and addressing. This is in contrast to the usual radio transmitter, where only modulation is required to send messages. To perform modulation, voice or other information is converted into a train of pulses which can be called modulation pulses. Then, to perform the addressing function, each modulation pulse is transmitted as several (usually three or four) short RF pulses (bursts of RF energy). Each RF pulse usually is sent on different frequency and in different time relation to other pulses. One can say that the combination of time and frequency slots in which RF pulses are transmitted—and which is selected from the so-called time-frequency address matrix (TF matrix) shown in Fig. 3—constitutes an address which is continuously recognized

FIG. 1 Narrowband address assignment.   FIG 2. RADA address assignment.

by the receiver of the particular user from the maze of other addresses. Only by transmitting this address can the particular user be reached.

In RADA each user is entirely independent and can transmit the messages to others at any time without delay, correlation, or synchronization with other users. Obviously some of the transmitted pulses will coincide with other pulses transmitted by other users and will result in a certain amount of system self-interference, which will be discussed later. By selection of a proper technique, this interference can be minimized.

2. Advantages of Address Communication (RADA)

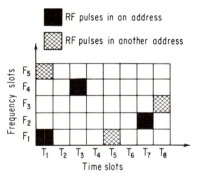

FIG. 3. Time-frequency address matrix.

Perhaps the most important advantage of RADA is that each user can contact any other user without delay by simply setting his transmitter to the address of the other user. This can be done any time without synchronization, without correlation with others or checking to see if the channel is free and without any central control. One can say that a private "hot line" is provided by RADA for every user to any other user.

Another advantage of this system is the so-called "graceful degradation" under an overloading condition. As more users utilize the common wideband channel, the self-interference of the system will gradually increase but will be noticeable only by those users who are talking at the maximum range and thus receive a very weak desired signal. This gradual degradation will permit perhaps twice as many talkers as the system was designed for before a very serious degradation of signal-to-noise ratio due to self-interference occurs but again only in the receivers receiving weak signals. And then again, many times more users may enter the system before the maximum range of communication, which is decreasing because of the self-interference, is seriously limited.

Figure 4 illustrates this property of the system by showing a system which is designed for N talkers with 20-db signal-to-interference ratio. It shows that twice as many talkers can use the system before the signal degrades to the point that it becomes unusable. Then the useful range of communication decreases, but it will not be cut in half until about eight times as many talkers enter the system. This gradual overloading characteristic of the system is contrasted to narrowband communication in which, when the assigned channel or channels are occupied, the next user cannot enter the system and has to wait. Thus, an abrupt system overloading is experienced in the narrowband approach.

Another advantage of RADA is that of better frequency spectrum utilization in systems of the mobile and emergency communication types. In such systems, delays cannot be tolerated, and if narrowband channels are used, they can be utilized only a small percentage of the time so as to permit free access to a channel without delay in case of emergency. Even then the emergency message could be delayed. In a voice-modulated RADA system, about 200 kc per talker will be required, as will be discussed later. Narrowband voice channels can be very narrow, only 4 kc in case of SSB.

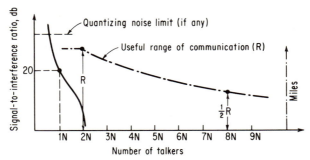

FIG. 4. RADA overloading characteristic.

However, sufficient guard band has to be provided between channels in order that the strong undesired signals on the nearby channels will not interfere with the reception. Also, in VHF and UHF, frequency stability is a problem requiring additional guard bands. If a system with a large dynamic range is contemplated, such as a mobile communication system, when a nearby user could be talking on an adjacent channel and could provide a very strong interfering signal, these guard bands have to be very large. In a practical system, 50- to 100-kc spacing between channels is generally provided. If one considers several such channels, which are used only a small fraction of the time because of emergency consideration, it turns out that a RADA system will provide much better service with no delays in less frequency spectrum. However, this will not be true for point-to-point systems or systems with small dynamic range of signals and in systems where a free channel can be selected out of many available and occasional delays in transmitting messages can be tolerated.

Another advantage of RADA is its antijamming quality, which is common to wideband systems, and also its equipment cost compared with highly stable tunable equipment requiring complex frequency synthesizers.

The RADA technique is very flexible and not only can provide an equivalent of a telephone service over the radio, including duplex service, but also can do much more than the telephone by providing conference calls or nets, busy-signal-overriding capabilities, general alarm provisions, etc. These are some of the reasons why RADA is a very promising technique for future VHF and UHF mobile communication radio systems, primarily military systems.

3. Modulation and Addressing Methods

In the RADA system, the transmitted waveform has to carry the address in addition to the modulation that is used in the other known communication systems. Thus, the possible RADA implementations may use different modulation techniques and also different addressing techniques. The modulation technique used has to be compatible with the addressing technique. The most promising addressing technique is the TF matrix technique as shown previously in Fig. 3. This type of addressing requires that the modulation be in the form of pulse train, because each of the modulation pulses is transmitted as a few (usually three or four) much shorter RF pulses selected from the TF matrix.

The possible modulation schemes that are compatible with the TF matrix addressing can be divided into two classes: analogue modulations and digital pulse modulations. In the analogue pulse modulation class, pulse position modulation (PPM) and pulse frequency modulation (PFM) are often considered for the RADA system. In PFM, the frequency of the transmitted pulses (RF bursts) varies with the modulation instead of the position in time, as in PPM. To mention other analogue pulse modulations, pulse amplitude modulation (PAM) is not advantageous for application to RADA because it is very sensitive to RADA self-interference. Even a relatively small interfering pulse, when added to or subtracted from the amplitude-modulated pulse, will cause a considerable amount of interference to the detected output. Pulse repetition rate modulation (PRM) requires a large average amount of pulses to be transmitted, thus increasing interference. Finally, pulse width modulation (PWM) is not compatible with the TF address matrix.

The advantage of the usable analogue pulse modulations is that a relatively small number of pulses (only one modulation pulse per voice sample) are transmitted. Sync pulses, which are sometimes used with PPM, do not need to be transmitted, since the center position of the PPM pulse can be determined by averaging in time. Also, the transmission of pulses during the pauses in the speech can be stopped, thus making the actual number of transmitted pulses in PPM and PFM smaller than the number of voice samples. One disadvantage of analogue modulation is its relatively high sensitivity to interfering pulses. Another disadvantage of analogue modulation shows up when repeaters are used. Not only is the design of pulse repeaters difficult in this case, but also each repeater adds noise and distortion to the modulation.

If the pulse modulation methods mentioned previously are quantized (which means that instead of continuous variations in pulse position, frequency, or amplitude, as the case may be, only a certain number of fixed levels is permitted), then the pulse modulation becomes digital. Out of all the previously discussed pulse modulations, only the quantized pulse position modulation (QPPM) is advantageous for RADA because a relatively large separation of quantizing levels in time is possible. In this case, an addition of a small interfering pulse cannot cause error by shifting the desired pulse sufficiently to jump over to the next quantized time slot.

In addition to these quantized pulse modulation methods, there are two binary digital modulations, namely, PCM pulse code modulation (PCM) and delta modulation (DM).[2] These are characterized by a continuous, uniformly spaced train of modulation pulses or no pulses. In PCM, each voice sample corresponds to a group of pulses (group synchronization is required), while in DM, the voice is sampled much more frequently and only one pulse per sample is sent to signify the increase in voice amplitude or omitted to signify the decrease.

The one disadvantage of these two methods in RADA application is that more pulses per second have to be transmitted. For example, in the case of PCM, voice may be sampled 8,000 times per second and for each voice sample a group of several pulses have to be transmitted. In case of 32 quantizing level PCM, there are up to five pulses in a group plus a sync pulse, making total 24,000 pps in average. In DM, while only one pulse per sample is transmitted, the sampling rate has to be increased about four to five times, to provide sufficient number of quantizing levels and equivalent voice quality to 32-level PCM. Thus, the number of transmitted pulses in DM is about 20,000 pps. In QPPM, only 8,000 pps can be transmitted. The increased pulse rate in PCM and DM increases the interference probability in the RADA system. However, this can be compensated by the use of time gates, which are very effective in rejecting interference by muting the receiver most of the time and energizing it only for short time periods, to receive the RF pulses (and interference) when such RF pulses are expected to arrive. The time gates are not effective in QPPM because the RF pulse will occupy only one out of many possible time slots, but the receiver has to be ready to receive the RF pulse in all of them, thus increasing the probability to receive interference.

A variation of DM called DDM (Derived Delta Modulation) was developed by Motorola Inc.[3] In this modulation method, the DM pulse train is converted to a different pulse train, containing approximately half as many pulses, mainly because no pulses are transmitted during the pauses in the speech. The DDM preserves all the advantages of DM yet provides much less interference because only half as many pulses are transmitted by each talker.

One advantage of all digital systems for RADA applications is that simple pulse-by-pulse repeaters can be used and the messages can be repeated several times over several repeaters without accumulating distortion and noise.

In comparing PCM with DM (or DDM), it can be said that for telephone-type speech quality, DM has several advantages over PCM for RADA applications. One is the simplicity of the DM modulators and demodulators, and another is the relative insensitivity to transmission errors caused by self-interference, which always exists in RADA systems. In DM, each error in the pulse train will cause only a two-quantizing level error at the receiver output, which will manifest itself as a faint click. On the other hand, an error in PCM may cause an output error of several quantizing levels, depending on its position in the pulse group.

Another advantage of the digital modulations is that they provide uniform and reliable output. Fixed signal-to-noise ratio at the output and fixed volume is obtained, because it is a sharp transmission threshold system. As long as the receiver distinguishes whether pulses were transmitted or not, the output signal-to-noise ratio is fixed and the only noise is the so-called quantizing noise, which results from converting analogue voice to digital form. Also, the audio volume will be constant and independent of variation in signal strength (as long as the signal is above threshold) and also independent of the number of repeaters employed.

For military applications, digital modulation is easily adaptable to secured or secret voice transmission by scrambling the train of pulses and unscrambling at the receiving point. Also, digital modulation is more difficult to jam and is particularly insensitive to the so-called partial or nuisance jamming because of the wide frequency band and sharp threshold. In addition, addressing means redundant transmission, which is still more difficult to jam.

So far, modulation methods compatible with TF matrix addressing were discussed. TF matrix addressing may be called redundant transmission because each modulation pulse is transmitted as several RF pulses, all of them carrying the same information. Another class of addressing method which could be applicable to RADA is the pseudo noise or quasi-random continuous-wave (CW) coding method. In this method, the transmitter continuously emits a waveform containing many sidebands and spread over a wide band. Different synchronization time, different pseudo-noise coding or different waveform sequences are used as addresses to distinguish one transmission from the other. All these methods, in addition to equipment complexity, have a common disadvantage, namely, a small dynamic range of distinguishable signals. This dynamic-range of signals is proportional to the so-called bandwidth expansion factor. For example, if a 3-kc voice-modulation band is expanded a thousand times (which means that it is coded to cover a band of 3 Mc), no more than 30 db of dynamic range can be tolerated. This means that an interfering signal which is 30 db stronger than the one desired will obscure reception of the latter. Because of the dynamic range limitation, these CW addressing methods are suitable only to specific RADA systems in which there will be no possibility of having a strong interfering transmitter near the receiver. CW addressing is not applicable to the mobile RADA systems in which this situation can frequently occur.

In addition to the true RADA systems, conventional narrowband techniques could be adopted to provide a service which is similar to RADA. For example, if each user is provided with the access to a common pool of narrowband channels, he could select a free channel, occupy it, and protect it by a suitable busy signal, then ring the desired user and talk to him over this channel. Likewise, it is possible that, instead of the narrowband channel, a free time slot can be selected by the user in a large centralized and synchronized time-division radio multiplexing system.   However, not only a very complex equipment has to be provided in these systems to achieve selection of the free channel or time slot, protection of it, and signaling, but these systems are not true RADA systems.   One difference is that they do not have the graceful overloading characteristics, so that, when all channels are occupied, the next user has to wait and considerable delay in message transmission may result.   Also, to provide positive signaling channel protection and/or synchronization, these systems have to be centralized, and therefore, the users are not capable of independent, delay-free communication with other users as in a true RADA system.   RADA-type service could be provided by narrowband technique if a separate channel with sufficient guard band could be assigned to each user to serve as his address.   Obviously, the available frequency spectrum is not sufficient to achieve this.   Figure 5 shows the possible RADA techniques which were briefly reviewed here.

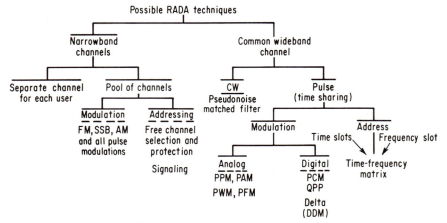

FIG. 5. Review of RADA techniques.

In conclusion, the most promising addressing technique is the TF address matrix, and the most promising modulation techniques are digital such as DM, PCM, or QPPM, although analogue types such as PPM or PFM could also be used.   Only these techniques will be further discussed.

4. Examples of RADA Implementation

Two examples of RADA implementation are illustrated by simplified block diagrams.   In Fig. 6 and 7, the RADA system, using binary modulation and time gates, is shown.   This is a RADEM (Random Access DElta Modulation) system developed by Motorola Inc.[3]

The transmitter includes a delta modulator, although a PCM modulator could be used to provide a train of binary pulses.   The pulses from the output flip-flop (FF) of the delta modulator are fed to the address TF matrix which controls the gates in accordance with the address-control setting.   The transmitter address control is set manually to the address of a desired user's receiver, and the address matrix is controlled in such a way that the appropriate gates will open in the appropriate time slots.   Thus, short pulses of different frequencies in different time slots will be fed through the gates to the common RF power amplifier and radiated from the antenna each time the modulation pulse is to be transmitted.   The design of the transmitter is relatively simple and noncritical.   The oscillators need not generate highly stable frequencies if the RF pulses used are short.

Figure 7 shows a receiver in which a single IF strip amplifier is used and the time gates are provided.

Time gates are shown here simply in the form of a switch connecting the antenna to the mixer and also as switches connecting the mixer to several local oscillators. Only one of these switches closes at the gate time, activating the receiver at the particular frequency which should be received in this particular time slot. Thus, the receiver is sensitive only to one frequency in one time slot at one time. To recognize the address,

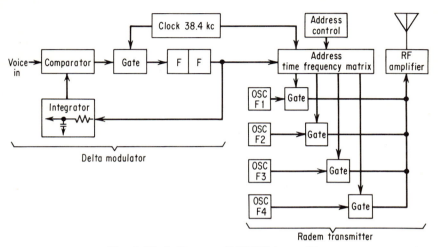

Fig. 6. Block diagram—RADEM transmitter.

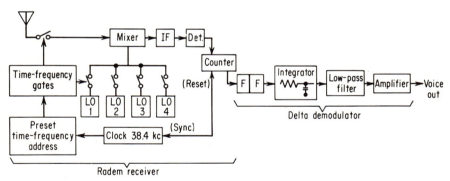

Fig. 7. Block diagram—RADEM receiver.

it is necessary only to determine (by counting) if all the desired pulses have been received in the desired time-frequency slots for which the time gates opened. For a four-pulse system, a counter to 4 will suffice, and if it counts to 4 during one modulation pulse period (26 μsecs), a pulse is sent to the flip-flop (FF) which signifies that the modulation pulse has been received.

In the delta demodulator, the flip-flop charges or discharges another integrating capacitor similar to the one in the transmitter. The reproduced voice curve, appearing across this capacitor, is then passed through the low-pass filter, amplified, and fed to the speaker or headphones. The time gates not only permit simplification of the equipment but also have a considerable advantage of eliminating interference, no matter how strong, if it falls outside the gates. The gates have to be synchronized with the transmitted address. To achieve this, the receiver searches continuously for the transmitted signal. To accelerate this process of finding and synchronizing with the transmitted signal, the receiver clock is adjusted to run slightly slower or slightly faster than the transmitter clock. The receiver clock continuously opens the time gates in time and frequency slots in accordance with the preset receiver address. It takes only a few milliseconds for the time gates to coincide with the transmitted

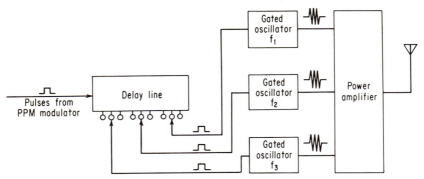

Fig. 8. Simplified RACEP transmitter. (*By permission from D. L. Haas, author of the paper entitled Satellite Communications Using RACEP delivered at the Air Force Association Convention in Washington in September, 1962.*)

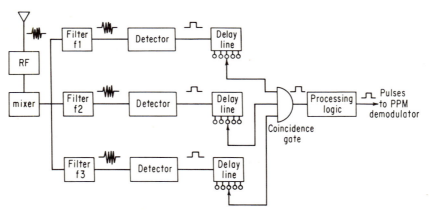

Fig. 9. Simplified RACEP receiver. (*By permission from D. L. Haas, author of the paper entitled Satellite Communications Using RACEP delivered at the Air Force Association Convention in Washington in September, 1962.*)

signals, then a few more milliseconds (some hundred or so of transmitted pulses) to synchronize the receiver clock and the gates. From that instant, the receiver will open the time gates only when the pulses are expected and keep them closed at all other times.

Another example of RADA implementation using analogue PPM is shown in Figs. 8 and 9. This is the RACEP (Random Access and Correlation for Extended Performance) system developed by Martin Co.[4] These figures are also representative of QPPM systems.

Figure 8 shows a transmitter in which pulses from the PPM modulator are sent through a delay line. Taps on this delay line permit selection of pulses with time separations corresponding to the time slots in the address of the desired user.   These pulses gate the appropriate oscillators which send the short RF pulses to the common RF amplifier and antenna.

Figure 9 shows the simplified RACEP receiver in which pulses received on different frequencies are fed to separate detectors and separate delay lines.   Again, taps on these delay lines correspond to the time slots of the preassigned receiver address.   They are so arranged that all the received RF pulses are fed at the same time to the coincidence gate if the time spacing between these RF pulses corresponds to the receiver address.   If coincidence occurs, a pulse is passed to the PPM demodulator and converted back to a voice sample.

5. Large RADA Systems with Repeaters

In the large RADA system with many users and/or long ranges, repeaters are often used to extend the range of operation of the user set.   Conversely, the use of repeaters

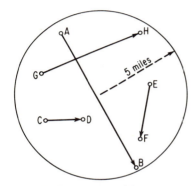

Fig. 10. RADA system without repeater.

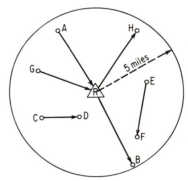

Fig. 11. RADA with centrally located repeater.

permits smaller and lighter user's equipment because the direct user-to-user range can then be smaller.   In addition, the judicious application of the repeaters permits better spectrum utilization in those systems in which conversations over short ranges are relatively much more frequent than long-range conversations.   To illustrate this, a simplified numerical example is provided.   In Fig. 10 a hypothetical RADA system is shown with a 5-mile-radius circle cut from a vast area over which many talkers are uniformly distributed.   The density of distribution is such that there are 25 talkers within this circle.

If no repeater is provided, the range of each user set has to be 10 miles in order that user A can reach user B at the opposite point within this circle.   With this 10-mile range, the receiver of user B will be exposed to interference from 100 talkers inside and outside the circle, and sufficient bandwidth has to be provided in such a RADA system to minimize this interference to a tolerable level.   Assuming that, if the band accommodates no more than five talkers per megacycle, the interference can be tolerated, the necessary bandwidth in this case will be 20 Mc.

Now suppose that a centrally located repeater R as shown in Fig. 11 is provided in this system with a separate frequency band for reception, so that only those conversations which exceed 5 miles in length are retransmitted by the repeater.

Further assume that 80 per cent of the conversations are shorter than 5 miles and that the remaining 20 per cent are from 5 to 10 miles.   In this case the user's equipment range can be reduced from 10 to 5 miles.   This can be done by providing about sixteen times less transmitter power for each user, assuming that this is a ground-based mobile system in which the signal attenuation increases 12 db when the range is doubled. Each user will have then much lighter equipment and battery pack, and his receiver will be exposed only to the interfering transmitters contained within a 5-mile circle because the others will be too weak to interfere. This means that only one-quarter, or 25, talkers could interfere with him in the worst case. Furthermore, according to our assumption 20 per cent (or 5) of these talkers are transmitting to the repeater over a separate frequency band because they are communicating over the range exceeding 5 miles. Thus only 80 per cent (or 20) talkers will interfere with the user receiver in the direct user-to-user band. Assuming again 5 talkers per megacycle, this band can be now only 4 Mc wide. However, the repeater will now receive calls from 20 per cent of all the talkers or 5 talkers, which will require also a 1-Mc repeater receiving bandwidth to accommodate them. Then this repeater will frequency-translate and retransmit these calls to other users, either in the separate repeater transmitting bandwidth or in the same user band. In either case, an additional 1 Mc of bandwidth will be required. The total system bandwidth will be 6 instead of 20 Mc, which means a considerable spectrum saving. Further saving of the spectrum can be obtained by using more sophisticated multiple repeater systems in which each repeater covers only a small portion of the area.

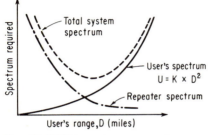

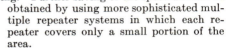

Fig. 12. RADA spectrum vs. user's range.

In general, one can say that, as the direct user-to-user range is decreased, the amount of spectrum required for this service will decrease in proportion to the user's range squared, which is shown on Fig. 12.

However, at the same time more sophisticated and numerous repeaters, consuming more bandwidth, will be required. The spectrum required for the repeaters is not a simple function and depends on the number, design, and distribution of repeaters. A second curve in Fig. 12 approximately illustrates the spectrum required for the repeaters. The third curve shows that the total spectrum required for a given system (which is the sum of the bandwidths required for the users and the repeaters) passes through a minimum.

RADA repeaters can be divided into two general classes: the simple pulse-by-pulse repeaters and more complex address-recognizing repeaters. In the first class, every pulse received by a repeater is retransmitted, with a short fixed delay, by the repeater transmitter. A block diagram of this simple repeater is shown in Fig. 13.

It consists of several receivers and pulse detectors, one for each frequency slot. If the pulse is detected by the receiver, it opens an appropriate gate which momentarily connects the oscillator to a common transmitter power amplifier. In this way, for each received pulse at a given frequency, the pulse is retransmitted at the corresponding frequency. Usually these repeaters are also frequency translators, which means that they receive pulses in one frequency band and retransmit them in another frequency band. Then the user has the choice either to transmit address pulses in the user band for direct user-to-user communication or transmit the same address structure in the repeater receiver band. Then his transmission will be retransmitted by the repeater either back to the user band or to a separate band if the user is equipped to receive calls in two bands.

It is possible to use this type of repeater without frequency translation, which means that the repeater is retransmitting in the same band as it is receiving. One possible use of such a repeater would be to overcome a natural obstacle dividing the RADA system into two areas. For example, if there is a mountain ridge dividing the RADA area in two, it would be possible to place such a nontranslating repeater on the top of the ridge so that the users on one side of the ridge could communicate with others on the other side of the ridge. However, such a repeater is not often used because it wastes spectrum, since all the RADA conversations are transmitted twice, once to the repeater and the second time by the repeater, whether it is necessary or not. This means that the interference level is increased approximately two times and additional bandwidth has to be provided.

The simple translating repeaters can be equipped with directive antennas, each covering different separate areas in the large RADA system.   In this way a kind of radio switchable system is obtained.   For example, a RADA system could be divided into three or more separate geographical areas.   A separate repeater having a different receiver frequency band is then used to connect each pair of areas.   The user, by selecting the frequency band over which he is transmitting, could direct his call to the one of the areas.   In this way the long-distance calls would not be broadcasted all over the RADA area, as with a single omnidirectional repeater, but would be limited to only one area.   A considerable spectrum conservation can be achieved this way at the expense of repeater complexity.

Such simple pulse-by-pulse repeaters could be airborne or even built into satellites, thus providing a RADA system with a small or large number of users (depending on the

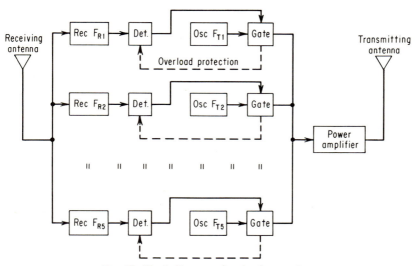

FIG. 13. Block diagram of a simple repeater.

spectrum utilized) with a long-range communication capability of the RADA type, which is characterized by independent system access (at random), no delays in communication, and graceful overloading degradation.

The other type of repeater is the so-called addressed or address-recognizing repeater.   It can be a very simple one consisting of a RADA receiver and transmitter wired back to back.   The receiver recognizes a single address and for each received modulation pulse retransmits one either with the same or with a different address. This type of simple repeater can be used to extend the range of the particular user or users group in some special situations.   For example, if a military patrol has to go outside the normal RADA communication range, it can leave behind this simple repeater, preferably equipped with directive antennas, making possible a considerable extension of the operating range.

In general, the addressed repeaters are not limited to single-address recognition. They can be very complex, operating as described below.   The repeater has its own specific receiving address, and it consists of several RADA transmitter/receiver pairs, each pair capable of being adjusted to any address, and of appropriate decoding address control and switching gear.   The user can contact the repeater by first sending his (user's) address, in a coded form, transmitting over the repeater address.   The repeater then decodes the user's address and adjusts one of the free RADA transmitter/ receivers to the address of the particular user. This corresponds in telephone technique to offering a free line to the user. Then the user sends over his own address a coded address of the desired party, which is decoded by the repeater and retransmitted to other repeaters in search of the desired party. Either a single repeater or repeater chains could be used. When the desired party is located, the two users can talk using two pairs of the transmitter/receiver units, which are located in the two interconnected repeaters each nearest to the talking user. The transmission from one repeater to the other is usually in the form of a modulation pulse train, modulating a separate channel. This arrangement is similar to a switchable telephone system except that radio transceivers and addressed communications are used between the users and the repeaters. These repeaters are very complex, with many RADA transceivers and all the address decoding and setting equipment, in addition to the switching gear, but considerable spectrum saving can be obtained with such a system, especially if many repeaters are used, each covering a relatively small area. One problem in such a system is avoiding the multiple transmission of messages in the overlapping areas. Overlapping areas are the areas which are within the range of more than one repeater.

In conclusion, when designing a practical large RADA system, one has to compromise between repeater system cost and complexity and spectrum saving.

DESIGN CONSIDERATIONS IN RADA SYSTEMS

6. Address Matrix

In the design of the addressing scheme for a given RADA system, the size of the TF matrix will determine the number of available addresses. Only the so-called "unique"

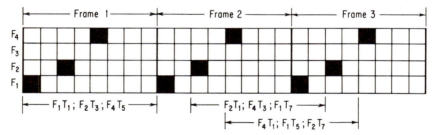

Fig. 14. Sequence of the TF matrix addresses.

addresses will be considered. An address transposed in time by one or several slots will not be a unique address because the RADA system is not synchronized and the receiver does not know which is the first time slot. For example, an address which has frequency 1 in time slot 1 (F_1T_1), frequency 2 in time slot 3 (F_2T_3), and frequency 4 in time slot 5 (F_4T_5) is exactly the same as the address (F_1T_4, F_2T_6, and F_4T_8), except that this last one is shifted by three time slots. In order to disregard these time-shifted addresses, only the addresses in which the first time slot is always occupied are considered. Furthermore, if the address is sequentially repeated several times, it may look like several other addresses to the receivers. These other addresses, again, will not be considered as unique. To illustrate this, see Fig. 14.

In this figure a simple address matrix consisting of four frequency and eight time slots is shown. The address F_1T_1, F_2T_3, and F_4T_5 is repeated sequentially three times. Because the receiver does not know which is the first time slot, it may recognize other addresses, such as F_2T_1, F_4T_3, and F_1T_7 or F_4T_1, F_1T_5, and F_2T_7. Those additional addresses again will not be counted as unique addresses. To eliminate them, only those addresses are considered in which one frequency is selected out of the frequency combination used ($F_1F_2F_4$ in our example) and is always put in the first time slot. An additional restriction on the address selection is that the same frequency and the same time slot are not used more than once in a given address.   This is done for practical reasons, such as to avoid the transmission of two frequencies in one time slot.

With these restrictions, the following formula permits the calculation of the number of available unique addresses A from a given TF matrix size:

$$A = C_N{}^F P_{N-1}{}^{T-1} = \frac{F!}{(F-N)!N!} \times \frac{(T-1)!}{(T-N)!}$$

In this formula F is the number of frequency slots, T is the number of time slots, N is the number of pulses transmitted in the address, C is the number of combinations, and P is the number of permutations.   The first part of the formula determines the number of available frequency combinations (triples, quadruples, etc., depending on the number of pulses in the address).   The second part determines the number of addresses for one frequency combination.

As example of address calculation, assume a simple three frequencies ($F = 3$), 12 time slots ($T = 12$) matrix and three pulses transmitted per address ($N = 3$).   In this case the formula shows the following number of addresses:

$$A = C_3{}^3 P_{3-1}{}^{12-1} = \frac{3!}{0!3!} \times \frac{11!}{9!} = 11 \times 10 = 110$$

In this simple case, since we have only three frequencies, all the frequencies have to be used in each address and the number of addresses can be calculated alternatively rather simply.   We always have to place one of the three frequencies, let us say F_1, in the first time slot, then the second frequency can be placed in one of the remaining 11 time slots, and the third frequency can be placed in one of the remaining 10 time slots, making the total number of addresses $11 \times 10 = 110$.

If one more frequency ($F = 4$) is added, there will be four frequency combinations available, which means that there will be $4 \times 110 = 440$ addresses.

This shows how rapidly the number of addresses increases with the size of the matrix. For further illustration, consider a large matrix with six frequencies and 16 time slots ($F = 6; T = 16; N = 3$).   In this case

$$A = \frac{6!}{3!3!} \times \frac{15!}{13!} = 30 \times 210 = 6,300$$

This is a sizable increase of the number of addresses over 440, although the matrix is only twice as big (6×16 vs. 4×12).

Another way of increasing substantially the number of addresses without changing the size of the matrix is to increase the number of pulses in the address.   Consider four pulses ($N = 4$) instead of three of the previous example.

$$A = \frac{6!}{2!4!} \times \frac{15!}{12!} = 15 \times 2,730 = 40,950$$

Note that in this case the number of frequency combinations was halved but the number of addresses for each combination increased over tenfold.

As can be seen from the above examples, the number of unique addresses from a relatively simple matrix can be very large and depends greatly on how many pulses are transmitted in each address.

This calculation is applicable only if the so-called N out of N address recognition logic is used, which means that all N pulses in the address have to be properly received to recognize the address.   For example, 4 out of 4 logic means that 4 pulses per address are transmitted and 4 have to be received to recognize the address.   Other address recognition logics are possible, for example, 3 out of 4, which means that the address will be recognized if any 3 out of 4 transmitted pulses are received.   This logic could be used to decrease the amount of bit cancellation errors, as will be discussed later.

When this logic is used, the number of unique addresses cannot be calculated from the formula given above and is much smaller.

In planning the RADA system each user is assigned a different unique address. In addition, in systems operating in a duplex mode, different transmitting and receiving addresses are assigned to each user. These are called the associated addresses. Sometimes a third associated address is assigned to each user for signaling purposes; then the number of unique addresses has to be at least three times larger than the number of users. Additional addresses may be needed for conference calls, repeaters, alarms, etc. On the other hand, some user groups may have a common address to operate in a net or conference mode only.

Distinction is made between the "user" and the "talker" in a RADA system. In order to evaluate the system self-interference, we define the talker as a user who is within the range of our receiver and who actually talks and transmits. For example, if there are 5,000 users in a large mobile RADA system, only 1,000 of them may be within the range of our receiver. Out of these 1,000, only 20 per cent (or 200) may be off hook, which means that there will be only 100 talkers because the other 100 will be listening. However, if the system uses a duplex mode arranged so that the signals are transmitted regardless of whether there is voice modulation or not, then the number of talkers is equal to the number of users off hook. On the other hand several users in a simplex conference mode (net) represent only one talker.

The matrix size determines not only the number of available addresses but also, under certain assumptions, the number of talkers and the bandwidth occupied by the RADA system. The length of the modulation pulse and of the time frame (address matrix length) is determined by a given modulation scheme. For example, it is 25 μsecs in DM with 40,000 voice samples per second. This modulation pulse length or time frame can be divided into several time slots. For example, dividing 25 μsec into 16 time slots will make each time slot approximately 1.56 μsec long. The transmitted RF pulses can be shorter than the time slots but not longer, because then the different possible addresses will be smeared and not unique. Assuming that the length of the RF pulses is equal to the time slot, the practical frequency separation (or frequency slots) is related to the pulse length. Usually, the frequency separation or the width of the frequency slot is equal to two times the inverse of pulse length. For example, if 1-μsec pulses are used, 2-Mc frequency slots will be about the optimum practical value. This is just "rule of thumb" and will depend on the dynamic range of signals received in a particular RADA system. Much larger frequency separations will require an unnecessarily wide frequency band, wasting spectrum. On the other hand, smaller frequency separations will lead to excessive interference caused by strong signals in adjacent frequency slots. With this rule assumed, the bandwidth occupied by the matrix is determined by the number of frequency slots used and the pulse length. One can say that the bandwidth is proportional to the area of the matrix if each time-frequency slot is shown as a square whose size is constant. If, for example, the 25-μsec frame is divided into 8 time slots instead of 16, each frequency slot will be half as big because pulses are twice as long and the total bandwidth occupied will be half as large. The matrix will have half as many squares, and its area will be also half as large.

The bandwidth occupied by the RADA system can be adjusted by changing the number of time slots or frequency slots in which case the number of available addresses is also changed. It can also be and usually is adjusted without changing the address structure by sending pulses shorter than the time slots. The shorter the pulses, the wider the frequency slots required. As discussed later, the bandwidth occupied by a RADA system is proportional to the number of talkers and may have to be quite large to accommodate very many talkers. The system bandwidth is not related to the number of addresses except through the matrix size, and usually there is a sufficient amount of addresses available from a given matrix size to satisfy the system requirements.

In conclusion the size of the TF matrix determines the number of unique addresses available and also the bandwidth occupied by the system if the frame length (modulation pulse length) is given. There is no fixed relation between number of addresses and the bandwidth occupied by the system, and each of these parameters can be independently adjusted. The number of addresses can be adjusted by changing the number of time slots and frequency slots used or by changing the number of transmitted pulses. The frequency band occupied is usually adjusted by changing the length of the transmitted pulse (this length has to be always shorter than the time slot) without changing the number of frequency or time slots and thus without changing the address structure.

7. RADA Self-interference

RADA systems are unsynchronized, time-sharing systems in which each user transmits, independently, RF pulses on different frequencies and at different times. The receiver is adjusted to recognize a particular time-frequency RF pulse pattern as his own address. The transmissions from other users may interfere with the desired transmission in two ways. First, the RF pulses from other talkers may occasionally fall into the particular time-frequency pattern that comprises the address which the receiver will recognize as his own, even though no desired address was sent to him at this particular time. This is a so-called false address or additive modulation bit error. The second way is the so-called cancellation error in which the desired address is sent but RF pulses from interfering talkers are added to one or more desired pulses in such an amplitude and phase relation (about equal in amplitude but opposite in phase) that the desired pulse is sufficiently canceled (its amplitude is dropped below the detection threshold level) and the reception of the desired address is not recognized by the receiver.

Both addition and cancellation bit errors will manifest themselves as background noise in the audio output. As the number of talkers in a given system increases, this background noise will become louder and eventually will make the reception of the desired signal impossible. Thus, to determine how many talkers a given system can tolerate, one has to define first what is the smallest tolerable output signal-to-noise (S/N) ratio in a system. Then, one must establish, either by calculation or by experiment, the number of talkers that will create the interference causing the decrease of the S/N ratio to this defined minimum. Usually, in voice systems, the output S/N ratio of either 13 or 20 db is considered a minimum.

The calculation of the relation between the number of talkers and the S/N ratio in analogue systems is very complex and depends much on system design, including pulse coincidence circuitry, delay lines, deviation used, etc. It would be difficult here to calculate interference in such a system, because the time, RF phase, and amplitude relations of numerous small interfering pulses to the desired RF pulse have to be considered. Even a relatively weak interfering pulse may change sufficiently the position and amplitude of the desired pulse to cause substantial output error. Therefore no analogue system will be further considered in this paragraph.

The case is different with digital systems in which the problem of interference calculation is simpler. In digital systems, individual RF pulses received are either accepted or not accepted and positive "yes" or "no" decisions are usually made at two levels: the individual RF pulse level and the whole address-recognition level.

The problem of estimating the interference in digital systems can be divided into two separate problems. One problem is to establish how the output S/N ratio will depend on the pulse error rate in the modulation pulse train. This relation will be different for different modulation schemes and can be easily established experimentally by introducing in the laboratory a given number of random pulse errors in the modulation pulse train and observing the output S/N ratio. These experiments were performed for DM,[3] and it was determined that, with DM having a 38.4-kc sampling rate, the 4 per cent pulse error rate corresponded to 20-db S/N ratio while at 15 per cent error rate 13-db S/N ratio was obtained. The maximum possible error rate is 50 per cent in this case, because, in a completely random pulse train, half of the pulses will correspond to the modulation pulses in any arbitrary modulation pulse train. This means that DM is very insensitive to errors, which makes it particularly suitable for RADA applications. Other modulation schemes such as PCM will not tolerate such a high error rate. For example, a 15 per cent error rate will mean an average of almost one error per group of six PCM pulses, which means that very seldom would a correct voice sample be obtained.

Once the relation between the S/N ratio and error rate is known, the second problem is to establish how many interfering talkers will actually cause a given error rate. It should be mentioned here that both problems have to be treated together to avoid wrong conclusions from the error rates alone. For example, in calculating a QPPM system, a lower error rate will be obtained from a given number of talkers because fewer pulses are transmitted, but at the same time, QPPM is more sensitive to errors and fewer errors can be tolerated for a given S/N ratio.

The problem of calculating error rate can be solved in two steps. The first step is to obtain, by calculation and/or simulation, the probability of a single RF pulse addition error p_a and cancellation error p_c. The second step is to calculate the modulation bit errors, assuming a given address matrix and address recognition logic.

Calculating addition errors p_a is easy when an assumption is made that all signals from the interfering talkers are equal in power to the desired signal (these are sometimes called laboratory talkers because they can be easily simulated in the laboratory). In this case the exact equation will be

$$p_a = 1 - (1 - d)^T \qquad d = K \frac{W}{L} \frac{N}{F}$$

where d is the average duty cycle contributed by each talker at each frequency and T is the number of talkers. The duty cycle d of each talker on each frequency can be calculated by multiplying the pulse length W by the average number of addresses per frame K and by the ratio of pulses transmitted per address N divided by the frame length L and by the number of frequency slots F available. For example, if there are four pulses in the address ($N = 4$) selected out of six frequency slots ($F = 6$), each frequency will have a pulse by each talker $\frac{4}{6}$ of the time on the average. Furthermore, if the pulse length is 1.5 μsec ($W = 1.5$), the frame length is 25 μsec ($L = 25$), and an address is sent an average of every second frame ($K = 0.5$), then the duty cycle will be

$$d = 0.5 \frac{1.5 \times 4}{25 \times 6} = \frac{1}{50} = 0.02$$

If the duty cycle is low and there are only a few talkers, a simplified equation can be used, which is shown below.

$$p_a = Td = KT \frac{W}{L} \frac{N}{F}$$

However, this equation is assuming that the interfering pulses never coincide, which is not valid, especially if each frequency slot is heavily loaded by many talkers (p_a is large). In this latter case, this simplified equation gives a pessimistic result and the exact equation should be used. For convenience, Table 1 shows the relation between p_a values calculated from the simplified equation and the actual p_a value from the exact equation under the assumption that the number of talkers is very large and each of them sends very short pulses.

Note that in Table 1, even if the probability $p_a = Td$ (calculated from a simple equation) is approaching or exceeding unity, the true probability p_a never exceeds unity because even if very many pulses are sent at random, many pulses will coincide and there will always be short, infrequent gaps when no pulses are transmitted. In practical RADA systems each frequency slot is usually loaded approximately half the time ($p_a = 0.3$ to 0.7), and the exact equation or Table 1 should be used. Table 1 is handy in converting the simple p_a calculation to exact p_a value.

Thus far p_a has been calculated by the rather impractical assumption that all the interfering talkers are equal in strength. The next paragraph (8) will show how this calculation of p_a can be used in practical systems with a large dynamic range of interfering signals by adjusting the number of talkers T. Also the next paragraph will indicate how to calculate p_c, the probability of cancellation error. Obviously, under the assumption that all talkers are equal, cancellation errors will be very frequent, making p_c about 0.3, since the amplitude is always correct and only the phase needs to be opposite.

Once the p_a and p_c are known for a given practical system, the calculation of modulation pulse errors or bit errors can be easily performed, knowing the address recognition logic. If a simple N out of N logic (as defined in paragraph 6 above) is used, then the total error rate E can be calculated from the following equation:

$$E = E_a + E_c = K_1 p_a{}^N + (1 - K_1) N p_c$$

In this equation, K_1 stands for the probability that the address was not sent in the frame (then only the addition error is possible) and $1 - K_1$ is the probability that the address was sent in the frame (only the cancellation error is then possible). In PCM and DM, $K_1 = 0.5$ because a modulation pulse is sent in every second frame as an average.

For example, assume that $N = 4$ and that 4 out of 4 logic is used (which means that all 4 pulses in the address have to be received to recognize the address and that, if any

Table 1. Probability of Pulse Addition (or Cancellation) vs. $d \times T$

$d \times T$.......	0.1	0.2	0.3	0.4	0.5	0.6	0.7	0.8	0.9	1	1.1	1.2	1.3	1.4	1.5
p_a or p_c.....	0.0952	0.181	0.260	0.329	0.393	0.451	0.503	0.550	0.593	0.632	0.667	0.699	0.727	0.753	0.777

one of them is canceled, the address will not be recognized), and further assume that p_a is equal to 0.5 and p_c is equal to 0.01. Then the error rate will be as follows:

$$E = 0.5 \times 0.5^4 + 0.5 \times 4 \times 0.01 = 0.031 + 0.02 = 5.1\%$$

One effective method of optimizing the system by decreasing the total number of errors E is to adjust the RF pulse detection threshold properly. In the average system this threshold usually is adjusted to about 6 to 8 db below the average level of the desired received pulse. If this threshold is adjusted upward, to -3 db for example, then the addition error will decrease because pulses 3 db or more weaker than the desired pulse will not be detected and cannot cause addition errors, but on the other hand, it will take a much weaker pulse in opposite phase relation to cause the cancellation error because the amplitude of the desired pulse must be decreased only by 3 db. If the threshold is lowered to below 6 or 8 db, then the addition errors will increase because quite weak pulses will now cause such errors, but at the same time cancellation errors will rapidly decrease because the interfering pulse will have to be almost equal in amplitude and opposite in phase to provide the necessary high degree of cancellation. After the optimum detection threshold is selected, the addition errors and cancellation errors are usually about equal in practical systems. While the bit addition and cancellation errors are about equal, the RF pulse addition errors are of an order of magnitude more frequent than the RF pulse cancellation errors.

8. RADA Bandwidth Consideration

In considering the bandwidth required for the given RADA system, it is important first to establish the so-called talker per megacycle (T/MC) number applicable to a given system. Then, when the largest number of talkers who may be within the range of the user's receiver in a given system is divided by this T/MC number, the bandwidth required for the user set is determined. Additional bandwidth may be required for repeater operation.

The T/MC number is also important for system design. Different system parameters, such as modulation and addressing scheme, pulse shape, pulse width vs. frequency separation, IF selectivity, detection threshold, time gate length and shape, etc., should be optimized to give the maximum T/MC number.

Before discussing how the T/MC number can be calculated, a more exact explanation of the term "talker" is necessary. The talker is defined from the viewpoint of the receiver receiving a desired signal at the maximum range of operation. Figure 15 shows a receiver R which is located in the center of the circle A whose radius corresponds to the maximum range of the user's set. The desired transmitter T_d is located on the circumference of this circle and is transmitting a message to the receiver R.

Thus the worst case is selected with the weakest desired signal. It is assumed that the interfering talkers are distributed uniformly everywhere in the plane of the circle, except perhaps inside a very small circle B which is determined by the minimum practical separation between the user sets and may be of an order of 50 to 100 ft. The active transmitters T within the large circle A are considered to be talkers, while active

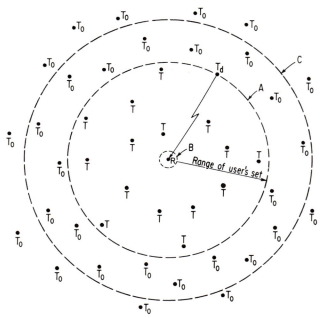

Fig. 15. Geographical presentation of RADA system.

transmitters outside the circle T_o are not counted as talkers, although they will cause some interference. A large ground-based system is assumed in this case in which the interference may come from other talkers who are outside the normal range of the receiver and who may or may not belong to the same RADA system. As both T and T_o talkers are added, the density of talkers per square mile increases until the interference causes an intolerable S/N ratio; then the talkers inside the circle are counted to obtain the T/MC number. If the detection threshold is adjusted to -6 db, for example, and ground-wave propagation is assumed (the signal decreases 12 db when the distance is doubled), then all talkers located within the circle C (with a radius 1.4 times larger) will have a sufficiently strong signal to cause addition errors and there will be twice as many of them than the talkers inside circle A who were counted as talkers in the T/MC number. On the other hand, cancellation errors will be caused mostly by the talkers located near the circumference A because either other talkers are too weak to cancel the desired pulse or, when too strong, they have little chance to cancel the pulse because the timing is critical.

From Fig. 15, the distribution of relative number of talkers vs. signal strength can be derived and is shown as a solid curve on Fig. 16.

Note from this curve that the number of strong talkers is relatively small; for example, talkers which have signal strength 40 db above the desired talker would be ten times less numerous.   On the other hand, talkers which have a signal strength 12 db weaker would be twice as numerous.   Assuming the number of talkers T inside circle A to be 100 per cent and by integrating this curve, the dotted curve on Fig. 16 indicates the total number of talkers having a signal equal or stronger than the abscissa value. It is shown that only 1 per cent of talkers will have a signal 40 db or stronger, for example.   On the other hand, 400 per cent of the talkers will have signals -12 db or stronger as compared with the desired talker.

In the previous paragraph, an equation was shown to calculate the probability of pulse addition p_a, assuming that all talkers have a signal strength equal to the desired talker.   A tacit assumption was made that each pulse from each such talker can occupy and interfere with only one time and frequency slot.   When the talker is stronger than the desired one, its pulse has sufficient power density over the longer time so that it can occupy (interfere with) more than one time slot and conversely a

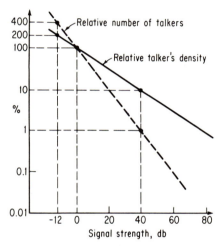

Fig. 16. Relative density and number of talkers vs. signal strength.

weaker talker will occupy less than one time slot.   If the time gates are used which are equal to the pulse width, then no matter how strong the pulse is, it will not interfere with more than two time slots.   However, if the interfering pulse is sufficiently strong, its frequency spectrum may have sufficient energy density to interfere with two adjacent frequency slots, and as it becomes still stronger, four adjacent frequency slots (two on each side) can be interfered with, and so on.   Knowing the pulse shape and frequency spectrum and the receiver selectivity and detection circuit, it is possible to evaluate for each pulse strength how many time and frequency slots such a pulse will interfere with, and therefore it is possible to determine the relation of interference capability, expressed in the form of number of time and frequency slots occupied vs. signal power.   An example of this relation is shown in Fig. 17.   For example, this curve may indicate that a talker which is 26 db stronger than the desired one may interfere with 1.9 time slots and in addition with 0.8 of time slots at two adjacent frequency slots.   This means that such a talker can interfere in $1.9 + 2 \times 0.8 = 3.5$ time-frequency slots, or one can say that such a talker is equivalent to 3.5 laboratory talkers.   By integrating the talker density shown on Fig. 16 multiplied by the interference capability of each such talker shown on Fig. 17, one can arrive at the total number of interfering "laboratory" talkers, which will be usually two or three times larger than the number of actual talkers.   Once this ratio is determined, it will suffice to multiply the number of actual talkers T by this number and use this new T' in the equation given in paragraph 7 to calculate the probability of addition errors p_a from Table 1.

The number of cancellation errors can also be calculated by evaluating the chance of pulse cancellation vs. signal strength.   A typical curve relating the two is shown on Fig. 18.

It can be seen from Fig. 18 that pulses equal to the desired pulse have the best chance to cancel the desired pulse, pulses much weaker have no chance at all, and pulses much stronger have only a very small chance because they have to arrive in proper phase and in a very short period of time (just before the decision on pulse reception is made) in order to cancel the accumulated energy from the desired pulse.

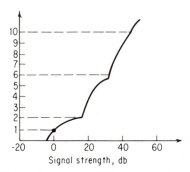

Fig. 17. Interference capability vs. signal strength.

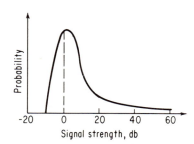

Fig. 18. Pulse cancellation probability vs. signal strength.

Again, when the curves of Figs. 16 and 18 are multiplied and integrated, the ratio of "canceling talkers" T'' to actual talkers T can be determined and the p_c calculated by substituting the corrected T'' number for T.   This completes the description of analytical method of calculation of interference and T/MC number for a given system, which is of an order of 2.5 to 7 in a well-designed system.

In optimizing the system, it is very important that the RF pulses will not be square in shape but will be shaped into cosine or cosine square in order to limit the pulse frequency spectrum spread.   It is well known that, if the square pulses are used, the frequency spectrum of a square pulse decays very slowly with frequency, and a square pulse only 15 or 20 db stronger than the desired one will interfere with many frequency slots.   Also if the time gates are used, the time gates should be likewise shaped in order to prevent an abrupt cutoff of the strong interfering pulse, which will then look like half square pulse and will have a spread spectrum.

The T/MC number not only will depend on the system design and selected parameters, such as pulse shaping, gate shaping, frequency spacing vs. pulse length, etc., but will also depend on the dynamic range of the signals. In the analytical approach above, the very large range of dynamic signals was assumed. This may not be the case with satellite or airborne repeater systems, for example, where all the talkers may be almost equally distant from the repeater or variations in distances will not be great. In such a system with small dynamic range of signals the T/MC number may be up to two or three times greater than indicated above. This higher number of T/MC will be also applicable in case of a repeater-to-user transmission if only one repeater retransmits all the messages in a separate band of frequencies. Then all the signals are equal in strength, and the dynamic range is zero.

In the calculation of the T/MC number, a uniform density talker distribution was assumed. In some systems, this approximation may not be correct. In such systems the receiver should be placed in the center of the area of highest talker density and then the T/MC number should be calculated from appropriate curve different from the one in Fig 16. The number of T/MC will be smaller in this case as there will be relatively more stronger talkers than the weaker ones. Also if the repeater is repeating messages in the user's band, the receiver should be placed as close to the repeater as it can be expected to find itself in a practical situation, and then a different T/MC number should be calculated, assuming that the repeater represents a certain fixed number of strong talkers. Usually in this case, the T/MC number will be small, and this is the reason why it is advantageous in a complex system with repeaters to use a separate frequency band for repeater-to-user communication.

In the T/MC number calculations so far discussed, the received pulse shape was assumed to be the same as the transmitted one. However, in actual pulse propagation, because of scatter reflections the received pulse does not decay so rapidly as the transmitted one. Usually the decay is approximately 20 db/octave of time. For example, if a received pulse decays 20 db in the first $\frac{1}{2}$ μsec after the transmission of the pulse ceased, it may decay another 20 db in the following 1 μsec and another 20 db in the following 2 μsec, and so on. This shape of pulse "tail" greatly depends on the terrain, antenna height, frequency used, etc. However, in general it is advantageous to use relatively long pulses, since the decaying tail of the pulse is almost independent of the pulse width. This means that address matrices with fewer time slots and more numerous frequency slots will yield a somewhat higher T/Mc number if pulse elongation due to the propagation is taken into consideration when calculating the T/MC number. This calculation can proceed as described above except that the interference by the actually received pulse shape should be evaluated instead of transmitted pulse shape. In digital RADA systems using pulses 1 μsec or longer the addition of the tail to the pulse usually results in additional error rate which decreases the T/MC number by 5 to 15 per cent.

In conclusion, it was indicated how to calculate the T/MC number for a practical RADA system as a function of system design parameters and dynamic range of signals. Usually in the well-designed system, the T/MC number is approximately 5, which permits an estimate of the bandwidth required for the system if the number of talkers within the range of the receiver in the worst possible situation is given.

9. Summary

Addressed communications, or RADA (Random Access Discrete Address), is a relatively new and promising communication system concept.

At the time of this writing, the RADA system concept is in the development state. Only a few experimental systems and equipments are in existence, but a relatively large military RADA system is in the planning and design stage. It is up to the future to confirm experimentally the calculations and predictions as to the system performance which has been discussed here, because no RADA operative system is now in existence. However, there are experimental indications that RADA will be successful and that RADA will be one of the most important communication systems concepts in the future.

REFERENCES

1. Hamsher, Donald H., System Concepts for Address Communications Systems, *IRE Trans. Vehicular Commun.* vol. VC-9, no. 3, December, 1960.
2. de Jaeger, F., Deltamodulation, A Method of PCM Transmission Using the One-Code, *Phillips Res. Rept.* 7, 1952.
3. Magnuski, Henry, Motorola's Contributions to Random Access Discrete Address Techniques, *Motorola Eng. Bull.*, vol. 11, Issues 1 and 2, 1963.
4. Haas, D. L., Satellite Communications Using RACEP, Martin Company. Paper delivered at the Air Force Association Convention in Washington in September, 1962.

Chapter 19

RADIO-FREQUENCY ALLOCATION AND ASSIGNMENT

PHILIP F. SILING, *Radio Corporation of America (retired)*

REGULATORY FACTORS

1. General

Selection of frequencies for radio communications depends upon administrative factors discussed in this chapter as well as technical factors discussed in Chaps. 6, 15, 16, and 17. Adequate knowledge of the administrative factors is essential to system design involving radio transmission to avoid a technical design in an unusable portion of the frequency spectrum. The present conditions depend upon:

Historical developments
International agreements
National or area policy
Extent of current usage
Elimination of interference

These topics and descriptions of the current organizations and procedures involved in obtaining a license or other appropriate authorizations are described in this chapter.

The radio-frequency spectrum is a national and international resource in the public domain. Therefore, the allocation and assignment* of radio frequencies and the national and international regulation of the radio spectrum are predicated upon the necessity to provide for efficient utilization of this finite and limited spectrum space under sound conservation principles. Theoretically, any user might employ any frequency for a particular purpose provided that it did not cause interference to another user. Practically, however, if this practice were followed, it would soon preclusively restrict the establishment of new and needed radio services as well as the expansion of existing services.

To promote an orderly development of this vital natural resource, it has been found necessary to have planned allocations to radio services so that operations within bands of frequencies are confined to particular types of services with comparable radiated powers and interference potential characteristics. In the main, this has been accom-

---

* Throughout this chapter, the term "allocation" is used to denote the subdivision of the radio spectrum into discrete bands of frequencies for use by one or more specific radio services whereas the term "assignment" is used for the designation of a specific frequency, within an allocated band, to be used by a particular station.

plished by international administrative radio conferences wherein the member nations agree to a table of international allocations by services.

Seven such international conferences have been held to promote orderly use of the spectrum and to adopt such international regulations as are necessary to accomplish this goal: 1906 Berlin; 1912 London; 1927 Washington, D.C.; 1932 Madrid; 1938 Cairo; 1947 Atlantic City; and 1959 Geneva.   The international table of frequency allocations now in force is that contained in the Radio Regulations adopted by the Geneva Conference in 1959.   Copies of these Regulations with annexed recommendations comprising a volume of 641 pages can be obtained, upon request, from the International Telecommunication Union, Geneva, Switzerland, at a cost of 19 Swiss francs. (Specify whether you desire the English, French, or Spanish edition.)

Figure 1 gives the international frequency allocation table in chart form.

It will be noted upon reference to Fig. 1 that the world has been subdivided into three regions, the area of each region also being shown.   Figure 1 shows the "tropical zone" where local broadcasting stations may use frequencies in the lower portion of the high-frequency spectrum to overcome high-level noise conditions prevalent in the lower frequency bands.   It will be noted also by reference to Fig. 1 that, with a few minor exceptions, the high-frequency portion of the spectrum (3 to 30 Mc/sec) is allocated on a world-wide basis.   This, of course, is because frequencies in this portion of the spectrum are those most capable of causing world-wide interference and, therefore, are those for which world-wide agreement as to allocations is essential.

The abbreviations primarily employed for the different bands of frequencies are shown in Table 1.

Table 1. Abbreviations of Frequency Bands

Band number	Band designations	Frequency range (lower limit exclusive, upper limit inclusive)		Corresponding metric subdivision
4	VLF (very low)	3–30 kc/sec	(kHz)*	Myriametric waves
5	LF (low)	30–300 kc/sec	(kHz)	Kilometric waves
6	MF (medium)	300–3,000 kc/sec	(kHz)	Hectometric waves
7	HF (high)	3–30 Mc/sec	(MHz)	Decametric waves
8	VHF (very high)	30–300 Mc/sec	(MHz)	Metric waves
9	UHF (ultra high)	300–3,000 Mc/sec	(MHz)	Decimetric waves
10	SHF (super high)	3–30 Gc/sec	(GHz)	Centimetric waves
11	EHF (extremely high)	30–300 Gc/sec	(GHz)	Millimetric waves
12	. . . .	300–3,000 Gc/sec or 3 Tc/sec	(GHz) (THz)	Decimillimetric waves

* Abbreviations generally used in other than English speaking countries.   Other abbreviations: c/s, cycles per second; Hz, hertz; k, kilo; M, Mega; G, Giga; T, Tera.

2. International Regulation

To accomplish the necessary international allocation of frequency bands, assignment of frequencies, and regulations of the use of the radio spectrum, the countries of the world have set up an International Telecommunication Union (hereafter referred to as the ITU).

The ITU was originally established in 1865 to handle the international regulation of wire telegraphy and was followed by the establishment of the Radio Telegraph Union at the Berlin Conference in 1906 for the regulation of radio.   These organizations were merged into the ITU by the Madrid Conference of 1932 and recast into its present form by the Atlantic City Conference of 1947.

The present organization of the ITU is shown in Fig. 2.

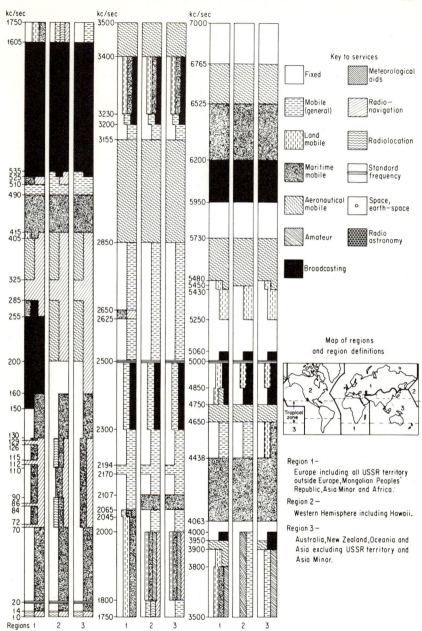

FIG. 1. Frequency allocations chart. (*By permission of Radio Corporation of America.*)

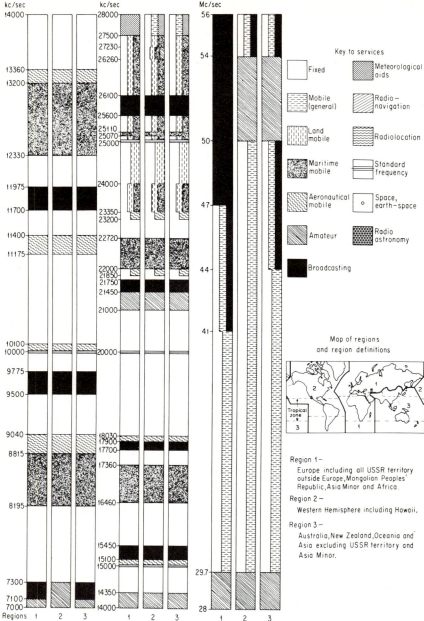

FIG. 1. Frequency allocations chart (*continued*).   (*By permission of Radio Corporation of America.*)

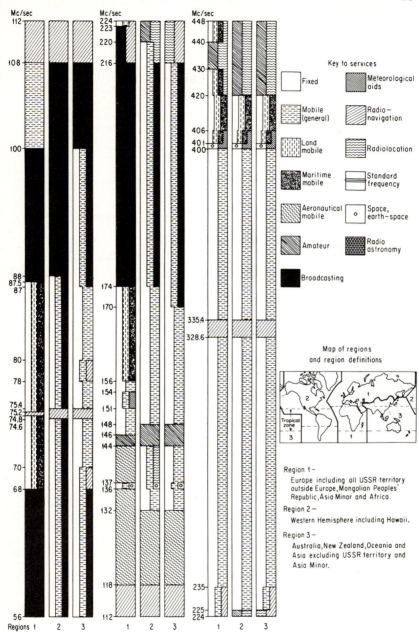

FIG. 1. Frequency allocations chart (*continued*). (*By permission of Radio Corporation of America.*)

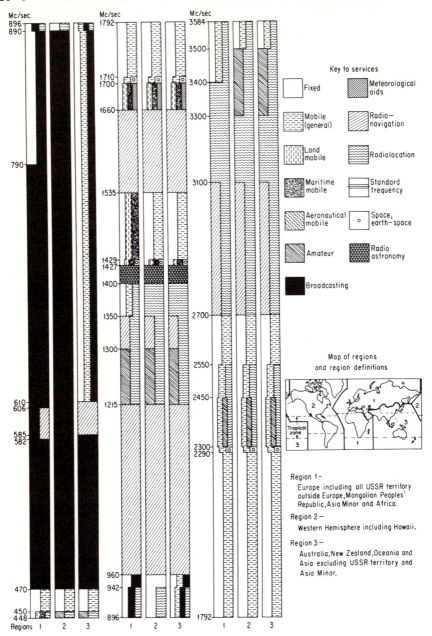

FIG. 1. Frequency allocations chart (*continued*).   (*By permission of Radio Corporation of America.*)

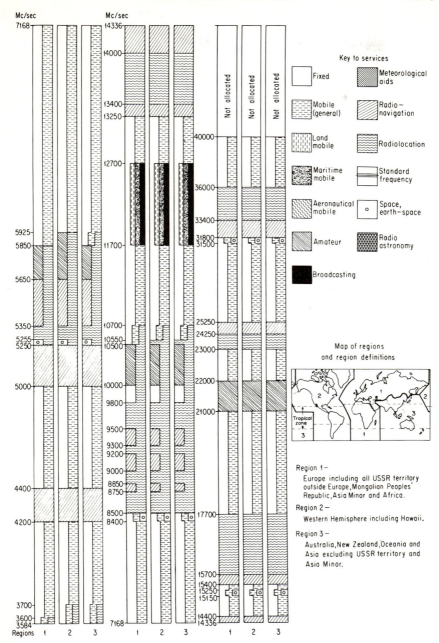

FIG. 1. Frequency allocations chart (*continued*). (*By permission of Radio Corporation of America*.)

The Plenipotentiary Conference, which normally meets about every five years, revises the basic convention for the ITU, establishes the general policies of the Union, determines the budgetary framework within which the Union must operate, and the basic salary scale of personnel on its permanent staff, and elects the Secretary General, the Deputy Secretary General, and the Members of the Union to serve on the Administrative Council.

The Administrative Council consists of 25 members chosen with due regard for equitable representation of all parts of the world.   It meets every year and acts for the Plenipotentiary Conference between sessions of that body in a manner similar to a board of directors serving under direction of the stockholders of a corporation.

The Administrative Radio Conferences revise the international radio regulations including the table of frequency allocations, meeting whenever advances in the art are such as to require this action, and also elect the 11 individuals who comprise the International Frequency Registration Board.

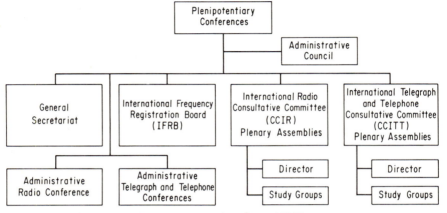

Fig. 2. Organization chart of ITU.

The Administrative Telegraph and Telephone Conferences in a similar way are responsible for the revision of the telegraph and telephone regulations.

There are four permanent organs of the ITU, namely, the General Secretariat, the International Frequency Registration Board, the International Radio Consultative Committee, and the International Telegraph and Telephone Consultative Committee.

The General Secretariat administers the affairs of the Union under the general direction of the Administrative Council and handles the Union's secretarial work, including the translation of documents and the interpretation necessary at all Conferences of the Union.   It also publishes the necessary documents of the Union, including the service documents and lists, which include the international frequency list, lists of fixed, broadcasting, coast, ship and radiodetermination and special service stations, and the alphabetical lists of call signs.   These documents can be obtained for a nominal price from the International Telecommunication Union, Place des Nations, Geneva, Switzerland.

The International Frequency Registration Board (IFRB) examines and records frequency assignments made by the different countries which are capable of causing harmful interference to the services of another administration, which are used for international communication, or for which it is desired to obtain international recognition of the use of the frequency.   It examines all such notices of assignments to determine the probability of these new assignments causing harmful interference to the service of a previously notified assignment of any other administration.   In case it finds such probability of interference, it returns the notice to the originating administration, explaining the reasons for the finding, with any suggestions that it may be able to offer for a satisfactory solution.

If, in fact, the originating administration has used this new assignment for 60 days without receiving any complaint of harmful interference and so advises the Board, the Board investigates the frequency assignments that contributed to the unfavorable finding in an endeavor to find the reasons why this incompatible assignment actually operates without harmful interference. As a result of its inquiries, the Board is then able to determine whether any of the listings should be deleted or modified, with the consent of the administrations involved, and in any case, records the new assignment with a note explaining the results of its findings.

The duties of the International Radio Consultative Committee (CCIR) are to study technical and operating questions relating specifically to radio communications and to issue recommendations on them.

In a similar way the duties of the International Telegraph and Telephone Consultative Committee (CCITT) are to study technical, operating, and tariff questions relating to telegraphy and telephony and to issue recommendations on them.

The questions to be studied by the Consultative Committees are those submitted to them by the Plenipotentiary Conferences, by an Administrative Conference, by the other consultative committees, by the International Frequency Registration Board, by the Plenary Assemblies of the Consultative Committees, or, in the interval between Plenary Assemblies, when requested or approved by correspondence by at least 12 members or associate members of the Union.

The recommendations approved by the Plenary Assemblies are for guidance and help to all administrations and, when appropriate, for consideration by the Administrative Conferences. In arriving at these recommendations, Study Groups, as set up by the Plenary Assemblies, carry on the work during the interval between Plenary Sessions, which normally are held about every 3 years. This work, usually accomplished by correspondence, may be the subject of interim meetings of a particular Study Group if this is found to be desirable.

Each Consultative Committee has a permanent Director, appointed by the Plenary Assembly, who coordinates and advises in the work of the Committee and the Study Groups.

3. National Regulation

Most countries exercise centralized control for the assignment and use of frequencies. In nearly every country this centralized control is vested in a Ministry of Communications or a Posts, Telegraph, and Telephone Administration. In Canada, the Department of Transport exercises this control function. Most of these government administrations also operate the communications services of the country, although these operations are conducted by private enterprise in some countries, particularly in the Western Hemisphere.

The United States is a notable exception to the above. Its centralizing control for *nongovernment* stations is vested in the Federal Communications Commission (FCC) set up by Congress by the Communications Act of 1934 to regulate nongovernment stations, to promote national and international public communications, and to provide for an orderly use of the radio spectrum. Similar control is exercised over *government* usage of the spectrum by the Director of Telecommunications Management, acting on behalf of the President with the advice of the Interdepartment Radio Advisory Committee (IRAC), a committee of representatives of those government agencies which are major users of radio. This rather complex system is shown graphically in Figure 3.

In any case, the procedures for securing authority to operate is along similar lines. Applications are submitted to the centralizing body in the form and including such relevant technical and other data as required by the centralizing authority. The centralizing body considers the applications from the standpoints of conformity to international rules and regulations, conformity to its own rules and regulations, and whether or not harmful interference would be caused to the service of any other station —national or international. If the finding is favorable in all aspects, a license or other

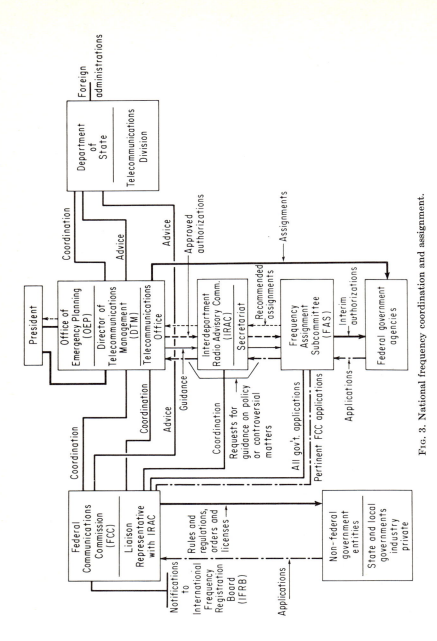

FIG. 3. National frequency coordination and assignment.

appropriate authorization is then issued.   This authorization may be qualified by instructions and restrictions as to areas of operation, hours of use, limitation of power, use of appropriate directional antennas, etc.   Therefore, it can be seen that the complexity of obtaining a license to operate will vary greatly with the type of service, the

Table 2. Volumes of FCC Rules and Regulations by Categories

Volume	Part	Price
Volume I (November, 1963)	Part 0, Commission Organization Part 1, Practice and Procedure Part 13, Commercial Radio Operators Part 17, Construction, Marking, and Lighting of Antenna Structures	$2.50 (3.50 foreign)
Volume II (January, 1964)	Part 2, Frequency Allocations and Radio Treaty Matters: General Rules and Regulations Part 5, Experimental Radio Services (other than Broadcast) Part 15, Radio Frequency Devices Part 18, Industrial, Scientific, and Medical Equipment	$2.00 (2.75 foreign)
Volume III (January, 1964)	Part 73, Radio Broadcast Services Part 74, Experimental, Auxiliary and Special Broadcast Services (Former part numbers 3 and 4, respectively)	$4.50 (5.75 foreign)
Volume IV (February, 1962)	Part 7, Stations on Land in Maritime Services Part 8, Stations on Shipboard in Maritime Services Part 14, Public Fixed Stations and Stations of the Maritime Services in Alaska (Now renumbered parts 81, 83, and 85, respectively)	$2.00 (3.00 foreign)
Volume V (January, 1964)	Part 87, Aviation Services Part 89, Public Safety Radio Services Part 91, Industrial Radio Services Part 93, Land Transportation Radio Services (Former part numbers 9, 10, 11, and 16, respectively)	$2.50 (3.50 foreign)
Volume VI (January, 1964)	Part 95, Citizens Radio Service Part 97, Amateur Radio Service Part 99, Disaster Communications Service (Former part numbers 12, 19, and 20, respectively)	$1.25 (1.75 foreign)
Volume VII (January, 1963)	Part 6, International Fixed Public Radio Communication Services (Now renumbered Part 23) Part 21, Domestic Public Radio Services (other than Maritime Mobile) Part 25, Satellite Communications	$2.00 (2.75 foreign)

congestion in the particular portion of the spectrum, and, most importantly, the rules and regulations set up by the control body itself.   For example, the FCC has a comprehensive set of Rules and Regulations which, where necessary, may require a formal hearing before the grant of a license.   This is particularly true in the case of broadcast applications.   It behooves a prospective nongovernment applicant in the United States, therefore, to consult the Commission's Rules involving the particular type of

Form of Notice*

For Use when Notifying to the International Frequency Registration Board a Frequency Assignment or a Change to an Assignment Recorded in the Master International Frequency Register

(see Article 9)

(a) Notifying administration

(e) Notice No. _____ Date _____

(b) New assignment [ ]

(c) Change of characteristics of a recorded assignment in the Master Register [ ]

(d) Deletion of an assignment [ ]

1 Assigned frequency _____ kc/s _____ Mc/s

For I.F.R.B. use

2c Date of putting into use

3 Call sign (Identification)

4a Name of transmitting station   4b Country   4c Longitude and latitude of the transmitter site

Locality(ies) or area(s) with which communication is established	Length of circuit (km)	Class of station and nature of service	Class of emission, necessary bandwidth and description of transmission	Power (in kW) P_c P_m P_p e	Transmitting antenna characteristics			Maximum hours of operation of the circuit to each locality or area (G.M.T.)	Megacycle order of the other frequencies normally utilized for the same circuit	Supplementary information
					Azimuth of maximum radiation	Angular width of radiation main lobe	Antenna gain (db)			
5a	5b	6	7	8	9a	9b	9c	10	11	

12a Operating Administration or Company

12b Name and postal address } of administration (Article 15)
Telegraphic address }

Regional or service agreement _____

Other information:

COORD/

* The actual size of the notice is a matter for individual administrations.

Fig. 4. Form of Notice-IFRB.

service requested. These regulations and rules, which, of course, are subject to frequent revision, can be obtained from the Government Printing Office.

The FCC acts in behalf of the United States in forwarding to the IFRB notice of all frequency assignments which should be recorded internationally. This activity by the FCC to provide the IFRB with the essential information required to conduct properly the international procedures applies to both government and nongovernment stations. The data required for such notification are usually prepared by the user on the form shown in Fig. 4.

A separate notice should be sent for each new frequency assignment requiring international recording, each deletion of such an assignment, or any change in the characteristics of a frequency assignment. While mentioned before, it should be repeated that only assignments which fall in one of the three following categories need be submitted to the IFRB:

1. Those capable of causing harmful international interference
2. Those in an international service
3. Those for which international recognition is desired

The basic characteristics which must be furnished on this form are shown in Table 3.

Table 3. Explanation of IFRB Notice

Column	Title
1	Assigned frequency
2c	Date of putting into use
3	Call sign
4a	Name of the transmitting station
4b	Country in which the transmitting station is located
4c	Longitude and latitude of transmitter site
5a	Localities or areas where communications are received or length of circuit for stations in the land mobile or space service
6	Class of station and nature of service
7	Class of emission, bandwidth, and description of transmission
8	Power in kilowatts
9a	Azimuth of maximum radiation
10	Maximum hours of operation of the circuit to each locality or area (GMT)
11	Megacycle order of the other frequencies normally utilized for the same circuit*
Supplementary information	Reference frequency or frequencies, if any†

* This information is required only for fixed point-to-point stations within the range of 4 to 28 Mc/sec.

† The assigned frequency is the center of the frequency band assigned to a station. A reference frequency is a frequency having a fixed and specified position with respect to the assigned which can be easily identified and measured, such as with multiplex, single sideband, independent sideband, etc., transmissions.

Furthermore, the FCC acts jointly with the Director of Telecommunication Management in development of the national table of frequency allocations. This table is generally within the framework of the international allocations table but gives such suballocations for each particular service as are necessary to meet the national needs. These suballocations are shown later in this chapter under the discussion of the individual services.

While we have dwelt at length on the centralized control mechanism in the United States, the procedures for securing radio licenses are, in broad essentials, the same for other countries through their respective control bodies.

The organization of the Federal Communications Commission is shown in Figure 5.

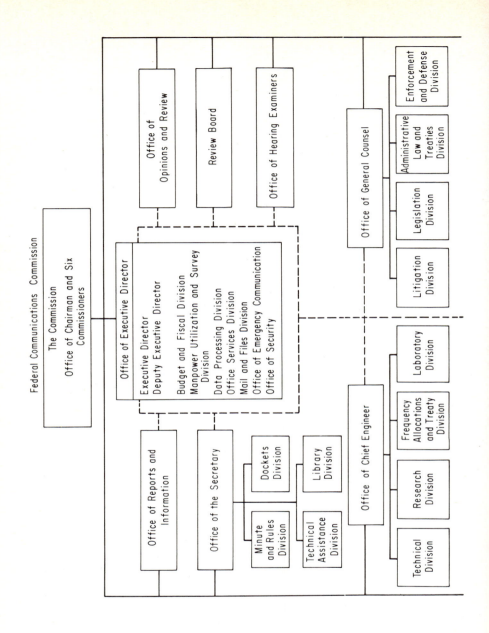

Federal Communications Commission

The Commission
Office of Chairman and Six Commissioners

Office of Executive Director

Executive Director
Deputy Executive Director

Budget and Fiscal Division
Manpower Utilization and Survey Division
Data Processing Division
Office Services Division
Mail and Files Division
Office of Emergency Communication
Office of Security

Office of Opinions and Review

Review Board

Office of Hearing Examiners

Office of General Counsel

Litigation Division

Legislation Division

Administrative Law and Treaties Division

Enforcement and Defense Division

Office of Reports and Information

Office of the Secretary

Minute and Rules Division

Dockets Division

Technical Assistance Division

Library Division

Office of Chief Engineer

Technical Division

Research Division

Frequency Allocations and Treaty Division

Laboratory Division

19–14

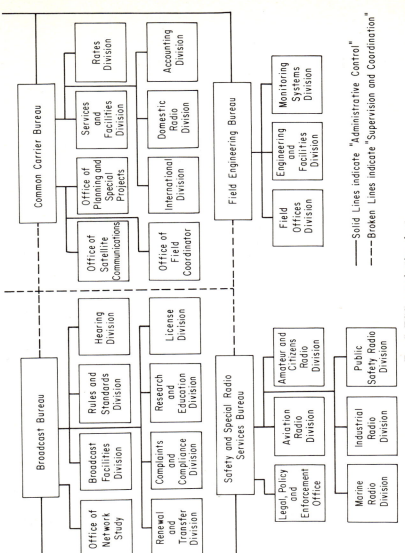

Fig. 5. FCC organization chart.

——— Solid Lines indicate "Administrative Control"
- - - Broken Lines indicate "Supervision and Coordination"

Common Carrier Bureau

Rates Division

Accounting Division

Services and Facilities Division

Domestic Radio Division

Office of Planning and Special Projects

International Division

Office of Satellite Communications

Office of Field Coordinator

Field Engineering Bureau

Monitoring Systems Division

Engineering and Facilities Division

Field Offices Division

Broadcast Bureau

Hearing Division

License Division

Rules and Standards Division

Research and Education Division

Broadcast Facilities Division

Complaints and Compliance Division

Office of Network Study

Renewal and Transfer Division

Safety and Special Radio Services Bureau

Amateur and Citizens Radio Division

Public Safety Radio Division

Aviation Radio Division

Industrial Radio Division

Legal, Policy and Enforcement Office

Marine Radio Division

Table 4. Fixed Service Frequencies

Region 1*	Region 2*	Region 3*	United States*,†
14–70 kc/sec	14–90 kc/sec	14–200 kc/sec	14–90 kc/sec
72–84 kc/sec	110–200 kc/sec		110–200 kc/sec
86–112 kc/sec			
115–126 kc/sec			
129–130 kc/sec			
1,605–2,170 kc/sec	1,605–2,065 kc/sec	1,605–2,065 kc/sec	1,605–1,800 kc/sec
2,194–2,625 kc/sec	2,107–2,170 kc/sec	2,107–2,170 kc/sec	2,107–2,170 kc/sec
2,650–2,850 kc/sec	2,194–2,850 kc/sec	2,194–2,850 kc/sec	2,194–2,850 kc/sec
3,155–3,400 kc/sec	3,155–3,400 kc/sec	3,155–3,400 kc/sec	3,155–3,400 kc/sec
3,500–3,900 kc/sec	3,500–4,063 kc/sec	3,500–3,900 kc/sec	
3,950–4,063 kc/sec		3,950–4,063 kc/sec	4,000–4,063 kc/sec
4,438–4,650 kc/sec	4,438–4,650 kc/sec	4,438–4,650 kc/sec	4,438–4,650 kc/sec
4,750–5,480 kc/sec	4,750–5,450 kc/sec	4,750–5,480 kc/sec	4,750–5,450 kc/sec
5,730–5,950 kc/sec	5,730–5,950 kc/sec	5,730–5,950 kc/sec	5,730–5,950 kc/sec
6,765–7,000 kc/sec	6,765–7,000 kc/sec	6,765–7,000 kc/sec	6,765–7,000 kc/sec
7,300–8,195 kc/sec	7,300–8,195 kc/sec	7,300–8,195 kc/sec	7,300–8,195 kc/sec
9,040–9,500 kc/sec	9,040–9,500 kc/sec	9,040–9,500 kc/sec	9,040–9,500 kc/sec
9,775–9,995 kc/sec	9,775–9,995 kc/sec	9,775–9,995 kc/sec	9,775–9,995 kc/sec
10,100–11,175 kc/sec	10,100–11,175 kc/sec	10,100–11,175 kc/sec	10,100–11,175 kc/sec
11,400–11,700 kc/sec	11,400–11,700 kc/sec	11,400–11,700 kc/sec	11,400–11,700 kc/sec
11,975–12,330 kc/sec	11,975–12,330 kc/sec	11,975–12,330 kc/sec	11,975–12,330 kc/sec
13,360–14,000 kc/sec	13,360–14,000 kc/sec	13,360–14,000 kc/sec	13,360–14,000 kc/sec
14,350–14,990 kc/sec	14,350–14,990 kc/sec	14,350–14,990 kc/sec	14,350–14,990 kc/sec
15,450–16,460 kc/sec	15,450–16,460 kc/sec	15,450–16,460 kc/sec	15,450–16,460 kc/sec
17,360–17,700 kc/sec	17,360–17,700 kc/sec	17,360–17,700 kc/sec	17,360–17,700 kc/sec
18,030–21,000 kc/sec	18,030–21,000 kc/sec	18,030–21,000 kc/sec	18,030–21,000 kc/sec
21,750–21,850 kc/sec	21,750–21,850 kc/sec	21,750–21,850 kc/sec	21,750–21,850 kc/sec
22,720–23,200 kc/sec	22,720–23,200 kc/sec	22,720–23,200 kc/sec	22,720–23,200 kc/sec
23,350–25,070 kc/sec	23,350–25,070 kc/sec	23,350–25,070 kc/sec	23,350–24,990 kc/sec
25,110–25,600 kc/sec	25,110–25,600 kc/sec	25,110–25,600 kc/sec	25,330–25,600 kc/sec
26,100–27,500 kc/sec	26,100–28,000 kc/sec	26,100–28,000 kc/sec	26,480–26,960 kc/sec
			27,230–27,280 kc/sec
			27,540–28,000 kc/sec
29.7–41 Mc/sec	29.7–50 Mc/sec	29.7–50 Mc/sec	29.8–30 Mc/sec
			32–33 Mc/sec
			34–35 Mc/sec
			36–37 Mc/sec
			38–39 Mc/sec
			40–42 Mc/sec
			49.6–50 Mc/sec
68–74.8 Mc/sec	54–74.6 Mc/sec	54–74.6 Mc/sec	72–73 Mc/sec
75.2–87.5 Mc/sec	75.4–88 Mc/sec	75.4–100 Mc/sec	75.4–76 Mc/sec
136–137 Mc/sec	132–144 Mc/sec	132–144 Mc/sec	137–144 Mc/sec
146–174 Mc/sec	148–220 Mc/sec	148–216 Mc/sec	148–150.8 Mc/sec
			162–174 Mc/sec
235–328.6 Mc/sec	225–328.6 Mc/sec	225–328.6 Mc/sec	225–328.6 Mc/sec
335.4–400 Mc/sec	335.4–400 Mc/sec	335.4–400 Mc/sec	335.4–400 Mc/sec
406–430 Mc/sec	406–420 Mc/sec	406–420 Mc/sec	406–420 Mc/sec
440–470 Mc/sec	450–470 Mc/sec	450–470 Mc/sec	
790–960 Mc/sec	890–960 Mc/sec	610–960 Mc/sec	942–960 Mc/sec

Table 4. Fixed Service Frequencies (*Continued*)

Region 1*	Region 2*	Region 3*	United States*·†
1,350–1,400 Mc/sec			
1,427–1,535 Mc/sec	1,427–1,435 Mc/sec	1,427–1,535 Mc/sec	1,427–1,435 Mc/sec
1,660–2,700 Mc/sec	1,660–2,300 Mc/sec	1,600–2,300 Mc/sec	1,660–1,670 Mc/sec
			1,710–2,290 Mc/sec
	2,450–2,700 Mc/sec	2,450–2,700 Mc/sec	2,450–2,690 Mc/sec
3,400–4,200 Mc/sec	3,500–4,200 Mc/sec	3,700–4,200 Mc/sec	3,700–4,200 Mc/sec
4,400–5,000 Mc/sec	4,400–5,000 Mc/sec	4,400–5,000 Mc/sec	4,400–4,990 Mc/sec
5,850–8,500 Mc/sec	5,925–8,500 Mc/sec	5,850–8,500 Mc/sec	5,925–6,425 Mc/sec
			6,575–8,500 Mc/sec
10.5–13.25 Gc/sec	10.55–13.25 Gc/sec	10.55–13.25 Gc/sec	10.55–10.68 Gc/sec
			10.7–11.7 Gc/sec
			12.2–13.25 Gc/sec
14.4–15.15 Gc/sec	14.4–15.15 Gc/sec	14.4–15.15 Gc/sec	14.4–15.15 Gc/sec
15.25–15.4 Gc/sec	15.25–15.4 Gc/sec	15.25–15.4 Gc/sec	15.25–15.35 Gc/sec
17.7–21 Gc/sec	17.7–21 Gc/sec	17.7–21 Gc/sec	17.7–19.3 Gc/sec
			19.4–21 Gc/sec
22–23 Gc/sec	22–23 Gc/sec	22–23 Gc/sec	22–23 Gc/sec
25.25–31.5 Gc/sec	25.25–31.5 Gc/sec	25.25–31.5 Gc/sec	25.25–31.3 Gc/sec
36–40 Gc/sec	36–40 Gc/sec	36–40 Gc/sec	36–40 Gc/sec

* For sake of brevity, exclusive provisions for standard frequency transmissions at 20, 2,500, 5,000, 20,000, and 25,000 kc/sec are not reflected in this table.

† This column reflects the combined government and nongovernment allocations. In some bands those allocations are not identical. For information as to bands restricted to government use, consult Part 2 of the Rules and Regulations of the Federal Communications Commission.

SELECTION OF FREQUENCIES FOR SPECIFIC ASSIGNMENTS

The other chapters of this handbook cover adequately the factors involved in the selection of the mode of transmission, i.e., cable, high-frequency radio, microwave link, etc., for the particular service desired. The propagation characteristics for the different parts of the spectrum and the propagation needs for each particular service are also discussed in the chapters describing the details of each service. Therefore, it is assumed that the prospective user knows the order of frequency or frequencies desired and has only to determine the specific frequencies available.

With this in mind, we shall discuss in the following paragraphs the problems involved in arriving at particular frequency assignments.

4. Fixed Service

The fixed service,* in general, is allocated frequency bands on either an exclusive or primary basis as shown in Table 4.

The use of frequencies below 200 kc/sec in the fixed service is constantly decreasing, and it would not seem unreasonable for it to disappear in the near future. The long wavelengths require the use of very large and expensive antennas and consequently extremely high power. Moreover, only narrow bandwidths can be obtained, limiting the information rate. Present use is largely of a backup nature for high-frequency long-distance circuits during conditions of ionospheric disturbances by taking advantage of the long-distance ground-wave coverage obtained. The Federal Communica-

* The fixed service is defined as a service of radiocommunication between specified fixed points.

tions Commission now assigns frequencies of this order in the International Fixed Public Service only in Alaska.

The fixed service use of frequencies between 1,600 and 4,000 kc/sec is generally confined to short-distance (in the order of 100 to 200 miles) communications circuits.   In the low end of the sunspot cycle, the frequencies in the high portion of the range may be used for long-distance radio communication at night.

Frequencies between 4,000 and 30,000 kc/sec are the backbone of the long-distance radio communication services.   As described in Chap. 15 on these services, a complement of frequencies is normally required to maintain service throughout the day, the year, and the sunspot cycle, the actual number of frequencies in the complement varying with the circuit path, the hours of operation, the type of modulation, etc.

The demand for frequencies in this range, not only for fixed services but also for long-distance mobile communications in the maritime and aeronautical services and for long-distance broadcasting, is so heavy that it makes this portion of the spectrum the most congested.   This coupled with the fact that these frequencies are capable of causing interference over wide areas makes it essential that the central control mechanism exercise the most rigid controls over their use.   These controls, as well as those exercised by the International Telecommunication Union, are fully discussed in other sections of this chapter.

In applying to the central control body for new frequencies or to increase the bandwidth of emission to permit the provision of more channels in a given bandwidth, it is usually the practice for the operating agency to determine the exact frequency and possible bandwidth by monitoring the particular portion of the spectrum involved. This monitoring should, if possible, be conducted at the desired point or area of reception by the correspondent of the operating agency and would consist of sweeping the particular portion of the spectrum over a period of several days to determine if there are any "holes" where signals of existing stations are not observed as well as the width of these holes.   This information is then transmitted back to the operating agency, which then makes tests to ensure that the frequency and bandwidth selected are satisfactory for reception purposes.   Usually the control body has to authorize such tests. If the tests indicate that the operations can be conducted satisfactorily, application is then made to the control body and the IFRB is notified in accordance with the procedures outlined in other portions of this chapter.   If, after such action has been taken, any interference with the operations of stations having prior use does develop and this interference cannot be adjusted, the agency establishing the new service must abandon the use of this particular frequency and try again, using the same procedures.

Frequencies in the 30- to 60-Mc/sec band have been used for ionospheric scatter transmissions in areas where the frequency usage density is low.   The high power and large antennas associated with this type of transmission give this method very high potential interference capabilities over wide areas.   Consequently, its use also must be subject to rigid control.

The main use in the fixed service for frequencies in the range 60 to 1,000 Mc/sec is for short distances using single or a few hops or for long-distance communications services by combining in tandem these essentially line-of-sight frequencies through use of repeater stations.   The same frequency can be repeated at frequent intervals, depending upon the spacing between repeaters, the terrain, and the care used in siting the repeater stations (i.e., if these stations are slightly off a straight line, the directional antennas employed permit the repetition of the frequency with lesser geographic spacing).   Fixed service systems in this range are particularly appropriate where relatively few channels are required and where the terrain affords extended line-of-sight paths. They have distinct advantages over HF radio circuits in that they are subject to less atmospheric and man-made noise, are less likely to encounter interference, use far less power, and have higher transmission quality and reliability.   Above all, they operate in a less congested portion of the spectrum.

The tropospheric scatter technique is also used in this portion of the spectrum. This mode of communication offers considerable merit where it is required to transmit wideband, multichannel communications beyond line of sight, particularly over inaccessible terrain or large bodies of water.   However, it requires very high power and highly directive transmitting and receiving antennas. Because of this high power, systems of this type are potential sources of interference to other users of the same portion of the spectrum, particularly the low-power line-of-sight relay systems, and the two cannot be accommodated as a rule in the same frequency bands and in the same areas.

Also operating in this range are international control circuits (i.e., links between transmitter and receiving stations and the operating center) and operational fixed circuits (i.e., links not open to the public correspondence operated by and for an operating agency to administer its own operations in a mobile service).

In the bands of frequencies above 1,000 Mc/sec, the principal use in the fixed service is by microwave relay systems which handle all forms of traffic both in systems open to the public correspondence and in private systems.

The systems have a capacity ranging from as few as 4 voice channels in the case of private systems to as many as 17,000 voice channels on heavy routes open to public correspondence. They use low power and highly directional antennas, which make it possible to re-use frequencies many times in a relatively small geographical area. Such radio relay systems provide better quality and reliability than most other modes of radio communications. They operate in a less congested portion of the spectrum, use frequencies which minimize interference, and are less susceptible to atmospheric and man-made noise.

In the future we can expect that these frequencies will be used for communications to and from satellites, sharing with the microwave relay service.

Also, tropospheric scatter, operational fixed and international control services previously described operate in this portion of the spectrum.

5. Aeronautical and Maritime Services Including Radio Navigational Aids

The aeronautical mobile services, in general, are allocated the bands shown in Table 5.

The aeronautical fixed service, in general, is allocated the same bands as shown in Table 4 for the fixed service. It may also operate in the bands designated by the fourth footnote in Table 5, sharing with the aeronautical mobile service.

Frequencies used in the aeronautical service in the band 1,605 to 2,850 kc/sec are primarily employed for short-distance fixed circuits. In general, the same bands of frequencies are used for aeronautical fixed services as are used for public-communication fixed services, in both this and higher ranges of frequencies. Similarly, the same rules and procedures apply as described under the section for fixed circuits.

Frequencies in the range from 2,850 to 4,063 kc/sec are available for short-distance aeronautical services, both for point-to-point and ground-air-ground communications. Frequency bands used in the aeronautical mobile R service, i.e., for aircraft flying scheduled routes, are 2,850 to 3,025 and 3,400 to 3,500 kc/sec. Those for aeronautical mobile OR service, i.e., for aircraft flying nonscheduled routes, are 3,025 to 3,155 kc/sec and, in Region 1, 3,800 to 3,950 kc/sec. In this range, the aeronautical fixed service employs frequencies in the band 3,155 to 3,400 and 4,000 to 4,063 kc/sec.

It should be noted that in the United States the band 3,500 to 4,000 kc/sec is reserved for amateurs, although in the remainder of the world the major portion of this band is used for fixed and mobile services.

In the high-frequency portion of the radio spectrum, approximately from 4 to 25 Mc/sec, the same problems of high congestion and necessity for rigid frequency control as outlined in the section under fixed service exist for the aeronautical services. It is on frequencies in this segment of the spectrum that reliance must be placed for the necessary long-distance point-to-point and transoceanic ground-to-air-to-ground communications. There is this difference, however: The aeronautical mobile services are covered by approved frequency-allotment plans for the world, copies of which can be obtained from the International Telecommunications Union for 9.15 Swiss francs. One of these plans covers the route service, and the other the off-route service.

Table 5. Aeronautical Frequencies

Region 1*,‡	Region 2*,‡	Region 3*,‡	United States*,†
490–510 kc/sec	490–535 kc/sec	490–510 kc/sec 525–535 kc/sec	490–535 kc/sec
	1,605–1,800 kc/sec	1,605–1,800 kc/sec	1,605–1,800 kc/sec
	2,000–2,065 kc/sec	2,000–2,065 kc/sec	2,000–2,065 kc/sec
2,065–2,170 kc/sec (OR)	2,107–2,850 kc/sec	2,107–2,850 kc/sec	2,107–2,850 kc/sec
2,170–2,194 kc/sec			
2,194–2,625 kc/sec (OR)			
2,650–2,850 kc/sec (OR)			
2,850–3,025 kc/sec (R)	2,850–3,025 kc/sec (R)	2,850–3,025 kc/sec (R)	2,850–3,025 kc/sec (R)
3,025–3,230 kc/sec (OR)	3,025–3,230 kc/sec (OR)	3,025–3,230 kc/sec (OR)	3,025–3,230 kc/sec (OR)
3,400–3,500 kc/sec (R)	3,400–3,500 kc/sec (R)	3,400–3,500 kc/sec (R)	3,400–3,500 kc/sec (R)
3,800–3,950 kc/sec (OR)	3,500–4,000 kc/sec (OR)	3,500–3,950 kc/sec	
4,438–4,650 kc/sec (OR)	4,438–4,650 kc/sec (OR)		4,438–4,650 kc/sec (OR)
4,650–4,700 kc/sec (OR)	4,650–4,700 kc/sec (OR)	4,650–4,700 kc/sec (R)	4,650–4,700 kc/sec (R)
4,700–4,850 kc/sec (OR)	4,700–4,750 kc/sec (OR)	4,700–4,750 kc/sec (OR)	4,700–4,750 kc/sec (OR)
5,430–5,480 kc/sec (OR)	5,450–5,680 kc/sec (R)	5,430–5,480 kc/sec (OR)	5,450–5,680 kc/sec (R)
5,480–5,680 kc/sec (R)		5,480–5,680 kc/sec (R)	
5,680–5,730 kc/sec (OR)	5,680–5,730 kc/sec (OR)	5,680–5,730 kc/sec (OR)	5,680–5,730 kc/sec (OR)
6,525–6,685 kc/sec (R)	6,525–6,685 kc/sec (R)	6,525–6,685 kc/sec (R)	6,525–6,685 kc/sec (R)
6,685–6,765 kc/sec (OR)	6,685–6,765 kc/sec (OR)	6,685–6,765 kc/sec (OR)	6,685–6,765 kc/sec (OR)
8,815–8,965 kc/sec (R)	8,815–8,965 kc/sec (R)	8,815–8,965 kc/sec (R)	8,815–8,965 kc/sec (R)
8,965–9,040 kc/sec (OR)	8,965–9,040 kc/sec (OR)	8,965–9,040 kc/sec (OR)	8,965–9,040 kc/sec (OR)
10,005–10,100 kc/sec (R)	10,005–10,100 kc/sec (R)	10,005–10,100 kc/sec (R)	10,005–10,100 kc/sec (R)
11,175–11,275 kc/sec (OR)	11,175–11,275 kc/sec (OR)	11,175–11,275 kc/sec (OR)	11,175–11,275 kc/sec (OR)
11,275–11,400 kc/sec (R)	11,275–11,400 kc/sec (R)	11,275–11,400 kc/sec (R)	11,275–11,400 kc/sec (R)
13,200–13,260 kc/sec (OR)	13,200–13,260 kc/sec (OR)	13,200–13,260 kc/sec (OR)	13,200–13,260 kc/sec (OR)
13,260–13,360 kc/sec (OR)	13,260–13,360 kc/sec (R)	13,260–13,360 kc/sec (R)	13,260–13,360 kc/sec (R)
15,010–15,100 kc/sec (OR)	15,010–15,100 kc/sec (OR)	15,010–15,100 kc/sec (OR)	15,010–15,100 kc/sec (OR)
17,900–17,970 kc/sec (R)	17,900–17,970 kc/sec (R)	17,900–17,970 kc/sec (R)	17,900–17,970 kc/sec (R)
17,970–18,030 kc/sec (OR)	17,970–18,030 kc/sec (OR)	17,970–18,030 kc/sec (OR)	17,970–18,030 kc/sec (OR)
21,850–22,000 kc/sec § (R)	21,850–22,000 kc/sec § (R)	21,850–22,000 kc/sec § (R)	21,850–22,000 kc/sec § (R)
23,200–23,350 kc/sec § (OR)	23,200–23,350 kc/sec § (OR)	23,200–23,350 kc/sec § (OR)	23,200–23,350 kc/sec § (OR)
	27,500–28,000 kc/sec	27,500–28,000 kc/sec	27,230–27,280 kc/sec
			27,540–28,000 kc/sec
29.7–41 Mc/sec	29.7–50 Mc/sec	29.7–50 Mc/sec	29.89–29.91 Mc/sec
			30–30.56 Mc/sec
			32–33 Mc/sec
			34–35 Mc/sec
			36–37 Mc/sec
			38–39 Mc/sec
			40–40.66 Mc/sec
			40.7–42 Mc/sec
			46.6–47 Mc/sec
			49.6–50 Mc/sec
	54–74.6 Mc/sec	54–74.6 Mc/sec	
	75.4–88 Mc/sec	75.4–100 Mc/sec	

* For sake of brevity, exclusive provisions for standard frequency transmissions at 2,500 kc/sec are not reflected in the table.

† This column reflects the combined government and nongovernment allocations. In some bands these are not identical. For information as to bands restricted to government use, consult Part 2 of the Rules and Regulations of the Federal Communications Commission.

‡ The letter "R" following a band of frequencies indicates that the aeronautical mobile use of this band is restricted to "Route" service, i.e., for aircraft flying scheduled routes. The letters "OR" following a band of frequencies indicates that the aeronautical mobile use of this band is restricted to "Off Route" service, i.e., for aircraft flying nonscheduled routes.

§ This band is also allocated to the aeronautical fixed service.

¶ Restricted to telemetering and flight tests.

Table 5. Aeronautical Frequencies (*Continued*)

Region 1*,‡	Region 2*,‡	Region 3*,‡	United States*,†
100–108 Mc/sec (OR)			
117.975–136 Mc/sec (R)	117.975–132 Mc/sec (R)	117.975–132 Mc/sec (R)	117.975–121.975 Mc/sec (R)
			121.975–128.825 Mc/sec
136–137 Mc/sec	132–144 Mc/sec	132–144 Mc/sec	128.825–136 Mc/sec (R)
137–144 Mc/sec (OR)			137–144 Mc/sec
146–156 Mc/sec (OR)	148–220 Mc/sec	148–216 Mc/sec	148–150.8 Mc/sec
			162–174 Mc/sec
235–328.6 Mc/sec	225–328.6 Mc/sec	225–328.6 Mc/sec	225–328.6 Mc/sec
335.4–400 Mc/sec	335.4–400 Mc/sec	335.4–400 Mc/sec	335.4–400 Mc/sec
450–470 Mc/sec	450–470 Mc/sec	450–470 Mc/sec	
		610–960 Mc/s	
1,350–1,400 Mc/sec	1,429–1,535 Mc/sec	1,429–1,535 Mc/sec	1,429–1,435 Mc/sec
			1,435–1,535 Mc/sec ¶
	1,700–2,300 Mc/sec	1,700–2,300 Mc/sec	1,710–1,850 Mc/sec
			1,990–2,110 Mc/sec
			2,220–2,290 Mc/sec
2,450–2,700 Mc/sec	2,450–2,700 Mc/sec	2,450–2,700 Mc/sec	2,450–2,500 Mc/sec
3,400–3,600 Mc/sec	3,500–4,200 Mc/sec	3,700–4,200 Mc/sec	
4,400–5,000 Mc/sec	4,400–5,000 Mc/sec	4,400–5,000 Mc/sec	4,400–4,990 Mc/sec
5,850–8,500 Mc/sec	5,925–8,500 Mc/sec	5,850–8,500 Mc/sec	6,425–6,575 Mc/sec
			6,875–8,500 Mc/sec
10.5–11.7 Gc/sec	10.55–11.7 Gc/sec	10.55–11.7 Gc/sec	10.55–10.68 Gc/sec
12.7–13.25 Gc/sec	12.7–13.25 Gc/sec	12.7–13.25 Gc/sec	13.2–13.25 Gc/sec
14.4–15.15 Gc/sec	14.4–15.15 Gc/sec	14.4–15.15 Gc/sec	14.4–15.15 Gc/sec
15.25–15.4	15.25–15.4 Gc/sec	15.25–15.4 Gc/sec	15.25–15.35 Gc/sec
17.7–21 Gc/sec	17.7–21 Gc/sec	17.7–21 Gc/sec	17.7–19.3 Gc/sec
			19.4–21 Gc/sec
22–23 Gc/sec	22–23 Gc/sec	22–23 Gc/sec	22–23 Gc/sec
25.25–31.5 Gc/sec	25.25–31.5 Gc/sec	25.25–31.5 Gc/sec	25.25–31.3 Gc/sec
36–40 Gc/sec	36–40 Gc/sec	36–40 Gc/sec	36–40 Gc/sec

Very limited use of the band 29.7 to 50 Mc/sec is made for short-distance communication links, and the band 162 to 174 Mc/sec is normally used for this purpose.

The world-wide common band for air-ground communications is the 118- to 136-Mc/sec band. Frequencies in this band are used in lieu of the HF bands previously mentioned wherever line-of-sight or slightly beyond horizon service will suffice. Other bands in this portion of the spectrum, primarily the 225- to 400-Mc/sec band, are also used for some off-route air-ground communications.

Frequencies above 1,000 Mc/sec are not at present normally used for air-ground communications, but experimental work in this portion of the spectrum is being conducted, particularly for data communications and supersonic aircraft. Wideband microwave link systems in the aeronautical fixed service operate in the 7,125- to 8,400-Mc/sec band.

Radionavigational aids, in general, are allocated the bands of frequencies shown in Table 6.

The frequency bands 10 to 14 and 90 to 110 kc/sec are used for long-range radio aids to navigation. In the frequency range 70 to 130 kc/sec, the Decca system is used to satisfy medium- and short-range aids to navigation and Loran C to satisfy long-range requirements.

The band 200 to 415 kc/sec is used for aeronautical aid systems, such as four-course ranges, nondirectional beacons, low-power voice facilities at range locations and towers, compass locators, and Consolan and Consol navigation systems. In the United States, the Federal Aviation Agency is responsible for the engineering of frequencies for all users in this band. That portion of this band between 285 and 325 kc/sec is used for marine radio beacons. The frequency 410 kc/sec is used for radio direction finding.

The band 1,800 to 2,000 kc/sec is used throughout the world by high-power Loran A systems for radio-navigation serving both maritime and aeronautical users.

The 74.6- to 75.4-, 108- to 118-, 328.6- to 335.4-, and 960- to 1,215-Mc/sec bands accommodate a group of radio aids to air navigation generally referred to as the world-wide common system of air navigation. These include marker beacons, localizers, omnidirectional ranges, glide slope, and distance-measuring devices.

Table 6. Radionavigational Aid Frequencies

Region 1*	Region 2*	Region 3*	United States*,†
10–14 kc/sec	10–14 kc/sec	10–14 kc/sec	10–14 kc/sec
70–130 kc/sec	70–90 kc/sec M	70–130 kc/sec	90–110 kc/sec
	90–110 kc/sec		
	110–130 kc/sec M		
255–285 kc/sec A	200–285 kc/sec A	200–285 kc/sec A	200–285 kc/sec A
285–315 kc/sec M	285–325 kc/sec M	285–325 kc/sec M	285–325 kc/sec M
315–405 kc/sec A	325–405 kc/sec A	325–405 kc/sec A	325–405 kc/sec A
405–415 kc/sec	405–415 kc/sec M	405–415 kc/sec	405–415 kc/sec A
	1,605–1,800 kc/sec A		1,605–1,750 kc/sec A
	1,800–2,000 kc/sec	1,800–2,000 kc/sec	1,800–2,000 kc/sec
2,625–2650 kc/sec M			
		68–70 Mc/sec A	
74.8–75.2 Mc/sec A	74.6–75.4 Mc/sec A	74.6–75.4 Mc/sec A	7.46–75.4 Mc/sec A
		78–80 Mc/sec A	
108–117.975 Mc/sec A	108–117.975 Mc/sec A	108–117.975 Mc/sec A	108–117.975 Mc/sec A
216–235 Mc/sec A		216–235 Mc/sec A	
328.6–335.4 Mc/sec A	328.6–335.4 Mc/sec A	328.6–335.4 Mc/sec A	328.6–335.4 Mc/sec A
582–606 Mc/sec		585–610 Mc/sec	
960–1,215 Mc/sec A	960–1,215 Mc/sec A	960–1,215 Mc/sec A	960–1,215 Mc/sec A
1,300–1,350 Mc/sec A	1,300–1,350 Mc/sec A	1,300–1,350 Mc/sec A	1,300–1,350 Mc/sec A
1,535–1,660 Mc/sec A	1,535–1,660 Mc/sec A	1,535–1,660 Mc/sec A	1,535–1,660 Mc/sec A
2,700–2,900 Mc/sec A	2,700–2,900 Mc/sec A	2,700–2,900 Mc/sec A	2,700–2,900 Mc/sec A
2,900–3,100 Mc/sec	2,900–3,100 Mc/sec	2,900–3,100 Mc/sec	2,900–3,100 Mc/sec M
4,200–4,400 Mc/sec A	4,200–4,400 Mc/sec A	4,200–4,400 Mc/sec A	4,200–4,400 Mc/sec A
5,000–5,250 Mc/sec A	5,000–5,250 Mc/sec A	5,000–5,250 Mc/sec A	5,000–5,250 Mc/sec A
5,350–5,460 Mc/sec A	5,350–5,460 Mc/sec A	5,350–5,460 Mc/sec A	5,350–5,460 Mc/sec A
5,460–5,470 Mc/sec	5,460–5,470 Mc/sec	5,460–5,470 Mc/sec	5,460–5,470 Mc/sec
5,470–5,650 Mc/sec M	5,470–5,650 Mc/sec M	5,470–5,650 Mc/sec M	5,470–5,650 Mc/sec M
9,000–9,200 Mc/sec A	9,000–9,200 Mc/sec A	9,000–9,200 Mc/sec A	9,000–9,200 Mc/sec A
9,300–9,500 Mc/sec	9,300–9,500 Mc/sec	9,300–9,500 Mc/sec	9,300–9,500 Mc/sec
13.25–13.4 Gc/sec A	13.25–13.4 Gc/sec A	13.25–13.4 Gc/sec A	13.25–13.4 Gc/sec A
14–14.4 Gc/sec	14–14.4 Gc/sec	14–14.4 Gc/sec	14–14.4 Gc/sec
15.4–15.7 Gc/sec A	15.4–15.7 Gc/sec A	15.4–15.7 Gc/sec A	15.4–15.7 Gc/sec A
24.25–25.25 Gc/sec	24.25–25.25 Gc/sec	24.25–25.25 Gc/sec	24.25–25.25 Gc/sec
31.8–33.4 Gc/sec	31.8–33.4 Gc/sec	31.8–33.4 Gc/sec	31.8–33.4 Gc/sec

* The letter "A" following an allocated band denotes that this band of frequencies is restricted to aeronautical navigation aids. The letter "M" following an allocated band denotes that this band is restricted to maritime navigational aids.
† This column reflects the combined government and nongovernment allocations. In some bands these are not identical. For information as to bands restricted to government use, consult Part 2 of the Rules and Regulations of the Federal Communications Commission.

Surveillance radar for air traffic control purposes is operated principally in the bands 1,300 to 1,350 and 2,700 to 2,900 Mc/sec. The bands 1,600 to 1,660 and 4,200 to 4,400 Mc/sec are used for radio altimeters. The bands 5,350 to 5,470 and 9,300 to 9,500 Mc/sec are used for airborne weather radar, and the bands 8,750 to 8,850 and 13,250 to 13,400 Mc/sec are utilized for airborne doppler radars. Some bands allocated for air navigation such as 5,000 to 5,250 and 15,400 to 15,700 Mc/sec are not yet in general use pending the development of suitable equipments for their use.

Table 7. Maritime Mobile Frequencies

Region 1*	Region 2*	Region 3*	United States*·†
14–70 kc/sec 72–84 kc/sec 86–112 kc/sec 115–126 kc/sec 129–160 kc/sec	14–90 kc/sec 110–160 kc/sec	14–160 kc/sec	14–90 kc/sec 110–160 kc/sec
255–285 kc/sec 405–525 kc/sec 1,605–2,850 kc/sec	415–535 kc/sec 1,605–2,850 kc/sec	415–535 kc/sec 1,605–2,850 kc/sec	415–535 kc/sec 1,605–1,750 kc/sec 2,000–2,850 kc/sec
3,155–3,400 kc/sec	3,155–3,400 kc/sec	3,155–3,400 kc/sec	3,155–3,400 kc/sec
3,500–3,800 kc/sec 4,063–4,650 kc/sec 6,200–6,525 kc/sec 8,195–8,815 kc/sec 12,330–13,200 kc/sec	3,500–4,000 kc/sec 4,063–4,650 kc/sec 6,200–6,525 kc/sec 8,195–8,815 kc/sec 12,330–13,200 kc/sec	3,500–3,900 kc/sec 4,063–4,650 kc/sec 6,200–6,525 kc/sec 8,195–8,815 kc/sec 12,330–13,200 kc/sec	4,063–4,650 kc/sec 6,200–6,525 kc/sec 8,195–8,815 kc/sec 12,330–13,200 kc/sec
16,460–17,360 kc/sec 22,000–22,720 kc/sec 25,010–25,600 kc/sec 26,100–27,500 kc/sec	16,460–17,360 kc/sec 22,000–22,720 kc/sec 25,010–25,600 kc/sec 26,100–28,000 kc/sec	16,460–17,360 kc/sec 22,000–22,720 kc/sec 25,010–25,600 kc/sec 26,100–28,000 kc/sec	16,460–17,360 kc/sec 22,000–22,720 kc/sec 25,330–25,600 kc/sec 26,480–26,950 kc/sec 27,230–27,280 kc/sec
			27,540–28,000 kc/sec
29.7–41 Mc/sec	29.7–50 Mc/sec	29.7–50 Mc/sec	29.80–29.91 Mc/sec 30–30.56 Mc/sec 32–33 Mc/sec 34–35 Mc/sec
			36–37 Mc/sec 38–39 Mc/sec 40–40.66 Mc/sec 40.7–42 Mc/sec
68–74.8 Mc/sec	54–74.6 Mc/sec	54–74.6 Mc/sec	
75.2–87.5 Mc/sec 100–108 Mc/sec 136–137 Mc/sec 146–174 Mc/sec	75.4–88 Mc/sec 132–144 Mc/sec 148–220 Mc/sec	75.4–100 Mc/sec 132–144 Mc/sec 148–216 Mc/sec	137–144 Mc/sec 148–150.8 Mc/sec 156.25–157.45 Mc/sec
			161.575–161.625 Mc/sec 161.775–174 Mc/sec
235–328.6 Mc/sec 335.4–400 Mc/sec 406–430 Mc/sec	225–328.6 Mc/sec 335.4–400 Mc/sec 406–420 Mc/sec	225–328.6 Mc/sec 335.4–400 Mc/sec 406–420 Mc/sec	225–328.6 Mc/sec 335.4–400 Mc/sec 406–420 Mc/sec
440–470 Mc/sec	450–470 Mc/sec	450–470 Mc/sec 610–960 Mc/sec	462.525–463.225 Mc/sec 464.725–466.475 Mc/sec
1,350–1,400 Mc/sec 1,427–1,535 Mc/sec 1,660–1,700 Mc/sec	1,427–1,535 Mc/sec 1,660–2,300 Mc/sec	1,427–1,535 Mc/sec 1,660–2,300 Mc/sec	1,427–1,435 Mc/sec 1,660–1,670 Mc/sec
			1,710–1,850 Mc/sec 2,200–2,290 Mc/sec
2,450–2,700 Mc/sec 3,400–3,600 Mc/sec 4,400–5,000 Mc/sec	2,450–2,700 Mc/sec 3,500–4,200 Mc/sec 4,400–5,000 Mc/sec	2,450–2,700 Mc/sec 3,700–4,200 Mc/sec 4,400–5,000 Mc/sec	2,450–2,500 Mc/sec 4,400–4,990 Mc/sec
5,850–8,500 Mc/sec	5,925–8,500 Mc/sec	5,850–8,500 Mc/sec	6,425–6,575 Mc/sec 7,125–8,500 Mc/sec
10.5–13.25 Gc/sec	10.55–13.25 Gc/sec	10.55–13.25 Gc/sec	10.55–10.68 Gc/sec 11.7–12.2 Gc/sec 13.2–13.25 Gc/sec
14.4–15.15 Gc/sec 15.25–15.4 Gc/sec 17.7–21 Gc/sec	14.4–15.15 Gc/sec 15.25–15.4 Gc/sec 17.7–21 Gc/sec	14.4–15.15 Gc/sec 15.25–15.4 Gc/sec 17.7–21 Gc/sec	14.4–15.15 Gc/sec 15.25–15.35 Gc/sec 17.7–19.3 Gc/sec 19.4–21 Gc/sec
22–23 Gc/sec	22–23 Gc/sec	22–23 Gc/sec	22–23 Gc/sec
25.25–31.5 Gc/sec 36–40 Gc/sec	25.25–31.5 Gc/sec 36–40 Gc/sec	25.25–31.5 Gc/sec 36–40 Gc/sec	25.25–31.3 Gc/sec 36–40 Gc/sec

* For sake of brevity, exclusive provisions for standard frequency transmissions at 20 and 2,500 kc/sec are not reflected in the table.

† This column reflects the combined government and nongovernment allocations. In some bands these are not identical. For information as to bands restricted to government use, consult part 2 of the Rules and Regulations of the Federal Communications Commission.

The bands 2,900 to 3,100, 5,470 to 5,650, and 9,300 to 9,500 Mc/sec of Table 6 are the bands in which all merchant ship radars and racon devices operate.

The maritime mobile services are, in general, allocated the bands of frequencies shown in Table 7.

The band 110 to 160 kc/sec is used primarily for broadcasts from coast stations to ships giving press transmissions, traffic lists, weather broadcasts, ice warnings, and other general marine information.

The band 415 to 535 kc/sec is the backbone of telegraph maritime mobile communications on the high seas for both the coast stations and ships. Ships operate on a selected frequency from among 425, 448, 454, 468, 480, or 512 kc/sec after calling on 500 kc/sec. The coast stations are usually assigned one working frequency in this band and one or more working frequencies in each of the frequency bands in the 2- to 22-Mc/sec range. The frequency 500 kc/sec is universally recognized as the distress and calling frequency for telegraphy.

The band 1,605 to 2,850 kc/sec is useful in some parts of the world for radio telephone communications between coast stations and relatively small ships, such as fishing fleets operating in the Alaskan or North Sea waters.

Table 8. Maritime Ship-Coast Frequencies

Calling band, telegraph, kc/sec	Ship station		Coast station	
	Telephone, kc/sec	Telegraph, kc/sec	Telephone, kc/sec	Telegraph, kc/sec
4,177–4,187	4,063–4,140	4,140–4,238	4,368–4,438	4,238–4,368
6,265.5–6,280.5	6,200–6,211	6,211–6,357		6,357–6,525
8,354–8,374	8,195–8,280	8,280–8,476	8,745–8,815	8,476–8,745
12,531–12,561	12,330–12,421	12,421–12,714	13,130–13,200	12,714–13,130
16,708–16,748	16,460–16,562	16,562–16,952	17,290–17,360	16,952–17,290
22,220–22,270	22,000–22,100	22,100–22,400	22,650–22,720	22,400–22,650
		25,070–25,110		

The band 2,000 to 2,035 kc/sec is used in the radio telephone service for both coast and ship stations, the band 2,035 to 2,065 kc/sec for coast stations in the radio telegraph service, and the band 2,065 to 2,107 kc/sec for ship stations in the radio telegraph service. The maritime mobile use of the band 2,107 to 2,495 kc/sec is for both coast and ship stations primarily in the radio telephone service. The frequency 2,182 kc/sec with a guard band of 12 kc/sec is the distress and calling frequency for the maritime mobile radio telephone service. The frequencies 2,638 and 2,738 kc/sec are intership frequencies for radio telephone communications. The band 3,155 to 3,400 kc/sec is used for both ship and coast stations and for both radio telephony and radio telegraphy.

The use of the high-frequency portion of the spectrum (4 to 27.5 Mc/sec) in the maritime mobile service is based upon a world-wide plan for coast stations developed some years ago. This plan continues to be followed with minor changes from time to time. The coast telegraph plan is now incorporated in the international frequency list and no longer is available as a separate document. The coast telephone portion of the plan can be obtained from the International Telecommunications Union, Geneva, Switzerland, at a cost of 17.25 Swiss francs. The maritime mobile use in this portion of the spectrum is subdivided as shown in Table 8.

The bands 156.3 to 157.4 and 161.6 to 162 Mc/sec are used for shore-ship-shore and intership telephony for communications over line-of-sight distances or slightly beyond.

Table 9. Land Mobile Frequencies

Region 1*	Region 2*	Region 3*	United States*,†
1,605–2,170 kc/sec	1,605–2,065 kc/sec	1,605–2,065 kc/sec	1,605–1,750 kc/sec
			2,000–2,065 kc/sec
	2,107–2,170 kc/sec	2,107–2,170 kc/sec	2,107–2,170 kc/sec
2,194–2,625 kc/sec	2,194–2,850 kc/sec	2,194–2,850 kc/sec	2,194–2,850 kc/sec
2,650–2,850 kc/sec			
3,155–3,400 kc/sec	3,155–3,400 kc/sec	3,155–3,400 kc/sec	3,155–3,400 kc/sec
3,500–3,900 kc/sec	3,500–4,000 kc/sec	3,500–3,900 kc/sec	
4,438–4,650 kc/sec	4,438–4,650 kc/sec	4,438–4,650 kc/sec	
4,750–4,995 kc/sec	4,850–4,995 kc/sec	4,850–4,995 kc/sec	4,850–4,995 kc/sec
5,250–5,480 kc/sec	5,250–5,450 kc/sec	5,250–5,480 kc/sec	5,250–5,450 kc/sec
23,350–25,070 kc/sec	23,350–25,070 kc/sec	23,350–25,070 kc/sec	23,350–25,070 kc/sec
25,110–25,600 kc/sec	25,110–25,600 kc/sec	25,110–25,600 kc/sec	25,110–25,600 kc/sec
26,100–27,500 kc/sec	26,100–28,000 kc/sec	26,100–28,000 kc/sec	26,100–26,950 kc/sec
			26,960–28,000 kc/sec
29.7–41 Mc/sec	29.7–50 Mc/sec	29.7–50 Mc/sec	29.7–29.8 Mc/sec
			29.89–29.91 Mc/sec
			30–40.66 Mc/sec
			40.7–50 Mc/sec
68–74.8 Mc/sec	54–74.6 Mc/sec	54–74.6 Mc/sec	
75.2–87.5 Mc/sec	75.4–88 Mc/sec	75.4–100 Mc/sec	
100–108 Mc/sec			
136–137 Mc/sec	132–144 Mc/sec	132–144 Mc/sec	137–144 Mc/sec
146–174 Mc/sec	148–220 Mc/sec	148–216 Mc/sec	148–156.25 Mc/sec
			157.45–161.575 Mc/sec
			161.625–161.775 Mc/sec
			162–174 Mc/sec
235–328.6 Mc/sec	225–328.6 Mc/sec	225–328.6 Mc/sec	225–328.6 Mc/sec
335.4–400 Mc/sec	335.4–400 Mc/sec	335.4–400 Mc/sec	335.4–400 Mc/sec
406–430 Mc/sec	406–420 Mc/sec	406–420 Mc/sec	406–420 Mc/sec
440–470 Mc/sec	450–470 Mc/sec	450–470 Mc/sec	450–470 Mc/sec
		610–960 Mc/sec	
1,350–1,400 Mc/sec			
1,427–1,535 Mc/sec	1,427–1,535 Mc/sec	1,427–1,535 Mc/sec	1,427–1,435 Mc/sec
1,660–1,700 Mc/sec	1,660–2,300 Mc/sec	1,660–2,300 Mc/sec	1,660–1,670 Mc/sec
			1,710–1,850 Mc/sec
			1,990–2,110 Mc/sec
			2,200–2,290 Mc/sec
2,450–2,700 Mc/sec	2,450–2,700 Mc/sec	2,450–2,700 Mc/sec	2,450–2,500 Mc/sec
3,400–3,600 Mc/sec	3,500–4,200 Mc/sec	3,700–4,200 Mc/sec	
4,400–5,000 Mc/sec	4,400–5,000 Mc/sec	4,400–5,000 Mc/sec	4,400–4,990 Mc/sec
5,850–8,500 Mc/sec	5,925–8,500 Mc/sec	5,850–8,500 Mc/sec	6,425–6,575 Mc/sec
			6,875–8,500 Mc/sec
10.5–13.25 Gc/sec	10.55–13.25 Gc/sec	10.55–13.25 Gc/sec	10.5–10.68 Gc/sec
			11.7–12.2 Gc/sec
			12.7–13.25 Gc/sec
14.4–15.15 Gc/sec	14.4–15.15 Gc/sec	14.4–15.15 Gc/sec	14.4–15.15 Gc/sec
15.25–15.4 Gc/sec	15.25–15.4 Gc/sec	15.25–15.4 Gc/sec	15.25–15.35 Gc/sec
17.7–21 Gc/sec	17.7–21 Gc/sec	17.7–21 Gc/sec	17.7–19.3 Gc/sec
			19.4–21 Gc/sec
22–23 Gc/sec	22–23 Gc/sec	22–23 Gc/sec	22–23 Gc/sec
25.25–31.5 Gc/sec	25.25–31.5 Gc/sec	25.25–31.5 Gc/sec	25.25–31.3 Gc/sec
36–40 Gc/sec	36–40 Gc/sec	36–40 Gc/sec	36–40 Gc/sec

* For sake of brevity, provisions for standard frequency transmissions at 2,500 and 25,000 kc/sec are not reflected in the table.

† This column reflects the combined government and nongovernment allocations. In some bands these are not identical. For information as to bands restricted to government use, consult Part 2 of the Rules and Regulations of the Federal Communications Commission.

6. Land Mobile Services

The land mobile services, in general, are allocated the bands shown in Table 9.

In order to secure maximum efficiency, the allocations are frequently subdivided, providing individual blocks of frequencies for each particular service in the land mobile category. In the United States there are the suballocations shown in Table 10.

Because of the congestion in certain areas in many of the bands of Table 10, most of the services have established an industry coordinating or advisory committee on either a local area or a national basis. It might be well, although not compulsory, for a prospective user to consult the service coordinating committee in his area before attempting to pick the exact frequency for his operation. In case he does not know how to get in touch with this coordinating committee, the manufacturers of his equipment can usually advise him or he can obtain this information from the Federal Communications Commission.

Frequency coordination requirements for mobile service frequencies, with a few exceptions, are similar in all the land mobile radio services. In those services requiring the submission of evidence of frequency coordination, two methods of coordination are offered applicants.

One method provides that the applicant may choose his own frequency and accompany his application with evidence, based on a field or engineering survey, that the selected frequency is the best choice and will cause the least interference to other stations in the area. In addition, a statement must be made that all existing licensees located within a radius of 75 miles of the proposed station and operating on frequencies

Table 10. Frequencies for Specific Services

Service	Frequency
Industrial:	
General............................	1,605–1,750 kc/sec
	2,107–2,170 kc/sec
	2,194–2,850 kc/sec
	3,155–3,400 kc/sec
	4,438–4,650 kc/sec
	27.23–27.28 Mc/sec
	460.5–460.95 Mc/sec
Business.........................	27.39–27.53 Mc/sec
	33.14 Mc/sec
	35.02–35.18 Mc/sec
	35.70–35.98 Mc/sec
	42.96–43 Mc/sec
	151.625–151.955 Mc/sec
	152.3–152.42 Mc/sec
	154.54–154.6 Mc/sec
	157.56–157.68 Mc/sec
	461.05–462 Mc/sec
	463.25–464.7 Mc/sec
	466.5–467.45 Mc/sec
	468–469.95 Mc/sec
Forest products...................	29.73–29.77 Mc/sec
	48.56–49.58 Mc/sec
	153.05–153.68 Mc/sec
	154.57 Mc/sec
	158.16–158.43 Mc/sec
	451.55–451.75 Mc/sec
	456.55–456.75 Mc/sec
Manufacturers....................	153.05–153.38 Mc/sec
	158.28–158.43 Mc/sec
	462.05–462.50 Mc/sec
	467.50–467.95 Mc/sec
Motion pictures..................	152.87–153.02 Mc/sec
	173.225–173.375 Mc/sec

Table 10. Frequencies for Specific Services (*Continued*)

Service	Frequency
Petroleum	25.02–25.3 Mc/sec
	30.66–30.82 Mc/sec
	38.18–33.38 Mc/sec
	48.56–49.50 Mc/sec
	153.05–153.68 Mc/sec
	158.16–158.43 Mc/sec
	173.25–173.35 Mc/sec
	451.55–451.75 Mc/sec
	456.55–456.75 Mc/sec
Power	37.46–37.86 Mc/sec
	47.70–48.54 Mc/sec
	153.41–153.71 Mc/sec
	158.13–158.25 Mc/sec
	173.25–173.35 Mc/sec
	451.05–451.25 Mc/sec
	456.05–456.25 Mc/sec
Relay press	173.225–173.375 Mc/sec
Special industrial	27.29–27.37 Mc/sec
	30.58 Mc/sec
	30.62 Mc/sec
	35.74–35.86 Mc/sec
	43.02–43.18 Mc/sec
	47.44–47.68 Mc/sec
	49.52–49.58 Mc/sec
	151.505–151.595 Mc/sec
	152.87–153.02 Mc/sec
	154.49 Mc/sec
	158.4 Mc/sec
	451.8–451.95 Mc/sec
	456.8–456.95 Mc/sec
Telephone maintenance	35.16 Mc/sec
	43.16 Mc/sec
	151.985 Mc/sec
	158.34 Mc/sec
	451.3–451.5 Mc/sec
	456.3–456.5 Mc/sec
Land transportation:	
General	27.23–27.28 Mc/sec
Auto emergency	150.815–150.875 Mc/sec
	150.905–150.965 Mc/sec
	157.45–157.5 Mc/sec
	452.55 Mc/sec
	452.60 Mc/sec
	457.55 Mc/sec
	457.60 Mc/sec
Motor carrier	43.70–44.60 Mc/sec
	159.495–160.2 Mc/sec
	452.65–452.85 Mc/sec
	457.65–457.85 Mc/sec
Railroad	160.215–161.565 Mc/sec
Taxicab	152.27–152.45 Mc/sec
	157.53–157.71 Mc/sec
	452.05–452.5 Mc/sec
	457.05–457.5 Mc/sec
Public safety:	
General	1,605–1,740 kc/sec
	2,107–2,170 kc/sec
	2,194–2,850 kc/sec
	3,155–3,400 kc/sec
	27.235–27.275 Mc/sec
	46.51–46.60 Mc/sec

Table 10. Frequencies for Specific Services (*Continued*)

Service	*Frequency*
Fire	33.42–33.98 Mc/sec
	45.88 Mc/sec
	46.06–46.5 Mc/sec
	153.77–154.445 Mc/sec
	166.25 Mc/sec
	170.15 Mc/sec
	453.05–453.95 Mc/sec
	458.05–458.95 Mc/sec
Forestry conservation	30.86–31.98 Mc/sec
	44.64–45.04 Mc/sec
	46.54–46.82 Mc/sec
	151.145–151.475 Mc/sec
	159.225–159.465 Mc/sec
	453.05–453.95 Mc/sec
	458.05–458.95 Mc/sec
Highway maintenance	33.02–33.10 Mc/sec
	37.90–37.98 Mc/sec
	45.68–45.84 Mc/sec
	46.86–47.40 Mc/sec
	150.995–151.130 Mc/sec
	156.045–156.24 Mc/sec
	157.05 Mc/sec
	157.11 Mc/sec
	158.985–159.195 Mc/sec
	453.05–453.95 Mc/sec
	458.05–458.95 Mc/sec
Local government	39.06 Mc/sec
	45.08–45.64 Mc/sec
	46.52–46.58 Mc/sec
	153.755–154.115 Mc/sec
	154.965–155.145 Mc/sec
	155.715–156.015 Mc/sec
	158.745–158.955 Mc/sec
	453.05–453.95 Mc/sec
	458.05–458.95 Mc/sec
Police	37.02–37.42 Mc/sec
	39.02–39.98 Mc/sec
	42.02–42.94 Mc/sec
	44.62–46.02 Mc/sec
	154.65–156.21 Mc/sec
	158.73–159.21 Mc/sec
	453.05–453.95 Mc/sec
	458.05–458.95 Mc/sec
Special emergency	33.02–33.10 Mc/sec
	37.90–37.98 Mc/sec
	45.92–46.04 Mc/sec
	47.42–47.66 Mc/sec
	155.16–155.4 Mc/sec
Domestic public:	
General	35.22–35.66 Mc/sec
	43.22–43.66 Mc/sec
	152.03–152.21 Mc/sec
	152.51–152.81 Mc/sec
	157.77–158.07 Mc/sec
	158.49–158.67 Mc/sec
	454.05–454.95 Mc/sec
	459.05–459.95 Mc/sec
Citizens' radio:	
General	26.96–27.23 Mc/sec
	27.255 Mc/sec
	462.55–463.20 Mc/sec
	464.70–466.45 Mc/sec

within 15 kc/sec of the requested frequency have been notified of his intention to request assignment of the particular frequency.

The second method provides that the applicant, if he so desires, may request a frequency recommendation from the frequency advisory committee for his radio service. These committees are voluntary and must be so organized as to be representative of all persons eligible for radio facilities in the radio service concerned in the area the committee purports to serve. Their recommendations are purely advisory and are not binding upon either the applicant or the Commission.

There are some radio services, however, which do not require the submission of any evidence of frequency coordination. These services include the Special Emergency Radio Service under Part 10 of the Rules, the Business Radio Service under Part 11 of the Rules and, the Citizens Radio Service under Part 19. In these services the applicant must choose his own frequency. In general, frequencies in these services are chosen by the equipment supplier, his choice being based on any available records or his own knowledge of frequency usage in the area.

Up to the present time there has been only experimental use in the mobile service for the frequencies above 10 Gc/sec.

7. Broadcast Services

The broadcast services, in general, are allocated the frequency bands shown in Table 11.

The frequencies from 150 to 1,605 kc/sec are used for AM broadcasting. Those between 2,300 and 5,060 kc/sec (with the exception of the band 3,900 to 4,000 kc/sec) are limited in the broadcasting service to AM national broadcasting in tropical zones. Tropical zones are roughly defined as those areas lying between the Tropic of Cancer

Table 11. Broadcast Service Frequencies

Region 1*	Region 2*	Region 3*	United States
150–285 kc/sec			
525–1,605 kc/sec	535–1,605 kc/sec	535–1,605 kc/sec	535–1,605 kc/sec
2,300–2,498 kc/sec	2,300–2,495 kc/sec	2,300–2,495 kc/sec	
3,200–3,400 kc/sec	3,200–3,400 kc/sec	3,200–3,400 kc/sec	
3,950–4,000 kc/sec		3,900–4,000 kc/sec	
4,750–5,060 kc/sec	4,750–5,060 kc/sec	4,750–5,060 kc/sec	
5,950–6,200 kc/sec	5,950–6,200 kc/sec	5,950–6,200 kc/sec	5,950–6,200 kc/sec
7,100–7,300 kc/sec		7,100–7,300 kc/sec	
9,500–9,775 kc/sec	9,500–9,775 kc/sec	9,500–9,775 kc/sec	9,500–9,775 kc/sec
11,700–11,975 kc/sec	11,700–11,975 kc/sec	11,700–11,975 kc/sec	11,700–11,975 kc/sec
15,100–15,450 kc/sec	15,100–15,450 kc/sec	15,100–15,450 kc/sec	15,100–15,450 kc/sec
17,700–17,900 kc/sec	17,700–17,900 kc/sec	17,700–17,900 kc/sec	17,700–17,900 kc/sec
21,450–21,750 kc/sec	21,450–21,750 kc/sec	21,450–21,750 kc/sec	21,450–21,750 kc/sec
25,600–26,100 kc/sec	25,600–26,100 kc/sec	25,600–26,100 kc/sec	25,600–26,100 kc/sec
		44–50 Mc/sec	
41–68 Mc/sec	54–74.8 Mc/sec	54–68 Mc/sec	54–72 Mc/sec
87.5–100 Mc/sec	75.4–108 Mc/sec	87–108 Mc/sec	76–108 Mc/sec
174–223 Mc/sec	174–216 Mc/sec	170–216 Mc/sec	174–216 Mc/sec
470–960 Mc/sec	470–890 Mc/sec	470–585 Mc/sec	470–890 Mc/sec
		610–960 Mc/sec	
11.7–12.7 Gc/sec	11.7–12.7 Gc/sec	11.7–12.7 Gc/sec	

* For the sake of brevity, exclusive provisions for standard frequency transmissions at 5,000 kc/sec are not reflected in the table.

and the Tropic of Capricorn in Region 2 and between 30 degrees North latitude and 35 degrees South latitude in Regions 1 and 3.   (For a more exact delineation of the tropical zones see Appendix 24 of the International Radio Regulations, Geneva, 1959.) In these zones atmospheric noise levels and solar absorption are high and a more satisfactory broadcasting service can be obtained by the use of these frequencies than by using low or medium frequencies.

The frequencies between 5,900 and 26,100 kc/sec plus the band 3,900 to 4,000 kc/sec in Region 3 and 3,950 to 4,000 kc/sec in Region 1 are used for long-distance AM broadcasting.   While the primary use of these frequencies is for international AM broadcasting, they are also used to a considerable extent for national coverage where it is

Table 12. Remote Pickup Frequencies

	Frequency
Remote pickup broadcasts......................	1,605–1,715 kc/sec
	26,110–26,470 kc/sec
	152.87–153.71 Mc/sec
	161.64–161.71 Mc/sec
	166.25–Mc/sec
	170.15–Mc/sec
	450.05–450.95 Mc/sec
	455.05–455.95 Mc/sec
FM and AM broadcast studio transmitter links and television studio transmitter links (audio only)...	942–952 Mc/sec
Television pickup and television studio transmitter links	1,999–2,101.5 Mc/sec
	2,450–2,500 Mc/sec
	6,887.5–7,112.5 Mc/sec
	10,550–10,675 Mc/sec
	12,700–13,250 Mc/sec

economically impracticable to use the more conventional methods of medium-frequency or VHF stations linked together by means of microwave or coaxial-cable systems.

The band from approximately 88 to 100 Mc/sec (88 to 108 Mc/sec in Region 2) is almost universally used for FM broadcasting, with the remainder of the frequencies in the broadcast service between 41 and 960 Mc/sec available for television broadcasting.

Throughout the spectrum, individual frequencies or bands of frequencies are set up for remote pickup broadcasts and station-to-transmitter links.   In the United States, these are shown in Table 12.

Because of the extreme congestion of the majority of the broadcast bands and because of the complicated nature of the installation, in order to avoid interference the prospective user should consult a specialized broadcasting engineer before making a definite selection of frequency.

8. Protection from Interference

If the above regulations are followed, in the majority of cases where frequencies not capable of international interference are employed, little or no interference with other operations will result.   If harmful interference should develop, however, the interfering signal should be identified if possible, and if it cannot be identified, it should be referred to the central control body for identification.   In any case, all pertinent data as to signal strength, times of interference, type of emission, etc., should be given the central control body so that it can resolve the problem.

In cases of international interference the situation is more complex.   If the interfering signal can be identified, its source and characteristics can be referred directly to the agency which operates the interfering station and, in general, through the cooperation of that agency, the interference can be cleared.   Otherwise, the problem, together with such data as to the source and characteristics of the interfering signal as can be obtained, should be submitted to the central control body so that it can verify the facts and make appropriate representations to the administration having jurisdiction over the interfering signal. This will usually take longer than the direct method but frequently will prove more effective. In any case, *prior use* of a frequency is the most important element in determining which of two stations has priority, and the newer service must make such adjustments as are necessary to resolve the interference.

Chapter 20

POWER FOR COMMUNICATION SYSTEMS

JAMES M. DUGUID, *Consulting Engineer**

INTRODUCTION

1. Scope and Functions

This chapter treats of the power necessary for the operation of the communication systems most frequently encountered. Both alternating and direct currents are involved, and although requirements for as much as 1,000 kw are encountered in centralized communication buildings in our largest cities, in general the capacities are small compared with public-service steam-generating stations. Except for special applications such as cable supplies, the voltages are low and generally classified by the Underwriters for signaling.

Since most communication apparatus, i.e., telephone receivers and transmitters, electron tubes, relays, etc., is designed for d-c operation and a-c power is commonly available, rectification is the most important function of these power supplies. As implied, primary power is purchased, and the communication power system rectifies, regulates, filters, and adds reliability as required by the load. After rectification, reliability is the most important function, because most commercial power sources are subject to unpredictable interruptions.

Alternating-current supplies are required in frequencies of the familiar 60 cps and in small quantities both above and below 60. For signaling purposes 20 cps is the common frequency of the current which rings the telephone bell, while tones such as dial and busy tones are in the order of 600 cps modulated by a lower frequency such as 120 cps.

2. Types of Plants

Two main types of power systems are involved; common or centralized and decentralized. After the introduction of the electrically rechargeable storage battery and the common battery switchboard in 1893, the centralized power plant became widely used because rectification and conversion required rotating machinery which could hardly be built into telephone switchboards and racks. Figure 1 shows a block diagram of a centralized power system. With the advent of small "dry" diode rectifiers beginning with copper oxide and continuing with selenium, germanium, and silicon types, as well as dry capacitors for noise filtering, it became economical and feasible in many applications to incorporate these components into the communications equipment and provide a cord or terminals for direct connection to a-c power. This is

* Retired from Bell Telephone Laboratories, Inc.

known as decentralized power in contrast to the older centralized power systems where the a-c purchased power was rectified, regulated, converted, and firmed up* on the wholesale basis for the entire load.   The trend has been to decentralized power, especially for transmission systems requiring high plate voltages for transmitting electron tubes.   While efficiency and economy in rectification may be sacrificed, the confinement of the high voltages to relatively small enclosures is of considerable advantage from the standpoint of safety and ease of distribution.   One pronounced hardship must be faced.   To provide firm power it is necessary to introduce switching in the a-c supply.   As shown in Fig. 2, this has to be carried out without the continuity of power provided by the storage battery shown in Fig. 1.   This disadvantage has continued the use of the centralized d-c power system with its storage battery in parallel

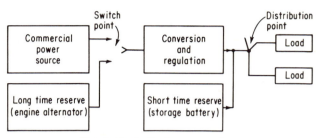

Fig. 1. Centralized power system.

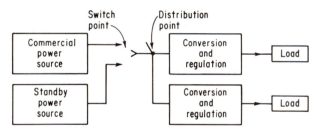

Fig. 2. Decentralized power system.

with the load far beyond predictions of a decade ago.   The introduction of the transistor, which is capable of operation on very low d-c voltages, will also tend to prolong the use of the centralized d-c power plant.

    Combinations of decentralized and centralized power systems have also come into use.   Coaxial-cable systems might be cited as an example.   Here kilovolt a-c power is sent out over the cable conductors and supplies the decentralized rectifiers in the amplifiers on a series basis.   At regular intervals along the cable route, common a-c power plants feeding both forward and rearward repeaters provide voltage conversion, regulation, and standby facilities.

MAJOR POWER COMPONENTS

3. Batteries

    Originally primary cells (nonrechargeable electrically) were used in telephone systems universally for talking current supply and were located at each subscriber's station.   With the advent of common central office batteries in 1893, the primary

---

* Firm power describes the deliberate increase in system reliability including redundance and switch-over characteristics.

cells were replaced with secondary (rechargeable electrically) cells which have been used ever since for energy-storage purposes in general. The primary battery has survived in its modern form, dry cells, principally for a source of potential rather than current. A few modern "wet" primary batteries known as air cells are also in use for pole-mounted power supplies such as line concentrators. They are useful in areas where no power lines are near by or where the drains do not justify a metered connection to the power lines.

Of the secondary batteries, the most important by far is the lead-acid battery, which is universally used in modern telephone plants for energy storage. Its low relative cost per ampere-hour reserve and long record of reliable operation have given it the preeminent position it occupies. Compared with the well-known automotive battery, it is much heavier per ampere-hour, but this is not important in a stationary battery. More to the point, in its particular application of a reserve battery to the telephone plant, it is much longer lived by a factor of 10 or 15 compared with the automobile battery which supplies the actual power and is therefore charged and discharged. The telephone lead-acid cell is designed for what is known as floating duty. This means that the load is carried by the charger or rectifier while the storage battery "floats"

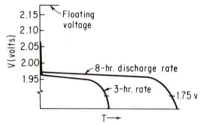

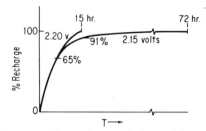

FIG. 3. Constant-current discharge dependence of voltage and capacity on rate of discharge. (*Reproduced by permission of Bell Telephone Laboratories, Inc.*)

FIG. 4. Recharge characteristics at different voltages. (*Reproduced by permission of Bell Telephone Laboratories, Inc.*)

on the line in parallel with the load, being neither charged nor discharged. Actually a very small current is taken by the battery just to make up for its internal losses. For this service an electrolyte of lower specific gravity (compared with the automotive battery) is used. The long life is attained by the quality of the original product, by careful maintenance over the years, and by the floating routine which avoids the destructive tendencies of high charging and deep discharge cycles.

Under the floating routine, the voltage of each lead-acid cell will vary from about 2.2 to 1.75 volts from charge to discharge as shown in Fig. 3. For an individual cell, a floating voltage as low as 2.13 volts would maintain the cell in good condition and retain the full reserve. However, to allow for individual cell variations it has been found necessary to use voltages of 2.15 to 2.17 when the battery consists of several cells in series. This will permit a voltage of at least 2.12 across the lowest cells in the string. These figures indicate the need for a closely regulated charging device. A capability of ±1 per cent regulation with a nominal setting of 2.17 volts results in a −20 per cent load regulation to get out the full reserve in the battery. Higher upper floating voltage limits tend to shorten the battery life. As shown in Fig. 4 after a discharge cycle, the total reserve can be restored in about 15 hr at a charge voltage of 2.20 per cell, so that the expense of providing for charging voltages of 2.5 and 3.0 volts is no longer warranted. Occasional high floating at 2.2 volts per cell, even though no discharge has taken place, will equalize the individual cell voltages without appreciably affecting the long-time life of the battery.

Assuming that the load is designed to operate under a 20 per cent power regulation, a very simple and inexpensive power plant can be used to utilize the complete reserve of the battery. Dial switching equipment, however, generally will tolerate only about

−10 per cent voltage regulation.   This may be met under relatively light load conditions by providing an extra large battery and only partially discharging it.   In most cases load regulation in the form of counter cells (CEMF) or emergency cells will be required.   The former are alkaline cells with nickel electrodes which develop a relatively constant voltage drop when put in series with the load current during normal floating conditions.   When the battery is on discharge, they can be short-circuited without damage.

Emergency cells are ordinary lead-acid cells which are arranged to be automatically switched in series with the load to add to the voltage under discharge conditions.   Generally they have to be introduced in steps to avoid exceeding the allowable upper voltage limit.   The size of the load that can be regulated by the counter cell method is limited by economic considerations.   Eventually the annual power losses through the CEMF cells will justify the use of the more expensive end cell system.

Manually operated telephone switchboards are generally operable over wide voltage limits.   As mentioned previously, the electromechanical dial switching equipments generally require the narrower (−10 per cent) regulation.   It is interesting to note that the electronic switching system designed by the Bell Telephone System will revert to the −20 per cent regulation to achieve a less complicated, cheaper power supply.

The smaller lead-acid cells, 180 to 1,680 amp-hr, are housed in transparent plastic jars so that the condition of the plates and separators can be watched.   The larger sizes, 4,000 to 7,000 amp-hr, are housed in rubber composition tanks.   Both types are enclosed and sealed with a special hydrogen vent to avoid the danger of explosion.   The jar batteries are usually mounted on racks, and the tanks placed directly on the floor.   One of the advantages of the floating routine, which does not vent much gas or acid vapor, is that there is no longer a need to enclose the cells in a separate battery room or treat the bus bars or cables with acid-resisting paint.   It is common practice to locate the enclosed cells in the power room with the charging equipment.

It is good practice to furnish a minimum of three busy-hours reserve in the battery so as to allow time to correct any trouble in starting a reserve engine-driven alternator.   Where no reserve alternator is furnished in the building or where an unattended telephone office is in a remote location from the maintenance center, much longer reserves such as 8 to 24 hr may be justified.

Batteries using the more expensive alkaline-nickel-cadmium system are used for special applications in communication equipments.   Engine-starting duty for standby alternators in unheated quarters is one typical use.   Here the load is very heavy, causing a deep discharge if the engine does not start quickly.   Then long periods of nonuse follow.   At temperatures below ordinary room levels, the nickel-cadmium batteries have marginally better performance.   In the smaller capacities, nickel-cadmium cells are now available completely sealed so that they can be used like a dry cell without thought of spilling.   Under this condition they can be subjected to many charge and discharge cycles, so they make ideal reserve-energy sources for decentralized power plants, such as line concentrators.   In the Telstar communications satellite, such cells were used in combination with the Bell Solar (battery) cell to provide continuous power when Telstar was in the earth's shadow.

4. Conversion Equipment

a. A-C/D-C Machines.   In order to convert 208-volt 60-cycle power, which is now practically universal in the United States, into direct current or alternating current at the voltages required and at higher or lower frequencies of alternating current, extensive use is made of converters in communication power systems.   While this field was originally occupied by rotating machines and vibrators, static converters (without moving parts) are gradually replacing them, with worthwhile savings in maintenance labor.   For convenience, this subject will discuss the various types which are important in old and new equipments, beginning with the a-c/d-c converters, which have the widest use.

Alternating-to-direct-current converters are used to convert the commercial 60-cycle alternating current to direct current at the required voltage.   At the same time the resulting direct-current ripple is suppressed and the output voltage is usually regulated. Motor-generator sets are still in wide use in existing offices in output ratings of 200 to 1,500 amp at 24 and 48 volts. Smaller machines for plate voltages are also common. Early machines known as the "M" type were especially built for charging talking batteries under load. Charging under load required surface-wound armatures, a greatly increased number of commutator bars, and copper gauze brushes to reduce the slot and commutator ripple so that the filters then available could produce quiet batteries for talking. With the introduction of first the wet and later the dry electrolytic capacitor, it was possible to use machines which were more nearly like those in ordinary commercial use at considerable economy. These are widely used, and only a very few of the special telephone machines are still operating. While the machines often compete favorably with static converters and rectifiers on a first-cost basis, lower maintenance costs, floor-space savings, and higher efficiency will generally result in reduced annual charges for the latter.

For battery-charging applications, d-c generators with a shunt field designed to work with a fast-response electronic voltage regulator which together produce a

Table 1. Characteristics of Copper Oxide, Selenium, Germanium, and Silicon Diode Rectifiers*

Type of diode	Copper oxide	Selenium	Germanium	Silicon
Efficiency 1ϕ full wave, %..............	75–90	70–85	90–95	89–94
Forward current density, amp/sq in......	0.10	0.32	300	1000
Peak inverse voltage per cell............	6–8	30	100	200 up
Maximum cell temperature, °C..........	85	75–125	55	150
Aging characteristics...................	Slight initial	Continuous	None observed	None observed

* Reproduced by permission of Bell Telephone Laboratories, Incorporated.

marked droop in the load-voltage characteristic of the generator at full load are now in general use. Formerly slow-acting mechanical regulators were used. In any event, the response of the regulator-generator combination must be much slower or faster than the battery and load; otherwise severe hunting will result. The abrupt droop in the output voltage on overload is necessary to permit the generator to be connected across a discharged battery without having its protection breaker or fuses operate.

b. Rectifiers. Early copper oxide rectifiers were limited in capacity and were useful only on special low-drain loads such as message registers, coin collectors, or ringing batteries. Mercury arcs were often used on central office batteries to support the light load during early morning hours. Tungar rectifiers in multiple were also helpful in replacing very small charging machines. Selenium and germanium solid-state diodes permitted the introduction of multiple 5- to 10-kw rectifiers during the past decade. To achieve a reasonable space factor forced-air cooling was often used, which reduced the reliability. In combination with passive circuit elements, such as zener diodes and magnetic amplifiers in the voltage-regulation and droop-control circuit, the selenium and germanium diodes offered a very satisfactory battery-charging unit in a single bay. Many are in use in the field for both central office switching and plate-supply batteries.

The introduction of silicon controlled rectifiers (SCR's) a few years ago was a real break-through in the power field. Table 1 illustrates the dramatic increase in forward current density, inverse peak voltage, and operating-temperature capabilities. The control terminal in the diode reduced space by eliminating the need for large saturable reactors in series with the input which were needed for regulation with selenium and germanium diodes. Silicon controlled rectifiers permit regulation similar to that of thyratron tubes, which would have been more widely used in communication power rectifiers except for their unpredictable failure characteristic. Two other very useful characteristics of SCR's are worth mentioning. Silicon rectifiers do not need forced-air cooling, and their efficiency remains high even when operating at low loads. This makes it economically feasible to carry variable loads with permanently connected parallel rectifiers in contrast to motor-generator sets and other rectifiers, which were switched in and out to save power as the load varied.

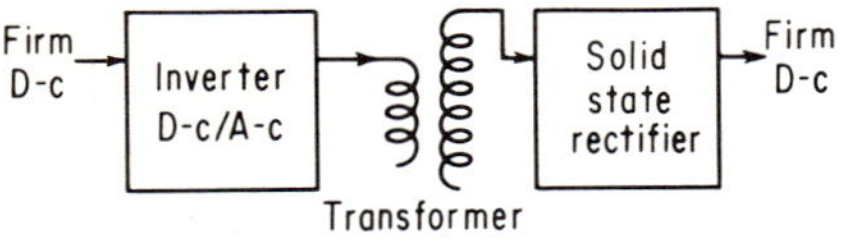

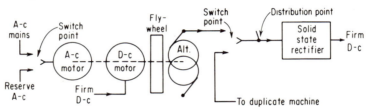

FIG. 5. Solid-state inverter (d-c transformer).

c. D-C/D-C Converters. The problem of obtaining *firm* high or low d-c voltages by conversion of the floated battery central office or filament supply has also been neatly solved by the solid-state silicon device used as a rectifier or inverter. Now it is possible and good practice to use a power inverter to convert firm battery low voltage to alternating current, step the alternating current up or down with transformers, and rectify and regulate to obtain the desired d-c supply at the voltage level required (see Fig. 5). Of course, this has been done (automobile radio) for some time using mechanical vibrators but only at very limited capacities. The once joked about d-c/d-c transformer has come into being with the solid-state rectifier!

This can be contrasted favorably to existing plants employing rotating machinery for inverters supplying individual rectifiers for the various plate voltages required as shown in Fig. 6.

FIG. 6. Block diagram of rotating machinery inverter.

5. Standby Equipment

Most communication services require firm power. The principle of redundancy in the power system is employed extensively to overcome equipment failures which cannot be predicted. The storage battery can be looked upon as an extra short-term supply for the rectifiers which normally carry the load. Before the reserve energy in the battery runs out, important power supplies are protected against long commercial power failures by standby alternators, powered by their own engines.

a. Diesel Alternators. Diesel-engine-driven alternators are the principal work horse to provide a standby for extended failures of outside power. They have been in use in telephone plants for about 25 years. Prior to that, gasoline and illuminating-gas engines were used to power standby generators. The latter were too bulky, and the storage and use of the highly volatile gasoline were hazardous. At present gasoline engines are still commonly supplied with the small 5- and 10-kw alternators. Diesel engines are furnished with 20- to 2,000-kw sets. The safest installations have the fuel in buried tanks with fuel pumps on the set equipped with return fuel lines. Gravity fuel feed eliminates the pump but adds the risk of putting the fuel pressure on the engines 24 hr every day, so that a leaky valve would fill the cylinders. Small sets are also offered to burn liquefied petroleum. Even with the LP tanks located outside the building and equipped with pressure-reducing valves, these are more dangerous than gasoline and have caused fires and explosions when the sets are indoors. Installation in their own metal housing with a grating for a floor so the heavy gas will not accumulate is much less hazardous.

Whatever the fuel used, automatic shutdown features are good practice and required in many cities. Among the most common safety features are overspeed trips, low-oil-pressure switches, high cooling-water temperature switch, and remote fuel shutoff valve or air shutdown. The air shutdown acts much more quickly than the fuel valve.

Modern diesel alternators are self-contained assemblies with the engine, alternator, and cooling radiator or heat exchanger all mounted on a common subbase. In the smaller sets and sometimes up to about 200 kw the control equipment is also mounted in a cabinet which straddles the alternator. This assembly is ready to be placed onto the building floor cushioned by vibration isolators. These permit the set to be located on upper floors of a building, if desired, without undue vibration interference to communication apparatus. With larger sets the control gear is mounted in separate floor-mounted cabinets.

For cooling, a set-mounted radiator with an engine-driven fan which pushes the air from the engine out through the radiator is common practice. This permits a short air duct to be fastened to a duct ledge on the outside of the radiator to an outside wall opening. A short, flexible section in the duct is necessary to avoid short-circuiting the vibration isolators. Flexible sections are also necessary in fuel, exhaust, and electrical lines. On large sets of several hundred kilowatt ratings, remote radiators are sometimes used to permit the air ducts to be short and simple. As an example, with large sets located in a subbasement it is easier to run water pipes and electric conduits to the motor-driven radiator on the roof than to run air ducts. Another variation is to use running-water cooling with a heat exchanger mounted on the set adjacent to the engine. This method subjects the standby equipment to the possibility of water-pressure failure unless a private driven well is used, which is expensive.

Electric starting motors with centrifugal disconnect devices similar to automobiles are almost universal for the smaller diesels and many of the larger ones. Compressed-air starting using the engine cylinders, which was once universal, is now confined to the larger engines. Hydraulic starting motors are a recent development to avoid the use of batteries. Unfortunately a battery-driven compressor motor should be available to avoid failure to start during power failures after the oil pressure is used up.

Diesels need no separate ignition system but ignite the fuel owing to the heat of compression in the cylinders. To start a cold engine in ambient temperatures lower than normal room temperature is not the easiest thing to do reliably. Therefore it is good practice to employ a thermostatically controlled electric heater in the cooling water set for at least the temperature of a heated room. Another "must" to assure reliable starting is periodic routine running of the set to relubricate the cylinder walls and provide system check-out.

b. Gas-turbine Sets. While the diesel set still occupies the center of the standby equipment stage, the gas-turbine set is coming into use. It is being called upon under special circumstances such as cramped quarters. Because of the high-speed operation of the turbine and compressor, the space and weight are dramatically less than the reciprocating engine. In the 750-kw size the turbine set can be one-fourth of the weight and occupy one-fourth the area of a diesel set. This is true even when the alternator is (for 60-cycle) run at 1,800 or 3,600 rpm through reduction gears. Ultimately a turbine with a small high-frequency alternator mounted on it will be used. If 60 cycles is required, a circuit made up of passive elements has been developed to convert frequencies in the kilocycle range up to the lower values. At present passive elements are available for low power loads, but larger capacities are being made available continuously.

There are two types of gas-turbine–compressor combinations. The fixed turbine is directly connected to the compressor in one type, while in the free turbine, the compressor is on a separate shaft and can be run independently of the turbine blade wheel. The flexibility of the free turbine has been found desirable for stationary turbo-alternator sets as contrasted to aviation turbines.

At present, the gas turbine can compete fairly well with the diesel in sizes of 750 kw and larger even when floor space is not at a premium. One handicap of the gas turbine is the deafening noise which results from the high-speed operation. This can be attenuated by enclosing the set in a sound-insulated case with suitable baffles. Partly because of such expensive features, the smaller sizes do not yet compete with diesel drives. The diesels also operate at greater thermal efficiency compared with the simple one-stage turbine. Compound stages which re-use the wasted hot gases and therefore have improved efficiency are much more expensive. The simple, inexpensive type is ideally suited to standby duty, since it runs for only relatively few lifetime hours.

POWER PLANTS

6. D-C Plants

a. Classification. Direct-current plants can be classified arbitrarily as follows:

Class A: low voltage, 150 or less volts to ground
Class B: intermediate voltage, 151 to 600 volts
Class C: high voltage, over 600 volts

This classification determines the protection measures necessary from the point of view of exposure of personnel to accidental voltage shock; insulation of equipment, both power and communication; and other important design characteristics such as contact separation. Development engineers try to utilize Class A low voltages as far as possible because this will result in the least expensive overall equipment design. Although low voltage results in higher currents for the same power, these are confined to the power system, which represents only a fraction of the cost of the overall system. From the point where the communications circuits are fused, out through the outside plant cables, the copper-wire size is generally determined by mechanical strength rather than carrying capacity.

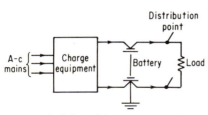

FIG. 7. Ideal floated battery power plant.

b. Low-voltage Centralized Plants. Centralized low-voltage d-c power plants in their simplest form are illustrated by Fig. 7. The storage battery is the heart of the plant, although as explained under paragraph 3 it is normally not used to support the load. It is available to do so, however, on a split second's notice without any switching if the load voltage drops below the "float" value for any reason. This makes the floated battery the ideal supply for communication systems where even milliseconds interruption to the power supply could mean dropping all calls in progress in a switching center with the ensuing unmanageable load peaks in trying to reestablish service. Indeed most telephone offices which have this type of power have never experienced such a failure during their entire life. When such failures do occur, they generally receive unfavorable publicity.

Three nominal voltages are in common use in the low-voltage class: 24, 48, and 130 volts. Twenty-four-volt power plants find application for supplying 6- and 10-volt heaters in electron tubes, transistor circuits, and manual switchboards. Forty-eight-volt power supplies dial switching offices of all kinds except the latest electronic type, and 130-volt power is very commonly used for electron-tube plate supplies except for the more powerful transmitting and traveling-wave tubes which must have higher voltages.

Where relatively light loads are involved, such as in 24-volt single-position switchboards and small 130-volt repeater concentrations, and voltage limits in the order of ±10 per cent can be tolerated, the simple power system typified by Fig. 7 may be feasible. Where closer voltage limits and heavier current drains are encountered, such as large electromagnetic dial offices, economics dictate the introduction of complications which compromise the ideal reliability as mentioned previously. Most dial switching office circuits cannot tolerate greater voltage variations than 10 or maybe 12 per cent. This means that provision must be made for switching into the circuit added emergency cells in a regular battery string in order to get the available reserve out of the battery. Nominal 48-volt plants are usually floated at about 50 volts with the use of 23 lead-acid cells. Under discharge the string voltage (see Fig. 3 and paragraph 3) drops to less than 46 volts without giving up an appreciable amount of its reserve power. To get the complete reserve the load would have to take voltages as low as 40 volts, which would cause failures of marginal equipment. However, 26 cells can be discharged to the complete discharge point of 1.75 volts per cell and maintain 45 volts. Therefore, three extra cells are switched in* one by one as the load voltage falls, as shown in Fig. 8. Provision also has to be made to charge and float these cells so that they will be in good condition when called upon under emergency conditions. Emergency cell switching has to be done without opening the load circuit, so the added complication of circuit-maintaining resistor grids is required, as shown in Fig. 8, to avoid a complete short circuit of the individual battery cells. Further

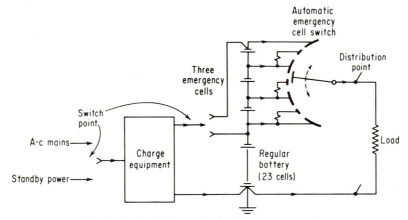

FIG. 8. Floated battery plant with emergency cells.

switching is generally necessary in the power service input circuit to the battery-charging equipment to provide for standby alternator power before the reserve in the battery runs out. Although different numbers of cells will be involved in both the regular and emergency batteries, the same principles are applied in 24- or 130-volt plants which require the added complication.

Twenty-four-volt supplies which are relatively light can be derived from the larger 48-volt supply by the use of series resistors for steady loads or CEMF cells for variable loads. It will be found, however, that when the 24-volt load approximates 100 amp, the power bill for the load wasted over the series lossers will justify the cost of a separate 24-volt battery and chargers.

c. High-voltage D-C Plants. Class B and C, intermediate and high voltage, d-c plants will be covered in paragraph 8. Although the a-c input and standby components are usually centralized, the high-voltage rectification and distribution tend to be decentralized with new constant-voltage systems being developed. The information given below is equally applicable to both low- and high-voltage plants.

d. Information Needed for Engineering Power Systems. The project engineer will need to know, among other items, data on the following:

1. Normal and emergency voltage limits
2. Initial and ultimate loads

* A further refinement has been introduced in recent plants to obtain more uniform discharge of the emergency cells. The first step introduces paralleled emergency cells into the regular string. This is followed as the overall string voltage drops further by reconnecting the same emergency cells to a series connection.

3. Type, voltage, frequency, and reliability of commercial power
4. Maximum elapsed time of operation without maintenance
5. Noise requirements for power
6. Centralized or decentralized power

Comments on the above items in sequence are as follows:

1. Normal voltage limits of 1 or 2 per cent and 15 per cent regulation in an emergency can usually be met easily and economically. Normal limits of $\frac{1}{4}$ of 1 per cent and 5 or 10 per cent in emergency will increase the cost of power enormously.

2. The initial load and up to the first 5 years largely determine the size of the battery and charging equipment needed initially. The ultimate loads determine the floor space needed, the size of the emergency cells and associated switching, the ultimate battery-charging equipment, and reserve engine. Since the life of the office may be as much as 30 years, it would not pay to put in the ultimate plant in the initial period.

3. Network power in the United States is slowly approaching standardization as three-phase 120/208 volts 60 cycles. Variants are 110, 115, and 190 up to 240 in steps. The frequencies are now pretty well confined to 60 cps in the United States and North and Central America and 50 cycles in the rest of the world, with exceptions for 60-cycle, such as Japan and some northern South American countries like Colombia and Venezuela. Reference to the U.S. government publication *Electric Current Abroad*, U.S. Department of Commerce, will be helpful in determining what frequency and voltage are available abroad.

The power reliability should be obtained in terms of the frequency and duration of interruptions to service. Network power with underground distribution averages less than one interruption per year. With overhead distribution the interruptions may average twice a year. Non-network systems in rural areas may be much worse. Conditions prevailing locally are needed to decide on the reserve in the battery and whether standby power is required.

4. The length of time it takes maintenance personnel to reach the equipment in an emergency has a direct bearing on the battery reserve and the need for a standby power source. If the installation is fully attended, including nights and week ends, and uninterrupted service is required, a 3-hr battery reserve and a manually started standby set may suffice. If reliable network power is available and a suburban office is involved, maybe the standby set can be omitted by having a little longer reserve in the battery and dependence on a portable standby. Partial maintenance may indicate the need for longer battery reserve and an automatically started standby.

5. The ripple or noise requirement is needed for the load. This will determine whether mop-up noise filters are needed in addition to the filters built into the charging equipment and the capacitor effect of the storage battery.

6. Decentralized power is often the answer for keeping the exposure of personnel to high voltages to a minimum. As covered in paragraph 2, this should be considered on an overall project basis rather than a strictly power-efficiency or conversion-cost basis. The availability of solid-state d-c voltage-conversion equipment will tend to encourage the use of a common plant providing a firm d-c voltage such as 48- or 130-volt storage battery which will supply decentralized voltage converters for the individual loads.

e. Centralized Battery Plants in Building-block Design. There are many advantages to engineer these plants on a building-block basis, adding blocks as the load grows over 5- or 10-year intervals. The batteries may be arranged in separate strings which are paralleled at their ends and will operate satisfactorily. This practice simplifies the replacement problem as compared with a single large battery. It is not good practice to put cells of different age and manufacture in the same series string.

Charging rectifiers are available in sizes of a few amperes to over one thousand. It is far better to pay the added cost of several smaller chargers as compared with one large one because the failure of one is then less important. This is especially true since one does not usually have to invest all the capital initially. Some manufacturers provide balancing connections when operating parallel rectifiers; others do not on the basis that the balance between the load each unit takes at light load is not important.

When the heavy load occurs, paralleled regulated rectifiers come closer to dividing the load equally.

The building-block principle is also applicable to standby alternator sets provided personnel can be available to parallel them after the first set takes the load automatically. Automatic paralleling equipment can be obtained but is usually quite expensive.

f. Battery Plants with Variable Loads. In planning the battery and charging capacity for a constant load, engineering is straightforward. The battery should be rated at the end of life to deliver the load for as many hours as the reserve required. The charge capacity should be distributed among at least three rectifiers, two of which can carry the load. The third is a spare and will permit high float voltage equalizing charges owing to the extra capacity. For load requirements, the estimated initial load plus maybe 5 years' growth should be used. After that, more charging equipment may be added.

With a variable load, such as found in dial switching offices, the same principle is involved except that the load is not quite so simple to determine. In this case what is known as the busy-hour load is used. Traffic surveys indicate that in a predominately business switching office there will be a traffic peak in the forenoon and another in the late afternoon. In residential offices, another may occur in the evening. The daily load in ampere-hours will be the area under the load curve plotted as amperes vs. hours. An approximation is 10 busy hours in a day. The load to be used to determine the battery and charging capacity will be the sum of any constant loads such as alarms or electron-tube heaters plus the busy-hour variable load. With variable loads as with constant ones, the battery should be sized to carry the heaviest (busy-hour) load because that may be the time it is called upon. The charging equipment may be somewhat under this peak because the battery will deliver the busy-hour load and the loss can be put back during the off-peak hours.

g. Power Board Growth. It was mentioned above that the emergency cell switching equipment is put in for the ultimate load. Other equipment such as meters and alarms, which are complete, should be located at the nongrowing end of the power switchboard, while the discharge fuses and charging rectifiers, which are added to as the load grows, should be at the opposite end. The nongrowing end is ordinarily adjacent to a wall or partition.

h. Existing Variable-load Power Plants with Machine Chargers. In existing power plants with charging motor-generator sets, the permanently paralleled pattern covered above for high, flat efficiency curve rectifiers is not followed. With these chargers the efficiency curve drops off steeply at light loads. Therefore, it is generally necessary to switch the chargers in and out as the load varies. Most of the sets are shut down during the light-load period from midnight to 5 or 6 A.M. A common practice is to equip the first two sets (one may be small for the light load) with voltage regulators. The others are arranged for constant-current operation and switch in and out as "slaves" under control of an ammeter relay which operates when the voltage-regulated sets approach their no or full loads. In this way, several sets are not sitting on the load at a fraction of their rating.

i. Batteryless Supplies. There are some communication facilities which do not require a standby reserve to cover power failures. An example is a business office which is completely mechanized. If the typewriters, adding machines, etc. are shut down, the telephone extensions may as well be, as calls to the outgoing lines can be made. This power comes from the central office battery. To supply the local lines and wiring plan equipment, a simple solid-state component rectifier with load-compensation voltage control is usually sufficient without even dry-cell reserve.

7. A-C Plants

a. Power Ringing and Signaling. Ringing power refers to the power supplies which provide alternating current to operate the subscribers' bells, and signaling refers to the supplies, such as interrupters, which control the ringing current, also audible tones such as dial tone, audible ringback tone, interrupters for busy tone, and lamps.

Except for harmonic ringing, the bells operate on 20-cycle alternating current with

±5 per cent frequency variation acceptable but normally controlled to within ±1 per cent. Local central office ringing in the Bell System is 80 to 90 volts ±1 or 2 per cent. Ringing supplied over PBX trunks is 100 volts ±10 per cent. Manual ringing is controlled by the operator's key. Machine ringing is controlled automatically by an interrupter and shut off by a circuit-breaker (tripping) relay when the called party answers. It usually consists of alternating current which is biased by low-voltage direct current (20 to 50 volts) to facilitate the adjustment of the tripping relays.

Direct Line Ringing. Private lines are equipped with ringers which are bridged across the line (in series with a capacitor to block the direct current).

Two-party Selective Ringing. Ringers are bridged from each side of a two-party line to ground so that only one can be rung when desired to give selective ringing.

Four-party Semiselective Ringing. Two ringers are bridged from each side of the line to ground; one party answers to one long ring and the other to two short rings. This provides means of ringing four parties on a semiselective basis. Both ringers on the one side sound.

Four-party Selective Ringing. This type uses alternating current which is biased with positive and negative direct current. As in four-party semiselective ringing, two ringers are bridged from each side of the line to ground. One is connected by a gas tube so that it will ring only on the positively biased current called superimposed positive, while the other will only ring on negative superimposed alternating current. This provides four-party selective ringing, since only one ringer can be rung at a time.

Eight-party Semiselective Ringing. This type is a combination of the last two types described, where four ringers are bridged to ground on each side of the line. Two ring at a time, since they are biased for positive and negative superimposed current. The selection between the parties is accomplished by long and short rings.

Code Ringing. Rural lines with up to 10 parties on each side of the line are rung by ringing codes made up of combinations of long and short rings.

Harmonic Ringing. While used very little in urban areas, it has been widely used in the rural offices to provide four-party selective ringing by the use of bells tuned to four frequencies. Historically, 20, 33, 50, and 66 cycles were used, but cross ringing has been a problem with this system, especially between the 50- and 66-cycle bells. This caused the introduction of frequencies such as 20, 30, 40, and 60 as well as others which are purposely selected so that the frequencies used do not contain harmonics of each other.

Tone Ringing. This is a form of harmonic ringing where frequencies in the audible range (300 to 1,000 cycles) are used. It has been used for special systems such as radio telephones in moving cars and for paging systems.

b. Ringing and Tone Generators. *Ringing.* Historically the original source of ringing current was Watson's hand-cranked generator, which was succeeded by vibrators and other d-c choppers, dynamotors, and rotary inverters. Large multiunit dial offices use 300- to 600-va. inverters driven by a-c or d-c motors. Inverter output is regulated 20-cycle alternating current, and positive and negative 120-volt d-c for operating coin-box magnets. Smaller offices and PBX's use static type subharmonic 60/20-cycle generators from 3 to 50 watts 20-cycle capacity.

Tones. Tones were originally generated by vibrators, buzzing relays, split segment drums, and other d-c choppers. Inductor-alternators and static-type harmonic generators are now in the field as well as a few transistor oscillators.

	Cycles
Bell System low tone for dial and busy signals	660/120*
Bell System audible ringing	420/40
Bell System high tone	500

* 660-cycle signal modulated at 120 cycles.

Voltages for these tones are low, in order of a volt or less.

For new offices such as the Bell System central office with electronic and touch tone transmission, new precise frequency and volume tones are being used. They are in the audible range (350 to 900 cycles). Harmonics are limited to 50 db below the fundamentals. Mixtures of precise tones will be used for various functions.

Interrupters. Slow-speed interrupters are used to interrupt the ringing current and place tripping battery on the line during the interruption (silent interval) and to produce busy tones and flashing lamp signals. The machine ringing interrupters are in the form of single-pole double-throw switches and produce long rings of 2-sec duration followed by a 4-sec silent interval. Also for party ringing, short rings of ½ to 1 sec with total repetitive cycles of 6 to 8 sec are produced. The busy signals are on a ½- or 1-sec cycle with approximately 50 per cent on and off periods and are SPST switches.

Relay, spring, and rotating cam; split segment drums; and mercury contact drums are all in use in the field. With 20-cycle rotating machines the springs were driven (or drums for that matter) through gears on a low-speed shaft. With solid-state ringing and tone generators, separate motor-driven interrupters have been used. Solid-state interrupters are in development. With the early solid-state switching devices the problem was to get adequate peak inverse voltage ratings (PIV's), but

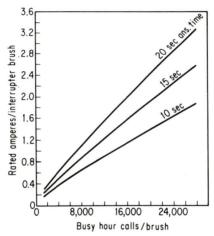

FIG. 9. Ringing drain curve. (*Reproduced by permission of Bell Telephone Laboratories, Inc.*)

these devices are now available. Economics still favor motor-driven multispring interrupters, which can be made for considerably less than an electronic wired unit.

The reliability of ringing equipment has to be high, since it is common to one or more central offices. Redundancy is always needed and is commonly offered by duplicate sources, at least one of which is battery driven and independent of commercial power. These sources must be quick starting because the loads do not tolerate warm-up periods.

Ringing Drains. Field readings and probability studies have established average 20-cycle ringing drains of 0.75 to 1.3 amp per 10,000 busy-hour calls. This drain will gradually be reduced, since high-impedance ringers are used in all new subscribers' sets. The variation is due to long (20 sec) or short average answering times. When a 6-sec machine ringing cycle with 2-sec ringing intervals is used, the load is divided among three interrupters which do not overlap, so that a given capacity generator can supply three times the load. Figure 9 shows ringing drain curves for an average ringer current of 0.020 amp.

Manual ringing, which is not very prevalent any more, will be in the same order, since a random distribution will usually obtain. When ringing is fed over numerous long cable pair feeders, however, the continuous capacitance to ground of the pairs puts an out-of-phase drain on the ringing source which has been known to measure over an ampere.

Tone drains are largely in the milliwatt range. Very large multiunit offices may approach a 1-watt or at least a fractional-watt range.

   c. Motor and Engine Alternator Plants for Continuous A-C Supplies. The use of alternating current for electromagnetic communication equipment has been held back by the less-than-satisfactory a-c magnet and the complications involved in providing power as firm as direct current backed by storage batteries. Light a-c loads such as electric clocks, flea-power motors, and a few fluorescent lamps have been switched to a standby vibrator driven by firm direct current. Fractional-kilowatt loads and greater have been switched on power failure to normally inert alternators driven by d-c motors or engines in cases where the start-up time delay could be tolerated. In many electronic loads interruptions of more than 25 or 50 msec cannot be accepted, since the plate currents do not persist beyond this period even when ballasted by electrolytic capacitors. To meet this short open interval requirement from regular to emergency power supplies, the energy stored in a rotating flywheel has been used.

   One system employs an alternator and flywheel driven normally by an a-c motor. If this motor fails, the drive is automatically shifted to a direct-connected d-c motor, the flywheel supporting the load during the switching interval. The set then continues on battery drive until the standby engine alternator is started and the a-c motor again

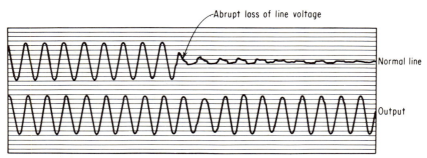

FIG. 10. Oscillogram of normal and inverter output voltage waves. (*Courtesy of Westinghouse Electric Corporation.*)

assumes the load. The a-c load is divided among several sets, and a spare one can also be substituted for any regular set should it fail (see Fig. 6).

   Another system substitutes for the d-c motor an internal-combustion engine which can carry on subject only to the fuel-tank capacity. In either case the flywheel on the set is kept running at all times to support the alternator load while the drive is being transferred. With the engine set, a clutch separates the engine from the flywheel and alternator so that the engine can be normally inert. Upon power failure the clutch is closed and the energy stored in the wheel starts the engine and supports the load.

   The all-electric system is somewhat more reliable than the engine system. It costs considerably more, however, when the cost of the battery for the d-c motor and charging equipment is taken into account. The engine units have been built in capacities up to 50 kw, and the motor-driven units to 25 kw. It is often the practice to distribute the load over more than one set, especially with the all-electric system.

   d. Static Inverters for Continuous A-C Supplies. Recent development activity has resulted in the availability of solid-state inverters using completely static (passive) circuit components. Alternating current is produced from a d-c battery supply using a bridge-type circuit gated by silicon controlled rectifiers (SCR's). A silicon transistor LC oscillator controls the output frequency. Ferroresonant-type regulating transformers maintain a 5 per cent output voltage regulation. With an instantaneous-type transfer circuit, the inverter takes over the supply without mutilating the a-c wave shape (see Fig. 10). As an optional system, the a-c supply is furnished continuously by the inverter. This involves the continuous power loss due to the efficiency* of the device.

Although long-time reliability has not been established from actual field experience at this writing, this development promises to offer the best solution to obtain firm a-c supplies for important electronic loads which cannot tolerate even millisecond interruptions. Because of its static elements it will permit extending the redundancy principle for important loads to the ultimate.

e. Engine Alternator Plants for Standby and Primary Power. In the majority of communication facilities the importance of maintaining continuity of service justifies the provision of a long-time reserve which is beyond that maintained in the storage battery. This reserve is commonly in the fuel-oil tank and may be of a week's time or more in outlying locations which are inaccessible during heavy snowfalls.

Speaking of standby sets, for ordinary offices one set is used sized to carry all or enough of the load so that the battery can be brought to full charge over a 24-hr period. Sometimes nonbattery loads such as lighting and elevator loads are also included. Where these loads are greater than 100 kw or so, it is good practice to use multiple standby sets to share the load even at some financial sacrifice so as to avoid having all the eggs in one basket.

The use of automatic starting and transfer is dictated by the need for instantaneous transfer and nonavailability at all times of maintenance personnel. If battery reserves will permit a manual start and warm-up of the engine, the simplicity of the arrangement has proved reliable. However, with nights and week ends generally partially uncovered by the maintenance forces, automatic operation is usually necessary. In order to avoid the expense and complication of automatic synchronizing equipment for multiple sets, one automatic transfer set can be arranged to take that part of the load which requires immediate backup. The rest of the load, including that being carried by the battery, can wait until the arrival of the maintenance personnel, who can then start and transfer the remaining sets, including paralleling and synchronizing them manually with one another or with the original automatic set. In case the majority of the load needs power within the starting time (a few seconds up to 20) of an engine, a single large set can be used. However, reliability is sacrificed compared with allowing for a 5-min engine warm-up period before taking the load.

The remarks about sharing the load are especially applicable to primary power plants. These are not often necessary in the continental United States. However, when a cost study shows an advantage over a period of years for a primary plant compared with the expense of extending the commercial power lines, diesel plants have proved successful and reliable. The study should include the load being alternately carried by at least two and in remote areas three sets. The last set will allow a team of two to carry the load alternately while one set is torn down for extended maintenance. Depending on the type of set, the period for major overhaul varies from 400 to 1,000 or more running hours. Where large (several hundred kilowatt) loads are involved and continuous maintenance is available, the operation assumes the proportions of an isolated commercial diesel plant. Here multiple sets are operated on the load bus, care being taken always to have enough capacity operating at any one time to continue to support the load even if the largest set suddenly fails and is taken out by its reverse power circuit breaker. In military systems this principle is extended to split load busses for essential and less important loads. In such cases, the essential load bus is supported as above by arranging to switch sets from the other bus in emergencies.

In complexes such as described above using multiple sets and busses, the switching becomes complicated. This justifies the preparation of a so-called "mimic" control panel which by means of different-colored lines and lights gives the operator the operating situation at any instant.

In both primary and standby plants it is well to have the engines in a separate room because of noise and the heating problem in cold weather. The cooling fans will draw all the warm air out of the room not only on power failure but on maintenance runs.

* About 70 per cent for inverter and 85 per cent for the rectifier.

In primary plants a major economy in space heating costs can be gained by using the heat in the exhaust gases from the engines. A convenient method to do this is to install "wet" mufflers in the exhaust pipe and use a forced-hot-water system for space heating. The engine heat can be used for part of the space or as a supplement to the regular heat source. This system was used successfully in the primary plants supplying the "DEW Line."

f. Transfer Equipment. A very important but sometimes neglected area is the equipment for transferring the load from normal to standby power. Failures here can block all return on capital invested in reliable emergency engine sets. Attention to a few basic principles will minimize the likelihood of such failures.

1. Follow strictly the National Electric Code which states, "Overcurrent protection devices shall be located at a point where the conductor receives its supply." Refer to Fig. 11. Circuit breaker (or fuse) *A* (Fig. 11) will protect the normal power conductor against a fault at point *X* no matter how long the lead. Then transfer to standby power will isolate the fault so that repairs can be made without affecting the

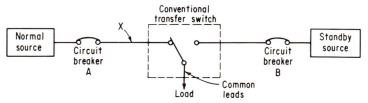

Fig. 11. Adequate overcurrent protection and transfer switching. (*Courtesy of Automatic Switch Company.*)

load. Likewise breaker *B* will protect the conductors from the standby source. It is also apparent that the less exposure to faults by long common load leads the better.

An unadvised short cut sometimes followed for economy reasons is to bring the *A* and *B* circuit breakers together physically so that they can be mechanically interlocked against accidental paralleling of the two sources. The transfer switch is then omitted, and breakers perform both functions. This leaves both normal and standby conductors unprotected against faults between their source and the breakers.

2. Locate the transfer equipment as near the load it supplies as practical. This will keep the common conductors (see Fig. 11) to a minimum so that faults, fires, or floods will more likely be in the individual leads. If the standby set can also be near the load, lead duplication can also be held to a minimum. It is preferable, however, to have lead duplication than common conductors which are longer than necessary.

3. The transfer switch itself should be capable of:

 a. Interrupting and carrying full load currents continuously as well as large overloads momentarily due to inrush currents on lighting loads, etc., without overheating
 b. Quick transfer and arc suppression*
 c. Double source control, i.e., operation from the power source to which the load is to be transferred by the automatic switches (see Fig. 12)
 d. Mechanically locking in place after operation
 e. Mechanically preventing the paralleling of the normal and standby sources

The transfer switch, being common to both power sources, cannot be easily isolated and therefore should be designed with special integrity. Furthermore, it has to carry the load continuously without down time. The double source requirement stems from the fact that the power supply from which the load is to be switched is usually out;

---

* There is one possible exception to the quick-transfer practice. Large induction motors are sometimes damaged mechanically by the application of an out-of-phase current before their existing residual field has collapsed. One way to avoid this is to arrange the motor-starting contactors to drop out for a definite time when power is interrupted.

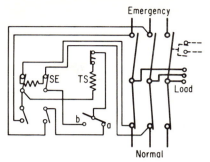

FIG. 12. Double power source automatic transfer switch. Momentary coil energization principle: On power failure upper contacts on relay *SE* open and lower contacts close. Main coil *TS* is momentarily energized from emergency source. Contact *a* immediately opens coil circuit. Normal power restoration again energizes *SE*; upper *SE* contacts close,* momentarily energizing *TS* through contact *b*. Switch returns load to normal, and *b* opens main coil circuit. As soon as the main contacts close in either position, they are mechanically locked by a toggle linkage until the coil is reenergized. (*Courtesy of Automatic Switch Company.*)

either the one to which it is to be connected is available or the transfer is of no value. The quick transfer and arc suppression, like the provision against paralleling sources, are to prevent opening one or both overcurrent devices. The mechanically locked-in requirement is to prevent load interruptions on line dips.

8. Specialty Power Plants

a. Constant-current Power Supply for Repeatered Submarine Telephone Cables. Repeaters for submarine cables, being inaccessible after the cable is laid, are supplied in series with a constant fractional-ampere direct current. (Direct current minimizes the filter equipment in the repeater.) The power is supplied at each land end, and the voltage, which depends on the length of the cable, is varied automatically to maintain the current constant. This provides regulated voltages across fixed resistances in the repeaters. To limit the voltage to ground or sea water at any point, the voltage necessary to maintain the current through the repeaters, cable wire resistance, and stray earth potentials is divided between the two shore stations at the ends of the cable. This limits the dielectric stress to the cable and repeater insulation. Single-ended supplies are used only on short-haul cables, where the penalty of doubling the dielectric stresses can be tolerated.

At each shore station, variable-voltage rectifiers and constant-current regulators deliver the current to the cable. The voltage regulators are twofold. An instantaneously acting series vacuum-tube regulator can limit the cable current even in the event of a sudden short circuit. This regulator is immediately unloaded by a slower acting variable a-c supply to the rectifiers. This method has been successful in preventing repeater tube heater burn-outs which would occur as a result of sudden short circuits or slower acting earth potentials in an aiding direction.

For reliability, the redundancy principle is used extensively in cable power plants. Two rectifiers and regulators supply each cable in parallel. They are arranged so that, if one fails, the other will immediately take over without any reaction in the cable current. A third set can replace either to permit maintenance operations. The a-c supply to the rectifiers and regulators is taken from two firm a-c distribution systems, so that, if one fails, one of the parallel regulating equipments will still function. The firm a-c bus bars are each supplied by an all-electric flywheel motor-

---

* The single coil is only momentarily energized—only the low-power-loss supervisory relays remain continuously energized. Accordingly, hum and vibration stay at an absolute minimum.

alternator set described in paragraph 7.   In this case a rotating spare set is supplied to take over in case the regular set fails.   In the original cable projects either 60- or 50-cycle alternating current was used depending on the local supply.   In the newer cables the motor-alternator sets are being replaced by solid-state inverters which generate 400 cycles.   The higher frequency is used only on the cable rectifiers and regulators, where its use results in smaller components.   The inverter takes direct current from a firm power-storage battery plant.   This plant or the motor alternators take power from a local source, when available, backed up by multiple standby sets.

As with all Class C voltage supplies, elaborate precautions are taken to protect personnel from the 2- to 4-kv cable voltages.   These include key interlocks and foolproof switching means to permit maintenance at lower, safer voltages.   To protect the inaccessible undersea repeaters from damage, elaborate overvoltage and overcurrent automatic shutdown and alarms are furnished.   The shutdowns do not take the cable current to zero, which would cool off the components, but to a lower than normal operating value.*

b. Microwave Long-haul Power.   At microwave repeater stations floated battery Class A and B power plants of 12, 130, and 250 volts supply vacuum-tube heaters, receiver tube plates, and transmitter tube plates, respectively.   These plants are backed up by automatically operated diesel alternator sets.   The 250-volt (Class B) power supply is fully enclosed to avoid accidental contact by the maintenance forces, but key interlocks are not required for voltages under 600 volts (see paragraph 8a).   In the latest Bell System routes which have traveling-wave transmitter tubes, 1- and 2-kv supplies are required and are built into the transmitting equipment.   Firm 60-cycle power to the rectifiers is supplied by all-electric flywheel sets using emergency 130-volt battery motors.   This battery also supplies the receiver tubes and other miscellaneous equipment.   In the not too far distant future, these rotating machinery plants will probably be replaced on new additions by solid-state converters which will transform the firm 130-volt direct current to the higher voltages.

c. Coaxial-cable Carrier.   Long-haul coaxial-cable systems are characterized by rather closely spaced repeaters which are supplied in series from main stations by a constant 60-cycle alternating current.   The current is regulated by variable voltage transformers, which in turn are energized by flywheel-type motor-alternator sets backed up for long failures by suitable diesel alternators.   The main stations generally feed several cables in both directions, each cable having an individual regulator.

* Detailed description is given in G. W. Meszaros and H. H. Spencer, Power Feed Equipment for the North Atlantic Link, *Bell System Tech. J.*, vol. 36, pp. 139–162, January, 1957.

Chapter 21

WIRE AND CABLE INSTALLATION PRACTICES

C. H. ELDER, *Reliable Electric Company*

The cable and wire facilities associated with the "outside plant" of a communication system comprise an appreciable segment of the total plant investment. From an economical viewpoint, the principal aim is to keep the long-term overall costs of the communication system as low as practicable. The facilities must be designed to meet the transmission-loss objectives of the various networks of wires and cables. Of equal importance, however, are the requirements for careful consideration of service integrity and safety which must also enter into any planned design of a plant. Both the *real* and *intangible* factors should be evaluated in order to provide what is considered the most economic communication facility.

The outside plant, in general, comprises all that portion of the communication network between the main frame of a telephone office and the station apparatus located on a customer's premises. It consists mainly of the cables and wires as placed aerially, in underground conduit, or buried directly in the ground. These circuits connect the switching equipment, repeaters, and other sophisticated apparatus designed to meet the variations in the grades of transmission demanded by the customer.

By reason of the out-of-door location of this network, the circuits are highly vulnerable. They are subject to the deteriorating effects of the weather, rodent damage, electrolytic action, abrasion, tensile and bending stresses, high voltages, and currents from lightning strokes and man-made damage. Consequently, it is most desirable that all these elements be carefully considered as an entity for each installation in order to maintain reasonable upkeep costs.

TYPES OF COMMUNICATION CABLES AND WIRES

Various types of cables with different sheath protections and conductor insulations have been developed over the course of years, and some types have become standard throughout the communication industry.

The predominant conductor material for cables is copper; however, aluminum conductors and copper-steel conductors are used in some specific types of cables. The major insulating materials for cable conductors are paper, paper pulp, polyethylene, and polyvinylchloride (PVC).

The popular types of communication cables are described in paragraphs 1 to 4.

1. Lead-sheathed Cable

The use of lead sheath for telephone cable dates back to the very early history of the telephone art. Although iron water pipe was used for some of the cable first

21–1

installed in the underground telephone plant, the lead-sheath extrusion methods were common to the outside plant prior to 1890.   Pure lead is subject to fatigue cracking in aerial installations and to rapid electrolytic action in underground plants.   As a result, the majority of cable sheaths without any outer protection contain a composition of 1 per cent antimony, which was found better to withstand the physical and chemical requirements expected of it (Fig. 1).

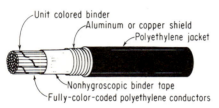

FIG. 1. Lead-covered cable.

As the lead sheath provides a hermetically sealed structure, the cable conductors are usually insulated with paper or paper pulp, a paper core belt, and an outer lead sheath.   In some high-lightning areas, a polyethylene jacket may be extruded over the cable core and under the lead sheath to provide an added high-voltage protection to the cable conductors.   This type of cable is designated as Lepeth cable—a contraction of lead-polyethylene.   Stubbing cables for various types of terminals are currently using lead-sheath structures.   However, to provide pressure dams and prevent conductor exposure to moisture associated with the apparatus, the conductors of these cables are generally insulated with PVC.

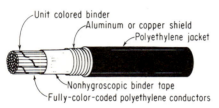

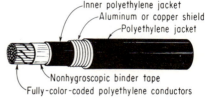

FIG. 2. Alpeth or Cupeth cable.   Used only on plastic-insulated conductors, mainly aerial plant or buried in minimum lightning exposure areas.

FIG. 3. Double-sheathed Alpeth or Cupeth cable.   Used only on plastic-insulated conductors.   Main use for direct burial in severe lightning areas and aerial plant in these areas.

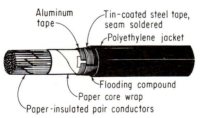

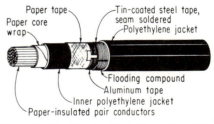

FIG. 4.   Stalpeth cable.   Used mainly on paper-pulp-insulated conductors in underground or aerial plant in low lightning exposure areas.   Not intended for direct burial.

FIG. 5.   Double-sheathed Stalpeth cable.   Used mainly on paper-pulp-insulated conductors in buried or aerial plant in high lightning areas.   Also in some possible gopher-infested areas.   Where full gopher protection is required, an outer sheath protection should be considered.

Lead-sheathed cables, however, are rapidly being superseded by the plastic-sheathed cables.   The availability of new protective materials to prevent the entrance of moisture to the cable core, higher dielectric strength of the conductors, and cable sheath with excellent weather resistance has furnished the means to provide more cable plant for fewer dollars than before.

2. Plastic-sheathed Cables

Plastic-sheathed cables are made up in generally accepted configurations of Alpeth or Cupeth cable (Fig. 2), double-sheathed Alpeth or Cupeth cable (Fig. 3), Stalpeth cable (Fig. 4), and double-sheathed Stalpeth cable (Fig. 5).

3. Self-supporting Cable and Wire (Figure-eight. Integrated Messenger, etc.)

Most manufacturers of cable will supply a cable with any specified number of pairs the engineer decides to order, but generally the manufacturers list catalogue

Table 1. Standard Cable Sizes (No. of Pairs)

Number of pairs		19 AWG		22 AWG		24 AWG		26 AWG	
Paper	Plastic	Paper	Plastic	Paper	Plastic	Paper	Plastic	Paper	Plastic
6	6	x	x	x	x	x	x	x	x
11	12	x	x	x	x	x	x	x	x
16	18	x	x	x	x	x	x	x	x
26	25	x	x	x	x	x	x	x	x
51	50	x	x	x	x	x	x	x	x
76	75	x	x	x	x	x	x	x	x
101	100	x	x	x	x	x	x	x	x
152	150	x	x	x	x	x	x	x	x
202	200	x	x	x	x	x	x	x	x
303	300	x	x	x	x	x	x	x	x
404	400	x	...	x	x	x	x	x	x
455	...	x	...	x	x				
606	600	...	...	x	x	x	x	x	x
909	900	...	...	x	...	x	x	x	x
1,212	...	...	...	x	...	x	...	x	
1,515	...	...	...	...	...	x	...	x	
1,818	...	...	...	...	...	...	...	x	
2,121	...	...	...	...	...	...	...	x	
2,424	...	...	...	...	...	...	...	x	

standard sizes of paper- and plastic-insulated conductors. The standard sizes vary by gage of the conductors and by type of insulation (Table 1) and consequently have different diameters (Tables 2 and 4) and weights (Tables 3 and 5).

The trend is toward greater use of plastic-insulated conductors for cables of 300 pairs and less.

An increasingly popular type of Alpeth cable structure for some installations, one that eliminates lashing of the cable to a separate messenger or strand, is the integrated or figure-eight cable as shown in Fig. 6. It is constructed in a manner to form a parallel configuration of the cable core and messenger. As a result, the messenger and cable sheath are simultaneously polyethylene jacketed and are separated by a supporting web of jacketing material. The messenger is generally 1/4-in. extra-high-strength steel (6,000-lb breaking strength). All interstices of the support strand are filled with a sealing compound to prevent strand corrosion.

4. Self-supporting Wire

For the self-supporting wire design, a single support wire of 0.109 or 0.134 in. is used in place of the heavier 1/4-in. strand. This structure does not contain a metal

Table 2. Paper-insulated Conductor Cable
(Cable diameters, inches)

Size	19 AWG			22 AWG			24 AWG			26 AWG		
	PL	STAL	DS STAL	PL	STAL	DS STAL	PL	STAL	DS STAL	PL	STAL	DS STAL
6	0.4	0.5	0.6									
11	0.5	0.6	0.7	0.4	0.5	0.65	0.4	0.5	0.5	0.3		
16	0.6	0.7	0.8	0.5	0.6	0.70	0.4	0.5	0.6	0.4		
26	0.7	0.8	0.9	0.6	0.7	0.78	0.5	0.6	0.7	0.4		
51	1.0	1.1	1.2	0.7	0.8	0.96	0.6	0.7	0.8	0.5	0.6	0.72
76	1.2	1.3	1.4	0.9	1.0	1.08	0.7	0.8	0.9	0.6	0.66	0.81
101	1.3	1.5	1.5	1.0	1.1	1.20	0.8	0.9	1.0	0.7	0.72	0.81
152	1.6	1.7	1.8	1.2	1.3	1.42	0.9	1.1	1.2	0.8	0.85	1.00
202	1.8	1.9	2.1	1.3	1.5	1.56	1.1	1.2	1.3	0.9	0.94	1.13
303	2.2	2.3	2.4	1.6	1.7	1.86	1.2	1.4	1.5	1.1	1.1	1.29
404	2.5	2.7	2.8	1.8	1.9	2.08	1.4	1.6	1.7	1.2	1.3	1.44
455	2.6	2.8	2.9	1.9	2.1	2.20						
606	...	...	...	2.2	2.3	2.47	1.7	1.9	2.0	1.4	1.5	1.68
909	...	...	...	2.6	2.8	2.94	2.0	2.2	2.3	1.7	1.8	1.99
1,212	...	...	...	...	...	....	2.4	2.5	2.7	2.0	2.0	2.25
1,515	...	...	...	...	...	....	2.6	2.8	2.9	2.2	2.3	2.47
1,818	...	...	...	...	...	....	...	...	...	2.4	2.5	2.69
2,121	...	...	...	...	...	....	...	...	...	2.6	2.6	2.85
2,424	...	...	...	...	...	....	...	...	...	2.7	2.8	3.01

NOTE: PL = plain lead, STAL = Stalpeth, DS STAL = double-sheathed Stalpeth.

Table 3. Paper-insulated Conductor Cable
(Cable weights, pound per foot)

Size	19 AWG			22 AWG			24 AWG			26 AWG		
	PL	STAL	DS STAL	PL	STAL	DS STAL	PL	STAL	DS STAL	PL	STAL	DS STAL
6	0.4	0.1	0.2									
11	0.6	0.2	0.2	0.4	0.1	0.18	0.3	0.1	0.1	0.3		
16	0.7	0.2	0.3	0.5	0.15	0.2	0.4	0.1	0.1	0.3		
26	0.9	0.3	0.4	0.6	0.2	0.3	0.5	0.15	0.2	0.4		
51	1.5	0.6	0.8	0.9	0.3	0.4	0.7	0.2	0.3	0.6	0.15	0.26
76	2.0	0.9	1.0	1.2	0.5	0.6	0.9	0.3	0.4	0.7	0.2	0.32
101	2.5	1.2	1.3	1.5	0.6	0.8	1.1	0.4	0.5	0.8	0.3	0.38
152	3.4	1.7	1.9	2.0	0.9	1.1	1.4	0.6	0.7	1.1	0.4	0.51
202	4.2	2.2	2.4	2.5	1.2	1.3	1.7	0.8	0.9	1.3	0.5	0.63
303	6.0	3.2	3.5	3.4	1.7	1.9	2.3	1.1	1.2	1.8	0.8	0.87
404	7.6	4.3	4.6	4.3	2.2	2.4	2.9	1.4	1.6	2.2	1.0	1.1
455	8.5	4.8	5.1	4.8	2.6	2.7						
606	...	...	...	6.0	3.2	3.5	4.0	2.1	2.3	2.9	1.4	1.55
909	...	...	...	8.5	4.7	5.0	5.5	3.0	3.3	4.0	2.0	2.24
1,212	...	...	...	...	....	....	7.1	3.9	4.3	5.1	2.6	2.87
1,515	...	...	...	...	....	....	8.6	4.9	5.2	6.1	3.3	3.51
1,818	...	...	...	...	....	....	...	....	...	7.2	3.9	4.14
2,121	...	...	...	...	....	....	...	....	...	8.2	4.5	4.75
2,424	...	...	...	...	....	....	...	....	...	8.9	5.0	5.33

Table 4. Polyethylene-insulated Conductor Cable
(Cable diameters, inches)

Size	19 AWG		22 AWG		24 AWG		26 AWG	
	AL	DS AL	AL	DS AL	AL	DS AL	AL	DS AL
6	0.49	0.59	0.41	0.51	0.36	0.46		
12	0.62	0.72	0.50	0.60	0.44	0.54	0.38	0.48
18	0.71	0.87	0.56	0.66	0.48	0.58	0.42	0.52
25	0.79	0.97	0.60	0.72	0.51	0.62	0.44	0.54
50	1.13	1.25	0.79	0.90	0.67	0.78	0.56	0.66
75	1.35	1.48	1.01	1.14	1.78	0.96	0.65	0.76
100	1.51	1.68	1.12	1.24	0.96	1.08	0.72	0.83
150	1.83	1.98	1.33	1.46	1.11	1.23	0.85	1.03
200	2.09	2.22	1.49	1.66	1.24	1.36	1.04	1.17
300	2.84	2.64	1.82	1.96	1.46	1.63	1.20	1.33
400	....	....	2.06	2.19	1.67	1.80	1.35	1.49
600	....	....	2.48	2.64	1.99	2.13	1.64	1.77
900	....	....	....	....	2.39	2.56	1.95	2.10

NOTE: AL = Alpeth, DS AL = double-sheathed Alpeth.

Table 5. Polyethylene-insulated Conductor Cable
(Cable weights, pound per foot)

Size	19 AWG		22 AWG		24 AWG		26 AWG	
	AL	DS AL	AL	DS AL	AL	DS AL	AL	DS AL
6	0.10	0.14	0.06	0.09	0.05	0.07		
12	0.17	0.22	0.10	0.14	0.08	0.11	0.06	0.09
18	0.24	0.31	0.14	0.18	0.10	0.14	0.07	0.10
25	0.31	0.40	0.18	0.22	0.12	0.16	0.09	0.12
50	0.64	0.71	0.31	0.40	0.21	0.26	0.15	0.19
75	0.91	1.00	0.51	0.57	0.30	0.39	0.20	0.25
100	1.18	1.30	0.65	0.72	0.45	0.50	0.26	0.31
150	1.73	1.87	0.92	1.02	0.62	0.69	0.36	0.46
200	2.27	2.44	1.19	1.36	0.79	0.88	0.54	0.60
300	3.32	3.55	1.76	1.90	1.14	1.26	0.76	0.84
400	....	....	2.30	2.47	1.50	1.62	0.97	1.07
600	....	....	3.37	3.60	2.18	2.34	1.42	1.54
900	....	....	....	....	3.19	3.41	2.08	2.23

shield, and the largest size is generally 18 pairs of 19 AWG or 22 AWG conductors. This design (Fig. 7) gives better electrical performance than the older multipair distribution wire.

5. Cable Sheath Protections

Various types of outer sheath protections for communication circuits have been developed over the period of years. Some of the hazards to overcome in aerial, underground, and buried systems include abrasion, electrolysis, lightning, tensile and bending stress, etc. Several outer sheath coverings that are considered standard are buried-tape armored (Figs. 8 and 9), aerial-tape armored (Fig. 10), wire armored submarine (Figs. 11 and 12), and light-wire armored (Fig. 13).

a. Buried-tape Armored. *Purpose.* Mechanical protection from abrasion and possible damage to circuits from rocky terrain and frost heaving. Also, some shielding from low-frequency induction.

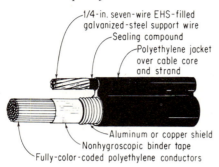

FIG. 6. Figure-eight self-supporting cable.

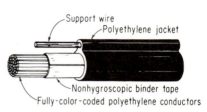

FIG. 7. Figure-eight Multipair distribution wire.

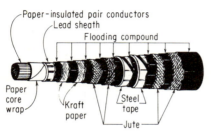

FIG. 8. Buried-tape armored cable.

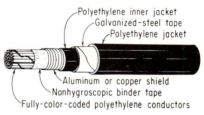

FIG. 9. Armored Alpeth cable.

Structure. Impregnated paper tape, cushion of jute roving, two layers of steel tape, and an outer covering of impregnated jute. Asphalt floodings are applied over each application of material. The outer covering is coated with a "whiting" or mica compound to prevent adjacent turns of cable from "sticking" on the reel. It was generally applied over lead-sheathed cables in the past. However, with the use of double-sheathed Alpeth types of cable, this type of cable protection is considered only in those cables containing toll and special circuits for signaling. The layer of jute cushioning and asphalt floodings are omitted, and the layer of steel tapes are "sandwiched" between the inner and outer polyethylene jackets of the PIC core type of cables as shown in Fig. 9.

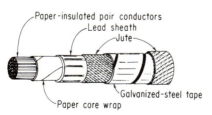

FIG. 10. Aerial-tape armored cable.

b. Aerial-tape Armored. *Purpose.* Mechanical protection and shielding from low-frequency induction.

Structure. Impregnated paper tape, cushion of jute roving, and galvanized steel tapes. For lead-sheathed cables only, the sheath and paper-tape application is coated with asphalt as a seal against moisture.

As noted for the buried-tape armored structures, the type of cable sheath for double-sheath Alpeth cables, with a layer of steel "sandwiched" between the polyethylene jackets, is the type of covering generally applied on important circuits where extra mechanical protection is required.

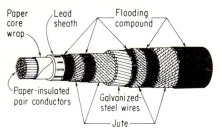

FIG. 11. Single-wire armored submarine cable.

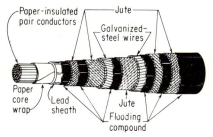

FIG. 12. Double-wire armored submarine cable.

c. Wire Armored Submarine. *Purpose.* High tensile strength and abrasion resistance for river inlet and bay crossings where excessive flexing and movement by rough water or tidal action are problems.

Structure. Jute bedding, galvanized-steel armor wires, and outer jute yarn. Asphalt floodings over protective materials when circuits are lead sheathed. The structure is modified for Alpeth-sheathed cables to omit the asphalt flooding compounds. Where the installation is especially severe, a double armored protection should be considered which basically consists of two layers of galvanized-steel armor wires with a jute cushion between the layer of armor wires.

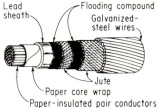

FIG. 13. Light-wire armored cable.

d. Light-wire Armored. *Purpose.* Lighter armor wire construction than the above to strengthen the cable structure for winching across marshy regions, gullies, and rocky ravines.

Structure. Same as single-armored submarine without an outer layer of jute. Uses smaller diameter galvanized armor wires.

6. Cable Conductor Color Coding

Most of the polyethylene-insulated conductor (PIC) cable manufactured today is fully color coded using the standard telephone conductor color scheme, and when possible, it is advisable to splice the conductor color-to-color in order to make full use of the color for pair identification. Some paper-insulated cables are also fully

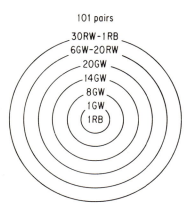

FIG. 14. Layer structure.

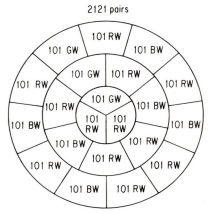

FIG. 15. Unit structure.

color coded, which is an aid to eliminate split and transposed pairs during splicing. Each color group is in either a binder or a layer group (Figs. 14 and 15).

7. Coaxial Cable

Coaxial-cable construction has provided the means to meet the requirements for larger blocks of channels on communication and UHF signals for closed-circuit networks. For broadband use on many long toll facilities, these coaxials are included in the same cable sheath with other voice and carrier circuits. Educational TV (ETV) and community antenna TV (CATV) use the closed-circuit network as the means to transmit their respective programs, and coaxial cables and short-haul microwave carry these signals. Although separate coaxials, generally, are placed along existing cable routes in many areas, there is growing interest by planning engineers to include coaxials in new cables in order to meet the anticipated demands of their customers along new cable routes.

The basic requirements for these systems are that the network of cables has uniform characteristic impedance, low losses or reflections, and proper shielding from extraneous electric fields. Relatively small tolerances in these characteristics are permitted, and the systems are designed accordingly. As there are many types of coaxials manufactured to meet these requirements, Fig. 16 is an example of one type of composite coaxial cable that incorporates the above features.

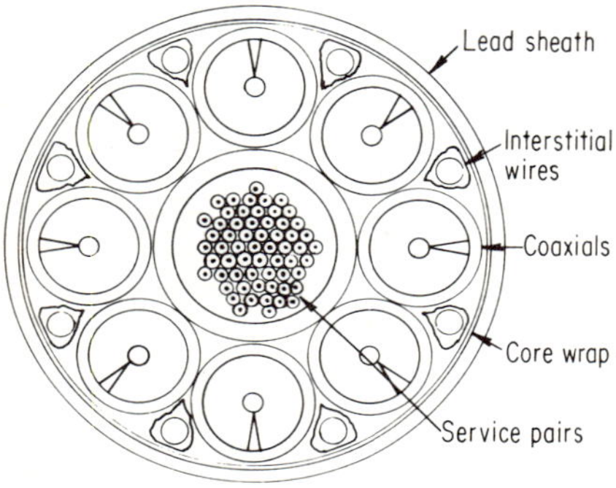

Fig. 16. Composite coaxial cable.

8. Cable Reels

Each manufacturer of cable will have standard reels on which cable will be furnished, and most companies will supply any length of cable up to the maximum capacity of the reel. In general, however, reels can be handled quite satisfactorily on standard reel trailers, which will carry 6,000 ft of cable up to 0.92 in. outside diameter and 1,000 ft of 3.18-in.-outside-diameter cable except for plain lead sheath. This reel will take 1,400 ft of 2.64-in.-outside-diameter plain lead sheath.

9. Communication Wire

Bare copper, bare copper steel, bare steel, and various compositions of insulations are used for line wires and service wires. Most of the insulated types are paired in either single- or multiple-paired formations. Owing to the effect of atmospheric conditions on the transmission quality of a bared wire circuit, most new construction uses insulated wires; however, there are many situations where a bared wire facility is desirable. The characteristic weights and equivalents of bare wire are shown in Table 6.

In places where long span construction is desirable, a single pair of 14 AWG extra-high-strength copper-steel conductors, insulated polyethylene, is being used extensively and can be placed in 350-ft spans in a heavy storm loading area. A single ridge is molded in the insulation to provide conductor identification. The wire has a minimum breaking strength of 1,100 lb and weighs approximately 35 lb per 1,000 ft.

Buried rural wire is usually made up of a single pair of 19 AWG polyethylene-insulated conductors wrapped with either a copper or aluminum spiral shield over which a polyethylene jacket has been extruded. In gopher-infested areas or where greater strength is desired, a steel mesh is embedded in the polyethylene sheath.

Buried service wire is usually a four-conductor wire of 20 AWG polyethylene-insulated conductor enclosed in a copper or aluminum spiral shield over which a

Table 6. Weights and Equivalents of Bare Wire

	Copper wire				Copper steel			Steel, regular			Steel, high strength			Steel, extra high strength
Diameter, in....	0.080	0.104	0.128	0.156	0.080	0.104	0.128	0.083	0.109	0.134	0.083	0.109	0.134	0.109
Lb per wire mile.	102	173	262	435	93	158	240	99	170	258	99	170	258	170
Min breaking strength, lb...	330	550	819	1,325	770	1,283	1,647	460	793	1,200	703	1,213	1,893	1,800

PVC sheath is extruded. Most drop wire is 18.5 AWG polyvinyl-insulated paired conductors.

SYSTEM ENGINEERING

10. Cable and Wire Facilities—General Design

Cable and wire plant is sized on the basis of growth forecasts not only to meet known requirements but to be adequate for some predetermined time. It is difficult to predict the growth pattern, the number of lines, and the types of service for a central-office area, and it is even more hazardous to estimate the growth along any given cable route. The exchange outside-plant cable network serves as a medium to connect the central office and station equipment in a manner which is compatible with signaling and transmission requirements. These requirements are usually stated in terms of circuit resistance and transmission limits. The cable networks are designed to keep the electrical losses within these limits regardless of the distance between the office and the customers. This is accomplished by planning the network around the several options of wire gages (19, 22, 24, and 26) for carrier systems and the various loading arrangements for voice frequency (H-88, H-44, etc.).

Multipling of cable pairs along the cable route is necessary to achieve high cable-pair utilization and to provide party-line association. "Multipling" indicates the appearance of the same cable pair at more than one location. Main feeder cables contain a preponderance of pairs that extend a considerable distance from the central office without dropping off circuits to a subscriber. Generally, feeder cables are large cables and consequently consist of strip-paper- or paper-pulp-insulated conductors. These cables, by nature of their construction, must be hermetically sealed at all splice closures. To gain access to these pairs by multipling or rearrangement can be costly. To offset this difficulty, distribution cables generally contain PIC insulations, and the use of ready-access closures provides the means to match the growth in many areas by reducing the costs to provide service to a new subscriber.

A further scheme that is now popularly viewed in many communication networks for new plant is the dedicated-plant concept which eliminates the need for rearrangement of cable pairs and handles the growth on shorter time intervals. This new concept in plant design involves the permanent assignment of a cable pair from the central office to each main station or customer's premises. Once a cable pair has been assigned to an address, it remains dedicated to that location whether the pair is working or idle and regardless of class of service. Any required bridging of party lines will, of necessity, be done at the central office, utilizing switchlike devices to remove the effect of other party stations during conversation. As expected, this scheme usually results in some advance in capital expenditures for additional feeder cable pairs and other apparatus. However, the savings in cost, in many instances, have been found to outweigh the carrying charges on the principal. As less total capital will be needed for future planning of a dedicated-plant network, the increased flexibility of the circuits and the overall elimination of the multipled portion of all circuits feeding a given area will provide a better transmission pattern in that area. Dedicated-plant wire, however, will not eliminate rearrangements and changes necessary to reroute customers' service to a different central office. Shortage of switching equipment and the recovery of coarse (19 and 22) gage cable plant are some of the reasons that limit the full concept of dedicated plant in some specific exchange areas.

As the multipling and rearrangements of circuits for old and new plants will exist for many years, the concept of cross connection is an important tool for the plant engineer. Cross connection is a method of semipermanently (or permanently) connecting a terminated cable pair to any one of a group of other cable pairs similarly connected. At a point of cross connection, the conductors of two cables are terminated in a common location in such a manner that jumper wires can be used to connect any pair of one terminated group to any pair of another group. A pole-mounted cross-connect terminal with a prewired stubbing cable is used to accomplish this circuit flexibility rather than the inflexible arrangement at splice closures. These weather-proof terminals with prewired and soldered conductors of plastic-insulated wires are designed to cross-connect circuits at any one location. They are available with lead-covered or Alpeth-type cable stubs in 100- to 600-pair sizes. Improvements in designs of these terminals over the years from the screw and nut type of terminals to the Type 66 Quick Clip terminals have increased the interest and economical aspects of this terminal. Cross connection serves four basic functions which have become more important over the years. They are:

1. Cable efficiency
2. Power circuit exposure
3. Transmission (removal of bridged tap)
4. Test point

Cross connection makes available a choice of feeder-cable pairs for those working circuits which are cross-connected. This flexibility promotes greater cable efficiency and higher cable fills (ratio of pairs in use to total pairs) than can be obtained otherwise. Usually it is possible to obtain up to 75 per cent without connection. With cross connection, the improvement in efficiency comes about through the elimination of unavailable pairs or those pairs not terminated at a location where they can be used or are impracticable to reallocate in order to serve a given area effectively.

Cross connection, by breaking the bridge between exposed and unexposed cable pairs, is one way to eliminate foreign power exposures. A branch cable not directly exposed to power circuits operating at high voltages but rigidly multipled to other circuits that are so exposed should be considered as exposed. Stations fed from terminals in the otherwise unexposed areas must be considered as exposed and require standard station protection. A cross-connect terminal, therefore, provides a method to control these two types of circuits.

The multipling of cable pairs on each working pair results, in some instances, in a "dead" bridge or nonworking multipled conductor. A "dead" bridge adds capacitance to the line which impairs transmission. Under some conditions, a certain amount of this loss can be tolerated. However, where transmission limits are exceeded owing to the losses introduced by bridged taps, corrective measures must be taken. The use of cross-connection points to drop off the bridge taps is one means of reducing these losses. Some of the advantages offered at these terminal locations are as a test point for trouble shooting and transmission-loss testing to select pairs for special circuits.

11. Route Selection

The route or path selection will be influenced by the cost of available right of way. Public right of way, such as highways, streets, and alleys, is usually less expensive than private-property rights of way or easements, but it may be less secure. Most governmental bodies which control public rights of way will insist that any cables, wires, or supporting structures placed therein must be relocated or removed at the owner's expense at any time the controlling agency wants use of the space the communication plant occupies. Private-property right of way may not be available at any price, or it may be so costly that public right of way, despite its insecurity, will still be the most desirable course. Both public and private right of way will be subject to political considerations, particularly the franchise terms, if any, local ordinances governing the use of public thoroughfares, traffic restrictions, and tree conditions. There will be different rules regarding the use of streets, alleys, main thoroughfares, and expressways, and these factors should be taken into account in choosing a route.

Certainly the terrain, width of the thoroughfare, and soil conditions are all factors in a decision. Probably in almost all cases the engineer will use both private and public rights of way in the design of a system and will have to make the best deal he can with both the owners of the private property and the people responsible for the public interest in the case of public property.

Before making a final decision, the engineer will have determined whether there is a single-cable or wire or a multicable requirement, for he must choose a supporting structure accordingly. One or two cables or wires may be buried direct, a pole line may suffice if two or three cables are required, but if more than this is needed, he will surely consider an underground conduit system.

Regardless of the choice of right of way a comprehensive survey is in order to locate property lines accurately and to locate the cable accurately, particularly if underground, so it can be found later on. There will probably be good reasons for locating poles precisely, and a survey by a competent surveyor will be needed. In all likelihood, the necessary permits and easements could not be acquired without an accurate survey having first been obtained. Since there will probably be other utilities in the public thoroughfare which will need to be protected from possible damage during the construction of the communication facility, a survey will be needed to locate them accurately.

Special permits are usually required to cross or occupy railroad property, and consequently, these permits should be secured well in advance of any scheduled construction work.

12. Systems Components

Most communication systems will have a main feeder which is the backbone cable or wire facility which, in turn, will split into branch or secondary feeders and which serve the distribution plant located on the customer's or subscriber's premises. The distributing terminal to which the customer's service wire is attached is an important and expensive part of the system.

The cable facilities which connect switching centers is usually referred to as trunk or toll cable.

The best main feeder facility will have the number of pairs for economical life depending upon any short-term or long-term limitations. It will have the proper gage to meet the resistance limit and transmission standards of the system, and it will be placed aerially, underground, or in conduit or buried as determined by studies of the future growth for that particular route or path. Whether or not it will have a complement for transmitting higher-frequency signals will depend on the need for ETV-CATV service or other services using frequencies beyond the voice range.

13. Aerial Construction Practices

a. General. A pole line for either aerial cable or wire must be designed to withstand wind and ice loads in addition to the cable or wire. Standard construction practices follow the storm loading map (Fig. 17), which shows the different zones for heavy, medium, and light loading and will determine the amount of anchoring, depth of pole set, and size of poles.

Non-self-supporting cable is suspended from a messenger and held by rings or lashed to the strand with a spirally wound stainless-steel lashing wire usually 0.045 in.

in diameter.   Lashing has become the most popular method because it is less expensive and does not require riding the messenger to place the rings.

There are many ways to anchor a pole line properly, and each manufacturer of anchoring devices has technical data to show the holding power of each one.   It is poor economy to anchor or brace a pole line inadequately, since the failure of one pole to support the load can endanger the entire lead.   For maximum benefit a guyed anchor should follow the standard lead-to-height ratio.   Where a suitable guyed anchor cannot be placed for reasons of space or terrain, a push brace or push-pull brace may suffice, and under extremely restrictive conditions or where only minor anchoring is required, a self-sustained pole can be used.   This consists of a block of concrete placed at the toe of the pole and another block on the side of the strain at the ground line or slightly below.   Sometimes the anchoring must be done at the next to

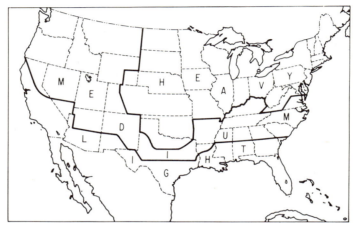

FIG. 17. Storm loading reference map.

the end pole in the lead, thus leaving a slack span between the next to the end and the end pole.

Usually there are governmental regulations, which are based on good safety practices, for determining minimum clearance for cable and wire which crosses streets, alleys, highways, railroads, etc.   These clearances should be carefully checked with the proper authorities before construction of a pole line is started.

For pole lines used exclusively by the communication facilities, the size and height of the poles will be set by the weight of the cable and suspension strand (Table 7), the storm loading and anchoring conditions, and the required clearances of the cables and wires.   Pole lines used jointly with electric power wires or cables will require poles high enough to give minimum safe clearance between the power and communication

Table 7. Galvanized-steel Suspension Strand Dimensions

Size, lb	Strand diameter, in.	Wire diameter, in.	Breaking strength, lb min	Weight, lb per 1,000 ft
2,200	$\frac{3}{16}$	0.065	2,400	77
6,000	$\frac{5}{16}$	0.109	6,000	225
10,000	$\frac{3}{8}$	0.120	11,500	270
16,000	$\frac{7}{16}$	0.144	18,000	390
25,000	$\frac{1}{2}$	0.165	25,000	510

facilities and space for the power wires, transformers, etc., as well as minimum clearance for the communications cables and wires. Safety code requirements in most places will be met by the span lengths and clearances as shown in Tables 8 to 12.

Clearances over waterways which are not classed as navigable are not prescribed by code; however, clearances over navigable streams in the United States are under the

Table 8. Horizontal Clearances—Poles and Stubs

Between nearest part of pole or stub	Minimum horizontal clearance	Remarks
Fire hydrant, signal pedestals...	3 ft	Obtain 4 ft or more if possible
Nearest rail of main track......	12 ft	
Nearest rail of sidings.........	7 ft	At loading siding leave room for driveway
Curb lines..................	6 in.	Measure to street side of curb
Power lines 0–750 volts.........	5 ft	Provided the nearest support for the power wires does not exceed 96 ft
Power wires 750–15,000 volts...	9 ft	Provided the nearest support for the power wires does not exceed 150 ft
Power cables.................	5 ft	Provided the nearest support for the power cable does not exceed 120 ft

Table 9. Clearances over Roads, Alleys, Driveways, Walkways, and Railroad Tracks

	Basic clearance at 60°F, ft	Remarks
Public street or roads..........	18	Clearance for drops or guys insulated against the highest voltage to which they are exposed up to 8,700 volts may be reduced to 16 ft at one side of the road
Driveways..................	10	
Alleys.....................	15	
Walkways—pedestrian only.....	8	10 ft for voltage to ground more than 160 volts
Railroad tracks handling freight cars on which men are permitted	27	Does not apply to overhead trolley electric railroads. May be reduced to 25 ft for guys and cable carried on messenger
Other railroads...............	18*	

* For crossing over railroads handling only cars lower than 16 ft, the clearance may be reduced by an amount equal to the difference between the highest car handled and the highest ordinary car but never below 18 ft.

jurisdiction of the U.S. Army Corps of Engineers. State and local authorities may exercise jurisdiction over other bodies of water. Clearances should be adequate for any special movement of vehicles and should be worked out with the property owners or other responsible persons. Clearances over rooftops should be at least 8 ft where the roof may be walked upon and a minimum of 2 ft otherwise.

In general there should be adequate separation between communication and power conductors at the pole (Table 13) to provide safe working areas for the linemen and

Table 10. Clearance for Facilities Running along Public Roads and Alleys

	Basic clearance at 60°F, ft	Remarks
Along highways in urban areas...............	18	
Along highways in rural areas................	14	
In both urban and rural areas where line is accessible to pedestrians only	8	10 ft required for conductors carrying over 160 volts (does not apply to cable)
In rural areas if no part of the line overhangs the traveled part of the road and it is unlikely that loaded vehicles will be crossing under the line	13	
Along alleys................................	15	

Table 11. Recommended Vertical Clearance at Span Crossings

Between communication facility and	Clearance, ft		Remarks
	Wires and cables	Guys	
Line wires 300 volts or less to ground.....	2	2	4 ft required if within 6 ft of pole (not applicable to guys)
Neutrals associated with systems of 22,000 volts or less to ground if effectively grounded	2	2	If within 5 ft of pole, wire-to-pole clearance controls
Line wires 301–750 volts to ground.......	4	2	
Line wires and service drops 750–8,700 volts to ground	4	4	If crossing is within 9 ft of pole, wire-to-pole clearance will control
Line wires 8,700–50,000 volts to ground...	6	4	
Neutrals of systems above 22Kv to ground and any neutral not effectively grounded	. . .	. . .	Same clearance as phase wires
Power cables having effectively grounded continuous metallic sheaths or messenger, any voltage	2	2	If within 5 ft of pole, wire-to-pole clearance will control
Other power cables......................	. . .	. . .	Same clearance as for open power wires of same voltage
Service drops 750 volts or less to ground..	2	2	4 ft required for cable crossing above open service drops*
Guys, span wires, lightning-protection wires	2	2	
Foreign communication cables, wires, and guys	2	2	

* Where 2 ft of clearance is permitted, if the sag increase of the upper conductor exceeds 1½ ft under storm loading (at the crossing point), the clearance should be increased by the actual sag increase less 1½ ft.

Table 12. Maximum* Span Lengths for 18-foot Highway

Kind of wire	Clearance at 60°F	
	Span lengths, ft	Extra initial clearance, ft
Multiple-line wire:		
120 Support..	165	
109 Support—min or recommended sag...........	155	
Drop wires:		
Single using min sag...........................	100	0.5
Single using recommended sag..................	130	0.5
Multiple wires using min sag...................	175	1.0
Multiple wires using recommended sag.............	175	0.5
Rural...	125	
Copper-line wires:		
104...	175	3.0
128...	175	1.5
165...	175	1.0
Copper-steel-line wires:		
080...	135	1.0
140...	155	
128...	170	0.5
Steel-line wires:		
109H..	140	
109F..	155	
109...	175	1.0

* Assumes midspan occurring somewhere above traveled part of road.

Table 13. Recommended Clearances between Communication and Power Conductors

Power conductor	Midspan clearance to communication conductor, in.	Min separation at the pole, in.
Effectively grounded neutrals associated with 22,000 volts and less.....................	30	40
Conductors of 750 volts and less..............	30 but not below line of sight	40
Conductors at 750–8,700 volts...............	30 above line of sight	40
Conductors of 8,700–50,000 volts and associated neutrals not effectively grounded	45 but at least 30 above line of sight	60
Effectively grounded neutrals associated with 22,000–50,000 volts	45	60
Any grounded power cable...................	30	40
Spacer-type power cable 750 volts and less....	30 but not below line of sight	40
Spacer-type power cable 750–8,700 volts......	30 above line of sight	40
Spacer-type power cable 8,700–50,000 volts....	45 but at least 30 above line of sight	60

Table 14. Recommended Span Lengths and Cable Sags
(No wind and 60°F)

Wt of cable, lb/ft	Size of strand	Span lengths, ft										
		Estimated sag, in.										
		100	125	150	175	200	225	250	275	300	325	350	375
0.2	6M	6	8	12	16	20	25	30	36	42	48	55	62
	10M	3	5	7	10	13	16	20	24	28	33	38	43
0.4	6M	8	12	16	21	26	32	38	45	52	60	67	75
	10M	5	7	10	14	18	22	27	32	37	43	49	55
0.6	6M	10	14	19	25	31	38	45	52	60	68	77	87
	10M	6	9	13	17	22	27	32	38	44	51	58	65
0.8	6M	11	16	22	28	35	42	50	58	67	76	85	94
	10M	7	11	15	20	25	31	37	44	51	58	66	73
1.0	6M	13	18	25	31	39	47	55	63	72	82	92	102
	10M	8	13	17	23	29	35	42	49	56	64	72	81
1.2	6M	14	20	27	34	42	50	59	68	78	87	98	
	10M	9	14	19	25	32	38	46	53	61	70	79	88
1.4	6M	15	22	29	36	45	53	62	72	82			
	10M	10	15	21	27	34	42	49	57	66	75	84	93
1.6	6M	16	23	30	39	47	56	66					
	10M	11	17	23	30	37	45	53	61	70	79	89	99
1.8	6M	17	24	32	40	49	59						
	10M	12	18	24	32	39	47	56	65	74	84	94	104
2.0	6M	18	25	34	42	52	61						
	10M	13	19	26	33	41	50	59	68	77	88	98	109
2.2	6M	19	27	35	44	54							
	10M	14	20	27	35	43	52	61	71	81	91	102	113
2.4	6M	20	28	36	40								
	10M	15	21	29	37	45	54	64	74	84	95	106	117
2.6	10M	15	22	30	38	47	56	66	76	87	98	109	121
	16M	11	16	22	29	36	44	52	61	70	79	89	99
2.8	10M	16	23	31	40	49	58	68	79	90	101	113	125
	16M	12	17	23	30	38	46	54	63	72	82	92	102

Table 14. Recommended Span Lengths and Cable Sags (*Continued*)

Wt of cable, lb/ft	Size of strand	Span lengths, ft — Estimated sag, in.											
		100	125	150	175	200	225	250	275	300	325	350	375
3.0	10M	17	24	32	41	50	60	70	81	92	104	116	
	16M	12	18	24	31	39	47	56	65	75	84	95	105
3.5	10M	19	27	35	44	54	65	75	87	99	111	124	
	16M	14	20	27	35	43	52	61	70	80	91	102	113
4.0	10M	20	28	37	47	57	68	80	92	104	117		
	16M	15	21	29	37	46	55	65	75	85	96	108	119
4.5	10M	21	30	40	50	61	72	84	96	109	123		
	16M	16	23	31	40	49	59	69	79	90	102	113	
5.0	10M	22	31	41	52	63	75	88	101	114			
	16M	17	24	33	42	51	61	72	83	94	106	119	
5.5	10M	24	33	44	55	66	79	92	105				
	16M	18	26	35	44	54	64	75	87	99	111	124	
6.0	10M	24	34	45	56	69	81	95					
	16M	19	27	36	46	56	67	78	90	102	115		
6.5	16M	20	28	38	48	58	70	81	94	106	119		
	25M	12	18	25	33	41	50	60	70	80	91	102	113
7.0	16M	21	29	39	49	60	72	84	96	109			
	25M	13	19	26	34	43	52	62	72	83	94	106	118
7.5	16M	22	31	40	51	62	74	87	100	113			
	25M	13	20	28	36	45	55	65	75	86	98	110	122
8.0	16M	22	31	41	52	64	76	89	102				
	25M	14	21	29	37	46	56	67	78	89	101	113	
8.5	16M	23	32	42	54	65	78	91					
	25M	15	22	30	38	48	58	69	80	91	104		

Table 15. Recommended Wire Sags
(Span length avg. 100 to 250 ft. All loading areas)

Span, ft	Multiple-line wire					Paired rural wire				
	Temperature, °F					Temperature, °F				
	120	90	60	30	0	120	90	60	30	0
	Sag, in.					Sag, in.				
100	9	9	8	7	6	4.5	3.5	3	2.5	2
110	11	10	9	8	8	5.5	4.5	3.5	3	2.5
120	13	12	11	10	9	6.5	5	4.5	3.5	3
130	16	14	13	12	10	7.5	6	5	4	4
140	18	17	15	13	12	8.5	7	6	5	4.5
150	21	19	17	15	14	10	8	6.5	5.5	5
160	24	22	19	17	16	11	9.5	7.5	6.5	6
170	27	24	22	20	18	12.5	10.5	8.5	7	6.5
180	30	27	25	22	20	14	12	9.5	8	7.5
190	34	30	27	25	22	16	13	11	9	8
200	37	34	30	27	24	17.5	14.5	12	10	9
210	41	37	33	30	27	19	16	13	11	10
220	45	41	36	33	29	21	17.5	14.5	12	11
230	49	45	40	36	32	23	19	16	13	12
240	53	49	43	39	35	25	21	17	14.5	13
250	58	52	47	42	38	27.5	22.5	19.5	15.5	14

for proper clearance at midspan. Proper tensioning of cable and wire is extremely important for good maintenance, and adequate sag (Tables 14 and 15) will allow for expansion and contraction without undue strain on the facility.

In general, the recommended spacings at the pole are adequate to meet midspan requirements for span lengths up to 150 ft and, in some cases, longer spans. However, for longer spans the engineer must allow increased separation at the pole in order to meet midspan requirements.

b. Classification of Pole Lines. It is common practice to classify pole lines supporting only communication circuits into five classes based on their relative service values, and appropriate structure and maintenance levels are recognized for each class (Table 16). The appropriate pole size (Table 17), guying requirements (Table 18),

Table 16. Pole-line Classification

Class of line	Designation of line	Relative strength factor
AA	Toll, open wire or cable	1.33
A	Toll, open wire or cable	1.00
B	Toll, open wire or cable	0.67
C	Toll, open wire only	0.50
C	Exchange open wire or cable	0.50
R	Open wire only—maximum of 10 wires	0.33
JB	Joint-use lines carrying communication wires or	1.77
JC	cable and electrical supply circuits	0.88

Table 17. Minimum Class of Treated Pole for Average Conditions

Equivalent wire load — Effective communication wires	Pole-line class	Class of new poles span, ft							Actual wire loads — Total communication wires
		100	130	150	175	200	225	250	
10 or less	A	7	7	7					1–15
	B	7	7	7					
	C	7	7	7	7	7	7	7	
	J	7	7	7	6				
11–14	A	7	6	6					16–20
	B	7	7	7					
	C	7	7	7	7	7	6	6	
	J	7	6	5	5				
15–20	A	5	4	4					21–30
	B	7	6	5					
	C	7	7	7	6	5	5	4	
	J	6	5	4	4				
21–27	A	4	3	3					31–40
	B	6	5	4					
	C	7	6	5	5	4	4	3	
	J	5	7	3	3				
28–34	A	4	2	2					41–50
	B	5	4	4					
	C	7	5	5	4	4	3	3	
	J	4	3	3	2				
35–40	A	3	2	1					51–60
	B	5	4	3					
	C	6	5	5	4	3	3	2	
	J	4	3	2	1				

Table 18. Minimum Sizes of Anchor-guyed Poles and Stubs

Ratio of lead/ height of anchor guys	Sum of anchor guys	Minimum class of anchor-guyed pole or stub. Any storm loading area or kind of timber — Length of pole					
		25	30	35	40	45	50
1	20M or less	7	6	6	5	5	4
	21M–30M	6	5	5	4	4	3
	31M–40M	5	4	4	3	3	2
	41M–50M	4	3	3	2	2	1
Less than 1	20M or less	6	5	5	4	4	3
	21M–30M	5	4	4	3	3	2
	31M–50M	4	3	3	2	2	1

Table 19. Recommended Pole Depth of Setting

Pole length, ft	Firm earth, ft	Solid rock, ft
18	4	3
20	4	3
22	4	3
25	5	3
30	5½	3½
35	6	4
40	6	4
45	6½	4½
50	7	4½
55	7½	5
60	8	5
65	8½	6
70	9	6

and depth of setting (Table 19) for various pole-line classes and loads are essential considerations for a properly designed communication pole lead.

The suppliers of pole products have standardized so that the class and length of each timber give the equivalent strength of all the other kinds of timber. The minimum circumference at 6 ft from the butt may be different for each kind. The engineer may select the kind of timber best suited to the area as the most economical at the moment, and if properly treated, all kinds will give good service. It will be advisable to study the specifications of each pole product supplier before specifying a particular kind of timber.

c. Guying. Pull is the distance in feet measured as shown in Fig. 18.

Guys may be omitted if the pull is less than 3 ft for 6M suspension strand and if less than 2 ft for 10M strand. Frequently a push-pull brace (Fig. 19) is used as a substitute for an anchor guy.

When the end pole cannot be properly guyed, it is good practice to anchor-guy to the next to last pole and dead-end the messenger at this pole and then slack span to the

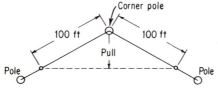

FIG. 18. Illustration of pull.

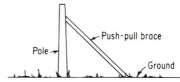

FIG. 19. Push-pull brace.

end pole, but even then, the end pole should be ground braced and the span should not exceed 100 ft for a 6M strand and 75 ft for a 10M strand.

The most used type of anchor is an expanding patent anchor designed to be used with specified guy strand and rod dimensions (Table 20).

d. Wire Equivalents. The customary method for determining the total load on a pole is to equate the load imposed by the several types of attachments (wires, cables, terminals, load pots, transformers, etc.) to a common basis of wire equivalents (Table 21). There are two sets of equivalents, one for equating attachments to equivalent 104 communication wires and one for equating them to equivalent No. 4 covered power wire. Communication engineers will generally use the 104 base.

To calculate the wire equivalents for sizes of wires and cables not shown in table, use the following method: For a heavy loading area, multiply the diameter in inches +

Table 20. Patent Anchors (Guy Strand and Rod Dimensions)

Type and size	Size of guy strand	Size of guy rod and marking	Size of hole, in.	Vertical depth of setting, ft	Augur re-quired	Special equipment required to place anchor
Expanding plate anchors: 3-way 6 in. 8-way 6 in.	2.2M or 6M	½ in. × 7 ft Guy rod 6M	6	About 5	Yes	Anchor tamping bar
3-way 8 in. 4-way 8 in. 8-way 8 in.	10M 2–6M	⅝ in. × 8 ft Guy rod 12M	8	About 5½	Yes	Anchor tamping bar
3-way 10 in. 4-way 10 in. 8-way 10 in.	1–16M or 1–6M and 1–10M	¾ in. × 9 ft Guy rod 16M	10	About 6½	Yes	Anchor tamping bar
4-way 12 in.	2–16M	1¼ × 10 ft Guy rod 32M	12	About 7	Yes	Anchor tamping bar

1.0 × 0.91 to obtain equivalent 104 wires or ×0.72 to obtain equivalent No. 4 covered power wires.

14. Underground Conduit

Underground conduit for communication cable is available in various kinds of materials: concrete, vitrified clay, asbestos cement, and several forms of fiber and plastic compounds. Concrete and clay are made in multiple-duct cross sections of one, two, three, four, six, eight, and nine holes. Almost any desired total cross section can be made by using combinations of pieces. The desired cross section of single-duct conduit can be made up by strapping the individual ducts together in accordance with the manufacturer's instructions.

Concrete conduit can be obtained with straight, mortise-and-tenon, or bell-and-spigot ends. Bell and spigot is the most popular type of concrete conduit. The approximate dimensions of vitrified-clay and bell-and-spigot conduit are given in Table 22.

Most standard communication cable conduit is made with a 3½-in. bore, which is large enough for all full-size cables up to 2,400 pairs.

Under reasonably firm soil conditions and average traffic loads, no protection is necessary around conduit having a cover of 24 in. of tamped earth. When crossing under a railroad track, it is important to consult with the railroad company for its specifications for depth and construction conditions. It is extremely important in the construction of a good conduit system to prepare the trench bed properly so that the conduit can be placed evenly and good joints made at all times.

No trouble should be experienced in pulling a full-size cable through a 1,000-ft section of conduit that is properly placed and has no bends. However, it is most difficult under actual field conditions to avoid either vertical or horizontal bends, and the section lengths between pulling joints will need to be adjusted accordingly (Fig. 20).

All the suppliers of conduit material provide various special fittings and pieces in order to make proper curves, to extend conduit of a different type, to allow a lateral connection, etc. To construct a proper bend, the makers of clay and concrete conduit provide mitered pieces and short, straight pieces while the makers of fiber and plastic usually provide bends with different radii to meet proper curve dimensions.

Table 21. Table of Wire Equivalents

Attachment	Telephone wire base (effective 104 tel. wires) Storm loading area			Power wire base (effective No. 4 covered power wires) Storm loading area		
	Heavy	Medium	Light	Heavy	Medium	Light
Communication plant:						
Bare open wire, 109, 104 or smaller, per wire	1	1	1	0.8	0.7	0.3
Bare open wire, 128, 134 or larger, per wire	1	1	1.3	0.8	0.7	0.4
Covered paired wire per pair or covered single wire per wire	1	1	2.5	0.8	0.7	0.7
Strand, all sizes	2	2	5	2	2	2
Cable and 6,000-lb ($^5/_{16}$ in.) strand	4	5	15	3	3	4
Cable and 10,000-lb ($^3/_8$ in.) strand	5	6	22	4	4	6
Cable and 16,000-lb ($^7/_{16}$ in.) strand	6	7	30	5	5	8
Cable and 25,000-lb ($^1/_2$ in.) strand	7	8	35	6	6	10
Cable terminal, "B" or "BB" type, 202 pair and less or "BD" type, all sizes	1	1	4	1	1	1
Cable terminal, "B" or "BB" type, more than 202 pair	1	1	6	1	1	2
Cable loading pot	1	1	4	1	1	1
Service drops, per unbalanced drop	1	1	2	1	0.5	0.5
Clothesline on Class C line poles per unbalanced clothesline	8	14	50	3	5	8
Clothesline on Class J line poles per unbalanced clothesline	4	7	25			
Power plant:						
Covered wire, No. 8 AWG (approx. 0.26 in. OD) or smaller per wire	1.1	1.3	2.5	0.9	0.9	0.7
Covered wire, No. 6 AWG (approx. 0.32 in. OD) per wire	1.2	1.4	3.1	1.0	1.0	0.9
Covered wire, No. 4 AWG (approx. 0.38 in. OD) per wire	1.3	1.5	3.7	1	1	1
Covered wire, No. 0000 AWG (approx. 0.65 in. OD) per wire	1.5	1.9	6.3	1.2	1.3	1.7
Covered wire, 500,000 cir mils (approx. 1.11 in. OD)	2	3	11	2	2	3
Covered wire, 1,000,000 cir mils (approx. 1.53 in. OD)	3	4	15	3	3	4
Covered wire, 2,000,000 cir mils, (approx. 2.15 in. OD) or larger	3	5	20	3	4	6
Power cable on strand (approx. diam of cable 2.56 in. or less)	6	7	30	5	5	8

Table 21. Table of Wire Equivalents (*Continued*)

Attachment	Telephone wire base (effective 104 tel. wires) Storm loading area			Power wire base (effective No. 4 covered power wires) Storm loading area		
	Heavy	Medium	Light	Heavy	Medium	Light
Suspension wire extending transversely between two pole lines and supporting trolley contact wires:						
One contact wire.................	6	10	25	5	7	7
Two contact wires................	12	20	50	10	14	14
Four contact wires...............	18	30	75	15	21	21
Bracket and one trolley contact wire on one side of pole line.....	2	2	8	2	2	2
Brackets and two trolley contact wires, one on each side of pole line.....	3	3	9	3	2	3
Bracket and two trolley contact wires, over tracks on same side of pole line.....	5	6	19	4	4	5
Transformers, 37½ kva or less.........	1	1	6	1	1	2
Transformers, over 37½ kva.........	1	2	9	1	2	3
Transverse clearance attachment for service drop above telephone attachments per wire.................	1	2	4	1	2	1
Service drops per unbalanced drop wire.....	1	1	3	1	1	1
Street lamp supported by mast arm (not bracket).................	1	1	3	1	1	1

Table 22. Approximate Conduit Dimensions

No. of ducts	Bell and spigot, avg. length 36-in. piece, in.				Clay, avg. length 40 in., in.	
	Bell end		Spigot end			
	W	H	W	H	W	H
4	11 1/16	11 1/16	9 1/8	9 1/8	9	9
6	15 3/16	11 1/16	13 1/4	9 1/8	13	9
8	19 1/4	11 1/16	17 3/8	9 1/8	17	9
9	15 3/16	15 3/16	13 1/4	13 1/4	13	13

The main purpose of a conduit system is to provide a means for placing cable underground at some future desired time without digging a trench each time a cable is to be placed. The secondary purpose is to protect the cable from damage from foreign sources. Therefore, it is most important that care is taken to make good joints so the ducts will remain clean for the next cable and so the conduit will stay in place. Trying to save cost at the expense of a good job is poor economy in the long run.

Subsidiary duct sections, the conduit from the main run to a pole, building, or underground pedestal, should be placed with equal care and with the minimum number of bends.

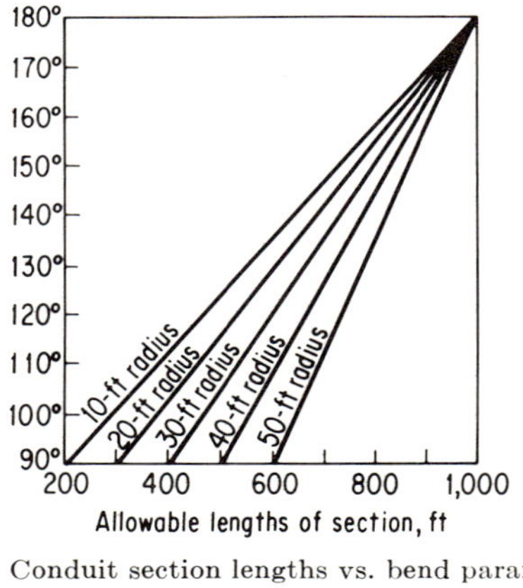

FIG. 20. Conduit section lengths vs. bend parameters.

The manholes in a conduit system are provided for the purpose of working on the cables which will be pulled into them and should be designed to allow adequate working room for the cable splicer and space for the cables, load pots, and other equipment which may be placed there. The headroom must be adequate for racking the cables plus a clear space of 12 in. between the floor and the lowest cable position and 18 in. between the roof and the highest racking position. Sufficient working space will be provided if the cables are racked vertically on 9-in. centers. For single racking, the length should be 8 ft 6 in. and width 4 ft 6 in., and when double racking, the width should be increased to 5 ft 0 in. and the length to 10 ft 6 in. in order to allow proper splice and working space. Where side-wall duct entrances are necessary, add 3 to 6 in. to the length.

Unless a larger opening is needed for ventilation or large loading coil cases, a 27-in. cover is usually provided. However, 30-in. covers are available for the unusual situation.

If the finished street or surface grade is raised after a manhole frame and cover are installed, it is common practice to place a manhole extension ring or rings to bring the cover up to proper grade.

15. Electrical Protection

Communication outside plant is subject to the hazards of voltages and currents in excess of normal operating quantities by virtue of exposure to lightning and energized power conductors. Without suitable protective devices and practices such exposure could create conditions where damage to property and injury to personnel would be severe. Therefore, electrical protection practices are designed to provide as much protection to life, equipment, and property as seems reasonable and practical.

Communication conductors are generally classified as exposed or unexposed, and protection must be provided accordingly. Conductors are considered to be exposed where they might come in contact with power circuits of greater than 300 volts or where they are subject to the effects of lightning. Conductors are unexposed under the following conditions:

1. Cables entirely underground from the switching center to the subscriber's premises and cables entirely underground from switching point to switching point.

2. Cables and wires carried on walls, fences, or poles entirely within a city block which is fed by an underground cable to the block, provided there is no open-wire power circuit with over 300 volts that may be contacted.

3. If the power wires are in well-grounded metallic sheath cables or in nonmetallic sheath cables supported by well-grounded conductive messengers or in fixed wiring either in or on buildings, the voltages may exceed 300 volts and be considered unexposed.

4. Aerial cables extending outside a block where the aerial construction of power circuits over 300 volts is prohibited.

All communication conductors which are not specifically classified as unexposed as in 1 to 4 shall be considered to be exposed.

It is permissible for a complement of cable pairs considered unexposed to be within the same cable sheath having a complement of exposed pairs, but it is not permissible to multiple pairs serving exposed with pairs serving unexposed areas.

The protection measures used to supplement the basic protective features of insulation and conductivity incorporated in the design of the facility itself employ the following basic principles of protection.

1. *Shielding:* diversion of foreign currents before contact with communication conductor

2. *Grounding:* diversion of excess current from communication conductor to ground

3. *Current control:* impedance coils or fuses

4. *Parallel conductivity:* provision of parallel paths thus reducing current flow

5. *Voltage limitation and equalization:* nonlinear resistances, discharge gaps, bonds, gas tubes, semiconductor diodes

Where grounding is employed, it is important to know the extent to which it will be effective, particularly if it is a ground to earth because of the differences in earth resistances.

There are no exact rules for the provision of protection devices for communication circuits, but certain recommended measures will generally suffice to meet safety codes for most situations encountered in practice.

RECOMMENDED PROTECTION MEASURES

16. Central-Office Equipment

Measures. Equipment connected to line wires should be protected by a fuse element between the line wires and the office and the point where exposure begins. Fuse elements shall be 7-amp fuses or 6-ft lengths of 24 AWG or smaller which will fuse on current values not high enough to overheat central-office protectors.

Protective Units. Use 0.003-in. open space cutouts on all entrance conductors. Use of heat coils or heat-coil fuses is optional.

Ground. No. 6 AWG copper ground wires are used for small offices, and No. 2 or larger size wire is used for large offices. Power-line ground and central-office ground should be equivalent and connected to local water mains or other low-resistance systems.

17. Aerial Cable

a. Joint Use with Power Line—Exposure More Than 1.5 Miles. *Measures.* Bond the cable sheath to ground at the start and end of a joint-use section. Place additional bonds at 1-mile intervals along the cable route. Bond the sheath to all uninsulated guy wires.

Protective Units. At cable terminals, use open space cutout protectors or use unprotected terminals and crossarm arrestors. At junctions with open wire, use fused terminals with open space cutouts. Aerial or buried wires may be used to provide added protection if required.

Ground. Ground the cable sheath at terminal location. Use No. 6 AWG copper wire to ground the rod or ground the system to establish 15 ohms or less resistance. Bond all grounds to the power company multiground neutral, to all underground cable, and to the central-office ground to complete a common ground system.

b. Lightning Protection for Non-joint-use Lines. *Measures.* Bond all cable sheaths to ground at $\frac{1}{4}$-mile intervals.

Protective Units. Install cable terminals with open space cutout protection, or use unprotected terminals and crossarm arrestors to connect to ground and to the cable sheath. Aerial or buried shield wires may be used for additional protection.

Ground. Use No. 6 AWG copper ground wire, and use a ground rod or electrode to approximate a resistance of 15 ohms or less.

18. Buried Cable

Measures. Select high-dielectric cables for high-lightning areas, preferably double-sheathed Alpeth or Stalpeth cables. Steel-taped sheath protection provides some low electrical induction and adds mechanical protection to the structure if required by the location. Placement of spaced shield wires above the cable lowers the instance of lightning damage to the cable in critical areas.

19. Underground Cable (Conduit Placement)

Measures. Bond metallic sheaths or shields of all cables together at each manhole.

20. Open Wire

a. Joint Use with Power Lines under 5 Kv. *Measures.* Use 0.006-in. open space cutouts or sawtooth arrestor blocks in a pole-mounted protector unit for one to three pairs. For four or more pairs of wires, use crossarm arrestors. Use No. 16 AWG bridal wire to connect open-wire lines to crossarm arrestor units at 20-ohm intervals along the line and 600 ft from the end of joint-use sections. Also, use protectors where circuits branch off the main line.

Ground. Use No. 6 AWG copper ground wire. Ground rods should measure 15 ohms or less resistance and should be tied to the separate multiground neutral of the power company to obtain an improved protection afforded by the rods in parallel.

b. *Joint Use with Power Lines over 5 Kv.* *Measures.* Use high-voltage protector units spaced at the same intervals as noted in *a* above. Use heavier No. 10 AWG bridal wire to connect open-wire lines to high-voltage arrestor units.

Ground. Use the same size of wire and grounding techniques as noted for *a* above.

c. *Non-joint-use Lines.* *Measures.* Use 0.006-in. cutouts or sawtooth arrestors as noted in *a* above. In areas subjected to severe lightning, crossarm arrestors are recommended for placement every 2 miles. Space cutouts or sawtooth arrestors are placed on every third or fourth pole from switching centers and every fourth or fifth pole from a fused terminal when used at the junction of aerial cable. In areas where there are few or no subscriber stations, a good practice is to place protection units every 2 miles along the route.

d. *Exposure to Inductive Interference of Power Line—Over 10 Miles.* *Measures.* Install drainage units in addition to standard protection equipment. Space protection equipment 3 miles from the start of the joint-use section. For single circuits, space units at every 7 miles and every 10 miles for two or more circuits. Only two circuits need to be drained on lines with two to five circuits.

Ground. Tie separate ground to power company multigrounded neutral.

e. *Frequent High-voltage Faults or Interference of Parallel Power Line.* *Measures.* Install short-circuiting relay protectors at 20-ohm intervals along the line, also at 600 ft from end of parallel section of power line.

Ground. Use No. 12 AWG copper wire for ground wire, and establish ground resistance of 15 ohms or less by a ground rod or electrode. Parallel ground rods by connection to the power company multigrounded neutral.

21. Subscriber Station

a. *Exposed Area.* *Measures.* Use station protectors equipped with 0.003-in. open space cutouts and 7-amp fuses, or use fuseless station protectors. For one protector, use No. 14 AWG ground wire; for two protectors, use No. 12 AWG ground wire; for three or more protectors, use No. 6 AWG ground wire.

Ground. Establish ground at a resistance of 25 ohms or less by using an electrode or ground rods. Connect the communication ground wire to the power company multigrounded neutral. If ground rods are used, allow a 6-ft separation between power and communication ground rods. Bond the rods together with No. 14 AWG copper wire.

b. *Located at Power Station or Substation.* *Measures.* Use a neutralizing transformer designed to match the impedance of the system.

c. *Exposed Remote Rural Location.* *Measures.* Use station protectors equipped with sawtooth arrestor blocks. Also, use high-voltage arrestors approximately $\frac{1}{2}$ mile from the station. Where a number of circuits require protection, a crossarm arrestor with 0.010-in. gap may be substituted for the high-voltage protector.

Ground. Establish 25 ohms or less ground resistance, and tie the ground to the power company multigrounded neutral by a parallel ground-rod connection.

d. *Radio and Television Transmitter Location.* *Drop Wire Feed.* *Measures.* Use crossarm arrestors equipped with sawtooth protector blocks located at cable terminals and drop-wire feed. Also, use standard station protectors on drop wires located at the station. At the cable terminal or not less than 500 ft from the station, establish 25 ohms or less ground resistance and connect to the cable sheath.

Ground. Ground all crossarm arrestor units to the cable sheath. Also, bond the station protector ground to the radio or television station ground.

Direct Cable Feed. *Measures.* Use standard station protectors on all working and nonworking cable pairs.

Ground. Bond the cable sheath to the radio or television station ground at the station location.

e. Explosive Area.   *Measures.*   All station wire must be placed in sealed conduit and extended beyond the explosive area.   When standard station protectors are used, such protectors must be placed outside the explosive area.   Specially designed station apparatus for explosive atmospheres are available and, when authorized, may be used in these locations.

22. Private Branch Exchange (Exposed Areas)

Measures.   Equip all lines with standard station protector units and low-voltage sneak current fuses, heat coils, or heat-coil fuses.   To connect three protectors or less, use No. 10 AWG copper wire and No. 6 AWG wire for four protectors or more.

Ground.   Establish a ground resistance of 25 ohms or less using ground rods or a ground electrode.

Chapter 22

CENTRAL-OFFICE CONFIGURATIONS

C. JULIAN GUSTAFSON,* *Consultant*

PLANNING FOR BUILDINGS AND EQUIPMENT

1. Long-range Fundamental Plan

In most telephone companies the engineering organization is responsible for preparing plans and arranging for the construction of outside plant, buildings, and central-office equipment necessary to meet the requirements for telephone service. The engineering organization is also responsible for the review and evaluation of various plans for meeting telephone-service requirements in the most practical and economical manner.

In view of the appreciable intervals usually prevailing between the time that the need for additional facilities is recognized and the time that these facilities become available for service, a long-range fundamental plan is necessary to establish the general program of procedure. The plan anticipates not only immediate requirements but also the needs of an ultimate period some 15 to 20 years hence.

The fundamental plan is generally prepared from data furnished by the traffic and commercial organizations pertaining to such considerations as estimated population growth, character of the neighborhood, telephone usage, and service requirements. The fundamental plan is usually developed in considerable detail for the immediate period of the next 5 years and on a broader basis for the 10- and 15-year periods leading up to the ultimate period.

2. Schedule[1]

A coordination schedule is developed on the basis of commercial surveys and fundamental plan studies. The schedule as set up in detail includes dates for starting and completing the various important steps which will be involved in the project from the time it is first considered until the office is placed in service. Here a PERT technique is useful.

The program of coordinating the various actions in connection with the establishment of a central office includes the following general procedures:

1. Determination of type of equipment to be used and the date when required for service. Normally, the completion of the building is timed to coordinate with the equipment shipping schedules to avoid carrying unused space in a new building for any appreciable length of time.

* Retired from Western Electric Company, Inc.

2. Recommendation for and purchase of lot.

3. Information from the commercial department on line and station estimates and classifications on which the traffic study is based.[2]

4. Data from the traffic department on the traffic study.[3]

5. Determination of equipment requirements.

6. Office-space requirements from various departments for other than equipment needs.

7. Preparation of rough floor-plan sketches.

8. Review of layouts with the departments involved.

9. Preparation of cost estimates.

10. Development of final plans after approval of management.

LAND AND BUILDINGS

3. Siting

a. Theoretical Wire Center. One of the prime factors to be considered in the selection of a site for a new central office is the location of the wire center. If, for appropriate reasons, it is not practical to procure a suitable site at the theoretically determined wire center, a selection is usually made by considering the advantages and disadvantages of alternate sites which are available. For example, a theoretical wire center may be ruled out because of flood hazards.

b. Size of Lot. The lot should be of ample size to meet the anticipated ultimate requirements of the building, taking into consideration any municipal regulations applying to the location with respect to property lines and setback from the street.

In addition to providing sufficient space for orderly expansion of the building, a liberal amount of land provides ample light and air, reduces fire exposure, permits flexibility in arranging the building on the lot, and allows for appropriate landscaping. The need for employee and company vehicle parking should be considered in view of increasing municipal requirements for off-street parking. Enough land should be purchased initially to provide for necessary parking even though not currently required by local ordinances. It is expected that such zoning will be in effect when subsequent additions are made.

c. Separation from Potential Hazards. Consideration should be given toward avoiding locations likely to be affected by hazards such as the following:

Flooding due to inadequate runoff or proximity to streams
Inadequacy of storm sewers
Proximity to airports
Proximity to industrial plants generating electrical interference or emitting dust, corrosive fumes, smoke, or dirt
Vibration due to nearby railroad or truck route
Proximity to oil or gas transmission lines
Proximity to highly congested areas where buildings are not of fire-resistant construction
Areas having establishments where explosives and flammable materials are manufactured, stored, or used
Areas having defense plants, air bases, oil refineries, etc.
Transmission interference from nearby electric power plants or transmission lines
Type of neighborhood through which employees must pass

d. Site Conditions. Conditions of the land must be considered, since they affect building costs, such as:

Rock excavation or blasting
Soft bottoms requiring pile foundations
Sheet piling and pumping in soft or wet excavation
Subsurface water necessitating pumping and waterproofing
Excessive cut or fill excavation

Excessive landscaping requirements

Slope and contour of lot with consideration of drainage problems, retaining walls, etc.

Natural and established grades in relation to proposed floor elevations

e. Availability of Utilities. The availability and adequacy of water and gas mains, electric distribution systems, and storm and sanitary sewers should be evaluated. Also, the accessibility to adequate public transportation should be checked.

4. Building Design Considerations

a. General. The primary considerations with respect to building design and construction are the lowest possible cost consistent with actual needs of equipment and personnel and the most efficient use of available space.

b. Size of Building. The proper size of the initial building and the capacity of the ultimate size to which it must be capable of being extended are determined in the consideration of the fundamental data, the availability and cost of land, and in the study of the floor plans. See Table 1 for typical requirements.

Table 1. Typical Space Requirements for Unattended Local Switching Centers
(Space for power but not operators or offices is included.)

Local lines	Ceiling height, ft	Length, ft	Width, ft
580	13	60	32
980	13	77	32
1,960	13	77	40
2,940	13	96	40

Careful planning is essential in order to achieve the most effective utilization of space in the initial building without affecting flexibility for future expansion. Beneficial occupancy of 90 per cent or better of the space available might be considered a desirable objective at which to aim.

c. Configuration of the Building. Floor-space requirements, type of equipment to be used, and office-space requirements will largely determine the size and the shape of the building as well as its facing and location on the lot.

The building design should be planned with the architect with the objective of spacing columns for maximum efficiency in the utilization of equipment floor space. To assist in attaining this objective, it is wise to develop standard equipment area floor plans which permit a certain amount of building flexibility in three dimensions (width, length, and height) to meet variations in lot dimensions and the amounts of equipment to be installed. Figure 1 shows how the three variables—length, width, and height—have been resolved into three basic equipment area plans whose bay dimensions can accommodate No. 5 crossbar, No. 1 crossbar, crossbar tandem, and No. 1 step-by-step equipment. The use of these standard equipment floor plans assists significantly in the preparation of layouts giving optimum arrangements for equipment, cabling, ironwork, and maintenance consistent with the building size and shape and column spacing. In all cases doors for equipment delivery must be large enough and well located.

Referring to Figs. 1 and 2 it will be noted that columns located in the direction perpendicular to equipment lines are spaced 20 ft between exterior walls and exterior columns and 18 ft 6 in. between interior columns. Columns located in a direction parallel with equipment lines may be spaced 16, 18, or 20 ft apart dependent on building size and telephone-company requirements. Figure 2 shows a typical equipment area layout with 16-ft column spacing.

Minimum ceiling clearances are required for central-office equipment rooms where 11-ft 6-in. frames are used in order to effect minimum building story heights. Generally a minimum clearance under all obstructions of 13 ft is necessary in Nos. 1 and 5 crossbar and No. 1 step-by-step local central offices and a minimum clearance of 13 ft 6 in. in toll terminal and No. 4 toll crossbar equipment rooms.

In community dial offices and in main and auxiliary repeater and carrier stations where 9-ft frames are used, a clearance of 10 ft 6 in. under all obstructions is usually adequate.

Expansion of an office to provide for growth is frequently desirable in the initial planning stages. In this case, floor loading and ceiling height should be included in

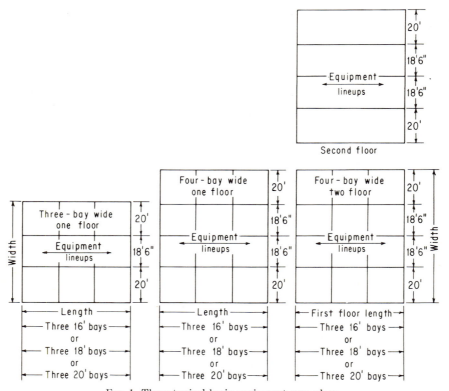

Fig. 1. Three typical basic equipment area plans.

the initial requirements. Because of continual increases in the amount of heat generated by newer equipment, air conditioning ducts probably will be added as equipment is added in a building. Therefore, the space over the heights indicated above should not be used for cable runs except in rare cases where the building engineers agree that such a procedure will not restrict their ability to achieve subsequent air duct installation. Occupancy of space in excess of immediate equipment requirements can be assigned to other activities, particularly those which can be moved to other locations when the space is needed for equipment.

d. Defense Considerations. In the overall planning of buildings it is suggested that defense aspects be given consideration, especially for buildings located in critical target areas and for those buildings which serve atomic energy and key military installations. Some of the factors to be considered in this connection are:

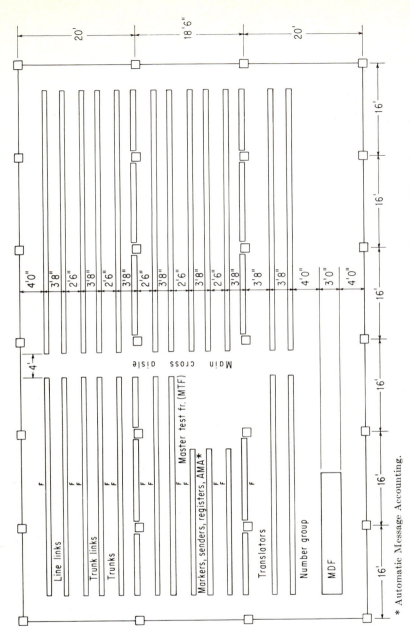

FIG. 2. Typical three-bay wide, one-floor equipment room accommodating typical crossbar equipment.

* Automatic Message Accounting.

1. Provision of a personnel shelter within the building. This may be dual-purpose space rather than space reserved solely for emergency use as a shelter area.

2. Allocation of protective space for emergency power equipment.

3. In laying out equipment on floor plans, it is desirable to locate common equipment toward the core of the building or in other comparable areas away from the exterior walls to provide some degree of protection from blast pressure and debris.

4. A radiation protection factor of 100 is recommended.

5. Important switching centers usually have poured concrete walls and floors.

6. Elimination of building features that may become hazardous in a disaster, such as the use of solid panels in lieu of glass in doors and the elimination of transoms, etc.

e. Building Requirements. In addition to space requirements for local and toll central-office facilities, space may be required for some of the following services:

Traffic and plant quarters
Dining service
Commercial
Accounting center and quarters
Coin-counting center

f. Heating, Ventilating, and Air Conditioning. Except for unattended repeater and carrier stations and small community dial offices, most central offices in the Northern states are provided with heating systems. Since a significant amount of heat is generated by the central-office equipment, due allowance for this supplemental heating should be made in the design of the heating system to avoid over-capacity. The amount of supplemental heating may be so small that it may be neglected, as in the case of dial offices having low traffic volumes. However, in repeater, carrier, or toll terminal offices having electron tubes operating continuously, an allowance as high as 35 per cent may be made.

To avoid excessively high temperatures in central-office equipment areas, ventilating fans and systems may be required. In some instances air conditioning will be necessary to keep the temperature of the equipment within optimum operating range. Toll and electronic switching systems equipment areas are likely candidates in some locations.

In hot climates air conditioning may be required for operating personnel comfort.

CENTRAL OFFICE EQUIPMENT[4,5]

5. Development of Switching Systems

a. Manual Switching. When Alexander Graham Bell obtained his patent for the telephone in 1876, there were no switchboards to interconnect a number of users. In 1877 the first experimental switchboard was provided by the Holmes Electric Company in Boston with about five lines. In 1878 the first commercial telephone exchange was opened in New Haven, Conn., with 21 subscribers. During the next 10 years manual switchboards were installed in most of the larger cities of the United States and Canada.

The early switchboards and the subscribers' sets were equipped with hand-operated magnetos for signaling. Each subscriber's telephone had its own set of batteries to energize the transmitter for talking. The frequent testing and replacement of batteries were sources of appreciable expense. To circumvent the expense of frequent battery replacement, Hammond V. Hayes, later chief engineer of the American Telephone and Telegraph Company, invented a common battery switchboard and obtained a patent in 1888. This design utilized a main battery at the central office to supply current to most of the telephones connected to that particular board. In 1894 the first complete common battery switchboard was put into service in Lexington, Mass.

In 1897 the first common battery multiple switchboard was placed in service in Louisville, Ky. This type of switchboard became known as the No. 1 relay switchboard. It used lamp supervision instead of drops and was widely used until the mid-twenties, when the dial systems were finding extensive application.

These earlier designs of switchboards generally used low, pedestal-type equipment frames or racks which could be installed in buildings having low ceiling heights such as were found in residences, stores, etc. Later designs of larger switchboards used 11-ft 6-in. frames for mounting the relay equipment, which usually necessitated equipment rooms having minimum clearances of 13 ft under all obstructions in buildings specially designed for telephone use.

b. Direct-dial Systems. Although manual switchboards were improved to the point where by 1915 they were capable of providing high-quality service, it was recognized that fully automatic service under control of the customer offered a number of advantages among which were (1) faster completion of calls particularly during periods of low traffic volume when manual switchboards were lightly manned, (2) ability to provide faster service during periods of temporary overloads, (3) freedom from the difficulties being experienced in large metropolitan areas in recruiting sufficient personnel, and (4) economies in building costs due to reductions in the amount of facilities required by operators.

In the development of dial telephone switching systems two basically different arrangements have been devised for controlling the operations of the switches. In one arrangement, such as in the step-by-step system as commonly used in the United States, the switches in each successive stage are directly responsive to the digit that is being dialed. This is known as the direct-dial or direct-control system.

As early as 1879 thought was being directed toward automatic switching systems, and the first patent for such a system was issued in 1879 to Daniel Connolly, T. A. Connolly, and T. J. McTighe. This system was not commercially successful, however.

In 1889 Almon B. Strowger devised an automatic system which was placed in service in La Porte, Ind., in 1892. This system used push buttons to interrupt the circuit between the subscriber's telephone and the central-office switch to cause the switch to step up and in to the desired number. This system, known as the Strowger or step-by-step system, was initially limited to 99 lines. To obtain a number such as "63," the subscriber would push the "vertical" pulsing button six times, pause, and then push the "horizontal" pulsing button three times. This arrangement was not entirely satisfactory from the subscribers' standpoint and besides had the disadvantage of requiring five conductors between the subscriber and the central office as compared with the two conductors required for a manual telephone.

By 1896, the Automatic Electric Company operating under the Strowger patents had improved the step-by-step system so that the number of conductors was reduced from five to two and the push buttons were replaced with a dial to send the proper number of pulses to the central office.

c. Common-control Rotary and Panel Systems.[6,7] With the tremendous gain in the number of telephone lines and the increasing use of the telephone, it became apparent around 1915 that the time was approaching when it would no longer be possible to hire enough operators to handle the ever-increasing volume of traffic in the larger cities.

Since the direct-dial system was limited to relatively small trunk groups, consideration was given to the design of switching systems having access to larger trunk groups which could be used more efficiently in large metropolitan centers. In a program beginning in 1906 the Western Electric Company undertook the design of two motor-driven switching systems providing larger trunk groups and more efficient access to available trunks than was available in existing direct-dial systems. Both of the motor-driven switching systems operated on the "Common-Control" principle, where pulses from the subscriber's dial were stored for a short time, in a device known as a register or sender, before being used for controlling the switching operations. One system, known as the "rotary" system, was first placed in service in England in 1914. In this system the switch wipers rotate horizontally across the multiple bank terminals but only the wipers of the desired level are connected through to complete the connection to the called line or trunk terminal. The rotary system was used quite extensively in Europe but did not find application in the Bell System because less expensive cabling arrangements were attainable in the panel system for use in large exchange areas.

The other motor-driven common-control system developed by the Western Electric Company was the panel system in which the trunks or lines appear on multiple banks or panels.

Each frame used for trunk selection consists of five multiple banks mounted one above the other. Each bank has 100 sets of terminals located vertically and multipled horizontally in such a manner that access to the terminals is obtained by means of 60 vertical moving selector rods, 30 of which are on each side of the frame. Each selector rod has five sets of brushes, one of which may be closed through to the bank in which the desired terminal appears.

The first installation of the panel system was in Newark, N.J., in 1915. In this office the customer gave the called number to an operator who in turn keyed the information into the common-control equipment which established the connection to the called line.

In 1921 the first full mechanical dial office having common-control equipment was cut into service in Omaha, Neb. Panel-type equipment was extensively installed throughout the United States during the 1920's and the 1930's in many of the larger cities such as New York, Chicago, Philadelphia, Boston, Detroit, St. Louis, Baltimore, Washington, Cincinnati, Cleveland, Milwaukee, and San Francisco. Most of these central offices are still in operation.

In both the rotary and the panel systems the selectors or switches, in moving to reach the desired trunk or line, send back revertive pulses to the sender or register. When the revertive pulses match those in storage, the sender or register stops the selector or switch on the desired terminal.

The Bell System selected the panel system for use in large exchange areas because (1) more efficient use of the trunks was possible through the use of the larger trunk groups, (2) trunking was divorced from the decimal numbering arrangement since it was no longer under control of the dial, and (3) the panel system provided a much greater degree of flexibility for growth and adaptability for the introduction of new types of service.

Common-control principles have been applied by the Automatic Electric Company to the step-by-step system, and such a system using a common-control device known as the director has been in use in London since the mid-twenties.

d. The Crossbar Switch. The basic element of the crossbar system is the crossbar switch, which was invented by J. H. Reynolds of the Western Electric Company in 1913 and on which a patent was issued in March, 1915. The original intent was to use the device as a line switch, but the plans were not formulated, and the switch was not used as either a line switch or a selector.

The 100-point crossbar switch is made up of 10 vertical units and 10 horizontal levels which provide 100 crosspoints, any one of which may be closed by the coordinated operation of one of 10 horizontal selector magnets and one of the vertical units. The crossbar switch is also available with 20 vertical units which in conjunction with the 10 horizontal levels provide 200 crosspoints.

During the twenties an improved version of the crossbar switch was developed by the Swedish Telephone Administration and initially used in crossbar offices for line switch and selector operations on the direct-dial principle.

e. Crossbar Systems.[4,5] Although the crossbar switch was first used in Sweden in 1926 in a direct-dial system, its first application in a common-control system, known as the No. 1 crossbar system, was in the Troy Avenue central office in Brooklyn, which was placed in service early in 1938.

The crossbar system offers many advantages over the earlier automatic systems, with the result that it has superseded the panel system in large exchange areas and is competitive with the step-by-step system in smaller exchange areas.

The biggest advantage of the crossbar over earlier systems is in the selection of trunks. In panel and step-by-step systems the size of a trunk group is limited by the number of terminals that a switch can hunt over in one sweep. In the crossbar system a marker or a comparable unit tests all the trunks in a group, selects and makes busy an idle trunk, and then establishes a connection to the selected trunk. Any idle trunk is immediately available to the marker, whereas in the panel or step-by-step system a released trunk is no longer available once the selector has swept by.   Another advantage of crossbar is in the reduction in maintenance effort due to the significantly decreased role of mechanics and the use of precious-metal contacts.   Crossbar offers improved transmission as well as a reduction in maintenance through the use of precious-metal contacts in the transmission switching paths in place of the base-metal contacts used in the earlier systems.

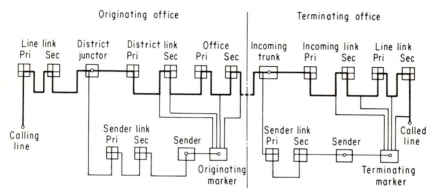

Fig. 3. Progress of a call through a No. 1 crossbar office with transmission path through principal equipment components shown in heavy lines.

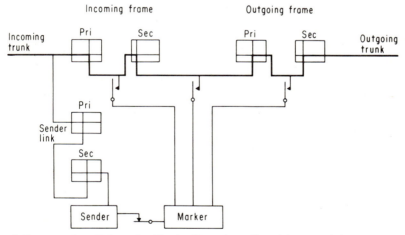

Fig. 4. Progress of a call through a No. 4 toll crossbar office with transmission path through principal equipment components shown in heavy lines.

Figure 3 shows the progress of a call through a No. 1 crossbar office with the transmission path through the principal equipment components shown in heavy lines.

The use of No. 1 crossbar proved to be so successful that its basic principles have been applied to other switching systems such as the No. 1 crossbar tandem system and the No. 4 toll crossbar system.   Figure 4 shows the progress of a call through a No. 4 toll crossbar office and the principal central-office equipment components.

In 1948 the first No. 5 crossbar system designed primarily to serve suburban areas was placed in service in Media, Pa.   The No. 5 system is now arranged to serve both suburban and large metropolitan areas, and it is being applied extensively.

f. Conclusion. The historical description of the development of switching centers and the reasons for conversion to new types of offices indicates a basic requirement in the design of long-life facilities. Advances in the state of the art led to provision for new services supplied sometimes by additional equipment and sometimes by replacement equipment. When new services to the customer are not features of new equipment, reduced building requirements or reduced labor requirements are generally the important features warranting the new equipment. In addition, compatability with new equipment installed in other offices may introduce a requirement for additional equipments in existing offices not being converted to the new system. Consideration of past developments can lead to a forecast of future possibilities.

6. A Typical Crossbar System[8–10]

a. General. This crossbar system, like the No. 1 crossbar system, uses the common-control method of operation in establishing connections between lines and trunks. In addition to handling local office traffic it is designed to serve as a tandem office or as a toll center switching point. It can also be equipped to function as a local automatic accounting center and a centralized automatic message accounting center, to identify automatically a calling party's number to a distant centralized automatic acounting center, and to act as a switching center for PBX subscribers who wish to have direct inward dialing and the use of automatic accounting facilities for outward-dialed calls.

Switchboards are provided in the crossbar system, as in other systems, to enable operators to handle certain classes of calls and to render assistance to subscribers. The switchboard may be located in the local office or in some other office at a distant location.

In a discussion of the customer services available in the system two categories are used, namely, "basic customer service" and "additional customer services."

Basic customer service comprises those facilities used for handling calls within the office, calls to other offices, reverting to other parties on the same party line, calls to lines in another marker group in the same building, and tandem and toll switched calls on a two-wire switched basis.

Additional customer services involve those facilities used in handling calls on a four-wire basis and in handling calls which require automatic message accounting,[11] Centrex,[12] or line link pulsing.

Figure 5 shows the progress of a call through the crossbar office and the principal equipment components involved in providing basic customer service.

b. Capacity of the Crossbar System. This system is made up of "marker groups" of equipment with each group containing a maximum of 60 line link frames, 30 trunk link frames, 4 dial-tone markers, and 8 completing markers along with appropriate quantities of senders, registers, connectors, traffic-measuring, and maintenance equipment.

A marker group can serve a maximum of 40,000 customer numbers assigned to 35,000 customer lines and can also serve 2,000 trunk numbers when the office is used as a tandem or toll switching center.

Equipment requirements for a typical crossbar central office up to 4,000 customer numbers are shown in Table 2. Variations from this typical office depend upon the service features being provided.

c. Framework. Formed sheet-metal frames are 11 ft 6 in. high with a base $10\frac{1}{2}$ in. deep. Single-bay connector, sender, and register frames are 2 ft $2\frac{1}{16}$ in. wide. Double-bay marker frames are 4 ft $4\frac{1}{16}$ in. wide, and link frames are 5 ft $7\frac{1}{8}$ in. wide.

d. Frames and Units. Frames and units used for mounting the apparatus required for system operation are of two types, namely, those used in the transmission path and those used in the control path.

e. Frames and Units Used in the Transmission Path. *Line Link Frames (LL).* The line link frame is used to give a customer's line, PBX line, or trunk a connection to a dial-tone marker through its line link marker connector and to request a connection to an originating register.

The lines terminate on verticals of line switches, and the junctors terminate on verticals of junctor switches. The horizontals of the line and junctor switches are connected in a standard crossbar link pattern.

The basic frame provides access to 100 junctors which serve 190 lines, and supplementary frames are available to increase the number of lines to a maximum of 590.

Trunk Link Frames. The trunk link frame provides to the trunks and originating registers full access over junctors to all lines and trunks terminated on the verticals of the line link frames.

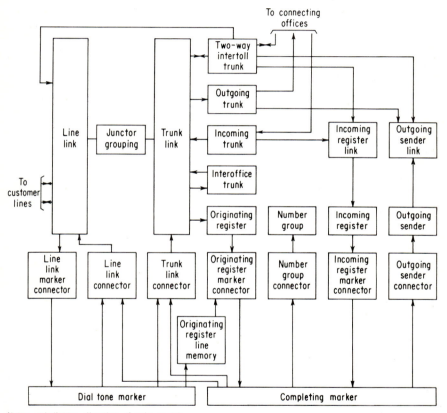

Note: ◄ indicates direction of seizure only

Fig. 5. Progress of a call through the typical crossbar office.

The trunks and originating registers terminate on the horizontals of trunk switches, and the junctors terminate on the horizontals of junctor switches. The verticals of the trunk and junctor switches are connected in the standard crossbar pattern, giving all junctors full access to all trunks.

The basic trunk link frame contains facilities for 160 trunks and 200 junctors which provide access to the customer lines or trunks terminated on from 2 to 20 line link frames. A basic frame with two extensions can serve a maximum of 60 line link frames.

Trunks. Trunks consist of two-way intertoll, outgoing, incoming, intraoffice, test, maintenance, and miscellaneous circuits. Trunks used in the transmission path pro-

Table 2. Number of Frames Required by the Typical Crossbar Central Office

Frames	Lines 580	980	1,960	2,940
	Numbers* 1,000	2,000	4,000	4,000
Line link..	2	2	4	6
Trunk link......................................	2	2	3	4
Junctor grouping.............................		1	1	1
Number group.................................	1	2	4	4
Originating register..........................	1	2	2	2
Originating register line memory...............	1	1	1	1
Coin supervisory link.........................	1	1	1	1
Originating line identifier.....................	2	2		
Dial-tone marker.............................	†	†	2	2
Completing marker...........................	2	2	2	2
Line link connector...........................	2	2	2	2
Trunk link connector.........................	2	2	8	8
Number group connector......................	2	2	4	4
Outgoing sender connector....................	2	2	2	2
Dial-tone marker connector...................			2	2
Completing marker connector.................			2	2
Main distributing frame verticals..............	15	15	25	28
Trunk—reverting call........................	4	4	6	8
Trunk—incoming.............................	8	8	12	16
Trunk—outgoing.............................	8	8	12	16

* Includes multiparty lines.
† An originating line identifier associated with a completing marker provides dial-tone functions in a small office.

vide the signaling and supervision required after the transmission path has been established.

Main Distributing Frame (MDF). The main distributing frame provides protectors on the vertical side and terminal strips on the horizontal side for cross connection between line or trunk cable pairs and the central-office equipment. Line numbers are assigned by these connections so that the load is evenly distributed over the line link frames. Trunks are similarly assigned to provide even distribution over the trunk link frames. A frame with 15 verticals is 10 ft long by 3 ft wide.

Junctor Grouping Frame. This frame provides a centralized location for cross-connecting junctor ends from the line link frames to the junctor ends from trunk link frames. The distribution patterns are arranged so that all line link frames have access to all trunk link frames. Junctor grouping frames are required to serve line and trunk link frames as follows:

JG frames required	To serve a maximum of	
	LL frames	TL frames
1	20	10 single
2	40	10 paired
3	60	10 triple

f. Frames and Units in the Control Path. *Dial-tone and Supplementary Dial-tone Markers.* The dial-tone marker selects a trunk link frame and an originating register and establishes the path for dial tone to and subsequent dialing from the customer's line through the line link, trunk link, and junctor grouping frames. The dial-tone marker also identifies the originating line by conveying the originating line location and the class of service indication to the originating register line memory frame, where the information is stored for subsequent use.

The supplementary dial-tone marker provides auxiliary facilities which enable the dial-tone marker to operate with more than 40 line link frames or when triple trunk link frame operation is desired. One supplementary dial-tone marker serves four dial-tone markers. Both the dial-tone and the supplementary dial-tone markers are mounted on single-bay frames.

Completing Marker. This marker comprises several frames functioning together to select and establish a path through the switching equipment for traffic originating in, completing to, or switching through a No. 5 crossbar office. The common equipment, translator and code treatment, and route relay frames are required for all types of service. The supplementary service treatment, FC relay, code conversion, and PBX allotter frames provide marker group capacities and types of service not commonly required in all offices.

Common Equipment Frame. This frame controls the overall operation required to select and establish a transmission path through the switching equipment. It receives the called number from the originating register, the calling party's line link frame location, and class of service from the originating register line memory frame. It also determines the type of call to be handled and with the aid of other frames in the marker complex selects and establishes the transmission path through the proper switching equipment.

Translator and Code Treatment Frame (T&CT).[13] This frame determines from the office code the identity of the trunk link frames which have trunk appearances to the called office, selects an outgoing sender of the proper type, and passes the last four digits of the called number to the sender. Also this frame indicates to the control equipment the class of service to which the customer is entitled.

Route Relay Frame (RR). This frame provides the route information necessary to establish the transmission path to the connecting office. It is a double-bay frame which mounts the multicontact relays for 100 routes for each of four completing markers.

Code-conversion Frame (CC). This frame converts the office code portion of the called number into arbitrary codes which are required when the actual office code is unsuitable for a call from this type of crossbar marker group through the same or another marker group.

PBX Allotter Frame. This frame allots numbers from two or more number group frames to lines from large PBX's which connect to this type of crossbar marker group.

Supplementary Service Treatment Frame (SST) and Auxiliary Service Treatment Frame (AST). The SST frame provides an additional 60 service treatment relays* when more than the 60 furnished on the T&CT frame are required. The AST frame is identical with the SST frame and increases the capacity of each of four markers to 180 service treatments.

FC Relay Frame (FCR). This frame provides facilities for determining whether trunks for a route on trunk link frames 20 to 29 are idle. This frame is used in offices having more than 20 trunk link frames. Its facilities enable the control equipment in the office to determine whether trunks, for a particular destination, are idle and available in frames 20 to 29.

Originating Line Identifier Unit. The originating line identifier unit with its associated completing marker establishes the dialing connection between the calling line and

---

* The service treatment relays indicate to the switching equipment the class of service to which a line originating a call is entitled (and the applicable rates for commercial service). The many different classes of service are necessary because of variations in calling charges, geographical boundaries, denied outward service, etc.

originating register.   The unit is used in small marker groups which do not warrant the use of a dial-tone marker.

Foreign-area Translator Frame (FAT).   This frame translates the office codes of certain six-digit direct distance dialed calls and furnishes the correct routing to the marker.

Number Group Frame (NG).   This frame translates the called number received from the marker and sends back to the marker the line link frame location of the line originating the call.

Originating Register (OR).   The originating register transmits dial tone to the customer, receives and stores the digits of the called line, and, after dialing is completed, transfers the digits to the completing marker.   The originating register also determines whether the calling party is on the tip or the ring side of two-party lines. The originating register can be arranged to accept both dial pulses from a rotary dial set or Touch-Tone® pulses from a Touch-Tone® set.

Originating Register Line Memory Frame (ORLM).   This frame registers the calling customer's line link frame location and class of service and stores this information until required by the completing marker.

Pretranslator, Pretranslator Connector, and Pretranslator Frame.   The pretranslator determines the total number of digits to be expected in the dialed number from three-digit service, office, or foreign-area codes.   The pretranslator connector gives three subgroups of originating registers access to each of a maximum of three pretranslators.

Outgoing Sender and Outgoing Sender Frames.   The outgoing sender receives the digits from the marker for an outgoing call and transmits them to the connecting office.   Dial pulsing, multifrequency pulsing, frequency-shift pulsing, and panel call indicator senders are available to provide the signaling required by the various types of connecting offices.

Intermarker Group Sender and Intermarker Group Sender Frame.   The intermarker group sender receives the digits of the called number and the call handling information from the originating marker group and passes this information on to the terminating marker groups when intermarker group operation is used.

Incoming Register and Incoming Register Frames.   The incoming register receives the digits of the called number from a connecting office and transmits them to the completing marker.   Dial pulsing, multifrequency pulsing, tandem incoming revertive pulsing, revertive pulsing, and central B-type incoming registers are available to match the signaling required by the various types of connecting offices.

Outgoing Sender Link Frame (OSL).   This frame provides a connection through crossbar switches between a trunk and an outgoing or intermarker group sender.

Incoming Register Link Frame (IRL).   This frame selects an idle incoming register and connects it through the crosspoints of a crossbar switch to the incoming trunk associated with the incoming office.

Coin Supervisory Link Frame (CSL).   This frame gives coin supervisory circuits access to coin trunks and junctors.

Connectors and Connector Frames—General.   Connectors are used to establish control paths between markers and switching frames.   Two basic types are used in this crossbar system.   The first type includes three kinds of connectors: line link, originating register, and incoming register marker connectors.   The second type includes four kinds of connectors: line link, trunk link, number group, and outgoing sender connectors.

The inclusion of the word "marker" in the name of a connector indicates that it is used by a switching frame for establishing a connection to a marker, and the exclusion of the word indicates that it is used by a marker for establishing a connection to a switching frame.

Line Link Connector Frame (LLC).   This frame contains the connector equipment which can give a maximum of 10 dial-tone and completing markers access to five line link frames.

Line Link Marker Connector Frame (LLMC).   This frame contains the connector equipment to give a maximum of 20 line link frames access to four dial-tone markers.

Incoming Register Marker Connector Frame (IRMC).   This frame contains the marker part of the connector equipment for connecting incoming registers and inter-marker group senders assigned within each of four connectors to a maximum of eight completing markers.

Originating Register Marker Connector Frame (ORMC). This frame contains the marker part of the connector equipment for connecting the originating registers assigned within each of four connectors to a maximum of eight completing markers.

Trunk Link Connector Frame (TLC). This frame normally contains the connector equipment required to give a maximum of 10 dial-tone and completing markers access to three trunk link frames. However, when the marker group is to include the ulti-mate of 12 markers, this frame contains the equipment for the 8 completing markers and a supplementary trunk link connector frame contains the equipment for giving the four dial-tone markers access to 10 trunk link frames.

Number Group Connector (NGC). This frame contains the connector equipment for connecting a maximum of eight completing markers to five number group frames.

Number Group Connector Control Frame (NGCC). This frame contains the general-purpose relay-control and preference units for a maximum of nine markers and 24 number group frames.

Outgoing Sender Connector Frame (OSC). This frame contains the connector equip-ment for connecting eight completing markers to 50 senders in five sender groups assigned to four connectors.

Foreign-area Translator Connector Frame (FATC). This frame connects a foreign-area translator to the marker at the request of the marker.

Line Link-Originating Register-Incoming Register Marker Connector Frame. This frame contains the connector and the associated connector control and marker prefer-ence units for four line link, two originating register, and two incoming register marker connectors with access to four dial-tone or completing markers.

Line Link-Trunk Link Connector Frame. This frame contains connector equipment for four line link and two trunk link connectors equipped for four dial-tone or com-pleting markers.

Number Group-Outgoing Sender Connector Frame. This frame contains equipment for four number group and two outgoing sender connectors equipped for four com-pleting markers.

Line Link-Trunk Link-Number Group-Outgoing Sender Connector Frame. This frame contains eight unitized connectors which serve as two line link, two trunk link, two number group, and two outgoing sender connectors—each with access to two com-pleting markers.

g. System Power Requirements. The principal power requirement for the operation of crossbar apparatus is provided by a 48-volt d-c storage battery charged by motor-generator sets and rectifiers. The system also requires $+130$ and -130 volts d-c provided by a d-c-d-c converter for coin supervisory circuits.

Twenty-two volts alternating current, derived from a 110- to 22-volt transformer, is required for timers on coin and message rate trunks.

In addition, various ringing and tone potentials are available for ringing, dial tone, busy tone, etc.

The storage batteries, charging equipment, ringing machines, and associated control equipment are usually located in the basement in the larger central offices and on the first floor in offices having less than 3,000 to 4,000 lines.

The ground supply and -48-volt battery are cabled from the power plant to cross-aisle power feeders. The -48-volt cross-aisle feeders are connected to 100-amp fuses on the end guard of each row of frames. Ground and battery feeders are run on the cable rack. The ground and battery feeders for each frame are tapped to the row feeders. Each frame is provided with fuse panels and fuses for protecting the equip-ment units on the frame. Also built into the structural portion of the frame are the battery filters, appliance outlets, testing and miscellaneous equipment required.

h. Maintenance Provisions. Testing and maintenance activities are controlled from a centrally located area in the switch room known as the "maintenance center." In this center is located the group of frames known as the "master test frame," consisting of the trouble recorder, the automatic monitor register and sender test, trunk test, master test connector, and associated circuits.

The trouble recorder automatically indicates, on a punched card, information pertaining to troubles encountered on service calls or on test calls originated from the master test frame.

The automatic monitor is used with the register and sender test bay or with manually operated register and sender test sets for testing the operation of registers and senders. It is also arranged to monitor the pulsing of certain registers or senders during normal operation to determine whether or not they are functioning properly.

The trunk test frame is used for testing automatically the operation of trunks. Failures are indicated by visual and audible alarms. In smaller offices, a less expensive manually operated outgoing test frame may be used.

The master test frame connector provides access to the various circuits for testing. Audible and visual alarms indicate the nature and the location of trouble conditions. When the office is unattended, the alarms may be transmitted to another office where there is a maintenance force.

In addition to the test frames located at the maintenance center, there are available a line insulation test frame, which is usually located near the line link frames, and a line verification unit, which is usually located near the main distributing frame (MDF).

The line insulation test frame tests customer lines for insulation defects. The line verification unit tests the cross connections, associated with line numbers, for both number groups and automatic message accounting translators. The line verification unit is also arranged to test the class of service assignment cross connections for line link frame terminations of customer lines.

i. Service Observing Facilities. These facilities are provided to enable the traffic department to make a cross-sectional check on customer lines to determine the quality of service being rendered by the switching equipment and toll and assistance operators.

The service observing desk may be in the local office or in another office. Lines selected for observation are connected to the service observing desk by connecting a cord from the line link vertical appearance of the customer line to a service observing jack on the line link frame. This line link service observing jack is cabled to an associated jack on the service observing jack bay. The jack on the service observing jack bay is then connected by a cord to a service observing line circuit which is terminated on the local or distant service observing board.

j. Traffic-measuring Facilities. In addition to the conventional traffic registers which indicate the number of times specific equipment is used for calls and the number of times that the equipment is busy, a traffic usage recorder is also available.

The traffic usage recorder provides a count of equipment usage on a CCS (100 call second basis). The recorder unit scans the equipment under study thirty-six times an hour at 100-sec intervals. The number of times that the equipment is in use during an hour is automatically recorded on traffic registers.

Readings on the traffic registers in both methods of operation may be recorded manually or photographed by a camera set up at the traffic register cabinet.

REFERENCES

1. Turrell, G. B., Creating a Central Office, *Bell Telephone Mag.*, Winter, 1955–1956.
2. Frost, G. R., William Keistler, and A. Ritchie, A Throwdown Machine for Telephone Traffic Studies, *Bell System Tech. J.*, vol. 32, March, 1953.
3. Pilliod, J. J., Fundamental Plans for Toll Telephone Plant, *Bell System Tech. J.*, vol. 31, September, 1952.
4. Pennoyer, D. H., Automatic Number Identification, Bell Telephone System, *Bell System Tech. J.*, vol. 37, September, 1958.
5. Clarke, A. B., and H. S. Osborne, Automatic Switching for Nation-wide Telephone Service, *Bell System Tech. J.*, vol. 31, September, 1952.
6. Brooks, C. E., Class of Service Markings, What and Why, *Bell Lab. Record*, October, 1962.

7. Hersey, R. E., Machine Memory in Telephone Switching, *Bell Lab. Record*, January, 1960.
8. Ferguson, J. G., and F. A. Korn, No. 5 Crossbar Switching System, *Trans. AIEE*, vol. 69, part 1, 1950.
9. Dehn, J. W., and R. E. Hersey, Recent New Features for the No. 5 Crossbar System, *Trans. AIEE*, part 1, Communications and Electronics, pp. 457–466, September, 1956.
10. Greene, J. E., For No. 5 Crossbar—A Packaged Central Office, *Bell Lab. Record*, October, 1962.
11. Hurst, G. A., Extending CAMA with No. 5 Crossbar, *Bell Lab. Record*, October, 1958.
12. Spiro, G., Centrex Service with No. 5 Crossbar, *Bell Lab. Record*, October, 1962.
13. Bishop, G. S., A New AMA Translator for No. 5 Crossbar, *Bell Lab. Record*, November, 1958.

Chapter 23

RADIO TERMINAL CONFIGURATIONS

EDWARD W. BROWN, JR., *Raytheon Company**

INTRODUCTION

The planning and implementation of radio communications station facilities for optimum performance at minimum cost are complex tasks involving many technical skills. In the usual case, where the stations under consideration are elements in a widespread network, the engineering of an individual station becomes only a facet of the overall system planning.

The present-day communications engineer has available as tools a variety and complexity of equipment and new techniques with which to design a system. The solution to a complex communications problem may rest with the use of a combination of several modes of operation including microwave or UHF line of sight, tropospheric scatter, ionospheric scatter, VHF diffraction or obstacle gain, high-frequency single sideband (SSB) or independent sideband (ISB). To assure a correct choice, a logical determination should be made, based on requirements for performance, projected expansion, cost, frequency allocations, interference, access, etc.[1-4]

The first step in the planning and engineering of a system is the preparation of a statement of the overall communications objectives and the attendant economic or strategic justification.[5-7] Consideration should be given to the use of existing commercial facilities with concern for performance, reliability, and cost. Future requirements for increased traffic or for television or data transmission must be weighed and treated with importance.

The second phase of importance in planning is normally a feasibility study made to determine traffic requirements and the optimum system configuration to meet these and future requirements. A field reconnaissance is essential to develop this information, to survey existing facilities, and to establish the many factors to be weighed in the choice of communications techniques and system routing. Overseas, in underdeveloped areas, maintenance and operating requirements must be heavily weighed, particularly if performed by indigenous personnel.

Finally comes the task of system planning, including the engineering and implementation of station facilities. This chapter is primarily concerned with the provision of station facilities. It begins with a discussion of the considerations governing the choice of station sites followed in logical order by sections covering site development, the planning and construction of facilities, and other factors of importance. Very little information exists in present-day literature on the planning and execution of station facilities. Most operating agencies, telephone companies, and communications systems contractors have developed standard engineering and installation prac-

* Now with Page Communications Engineers.

tices based on acquired experience. For the facilities planning engineer, experience is the best teacher.

STATION SITING

1. General Considerations

The proper selection of sites for terminal and repeater stations is an essential and exacting task in the planning of any radio communications system. The cost of site development and the installation of facilities is high and warrants the utmost care. Cognizance of the factors involved in the correct choice of equipment sites is essential in order to avoid costly mistakes resulting in poor performance or even total system failure.

There are many documents and publications useful in the selection of proper sites. Among the most valuable are *accurate contour maps* of the site area and the terrain between proposed site locations. *Aerial photographs* are also useful in showing the proximity of proposed sites to sources of possible interference such as industrial plants, cities, heavily traveled highways, etc. Commercial *air schedules* may be used to check proposed propagation paths against air routes and their traffic densities, and *local weather maps* to reveal any unusual climatological features such as ice loading, sandstorms, prevalent wind velocities, etc.

Because of many useful documents available, selection of sites and propagation paths will generally be undertaken in two phases. Much of the initial work required to choose feasible sites can be accomplished in a *preliminary study* with the aid of maps, reports, and other documents. The second phase is a *field survey*, including visits to sites proposed in the preliminary study, for the final site selection. The field survey should also include collection of data to be used in estimating installation and operation costs, obtaining land rights, and preliminary system design.

The overall station siting requirements for different types of service will vary, for example, between microwave line-of-sight, tropospheric-scatter, and high-frequency transmitting and receiving stations. There are, however, certain common considerations of importance which include the following:

Site accessibility for construction and operation.

Proximity to toll centers or other user facilities.

Availability of reliable commercial power.

Propagation characteristics suitable for the intended service. Sometimes two or more different services will terminate in the same station.

Interference, natural and man-made.

Geological formation.

Site security.

Environmental conditions.

2. Economic Considerations

In the overall planning of single stations or of networks of stations, compromises must be made between the actual costs of facilities and the costs associated with the interconnection to communications centers and with operation. Two important parameters to be considered are the site location and on-site topography and facilities.

The selection of remote sites is often based on the desire to minimize antenna structure height and, in some instances, the number of stations required. Done without regard to other considerations, this approach can result in prohibitive costs for site development, construction, access roads, communications center interconnection, commercial power lines, and inaccessibility in inclement weather.

On-site conflicts may occur between requirements for satisfying the rigid conditions for the highest quality of facilities and performance on one hand and conditions for economy in structural designs of buildings, antenna structures, suitable topography, and soil load-bearing characteristics.

All these factors must be properly weighed and applied to select the optimum in facilities, performance, and accessibility for operation and maintenance at reasonable cost.

3. Preliminary Study

The aim of the preliminary study is to determine feasible system routing and possible station site locations. The study would be based on available maps, reports, and other relevant documents and should result in several equally feasible alternative site locations suitable for integration into the system should the prime site be unobtainable or inaccessible. The choice of possible sites should be made in terms of their ability to satisfy the conditions itemized below. These conditions are listed in suggested order of importance, and their relevance to the selection should be weighed accordingly.

a. Propagation. A site to be acceptable must satisfy requirements necessary for optimum transmission for the planned mode of propagation. For example, a microwave site must provide adequate clearance over intervening terrain to the adjacent sites, with reasonable tower heights. Paths over large bodies of water should be avoided if economically possible. At troposcatter sites, the on-path optical horizon must be at as low an angle of elevation as possible.

b. Interference. Receiving sites should not be located close to intense sources of interference. The proposed antenna patterns will, of course, affect the relative importance of any sources. Possible sources of interference are cities, large industrial plants, powerful radio transmitters, radar equipment, heavily traveled highways, and high-tension lines.

c. Airports and Air Routes. Because of tower height restrictions, sites preferably should not be chosen close to major airports. Troposcatter paths which are closely parallel to or which cross heavily traveled air routes should be avoided, since aircraft can cause severe multipath interference.

d. Terrain. The terrain must be suitable for the installation of transmitting/receiving station buildings and antenna structures of the size determined by the mode of propagation and desired performance. This requires adequate ground area and soil-bearing characteristics appropriate to the necessary structures.

e. Communication Centers. Distances from existing or proposed user communications centers should be accurately determined, and an estimate made of the difficulties and cost involved in establishing wire or radio links to them.

f. Accessibility. A determination should be made of the accessibility of a proposed site for construction and for supply and maintenance during operation.

4. Field Survey

The field survey should confirm the results of the preliminary study and provide a more detailed and comprehensive description of the local environmental features which may influence transmission and circuit quality. Additionally, it should provide descriptions of various climatological and topographical features at the site to be considered in site preparation, design of structures, foundations, and access roads.

The field survey should generally include the elements listed in the program below. Data so accumulated will make possible the final selection of sites, specification of system parameters, a reasonable prediction of performance, and an approximate estimate of construction and installation costs. The survey team should include a system planning engineer, construction engineer, steel rigger, and other applicable specialties.

1. Check the latitude and longitude of sites by celestial sightings if necessary.
2. Compute the path azimuth with respect to celestial north.
3. Photograph the site area.
4. Check availability and reliability of commercial power. Determine voltage, frequency, and regulation.
5. Where necessary, path profiles must be verified by a physical survey with determinations of site elevations and the elevations and nature of major on-path obstruc-

tions. For troposcatter sites, the angle to the horizon should be measured and photographed over a range of 45° either side of the path azimuth.

6. Check route distances and possible difficulties in running wire lines or radio
links to appropriate communications centers.

7. Study site geographical formations relative to the types of buildings necessary
and the footings required for suitable antenna structures. Make test borings if
required.

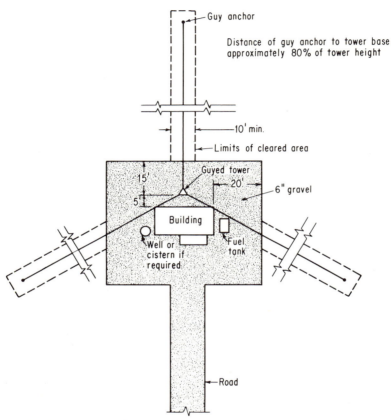

FIG. 1. Typical site plan for line-of-sight microwave station. This typical plot plan to be
varied as required, depending on conditions at each site.

8. Note any climatological features influencing the design of structures and equipment or radio propagation characteristics.

9. Determine the requirements for access roads and general site accessibility for
construction and station maintenance and supply.

10. Determine possible sources of interference. Check with a receiver around the
proposed operating frequency where indicated.

11. Check availability of drinking water and cooling water if required. Note the
temperature of cooling water and estimate Btu per hour heat sink capacity.

12. Investigate the availability of local contractors and building materials. Obtain
cost estimates.

13. Determine the location of the nearest shipping point and warehousing facilities with cost estimates.

14. For remote sites, the availability and cost of living and feeding accommodations, medical care, communications, etc., must be investigated.

A typical site plan for a microwave line-of-sight station is depicted in Fig. 1.

SITE DEVELOPMENT

5. Description

The task of site development encompasses all the work necessary to provide access to the site and to prepare it for the installation of foundations for buildings and antenna structures and for other outside-plant facilities. The major activities comprising this work are outlined below in their usual order of accomplishment.

6. Field Survey

Survey parties made up of civil engineers, construction engineers, communications equipment engineers, antenna structure riggers, and other applicable specialties should be dispatched to field sites as soon as possible, prior to station planning activities. This team will be responsible for the accumulation of all field data to facilitate the design of access roads, drainage facilities, structure foundations, etc. Additionally, it will also determine the location of necessary natural materials, including road surfacing materials, concrete aggregate, water, etc. For those sites where the soil conditions are known beforehand to be poor, proper earth borings should be made, vane tests performed, and undisturbed soil samples taken for laboratory analysis. Based on the results obtained and on the expected loads, the building and antenna structure foundation will be designed. For sites where the bearing capacity of the subsoil is questionable, weighted drill-point tests will be made, later to be supplemented by sample borings if found necessary.

Foundation for towers located on hilltops and shoulders will not usually cause problems, as rock or hard soil can be expected in practically all cases. However, for the towers located on low-lying wet or swampy areas requiring heavy filling of earth, it must be determined whether a direct foundation is sufficient or piling will have to be used.

Where required, the survey team should establish boresites to the necessary accuracy for antenna construction and aiming.

7. Access Roads

Economics and the small traffic volume will usually dictate that access roads be inexpensive and of minimum standards. The design requirements, i.e., grade, curve radius, drainage culverts, surfacing, etc., should be predicted on weather and other environmental considerations and on the classes of vehicles requiring access during operation and maintenance of the station. This may include heavy fuel-delivery trucks. During the construction period, access may be required for large and heavy vehicles. In such cases, it may be advisable to construct a temporary road and to delay completion and surfacing until the station is completely installed.

Roadways and parking areas must be kept away from the foreground of ground-mounted troposcatter or other antennas where interference or radiation hazard might result.

8. Site Grading

Before construction of structures may commence, the site must be properly leveled or graded and drained. The controlled area should be adequate to provide for all structures, including the anchors for guyed towers. Possible expansion of facilities should be allowed for in the overall planning. Adequate foreground clearance, possibly off-site, must be provided for microwave line-of-sight and for troposcatter stations. For the latter, fencing or other protective measures are required to define the perimeter of the hazardous zone.

9. Excavations

Following the completion of site grading, excavations will be made for all foundations and footings for buildings and other structures. The presence of rock will present special problems and will influence the design of foundations and footings as will the soil tests. Additional excavations may be necessary for grounding systems, fuel-storage tanks, wells or cisterns, septic tanks and seepage fields, pole lines, transmission-line support structures, etc.

10. Completion

The final phases of site development are normally completed after major construction activities are completed. These will include the erection of permanent fencing, finished grading, planting, and surfacing.

BUILDINGS

11. General

The planning and construction of buildings for radio communications stations present a host of peculiar requirements which vary widely with the nature and intended use of the facilities to be installed. In many instances, the stations comprising a system are widely separated, relatively inaccessible, and unattended and are required to operate continuously for long periods without maintenance. Full consideration must be given to the conditions and requirements at each site during planning activities to assure the achievement of optimum performance and reliability.[8,9]

At many large manned facilities, the radio terminal equipment may be only one small element of a major complex, and overall building design will be governed by

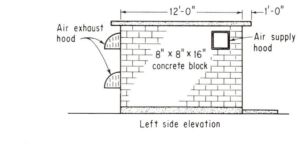

Left side elevation

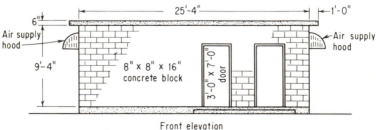

Front elevation

Fig. 2. Typical building elevations for line-of-sight microwave station.

other requirements. However, in such cases, the rooms or areas designated for the communications equipment must satisfy certain basic requirements.

Buildings for radio terminals and repeaters should be fireproof, requiring that they be of masonry or metallic construction. The use of steel prefabricated buildings has met with considerable popularity by many users. With due consideration for their disadvantages, they offer an economical and expeditious solution at unattended stations in many systems. These buildings are usually constructed of modular sections and can be easily erected. In tactical systems or in some fixed systems, consideration can be given to the use of transportable "huts" or modified truck trailers. This approach allows the complete station installation and checkout to be made at a central location before being moved to the site.

Figures 2 and 3 depict the front and side elevations and floor plan for a typical unattended line-of-sight microwave relay station.

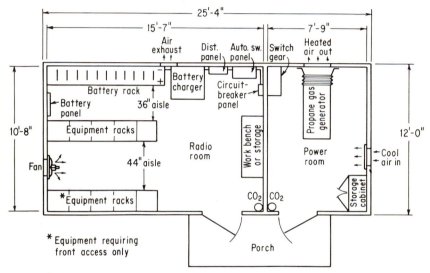

Fig. 3. Typical floor plan and equipment layout for line-of-sight microwave station.

12. Design Considerations

In the planning of buildings to house radio and multiplex equipment together with power and ancillary facilities, the general characteristics and peculiar requirements of each must be considered. The physical dimensions, access for servicing and adjustment, and relationship to other facilities are factors which largely dictate the space requirements and placement of each piece of equipment. The headroom required by all equipments with allowance for cable ladders and trunking; entrance openings for RF feeders, power supply, telephone cables, and other services; and methods for attaching equipment to the floor are additional important design considerations. In many remote areas, experience has shown that station buildings should be reasonably bulletproof. Due allowance must be made for future expansion requirements with space allowed or provisions made for enlargement of facilities in the building planning.

The choice of the type of construction and of building materials should be largely based, for reasons of economy, on factors determined by the locations. The availability of required building materials and local skills, local building codes, and zoning regulations are important factors. Where site accessibility is poor, the use of light-weight and easily transported materials may be dictated. Some of the building materials to be considered are concrete block, brick, poured concrete, steel panels, transite panels, and prefabricated buildings.

The climatic environment of each station site must be carefully considered in the planning of buildings. Temperature, humidity, rainfall, wind, dust, ice, and snow are factors of primary importance. The ambient temperature must be maintained within the limits dictated by each type of equipment. In dusty locations, air filters must be provided. Adequate protection and drainage are necessary in areas of extreme rainfall.

13. Facilities Required

In addition to the normal communications equipment, station buildings must also provide for other necessary facilities. Power generation, heating and ventilation, and transmission-line pressurization are important considerations. Adequate planning must be made to facilitate the routing and distribution of fuel piping, electric power, RF transmission lines, baseband, and multiplex cables and telephone cables. The use of cable ladders, trunking, and floor troughs will provide for ease of installation and a high degree of flexibility.

Adequate lighting must be provided to facilitate maintenance and adjustment of installed equipment. A workshop, lavatory and kitchen facilities, and sleeping quarters should be provided to the degree necessary at each station. Fire-extinguishing equipment in the form of portable extinguishers or automatic built-in equipment is an essential adjunct to any station.

14. Heating and Ventilation

The importance of an adequate heating and ventilating system cannot be too strongly emphasized. Electronic equipment to provide optimum performance and long life must be maintained within prescribed temperature limits. High temperature extremes must be controlled by the use of ventilating fans or, in some cases, by mechanical cooling. Air conditioning, for operating-personnel comfort, may be required in warm or hot climates.

The simplest system can employ thermostatically controlled exhaust fans with automatic louvers, relying on heat dissipation from equipment for some degree of cold-weather heating. With continuously running internal-combustion power plants, waste heat is usually adequate for station heating. Storage battery banks used in "no-break" systems or for direct operation of equipment must be maintained at relatively warm temperatures and may in some stations require supplementary electric or fuel-fired heating.

Insulation properly utilized will greatly facilitate temperature control of station buildings. Equipment which dissipates a large quantity of heat may be fitted with ducts to carry it directly to the outside or with heat exchangers. In dusty environments, the use of air filters on all air intakes is recommended. These may ordinarily be of the replaceable or washable type, but, in extreme cases, consideration should be given to the use of traveling-screen oil bath filters.

STATION PLANNING

15. Introduction

The task of station planning must take into detailed consideration all pertinent requirements related to proposed equipment, facilities, site parameters, utilization, and other factors to be covered in detail below.

We are concerned here with much more than simply designing a station building and determining the location of equipment and power and signal distribution. Thorough planning is essential to provide for ease of installation, maximum equipment accessibility, reliability, and expansion of facilities, all at minimum practical cost. An adequate job of planning engineering will assure the provision of all necessary equipment and installation materials on site and will obviate the need for field changes and procurement of materials. The savings in installation time, particularly for remote and foreign locations, will pay many times over for the planning costs and will result in a well-executed facility.

It is customary and convenient to treat separately the planning for *inside-plant* and *outside-plant* requirements. By definition, these refer respectively to the equipment and facilities located within the station building and to those located external to the building.

Figures 1 to 3 depict site, building, and equipment configurations for a typical microwave line-of-sight station, embodying many of the principles outlined in this chapter.

16. Factors to Be Considered

a. Site Parameters. Site topography and environmental requirements must be fully satisfied in station planning if operational reliability and performance objectives are to be met. Site topography will govern the design of buildings and antenna support structures. The relative remoteness and accessibility may limit the size and weight of equipment and materials to be transported to the site as well as fuel delivery when the station is operational.

Environmental considerations will include temperature, humidity, wind, rain, snow, ice, and elevation. These bear importantly with respect to the design of antennas, towers, buildings, and site development.

b. Degree of Permanence. The relative permanence of a station configuration, particularly in military systems, is an important parameter. The use of prefabricated buildings, trailers, or transportable shelters will facilitate the installation and withdrawal of facilities from temporary sites. Portable aluminum-alloy towers are available which can be quickly erected or dismantled and are easily transported.

c. Cost. Most commercial and military communications equipment, systems, and facilities are procured through competitive bidding. To this end, the system planner, equipment manufacturers, the systems contractor, and the operating company or agency must all aim at achieving the required specifications for equipment, performance, and reliability at minimum cost.[10]

On the part of the customer, this dictates the necessity for exacting technical specifications and other procurement and planning definitives. For equipment suppliers and systems contractors, emphasis must of necessity be placed on meeting stated requirements without substantially exceeding them and without frills.

d. Mode of Operation. The mode of operation of the system or station, i.e., line-of-sight, tropospheric-scatter, ionospheric-scatter, high frequency, very low frequency, etc., will dictate the overall station configuration. Long-distance HF facilities will require separate isolated transmitting and receiving sites to accommodate the large antenna arrays. Tropospheric- or ionospheric-scatter stations require large, level sites to provide for buildings and antenna structures. Large stations of any category may utilize separate buildings for power plants and communications equipment.

The manning requirements at any station will influence building planning and site access needs. A station which will be fully manned with operating personnel or which may function as a maintenance center for a section of a system must provide the requisite amenities for personnel and appropriate workshop and storage facilities. The provision of housing and logistic support will present special problems in remote areas. Unmanned repeater stations need to provide a minimum of building space and personnel comforts. Relative accessibility must satisfy requirements for installation and maintenance.

e. Other Factors. Planned future expansion of inside- and outside-plant facilities is an important consideration in station engineering. Power plants must possess sufficient reserve capacity or other provision to meet future needs. The required space for additional equipment should be provided or must be easily added.

Equipment accessibility for operation and maintenance must be adequate and should not be inhibited by later additions. Reliability of operation will be influenced by this and by considerations of protection against temperatures, humidity, and other ambient conditions.

17. Inside-plant Planning Requirements

a. Equipment Layout. The proper physical placement of communications and ancillary equipment within the station buildings must satisfy the important requirements listed below:

Accessibility for maintenance with provision for removing major assemblies from front, back, and sides is required.
In large stations, similar equipments may be conveniently grouped together for ease of installation, operation, maintenance, power and signal connections.
The operation of high-power transmitters in the presence of receivers and other low-level equipment may require physical separation to minimize interference effects.
Equipment having a high heat dissipation should be located away from other equipments, with adequate facilities for heat removal.

b. Power Plant. Internal-combustion or other rotating power-generation equipment should be completely isolated from the space occupied by communications equipment to minimize fire hazard. In small stations, power equipment may be located in a separate room. In large station complexes, the power plant is preferably located in a separate building. Vibration isolation is a requisite in all cases.

Space is required for maintenance of rotating machinery. For heavy units, chain hoist rails fixed to the ceiling together with outside openings are recommended to allow removal of units from the building for overhaul and replacement.

The present trend toward direct battery operation of microwave relay equipment presents special problems. Owing to the use of relatively low voltages and high currents, power distribution circuits must be kept short. The maintenance of batteries at full reserve capability requires a relatively warm temperature environment.

c. Power and Signal Distribution. Facilities for power and signal distribution should be kept extremely flexible to accommodate wiring revisions and the addition of new equipment. In general, circuits should be planned for minimum length to minimize voltage drop and exposure. The required flexibility for all circuit routing may be achieved through the use of wall-mounted duct and conduit and cable ladders suspended from the ceiling. Interface requirements are treated later in this chapter.

It is advisable, where possible, to run power-distribution circuits under the floor in duct, conduit, or trenches with removable cover plates. Power-distribution centers should be located in the main equipment areas for accessibility and to minimize circuit length. Each major equipment item should be supplied by an independent circuit.

d. RF Feeders and Branching. Thorough planning of RF feeder and branching facilities is essential to minimize losses and cost. Maximum use should be of straight runs, as waveguide bends are expensive and large cable is difficult to bend in a confined space. Cable ladder is recommended for the support of heavy cables. Large waveguide and branching networks may be suspended from the ceiling. Proper provision must be made for the exit of feeders from the station building to the outside.

e. Line Pressurizing System. A line pressurizing system capable of maintaining transmission lines, branching networks, and antenna feeds with dry air or gas under slight pressure is a necessary adjunct to any large station configuration. Either a mechanical system or a static system with dry nitrogen gas may be employed.

f. Other Requirements. The provision of facilities for other requirements, including heating and ventilating, fire fighting, workshops, parts storage, sanitary facilities, personnel amenities, etc., is discussed elsewhere in this chapter.

18. Outside-plant Planning Requirements

a. Site Development. The problems of site development, access roads, etc., are discussed in paragraphs 5 to 10 of this chapter.

b. Antennas and Antenna Support Structure. *General.* In radio terminal configurations, the required antennas and support structures generally constitute the major outside-plant facility. Thorough planning and proper installation are essential to assure optimum performance and stability and even survival against the ravages of time and the elements. The manufacturers and fabricators of antenna facilities represent the most reliable source of planning information and technical specifications.

Towers. Vertical towers, generally of galvanized steel construction, are extensively used as active antenna elements (radiators) or as support structures for antennas or to serve both tasks simultaneously. Guyed towers are more commonly employed in preference to self-supporting towers because of their inherently lower cost. They do, however, need relatively large areas to accommodate the guying arrangement, and where large land plots are not available or are relatively costly, the use of self-supporting towers may be dictated. Below heights of 60 to 80 ft, the use of self-supporting towers is generally economical.

Expected maximum wind conditions with or without ice form the basic criteria for the design of any tower structure. The two main conditions to be met are the required directional stability of the antenna to provide reliable communications and ultimate survival of the structure. Typical requirements for a tower supporting a microwave radio relay antenna may call for a maximum deflection or twist of the antenna of $\pm 1°$ for a 70-mph wind with survival at 100 mph. Ice loading conditions will be specified according to requirements. In some areas of the world, the survival wind velocity may be in excess of 200 mph. The effect of tornadoes is rarely considered in the design of structures, and a calculated risk is taken whenever they are installed in areas where tornadoes occur. Wind loading is generally expressed in pounds per square foot, which is proportional to the square of the wind velocity. A velocity of 70 mph is equivalent to a loading of 20 psf, while 100 mph is equivalent to 40 psf.[11,12]

Obstruction lighting and painting are required for most tower structures to conform to United States or international standards. Lighting kits are available to conform to requirements for specified tower heights. Tower-climbing facilities are essential in the form of simple steps or through the use of a ladder mounted inside or outside the tower.

In radio relay systems, antennas are often conveniently installed atop station buildings on simple mounts or on short, self-supporting towers. For tactical military equipment or in other temporary installations, portable, quick-erect towers of aluminum construction are available.

Antennas. For microwave radio relay systems, relatively simple antennas are employed, mounted atop towers, buildings, or other supporting structures. Above 4,000 Mc, it is often convenient to employ tower-mounted passive reflectors in the familiar "periscope" (fly-swatter) configuration. Requirements for channel allocations or interference considerations may dictate the use of tower-mounted, high-discrimination parabolic antennas. Passive repeaters are sometimes used to route microwave beams around or over obstacles, eliminating the need for an active repeater station.[13]

Large radiating antenna structures, such as the familiar billboards or dishes used for tropospheric-scatter systems, must be designed and implemented to the same rigid requirements as tower structures for stability and survival, with due allowances for environmental conditions. The somewhat higher directional characteristics of such antennas dictate more stringent requirements for operational stability and accuracy in installation. Other large antenna structures, as used in ionospheric-scatter or in high-frequency systems, must meet similar, though less exacting, standards.

Radomes, heated or unheated, may be employed to prevent icing or to minimize its effects. In the case of large tropospheric-scatter antennas, only the feed horn will be so protected. Electrically heated radomes consume a great deal of power and should be employed with discretion.

Transmission Lines. Antenna feeders, whether open wire, coaxial cable, or waveguide, must be properly installed and protected for maximum efficiency and reliability.

They may be run above ground, supported by wood or steel poles, or may be run underground in ducts. Protection should be provided for aboveground runs from falling ice. Protection from the direct rays of the sun may be advisable, particularly for large waveguides in hot climates. Use must be made of flexible waveguide sections to allow for expansion and for installation tolerances. Care should be exercised in assembling waveguide runs to minimize reflections occurring from joints.[15]

Structure Foundations. The foundations of a guyed or self-supporting tower or antenna structure must be properly designed to accommodate the structure load under expected soil conditions. For planning purposes, normal soil is considered to have a load-bearing capacity of 4,000 psf. Special designs are necessary for swampy or rocky soil. Antenna structure bases are constructed of concrete, steel reinforced. The design should provide for proper load distribution to prevent excessive stresses at any point. Guy anchors are usually of concrete and steel construction, designed to meet soil conditions and loading requirements.

Siting. The accuracy required in siting antenna structures is related to the intended utilization. Tower-mounted microwave antennas are usually provided with a wide adjustment range, making extreme accuracy unnecessary. Large tropospheric-scatter antennas or microwave passive repeaters will provide only a very narrow adjustment range. For these structures, an accurate boresite with concrete markers must be established from a celestial survey. With this reference line, the locations of concrete piers and anchor bolts can be accurately fixed.

c. Grounding. The station protective grounding system comprises two or three major elements; the equipment grounding system, the antenna structure grounding system, and the underground fuel supply tank cathodic protection system (not connected to other grounds).

The overall grounding system bonds together and connects to earth all equipment assemblies, power-generating equipment, the electrical system, transmission lines, antenna structures, steel framework or buildings, etc. Protection is provided against lightning strikes and the possibility of appreciable voltage differences between equipments and above ground.

Station ground systems which form a part of the antenna radiating system require exacting engineering to meet required performance and prevalent soil conditions.

Soil-resistivity measurements should be made at each station site as an aid in the design of the grounding system.

d. Fuel Storage. Fuel-storage tanks for operation of the power plant may be installed above or below ground as required. Storage capacity is predicated on the planned interval between fuel deliveries, with due allowance for inclement weather, and on the load factor of the plant. The latter is governed by expected power outage if the plant is of the standby type or by operating requirements if it is the primary power source.

In warm or hot climates, it is advisable to protect aboveground tanks with a sunshade of corrugated asbestos or similar material. Buried tanks should be provided with cathodic protection and should be completely isolated from the station grounding system. This will necessitate the use of a section of neoprene tubing (or substitute) in the fuel line between the tank and the power plant.

19. Sequence of Planning

An orderly sequence of station planning is essential to achieve optimum results with minimum effort. While each individual planner will approach the problem differently, the following schedule is suggested for essential tasks in the order listed.

  Site layout
  Preliminary building design
  Equipment placement
  Power-system layout
  Power distribution and interfaces
  RF distribution and interfaces

Traffic distribution and interfaces
Final building design
Outside-plant engineering

POWER PLANTS

20. General

The subject of power systems for communications equipment has been covered in a separate chapter and will be treated here only with regard to special considerations for radio communications stations.

Power plants for radio communications stations are of two main categories: those supplying prime power and those serving as standby to a normal commercial power source. The present stringent performance objectives for military and commercial systems and the high degree of available equipment reliability often dictate the requirement for uninterrupted power. This has led to the prevalent use of "no-break" types of power systems of the mechanical (flywheel) or all-electric (storage-battery) types. In many systems, the trend to solid-state equipment and the attendant low-power drain encourages d-c operation from storage batteries capable of satisfying the requirements for standby and no-break operation.

21. Design Considerations

In general practice, internal-combustion engine-generator equipment is located in separate rooms or, in the case of large installations, in separate buildings from the electronic equipment. This is dictated by requirements for protection against vibration, noise, fumes, heat, fire hazard, etc. Waste heat given off by the radiator or air cooling may be utilized and controlled to supply cold-weather heating for electronic equipment and power-plant areas. Outlets for cooling air and exhaust gases must be well isolated from ventilating-air intakes to prevent contamination.

The transmission of vibration from operating engine-generators to other equipment must be minimized. Generally, this is done by mounting the unit on massive concrete pads which are isolated from the building floor and foundation. The use of skid-mounted engine-generators simplifies installation and, for the smaller sizes, facilitates their removal and replacement for overhaul.

For prime-source power plants, the use of relatively slow-speed diesel electric engine-generators is mandatory for reliability, economy, and long maintenance intervals. For standby requirements, with intermittent operation, the use of gasoline or propane gas units offers the advantages of simplicity and ease of maintenance and starting.[14] In each case, the relative merits of diesel vs. gas operation must be weighed against initial and operating costs, maintenance requirements and cost, and the advantages previously cited for both. In any network or system, compatibility in equipment is desirable at all stations through the use of identical engine-generators or units using common replacement parts.

Figure 3 shows a suggested standby power plant with a propane gas generator for a typical microwave line-of-sight relay station.

22. Special Considerations for Battery Plants

Lead-acid storage batteries require a relatively warm environment for efficient operation. Rated ampere-hour capacity is ordinarily predicated on an ambient temperature of 77°F. On a 3-hr discharge cycle (to 1.75 volts per cell), the capacity is reduced approximately to 77 per cent at 40°F and to 52 per cent at 0°F. To take advantage of the heat usually dissipated by the operating electronic equipment, the battery bank may be located in the equipment room. Proper insulation in this room with a thermostatically controlled ventilating system may provide a favorable temperature environment without requiring supplementary heat except in the most extreme cases. Additionally, the length of d-c power feeders is minimized together with voltage drops resulting from the relatively high currents and low voltages (24 or 48 volts) employed.

The reserve operating time for station electronic equipment will vary according to the reliability of the prime power source and the availability of standby power. In practice, reserve capacities of 4 to 24 hr are provided. The battery charger must be of a rating which is adequate to carry the equipment load and to recharge the battery in a maximum of 24 hr.

INTERFACE REQUIREMENTS

23. Description

Interface points must be generally established within or without a station for interconnection with external facilities and services, whether those of the user, customer, supplier, or contractor. The form and location of the interface point and responsibility for its supply and installation will be established by agreement of the parties concerned, usually in accordance with prevailing codes and practices. The most common requirements are for telephone (signal) and power interconnections.

24. Telephone

The interface point between the radio terminal facilities and external cable or wire plant is usually provided in the form of protected and unprotected distribution frames, along with the proper number of protectors and terminal blocks to meet requirements. Building entrance provisions should be according to governing codes.

25. Power

Externally supplied electric power facilities will be terminated in a junction box or protected disconnect switch to be located within or outside the station building. This facility and building entrance provisions will be governed by established codes.

26. Other

Examples of other facilities which may require the establishment of an interface point include the following:

Antenna transmission lines
Grounding systems
Fuel-supply piping
Tower or other outside lighting
Control circuits

ENVIRONMENTAL CONSIDERATIONS

27. Planning Considerations

The total characteristics of environment weigh very heavily in the overall engineering of radio communications station facilities. In particular, attention must be given to the problems presented relative to planning for the following:

Site location
Access road
Site development
Outside-plant facilities
Building design
Equipment operation
Access for operation and maintenance
Fuel-storage facilities

28. Specific Environmental Parameters

a. Temperature. All inside-plant and outside-plant facilities must be designed for proper operation and stability throughout the expected range of temperature variations. Heating, ventilating, and air-conditioning facilities will be planned to provide for personnel comfort, equipment life, and operational efficiency and stability.

b. Humidity. Extreme humidity conditions may necessitate the use of air-conditioning or dehumidifying equipment for equipment protection and operating-personnel comfort. In tropical locations, problems of fungus growth may be encountered.

c. Rainfall. In areas of unusually heavy rainfall, precautions must be taken in planning the design of access roads, site drainage, and building design. Microwave line-of-sight systems operating above 6,000 Mc may encounter excessive signal attenuation in tropical downpours.

d. Snow. Heavy snow accumulation can present problems in site access for operation and maintenance, particularly in remote areas. Horizontally mounted microwave antennas may require radomes or heaters for protection against snow. Special building design may be required to combat rooftop snow accumulation.

e. Ice and Wind. Antenna structures and towers must be designed for required stability and survival for expected icing and wind conditions. Protection from falling ice is essential for antennas, transmission lines, and personnel. Radomes or other de-icing facilities may be necessary for UHF or SHF antennas or feeds.

f. Dust, Sand, Salt Atmosphere. In desert or seacoast areas, protection may be required against airborne dust, sand, and salt. This can be accomplished by the use of simple air filters in all building air intakes. In extreme environments, buildings may require slight pressurization along with traveling screen, oil bath air filters.

g. Storms. Overall station planning must assure continuing operation and survival in the face of expected local storms, including hurricanes, typhoons, etc. Protection against tornado-type winds for towers and other antenna structures is not normally provided.

h. Elevation. Extreme site elevation may pose special problems for operating personnel and electronic equipment design to assure performance objectives and reliability. Engine-driven power generators must be derated in operation at increased elevations.

CONSTRUCTION AND SUPPORT LOGISTICS

29. Requirements

For the implementation and operation of a single station or a complex system it is necessary to develop a comprehensive logistical support plan. Three important areas of consideration are discussed in the paragraphs which follow. The relative importance of each point will vary between the phases of installation and operation, but all are important. Special considerations for foreign and remote areas are treated in more detail later in this chapter.

30. Equipment and Material Handling

The shipping, warehousing, transport to the station site, and on-site storage of station equipment and building materials represent a major logistical task in the implementation of a facility. Proper planning, as a result of field investigation, can effect a significant saving of time and money. The field survey should include checks on rail or harbor facilities, storage warehouses, trucking availability, etc. The routes to be traversed by trucks must be inspected for width, grade, curves, overhead obstructions, and weather restrictions. Requirements for special loading and unloading equipment must be determined.

31. Personnel Support

In remote and foreign areas, the support of installation and operating personnel must be adequately provided for. Consideration is required of problems associated with personnel transport, housing, messing, medical attention, rest and recreation, mail deliveries, etc. Plans must exist for hurried evacuation of sick or injured persons.

32. Facilities and Supplies

An advance determination of available facilities, services, and supplies is necessary prior to commencing field activities. Commercial communications facilities, if they exist at all, may require supplementing. Electric power for construction may not be commercially available. Water for construction purposes may need to be trucked over long distances. Fuel deliveries during construction and operation must be arranged.

SYSTEM OR STATION IMPLEMENTATION

33. Introduction

Previous discussions have been concerned with the important tasks and considerations involved in the planning and execution of a system or station facility. The overall job of implementation is complex and necessitates careful and thorough planning and scheduling. The major divisions of this chapter define the main tasks and considerations in order of execution and relative importance and can serve as a guide. In staffing for planning and implementation, a variety of skills and specialties is required, to a degree dependent on the size and complexity of the facility. From consideration of cost, scheduling, and logistical support, it is prudent to keep manpower at a minimum by utilizing well-rounded and *experienced* planning and field engineers.

34. Equipment Installation and Commissioning

Other sections of this chapter have outlined the major requirements for planning and implementation activities. The job of installing and testing equipment and commissioning the facility is the culmination of the program and its measure of accomplishment.

The installation and commissioning of equipment in an unsophisticated system of a few stations are most expeditiously performed by a small crew of highly experienced technicians, each familiar with all or most of the requirements.

In a large and complex system configuration, the utilization of teams of specialists provides the optimum in economy, scheduling, quality, and performance. Individual teams would, for example, be established to carry out one or more of the following phases of the program:

    Physical equipment installation
    Electrical wiring
    Fuel- and lubricating-system piping
    Equipment commissioning
    Subsystem alignment and testing
    System alignment and testing
    Acceptance and proof-of-performance tests

Each team consists of a qualified supervising engineer and a staff of required skills. These teams operate independently on a schedule planned to expedite the program and to prevent interference at stations between different teams. The use of small, well-organized teams for specific tasks results in a high degree of efficiency, as each team will be doing repetitive work. The equipment-installation team moves into each site first, to be followed in succession by the other teams. The use of portable radio communications sets in the field may facilitate all operations relating to scheduling of work and of material and equipment to the field.

Individual teams should be staffed with qualified tradesmen such as riggers, electricians, pipe fitters, etc. Except for overall supervision, the need for communications engineers and technicians arises only in the equipment and system commissioning, alignment, and testing phases. Attempts to utilize communications specialists for tradesmen's tasks is a false economy and can result in schedule delays and in poor quality of installation and performance.

FOREIGN AND REMOTE AREA CONSIDERATIONS

35. General

The planning, construction, and operation of radio communications facilities in foreign areas present a host of new problems and engineering considerations to even the most experienced designers and contractors. Many of these same problems are also met in the implementation and operation of domestic systems in remote areas. Proper recognition of these factors and the planned development of solutions to meet them will avoid costly mistakes and delays in the execution of required facilities.

For convenience, it is possible to group most of these special problems and considerations into three general categories: (1) logistics, availability of needed materials, services, and facilities; (2) environmental and topographical problems; (3) social, legal, and technical problems peculiar to the host country. These are treated separately below.

36. Logistics, Availability of Materials, Services, and Facilities

In many parts of the world, the availability of qualified contractors, technical skills, and crafts may be limited in quantity and quality or may be nonexistent. Construction machinery may be available in limited quantity, possibly of an obsolete or rudimentary design. It may be necessary to import all or part of the required building materials. The host country governments in many developing countries will encourage the maximum use of manual labor, replacing machinery in construction activities to aid the indigenous labor force and economy.

Support for personnel working in the field must be provided, including billeting, messing, medical services, etc. Communications will be needed between sites and to the contractor's operating bases. This will often require the use of HF or VHF radio. The problems associated with material handling, including shipping, warehousing, and overland transport to the site, warrant thorough investigating and planning.

37. Environmental and Topographical Problems

The subject of environmental considerations has been treated earlier in this chapter and is mentioned again here for the sake of emphasis. Special consideration must be given the design of facilities in areas subject to earthquakes, extremely high winds, typhoons, and other forms of tropical storms. Unmanned stations must be planned to operate for long periods without attention in these environments. Manned stations must possess the capabilities of remaining self-sufficient without outside support for prolonged periods.

Adequate topographical mapping of the area of construction may be unavailable, requiring a field survey by the contractor to determine site topography and access. In an extensive system, where path topography between sites must be established, aerial photogrammetric or radar surveys may be required.

38. Legal, Social, and Technical Problems

In many foreign areas, particularly in underdeveloped countries, the problems arising out of regulations pertaining to customs, immigration, and other legal requirements can be costly and time consuming. It is mandatory to employ a local business agent or a host country business manager and staff to expedite matters concerned with employing local nationals, immigration, customs, work permits for foreign nationals, local subcontract administration, taxes, cost accounting, and a multitude of other problems.

Living conditions will be alien to foreign nationals working in many countries. Housing accommodations, sanitary facilities, medical standards, and services may seem primitive. Satisfactory foodstuffs and potable water may be scarce, requiring transporting over long distances. Leave and facilities for rest and recreation are recommended for foreign nationals working in extremely remote or primitive situations.

Differences in construction and engineering codes and practices existing in foreign countries usually must be taken into account in overseas system planning. Architectural and construction standards differ markedly, depending to some degree on availability and class of construction materials. In the case of electric power supply, voltage and frequency standards may be different along with the practices followed and hardware utilized.

REFERENCES

1. Egli, J. J., UHF Radio Relay System Engineering, *Proc. IRE*, January, 1953.
2. Hancock, E. G., National Telecommunications Network Planning, *Point to Point Telecommunications*, June, 1961.
3. Marsh, D. R., How to Plan a Microwave System, *Oil Gas J.*, Nov. 26, Dec. 3, 1956.
4. Nexon, V. J., and G. Karger, A Modern Approach to Communications Systems Planning and Engineering, *Pipe Line News*, April, 1962.
5. Bray, W. J., The Standardization of International Microwave Radio-relay Systems, *Proc. IEE*, Paper 3412E, March, 1961.
6. Microwave Transmission Systems, *EIA Standard* RS-203.
7. *Transmission Systems for Communications*, vols. I and II, by Members of the Technical Staff, Bell Telephone Laboratories, 463 West Street, New York.
8. Microwave Housing Facilities, *EIA Standard* TR-142.
9. *Standardized Buildings for RCA Microwave Radio Stations*, Radio Corporation of America, Camden, N.J.
10. How to Estimate the Cost of a Private Microwave System, *Railway Signaling and Communications*, May, 1962.
11. Structural Standards for Steel Transmitting Antennas, Supporting Steel Towers, *EIA Standard* RS-222.
12. Wells, H. A., Microwave Radio Relay Towers, *Bell Lab. Record*, August, 1959.
13. Mechanical Characteristics for Microwave Relay System Antennas and Passive Reflectors, *EIA Standard* RS-195.
14. Emergency Stand-by Power Generators and Accessories for Microwave Systems, *EIA Standard* RS-173.
15. Mechanical Considerations for Transmission Lines in Microwave Relay Applications, *EIA Standard* RS-158.

Chapter 24

ADMINISTRATION FOR COMMUNICATION SYSTEMS

KENNETH D. HOWATT, *American Telephone and Telegraph Company*

INTRODUCTION

Administration as outlined in this chapter refers to the direction, management, and supervision of a going communications system. Organization, practices, and the duties and responsibilities of the various units of an organization are all implicit in the discussion of such a topic. Just as it is the responsibility of the engineer to design and to install a communications system in a manner which will provide the most practical

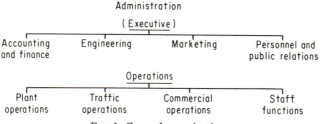

FIG. 1. General organization.

and economical results, so there is a continuing responsibility by the owners and managers to operate the system as efficiently as possible. Certain aids to efficient operation involve planning, methods, the maintenance of records, and results measurements which will further the quality of operating performance. The importance of the different facets of proper administration will vary in accordance with the size of the system, its objectives and goals, and the opinions of the men in charge.

The organizational arrangement of most large communications systems generally indicates at least the functions or departments shown in Fig. 1. In a commercial firm, the chief executive officer is responsible to the board of directors, and in a government-owned system he is responsible to a bureau head or other government official or body.

In a smaller system, the executive head might be the owner or his appointee. If the system is not privately owned, the chief administrator would probably be a member of a Federal, state, or local governmental bureau or division. In a small organization, the overall operation would most likely be arranged along nonfunctional lines, so that a given manager or supervisor would be responsible for several functions. Irrespective of the precise form of organization, all the functions would be performed to some degree in the day-to-day operations of the system. This chapter will primarily discuss administrative subjects influencing operations of a communications system directly: accounting, engineering, and operations.

ACCOUNTING

The American Institute of Certified Public Accountants defines the accounting function as follows:[1] "Accounting is the art of recording, classifying and summarizing in a significant manner and in terms of money, transactions and events which are, in part at least, of a financial character, and interpreting the results thereof."

Accounting is the tool upon which management relies for direction in the overall operations of the business. It assists in the control of costs, the preparation of budgets, and the maximizing of profits. Many decisions which are fundamental to the successful operation of the business could not be made intelligently without the information made available by a modern accounting system.

The accounting process measures and records input and output, and the raw material of the accounting process is money. Money flows into a business from many sources and flows out from a business to many destinations.

1. Examples of sources of cash (cash inflow):

    a. Revenues from customers for services rendered
    b. New capital from investors in business
    c. Amounts from sales of property (including junk)
    d. Interest from cash deposits and short-term securities
    e. Rent income from building space or communications facilities leased to other carriers

2. Examples of destinations of cash (cash outflow):

    a. Construction of communications plant and facilities
    b. Salaries and wages of employees
    c. Payments for services rendered by others including light and power, technical services, and legal fees
    d. Cost of maintenance materials
    e. Rent payments for office space or plant and facilities leased from other carriers
    f. Local, state, and Federal taxes
    g. Interest on notes and bonds
    h. Dividend payments to shareholders

In a government-owned operation, the cash inflow would consist of taxes paid into the government, income from property leased to others, and proceeds from debt incurred by the government. Cash outflow, on the other hand, would be composed primarily of amounts for payrolls, construction and maintenance materials, property leased from others, and technical assistance.

Table 1. Income Statement*

Operating revenues............	$100,000
Operating expenses..............	60,000
Net operating revenues.........	$40,000
Operating taxes.................	23,500
Net operating income...........	$16,500
Other income..................	300
Total income..................	$16,800
Interest deductions..............	2,500
Net income...................	$14,300
Dividends.....................	10,800
Retained earnings..............	$3,500

* Amounts are hypothetical.

The faithful recording of the monetary transactions affecting the financial status of a communications system is of paramount importance in the efficient operation of a communications system. The accumulated totals of income and outgo for all the business transactions which take place during the accounting period, e.g., one month or one year, are shown on the income statement. A typical income statement showing the financial progress over a given period would read as shown in Table 1.

The other part of the financial picture is the balance sheet, which depicts the financial status of a going enterprise as of a given point of time and is prepared as of the end of the accounting period, e.g., end of month or year. A typical balance sheet would be as shown in Table 2.

Table 2. Balance Sheet*

Assets		*Liabilities*	
Plant......................	300,000	Common stock................	145,000
Depreciation reserve...........	70,000	Funded debt..................	60,000
Net plant....................	$230,000	Bank loans...................	15,000
Current assets................	29,000	Surplus......................	25,000
Other assets.................	1,000	Total capital...............	$245,000
		Current liabilities.............	15,000
Total assets................	260,000	Total liabilities.............	$260,000

*Amounts are hypothetical.

1. Accounting Records and Their Uses

In addition to its use for internal administrative purposes, it is often necessary to maintain a complete set of books, i.e., accounts, in order to comply with local, state, or Federal regulations and with tax laws. In the United States, communications common carriers rendering interstate service are under the jurisdiction of the Federal Communications Commission, and the latter has prescribed the accounts by name and number which shall be maintained by each such carrier. A communications system subject to local, county, state, or Federal taxes will need sufficient records to satisfy the taxing authorities. Communications systems restricting their operations within the boundaries of a given state may, unless specifically exempted, come under the jurisdiction of a state utility commission or similar body.

Ordinarily it is desirable to accumulate the different kinds of revenues and the various types of expenses in separate accounts, each such account bearing an appropriate name. Such segregations of revenues and expenses provide a positive basis for administering the business with respect to revenue requirements, cost controls, operating efficiency, budget requirements, and financing which may be necessary from time to time. These terms may be described as follows:

Revenue Requirements. If a communications system is to prosper, remain financially sound, and provide the services for which it was chartered, the income from revenues must be sufficient to meet all current expenses such as those for wages, taxes, maintenance materials, depreciation, and all sundry purposes. In addition, amounts must be available to meet scheduled interest payments on bonds or notes. Furthermore, if the system is publicly or privately owned, there should be a remainder after expenses, taxes, and interest to provide a reasonable return to those who invested capital in the system.

Cost Controls. The record of expenses as shown by the accounts will indicate the level and direction of costs and provide a warning signal with respect to those costs experiencing unusual fluctuations or trending toward unjustifiably high levels. Prompt administrative action is possible only where reliable controls exist.

Operating Efficiency. The volume and quality of services furnished may be compared with the costs of rendering such services by the system. This relationship may then be compared with standard objectives or other systems to determine the relative efficiency of the system being analyzed. Without ample cost records, operating efficiency cannot be adequately controlled.

Budget Requirements. Sound planning dictates the preparation of a budget for the coming year's operations and entails an estimate of plant additions or changes, revenues, expenses, taxes, interest, and profit or return. The accounting records are invaluable in assisting in such forecasts.

Financing. Expansion, major rearrangements, and changes or unforeseen disasters may require additional financing. Without reliable financial statements to show the

Table 3. Account Titles for Revenues and Expenses[2]

Revenues:
Local service revenues
Toll service revenues
Rent revenues
Other operating revenues
Uncollectable revenues (Dr.)

Expenses:
Maintenance expenses:
Repairs of pole lines
Repairs of aerial cable
Repairs of underground cable
Repairs of buried cable
Repairs of submarine cable
Repairs of aerial wire
Repairs of underground conduit
Test desk work (circuit testing and trouble location)
Repairs of central-office equipment
Repairs of subscribers' station equipment
Repairs of building and grounds
Costs of power:
Commercial power purchased
Maintaining generators and batteries
Other miscellaneous maintenance expenses
Depreciation expenses
Traffic expenses:
General traffic supervision
Operators' wages
Stationery and printing
House service expense
Other traffic expenses
Commercial expenses:
General commercial administration
Advertising
Sales expense
Local commercial operations
Directory expenses
Other commercial expenses
General expenses:
Executive or administration
Accounting department
Treasury department
Engineering expense
Insurance
Accidents and damages
Operating rents
Medical and pension plans
Other general expenses

condition of the business, past and present, it would probably be impossible to raise the required capital at a reasonable cost, if at all.

Representative names of typical accounts[2] employed by communications systems for income statement purposes are indicated in Table 3. The accounts listed represent the major sources of revenues and the major items for which expenses are incurred. From a marketing standpoint and for purposes of earnings requirements it is important that the management be able to analyze on a continuing and current basis the source of its revenues. Such information indicates, for example, where more sales effort should be applied and where rate adjustments may be required.

Similarly, with respect to cost controls, it is necessary that expenses of a particular type be segregated from all other expenses in order that effective action may be taken immediately upon the occurrence of an unusual fluctuation in the level of expense for a given classification of plant or activity. If, for example, maintenance expenses for wire and cable were combined in one account and the level suddenly peaked up, it would not be apparent whether it was one or more kinds of cable or the aerial wire which was causing the increased outlays of money.

Oftentimes it is desirable to establish subaccounts under the major accounts in order to obtain a finer division of revenues or expenses. For example, under repairs

Table 4. Account Titles for Assets and Liabilities[2]

Assets:
- Telephone plant:
  - Land
  - Buildings
  - Central-office equipment
  - Subscribers' station apparatus
  - Station connections
  - Private branch exchanges
  - Pole lines
  - Aerial cable
  - Underground cable
  - Buried cable
  - Submarine cable
  - Aerial wire
  - Underground conduit
  - Furniture and office equipment
  - Vehicles and other work equipment
- Other investments:
  - Sinking funds
  - Cash
  - Notes receivable
  - Amounts due from customers and agents
  - Other accounts receivable
  - Material and supplies
  - Other current assets
  - Prepaid rents
  - Prepaid taxes
  - Prepaid insurance
  - Insurance and other funds

Liabilities:
- Capital stock
- Premium on capital stock
- Funded debt
- Other long-term debt
- Notes payable
- Accounts payable
- Other current liabilities
- Taxes accrued
- Depreciation reserve
- Surplus

of central office equipment it might be beneficial to the managers of an operating enterprise to have separate subaccounts in which to record expenses associated with switching equipment, microwave equipment, mobile radio base-station equipment, etc. Similarly under the repairs of subscribers' station equipment it might be desirable to segregate maintenance costs for telephone equipment, teletypewriter equipment, mobile radio apparatus, etc.

In addition to the income and expense accounts, it is necessary to establish and maintain a set of accounts for the balance sheet. As previously indicated, the balance sheet is comprised of two sections, viz., assets and liabilities. Representative names of typical balance sheet accounts[2] for communications systems are illustrated in Table 4. It is necessary to segregate the various kinds of assets and liabilities in order to manage the operation properly as indicated for the accounts on the income statement.   Different kinds of plant, for example, have different characteristics and different unit costs, and amounts expended for construction are relatively unimportant for some kinds of plant but of very large proportions for other kinds.   The amount of cash on hand is an important indication of liquidity and solvency.   If cash were lumped with other assets, the cash position could not be readily ascertained. The different kinds of liabilities have individual and distinct characteristics and therefore are segregated to the extent required to present the clearest possible corporate picture.

Subaccounts of the main asset and liability accounts are also established where a finer breakdown in the amounts accounted for is desirable.   For example, under central office equipment subaccounts may be established for the various kinds of this plant such as dial switching, microwave, mobile radio, etc.

The dollar amounts recorded in the revenue accounts consist primarily of amounts paid to the communications system by customers for services furnished.   If the system is not a common carrier offering services to the public but rather is a private or government operation, a suitably captioned account or accounts are established to record the receipt of funds used in operating the system.

Table 5. Cost Elements of Plant[2]

Material
Installation labor
Prorate of motor vehicle and special tools expense
Supply expense
Engineering costs
Architects' fees
Administrative costs
Interest costs during construction
Taxes incurred during construction
Any other costs directly related to construction

Except for depreciation expense, the amounts expended for operating the system as recorded in the expense accounts will represent to a large extent amounts paid for wages and salaries of employees and others.   Maintenance expenses will also include amounts spent for materials, freight, motor-vehicle expense, supply costs, and all the sundry expenditures made in the day-to-day maintenance job.   Under some situations, expenditures for facilities leased from others may be relatively large.   Certain accounts such as insurance and operating rents, which obviously cover payments for insurance premiums and rented space, respectively, do not involve elements of wages and sundry costs.

The periodic accrual of depreciation expense is the method by which the cost of the plant in service (included on the balance sheet) is charged off against operations over the life of the plant.   Generally speaking, communications plant and equipment have a life of more than one year.   When the system is constructed, the cost of the plant represents a prepaid cost, and depreciation accounting is the method by which these prepaid costs are assigned to the various accounting periods during the life of the plant.   The rate at which the plant costs are amortized depends upon the estimated life of the plant.   The studies required as well as representative lives and depreciation rates are discussed later in the Engineering section of this chapter.

As indicated by the sample balance sheet, the dollars invested in plant represent the major portion of the assets of any communications system.   Plant may involve many kinds of physical property such as buildings, central office equipment, buried cable, etc.   The cost of plant reported in a balance sheet consists of several elements and typically includes the costs illustrated in Table 5.

Underlying the dollars shown on the balance sheet for plant, there should be a rather complete record showing the quantities and locations of the plant for which the dollars were spent.   This record is sometimes referred to as the continuing property record and should be maintained in such a way as to meet the following basic objectives:

1. An inventory of property record units which may be readily spot-checked for proof of physical existence

2. The association of costs with such property record units to assure accurate accounting for retirements

3. The determination of dates of installation and removal of plant retired to provide data for use in connection with depreciation studies

Such a record fills the need for an inventory of plant which may be used for appraisals required for the determination of rate schedules, sales of plant, property tax assessments, or other studies of a similar nature.   Typical entries in the continuing property record for central office equipment at a long-distance terminal office are illustrated in Table 6.   Additional information may be shown as desired.   The installed cost includes the material cost and a prorate of labor, supervision, and all other overheads

Table 6. Continuing Property Record

Bay loca-tion	Name of unit	Specification or drawing No.	Quan-tity	Installed cost	Year placed
2101	O–N terminal mounting	J98705A	4	$1,055.95	1956
2101	O–N channel terminal unit	J98705D	16	3,699.02	1956
2101	O–N twin-channel unit	J98705E	4	1,011.15	1956
2101	O–N twin-channel unit	J98705E	4	1,017.55	1956
2101	O–N miscellaneous oscillator	J98705W	1	62.40	1956
2102	O–N channel terminal unit	J98705D	16	4,387.71	1957
2102	O–N twin-channel unit	J98705E	8	1,799.97	1957
2103	O–N combined net and oscillator panel.	J98706G	1	52.04	1958
2105	O–N LH-LL repeater unit	J98706E	2	544.05	1958
2106	O–N channel terminal unit	J98705D	16	4,962.95	1958

associated with the installation of the particular equipment itemized.   The information shown is generally recorded on punched cards or magnetic tape and is printed out on summary tabulations periodically as may be required.

The detail shown in the continuing property records for each class of property would be about as follows:

Class of property	*Inventory detail*
Land.	Original purchase cost by plots with sufficient legal description to show location
Buildings.	Original construction costs for each building. Subsidiary records generally maintained by engineering would show floor layout, square feet of space, and other pertinent data
Central office equipment.	Original installed costs for each building location. Related engineering records would show equipment layouts, bay numbers, and various drawings and diagrams
Subscribers' station equipment. . .	An inventory of the numbers of telephones, teletypewriters, and radiotelephones in service and in supplies
Pole lines.	An inventory of poles by length
Cable.	An inventory of the sheath feet of cable with separate quantities recorded for different sizes and make-ups. Separate counts are made for cable terminals and loading coils
Conduit	An inventory of duct feet by duct groups

Broad gage unit costs as well as more finite average unit costs may be determined from the numbers of units reflected in the continuing property record and from the dollars of construction costs shown in the accounts and in the property record.

From the amounts shown on the income statement and on the balance sheet, several control ratios may be developed periodically which will show trends in operating efficiency and other characteristics of the system:

1. *Revenue-to-plant Ratio.* The ratio obtained by dividing the amount of total operating revenues (annual basis) by the amount of average telephone plant in service.

2. *Expense-to-revenue Ratio.* Sometimes called the operating ratio and is obtained by dividing the amount of total operating expenses by the amount of total operating revenues.

3. *Maintenance Ratio.* The ratio obtained by dividing the amount of total maintenance expenses (annual basis) by the amount of average telephone plant in service.

4. *Depreciation Ratio.* The ratio obtained by dividing the amount of depreciation expense (annual basis) by the amount of average depreciable telephone plant.

5. *Earnings Ratio.* The ratio of the amount of total income (or net operating income) to the amount of average net telephone plant. Typical ratios experienced by large communications companies in the United States during 1962 fell within the ranges shown on Table 7.

Table 7. Control Ratios

	From	To
Revenue to plant.............	32%	36%
Expense to revenue...........	57%	62%
Maintenance ratio............	5.2%	6.2%
Depreciation ratio...........	4.3%	5.3%
Earnings ratio...............	7.0%	8.5%

ENGINEERING

When it is considered that the engineering forces plan the facilities which ultimately result in large outlays of cash for relatively long-term investments, it becomes evident that engineering functions are of major importance. The chief activities of the engineering section of any going communications system are generally in connection with:

Planning and recommending new facilities to meet the growing needs of the system

Furnishing advice and assistance in the application of engineering methods to the construction, operation, and maintenance of plant

Preparing specific plans and specifications for construction or rearrangements and changes of plant

Improvement of facilities and introduction of instrumentalities for improving service or reducing costs

Maintaining records, maps, specifications, diagrams, and location records relative to equipment installations, cable routes, microwave routes, and other plant and facilities as required

Providing the accounting section with information needed to maintain reliable and complete accounting records, also developing depreciation rates for the accountants' use in accruing depreciation

2. Fundamental Plans

Proper management of the capital invested in the plant of a communications system dictates the existence of fundamental plans. These consist principally of requirements for additional circuits and broad, long-range plans for line sections where major growth is expected. General requirements for switching systems, where involved, would also be included in fundamental plans as would standard building designs and new engineering methods.

Periodically, tabulations of messages over voice and telegraph circuits should be prepared as a means of determining the volume of traffic handled between pairs of points and over specific circuit or trunk groups.   From the trends in these volumes and from other information which is known to have an effect on the volume of traffic handled (e.g., economic trends, population growth, industrial activity), it is then possible to forecast future requirements for additional circuits and related facilities. Usage of trunk groups may also be measured by holding time-recording devices which will record the total minutes of use by hours of the day for the specific circuits being measured.   Immediate or imminent requirements for additional circuits may become apparent from an inordinate number of no-circuit conditions which result in too many busy signals and delayed call attempts.   Sound fundamental plans accompanied by timely execution are intended to avoid retrogression in service of this nature (see Chap. 6).

Tabulations of holding times and message volumes should be analyzed by engineers, and the results used to determine the most practicable and economical type of construction required to obtain the additional circuits needed over the ensuing years. Depending upon circumstances, the solution may rest anywhere from adding channel banks to an existing backbone route to developing another route which, besides providing additional capacity, would furnish an alternate route for service protection purposes.

Other general considerations are generally required in the development of fundamental plans.   For example, it is necessary to ascertain that the proposed arrangements in the plans will provide the desired transmission capabilities.   Furthermore, it is necessary to know that the funds required for the new construction will be available as the plan comes into fruition.

3. Construction Program

The requirements of the communications system should be under more or less continuous analysis with respect to circuit requirements and service conditions.   The quantities of circuits to be added under a given construction program should be anticipated sufficiently in advance of actual needs to provide practical and economical intervals in which to carry out the construction work.   The interval will depend upon the kind of plant or facilities to be constructed.   For example, it may take from 12 to 18 months to engineer, install, and cut into service additional circuits on existing microwave or coaxial-cable systems.   On projects which require major outside-plant, building, and equipment installations, an interval of 24 to 30 months may be needed with even longer intervals for switching systems.

Accordingly, a given construction program will cover the addition of facilities which will become available during a period beginning about 1 year after and stretching to about 2 years after the plans were prepared.   The preparation of plans for a specific construction program begins after a forecast based on continuing analysis of the traffic over the system indicates the need for more circuits or other types of plant.   Assuming that broadband carrier facilities are to be used to meet the additional circuits needed, the requirements may be broken down into several projects and the priority for each project determined.   Construction would then be carried out in the order in which the respective facilities are needed for service.

This obviously leads to the next step of formulating the construction program per se, with schedules established for performing the work.   At this juncture it is necessary to prepare detailed plans and specifications and obtain whatever authorization is required to proceed with the work.   The schedules established are effectively timetables for the various groups or activities involved and show the periods in which the details of engineering and planning must be prepared and when construction must be started and completed.   The schedule should be prepared in such a manner that the work will be distributed among the various projects to avoid peaks of work loads and yet meet the completion dates in so far as practicable.

In the preparation of the detailed plans, several possible specific arrangements for constructing the proposed plant additions should be studied.   The arrangements which will provide the most practicable and economical construction are then selected and outlined. The outlines, in effect, become project sheets and statements and are used for preparation of the construction specifications and monetary estimates of expenditures for the construction program. The project sheets and statements tabulate major items of material, equipment, and buildings to be constructed, and where plant is removed in connection with the project, it is also itemized. In preparing the outlines it is necessary to determine specific transmission requirements, switching requirements if involved, and also the requirements for signaling and terminating equipment for circuits and the amount of power and floor space which will be needed as a result of expansion. If outside plant is involved, such as cable or open wire, details must be shown with respect to cable placing and splicing and for wire-transposing arrangements. Provision must also be made for the normal complement of alarm, control, and order wires needed adequately to maintain and operate the new facilities.

4. Economic Engineering Interval

Planning, providing, and utilizing the facilities needed to satisfy the demands for communications service require an exacting solution in order that the service-cost balance be maintained. The amount and timing of additions must be carefully gaged to avoid facility shortages on the one hand and excessive margins which unnecessarily increase costs of operation on the other hand. It may appear that optimum results would be obtained by frequent small additions. Actually, the size of additions are governed by economic considerations which determine the economic engineering interval. Small additions will result in a reduction of the average amount of unused or spare facilities as compared with one large addition. However, the cost per unit of plant will be greater with the small additions than with the larger one. This results from the fact that the increased engineering time to design the larger job is not proportionate to the increased size. Other costs such as those for ordering and getting delivery of material and readying a construction crew for installation of the material and equipment are little if any more for the large job than for the small job (see Chap. 6).

An engineering economy study, explained in Chap. 25, is required to determine the size of addition or engineering interval which will result in the proper balance between higher excess capacity with lower unit cost and lower excess capacity with higher unit cost. The economic engineering interval will be different for each category of plant and for different rates of growth within a given category of plant. It will be shorter with higher rates of growth because the jobs are larger, resulting in lower unit costs but increasing the amount and, therefore, the cost of excess capacity over and above requirements. The amount of plant provided in accordance with an economic engineering interval is, by definition, that which results in the least cost for the capacity produced. Any increase or decrease in the interval would raise the cost.

In the preceding paragraphs, no reference has been made to station equipment, i.e., telephones, teletypewriters, and auxiliary equipment. This equipment is highly portable and can be obtained from the manufacturer or supplier on fairly short order. Furthermore, the amount required is directly related to the number of actual or potential subscribers or locations to be served within the communications system. The turnover in such apparatus is relatively high as compared with the other plant and equipment, and therefore its location life is relatively short. Generally, an ample supply of various types of station apparatus is carried in stock in order to meet any immediate needs for such equipment. Because of all these factors, the problem of planning and providing station equipment does not play a large role in the overall construction program even though the investment in such plant might be sizable.

5. Depreciation Studies

As used in everyday speech, depreciation generally means a reduction in value and is the difference between the monetary value of an asset determined as of two different dates. As used in accounting, depreciation expense on the income statement refers to the estimated amount of capital consumed during each accounting period. The depreciation reserve on the balance sheet represents the accumulated capital consumption as related to capital invested in existing plant. As mentioned before, the cost of plant in a communications system as well as in other industries is viewed as a prepaid asset, a cost which is allocated proportionately to each of the accounting periods during which the plant is used. The accounting concept is strictly one of value, and the official[1] accounting definition states, "Depreciation accounting is a system of accounting which aims to distribute cost or other basic value of tangible capital assets, less salvage (if any), over the estimated useful life of the unit (which may be a group of assets) in a systematic and rational manner. It is a process of allocation, not of valuation."

Table 8. Determination of Average Life*

Year	Unit retired	Service years	Unit years
1	5	0.5	2.5
2	7	1.5	10.5
3	8	2.5	20.0
4	10	3.5	35.0
5	15	4.5	67.5
6	35	5.5	192.5
7	60	6.5	390.0
8	90	7.5	675.0
9	120	8.5	1,020.0
10	150	9.5	1,425.0
11	150	10.5	1,575.0
12	120	11.5	1,380.0
13	90	12.5	1,125.0
14	60	13.5	810.0
15	35	14.5	507.5
16	15	15.5	232.5
17	10	16.5	165.0
18	8	17.5	140.0
19	7	18.5	129.5
20	5	19.5	97.5
Total	1,000	....	10,000.0

* Average service life = 10,000 ÷ 1000 = 10 years.

Engineers who deal with capital expenditures for plant and those who are concerned with the management of investment need a good understanding of the behavior of plant.[3] Because the major question in connection with the determination of annual depreciation rates is, "how long will the plant live?", the engineer should understand its life characteristics which determine the patterns of capital repayment from revenue and financial standpoints, the significant effects of growth on revenue requirements and financing, and the relationship between plant and capital as affected by capital repayment practices.

Table 8 illustrates how 1,000 units placed in service on January 1 of a given year might be retired over a 20-year period. In this example, 5 units were retired during year 1. Assuming these retirements to be distributed evenly throughout the year, the average life of these 5 units is 0.5 year. The 7 units retired in the second calendar year then had an average life of 1.5 years, and so on through the entire life span of the entire group. It is found that, although the life span of this vintage group covered 20 years, the arithmetical average of the lives of the units was only 10 years.

The straight-line depreciation rate for a group of property is calculated by the formula

$$\text{Depreciation rate } \% = \frac{100\% - \text{average net salvage } \%}{\text{average service life}}$$

Referring to Table 8, if it is assumed that the net salvage (gross salvage less cost of removal) for each of the units as they are retired is approximately 5 per cent of their original cost, the resulting annual depreciation rate would be

$$\text{Rate} = \frac{100\% - 5\%}{10 \text{ years}} = 9.5\%$$

It is obvious that the experience in its entirety as shown in the simplified illustration of Table 8 for a particular type of plant could not be tabulated until the last item of plant had expired and been removed from service. In actual practice, the average life of each vintage group (each year's new construction is a vintage) is a matter of estimate until all the units in that group have been retired. In order to make the maximum possible use of actual past experience, the average life that has been realized to date is calculated for each vintage group. Then an estimate is made of the life expectancy of those units of the group remaining in service. The life realized in the past is merged with the expectation for the future to determine the average life for the vintage group. The average lives of the several vintage groups are then composited to obtain the average service life for the plant category. Different kinds of

Table 9. Typical Depreciation Rates[4]

	Average service life, years		Average net salvage, %		Annual rate, %	
	High	Low	High	Low	High	Low
Buildings..........................	51	38	8	0	2.6	2.0
Dial switching.....................	33	20	10	−2	4.4	3.0
Switchboards.......................	29	22	7	−3	8.5	4.8
Microwave*........................	14	10	20	5	10.7	6.6
Station apparatus..................	16	13	8	1	7.2	6.0
Pole lines.........................	42	19	−9	−50	6.5	2.8
Aerial cable.......................	37	26	32	−21	4.5	1.9
Underground cable..................	47	37	23	2	2.6	1.8
Buried cable.......................	42	16	0	−1	4.8	2.4
Aerial wire........................	43	8	34	−22	14.1	2.0
Underground conduit................	75	60	0	−5	1.7	1.4
Furniture and office equipment......	24	16	19	3	6.1	3.7
Vehicles and work equipment........	9	6	29	6	13.2	9.6

* Includes radio-telephone station apparatus and equipment.

plant such as poles, cable, station apparatus, radio-telephone equipment, etc., are studied separately. In actual practice the original cost of plant (monetary value of the units) is generally used as depicted in Table 6 instead of the count of physical units.

Table 9 shows ranges of average service lives, average net salvage percentages, and annual depreciation rates[4] based on the experience of large communications systems in the United States. The data are shown separately for various categories of communications plant and equipment. The "High" average service life shown in the first column is not associated with the "High" average net salvage given in the third column. The same is true of the figures shown in the second and fourth columns. Therefore the "High" and "Low" annual rates in the last two columns cannot be computed from data in the other columns. Although the annual depreciation rates shown are the highs and lows for each category of plant for a large number of companies sampled, the underlying service lives and net salvage for each may fall within or at the other end of the range for the highs and lows shown in the service life and net salvage columns. For example, the company with the high average service life of 51 years for buildings experienced low net salvage of zero. The resulting annual rate of depreciation (100 per cent − 0 per cent divided by 51 years) is 2 per cent (rounded). Therefore, in this case, a combination of the high average service life and low net salvage resulted in a low annual rate of depreciation. In the case of aerial cable, the high service life of 37 years was accompanied by average net salvage of 17 per cent, resulting in an annual depreciation rate of 2.2 per cent.

The retirement of individual items of plant may be brought about by a variety of causes. These causes may be physical—wear and tear; rust, rot, and decay; or casualties. They may be functional—obsolescence, inadequacy, customers' requirements, or orders from public authority. The combined action of all causes results in some items having a very short life and other similar items a much longer life. If a large number of similar items is involved, the retirements by ages generally tend to follow a consistent statistical pattern known as a "life table" or survivor curve. Such curves are plotted to show the percentage of plant surviving against the age of the plant (or against the age in per cent of average life of the plant). The survivor curve in Fig. 2 is for central office equipment associated with intercity circuits. The estimated average life of the equipment in this study is 27 years. The curve indicates, based on retirements up to the date of the study, that for a given vintage of plant about 80 per cent will remain in service at the end of 13 years and 20 per cent will still remain in service at the end of 40 years. The introduction of new instrumentalities resulting in the obsolescence of existing equipment, and therefore in heavy retirements of such plant, could change this picture quite drastically and shorten the estimated life of the equipment.

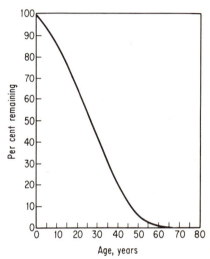

Fig. 2. Life table.

The time pattern for recovering capital through depreciation accruals is subject to wide variation. In addition to the straight-line method whereby the capital is recovered as nearly as possible in equal increments over the life of the plant, it could be recovered immediately by expense accounting or at the end of the plant's life by retirement accounting. Furthermore, a schedule of accruals could be established whereby relatively more is recovered during the early years (accelerated depreciation) or the process could be reversed. This is of interest to the engineer, since in engineering economy studies, alternate plans which involve expenditures at different times are compared by the use of compound-interest factors to equate all expenditures to present worth as of the same time or to equivalent annual expenditures over the same period of time. It is useful to note that in the simple case in which (1) the entire life of the plant is included in the study period and (2) income taxes are not considered, the present worth of the sum of interest (return on capital) and depreciation (repayment of capital) is not affected by the time pattern of repayment. This is illustrated in Table 10, which considers a single item of plant with a life of 5 years and assumes that all financial transactions occur on December 31. Plant costing $100 is installed on December 31 of year zero.

In summary, the engineering department in a communications enterprise should be organized to perform many functions. In addition to planning, designing, and assisting in the installation and operation of plant facilities, the department should have the capability of keeping important records and using them in connection with

Table 10. Capital Recovery Timing

Year	Gross plant	Repay-ment of capital*	Net plant	Return on capital†	Total revenue require-ment	8% present worth factor	Present worth of revenue require-ment
a	b	c	d	e 8% of d	f c + e	g	h f × g
			Expense Accounting (Immediate Repayment)				
0	. . . . . .	100.00	. . . . . .	. . . . .	100.00	1.0000	100.00
			Sum-of-the-years Digits Depreciation (Accelerated)				
1	100.00	33.33	100.00	8.00	41.33	0.9259	38.27
2	100.00	26.67	66.67	5.33	32.00	0.8537	27.44
3	100.00	20.00	40.00	3.20	23.20	0.7938	18.42
4	100.00	13.33	20.00	1.60	14.93	0.7350	10.97
5	100.00	6.67	6.67	.53	7.20	0.6806	4.90
Total		100.00	. . . . . .	18.66	118.66	. . . . . .	100.00
			Straight-line Depreciation				
1	100.00	20.00	100.00	8.00	28.00	0.9259	25.93
2	100.00	20.00	80.00	6.40	26.40	0.8573	22.63
3	100.00	20.00	60.00	4.80	24.80	0.7938	19.69
4	100.00	20.00	40.00	3.20	23.20	0.7350	17.05
5	100.00	20.00	20.00	1.60	21.60	0.6806	14.70
Total		. . . . . .	. . . . . .	24.00	124.00	. . . . . .	100.00
			Retirement Accounting (Repayment at End of Life)				
1	100.00	. . . . . .	100.00	8.00	8.00	0.9259	7.41
2	100.00	. . . . . .	100.00	8.00	8.00	0.8573	6.86
3	100.00	. . . . . .	100.00	8.00	8.00	0.7938	6.35
4	100.00	. . . . . .	100.00	8.00	8.00	0.7350	5.88
5	100.00	100.00	100.00	8.00	108.00	0.6806	73.50
Total		100.00	. . . . . .	40.00	140.00	. . . . . .	100.00

* That is, amount of depreciation expense.
† Assumed objective return (earnings) on net plant.

cost analyses, plant retirements, depreciation studies, and other activities required in managing the business.

PLANT OPERATIONS

The major responsibility of the plant operating people is the maintenance of communications facilities.   The physical upkeep of transmission lines and associated equipment as well as the continuing maintenance of complete communications systems forms a large part of their task.   Rearrangements and changes resulting from the continual change in service requirements will also add to the work load of the plant forces.

Construction of new outside plant, such as open-wire and cable lines, is typically a function of the plant department.   Plant forces install telephone and teletypewriter station apparatus. They also install central office switching, toll terminal, and radio equipment where such work is not contracted to the manufacturer. In addition, plant employees are generally responsible for procuring rights of way and real estate that may be required in connection with new construction projects. In order to carry out their responsibilities, plant forces must be deployed at strategic locations throughout the territory served by the communications system. A typical organizational arrangement for a particular district or operating area is shown in Fig. 3.

The plant supervisor performs the staff functions pertaining to equipment maintenance, transmission, service, and outside plant. The organization may or may not include cable and line supervisors and their forces, depending upon the size of the operation and the kind of plant involved. In the larger central offices, the chief or his equivalent may have under him other supervisors in charge of forces responsible for transmission, telegraph service, equipment maintenance, etc.

If communications plant is to be kept in first-class operating condition at reasonable cost levels, the proper proportions of two kinds of maintenance work must be sustained. The first type is restorative or trouble-clearing maintenance, which is done whenever a service interruption occurs and the trouble must be located promptly and defective units replaced if necessary. Equally important, of course, is preventive maintenance, which is carried on in order to discover potential troubles before they can cause actual failures. This avoids service interruptions and eliminates subsequent trouble maintenance work. As a practical matter, maintenance work falls into three main categories, viz., outside-plant maintenance, inside-plant maintenance, and transmission maintenance.

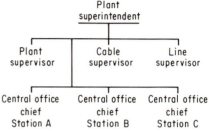

Fig. 3. Plant organization.

6. Outside-plant Maintenance

Open-wire lines are maintained by section linemen who are responsible for a specified section of open-wire line. The length of section will depend upon such factors as the number of wires, characteristics of the terrain, nature of service furnished, trouble experience with the line, etc. The duties of the section linemen consist primarily of:

Patrolling and inspecting the line periodically
Effecting minor repairs and line changes
Clearing brush and trimming trees
Working with test-board forces at adjacent central offices in the location and the clearing of reported or found troubles
Representing his employer in day-to-day contacts with the property owners, local officials, and public in general within his territory

Major jobs requiring pole replacements, rearrangements and changes, and extensive wire stringing or cable placing would normally be performed by a construction crew rather than the maintenance forces.

A section cableman has duties similar to those of a section lineman except that he is largely concerned with cables. The length of section is influenced by the factors stated above and in addition depends upon the size of cables, whether the cables are aerial or underground, and whether there are coaxial conductors involved.

In modern systems, cables are kept under continuous pressure of about 6 psi. The gas may consist of dry nitrogen applied from tanks at intervals of about 17 miles (for intercity cables), or dry air may be applied by special compressors located in central offices and repeater stations. This important development has assisted considerably in the maintenance of cables, and it is the section cableman who is responsible for checking gas pressure valves located approximately every 3,000 ft along the cable sheath. He analyzes the pressure readings for indications of leaks in the cable sheath which could result in conductor troubles if not repaired.

A complete maintenance procedure provides for the recording and analysis of all trouble reports. Trouble tickets illustrated in Fig. 4, may be designed for use in recording each individual trouble and for showing the disposition and cause of the trouble. Periodically the information is summarized and analyzed from the standpoint of where more or less maintenance attention should be directed in minimizing service interruptions. Troubles would be reported to a central location such as the nearest test board. A craftsman would be dispatched or alerted to the trouble, and after restoration had been completed, the test board would be informed of the disposition made and pertinent information would be entered on the trouble ticket.

Troubles reported by persons using the communications system may be classified as illustrated in Table 11.

Disposition of troubles with respect to where faults were located may be classified as illustrated in Table 12.

The cause of the trouble may be classified in one of several categories as shown on Table 13.

The work of reporting, summarizing, and analyzing reported troubles may be greatly simplified by the adoption of abbreviations and the numerical coding of the various categories illustrated above. Troubles found in connection with routine preventive maintenance should be recorded and analyzed in the same manner as the reported troubles. Obviously, more detailed classifications may be established where warranted. For example, with regard to outside plant, cable troubles could be separated between those in pressurized and nonpressurized cables and troubles could be pin-pointed, such as sheath breaks due to cuts, bullet holes, fatigue, splice failures, etc.

Trouble report: *Pairs 1+2, 5+6, 9+10 short*
Trouble found: *Wet splice*
Work done: *Opened splice, dried out cable, closed splice*

Reported		Cleared
8/6	Date	*8/6*
7³⁰ AM	Time	*3¹⁵ PM*
JRO	By	*NLR*
X X	Time spent	*4 hours*

Summary for analysis

Aerial *x*   Underground

Moisture *x*

Lightning

Open sheath *x*

Cable terminal

Other

FIG. 4. Trouble ticket for cable circuits.

Table 11. Types of Trouble

Can't call—no dial tone	Cuts out
Can't raise operator	Garbled
Double connection	Bell doesn't ring
No audible ringing	Doesn't get calls
Wrong party reached	Bell rings—can't answer
All trunks busy	Can't trip ringing
Don't answer	Bell interference
Get false busy	Worn equipment
Can't hear	Damaged equipment
Can't be heard	Loose equipment
Noisy	Out of adjustment
Crosstalk	Other
Clicking on line	

7. Inside-plant Maintenance

In the modern-day communications systems, inside-plant maintenance has grown to be both more complex and more extensive than outside-plant maintenance. This situation has resulted from the rapid growth of microwave radio systems and cable systems involving coaxial cable. In both these systems, large amounts of equipment are used for the basic carrier system as well as for channelizing, amplification, power, and testing. As compared with the capacity in numbers of circuits, the amount of outside plant required is minimized. The introduction of complex switching systems has also increased the importance of inside maintenance. It is important that the inside equipment be maintained properly, since the failure of a single component may interrupt services on many circuits.

Table 12. Locations of Trouble

Station set:

Teletypewriter machine	Cords
Transmitter unit	Key and lamp
Receiver unit	Internal circuit
Ringer	Other
Dial	

Station wiring:

Drop wire	Inside wire
Block wire	Inside cable
Protector	Ground wire

Outside plant:

Cable conductors	Protector
Cable terminal	Carrier
Line wire	Other

Central office:

Frame	Voice repeater
Line circuit	Carrier
Common equipment	Other

Other:

Test OK	Referred to others
Found OK	

Table 13. Causes of Trouble

Man-made:

Repair	Foreign workmen
Installation	User action
Construction	Malicious damage

Plant or equipment:

Dirt	Overload
Deterioration	Other
Adjustment	

Weather:

Lightning	Wind
Flood	Ice and sleet
Tornado	Other

Other:

Fire	Birds
Insects	Trees and foliage
Rodents	Other

Trouble on inside plant is caused primarily from dust and dirt, wear, aging, deterioration of vacuum tubes, need for readjustment of mechanical and electronic components, opens and shorts on open terminal strips caused by craftsmen, etc. Routine testing and inspection in accordance with established schedules are the best forms of preventive maintenance and when administered properly will keep troubles at a minimum. Despite a good program of preventive maintenance, troubles will occur and some part of total maintenance time will be spent in locating and clearing such faults.

A large part of the total investment in central office equipment will be at locations which are unattended, such as an unattended radio relay point. The only time plant forces will be present will be in connection with the performance of routine equipment tests for the purpose of clearing reported troubles or because of some necessary equipment changes or rearrangements. The existence of unattended locations presents certain problems.   Motor vehicles must be available at all times, and in some climates heavy equipment such as snow tractors may be required to gain access to isolated places.   A trained force of equipment men must be available and located within a reasonable distance of all such sites.   A variety of portable and stationary test gear must be kept in good shape and ready for use.   One or more maintenance centers will generally be required, and their locations will be dictated by economy and service protection.   While much of the routine maintenance and trouble clearing will be done on the premises of the attended and unattended central offices, occasions will arise where it will be necessary to substitute spare equipment for a component in trouble and return the defective unit to the centralized maintenance center for repair.

In the management of central office maintenance, it may be desirable to establish a procedure for work measurement whereby the efficiency of the maintenance force may be measured and also the requirements for additional manpower may be estimated in connection with the expansion of a communications system or the establishment of a

Table 14. Work Measurements

Plant unit (a)	No. (b)	Maintenance time, min (c)	Work units per month (d)
Radio equipment:			
1. Microwave receiver.....................	25	60,000	6.67
2. Microwave transmitter.................	25	54,000	6.00
3. Batteries (per cell).....................	1,250	6,250	0.014
4. Rectifiers...............................	30	4,200	0.39
5. Emergency alternators.................	Etc.	Etc.	Etc.
6. Alarm equipment.......................			
7. Amplifiers..............................			
8. Channel switching equipment...........			
9. Miscellaneous equipment...............			
Terminal equipment:			
10. Terminating sets......................			
11. Test boards (positions)................			
12. Amplifiers.............................			
13. Channel bank equipment..............			
14. Master group equipment..............			
15. Submaster group equipment...........			
16. Supergroup equipment.................			
17. Group equipment......................			

new system.   The initial step in setting up such a procedure involves (1) an inventory of all principal plant units at each location where equipment is installed, (2) the record-ing of actual time spent in maintaining all such units for a fixed study period, and (3) the development of a maintenance time per plant unit for each unit of plant inventoried.

Table 14 illustrates the steps to be followed in establishing the procedure.   The quantities of various kinds of plant units as illustrated in column a would be entered in column b.   Over a period of 6 to 12 months, the actual minutes of time spent in maintaining these units would be entered in column c.   The work units per month in column d are the result of (1) finding the quotient of column c divided by column b and (2) dividing this quotient by the product of the number of months in the study times the number of minutes established as one work unit.   The information shown in Table 14 would be prepared for each site where equipment is located.   The minutes spent on maintenance effort entered in column c for each site would be the sum of time spent on routine tests (as prescribed by the equipment manufacturers or in accordance with an established schedule based on experience) plus time spent in locating and clearing troubles plus time spent in repairing equipment at centralized maintenance centers. Travel time associated with repairs would be included. For purposes of establishing average maintenance time per plant unit to be used in the future, amounts in columns *b* and *c* should be accumulated for all locations and average or composite ratios would then be computed in column *d* for each kind of plant unit in column *a*.

Assuming that the work force was reasonably in balance during the 6- to 12-month study period and that the maintenance procedures had been shaken down to the point where the objective quality of service was being sustained, the ratios in column *d* of the table could be applied as follows:

Assume that line 1 for microwave receivers for the entire system showed 25 units in column *b* and 60,000 min for column *c* actually recorded over a 6-month period for maintaining microwave receivers. The average "work per unit" ratio for 1 month would be (60,000 ÷ 25) ÷ 6, or 400 min.

If 60 min of work were arbitrarily established as one work unit, then the study indicates that there are 6.67 "work units" required on the average each month for each microwave receiver in the communications system under study. Similar work units would be established for each plant item listed. The work units in column *d* are illustrative only because they will vary considerably between systems, depending upon type of equipment, service objectives, experience of force, territory, assumed minutes per work unit, etc.

In future months, the total work units required for the system can be computed by applying the work-unit factors to the inventory of plant items which may be expected to increase as the system expands. The total work units thus obtained may be compared with the actual maintenance hours paid for to determine whether efficiency is increasing or decreasing. Similarly, the number of additional man-hours required as the result of a major expansion can be estimated by multiplying the units of plant to be added by the work-unit factors and converting these hours to additional workmen.

Cct. No. *104*   From *Rankin* To *Heclin*

Date placed in service *7/8/61*

Method of operation *Auto.*

Class of circuit *Common*

Facilities and mileage

Facility

From	To	Type	Unit No.	Miles
Rankin	*Heclin*	*R*	*100*	*34*

Total miles *34*

Equipment

Location	Type	No.
Rankin	*Chan Modem*	*4-2*
Heclin	*" "*	*2-2*

Other data:

Cct. net loss 6

FIG. 5. Circuit or trunk card.

Within a communications system, one or more central office locations will be designated as service centers. Ordinarily, a test board will be located at each service center for purposes of testing local and intercity circuits and equipment. At each service center it is necessary to maintain line cards for each of the local lines serving telephone, teletypewriter, or other stations and for each of the intercity or interoffice circuits. Such line and circuit cards show information which will assist in maintaining service throughout the system.

The line cards for each line serving a local station give the following information:

Telephone number or other station designation
Name of person to whom telephone number is assigned (or name of department, etc.)
Address or other location designation
Cable number and pair number or other line information
Type and quantity of equipment at station premises, including key equipment, intercommunicating equipment, and extensions

A circuit card, as shown in Fig. 5, would be prepared for each intercity (or interoffice) circuit and the following information would be shown:

Circuit number and name of terminal locations
Method of operation such as automatic, dial pulse, ringdown, call announcer, etc.
Class of circuit, e.g., common trunk, private wire, via trunk, terminal trunk, etc.
Mileage—broken down by types of facilities involved such as open wire, cable, radio
Cable number and pair designation (if cable)
Carrier unit (if carrier circuit)
Special equipment involved
Transmission characteristics, as required
Any other information which would assist in trouble location and clearance

8. Transmission Maintenance

Transmission maintenance is concerned with the overall operations of communications systems. This work is the responsibility of plant central office forces working at test boards, in the repeater rooms, carrier terminal rooms, and microwave terminals. This work is partly preventive-maintenance work and partly corrective or trouble maintenance. The personnel involved check voice repeaters and other amplifiers and maintain them at proper gains and frequency characteristics. They also measure and adjust the transmission equivalent and the transmission—frequency characteristics of circuit units, carrier systems, and overall circuits. The transmission personnel may also be responsible for making switches in the communications network at test boards and patch bays in accordance with predetermined schedules.

Table 15. Routine Test Schedule

Test	Station to station*	Serving test center to	
		Serving test center	Station
Net loss...............	Quarterly	Monthly	Quarterly
Noise..................	Annually	Annually	Annually
Frequency response......	Annually	Annually	Annually
Station inspection.......	Annually	Annually	Annually

* Tests are made station to station where practical, otherwise as indicated.

Transmission-maintenance forces are responsible for certain line and system tests, the intervals for which are based on recommendations of the equipment manufacturer and trouble incidence. Net loss measurements should be made in both directions on all repeatered circuits with appropriate transmission-measuring sets. Frequency-response measurements are also required periodically using transmission-measuring sets equipped with an oscillator which will produce frequencies of 500 to 2,700 cycles. In addition, noise-measurement tests are made with a noise-measuring set, and the overall performance of the circuits are checked by making talking and signaling tests between stations. A typical schedule for routine tests is shown in Table 15.

Tests will be required at other times in connection with reported troubles and trouble location. Furthermore, special nonvoice services may require special tests such as those requiring envelope delay measuring sets.

TRAFFIC OPERATIONS

In an extensive telephone system, the traffic department is responsible for three general functions, viz., telephone operation, administration, and certain engineering. The first classification consists of the selection and training of operating forces and the handling of telephone calls together with the supervision required. Traffic administration involves studies and measurements of operating results and of the proper distribution of circuit and operator loads, as well as the development of improved operating methods and practices. The basic responsibility of the traffic engineers is the determination of the numbers of circuits required currently and in the future. Coupled with this is the responsibility for the determination of the circuit routes to be used between any two points in the communications network. In a large system, every point on the network is not connected by direct circuits with every other point, and intermediate switching is required to connect points not directly connected. Traffic engineers are also responsible for the determination of the requirements for switchboard positions and other switching equipment.

9. Operation and Administration

Efficient switchboard operating entails considerable knowledge and skill on the part of the operator. Because of this, the selection and training of personnel for operating work are important functions of the traffic organization. Extensive training facilities in the form of classrooms and practice switchboards may be required. It goes without saying that capable instructors are a prime necessity. Good operating is dependent upon cooperative effort and "teamwork," and in the interests of standardizing the work it is necessary to develop detailed procedures for use within the system so that the operating job will be handled uniformly in all places.

The traffic volume at a switchboard location generally varies considerably from hour to hour during the 24 hr of the day. It is ordinarily greatest during the normal business hours, with morning and afternoon peaks. It is desirable that the number of operators assigned throughout the day be sufficient to answer the calls promptly and complete them without delay. On the other hand it is obvious that more than the required number is uneconomical. A tool has been developed to aid in force adjustment work and in the determination of work loads. This tool is known as a traffic coefficient.

Over the years, studies have been made in the telephone industry to determine the average amounts of operating work required to handle all the various types of calls, e.g., different kinds of local calls, information calls, assistance calls, and various types of long-distance calls. Many years ago it was determined that, on the average, a group of fully experienced operators of average ability working at a local manual switchboard, with all positions covered, could handle 230 local calls per operator-hour and give the required quality of service. The amount of work required to handle one such call was adopted as "one unit" of work, since the great majority of operating work done at that time consisted of handling local calls.

Traffic coefficients are used to express the labor required to handle the various kinds of calls and other kinds of switchboard work in terms of a common denominator, traffic units. The coefficient is a multiplier which, applied to the different kinds of calls, reduces all to the common denominator of work units. The coefficient is a factor which states the number of traffic units per call for a specific type of call. The formula for calculating a traffic coefficient is as follows:

$$\text{Coefficient} = 230 \div \frac{3{,}600 \times \% \text{ occupied time}}{\text{work time per call}}$$

where 230 represents the number of local calls which can be made in 1 hr under the conditions described heretofor, 3,600 represents the seconds in 1 hr, % occupied time represents the proportion of the time the operators handling such calls are actually doing operating work. In order to answer calls promptly, some portion of an operating team must be idle, as of any given moment, awaiting the next calls to be answered. Work time per call represents the average time, in seconds, required to complete a call based on a reasonable sample over a given period of time. Assume that studies indicate that it takes 150 sec, on the average, to complete the operating work on long-distance calls originating at a particular switchboard location. Also, occupied time is assumed to be 75 per cent.   The traffic coefficient would be

$$230 \div \frac{3,600 \times 75\%}{150} = 12.8 \text{ units per call}$$

The factors affecting per cent occupied time are:

The grade of service to be maintained
The size of the operating team
The work time per call

The factors affecting the work time are:

The type of call
The operating method
The equipment being used
The ability of the operating force
The training of the operating force
The experience of the operating force

In connection with cost controls, the total traffic units handled at a given switch-board may be computed each month by multiplying the volumes of calls handled by the appropriate coefficients.   The total wages paid to traffic operators may be divided by the total traffic units to determine the operating costs per thousand traffic units. A record of the cost per thousand traffic units maintained from month to month may be used to establish cost trends.   Also, the costs incurred at various operator locations may be compared, one with another.

10. Traffic Engineering

Generally, the traffic engineers are responsible for recommending to the engineering department the numbers and kinds of circuits required and the quantities of switch-board positions and related terminating and dial switching equipment needed to meet anticipated traffic loads throughout the system.   Each circuit group has character-istics of its own with respect to busy hour, percentage first routed traffic, short or long average length of conversation, a fairly steady flow of traffic, or a seasonal fluctuation.

Circuit requirements should be determined 1 to 2 years in advance, and more lead time is needed with respect to switching equipment.   Periodic counts of messages which can be summarized in various ways and circuit-usage records form the basis of the traffic engineering approach.   A "base load record" is established by counting the number of average business-day messages over circuit groups between each intercity terminal and all other terminals.   These base loads may then be projected to levels expected at future periods to determine the future needs for circuits and switching equipment.

For switchboards, the total operating work units per call are determined by using standard work units for different types of calls and for different operating steps performed in completing these calls.   The busy-hour calls multiplied by the total work units per call give the busy-hour units for the office under study.   The appropri-ate operator work load per position is then selected and is divided into the units to determine the number of switchboard positions required.

REFERENCES

1. American Institute of Certified Public Accountants, *Accounting Terminology Bull.* 1, New York, 1953.
2. Uniform System of Accounts, Part 31, *Rules and Regulations of Federal Communications Commission*, revised to Sept. 1, 1962.
3. *Nat. Assoc. Railroad Utility Comm. Rep. 72d Ann. Conven.*, 1960, pp. 305–318.
4. *Annual Report Form M*, Federal Communications Commission, Class A and B Telephone Companies, Year ended Dec. 31, 1962.

Chapter 25

COST STUDY METHODS

M. A. LISH, *American Telephone and Telegraph Company*

COST STUDY TYPES[1]

1. Types of Cost Studies

Cost studies which reflect all pertinent elements of costs are essential to the successful provision and management of any communications system, regardless of the purpose for which it may have been or is being constructed. While cost studies relating to communications systems vary quite widely in scope and complexity, they are essentially of two main types, namely:

a. Engineering Economy or Economic Selection Studies. These are designed to provide management with cost information to assist it in appraising the relative merits of alternate projects. These projects generally relate to the construction of plant for growth or modernization and are designed to answer three very fundamental questions:

1. Why undertake this project at all?
2. Why do it now?
3. Which of the various alternative ways it may be done will prove to be the most economical in the long run? In brief, why do it this way?

b. Revenue Requirement and Profitability Studies. These are designed to provide management with cost information to assist it in establishing or revising prices or rates for communication services provided to the public or to others.

In the preparation of these types of cost studies, a thorough understanding of what constitutes costs is essential; i.e., it is necessary to know the nature and kinds of costs.

NATURE OF COSTS[1]

2. Major Kinds of Costs

In cost studies, there are two major kinds of costs which must be recognized because they require different treatment. These major kinds are:

1. Operating costs
2. Capital costs

3. Operating Costs

Operating costs comprise those expenditures, usually of a recurring nature, which are made or will be incurred because a company operates a functioning plant and which

25–1

are not a direct function of the capital costs of the plant.* The level of these expenditures is a function of the numbers and kinds of plant units in operation (although not necessarily a direct function), how these units are designed and assembled, where the plant is located, methods used to operate it, amount and methods of routine upkeep, the extent to which it is rearranged and relocated, whether the plant is used for communications furnished to the public and, therefore, requiring customer services such as sales and billing services or the communications plant is subsidiary to a larger operation, such as a private microwave system owned by a gas-transmission pipeline company, etc. Under most arrangements for plant operation, the major portion of operating costs consists of payroll costs.

4. Kinds of Operating Costs[2]

The principal kinds of operating costs for a communications system provided by a common carrier company, such as a telephone company, are as follows:

a. Maintenance Expenses. These expenses include (1) the cost of material and labor associated with the routine repair of plant; (2) training of maintenance forces; (3) testing of equipment and facilities for transmission imperfections; (4) rearrangements and modifications of plant of a minor nature; (5) miscellaneous expenses such as purchases of small tools, routine repairs of maintenance tools, and house service associated with space occupied by forces who maintain plant; (6) fuel and purchased power to operate power-plant equipment; and (7) general supervision and engineering work associated with maintenance work. Maintenance expenses, while frequently estimated in certain types of studies by the use of ratios of maintenance expenses to initial costs of plant, actually have no inherent relationship to the cost of plant. Frequently reductions in maintenance expenses can be effected through the expenditure of higher initial amounts for plant. Therefore, in specific problems, particularly those involving economic selection among alternate plans or determination of revenue requirements for new instrumentalities, full cognizance of the types of plant, conditions of use, testing equipment and procedures, and any other factors affecting maintenance costs, must be given.

b. Traffic Expenses. Traffic expenses consist, for the most part, of operators' salaries and related operator training expenses. They also include expenses related to service observing; customer instruction; service inspection; operation of rest rooms and lunchrooms for operators; house service associated with operating, rest, and lunchrooms; miscellaneous supplies, such as toll tickets and other central-office stationery; general supervision of the operating forces; and certain engineering work related to the provision of switching equipment and circuit facilities. The amount of traffic expenses involved in cost studies may vary from none in many types of studies to a very significant amount in such studies as an engineering economy study relating to conversion from toll to extended area service between exchanges.

c. Other Customer Servicing Expenses. Other customer servicing expenses cover three main types of work namely, marketing, commercial, and billing and customer account work. Marketing expenses cover the costs of advertising, sales work, production of directories, and related general staff work. Commercial expenses include the costs of (1) operating local business offices, which handle most customer contact work and are responsible for the preparation of service orders and the collection of past due accounts; (2) arranging for and paying commissions for coin telephones; and (3) related general staff and supervision work. Billing and customer account expenses include the costs incurred in the preparation and mailing of customers' bills and in the maintaining of customer accounts. Customer servicing expenses, like traffic expenses, are present in studies in widely varying degrees dependent upon the

---

* While operating costs are not a direct function of capital costs of plant, certain operating costs, e.g., maintenance expenses and ad valorem taxes, frequently are estimated in certain types of cost studies by the application of ratios of expenses or taxes to initial capital costs. The use of such ratios, however, are merely empirical means of estimating the operating costs involved, and their use should not be construed as implying any fundamental relationship between initial capital costs and the level of such operating costs.

nature of the study.   They are always present, for example, in a revenue-requirement study relating to a new instrumentality.

d. General Administrative Expenses.   The term "general administrative expenses" refers to those expenses which are present in all businesses but which are seldom directly associable with any specific project or service.   These expenses include the expenses of (1) the executive department; (2) law department; (3) treasury department; (4) accounting department, other than billing and customer account maintenance; and (5) other general staff groups such as personnel and public relations groups. In this classification may be included costs of insurance, accident and damage claims, and miscellaneous expenses not includible in other classifications.   Sometimes operating rents and employee welfare costs, such as employee contributions to pension funds, sickness benefits, and death benefits, are treated as general administrative expenses, but usually they can be directly associated with a service or project.   Some corporations in their accounting processes allocate some of the individual general administrative expenses, such as executive department, law department, treasury department, and accounting department general administrative expenses, between functions relating to the operating plant and functions relating to the acquisition of plant, treating the former as operating costs and the latter as capital costs.

e. Operating Taxes.   The term "operating taxes," as used herein, refers to taxes other than taxes based on net income (net income taxes are directly related to earnings and, therefore, are treated in most cost studies as a part of capital costs).   Operating taxes include the following broad types of taxes:

1. *Ad Valorem Taxes.*   These are the taxes which are based on assessed valuations of property.

2. *Gross Receipts, Gross Revenue, and Gross Income Taxes.*   These taxes are based on total receipts, revenues, or income, which, except for minor variations in tax laws, generally have the same meaning.   Gross receipts, gross revenue, and gross income taxes are generally levied on businesses only.

3. *Sales Taxes.*   These taxes are levied on the purchase price of all or selected types of materials but may in some states apply to certain kinds of services.   These taxes may be classified either as capital costs or as operating costs dependent upon whether the related purchased material or service is to be used in connection with the construction of plant or with operation or maintenance activities.

4. *Social Security and Unemployment Taxes.*   In general, these taxes are based on wages and, for cost study purposes, are treated as loadings on labor costs.   Accordingly, they are classified as capital costs when the related labor is a capital cost and as operating costs when the related labor is an operating cost.

5. *Miscellaneous Taxes.*   In addition to the aforementioned types of taxes, there are a number of miscellaneous taxes, such as capital stock taxes, occupation or use taxes, and franchise taxes.   Because of their minor nature, these taxes are combined with ad valorem taxes and treated as operating costs in most cost studies.

5. Capital Costs

Capital costs are those costs which are incurred because property, either tangible or intangible, is owned.*   They are costs which are created because an investment has been made in property and not just because the property physically exists.   Capital costs have two distinct aspects, namely, (*a*) initial capital costs and (*b*) recurring capital costs.

a. Initial Capital Costs.   The first aspect of capital costs is related to the initial investments made when ownership in a plant unit or units is acquired or when other assets of a business are purchased.   These initial investments or costs, referred to hereinafter as *initial costs,* are the nonrecurring costs capitalized as a result of acquisition of plant or other assets, e.g., working capital.   In general, initial costs comprise

---

* Small units of property, such as small tools, and property which is used up rapidly, such as fuel and office supplies, frequently are charged to operating expenses when they are purchased, and therefore, because they are not capitalized, they are considered to be operating costs rather than capital costs and are so treated herein.

amounts paid for material and apparatus, for installation labor, for engineering labor directly associated with the project, and for contractors' bills if any part of the construction work is done under a contract. Initial capital costs also include amounts of cash or short-term investments maintained to pay bills for new construction and to meet current expenses in advance of the receipt of revenues from the business.

b. Recurring Capital Costs. The second aspect of capital costs relates to the recurring costs, usually expressed as an amount, which results from the facts that money has been invested in the business, that this investment must be kept whole, and that the business must pay the investor something, usually referred to as a return on capital, for the use of his money in the business. Recurring capital costs consist of three elements:

1. Capital repayment, more commonly referred to as depreciation expense.
2. Return on capital.
3. Income taxes. Income taxes are treated as capital costs because they are directly related to amounts capitalized and vary with both the amounts of return on capital and the method of capital repayment.

c. Capital Repayment. Capital repayment is the term used herein to reflect the costs which are incurred because of the fact that each physical property unit eventually comes to the end of its useful life for one or more of a number of reasons. These reasons may be wear and tear, obsolence due to technological improvements, storm or accident damage, major plant relocations, and major plant expansions.

In actual practice, capital repayment normally is accomplished by means of depreciation accounting in which annual amounts of "depreciation expense" are deducted from current revenues to repay the initial capital costs over the service life of the property involved. Depreciation accounting generally is used only for physical property having a useful life expectancy of more than one year.

Although there are a number of methods used in depreciation accounting, the most common practice, particularly among common-carrier telephone companies, is to repay the cost of the property on a straight-line basis. On this basis the cost of the property is allocated as evenly as possible over its useful life on a year-by-year or month-by-month basis, and an amount equal to such even proportion is set aside in a reserve, i.e., repaid each year or each month.

Where straight-line depreciation accounting is used, the depreciation-expense amount entered in the accounts is determined by multiplying the initial capital cost for all property of a given class (e.g., conduit, pole lines, radio central-office equipment) by an annual or monthly depreciation rate. This depreciation rate, which is usually expressed as a percentage, is determined by the use of the following formula:

$$\% \text{ depreciation} = \frac{100\% - \% \text{ that net salvage is of initial cost}}{\text{average service life in years (or months)}}$$

where　　　　Net salvage = gross salvage − cost of removal

Other methods of depreciation accounting are:

1. *Declining-balance Depreciation Accounting.* In this method of depreciation accounting, a specified rate is applied each year to the remaining book value, i.e., the initial cost of the asset being depreciated less the sum of the depreciation amounts determined in prior years. For example, assuming an asset having an initial cost of $10,000 and an annual depreciation rate of 10 per cent, the depreciation expense during the first year would be $1,000, $900 during the second year, i.e.,

$$0.10 (\$10,000 - \$1,000) = \$900$$

and so on. This method frequently is used in cases where it is judged that the income from an asset is likely to be greater in the earlier years of its life than in its later years, such as in the case where the asset is likely to be used as a standby rather than as a productive facility in its later years. A declining-balance rate which recovers two-thirds of the initial costs during the first half of the service life is usually acceptable for Federal income tax purposes.

2. *Sum-of-the-years-digit Depreciation Accounting.* This method is similar in principle to the declining-balance method. The digits for the number of years estimated as the service life are added together, and the amount of depreciation expense is determined as a declining fraction of the initial cost less net salvage computed as follows:

$$\frac{\text{Estimated service life} - \text{realized life} + 1}{\text{Sum-of-the-years in estimated service life}}$$

For example, if an asset having an initial cost of $12,000 is estimated to have a service life of 10 years with a net salvage of $1,000, the depreciation expenses would be 10/55 ($12,000 − $1,000 = $11,000) = $2,000 during the first year, 9/55 ($11,000) = $1,800 during the second year, 8/55 ($11,000) = $1,600 during the third year, and so on. This method recovers about three-fourths of the initial costs less net salvage during the first half of the service life.

3. *Sinking-fund Depreciation Accounting.* This method can be visualized best by assuming that a savings account is established in which an equal amount will be deposited each year during the estimated service life of the asset. The annual amounts so deposited together with accrued interest at a specific interest rate compounded annually will accumulate during the estimated service life to an amount equal to the initial cost of the asset less estimated net salvage. Under this plan, the amount of depreciation expenses incurred each year is the equal annual amount plus the interest credited to the account during the year. While this method was used extensively in the early decades of this century, it is seldom used today. It is mentioned here primarily because, as discussed hereinafter, this method is used extensively in engineering economy studies and will be described in greater detail in paragraph 9e.

In considering capital repayment costs from an engineering economy study viewpoint, it is important to remember that the first or initial capital cost of plant is just as much an expense of a business as are such current operating expenses as maintenance, commercial, and traffic expenses; the only difference is that the plant is used over a number of years and, therefore, its entire cost is not properly chargeable to just one or a few of those useful years. Also, the engineer in making an engineering economy study deals with the same phenomena as the accountant in depreciation accounting, but whereas the accountant is dealing with broad average situations pertaining to existing plant and to current periods, the engineer is concerned with more specific situations involving only future periods. The accountant is interested in the total costs; the engineer is interested only in those parts which are susceptible to future control. Depreciation expense in the accounting sense enters into engineering economy studies only for the purpose of estimating income taxes.

d. Return on Capital. Because an investor will not knowingly furnish any of his funds to a business unless he expects its total income to be sufficient to cover all its current costs and, in addition, to provide a return on his investment, return on capital becomes a necessary element in any engineering economy or revenue-requirement cost study. For this reason, return on capital is considered to be a cost element for the purposes of this chapter.

Determination of the level of return on capital to be used in a study must be based on a combination of factors. These include the nature of the study, e.g., engineering economy or revenue requirements; past experience; the current situation as to the cost of raising new capital, which in turn is affected by the current credit rating of the business; and current economic conditions and climate.

For engineering economy studies, an expected long-term average rate of return rounded to the nearest whole per cent generally is appropriate. For a revenue-requirement study, however, it may be desirable to use a rate of return more nearly related to current earnings levels. In examples which follow a rate of 8 per cent for return on capital has been used.

e. Income Taxes. Since a communications company, and in particular a regulated public utility, receives net income only because it has capital invested in property on which a return must be earned, income taxes, which are a direct function of net income, are also treated as capital costs.

There are two types of net income taxes generally encountered, namely, income taxes levied by the Federal government and income taxes levied by state governments. In relatively rare instances, some local governments also have established net income taxes. State and local income taxes usually are deductible in determining Federal income taxes, but in general, Federal income taxes are not deductible in determining state and local income taxes, although there are a few exceptions.

The amount of net income taxes associable with a given return on capital is not only a function of the amount of the return and the income tax rate but also a function of the composition of the capital as to proportions of debt and equity. This is because interest on debt is not subject to income taxes and income taxes, therefore, are a direct function only of the equity portion of return on capital.

The equity portion of return on capital can be determined by the use of the following formula:

$$E = 1 - d \frac{i_d}{r} \tag{1}$$

where E = equity portion of return
d = proportion that debt is of total capital = debt ratio
r = overall rate of return
i_d = interest rate on debt

Example:

Assume:
$$d = 40\%$$
$$i_d = 5.0\%$$
$$r = 8.0\%$$

Then
$$E = 100\% - 40\% \frac{5.0\%}{8.0\%} = 100\% - (40\% \times 0.625)$$
$$= 75\%$$

Using this formula for the equity portion of return, income taxes can be determined by means of the following formula:

$$T = \frac{t}{1-t} ER = \frac{t}{1-t}\left(1 - d\frac{i_d}{r}\right) R \tag{2}$$

where T = income taxes
t = income tax rate
R = total return
E, d, i_d, and r are the same as defined above.

Example:

Assume:
$$t = 48\%$$
$$R = \$1,000$$

E, d, i_d, and r are the same as in the above example. Then

$$T = \frac{t}{1-t} ER = \frac{48\%}{52\%} \times 0.75 \times \$1,000$$
$$= 0.923 \times \$750$$
$$= \$692$$

In the foregoing formula for determining income taxes, the factor t in states which have a state income tax is a composite of the Federal and state income tax rates determined as follows:

1. In states where Federal income taxes *are not deductible* for purposes of computing state income taxes, the composite tax factor t is determined by means of the following formula:

$$t = s + f(1 - s) \tag{3}$$

where f = Federal income tax rate

$\quad s$ = state income tax rate

2. In states where Federal income taxes *are deductible* for purposes of computing state income taxes, the composite tax factor t is determined by use of the following formula in which the f and s factors are the same as defined above:

$$t = \frac{f + s - 2fs}{1 - fs} \tag{4}$$

In all the foregoing formulas, it is assumed that, for the purposes of computing income taxes, the only significant deductions from revenues are expenses, taxes, and interest payments on debt. In certain studies where greater refinement in income tax determinations is deemed desirable, there are a number of other cost elements which may be present and which would have to be considered in income tax computations. Examples of these other elements are:

1. Capitalized interest during construction*
2. Capitalized employee welfare costs, such as pensions expenses related to installation labor*
3. Capitalized taxes, such as social security taxes, associated with installation labor and sales taxes*
4. Debt refunding costs
5. Dividends on preferred stocks
6. Investment incentive tax credit

Where it is deemed desirable to consider these items, it is suggested that their treatment be discussed with income tax experts.

TIME VALUE OF MONEY[1]

6. Fundamental Time-value Concept

The fundamental time-value concept of money originates from a recognition of the fact that all money has potential earning power. This fundamental concept, which must be recognized in cost studies, is that the use of money entails a cost,† a cost which is a function of the length of time the money is used, and that the value of monetary transactions depends upon when they occur. Hence, the longer an expenditure can be deferred, the less is its impact on the business, and conversely, the earlier a receipt or cost savings can be realized, the greater is its value to the business.

7. Equivalence

Since the value of money is related to time, all expenditures and receipts in a comparative cost study must be translated to a common date or period for a meaningful comparison; that is, they must be expressed on an equivalent basis.

For example, consider a situation involving two alternatives both of which will accomplish the desired results equally well. Alternative A would cost $100 now, and alternative B would cost $100 one year from now. Money can be invested elsewhere by the firm to earn 8 per cent per year.

---

* These capitalized costs may be claimed as income tax deductions in the year in which they are incurred. However, where this is done, such capitalized costs are not depreciable for tax purposes. Hence, in studies involving the entire life span of plant items, whether or not they are considered generally has no significant effect.

† Perhaps, the most readily understood illustration of the "cost of money" is that of interest on a loan. However, the term "cost" as used in this discussion does not necessarily refer to an obligation to pay; it may refer to the loss of an opportunity to earn. For example, if an individual must borrow money to purchase a desired item, there is clearly a cost of money involved in the purchase, namely, interest on the loan he has made. But even if the individual has sufficient cash to make the purchase without borrowing, there is still a cost-of-money aspect to the purchase, for if the purchase were not made, the money could be used productively, perhaps deposited in a savings bank to earn interest.

If there were no cost associated with the use of money, that is, if money had no earning potential, the two alternatives would be equally attractive. However, in reality, money does have earning power, and therefore, alternative B is the more desirable.

But while it is apparent that it would be misleading to consider these expenditures as being equally desirable, a method for measuring the difference in the economic impact of alternatives A and B is essential. One way to do this is to express the $100 expenditure under alternative B, which is to be made one year from now, in terms of an equivalent expenditure now. This can be done by determining how much money would be required at the present time in order to have, with 8 per cent earnings, $100 at the end of the year. The table of factors in Table 1 indicates that this amount is $92.59.

Selecting alternative A would require $100 now; however, if B is chosen, only $92.59 would be required now. Thus, the cost of alternative B has been expressed in terms equivalent to those of alternative A, and the measurable difference in economic impact is $7.41.

The concept of calculating an equivalent expenditure, illustrated in the preceding example, is an important one in comparative cost study work. Usually such a cost study will involve expenditures and receipts occurring at different times, and a proper evaluation can be made only by translating these expenditures and receipts to equivalent transactions as of some common date.

As well as translating a single sum of money to an equivalent amount as of another date, equivalence for a single sum of money may be expressed in terms of an equivalent series of equal amounts expended at specific intervals over any specified time span and, conversely, the equivalent of a uniform series of expenditures may be expressed as a single equivalent amount occurring at any specified point in time. Such a uniform series is called an "annuity." An example of an equivalent relationship between a single amount and a uniform series is evident in most house mortgages, which generally provide for equal monthly payments covering amortization of the principal plus interest on the unpaid balance.

8. Time-value Factors

The following are the principal types of factors, referred to hereinafter as "time-value factors," used in translating given amounts of money to equivalent amounts at some other specified date or period.

Factor	Symbol*	Mathematical expression
Single amount to single amount:		
Future worth of present amount.....	$\left(\dfrac{f}{p}\right)_n^i$	$(1 + i)^n$
Present worth of future amount......	$\left(\dfrac{p}{f}\right)_n^i$	$\dfrac{1}{(1 + i)^n}$
Single amount from a uniform series:		
Future worth of an annuity..........	$\left(\dfrac{f}{a}\right)_n^i$	$\dfrac{(1 + i)^n - 1}{i}$
Present worth of an annuity........	$\left(\dfrac{p}{a}\right)_n^i$	$\dfrac{(1 + i)^n - 1}{i(1 + i)^n}$
Uniform series from a single amount:		
Annuity for a future amount........	$\left(\dfrac{a}{f}\right)_n^i$	$\dfrac{i}{(1 + i)^n - 1}$
Annuity from a present amount......	$\left(\dfrac{a}{p}\right)_n^i$	$\dfrac{i(1 + i)^n}{(1 + i)^n - 1}$

* The symbols used here are the same as those used in *Engineering Economy*, a copyrighted publication of the American Telephone and Telegraph Company issued in 1963.

where: i = interest or earnings rate per compounding period*

n = number of compounding periods

f = future worth

p = present worth

a = annuity

In the use of these time-value factor symbols and mathematical expressions, the following are worthy of note:

1. In each time-value symbol, the numerator of the factor represents the state *to* which the given amount is being converted and the denominator represents the state *from* which the given amount is being converted. As an illustration, the factor p/a symbolizes the conversion of an annuity to an equivalent present amount.

2. The symbols are stated as fractions, and they may be treated as such in expressing relationships among the factors. For instance, $a/p = \dfrac{1}{p/a}$. Similarly, the mathematical expression for a/p is the reciprocal of that for p/a.

3. Usually i is given as an annual rate and n is expressed in years.

4. It is assumed that all transactions occur at the end of the compounding period, i.e., in the usual case, at the end of the year.

9. Definitions and Illustrative Computations

The following tabulation of time-value factors is provided for use in subsequent illustrative problems of determining equivalent values for single amounts or series of uniform amounts. Each of these factors assumes that r is 8 per cent per annum compounded annually, and that n is 5 years.

f/p	p/f	f/a	p/a	a/f	a/p
1.469	0.6806	5.867	3.993	0.17046	0.25046

For other values of r (rates of return or earnings) and values of n (time periods) see Table 1.

a. Future Worth of a Present Amount f/p. This is an expression denoting the value, n periods in the future, to which a specified amount of money will accrue with return of r per cent per period.

Example: If $100 is deposited in a bank paying 8 per cent per year, how much would be in the account at the end of 5 years?

Answer: $100 \times f/p = \$100 \times 1.469 = \$146.90$.

b. Present Worth of a Future Amount p/f. This is an expression denoting the amount required at present to accrue a specified amount n-periods in the future with a return of r per cent per period.

Example: To have $146.90 in a bank account at the end of 5 years, it would be necessary to deposit how much at present if the bank pays 8 per cent per year?

Answer: $146.90 \times p/f = \$146.90 \times 0.6806 = \$100$.

c. Future Worth of an Annuity f/a. This is an expression describing the amount to which a series of equal payments, made one per period, will have accrued at the end of n periods, with return at r per cent per period.

* In the tables of time-value factors and in much of the literature in the field of engineering economics, the term "interest" is used, as it is in this definition, to mean the cost attributed to the use of money. Throughout this chapter, however, the word "interest" generally is used in the more restricted sense of meaning interest payments on bonded and other indebtedness, and the term "return" or "earnings" is used as "interest" is in these symbols, as being synonymous with the "cost of money." Accordingly, elsewhere in this chapter, the letter r is used for return and the letter i for interest on indebtedness.

Example: If deposits of $100 each are made annually in a bank account paying 8 per cent per year, how much will be in the account at the end of the fifth year?

Answer: $100 $\times f/a$ = $100 $\times$ 5.867 = $586.70.

d. Present Worth of an Annuity p/a. This is a term denoting the amount required at present to provide for a series of specified equal amounts, one per period over n periods, with a return rate of r per cent per period.

Example: How much would be required in a fund earning at a rate of 8 per cent to support five annual payments of $100 each?

Answer: $100 $\times p/a$ = $100 $\times$ 3.993 = $399.30.

e. Annuity for a Future Amount a/f. This is an expression describing the amount of each of a series of equal payments, made one per period and earning at a rate of r per cent per period, needed to accumulate to a specified amount at the end of n periods.

Example: How large an annual deposit will be necessary for five such deposits, earning at a rate of 8 per cent, to accumulate to $586.70?

Answer: $586.70 $\times a/f$ = $586.70 $\times$ 0.17046 = $100.

The annuity for a future amount factor is also known both as the "sinking-fund" factor and as the "annuity-depreciation" factor.

f. Annuity from a Present Amount a/p. This is a term describing the amount of each of a series of equal payments made one per period, for n periods, which would be provided by a specified present amount earning at a rate of r per cent per period.

Example: How much could be withdrawn at the end of each of 5 years from an initial fund of $399.30 earning at 8 per cent per year?

Answer: $399.30 $\times a/p$ = $399.30 $\times$ 0.25046 = $100.

The annuity from a present amount factor is also known both as the "capital-recovery" factor and as the "amortization" factor, since the factor may be used to compute the size of the level payments necessary to repay a loan (present amount) and pay an r per cent return on the unpaid portion of the loan.

g. Time-value Factor Relationships. To illustrate the algebraic character of the symbols, it is clear that algebraically $p/a \times f/p = f/a$; now, if the time-value factor p/a, used in Example *d*, is multiplied by the time-value factor f/p, used in Example *a*, i.e., multiply 3.993 by 1.469, the result is 5.867, which is the time-value factor f/a, used in Example *c*.

ECONOMIC SELECTION STUDIES[1]

10. Definition and Principles

The term "economic selection studies" is used in this chapter to denote those types of economic studies which are made for the purpose of determining which of two or more alternatives is the least costly and, hence, the more attractive to the business. Some of the fundamentals for such studies are as follows:

1. Alternatives being studied must be comparable; that is, they should be equally satisfactory ways to accomplish a desired result.*

2. The time value of money must be considered.

3. All pertinent expenditures and receipts should be included. However, those expenditures which are common to all alternatives may be ignored. It is important, on the other hand, to recognize costs which may be generated by a plan but which may not be directly associated with it.

---

* Subjective evaluation of a problem such as comparing less service at lower cost vs. better service at higher cost is left to the ingenuity of the reader. The principles described herein will apply to the objective aspects of such a problem.

4. Only *future* transactions should be considered. Past expenditures cannot be changed by a decision made now, and the accounting records of past expenditures are irrelevant when economic effects of future action are being studied.

5. Expenditures and receipts should be evaluated in terms of cash effect. For instance, in an economic selection study to determine whether to retain or replace an item of plant, the "cost" of that item of plant for the purposes of the study is its net salvage value. This is true regardless of what its investment value on the books of the company may happen to be, because the net salvage value is the true measure of the cash differential between the two alternatives, retention or nonretention, which are being evaluated.

11. Study Method

The study method generally considered the most desirable in economic selection studies is known as the revenue-requirements method. This method compares alternatives by finding which requires the least revenue to support its costs. Since an adequate return is an essential part of revenues if a business is to be healthy, this method includes a return on capital as a cost of each investment proposal.

There is an important distinction, however, between the revenue-requirement study method used in an economic selection study and a revenue-requirements study. A revenue-requirements study computes the total amount of revenue required to support all costs associated with a single investment proposal. In contrast, in an economic selection study, costs are measured in terms of revenue requirements but, since only the differences among alternatives are of interest, identical costs common to all alternatives being studied may be omitted. As a consequence certain overhead costs, such as executives' salaries and accounting expenses, usually are omitted in the revenue requirements developed for economic selection studies. Therefore, the economic selection study measures only certain revenue requirements and usually not *all* revenue requirements.

12. Revenue Requirements of Current Expenses and Capital Expenditures

The nature of capital costs and of operating expenses and their essential differences are discussed in paragraphs 3 through 5. In an economic selection study, it is necessary to distinguish between the two types of costs only because the way in which they are recovered affects the treatment of income taxes in the study.

Capital expenditures, such as plant investments, usually provide service over a number of years and are not treated as expenses when they occur. Rather, capital expenditures, which usually are obtained from funds provided by investors, are repaid through the medium of depreciation over the life of the plant. Furthermore, until the investors' money is recovered from the operations of the business through depreciation, return on the unrecovered (i.e., undepreciated) investment must be paid out of current revenues.

However, not all the money which remains out of current revenues after paying operating expenses, including depreciation, is available for return; some of it must be used to pay income taxes. For this reason, the revenue requirement resulting from a capital expenditure comprises not just depreciation expenses and return on investment but an income tax element as well.

Operating expenses, on the other hand, are paid out of revenues during the accounting period in which the expenses occur, and since there is no investment to be repaid and no obligation on which a return must be earned, there is no associated revenue requirement for income taxes. Hence, the revenue requirement resulting from a current expense is only the expense itself.

13. Computing Revenue Requirements for Capital Expenditures

a. Return and Depreciation. An annual amount providing for the sum of return on and repayment of an investment can be derived from the amortization factor *a/p*, and such combined amounts are usually determined for use in economic selection studies where equal annual-depreciation expense amounts are assumed. That such annual amounts are equivalent in time value to the straight-line depreciation pattern of capital recovery used in most communications company accounting is shown in the following demonstration involving a $100 investment with a 5-year life and return at 8 per cent per anum:

(1) *Straight-line Depreciation*

Year	Unre-covered invest-ment	Straight-line depre-ciation	Return at 8 per cent on unre-covered invest-ment	Total annual amount	Present worth of annual amounts		
					Depre-ciation	Return	Total*
1	$100.00	$20.00	$8.00	$28.00	$18.52	$7.41	$25.93
2	80.00	20.00	6.40	26.40	17.14	5.49	22.63
3	60.00	20.00	4.80	24.80	15.88	3.81	19.69
4	40.00	20.00	3.20	23.20	14.70	2.35	17.05
5	20.00	20.00	1.60	21.60	13.61	1.09	14.70
Total					$79.85	$20.15	$100.00

* The present worth of return and the present worth of depreciation will vary according to the assumed depreciation pattern, but their sum will always equal the original investment. The slower the recovery of capital, the greater the present worth of the return component.

(2) *Amortization Annuity*

Year	Annual amount	Present worth of annual amount
1	$25.05	$23.19
2	25.05	21.47
3	25.05	19.88
4	25.05	18.41
5	25.05	17.05
Total		$100.00

Since the present worths of payments under the two series are the same, the two series obviously are equivalent to each other.

b. Income Taxes. For the purpose of determining the income tax portion of the revenue requirements associated with capital expenditures, it is necessary to isolate the return portion from the depreciation-expense portion. In calculations such as that in paragraph 13a (1) above, this is no problem. Where a combined amount for return and depreciation expense is determined by use of the amortization annuity method, it is practicable to determine a level annual amount for return by subtracting a straight-line depreciation accrual amount from the annual amortization annuity amount. Using figures from the examples in paragraph 13a, this may be illustrated as follows: Amortization annuity ($25.05) − straight-line depreciation ($20.) = level annual amount for return ($5.05).

The level amount for return also can be computed from the example in paragraph 13a by determining the sum of the present worths of the return amounts computed in the straight-line depreciation example in Paragraph 13a (1) and by determining the annuity such a sum will produce; i.e.,

$$\text{Present worth of return} \times \left(\frac{a}{p}\right)_5^{8\%} = \$20.15 \times 0.25046 = \$5.05$$

The level annual amount for return having been determined as indicated in the foregoing, the income tax portion of the annual-revenue requirement associated with capital expenditures is determined by multiplying the level annual amount by an appropriate income tax factor.   The income tax factor may be determined by the use of the formulas (1), (2), and (3) or (4) set forth in paragraph 5e of this chapter.

   c. *Salvage.*   Frequently, some of the money originally invested in an item of plant may be recovered from salvage obtained when the item is disposed of at the end of the life being studied.   This would reduce the amount which must be amortized.   On the other hand, the cost of removing the item will tend to increase the amount which must be amortized, so that it is the "net salvage," the difference between the cash received for the item and the cost of removing it, which requires consideration.

   In a present-worth-type study, the present worth of the net salvage may be subtracted from the first cost of the item to determine the present worth of the revenue requirement subject to amortization treatment.   In a study using annual costs, the annual equivalent of the future net salvage is subtracted from the annual equivalent of the first cost.   The algebraic expressions for amortization reflecting net salvage are

$$\text{Present worth of amortization} = \text{first cost} - \text{net salvage} \times \left(\frac{p}{f}\right)$$

$$\text{Amortization annuity} = \text{first cost} \times \left(\frac{a}{p}\right) - \text{net salvage} \times \left(\frac{a}{f}\right)$$

In a computation of the revenue requirement for income taxes, net salvage is reflected in both the amortization and the depreciation amounts.

   Example: Compute the capital costs associated with the assumptions outlined below on both an annual-revenue-requirement basis and on a present-worth-of-revenue requirements basis:

   Assumptions:

First cost = $1,000	$t = 48\%$	$r = 8\%$
Life = 5 years	$d = 40\%$	$i_d = 5.0\%$
Net salvage = $100		

$$\text{Income tax factor} = \frac{t}{1-t} \cdot \frac{r - di_d}{r} = 0.69231$$

1. *Annual Revenue Requirements Basis*

   a. Amortization = $\$1,000 \times \left(\frac{a}{p}\right)_{5\text{ years}}^{8\%} - \$100 \times \left(\frac{a}{f}\right)_{5\text{ years}}^{8\%} =$ $233

   b. Depreciation = $\dfrac{1{,}000 - 100}{5} =$ $180

   c. Level annual return (a) − (b) = $ 53

   d. Income taxes = (c) × 0.69231 = $ 37

   e. Amortization + income taxes = (a) + (d) = $270

2. *Present Worth of Revenue Requirements Basis*

   a. Amortization = $\$1,000 - \$100 \times \left(\frac{p}{f}\right)_5^{8\%} =$ $932

   b. Depreciation = $\dfrac{1{,}000 - 100}{5} \times \left(\frac{p}{a}\right)_5^{8\%} =$ $719

   c. Return = (a) − (b) = $213

   d. Income taxes = (c) × 0.69231 = $147

   e. Amortization + income taxes = (a) + (d) = $1,079

   Proof: $\$270 \times \left(\frac{p}{a}\right)_5^{8\%} = \$270 \times 3.993 =$ $1,079

14. Economic Selection Study Techniques

One of the principles to be observed in making an economic selection study is that the alternatives must be comparable (see paragraph 10). One facet of comparability is provision of service over the same periods of time.

Frequently, the alternative plans being studied involve investments in plant having different lives or different dates of installation. It then is necessary to structure the study in such a way that equal service periods will be involved. This is usually done by considering not just the first plant placements but subsequent placements as well, since typically the problem is one of making the final dates of the alternatives coincident. (If the initial placements occur at different times, some provision for service during the period prior to the later placement will be evident. Commonly, it will be the retention of existing plant.)

The conditions and assumptions associated with the alternatives influence the choice of study techniques. In the simple comparison of two items, to be installed at the same time and retired at the same time, either a present-worth or an annual-revenue-requirement type of study would be suitable.

But consider two more complex alternatives, one of which would install new plant now to last for 20 years and the second of which would continue to use existing plant for the first 5 years and then install sufficient new plant to serve the remaining 15 years. The second alternative would have two different series of annual-revenue requirements, one for the existing plant during the first 5 years and a dissimilar one, associated with the replacing plant, during the next 15 years. Neither of these two series of annual requirements is necessarily typical of the whole 20-year period. In cases of this type, it usually is more appropriate to compare the present worths of the alternative expenditures involved, both capital and expenses.

By contrast, consider two still different alternatives each of which involves concurrent installations of plant. The plant will last 4 years under one alternative, 5 years under the other, and there is no reason to believe that either type of plant would not be replaced with the same type upon retirement. Therefore, to make the periods of service equal, it is decided to study the relative costs of five successive installations of the 4-year plant vs. four successive installations of the 5-year plant. This example could be analyzed on a present-worth basis, but it would involve numerous calculations to translate all the successive expenditures to present worths at the initial data. On the other hand, the annual-revenue requirements associated with the first installations under either alternative would be typical of the annual requirements for each alternative during the entire 20-year study period. Thus, for this case, an annual-revenue-requirement study would portray the relative economics of the alternatives with greater facility.

In a decision of what study technique to use, it should be remembered that the present-worth approach totals, on a time-value basis, all pertinent costs while the annual-revenue-requirements study compares cross sections of the costs. Thus an annual-revenue-requirement study must be able to depict, with a single annual requirement figure, the economic impact of the whole study period. When an alternative is composed of a succession of dissimilar annual-revenue-requirement streams, some form of present-worth study is suggested.

Generally, studies involving coincident placements of plant seem to favor annual-revenue-requirement analysis while those with coincident retirements suggest the present-worth approach. Either method may be used in studies with both coincident placements and coincident retirements.

In the problems encountered in the communications business, however, it is not uncommon to find that the alternatives have neither coincident placements nor coincident retirements and that the alternatives involve additions of several units of plant, making it difficult to compensate for an irregular pattern simply by assuming successive additions of similar plant. For this type of study, a combination of the annual-revenue requirement and present-worth methods is used. It is called the present worth of annual revenue requirements (PWRR) method. In the PWRR method, the present worths, at the initial date of the study, are totaled for all the annual-revenue requirements (annual charges) generated under the alternatives during a period of time which is judged to be sufficiently long to indicate the significant economic effects. In judging how long a period to select, it should be borne in mind that the influence of costs occurring at the end of a period will be diminished because of time-value weighting. For instance, with an 8 per cent rate of return, costs occurring at the beginning of a study period will have slightly more than ten times the effect of the same costs occurring 30 years later.

15. Examples of Economic Selection Study Techniques

The following examples are illustrative of some of the economic selection study principles and techniques which have been discussed. The figures used are not necessarily typical of any actual situations.

In each of the examples, the following assumptions are made:

$$\text{Income tax rate } t = 48\%$$
$$\text{Composite rate of return } r = 8\%$$
$$\text{Rate of interest on debt } i_d = 5.0\%$$
$$\text{Debt ratio } d = 40\%$$
$$\text{Income tax factor} = \left(\frac{t}{1-t}\right)\left(\frac{r - di_d}{r}\right) = 0.69231$$

PW is the abbreviation for "present worth."

a. Example 1—Annual-revenue-requirement-study Technique. The purchase of an automatic toll ticket sorter at a cost of \$10,000 will reduce clerical expenses by \$1,500 per year. The machine will be serviceable for 10 years, at which time its net salvage value will be \$200.

(1) *Plan A—Purchase Sorter*

		Annual-revenue requirement
(a) Amortization $= \$10,000 \times \left(\dfrac{a}{p}\right)^{8\%}_{10} - \$200 \times \left(\dfrac{a}{f}\right)^{8\%}_{10}$		\$1,476
(b) Depreciation $= \dfrac{10,000 - 200}{10} =$		\$980
(c) Level annual return $= (a) - (b) =$		\$496
(d) Annual requirement for income taxes $= (c) \times 0.69231 =$		\$343
(e) Total annual revenue requirement $(a) + (d) =$		\$1,819

(2) *Plan B—Continue Present Clerical Operation*

(f) Operating expense $=$		\$1,500
(3) *Difference in favor of Plan B* $= (e) - (f) =$		\$319

(4) *Comments on Example 1.* This example could have been studied on a present-worth-study basis equally as well.

b. Example 2—Present-worth-study Technique. It is necessary that a certain service be continued for 10 more years. However, the station apparatus now providing that service will be usable for only 5 more years. It is possible to replace the existing apparatus unit now with a new unit costing \$10,000 or to replace it at the end of 5 years with a new unit costing \$8,000. The existing equipment unit will have a \$500 net salvage value if replaced now or a \$50 net salvage if replaced at the end of 5 years. The \$10,000 replacement unit will have zero net salvage value at the end of its 10 years of service; the \$8,000 replacement will have a \$400 net salvage value at the end of its 5 years of service. Other factors to be considered are tabulated below:

Annual Costs

	Existing unit	$10,000 replacement unit	$8,000 replacement unit
Maintenance............	$700	$500	$400
Administration.........	100	100	80
Ad valorem taxes........	400	400	320
Total.................	$1,200	$1,000	$800

(1) *Plan A—Replace Now—PW of Revenue Requirements at Year 0 Associated with $10,000 Replacement Unit*

(a) Amortization $= \$10,000 - 0 \times \left(\dfrac{p}{f}\right)_{10}^{8\%} = $ $10,000

(b) Depreciation $= \dfrac{10,000 - 0}{10} \times \left(\dfrac{p}{a}\right)_{10}^{8\%} = $ $6,710

(c) Return $= (a) - (b) = $ $3,290

(d) Income tax $= (c) \times 0.69231 = $ $2,278

(e) Total: PW capital costs $= (a) + (d) = $ $12,278

(f) Total PW annual expenses $= \$1,000 \times \left(\dfrac{p}{a}\right)_{10}^{8\%} = $ $6,710

(g) Total PW revenue requirements for plan A at year 0 $= (e) + (f) = $ $18,988

(2) *Plan B—Replace at the End of Five Years*

  (2.1) *PW of Revenue Requirements at Year 0 Associated with Existing Equipment*

(h) Amortization $= \$500 - \$50 \times \left(\dfrac{p}{f}\right)_{5}^{8\%} = $ $466

(i) Depreciation $= \dfrac{\$500 - \$50}{5} \times \left(\dfrac{p}{a}\right)_{5}^{8\%} = $ $359

(j) Return $= (h) - (i) = $ $107

(k) Income tax $= (j) \times 0.69231 = $ $74

(l) Total PW capital costs $= (h) + (k) = $ $540

(m) Total PW annual expenses $= \$1,200 \times \left(\dfrac{p}{a}\right)_{5}^{8\%} = $ $4,792

(n) Total PW at year 0 of revenue requirements $= (l) + (m) = $ $5,332

  (2.2) *PW of Revenue Requirements at Year 5 Associated with $8,000 Replacement Unit*

(o) Amortization $= \$8,000 - 400 \times \left(\dfrac{p}{f}\right)_{5}^{8\%} = $ $7,728

(p) Depreciation $= \dfrac{8,000 - 400}{5} \times \left(\dfrac{p}{a}\right)_{5}^{8\%} = $ $6,069

(q) Return $= (o) - (p) = $ $1,659

(r) Income tax $= (q) \times 0.69231 = $ $1,149

(s) Total PW at year 5 of revenue requirements for capital costs $= (o) + (r) = $ $8,877

(t) Total PW at year 5 of revenue requirements for annual expenses $=$

    $\$800 \times \left(\dfrac{p}{a}\right)_{5}^{8\%} = $ $3,194

(u) Total PW at year 5 of revenue requirements $= (s) + (t) = $ $12,071

  (2.3) *PW of Total Revenue Requirements at Year 0 Associated with $8,000*

    Replacement Unit $= (v) = (u) \times \left(\dfrac{p}{f}\right)_{5}^{8\%} = $ $8,216

(2.4) *PW of Total Revenue Requirements at Year* 0 *Associated with*
Plan B = (*w*) = (*n*) + (*v*) = $13,548

(3) *Comments on Example 2*

(*x*) Plan B has two successive streams of annual charges neither of which would offer a cross-sectional view representative of all of Plan B. Therefore, a present-worth study is more appropriate.

(*y*) The cost of retaining the existing plant in Plan B includes the present net salvage value of the plant as a "first cost," since this value must be foregone if the plant is kept and not sold. The net difference between Plans A and B would be the same if, instead, this first cost were shown as a credit to Plan A, analogous to a "trade-in allowance." Associated with this first cost is the ultimate net salvage which would be realized if the existing machine were retained for the next 5 years.

(*z*) Two present-worth operations are involved in connection with the $8,000 replacement equipment unit. One finds the present worth of annual charges at the date of installation; the next translates this to a present worth as of the initial date of the study.

c. Example 3—Present Worth of Annual-revenue Requirements (PWRR)—
Single Time Period. A 5-year-old construction truck will be replaced by a new truck estimated to cost $10,000. The new truck is estimated to have an 8-year service life and a net salvage value of $600 set at the end of that life. The replacement either must be made now or can be delayed 5 years by improvements costing $1,500 (to be capitalized). The existing truck, if replaced now, is estimated to have a net salvage value of $600; if replaced later, it would have a net salvage value of $100. Under either plan, it may be assumed that the $10,000 truck would be followed at the end of its service by an indefinite succession of similar trucks.

Maintenance of the new truck would cost about $300 per year, while maintenance of the existing truck would cost about $500 per year. All other costs may be assumed to be common to both plans.

(1) *Comments on Example 3.* A present-worth study does not seem appropriate for this type of case, since, with any finite number of successive purchases or installations, one plan will always provide a longer service life than the other. However, under the conditions of the problem, it does not appear unreasonable to assume that successive purchases or installations will occur under either plan at least long enough so that, with time-value weighting, subsequent additions will have negligible influence. These successive purchases or installations will each renew the same stream of annual-revenue requirements. Therefore, the problem can be restated as follows:

Plan A. Start generating annual-revenue requirements associated with a new truck immediately.

Plan B. Start generating annual-revenue requirements associated with a new truck 5 years from now by repairing the existing truck now.

It can be seen that, beyond the 5-year point, the annual-revenue requirements under both plans will be identical and the only difference between the two plans is the difference in annual-revenue requirements during the first 5 years. However, an annual-revenue-requirement format might be misleading because of the implication that its results are representative of all years, not just a few. Therefore, it would be most appropriate to compare the present worth of the annual-revenue requirements for the two plans.

(2) *Plan A—Replace Immediately*

(2.1) *Annual-revenue Requirement for New Truck*

(*a*) Amortization = $10,000 \times \left(\dfrac{a}{p}\right)_8^{8\%} - \$600 \times \left(\dfrac{a}{f}\right)_8^{8\%} = $1,684

(*b*) Depreciation = $\dfrac{\$10,000 - \$600}{8}$ = $1,175

(c) Level annual return = (a) − (b) = $ 509
(d) Annual income taxes = (c) × 0.69231 = $ 352
(e) Total annual capital cost = (a) + (d) = $2,036
(f) Annual expenses = $ 300
(g) Total annual-revenue requirement = (e) + (f) = $2,336

(2.2) *Present worth of annual revenue requirements for new truck* =

$$(g) \times \left(\frac{p}{a}\right)^{8\%}_{5} = \qquad \$9,328$$

(3) *Plan B—Improve Old Truck and Defer Replacement for 5 Years*
    (3.1) Annual Revenue Requirements for Existing Truck

$$(h)\ \text{Amortization} = (\$600 + \$1,500) \times \left(\frac{a}{p}\right)^{8\%}_{5} - 100 \times \left(\frac{a}{f}\right)^{8\%}_{5} = \qquad \$\ 509$$

$$(i)\ \text{Depreciation} = \frac{2,100 - 100}{5} = \qquad \$\ 400$$

(j) Level annual return = (h) − (i) = $ 109
(k) Annual income taxes = (j) × 0.69231 = $ 75
(l) Total annual capital cost = (h) + (k) = $ 584
(m) Annual expenses = $ 500
(n) Total annual revenue requirements = (l) + (m) = $1,084

(3.2) *PW of annual-revenue requirements for existing truck* =

$$(n) \times \left(\frac{p}{a}\right)^{8\%}_{5} = \qquad \$4,328$$

(3.3) *Difference in favor of Plan B* = (2.2) − (3.2) = $5,000

(4) *Additional Comments on Example* 3. The present worth of annual-revenue-requirements format is useful, as illustrated in this example, to determine whether a proposal for temporary relief is justified by the savings resulting from deferring a large planned expenditure. The PWRR analysis can also be adapted to evaluating the use of spare capacity. To find the economic penalty resulting from the use of spare plant, a PWRR study is used to measure the effect of advancing the next scheduled relief of that plant.

d. Example 4—Present Worth of Annual-revenue Requirements (PWRR)— Multiple Time Periods. It is estimated that the future demand for a particular service will require a growth rate in equipment capacity of 100 lines per year for 15 years. This requirement can be met by either of the following:
Plan A. Install a 1,500-line unit now.
Plan B. Install a 300-line unit now and at the end of the third, sixth, ninth, and twelfth years.
The following data should be assumed:

	1,500-*line* unit	300-*line* unit
First cost	$10,000	$3,500
Net salvage	−10%	−10%
Life	20 years	20 years
Annual maintenance	$500	$200
Annual ad valorem taxes	$200	$ 60

Since the plans involve noncoincident placements and noncoincident retirements, a present-worth-of-revenue-requirements study is indicated. The first step is to develop annual-revenue requirements for each of the two unit sizes. In this example, the initial stage of the computation employs percentages of first cost rather than dollars, as follows:

	1,500-*line* unit	300-*line* unit
(a) Amortization % =		
$100\% \times \left(\frac{a}{p}\right)^{8\%}_{20} + 10\% \times \left(\frac{a}{f}\right)^{8\%}_{20} =$	10.40%	10.40%

(b) Depreciation $\% = \dfrac{100\% + 10\%}{20} =$ 5.50%   5.50%

(c) Level annual return $\% = (a) - (b) =$ 4.90%   4.90%

(d) Income tax $\% = (c) \times 0.69231 =$ 3.39%   3.39%

(e) Total capital annual-revenue requirements =
   (a) + (d) =   13.79%   13.79%

(f) Total capital annual-revenue requirements =
   (e) % $\times$ first cost =   \$1,379   \$483

Annual expenses:

(g) Maintenance   \$ 500   \$200

(h) Ad valorem taxes   \$ 200   \$ 60

(i) Total annual-revenue requirements =
   (f) + (g) + (h) =   \$2,079   \$743

The next step is to develop the present worth of the annual-revenue requirements associated with the two unit sizes for a period long enough to include the entire life of each of the 300-line units being studied. This period extends from the initial installation to the final retirement. In this example, the last unit is placed in year 12 and survives 20 years, so that the period is 32 years. For the purposes of this study, it is assumed that the units which have been retired previous to the end of the 32-year period will have been replaced by like units.

The present worths of the annual-revenue requirements are computed as follows:

Plan A	Annual-revenue requirements	Factors	PWRR year 0
1,500-line unit installed year 0	\$2,079	$\left(\dfrac{p}{a}\right)^{8\%}_{32}$	\$23,773
Plan B			
300-line unit installed year 0	743	$\left(\dfrac{p}{a}\right)^{8\%}_{32}$	8,496
300-line unit installed year 3	743	$\left(\dfrac{p}{a}\right)^{8\%}_{29} \times \left(\dfrac{p}{f}\right)^{8\%}_{3}$	6,581
300-line unit installed year 6	743	$\left(\dfrac{p}{a}\right)^{8\%}_{26} \times \left(\dfrac{p}{f}\right)^{8\%}_{6}$	5,062
300-line unit installed year 9	743	$\left(\dfrac{p}{a}\right)^{8\%}_{23} \times \left(\dfrac{p}{f}\right)^{8\%}_{9}$	3,854
300-line unit installed year 12	743	$\left(\dfrac{p}{a}\right)^{8\%}_{20} \times \left(\dfrac{p}{f}\right)^{8\%}_{12}$	2,897
Total Plan B			\$26,890

(3) *Comments on Example* 4

(a) Over the selected study period, the present-worth treatment of the annual-revenue requirements results in changing from a condition appearing to favor the small increments of plant during the early years to one in which the installation of single large units clearly is the more economic plan.

(b) Net salvage is not always positive. Sometimes the cost of removal exceeds the gross salvage value of the equipment, thereby necessitating the amortization of a higher amount than first cost. A negative net salvage value has been used in this example as an illustration of this condition.

(c) This example, as have all previous examples, assumes that all units will be retired at exactly the estimated average life. For plant composed of a large number of units, this estimated average results from the life experience of many units, some being retired with shorter lives than "average," some with longer lives.　This distribution of lives is called a "dispersion of retirements." Because of time-value weighting, the influence of those units with shorter-than-average lives is greater than the influence of units with longer-than-average lives.　Therefore, an amortization factor which reflects the dispersion of retirements is larger than one which, in effect, assumes that all units will retire at exactly the average life.　Tables of factors are available which reflect some of the patterns of dispersed retirements associated with various types of telephone plant.　It is appropriate to use such dispersed annuity factors in some types of economic selection studies but their use is more common in revenue requirements studies of the type illustrated in Example 4.

REFERENCES

1. *Engineering Economy*, 2d ed., American Telephone and Telegraph Company, 1963. Grant, Eugene L., and W. Grant Ireson, *Principles of Engineering Economy*, 4th ed., The Ronald Press Company, New York, 1960.
2. Part 31—Uniform System of Accounts for Class A and Class B Telephone Companies, Federal Communication Commission, *Rules and Regulations*.

Table 1

4%

ANNUAL COMPOUNDING AT ABOVE EFFECTIVE (i) RATE PER YEAR

FACTORS APPLICABLE TO END OF YEAR AMOUNTS

	SINGLE AMOUNT		SERIES OF UNIFORM AMOUNTS				
	PRESENT WORTH OF A FUTURE AMT.	FUTURE WORTH OF A PRESENT AMT.	ANNUITY FROM A PRESENT AMOUNT	ANNUITY FOR A FUTURE AMOUNT	PRESENT WORTH OF AN ANNUITY	FUTURE WORTH OF AN ANNUITY	
n	$(p/f)_n$	$(f/p)_n$	$(a/p)_n$	$(a/f)_n$	$(p/a)_n$	$(f/a)_n$	n
	$\dfrac{1}{(1+i)^n}$	$(1+i)^n$	$\dfrac{i(1+i)^n}{(1+i)^n - 1}$	$\dfrac{i}{(1+i)^n - 1}$	$\dfrac{(1+i)^n - 1}{i(1+i)^n}$	$\dfrac{(1+i)^n - 1}{i}$	
1	0.9615	1.040	1.04000	1.00000	0.962	1.000	1
2	0.9246	1.082	0.53020	0.49020	1.886	2.040	2
3	0.8890	1.125	0.36035	0.32035	2.775	3.122	3
4	0.8548	1.170	0.27549	0.23549	3.630	4.246	4
5	0.8219	1.217	0.22463	0.18463	4.452	5.416	5
6	0.7903	1.265	0.19076	0.15076	5.242	6.633	6
7	0.7599	1.316	0.16661	0.12661	6.002	7.898	7
8	0.7307	1.369	0.14853	0.10853	6.733	9.214	8
9	0.7026	1.423	0.13449	0.09449	7.435	10.583	9
10	0.6756	1.480	0.12329	0.08329	8.111	12.006	10
11	0.6496	1.539	0.11415	0.07415	8.760	13.486	11
12	0.6246	1.601	0.10655	0.06655	9.385	15.026	12
13	0.6006	1.665	0.10014	0.06014	9.986	16.627	13
14	0.5775	1.732	0.09467	0.05467	10.563	18.292	14
15	0.5553	1.801	0.08994	0.04994	11.118	20.024	15
16	0.5339	1.873	0.08582	0.04582	11.652	21.825	16
17	0.5134	1.948	0.08220	0.04220	12.166	23.698	17
18	0.4936	2.026	0.07899	0.03899	12.659	25.645	18
19	0.4746	2.107	0.07614	0.03614	13.134	27.671	19
20	0.4564	2.191	0.07358	0.03358	13.590	29.778	20
21	0.4388	2.279	0.07128	0.03128	14.029	31.969	21
22	0.4220	2.370	0.06920	0.02920	14.451	34.248	22
23	0.4057	2.465	0.06731	0.02731	14.857	36.618	23
24	0.3901	2.563	0.06559	0.02559	15.247	39.083	24
25	0.3751	2.666	0.06401	0.02401	15.622	41.646	25
26	0.3607	2.772	0.06257	0.02257	15.983	44.312	26
27	0.3468	2.883	0.06124	0.02124	16.330	47.084	27
28	0.3335	2.999	0.06001	0.02001	16.663	49.968	28
29	0.3207	3.119	0.05888	0.01888	16.984	52.966	29
30	0.3083	3.243	0.05783	0.01783	17.292	56.085	30
31	0.2965	3.373	0.05686	0.01686	17.588	59.328	31
32	0.2851	3.508	0.05595	0.01595	17.874	62.701	32
33	0.2741	3.648	0.05510	0.01510	18.148	66.210	33
34	0.2636	3.794	0.05431	0.01431	18.411	69.858	34
35	0.2534	3.946	0.05358	0.01358	18.665	73.652	35
40	0.2083	4.801	0.05052	0.01052	19.793	95.026	40
45	0.1712	5.841	0.04826	0.00826	20.720	121.029	45
50	0.1407	7.107	0.04655	0.00655	21.482	152.667	50
55	0.1157	8.646	0.04523	0.00523	22.109	191.159	55
60	0.0951	10.520	0.04420	0.00420	22.623	237.991	60
65	0.0781	12.799	0.04339	0.00339	23.047	294.968	65
70	0.0642	15.572	0.04275	0.00275	23.395	364.290	70
75	0.0528	18.945	0.04223	0.00223	23.680	448.631	75
∞			0.04000	0.00000	25.000		∞

From *Engineering Economy*, 2d ed., copyright by the American Telephone and Telegraph Company, 1963. By permission.

Table 1. (*Continued*)

5%

ANNUAL COMPOUNDING AT ABOVE EFFECTIVE (i) RATE PER YEAR

FACTORS APPLICABLE TO END OF YEAR AMOUNTS

n	SINGLE AMOUNT		SERIES OF UNIFORM AMOUNTS				n
	PRESENT WORTH OF A FUTURE AMT. $(p/f)_n$ $\dfrac{1}{(1+i)^n}$	FUTURE WORTH OF A PRESENT AMT. $(f/p)_n$ $(1+i)^n$	ANNUITY FROM A PRESENT AMOUNT $(a/p)_n$ $\dfrac{i(1+i)^n}{(1+i)^n-1}$	ANNUITY FOR A FUTURE AMOUNT $(a/f)_n$ $\dfrac{i}{(1+i)^n-1}$	PRESENT WORTH OF AN ANNUITY $(p/a)_n$ $\dfrac{(1+i)^n-1}{i(1+i)^n}$	FUTURE WORTH OF AN ANNUITY $(f/a)_n$ $\dfrac{(1+i)^n-1}{i}$	
1	0.9524	1.050	1.05000	1.00000	0.952	1.000	1
2	0.9070	1.103	0.53780	0.48780	1.859	2.050	2
3	0.8638	1.158	0.36721	0.31721	2.723	3.153	3
4	0.8227	1.216	0.28201	0.23201	3.546	4.310	4
5	0.7835	1.276	0.23097	0.18097	4.329	5.526	5
6	0.7462	1.340	0.19702	0.14702	5.076	6.802	6
7	0.7107	1.407	0.17282	0.12282	5.786	8.142	7
8	0.6768	1.477	0.15472	0.10472	6.463	9.549	8
9	0.6446	1.551	0.14069	0.09069	7.108	11.027	9
10	0.6139	1.629	0.12950	0.07950	7.722	12.578	10
11	0.5847	1.710	0.12039	0.07039	8.306	14.207	11
12	0.5568	1.796	0.11283	0.06283	8.863	15.917	12
13	0.5303	1.886	0.10646	0.05646	9.394	17.713	13
14	0.5051	1.980	0.10102	0.05102	9.899	19.599	14
15	0.4810	2.079	0.09634	0.04634	10.380	21.579	15
16	0.4581	2.183	0.09227	0.04227	10.838	23.657	16
17	0.4363	2.292	0.08870	0.03870	11.274	25.840	17
18	0.4155	2.407	0.08555	0.03555	11.690	28.132	18
19	0.3957	2.527	0.08275	0.03275	12.085	30.539	19
20	0.3769	2.653	0.08024	0.03024	12.462	33.066	20
21	0.3589	2.786	0.07800	0.02800	12.821	35.719	21
22	0.3418	2.925	0.07597	0.02597	13.163	38.505	22
23	0.3256	3.072	0.07414	0.02414	13.489	41.430	23
24	0.3101	3.225	0.07247	0.02247	13.799	44.502	24
25	0.2953	3.386	0.07095	0.02095	14.094	47.727	25
26	0.2812	3.556	0.06956	0.01956	14.375	51.113	26
27	0.2678	3.733	0.06829	0.01829	14.643	54.669	27
28	0.2551	3.920	0.06712	0.01712	14.898	58.403	28
29	0.2429	4.116	0.06605	0.01605	15.141	62.323	29
30	0.2314	4.322	0.06505	0.01505	15.372	66.439	30
31	0.2204	4.538	0.06413	0.01413	15.593	70.761	31
32	0.2099	4.765	0.06328	0.01328	15.803	75.299	32
33	0.1999	5.003	0.06249	0.01249	16.003	80.064	33
34	0.1904	5.253	0.06176	0.01176	16.193	85.067	34
35	0.1813	5.516	0.06107	0.01107	16.374	90.320	35
40	0.1420	7.040	0.05828	0.00828	17.159	120.800	40
45	0.1113	8.985	0.05626	0.00626	17.774	159.700	45
50	0.0872	11.467	0.05478	0.00478	18.256	209.348	50
55	0.0683	14.636	0.05367	0.00367	18.633	272.713	55
60	0.0535	18.679	0.05283	0.00283	18.929	353.584	60
65	0.0419	23.840	0.05219	0.00219	19.161	456.798	65
70	0.0329	30.426	0.05170	0.00170	19.343	588.529	70
75	0.0258	38.833	0.05132	0.00132	19.485	756.654	75
∞			0.05000	0.00000	20.000		∞

From *Engineering Economy*, 2d ed., copyright by the American Telephone and Telegraph Company, 1963. By permission.

TABLE 1   25–23

Table 1. (*Continued*)

6%

ANNUAL COMPOUNDING AT ABOVE EFFECTIVE (i) RATE PER YEAR

FACTORS APPLICABLE TO END OF YEAR AMOUNTS

n	SINGLE AMOUNT		SERIES OF UNIFORM AMOUNTS				n
	PRESENT WORTH OF A FUTURE AMT. $(p/f)_n$	FUTURE WORTH OF A PRESENT AMT $(f/p)_n$	ANNUITY FROM A PRESENT AMOUNT $(a/p)_n$	ANNUITY FOR A FUTURE AMOUNT $(a/f)_n$	PRESENT WORTH OF AN ANNUITY $(p/a)_n$	FUTURE WORTH OF AN ANNUITY $(f/a)_n$	
	$\dfrac{1}{(1+i)^n}$	$(1+i)^n$	$\dfrac{i(1+i)^n}{(1+i)^n-1}$	$\dfrac{i}{(1+i)^n-1}$	$\dfrac{(1+i)^n-1}{i(1+i)^n}$	$\dfrac{(1+i)^n-1}{i}$	
1	0.9434	1.060	1.06000	1.00000	0.943	1.000	1
2	0.8900	1.124	0.54544	0.48544	1.833	2.060	2
3	0.8396	1.191	0.37411	0.31411	2.673	3.184	3
4	0.7921	1.262	0.28859	0.22859	3.465	4.375	4
5	0.7473	1.338	0.23740	0.17740	4.212	5.637	5
6	0.7050	1.419	0.20336	0.14336	4.917	6.975	6
7	0.6651	1.504	0.17914	0.11914	5.582	8.394	7
8	0.6274	1.594	0.16104	0.10104	6.210	9.897	8
9	0.5919	1.689	0.14702	0.08702	6.802	11.491	9
10	0.5584	1.791	0.13587	0.07587	7.360	13.181	10
11	0.5268	1.898	0.12679	0.06679	7.887	14.972	11
12	0.4970	2.012	0.11928	0.05928	8.384	16.870	12
13	0.4688	2.133	0.11296	0.05296	8.853	18.882	13
14	0.4423	2.261	0.10758	0.04758	9.295	21.015	14
15	0.4173	2.397	0.10296	0.04296	9.712	23.276	15
16	0.3936	2.540	0.09895	0.03895	10.106	25.673	16
17	0.3714	2.693	0.09544	0.03544	10.477	28.213	17
18	0.3503	2.854	0.09236	0.03236	10.828	30.906	18
19	0.3305	3.026	0.08962	0.02962	11.158	33.760	19
20	0.3118	3.207	0.08718	0.02718	11.470	36.786	20
21	0.2942	3.400	0.08500	0.02500	11.764	39.993	21
22	0.2775	3.604	0.08305	0.02305	12.042	43.392	22
23	0.2618	3.820	0.08128	0.02128	12.303	46.996	23
24	0.2470	4.049	0.07968	0.01968	12.550	50.816	24
25	0.2330	4.292	0.07823	0.01823	12.783	54.865	25
26	0.2198	4.549	0.07690	0.01690	13.003	59.156	26
27	0.2074	4.822	0.07570	0.01570	13.211	63.706	27
28	0.1956	5.112	0.07459	0.01459	13.406	68.528	28
29	0.1846	5.418	0.07358	0.01358	13.591	73.640	29
30	0.1741	5.743	0.07265	0.01265	13.765	79.058	30
31	0.1643	6.088	0.07179	0.01179	13.929	84.802	31
32	0.1550	6.453	0.07100	0.01100	14.084	90.890	32
33	0.1462	6.841	0.07027	0.01027	14.230	97.343	33
34	0.1379	7.251	0.06960	0.00960	14.368	104.184	34
35	0.1301	7.686	0.06897	0.00897	14.498	111.435	35
40	0.0972	10.286	0.06646	0.00646	15.046	154.762	40
45	0.0727	13.765	0.06470	0.00470	15.456	212.744	45
50	0.0543	18.420	0.06344	0.00344	15.762	290.336	50
55	0.0406	24.650	0.06254	0.00254	15.991	394.172	55
60	0.0303	32.988	0.06188	0.00188	16.161	533.128	60
65	0.0227	44.145	0.06139	0.00139	16.289	719.083	65
70	0.0169	59.076	0.06103	0.00103	16.385	967.932	70
75	0.0126	79.057	0.06077	0.00077	16.456	1300.949	75
∞			0.06000	0.00000	16.667		∞

Table 1. (*Continued*)

7%

ANNUAL COMPOUNDING AT ABOVE EFFECTIVE (i) RATE PER YEAR

FACTORS APPLICABLE TO END OF YEAR AMOUNTS

	SINGLE AMOUNT		SERIES OF UNIFORM AMOUNTS				
n	PRESENT WORTH OF A FUTURE AMT. $(p/f)_n$	FUTURE WORTH OF A PRESENT AMT. $(f/p)_n$	ANNUITY FROM A PRESENT AMOUNT $(a/p)_n$	ANNUITY FOR A FUTURE AMOUNT $(a/f)_n$	PRESENT WORTH OF AN ANNUITY $(p/a)_n$	FUTURE WORTH OF AN ANNUITY $(f/a)_n$	n
	$\dfrac{1}{(1+i)^n}$	$(1+i)^n$	$\dfrac{i(1+i)^n}{(1+i)^n-1}$	$\dfrac{i}{(1+i)^n-1}$	$\dfrac{(1+i)^n-1}{i(1+i)^n}$	$\dfrac{(1+i)^n-1}{i}$	
1	0.9346	1.070	1.07000	1.00000	0.935	1.000	1
2	0.8734	1.145	0.55309	0.48309	1.808	2.070	2
3	0.8163	1.225	0.38105	0.31105	2.624	3.215	3
4	0.7629	1.311	0.29523	0.22523	3.387	4.440	4
5	0.7130	1.403	0.24389	0.17389	4.100	5.751	5
6	0.6663	1.501	0.20980	0.13980	4.767	7.153	6
7	0.6227	1.606	0.18555	0.11555	5.389	8.654	7
8	0.5820	1.718	0.16747	0.09747	5.971	10.260	8
9	0.5439	1.838	0.15349	0.08349	6.515	11.978	9
10	0.5083	1.967	0.14238	0.07238	7.024	13.816	10
11	0.4751	2.105	0.13336	0.06336	7.499	15.784	11
12	0.4440	2.252	0.12590	0.05590	7.943	17.888	12
13	0.4150	2.410	0.11965	0.04965	8.358	20.141	13
14	0.3878	2.579	0.11434	0.04434	8.745	22.550	14
15	0.3624	2.759	0.10979	0.03979	9.108	25.129	15
16	0.3387	2.952	0.10586	0.03586	9.447	27.888	16
17	0.3166	3.159	0.10243	0.03243	9.763	30.840	17
18	0.2959	3.380	0.09941	0.02941	10.059	33.999	18
19	0.2765	3.617	0.09675	0.02675	10.336	37.379	19
20	0.2584	3.870	0.09439	0.02439	10.594	40.995	20
21	0.2415	4.141	0.09229	0.02229	10.836	44.865	21
22	0.2257	4.430	0.09041	0.02041	11.061	49.006	22
23	0.2109	4.741	0.08871	0.01871	11.272	53.436	23
24	0.1971	5.072	0.08719	0.01719	11.469	58.177	24
25	0.1842	5.427	0.08581	0.01581	11.654	63.249	25
26	0.1722	5.807	0.08456	0.01456	11.826	68.676	26
27	0.1609	6.214	0.08343	0.01343	11.987	74.484	27
28	0.1504	6.649	0.08239	0.01239	12.137	80.698	28
29	0.1406	7.114	0.08145	0.01145	12.278	87.347	29
30	0.1314	7.612	0.08059	0.01059	12.409	94.461	30
31	0.1228	8.145	0.07980	0.00980	12.532	102.073	31
32	0.1147	8.715	0.07907	0.00907	12.647	110.218	32
33	0.1072	9.325	0.07841	0.00841	12.754	118.933	33
34	0.1002	9.978	0.07780	0.00780	12.854	128.259	34
35	0.0937	10.677	0.07723	0.00723	12.948	138.237	35
40	0.0668	14.974	0.07501	0.00501	13.332	199.635	40
45	0.0476	21.002	0.07350	0.00350	13.606	285.749	45
50	0.0339	29.457	0.07246	0.00246	13.801	406.529	50
55	0.0242	41.315	0.07174	0.00174	13.940	575.929	55
60	0.0173	57.946	0.07123	0.00123	14.039	813.520	60
65	0.0123	81.273	0.07087	0.00087	14.110	1146.755	65
70	0.0088	113.989	0.07062	0.00062	14.160	1614.134	70
75	0.0063	159.876	0.07044	0.00044	14.196	2269.657	75
∞			0.07000	0.00000	14.286		∞

From *Engineering Economy*, 2d ed., copyright by the American Telephone and Telegraph Company, 1963. By permission.

TABLE 1   25–25

Table 1. *(Continued)*

8%

ANNUAL COMPOUNDING AT ABOVE EFFECTIVE (i) RATE PER YEAR

FACTORS APPLICABLE TO END OF YEAR AMOUNTS

	SINGLE AMOUNT		SERIES OF UNIFORM AMOUNTS					
n	PRESENT WORTH OF A FUTURE AMT. $(p/f)_n$	FUTURE WORTH OF A PRESENT AMT. $(f/p)_n$	ANNUITY FROM A PRESENT AMOUNT $(a/p)_n$	ANNUITY FOR A FUTURE AMOUNT $(a/f)_n$	PRESENT WORTH OF AN ANNUITY $(p/a)_n$	FUTURE WORTH OF AN ANNUITY $(f/a)_n$	n
	$\dfrac{1}{(1+i)^n}$	$(1+i)^n$	$\dfrac{i(1+i)^n}{(1+i)^n - 1}$	$\dfrac{i}{(1+i)^n - 1}$	$\dfrac{(1+i)^n - 1}{i(1+i)^n}$	$\dfrac{(1+i)^n - 1}{i}$	
1	0.9259	1.080	1.08000	1.00000	0.926	1.000	1
2	0.8573	1.166	0.56077	0.48077	1.783	2.080	2
3	0.7938	1.260	0.38803	0.30803	2.577	3.246	3
4	0.7350	1.360	0.30192	0.22192	3.312	4.506	4
5	0.6806	1.469	0.25046	0.17046	3.993	5.867	5
6	0.6302	1.587	0.21632	0.13632	4.623	7.336	6
7	0.5835	1.714	0.19207	0.11207	5.206	8.923	7
8	0.5403	1.851	0.17401	0.09401	5.747	10.637	8
9	0.5002	1.999	0.16008	0.08008	6.247	12.488	9
10	0.4632	2.159	0.14903	0.06903	6.710	14.487	10
11	0.4289	2.332	0.14008	0.06008	7.139	16.645	11
12	0.3971	2.518	0.13270	0.05270	7.536	18.977	12
13	0.3677	2.720	0.12652	0.04652	7.904	21.495	13
14	0.3405	2.937	0.12130	0.04130	8.244	24.215	14
15	0.3152	3.172	0.11683	0.03683	8.559	27.152	15
16	0.2919	3.426	0.11298	0.03298	8.851	30.324	16
17	0.2703	3.700	0.10963	0.02963	9.122	33.750	17
18	0.2502	3.996	0.10670	0.02670	9.372	37.450	18
19	0.2317	4.316	0.10413	0.02413	9.604	41.446	19
20	0.2145	4.661	0.10185	0.02185	9.818	45.762	20
21	0.1987	5.034	0.09983	0.01983	10.017	50.423	21
22	0.1839	5.437	0.09803	0.01803	10.201	55.457	22
23	0.1703	5.871	0.09642	0.01642	10.371	60.893	23
24	0.1577	6.341	0.09498	0.01498	10.529	66.765	24
25	0.1460	6.848	0.09368	0.01368	10.675	73.106	25
26	0.1352	7.396	0.09251	0.01251	10.810	79.954	26
27	0.1252	7.988	0.09145	0.01145	10.935	87.351	27
28	0.1159	8.627	0.09049	0.01049	11.051	95.339	28
29	0.1073	9.317	0.08962	0.00962	11.158	103.966	29
30	0.0994	10.063	0.08883	0.00883	11.258	113.283	30
31	0.0920	10.868	0.08811	0.00811	11.350	123.346	31
32	0.0852	11.737	0.08745	0.00745	11.435	134.214	32
33	0.0789	12.676	0.08685	0.00685	11.514	145.951	33
34	0.0730	13.690	0.08630	0.00630	11.587	158.627	34
35	0.0676	14.785	0.08580	0.00580	11.655	172.317	35
40	0.0460	21.725	0.08386	0.00386	11.925	259.057	40
45	0.0313	31.920	0.08259	0.00259	12.108	386.506	45
50	0.0213	46.902	0.08174	0.00174	12.233	573.770	50
55	0.0145	68.914	0.08118	0.00118	12.319	848.923	55
60	0.0099	101.257	0.08080	0.00080	12.377	1253.213	60
65	0.0067	148.780	0.08054	0.00054	12.416	1847.248	65
70	0.0046	218.606	0.08037	0.00037	12.443	2720.080	70
75	0.0031	321.205	0.08025	0.00025	12.461	4002.557	75
∞			0.08000	0.00000	12.500		∞

From *Engineering Economy*, 2d ed., copyright by the American Telephone and Telegraph Company, 1963.   By permission.

COST STUDY METHODS

Table 1. (*Continued*)

9%

ANNUAL COMPOUNDING AT ABOVE EFFECTIVE (i) RATE PER YEAR

FACTORS APPLICABLE TO END OF YEAR AMOUNTS

	SINGLE AMOUNT		SERIES OF UNIFORM AMOUNTS				
	PRESENT WORTH OF A FUTURE AMT. $(p/f)_n$	FUTURE WORTH OF A PRESENT AMT. $(f/p)_n$	ANNUITY FROM A PRESENT AMOUNT $(a/p)_n$	ANNUITY FOR A FUTURE AMOUNT $(a/f)_n$	PRESENT WORTH OF AN ANNUITY $(p/a)_n$	FUTURE WORTH OF AN ANNUITY $(f/a)_n$	
n	$\dfrac{1}{(1+i)^n}$	$(1+i)^n$	$\dfrac{i(1+i)^n}{(1+i)^n - 1}$	$\dfrac{i}{(1+i)^n - 1}$	$\dfrac{(1+i)^n - 1}{i(1+i)^n}$	$\dfrac{(1+i)^n - 1}{i}$	n
1	0.9174	1.090	1.09000	1.00000	0.917	1.000	1
2	0.8417	1.188	0.56847	0.47847	1.759	2.090	2
3	0.7722	1.295	0.39505	0.30505	2.531	3.278	3
4	0.7084	1.412	0.30867	0.21867	3.240	4.573	4
5	0.6499	1.539	0.25709	0.16709	3.890	5.985	5
6	0.5963	1.677	0.22292	0.13292	4.486	7.523	6
7	0.5470	1.878	0.19869	0.10869	5.033	9.200	7
8	0.5019	1.993	0.18067	0.09067	5.535	11.028	8
9	0.4604	2.172	0.16680	0.07680	5.995	13.021	9
10	0.4224	2.367	0.15582	0.06582	6.418	15.193	10
11	0.3875	2.580	0.14695	0.05695	6.805	17.560	11
12	0.3555	2.813	0.13965	0.04965	7.161	20.141	12
13	0.3262	3.066	0.13357	0.04357	7.487	22.953	13
14	0.2992	3.342	0.12843	0.03843	7.786	26.019	14
15	0.2745	3.642	0.12406	0.03406	8.061	29.361	15
16	0.2519	3.970	0.12030	0.03030	8.313	33.003	16
17	0.2311	4.328	0.11705	0.02705	8.544	36.974	17
18	0.2120	4.717	0.11421	0.02421	8.756	41.301	18
19	0.1945	5.142	0.11173	0.02173	8.950	46.018	19
20	0.1784	5.604	0.10955	0.01955	9.129	51.160	20
21	0.1637	6.109	0.10762	0.01762	9.292	56.765	21
22	0.1502	6.659	0.10590	0.01590	9.442	62.873	22
23	0.1378	7.258	0.10438	0.01438	9.580	69.532	23
24	0.1264	7.911	0.10302	0.01302	9.707	76.790	24
25	0.1160	8.623	0.10181	0.01180	9.823	84.701	25
26	0.1064	9.399	0.10072	0.01072	9.929	93.324	26
27	0.0976	10.245	0.09973	0.00973	10.027	102.723	27
28	0.0895	11.167	0.09885	0.00885	10.116	112.968	28
29	0.0822	12.172	0.09806	0.00806	10.198	124.135	29
30	0.0754	13.268	0.09734	0.00734	10.274	136.308	30
31	0.0691	14.462	0.09669	0.00669	10.343	149.575	31
32	0.0634	15.763	0.09610	0.00610	10.406	164.037	32
33	0.0582	17.182	0.09556	0.00556	10.464	179.800	33
34	0.0534	18.728	0.09508	0.00508	10.518	196.982	34
35	0.0490	20.414	0.09464	0.00464	10.567	215.711	35
40	0.0318	31.409	0.09296	0.00296	10.757	337.882	40
45	0.0207	48.327	0.09190	0.00190	10.881	525.859	45
50	0.0134	74.358	0.09123	0.00123	10.962	815.084	50
55	0.0087	114.408	0.09079	0.00079	11.014	1260.092	55
60	0.0057	176.031	0.09051	0.00051	11.048	1944.792	60
65	0.0037	270.846	0.09033	0.00033	11.070	2998.288	65
70	0.0024	416.730	0.09022	0.00022	11.084	4619.223	70
75	0.0016	641.191	0.09014	0.00014	11.094	7113.232	75
∞			0.09000	0.00000	11.111		∞

From *Engineering Economy*, 2d ed., copyright by the American Telephone and Telegraph Company, 1963. By permission.

TABLE 1   25–27

Table 1. (*Continued*)

10%

ANNUAL COMPOUNDING AT ABOVE EFFECTIVE (i) RATE PER YEAR

FACTORS APPLICABLE TO END OF YEAR AMOUNTS

	SINGLE AMOUNT		SERIES OF UNIFORM AMOUNTS				
n	PRESENT WORTH OF A FUTURE AMT. $(p/f)_n$	FUTURE WORTH OF A PRESENT AMT. $(f/p)_n$	ANNUITY FROM A PRESENT AMOUNT $(a/p)_n$	ANNUITY FOR A FUTURE AMOUNT $(a/f)_n$	PRESENT WORTH OF AN ANNUITY $(p/a)_n$	FUTURE WORTH OF AN ANNUITY $(f/a)_n$	n
	$\dfrac{1}{(1+i)^n}$	$(1+i)^n$	$\dfrac{i(1+i)^n}{(1+i)^n - 1}$	$\dfrac{i}{(1+i)^n - 1}$	$\dfrac{(1+i)^n - 1}{i(1+i)^n}$	$\dfrac{(1+i)^n - 1}{i}$	
1	0.9091	1.100	1.10000	1.00000	0.909	1.000	1
2	0.8264	1.210	0.57619	0.47619	1.736	2.100	2
3	0.7513	1.331	0.40211	0.30211	2.487	3.310	3
4	0.6830	1.464	0.31547	0.21547	3.170	4.641	4
5	0.6209	1.611	0.26380	0.16380	3.791	6.105	5
6	0.5645	1.772	0.22961	0.12961	4.355	7.716	6
7	0.5132	1.949	0.20541	0.10541	4.868	9.487	7
8	0.4665	2.144	0.18744	0.08744	5.335	11.436	8
9	0.4241	2.358	0.17364	0.07364	5.759	13.579	9
10	0.3855	2.594	0.16275	0.06275	6.144	15.937	10
11	0.3505	2.853	0.15396	0.05396	6.495	18.531	11
12	0.3186	3.138	0.14676	0.04676	6.814	21.384	12
13	0.2897	3.452	0.14078	0.04078	7.103	24.523	13
14	0.2633	3.797	0.13575	0.03575	7.367	27.975	14
15	0.2394	4.177	0.13147	0.03147	7.606	31.772	15
16	0.2176	4.595	0.12782	0.02782	7.824	35.950	16
17	0.1978	5.054	0.12466	0.02466	8.022	40.545	17
18	0.1799	5.560	0.12193	0.02193	8.201	45.599	18
19	0.1635	6.116	0.11955	0.01955	8.365	51.159	19
20	0.1486	6.727	0.11746	0.01746	8.514	57.275	20
21	0.1351	7.400	0.11562	0.01562	8.649	64.002	21
22	0.1228	8.140	0.11401	0.01401	8.772	71.403	22
23	0.1117	8.954	0.11257	0.01257	8.883	79.543	23
24	0.1015	9.850	0.11130	0.01130	8.985	88.497	24
25	0.0923	10.835	0.11017	0.01017	9.077	98.347	25
26	0.0839	11.918	0.10916	0.00916	9.161	109.182	26
27	0.0763	13.110	0.10826	0.00826	9.237	121.100	27
28	0.0693	14.421	0.10745	0.00745	9.307	134.210	28
29	0.0630	15.863	0.10673	0.00673	9.370	148.631	29
30	0.0573	17.449	0.10608	0.00608	9.427	164.494	30
31	0.0521	19.194	0.10550	0.00550	9.479	181.943	31
32	0.0474	21.114	0.10497	0.00497	9.526	201.138	32
33	0.0431	23.225	0.10450	0.00450	9.569	222.252	33
34	0.0391	25.548	0.10407	0.00407	9.609	245.477	34
35	0.0356	28.102	0.10369	0.00369	9.644	271.024	35
40	0.0221	45.259	0.10226	0.00226	9.779	442.593	40
45	0.0137	72.890	0.10139	0.00139	9.863	718.905	45
50	0.0085	117.391	0.10086	0.00086	9.915	1163.909	50
55	0.0053	189.059	0.10053	0.00053	9.947	1880.591	55
60	0.0033	304.482	0.10033	0.00033	9.967	3034.816	60
65	0.0020	490.371	0.10020	0.00020	9.980	4893.707	65
70	0.0013	789.747	0.10013	0.00013	9.987	7887.470	70
75	0.0008	1271.895	0.10008	0.00008	9.992	12708.954	75
∞			0.10000	0.00000	10.000		∞

INDEX

1

baud

INDEX

Carson's Rule - P. 16-14

Multi-carrier power sharing — 16-13

Vocoder